ANNUAL REVIEW OF ASTRONOMY AND ASTROPHYSICS

ANNUAL REVIEW OF ASTRONOMY AND ASTROPHYSICS

VOLUME 34, 1996
(revised edition)

GEOFFREY BURBIDGE, *Editor*
University of California, San Diego

ALLAN SANDAGE, *Associate Editor*
Observatories of the Carnegie Institution of Washington

http://annurev.org science@annurev.org 415-493-4400

ANNUAL REVIEWS INC. 4139 EL CAMINO WAY P.O. BOX 10139 PALO ALTO, CALIFORNIA 94303-0139

ANNUAL REVIEWS INC.
Palo Alto, California, USA

International Standard Serial Number: 0066-4146
International Standard Book Number: 0-8243-0934-0
Library of Congress Catalog Card Number: 63-8846

♾ The paper used in this publication meets the minimum requirements of American National Standard for Information Sciences—Permanence of Paper for Printed Library Materials, ANSI Z39.48-1984.

TYPESET BY TECHBOOKS, FAIRFAX, VA
PRINTED AND BOUND IN THE UNITED STATES OF AMERICA

PREFACE

This is a revised printing of Volume 34. The volume was originally printed and distributed starting in September 1996. However, the printing and production were not satisfactory and there were a number of errors and omissions, particularly in the last article. The problems all arose during the final stages of production in the summer of 1996. We have reprinted the volume in its entirety.

In the first version, my preface was accidentally omitted. Here it is as I wrote it last spring:

This volume was planned at a meeting held on April 30, 1994, in La Jolla, California. Those who attended the meeting included Geoffrey Burbidge (Editor), Allan Sandage (Associate Editor), Anne Cowley, John Leibacher, and Anneila Sargent (Editorial Committee Members), and DA Mendis, B Rickett, and AM Wolfe, who were guests. David Couzens (Production Editor) also attended.

In the preface to Volume 33, I pointed out that 29 articles were scheduled for this volume. Eighteen articles are contained here. For volume 35, 25 articles are presently scheduled.

David Couzens, who has been Production Editor since early 1991, is leaving us as I write this preface (March 1996). Since 1973, I have worked with a number of Production Editors and David has been one of the best. He has done the bulk of the work involved in the production of this volume, and as usual the work has been excellent. We are all sorry to see him go.

GEOFFREY BURBIDGE
EDITOR
December 1996

Annual Review of Astronomy and Astrophysics
Volume 34 (1996)

CONTENTS

 (continued)

SOME RELATED ARTICLES IN OTHER *ANNUAL REVIEWS*

From the *Annual Review of Earth and Planetary Sciences*, Volume 23 (1995):

Radar Observations of Mars, Mercury, and Titan, DO Muhleman, AW Grossman, and BJ Butler

The Origin of Life in the Solar System: Current Issues, CF Chyba and GD McDonald

Volume 24 (1996)

Io on the Eve of the Galileo Mission, JR Spencer and NM Schneider

Probing Planetary Atmospheres with Stellar Occulations, JL Elliot and CB Olkin

From the *Annual Review of Nuclear and Particle Science,*

Volume 44 (1994)

Astrophysical Sources of Gravitational Radiation, S Bonazzola and J-A Marck

Volume 45 (1995)

Matter at Large Neutron Excess and the Physics of Neutron Star Crusts, CJ Pethick and G Ravenhill

ANNUAL REVIEWS INC. is a nonprofit scientific publisher established to promote the advancement of the sciences. Beginning in 1932 with the *Annual Review of Biochemistry*, the Company has pursued as its principal function the publication of high-quality, reasonably priced *Annual Review* volumes. The volumes are organized by Editors and Editorial Committees who invite qualified authors to contribute critical articles reviewing significant developments within each major discipline. The Editor-in-Chief invites those interested in serving as future Editorial Committee members to communicate directly with him. Annual Reviews Inc. is administered by a Board of Directors, whose members serve without compensation.

ANNUAL REVIEWS OF

Anthropology
Astronomy and Astrophysics
Biochemistry
Biophysics and Biomolecular Structure
Cell Biology
Computer Science
Earth and Planetary Sciences
Ecology and Systematics
Energy and the Environment
Entomology
Fluid Mechanics
Genetics
Immunology
Materials Science
Medicine
Microbiology
Neuroscience
Nuclear and Particle Science
Nutrition
Pharmacology and Toxicology
Physical Chemistry
Physiology
Phytopathology
Plant Physiology and Plant Molecular Biology
Psychology
Public Health
Sociology

SPECIAL PUBLICATIONS

Excitement and Fascination of Science, Vols. 1, 2, 3, and 4
Intelligence and Affectivity, by Jean Piaget

For the convenience of readers, a detachable order form/envelope is bound into the back of this volume.

Evry Schatzman

Annu. Rev. Astron. Astrophys. 1996. 34:1–34

THE DESIRE TO UNDERSTAND THE WORLD

Evry Schatzman

Observatoire de Paris, Section de Meudon, Département d'Astrophysique Stellaire et Galactique, 92195 Meudon-Cedex, France

---

THE WAY TO ASTROPHYSICS[1]

Saving My Life

July 1, 1943: I was joining the Observatory of Haute-Provence. I was full of emotions and feelings, which certainly had, in a very subtle way, an influence on my scientific life. It was the beginning of an illegal life. I had a complete set of papers (false papers), identity card, food card, and most important the *Carte du Service du Travail Obligatoire* (the government of Pierre Laval in Vichy had negotiated an agreement with the German Government: boys born between 1920 and 1922 had to go to work in Germany), with the notation "*trente-quatre mois de captivité*."[2] When I went to Digne a few days later to meet my young wife Ruth who was coming from Nice, these papers demonstrated their validity. Just as I got off the bus, I had an identity check by a gendarme[3] and he let me go without any problem. I learned much later (after the liberation) that the gendarmerie in this part of France was closely connected with the resistance movement. Did the gendarme simply feel comfortable seeing a young man, who was of an age to work in Germany, carrying the proof that he was exempted from this kind of constraint, or was he connected with the resistance movement and supporting illegal activities? I shall never know.

I had stopped in St. Michel, a few days before my marriage, in order to register in due time at the butcher shop, the bakery, and the grocery store, with my fresh food tickets. The tickets were "legal," as well as the food card, even

[1] It is impossible to follow the chronological order! The logic of things has determined the presentation of the biography.

[2] Prisoner thirty-four months.

[3] There are three kinds of police in France: the national police, which is under the orders of the Interior Ministry; the city police, which in all big cities is under the orders of the city council; and in the countryside, the gendarmerie, which is a special part of the national army.

0066-4146/96/0915-0001$08.00

if they had been obtained in a complex way where false documents had been used to obtain legal documents. I was settling in a quiet place, in a region that was still under Italian occupation, with the feeling that I had greater chances to survive than in any other place. I wished to do science, but at the time of my arrival, this was far from my mind. To survive: Survival was my only worry.

Coming from Nice by this extraordinary 130 kilometer railroad and going through the valleys and the hills of the Alps of Provence with an engine working with wood instead of coal took seven hours. Then came the bus, and, after the St. Michel stop on the main road, I rode five kilometers, with my suitcase on my bicycle, to the Observatory. I still remember the extraordinary smell coming from the fields and from the trees, and this characteristic noise filling the landscape: the buzzing of the cicadas. I had "gone underground" (metaphoric expression for illegality); I was feeling free, even if the fear of being arrested and sent to a concentration camp in Germany, like my father nine months before, had not left my mind.

I knew that my only duty was to work as a night assistant to the assistant director of the Observatory, Ch. Fehrenbach. Otherwise, I was a guest of the Observatory, and I was deeply grateful to Jean Dufay, Director of the Lyons Observatory and the *Observatoire de Haute Provence*, who had given me a place to hide, a place that was incredibly safe. And in fact, even after the collapse of Fascist Italy and the complete occupation of the Italian occupation zone by the German troops, the Observatory of Haute-Provence remained a safe place. The nearest German soldiers were the four men working in the railroad station of Manosque, 25 kilometers away, and there was a larger garrison in Digne, at a distance of 62 kilometers. Why do I give all these details? I want my readers to understand what was going on in the mind of a young man, who had lost his civil rights, who was under the threat of the racist decisions taken by the Nazis and the Vichy government.

To survive: Survival came first.

Student Life: 1939–1941

I was already attracted by science, during my younger years, perhaps influenced by my father, who very much liked reading popular science publications. I was educated in school without any problem and studied for one of those competitive exams that are so characteristic of the French educational system (Ecole Normale Supérieure or in short ENS or l'École).[4] I passed it in July 1939.

[4]The ENS was created by a decree of the parliament *(Convention)* in 1794, as were many other Écoles the same year. It was only at the beginning of the twentieth century that a large fraction of ENS science students obtained academic and research positions. For example, Eugène Bloch, Director of the physics lab (ENS Class of 1897) was a high school teacher for 19 years before getting a University position. He used all his free time to do research.

The war started soon afterwards, with the declaration of war on Germany by France on September 2, 1939. At that time, the academic year started the first week in November. I was just 19 years old—too young to be drafted. I remember reading before the start of the academic year *Men Like Gods* by HG Wells. I remained seduced during all my life by this view of a free society, where the main aim was knowing and understanding the world better and better.

Because of the mobilization, the École Normale was almost empty. The boarding school was closed except for non-Parisian students. I had dreamed of leaving the family, but it was not possible. Only the few *Élèves* who were discharged from military service, or the youngest ones, who were not drafted, about 40 altogether, were there. Élèves of the second, third, and fourth year of studies were in the army. Once, during the year, a large fraction of these Élèves, most of them officers, succeeded in getting leave on the same day and organized a banquet at the ENS, with the first-year students taking part in this enjoyable meeting.

I went to the École everyday, attending lectures on mathematics and physics at the Sorbonne. Twice a week, École students attended premilitary officer training: artillery for sciences students and infantry for humanities students.[5] The material of the University exams appeared so easy compared to the level of the competitive exam that little work was required to be successful. This left time for other occupations.

One of the ENS students who entered the school with me, Louis Le Blan, was very fond of experimental work. He had found it possible to build electronic systems in the physics laboratory. He told me, a couple of weeks after the opening of the University (the first Monday in November at that time), that the Director of the lab, Eugène Bloch, wanted a student to help him set up his lab. Eugène Bloch was a specialist in spectroscopy, and the laboratory had just moved from the old building of the ENS, on Rue d'Ulm, to the new building, on Rue Lhomond. He wanted to use his Rowland grating again, in a completely new room, in order to begin his research again. There was a circular concrete base, completely isolated from ground vibrations. In the old building, it had been noticed that sometimes all the spectral lines on the photographic plate, after a long exposure, were double: The vibrations produced by a truck going by on the nearby street had displaced the grating by a small amount, large enough to modify the recorded spectrum. The aim was to put the grating and the support of the photographic plates in the correct position to take spectra. This was done

[5]Usually the military preparation for an ENS student lasted two years, but in war time, in order to have more officers, the density of preparation had doubled, and we were supposed to pass the exam giving us the grade of sous-lieutenant in June. We learned about the 75-mm gun, balistics, and how to hit the target successfully.

with an iron arc as a light source. I was asked to measure the wavelengths and to identify spectral lines with the help of a catalog of iron lines. I measured hundreds of wavelengths but could identify only a very small number of them. Many years later, I realized that a large number of faint molecular lines appeared on these spectrograms.

I do not know whether I really contributed to setting up the large grating spectroscopy room, or whether what I was doing was just a way of initiating me to laboratory work and experimental research. All I remember is that doing this experimental work in complete freedom made me very enthusiastic. I was very shy, but during the year, having looked here or there at various books, I asked Eugène Bloch about a theoretical problem. I do not remember his exact answer, but his meaning was that I should first be concerned with experiments.

At that time, there was no "modern physics" in the physics program of the University. The famous textbooks of Georges Bruhat contained practically no quantum mechanics; the volume on optics had some development on spectral lines. The first books of Louis de Broglie were easy to find in bookstores, and I bought a couple of them. But these books were devoted to the basis of quantum mechanics, and there were no applications to practical physical problems, like those found in the old classical book of Mott & Massey. Although reading Louis de Broglie's books was fascinating, there was no possible practical use for them. In fact, walking around in the second-hand area of the well-known Gibert bookstore, I had found a very attractive theoretical book on atomic structure, which I judged a few years later to have been written by one of these crazy crackpots whose work winds up on the second-hand shelves. This gives the reader an idea of the intellectual landscape facing a young student.

Disorderly Flight

Real war in France began during the night of May 9, 1940. Very soon afterwards troops and civilians began to flee in the face of the German army. At the beginning of June, it appeared that Paris was going to be taken. I asked the assistant director of the École Normale, Georges Bruhat, who was in charge of the science section, what to do. At that time he was certainly much worried. Because the physics lab had been doing research for the army during the *drôle de guerre,* it was necessary, as I learned years later, to destroy documents. He just gave me an appointment in Bordeaux (reminding me of historical memories of the Franco-Prussian War of 1870—would the government be set up in Bordeaux like that of Gambetta?), telling me that I had to leave Paris by bicycle by the Porte d'Orléans at the time German troops would be entering Paris by the Porte de Clignancourt. I left Paris, riding on a bicycle with the smallest possible amount of luggage, early on the morning of June 10th. I avoided an incredible traffic jam, left Etampes just a few minutes before a terrible bombing, finally took a small road, and after a few kilometers found a small village. I went to the railroad station, asking for a place to sleep. The station master gave me hospitality, opening a folding bed in the waiting room, with perfectly fresh white sheets. He did not want me to pay anything. This peaceful image, at the time of this disaster, is still very vivid in my mind. Early the next morning, I started riding again in the direction of Blois. On the way I was joined by a 16-year-old factory worker who had been ordered to go to his factory, which had moved to the southwest of France. The road was less boring with him. There were not so many cars as before, as people were crossing the Loire river. In their mind a great battle could stop the German army there! At the end of the day, we got off the main road. We found a farm, where the farmers offered us a peasant dinner. We shared a large bed for the night, and next morning I went to Blois. I expected to find a train going to Bordeaux, got on one, and after 25 hours, including several stops in the country because of bombing alarms, arrived at the Bordeaux railroad station. Cousins gave me hospitality.

The next day I went to the University and found the École Normale Supérieure professors gathered there. In the University courtyard, the chemistry professor Georges Dupont was demonstrating the possible use against tanks of what was going to be called a few years later "Molotov cocktails." I did not know what to do. Alfred Kastler, who at that time was a professor at Bordeaux University, gave me physics papers of bachelor candidates to grade. In Bordeaux I met one of the ENS students who had taken the competitive exam with me, Jacques Polonowski, and another one, Jean Mayer, who had been drafted in April and had escaped from his barracks not far from Bordeaux. The father of Jean Mayer was a well-known professor at the Collège de France, and Jean Mayer was probably much more aware than I of what could happen. He was terribly nervous, was frightened that terrible antisemitic decisions of the Nazis would fall on us. He took me to the harbor to try to board an English boat for England. It was not successful! His mother and sister were staying in Valence d'Agen, where his family had a property. He convinced me to go there, where I spent a few days. Jean Mayer was already dreaming of hiding in the country and taking part in resistance actions against the Germans. In fact, he left France with his family for the States later. In the States he became a well-known professor of biology, member of the American Academy of Sciences, and member of the French Academy of Sciences. It was during the short time I spent in Valence d'Agen that I first heard the famous appeal of de Gaulle to the French to reject the capitulation by Petain and to keep fighting against the Germans.

I do not remember how I got a message from Albert Lejeune, another ENS student who had reached Montpellier, suggesting that I should come to take my physics and math exams there. This seemed the most important thing to do at that time. I went to Montpellier, got a bed there at the boarding high school for boys and, as a student of the École Normale Supérieure, obtained some money from the Rector, Henry Pariselle. I met other students from ENS. I registered for the special exams organized for refugees and was able to get in touch with my parents.

During this time in Montpellier, I spent a long time in the library, reading some of the volumes of the physics course of Bouasse,[6] but I also went walking in the *garrigue* (sun-drenched hills). Maurice Pariselle (same ENS class), who had been drafted in April, was in Montpellier barracks. Visiting him, I discovered that I could take the military preparation exam, but the collapse of the French army had raised in my mind such a lack of trust in the military that I did not want to have any connection with them anymore. The great speech of de Gaulle was known, but London seemed so far away, and I had so little taste for arms and their use that after the failure with Jean Mayer in Bordeaux it did not enter my mind that I could begin to look for an escape from defeated and occupied France. The other students of the *ENS* whom I met in Montpellier were only thinking about continuing their studies and, in the end, I did as they did. I finally succeeded in passing the two exams for which I had prepared during the year.[7]

At the end of July 1940, I did not know what to do. I left Montpellier and went to Martres-Tolosanes, a small village near Toulouse, where I could be lodged by a niece of my father. At the end of September, I learned from an exchange of letters with my mother that Bruhat had decided that I had to come back to Paris and attend the normal activities of the School.

Return to Paris

It remained to organize my trip back to Paris. I went to the railroad station of Toulouse to buy a ticket. A poster, above the ticket office, was frightening: "*Le passage de la ligne de démarcation est interdit aux nègres, aux chiens et aux juifs.*"[8] I nevertheless bought the ticket. I left Toulouse on September 19. The train stopped in Chateauroux, for a check of the situation of the passengers. The French officer in charge asked me if I was Jewish. Surprised, and not prepared for that question, I just said "no," shaking my head. His comment was:"*Tenez-bon!*".[9] I prepared myself for the border crossing. But my name,

[6]Bouasse did not believe in quantum physics, but he wrote a physics treatise of 48 volumes, which had the advantage of going into details that are usually skipped. These books were present in all University libraries. Easy reading! Completely obsolete now!

[7]The French university system is such that degrees delivered by one university are valid all over the country.

[8]Crossing the border of the occupied zone is forbidden to negroes, dogs and Jews.

[9]Stand firm!

which looked like a German name,[10] did not raise any question and I could renew breathing when the train started.

The École was almost full. There were several Élèves who were war prisoners. After the demonstration supporting de Gaulle on November 11, at the Place de l'Etoile, the University was closed for several weeks and we had to sign in everyday at the nearest police station, on the rue Vauquelin.

Second-year students had to study chemistry and classical mechanics. I liked mechanics, but I did not like the courses in chemistry, except those in physical chemistry. Here I took my first step in research. In Spring 1941, studying binary mixtures, I was curious to examine ways of measuring the chemical composition of the solid and liquid phases obtained during solidification or melting. It appeared to me that fast cooling, or tempering, would produce a wrong relation between the chemical composition of the solid or liquid phase and the temperature of solidification or melting, by some departure from thermodynamic equilibrium. With a simple picture of the chemical composition of the layer separating the liquid from the solid phase I tried to estimate the deformation of the classical diagram describing binary mixtures.

That this work was published is due to chance. I was sent by Bruhat to visit Paul Pascal,[11] President of the Jury of chemistry, to get in advance the marks of the written part of the chemistry exam of ENS students. Pascal, after giving me the marks, and noticing that I was not happy with the mark he had assigned to me, appeared to be pleased to talk with me and asked me about my interests. I mentioned this computation on tempering of binary mixtures. He offered to let me present a short paper on the subject in the *Comptes Rendus de l'Académie des Sciences.* I brought him the paper one week later. It was published in the *Comptes Rendus* in June 1941. Later I tried to publish a complete paper in *Cahiers de Physique.* Was it delayed because Jews were not allowed to publish,[12] or because of the slowness of the publication? It appeared only in 1945.

That same year I studied for an exam on probabilities. The main course was given by the famous Professor Emile Borel. It was also necessary to make a choice among several optional subjects, and I chose statistical mechanics. The

[10]It is definitely not a German name, but a Jewish name. Fortunately, this was apparently not known by German officers! In the spring of 1943, as I was going from Lyon to visit friends of my father in Annonay, it happened that there was an indentity card check. The German officer asked me kindly: "Sprechen Sie deutsch?" I said "nein" instead of saying "non." The officer smiled!

[11]Pascal was a well-known specialist in physical chemistry and Member of the French Academy of Sciences.

[12]Obviously, Academician Paul Pascal did not care about it. I do not think that there was any checking of the *Compte Rendus.*

course was delivered by Francis Perrin. The introduction to Fermi-Dirac and Bose-Einstein statistics was absolutely exciting.

During the summer of 1941, students were called up to help harvesting. To avoid that, I studied for an exam of mathematics: *mécanique analytique*. There was a very large program from Lagrange-Hamilton theorems to elasticity, hydrodynamics, and shock waves. At home I enjoyed studying elasticity, fluid mechanics, Lagrangians, and a terrible course of Gaston Julia on matrices. In October, I discovered that the professor responsible for the exam was no longer Gaston Julia, but Jean Chazy. The subject of the exam was not matrices, but an elementary application of Hamiltonians to eastward deviation of projectiles. I was the only candidate, and Jean Chazy asked me to come to his apartment for the oral part of the exam. There was no heating at that time, and Chazy was wearing a blanket on his shoulders. He sharply criticized my solution as being not very elegant and asked me a few questions on elasticity. I was awarded the degree of *Certificat de Mécanique Analytique.*

Deportation of My Father[13]

It is perhaps difficult today to make the reader appreciate the emotions and feelings of a young Jewish student in the fall of 1942. I had lost my civil rights. In October 1940, Jewish teachers were forbidden to teach: They should not face Aryan pupils; in June 1941 *numerus clausus* of Jewish students was set up. Becoming a student of ENS in 1939 was the opening of a career as a science teacher, and this prospect had vanished. The first arrest of Jews in Paris took place in May and the first roundup in August 1941. They were sent to a concentration camp. That camp was in France, but they were going to be deported later. After the murder of a German colonel in Paris by the resistance, 742 Jewish intellectuals (physicians, graduate engineers, writers, lawyers, dentists: my father was one of them) were arrested and sent to the Compiègne *lager* on December 12, 1941. At that time, only men were arrested, either individually or in roundups. With the stamp "JUIF" on my identity card I was facing a real danger. After consulting several people, it appeared that the best thing to do was to flee to the south, to the nonoccupied zone, to Lyons. I obtained the agreement of Georges Bruhat, who took the necessary administrative decisions, especially transmitting my student fellowship from Paris to Lyons. It was a long trip; I received unexpected help crossing the border illegally between north and south zones and arrived in Lyons on January 16, 1942. I felt safer in Lyons than in Paris, but I was worried about my mother,

[13] An excellent book concerning the situation of the French Jews is *Vichy France and the Jews,* by Michael R Marrus and Robert O Paxton (New York: Basic Books, originally published as "Vichy et les Juifs" by Calmann-Lévy, Paris, 1981).

left alone in Paris. Drastic decisions about Jews in prison camps had not been taken yet and she could send letters and food packages to my father; she could receive letters from him as well. I had the feeling of a rope around my neck, strangling me little by little.

In January 1942, the war in Russia was terribly impressive. London was broadcasting information, and even with their optimistic presentation of the news and the extraordinary firmness of England, the deep penetration of the German army into the Soviet Union looked like an irreversible process, raising the fear of a possible final victory for Hitler. In that case, what kind of slave would I become? The decision of the *final solution* was not known at that time.

September 23, 1942 is the day of deportation of my father. He died in Auschwitz, probably on September 25, at the age of 65. He had dropped a letter to my mother on the railway tracks, thrown through the opening of the cattle wagon, where he had written: "nous sommes quarante cinq dans un wagon à bestiaux..."[14]. The letter was picked up by a railroad employee who sent it to my mother. As soon as she received it, she sent me a short note giving me the horrible news, which I received only on October 2. My father had been sent to the camp of Pithiviers in August. On September 21, the Prefect of Loiret, noticing his First World War services, tried to save my father by moving him to the camp of Beaune-la-Rolande. But he was not French born and the Prefect was obliged to move him again to Pithiviers, from where he was deported.

Protection from Racism

My parents believed that French Jews were safe. In fact, my father, a dentist, did not imagine how he could make a living in the southern zone[15] and never mentioned the idea that perhaps it was safer there than in the north.

As already mentionned, after the arrest of my father on December 12, 1941, I left Paris and arrived in Lyon on January 16, 1942. It was the beginning of a new life.

By May 1943, I was wondering how to escape from the Compulsory Working Service (STO: *Service du Travail Obligatoire*). In the physics lab of Lyon, there was an assistant lecturer, Cavassillas, who sometimes went to St. Michel, to help Jean Dufay observe the night sky. He suggested that I hide in the Observatory and advised me to see Dufay[16] and ask him if he would agree. I waited for Dufay to come out from lecturing at the University and began explaining who I was: a student of the Ecole Normale Supérieure, whose father was deported....

[14]We are forty five in a cattle wagon, 25 women and children, 15 children, 9 without parents....

[15]Until November 1942, the German Occupation forces were not present in the southern part of France.

[16]I learned after the war that the "Conseil national de la Résistance" had been holding meetings at Lyons Observatory.

I asked him if he would accept me under a false identity at the Observatory. Before I had finished my sentence, he had said yes. Furthermore, he obtained a fellowship for me.[17]

Eugène Cotton, who entered ENS in 1934, was also working in the physics lab. Since we both came from the École we had a feeling of friendship. He came once to see me in the winter of 1942 and gave me a small paper from an underground resistance movement. I did not want it. It is true that I was depressed by the fact that my father was in a concentration camp, but at that time I did not believe that doing something against the occupation forces was possible. At the time of this Service du Travail Obligatoire story, I knew Eugène quite well. He introduced me to another former *Élève* of ENS, of the same class as he, who was also a physicist. His name was Vigneron, and he was already deeply involved in the undergound resistance movement. Vigneron provided me with a paper attesting liberation from *Stalag 1A* (in fact, I never used it!). I got in touch with David Donnoff, who was working in the Jewish resistance. He provided me with a blank birth certificate, which I later filled in with the birth on January 5, 1920 of Antoine Emile Louis Sellier. I went to the "Prisoners House" to obtain an official paper concerning the time spent in a stalag; I was very much afraid of not being able to answer questions. I arrived in an office on the first floor of a buiding and found a man of about 50, to whom I explained that I had been liberated from Stalag 1A for health reasons. The man joked: "You know well that prisoners from Stalag 1A are not liberated for health reasons"; he asked for my name and filled out the papers; a few minutes later I was in the street with this precious document. There remained the problem of getting an identity card. I had the address of a police station in Grenoble, connected with the resistance, where I could obtain it. But when I arrived in front of the police station, I read a posted notice that listed the documents necessary for an identity card and found that I had only one instead of two. I needed a witness. I asked a grocer, in front of his shop, if he would agree to testify for me. He looked at me, asked me to give him my document (the blank identity card with just the fiscal stamp on it), told me to wait a few minutes, went to the police station, and indeed came back within a short time with my false-true identity card. With it, and the document from the Prisoners House, I got the card and tickets concerning food products the next day. When registration of young men for compulsory work was open, I got the card with the words already mentioned:

[17]Not being registered in Paris, I had lost my ENS student fellowship in the Fall of 1942. Jean Dufay procured the help of Bruhat to obtain a fellowship of the *Aide à la recherche scientifique.* Breaking racist discrimination rules, Champetier, assistant director of the CNRS, sent a check to Dufay without stating the name of the payee. I received the money regularly through the St. Michel post office.

"prisoner thirty-four months" without any trouble. I wish to express here my gratitude:

> to Georges Bruhat, dead in deportation, December 31, 1944, for his constant support;
>
> to Jean Dufay, who provided me hospitality at the *Observatoire de Haute Provence*;
>
> to Charles Fehrenbach, who protected us during our stay at the Observatory; and
>
> to those members of the communist resistance movement, Léopold Vigneron, Eugène Cotton, Cavassilas, and to the member of the Jewish resistance, David Donnoff, killed later in Lyon by the Gestapo, who helped me to go under cover.

Research: Second Step

Beginning with my third year of the ENS (Fall 1941), we had to practice some elementary research, which for physicists was preferably experimental. My little knowledge of statistical mechanics and chemical physics led me to imagine that water molecules would not have their electric dipole isotropically distributed in the vicinity of ions in an ionic solution. Because of the scale of the process, it seemed to me that it would be possible, using X-ray diffraction, to test the effect. Georges Bruhat sent me to the laboratory of crystallography of Jean Wyart to present my project. It was accepted, and I started immediately, trying to accomplish the experiment. The idea was to put the ionic solution in between two thin mica plates. The beam of X rays would be limited by a slit of variable width. However, by the time I had completed the first items in the experiment, I had decided, as mentioned before, to go to Lyons to find a refuge.

Lyons had an important laboratory of mineralogy and crystallography, headed by Henri Longchambon. I was immediately accepted there. I was given a small lab, where there was a relatively old fashioned X-ray source, which had not been used for several years. The first thing I had to do was to produce X rays. There was a Holweck vacuum pump, which could only be started after preliminary pumping. I failed.

Visiting the physics lab, I became acquainted with Professor Max Morand, also a former student of the ENS (class of 1920), who had been a professor of physics at the University of Liège. After the bombing of Liège in May 1940, with all his instruments destroyed, he left Belgium and found a refuge at the Faculté des Sciences of Lyons. He was trying to rebuild a Van de Graaf generator. I told him the trouble I was having with my instruments. He came to my lab and showed me that the entire instrument was dirty and had to be cleaned carefully, in order to avoid dirty vapors preventing a good vacuum from taking place. I complained about the fact that I had to obtain my degree, the Diplôme d'Études supérieures de Physique, before the end of the academic year. He then suggested I should build a hot-wire manometer and make measurements of gas pressure with it. The hot wire was in fact a gold foil. When measuring its electric resistance, I found that it had a ten times larger resistance than expected from its cross section. When looked at with a microscope, it appeared to be full of holes. Following a complex transform that I had found in Bouasse during my stay in Montpellier, I calculated the actual resistance, due to a large number of holes in the gold foil.

Usually, to get the degree, students would present the results of their experiment and give a short lecture on a subject chosen by the examining board. However, having explained to Prettre, professor of chemical physics, my work on mixtures tempering, he asked me to present it. For the degree I presented two original subjects: the hot gold-foil manometer and the quenching of binary mixtures.

The audience was quite limited. However, I had the honor of the presence of Eugène Bloch and his brother Léon, who had left the northern zone of occupation for Lyons. I spoke almost two hours, as no one had told me to prepare a shorter presentation of my work. At the end, Eugène Bloch, who was famous for his teaching of didactics of physics to the Élèves[18] first congratulated me, and then said that, for such an elementary subject, my talk was perfectly incomprehensible. I never forgot the remark. Eugène Bloch was certainly a very important person in French academia. In 1943 he had a visit from Jean Chazy, and I went to his office to meet him. When I arrived there, Jean Chazy was making surprising comments about the migration of a large number of Jews from the occupied norther zone to the southern zone. He was in fact comparing the system of the two zones to communicating vessels and the number of Jews in each zone to fluid heights in communicating vessels. His statement was that the height of the "jewish liquid" in the southern zone being too large, the situation was incompatible with the properties of communicating vessels. Eugène Bloch was smiling! Obviously, having been a University Professor of Rational Mechanics did not enable him to have a good view of the actual situation. In 1944, Eugène Bloch, 66 years old, was deported and died in Auschwitz. His wife and his brother escaped.

[18]During the last year of ENS, students prepared a difficult competitive exam called the *agrégation*, with an important test consisting of a lesson in physics and a lesson in chemistry for high school students.

During the next year, 1942–1943, I kept my office in the physics lab of Max Morand. With the help of Longchambon, I found a small job, using polarized light to test stresses in a plexiglass model of a railroad bridge over the Rhône. This lasted until my departure to Saint Michel.

IDEOLOGY, POLITICS, AND SOCIETY

Introduction

Having taken a small part in the resistance and having been seduced by an ideology, dreaming of a new and better society, I felt it necessary after the war to become active in politics and trade unions. I felt the need to write impassioned articles in periodicals. This has been known in France and abroad, including the States. It played a part on my relationship with the scientific community. I find it necessary to tell that story, a story that goes from an enthusiastic attitude at the beginning to pain, regrets, and melancholy, feelings that have not passed away.

Resistance and Belief

To understand what follows I have to say a few words about my father. Born in Tulcea (Rumania) in 1877, he was taken by his parents to Palestine in 1882, participating in the emigration movement "Zion lovers." A good pupil, he was sent in 1896 by *Alliance Israëlite Universelle*, under the sponsorship of the Baron Rothschlid, to study agriculture in Grignon.[19] This was at the time of the Dreyfus affair, and he discovered in France the battle between right and left, the fight of the famous *Ligue des Droits de l'Homme* in favor of Dreyfus and against antisemitism. Although brought up religiously he became completely atheistic. Back in Palestine in 1899 he was sent to work in the region of Metula. At that time this area was full of swamps and he caught malaria. Because he had learned modern agriculture he got into trouble with the old Jewish peasants. He finally went to New Zealand in 1902. There, he discovered the Labour Party and appreciated the quality of its politics. As he could not settle in New Zealand under the conditions he had hoped for, he came back to France in 1905. He would have liked to study medicine, but he did not have the proper degree to do it. However, he was able to register in the best dentistry school and got his degree in 1908. That same year he obtained French nationality. He married one of the daughters of the Secretary of the Jewish Paris Consistory, Léon Kahn, who had died in 1900. Despite his atheistic attitude, he was very respectful of the faith of my grandmother. I was educated with a mixture of atheism, skepticism, and Jewish tradition.

[19]There is still a high level school in Grignon, training graduate engineers in agriculture.

My father, being a great admirer of Jaurès, became a member of the French Socialist Party. During the First World War, he had to treat disfigured soldiers and found that work difficult to bear. When the "Third Internationale" was created he joined the French Communist Party, but he resigned when the Organization asked him to leave the *Ligue des Droits de l'Homme.* He was so grateful to the *Ligue* for its fight in favor of Dreyfus that he could not stand such a requirement.

I became acquainted with the political ideas of my father essentially at the time of the "Front Populaire" and the fight against French fascists. The political victory of the left in the 1936 elections was very exciting, and at that time I certainly did not see the difference between socialists and communists. During the last two years before the beginning of World War II, I was so busy studying for the competitive exam that I did not closely follow international events. My father was certainly very much worried by the great trials in the USSR, but I did not feel concerned. Then came the Münich agreement and the victory of Franco in Spain—all that had the smell of the coming war.

A short time after the terrible news of the deportation of my father, I went to see Eugène Cotton, telling him that I was ready to join the Communist Party. I needed to do something to give me the conviction that I was contributing to the fight against the Nazis. A few days later, I started to stencil at night, in the lab, packages of 50 or a 100 copies of short leaflets calling students to resistance. Leaving the lab, I had to walk just 20 meters in order to drop the stencil in the Rhône river; the following day, the leaflets were given to someone I did not know, to be quickly thrown in a lecture room. Once I made a photocopy of a document about the use of plastic bombs.

Obviously, Eugène Cotton, acting like the intellectual he was, wanted to let me know not only the 1942 slogans, but also Marxist theories. He lent me Engels's *Origin of Family, Property and State.* The student of the underground communist organization with whom I had a direct connection lent me *Materialism and Empiriocriticism* by Lenin, but I was not ready to read that kind of philosophy. On the contrary, Engels seemed to me amazingly clear. I was seduced, as if I had an unknown need for an ideology, which at last was fullfilled. In fact, looking today at my intellectual behavior at that time, it looks as if I had taken holy orders. Suddenly it was like sunshine enlightening my life. The theories of Marx and Engels, explaining human history completely and definitely, the success of the Communist Revolution in Russia in 1917, the idea that practical applications of the knowledge of history's laws would lead to a new society, were incredibly attractive. This kind of behavior, a sort of faith, with a complete respect for these written "scientific" laws of history, looks today just like the respect of dogmas, and at that time touching these dogmas was sacriligeous.

The situation was so desperate; the fear of being arrested as a Jew and sent to a camp was turning over in my head and in my stomach. The help I received from the Communist organization certainly brought me to a feeling of gratitude. Coming back to Paris in October 1944, I did not join the Party again, although I was feeling guilty for not expressing my thanks and not being a communist militant.

At the end of 1945 or beginning of 1946, I joined the Party again. Communist publishers were publishing the most important Marxist writings, little by little. The ideology in these books went much beyond the daily political attitude of the Communist Party. The idea that a "scientific" knowledge of the laws of history, economy, and production could lead to a better society was very appealing. Becoming known among scientists, I joined the trade union Syndicat de l'Enseignement Supérieur et de la Recherche Scientifique. It was at that time a very small union, which was actually in the hands of the communists. I quickly went down the path of a more intellectual Marxist activity; I published several ideological articles, completely in agreement with the dogmas of that time in *La Pensée,* founded by Paul Langevin in 1939. I became known as an active Communist in the French astronomical community, especially when I became general secretary of the Union in 1949. I kept this responsibility for eight years.

In the spring of 1950, A Danjon organized an international meeting on astrometry in Paris. There was a Soviet delegation, which included Zverev (chief of the delegation), Ambartsumian (for whom it was sort of an award), Nemiro and Orlov from Pulkovo, and Batruchevitch (a "secretary" of the Academy). It was well known at that time that Soviet people were not allowed by the KGB to have private relationships in the country they were visiting. Nevertheless, I dared to invite them for dinner at home. We were expecting them at 8 pm, but would they come? Three of them, Ambartsumian and the two others arrived at 9:30 pm, explaining that Zverev was sick. Batruchevitch was not there either. Since both Ruth and my mother-in-law spoke Russian, the guests enjoyed the dinner and it was a pleasant evening.

Through the channel of something akin to cultural exchange sponsored by the Soviet Embassy, I was invited to go to Moscow for the celebration in November of the anniversary of the Revolution. When I arrived in Moscow, I discovered that it was essentially a tourist visit, but I asked again and again to see astronomers. Finally, I had a meeting in the building of the Soviet-French Association with Kukarkin and Kulikovsky; Kulikovsky who spoke French also acted as translator. I was also invited to meet the Dean of Sciences. Because I was not a mathematician, he looked for an astrophysicist and found Alla Massevitch, who came to his office. Meeting Alla was a great pleasure, for I was familiar with several of her publications on the internal constitution of the stars. As we came out of the University (which at that time was in an old building close to Red Square), I left her at the bus stop; we both hoped to see each other again. Many years later, she told me that she desperately tried to see me again without success. When I insisted on seing Alla Massevitch again, I was told that she was sick, then that she was in her "datcha," and finally, I had another meeting with Kukarkin and Kulikovsky. It is difficult, after 45 years, to estimate the results of these meetings. Perhaps the simplest thing is to say that I did get acquainted with astronomical research in the USSR.

Stalinist Regime, Stalinist Antisemitism

I had a painful experience in 1956. In order to make it understandable, I have to give some information on my family. My father-in-law, Joseph Ariel (Fisher), had been expelled from the USSR in 1924 for Zionist activities. He worked in the Keren Kayemeth Leisrael in France. He had always closely followed the situation, events, and politics of the USSR. He settled in Israel in 1950 and became the Israeli ambassador in Brussels (1952–1956). He was terribly affected by the antisemitic campaign in the USSR at the end of 1952, the accusation against the Jewish physicians, and its consequences at the level of the French Communist press.

In 1956 I was invited to take part in a colloquium in Burakan. This took place a few months after the famous Khrushchev speech. Because of the softening of the Communist dictatorship I was able to stop for two days in Moscow before the colloquium and visit the two living sisters of my father-in-law. I stayed for these two days at the home of the oldest sister. She had had a French nanny when she was a young child, before the First World War, and spoke relatively good French.

I am still under the shock of that visit.

I heard stories of antisemitism. Israel Gradsztajn, who was the brother of my mother-in-law, had been expelled in 1952 from his Institute of Mathematics, but was accepted as a teacher of mathematics in a high school. Gradsztajn! The author of the famous book of mathematical formulae. He apologized for giving me only a Rumanian translation of an old edition of his book.

I heard stories about arrests and deportations. That was the way in which these two sisters had lost their husbands, one deported in 1937, the other in 1945. The last one was freed in the summer of 1953, came back to Moscow by his own means, and died six weeks later of dysentery.

I want to be clear. They were deported *not* because they were Jewish. They were just carried along to the camps with millions and millions of people. The existence of *gulags* was known officially at that time, but it was deeply moving to be in touch with people who experienced the Stalinist dictatorship. This was much more moving than the events surrounding Lysenko. French biologists never accepted the claims of Lysenko, and I was aware of this early, due to the relationships I had with some of them in the Trade-Union of University Professors, especially the geneticist Georges Teissier and the biologist Marcel Prenant.

Coming back to western Europe, I stopped in Brussels to see my father-in-law and my mother-in-law. My father-in-law, who had a diplomatic passport, was waiting for me in the arrival hall before the police check point. I began immediately to tell him what I had heard and what I had learned. Despite his continuous interest in events in the USSR, this was far beyond what he actually knew, and his first reaction in this public place was to tell me to keep silent.

Sorrow

This contact with Soviet reality was painful. I had given time and thought to this political activity, I had run risks, and I had been mystified, deceived, and betrayed. It was unbearable. Back in Paris, I visited officials of the Party to tell them that it was necessary to tell the truth, to have a critical look at the Stalinist dictatorship. I believe now that these minor officials were frightened. They rejected my arguments gently, as if I had been the victim of a lie and was some kind of irresponsible child. That is exactly it: What I was saying was childish, and they had only to reprove me in order to put me back on the right path.

All of that was useless. In fact, when I began to read the classics mentioned above, I had already considered that Marxist theories were just a step in the direction of a better understanding of history and of present society. Just a minor incident had stopped this thinking for a while. In 1948, I was giving a lecture in one of suburbs of Paris. I was trying to explain the *Dialectical Materialism* of Engels. One finds there a comparison between the discontinuous change from liquid to vapor and the discontinuous change of a society through a revolution. Then I explained that, just as we can bring liquid water to vapor water by going around the critical point, it was possible to avoid a revolution and to go directly to the new society. The result, during the discussion, was an acid criticism of this metaphorical description of historical changes in society, as if it had been against the Revolution. At that time, I concluded that I had to be respectful of the dogmas.

I have already mentioned the origin of my emotional link to communism. But in 1956 I was beginning to judge what happened in the USSR. It took me two years to decide to leave the Party. It was then 1959. It is perhaps difficult for the reader to understand this slow, too slow change of mind. I was able to reject Stalinism, but I still believed in the basic dogmas of Marx and Engels. The horrible dictatorship of the Communist Party in the USSR seemed to me an inheritance of Russian history, a deviation from the path drawn by Marx and

Engels. It was only in the 1970s that it became clear to me that knowledge of society is not of the same nature as science.

At that time, in the 1970s, it became apparent to me that I had not only been betrayed by the Communist Party, but that I had betrayed myself by the belief that it was possible to use a social, historical, and economic theory to build up a new society, exactly as it is possible to use a new physical theory to achieve a successful new physical experiment. I had the feeling that I had been a supporter of an unacceptable dictatorship, that I had been an accomplice, even if only in a very remote way, in unbearable crimes.

Philosophy

What remains on one hand of this ideological experience is just a philosophy. If I use different words to speak of the physical, chemical, ... sciences on the one hand, and the social sciences on the other hand, I would say first that it is not possible to give to "science" and to other "knowledge" the same basic epistemological meaning. Next, Nature can be known and understood, and the most fascinating research subject of a theoretician is to look at the contradictions between theory and observational and experimental data. Although this seems elementary (*elementary, my dear Watson!*), it nevertheless is a powerful research motivation.

On the other hand, I remain worried about the future of science. Education does not provide the majority of citizens with the understanding of the aim of science: knowing the laws of Nature. People confuse science and technique. I have the feeling that science is not really accepted by society[20] and that elementary teaching of science is not satisfactory and has not been adjusted to serve the future of mankind. [21] It is very difficult to transmit such a message. Can I dare to say that it is not only a question of understanding, but that all over the world most scientists, simply, do not care?

RESEARCH

White Dwarfs in Saint Michel

On July 1, 1943, I was in the hills of Haute-Provence, facing wild sunlight and heat. There were four families living in the Observatory: Fehrenbach, the assistant director; two technicians, Pissavin and Blanc; their wives and children; and Sellier (myself!) and my wife, Ruth. Later on, David Belorizky, an astronomer at the Marseille Observatory, and his small son, came to find

[20]There was a remarkable editorial on this subject in *Physics Today* (September 1970).

[21]I have tried to approach these questions in two books: *Science et Société* (1971) and *La science menacée* (1988).

a refuge from antisemitic roundups. Jean Daudin, a specialist in cosmic rays and a very active member of the resistance also found a temporary refuge there, with his wife Alice Daudin, who wanted to become an astrophysicist and finally remained after the war working in the field of nuclear physics. I was committed to help Fehrenbach as a night assistant. The library of the Observatory consisted at that time of just one cupboard of books. It included the first edition of the *Handbuch der Astrophysik,* the first edition of Unsöld's *Physik der Sternatmosphären*, the proceedings of the 1938 Paris colloquium on *Novae, Supernovae and White dwarfs,* the complete collection of *Zeitschrift für Astrophysik,* and a few other books. Fehrenbach advised me to read Unsöld. My knowledge of German was not good, and I had to translate what I was trying to read into French. I translated perhaps one hundred pages, but I did not understand their meaning until I had to practice radiative transfer three years later. In order to give a scientific life to the Observatory, Fehrenbach organized a seminar for the five scientists who were there. He asked me to give an account of the 1938 meeting. It turned out that, having attended the lectures of Francis Perrin in 1941 on statistical mechanics, I knew Fermi-Dirac statistics and I could easily understand the problem of the internal structure of white dwarfs. It also happened that some time before I had read in a popular science periodical, *Science et Vie,* a paper written by Paul Couderc on stellar energy sources in which he cited the name of Bethe. I gave the CNRS an order for a microfilm of Bethe's paper. I found an older paper of Gamow's in *Z. Astrophys.* on thermonuclear reactions and discovered immediately that a white dwarf having the same chemical composition as the Sun would explode. Hydrogen should not be present inside white dwarfs. I also found that, due to the high gravity field in white dwarfs, hydrogen should float to the surface, and that the size of the transition region from pure hydrogen to metals had to be very thin.

Can I mention the origin of my passion? Doing research gave me great pleasure. I had the impression of exploring unknown countries, discovering one after the other extraordinary things, and bringing them piece by piece into the realm of knowledge.

When I arrived in Paris, in October 1944, I had to study for the competitive exam called the *Agrégation,* which at that time was compulsory for ENS Élèves and was a prerequisite for a career as a high school teacher. On the other hand I wanted to finish the first step of a research project leading to the doctoral degree. However, the date at which the competitive exam was to take place was not known, and it was not even known if there was going to be one during war time. This gave me the opportunity to continue my research.

However, it was also necessary to attend lectures and perform student duties. Starting the study of "classical" physics and chemistry again was difficult after having experienced research activity. It was necessary to know all classical physics and all chemistry to be ready for the exam. And both fields were much less attractive than quantum physics and statistical mechanics. I remember the difficult conditions of the winter of 1944–1945: not much heating, just a very small stove. I made myself a vest with several layers of paper cut out of newspapers (good thermal insulation!) but had the greatest difficulty in using it, due to the noise produced by any motion.

Finishing my work on white dwarfs was very absorbing, and I visited Louis de Broglie several times, bringing him manuscripts to be published in the *Compte Rendus de l'Académie des Sciences.* These short papers, called *Notes,* were published during the winter of 1944–1945.

It was during these months that I became acquainted with Jean-Claude Pecker, who was also finishing his studies at the ENS. His parents had been deported in May 1944 and he was deeply affected by this. I remember walking with him in the corridors of the school and talking, talking about everything, but essentially about astrophysics. From that time on we have had a friendship, a brotherhood that has never weakened.

Finally, we learned in March that the written part of the competitive exam was going to take place in April. I hurried up, desperately reading hundreds of pages of physics and chemistry. The oral part took place three months later. Thirteen young physicists were successful. I was the thirteenth. Undoubtedly, initiation in research work was not a good preparation for a competitive exam!

Many years later, in 1967, I was elected as a member of the CNRS Scientific Committee of Astronomy, Astrophysics and Geophysics, and elected immediately afterwards as a member of the Scientific Council of the CNRS. There I met the well-known physicist, Louis Néel, a Nobel prize winner, who had been the chairman of the *Jury d'Agregation* in 1944. The problem he had given in 1944 consisted in building up, step by step, the theory of the Lyot monochromator. He told me what I felt was perhaps the best compliment I ever had: "I have recently been sorting out old papers. I found my notes concerning your part in the competitive *agregation.* And it was clear that I was not mistaken about you." After the exam, successful candidates were received by the chairman, and I was asked which kind of position I wanted. I said that I was expecting a fellowship in the States. This was not true, but it was not wrong either, as GP Kuiper, in the fall of 1944, had offered me an invitation to the States. Because I had experienced illegal life, escaping from the compulsory working duty, I was considered a special case, and I avoided having to take a teaching position in a high school.

I still had to complete the writing of my thesis and I worked hard to have it published in due time in the *Annales d' Astrophysique.* I presented it in March 1946 to the examining board that had Louis de Broglie as chairman.

Chargé de Recherche

I applied for a position in the CNRS. My file happened to be supported by Francis Perrin and I finally got a position in physics in the fall of 1945, although it was not until February 1946 that I began to get the salary of *Chargé de Recherche*. The rank of the position was very exceptional, as I was entering the CNRS not at the lowest rank of *Attaché de Recherche*, but at the rank above. In fact, I think that for a beginner who had not yet obtained the doctoral degree (I got it in March 1946), this has never been done again.

A relatively short time after publication of my thesis in *Annales d'Astrophysique*, I received a very nice letter from L Biermann in Munich, drawing my attention to papers I had not quoted. I gave a curt answer, mentioning the deportation of my father and the difficult conditions under which I had been working at St. Michel. I met Biermann much later at a meeting on the edge of the lake of Tägernsee in 1959. He wanted to talk with me; we entertained Biermann and his wife at home in Paris a couple of weeks later. Biermann said that he had no knowledge of the extermination camps during the war. We met the Biermanns again in Boulder in 1968 and in Copenhagen in 1978, on the occasion of the seventieth birthday of Bengt Strömgren. Our relationship with Biermann has been rare but friendly.

Institut d'Astrophysique (IAP)

I became a member of the Institut d'Astrophysique in Paris. During my stay in St. Michel, I had known Daniel Chalonge and Daniel Barbier. At that time the IAP building was not finished, and I did not get an office. I was working at home, often visiting the IAP. The library was very small—it was in a room that later became one of the guest rooms for IAP visitors. Henri Mineur was the Director. Chalonge and Barbier had offices. Volodia Kourganoff was working in a tiny space, which had become the kitchen of the guest area. I had lots of discussions with all of them. Kourganoff was beginning his work with Chalonge on the role of the negative hydrogen ion in solar limb darkening. Several times I met Rupert Wildt who often visited Chalonge. He was famous for his paper on the negative hydrogen ion. Tuominen arrived in 1946; he got a position in Paris for two years and an office at the IAP. I talked with him almost every day.

I had constantly in mind the proceedings of the 1938 Paris colloquium, and taking into account a remark of Biermann on instabilities, I imagined a model of collapsing white dwarfs, leading to a supernova explosion. In 1939 Zwicky had considered the collapse of a white dwarf into a neutron star. I had no knowledge of Landau's model of neutron stars, and my theory of supernovae was based on the idea of an instability leading to an explosion and the ejection of matter by a shock wave. Strangely enough, the orders of magnitude of radius, stellar density, and velocity of ejection were not so far from what we presently know, even if our knowledge of SN maximum luminosity was wrong by six or seven magnitudes. Altogether the theory was incredibly naive. Barbier, who was editor of *Annales d'Astrophysique* submitted my paper to Strömgren. Strömgren turned out to be very laudatory, and my paper was published in 1946. In July 1948, a short time before the IAU meeting in Zürich, I received a very kind letter from Paul Ledoux, criticizing my view on instability. There was some kind of heroism in building up theories based on an incomplete knowledge of physical processes with the aim of explaining poor observational data.

I considered thermonuclear reactions as having a great astrophysical importance, and I continued several years to work on the subject.

Barbier wanted to keep me in astrophysics. At that time, I often visited Edmond Bauer, formerly a Professor in Strasbourg and later Professor in Paris. He had his office at the Institut Curie, and I had him as a professor for the preparation of the *Agregation.* He told me once that he had convinced Danjon, during the time he was in Strasbourg, of the reality of quantum mechanics. These ideas of modern physics were seductive. Bauer persuaded me to apply for a position in Dublin, to study quantum mechanics with Walter Heitler. I began to exchange letters with Cecile Morette (now DeWitt) who was working there, but before any conclusion was reached, Barbier obtained the agreement of Bengt Strömgren to receive me in Copenhagen. I was sufficiently involved in astrophysics to accept the invitation immediately.

My first initiation into astrophysics was in St. Michel. For my second initiation I was to have the guidance of a prestigious astrophysicist and so did not to have to use books alone.

When I arrived in Copenhagen on April 9, 1947, the anniversary of the invasion of Denmark in 1940, the street were almost empty. I had little money, was afraid of boarding a streetcar with my small knowledge of Danish, and did not take a cab. I walked from the railroad station to the Observatory, appreciating the beauty of the city.

Copenhagen

Strömgren was waiting for me. I asked him: "What shall I do?". His answer was "You have worked on the inside of white dwarfs; now you should work on white dwarf atmospheres." Rudkjöbing helped me to find my way. I had no doubt that I had to study a pure hydrogen atmosphere. I started reading Unsöld, having completely forgotten my reading of *Physik der Sternatmosphären* four years before. I discovered the limitation of the number of excited states, due to the presence of a microscopic random electric field and pressure ionization, and I finally computed atmosphere models and line profiles with a small desk hand-computer. I was so much seduced and attracted by the publications of the

Copenhagen Observatory that I asked Strömgren to have my work published in this series. I did not know at that time that it is better to publish in international periodicals. That work, which was published only in 1950 because the Copenhagen Observatory did not have the money to publish it earlier, remained entirely unknown. It was the first complete description of pressure ionization and the Stark broadening of hydrogen lines in white dwarfs. The MHD (Mihalas, Hummer, Däppen) paper, published in 1989, was certainly much more elaborate than my 1950 one. But I confess that I was not pleased at not being quoted! I wrote to Mihalas and discovered that he was completely unaware of the existence of that paper. I sent him a photocopy of it. Many thanks to Mihalas for his apologies.

I regularly attended the seminar at the Bohr Institute and enjoyed having a direct view of the most recent developments in physics. I had an appointment with Niels Bohr, gave a lecture on white dwarfs, and became acquainted with a number of young physicists, including Cecile Morette, with whom I had the exchange of letters about Dublin.

Beginning an Astrophysicist's Life

I was still in Copenhagen when I received an invitation from Lyman Spitzer for a stay in Princeton in 1948–1949. I was going to have a fellowship, a sum that seemed large to me. The aura of Princeton was such that I felt much flattered and honored. I accepted immediately.

In the spring of 1948. Danjon asked me if I would be willing to teach astrophysics at the University of Paris. Danjon was teaching fundamental astronomy and Jacques Lévy, an astronomer at the Paris Observatory, was taking care of training students. These courses were given under the common title *Astronomie Approfondie* and each of us had to lecture two hours a week, during the first semester, from November 1 to February 28. As far as I know, this offer was made on the advice of Otto Struve. I felt very flattered. I accepted, but as I was going to the States, this appointment had to be postponed until the fall of 1949. When I came back from Princeton in June 1949, I went to see Danjon and asked him if he was still willing to give me this lecturing opportunity. It was the beginning of a long story. I taught for 27 years, but I shall tell more later about this extraordinary experience.

In spring of 1948, I was invited by Jacques Cox, former Rector of Brussels University and a professor of astronomy, to give a few lectures at the University. I remember speaking about supernovae. There were in the audience several well-known professors: Émile Picard, who went up in balloons to measure cosmic rays; Bourgeois, the Director of the National Belgian Observatory; and naturally Jacques Cox. At the end of my stay, Cox asked me if I would agree to teach astrophysics at Brussels University. I was not going to have a full position there, just a small salary, enough to pay for my trips from Paris and my hotel expenses. But this gave me a feeling of recognition, and I accepted. I started this teaching also in 1949. I kept this position for 15 years.

RESEARCH AND UNIVERSITY EDUCATION

Learning New Fields of Physics

Arriving in Princeton, I had no precise subject in mind, except perhaps some concern about solar granulation, resulting from discussions with Jean-Claude Pecker. It was through discussions with Martin Schwarzschild and Lyman Spitzer that I came to the two subjects I dealt with during my stay in Princeton: heating of the solar corona by shock waves and the effect of a μ-gradient due to gravitational settling on a white dwarf structure. I talked much with Martin and Lyman and discovered the way of thinking of theoreticians who had a great knowledge of observational data. They were critical and direct in the discussion. I had the feeling of my great ignorance: The trouble is that all possible physical processes can be important in astrophysics.

Let me give an example here. During June 1950 I spent a week at Leiden Observatory, gave a lecture on white dwarfs, and talked much with Oort and Van de Hulst. Oort asked me if I could give an explanation of some spectroscopic properties of the Crab Nebula. After my visit, I spent a couple of months working on the problem of radiative transfer in the nebula and did not find any explanation, so I finally dropped the subject. Schklovsky, who had many more relationships with physicists than I (as he told me many years later), concluded, as did others, that the continuum of the Crab was due to synchrotron radiation.

When I started lecturing on astrophysics in Paris, in the fall of 1949, Danjon left me free to choose the program. I felt that I had to cover the whole field, from stars to galaxies and cosmology. I had to learn how to present basic data from the Hertzsprung-Russell diagram and binaries to the structure of galaxies. This experience, which lasted 27 years, was also the origin of many publications, owing to the number of cases presenting unexplained facts, which actually turned out to be easily explainable. This is the reason for the great variety of subjects that I approached, even if they remained essentially in the field of stellar physics: the structure, evolution, and origin of stars.

Scientific Publications and Reputation

If I remember correctly, I once heard Otto Struve saying that Soviet astronomers had been clever enough to avoid something similar to the famous catastrophic Lysenko affair. Their solution had been to organize meetings on cosmogony and to publish yearly *Questions of Cosmogony*. Kukarkin, during the IAU meeting in Rome, asked me to write a report on *Cosmogony and Cosmology in Western*

Countries. I collected about four hundred microfilms during the academic year 1952–1953, and during summer vacation I wrote a long paper, which was published in this Soviet periodical in 1954. I felt proud of that analysis, which was the result of quite hard work, but I did not like having it published only in Russian. The French text, after undergoing a few modifications to make it sufficiently popular, was published in France by the publisher Albin Michel (1957), under the title *Origine et évolution des mondes.* A Spanish translation by Raquiel Rabiela de Gortari and Arcadio Poveda was published in 1960. In between, I had been asked in 1959 to bring the book up to date, in order to have it published in English. Spending one month in Pasadena in September 1959, I discovered a fantastic amount of literature on cosmogony and cosmology published between 1952 and 1959. I felt unable to achieve a complete updating and finally wrote just a few adjustments. It is true that, at that time, I was working on the loss of angular momentum in stars, which was the continuation of the short paper I had given during the IAU meeting in Moscow in 1958. I was more eager to complete a paper on this subject than to write corrections to the book on cosmogony and cosmology. Anyhow, I sent these few corrections to the publisher and the English translation, by Annabel and Bernard Pagel, of *The Origin and Evolution of the Universe* was published in 1966.

The draft of the paper on loss of angular momentum was finished when I left Pasadena at the end of September 1959, and a preprint was available.

It is true that my publishing *Questions of Cosmogony* confirmed the rumor about my membership in the French Communist Party. During an IAU meeting, I was trapped by GP Kuiper and W Iwanowska; the latter was Director of Torun Observatory, whose fiancé had been killed in Katyn. They asked me to give my opinion about this horrible murder of 15,000 Polish officers in Katyn. Was it the Nazis or the KGB?

In 1967, after the death of André Danjon, I decided to be a candidate for membership in the Academie des Sciences. At that time, the rule was to pay a visit to academicians, to introduce oneself. Francis Perrin who had helped me in 1945 to get my position in the CNRS asked me: "Is it true, as Danjon told me, that you stopped working on the internal constitution of the stars, because it was condemned by Russian communists?". This was incredibly surprising as I had never stopped working on this subject. I still remember the bright look in Perrin's eyes. Can I mention here the fact that several people never forgave me for my former membership of the Communist Party? They did not want to see me joining the highest institutions.

In 1954 I received the offer of a professorship in Manchester. I did not want to leave France. Continuing to contribute to the development of astrophysics in France was a way of expressing my gratitude for the help I had received during the war. Because of the political situation and the statutes of CNRS research workers, was there also a threat of losing my CNRS position? There were rumors about it. I asked Danjon if I could become a university professor in Paris (actually, an associate professsor or something equivalent). He agreed to support my application and I secured the new position in October 1954. As noted in the last edition of *Petit Larousse,* this was the creation of the the first Astrophysics Chair in France.

It was in these years that I asked JC Pecker if he would be willing to write an astrophysics textbook with me. On the advice of Alfred Kastler we approached the publisher Masson, and *Astrophysique Générale* was published in 1957.

Students

Almost immediately after starting my astrophysics lectures, I began to have contact with students who wanted to prepare their PhD with me. The French doctoral degree, at that time, was obtained after several years of research: Four or five years was considered normal. People preparing a doctoral degree already had jobs, such as *Attaché de Recherche* (CNRS position), assistant lecturer at the University, or assistant astronomer at the Observatory.[22]

When discussing topics with a beginner, I would mention a variety of unsolved theoretical problems until enthusiasm sparked in the mind of the student. Some of those attending my lectures at the University got interested in important questions. They found their research subject by themselves, asking me only to be their sponsor and the person to talk with. I have sponsored many doctoral students; I do not remember how many. Some of them were remarkable. This situation opened up the possibility of developing theoretical astrophysics in France.

A symposium on plasma physics in Stockholm (1955) gave me the firm belief that it was absolutely necessary to develop this field in France. With Jean-Loup Delcroix, Jean-François Denisse, and Theo Kahan, I organized in the fall of 1955 a graduate course in plasma physics, which immediately met with great success. I taught magnetohydrodynamics. This interest in plasma physics and magnetohydrodynamics led me to meet R Lüst, A Schlüter, E Parker, and many others, and I kept in touch with them for several years.

There was a great change in the physics programs in 1957, and I succeeded in having astrophysics become one of the optional subjects of the undergraduate courses for the *Licence.* Next I found that it was necessary to create a graduate course in astrophysics (1961). As I could not deliver courses on all subjects by myself, I solicited the help of many fellow astrophysicists. Some of these

[22]The present situation is completely different. Essentially, students have to complete their doctorates in two or three years. They have a fellowship.

courses were very good. This graduate course in astrophysics was the origin of the training and recruiting of many astrophysicists, both French and foreign.

After creating the graduate course in astrophysics in 1961, I found that leading these two teams of graduate studies was too heavy a load, so I gave up plasma physics in 1969. This experience gave me some training in both plasma physics and MHD and is the origin of several applications of this knowledge to astrophysics.

University studies were reorganized again in 1967. This afforded the chance to teach astrophysics at three levels: beginning, for undergraduate studies (optional), advanced, for undergraduate studies (optional), and, as mentioned above, graduate studies.

School of Astrophysics

Graduate courses in astrophysics and plasma physics attracted students. Whereas the numbers of girls and boys attending the courses in plasma physics were respectively 1 and 20, or something like that, there was roughly an equal number of girls and boys attending the astrophysics courses. Many of them remained in the field and had brilliant careers. There was little space in the Institut d'Astrophysique, and at the beginning, despite the fact that the number of young research workers was small, they were squeezed together on the second floor. In 1964 I started the necessary steps that led to the building of the Laboratoire d'Astrophysique de Meudon or LAM, inaugurated in 1972.

In Princeton I discovered the value of a weekly seminar. Nothing like that had been put into practice in the French astronomical community. I introduced it at the IAP as soon as I came back from the States in 1949. This gave a new life to the IAP.

I never noticed any differences between male and female students, except perhaps that, at the beginning of their careers, girls were more shy than boys. I cannot give names here. I do not remember all of them, and naming some might make the others jealous.

Today I find myself surrounded by the product of what I can call the "astrophysics school." In other words, there are now in France many people with whom I can talk about astrophysics; many of them have reached a level that I consider higher than mine at the same age. When I was young, traveling mainly to the States and taking part in meetings was not only a way of becoming known to foreign colleagues, but was necessary to find people to talk with. This time has passed.

About 250 students have taken my graduate courses. Perhaps two thousand students have taken my undergraduate course. I have often discovered that some of my younger colleagues have attented my lectures. Many of my former students are now in charge of astronomical life in different places. There are now groups of astrophysical theoreticians in Nice, Toulouse, Lyons, and Strasbourg (chronological order). Ramon Canal is at the origin of such a group in Barcelona.

This clearly shows that theoretical astrophysics now occupies its proper place in France.

Publications

Beginning in the fall of 1949, in order to prepare my lectures at the University, I had to get acquainted with the most recent literature. And it was also necessary, for the consistency of the lectures, to have at least some knowledge of the large variety of fields of physics involved in astrophysics. It was easy, years ago, when considering a physical property, to notice references to an obvious physical process. This is essentially the origin of the large number of papers (more than 200) that I have published. When I talked with Van de Hulst in the 1960s, he already noticed that I had not specialized in a narrow domain of astrophysics. In recent years, this rate of publishing has decreased. Does the writing of a paper at the present international level take more time? I would like to mention here what I consider my most important papers.

I first recall some of the work mentioned previously.

The obvious fact that the vertical scale of gravitational separation was much smaller than the radius of a white dwarf led me to the description of white dwarf structure (Schatzman 1946).

Similarly, to give a proof of the absence of hydrogen inside white dwarfs I found it possible to take into account the effect of screening on thermonuclear reaction rates (Schatzman 1948b). For this, I used knowledge of the Debye field that I had acquired in 1941. Great improvements in this theory were made by Van Horn & Salpeter (1969) and by DeWitt, Graboske & Cooper (1973). Many years later, I convinced Bernard Jancovici, from Orsay University, to apply better statistical methods to the treatment of the thermonuclear rate at high density. The result was an important paper by Alastuey & Jancovici (1978).

The conditions of wave propagation in the solar atmosphere, and my knowledge of the properties of shock waves, led me to a theory of the heating of the solar chromosphere and corona (Schatzman 1949). I knew nothing at that time about plasma physics and was unable to take into account the presence of magnetic fields.

The fact that after a nova outburst a star is still visible means that internal energy of the star is certainly larger than energy of the explosion. This means that the star has a small radius and is necessarily a white dwarf. I already had proposed the idea of a shock wave in novae (Schatzman 1948a). Reading a paper by Alpher on the origin of the elements, which included cross section estimates, led me to the description of the pp I cycle closing by the ^{3}He(^{3}He, pp)^{4}He reaction, and I considered the possibility of a detonation wave due to that reaction (Schatzman 1951). However, the difficulty of starting the detonation led me to a theory of novae based on the idea of instabilities generated by tidal effects in a white dwarf companion in a binary (Schatzman 1958). Furthermore, it raised the possibility of explaining the presence of rings in the gas cloud ejected by a nova. I greatly appreciate the fact that Sumner Starrfield studied these papers carefully before beginning his work on novae. It is well known that he studied the role of accretion. Why did I not consider that possibility? Herman Bondi and Fred Hoyle had produced an accretion theory that ignored the role of radiation. I have shown, at the 1952 Cambridge IAU meeting, that because of ionization due to radiation the rate of accretion was much smaller (Schatzman 1955). The consequence of this result is that I had in mind the idea that accretion was not important.

The polarization of light by interstellar matter was supposed to be due to the presence of anisotropic particles. However, there was no model of formation of such particles, and there was the question of the magnitude of the interstellar magnetic field. In a lecture at the IAP, Pol Swings had mentioned the presence of soot particles in the atmosphere of N-dwarfs. Subsequently, Cayrel and I considered the possibility of polarization of light by interstellar graphite particles (Schatzman & Cayrel 1954). May I mention again a case of nostalgia? In 1962 Hoyle and Wickramasinge made the same assumption but ignored my paper. We know today that this model is too simple and disagrees with the observed wavelength dependance of polarization.

After attending a remarkable lecture by O Struve in the 1950s at the Institut d'Astrophysique and reading his book on stellar evolution, I got the idea of looking at the question of loss of angular momentum. But losing angular momentum from stellar surfaces meant too much mass loss and was incompatible with our knowledge of stellar evolution and the absence of a big cloud around main sequence stars. When I began to learn plasma physics in 1956, I proposed the idea of spindown due to mass loss in the presence of a magnetic field (Schatzman 1959, 1962). I still think that the relatively large magnetic field above sunspots is more important than the weak one around the whole star.

Another product of my teaching of magnetohydrodynamics is the idea of an acceleration mechanism of cosmic rays by shock waves (Schatzman 1963).

Since my early years at the Observatory of Haute-Provence, I have been skeptical about the idea of complete quiescence inside the stars. I began to look at the effect of macroscopic diffusion on gravitational separation (Schatzman 1969), then I looked at the generation of turbulent diffusion by differential rotation (Schatzman 1970), considered the effect on lithium (Schatzman 1977)—but unfortunately used a bad model of giants, and finally got the idea of applying turbulent diffusion to give an explanation of the solar neutrino deficiency (Schatzman et al 1981). The oversimplified diffusion model led to results in disagreement with helioseismology data.

Talking with Roland Omnès in the late 1960s about his idea of annihilation in the early Universe, I gained at least some knowledge of the annihilation process. In the early 1980s, I was approached by Jean-Marie Souriau about his result (presently not accepted by cosmologists) concerning the presence of a missing zone of quasars. He found that to explain the space distribution of quasars it was necessary to assume a spherical, closed Universe, with a nonzero cosmological constant. His main idea was that half of the sphere was made of matter and half made of antimatter. From this picture I tried to see if the order of magnitude of the annihilation processes and its effect before galaxy formation would explain the size of the missing zone. I was surprised to find the proper order of magnitude. Why not try to do better? With Xavier Desert, we published a paper on " Physical constraints on the bi-partition of the Universe" (Desert & Schatzman 1986), where the theory of the missing zone was in agreement within 10% with the results of the group of Marseille-Luminy.

The difficulty of having a consistent model of lithium depletion (Schatzman & Baglin 1991), and discussions with Zahn, led me to take into account another physical process. Random gravity waves can generate a diffusion process. The first step is to obtain an estimate of the amplitude of gravity waves generated in the radiative zone by turbulent flow in the convective zone. The second step is to give an estimate of the diffusion coefficient. I started this work in 1991 and I am not yet through. This diffusion process provides a good contribution to the understanding of lithium depletion. But the problem of neutrino deficiency is much more difficult. Being a member of the GALLEX group since 1984, I have signed (with the other members) several papers concerning the results of neutrino counting. The first papers of the series were published in 1992 (GALLEX 1992a,b). The GALLium EXperiment uses a tank containing 30 tons of gallium, including 12 tons of ^{71}Ga (exactly 1.044×10^{29} ^{71}Ga atoms). The 1995 result is an average solar neutrino flux of 87 SNUs (SNU = solar neutrino unit = 10^{-36} ν cm^{-2} s^{-1}). This is in agreement with the results of the Russian-American experiment (SAGE), but in complete disagreement with the predictions of the Standard Solar Model (SMM) of about 130 SNUs. There is presently a preference for explaining the discrepancy between the experimental results [Homestake goldmine chlorine experiment, GALLEX and Kamiokande, the direct counting of high-energy neutrinos (above 7 MeV)] and the predictions of SSM by matter enhanced neutrino oscillations: the so-called MSW effect (Mikheyev, Smirnov, Wolfenstein effect; e.g. GALLEX 1992b). Considering that there are still many uncertainties in solar models, I think that, before taking into account the MSW effect, it is necessary to assume massless neutrinos and to try taking into account more physical processes and their effect on stellar structure, to explain the deficiency.

Life of a Scientist

Since 1957 I have taken part in a large number of colloquia, symposia, and meetings, and I have been invited abroad to give lectures or courses in summer or winter sessions. I was invited to give astrophysics lectures at the 1959 Summer Course of Les Houches and contributed to the organization of the 1966 Summer Course on High Energy Astrophysics. I was invited by Robert Oppenheimer to take part in the 1958 and 1964 Congrès Solvay, and I was invited again to the 1974 one. Leon Van Hove invited me to spend one year at CERN, and I came for the academic year from October 1969 to September 1970. Because of these opportunities, I feel that I have had a better relationship with the international scientific community than with the French one.

My ideas and my work have been appreciated by H Mineur, D Chalonge, D Barbier, B Strömgren, M Schwarzschild, and L Spitzer, and A Danjon gave me the task of teaching astrophysics at the university level. But because I was kept away from scientific responsibility by French astronomers of an older generation than mine, and felt that most of them never gave me proper credit for my work, I have always had the impression of being kept at the border of the French scientific community. This explains why it was not until 1971 that I had for the first time a real feeling of recognition in France. I got the Robin award from the French Physical Society and delivered a speech at the meeting of the Society in Evian. Although in 1966 I had received from the Académie des Sciences the award in memory of the French scientists killed by the Nazis during the war, I could not forgive the fact that the short description of my scientific work, giving the reasons for the award, was completely wrong: It referred to a research field that I had never touched! I felt misunderstood. In 1975, I got the Holweck award given both by the Institute of Physics of Great Britain and the French Physical Society. I gave a lecture in Cambridge and, following what I considered as a great honor, I was elected President of the French Physical Society (1976–1977). That same year, I founded the French Society of Professional Astronomers (SFSA or *Société Française des Spécialistes d' Astronomie*), which is now an effective association.

After the student revolution of May 1968 (in which I took much part) the number of meetings at the University increased so much that I felt that my duties were becoming more and more heavy. On January 1, 1976, I again became a member of CNRS as *Directeur de Recherches.* I gave up the direction of my small laboratory of theoretical astrophysics and moved part time to the Nice Observatory. It gave me the chance to be close to my four-year-old son and his mother Annie Baglin. I appreciated the working conditions in Nice. When he was Director of Nice Observatory, JC Pecker had organized an excellent library there. I was delivered from all obligations, which meant being a full-time research worker. This marked the end of lecturing and the beginning of more efficient research activity again.

With the new generation, I had other relationships with the French scientific community, and it is under these conditions that I got the Gold Medal of the CNRS in 1983. The situation in the Académie des Sciences had changed and I was presented as a candidate by JC Pecker; I have been told that I received excellent support from several physicists, and I was elected in 1985. I was named a member of the High Scientific Council of Paris Observatory for four years in 1989 and named a member of the Advisory Commitee of the CNRS High Administration of Astronomy and Geophysics (INSU or Institut National des Sciences de l'Univers) in 1994. It was the "young" generation who organized a large meeting in 1990 to celebrate my seventieth aniversary. The proceedings have been published by *Union Rationaliste.*[23] Jean-Claude Pecker (1991) gave a complete description of my scientific career; Jean-Paul Zahn (1991) gave the impression of beginners in the 1950s to 1960s. I want to quote his testimony concerning my relationship with the astronomical community:

> We were admiring him also for his commitments in politics, trade unionism, and philosophy. We knew these commitments only indirectly, as he had not allowed himself to proceed to any kind of proselytism in the working place: Commitments that have not always been understood and have produced strong enmities. Should I confess that we are still deeply shocked by the late recognition of his role by the official authorities of our community?

This confirms the fact that professionally I never mixed science and politics, and it justifies my feelings concerning my relationship with the community. I conclude with the 1993 Jubilee colloquium "Physical Processes in Astrophysics," organized by Ian Roxburgh with the chairmanship of Lyman Spitzer, and, to my great surprise, announced by a wonderful poster drawn by Pecker. It was terribly moving.

Things have changed since the old times!

I retired in the fall of 1989. I still have an office at Meudon Observatory, I am still the leader of young beginners, and I am trying to remain as active a scientist as possible!

[23]The *Union Rationaliste* was founded by Paul Langevin in 1932. I am presently its president. The *UR* is a small association that bears some similarity with the American Association, *CSICOP,* whose president is Paul Kurtz, Professor emeritus of Philosophy, State University of New York at Buffalo.

Returning to Philosophy

We have in France the philosophical heritage of Auguste Comte's positivism. We can smile now when recalling that Comte said that we shall never know the chemical composition of stars, as we shall never be able to go there and take a sample. It is already remarkable, after all, that he was assuming that stars were made of matter! But this idea that we can know only what we can touch has not disappeared. There are still many scientists in France who are skeptical about any kind of theory. We can naturally smile again when we look at the book of Georges Bruhat, *Les Etoiles* (1937), and notice in the foreword that he was not going to write about the interiors of the stars, as we shall never go there to touch them. Remember that Eddington's book was published in 1926 and that there were many papers on the subject in *Ap. J., MNRAS,* and *Z. Astrophys.* But in 1971, a few months before moving from the Institut d'Astrophysique to the new laboratory that I created in Meudon (LAM, now world renowned), I was talking with André Lallemand, then Director of the IAP, about people who were moving with me to Meudon. Suddenly he said : "You, theoreticians, you are not concerned with the important things." I was shocked and I replied by mentioning some of my subjects of research. He stopped me saying: "Oh! You are different, but the others. . . ". That sentence was painful for me. It expressed such a strong prejudice against theory that I find it necessary to emphasize this. A large fraction of French astronomers are "instrumentalists." Some of them are remarkable inventors and are essentially interested in instruments that bring in new data. But they do not understand that this is not enough. We must also be concerned equally with the physical problems that can be solved by obtaining proper data. Of course not every one is like this. We can consider, for example, the case of Lacroute, a man of the same generation as Lallemand. When I met Lacroute for the first time in 1946, he questioned me with great interest about the work of Bethe on thermonuclear energy sources. He is behind the *HIPPARCOS* space program , which is producing an incredible amount of data (luminosity and, for binaries, better values of masses) important for the understanding of stars.

Message

I have always tried to transmit my message. Astrophysical objects are the seat of a large variety of physical processes. Understanding the phenomena rests on the knowledge and the understanding of those physical processes. Philosophical ideas play an important role. The philosophical conviction of the reality of the world surrounding us is important; it inspires the search for these processes that we do not see, but which are the origin of what we observe. This intellectual behavior is so deeply rooted in me that it has always been working unconsciously. I can even say that it was more by an instinctive materialistic attitude (Louis de Broglie used the word *realistic*, but with the same meaning) than by any philosophical reasoning that I considered gravitational separation in white dwarfs or electron screening in thermonuclear reactions. I think that it is important to keep in mind the occasional contradictions between models and observational data. It is a glory to improve the fit between theory and observations, but there always remains some discrepancy. I think that the most important attitude is to find which forgotten physical processes are responsible for something we do not understand. I have tried to transmit this message to the scientific community.

ACKNOWLEDGMENTS

Janet Rountree was kind enough to translate my "Frenglish" into English.

Literature Cited

Alastuey A, Jancovici B. 1978 *Ap. J.*
Desert X, Schatzman E. 1986. *Astron. Astrophys.* 158:135
DeWitt HE, Graboske HC, Cooper MJ. 1973. *Astrophys. J.* 181:439
GALLEX Collab., Anselmann P, et al. 1992a. *Phys. Lett. B* 285:376
GALLEX Collab., Anselmann P. et al. 1992b. *Phys. Lett. B* 285:390
Pecker JC. 1991. *Les Cahiers Rationalistes.* 459–60:256
Pecker JC, Schatzman E. 1957. *Astrophysique Générale.* Paris: Masson
Schatzman E. 1946. *Ann. Astrophys.* 8:143
Schatzman E. 1948a. *Bull. Cl. Sci. Acad. R. Bel.* 34:828
Schatzman E. 1948b. *J. Phys.* 9:46
Schatzman E. 1949. *Ann. Astrophys.* 12:203
Schatzman E. 1951. *C. R. Acad. Sci.* 232: 1740
Schatzman E. 1955. In *Gas Dynamics of Cosmic Clouds,*, ed. HC Van de Hulst, JM Burgers, p. 193. Amsterdam: North-Holland
Schatzman E. 1958. *Ann. Astrophys.* 21:1
Schatzman E. 1959. *Ann. Astrophys. Suppl.* 8:129
Schatzman E. 1962. *Ann. Astrophys.* 25:18
Schatzman E. 1963. *Ann. Astrophys.* 26:234
Schatzman E. 1969. *Astron. Astrophys.* 3:231
Schatzman E. 1970. *CERN 70-31, Lectures given in the Academic training of CERN,* p. 50
Schatzman E. 1977. *Astron. Astrophys.* 55:151
Schatzman E. 1991. In *Solar Interior and Atmosphere,* ed. AN Cox, WC Livingson, MS Matthews, p. 192. Tucson: Univ. Ariz. Press, p. 192
Schatzman E, Baglin A. 1991. *Astron. Astrophys.* 249:125
Schatzman E, Cayrel R. 1954. *Ann. Astrophys.* 17:555
Schatzman E, Maeder A, Angrand F, Glowinski R. 1981. *Astron. Astrophys.* 96:1
Van Horn H, Salpeter EE. 1969. *Ap. J.* 155:183
Zahn JP. 1991. *Les Cahiers Rationalistes.* 459–60:285

Annu. Rev. Astron. Astrophys. 1996. 34:35–73

COROTATING AND TRANSIENT SOLAR WIND FLOWS IN THREE DIMENSIONS

J. T. Gosling

Los Alamos National Laboratory, Los Alamos, New Mexico 87545

KEY WORDS: corotating interaction regions, shock disturbances, coronal mass ejections, magnetic reconnection, *Ulysses*

ABSTRACT

Two types of flows dominate the large-scale structure of the solar wind: corotating flows and transient disturbances. Corotating flows are associated with spatial variability in the coronal expansion and solar rotation, whereas transient disturbances are associated with episodic ejections of material into interplanetary space from coronal regions not previously participating in the solar wind expansion. *Ulysses'* recent epic journey over the poles of the Sun has provided new insights on the three-dimensional nature of both corotating flows and transient disturbances in the solar wind and their evolution with heliocentric distance and latitude. This paper provides a simple physical description of the origins and dynamics of both of these types of solar wind flows, highlighting new understanding gained from the unique *Ulysses* high-latitude observations.

INTRODUCTION

The coronal expansion that forms the solar wind is nonuniform because it is modulated by the Sun's complex magnetic field. High-speed flows in the quiescent (i.e. nontransient) wind usually originate in coronal holes that extend equatorward from the magnetic poles of the Sun (e.g. Krieger et al 1973), whereas low-speed flows tend to originate within the relatively dense coronal streamers that straddle magnetic neutral lines in the solar atmosphere (e.g. Gosling et al

0066-4146/96/0915-0035$08.00

1981). Solar rotation causes high- and low-speed flows to interact with one another in interplanetary space at low heliographic latitudes, producing large-scale compressive structures that corotate with the Sun (e.g. Hundhausen 1972).

As the coronal magnetic field evolves during the $\sim$11-year solar activity cycle the pattern of coronal expansion varies. The most dramatic changes in the coronal expansion occur during coronal mass ejections (CMEs) in which solar material is propelled outward into interplanetary space from closed magnetic field regions in the solar atmosphere not previously participating in the solar wind expansion [see reviews by Hundhausen (1988, 1996) and by Kahler (1987, 1988)]. CMEs produce transient disturbances in the solar wind that propagate to the outer boundary of the heliosphere.

The solar wind at low heliographic latitudes is dominated by the above two types of flows: 1. large-scale, corotating structures associated with spatial variability in the coronal expansion and solar rotation and 2. transient disturbances associated with CMEs. In his treatise on the solar wind, Parker (1963) anticipated both corotating structures and transient solar wind flows. Direct observations of the solar wind at low latitudes over a wide range of heliocentric distances in the intervening 32 years have added considerably to our knowledge of these types of flows. These in situ measurements of the solar wind have been complemented by new information on coronal structure obtained from white-light coronagraphs and soft x-ray telescopes flown on satellites. In concert with this growth in observational knowledge, gas dynamic and magnetohydrodynamic (MHD) numerical codes of varying degrees of sophistication have been developed to model both corotating and transient solar wind flows. By 1990 a reasonably good physical understanding had been obtained of both types of flows at low heliographic latitudes, although not all of the details had been worked out or agreed upon.

Until recently, however, our understanding of corotating and transient flows was essentially two dimensional, i.e. we basically understood how these flows evolve as functions of radius and solar longitude, but we lacked direct information on their three-dimensional (3-D) structure and their evolution as a function of solar latitude. *Ulysses*' epic journey over the poles of the Sun has provided direct observations of the solar wind at heliographic latitudes well out of the ecliptic plane for the first time (e.g. Smith et al 1995). These observations have given us new insights on the physical nature of both corotating structures and transient disturbances in the solar wind. In this paper we present simple physical descriptions of both corotating flows and CME-driven disturbances in the solar wind, highlighting the new understanding of 3-D structure provided by the *Ulysses* observations at high latitudes. Our goal is to place the new *Ulysses* observations into context by furnishing short tutorials on these subjects.

COROTATING FLOWS IN THE SOLAR WIND

Solar Wind Stream Structure

Large excursions in solar wind speed, density, and pressure in the quiescent wind near Earth generally conform to a characteristic pattern of variability that has come to be known as solar wind stream structure (e.g. Snyder et al 1963). This pattern is illustrated in Figure 1, which shows the result of superposing data from 25 streams of moderate amplitude, keying on the peaks in solar wind density on their leading edges. Speed rises rapidly on the leading edges of the streams and falls back to average values much more slowly. The density peaks as the speed rises, and it falls to lower than average values as the speed declines. The gas pressure also maximizes as the speed rises, and the flow is deflected there, first to the west (i.e. in the sense of planetary motion about the Sun) and

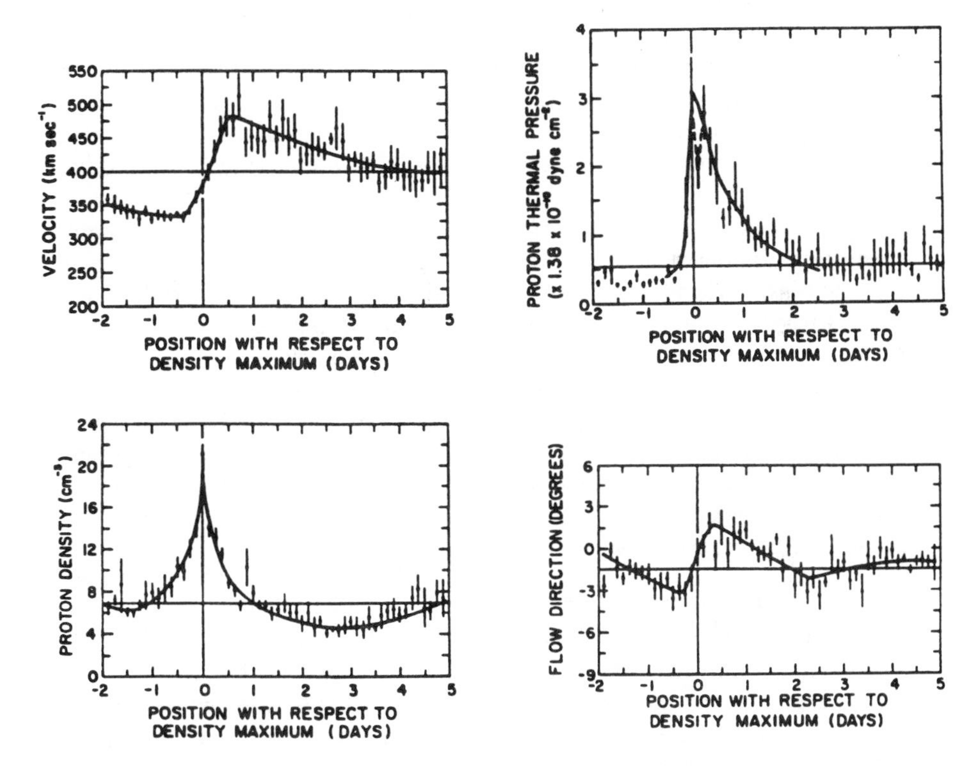

Figure 1 Characteristic temporal variations in flow speed, density, pressure, and flow azimuth (negative is in sense of corotation with Sun) associated with solar wind streams at Earth's orbit. These are average profiles of 25 streams that contained relatively large density compressions. (Adapted from Gosling et al 1972.)

then to the east. This pattern of variability is the inevitable consequence of the evolution of a stream as it progresses outward from the Sun, as discussed below.

Stream Steepening and Interaction Regions

Flows of different speeds become radially aligned at low heliographic latitudes as the Sun rotates (with a period of ~ 27 days as observed from Earth). A snapshot of the speed of the wind as a function of heliocentric distance at any particular longitude in the equatorial plane at some initial time might appear as the profile at the left in the upper portion of Figure 2. Parcels of plasma within the "stream" originate from different longitudes on the Sun at different times. These parcels cannot interpenetrate by virtue of the inclined magnetic field that is frozen into the flow. At later times, the faster-moving plasma at the crest of the stream overtakes and collides with slower plasma ahead while outrunning slower-moving plasma from behind. Thus high-speed streams steepen with heliocentric distance. Material within the stream is rearranged as the stream steepens; parcels of plasma on the rising-speed portion of the stream are compressed, causing an increase in pressure there (lower portion of Figure 2), while parcels on the falling-speed portion of the stream are increasingly separated, producing a rarefaction. The temporal variations of solar wind speed, density, and pressure observed near Earth and shown in Figure 1 are readily identified with such evolved streams.

It is common to refer to the compression on the leading edge of a high-speed stream as an interaction region. Being a region of high pressure it expands into the plasma ahead and behind at the fast mode speed, c_f. The leading edge of the interaction region is called a forward wave because it propagates through the plasma in the same sense as the flow (i.e. antisunward); the trailing edge is called a reverse wave because it propagates back toward the Sun in the solar wind rest frame. The propagation of these waves and the overall pressure gradients behind them produce an acceleration of the slow wind ahead of the stream and a deceleration of the fast wind within the stream itself. The net result of the interaction is to limit the steepening of the stream and to transfer momentum and energy from the fast wind to the slow wind.

Shock Formation Associated with Interaction Regions

When the amplitude of a high-speed stream is less than about $2c_f$, it will gradually damp out with increasing heliocentric distance as just described. However, when the amplitude is greater than $2c_f$, the stream initially steepens faster than the region of high pressure can expand so that the latter at first squeezes down with increasing heliocentric distance (Hundhausen 1973). The nonlinear rise in pressure associated with this squeezing eventually causes the forward and reverse waves bounding the interaction region to steepen into shocks. The shocks

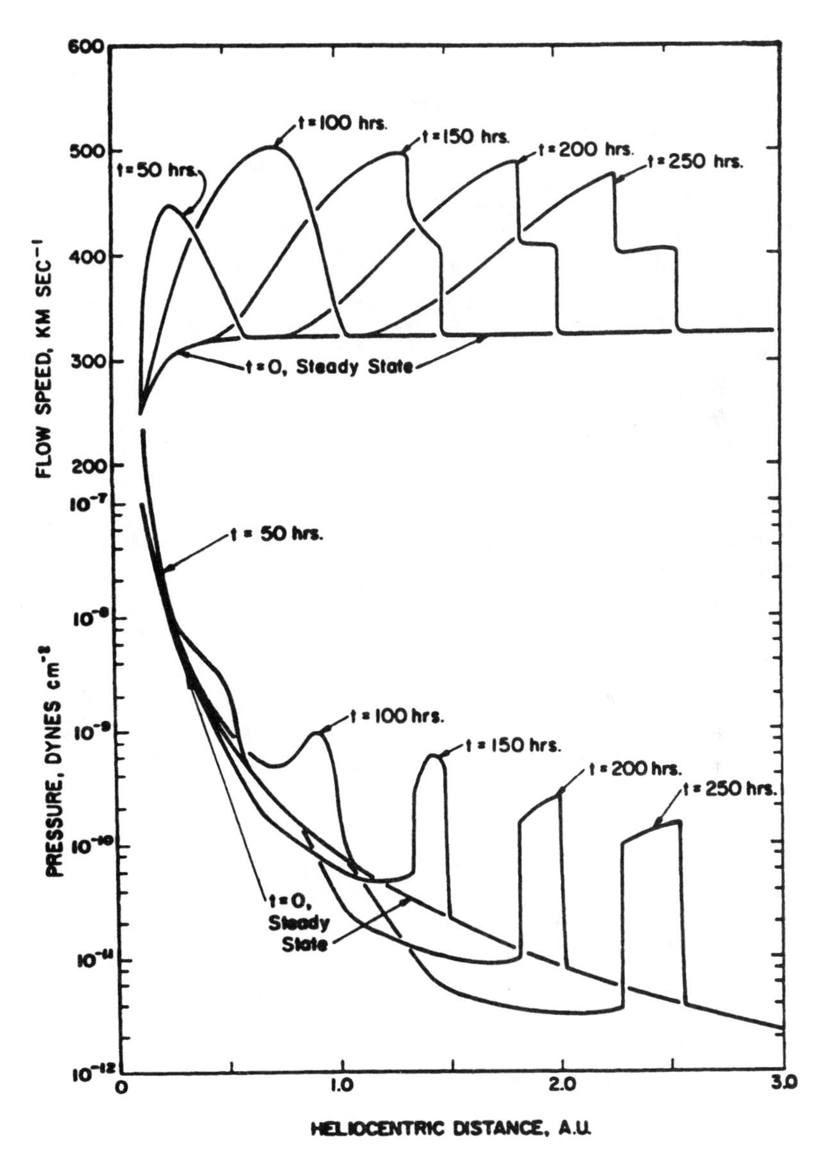

Figure 2 Snapshots of solar wind flow speed and pressure as functions of heliocentric distance at different times during outward evolution of a high-speed stream as calculated from a simple 1-D gas-dynamic code. (Adapted from Hundhausen 1973.)

can propagate into the surrounding plasma at speeds considerably greater than c_f, enabling the interaction region to expand once shock formation occurs. The major accelerations and decelerations associated with stream evolution now occur discontinuously at the shock surfaces, giving the stream/speed profile the appearance of a double sawtooth wave. Observations indicate that few solar wind streams steepen sufficiently inside 1 AU to cause shock formation by the time the streams cross Earth's orbit (e.g. Gosling et al 1972, Ogilvie 1972). Nevertheless, because c_f generally decreases with increasing heliocentric distance, most stream interaction regions are bounded by a pair of shocks at heliocentric distances beyond about 2 AU (e.g. Hundhausen & Gosling 1976, Gosling et al 1976b, Smith & Wolfe 1976). At low latitudes in the very distant heliosphere, stream amplitudes are greatly reduced, and the dominant structures are expanding compression regions that merge with one another (e.g. Burlaga 1983, 1984).

Stream Evolution in Two Dimensions

To this point we have considered stream evolution along a fixed radius extending outward from the Sun in the equatorial plane. When the coronal expansion is time stationary but spatially variable, stream evolution proceeds in the same manner at all longitudes; however, the state of evolution is a function of longitude. As a result, as illustrated in Figure 3, an interaction region in the equatorial plane becomes aligned with an Archimedean spiral, having a pitch intermediate between that of the fast and slow flows that produce it. The entire pattern of interaction corotates with the Sun, and the region of compression is known as a corotating interaction region or CIR (e.g. Smith & Wolfe 1976). It is worth emphasizing that only the pattern actually corotates—each parcel of solar wind plasma moves outward nearly radially as indicated by the arrows in the figure. Because CIRs are inclined relative to the radial direction, the forward and reverse waves bounding the CIRs have both radial and azimuthal components of propagation. With increasing heliocentric distance the forward waves propagate antisunward and in the direction of planetary motion about the Sun (westward), while the reverse waves propagate sunward (in the plasma rest frame) and eastward. Thus, slow wind is accelerated and deflected westward as it is overtaken by the forward wave bounding a CIR, and the fast wind is decelerated and deflected eastward as it encounters the reverse wave. This accounts for the small west-east deflections commonly observed on the leading edges of high-speed streams in the ecliptic plane (Figure 1). Indeed, those deflections have long been used to infer CIR orientations in the ecliptic plane (e.g. Siscoe et al 1969, Gosling et al 1978). One consequence of the transverse deflections is that the plasma partially relieves the stresses induced by kinematic steepening by simply slipping aside. Hence, real solar wind streams steepen less rapidly than predicted by the simple one-dimensional (1-D) calculation shown in Figure 2.

The Heliospheric Current Sheet

High-speed streams in the solar wind tend to be unipolar in the sense that the interplanetary magnetic field (IMF) within a stream points either entirely away from or entirely toward the Sun along an Archimedean spiral (e.g. Wilcox & Ness 1965). The magnetic polarity of a high-speed stream reflects the polarity of the coronal hole from which it originates. Field polarity reversals occur within the dense low-speed flows between streams that originate within coronal streamers. These reversals correspond to crossings of the heliospheric current sheet (HCS) that wraps around the Sun and that maps back to the solar magnetic equator. To zeroth order the Sun's large-scale magnetic field is that of a dipole, tilted with respect to its rotation axis. The tilt varies with the phase of the solar activity cycle. As illustrated in the left portion of Figure 4, at solar minimum the dipole tends to be aligned nearly with the rotation axis, whereas on the declining phase of the solar cycle it tends to be inclined substantially to the

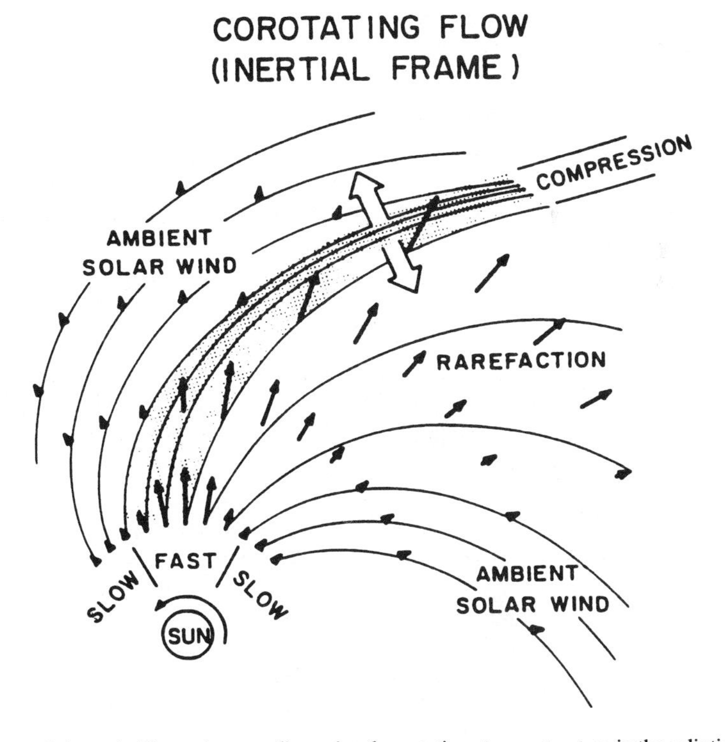

Figure 3 Schematic illustrating two-dimensional corotating stream structure in the ecliptic plane in the inner heliosphere. (From Pizzo 1978.)

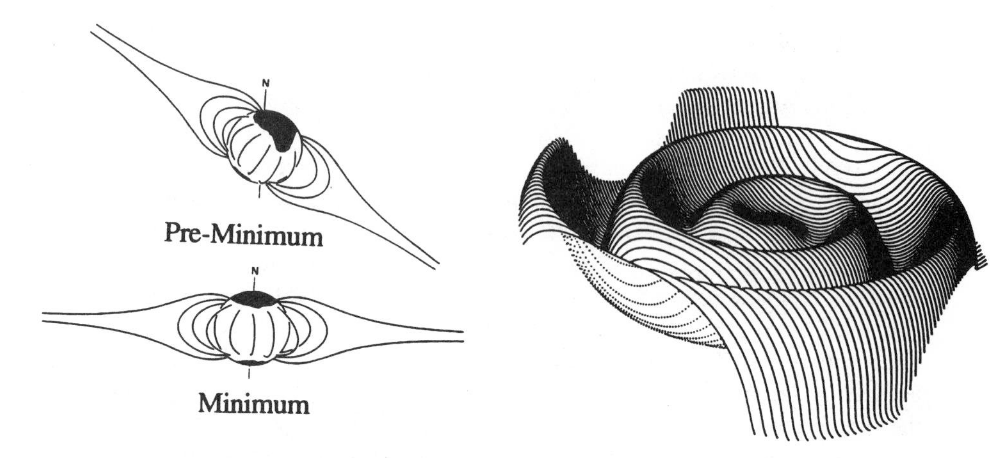

Figure 4 Schematics illustrating the changing tilt of the solar magnetic dipole and coronal structure relative to the rotation axis of the Sun (*left*) and the configuration of the heliospheric current sheet in interplanetary space when the solar magnetic dipole is inclined substantially relative to the solar rotation axis (*right*). (Adapted from Hundhausen 1977 and Jokipii & Thomas 1981.)

rotation axis. Near solar maximum the solar field is sufficiently complex that the dipole concept is probably not useful.

Near solar minimum, when the solar magnetic dipole and the rotation axis are nearly aligned, the HCS in interplanetary space is often nearly planar and roughly coincides with the solar equatorial plane. However, at times when the dipole tilt is substantial, the HCS becomes warped into a structure resembling a twirling ballerina's skirt (e.g. Jokipii & Thomas 1981), as illustrated by the sketch in Figure 4. Successive outward folds in the HCS correspond to successive solar rotations and are separated by about 4.7 AU when the flow speed at the current sheet is 300 km s^{-1}. The maximum solar latitude attained by the current sheet in this simple picture is the same as the tilt of the dipole to the rotation axis. As is seen below, CIRs tend to be tilted in the same sense as is the HCS.

The Ulysses *Mission*

Ulysses, which is a joint European/United States space mission, was launched on October 6, 1990 into an orbit that used a Jupiter swingby in February 1992 to take the spacecraft nearly over the south and north poles of the Sun in September 1994 and July 1995, respectively. The launch occurred shortly after solar activity maximum, and solar activity remained relatively high throughout the near-ecliptic cruise to Jupiter. Solar activity declined noticeably as *Ulysses* made its initial traversal to high southern heliographic latitudes. The near-polar passes occurred just prior to solar activity minimum, expected sometime in 1996.

Plasma Observations During the First Five Years of the Ulysses *Mission*

Figure 5 provides an overview of selected solar wind plasma parameters from the Los Alamos/Jet Propulsion Laboratory plasma experiment on *Ulysses* (Bame et al 1992) from instrument turn-on on November 17, 1990 until shortly after the north polar passage almost five years later. During the near-ecliptic cruise to Jupiter and shortly thereafter the experiment observed a complex stream structure that was a mixture of corotating flows associated with coronal holes and streamers and transient flows associated with CMEs. However, beginning in

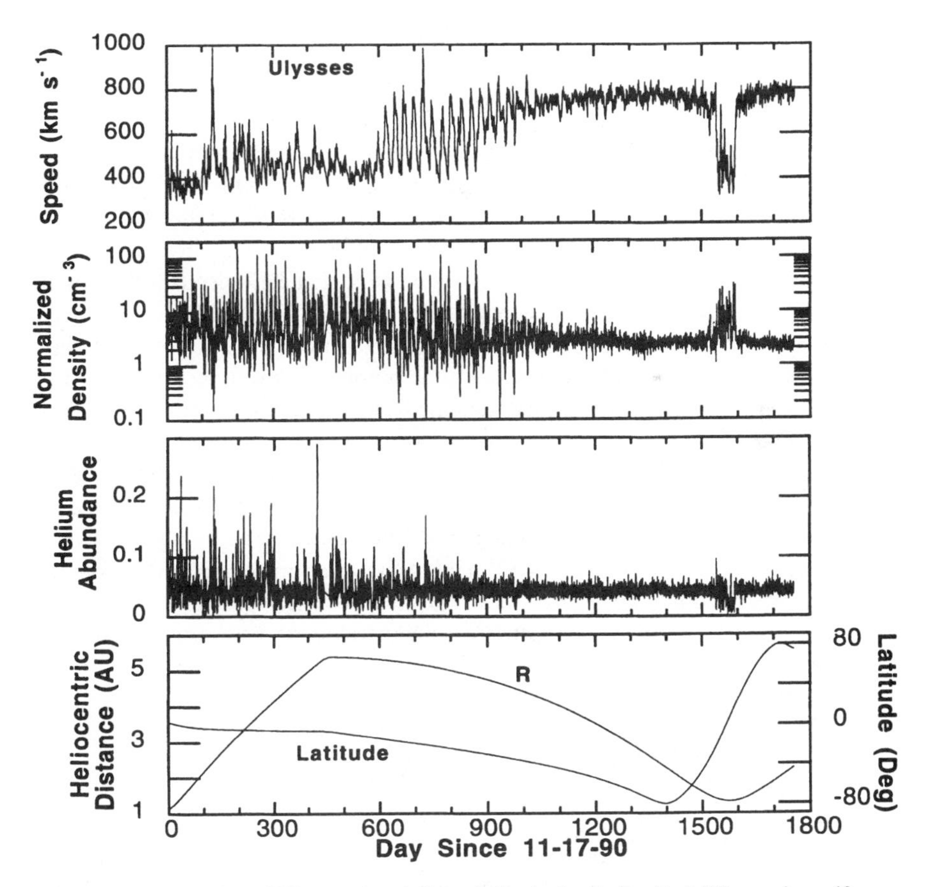

Figure 5 An overview of *Ulysses* solar wind speed, density (normalized to 1 AU assuming uniform spherical expansion), and helium abundance (relative to H^+) from instrument turn-on on November 17, 1990 through September 6, 1995. The bottom panel shows the changing radius and latitude of the spacecraft.

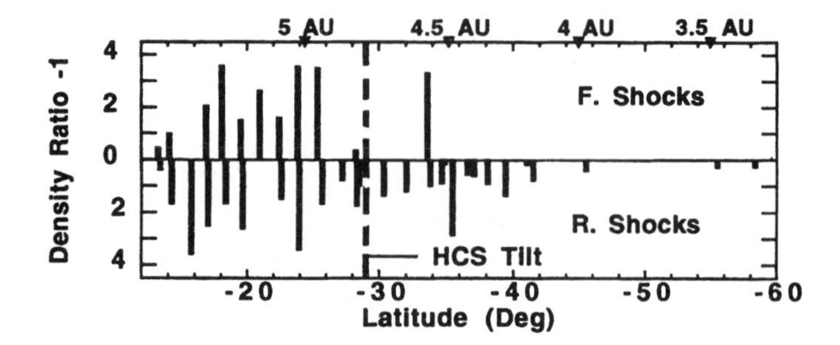

Figure 6 Bar diagram of shock "strength" vs latitude for corotating shocks associated with the stream from the southern coronal hole observed by the plasma experiment on *Ulysses* during its initial transit to high southern latitudes. Heliocentric distance of the observations is indicated by the triangles at the top of the plot.

July 1992 (near day 600), as *Ulysses* moved poleward of S14°, the instrument detected a single, broad, high-speed stream interleaved with low-speed flow each solar rotation for 15 successive solar rotations (Bame et al 1993). At a latitude of ~S36° the low-speed flows vanished and a comparatively uniform high-speed flow was detected that persisted to higher latitudes (Phillips et al 1994). Magnetic field measurements (Smith et al 1993) establish conclusively that both the recurring stream and the high-speed flow above S36° originated in a large coronal hole that covered the southern polar region. In contrast, the interleaved low-speed flows originated in the band of coronal streamers centered on the magnetic equator. The combined plasma and field observations have been used to infer that the the coronal streamer belt and the embedded heliospheric current sheet were inclined ~29° to the equatorial plane at the time of *Ulysses*' southward excursion (Bame et al 1993, Smith et al 1993), consistent with measurements of the magnetic field in the solar photosphere (Hoeksema 1995). By the time of *Ulysses*' rapid traversal back across the solar equator to high northern latitudes early in 1995, the inclination of the streamer belt to the equator was ~10°, and variable wind speed was observed in a band ~43° wide, centered roughly on the solar equator.

Corotating Shock Observations During Ulysses' *Transit to High Latitudes*

Ulysses' initial transit to high southern latitudes occurred at a heliocentric distance of ~5 AU, where the waves bounding corotating interaction regions are usually shocks. Indeed, at moderately low heliographic latitudes the CIR associated with the recurrent stream discussed above was usually bounded by a forward-reverse shock pair. This is illustrated in Figure 6, which summarizes corotating shocks observed by *Ulysses* from July 1992 when the spacecraft was at S12° through April 17, 1994 when the spacecraft was at S60°. The figure shows the ratio of downstream to upstream density minus one as a function of heliographic latitude. (The density ratio is a measure of shock strength and is inversely proportional to the ratio of upstream to downstream normal flow speeds in the shock frame.) Forward shocks are plotted above the horizontal axis and reverse shocks are plotted below the axis. Of the 10 CIRs observed equatorward of S26°, which was slightly less than the tilt of the solar magnetic dipole at this time, 8 were bounded by forward-reverse shock pairs. The forward wave was not a shock in one of the other events, while the reverse wave was not a shock in the other. The most surprising results demonstrated in the figure, however, are that (*a*) *Ulysses* encountered only two CIR forward shocks poleward of S26°, while (*b*) it continued to encounter reverse shocks regularly up to a latitude of S42° and sporadically thereafter up to S58°, well poleward of the latitudes where low-speed flow was observed (Gosling et al 1993). These results were not specifically predicted prior to *Ulysses*' journey; in fact, exactly the opposite was predicted in one case (Burlaga 1986). Nevertheless, as we show below, the *Ulysses* results have a natural explanation in terms of an existing 3-D model of stream evolution.

Flow Deflections and Shock Propagation in Three Dimensions

Flow deflections observed downstream from the corotating shocks are crucial to understanding the physical origin of the above effects. The left panel of Figure 7 shows selected solar wind plasma parameters for a 6-day interval encompassing the leading edge of the stream in November, 1992 when *Ulysses* was at 5.12 AU and S19.5°. The forward shock that bounds the leading edge of the CIR is distinguished in the figure by abrupt increases in speed and pressure on DOY 307; the reverse shock that bounds its trailing edge is distinguished by an abrupt increase in speed and an abrupt decrease in pressure on DOY 310. Because both shocks were convected away from the Sun by the supersonic flow of the wind, *Ulysses* sampled the region downstream of the forward shock after its passage and the region downstream of the reverse shock prior to its passage. Of particular interest here are changes in flow direction associated with the CIR. Downstream of the forward shock the flow turned northward and westward, while downstream of the reverse shock the flow turned eastward and southward. These flow changes indicate that not only was the forward shock propagating antisunward and westward as expected, but also equatorward; likewise, not only was the reverse shock propagating sunward (in the plasma rest frame) and eastward as expected, but also poleward. Flow deflections observed for other recurrences of the CIR equatorward of S26° demonstrate that this result was general: The forward waves were preferentially propagating antisunward, westward, and equatorward, while the reverse waves were preferentially propagating sunward, eastward, and poleward.

The reverse shocks tended to get weaker as *Ulysses* moved poleward of S26°, and many of them had no obvious associations with forward waves of any sort. The *right* panel of Figure 7 shows plasma parameters for an interval encompassing the last regularly observed reverse shock at S41.5°. Not only was there no accompanying forward shock, but the reverse shock itself was relatively poorly distinguished in the data (compare this with the low-latitude event on the *left*). Nevertheless, small eastward and poleward deflections of the flow were observed downstream (earlier in time) of the shock. Similar and larger poleward

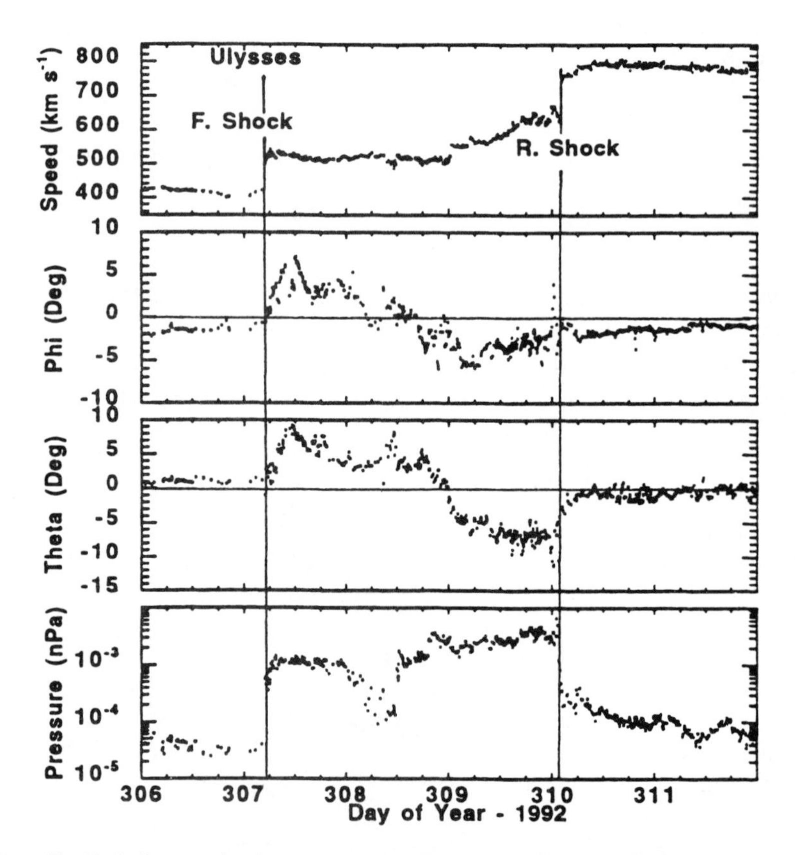

Figure 7 (*Left*) A corotating interaction region observed by *Ulysses* at 5.12 AU and S19.5°. Parameters plotted from top to bottom are solar wind flow speed, azimuthal and meridional flow angles (positive for flow in the direction of planetary motion about the Sun and equatorward, respectively), and proton thermal pressure. (*Right*) A weak reverse shock observed at 4.18 AU and S41.5°. Parameter scaling is as on the left. (From Gosling et al 1995b.)

deflections were measured downstream of all but two of the reverse shocks observed poleward of S26°, indicating that most of the reverse shocks at high latitudes were propagating poleward, as at lower latitudes (Gosling et al 1993).

North-South CIR Tilts

It is thus clear what caused the shock pattern apparent in Figure 6: The corotating interaction region formed at low and intermediate southern latitudes had a substantial north-south tilt. The tilt was such that the forward shock associated with the CIR was usually propagating equatorward and thus was not usually observed at latitudes higher than where the interaction between fast and slow wind was occurring. On the other hand, the reverse shock associated with the CIR was usually propagating poleward and thus was able to propagate to latitudes considerably poleward of where it was originally formed. The shocks

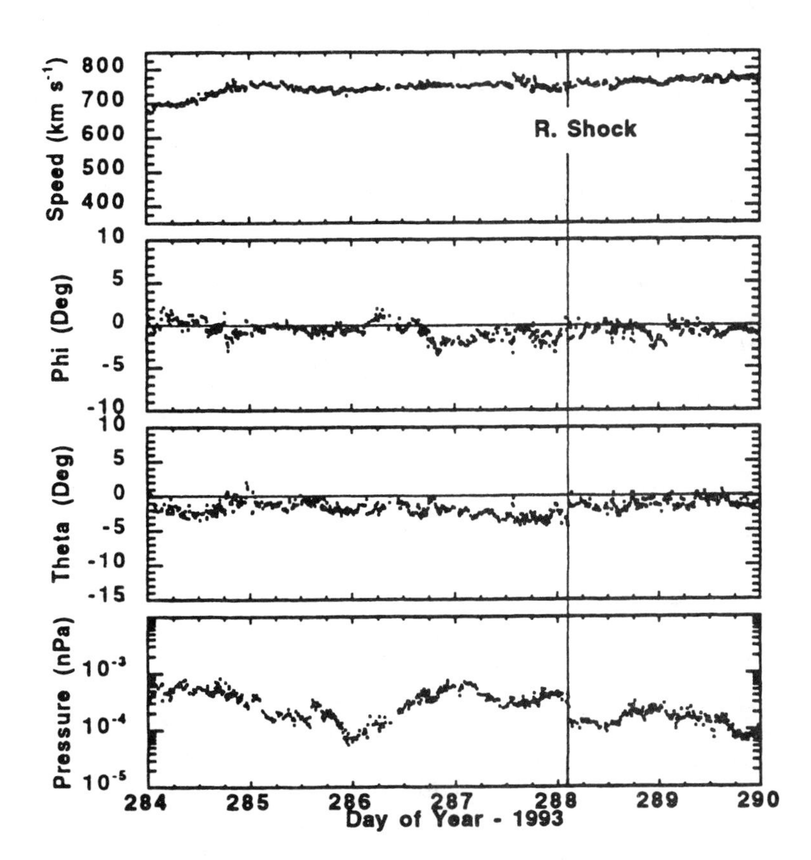

Figure 7 (*continued*)

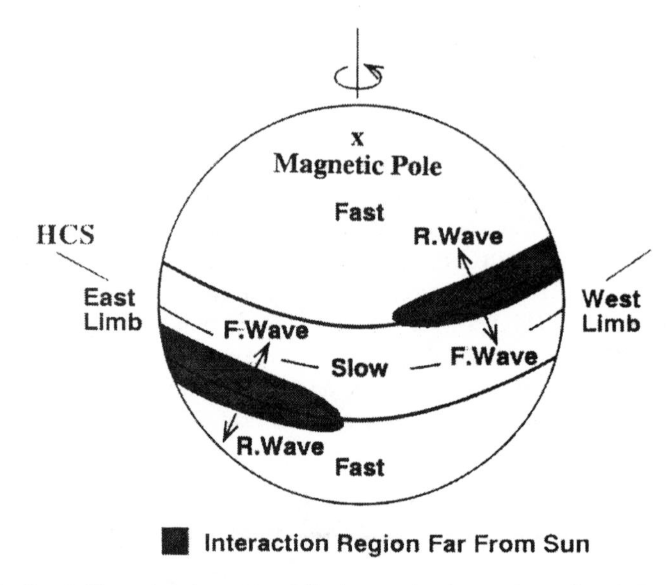

Figure 8 Sketch illustrating the origin of tilted corotating interaction regions in interplanetary space in terms of a tilted magnetic dipole and solar wind structure close to the Sun. The interaction regions form well away from the Sun. (From Gosling et al 1993.)

were strongest at latitudes where the interaction was directly driven, and they weakened as they propagated to higher and lower latitudes.

It had previously been suggested that CIRs should have north-south tilts that are opposed in the northern and southern hemispheres (McNutt 1988; Pizzo 1991, 1994a, 1994b). Indeed, evidence for opposed CIR tilts can be found in systematic north-south flow variations seen in the distant heliosphere between 20 and 25 AU by *Voyager 2* (Lazarus et al 1988), although the *Voyager* flow observations have also been interpreted in terms of large-scale vortex streets (Burlaga 1990; Siregar et al 1992, 1993), an interpretation the present author does not favor. More recently, we have found evidence for opposed CIR tilts in the opposite solar hemispheres in *Ulysses* data obtained during its rapid northward traverse across the solar equator (Gosling et al 1995d). These opposed tilts arise because, as already noted, 1. the solar wind expansion near the Sun is controlled by the solar magnetic field, 2. the solar magnetic dipole is often tilted relative to the rotation axis of the Sun, and 3. solar rotation drives CIRs. The simple sketch shown in Figure 8 illustrates how these factors combine to produce opposed north-south CIR tilts in the opposite solar hemispheres. The sketch shows a band of slow wind girding the Sun and inclined relative to the solar equator. This band is surrounded on either side by regions of fast wind. We associate the slow wind with the coronal streamer belt that is centered roughly on the magnetic equator and the fast wind with coronal holes that are centered on the magnetic poles. The slow wind band is inclined relative to the solar equator because of the tilt of the solar magnetic dipole relative to the rotation axis of the Sun; its inclination is the same as that of the HCS that is embedded near its center. As the Sun rotates, the fast wind overtakes the slow wind in interplanetary space along interfaces that are inclined relative to the solar equator in the same sense as is the band of slow wind. The interfaces, and the CIRs in which they are embedded, have opposite north-south tilts in the northern and southern solar hemispheres. With increasing heliocentric distance in both hemispheres, the forward waves bounding the CIR propagate westward and equatorward and eventually across the equator into the opposite hemisphere, while the reverse waves propagate eastward and poleward and eventually to latitudes above the band of slow solar wind flow.

Modeling CIRs in Three Dimensions

A variety of gas dynamic and MHD codes have been developed to model corotating flows (e.g. Hundhausen 1973; Whang 1980; Pizzo 1980, 1982, 1991, 1994a; Hu & Habbal 1993; Kota 1992). At present, the most sophisticated appears to be the recent 3-D MHD code developed by Pizzo (1994a). The geometry illustrated in Figure 8 has been used to specify the inner boundary conditions for a calculation using this code (Pizzo & Gosling 1994). The input flow configuration was that of a dipole tilted 30° to the solar rotation axis (similar to that prevailing during *Ulysses*' initial transit to high solar latitudes), with fast, hot, low-density flow emanating radially from regions centered on the dipole axis and slow, cold, dense flow emanating from a belt girding the dipole equator. Parameter values were chosen at the inner boundary of the calculation at 0.15 AU to produce a 310–750 km s^{-1} speed range at 1 AU; the calculation was carried out over longitude and latitude ranges of 360° and ±64° with half-degree resolution in both angles and out to a heliocentric distance of 10 AU with radial steps that nowhere exceeded .01 AU.

Figure 9 shows resulting longitudinal traces of flow speed and gas pressure at 5 AU in one degree intervals in latitude (λ) from the equator up to λ = N50° (the pattern is identical at southern latitudes although shifted in longitude). Since the pattern rotates with the sun, these longitudinal traces from right to left represent the time profiles a spacecraft would measure at various latitudes at this distance from the Sun. The simulation reproduces the basic observational effects shown in Figures 6 and 7. The trace at the equator shows a symmetric structure for streams from the opposite solar hemispheres, with classic signatures of forward shock, stream interface (separating what were originally the fast and slow flows), and reverse shock evident. The northern stream (the one near a longitude of 90°)

retains this basic structure up to a latitude near that associated with the tilt of the dipole axis to the solar rotation axis ($\lambda = 30°$). Poleward of that point, however, the forward shock decays rapidly into a broad forward wave, while the reverse shock persists with ever weakening intensity to latitudes 10–15° higher, as in the observations. Although not reproduced here, the simulation also accurately reproduces the sense and magnitude of the average flow deflections observed downstream of the shocks and thus also the average east-west orientation and north-south tilt of the CIR.

Summary

Our present understanding of corotating flows in the solar wind is as follows. A time stationary, but spatially variable, coronal expansion when coupled with solar rotation causes flows of different speed to be radially aligned at low heliographic latitudes. High-speed streams steepen with increasing heliocentric distance because the crests of the streams travel out from the Sun faster than the troughs between the streams. Stream steepening produces a buildup of pressure on the leading edge of a stream that acts to resist the steepening of the stream by accelerating low-speed plasma ahead of the stream and decelerating high-

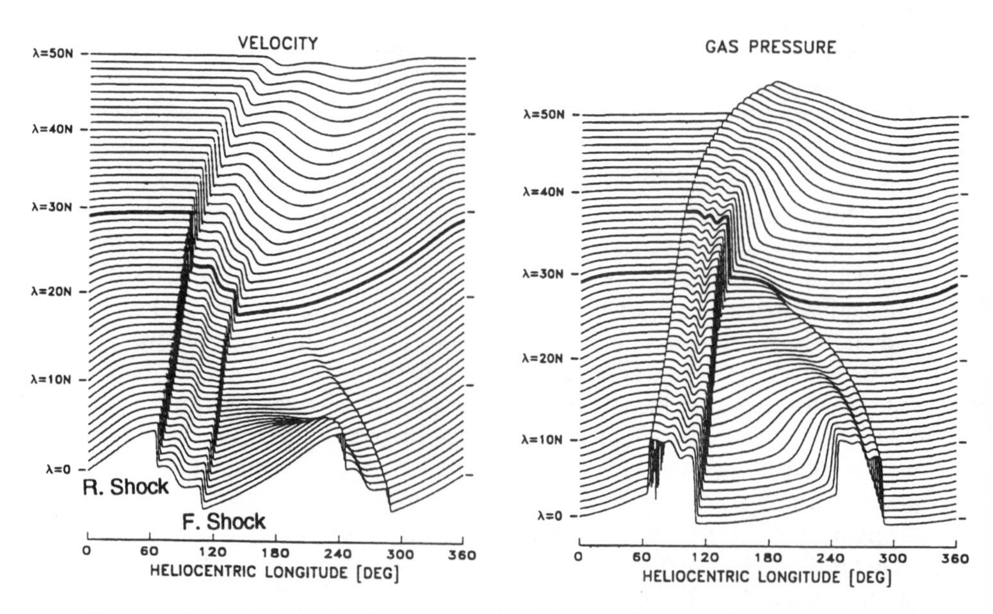

Figure 9 Solar wind speed and gas pressure at 5 AU vs longitude in 1° intervals in latitude predicted by a 3-D MHD solar wind stream model. The heavy trace at 29° indicates the inferred tilt of the current sheet at the time of *Ulysses*' transit to high southern latitudes. Longitudinal traces from right to left mimic the time profiles that a stationary spacecraft would measure at various latitudes at 5 AU. (Adapted from Pizzo & Gosling 1994.)

speed plasma within the stream itself. By this process high-speed streams damp with increasing distance from the Sun. When the stream amplitude is greater than about twice the fast-mode speed, a pair of shocks form on the leading and trailing edges of the high-pressure interaction region. The forward shock propagates into the low-speed plasma and the reverse shock propagates back into the high-speed plasma. Because the state of evolution of a high-speed stream is a function of solar longitude, the interaction region and its bounding waves are roughly aligned with Archimedean spirals in the solar equatorial plane, and these appear to corotate with the Sun. In addition, these corotating interaction regions commonly have substantial north-south tilts that are opposed in the opposite solar hemispheres. The tilts are such that with increasing heliocentric distance the forward shocks bounding corotating interaction regions in both hemispheres propagate toward and eventually across the equator, while the reverse shocks propagate poleward to latitudes well above where the interaction between fast and slow wind originally occurs. The latter effects fundamentally are a consequence of the fact that the solar magnetic dipole is commonly tilted relative to the rotation axis of the Sun.

CORONAL MASS EJECTIONS AND TRANSIENT DISTURBANCES IN THE SOLAR WIND

CMEs Close to the Sun

The most rapid and dramatic changes in the corona occur during coronal mass ejections. CME events originate in closed magnetic field regions in the corona where the field normally is sufficiently strong to constrain the plasma from expanding outward (e.g. Hundhausen 1988, 1995). Usually these closed field regions are found in the coronal streamer belt surrounding the solar magnetic equator. During a typical CME, somewhere between 10^{15} and 10^{16} g of solar material are ejected into interplanetary space. Ejection speeds within 5 solar radii of the Sun's surface range from less than 50 km s^{-1} in some of the slower events to as high as 2000 km s^{-1} in some of the faster ones (Gosling et al 1976a, Howard et al 1985, Hundhausen et al 1994). Many coronal mass ejections accelerate as they move out through the corona; others appear to move with nearly constant speed. The average CME speed in the corona is close to the overall average speed of the solar wind ($\sim$470 km s^{-1}) observed near Earth. CME speeds in the corona are roughly independent of heliographic latitude on average (Hundhausen et al 1994); however, CMEs occur much more frequently at low than at high latitudes.

Coronal mass ejections are commonly, but not always, observed in association with other forms of solar activity such as eruptive prominences, metric-wavelength radio bursts, and both impulsive and gradual solar flares (e.g. Gosling et al 1974, Munro et al 1979, Webb 1992). The most common association is with eruptive prominences, which often are embedded within the interiors of CMEs, and with gradual flares (Webb et al 1976, Kahler 1977, Sheeley et al 1983). Like other forms of solar activity, CMEs occur with a frequency that varies in a cycle of ~ 11 years. On average, the Sun emits about 3.5 CMEs day^{-1} near the peak of the solar activity cycle, but only about 0.2 CMEs day^{-1} near solar activity minimum (Webb & Howard 1994). The physical origin of CMEs remains obscure; however, there is good evidence indicating that flares are not the fundamental physical cause of CMEs (e.g. Kahler 1992, Gosling 1993, Hundhausen 1996). Many CMEs can be associated with long-duration (many hours) soft X-ray events that commence close to the time that CMEs lift off from the Sun. These long-lived events, sometimes known as gradual flares, typically are associated with a restructuring of the solar corona beneath the departing CMEs. It is commonly believed that the newly formed coronal loops visible in gradual flares result from reconnection within the magnetic legs of the rising CMEs (e.g. Kopp & Pneuman 1976, Hiei et al 1993).

Interplanetary Disturbances Driven by Fast CMEs at Low Heliographic Latitudes

The leading edges of the faster CMEs have outward speeds considerably greater than that associated with the normal solar wind expansion at low heliographic latitudes, and thus they commonly drive transient shock wave disturbances in the solar wind at low latitudes (e.g. Sheeley et al 1985). Figure 10 illustrates how shocks are normally formed in association with fast CMEs at low latitudes. The figure shows calculated radial speed and pressure profiles of a simulated solar wind disturbance driven by a fast CME at the time the disturbance first reaches 1 AU. As indicated by the insert in the top panel, the disturbance was initiated at 0.14 AU (the inner boundary of the one-dimensional gas dynamic calculation) by abruptly raising the flow speed from 275 to 980 km s^{-1} in a square wave pulse lasting 6 hours. The initial disturbance thus mimics a uniformly fast, spatially limited CME with an internal pressure equal to that of the surrounding solar wind plasma. A region of high pressure (an interaction region) develops on the leading edge of the disturbance as the CME overtakes the slower ambient wind ahead. This region of high pressure is bounded by a forward shock on its leading edge that propagates into the ambient wind ahead and on its trailing edge by a reverse shock that propagates backward through the CME. Both observations (e.g. Gosling et al 1988) and simulations (e.g. Pizzo 1985) indicate, however, that at low latitudes the reverse shock is ordinarily present only near the middle of CME-driven disturbances where the

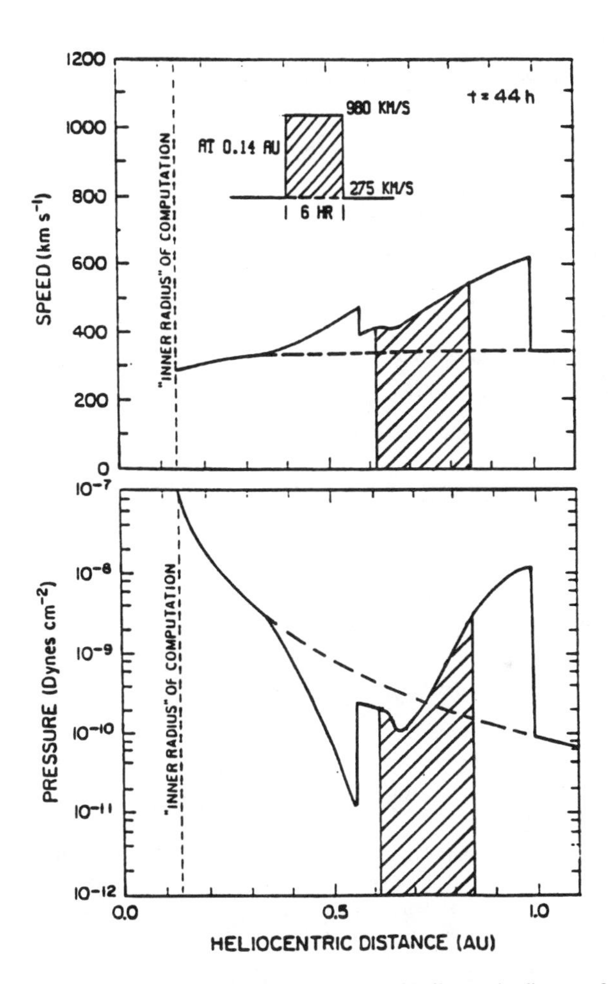

Figure 10 Solar wind speed and pressure as functions of heliocentric distance for a 1-D gas-dynamic simulation of a CME-driven disturbance. The dashed curve indicates the steady state prior to introduction of the temporal variation in flow speed imposed at the inner boundary of 0.14 AU, which is shown in the insert in the top panel. Hatching identifies material that was introduced with a speed of 980 km s^{-1} at the inner boundary and therefore identifies the CME in the simulation. (Adapted from Hundhausen 1985.)

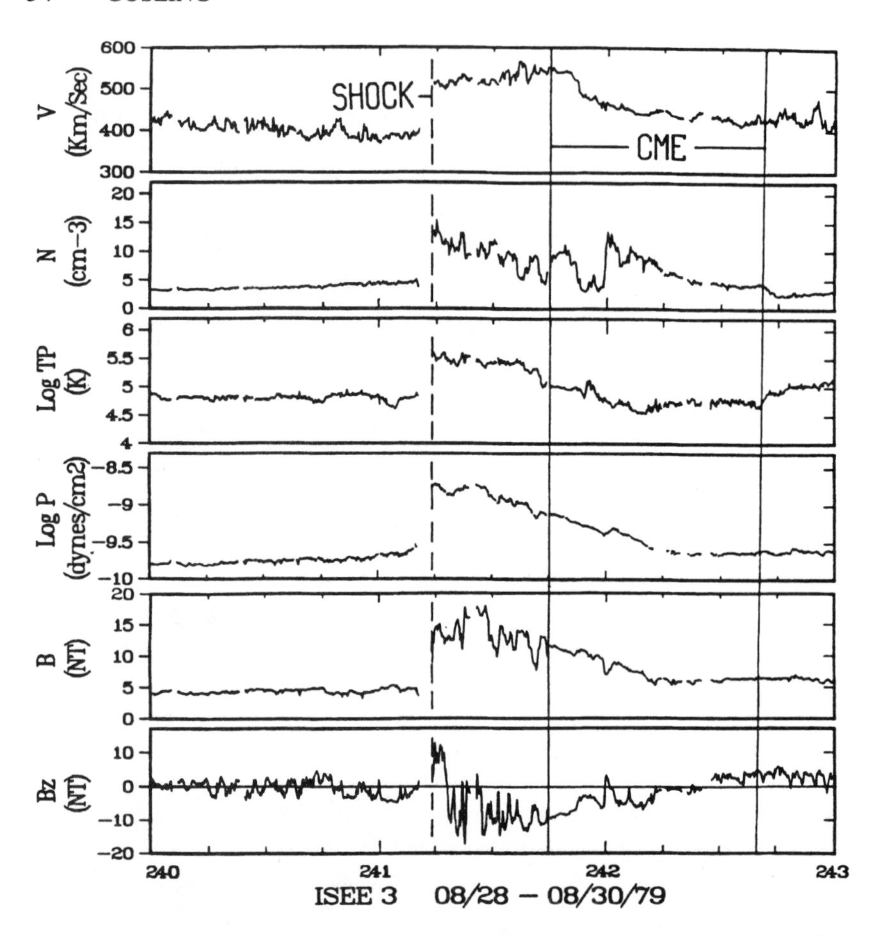

Figure 11 Selected solar wind plasma and magnetic field parameters measured near Earth encompassing a CME-driven shock disturbance. From top to bottom the parameters plotted are flow speed; proton density and temperature; combined proton, electron, and field pressure; magnetic field strength; and field component perpendicular to the ecliptic. (From Gosling & McComas 1987.)

interaction is most nearly one dimensional in nature. The CME slows as a result of its interaction with the plasma ahead and behind (it interacts with the plasma behind via the rarefaction wave produced as the CME outruns the slower trailing plasma), but still has a higher speed near 1 AU than the ambient wind ahead in this simple calculation.

Figure 11 displays selected plasma and magnetic field data from a CME-driven shock wave disturbance observed in the solar wind near Earth by the *ISEE 3* spacecraft. The shock is distinguished in the plot by discontinuous increases in flow speed density, temperature, magnetic field strength, and pressure. As would be expected, the plasma volume identified as the CME resided well behind the shock and had a higher speed than the ambient unshocked wind ahead. In this case the CME was also distinguished by a somewhat lower proton temperature than in the surrounding plasma, by a moderately strong and smoothly varying field strength, and by a smooth rotation in the vertical (out of the ecliptic) component of the magnetic field.

CMEs in the Solar Wind Near Earth

The identification of CMEs in solar wind plasma and magnetic field data is still something of an art. In this regard, shocks serve as useful fiducials for identifying fast CMEs in the solar wind at low latitudes. A number of plasma and field signatures have been recognized in low-latitude solar wind plasma and magnetic field data near 1 AU that qualify as unusual compared to the normal solar wind, but that are commonly observed a number of hours after shock passage. These signatures, some of which are apparent in Figure 11, are commonly used to identify fast CMEs (for reviews, see Gosling 1990, 1992) and include the following: 1. counterstreaming (along the field) suprathermal (> 60 eV) electrons, 2. counterstreaming energetic ($> \sim 20$ keV) protons, 3. helium abundance enhancements ($He^{++}/H^{+} > \sim .08$), 4. ion and electron temperature depressions, 5. strong magnetic fields ($> \sim 8$ nT), 6. low plasma beta (< 1.0), 7. low magnetic field variance, 8. anomalous field rotations (flux ropes), and 9. unusual plasma ionization states (e.g. Fe^{+16}, He^{+}).

Many of the above signatures are also observed in the absence of shocks where, presumably, they serve to identify the relatively low-speed CMEs that do not drive shock disturbances. However, few CMEs in the solar wind exhibit all of these signatures, and there is no one signature that is 100% reliable (e.g. Zwickl et al 1983, Richardson & Cane 1993). It is our experience though that CMEs can usually be identified in the solar wind at low latitudes near 1 AU by counterstreaming fluxes of suprathermal electrons, and that is how the original CME identification was made for the event shown in Figure 11. As illustrated in Figure 12, suprathermal electron fluxes can be used to distinguish different magnetic field topologies in interplanetary space. These electrons carry the solar wind electron heat flux and are beamed, almost scatter-free, outward from the Sun along the IMF (e.g. Feldman et al 1975). In the normal solar wind suprathermal electron beams are unidirectional, reflecting the fact that field lines there are open and effectively connected to the hot corona at only one end. Coronal mass ejections, on the other hand, originate in closed field regions in the corona, and field lines threading CMEs are initially connected to the corona at both ends. Such a field topology leads to suprathermal electron fluxes that move outward from the corona from both

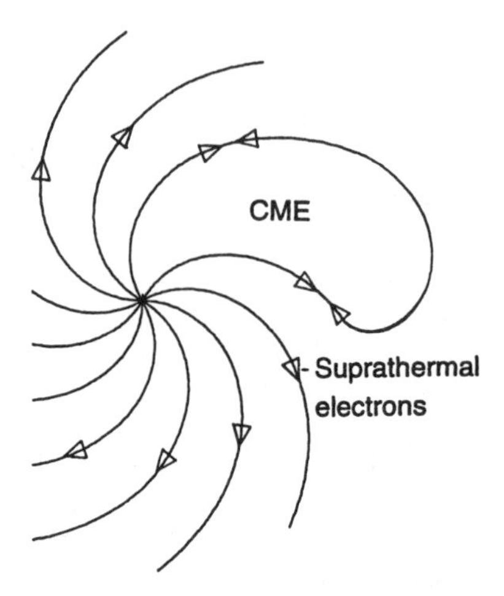

Figure 12 Sketch illustrating open and closed magnetic field topologies in the solar wind and the corresponding types of suprathermal electron streaming observed. Such streaming is observed whether or not the field lines are planar as drawn or are three-dimensional helixes.

footpoints, producing a counterstreaming flux of these electrons in interplanetary space.

CMEs, as identified by the counterstreaming suprathermal electron signature, have variable radial thicknesses near Earth. Their average radial thickness near 1 AU is close to 0.2 AU, and they usually are expanding as they propagate farther out into the heliosphere (e.g. Burlaga & Behannon 1982, Farrugia et al 1993). During the past sunspot cycle, CMEs accounted for about 15% of all the solar wind observed near Earth near solar activity maximum, whereas they accounted for less than 1% of all measurements near solar minimum (Gosling et al 1992). The Earth intercepted $\sim$72 CMEs year^{-1} near solar activity maximum and $\sim$8 CMEs year^{-1} near solar minimum. Solar wind disturbances driven by fast CMEs are the cause of virtually all large, nonrecurrent geomagnetic storms (Gosling et al 1991).

Speeds of CMEs in the Ecliptic Plane Near 1 AU

Coronal mass ejection bulk speeds in the ecliptic plane near 1 AU range from about 280 km s^{-1} to greater than 1000 km s^{-1}, with the average speed being comparable to that of the normal solar wind ($\sim$470 km s^{-1}) (Gosling et al 1987). Approximately one third of all CMEs in the ecliptic have sufficiently high speeds relative to the ambient wind ahead to drive shock disturbances; the remainder simply ride along with the rest of the wind. It is of particular interest that the minimum CME speed in the ecliptic plane at 1 AU is essentially the same as that of the normal solar wind there. That is, even the slowest CMEs observed in the corona eventually get accelerated up to at least the same minimum speed as the rest of the wind in the ecliptic plane. Because slow CMEs near Earth are usually found at times of declining or constant flow speed and in the absence of any sizable local pressure gradients, they do not appear to attain their observed speed by virtue of being pushed outward by faster plasma from behind. On the contrary, slow CMEs simply seem to experience the same basic outward acceleration as does the rest of the wind.

CMEs with Magnetic Flux Rope Topologies

Approximately one third of all CMEs identified in the solar wind near Earth exhibit internal field rotations characteristic of nearly force-free magnetic flux ropes (Gosling 1990), which consist of a series of helical field lines of ever-increasing pitch wrapped about a central axis. The CME event shown in Figure 11 exhibits such a rotation and is a flux rope. When the field strength is high and the plasma beta is low, such CMEs are known as magnetic clouds (e.g. Burlaga 1991). It has been suggested (e.g. Burlaga 1991, Rust 1994) that magnetic clouds may be interplanetary extensions of solar filaments (prominences), which often have helical magnetic field topologies and often are embedded within CMES observed by coronagraphs. We believe, however, that this suggestion is unlikely to be correct: Filaments typically comprise only small fractions of the volumes of CMEs when viewed by coronagraphs and thus would be expected to be found either deep within the interiors of CMEs in the solar wind or on their trailing edges. On the other hand, the events identified as flux ropes in the solar wind typically, as in Figure 11, are the entire CMEs, not small portions thereof.

An alternative suggestion is that the flux rope topology characteristic of some CMEs in the solar wind is a consequence of 3-D magnetic reconnection (Gosling 1990, 1993). As already noted, new magnetic loops formed in the corona beneath CMEs during gradual flares provide strong evidence that magnetic reconnection commonly occurs within the magnetic legs of the departing CMEs (e.g. Kopp & Pneuman 1976). As visualized in two dimensions, the magnetic loops threading a CME reconnect with themselves to produce detached magnetic loops within the CME and new magnetic loops statically confined to the corona below. However, because of the high field symmetry required, it is unlikely that reconnection ever actually occurs in such a manner. Any skewing or shearing of the field results in reconnection between neighboring CME loops. As illustrated in Figure 13, reconnection in the presence of skew or shear produces helical field lines threading the CME as well as new magnetic loops in the corona below. When the plasma beta is low the helical field lines within the CME eventually relax to the nearly force-free structure characteristic of a magnetic cloud. In contrast to the two-dimensional case, reconnection in three dimensions does not initially result in magnetic disconnection from the Sun.

CMEs at High Solar Latitudes

A number of CMEs were observed in the solar wind by *Ulysses* during its low-latitude cruise to Jupiter and in the year thereafter (e.g. Phillips et al 1992). These CMEs were partly responsible for the jumbled character of the stream structure observed at low latitudes evident in Figure 5; moreover, CMEs observed in March 1991 and in November 1992 produced the fastest flows yet detected by *Ulysses,* although many of the low-latitude CMEs also had low or average speeds. Many of the low-latitude CME events can be recognized in the third panel of Figure 5 as intervals of enhanced helium abundance, which

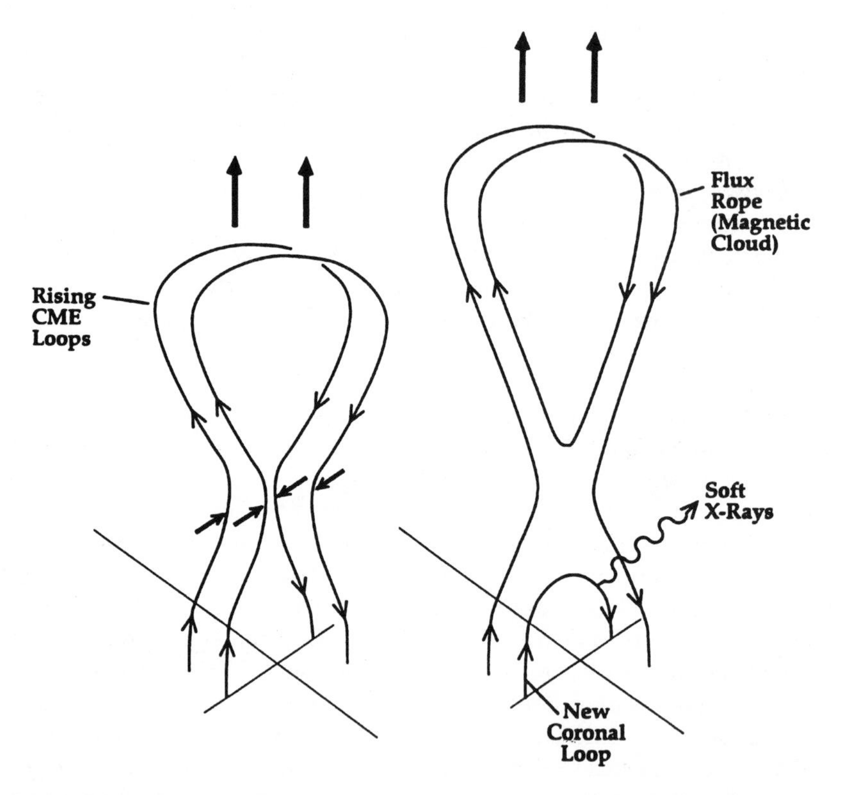

Figure 13 Sketches illustrating the pinching off (reconnection) of magnetic loops in a rising CME whose legs are sheared relative to one another. Reconnection produces a magnetic flux rope topology within the CME and new loops in the corona beneath. (From Gosling 1993.)

is a common characteristic of CMEs in the ecliptic plane. Fewer CMEs were observed at high latitudes, owing both to the overall decline in solar activity during the high-latitude phase of the mission and to the fact that CMEs occur less frequently at high than at low solar latitudes. Only six CMEs have been positively identified in the *Ulysses* data poleward of S30°; these identifications were made using a combination of counterstreaming electron measurements and magnetic field data. It is of considerable interest that none of those six CMEs can be distinguished easily in Figure 5 on the basis of their speed or helium content.

The Speeds of High-Latitude CMEs in the Solar Wind

Figure 14 shows selected plasma and magnetic field parameters for intervals encompassing the first and last of the high-latitude CMEs detected by *Ulysses* at 4.6 AU and S33° and at 3.2 AU and S61°, respectively. Both of these CME events had clear associations with coronal loop formation events (gradual flares) observed by the soft X-ray experiment on *Yohkoh.* One striking aspect of these two CMEs, characteristic of all the high-latitude CMEs observed to date (Gosling et al 1994a), was their high speeds. These speeds were comparable to that of the rest of the wind observed at high latitudes and were greater than about 95% of all CME speeds observed in the ecliptic plane near Earth. The average speed of the six certain high-latitude CME events was ~ 730 km s^{-1}, ~ 250 km s^{-1} higher than the average speed of CMEs detected in the solar wind near Earth. All of the high-latitude CMEs were observed at times of nearly constant or declining flow speed, as for the events shown in Figure 14. This indicates that these CMEs did not attain their high speeds as a result of interaction with trailing solar wind plasma.

It is useful to compare these results with observations of CME speeds in the corona as measured during 1980 and 1984–1989 by the coronagraph on the *Solar Maximum Mission, SMM* (Hundhausen et al 1994). Although the scatter of speeds observed in the corona by the *SMM* coronagraph was large at all latitudes, the average CME speed within ~ 6 solar radii of Sun center was ~ 350 km s^{-1}, nearly independent of latitude. Moreover, few CMEs in the corona had speeds as high as have been observed in the solar wind at high latitudes by *Ulysses.* Because only six high-latitude events have been identified to date, our result concerning the high speeds of CMEs in the high-latitude solar wind is perhaps only a statistical anomaly. That is, it is possible that the Sun emitted a set of unusually fast CMEs toward *Ulysses* when the spacecraft was at high latitudes. We suspect, however, that the observations are not anomalous, but rather provide basic insight about the forces that accelerate CMEs.

It is often assumed that the forces accelerating CMEs outward from the Sun are different from those responsible for accelerating the normal solar wind

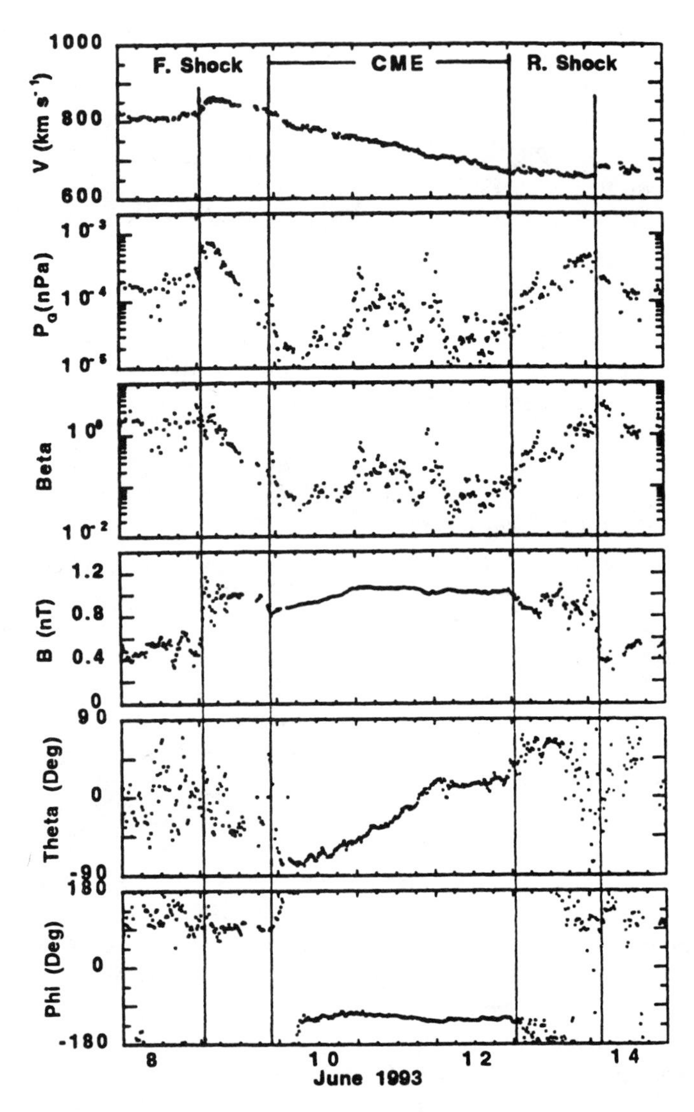

Figure 14 Selected plasma and magnetic field parameters for two CME events observed by *Ulysses* at 4.6 AU and S33° (*left*) and at 3.2 AU and S61° (*right*). Parameters plotted from top to bottom are flow speed, gas pressure, plasma beta, and magnetic field strength, polar angle, and azimuthal angle. (Adapted from Gosling et al 1994b,c.)

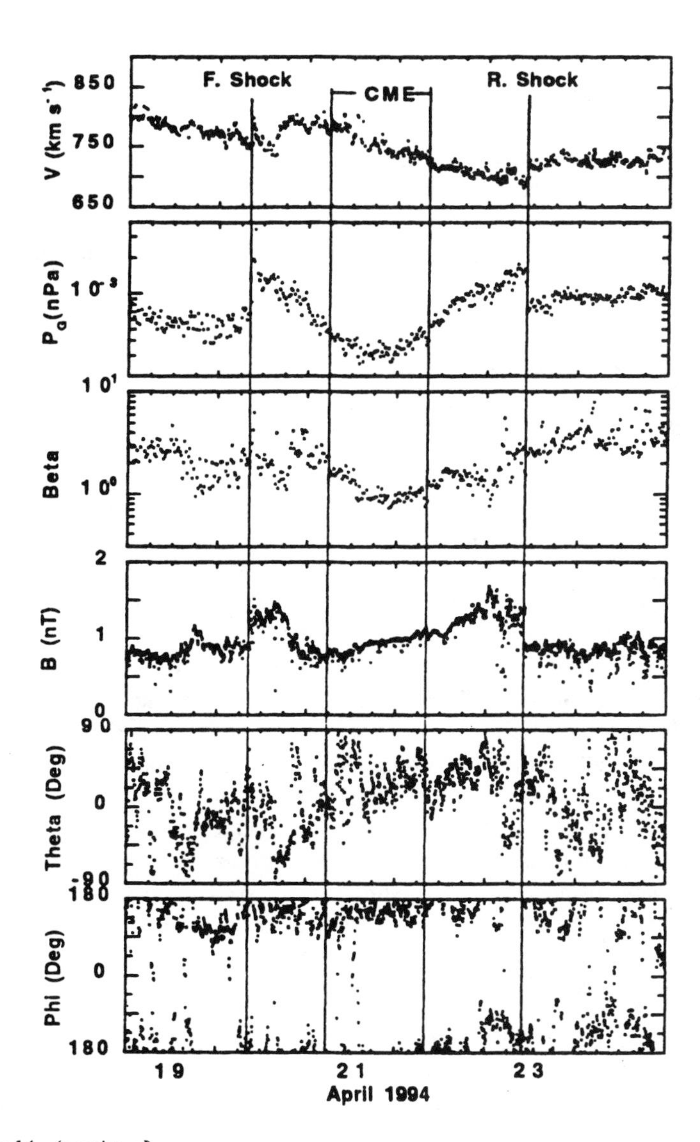

Figure 14 (*continued*)

(e.g. Chen & Garren 1993). The *Ulysses* observations, on the other hand, when coupled with measurements of minimum CME speeds in the ecliptic plane at 1 AU, suggest that slow CMEs experience the same basic acceleration as does the normal solar wind. It may be that an instability in the corona is responsible for the initial release of a slow CME (e.g. Low 1993, 1994; Priest 1988), but thereafter the CME experiences the same overall pressure forces that accelerate the surrounding solar wind. The prime difference between the low- and high-latitude measurements in this respect is that the minimum speed of the normal wind at high latitudes is much greater than it is at low latitudes, so the CMEs reach much higher minimum speeds at high latitudes.

Forward-Reverse Shock Pairs Associated with CMEs in the Solar Wind at High Heliographic Latitudes

A forward-reverse shock pair was associated with each of the CME events shown in Figure 14. In both events the shocks were offset by approximately equal distances from the edges of the CME, and in both events the pressure maximized immediately downstream from the shocks and reached a minimum roughly in the center of the CME. Moreover, in both events the speed declined from the forward shock to the reverse shock. Although both CMEs had high speeds, they were not traveling faster than the ambient solar wind ahead of the forward shocks. Thus, the shock pairs could not have been produced by relative motion between the CMEs and the ambient wind ahead (compare, for example the event profiles in Figure 14 with those of the numerical simulation in Figure 10). Rather, the observations suggest that the shock pairs in these events were produced by overexpansion (i.e. expansion driven by a high internal pressure) of the CMEs. To the best of our knowledge, shock pairs associated with overexpansion of CMEs have never been observed at low heliographic latitudes at any distance from the Sun. Yet of six certain CMEs observed poleward of S30°, three had associated shock pairs of this nature. These events thus form a new class of forward-reverse shock pairs that appears to be restricted to high solar latitudes (Gosling et al 1994c).

Simulation of an Overexpanding CME in the Solar Wind

The left panel of Figure 15, which shows the result of a simple numerical simulation of an overexpanding CME, illustrates what we believe is the basic physics behind these events. The figure shows a disturbance arriving at 5 AU that was initiated at the inner boundary of the 1-D gas-dynamic calculation at 0.14 AU by increasing the density by a factor of four in a bell-shaped pulse 10 hours wide while simultaneously holding the temperature and speed constant. This mimics the ejection of a dense CME from the Sun, whose internal pressure is higher than that of the surrounding solar wind. The radial width of the simulated CME at the inner boundary, which lies well outside the critical point where the solar wind goes supersonic, was ~ 0.17 AU. Owing to its high internal pressure, the CME expands as it travels out from the Sun, so that at ~ 4.6 AU its radial width is 0.5 AU. Because the expansion drives a forward wave that propagates ahead of the CME and a reverse wave that propagates back into the trailing solar wind plasma, the overall disturbance is ~ 0.95 AU wide. These relatively modest-strength pressure waves steepen into shocks before the disturbance arrives at 3 AU. The expansion causes the density and temperature (and hence also the pressure) within the CME to be lower than that in the plasma immediately surrounding the disturbance at large distances from the Sun. The temporal signature produced at a fixed point in the outer heliosphere by the above simulated disturbance bears a remarkable resemblance to the overall appearance of the events shown in Figure 14. Quantitative differences between the observations and the calculations can probably be ascribed to differences between the initial conditions assumed in the simulation and in the real solar wind in these events and to limitations inherent in the 1-D gas-dynamic code employed in the calculation.

The sketch shown in the right panel of Figure 15 helps summarize our interpretation of these high-latitude shock-pair events. CMEs having internal pressures higher than and speeds comparable to that of the surrounding solar wind are sometimes ejected into the high-latitude wind. Over-pressure causes the CMEs to expand relative to the surrounding solar wind as they travel out from the Sun, ultimately producing an expansion shock that wraps around the CME. The expansion shock propagates into the ambient wind with a speed V_e that is considerably less than the bulk speed V_{cme} of the CME; thus the entire structure is convected away from the Sun. In this simple picture, the forward and reverse shocks are actually both part of the same expansion shock that is sampled twice as the structure convects past a spacecraft. With increasing heliocentric distance the disturbance evolves from one where the pressure is enhanced within the CME relative to the ambient wind to one where high pressure is concentrated in the region immediately downstream from the expansion shock, and the region interior to the CME has lower than average pressure. Since the background wind pressure continues to decrease with increasing distance from the Sun, there is nothing to prevent the CME from continuing to expand as it travels into the far outer reaches of the heliosphere. This contrasts, for example, with the case of overexpansion in the Earth's atmosphere produced by an explosion. In that case, back pressure from the surrounding atmosphere eventually causes an implosion at the center of the disturbance.

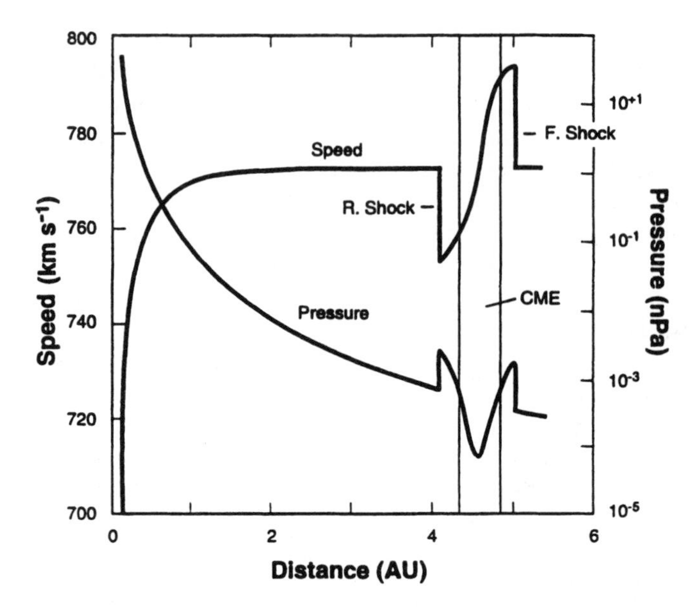

Figure 15 (*Left*) Solar wind speed and pressure as functions of heliocentric distance for a simulated disturbance that has just arrived at 5 AU. The simulation utilizes a 1-D gas-dynamic code and was initiated at 0.14 AU by increasing the density by a factor of four in a bell-shaped pulse 10-hours wide. (*Right*) Artist's sketch illustrating the evolution with increasing heliocentric distance of a CME ejected into interplanetary space with an internal pressure higher than and a speed comparable to that of the surrounding solar wind. Shading indicates the distribution of pressure within the disturbance. (Adapted from Gosling et al 1994c.)

Simultaneous Low- and High-Latitude Observations of a CME-Driven Disturbance

Despite relatively good agreement between observation and simulation noted above, there remain a number of unanswered questions relative to these forward-reverse shock-pair events. Among them are questions of why they are restricted to high heliographic latitudes and how the reverse wave avoids running back into the Sun in the real solar wind. (For the simulation shown in Figure 15 the inner boundary lies well outside the critical point where the solar wind goes supersonic, so the reverse wave does not encounter the Sun).

Some insight into resolving these questions can be found in observations of an event observed simultaneously by both *Ulysses* at high latitudes and by *IMP 8* in orbit about Earth. Figure 16 contrasts the high- and low-latitude plasma and magnetic field observations of this event associated with a disturbance near

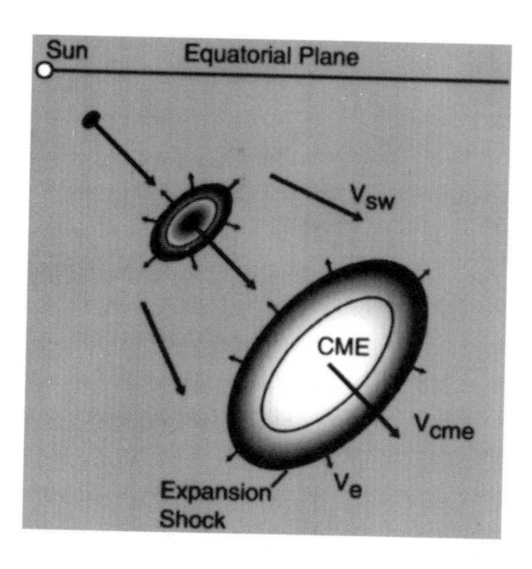

Figure 15 (*continued*)

the center of the Sun on February 20, 1994. At the time of these observations *Ulysses* was at 3.53 AU (which accounts for the time delay between *IMP 8* and *Ulysses'* detection of the event) and 11.4° west and 47.2° south of Earth. It is of interest that the CME itself had a somewhat different character at low and high latitudes; for example, *IMP 8* observed a substantial helium abundance enhancement while *Ulysses* did not. More to the point, the disturbance in the ecliptic included a strong forward shock but no reverse shock, while at high latitudes the disturbance was bounded by a relatively weak forward-reverse shock pair centered on the CME. Clearly, the shock disturbance in the ecliptic was driven by the relative speed between the CME and a slower ambient wind ahead, whereas at *Ulysses* the disturbance was driven by overexpansion of the CME. This suggests that a high CME speed close to the Sun may be crucial in allowing the reverse wave associated with overexpansion to escape from the Sun. If the initial outward speed of a CME having a high internal pressure exceeds the fast-mode speed, the reverse wave associated with overexpansion does not run back into the Sun. In contrast, at low latitudes the disturbance resulting from such a CME is dominated by compressional effects associated with the CME overtaking slower wind ahead rather than by overexpansion. This may explain why overexpansion shock pairs have not yet been observed at low latitudes.

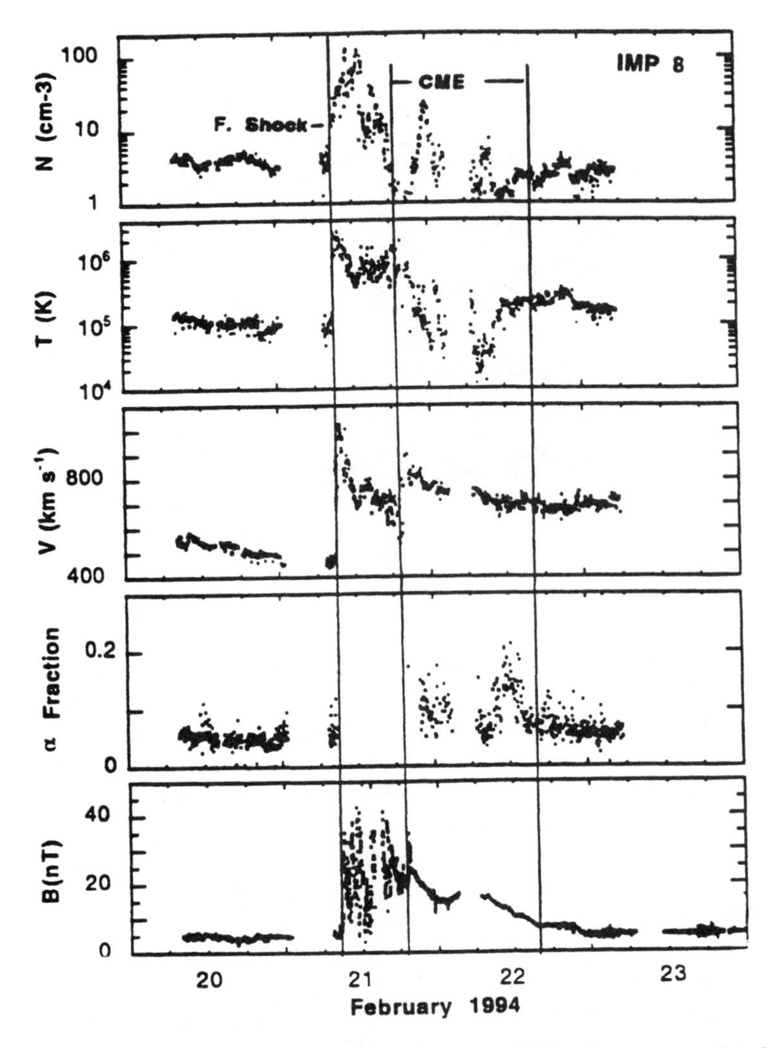

Figure 16 Selected solar wind plasma and magnetic field parameters for a solar wind disturbance observed near Earth by *IMP 8* (*left*) and at 3.53 AU and S54.3° by *Ulysses* (*right*) in February 1994. Parameters plotted from top to bottom are proton density, proton temperature, bulk flow speed, He^{++}/H^{+} fraction, and magnetic field strength. *Ulysses* was 11.4° west and 47.2° south of Earth at the time of these observations. (Adapted from Gosling et al 1995e.)

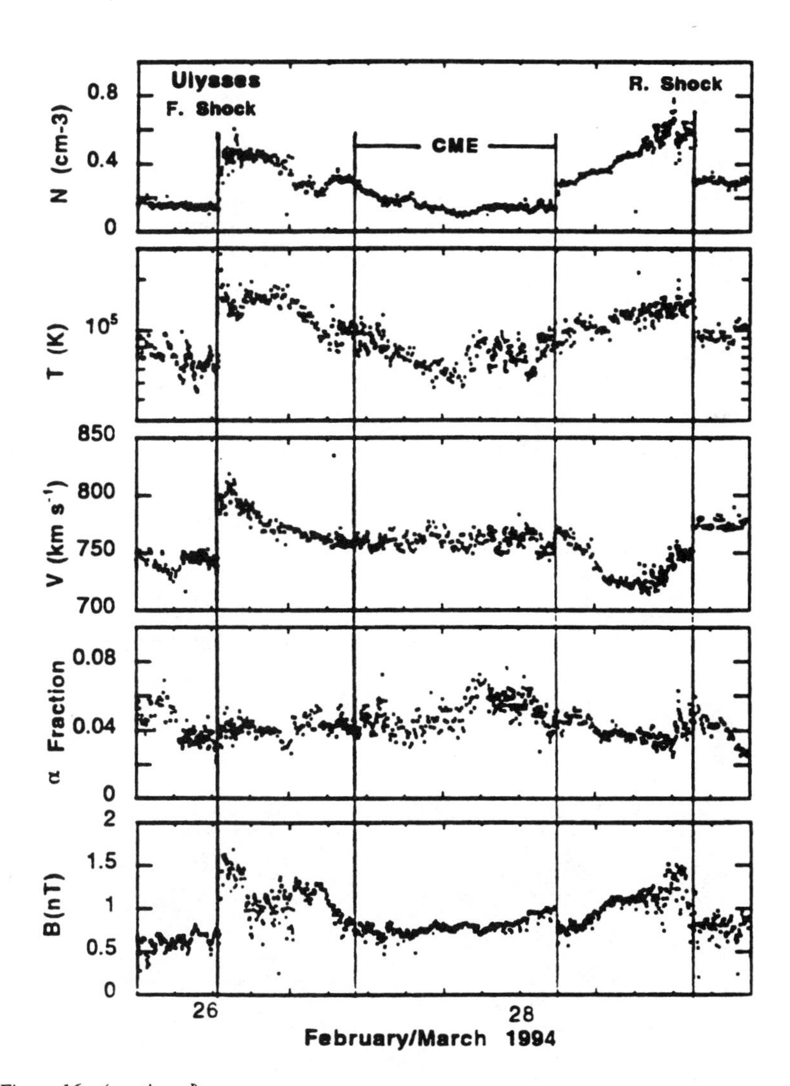

Figure 16 (*continued*)

Ulysses *Observations Related to the Origin of Magnetic Flux Ropes*

The best example of a nearly force-free magnetic flux rope in the *Ulysses* high-latitude data is the event of June 1993, shown in the left portion of Figure 14 (p. 60). At the time of these observations *Ulysses* was at 4.64 AU and S32.5° and was located slightly behind the southeast limb of the Sun as viewed from Earth. Because the center of that CME was neither overtaking slower plasma ahead nor being pushed outward by faster plasma from behind, it should have traveled out to *Ulysses* at a nearly constant speed of 740 km s^{-1}. It thus should have left the Sun $\sim$ 11 days earlier, near the middle of the day on May 31, 1993. Figure 17 shows two images of the Sun obtained by the soft X-ray telescope on *Yohkoh* on May 31. These images reveal that large coronal loops, which persisted for many hours, formed on and probably slightly behind the southeast solar limb (lower left in the photographs) almost directly beneath *Ulysses* between 08:26 and 17:51 UT, near the time when we calculate that the CME should have departed from the Sun.

The combined *Ulysses* and *Yohkoh* May/June 1993 observations relating a nearly force-free magnetic flux rope CME with the formation of large magnetic loops in the corona provide support for the idea that the flux rope topology is a result of 3-D reconnection occurring within the magnetic legs of CMEs close to the Sun, as sketched in Figure 13. The April 1994 event, however (shown in the *right* panel of Figure 14), was also clearly related to a large coronal loop formation event (on April 14, 1994), but it did not exhibit the internal rotation

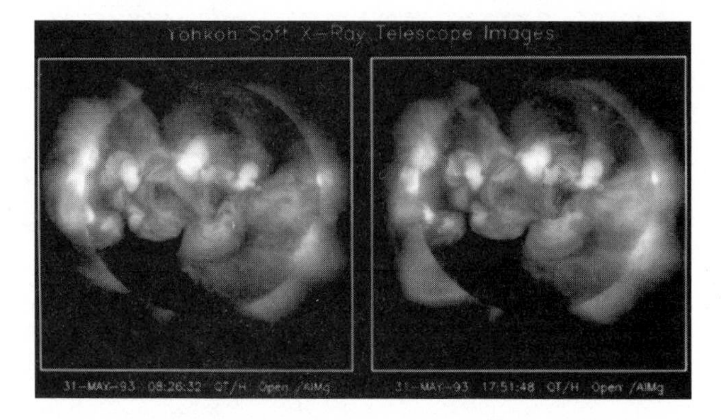

Figure 17 Images of the Sun taken by the soft x ray telescope on *Yohkoh* on May 31, 1993. Large coronal loops formed on the southeast limb (lower left portion of photographs) directly beneath *Ulysses* in the interval between 08:26 and 17:51 UT. (From Gosling et al 1995a.)

characteristic of a force-free magnetic flux rope. The essential difference between the June 1993 and April 1994 events appears to be associated with the plasma beta. The June 1993 event was magnetically dominated (low beta), so the helical fields relaxed to a nearly force-free state, whereas the April 1994 event was plasma dominated (high beta) and no such relaxation of the helical fields occurred.

Sustained Magnetic Reconnection and the Creation of Open and Disconnected Field Lines within CMEs

Despite the usefulness of the counterstreaming suprathermal electron signature for identifying CMEs in the solar wind, it has long been apparent that not all portions of CMEs and perhaps not even all CMEs can be identified in solar wind data by that signature (e.g. Zwickl et al 1983, Crooker et al 1990, Gosling et al 1992, Richardson & Cane 1993, Kahler 1994). This absence suggests that some field lines within CMEs are occasionally open. For example, counterstreaming was not observed continuously throughout the June 1993 CME event shown in Figure 14. As noted below, there is reason to believe that open field lines within the interior of a flux rope CME are caused by sustained 3-D reconnection at the footpoints of the CME close to the Sun (Gosling et al 1995c).

The *left* portion of Figure 18 illustrates the topologies that may result from sustained 3-D reconnection in the corona behind a CME. The sketches are based upon results of numerical simulations of 3-D reconnection in the qualitatively similar geometry in the Earth's magnetic tail in the presence of a cross-tail field component (Birn & Hesse 1991, Hesse & Birn 1991). Each panel in the figure represents a successive stage in the reconnection process ranging from panel *a,* which represents the case already illustrated in Figure 13, to panel *d,* which represents the reconnection of open field lines of the normal solar wind behind a disconnected CME. Note that in three dimensions disconnected field lines produced by reconnection are attached to the outer heliosphere at both ends.

Coronal mass ejections open new magnetic flux into interplanetary space (see the review by McComas 1995). The amount of new flux opened up by a CME is reduced by the process of reconnection. When reconnection proceeds to the point illustrated in panel *b* (Figure 18), that portion of the CME adds no net open magnetic flux to interplanetary space. When it proceeds to the point illustrated in panel *c* and beyond, that portion of the CME produces a net decrease in the open magnetic flux in interplanetary space and an increase in the amount of magnetic flux closed off at the Sun.

Helical field lines threading the center of a CME are exposed to ordinary open field lines of opposite polarity at the base of the CME. In simulations

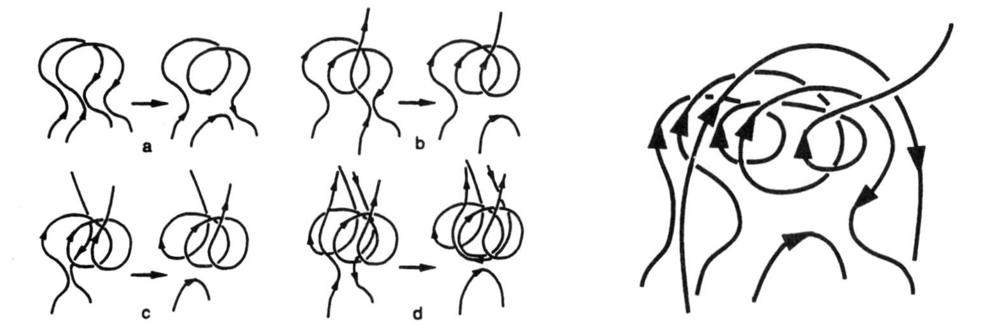

Figure 18 (*Left*) Sketches illustrating successive steps in 3-D reconnection within the magnetic legs of a CME and the changing magnetic topologies that result. (*Right*) Sketch illustrating several different magnetic topologies possible in a CME that has undergone 3-D reconnection. Not all possible topologies are represented. None of the sketches reproduced here are to scale. (Adapted from Gosling et al 1995c.)

in the qualitatively similar geometry in the geomagnetic tail, these tend to be the helical field lines that are opened up first. As a result, a variety of field line topologies can simultaneously be present within a flux rope. The sketch shown in the right portion of Figure 18 illustrates an example of mixed magnetic topologies within a flux rope CME. The crucial point is that reconnection can produce open or disconnected field lines within the heart of a CME, even though the CME originates entirely as closed field lines attached to the Sun at both ends.

Summary

Coronal mass ejections in the solar wind usually have distinct plasma and magnetic field signatures that distinguish them from the normal solar wind. The most reliable of these signatures appears to be a flux of counterstreaming suprathermal electrons, reflecting the fact that CMEs originate in closed field regions in the solar corona. At all heliographic latitudes CMEs attain at least the minimum speed of the ambient solar wind, which indicates that slow CMEs experience the same outward forces as the rest of the wind (fast CMEs receive an additional acceleration from as yet undetermined sources). At low latitudes, about one third of all CMEs drive transient shock wave disturbances in the solar wind because of their high speeds relative to the ambient wind ahead; reverse shocks are relatively rare in such disturbances. At high latitudes, overexpansion often produces an expansion shock that surrounds a CME and that is observed as a forward-reverse shock pair by a spacecraft. Approximately a third of all CMEs in the solar wind are nearly force-free flux ropes, characterized by a series of helical field lines of ever-increasing pitch wrapped about a central axis.

There is substantial evidence to suggest that these flux ropes are a consequence of a low plasma beta and three-dimensional magnetic reconnection within the magnetic legs of the CMEs close to the Sun. Sustained magnetic reconnection behind a CME reduces the amount of open magnetic flux dragged into interplanetary space by a CME and can produce both open and disconnected field lines in its interior.

Ulysses' initial excursion to high solar latitudes occurred at a time when the magnetic structure of the Sun (and the resulting coronal expansion) was relatively simple and when transient disturbances in the solar wind were relatively rare. During *Ulysses'* next transit to high latitudes, which will occur on the rising phase of the solar activity cycle, the magnetic structure of the Sun should be more complex, corotating structure should be less pronounced, and transient disturbances should be more common in the solar wind. It will be interesting to see how well our conclusions on the 3-D nature of corotating and transient solar wind flows hold true for a more complex Sun and interplanetary medium.

ACKNOWLEDGMENTS

Much of the *Ulysses* work synthesized here was done in collaboration with a large number of individuals on the *Ulysses* plasma and magnetic field experiment teams at Los Alamos National Laboratory, Jet Propulsion Laboratory, and Imperial College. Sam Bame was the original Principal Investigator (PI) for the *Ulysses* plasma experiment and the present PI is John Phillips, while Andre Balogh is the PI for the magnetic field experiment. I thank Art Hundhausen and Vic Pizzo who, over the years, have contributed much to my understanding of both corotating and transient solar wind flows. This work was supported by a Los Alamos National Laboratory Directed Research and Development grant.

Literature Cited

Bame SJ, Goldstein BE, Gosling JT, Harvey JW, McComas DJ, et al. 1993. *Geophys. Res. Lett.* 20:2323–26

Bame SJ, McComas DJ, Barraclough, BL, Phillips JL, Sofaly KJ, et al. 1992. *Astron. Astrophys. Suppl. Ser.* 92:237–65

Birn J, Hesse M. 1991. *J. Geophys. Res.* 96:23–34

Burlaga LF. 1983. *J. Geophys. Res.* 88:6085–94

Burlaga LF. 1984. *Space Sci. Rev.* 39:255–316

Burlaga LF. 1986. In *The Sun and the Heliosphere in Three Dimensions,* ed. RG Marsden, pp. 191–204. Dordrecht: Reidel. 525 pp.

Burlaga LF. 1990. *J. Geophys. Res.* 95:4333–36

Burlaga LF. 1991. In *Physics of the Inner Heliosphere II,* ed. R Schwenn, E Marsch, pp. 1–22. Berlin: Springer-Verlag. 352 pp.

Burlaga LF, Behannon KW. 1982. *Sol. Phys.* 81:181–91

Chen J, Garren DA. 1993. *Geophys. Res. Lett.*

20:2319–22
Crooker NU, Gosling JT, Smith EJ, Russell CT. 1990. In *Physics of Magnetic Flux Ropes,* ed. CT Russell, ER Priest, LC Lee, *Geophys. Monogr. 58,* pp. 365–71. Washington, DC: Am. Geophys. Union. 685 pp.
Farrugia CJ, Burlaga LF, Osherovich VA, Richardson IG, Freeman MP, et al. 1993. *J. Geophys. Res.* 98:7621–32
Feldman WC, Asbridge JR, Bame SJ, Montgomery MD, Gary SP. 1975. *J. Geophys Res.* 80:4181–96
Gosling JT. 1990. In *Physics of Magnetic Flux Ropes,* ed. CT Russell, ER Priest, LC Lee, *Geophys. Monogr. 58,* pp. 343–64. Washington, DC: Am. Geophys. Union. 685 pp.
Gosling JT. 1992. In *Eruptive Solar Flares,* ed Z Svestka, BV Jackson, ME Machado, *Lect. Notes Phys. 399,* pp. 258–67. New York: Springer-Verlag. 409 pp.
Gosling JT. 1993. *J. Geophys Res.* 98:18,937–49
Gosling JT, Asbridge JR, Bame SJ, Feldman WC. 1978. *J. Geophys Res.* 83:1401–12
Gosling JT, Baker DN, Bame SJ, Feldman WC, Zwickl RD, Smith EJ. 1987. *J. Geophys. Res.* 92:8519–35
Gosling JT, Bame SJ, McComas DJ, Phillips JL, Balogh A, Strong KT. 1995a. *Space Sci. Rev.* 72:133–36
Gosling JT, Bame SJ, McComas DJ, Phillips JL, Goldstein BE, Neugebauer M. 1994a. *Geophys. Res. Lett.* 21:1109–12
Gosling JT, Bame SJ, McComas DJ, Phillips JL, Pizzo VJ, et al. 1993. *Geophys. Res. Lett.* 20:2789–92
Gosling JT, Bame SJ, McComas DJ, Phillips JL, Pizzo VJ, et al. 1995b. *Space Sci. Rev.* 72:99–104
Gosling JT, Bame SJ, McComas DJ, Phillips JL, Scime EE, et al. 1994b. *Geophys Res. Lett.* 21:237–40
Gosling JT, Bame SJ, Smith EJ, Burton ME. 1988. *J. Geophys. Res.* 93:8741–48
Gosling JT, Birn J, Hesse M. 1995c. *Geophys. Res. Lett.* 22:869–72
Gosling JT, Borrini G, Asbridge JR, Bame SJ, Feldman WC, Hansen RT. 1981. *J. Geophys. Res.* 86:5438–48
Gosling JT, Feldman WC, McComas DJ, Phillips JL, Pizzo VJ, Forsyth RJ. 1995d. *Geophys. Res. Lett.* 22:3333–36
Gosling JT, Hildner E, MacQueen RM, Munro RH, Poland AI, Ross CL. 1974. *J. Geophys. Res.* 79:4581–87
Gosling JT, Hildner E, MacQueen RM, Munro RH, Poland AI, Ross CL. 1976a. *Sol. Phys.* 48:389–97
Gosling JT, Hundhausen AJ, Bame SJ. 1976b. *J. Geophys. Res.* 81:2111–22
Gosling JT, Hundhausen AJ, Pizzo V, Asbridge JR. 1972. *J. Geophys. Res.* 77:5442–54
Gosling JT, McComas DJ. 1987. *Geophys Res. Lett.* 14:355–58
Gosling JT, McComas DJ, Phillips JL, Bame SJ. 1991. *J. Geophys. Res.* 96:7831–33
Gosling JT, McComas DJ, Phillips JL, Bame SJ. 1992. *J. Geophys Res.* 97:6531–35
Gosling JT, McComas DJ, Phillips JL, Pizzo VJ, Goldstein BE, et al. 1995e. *Geophys. Res. Lett.* 22:1753–56
Gosling JT, McComas DJ, Phillips JL, Weiss LA, Pizzo VJ, et al. 1994c. *Geophys. Res. Lett.* 21:2271–74
Hesse M, Birn J. 1991. *J. Geophys. Res.* 96:5683–96
Hiei E, Hundhausen AJ, Sime DG. 1993. *Geophys. Res. Lett.* 20:2785–88
Hoeksema JT. 1995. *Space Sci. Rev.* 72:137–48
Howard R, Sheeley NR, Koomen MJ, Michels DJ. 1985. *J. Geophys. Res.* 90:8173–91
Hu YQ, Habbal SR. 1993. *J. Geophys. Res.* 98:3551–61
Hundhausen AJ. 1972. *Coronal Expansion and Solar Wind.* New York: Springer-Verlag. 238 pp.
Hundhausen AJ. 1973. *J. Geophys. Res.* 78:1528–42
Hundhausen AJ. 1977. In *Coronal Holes and High Speed Wind Streams,* ed J. Zirker, pp. 225–329. Boulder: Colo. Assoc. Univ. Press. 454 pp.
Hundhausen AJ. 1985. In *Collisionless Shocks in the Heliosphere: A Tutorial Review,* ed. RG Stone, BT Tsurutani, *Geophys. Monogr. 34,* pp. 37–58. Washington, DC: Am. Geophys. Union. 115 pp.
Hundhausen AJ. 1988. In *Proc. Sixth Int. Solar Wind Conf., Tech. Note 306+Proc,* ed. V Pizzo, TE Holzer, DG Sime, Vol 1, pp. 181–214. Boulder: Natl. Cent.. Atmos. Res. 356 pp.
Hundhausen AJ. 1996. In *The Many Faces of the Sun,* ed K Strong, J Saba, B Haisch. New York: Springer-Verlag. In press
Hundhausen AJ, Burkepile JT, St. Cyr OC. 1994. *J. Geophys. Res.* 99:6543–52
Hundhausen AJ, Gosling JT. 1976. *J. Geophys. Res.* 81:1436–40
Jokipii RJ, Thomas B. 1981. *Ap. J.* 243:1115–22
Kahler SW. 1977. *Ap J.* 214:891–97
Kahler SW. 1987. *Rev. Geophys.* 25:663–75
Kahler SW. 1988. In *Proc. Sixth Int. Solar Wind Conf., Tech. Note 306+Proc,* ed. V Pizzo, TE Holzer, DG Sime, Vol 1, pp. 215–31. Boulder: Natl. Cent. Atmos. Res. 356 pp.
Kahler SW. 1992. *Annu. Rev. Astron. Astrophys.* 30:113–41
Kahler SW. 1994. In *Solar Dynamic Phenomena and Solar Wind Consequences,* ed JJ

Hunt, pp. 253–61. Noordwijk: ESA Publ. SP-373. 469 pp.
Kopp RA, Pneuman GW. 1976. *Sol. Phys.* 50:85–98
Kota J. 1992. In *Solar Wind Seven,* ed E Marsch, R Schwenn, pp. 205–8. New York: Pergamon. 711 pp.
Krieger AS, Timothy AF, Roelof EC. 1973, *Sol. Phys.* 29:505–25
Lazarus AJ, Yedidia B, Villanueva L, McNutt RL, Belcher JW, et al. 1988. *Geophys Res. Lett.* 15:1519–22
Low BC. 1993. *Adv. Space Res.* 13:63–69
Low BC. 1994. In *Solar Dynamic Phenomena and Solar Wind Consequences,* ed JJ Hunt, pp. 123–32. Noordwijk: *ESA Publ. SP-373.* 469 pp.
McComas DJ. 1995. *Rev. Geophys. Suppl., US Natl. Rep. to IUGG 1991–1994,* pp. 603–7
McNutt RL. 1988. *Geophys. Res. Lett.* 15:1523–26
Munro RH, Gosling JT, Hildner E, MacQueen RM, Poland AI, Ross CL. 1979. *Sol. Phys.* 61:201–15
Ogilvie KW. 1972. In *Solar Wind,* ed CP Sonett, PJ Coleman, JM Wilcox, pp. 430–34. Washington, DC: *NASA Spec. Rep. SP 308.* 717 pp.
Parker EN. 1963. *Interplanetary Dynamical Processes.* New York: Wiley. 272 pp.
Phillips JL, Balogh A, Bame SJ, Goldstein BE, Gosling JT, et al. 1994. *Geophys. Res. Lett.* 21:1105–8
Phillips JL, Bame SJ, Gosling JT, McComas DJ, Goldstein BE, et al. 1992. *Geophys. Res. Lett.* 19:1239–42
Pizzo VJ. 1978. *J. Geophys. Res.* 83:5563–72
Pizzo VJ. 1980. *J. Geophys. Res.* 85:727–43
Pizzo VJ. 1982. *J. Geophys. Res.* 87:4374–94
Pizzo VJ. 1985. In *Collisionless Shocks in the Heliosphere: Reviews of Current Research,* ed. BT Tsurutani, RG Stone, *Geophys. Monogr. 35,* pp. 51–69. Washington, DC: Am. Geophys. Union. 303 pp.
Pizzo VJ. 1991. *J. Geophys. Res.* 96:5405–20
Pizzo VJ. 1994a. *J. Geophys Res.* 99:4173–83
Pizzo VJ. 1994b. *J. Geophys. Res.* 99:4185–91
Pizzo VJ, Gosling JT. 1994. *Geophys. Res. Lett.* 21:2063–66
Priest ER. 1988. *Ap. J.* 328:848–55
Richardson IG, Cane HV. 1993. *J. Geophys. Res.* 98:15,295–304
Rust DW. 1994. *Geophys Res. Lett.* 21:241–44
Sheeley NR, Howard RA, Koomen MJ, Michels DJ. 1983. *Ap. J.* 272:349–54
Sheeley NR, Howard RA, Koomen MJ, Michels DJ, Schwenn R, et al. 1985. *J. Geophys. Res.* 90:163–75
Siregar E, Roberts DA, Goldstein ML. 1992. *Geophys. Res. Lett.* 19:1427–30
Siregar E, Roberts DA, Goldstein ML. 1993. *J. Geophys. Res.* 98:13,233–46
Siscoe GL, Goldstein BL, Lazarus AL. 1969. *J. Geophys. Res.* 74:1759–62
Smith EJ, Marsden RG, Page DE. 1995. *Science* 268:1005–7
Smith EJ, Neugebauer M, Balogh A, Bame SJ, Erdos G, et al. 1993. *Geophys. Res. Lett.* 20:2327–30
Smith EJ, Wolfe JH. 1976. *Geophys Res. Lett.* 3:137–40
Snyder CW, Neugebauer M, Rao UR. 1963. *J. Geophys. Res.* 63:6361–70
Webb DF. 1992. In *Eruptive Solar Flares,* ed Z Svestka, BV Jackson, ME Machado, *Lect. Notes Phys. 399,* pp. 234–47. New York: Springer-Verlag. 409 pp.
Webb DF, Howard RA. 1994. *J. Geophys. Res.* 99:4201–20
Webb DF, Krieger AS, Rust DM. 1976. *Sol. Phys.* 48:159–86
Whang YC. 1980. *J. Geophys. Res.* 85:2285–95
Wilcox JM, Ness NF. 1965. *J. Geophys. Res.* 70:5793–805
Zwickl RD, Asbridge JR, Bame SJ, Feldman WC, Gosling JT, Smith EJ. 1983. In *Solar Wind Five,* ed M Neugebauer, pp. 711–17. *NASA Conf. Publ.* 2280. 742 pp.

Annu. Rev. Astron. Astrophys. 1996. 34:75–109

SOLAR ACTIVE REGIONS AS DIAGNOSTICS OF SUBSURFACE CONDITIONS

Robert F. Howard

National Solar Observatory, National Optical Astronomy Observatories,[1]
Tucson, Arizona 85726

KEY WORDS: sunspots, plages, magnetic fields, solar activity

ABSTRACT

In the past decade a number of observational and theoretical studies have appeared that address the problem of how both the physical conditions in subsurface layers of the Sun and the nature of the magnetic flux tubes of active regions are reflected in the structure and behavior of these regions at the surface. This review discusses work in this area. Many characteristics of plages and sunspot groups are shown to be related to the conditions encountered by the region flux tube as it rises through the convective zone of the Sun to the surface. Size distributions, rotation and meridional flow rates and their covariances, and characteristics of growth and decay are among the factors that have been shown to depend on the nature of the source magnetic flux tube and the physical effects, such as the Coriolis force and magnetic tension, that act deep in the convection zone.

---

1. INTRODUCTION

Solar active regions have provided clues to the physical nature of the Sun starting with the first serious observations of sunspots in the early seventeenth century. For a long time, work of this sort was concentrated on the determination of solar rotation, and later, in the nineteenth century, sunspot observations were

[1]Operated by the Association of Universities for Research in Astronomy, Inc., under Cooperative Agreement with the National Science Foundation.

0066-4146/96/0915-0075$08.00

employed to demonstrate the existence of the solar activity cycle. In recent decades, the study of active regions on the solar surface has been concerned largely with active phenomena associated with these regions, seen in the photosphere and above. A great many interesting results in the areas of plasma physics and MHD have been obtained in relation to many types of structures and processes observed in and around active regions. This remains a vigorous and exciting field of solar research. [For recent reviews, see Moore & Rabin (1985), Lites (1994), Keil et al (1994), Keller (1993), and Semel et al (1991).] But another equally important area of research involves generally large-scale statistical studies of the characteristics of active regions and what constraints these regularities set on the behavior of the subsurface flux tubes from which the regions are formed. This broad area is the topic of this review. Many of the clues that are seen in the behavior of active regions have, as yet, no clear explanation in the physics of subsurface flux tubes. This may be primarily due to our lack of understanding of the details of the physical processes that take place beneath the surface. Nevertheless, we are beginning to get important clues to these processes, and we are beginning to set limits on some of the physical parameters of the flux tubes. This review highlights some recent progress in this area.

In recent years there has been a considerable rebirth of interest in this field of solar study, as we begin to develop some insights into the nature of the solar activity cycle. We are still not close to an understanding of the cycle mechanism, but observational constraints, many of which come from close study of the gross properties of active regions, such as are discussed here, are beginning to narrow the alternatives open to us in answering one of the major puzzles in the astrophysics of stars: What causes the activity cycles that we observe?

Active regions—which in this review are interpreted to mean the surface magnetic fields of plages and sunspots and the various solar optical and radio features that they are responsible for—are now generally thought to arise from toroidal flux tubes near the base of the convection zone, most likely in the overshoot layer just beneath the base of the convection zone (Parker 1955; Rosner 1980; Golub et al 1981; van Ballegooijen 1982; Schmitt & Rosner 1983; DeLuca & Gilman 1986, 1991; Choudhuri 1990). The active region flux tubes are believed to rise to the surface by buoyant forces (Parker 1979). They are Ω-shaped magnetic flux loops that have their lower ends linked to the subsurface toroidal tubes. During their buoyant rise to the surface, their motions may be influenced by several physical effects, such as the Coriolis force, magnetic tension, drag, and large-scale convective motions (Choudhuri & Gilman 1987; Choudhuri & D'Silva 1990; D'Silva & Choudhuri 1991, 1993; Fan et al 1993, 1994; D'Silva 1993;

D'Silva & Howard 1993). Studies of the motions and orientations of the regions at the surface will give us information about the relative importance of these forces, and thus about the physical conditions beneath the surface. For example, one can explain the morphological asymmetry of the regions that form at the surface from the physical forces mentioned above (Fan et al 1993). Information also can be gained about the flux and field strengths of the flux tubes. All of this information promises to aid in our understanding of the cycle mechanism. Recent reviews of progress in our understanding of the solar dynamo may be found in the papers by Rabin et al (1991), Kontor (1993), and Rozelot (1993), among others. Recent theoretical work suggests that there may be problems with the earlier interpretation of dynamo action on the Sun (Vainshtein et al 1993).

One area of investigation that has received recent attention and is not covered in this review is that of the distribution of active regions in longitude. Although there is a very long and somewhat confusing literature on this subject (e.g. Bogart 1982, Tabaoda & Moreno 1993), I omit mention of any of this work in this overview.

It is shown here that the magnetic fields of plages and sunspots give evidence at times of different dynamic behavior. This most likely reflects the fact that these two types of features arise from flux tubes with different field strengths and/or different fluxes. It is hard to imagine that these features can be very separate in their origins, because sunspots are never seen without accompanying plage magnetic fields.

The plage results, cited here from my own work, come from the analysis of magnetic field measurements with the Mount Wilson magnetograph covering the interval 1967–1995 (Howard 1989). The Mount Wilson sunspot results come from the digitized, daily white-light photographs (Howard et al 1984). These data cover the interval 1917–1985 and represent only sunspot umbrae. The progress made in these areas in the past decade or so, both from these and other compilations of synoptic observations, illustrates the great importance of long synoptic data sets in the study of solar and stellar activity.

2. THE SIZES OF ACTIVE REGIONS

The distributions of active region sizes and total magnetic fluxes are likely to give us information regarding the nature of the subsurface sources of magnetic flux from which the regions arise (Wentzel & Seiden 1992). In this regard it should be noted that a fundamental difference between sunspots and plages may be found in the manner in which they decay. Sunspot evolution is characterized by a generally monotonic decrease in area with time, usually over a relatively short interval—often only a few days (Kiepenheuer 1953, Bray & Loughhead 1965, Martínez Pillet et al 1993). Plages, on the other hand, decay by expanding and weakening over a relatively long interval, i.e. a large fraction of the lifetime of the plage (Leighton 1964, Howard 1965, Harvey 1996). This decay can last for many weeks or even months. In fact, this weakening is such a gradual process that there is often some disagreement about when a plage ceases to be a plage and becomes, instead, enhanced network emission in chromospheric lines. In the case of the Mount Wilson data discussed in this review, the criteria, listed in detail in the earlier paper (Howard 1989) at least have the advantage of being applied consistently and impartially over the full interval covered by the data set. In the past this has not always been the case, and in addition, some studies have relied on eye estimates of plage outlines, which can be uncertain. The positional data given here are determined by a magnetic flux–weighted calculation. It is clear that the size distributions and derived motions of these two types of features will reflect this difference in the process of their decay, and to some extent they will also reflect the difference in the sharpness of these two types of features. Evidence of these effects may be seen in the size distributions and in some of the other results that follow.

2.1 *Plages*

Plages cover a wide range of sizes, from the smallest ephemeral regions, requiring fairly high-resolution observations to detect them, to large, weak features, expanded by the "erosion" effects of local supergranular motions. The ephemeral regions (Golub et al 1979, Martin & Harvey 1979) form a subset of active regions that has been studied as a rather separate entity. However, from an analysis of their size distributions, it is clear that they merely represent the small end of the size distribution of all active regions (Harvey 1993, Harvey & Zwaan 1993).

In recent years, several studies of the area distribution of sunspots, plages, or surface magnetic fields have been published (Tang et al 1984, Schrijver 1988, Bogdan et al 1988, Wang & Sheeley 1989, Howard 1989, Harvey & Zwaan 1993, Ograpishvili 1994). In the Harvey & Zwaan study, each plage is represented only once in the statistics, an advantage not shared by all of the other studies referenced above. Careful corrections were made in the Harvey & Zwaan study for the variation of visibility of plages, especially small plages, across the solar surface, for gaps in the data coverage, for the east-west bias in the visibility of the maximum phase of development, and for the reduced visibility of small regions arising in existing active regions. Harvey & Zwaan analyzed separately those "regions" that arise within previously existing regions, thereby providing an important insight into the nature of magnetic flux eruption at the solar surface.

Harvey & Zwaan (1993) found that the shape of the size distribution of regions that form within existing regions is virtually the same as that for regions that form where there are no preexisting regions. These regions that form within regions make up at least 44% of all regions (larger than 3.5 square degrees). If one takes account of the total area occupied by regions at any one time, then the rate of formation of regions within an existing region must be much higher than that of outside regions (in the active latitudes). For regions arising outside of existing regions, this rate increases by a factor of 3.5 from minimum to maximum of the activity cycle, whereas for regions appearing within existing regions, the rate varies with a significantly lower amplitude.

The plage area distribution from the Harvey & Zwaan (1993) study is shown in Figure 1. The ordinate $n(A)$ is regions per day, normalized to counts per

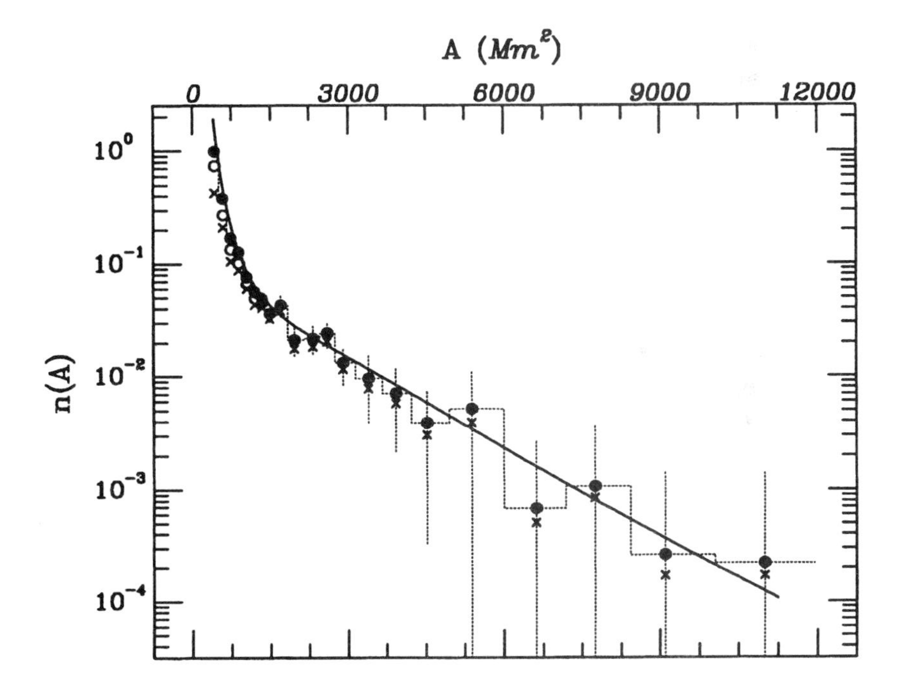

Figure 1 Size distribution of 978 individual active regions selected from National Solar Observatory/Kitt Peak full-disk magnetograms during 29 solar rotations in the interval 1975 through 1986. The ordinate is the number of regions normalized to counts per unit region area (1 square degree) and per one day of observation. The solid line is the fit to Equation 1. The × symbols represent the observed counts; the ∘ symbols represent the number of regions corrected for visibility function, data gaps, and observing the maximum development; and the • symbols and the histogram represent the number of regions corrected also for an underestimate of smaller regions emerging in existing active regions. Vertical bars represent estimated errors of the fully corrected data. (From Harvey & Zwaan 1993.)

unit area (1 square degree) and per day of observation. Clearly the number of regions falls off very rapidly with increasing area. These corrected counts refer to the values corrected as described briefly above. The solid line represents the following polynomial fit to the points:

$$n(A) = \frac{S(A)}{A}\left[c_2 + 2c_3(\log A) + 3c_4(\log A)^2\right], \tag{1}$$

where the cumulative size distribution is represented by

$$\log S(A) = c_1 + c_2(\log A) + c_3(\log A)^2 + c_4(\log A)^3$$

and $c_1 = 1.952 \pm 1.319$, $c_2 = -5.444 \pm 2.908$, $c_3 = 4.900 \pm 2.100$, and $c_4 = -1.798 \pm 0.497$.

The fact that regions are much more likely (per square degree in the active latitudes) to erupt within existing active regions than outside them suggests that at least some of the regions that arise within preexisting regions have an origin that is not similar to that of other regions. That is, they arise as a result of the preexisting magnetic flux tubes at the location of the plage. On the other hand, the fact that the size distribution of the regions that are born within regions is quite similar to that of independent regions tends to suggest a similar origin for all regions. Thus this presents us with a dilemma concerning the origin of magnetic flux that appears within preexisting flux. Is it just flux that comes from the same subsurface flux tubes as other regions and that just happens to arise in a preexisting region, or is the new flux somehow associated with the older flux of the region? Is it, perhaps, a part of the same flux tube that has simply separated from the flux of the original region near the surface? At the moment we have no answers to these questions. More careful studies of the eruption of magnetic flux in existing regions will be needed to resolve this problem. For example, differences in the evolution of these two types of flux might be found.

2.2 *Sunspot Groups*

The area of a sunspot group is defined to be the sum of the areas of all the individual sunspot umbrae that make up the group. One might expect that the size distribution of sunspot groups should be similar to that of plages, since both represent the full magnetic flux of each type of feature. In Figure 2, however, it may be seen that this is not the case. The sunspot group distribution from the Mount Wilson data set (Howard et al 1984) (denoted by *crosses* and the *dashed line*) does not show a sharp increase at small areas. Compare this with Figure 1. This is another difference between the characteristics of plages and sunspots, and it undoubtedly represents a difference either in the characteristics of the subsurface magnetic flux tubes from which these two types of features originate or in the process of their formation at the surface.

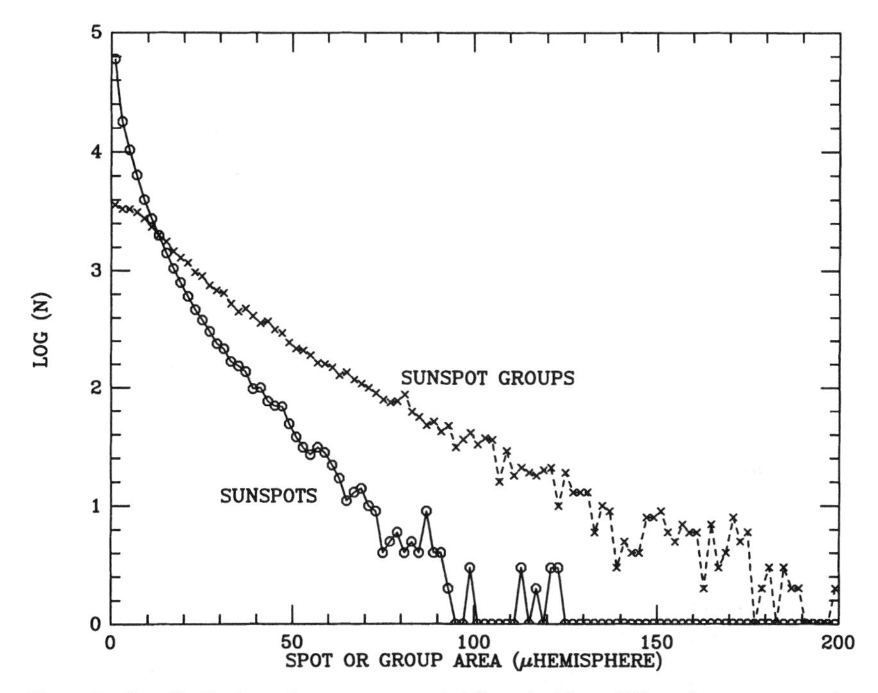

Figure 2 Size distributions of sunspot groups (×) from the Mount Wilson data set representing the years 1917–1985 and individual sunspots (○) from the same data set. There are 36,577 spot groups and 110,100 individual spots represented in this plot. The bin size is 2 μhemisphere.

It should be noted that both the Mount Wilson plage data set and the spot group data set represent the features on each day on which they are observed. Thus, a single feature—a plage or spot group—may be represented several times on a single disk passage. Because this is a feature of both of these data sets, this characteristic cannot affect the differences seen in their distribution functions.

2.3 *Individual Sunspots*

Bogdan et al (1988) investigated the size distribution of individual sunspots (umbrae), using the Mount Wilson data set (Howard et al 1984). Through nearly the entire size range, the distribution is shown to be log normal. All phases of the activity cycle and the individual full cycles within the data set (between 1917 and 1982) show the same log-normal distribution. The mean and the geometric logarithmic standard deviations of the distribution remain essentially constant over the whole time interval; only the number of spots varies in a well-known way over each activity cycle. The log-normal nature of the size distribution and its invariability over the cycle both suggest that the distribution arises from fragmentation of larger flux tubes somewhere beneath the surface during, or more likely before, the birth of the spot group in the photosphere. It should be noted that the Harvey & Zwaan (1993) study did not find a log-normal distribution of plage sizes. This, then, represents another difference between sunspots and plages.

The size distributions of both individual sunspots (*circles*) and sunspot groups (*crosses*) are shown in Figure 2. The bin size for these distributions is 2 μ hemisphere (where 1 μhemisphere is 0.5×10^{-6} of the area of the spherical solar surface). The difference between the two distributions is striking. Clearly, although the sizes of individual spots may result from the fragmentation of a larger flux tube, as noted above, the same certainly cannot be said for sunspot groups, even though some fraction of sunspot groups consists of only one sunspot.

3. ROTATION

One of the first physical properties noted about the Sun centuries ago was its rotation. We now know that the Sun rotates differentially and that sunspots and other magnetic features rotate faster than the surface plasma (Howard & Harvey 1970). Early explanations for this difference centered on possible magnetic linkage to faster-rotating subsurface layers (Howard 1984), but a more recent study (D'Silva & Howard 1994) provides a different explanation: that the effects of buoyancy and drag, coupled with the Coriolis force can lead to a faster rotation rate for emerged magnetic flux tubes. D'Silva & Howard (1994) also provide an explanation for other effects, which are discussed below.

Gilman & Howard (1986a) found that differential rotation (with latitude) of individual sunspots within sunspot groups showed very much the same amplitude as was determined for all sunspots, both within groups and as single-spot groups. This implies that the individual flux tubes that form a sunspot group can be influenced by surface plasma rotation, even though they may originate from a deep subsurface flux tube that might be expected to have a uniform rotation rate.

A north-south difference in sunspot group rotation rates (and in some other parameters as well) has been reported by Antonucci et al (1990), Verma (1993), and Carbonell et al (1993). Slight hemispheric differences may also be seen in individual sunspot rotation (Howard et al 1984), although this may not be statistically significant.

A study of a pair of active regions located close to each other showed that the regions demonstrated similar dynamic behavior (van Driel-Gesztelyi et al 1993). This similarity might be due to some sort of subsurface connection of the flux tubes or to a common subsurface origin for adjacent regions.

A torsional oscillation of the Sun has been observed in the form of a traveling shear wave with an amplitude of a few m s^{-1} from high to low latitudes in each hemisphere as the cycle progresses (Howard & LaBonte 1980, Scherrer & Wilcox 1980, LaBonte & Howard 1982a, Godoli et al 1993a,b). These results come from surface Doppler measurements. The same velocity field has been detected, although just barely, in sunspot rotation results (Godoli & Mazzucconi 1982, Tuominen et al 1983). These motions have also been seen in the rotation of small magnetic features (Komm et al 1993a).

3.1 *Plages*

Among the earliest active region (as opposed to sunspot group) rotation measurements were those of faculae by Newton (1924). More recent determinations have been made by Belvedere et al (1978) and by Donahue & Keil (1995), using disk-integrated observations. Snodgrass (1983), Hejna (1986), and Stenflo (1989) have carried out studies of the rotation of weak magnetic features. Plage motion measurements suffer from the fact that plages are rather ill-defined features that evolve and change their shapes from day to day. After an initial period of rapid growth, they decay slowly by expanding and weakening, with the constant erosion of small flux elements. These factors, combined with generally poor photometric accuracy on a continuing basis, make it very difficult to get an accurate picture of the motions of plages. These difficulties were overcome to some extent with the use of the magnetic field measurements as a substitute for the plage emission (Howard 1989). In this recent work (and some subsequent studies by the same author, which are discussed elsewhere in this review), the use of magnetic flux–weighted positions provided a fairly accurate and objective means of calculating the positions of plages and thus their daily motions. An earlier review of solar rotation gives more details about still earlier work (Howard 1984). This area has seen many contributions from many research groups over the past decade or so.

Plage sidereal rotation results from the Mount Wilson magnetic data were published several years ago (Howard et al 1984, Howard 1990). Figure 3 shows the differential rotation of plages averaged in 5° latitude zones (*dashed lines*). There are 11,909 plages in the interval Jaunuary 1967 to July 1995 represented in this plot. The full error bars for this and all the plots that follow in this review represent ± one standard deviation. Any variations between the north and south hemispheres seen in Figure 3 are probably not real. The scatter in the observed points is too great to show any such possible variations. The smooth dashed line and the smoothed fits in the figures that follow are fits to all the original data points (not to the averages over 5° latitude zones) using the formula

$$\omega = A + B \sin^2 \theta, \tag{2}$$

where θ is the latitude. A least-squares fit to the plage data represented in Figure 3 gives the following values for the coefficients: $A = 14.099 \pm 0.022$ and $B = -2.619 \pm 0.203$. The units are degrees day^{-1} sidereal.

The leading and following portions of plages (as defined by their polarities) show no significant difference in equatorial rotation rates, as is illustrated in Figure 4. Here, the number of plages is the same as is represented in Figure 3 (11,909) from the same data set. The values of the coefficients from least-squares solutions of fits to Equation 2 are, for leading portions of plages, $A = 14.090 \pm 0.027$ and $B = -2.809 \pm 0.242$, and for following portions, $A = 14.060 \pm 0.026$ and $B = -1.909 \pm 0.241$. This result is discussed in more detail below in connection with the same plot for sunspots, but it should be

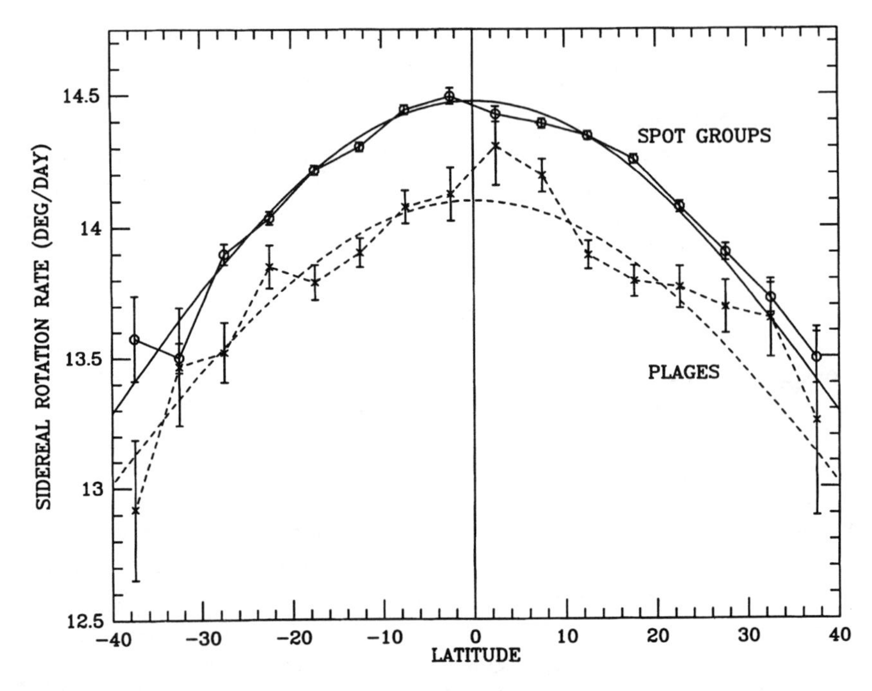

Figure 3 Sidereal rotation as a function of latitude for the same spot group data set represented in Figure 2 and for the Mount Wilson plage data set (1967–1995). Here the $\times$ symbols represent the plages and the $\circ$ symbols represent the sunspot groups. The averages are over 5° latitude zones. The error bars in this and the following figures in this paper represent ± 1 standard deviation.

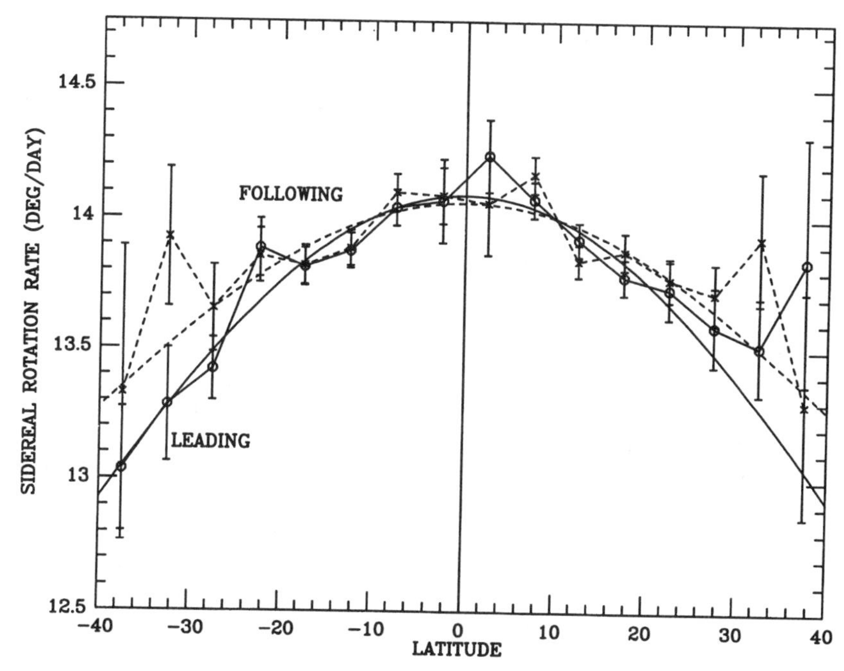

Figure 4 Sidereal rotation as a function of latitude for the leading (○) and following (×) portions of plages averaged over 5° latitude zones. This plot incorporates the same Mount Wilson plage data set represented in Figure 3.

noted here that the manner in which plages decay may affect their apparent dynamic properties. Magnetic flux, in a process analogous to the erosion of a beach, is carried away in small, approximately vertical magnetic flux tubes by the random-walk nature of supergranular, and perhaps granular, motions. This gradual decay, combined with the effects of differential rotation, generally tends to elongate the decaying plage in the east-west direction, especially toward the east because of the direction of the usual tilt of the magnetic axes of the regions. The elongation tends to spread out and dissipate the following portions of the plage more rapidly than the leading portions of the plage, and this will give the appearance of faster-moving following polarity fields (as the magnetic flux to the east disappears from the plage). This effect is clearly seen for the following fields in Figure 4 as a *lower* differential rotation (i.e. a faster rotation rate at high latitudes). It is difficult to estimate the amplitude of this observational bias.

Howard et al (1990) and Komm et al (1993a,b) have examined the rotation rates of small magnetic features that are presumed to be the dispersed remnants of old active regions. They find rotation values close to those of individual sunspots but faster than sunspot groups.

3.2 *Sunspot Groups*

The rotation rate of sunspot groups has been measured and thought about for nearly four centuries. Modern study of this quantity began with Carrington (1863), but the most accurate determination until recent years was that of Newton & Nunn (1951). Other later determinations include those of Ward (1966), Godoli & Mazzuconi (1979), Arévalo et al (1982), Balthasar et al (1986), Ternullo (1990), and Nesme-Ribes et al (1993a) (also see the review by Howard 1984). Most of these measurements were made using the extensive Greenwich sunspot group data set.

One advantage of the more recent analyses of the Mount Wilson sunspot data set (Howard et al 1984, Howard 1991a) and the Kodaikanal (India) data set (Sivaraman et al 1993, Gupta 1994), which use the same reduction codes, is that the daily positions of the sunspot groups are determined by area-weighted positions of the individual sunspots in the group. Within a group, sunspots appear and disappear from time to time, and this makes the definition of the position of the group (and thus its rate of motion in longitude and latitude) somewhat uncertain. Using area-weighted positions of groups allows for a more objective determination of rates of motion, although the ephemeral nature of the individual spots does introduce a scatter, and possibly even a systematic bias, that does not reflect accurately the motion of the basic, subsurface magnetic flux tube(s) forming the group, no matter which method is used to determine the group position. However, most of the time the spots that form and disappear are small spots, and generally they do not strongly influence the group positions.

Figure 3 shows the differential rotation of spot groups from the latest analysis of the Mount Wilson data (*circles* and *solid lines*). Differential rotation in the usual sense is clearly visible and is well above the random errors. The coefficients in the least-squares fit to Equation 2 for these data are $A = 14.476 \pm 0.006$ and $B = -2.875 \pm 0.058$. This solution is represented by the smooth (*solid*) curve seen for the groups in Figure 3. There are 36,577 groups in the interval 1917–1985 represented here.

The rotation of sunspot groups varies with the polarity separation of the group (Howard 1992a). Polarity separations in this study were defined to be the distances between the (area-weighted) positions of spots leading and trailing the area-weighted centroid in longitude. (No magnetic information is available for this data set.) Smaller polarity separations correspond to slower rotation rates. This may be because the magnetic tension in the subsurface flux loop is greater for groups with smaller polarity separations. This relationship, however, is somewhat complicated by the correlation between spot group size and rotation rate and also by the relationship between individual spot size and rotation rate, both of which are discussed below. It is clear that the fundamental association is between individual spot size and rotation rate.

The possibility that rotation rate varies with activity cycle phase is interesting because any such correlation might give us important information about a possible activity cycle mechanism. There is considerable evidence for a significantly faster rotation rate of spot groups near activity minimum and a less convincing case for a slight increase in rotation rate near the maximum phase (Balthasar & Wöhl 1980; Arévalo et al 1982; Lustig 1983; Gilman & Howard 1984, 1990; Hathaway & Wilson 1990; Kambry & Nishikawa 1990). The amplitude of the effect is only around 0.7%, or around 15 m s^{-1}. Such a rotation variation must be associated with the subsurface dynamo in a manner that is not obvious at this time. There is also some evidence for a secular variation of sunspot group rotation (Howard & Harvey 1970, Yoshimura & Kambry 1993).

The tilt angle of an active region is a parameter that is used frequently in this review. It is defined in general as the angle of the magnetic axis of the region on the solar surface, referenced to the local circle of latitude. The magnetic axis may differ from the axis defined visually because magnetic flux is not distributed uniformly within a region. The average tilt angle is roughly 5° in the direction represented by the leading portion of the region located closer to the equator than the following portion. There is a considerable dispersion in this distribution and a tendency for the angle to increase with increasing latitude (Joy's law), which is discussed below. Here we note that the rotation rate of sunspot groups shows a significant variation with tilt angle, as shown in Figure 5. The data come from the Mount Wilson sunspot data set (Howard et al 1984). There are 24,597 multi-spot sunspot groups represented in this plot, which covers the interval 1917 through 1985. The ordinate of Figure 5 is the average residual rotation rate, which is the rotation rate of the spots minus the average rotation rate of all spots for the full interval of the observations for the appropriate latitude. Using the residual rotation rate instead of the absolute rotation rate eliminates the effects of differential rotation, which would result from the latitude-tilt angle relationship (Joy's law). Clearly, from Figure 5, regions with tilt angles near the average value have larger than average residual rotation rates by about 0.1° day^{-1}, or about 0.7%. It is tempting to think that this represents the effect of variations in the magnetic tension of the subsurface field lines connecting the region at the surface with the subsurface source flux tubes. When the tilt angle differs from the average value, this results in a twist in the magnetic flux loop beneath the surface. The resulting increase in magnetic tension may act to slow slightly the rotation rate of the region. Note that in

Figure 5 the peak in the residual rotation rate curve is obviously close to the average tilt angle of spot groups (near 5°) and is not at 0°. This makes the magnetic tension argument given below seem more reasonable.

3.3 *Individual Sunspots*

A distinct advantage of the Mount Wilson (Howard et al 1984) and Kodaikanal (Sivaraman et al 1993) sunspot data sets is that individual sunspot positions and areas have been measured, and thus the motions of these features can be studied for the first time for a large data set. Earlier studies of individual spots were generally confined to single-spot sunspot groups, which are almost always large spots and may represent a special case.

The identification of individual sunspots from one day to the next poses a challenging problem, but is necessary for examining their dynamics (Howard et al 1984). Spot lifetimes are short and proper motions of individual spots

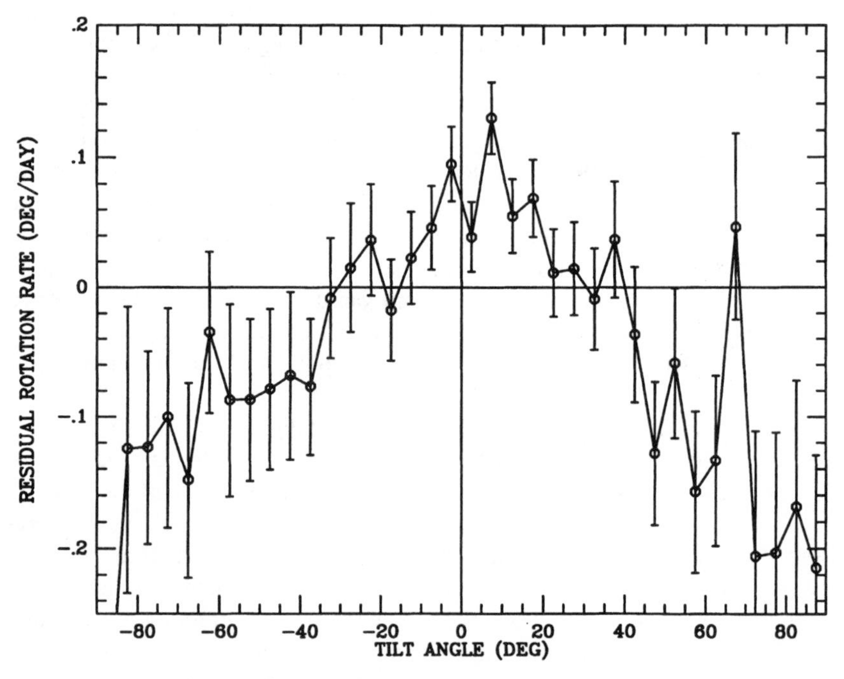

Figure 5 Average residual rotation rates averaged over 5° zones of axial tilt angle for all multi-spot sunspot groups in the Mount Wilson sunspot data set. The residual rotation is the rotation of a group minus the average rotation of groups for the whole data set at the latitude of the group. There are 24,597 sunspot groups represented here. Note that the peak of this curve occurs near the average tilt angle (5–10°), not at 0°.

often make it difficult to be certain which spot is which within a large, or even a medium-sized, group after an interval of one day. This is often true even when one has the luxury of examining daily spot observations by eye. Designing software to carry out this task for a very large data set is a particularly complex problem, involving pattern-recognition logic.

Combining obervations from two observatories widely separated in longitude is a means of shortening the interval between observations, as has been done successfully for data from Mount Wilson and Kodaikanal (Sivaraman et al 1993). Increasing the cadence even more by means of a network of several observatories separated in longitude would further improve the quality of such data and the reliability of identification of sunspots. In order to be certain of sunspot identifications from one observation to the next, it would probably be necessary to take observations with a cadence of two or three hours.

The details of the technique used to match sunspots from one observation to the next are given in the paper by Howard et al (1984) and will not be repeated here. It is a complicated procedure, and at best it is an approximation. Even matching spots by direct examination is an uncertain process in many cases for observations obtained once per day. The lifetimes of small sunspots are sufficiently short (of the order of a day) that ambiguities in identification of spots on the next day arise frequently. Larger, stable spots, of course, are generally not a problem.

Leading and following spots show significantly different rotation rates (Gilman & Howard 1985), as illustrated in Figure 6. Here the coefficients from Equation 2 for leading spots (*circles* and *solid lines*) are $A = 14.513 \pm 0.005$ and $B = -2.660 \pm 0.046$. For following spots (*crosses* and *dashed lines*), the coefficients are $A = 14.421 \pm 0.004$ and $B = -2.736 \pm 0.042$. The difference in the equatorial rates between these two data sets is about 0.6%. At all latitudes, leading spots rotate faster than following spots. This represents, on average, a slow separation of spots of opposite polarities; however, this is only an average. If one constructs a plot similar to Figure 6 but for growing spots only, the result is that leading spots rotate much faster (by about 3%) than following spots. On the other hand, if only decaying spots are analyzed, then following spots rotate faster than leading spots by about the same amount (Howard 1992b). Spots spend more time decaying than they do growing, on average, so the consideration of all spots, as in Figure 6, shows a slight preference for a net expansion of the group, in other words, for leading spots rotating faster than following spots. It has been known for some time that spot groups expand along their magnetic axes early in their lifetimes and contract in the later stage of their existence (Kiepenheuer 1953).

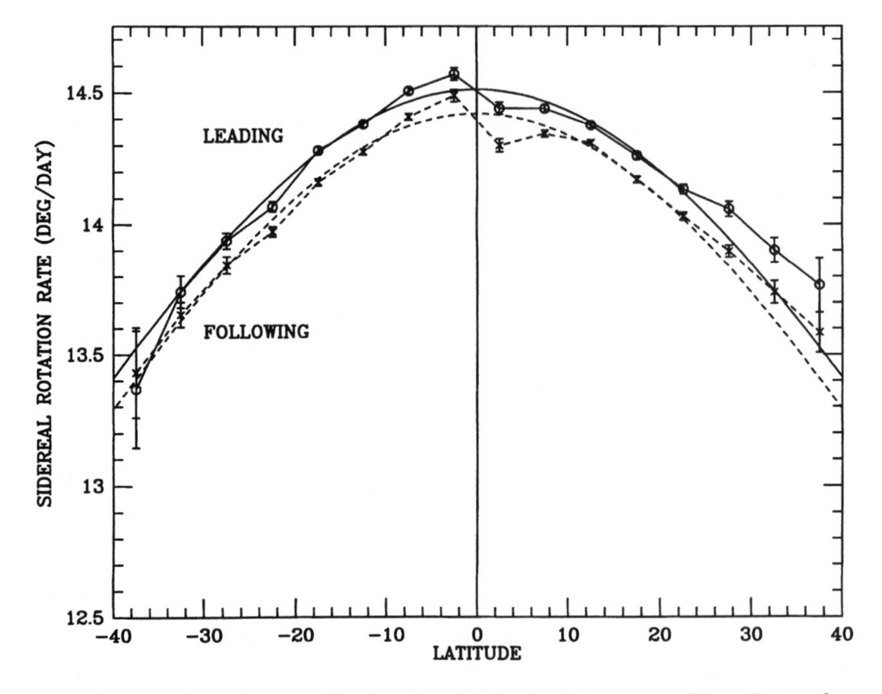

Figure 6 Sidereal rotation rate of leading (○) and following (×) sunspots. These data are from the multi-spot groups in the Mount Wilson data set. There are 39,020 leading spots and 46,075 following spots represented here. Rotation averages are over 5° latitude zones.

A recent study of the growth and decay of active regions (Howard 1992d) combined studies of the correlation of the separation of the leading and following portions of regions with the inclinations of their magnetic field lines to the vertical and the rotation rates of leading and following portions. The resulting patterns of behavior are consistent with the hypothesis that growing regions are represented by a rising magnetic flux loop and that decaying regions are represented by a descending flux loop. Of course, some of the magnetic flux of a decaying region expands and weakens to form the large-scale pattern of weak background magnetic fields that is seen to cover a large fraction of the solar surface at all latitudes (Bumba & Howard 1965), but Howard's (1992d) recent study indicates that it is likely that some of the flux sinks back below the surface, confirming the earlier results of Rabin et al (1984).

Large sunspots rotate at a slower rate than small sunspots. This is illustrated in Figure 7, where the result is clearly quite significant, especially for spots with umbral areas less than about 20 μhemisphere. (A circular sunspot umbra with area 20 μhemisphere has a diameter of about 12 arcsec at disk center, or about 8700 km.) It has also been established that old spots rotate slower than young spots (Ternullo et al 1981; Balthasar et al 1986; Godoli et al 1993a,b; Zuccarello 1993). Note that the ordinate in Figure 7 is the residual rotation rate, which is the difference between the rotation rate of a spot and the average rate for all spots at that latitude for the whole data set. Using the residual rotation rate eliminates any possible contribution from the variation of the average latitude of spots with the phase of the activity cycle.

In a recent study, D'Silva & Howard (1994) have demonstrated that the difference in the rotation rates of large and small flux tubes (sunspots) may be explained by the fact that large flux tubes rise through the convection zone faster than small flux tubes because of the lower drag on the larger flux tubes, and thus they have less time for the Coriolis effect to drain their preceding portions of material and to increase their rotational speed. A similar argument can explain the fact that spots rotate faster than the surrounding plasma.

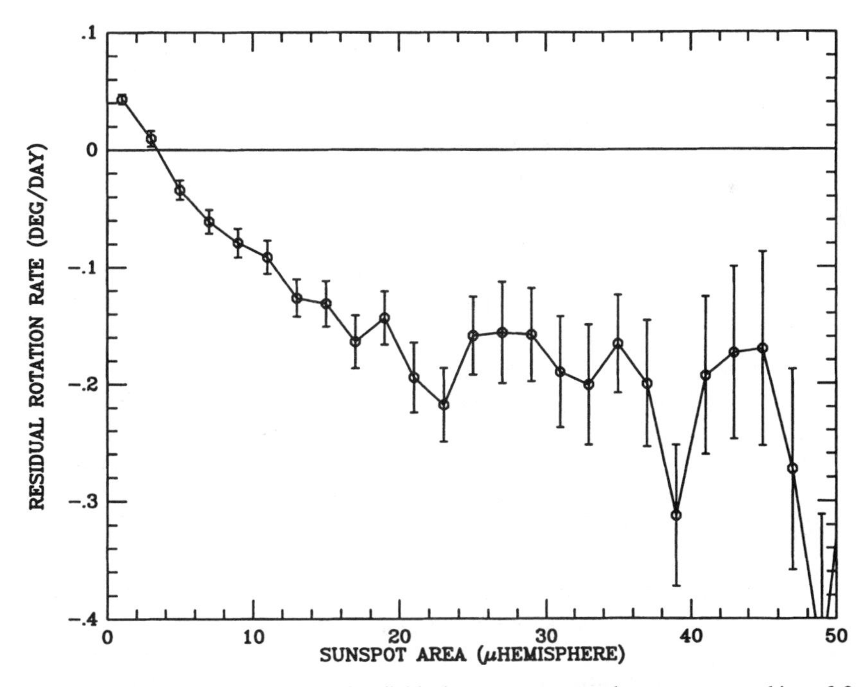

Figure 7 Residual rotation rates of individual sunspots averaged over spot area bins of 2 μhemisphere (1 μhemisphere is the area of a spot with a diameter of about 2 arcsec). The residual rotation is the rotation of a spot minus the average rotation of spots for the whole data set at the latitude of the spot.

4. MERIDIONAL MOTIONS

Studying motions in the north-south direction on the Sun has proved to be a considerable observational challenge, using either Doppler or tracer techniques. Meridional motions are observed to be smaller by at least two orders of magnitude than rotational motions, and they are correspondingly more difficult to detect. Amplitudes for plages are in the range of 15 m s^{-1}, which translates to about 0.1° day^{-1}, which is roughly 1.5 arcsec day^{-1} at low latitudes. This is close to the practical limit of individual positional determinations on the solar surface in conditions of good seeing. Meridional motion amplitudes for spot groups are about an order of magnitude smaller.

In spite of the observational difficulties, there is a rich literature in the area of sunspot group meridional motions (Tuominen 1941, 1955, 1976; Richardson & Schwarzschild 1953; Tuominin & Kyröläinen 1982; Tuominen et al 1983; Balthasar & Wöhl 1980; Howard et al 1984; Balthasar et al 1986; Gilman & Howard 1986b; Kambry et al 1991; Lustig & Wöhl 1991). These studies all give roughly the same result, differing only somewhat in amplitude. A study by Ribes et al (1985) indicates a latitude structure in this flow. Latushko (1994) recently found a similar result. None of the other studies referenced above confirm this. For features as ill-defined as plages, this is a very difficult measurement to make with confidence, even though the amplitudes are observed to be larger. The meridional motion of prominences, on the other hand, can be measured with some accuracy because of their long lifetimes and sharp boundaries (e.g. Waldmeier 1973, Makarov 1984). Komm et al (1993c) measured the meridional motion of small, dispersed magnetic features—the remnants of old active regions. They found a meridional flow that was poleward up to latitudes of about 40° in each hemisphere and that had an amplitude comparable to that of spot results; however, the flow differed from the plage results, suggesting that these weak, widely dispersed features have different subsurface magnetic connections than do the younger, stronger active region fields. Using the Mount Wilson magnetic data, which measures magnetic flux–weighted positions, it is possible to obtain relatively high accuracies for plages when averaging over many regions (Howard 1989). A good review of meridional observations of all kinds was given by Bogart (1987). The importance of meridional circulation in dynamo theories has been emphasized for many years (e.g. Durney 1974).

Doppler measurements of meridional flows suffer from the fact that at low latitudes a line-of-sight velocity measurement will contain only a small component of any north-south motions. Nevertheless, a number of determinations have been made. Probably the most precise measures are those of Duvall (1979) and LaBonte & Howard (1982b), who found poleward motions in the range 10–20 m s^{-1}. Other published studies have produced somewhat disparate results,

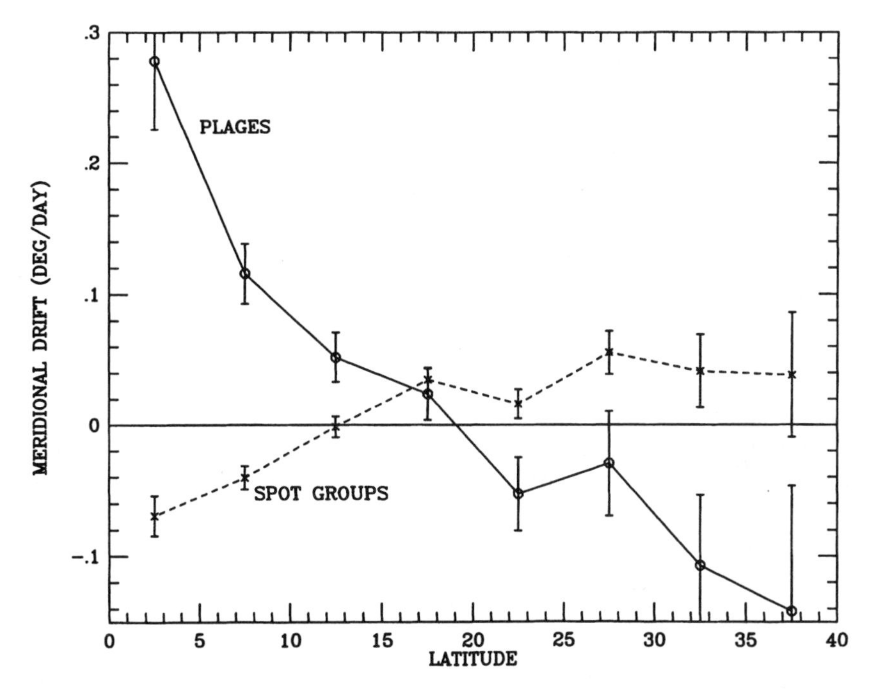

Figure 8 Meridional (north/south) drift of sunspot groups (×) and plages (∘) averaged over zones of 5° in latitude. The data sets are the same as in Figure 3. Positive drift is poleward.

in one case even in the equatorward direction (Pérez-Garde et al 1981, Andersen 1984, Pierce & LoPresto 1987, Lustig & Wöhl 1990, Cavallini et al 1992, Hathaway 1993, Ulrich 1993).

4.1 *Plages*

Figure 8 shows the average daily latitudinal motion of plages, averaged over 5° latitude bins (Howard 1991e). Spot group data (*crosses* and *dashed lines*) are included as well and are discussed below. The units are degrees day^{-1} (1° day^{-1} $\approx$ 140 m s^{-1}). Clearly, the latitude dependence of meridional motion is in the opposite sense for these two types of features.

It is of interest to examine the magnitude of the latitudinal motion of plages as a function of the distance of the plages from the average latitude of plages in each hemisphere at any epoch (Howard 1991b). We will call this quantity ξ, and the average latitude of activity (at any time) will be referred to as ξ_0, which, of course, varies with time, starting at high latitudes early in the cycle and ending near the equator at the time of activity minimum. Figure 9 shows the relationship between plage meridional motion and ξ. Plages poleward of ξ_0 tend to drift to lower latitudes; plages equatorward of ξ_0 tend to drift to higher latitudes. That is, plages move, on average, in the latitudinal direction toward the average latitude of activity. The amplitude of the effect in ξ (cf Figure 9) is larger than that seen in the latitude drift versus latitude plot for plages shown in Figure 8 by about a factor of 2, which suggests that this effect is principally a ξ relationship and not a latitude relationship. That is, the strong ξ dependence of the meridional drift is smeared by the variation of ξ_0 during the cycle, when viewed as a latitude dependence.

Other results from the earlier study (Howard 1991b) regarding the ξ dependence of region parameters are that the distribution of plages is asymmetric about ξ_0 in the direction of a broader distribution on the poleward side than on the equatorward side, with the median of the distribution located about 1° equatorward of ξ_0. Reversed polarity regions show no such asymmetry. The distribution of region areas or magnetic fluxes is peaked at ξ_0 and decreases by about a factor of 2 within about 10° on either side of ξ_0. The relationship

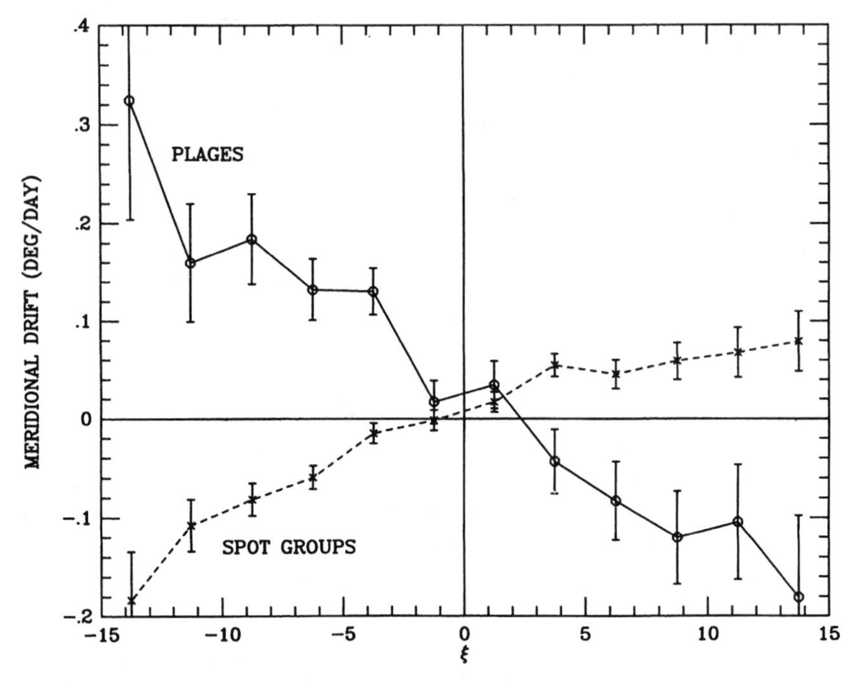

Figure 9 Similar to Figure 8 except that the abcissa is ξ, not latitude. ξ is the latitude distance from the average latitude of activity for the appropriate epoch. Positive drift is poleward. Note that the slope of this variation for sunspot groups is about twice that in Figure 8. (See text.)

between meridional drift and ξ mentioned above applies in general to younger (growing) plages and not to older (decaying) plages. This is confirmed by the work of Lustig & Wöhl (1991). Thus this may represent an effect of the flux tubes rising through the convective zone, and perhaps it does not represent the surface conditions, which is something we already suspect from the Doppler meridional results discussed above.

4.2 *Sunspot Groups*

There is a relationship that has been known for years between meridional drift of sunspot groups and latitude (Richardson & Schwarzschild 1953, Becker 1954, Coffey & Gilman 1969, Schröter & Wöhl 1975, Tuominen 1976, Duvall 1979, Howard 1979, Topka et al 1982, Tuominen & Kyröläinen 1982, Ribes et al 1985, Ribes 1986, Howard & Gilman 1986, Nesme-Ribes et al 1993b). However, in a recent study, Lustig & Wöhl (1994) found no significant drift in sunspot motions. But in general it has been found that meridional motions of spot groups at low latitudes are equatorward in both hemispheres, and at higher latitudes they are poleward. Amplitudes are in the range of 1–2 m s^{-1}. This is shown in Figure 8, where the circles and solid lines represent plage data.

Much of what is discussed in this section is taken from papers by Howard & Gilman (1986) and Howard (1991b). Many of the earlier papers use the extensive Greenwich data set, but these two use the Mount Wilson measurements discussed above (Howard et al 1984). As for the rotation results, the group positions are area-weighted by the individual sunspots.

As mentioned above, spot groups show average meridional motions that are equatorward at low latitudes and poleward at high latitudes. This is also illustrated in Figure 9, which shows the ξ dependence (see above). From Figures 8 and 9 and from the discussion above, one notes that the meridional motions of spots and of plages are in exactly opposite directions. Plages drift toward ξ_0, and spot groups drift away from ξ_0.

For sunspot groups, this effect is most evident for decaying groups, as opposed to the situation for plages where the (opposite) effect is most evident for growing plages. In fact, growing spot groups show a tendency for the opposite slope in a plot such as Figure 9, and so the slope for decaying groups is much greater than that seen in Figure 9 (Howard 1991b).

In the case of sunspot groups, the amplitudes of the effect seen in Figures 8 and 9 are roughly the same. Thus, unlike the situation for plages, where clearly the ξ relationship is dominant, here we cannot decide easily which is the fundamental correlation.

It is difficult to reconcile such different behavior and different amplitudes of these two features—plages and sunspot groups—that are so intimately connected. As was mentioned above, measurements of the motions of these two different types of features could suffer from systematic errors that differ for each feature. For example, as noted above, plages decay by expanding and stretching in longitude. This is especially true for the following portions of plages, which stretch eastward and poleward more rapidly than the leading portions stretch westward and equatorward (Bumba & Howard 1965). This may explain the equatorward motion of plages, especially at high latitudes, although the fact, as noted above, that this effect has a higher amplitude for growing plages than for decaying plages, tends to make this explanation seem less reasonable.

5. COVARIANCE OF LATITUDE AND LONGITUDE MOTIONS

5.1 *Plages*

Belvedere et al (1976) studied the covariance of latitudinal and longitudinal motions using facular data covering a four-year interval. They found results similar to those found previously by Ward (1965) for sunspot groups, which suggested that the differential rotation of the Sun could be maintained using only Reynolds stresses near the surface (Gilman 1986). (Reynolds stresses are nonaxisymmetric motions, having correlations between east-west and either north-south or radial motions, which provide angular momentum transport.) Komm et al (1994) studied the covariance of small magnetic features, presumably the remnants of decayed active regions. They found covariance values one to two orders of magnitude smaller than those for sunspot groups.

Plage rotation–meridional drift covariances have been studied by Howard (1991e). In this study, plages were separated into those that were growing and those that were decaying. Growing plages showed a very strong correlation between meridional drift and residual rotation. These are presumably young plages. Decaying plages, on the other hand, which may be considered to be old and perhaps somewhat disconnected magnetically from lower layers, showed no correlation at all. This makes it appear that the covariance that is found (at least for plages) is not a basic property of the convective motions in the solar surface plasma but is instead a result of the physical forces on the rising flux tube. This possibility is discussed in the next section in connection with the work of D'Silva & Howard (1995).

Figure 10 shows the correlation between meridional drift and residual rotation velocity for plages [*circles* and *solid lines* representing 9595 plages from the Mount Wilson data set (Howard 1989)] covering the years 1967–1995 and sunspot groups [*crosses* and *dashed lines* representing 34,825 sunspot groups from 1917–1985 (Howard et al 1984)]. The correlation is very strong. The slopes are 0.0927 ± 0.0058 for the plage data and 0.0789 ± 0.0027 for the spot

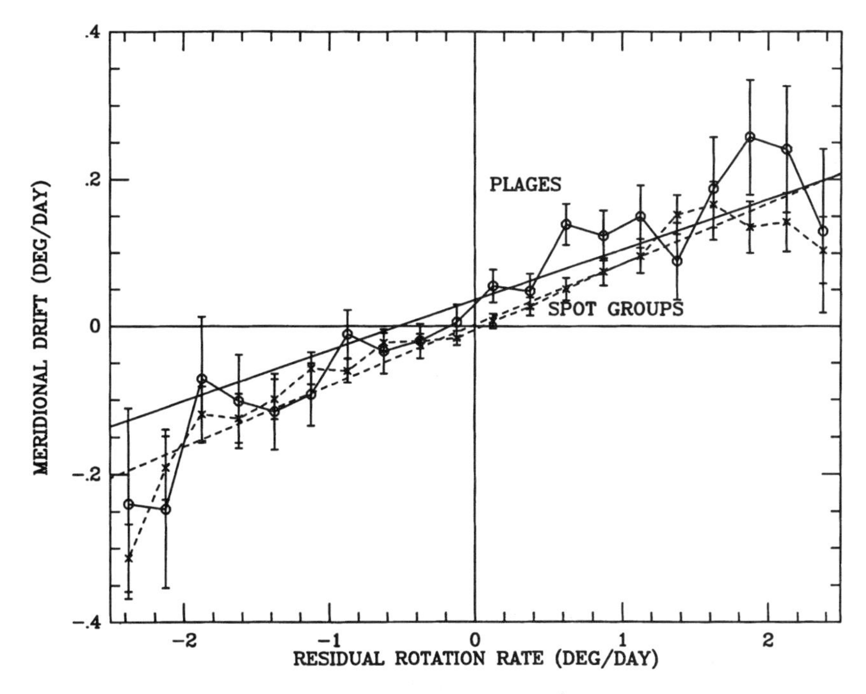

Figure 10 Average values of meridional drift over 0.25° day^{-1} zones of residual rotation rate for plages (○) and spot groups (×). These are the same data sets represented in Figure 3. Positive drift is poleward.

groups. Here the plage results represent the average of all plages, growing and decaying.

5.2 *Sunspot Groups*

The first observational study to reveal a positive correlation of latitudinal and longitudinal motions of sunspot groups was that of Ward (1965), using Greenwich data. Coffey & Gilman (1969) and Paternò et al (1991) found a similar correlation. A recent observational study (Nesme-Ribes et al 1993b) finds no significant covariance from sunspot data, using Meudon observations covering cycle 21. Theoretical models have sought to explain the equatorial acceleration of the Sun as being due to equatorial angular momentum transport from Reynolds stresses that are reflected in the correlations shown in the surface motion data.

Almost from the start there have been criticisms of these observational conclusions on the basis that the observed motions may result from the well-known expansion and contraction of sunspots along the tilted magnetic axes of the sunspot groups. These axes are inclined in such a way (leading spots equatorward of following spots) as to lead to the correlation that is seen (RB Leighton 1965, unpublished manuscript). Gilman & Howard (1984), however, argued that because the effect could be observed for whole sunspot groups, at least some fraction of the observed correlation must be due to Reynolds stresses near the solar surface and that this amount was sufficient to account for the angular momentum transport required to maintain the solar differential rotation. However, D'Silva & Howard (1995) have recently pointed out that the effects of the Coriolis force on rising flux tubes can also account for this surface velocity correlation, with no need for a convective explanation. Thus, this problem remains unresolved. If the covariance of the group motions does not arise from Reynolds stresses in the convective plasma, then there remains the problem of what *is* responsible for the differential rotation.

6. CHARACTERISTICS OF GROWTH AND DECAY OF REGIONS

Some interesting characteristics in the growth and decay of plages and sunspot groups have been found in recent years (Howard 1991d, 1992b). As all observers know, plages grow faster than they decay (Kiepenheuer 1953). The same is generally true of sunspots (Bray & Loughhead 1965) but with the qualifications discussed below. In addition, leading polarity fields grow and decay faster than following polarity fields (Howard 1991d). A further plage result is that reversed-polarity regions grow faster than normal-polarity regions, but this may be a size effect—reversed polarity regions are on average smaller than normal regions, and smaller regions grow faster than bigger regions (Howard 1991d). Regions with rotation rates that deviate from the average rate (for the appropriate latitude) grow and decay faster than regions near the average rotation rate (Howard 1991d). The effect for decaying plages might represent the enhanced cancellation of magnetic flux for field elements that are rotating at a different rate than the surrounding background fields, brought about simply because the regions encounter opposite polarity magnetic flux more rapidly as they move through the surrounding weak-field magnetic patterns. The same effect holds for meridional motion: The larger the deviation from zero meridional flow, the faster the growth and decay rates of the region (Howard 1991d), and again the explanation may be that this (at least the decay) is due to a faster motion of the region fields through the existing weak-field patterns.

In the case of spot groups, Howard (1992b) found that small groups grow faster than they decay, and large groups decay faster than they grow. As in the case of plages, groups with residual rotation rates and meridional drifts close to the average values show slower growth and decay rates than do groups that deviate from these average values. Spot groups show an increase in growth rates near solar activity minimum, followed by a sharp decline.

It is suggested that some of the differences in growth and decay rates between fields of opposite polarities are due to the fact that leading fields are inclined at a greater angle to the local vertical than are following fields (Howard 1991c). This is difficult to confirm from theoretical analysis or simulations because the inclination angles depend to some extent on initial conditions and the details of the model (D'Silva & Howard 1993, 1994). A recent model gives a steeper inclination for following than for leading fields (Moreno-Insertis et al 1992). For a recent review of many topics in this field see Moreno-Insertis (1994).

7. MAGNETIC FIELD ORIENTATIONS

7.1 *Tilt Angles*

One of the more interesting aspects of active region formation and evolution is that of the orientation of the magnetic field lines of the regions. The easiest component to measure is the *tilt angle,* which is defined to be the angle between the line joining the centroids of the two polarities in a region and the local parallel of latitude. Angles with the leading polarity fields equatorward of the following fields are generally considered to be positive. The average tilt angle of regions is roughly $+5°$. Figure 11 shows the tilt angle distribution of plages and sunspot groups separately. The difference seen there of about 5° between the tilt angles of these two features is artificially exaggerated by the low angular resolution of the plage data set. This causes a sharp increase in small region tilt angles near 0°. These tilt angles actually should be considered to extend over a larger range of angles.

The tilt angle distribution at the surface, when the regions first erupt, will be caused to some extent by the effects of the Coriolis force on the rising flux tube (Schmidt 1968, Wang & Sheeley 1989, D'Silva & Howard 1993). The slower the rise to the surface, the longer time this force will have to alter the original flux tube orientation. D'Silva & Howard (1993) showed that the rise time is greater than the rotation period. If the magnetic fields of sunspots are stronger than those of plages, the resulting increased magnetic tension will tend to counteract the Coriolis force, and thus sunspot groups will have lower tilt angles than plages. This mechanism provides a possible explanation of this effect.

One of the more interesting aspects of region tilt angles is the latitude dependence, which lately has been called Joy's law (Hale et al 1919, Wang & Sheeley 1989). There is a clear increase of tilt angle with latitude in both hemispheres.

Figure 12 shows this relationship for both plages and sunspot groups. This latitude dependence and the tilt itself have been attributed to the effect of the Coriolis force on the rising flux tubes (Choudhuri & Gilman 1987, Choudhuri 1989, Wang & Sheeley 1991, D'Silva & Choudhuri 1993, Fisher et al 1995); however, convective motions have also been suggested as a cause of this effect (Baranyi & Ludmány 1992).

This tilt angle–latitude dependence can be reproduced in numerical simulations of thin magnetic flux tubes rising in the convection zone. Reproducing this effect together with the tilt angle–polarity separation effect (Howard 1993), the simulations suggest that the field strengths of the subsurface source toroidal flux tubes should be in the range of 40 to 150 KG (D'Silva & Howard 1993). These results are confirmed by a similar analysis by Fisher et al (1995).

There is a clear relationship between the sizes of spot groups and their tilt angles. Groups with tilt angles near the average values show the largest areas,

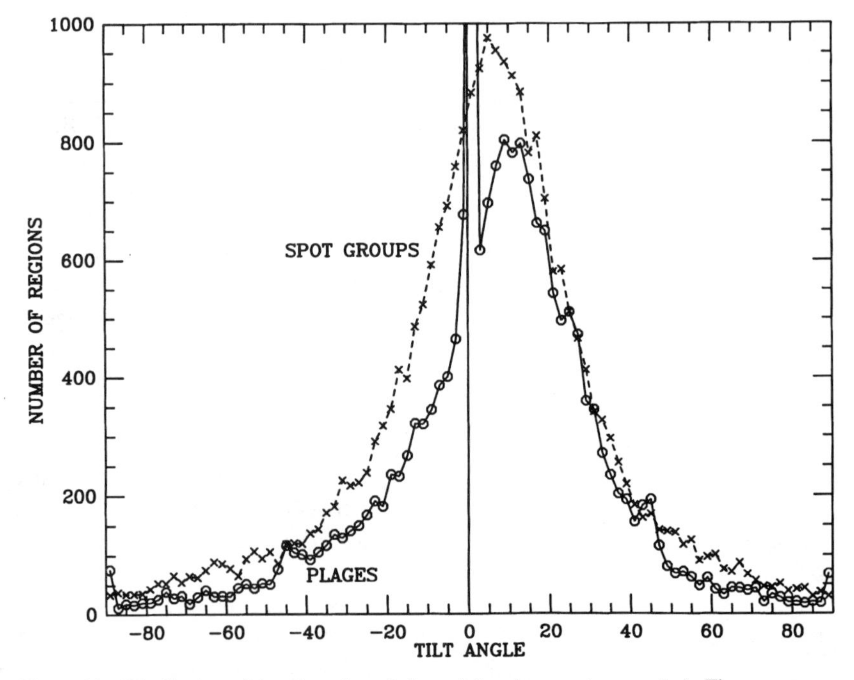

Figure 11 Distribution of the tilt angles of plages (o) and sunspot groups (×). The averages are over zones of 2° in tilt angle. The sunspot data set is the same as in Figure 2, but the Mount Wilson plage data set is larger than the one in Figure 3 because it includes many plages that were not observed on two consecutive days ($N = 22,243$). The large value, off the plot, for the plage tilt angle near 0° (2270 regions) is caused by the low angular resolution of the original data set. Most of these zero-tilt-angle regions are those that consist of two pixels, oriented exactly east-west.

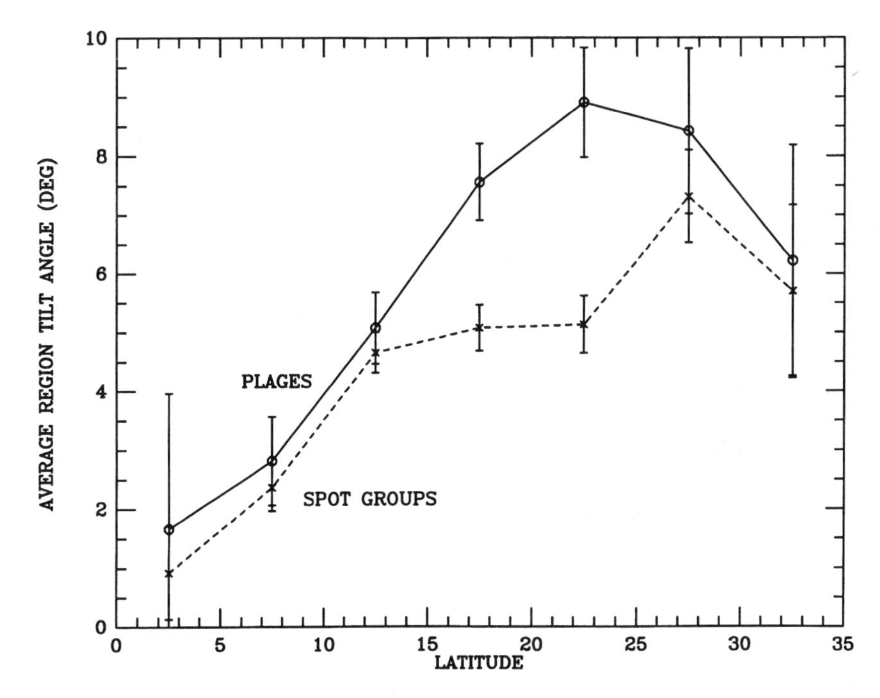

Figure 12 Average tilt angles over 5° zones of latitude (Joy's Law) for plages (o) and sunspot groups (×). The sunspot data set is the same as that in Figure 2, and the plage data set is the same as that in Figure 11.

on average. This effect is quite obvious in Figure 13. Groups with tilt angles that differ from the average by 40° in either direction have average areas that are roughly half those of groups with the usual orientation. Note that the peak of the curve is definitely at the average tilt angle, not at 0°. One explanation for this effect is that random, large-scale convective motions affect the resulting group tilt angles by the time the groups reach the solar surface. The larger the groups, the smaller will be the effects of these motions.

A recent observational study (Howard 1994) has demonstrated that the effects of the Coriolis force can be seen in surface motions of portions of sunspot groups. It should be noted that the leading and following polarities of regions tend to move apart in the early stages of the evolution of a region and move together again later in the lifetime of a region (Kiepenheuer 1953, Zwaan 1985). The correlation between this polarity separation change and magnetic axis rotation is illustrated in Figure 14, where the average daily tilt angle change is given for various values of daily polarity separation change. A positive tilt angle change represents clockwise motion of the group in the north and counterclockwise motion in the south. Negative separation change represents a contraction of the group, i.e. the leading and following spots move toward each other. From Figure 14 we see that a contracting group rotates, on average, counterclockwise in the north and clockwise in the south. This is the sense expected from the Coriolis force. Further analysis (Howard 1994) indicates that the amplitude of the effect is greater at higher latitudes, which is what is expected from the Coriolis effect. Moreover, the total amplitude of the rotational motion is in the correct range for this effect.

7.2 *Field Line Inclinations (to the Local Vertical)*

Field line inclinations have been estimated for a few regions from magnetic measurements, but in general it is difficult to derive such information accurately outside sunspots because of the low sensitivity of the transverse Zeeman effect. Another technique, first used by Minneart (1946) and Gleissberg (1947), is to examine statistically the field strengths or areas of features as a function of distance from the central meridian of the Sun. This technique was first

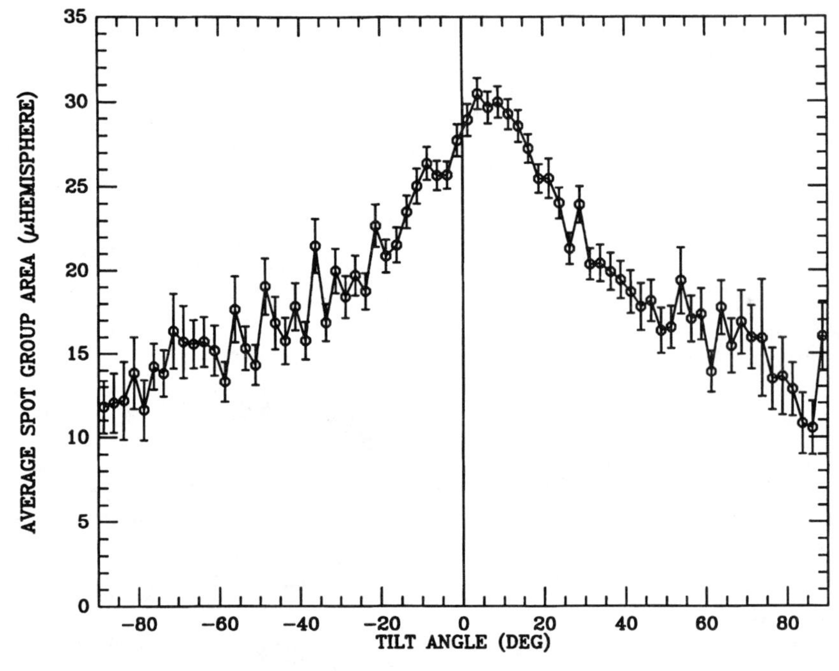

Figure 13 Average sunspot group areas for (multi-spot) groups with different tilt angles. The size of the tilt angle zones over which the averages were taken is 2.5°. There are 24,597 groups represented here from the Mount Wilson sunspot data set.

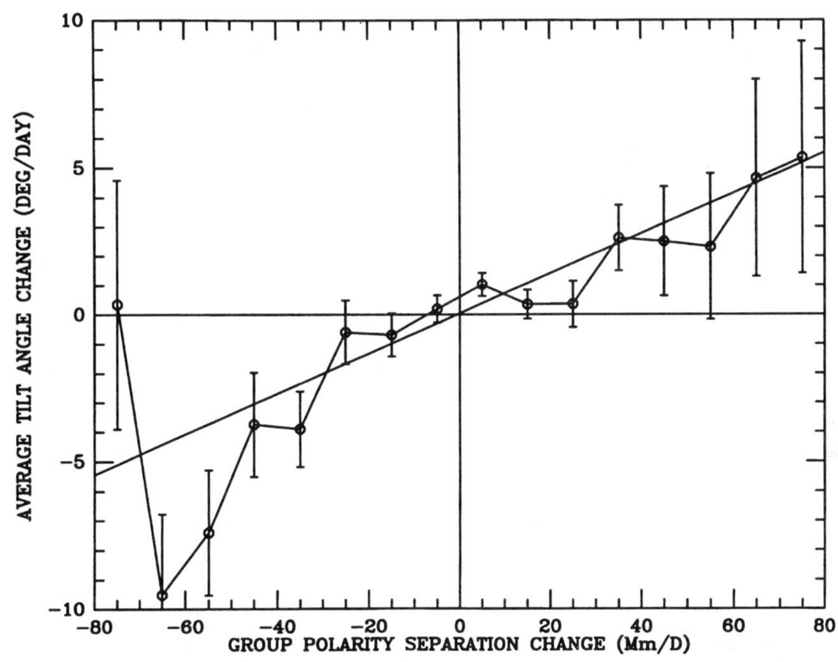

Figure 14 Tilt angle change in degrees day^{-1} for the Mount Wilson multi-spot sunspot data set, averaged over 10° zones in daily sunspot group polarity separation change in Mm day^{-1}.

applied to large-scale magnetograph data by Howard (1974). In more recent studies (Howard 1991c, 1992c), the method has been applied to plages and to sunspot areas (the Mount Wilson sunspot data set contains no magnetic field information). Other recent studies, using the same technique, were carried out by Murray (1992) and Shrauner & Scherrer (1994).

Field lines of plages, seen in the photosphere, are inclined (to the vertical) in such a way as to trail (i.e. they are inclined away from) the rotation by about 2°. Leading and following polarities are inclined toward each other by about 16° (Howard 1991c). For sunspot groups, as in the case of plages, the field lines trail the rotation by several degrees, but, in contrast to the plages, the leading and following spot field lines are inclined away from each other by a degree or so (Minneart 1946, Howard 1992c).

It is interesting to note that there is a very large difference in the inclination angles of plage magnetic field lines between growing and decaying plages (Howard 1991c). Growing regions are sharply inclined toward the west (to lead the rotation). This average angle is nearly 25°. The leading and following fields maintain roughly the same relative inclinations as are derived for all regions as discussed above, i.e. about 16°. Decaying plages are inclined to trail the rotation by about 5° on average, and again the same relative inclinations of the leading and following fields are maintained. This strong inclination of growing plages must be related in some as yet unknown fashion to the forces on the rising flux tubes. The fact that later the inclinations of the plage fields change sign and trail the rotation may suggest that they are affected over time by a differential rotation with depth.

Shrauner & Scherrer (1994) studied large-scale weak fields from the Stanford magnetograph. They demonstrated that the tendency for leading and following fields to be inclined toward each other starts at quite high latitudes in each hemisphere early in an activity cycle and moves slowly equatorward as the cycle progresses. This suggests that an equatorward-moving, subsurface dynamo may affect the orientation of magnetic field lines at the surface, even for weak fields, most of which are generally considered to be the remnants of old active regions. Thus the surface fields may retain a memory of their orientations long after they appear at the solar surface.

The east-west inclinations of magnetic field lines in plages (Howard 1991c) show, on average, values that trail the rotation poleward of ξ_0 (the average latitude of activity at any epoch) and values that lead the rotation (inclined toward the west) equatorward of ξ_0. The average differences in these values are found to be about 4° greater in the north than in the south, which is a large and statistically significant difference. It is difficult to be sure, however, if the basic parameter here is ξ or latitude.

8. DIFFERENCES BETWEEN SUNSPOT GROUPS AND PLAGES

One of the more interesting generalizations that one can make, following on some of the research done in the past decade or so and discussed in this review, is that sunspot groups and plages do not always show the same behavior in their velocities or other characteristics. This is a bit surprising because these two features have always been closely associated in the minds of observers and theorists, and generally they are treated as the same phenomenon. (There is, of course, the obvious difference that sunspots are dark and plages are bright in chromospheric lines, and we see them at slightly different heights in the solar atmosphere, but here the differences are taken to mean phenomenological differences.) Certainly one never sees a sunspot group without an accompanying plage, although the opposite situation can and does at times occur. A brief discussion of some of the more striking differences follows.

1. Sunspots (and sunspot groups) decay in a strikingly different manner than do plages. Sunspots disappear relatively rapidly, in situ, by a simple shrinking of their outer boundary (Bray & Loughhead 1965). High-resolution

observations of decaying spots (Sheeley 1969, Sheeley & Bhatnagar 1971, Harvey & Harvey 1973, Vrabec 1974, Brickhouse & LaBonte 1988) indicate that the decay takes place by means of directed outward flow of small magnetic flux elements. Plages, on the other hand, decay by the expansion and weakening of the plage as it ages. Supergranular motions erode the plage, removing flux elements from its boundary and distributing them widely in the surrounding atmosphere (Leighton 1964, Howard 1965, Lawrence & Schrijver 1993), which results in an enhanced supergranular network magnetic pattern surrounding decaying plages.

2. Sunspot groups and plages differ markedly in their size distributions, as is seen in Figures 1 and 2. The discrepancy here cannot be due to the limited angular resolution of the plage data set, because, if anything, that would lead to a discrepancy in the opposite direction. Because of this limitation, the smallest ephemeral regions are not completely represented in Figure 2. The size distribution of individual sunspots is closer to that of plages, but still the comparison is not good.

3. The rotation rates of plages and sunspot groups differ, as can be seen in Figure 3. The difference is rather large—about 2.5%, or 50 m s^{-1}—and it is clearly significant. It is difficult to draw conclusions from this comparison, because many factors play a role in these averages. Thus it is not possible to say that for any individual active region the spots may rotate slower than the plage. The reason is primarily that the rotation rates of both spot groups and plages depend on the size of the feature (cf Figure 7); in particular, large plages rotate slower than small plages. Many large plages are in the later stages of their development, without sunspots, and thus they will slant the plage rotation average to smaller values that are not related to spot group rotation rates. In addition, as mentioned above, the nature of the decay process in plages may affect the measured daily positions and thus the rotation rates.

4. Meridional motions of plages and sunspot groups are in the opposite direction. Plages drift toward the average latitude of activity (ξ_0), and spot groups drift away from this latitude. Again this is a relatively large and significant result, as may be seen in Figures 8 and 9. Actually the tendencies are in some sense the same. Growing plages or spot groups show a negative slope (such as is seen for all plages) in a plot such as Figure 9, and decaying plages or spot groups show a positive (opposite) slope. But for the plages, the amplitude of the negative slope is much greater than that of the positive slope, and for sunspot groups the situation is reversed. The result for all plages or spot groups is that the slopes are reversed for these two types of features.

5. Field line inclinations to the vertical present a particularly interesting and significant difference between plages and sunspot groups. Plages leading and following polarity fields are inclined toward each other by about 16°, whereas the inclinations of leading and following spots are *away* from each other by a few degrees. In both cases, the overall inclinations are the same—to trail the rotation by a few degrees. Growing plages are strongly inclined to lead (i.e. are inclined toward) the rotation, compared to decaying spots (on an absolute scale), whereas growing spot groups and spots are slightly inclined to trail the rotation, compared to decaying features.

6. Sunspot groups and faculae show activity cycles with slightly different phases (Foukal 1993).

9. CONCLUSIONS

It is possible to study the physical processes acting on the rising magnetic flux tubes in the convective zone by examining average characteristics of the active regions formed by these flux tubes. These characteristics—size distributions, motions, orientations—in some instances can set constraints on the physical effects that act on the rising flux tubes. A number of these observational results have been discussed in this review. Here the emphasis has been on the observations, largely because the theoretical conclusions are not always obvious and require extensive study, including detailed simulations with MHD codes. It seems clear that this field is just now developing to the point where we may begin to draw useful conclusions. The bulk of the results discussed in this review have come in the past five or ten years. Continued work along these lines, combined with the powerful helioseismic techniques now being perfected (Harvey 1995), promise within the next few years to greatly expand our knowledge of the conditions within the solar convection zone and to give us valuable clues about the nature of the solar dynamo. These are exciting times.

ACKNOWLEDGMENTS

The Mount Wilson observations discussed in this review were made possible through the dedication and hard work of several generations of observers. I am indebted to Dr. Roger K Ulrich for the use of these observations and to Dr. John W Harvey for some suggestions on an earlier draft of this paper.

Literature Cited

Andersen BN. 1984. *Sol. Phys.* 94:49–56

Antonucci E, Hoeksema JT, Scherrer PH. 1990. *Ap. J.* 360:296–304

Arévalo MJ, Gomez R, Vázquez M, Balthasar M, Wöhl H. 1982. *Astron. Astrophys.* 111:266–71

Balthasar H, Vázquez M, Wöhl H. 1986. *Astron. Astrophys.* 155:87–98

Balthasar H, Wöhl H. 1980. *Astron. Astrophys.* 92:111–16

Baranyi T, Ludmány A. 1992. *Sol. Phys.* 139:247–54

Becker U. 1954. *Z. Astrophys.* 34:129–36

Belvedere G, Godoli G, Motta S, Paternò L, Zappalà RA. 1976. *Sol. Phys.* 46:23–28

Belvedere G, Zappalà RA, D'Arrigo C, Motta S, Pirronello V, et al. 1978. *Proc. Workshop on Solar Rotation*, ed. G Belvedere, L Paternò. Univ. Catania Publ. No. 162, pp. 189–203. Catania: Univ. Catania

Bogart RS. 1982. *Sol. Phys.* 76:155–65

Bogart RS. 1987. *Sol. Phys.* 110:23–34

Bogdan TJ, Gilman PA, Lerche I, Howard RF. 1988. *Ap. J.* 327:451–56

Bray RJ, Loughhead RE. 1965. *Sunspots.* New York: Wiley

Brickhouse NS, LaBonte BJ. 1988. *Sol. Phys.* 115:43–60

Bumba V, Howard RF. 1965. *Ap. J.* 141:1502–12

Carbonell M, Oliver R, Ballester JLB. 1993. *Astron. Astrophys.* 274:497–504

Carrington RC. 1863. *Observations of the Spots on the Sun.* London: Williams & Norgate. 412 pp.

Cavallini F, Ceppatelli G, Righini A. 1992. *Astron. Astrophys.* 254:381–86

Choudhuri AR. 1989. *Sol. Phys.* 123:217–39

Choudhuri AR. 1990. *Ap. J.* 355:733–44

Choudhuri AR, D'Silva S. 1990. *Astron. Astrophys.* 239:326–34

Choudhuri AR, Gilman PA. 1987. *Ap. J.* 316:788–800

Coffey HE, Gilman PA. 1969. *Sol. Phys.* 9:423–26

DeLuca EE, Gilman PA. 1986. *Geophys. Astrophys. Fluid Dyn.* 37:85–127

DeLuca EE, Gilman PA. 1991. In *Solar Interior and Atmosphere,* ed. AN Cox, WC Livingston, MS Matthews, pp. 275–303. Tucson: Univ. Ariz. Press

Donahue RA, Keil SL. 1995. *Sol. Phys.* 159:53–62

D'Silva S. 1993. *Ap. J.* 407:385–97

D'Silva S, Choudhuri AR. 1991. *Sol. Phys.* 136:201–19

D'Silva S, Choudhuri AR. 1993. *Astron. Astrophys.* 272:621–33

D'Silva S, Howard RF. 1993. *Sol. Phys.* 148:1–9

D'Silva S, Howard RF. 1994. *Sol. Phys.* 151:213–30

D'Silva S, Howard RF. 1995. *Sol. Phys.* 159:63–88

Durney BR. 1974. *Ap. J.* 190:211–21

Duvall TL Jr. 1979. *Sol. Phys.* 63:3–15

Fan Y, Fisher GH, DeLuca EE. 1993. *Ap. J.* 405:390–401

Fan Y, Fisher GH, McClymont AN. 1994. *Ap. J.* 436:907–28

Fisher GH, Fan Y, Howard RF. 1995. *Ap. J.* 438:463–71

Foukal P. 1993. *Sol. Phys.* 148:219–32

Gilman PA. 1968. In *Physics of the Sun,* Vol. 1, ed. PA Sturrock, TE Holzer, DM Mihalis, RK Ulrich, pp. 95–160. Dordrecht: Reidel

Gilman PA, Howard RF. 1984. *Sol. Phys.* 93:171–75

Gilman PA, Howard RF. 1985. *Ap. J.* 295:233–40

Gilman PA, Howard RF. 1986a. *Ap. J.* 303:480–85

Gilman PA, Howard RF. 1986b. *Ap. J.* 307:389–94

Gilman PA, Howard RF. 1990. *Ap. J.* 357:271–74

Gleissberg W. 1947. *Observatory* 67:60–62

Godoli G, Mazzucconi F. 1979. *Sol. Phys.* 64:247–54

Godoli G, Mazzucconi F. 1982. *Astron. Astrophys.* 116:188–89

Godoli G, Mazzucconi F, Piergianni I. 1993a. *Mem. Soc. Astron. Ital.* 64:787–89

Godoli G, Mazzucconi F, Piergianni I. 1993b. *Sol. Phys.* 148:195–200

Golub L, Davis JM, Krieger AS. 1979. *Ap. J. Lett.* 229:L145–50

Golub L, Rosner R, Vaiana GS, Weiss NO. 1981. *Ap. J.* 243:309–16

Gupta SS. 1994. *Studies on the variation of rotation on the surface and in depth on the sun in relation to the photospheric magnetic fields.* Thesis. Ravishankar Univ., Raipur, India

Hale GE, Ellerman F, Nicholson SB, Joy AH. 1919. *Ap. J.* 49:153–78

Harvey J. 1995. *Science* 48:32–38

Harvey KL. 1993. See Zirin et al 1993, pp. 488–91

Harvey KL. 1996. *Sol. Phys.* In press

Harvey KL, Harvey JW. 1973. *Sol. Phys.* 28:61–71

Harvey KL, Zwaan C. 1993. *Sol. Phys.* 148:85–118

Hathaway DH. 1993. *Global Oscillation Net-*

work Group (GONG) Conf. on Seismic Investigation of the Sun and Stars, ed. TM Brown. Astron. Soc. Pac. Conf. Ser. 42:265–68
Hathaway DH, Wilson RM. 1990. *Ap. J.* 357:271–74
Hejna L. 1986. *Bull. Astron. Inst. Czech.* 37:175–79
Howard RF. 1965. In *Stellar and Solar Magnetic Fields,* ed. R Lüst, pp. 129–43. Amsterdam: North Holland
Howard RF. 1974. *Sol. Phys.* 39:275–87
Howard RF. 1979. *Ap. J. Lett.* 228:L45–50
Howard RF. 1984. *Annu. Rev. Astron. Astrophys.* 22:131–55
Howard RF. 1989. *Sol. Phys.* 123:271–84
Howard RF. 1990. *Sol. Phys.* 126:299–309
Howard RF. 1991a. *Sol. Phys.* 135:327–37
Howard RF. 1991b. *Sol. Phys.* 135:43–55
Howard RF. 1991c. *Sol. Phys.* 134:233–46
Howard RF. 1991d. *Sol. Phys.* 131:239–57
Howard RF. 1991e. *Sol. Phys.* 131:259–68
Howard RF. 1992a. *Sol. Phys.* 142:233–48
Howard RF. 1992b. *Sol. Phys.* 137:51–65
Howard RF. 1992c. *Sol. Phys.* 137:205–13
Howard RF. 1992d. *Sol. Phys.* 142:47–65
Howard RF. 1993. *Sol. Phys.* 145:105–09
Howard RF. 1994. *Sol. Phys.* 149:23–29
Howard RF, Gilman PI, Gilman PA. 1984. *Ap. J.* 283:373–84
Howard RF, Gilman PA. 1986. *Ap. J.* 307:389–94
Howard RF, Harvey JW. 1970. *Sol. Phys.* 12:23–51
Howard RF, Harvey JW, Forgach S. 1990. *Sol. Phys.* 130:295–311
Howard RF, LaBonte BJ. 1980. *Ap. J. Lett.* 239:L33–36
Kambry MA, Nishikawa J. 1990. *Sol. Phys.* 126:89–100
Kambry MA, Nishikawa J, Sakurai T, Ichimoto K, Hiei E. 1991. *Sol. Phys.* 132:41–48
Keil SL, Balasubramaniam KS, Bernasconi P, Smaldone LA, Cauzzi G. 1994. In *Solar Active Region Evolution: Comparing Models with Observations,* ed. KS Balasubramaniam, GW Simon, ASP Conf. Ser. 68:265–82. San Francisco: Astron. Soc. Pac.
Keller C. 1993. See Zirin et al 1993, pp. 3–10
Kiepenheuer KO. 1953. In *The Sun,* ed. GP Kuiper, pp. 322–465. Chicago: Univ. Chicago Press
Komm RW, Howard RF, Harvey JW. 1993a. *Sol. Phys.* 143:19–39
Komm RW, Howard RF, Harvey JW. 1993b. *Sol. Phys.* 145:1–10
Komm RW, Howard RF, Harvey JW. 1993c. *Sol. Phys.* 147:207–23
Komm RW, Howard RF, Harvey JW. 1994. *Sol. Phys.* 151:15–28
Kontor NN. 1993. *Adv. Space Res.* 13:417–27
LaBonte BJ, Howard RF. 1982a. *Sol. Phys.* 75:161–78
LaBonte BJ, Howard RF. 1982b. *Sol. Phys.* 80:361–72
Latushko S. 1994. *Sol. Phys.* 149:231–41
Lawrence JK, Schrijver CJ. 1993. *Ap. J.* 411:402–05
Leighton RB. 1964. *Ap. J.* 140:1547–62
Lites BW. 1994. In *Chromospheric Dynamics,* ed. M Carlsson, pp. 1–23. Oslo: Inst. Theor. Phys., Univ. Oslo, Norway
Lustig G. 1983. *Astron. Astrophys.* 125:355–58
Lustig G, Wöhl H. 1990. *Astron. Astrophys.* 229:224–27
Lustig G, Wöhl H. 1991. *Astron. Astrophys.* 249:528–32
Lustig G, Wöhl H. 1994. *Sol. Phys.* 152:221–26
Makarov VI. 1984. *Sol. Phys.* 93:393–96
Martin SF, Harvey KL. 1979. *Sol. Phys.* 64:93–108
Martínez Pillet V, Moreno-Insertis M, Vázquez M. 1993. See Zirin et al 1993, pp. 67–70
Minneart MGJ. 1946. *MNRAS* 106:98–100
Moore R, Rabin D. 1985. *Annu. Rev. Astron. Astrophys.* 23:239–66
Moreno-Insertis F. 1994. In *Solar Magnetic Fields,* ed. M Schüssler, W Schmidt, pp. 117–35. Cambridge: Cambridge Univ. Press
Moreno-Insertis F, Schüssler M, Ferriz-Mas A. 1992. *Astron. Astrophys.* 264:686–700
Murray N. 1992. *Ap. J.* 401:386–97
Nesme-Ribes E, Ferriera EN, Mein P. 1993a. *Astron. Astrophys.* 274:521–33
Nesme-Ribes E, Ferriera EN, Mein P. 1993b. *Astron. Astrophys.* 276:211–18
Newton HW. 1924. *MNRAS* 84:431–42
Newton HW, Nunn ML. 1951. *MNRAS* 111:413–21
Ograpishvili N. 1994. *Sol. Phys.* 149:93–104
Parker EN. 1955. *Ap. J.* 121:491–507
Parker EN. 1979. *Cosmical Magnetic Fields.* Oxford: Clarendon
Paternò L, Spadero D, Zappalà RA, Zuccarello F. 1991. *Astron. Astrophys.* 252:337–42
Pérez-Garde M, Vázquez M, Schwan H, Wöhl H. 1981. *Astron. Astrophys.* 93:67–70
Pierce AK, LoPresto JC. 1987. *Bull. Am. Astron. Soc.* 19:935
Rabin D, Moore R, Hagyard MJ. 1984. *Ap. J.* 287:404–11
Rabin DM, DeVore CR, Sheeley NR Jr, Harvey KL, Hoeksema JT. 1991. In *Solar Interior and Atmosphere,* ed. AN Cox, WC Livingston, MS Matthews, pp. 781–843. Tucson: Univ. Ariz. Press
Ribes E. 1986. *C. R. Acad. Sci. Paris* 302:871–73
Ribes E, Mein P, Mangeney A. 1985. *Nature* 318:170–71
Richardson RS, Schwarzschild M. 1953. *Prob-*

lemi della Fisica Solare, Fond. Alessandro Volta. Atti. Conv. 11:228–49. Rome: Accad. Naz. Lincei

Roselot JP. 1993. *Adv. Space Res.* 13:439–42

Rosner R. 1980. In *Cool Stars, Stellar Systems, and the Sun, Smithsonian Astrophys. Obs. Rep. No. 389,* ed. AK Dupree, pp. 79–96

Scherrer PH, Wilcox JM. 1980. *Ap. J. Lett.* 239:L89–90

Schmidt HU. 1968. *Structure and Development of Solar Active Regions, IAU Symp. No. 35,* ed. KO Kiepenheuer, pp. 95–107. Dordrecht: Reidel

Schmitt JHMM, Rosner R. 1983. *Ap. J.* 265:901–24

Schrijver CJ. 1988. *Astron. Astrophys.* 189:163–72

Schröter EH, Wöhl H. 1975. *Sol. Phys.* 42:3–16

Semel M, Mouradian Z, Soru-Escaut I, Maltby P, Rees D, et al. 1991. In *Solar Interior and Atmosphere,* ed. AN Cox, WC Livingston, MS Matthews, pp. 844–89. Tucson: Univ. Ariz. Press

Sheeley NR Jr. 1969. *Sol. Phys.* 9:347–57

Sheeley NR Jr, Bhatnagar A. 1971. *Sol. Phys.* 19:338–46

Shrauner JA, Scherrer PH. 1994. *Sol. Phys.* 153:131–41

Sivaraman KR, Gupta SS, Howard RF. 1993. *Sol. Phys.* 146:27–47

Snodgrass HB. 1983. *Ap. J.* 270:288–99

Stenflo JO. 1989. *Astron. Astrophys.* 210:403–09

Tabaoda RER, Moreno GG. 1993. *Sol. Phys.* 144:399–402

Tang F, Howard RF, Adkins JM. 1984. *Sol. Phys.* 91:75–86

Ternullo M. 1990. *Sol. Phys.* 127:29–50

Ternullo M, Zappalà RA, Zucarello F. 1981. *Sol. Phys.* 74:111–15

Topka K, Moore R, LaBonte BJ, Howard R. 1982. *Sol. Phys.* 79:231–45

Tuominen J. 1941. *Z. Astrophys.* 21:96–108

Tuominen J. 1955. *Z. Astrophys.* 37:145–48

Tuominen J. 1976. *Sol. Phys.* 47:541–50

Tuominen J, Kyröläinen J. 1982. *Sol. Phys.* 79:161–72

Tuominen J, Tuominen I, Kyröläinen J. 1983. *MNRAS* 205:691–704

Ulrich RK. 1993. *Inside the Stars, IAU Colloq. No. 137,* ed. WW Weiss, A Baglin, ASP Conf. Ser. 40:26–42. San Francisco: Astron. Soc. Pac.

Vainshtein SI, Parker EN, Rosner R. 1993. *Ap. J.* 404:773–80

van Ballegooijen AA. 1982. *Astron. Astrophys.* 113:99–112

van Driel-Gesztelyi L, Csepura G, Nagy I, Gerlei O, Schmieder B, et al. 1993. *Sol. Phys.* 145:77–94

Verma VK. 1993. *Ap. J.* 403:797–800

Vrabec D. 1974. *Chromospheric Fine Structure, IAU Symp. No. 56,* ed. RG Athay, pp. 201–31. Dordrecht: Reidel

Waldmeier M. 1973. *Sol. Phys.* 28:389–98

Wang Y-M, Sheeley NR Jr. 1989. *Sol. Phys.* 124:81–100

Wang Y-M, Sheeley NR Jr. 1991. *Ap. J.* 375:761–70

Ward F. 1965. *Ap. J.* 141:534–47

Ward F. 1966. *Ap. J.* 145:416–25

Wentzel DG, Seiden PE. 1992. *Ap. J.* 390:280–89

Yoshimura H, Kambry MA. 1993. *Sol. Phys.* 143:205–14

Zirin H, Ai G, Wang H, eds. 1993. *The Magnetic and Velocity Fields of Solar Active Regions,* ASP Conf. Ser. Vol. 41. San Francisco: Astron. Soc. Pac.

Zuccarello F. 1993. *Astron. Astrophys.* 272:587–94

Zwaan C. 1985. *Sol. Phys.* 100:397–414

Annu. Rev. Astron. Astrophys. 1996. 34:111–54

BIPOLAR MOLECULAR OUTFLOWS FROM YOUNG STARS AND PROTOSTARS

Rafael Bachiller

Observatorio Astronómico Nacional (IGN), Campus Universitario, Apartado 1143, E–28800 Alcalá de Henares (Madrid), Spain

KEY WORDS: star formation, interstellar medium, interstellar molecules, jets

ABSTRACT

A violent outflow of high-velocity gas is one of the first manifestations of the formation of a new star. Such outflows emerge bipolarly from the young object and involve amounts of energy similar to those involved in accretion processes. The youngest (proto-)stellar low-mass objects known to date (the Class 0 protostars) present a particularly efficient outflow activity, indicating that outflow and infall motions happen simultaneously and are closely linked since the very first stages of the star formation processes.

This article reviews the wealth of information being provided by large millimeter-wave telescopes and interferometers on the small-scale structure of molecular outflows, as well as the most recent theories about their origin. The observations of highly collimated CO outflows, extremely high velocity (EHV) flows, and molecular "bullets" are examined in detail, since they provide key information on the origin and propagation of outflows. The peculiar chemistry operating in the associated shocked molecular regions is discussed, highlighting the recent high-sensitivity observations of low-luminosity sources. The classification schemes and the properties of the driving sources of bipolar outflows are summarized with special attention devoted to the recently identified Class 0 protostars. All these issues are crucial for building a unified theory on the mass-loss phenomena in young stars.

0066-4146/96/0915-0111$08.00

1. INTRODUCTION

The study of mass-loss phenomena from young stars started in the early 1950s with the discovery by Herbig (1951) and Haro (1952) of small nebulosities with peculiar emission line spectra. The so-called Herbig-Haro (HH) objects were soon associated with stellar winds (Osterbrock 1958) and later found to be due to the interaction of a highly supersonic stellar wind with the ambient surrounding material (Schwartz 1975). Measurements of proper motions (Cudworth & Herbig 1979) confirmed that the ejection originates from a newly formed star. Moreover, the rapidly moving highly collimated HH jets, discovered in the visible by Mundt & Fried (1983), also originate from young star positions. On the other hand, the presence of winds around young T Tauri stars was recognized in their P Cygni profiles (Herbig 1962, Kuhi 1964) and in centimeter wavelength continuum observations (Cohen et al 1982).

Broad lines of CO at millimeter wavelengths generated by high-velocity molecular gas were discovered toward the Orion A molecular cloud in the mid-1970s (Kwan & Scoville 1976, Zuckerman et al 1976). High-velocity CO emission was soon detected toward other objects, and the structure of the outflowing material was found to be bipolar (Snell et al 1980, Rodríguez et al 1980). The first surveys revealed that these bipolar outflows are extraordinarily common around young stars (Bally & Lada 1983; Edwards & Snell 1982, 1983, 1984). Lada (1985) compiled the first catalog, which contained 68 outflow sources. Further searches carried out with unbiased selection criteria by using, for example, the *IRAS* data base, or the systematic observation of a full molecular cloud in CO lines, led to the detection of many more outflows. Fukui et al (1993) listed 157 outflows confirmed through complete or partial mapping. Observations since then have increased the number of presently known molecular outflows to nearly 200.

Outflows from young stars are a ubiquitous and energetic phenomenon; they have spectacular observational manifestations over a wide range of wavelengths from the ultraviolet to the radio. In general terms, we are now confident that virtually all young stellar objects (YSOs) undergo periods of copious mass loss. The highest resolution observations available show that the flows emerge bipolarly from a stellar or circumstellar region. The fast well-collimated stellar wind sweeps up the ambient molecular gas in its vicinity, forming two cavities oriented in opposite directions with respect to the central star. The molecular gas displaced from the cavities expands in the form of irregular lobes and incomplete shells and constitutes the CO outflow. However, even the most basic questions about the outflow phenomenon are still a matter of debate. It is not clear yet what physical mechanism produces the outflows, and the underlying stellar or protostellar wind that should sweep up the fast moving molecular gas is proving to be extremely hard to detect.

The new generation of large radiotelescopes and interferometers working at millimeter and submillimeter wavelengths is providing a wealth of information on the small-scale structure of bipolar molecular outflows. In addition to the classical outflows at standard high velocities (SHV, i.e. velocities ranging from a few kilometers per second to about 20 km s^{-1}) whose properties were summarized in the excellent review of Lada (1985), weak CO components that have extremely high velocities (EHV) have been discovered and mapped toward some outflows (e.g. Figure 1). The EHV CO components are reminiscent of the HH jets observed in the visible and seem to be of a different nature than the SHV components (Bachiller & Gómez-González 1992).

The purpose of this article is to review the progress in outflow research since Lada's (1985) review, by taking into account the observations carried out during this ten-year period with millimeter telescopes of high resolution and sensitivity. Special attention is devoted to the extraordinary outflow activity of "Class 0" sources, possibly the youngest (proto-)stellar low-mass objects known to date. Recent theoretical models for the origin of flows and for the interaction of the winds with the molecular surrounding medium are also discussed. Other recent review papers on closely related issues include those by Lada (1991), Bachiller & Gómez-González (1992), Fukui et al (1993), Königl & Ruden (1993), Sargent & Welch (1993), Edwards et al (1993), Cabrit (1993), and Reipurth & Bachiller (1995).

2. THE DIFFERENT COMPONENTS OF BIPOLAR OUTFLOWS

Bipolar outflows from YSOs contain ionized, atomic, and molecular gas in a wide range of excitation conditions. We next describe each of these components in turn and discuss their close relationship. We start by briefly summarizing the properties of the relatively cold molecular gas traced by the classical (SHV) CO outflows. (For a more complete description of their characteristics, see Lada 1985.) This SHV component is usually the most massive, since it consists of a large amount of ambient material that has been swept up during the full period of mass-loss. In contrast, the EHV CO component found in some outflows has different characteristics, and it is separately discussed in Section 3.

2.1 *Molecular Component: Standard CO Outflows*

Molecular outflows with standard high velocities have been extensively studied during the past 15 years (e.g. Bally & Lada 1983; Edwards & Snell 1982, 1983, 1984; Snell et al 1984; Goldsmith et al 1984; Parker et al 1991). Following a first suggestion of Snell et al (1980), there is a wide consensus now that these outflows consist of ambient gas swept up by an underlying wind. These SHV outflows are observed around young stellar objects of very different masses and

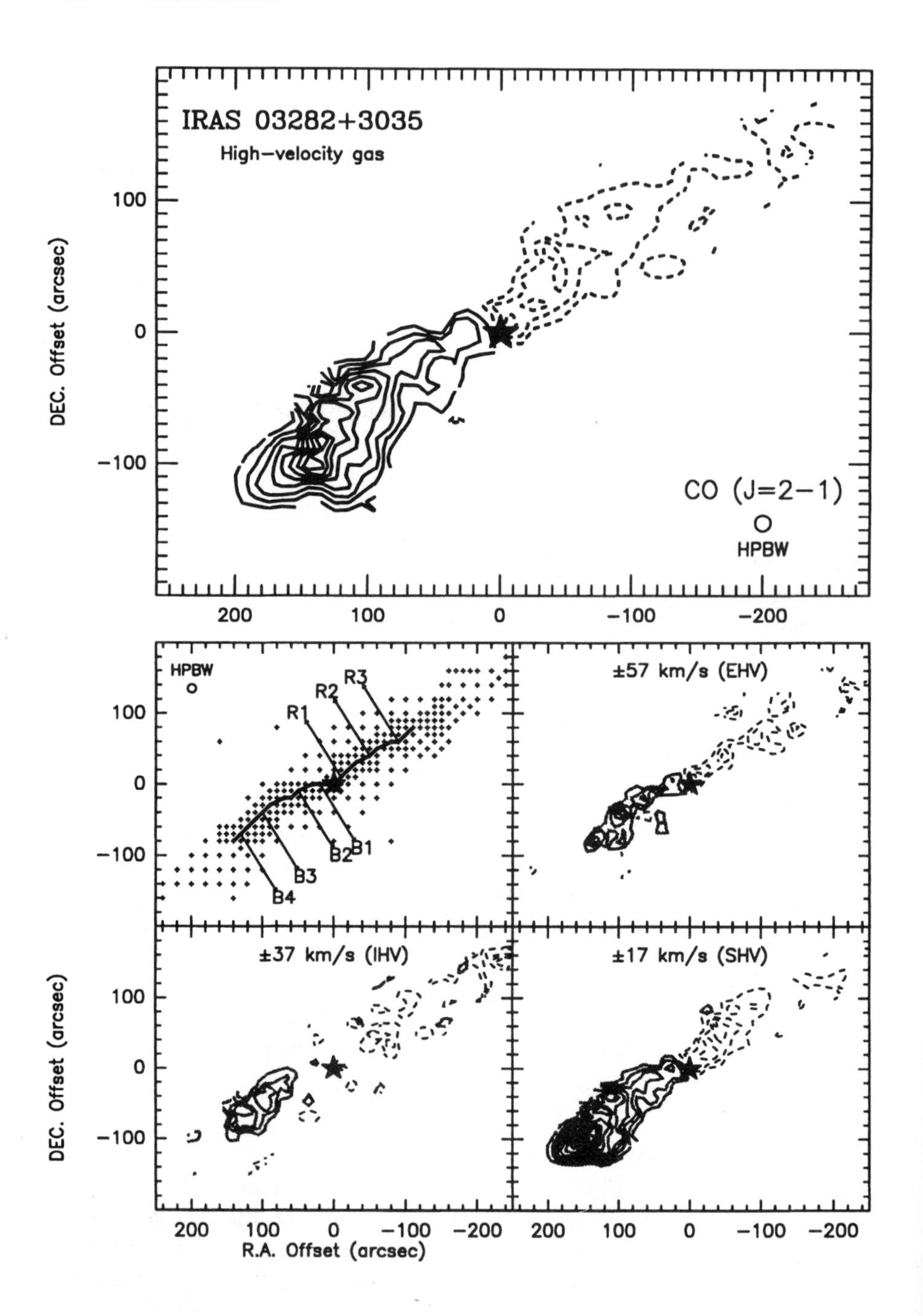
IRAS 03282+3035
High-velocity gas
CO (J=2−1)
HPBW
DEC. Offset (arcsec)
R.A. Offset (arcsec)
R1
R2
R3
B1
B2
B3
B4
±57 km/s (EHV)
±37 km/s (IHV)
±17 km/s (SHV)

luminosities, with low collimation factors (i.e. the ratio of the outflow length to its width) in the range of 2 to 5.

Most molecular outflows are bipolar, though some monopolar outflows are also reported in the literature (e.g. MWC1080, Bally & Lada 1983). In addition, there is an increasing number of multipolar outflows (e.g.: VLA 16293, Walker et al 1988; 723, Avery et al 1990; HH 111, Cernicharo et al 1996) that could result from the superposition of distinct bipolar outflows. In fact, the widely observed multiplicity of young stars (Mathieu 1994) seems to result in a correspondingly high number of multipolar molecular outflows.

The usual procedures used to estimate the physical parameters of bipolar outflows from CO observations have been summarized by Bachiller & Gómez-González (1992). Estimating the CO optical depth and excitation temperature requires the observation of at least two rotational lines of CO and one line of ^{13}CO. From this, it is then possible to estimate the mass of the outflowing gas (by assuming a CO/H_2 ratio). Estimates of the flow momentum and energy depend critically on knowledge of the inclination of the flow axis to the line of sight, and these can be subject to serious uncertainties (Margulis & Lada 1985, Cabrit & Bertout 1990). Some attempts have been made to model the 3-D kinematic structure of bipolar outflows (Cabrit & Bertout 1986, 1990; Cabrit et al 1988), and the kinematic structure of some particular outflows has been successfully accounted for (RNO43 and B335, Cabrit et al 1988; MonR2, Meyers-Rice & Lada 1991).

The amount of mass in a given molecular flow can range from less than $10^{-2}\,M_\odot$ (e.g. HH 34; Chernin & Masson 1995a, Terebey et al 1989) to about $200\,M_\odot$ (e.g. Mon R2, Wolf et al 1990; DR21, Russell et al 1992). The flow sizes go from less than 0.1 pc (e.g. Ori-I-2, Cernicharo et al 1992) to about 5 pc. The energy deposited in the CO outflow can reach 10^{47}–10^{48} erg (e.g.

←

Figure 1 (*Top*) Map of CO 2–1 intensity integrated in the line wings from the IRAS 03282+3035 outflow. Solid contours represent the blueshifted emission (from −60 km s^{-1} to 0 km s^{-1}), while dashed contours are for the redshifted emission (integrated from 14 km s^{-1} to 74 km s^{-1}). First contour and contour interval are 6 K km s^{-1}. (*Lower panels*) Maps of CO 2–1 line intensity integrated in different velocity intervals. The different panels are for the blueshifted (*solid contours*) and redshifted (*dashed*) emission in a velocity interval of 20 km s^{-1}centered at a velocity offset of ± 57 (*upper right panel*), ± 37 (*lower left*), and ± 17 km s^{-1} (*lower right*), with respect to the ambient cloud velocity, 7 km s^{-1}. In the two maps corresponding to the highest velocity emission, the first contour and the contour interval are 2 and 1 K km s^{-1}, respectively. In the map of lower velocity emission, the first contour and the contour interval are 4 K km s^{-1}. The positions of some high-velocity molecular bullets are indicated on the solid line in the upper left panel. (Adapted from Bachiller et al 1991c.)

DR21, Garden et al 1991). The measured kinematic time scales range from 10^3 to a few 10^5 years.

Clumping is a common characteristic of molecular outflows. From multiline observations of CO, Plambeck et al (1983) derived typical values of the filling factor of 0.1–0.2 in several outflows. Snell et al (1984) found that the filling factor approaches unity at the lowest outflow velocities. Clumps are directly observed in some nearby massive outflows as spatially localized velocity components superimposed to the SHV wings. Multiline CS observations have been used to derive the physical properties of such clumps in NGC 2071 (Kitamura et al 1990) and in Mon R2 (Tafalla et al 1994). In Mon R2, the observed SHV clumps are as dense as the ambient cloud (density of a few 10^5 cm^{-3}), and they have masses of up to several $M_\odot$. Most of the SHV outflowing gas could be in the form of clumps of a wide range of size and mass. The decrease of the filling factor at progressively high velocities seems to indicate that the smaller clumps move faster than the larger ones.

Outflow activity can vary over a wide range of time scales. The case of the L1551 outflow is particularly interesting because this is considered the CO outflow prototype. Recent observations of CO with high angular resolution have revealed that the outflow has passed through at least four periods of copious mass-loss (Figure 2; Bachiller et al 1994a). The duration of each mass-loss epoch and the time elapsed between two of them are both about a few 10^4 years. The mass of molecular material associated with each eruptive event ranges from a few tenths to 1 $M_\odot$. Other signs of variability in L1551 are observed in the visible (Neckel & Staude 1987, Campbell et al 1988, Davis et al 1995), and IRS5—the star driving the outflow—is probably a FU Ori star (Mundt et al 1985, Carr et al 1987). However, the time scales involved in the optical variability are of the order of 10^2–10^3 yr, much shorter than the time scales involved in the CO periodicity.

2.2 *Ionized Component: HH Objects and Radio Jets*

In addition to the molecular emission, bipolar outflows from young stars are also observed in the form of optical and centimeter-wavelength jets of ionized material. Particularly relevant are the optical HH jets (Mundt & Fried 1983, Dopita et al 1982, Mundt et al 1987, Reipurth 1991) observed to emerge from a wide variety of YSOs. One of the best examples is the HH 34 system, which contains multiple bow shocks (Reipurth et al 1986, Bührke et al 1988, Reipurth & Heathcote 1992, Morse et al 1992) and extends up to 1.5 pc (Bally & Devine 1994). Many HH jets have associated CO outflows, and in such cases the HH jet and the corresponding CO outflow have the same orientation, similar extension, and compatible kinematics. This is the case for HH 34 (Chernin & Masson 1995a), HH 111 (Reipurth & Cernicharo 1995, Cernicharo & Reipurth

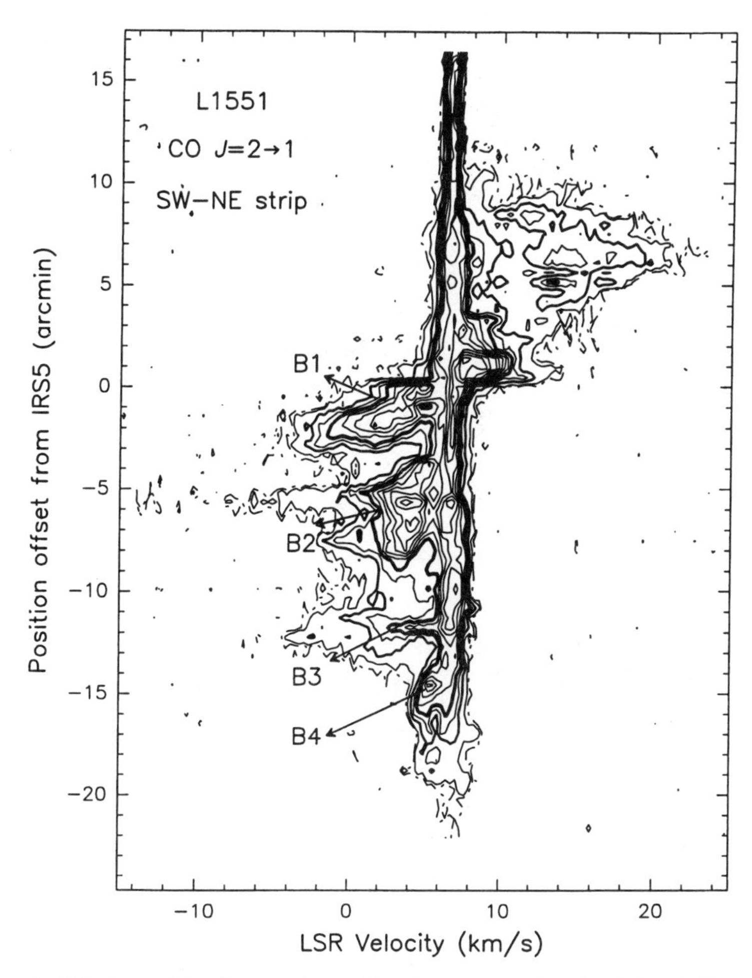

Figure 2 Velocity-position diagram along a line close to the main axis of the L1551 outflow. Position offsets are with respect to IRS5, the outflow exciting source. B1, B2, B3, and B4 are high-velocity features, corresponding to four different lumps of material in the blueshifted lobe of the outflow, which are likely associated with four successive ejection events. (From Bachiller et al 1994a.)

1996), and the HH complexes in L1551 (Mundt & Fried 1983, Snell et al 1980). Surprisingly, some conspicuous HH jets such as HH 34, HH 1/2, HH 46/47, and HH 83, are known to be associated with particularly weak molecular outflows (Chernin & Masson 1991, 1995a; Olberg et al 1992; Bally et al 1994), perhaps because HH jets become optically bright in regions of low visual extinction, in which most of the ambient molecular material has been already dispersed. Recent interferometric images of the HH 111 molecular outflow show that the HH jet lies in a hole of molecular emission, indicating that the jet has cleared out a narrow cylinder in the ambient molecular cloud (Figure 3; Cernicharo et al 1996).

Optical forbidden emission lines also provide powerful diagnostics of bipolar outflows. Observations of the [OI] λλ 6300, 6363; [NII] 6583; and [SII] 6716, 6731 Å lines show profiles that are blueshifted with respect to stellar velocity, probably because a thick circumstellar condensation obscures the receding part of the outflow (Mundt 1984; Edwards et al 1987; Appenzeller & Mundt 1989; Cabrit et al 1990; Hirth et al 1994a,b). In addition, the profiles are often double

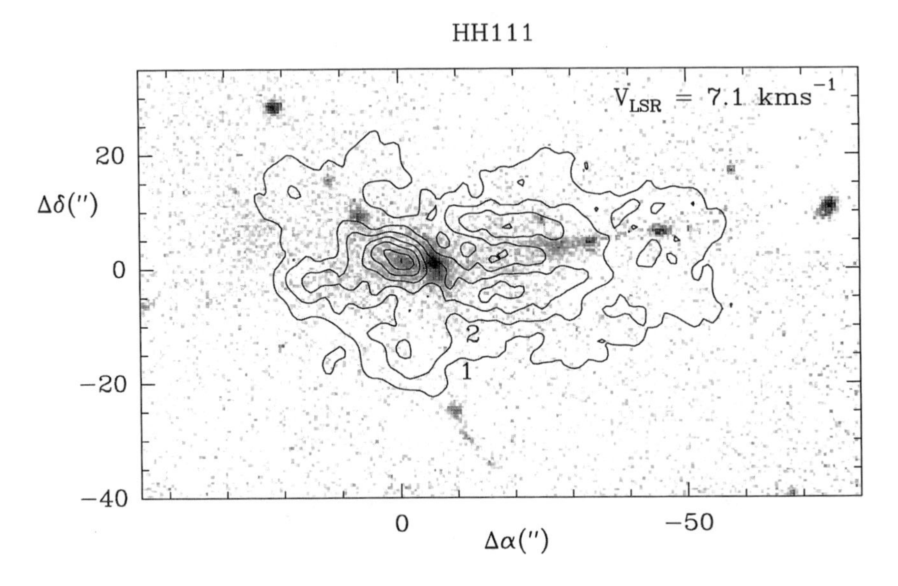

Figure 3 CO 1–0 emission contours superimposed on an H_2 image of the HH 111 region. The CO emission, observed at the Plateau de Bure interferometer, has been integrated in a velocity interval of 0.4 km s^{-1} centered at 7.1 km s^{-1} and traces quiescent ambient material. The H_2 image of the jet was obtained at Calar Alto by J Eislöffel. It appears that the HH 111 jet is clearing up a narrow cylinder in the ambient medium. (From Cernicharo et al 1996.)

peaked with a high-velocity component (HVC, at velocities < -100 km s^{-1}), and a low-velocity component (LVC, at > -30 km s^{-1}). There is observational evidence that the HVC arises from a highly collimated jet from the very vicinity of the star, while the LVC could originate in a wind at the circumstellar disk surface (Kwan & Tademaru 1988; Hirth et al 1994a,b).

Continuum emission at centimeter wavelengths was detected in the early 1980s towards the energy sources of some outflows (Cohen et al 1982, Bieging et al 1984, Bieging & Cohen 1985). Presently about 40 sources have been detected, 10 of which have been imaged (Rodríguez 1995). The maps reveal weak, well-collimated jets emerging from the YSOs. For instance, in the HH 80/81 system in Sagittarius (Reipurth & Graham 1988), a very narrow jet of 5 pc in length is found to be centered on the exciting source (Rodríguez & Reipurth 1989; Martí et al 1993, 1995). In the case of HH 1/2 the HH objects are well detected, and the source spectral index is characteristic of an ionized wind (Pravdo et al 1985). In most cases, the cm-wavelength emission is interpreted as free-free emission from a thermal jet (Reynolds 1986). In addition, spectral indices characteristic of nonthermal synchrotron emission have been derived toward the lobes of the YSO Serpens/FIR1 (Curiel et al 1993) and in the large arcs known as "Orion streamers" (HH 222) emerging from a faint near-IR source (Yusef-Zadeh et al 1990). The coexistence of thermal and nonthermal radio emission in YSO jets has been modeled by Henriksen et al (1991), who suggest that the nonthermal emission is due to relativistic electrons possibly accelerated by a diffusive shock at the region of interaction between the jet and the ambient cloud material.

It is unclear whether the observed (optical and/or radio) jets can drive the associated molecular outflows. Cabrit & Bertout (1992) found a good correlation of the 6-cm luminosity with the force and the luminosity of the associated CO outflows, which in principle argues in favor of the jets driving the outflows. However, Mundt et al (1987) claimed that the momentum of HH jets is not large enough, but momentum estimates are very uncertain because the optical jet densities are difficult to determine (e.g. Raga 1991, Ray 1993). The ejection velocity of the HH jets may also be time-variable (e.g. Raga & Kofman 1992). Thus estimates of the total HH jet momentum might need to be revised. Moreover, as suggested by Parker et al (1991), the lifetimes of the CO outflows could be much greater than their kinematic time scales, and the required momentum injection rate from a possible driving jet could be reduced accordingly. Finally, the possibility remains that a significant neutral (atomic or molecular) component coexists with the ionized jet, helping to drive the molecular SHV outflow. In such a case, the correlation found by Cabrit & Bertout (1992) could be due to a nearly constant ionization fraction in the winds of their YSO sample.

2.3 *Atomic Neutral Component*

The possibility of there being a high fraction of neutral matter in the primary wind appears as one of the most appealing recent suggestions (e.g. Natta et al 1988). Observations of the HI 21-cm line around a few low-mass YSOs such as HH 7–11/IRS, L1551/IRS, and T Tau (Lizano et al 1988, Giovanardi et al 1992, Rodríguez et al 1990, Ruiz et al 1992) have revealed broad wings indicative of winds of up to 200 km s^{-1} and mass-loss rates of 10^{-6} to 10^{-5} $M_{\odot}$ yr^{-1}. HI emission has also been detected in two high-mass bipolar outflows (NGC 2071, Bally & Stark 1983; DR21, Russell et al 1992). However, other searches for high-velocity HI emission have failed in some important objects such as L1448 (LM Chernin, private communication, 1994). Such HI observations are always hampered by the relatively poor angular resolution of the cm single-dish telescopes and the confusion of the background Galactic emission.

Certainly, some HI could be created from the dissociation of ambient molecular gas in the shocked regions. But the HI emission could also trace fast and mostly neutral winds, which could in principle drive the CO outflows by entraining ambient material in a mixing layer similar to that modeled by Cantó & Raga (1991) in the context of HH jets. Models of mixing layers more suited for the CO outflows have recently been considered by Lizano & Giovanardi (1995), who estimated the temperature in the mixing layers to be around 4000 K. H_2 is expected to be the main coolant in the layer, and its near-IR line emission is in fact detected toward a high number of CO outflows (see next subsection).

Unfortunately, from the existing observations it is impossible to know whether the high-velocity HI emission arises from a jet, because the poor angular resolution single-dish observations of the HI 21-cm line do not reveal the structure of the neutral atomic component. Lizano & Giovanardi (1995) proposed that the primary neutral wind in L1551 has the form of a wide-angle, radially directed flow, similar to the model of Shu et al (1991). However, this type of model has been found to be unable to explain some basic characteristics of bipolar CO outflows such as the observed distributions of mass and momentum within the flows (Masson & Chernin 1992, Chernin & Masson 1995b). Thus, if the driving agent of molecular outflows is a neutral wind, it has to be highly collimated. Moreover, the possibility that the CO outflows are driven by jets presents the advantage that the two phenomena of highly collimated jets and poorly collimated CO outflows are unified.

As was mentioned above, the forbidden lines of [OI] near 6300 Å also exhibit broad wings, and the highest velocity part of the emission seems to come from a highly collimated jet (e.g. Hirth et al 1994a). The NaI D line is also detected in T Tauri winds (Mundt 1984, Natta & Giovanardi 1990) However, it is difficult to obtain accurate estimates of the momentum rate from the [OI] and

NaI observations. Finally, a neutral wind is expected to contain a substantial fraction of molecules. Even if the jet were initially atomic, some molecules such as CO and SiO could form in relatively short times (Glassgold et al 1989, 1991), and observations of CO and SiO lines could then help in elucidating the structure of the wind. In fact, Lizano et al (1988) detected EHV emission in CO lines around HH 7–11, and this emission was subsequently found to be in the form of a highly collimated jet (Bachiller & Cernicharo 1990, Masson et al 1990). Similar EHV jets have been detected in other objects (see Section 3), suggesting that these EHV CO jets could be the neutral winds driving the standard CO outflows. Clearly, it is very difficult to distinguish the primary wind actually ejected from the central star/disk system from the high-velocity gas accelerated and processed by shocks. The observations reviewed in Section 3 show that, if not the primary wind itself, the EHV CO outflow is very intimately related to the primary driving agent.

2.4 *Molecular Component: High-Excitation H_2 Emitting Gas*

The vibrational transitions of H_2 arise from energy levels > 6000 K above the ground state; thus H_2 molecules become collisionally excited in dense regions at temperatures of a few 10^3 K. As a consequence, such transitions are good potential tracers of shocked molecular gas. In particular, the v = 1–0 and v = 2–1 S(1) lines of H_2 at $\lambda\lambda$ 2.122 and 2.247 μm are excited in shocks with velocities in the range 10–50 km s^{-1} (Shull & Beckwith 1982, Draine et al 1983, Smith 1994). At higher shock velocities, H_2 molecules are dissociated. Thanks to the recent development of sensitive detector arrays, it is now possible to explore large regions of the sky in the near-infrared at arcsec resolution. The v = 1–0 S(1) line has been observed toward the most conspicuous HH flows, including HH 43 (Schwartz et al 1988), HH 1/2 (Davis et al 1994b), HH 7–11 and 12 (Stapelfeldt et al 1991), CepA/GGD37 (Lane 1989), HH 111 (Gredel & Reipurth 1993, 1994; Davis et al 1994c), OriA (Taylor et al 1984), and many others (e.g. Hodapp & Ladd 1995). The extensive survey by Hodapp (1994) in the K' band, which contains the v = 1–0 line, reveals a variety of complex morphologies. In general, the H_2 emission is somewhat correlated with the lower-excitation optical features and can trace weaker shocks not visible in the optical (e.g. HH 46/47, Eislöffel et al 1994). Bow shock morphologies are observed at the head of some outflows. In OriA/IRc2, the H_2 jet-like filaments or "fingers" (Taylor et al 1984, Allen & Burton 1993) also terminate in bow shocks.

One of the main advantages of H_2 observations is that they allow the study of particularly young, optically invisible outflows, which are still deeply embedded within dense cores. Strong H_2 emission is observed around several very young Class 0 YSOs (see Section 4), including L1448-mm (Terebey 1991, Bally et al

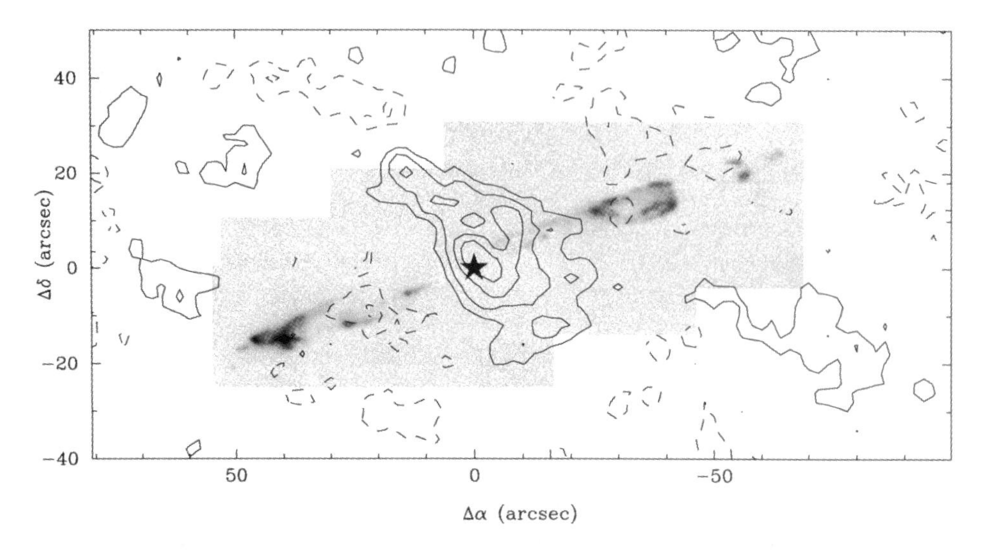

Figure 4 Superposition of a gray-scaled image of the HH 211 jet taken in the H_2 $v = 1$–0 S(1) line at 2.122 μm (from McCaughrean et al 1994) with a NH_3 (1,1) image obtained with the VLA at its D configuration (6″ angular resolution) (R Bachiller & M Tafalla 1995, unpublished data). The star marks the position of the jet source HH 211-mm (see also Table 1).

1993b, Davis et al 1994a), IRAS 03282 (Bally et al 1993a, Bachiller et al 1994b), IC 348-mm (McCaughrean et al 1994), VLA 1623 (Dent et al 1995), and L1157-mm (Hodapp 1994, Davis & Eislöffel 1995). As an example, we show in Figure 4 the jet emerging from IC 348-mm (also called HH 211). The Class 0 source is embedded in a dense NH_3 clump of a few $M_\odot$ (Bachiller et al 1987; R Bachiller & M Tafalla 1996, in preparation). The jet has a kinematical age of < 1000 yr (McCaughrean et al 1994), and the H_2 emission arises in a kind of cocoon around the true jet. This behavior is observed in most of the jets from Class 0 sources: The H_2 emission forms long filaments, but these filaments are not strictly coincident with the axes of the jets. The observations thus confirm that the H_2 line emission arises in the mixing layer where ambient material is entrained. The observation of bow shocks in several sources underscores the importance of the "prompt" entrainment at the jet head (Davis & Eislöffel 1995).

3. HIGHLY COLLIMATED CO OUTFLOWS

The high resolution and sensitivity provided by large millimeter-wave radiotelescopes has resulted in important developments in the study of bipolar outflows. In particular, highly collimated molecular outflows (with collimation factors

> 10) have been recognized as a distinct important class within molecular flows. Well-documented examples are L1448 (Bachiller et al 1990), IRAS 03282 (Figure 1, Bachiller et al 1991c), NGC 2024/FIR5 (Richer et al 1989, 1992), OMC 1/FIR4 (Schmid-Burgk et al 1990), NGC 2264G (Lada & Fich 1996), VLA 1623 (André et al 1990a, Dent et al 1995), and IC 348/HH 211 (McCaughrean et al 1994). Most of these highly collimated flows exhibit extremely high velocity components (in excess of 40 km s^{-1}) concentrated toward the flow axis, and the slower gas is less collimated, similar to the standard CO flows.

3.1 *EHV Components and Molecular Bullets*

In some highly collimated outflows, the CO component on the outflow axis is a jet-like structure flowing at extremely high velocities (i.e. velocities ~ 100 km s^{-1}). Good examples are IRAS 03282 and L1448. The momentum in the EHV jet-like component is large, generally sufficient to put into motion the standard SHV bipolar outflows. In addition, in some particularly clear cases (such as IRAS 03282 and L1448), the terminal velocity of the EHV jet is observed to decrease with distance from the outflow origin, whereas the terminal velocity of the SHV component is observed to increase. This behavior strongly suggests that the EHV jet-like component is injecting momentum into the ambient gas to produce the SHV outflow.

Rather than being a continuous jet, the EHV component presents discrete peaks that are well defined in space and in velocity. Such peaks are referred to as "molecular bullets" (Bachiller et al 1990). An illustrative example is provided by the outflow in IRAS 03282+3035 (Bachiller et al 1991c). Figure 1 shows the structure of the outflow. The EHV jet consists of a chain of molecular bullets interconnected by weaker emission. The standard (SHV) outflow is observed as extended lobes surrounding the EHV jet. Figure 5 shows a few spectra, obtained toward the outflow axis, in which the EHV features are well observed.

Molecular bullets are observed in the majority of highly collimated CO outflows. A remarkable example is that of the HH 7–11 flow. The observed radial velocities of the bullets in this outflow exceed 100 km s^{-1} with respect to the ambient cloud, and their CO linewidths are about 20 km s^{-1} (Bachiller & Cernicharo 1990, Masson et al 1990). Not all molecular bullets in highly collimated jets present such extreme radial velocities. For instance, the CO jet around VLA 1623 (André et al 1990a) also exhibits a clear structure in clumps, but the radial velocities observed toward this jet are ≤ 30 km s^{-1}, probably due to a very high inclination of the outflow with respect to the line of sight. Other highly collimated CO outflows presenting a clear structure with molecular bullets include NGC 2024/FIR5 (Richer et al 1989, 1992), HH 111 (Cernicharo & Reipurth 1996), and IRAS 2005 (Bachiller et al 1995a). The typical sizes of

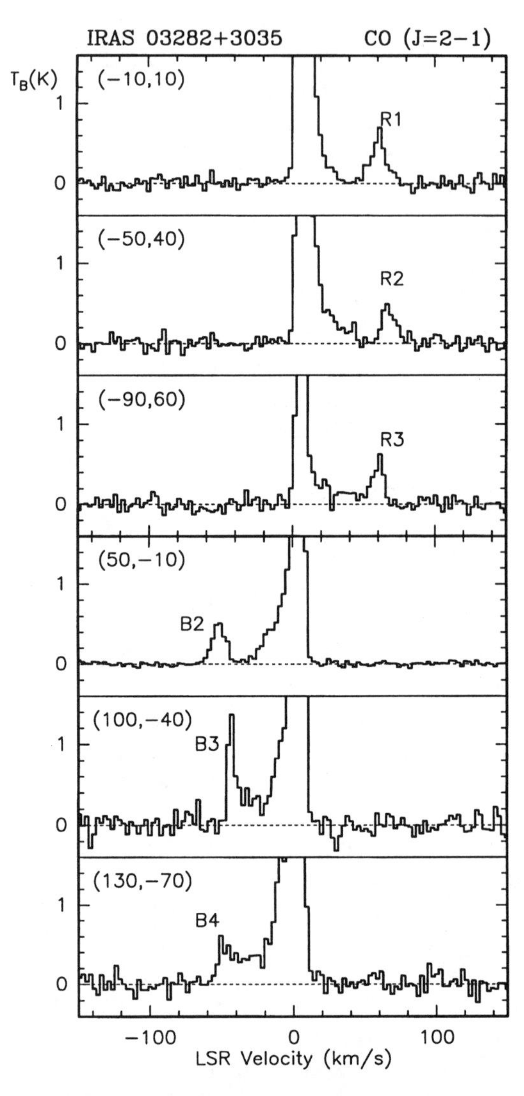

Figure 5 CO 2–1 spectra observed toward selected positions in the IRAS 03282+3035 outflow (from Bachiller et al 1991c). The position offsets (marked in the upper right corner of each panel) are relative to the position of the exciting source (see Table 1). High-velocity molecular bullets are denoted by R1, R2, and R3 (redshifted) and B2, B3, and B4 (blueshifted). Note that the emission extends over a velocity range of about 140 km s^{-1}.

such bullets are a few 10^{-2} pc, and their masses are a few $10^{-4}\ M_{\odot}$, as derived from multiline CO observations (e.g. Bachiller et al 1990). The kinematic time scales range from a few hundred to a few thousand years.

Molecular bullets tend to be regularly spaced along the axis of the highly collimated jets. In the IRAS 03282 and L1448 outflows (Bachiller et al 1991c, 1990), the bullets can be grouped in symmetrical pairs. The striking symmetry found in position and velocity between the redshifted and the blueshifted bullets indicates that, rather than produced in situ, each pair of bullets corresponds to a bipolar ejection taking place near the central YSO. Such ejection events are quasi-periodic, and the time elapsed between two successive outbursts is of the order of 10^3 yr. Intermittency seems to be a relatively common characteristic of highly collimated outflows.

Discrete high-velocity CO components are also observed in the immediate vicinity of some YSOs with near-infrared spectroscopy (e.g. Mitchell et al 1991), but the relationship of these features to the molecular bullets observed in the rotational CO transitions is unclear. Another phenomenon that could be related with molecular bullets are the successive shocked knots observed in optical and near-infrared jets (Reipurth 1989; Reipurth & Heathcote 1991, 1992; Hartigan et al 1993; Bally et al 1993b; Eislöffel & Mundt 1994). Such shocked knots are also believed to be associated with recurrent ejection events (Reipurth 1989).

The precise origin of the eruptions that cause molecular bullets is not well understood, but it is noteworthy that the masses and time scales of bullets are similar to those of the "optical" outburst observed in FU Ori stars. The FU Ori eruptions can be well explained by a large increase in the accretion rate through a circumstellar disk (Hartmann & Kenyon 1985), up to $10^{-4}\ M_{\odot}\mathrm{yr}^{-1}$. For an average duration of about 100 yr, this yields a total accreted mass up to $10^{-2}\ M_{\odot}$. The masses of the molecular bullets would thus be consistent with a ratio of accretion rate to mass outflow of $\sim$ 10 to 100.

The EHV molecular bullets present extraordinary chemical characteristics. For instance, the abundance of the SiO molecules can be enhanced by several orders of magnitude with respect to the quiescent ambient cloud (Bachiller et al 1991b, Guilloteau et al 1992). Chemical reactions in an initially neutral atomic wind seem able to produce significant amounts of SiO (Glassgold et al 1989, 1991). However, the large SiO abundance and its observed spatial distribution in different flows suggest that an important part of the SiO enhancement is produced by shocks associated with the highly collimated outflow (see Section 5).

3.2 *Molecular Bow Shocks*

One of the most intriguing aspects of bipolar outflows is that, in many cases, highly collimated jets coexist with poorly collimated molecular flows. Jets are usually observed along the axis of wide CO flow lobes that resemble ovoidal or ellipsoidal cavities. How a narrow highly supersonic jet can generate so wide a cavity is a fundamental question addressed by recent observations and models. Interferometry at millimeter wavelengths now makes it possible to image the CO outflows with a resolution close to an arcsec (Sargent & Welch 1993); these data can shed light on this issue. The most detailed observations available so far are those of the L1448 and L1157 outflows. Figure 6 shows a comparison of the H_2 image of L1448 (from Bally et al 1993b) with the images of the slow-moving CO obtained with the IRAM interferometer (Bachiller et al 1995b). The H_2 images reveal a series of well-defined bow shocks in the blueshifted lobe, whereas the slow-moving molecular gas traces the edges of a biconical limb-brightened cavity. The blueshifted part of the cavity is also seen in the continuum near-infrared emission, which is scattered at the cavity walls (Bally et al 1993b). It is remarkable that the walls of the CO cavity seen at blueshifted velocities (top left panel in Figure 6) seem to be complementary to the bow shock H_2 structure, which is the closest to the exciting source, L1448-mm. The arc structure delineated by the H_2 emission seems to close the conical CO cavity. This configuration strongly suggests that the large opening angle of the SHV CO outflow results from the entrainment of ambient material through the large bow shocks traced by the H_2 line emission.

Another example of a bow shock–driven outflow is that in L1157 (Umemoto et al 1992, Gueth et al 1996). Figure 7 shows velocity channel CO images of the blueshifted lobe. The images reveal at least two prominent limb-brightened cavities, which also seem to be created by the propagation of large bow shocks. These observations also illustrate the importance of combining single-dish data with the interferometric data. In fact, with only the purely interferometric images (top row of the figure), one could think that the cavities are empty structures. When one adds the zero-spacing information (middle row), significant CO emission arising from the inner part of the cavities becomes evident. The bow shocks at the head of the cavities are also well observed in NH_3 emission (Bachiller et al 1993), and VLA images reveal a structure similar to that seen in CO (M Tafalla & R Bachiller 1995; 1996, in preparation; see below).

Figure 6 Superposition of a gray-scaled H_2 image of the L1448 jet (from Bally et al 1993b) and the interferometric images of the CO 1–0 emission integrated over four intervals at low velocities (from Bachiller et al 1995b). The central LSR velocity for each interval is given at the upper left corner of each panel. First contour and step are 0.92 K km s^{-1}. The jet direction, defined by SiO observations, is indicated by the solid line. The positions of L1448-mm and IRS3 are marked with stars. A part of the CO outflow from IRS3 is visible in the 7 km s^{-1} panel. The H_2 emission traces large bow shocks, whereas the CO delineates the walls of a biconical cavity. Note that the walls of the CO cavity in the blueshifted lobe are complementary of the first H_2 bow shock. This morphology strongly suggests that the CO bipolar outflow results from entrainment of ambient material through the propagation of large bow shocks.

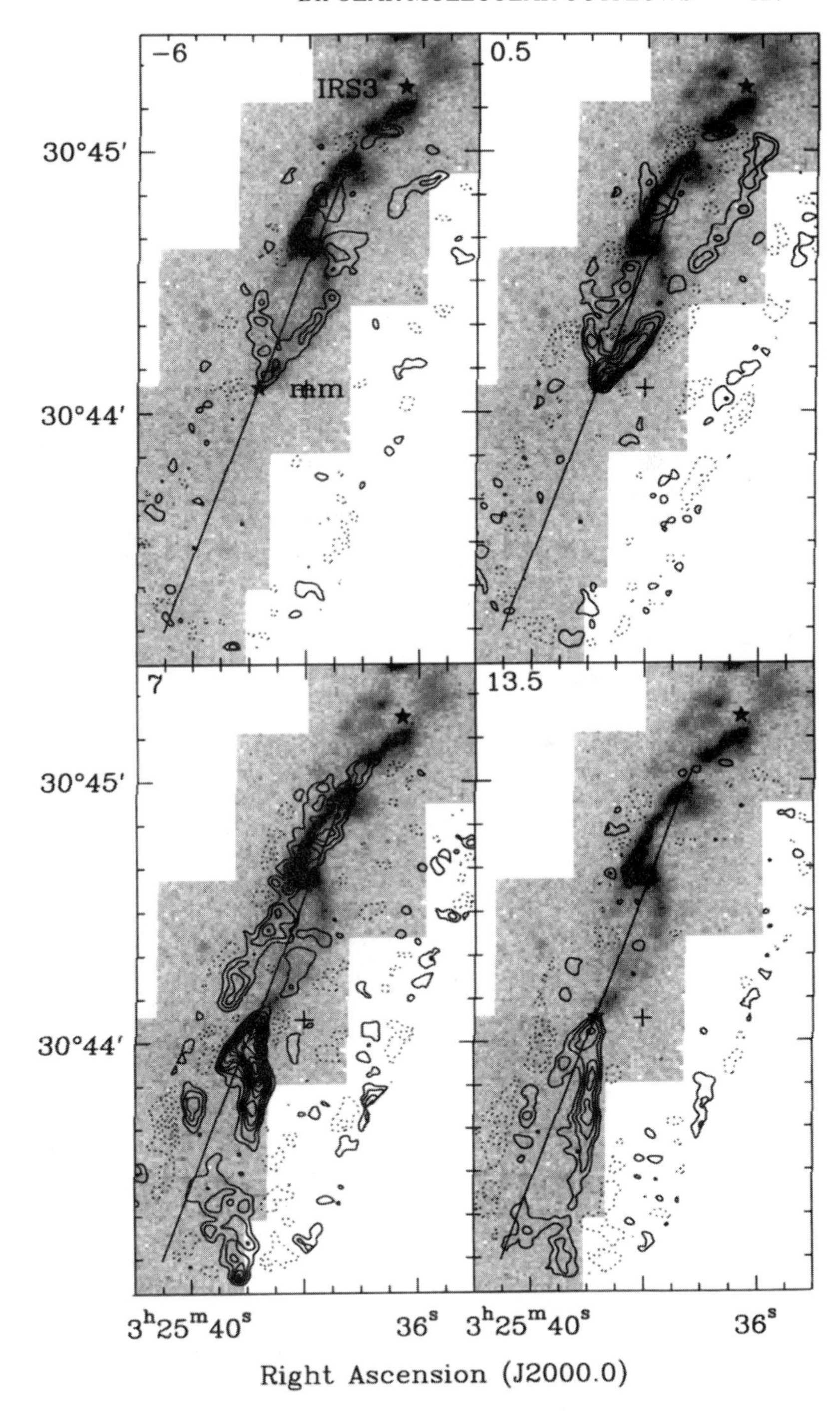
-6
IRS3
mm
0.5
7
13.5
30°45′
30°44′
Declination (J2000.0)
3h25m40s
36s
Right Ascension (J2000.0)

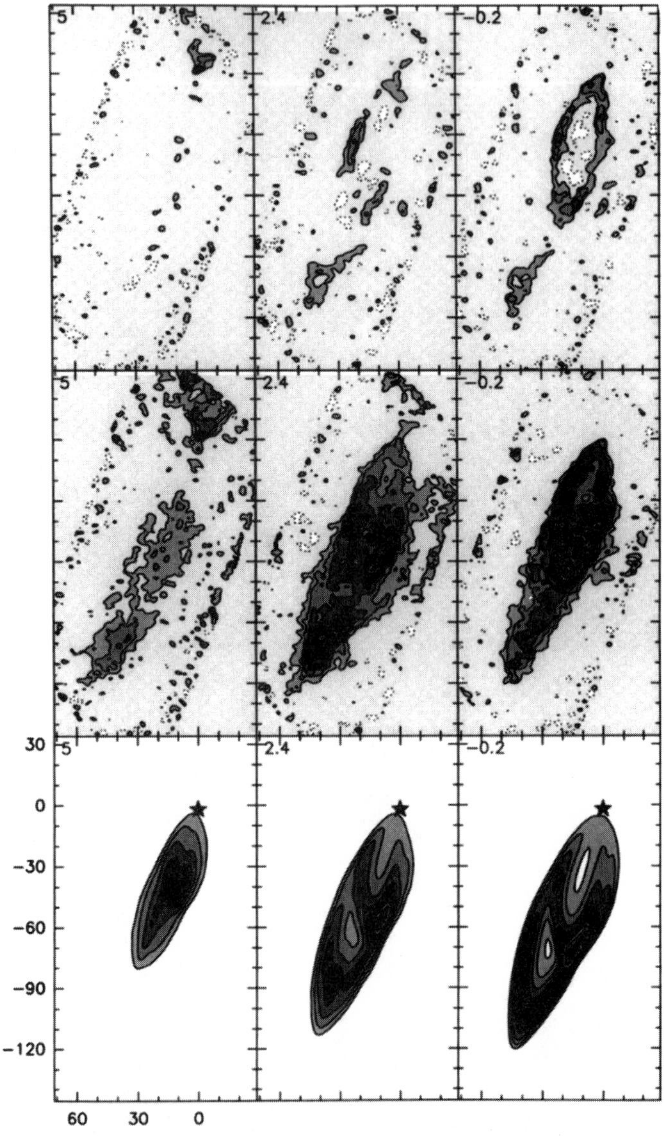

Figure 7 CO emission from the blueshifted lobe of the L1157 outflow integrated over velocity intervals of 2.6 km s^{-1}. The central LSR velocity for each interval is given at the upper left corner of each panel. The LSR velocity of the ambient gas is 2.75 km s^{-1}. Position offsets are in arcsec with respect to the L1157-mm (see Table 1), whose position is indicated with a star. First contour and step are 155 mJy/beam (1.3 K). The beam size is 3.6″ × 3″ at P.A. 90°. (*Top row*) CO 1–0 maps reconstructed from purely interferometric IRAM data. (*Middle row*) Images obtained after inclusion of the short spacing information obtained at the IRAM 30-m telescope. (*Bottom row*) Synthetic maps obtained with a precessing, episodic jet model smoothed to the resolution of the observations. (Adapted from Gueth et al 1996.)

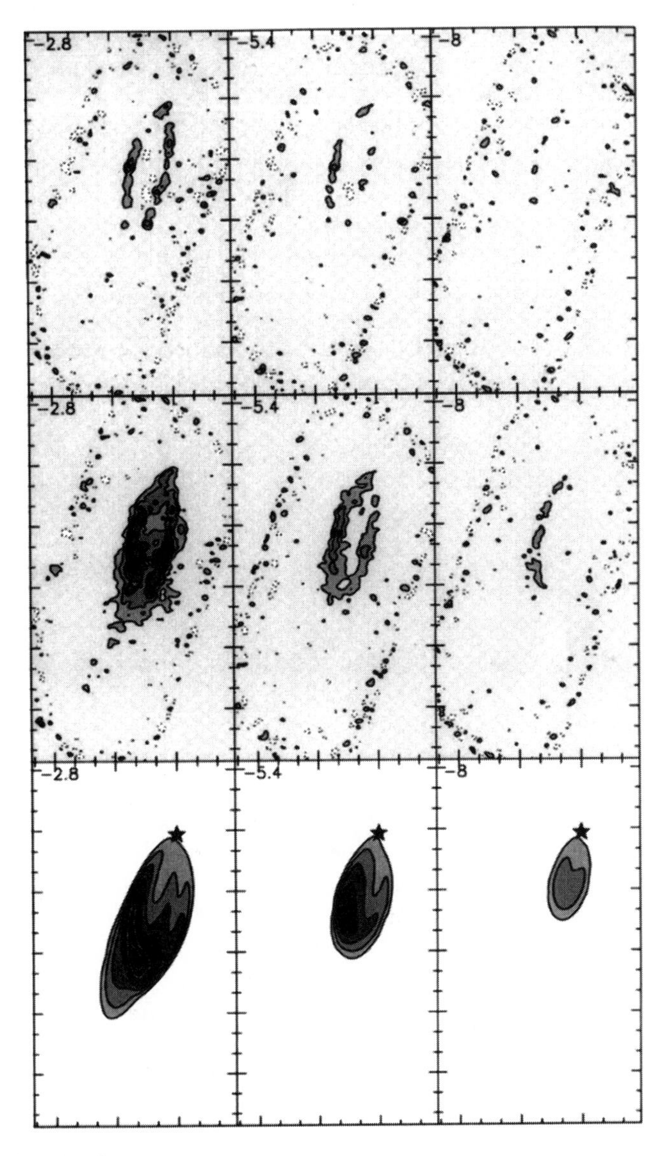

Figure 7 (*continued*)

Interestingly, the two cavities are not well aligned on a single line passing through the exciting source, L1157-mm, as if the axis of the underlying jet had precessed from the first ejection event to the second one. A simple spatio-kinematic model in which the jet precesses on a narrow cone (of opening angle close to 6°) provides an accurate description of the observations (bottom row of the figure). Thus, the large opening angle observed at the base of the CO outflow is very likely determined by the large size of the propagating bow shocks, rather than to the jet precession, which happens in a very narrow cone.

In more massive objects, higher confusion makes the identification of molecular bow shocks more difficult and also complicates the association of the bow shocks with the CO outflows. In the case of Orion/IRc2, multiple H_2 bow shocks are observed to emerge from the central source (Taylor et al 1984, Lane 1989, Allen & Burton 1993), forming a wide fan. However, the comparison with the SHV CO outflow is difficult in this case because of the confusion with the possible interaction of the molecular cloud with the extended HII region in the background (Rodríguez-Franco 1995). Finally, it is interesting to note that the H_2O maser features in W49 (Gwinn et al 1992) seem to delineate the surfaces of an elongated cocoon produced by a jet (Mac Low & Elitzur 1992, Mac Low et al 1994).

3.3 *What Drives the High-Velocity Gas?*

As mentioned above, a wind emanating from the central star/disk system was soon proposed as the primary physical agent that could drive the observed CO outflows (Snell et al 1980, Lizano et al 1988, Shu et al 1991). However, wide-angle winds fail to explain some important properties of molecular outflows, such as the observed amount of mass as a function of velocity (Masson & Chernin 1992). In fact, low-collimation winds would contain more material at extreme velocities than observed in actual outflows. Jets, on the other hand, can sweep the gas aside by means of bow shocks, and most of the gas would be at low velocities, in agreement with observations. Wide-angle winds are also unable to explain the observed spatial distribution of momentum in molecular outflows (Chernin & Masson 1995b) and the morphology of the highly collimated CO outflows (see discussion above). Thus, the primary driving agent of molecular outflows is very likely a jet. As mentioned above, a great advantage of the jet-driven picture is that it unifies two phenomena, HH jets and CO outflows, that were initially considered intrinsically different.

When a jet interacts with the ambient molecular medium, the shock is unlikely to be adiabatic (energy conserving), owing to fast cooling. In fact, diffuse far-infrared emission is observed in L1551 with a morphology similar to that of the outflow, and the far-infrared luminosity is about 18% of the bolometric luminosity of the central YSO (Edwards et al 1986, Clark & Laurejis 1986).

In the case of HH 46/47, the luminosity in the H_2 lines is comparable to the mechanical power of the CO outflow (Eislöffel et al 1994). In addition, the high bipolarity observed in some outflows makes the energy-conserving winds implausible for producing outflows (Meyers-Rice & Lada 1991, Lada & Fich 1996). Thus, outflows are likely driven in a momentum-conserving fashion. The transfer of momentum from the jet to the ambient medium can be achieved in different ways (see e.g. Dyson 1984). From numerical hydrodynamical simulations, De Young (1986) distinguished two basic processes: 1. the "prompt entrainment" happening at the head (bow shock) of the jet and 2. the "steady-state entrainment" taking place along the sides of the jet by turbulent mixing of the ambient material through Kelvin-Helmholtz instabilities. The numerical simulations (De Young 1986) show that the first process dominates in the case of intermediate Mach number jets ($M = 5$–10), with internal densities comparable to that of the external medium. Turbulent steady-state entrainment is the dominant process in low-velocity jets ($M \sim 1$).

Specific models for stellar jets were first developed to account for the observations of HH jets (Raga 1988, Tenorio-Tagle et al 1988, Blondin et al 1990, Gouveia dal Pino & Benz 1993, Hartigan & Raymond 1993, Stone & Norman 1994). These models are mainly concerned with the propagation of the jet itself, and they do not consider the possible generation of molecular flows through the entrainment of quiescent ambient material. Recently, attempts have been made to specifically model the creation of CO outflows by jets. Such jet-driven outflow models are still very approximate and do not intend to provide a full description of all the complex hydrodynamical phenomena involved. The models can be classified in two families, depending on the entrainment process assumed to be mainly responsible for the CO outflow.

Turbulent entrained outflow models (Stahler 1993, 1994; Raga et al 1993b) emphasize the steady-state turbulent entrainment. Such models seem to explain the velocity shear structure observed at the base of some outflows such as IRAS 03282 (Tafalla et al 1993b); they could also explain the "Hubble law" observed in many outflows (Stahler 1994; but see also Lada & Fich 1996). However, the observed low collimation of CO outflows in not well understood in this model, since the width of the boundary mixing layer in which the entrainment of ambient material (the CO outflow) occurs is expected to be very thin, comparable to the jet radius (Cantó & Raga 1991, Raga et al 1993b).

Bow shock outflow models aim to explain the production of the CO flows by prompt entrainment (Masson & Chernin 1993, Raga & Cabrit 1993, Chernin et al 1994b). In fact, prompt entrainment through bow shocks is the expected dominant process in outflows, owing to the high Mach numbers ($M > 10$). Narrow jets could, in principle, produce much larger bow structures, which would explain the coexistence of the highly collimated jets with the poorly collimated CO flows. In fact, as discussed above, large bow shocks are directly observed by their line emission in many sources. Bow shocks are able to sweep up large amounts of ambient material (Masson & Chernin 1993), conserving momentum. However, not all the observed details are well accounted for by the existing bow shock outflow models. For instance, the existing models (e.g. Raga & Cabrit 1993) fail to reproduce the shapes of the cavities excavated by the propagation of the bow shocks (Gueth et al 1996), and the observed spatial distribution of momentum cannot be accounted for (Chernin & Masson 1995b). But it is important to note that the Raga & Cabrit model is made for an "internal working surface," and not for a "leading jet head." This model also assumed that the passage of the bow shock will be followed by a turbulent wake, but at the present time it is unclear what kind of velocity field is expected in this complex region. Finally, the existing numerical models of bow shocks (Stone & Norman 1994, Chernin et al 1994b) predict bow structures that are narrower than what is necessary to explain the wide opening angles of CO outflows. Jet precession has been argued as a possible mechanism to broaden the outflow lobes (Masson & Chernin 1993), but the observational studies of precession in CO flows are difficult due to the complex morphologies of the lobes, and in the most promising cases for precession, the width of the CO lobes seems to be caused by large bow shocks (Figure 7, Gueth et al 1996).

In conclusion, bow shock outflow models seem to be the most promising ones for explaining most of the observational characteristics of CO outflows, but future models should address the problem of the large transverse sizes of the observed CO lobes and bow shocks. In addition, turbulent entrainment could still efficiently operate in some regions (or at some evolutionary stages) of the jet/outflow system. As discussed above, the observed jets are clearly eruptive, and first attempts have been made to model time-variations in velocity, mass-loss rate, and angle of ejection in bipolar outflows (Raga & Kofman 1992, Hartigan & Raymond 1993, Biro & Raga 1994, Raga et al 1993a). Clearly, time-variability is a dominant aspect of molecular outflows, and future models should take it into account.

4. CHEMISTRY

Shocks are the natural result of the propagation of high-Mach-number outflows within molecular clouds. Shock waves compress and heat the gas, triggering chemical reactions that do not operate in quiescent environments. In addition, shock processing of the dust grains results in the injection of some particular atoms and molecules into the gas phase. Thus the molecular gas in the vicinity of YSOs is expected to present a distinct and unusual chemical composition.

Some molecular abundances are known to be enhanced as a result of the action of bipolar outflows on the surrounding gas. Observations of the outflow around OriA/IRc2 have provided valuable information about the chemical processes activated by the birth of a high-mass star in the surrounding medium. Evaporation of molecules from dust grains and high-temperature shock chemistry have been found to be important processes in increasing the abundances of some sulfur, oxygen, nitrogen, and deuterium compounds (Plambeck et al 1982; Blake et al 1987; Plambeck & Wright 1987; Walmsley et al 1987; see Rodríguez-Franco 1995, for a complete review). Unfortunately, the case of OriA/IRc2 is particularly complex owing to the presence of at least five gas components (the outflow, the expanding "doughnut," the hot core, the ridge, and the compact ridge) of different physical conditions along the line of sight.

Important chemical effects have also been observed in some outflows from low-mass YSOs, which usually present much less confusion than OriA/IRc2. One of the most extreme examples is SiO, whose abundance is enhanced by several orders of magnitude at the heads and along the axes of some molecular outflows (Bachiller et al 1991b, Martín-Pintado et al 1992, McMullin et al 1994a). Other molecules with well-documented outflow enhancements include SO (Martín-Pintado et al 1992, Schmid-Burgk & Muders 1995, Chernin et al 1994a), NH_3 (Bachiller et al 1993, Tafalla & Bachiller 1995), and CH_3OH (Bachiller et al 1995c, Sandell et al 1994). As an example, the NH_3 (3, 3) line near λ 1.3 cm is dominated by broad emission around the L1157 outflow, and it becomes possible to image the flow with the VLA. Recent images obtained in its D configuration (5″ angular resolution) reveal a structure of successive bow shocks along the outflow axis (Figure 8; M Tafalla & R Bachiller 1995; 1996, in preparation).

Interest in detecting molecular lines from shocked regions not only comes from the chemistry involved but also from the fact that molecular lines provide a very useful tool for estimating the physical conditions of the shocked component. The case of ammonia is particularly important, since this is the best interstellar thermometer. Additionally, multiline studies of SiO and CH_3OH allow reliable estimates of the volume densities. For instance, in the bow shock associated with the L1157 blueshifted outflow, the kinetic temperature has been estimated to be 80 K and the volume density to be a few 10^6 cm^{-3} (Bachiller et al 1993, 1995c). It thus appears that the shocked gas traced by the mm molecular lines is not as hot as the gas traced by the near-infrared lines of H_2 (which is at a temperature of a few 10^3 K).

Modeling the complex shock chemistry operating in the vicinity of YSOs requires estimating a relatively large number of molecular abundances. Such estimates should be done through extensive molecular line surveys at millimeter

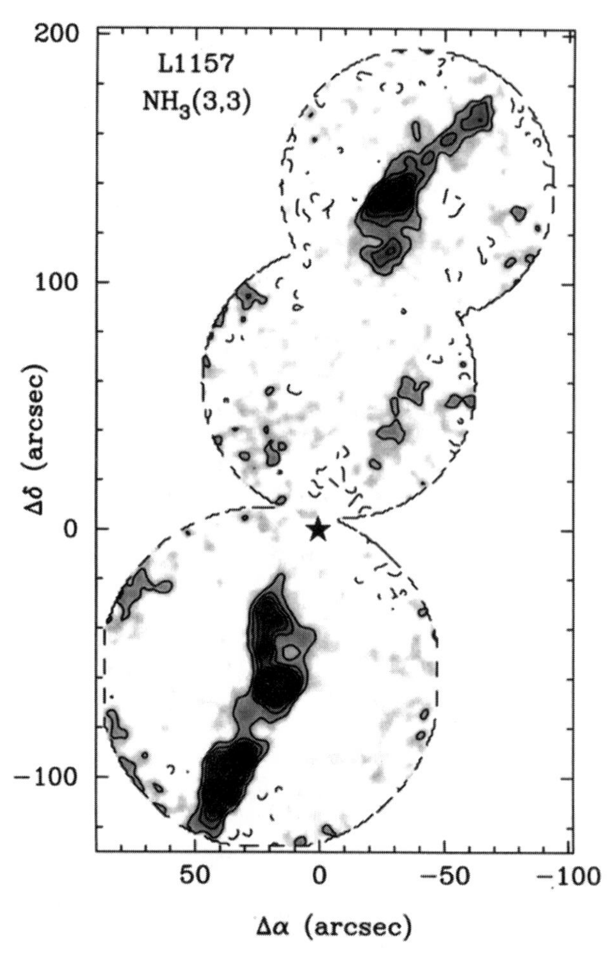

Figure 8 Integrated intensity map of the NH_3 (3, 3) line emission in the L1157 outflow obtained with the VLA (D configuration, 5″ resolution). The northern lobe is redshifted emission; the southern one is blueshifted. The map center is on L1157-mm (see Table 1), the outflow exciting source. The NH_3 (3, 3) line traces warm gas ($T \sim 70$ K) associated with bow shocks. Data are from Tafalla & Bachiller (1995), augmented with recent unpublished observations by the same authors.

wavelengths. The vicinity of low-mass YSOs are the best targets for these kinds of studies, because the confusion in these regions is expected to be less severe than in more massive clouds. The first extensive millimeter surveys have been carried out toward the Class 0 sources NGC 1333/IRAS 4 (Blake et al 1995) and IRAS 16293 (Blake et al 1994, van Dishoeck et al 1995). Figure 9 shows some results from one of these surveys toward the L1157 outflow. The chemical segregation in this cloud is particularly illustrative. The narrow line profiles observed toward the position of the source arise from cold quiescent gas, whereas toward the bow shock region the profiles are dominated by the broad lines associated with the shock. Some molecular lines such as those of DCO^+ and N_2H^+ are only observed toward the cold gas condensation around the exciting source, whereas some other molecules such SiO and methanol (CH_3OH) only trace the hot warm gas in the shock. CS and H_2CO lines are observed in both gas components (R Bachiller et al 1996, in preparation).

The detailed chemical processes induced by the action of shocks are poorly understood. In most cases, the emission of shock-chemistry molecules is seen at the position of the bow shocks [e.g. IRAS 03282 and L1157 (R Bachiller et al 1994b; 1996, in preparation)], but SiO emission is also seen arising from shocks *along* the highly collimated molecular outflow in L1448 (Bachiller et al 1991b, Guilloteau et al 1992, Dutrey et al 1996). It seems clear that SiO is a result of the shock chemistry following the destruction of the refractory grain cores. However, other molecules such as ammonia and methanol, which are known to be abundant in the ice dust mantles (e.g Allamandola et al 1992), could be directly desorbed from them. Deuterated species could also be removed from the grains by grain-grain collisions (van Dishoeck et al 1995). The origin of other molecules such as SO and HCO^+ is even less clear. Theoretical studies of the outflow chemistry should include a high number of processes, namely the effect of the UV near HH objects (Wolfire & Königl 1993), the chemistry of the atomic component in the jet (Glassgold et al 1989, 1991), the mixing layer chemistry (Taylor & Raga 1995), as well as the specific processes related to the shocks (e.g. Iglesias & Silk 1978, Neufeld & Dalgarno 1989, Millar et al 1991, Pineau des Forêts et al 1993).

5. DRIVING SOURCES

Most of the driving sources of bipolar molecular outflows are YSOs that are deeply embedded within the molecular cores in which they were born. YSOs are difficult to classify in a scheme such as the H-R diagram because they do not radiate as single blackbodies. Rather, their spectral energy distributions (SED) are often broad, resulting from the wide range of temperatures in their dusty envelopes (e.g. Scoville & Kwan 1976). Spectral types are also difficult to assign due to the strong obscuration, although near-infrared spectroscopy seems to be a promising tool (Hodapp & Deane 1993, Casali & Eiroa 1995). However, as protostellar outflows disperse the material surrounding the YSO, a systematic change in the SED is produced (see, e.g. Shu et al 1987), and this evolution of the SED shape can be used to classify the YSOs in different evolutionary classes (see below). Understanding the mechanisms of dispersion of the dense circumstellar gas around YSOs is also of crucial importance because this dispersion will probably determine the final mass of the star/disk system that is under formation.

5.1 *Core Disruption and the Classification of YSOs*

Dense cores harboring outflows are known to present broader NH_3 lines than nonoutflow cores (Myers et al 1988), but the nature of the line broadening mechanism is not well understood. Turbulence and systematic motions such as rotation, infall, and expansion have been proposed as possible causes of line broadening. However, in most cases of outflow cores that have been properly mapped in lines tracing high-density material, the velocity fields traced by these lines clearly reflect the outflow motions. Some examples are NGC 6334I (Bachiller & Cernicharo 1990), CepA (Bally & Lane 1990, Torrelles et al 1987), NGC 2071-N (Iwata et al 1988), L1551 (Menten & Walmsley 1985; M Tafalla & PC Myers 1996, in preparation), and the dense cores in the L1204/S140 (Tafalla et al 1993a). In some of these cases the velocity field, when studied with low resolution or sensitivity, was first interpreted as rotation, but further high-sensitivity observations revealed the association of the core velocity field with the outflow. In Figure 10 we summarize the situation observed in the L1228 dense core (from Tafalla et al 1995). In this core, the embedded source IRAS 20528+7724 drives a powerful CO outflow of 1.3 pc size (Bally et al 1995). The dense gas exhibits bright lines of C_3H_2 in a region of 0.1 pc around the IRAS source. The middle panels in Figure 10 show the velocity structure within the dense core by using three velocity channel maps of 0.5 km s^{-1} width. The blueshifted and redshifted emissions show no overlap and clearly reflect the outflow motion. The spectra observed at the positions B, R, and C (bottom panel of the figure) show that the entire line profile is shifted, by a full linewidth,

Figure 9 Spectra obtained toward two representative positions of the L1157 outflow. The (0, 0) position is that of the exciting source L1157-mm (see Table 1), whereas the (20″–60″) position is one of the prominent bow shocks in the blueshifted lobe (see e.g. Figure 8). Note the differences in the behavior of the different molecular lines. Some molecules (such as N_2H^+ and DCO^+) only trace the quiescent gas associated with the dense core, whereas other molecules (e.g. SiO, CS, H_2CO, CH_3OH) show broad emission associated with the outflow. [From R Bachiller et al (1996, in preparation).]

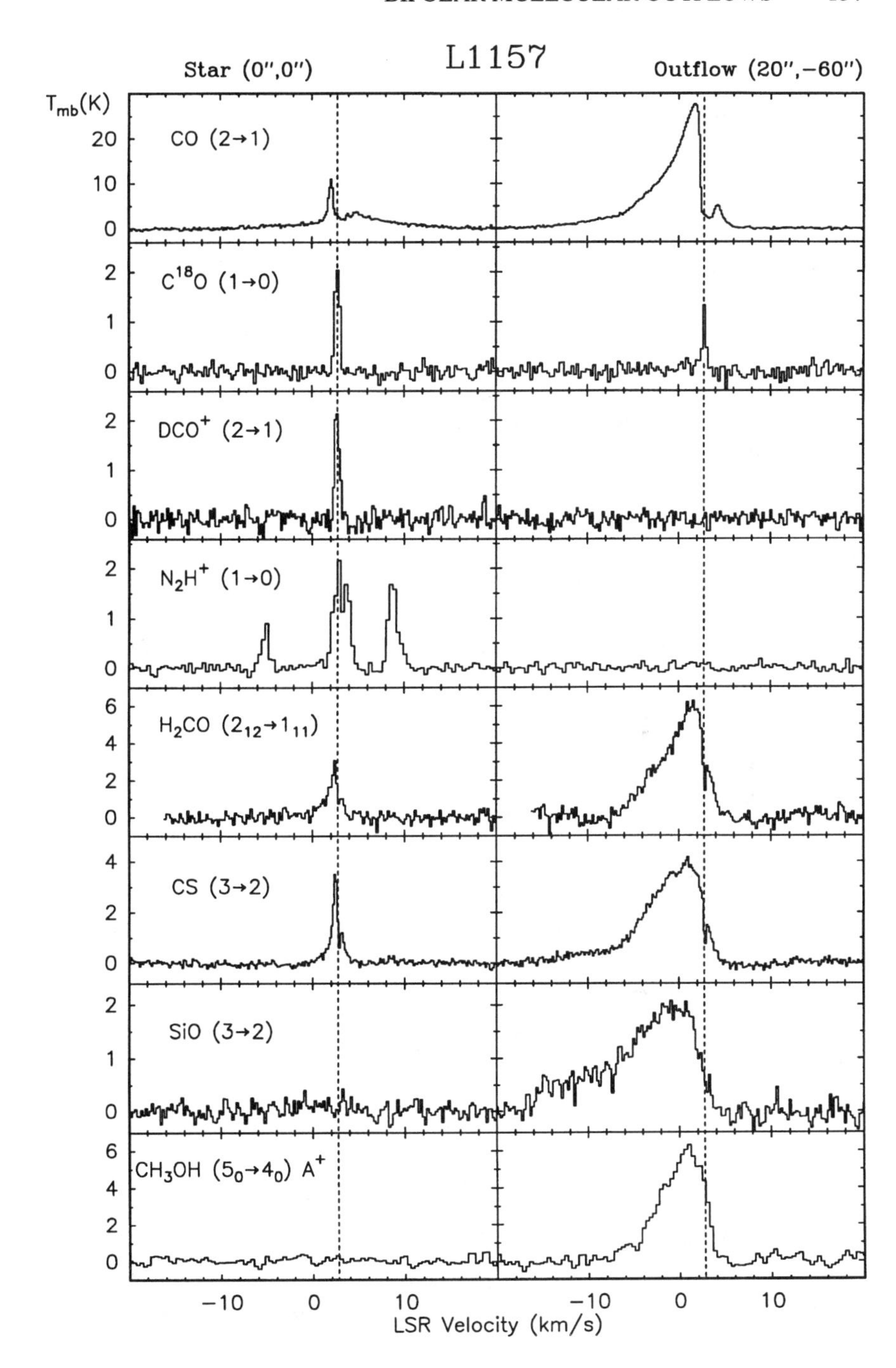
L1157
Star (0",0")
Outflow (20",-60")
T_{mb}(K)
CO (2→1)
$C^{18}O$ (1→0)
DCO^+ (2→1)
N_2H^+ (1→0)
H_2CO (2_{12}→1_{11})
CS (3→2)
SiO (3→2)
CH_3OH (5_0→4_0) A^+
LSR Velocity (km/s)

from one position to the other. The outflow seems to accelerate fragments of the dense core to velocities higher than the internal velocity dispersion. In this way, outflows can excavate cavities in dense cores, producing bipolar nebulae that will successively become visible in the near-IR [as in the case of L1448 (Bally et al 1993b, Bachiller et al 1995b)] and finally in the optical (such as NGC 2261, Hubble's variable nebula around R Mon).

As the circumstellar material is dispersed around a YSO by the action of the outflows, the SED of the YSO will systematically evolve. This process allows one to classify the YSOs in Classes I, II, and III, depending upon the value of the infrared spectral index [$\alpha_{IR} = -d \log(\nu F_\nu)/d \log \nu$] evaluated longward of 2.2 μm (Lada & Wilking 1984, Adams et al 1987, Lada 1991). Class I sources have $\alpha_{IR} > 0$ and SEDs broader than single blackbody functions, probably resulting from warm (300–1000 K) dusty envelopes around a hot (3000–5000 K) stellar-like object. These sources are associated with dense molecular cores (Myers et al 1987). Class II sources have $\alpha_{IR} < 0$ and again SEDs broader than a single temperature blackbody. They are optically visible and exhibit spectra similar to those of cool photospheres, i.e. they are probably classical T Tauri stars surrounded by dusty disks. Class III sources have $\alpha_{IR} < 0$, SEDs similar to those of single blackbodies, are visible, and do not exhibit large infrared excess. They include pre–main sequence stars surrounded by optically thin disks, very young stars of the main sequence, and "naked" T Tauri stars (Walter et al 1988).

In addition to these sources, recent observations with bolometers at millimeter wavelengths have revealed the existence of colder sources that do not fit in the classification scheme depicted above. Such objects, called "Class 0" sources, are thought to be in an evolutionary stage prior to Class I; these are described in the next subsection. The SEDs of these different classes of sources have been successfully modeled by assuming a systematic dispersal of the total circumstellar mass from Class 0 to Class III (Adams et al 1987, Kenyon et al 1993), and it is believed that the circumstellar mass decreases by a factor of 5–10 from one class to the next (André & Montmerle 1994).

In an attempt to describe the evolution of YSOs and main-sequence stars in a unified way, Myers & Ladd (1993) introduced a parameter called bolometric

---

Figure 10 Interaction of the molecular outflow with the dense core in L1228. (*Top panel*) Integrated intensity of the CO high-velocity gas superimposed to the C_3H_2 emission, which traces the core. (*Middle panel*) C_3H_2 emission from within the box marked in the upper panel, integrated over three velocity intervals 0.5 km s^{-1} wide. A clear bipolarity is observed in the kinematics of the core. (*Bottom panel*) C_3H_2 spectra from the positions B, C, R indicated in the top panel. A velocity shift in the line profiles is observed from each position to the next. (Adapted from Tafalla et al 1995.)

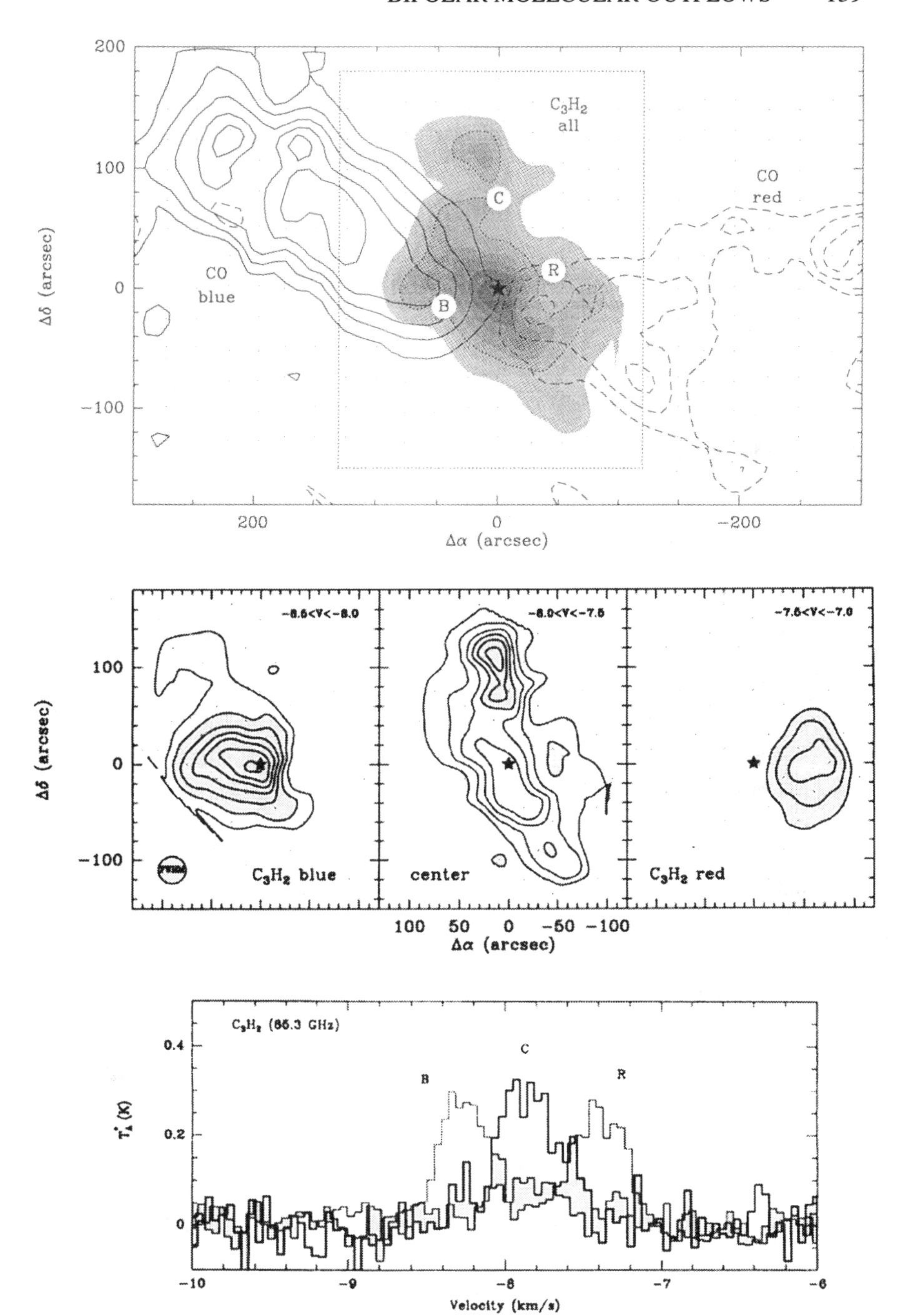

C_3H_2 all
CO red
CO blue
C
R
B
$\Delta\delta$ (arcsec)
$\Delta\alpha$ (arcsec)
C_3H_2 blue
center
C_3H_2 red
C_3H_2 (85.3 GHz)
C
B
R
T_A^* (K)
Velocity (km/s)

temperature, $T_{\rm bol}$, the temperature of a blackbody having the same mean frequency as the observed SED. $T_{\rm bol}$ increases monotonically from Class 0 objects to classes I, II, and III, corresponding to the SED evolution. Class 0 sources have $T_{\rm bol} < 100$ K, whereas typically $T_{\rm bol} \sim 700$ K in Class I sources, ~ 2000 K in Class II, and ~ 3500 K in Class III. The bolometric temperature seems to be an appropriate parameter to describe the different kinds of YSOs, because it uses all the available SED information, and it can be used for the whole range of YSO classes. In addition, the log-log diagram of $L_{\rm bol}$ vs $T_{\rm bol}$ (the BLT diagram) is the analog for YSOs to the H-R diagram, with the advantage that both diagrams have the same main sequence. The BLT diagram can be used to compare observations with theoretical evolutionary models and for comparative studies of different star-forming regions (Chen et al 1995).

5.2 *Class 0 Protostars*

One of the most interesting issues addressed in the study of YSOs is the identification of protostars. Recent observations at millimeteter wavelengths (André & Montmerle 1994) have shown that a YSO at the Class I stage has already assembled most of its stellar mass (i.e. the mass of its circumstellar envelope is well below its stellar mass: $M_{\rm CE} < M_*$). However, in protostars, i.e. objects in which the luminosity is generated from the gravitational accretion, one expects that the stellar mass has not been fully accumulated ($M_{\rm CE} > M_*$). Such pre–Class I objects are referred to as "extreme Class I" sources (Lada 1991) or "Class 0" protostars (André et al 1993). The evolutionary sequence from Class 0 to III is summarized in Figure 11. Pre-protostellar cores prior to the start of gravitational collapse can be identified by mapping the millimeter/submillimeter emission in dark clouds (Benson & Myers 1989, Ward-Thompson et al 1994).

Phenomenologically, a Class 0 protostar is defined as a submillimeter source with the three following attributes (André et al 1993, Barsony 1995): 1. At most, weak emission at $\lambda < 10\ \mu$m, 2. a spectral energy distribution similar to a blackbody at 15–30 K, and 3. $L_{\rm submm}/L_{\rm bol} > 5 \times 10^{-3}$, where $L_{\rm submm}$ is the luminosity measured at $\lambda > 350\ \mu$m and $L_{\rm bol}$ is the bolometric luminosity. In addition, Class 0 protostars can be observationally distinguished from pre-protostellar cores by the presence of a centimeter source (ionized gas) or an outflow. Table 1 lists most of the Class 0 sources identified at present, together with some of their characteristics.

Gravitational infall is expected to occur in Class 0 sources, and spectral lines of moderate optical depth should show the redshifted self-absorption asymmetry characteristic of infall motions. However, the velocity field around YSOs is often complicated by the presence of outflow motions, and the observed line profiles are complex. In some cases, nevertheless, the evidence for infall seems well founded, namely in B335 (Zhou et al 1993), IRAS 16293 (Walker et al

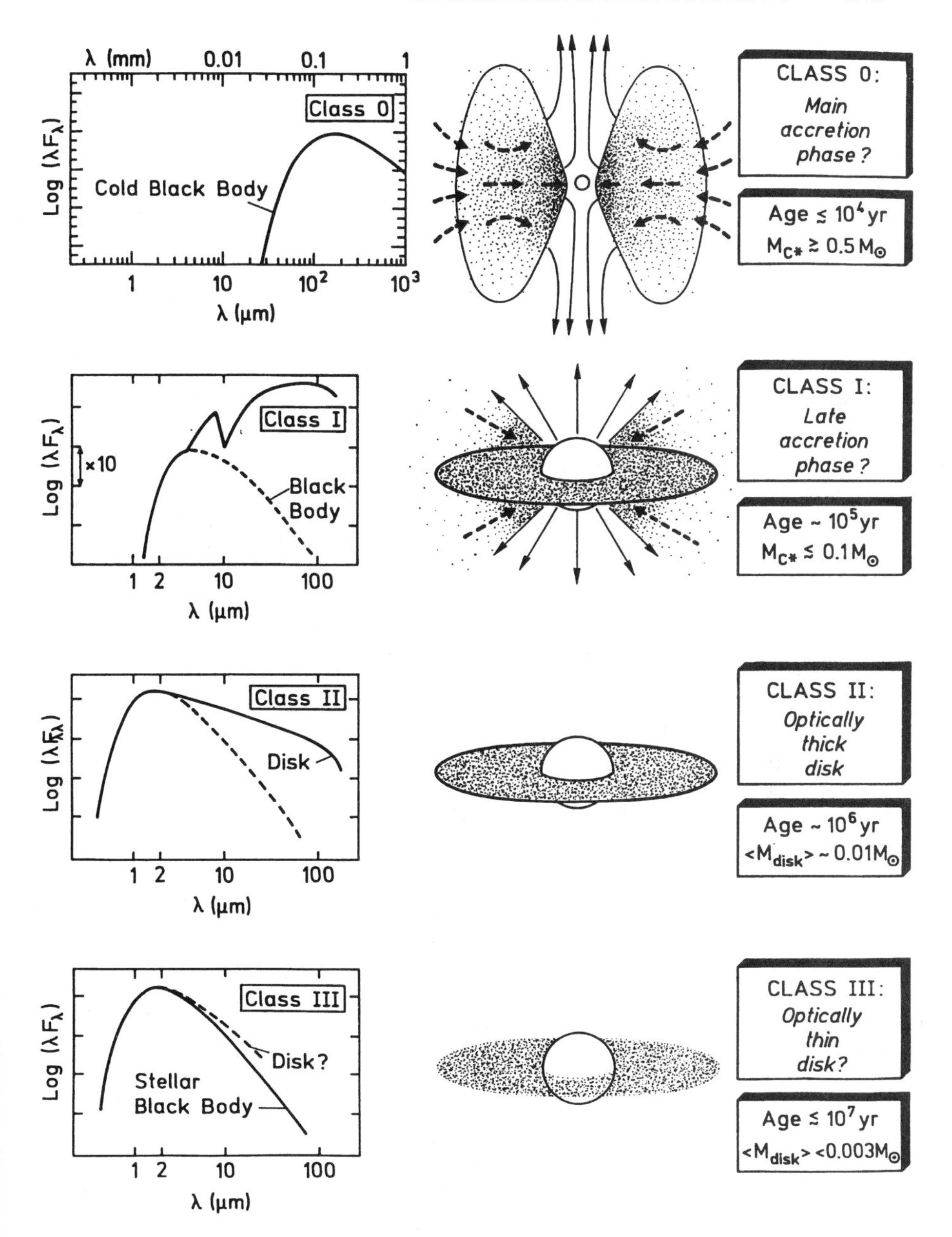

Figure 11 Evolutionary sequence of the spectral energy distributions for low-mass YSOs as proposed by André (1994). The four classes 0, I, II, and III correspond to successive stages of evolution.

Table 1 Properties of candidate Class 0 protostars

Source	IRAS name	α(1950.0)	δ(1950.0)	Dist. (pc)	$L_{\rm bol}$ ($L_\odot$)	$M_{\rm env}$ ($M_\odot$)	T_d (K)	Outflow[a]	Ref.[c]
L1448-mm		03 22 34.3	30 33 35	300	10	2	30	hc	1, 2, 3
L1448/IRS3	03225+3034	03 22 31.5	30 34 49	300	9	1.4		hc	1, 3
NGC1333/IRAS2*		03 25 49.9	31 04 16	350	40			hc,mu	4, 5
NGC1333/IRAS4A		03 26 04.8	31 03 13	350	14	7	37	hc	5, 6
NGC1333/IRAS4B		03 26 06.5	31 02 51	350	14	3		bip	5, 6
IRAS03282	03282+3035	03 28 15.2	30 35 14	300	2	0.6	26	hc	7, 8
HH211-mm		03 40 48.7	31 51 24	300	~5			hc	9
IRAS04166	04166+2706	04 16 37.8	27 06 29	140	0.4	0.2		bip	10, 11
L1527	04368+2557	04 36 49.5	25 57 16	140	2	0.4	59[b]	hc	12, 13
RNO43-mm	05295+1247	05 29 30.6	12 47 35	400	6	0.6	33	hc	14
NGC2024/FIR5		05 39 13.0	–01 57 08	400	> 10	15	20	hc,mo	15, 16, 17
NGC2024/FIR6		05 39 13.7	–01 57 30	400	> 15	6	20	hc,c	15, 16, 18
HH24-mm		05 43 34.8	–00 11 49	460	5	6	25		19, 20, 21
NGC2264G	06384+0958	06 38 25.8	09 58 52	800	5			hc	22, 23, 24, 25
IRAS08076*	08076–3556	08 07 40.3	–35 56 06	450	11		74[a]	hc	5
BHR71-mm*	11590–6452	11 59 03.1	–64 52 11	200	12			hc	26, 27
VLA1623		16 23 25.0	–24 17 47	160	1	0.6	20	hc	28, 29
IRAS16293	16293–2422	16 29 21.0	–24 22 16	160	23	2	39	hc,mu	30
L483*	18148–0440	18 14 50.6	–04 40 49	200	14		49[b]	hc	12, 13
Serp/S68N		18 27 15.2	01 14 57	310			20	w	31, 32, 33, 34
Serp/FIRS1		18 27 17.3	01 13 23	310	50		35		31, 32, 33, 34
Serp/smm3		18 27 27.3	01 11 55	310	< 11		< 20		31, 33, 34
Serp/smm4		18 27 24.7	01 11 10	310	< 11		20		31, 33, 34
L723-mm	19156+1906	19 15 42.0	19 06 55	300	3	0.6	—	mu?	35, 36
B335	19345+0727	19 34 35.1	07 27 24	250	3	0.8	30	hc	37
IRAS20050*	20050+2720	20 05 02.5	27 20 09	700	260		40	hc,mu	38, 39
S106-smm		20 25 32.4	37 12 48	600	> 24	<10	> 20		40
L1157-mm	20386+6751	20 38 39.3	67 51 36	440	11	0.5	—	hc	41, 42

* Firm classification as Class 0 source needs confirmation.

[a] Outflow characteristics: hc = highly collimated CO outflow, bip = bipolar, c = compact, mu = multiple, mo = monopolar, w = wings.

[b] Bolometric temperature in the sense of Myers & Ladd (1993).

[c] References: 1. Bachiller et al (1990), 2. Bachiller et al (1991a), 3. Bachiller et al (1995b), 4. Sandell et al (1994), 5. Hodapp & Ladd (1995), 6. Sandell et al (1991), 7. Bachiller et al (1991c), 8. Bachiller et al (1994b), 9. McCaughrean et al (1994), 10. Kenyon et al (1993), 11. Bontemps et al (1996), 12. Ladd et al (1991), 13. Myers et al (1995), 14. Zinnecker et al (1992), 15. Mezger et al (1992), 16. Mauersberger et al (1992), 17. Richer et al (1992), 18. Richer (1990), 19. Chini et al (1993), 20. Ward-Thompson et al (1995), 21. Bontemps et al (1995), 22. Margulis et al (1990), 23. Gómez et al (1994), 24. Ward-Thompson et al (1995), 25. Lada & Fich (1996), 26. Bourke & Lehtinen (1995), 27. Bourke & Lehtinen (1996), 28. André et al (1990a), 29. André et al (1993), 30. Walker et al (1986), 31. Casali et al (1993), 32. McMullin et al (1994b), 33. Hurt et al (1996), 34. White et al (1995), 35. Cabrit & André (1991), 36. Avery et al (1990), 37. Chandler et al (1990), 38. Wilking et al (1989), 39. Bachiller et al (1995a), 40. Richer et al (1993), 41. Gueth et al (1995), 42. Tafalla & Bachiller (1995).

1994), L1527 (Zhou et al 1994, Myers et al 1995), and L483 (Myers et al 1995). Note that all these are Class 0 sources and that apparently more evolved (Class I) objects exhibit no sign of infall (e.g. Zhou et al 1994).

The concept of Class 0 protostars has been introduced for low-mass YSOs, whereas high-mass counterparts of so young objects are more difficult to study because of their faster evolution and the higher confusion in the more massive and turbulent surrounding clouds. High-mass protostars could, however, have been detected as mm peaks near ultracompact HII regions (e.g. Cesaroni et al 1994), and evidence for infall has been claimed around some ultracompact HII regions (Welch et al 1987, Keto et al 1988, Rudolph et al 1990, Wilner et al 1995). We caution, however, that the interpretation in terms of infall in these objects is always complicated by the presence of strong outflows.

6. ORIGIN OF BIPOLAR MOLECULAR OUTFLOWS

6.1 *Disks*

Observations of molecular lines and of millimeter and infrared continua have provided evidence that YSOs are surrounded by circumstellar structures of about 10^2 AU and masses in the range of a few 10^{-3} to 1 $M_\odot$ (see Sargent 1996 for a recent review). Millimeter-wave continuum surveys are providing information on disk properties and on disk frequency rate in different cloud complexes (André et al 1990b; Beckwith & Sargent 1991, 1993; Reipurth et al 1993; Osterloh & Beckwith 1995). The *HST* has observed externally illuminated disks in the Orion complex and disks in absorption against the nebular background of the Orion clouds (O'Dell et al 1993, O'Dell & Wen 1994).

Disks are also necessary to explain a number of observational facts: 1. the asymmetries of the forbidden line profiles observed around YSOs, because the preferentially blueshifted profiles are assumed to result from the disk occultation of the redshifted emission (Appenzeller et al 1984, Edwards et al 1987), 2. the SEDs of the classical T Tau stars, since the typical spectral index of viscous disks explains quite accurately the IR observations (Bertout et al 1988), 3. the excess observed in the optical and UV ranges, which is responsible for the veiling of the photospheric absorption lines (Bertout 1989), and 4. the eruptions of FU Ori stars that can be understood as resulting from activity in the accretion disks (e.g. Hartmann & Kenyon 1985, Croswell et al 1987).

The physical parameters of the disks are difficult to obtain. Infrared observations provide the total spectrum of the star/disk system, and sophisticated models are necessary to extract disk properties (e.g. Bertout et al 1988). Interferometric observations at millimeter wavelengths currently provide the needed resolution to directly observe the gas and dust emission from some disks (Sargent & Welch 1993), but these disks may be surrounded by more extended envelopes that introduce confusion in the observations. In particular, scattering of the disk radiation in the envelope can produce a far-infrared or submillimeter excess (Natta 1993), distorting the SED of the disk (Butner et al 1994). These envelopes can also bias the interpretation of mm-wave continuum emission (Terebey et al 1993).

The interpretation of molecular line observations is also complicated by the coexistence of the disk rotation with other systematic motions such as infall and outflow. Convincing cases of disks have been found in HL Tau (Sargent & Beckwith 1987, 1991), T Tau (Weintraub et al 1989), and GG Tau (Dutrey et al 1994). In the case of HL Tau, the presence of outflow motions makes it difficult to obtain accurate estimates of the disk parameters (Cabrit et al 1996). However, the structure of the circumbinary disk around GG Tau is well revealed by the 2″ angular resolution images of Dutrey et al (1994). It appears that the material within 180 AU of the binary has been cleared up, and the disk has a radius of 900 AU.

6.2 *Relationship of Jets and Disks*

There is increasing observational evidence for the existence of a close link between jets and disks around YSOs. For instance, in T Tau stars, forbidden line emission, which is thought to arise from an outflow, is only seen in objects presenting near-IR excesses attributed to disks (Edwards 1995). In addition, the intensity of the high-velocity emission seen in the [OI] 6300 Å line is correlated with the near-IR color excess (Edwards et al 1993). In younger objects, such relations are also well observed. In particular, Cabrit & André (1991) found the momentum flux of molecular outflows to be well correlated with the mass of the circumstellar YSO envelope as determined from mm-wave observations. This correlation was recently improved by Bontemps et al (1996), demonstrating that there is a good continuity from Class 0 to Class I sources (Figure 12). We finally note that the ubiquity of high-velocity outflows around YSOs is accompanied by corresponding very high frequency rates in the observations of disks. For instance, in embedded clusters, the fraction of young low-mass stars that have circumstellar disks exceeds 80% (Strom 1995, Dougados et al 1996).

The observed relationship between the properties of jets and disks suggests that disks are necessary to drive winds. Hartmann & McGregor (1982) showed that purely stellar winds could not explain the observed mass-loss rates in typical T Tau stars, since this would require the stars to rotate near to breakup, in contrast to the observations showing T Tau stars rotating an order of magnitude slower than breakup velocities (Vogel & Kuhi 1981, Bouvier et al 1986). The formation of disks provides a powerful mechanism to store the angular momentum during the YSO evolution, offering a natural explanation for the slow T Tau rotation velocities. On the other hand, the neutral species (Na, HI, CO) observed at very high velocities around some YSOs suggest that the wind has a significant neutral component, supporting the idea that the wind arises from the disk, and not from the hot stellar surface (Königl & Ruden 1993).

In summary, accretion disks appear as the reservoir of momentum and energy that can potentially account for the enormous mechanical power observed in bipolar molecular outflows. In order to understand how the momentum is transferred from the disk to the wind, a series of theoretical models have been constructed. These are discussed in the next subsection.

6.3 *Models for the Wind Origin*

Purely hydrodynamical models were first constructed to explain the origin of bipolar outflows. In such models the wind was collimated externally by the

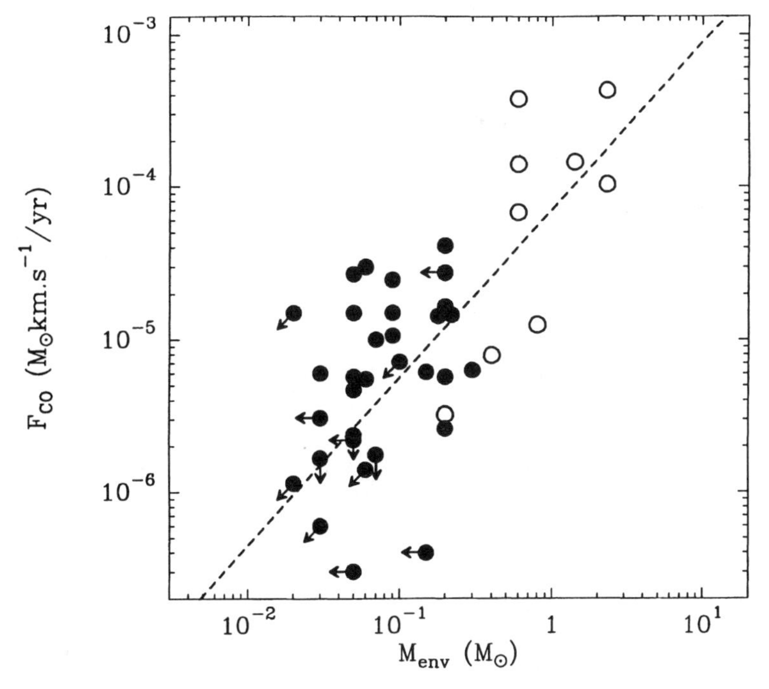

Figure 12 Momentum flux in the CO outflow vs circumstellar envelope mass for a sample of Class 0 (*open circles*) and Class I (*filled circles*) YSOs. The most powerful outflows emanate from the youngest objects with most massive envelopes. The dashed line is a fit to the observed correlation. (From Bontemps et al 1996.)

ambient medium. Barral & Cantó (1981) first proposed that an isotropic wind could be collimated by the thermal pressure from a surrounding large-scale flattened structure or disk. In fact, the wind could be collimated by the formation of de Laval nozzles when it expands through the decreasing density of the core. Such purely hydrodynamical models present the serious drawback that the jet acceleration depends strongly on the shape of the nozzle generated by the external pressure, i.e. the structure of the external medium critically determines the jet properties. Furthermore, as Königl (1982) concluded, a magnetic field is necessary to create the initial anisotropy, determining for instance the orientation of the protostellar disk by means of magnetic braking (Mouschovias & Paleologou 1980), which will eventually determine the jet orientation.

Magnetic fields seem also necessary to launch the wind. As De Campli (1981) pointed out, thermal pressure alone cannot generate the observed outflows, because the temperatures implied at the base of the flow would be very high, and the radiative losses (e.g. by X rays) would be several orders of magnitude higher than the stellar luminosity (Königl & Ruden 1993). However, a magnetic field coupled to the rotating disk surrounding the YSO provides a potentially powerful engine to explain the production of jets. It thus follows that the most suitable models for the origin of protostellar jets are those of magnetohydrodynamical (MHD) disk-driven winds. These models can be divided into two categories, depending on whether the jet arises at the disk surface or at the star/disk boundary layer.

6.3.1 DISK-DRIVEN WINDS The model of Blandford & Payne (1982) for extragalactic jets was soon applied to the case of the YSO jets (Pudritz & Norman 1983, 1986; Pudritz 1985; Königl 1989). In these models accretion and ejection are interdependent processes, and the wind is centrifugally driven by the poloidal magnetic fields threading the disk. Lovelace et al (1991, 1993) have tried to simplify the problem by averaging the different physical variables over the cross section of the jet at a given distance from the equatorial plane. Further refinements of MHD analytical models have been recently done by Appl & Camenzind (1992), Pudritz et al (1991), Pelletier & Pudritz (1992), Contopoulos & Lovelace (1994), and Rosso & Pelletier (1994). Numerical simulations have been carried out by Uchida & Shibata (1985) and Shibata & Uchida (1986). Most of these models do not consider the detailed structure of the disk at the base of the jet. However, as discussed above, observations suggest that the structures of the jet and disk are intimately related, so that accretion and ejection should be modeled together. Königl (1989) was the first to consider the disk structure with realistic magnetic fields. Wardle & Königl (1993) refined this model for the disk structure. Recent self-consistent models

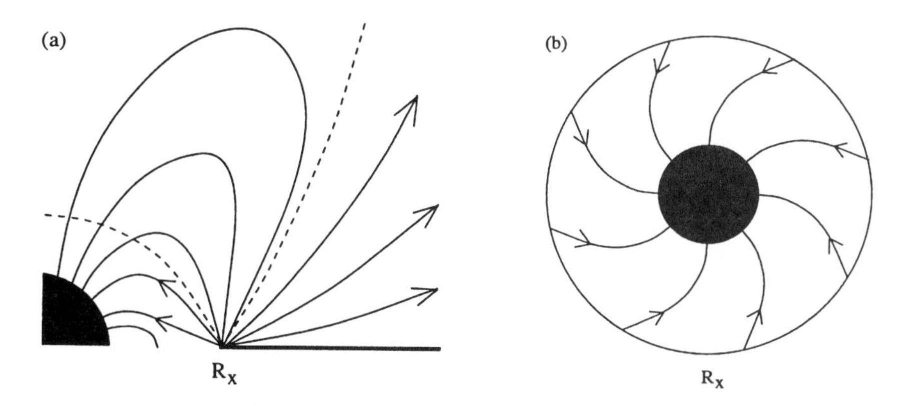

Figure 13 Schematic views of the (*a*) meridional plane and (*b*) equatorial plane of the configuration modeled by Shu et al (1994a,b) for the origin of bipolar outflows. The circumstellar disk is truncated at a distance R_X from the star. Both energetic outflows and funnel flows emerge from the disk truncation region. Gas accreting from the disk onto the star in a funnel flow drags the stellar field into a trailing spiral pattern. (From Najita 1995.)

of magnetized accretion-ejection structures have been developed by Ferreira & Pelletier (1993a,b, 1995).

Some problems of the disk-wind models remain poorly understood. The required magnetic field in the disk has to be maintained by a turbulent dynamo process, but the fields generated by the dynamo process are expected to be too weak (Stepinski & Levy 1990, 1991). Also, as emphasized by Shu (1995), the external magnetic fields retained in viscous disks are probably insufficient to launch the wind, owing to the likely low ionization of the disks.

6.3.2 BOUNDARY LAYER–DRIVEN WINDS In the boundary layer between the accretion disk and the star the rotational velocity of the disk is regulated to the star rotational velocity. This region is expected to be an important reservoir of energy where jets could potentially be efficiently formed. Torbett (1984) proposed that the shock developed by the effect of the accretion combined with the thermal instabilities were responsible for the generation of jets. This model, however, presents the same problems as thermal stellar winds (De Campli 1981). Pringle (1989) suggested that a strong toroidal magnetic field could be produced at the boundary layer by a dynamo effect, but the details of the ejection process were not modeled. Other models that place the origin of the jet at the boundary layer have been proposed by Camenzind (1990) and Bertout & Regev (1992).

The most popular model of this category is that of the X-celerator. This model assumed initially (Shu et al 1988) that the jet was generated at the region of the stellar equator where centrifugal and gravitational forces are compensated (the X-point). The YSO was assumed to have a strong magnetic field and was able to continue accreting by ejecting a strong outflow at a significant fraction of the infall rate. Since the mass-loss happens on the equatorial plane, the optical jets are produced by the expansion of the flow toward the rotational poles. The main difficulty of this model is that the star needs to rotate at breakup at its equator, while actual T Tauri stars are known to rotate at about a tenth of this velocity (Bouvier et al 1986). Recently, the X-celerator model has been modified (Shu et al 1994a,b; Najita & Shu 1994; Ostriker & Shu 1995) to allow for the star rotating below breakup. Following a suggestion by Königl (1991), it is assumed that the stellar magnetic field is strong enough to truncate the disk at a radius R_X from the star (Figure 13). The rapid rotation of the material in a small region around this radius seems able to drive a funnel inflow into the star together with a X-type outflow.

7. CONCLUSION

Bipolar outflows are ubiquitous around young stars and involve amounts of energy similar to those involved in the accretion processes. Thus, outflows are a dominant ingredient in the formation of stars. Outflows probably limit the mass of the star/disk system under formation and are indispensable for transporting away the excess angular momentum of accretion disks. Outflows seem able to perturb the dense gas within the cores where stars are born, and they could determine the dense core evolution after the first stellar generation. The youngest stellar objects presently known (the so-called Class 0 protostars) are sources of energetic outflows, implying that outflow and infall motions happen simultaneously and are closely linked from the very beginning stages of the star formation process. The idea of a new star forming from relatively simple hydrodynamic infall is giving way to a picture in which magnetic fields play a crucial role and stars are born through the formation of complex engines of accretion/ejection. The next generation of millimeter-wave interferometers will be decisive in elucidating the structure of such engines and will probably reveal unexpected phenomena related to the origin of outflows. It seems inevitable that future theories of star formation will have to take into account, together with the structure of the protostar and its surrounding accretion disk, the processes of infall and outflow in a unified manner.

ACKNOWLEDGMENTS

It is a pleasure to acknowledge Drs. André, Cabrit, Cernicharo, Dutrey, Eislöffel, Fuente, Gómez-González, Gueth, Guilloteau, Mardones, Martín-Pintado, Myers,

Reipurth, Richer, Tafalla, Pérez-Gutiérrez, and Planesas for interesting discussions and fruitful collaborative work on outflows during the past years. Drs. Eislöffel, Reipurth, and Tafalla gave me many comments and suggestions that helped to improve the manuscript. I also thank Drs. Bontemps and McCaughrean for providing figures and the numerous colleagues who provided manuscripts in advance of publication. A part of this article was written while I was guest professor at the Observatoire de Grenoble (France), where I enjoyed a friendly and stimulating atmosphere. Funding support from Spanish DGICYT (through grant PB93–48) is gratefully acknowledged.

Literature Cited

Adams FC, Lada CJ, Shu FH. 1987. *Ap. J.* 312:788–806

Allamandola LJ, Sandford SA, Tielens AGGM, Herbst T. 1992. *Ap. J.* 399:134–46

Allen DA, Burton MG. 1993. *Nature* 363:54–56

André P. 1994. In *The Cold Universe,* ed. T Montmerle, CJ Lada, IF Mirabel, J Trân Thanh Vân, pp. 179–92. Gif-sur-Yvette, France: Frontières.

André P. 1995. In *Circumstellar Matter,* ed. G Watt, P Williams, pp. 29–42. Dordrecht: Kluwer.

André P, Martín-Pintado J, Despois D, Montmerle T. 1990a. *Astron. Astrophys.* 236:180–92

André P, Montmerle T. 1994. *Ap. J.* 420:837–62

André P, Montmerle T, Feigelson ED, Steppe H. 1990b. *Astron. Astrophys.* 240:321–30

André P, Ward-Thompson D, Barsony M. 1993. *Ap. J.* 406:122–41

Appenzeller I, Jankovics I, Östreicher R. 1984. *Astron. Astrophys.* 141:108–15

Appenzeller I, Mundt R. 1989. *Astron. Astrophys. Rev.* 1:191–234

Appl S, Camenzind M. 1992. *Astron. Astrophys.* 256:354–70

Avery LW, Hayashi SS, White GJ. 1990. *Ap. J.* 357:524–30

Bachiller R, André P, Cabrit S. 1991a. *Astron. Astrophys.* 241:L43–46

Bachiller R, Cernicharo J. 1990. *Astron. Astrophys.* 239:276–86

Bachiller R, Cernicharo J, Martín-Pintado J, Tafalla M, Lazareff B. 1990. *Astron. Astrophys.* 231:174–86

Bachiller R, Fuente A, Tafalla M. 1995a. *Ap. J. Lett.* 445:L51–54

Bachiller R, Gómez-González J. 1992. *Astron. Astrophys. Rev.* 3:257–87

Bachiller R, Guilloteau S, Dutrey A, Planesas P, Martín-Pintado J. 1995b. *Astron. Astrophys.* 299:857–68

Bachiller R, Guilloteau S, Kahane C. 1987. *Astron. Astrophys.* 173:324–36

Bachiller R, Liechti S, Walmsley CM, Colomer F. 1995c. *Astron. Astrophys.* 295:L51–54

Bachiller R, Martín-Pintado J, Fuente A. 1991b. *Astron. Astrophys.* 243:L21–24

Bachiller R, Martín-Pintado J, Fuente A. 1993. *Ap. J. Lett.* 417:L45–48

Bachiller R, Martín-Pintado J, Planesas P. 1991c. *Astron. Astrophys.* 251:639–48

Bachiller R, Tafalla M, Cernicharo J. 1994a. *Ap. J. Lett.* 425:L93–96

Bachiller R, Terebey S, Jarrett T, Martín-Pintado J, Beichman CA, van Buren D. 1994b. *Ap. J.* 437:296–304

Bally J, Castets A, Duvert D. 1994. *Ap. J.* 423:310–19

Bally J, Devine D. 1994. *Ap. J. Lett.* 428:L65–68

Bally J, Devine D, Fesen RA. 1995. *Ap. J.* 454:345–60

Bally J, Devine D, Hereld M, Rauscher BJ. 1993a. *Ap. J. Lett.* 418:L75–78

Bally J, Lada CJ. 1983. *Ap. J.* 265:824–47

Bally J, Lada CJ, Lane, AP. 1993b. *Ap. J.* 418:322–27

Bally J, Lane AP. 1990. In *Astrophysics with Infrared Arrays,* ed. R Elston, p. 273. San Francisco: Astron. Soc. Pac.

Bally J, Stark AA. 1983. *Ap. J. Lett.* 266:L61–64

Barral P, Cantó J. 1981. *Rev. Mex. Astron. Astrofis.* 5:101–8

Barsony M. 1995. In *Clouds, Cores, and Low Mass Stars,* ed. DP Clemens, R Barvainis, pp. 197–206. San Francisco: Astron. Soc. Pac.
Beckwith SVW, Sargent AI. 1991. *Ap. J.* 381:250–58
Beckwith SVW, Sargent AI. 1993. *Ap. J.* 402:280–91
Benson PC, Myers PC. 1989. *Ap. J. Suppl.* 71:89–108
Bertout C. 1989. *Annu. Rev. Astron. Astrophys.* 27:351–95
Bertout C, Basri G, Bouvier J. 1988. *Ap. J.* 330:350–73
Bertout C, Regev O. 1992. *Ap. J. Lett.* 399:L163–66
Bieging JH, Cohen M. 1985. *Ap. J. Lett.* 289:L5–8
Bieging JH, Cohen M, Schwartz PR. 1984. *Ap. J.* 282:699–708
Biro S, Raga AC. 1994. *Ap. J.* 434:221–31
Blake GA, Sandell G, van Dishoeck EF, Groesbeck TD, Mundy LG. 1995. *Ap. J.* 441:689–701
Blake GA, Sutton EC, Masson CR, Phillips TG. 1987. *Ap. J.* 315:621–45
Blake GA, van Dishoeck EF, Jansen DJ, Groesbeck TD, Mundy LG. 1994. *Ap. J.* 428:680–92
Blandford RD, Payne DG. 1982. *MNRAS* 199:883–903
Blondin JM, Fryxell BA, Königl A. 1990. *Ap. J.* 360:370–86
Bontemps S, André P, Terebey S, Cabrit S. 1996. *Astron. Astrophys.* In Press
Bontemps S, André P, Ward-Thompson D. 1995. *Astron. Astrophys.* 297:98–102
Bourke TL, Lehtinen KK. 1995. In *CO: Twenty-Five Years of Millimeter-Wave Spectroscopy. IAU Symp. 170.* In press
Bourke TL, Lehtinen KK. 1996. Preprint
Bouvier J, Bertout C, Benz W, Mayor M. 1986. *Astron. Astrophys.* 165:110–19
Bührke Th, Mundt R, Ray TP. 1988. *Astron. Astrophys.* 200:99–119
Butner HM, Natta A, Evans NJ II. 1994. *Ap. J.* 420:326–35
Cabrit S. 1993. In *Stellar Jets and Bipolar Outflows,* ed. L Errico, A Vittone, pp. 1–14. Dordrecht: Kluwer
Cabrit S, André P. 1991. *Ap. J. Lett.* 379:L25–28
Cabrit S, Bertout C. 1986. *Ap. J.* 307:313–23
Cabrit S, Bertout C. 1990. *Ap. J.* 348:530–41
Cabrit S, Bertout C. 1992. *Astron. Astrophys.* 261:274–84
Cabrit S, Edwards S, Strom SE, Strom KM. 1990. *Ap. J.* 354:687–700
Cabrit S, Goldsmith PF, Snell RL. 1988. *Ap. J.* 334:196–208
Cabrit S, Guilloteau S, André P, Bertout C, Montmerle T, Schuster K. 1996. *Astron. Astrophys.* 305:527–40
Camenzind M. 1990. *Rev. Mod. Astron.* 3:234–65
Campbell B, Person SE, Strom SE, Grasdalen GL. 1988. *Astron. J.* 95:1173–84
Cantó J, Raga A. 1991. *Ap. J.* 372:646–58
Carr JS, Harvey PM, Lester DF. 1987. *Ap. J. Lett.* 321:L71–74
Casali MM, Eiroa C. 1995. *Rev. Mex. Astron. Astrof (Ser. Conf.)* 1:303–7
Casali MM, Eiroa C, Duncan WD. 1993. *Astron. Astrophys.* 275:195–200
Cernicharo J, Bachiller R, Duvert G, González-Alfonso E, Gómez-González J. 1992. *Astron. Astrophys.* 261:589–601
Cernicharo J, Neri R, Reipurth B, Bachiller R. 1996. Preprint
Cernicharo J, Reipurth B. 1996. Preprint
Cesaroni R, Olmi L, Walmsley CM, Churchwell E, Hofner P. 1994. *Ap. J. Lett.* 435:L137–41
Chandler CJ, Gear WK, Sandell G, Hayashi S, Duncan WD, et al. 1990. *MNRAS* 243:330–35
Chen H, Myers PC, Ladd EF, Wood DO. 1995. *Ap. J.* 445:377–92
Chernin LM, Masson CR. 1991. *Ap. J. Lett.* 382:L93–96
Chernin LM, Masson CR. 1995a. *Ap. J.* 443:181–86
Chernin LM, Masson CR. 1995b. *Ap. J.* 455:182–89
Chernin LM, Masson CR, Fuller G. 1994a. *Ap. J.* 436:741–48
Chernin LM, Masson CR, Gouveia Dal Pino EM, Benz W. 1994b. *Ap. J.* 426:204–14
Chini R, Krügel E, Haslam CGT, Kreysa E, Lemke R, et al. 1993. *Astron. Astrophys.* 272:L5–8
Clark FO, Laurejis RJ. 1986. *Astron. Astrophys.* 154:L26–29
Cohen M, Bieging JH, Schwartz PR. 1982. *Ap. J.* 253:707–15
Contopoulos J, Lovelace RVE. 1994. *Ap. J.* 429:139–52
Croswell K, Hartmann L, Avret EH. 1987. *Ap. J.* 312:227–42
Cudworth KM, Herbig GH. 1979. *Astron. J.* 84:548–51
Curiel S, Rodríguez LF, Moran JM, Cantó J. 1993. *Ap. J.* 415:191–203
Davis CJ, Dent WRF, Matthews HE, Aspin C, Lightfoot JF. 1994a. *MNRAS* 266:933–44
Davis CJ, Eislöffel J. 1995. *Astron. Astrophys.* 300:851–69
Davis CJ, Eislöffel J, Ray TP. 1994b. *Ap. J. Lett.* 426:L93–95
Davis CJ, Mundt R, Eislöffel J. 1994c. *Ap. J. Lett.* 437:L55–58
Davis CJ, Mundt R, Eislöffel J, Ray TP. 1995. *Astron. J.* 110:766–75

De Campli WM. 1981. *Ap. J.* 244:124–46
Dent WRF, Matthews HE, Walther DM. 1995. *MNRAS* 277:193–209
De Young DS. 1986. *Ap. J.* 307:62–72
Dopita MA, Schwartz R, Evans I. 1982. *Ap. J.* 263:L73–76
Dougados C, Carpenter J, Meyer M, Strom SE. 1996. Preprint
Draine BT, Roberge WG, Dalgarno A. 1983. *Ap. J.* 264:485–507
Dutrey A, Guilloteau S, Bachiller R. 1996. Preprint
Dutrey A, Guilloteau S, Simon M. 1994. *Astron. Astrophys.* 286:149–59
Dyson JE. 1984. *Astrophys. Space Sci.* 106:181–97
Edwards S. 1995. *Rev. Mex. Astron. Astrof. (Ser. Conf.)* 1:309–16
Edwards S, Cabrit S, Strom SE, Heyer I, Strom KM, Andrson E. 1987. *Ap. J.* 321:473–95
Edwards S, Ray T, Mundt R. 1993. In *Protostars and Planets III,* ed. EH Levy, JI Lunine, pp. 567–602. Tucson: Univ. Ariz. Press
Edwards S, Snell RL. 1982. *Ap. J.* 261:151–60
Edwards S, Snell RL. 1983. *Ap. J.* 270:605–19
Edwards S, Snell RL. 1984. *Ap. J.* 281:237–49
Edwards S, Strom SE, Snell RL, Jarrett TH, Beichman CA, Strom KM. 1986. *Ap. J. Lett.* 307:L65–68
Eislöffel J, Davis CJ, Ray TP, Mundt R. 1994. *Ap. J. Lett.* 422:L91–93
Eislöffel J, Mundt R. 1994. *Astron. Astrophys.* 284:530–44
Ferreira J, Pelletier G. 1993a. *Astron. Astrophys.* 276:625–36
Ferreira J, Pelletier G. 1993b. *Astron. Astrophys.* 276:637–47
Ferreira J, Pelletier G. 1995. *Astron. Astrophys.* 295:807–32
Fukui Y, Iwata T, Mizuno A, Bally J, Lane AP. 1993. In *Protostars and Planets III,* ed. EH Levy, JI Lunine, pp. 603–39. Tucson: Univ. Ariz. Press
Garden RP, Hayashi M, Gatley I, Hasegawa T, Kaifu N. 1991. *Ap. J.* 374:540–54
Giovanardi C, Lizano S, Natta A, Evans NJ II, Heiles C. 1992. *Ap. J.* 397:214–24
Glassgold AE, Mamon GA, Huggins PJ. 1989. *Ap. J. Lett.* 336:L29–32
Glassgold AE, Mamon GA, Huggins PJ. 1991. *Ap. J.* 373:254–65
Goldsmith PF, Snell RL, Hemeon-Heyer M, Langer WD. 1984. *Ap. J.* 286:599–608
Gómez JF, Curiel S, Torrelles JM, Rodríguez LF, Anglada G, Girart JM. 1994. *Ap. J.* 436:749–53
Gouveia Dal Pino EM, Benz W. 1993. *Ap. J.* 410:686–95
Gredel R, Reipurth B. 1993. *Ap. J. Lett.* 407:L29–32
Gredel R, Reipurth B. 1994. *Astron. Astrophys.* 289:L19–22
Gueth F, Guilloteau S, Bachiller R. 1996. *Astron. Astrophys.* In press
Guilloteau S, Bachiller R, Lucas R, Fuente A. 1992. *Astron. Astrophys.* 265:L49–52
Gwinn CR, Moran JM, Reid MJ. 1992. *Ap. J.* 393:149–64
Haro G. 1952. *Ap. J.* 115:572–73
Hartigan P, Morse JA, Heathcote S, Cecil G. 1993. *Ap. J. Lett.* 414:L121–24
Hartigan P, Raymond PC. 1993. *Ap. J.* 409:705–19
Hartmann L, Kenyon SJ. 1985. *Ap. J.* 299:462–78
Hartmann L, McGregor KB. 1982. *Ap. J.* 259:180–92
Henriksen RN, Ptuskin VS, Mirabel IF. 1991. *Astron. Astrophys.* 248:221–26
Herbig G. 1951. *Ap. J.* 113:697–99
Herbig G. 1962. *Adv. Astron. Astrophys.* 1:47–103
Hirth GA, Mundt R, Solf J. 1994a. *Astron. Astrophys.* 285:929–42
Hirth GA, Mundt R, Solf J, Ray TP. 1994b. *Ap. J. Lett.* 427:L99–102
Hodapp KW. 1994. *Ap. J. Suppl.* 94:615–49
Hodapp KW, Deane J. 1993. *Ap. J. Suppl.* 88:119–35
Hodapp KW, Ladd EF. 1995. *Ap. J.* 453:715–20
Hurt RL, Barsony M, Wootten A. 1996. *Ap. J.* 456:686–95
Iglesias ER, Silk J. 1978. *Ap. J.* 226:851–57
Iwata T, Fukui Y, Ogawa H. 1988. *Ap. J.* 325:372–81
Kenyon SJ, Calvet N, Hartmann L. 1993. *Ap. J.* 414:676–94
Keto ER, Ho PTP, Haschick AD. 1988. *Ap. J.* 324:920–30
Kitamura Y, Kawabe R, Yamashita T, Hayashi M. 1990. *Ap. J.* 363:180–91
Königl A. 1982. *Ap. J.* 261:115–34
Königl A. 1989. *Ap. J.* 342:208–23
Königl A. 1991. *Ap. J. Lett.* 370:L39–43
Königl A, Ruden SP. 1993. In *Protostars and Planets III,* ed. EH Levy, JI Lunine, pp. 641–87. Tucson: Univ. Ariz. Press
Kuhi LV. 1964. *Ap. J.* 140:1409–33
Kwan J, Scoville N. 1976. *Ap. J. Lett.* 210:L39–42
Kwan J, Tademaru E. 1988. *Ap. J. Lett.* 332:L41–44
Lada CJ. 1985. *Annu. Rev. Astron. Astrophys.* 23:267–317
Lada CJ. 1991. In *The Physics of Star Formation and Early Evolution,* ed. CJ Lada, ND Kylafis, pp. 329–63. Kluwer: Dordrecht
Lada CJ, Fich M. 1996. *Ap. J.* 459:638–52
Lada CJ, Wilking BA. 1984. *Ap. J.* 287:610–21
Ladd EF, Adams FC, Casey S, Davidson JA,

Fuller GA, et al. 1991. *Ap. J.* 366:203–20
Lane AP. 1989. In *ESO Workshop on Low Mass Star Formation and Pre-Main Sequence Objects,* ed. B Reipurth, pp. 331–48. Garching: ESO
Lizano S, Giovanardi C. 1995. *Ap. J.* 447:742–51
Lizano S, Heiles C, Rodríguez LF, Koo BC, Shu FH, et al. 1988. *Ap. J.* 328:763–76
Lovelace RVE, Berk HL, Contopoulos J. 1991. *Ap. J.* 376:696–705
Lovelace RVE, Romanova MM, Contopoulos J. 1993. *Ap. J.* 403:158–63
Mac Low MM, Elitzur M. 1992. *Ap. J. Lett.* 393:L33–36
Mac Low MM, Elitzur M, Stone JM, Königl A. 1994. *Ap. J.* 427:914–18
Margulis M, Lada CJ. 1985. *Ap. J.* 299:925–38
Margulis M, Lada CJ, Hasegawa T, Hayashi SS, Hayashi M, et al. 1990. *Ap. J.* 352:615–24
Martí J, Rodríguez LF, Reipurth B. 1993. *Ap. J.* 416:208–17
Martí J, Rodríguez LF, Reipurth B. 1995. *Ap. J.* 449:184–87
Martín-Pintado J, Bachiller R, Fuente A. 1992. *Astron. Astrophys.* 254:315–26
Masson CR, Chernin LM. 1992. *Ap. J. Lett.* 387:L47–50
Masson CR, Chernin LM. 1993. *Ap. J.* 414:230–41
Masson CR, Mundy LG, Keene J. 1990. *Ap. J. Lett.* 357:L25–28
Mathieu RD. 1994. *Annu. Rev. Astron. Astrophys.* 32:465–530
Mauersberger R, Wilson TL, Mezger PG, Gaume R, Johnston KJ. 1992. *Astron. Astrophys.* 256:640–51
McCaughrean MJ, Rayner JT, Zinnecker H. 1994. *Ap. J. Lett.* 436:L189–93
McMullin JP, Mundy LG, Blake GA. 1994a. *Ap. J.* 437:305–16
McMullin JP, Mundy LG, Wilking BA, Hetzel T, Blake GA. 1994b. *Ap. J.* 424:222–36
Menten KM, Walmsley CM. 1985. *Ap. J.* 146:369–74
Meyers-Rice BA, Lada CJ. 1991. *Ap. J.* 368:445–62
Mezger PG, Sievers AW, Haslam CGT, Kreysa E, Lemke R, et al. 1992. *Astron. Astrophys.* 256:631–39
Millar TJ, Herbst E, Charnley SB. 1991. *Ap. J.* 369:147–56
Mitchell GF, Maillard JP, Hasegawa TI. 1991. *Ap. J.* 371:342–56
Morse JA, Hartigan P, Cecil G, Raymond JC, Heathcote S. 1992. *Ap. J.* 399:231–45
Mouschovias T, Paleologou EV. 1980. *Ap. J.* 237:877–99
Mundt R. 1984. *Ap. J.* 280:749–70
Mundt R, Brugel EW, Bührke T. 1987. *Ap. J.* 319:275–303
Mundt R, Fried JW. 1983. *Ap. J. Lett.* 274:L83–86
Mundt R, Stocke J, Strom SE, Strom KM, Anderson ER. 1985. *Ap. J. Lett.* 297:L41–44
Myers PC, Bachiller RB, Caselli P, Fuller GA, Mardones D, et al. 1995. *Ap. J. Lett.* 449:L65–68
Myers PC, Fuller GA, Mathieu RD, Beichman CA, Benson PJ, et al. 1987. *Ap. J.* 319:340–57
Myers PC, Heyer M, Snell RL, Goldsmith PF. 1988. *Ap. J.* 324:907–19
Myers PC, Ladd EF. 1993. *Ap. J. Lett.* 413:L47–50
Najita JR. 1995. *Rev. Mex. Astron. Astrof. (Ser. Conf.)* 1:293–301
Najita JR, Shu FH. 1994. *Ap. J.* 429:808–25
Natta A. 1993. *Ap. J.* 412:761–70
Natta A, Giovanardi C. 1990. *Ap. J.* 356:646–61
Natta A, Giovanardi C, Palla F, Evans NJ Jr. 1988. *Ap. J.* 327:817–21
Neckel T, Staude J. 1987. *Ap. J. Lett.* 322:L27–30
Neufeld DA, Dalgarno A. 1989. *Ap. J.* 340:869–93
O'Dell CR, Wen Z. 1994. *Ap. J.* 436:194–202
O'Dell CR, Wen Z, Hu X. 1993. *Ap. J.* 410:696–700
Olberg M, Reipurth B, Booth R. 1992. *Astron. Astrophys.* 259:252–56
Osterbrock DE. 1958. *Publ. Astron. Soc. Pac.* 70:399–403
Osterloh M, Beckwith SVW. 1995. *Ap. J.* 439:288–302
Ostriker EC, Shu FH. 1995. *Ap. J.* 447:813–28
Parker ND, Padman R, Scott PF. 1991. *MNRAS* 252:442–61
Pelletier G, Pudritz RE. 1992. *Ap. J.* 394:117–38
Pineau des Forêts G, Roueff E, Schilke P, Flower DR. 1993. *MNRAS* 262:915–28
Plambeck RL, Snell RL, Loren RB. 1983. *Ap. J.* 266:321–30
Plambeck RL, Wright MCH. 1987. *Ap. J. Lett.* 317:L101–5
Plambeck RL, Wright MCH, Welch WJ, Bieging JH, Baud B, et al. 1982. *Ap. J.* 259:617–24
Pravdo SH, Rodríguez LF, Curiel S, Cantó J, Torrelles JM, et al. 1985. *Ap. J. Lett.* 293:L35–38
Pringle JE. 1989. *MNRAS* 236:107–15
Pudritz RE. 1985. *Ap. J.* 293:216–29
Pudritz RE, Norman CA. 1983. *Ap. J.* 274:677–97
Pudritz RE, Norman CA. 1986. *Ap. J.* 301:571–86
Pudritz RE, Pelletier G, Gómez de Castro AI. 1991. In *The Physics of Star Formation and Early Evolution,* ed. CJ Lada, ND Kylafis,

pp. 539–64. Kluwer: Dordrecht
Raga AC. 1988. *Ap. J.* 335:820–28
Raga AC. 1991. *Astron. J.* 101:1472–75
Raga AC, Cabrit S. 1993. *Astron. Astrophys.* 278:267–78
Raga AC, Cantó J, Biro S. 1993a. *MNRAS* 260:163–70
Raga AC, Cantó J, Calvet N, Rodríguez LF, Torrelles JM. 1993b. *Astron. Astrophys.* 276:539–48
Raga AC, Kofman L. 1992. *Ap. J.* 386:222–28
Ray TP. 1993. In *Stellar Jets and Bipolar Outflows,* ed. L Errico, A Vittone, pp 241–56. Dordrecht: Kluwer
Reipurth B. 1989. *Nature* 340:42–44
Reipurth B. 1991. In *The Physics of Star Formation and Early Evolution,* ed. CJ Lada, ND Kylafis, pp. 497–538. Kluwer: Dordrecht
Reipurth B, Bachiller R. 1996. In *CO:Twenty-Five Years of Millimeter-Wave Spectroscopy. IAU Symp. 170.* In press
Reipurth B, Bally J, Graham JA, Lane AP, Zealey WJ. 1986. *Astron. Astrophys.* 164:51–66
Reipurth B, Cernicharo C. 1995. *Rev. Mex. Astron. Astrof. (Ser. Conf.)* 1:43–58
Reipurth B, Chini R, Krügel E, Kreysa E, Sievers A. 1993. *Astron. Astrophys.* 273:221–38
Reipurth B, Graham JA. 1988. *Astron. Astrophys.* 202:219–39
Reipurth B, Heathcote S. 1991. *Astron. Astrophys.* 246:511–34
Reipurth B, Heathcote S. 1992. *Astron. Astrophys.* 257:693–700
Reynolds SP. 1986. *Ap. J.* 304:713–20
Richer JS. 1990. *MNRAS* 245:24p–27p
Richer JS, Hills RE, Padman R. 1992. *MNRAS* 254:525–38
Richer JS, Hills RE, Padman R, Russell APG. 1989. *MNRAS* 241:231–46
Richer JS, Padman R, Ward-Thompson D, Hills RE, Harris AI. 1993. *MNRAS* 262:839–54
Rodríguez LF. 1995. *Rev. Mex. Astron. Astrof. (Ser. Conf.)* 1:1–10
Rodríguez LF, Ho PTP, Moran JM. 1980. *Ap. J. Lett.* 240:L149–52
Rodríguez LF, Lizano S, Cantó J, Escalante V, Mirabel IF. 1990. *Ap. J.* 365:261–68
Rodríguez LF, Reipurth B. 1989. *Rev. Mex. Astron. Astrof.* 17:59–63
Rodríguez-Franco A. 1995. *Condiciones físicas y químicas en la nube molecular Orión A.* PhD thesis. Univ. Complutense, Madrid
Rosso F, Pelletier G. 1994. *Astron. Astrophys.* 287:325–37
Rudolph A, Welch WJ, Palmer P, Dubrulle B. 1990. *Ap. J.* 363:528–46
Ruiz A, Alonso JL, Mirabel IF. 1992. *Ap. J. Lett.* 394:L57–60
Russell APG, Bally J, Padman R, Hills RE. 1992. *Ap. J.* 387:219–28
Sandell G, Aspin C, Duncan WD, Russell APG, Robson IE. 1991. *Ap. J. Lett.* 376:L17–20
Sandell G, Knee LBG, Aspin C, Robson IE, Russell APG. 1994. *Astron. Astrophys.* 285:L1–4
Sargent AI. 1996. In *Disks and Outflows around Young Stars,* ed. SVW Beckwith, A Natta, J Staude. Berlin: Springer-Verlag. In press
Sargent AI, Beckwith SVW. 1987. *Ap. J.* 323:294–305
Sargent AI, Beckwith SVW. 1991. *Ap. J. Lett.* 382:L31–34
Sargent AI, Welch WJ. 1993. *Annu. Rev. Astron. Astrophys.* 31:297–343
Schmid-Burgk J, Güsten R, Mauersberger R, Schulz A, Wilson TL. 1990. *Ap. J. Lett.* 362:L25–28
Schmid-Burgk J, Muders D. 1995. In *Stellar and Circumstellar Astrophysics,* ed. G Wallerstein, A Noriega-Crespo. San Francisco: ASP Conf. Ser.
Schwartz RD. 1975. *Ap. J.* 195:631–42
Schwartz RD, Williams PM, Cohen M, Jennings DG. 1988. *Ap. J.* 334:99–102
Scoville NZ, Kwan J. 1976. *Ap. J.* 206:718–27
Shibata K, Uchida Y. 1986. *Publ. Astron. Soc. Jpn.* 38:631–60
Shu FH. 1995. *Rev. Mex. Astron. Astrof. (Ser. Conf.)* 1:375–86
Shu FH, Adams FC, Lizano S. 1987. *Annu. Rev. Astron. Astrophys.* 25:23–81
Shu FH, Lizano S, Ruden SP, Najita J. 1988. *Ap. J. Lett.* 328:L19–23
Shu FH, Najita J, Ostriker E, Wilkin F, Ruden S, Lizano S. 1994a. *Ap. J.* 429:781–96
Shu FH, Najita J, Ruden S, Lizano S. 1994b. *Ap. J.* 429:797–807
Shu FH, Ruden SP, Lada CJ, Lizano S. 1991. *Ap. J. Lett.* 370:L31–34
Shull JM, Beckwith S. 1982. *Annu. Rev. Astron. Astrophys.* 20:163–90
Smith MD. 1994. *MNRAS* 266:238–46
Snell RL, Loren RB, Plambeck RL. 1980. *Ap. J. Lett.* 239:L17–20
Snell RL, Scoville NZ, Sanders DB, Erickson NR. 1984. *Ap. J.* 284:176–93
Stahler SW. 1993. In *Astrophysical Jets,* ed. M Livio, C O'Dea, D Burgarella, p. 183–209. Cambridge: Cambridge Univ. Press
Stahler SW. 1994. *Ap. J.* 422:616–20
Stapelfeldt KR, Beichman CA, Hester JJ, Scoville NZ, Gautier TN. 1991. *Ap. J.* 371:226–36
Stepinski TF, Levy EH. 1990. *Ap. J.* 350:819–26
Stepinski TF, Levy EH. 1991. *Ap. J.* 379:343–55
Stone JM, Norman ML. 1994. *Ap. J.* 420:237–46
Strom SE. 1995. *Rev. Mex. Astron. Astrof. (Ser. Conf.)* 1:317–28
Tafalla M, Bachiller R. 1995. *Ap. J. Lett.*

443:L37–40
Tafalla M, Bachiller R, Martín-Pintado J. 1993a. *Ap. J.* 403:175–86
Tafalla M, Bachiller R, Martín-Pintado J, Wright MCH. 1993b. *Ap. J. Lett.* 415:L139–42
Tafalla M, Bachiller R, Wright MCH. 1994. *Ap. J. Lett.* 425:L93–96
Tafalla M, Myers PC, Wilner DJ. 1995. In *Clouds, Cores, and Low Mass Stars,* ed. DP Clemens, R Barvainis, pp. 391–95. San Francisco: Astron. Soc. Pac. In press
Taylor S, Raga AC. 1995. *Astron. Astrophys.* 296:823–32
Taylor KNR, Storey JWV, Sandell G, Williams PM, Zealey WJ. 1984. *Nature* 311:236
Tenorio-Tagle G, Cantó J, Rózyczka M. 1988. *Astron. Astrophys.* 202:256–66
Terebey S. 1991. *Mem. Soc. Astron. Ital.* 62(4): 823–28
Terebey S, Chandler CJ, André P. 1993. *Ap. J.* 414:759–72
Terebey S, Vogel SN, Myers PC. 1989. *Ap. J.* 340:472–78
Torbett MV. 1984. *Ap. J.* 278:318–25
Torrelles JM, Ho PTP, Rodríguez LF, Cantó J, Moran JM. 1987. *Ap. J.* 321:884–87
Uchida Y, Shibata K. 1985. *Publ. Astron. Soc. Jpn.* 37:515–35
Umemoto T, Iwata T, Fukui Y, Mikami H, Yamamoto S, et al. 1992. *Ap. J. Lett.* 392:L83–86
van Dishoeck EF, Blake GA, Jansen DJ, Groesbeck TD. 1995. *Ap. J.* 447:760–82
Vogel SN, Kuhi LV. 1981. *Ap. J.* 245:960–76
Walker CK, Lada CJ, Young ET, Maloney PR, Wilking BA. 1986. *Ap. J. Lett.* 309:L47–51
Walker CK, Lada CJ, Young ET, Margulis M. 1988. *Ap. J.* 332:335–45
Walker CK, Narayanan G, Boss A. 1994. *Ap. J.* 431:767–82
Walmsley CM, Hermsen W, Henkel C, Mauersberger R, Wilson TL. 1987. *Astron. Astrophys.* 172:311–15
Walter FM, Brown A, Mathieu RD, Myers PC, Vrba FJ. 1988. *Astron. J.* 96:297–325
Ward-Thompson D, Chini R, Krügel E, André P, Bontemps S. 1995. *MNRAS* 274:1219–24
Ward-Thompson D, Scott PF, Hills RE, André P. 1994. *MNRAS* 268:276–90
Wardle M, Königl A. 1993. *Ap. J.* 410:218–38
Weintraub DA, Masson CR, Zuckerman B. 1989. *Ap. J.* 320:336–43
Welch WJW, Dreher JW, Jackson JM, Terebey S, Vogel SN. 1987. *Science* 238:1550–55
White GJ, Casali MM, Eiroa C. 1995. *Astron. Astrophys.* 298:594–605
Wilking BA, Mundy LG, Blackwell JH, Howe JE. 1989. *Ap. J.* 345:257–64
Wilner DJ, Welch WJ, Forster JR, Murata Y. 1995. Preprint
Wolf G, Lada CJ, Bally J. 1990. *Astron. J.* 100:1892–902
Wolfire MG, Königl A. 1993. *Ap. J.* 415:204–17
Yusef-Zadeh F, Cornwell TJ, Reipurth B, Roth M. 1990. *Ap. J. Lett.* 348:L61–64
Zhou S, Evans NJ II, Kömpe C, Walmsley CM. 1993. *Ap. J.* 404:232–46
Zhou S, Evans NJ II, Wang Y, Peng R, Lo KY. 1994. *Ap. J.* 433:131–48
Zinnecker H, Bastien P, Arcoragi JP, Yorke HW. 1992. *Astron. Astrophys.* 265:726–32
Zuckerman B, Kuiper TBH, Kuiper ENR. 1976. *Ap. J. Lett.* 209:L137–42

Annu. Rev. Astron. Astrophys. 1996. 34:155–206

GALACTIC MAGNETISM: Recent Developments and Perspectives

Rainer Beck

Max Planck Institute for Radioastronomy, Auf dem Hügel 69, D-53121 Bonn, Germany

Axel Brandenburg[1]

Nordita, Blegdamsvej 17, DK-2100 Copenhagen Ø, Denmark

David Moss

Mathematics Department, The University, Manchester M13 9PL, United Kingdom

Anvar Shukurov

Computing Center, Moscow University, 119899 Moscow, Russia

Dmitry Sokoloff

Department of Physics, Moscow University, 119899 Moscow, Russia

KEY WORDS: radio syncrotron emission, radio polarization, spiral arms, galactic halos, dynamos

ABSTRACT

We discuss current observational and theoretical knowledge of magnetic fields, especially the large-scale structure in the disks and halos of spiral galaxies. Among other topics, we consider the enhancement of global magnetic fields in the interarm regions, magnetic spiral arms, and representations as superpositions of azimuthal modes, emphasizing a number of unresolved questions. It is argued that a turbulent hydromagnetic dynamo of some kind and an inverse cascade

[1]Now at Department of Mathematics and Statistics, University of Newcastle upon Tyne, NE1 7RU, United Kingdom.

0066-4146/96/0915-0155$08.00

of magnetic energy gives the most plausible explanation for the regular galactic magnetic fields. Primordial theory is found to be unsatisfactory, and fields of cosmological origin may not even be able to provide a seed field for a dynamo. Although dynamo theory has its own problems, the general form of the dynamo equations appears quite robust. Finally, detailed models of magnetic field generation in galaxies, allowing for factors such as spiral structure, starbursts, galactic winds, and fountains, are discussed and confronted with observations.

1. INTRODUCTION

The magnetic field of the Milky Way has been investigated for about 40 years, and those of external spiral galaxies for about 20 years. It now seems clear that spiral galaxies generally possess large-scale magnetic fields whose evolution and, possibly, their origins are controlled by induction effects in the partially ionized interstellar gas. Turbulent motions with scales below about 100 pc are present in this gas, and so the observed ubiquity of the large-scale galactic magnetic fields, coherent over scales of at least 1 kpc, requires special explanation. In fact, the theory of the galactic magnetic fields discussed in this review (known as mean-field magnetohydrodynamics) represents one of the earliest examples of synergetic theories describing how order can arise from chaos.

Our main emphasis is on magnetic fields whose scales exceed that of the interstellar turbulence. These are the fields—known as the mean, average, large-scale, global, or regular magnetic fields—that produce polarized radio emission in nearby spiral galaxies when observed at resolutions of 0.1–3 kpc. We also stress unresolved problems concerning the random (turbulent) magnetic fields in the interstellar medium (ISM), but we do not extend this discussion to the fields present in elliptical galaxies. Neither do we discuss phenomena connected with the central regions of the Milky Way.

The regular magnetic fields in the disks of spiral galaxies are usually considered to be the result of large-scale dynamo action, involving a collective inductive effect of turbulence (the α-effect) and differential rotation. Even though alternatives to dynamo theory have been proposed, we believe that something resembling an $\alpha\Omega$-dynamo is the dominant mechanism, possibly sometimes modified by other hydromagnetic effects such as induction by streaming motions associated with spiral arms, other noncircular motions, and galactic fountains. The dynamo is the key ingredient of the theory: Other mechanisms by themselves are unable to maintain the observed large-scale galactic magnetic fields over galactic lifetimes.

The main rival of the dynamo theory is the primordial field theory. In this theory, one assumes that the observed magnetic patterns arise directly from a pregalactic magnetic field, distorted by the galactic differential rotation. We discuss why we believe that this theory, in spite of its appealing simplicity, cannot by itself give a detailed explanation of the range of field structures observed in spiral galaxies. A great conceptual advantage of the dynamo theory is that it can provide a universal explanation for the varied field configurations observed in spiral galaxies: axisymmetric and bisymmetric in azimuth; odd, even, and mixed parity vertically; etc. Of course, a primordial field may influence subsequent dynamo action, or it may be amplified by a dynamo.

The dynamo theory has its own difficulties. The linear version, which is valid when the magnetic field is too weak to significantly affect the velocity field, is relatively well developed and agrees favorably with observations wherever such a comparison is meaningful. However, the nonlinear saturation of the dynamo is not well understood and the conventional ideas were recently strongly criticized. They certainly need substantial improvement (Section 4). We argue, however, that the mathematical form of the mean-field dynamo equations is rather generic and robust, so that the available results are expected to be at least qualitatively correct, even though the details and the physical meaning of the coefficients of the dynamo equations may need to be revised.

The topics of this article have recently been reviewed by Wielebinski & Krause (1993) and Kronberg (1994). We have attempted to avoid unnecessary repetition of their material.

2. INTERPRETATION OF RADIO OBSERVATIONS

Interstellar magnetic fields can be observed indirectly at optical and radio wavelengths. Heiles (1976), Verschuur (1979), and Tinbergen (1996) provide extensive reviews of observational methods. In recent years, observations of the linearly polarized radio continuum emission have improved significantly; these provide the most extensive and reliable information about galactic magnetic fields. We thus concentrate on results based on radio continuum data. Zeeman splitting measurements are discussed by Heiles et al (1993). For optical and infrared polarization data, see Roberge & Whittet (1996).

2.1 *Field Strength Estimates*

The strengths of the projections of the total (**B**) and regular ($\overline{\mathbf{B}}$) magnetic fields onto the plane of the sky ($\mathbf{B}_\perp$ and $\overline{\mathbf{B}}_\perp$) can be determined from the intensity of the total and linearly polarized synchrotron emission (e.g. Rybicki & Lightman 1979, p. 180). However, a relation between the energy densities of relativistic electrons, ϵ_e, and the total magnetic field, ϵ_B, has to be assumed. Direct measurements of cosmic rays are possible only near the Earth. The local cosmic-ray energy density ϵ_{CR} is comparable to ϵ_B, and $K = \epsilon_{CR}/\epsilon_e \simeq 100$ locally, but is possibly lower in other galaxies (Pohl 1993).

It is plausible to assume $\epsilon_{CR} = a\epsilon_B$, where a depends on the detailed model: pressure equilibrium, minimum total energy, or energy density equipartition. Although the validity of these assumptions may be questioned (Longair 1994, Urbanik et al 1994, Heiles 1996), they generally provide reasonable estimates.

Gamma-ray observations have been used to obtain indirect data about the distribution of cosmic-ray electrons in the Galaxy (Bloemen et al 1986) and in the Magellanic Clouds (Chi & Wolfendale 1993). Comparing radio and γ-ray data for the Magellanic Clouds, Chi & Wolfendale claimed that energy equipartition is not valid (see, however, Pohl 1993). Their arguments would not apply if γ and radio emissions originate from different regions.

The standard minimum-energy formulae generally use a fixed integration interval in frequency to determine the total energy density of cosmic-ray electrons. This procedure makes it difficult to compare minimum-energy field strengths between galaxies because a fixed frequency interval corresponds to different electron energy intervals, depending on the field strength itself. When a fixed integration interval in electron energy is used, the minimum-energy and energy equipartition estimates give similar values for $\langle B^2 B_\perp^{1+\alpha_s}\rangle \simeq \langle B_\perp^{3+\alpha_s}\rangle$, where α_s is the synchrotron spectral index (typically $\simeq 0.9$). The resulting estimate $\langle B_\perp^{3+\alpha_s}\rangle^{1/(3+\alpha_s)}$ is larger than the mean field $\langle B_\perp\rangle$ if the field strength varies along the path length, since $\langle B_\perp\rangle^{3+\alpha_s} \leq \langle B_\perp^{3+\alpha_s}\rangle$. (Here and elsewhere we denote the magnitude of a vector by $B = |\mathbf{B}|$.)

If the field is concentrated in filaments with a volume filling factor f, the equipartition estimate is smaller than the field strength in the filaments by a factor $f^{1/(3+\alpha_s)}$. The derived field strength depends on the power $(3+\alpha_s)^{-1} \simeq 1/4$ of any of the input values, so that even large uncertainties cause only a moderate error in field strength. For example, a probable uncertainty in K of 50% gives an error in magnetic field strength of $\simeq 15\%$, with the total uncertainty perhaps reaching 30%.

An estimate of the regular field strength $\overline{B}_\perp$ can be obtained by using the observed degree of polarization P, from $P \simeq P_0(\overline{B}_\perp/B_\perp)^2$, where $P_0 \simeq 75\%$ (Burn 1966). Note that regular field strengths are always lower limits because of limited instrumental resolution.

2.2 *Large-Scale Field Patterns*

The plane of polarization of a linearly polarized radio wave rotates when the wave passes through a plasma with a regular magnetic field. The rotation angle $\Delta\psi$ increases with the integral of $n_e B_\parallel$ along the line of sight (where n_e is the thermal electron density and $B_\parallel$ is the component of the total magnetic field along the line of sight) and with λ^2 (where λ is the wavelength of observation). The quantity $\Delta\psi/\Delta\lambda^2$ is called the rotation measure, RM. The observed RM is sensitive to the regular magnetic field $\overline{B}_\parallel$ because the random fields $b_\parallel$ mostly cancel. The sign of RM allows the two opposite directions of $\overline{B}_\parallel$ to be distinguished. An accurate determination of RM requires observations at (at least) three wavelengths because the observed orientation of the polarization plane is ambiguous by a multiple of $\pm 180°$ (see Ruzmaikin & Sokoloff 1979). Unlike equipartition estimates, which are insensitive to the presence of field reversals within the volume observed by the telescope beam, the observed value of Faraday rotation will decrease with increasing number of reversals.

Although the filled apertures of single-dish telescopes are sensitive to all spatial structures above the resolution limit, synthesis instruments such as the VLA cannot provide interferometric data at short spacings. This shortcoming results in some blindness to extended emission. Missing large-scale structures in maps of Stokes parameters Q and U can systematically distort the polarization angles and hence the RM distribution, so that the inclusion of additional data from single-dish telescopes in all Stokes parameters is required. In Section 3.4 (Figure 3) we show the result of such a successful combination by using a maximum-entropy method.

A convenient general way to parameterize the global magnetic field (irrespective of its origin) is by Fourier decomposition in terms of the azimuthal angle ϕ measured in the plane of the galaxy, $\overline{\mathbf{B}} = \sum_m \overline{\mathbf{B}}_m \exp(im\phi)$. In practice, observations are analyzed within rings (centered at the galaxy's center) whose width is chosen to be consistent with the resolution of the observations. The result is a set of Fourier coefficients of the large-scale magnetic field for each ring. Usually, a combination of $m = 0$ and $m = 1$ modes is enough to provide a statistically satisfactory fit to the data. This is a remarkable indication of the presence of genuine global magnetic structures in spiral galaxies.

All observed magnetic fields have significant radial and azimuthal components: The magnetic lines of the regular field are spirals (Section 8.3). We distinguish between spiral structures that can be considered as basically axisymmetric (ASS), and basically antisymmetric or bisymmetric (BSS), with respect to rotation by 180°. Note that higher azimuthal Fourier modes are expected to be superimposed on these dominant ones, but these should have relatively small amplitudes. Fields containing several Fourier components of significant amplitude have mixed spiral structure (MSS); this might be considered to be a combination of ASS and BSS.

A further classification of magnetic structures according to their symmetry with respect to the galaxy's midplane distinguishes symmetric S (i.e. even parity or quadrupole) from antisymmetric (odd parity or dipole) modes A.

Table 1 A two-dimensional classification of global magnetic structures listing the dominant azimuthal and vertical modes

Vertical	Azimuthal structure		
structure	ASS	BSS	MSS
Even	S0	S1	S0 + S1
Odd	A0	A1	A0 + A1
Mixed	M0	M1	M0 + M1

Mixed-parity distributions (M), in which the magnetic fields are neither even nor odd but are superpositions, are also possible. This notation is supplemented with a value of the azimuthal wave number m, e.g. S0 means a quadrupole axisymmetric field. The notation used in discussions of global magnetic structures in spiral galaxies in presented in Table 1.

An ASS (BSS) field produces a 2π-periodic (π-periodic) distribution of RM along ϕ (Sofue et al 1986, Krause 1990, Wielebinski & Krause 1993). For the $m = 0$ mode, the phase of the variation of RM with ϕ is equal to the magnetic pitch angle, $p = \arctan(\overline{B}_r/\overline{B}_\phi)$. Using the observed azimuthal distribution of RM in a galaxy, the structure of the line-of-sight component of a large-scale magnetic field can be studied. This method is difficult to apply if the data suffer from Faraday depolarization, if the regular field is not parallel to the plane of the galaxy, if its pitch angle in the disk is not constant, or if the disk is surrounded by a halo with magnetic fields of comparable strengths.

A more direct method of analysis considers polarization angles ψ without converting them into Faraday rotation measures (Ruzmaikin et al 1990; Sokoloff et al 1992; EM Berkhuijsen et al, in preparation). There are three main contributions to the observed polarization angle: $\psi = \psi_0 + \mathrm{RM}\,\lambda^2 + \mathrm{RM}_{\mathrm{fg}}\,\lambda^2$, where ψ_0 is determined by the transverse magnetic field in the galaxy, RM is associated with Faraday rotation by the line-of-sight magnetic field in the galaxy, and $\mathrm{RM}_{\mathrm{fg}}$ is the foreground rotation measure. Thus, a direct analysis of ψ patterns at several wavelengths allows a self-consistent study of all three components of the regular magnetic field. Another advantage of this method is that complicated magnetic structures along the line of sight can be studied. Implementations of this method employ consistent statistical tests such as the χ^2 and Fisher criteria, thereby allowing the reliability of the results to be assessed.

Note that Faraday rotation analysis yields an average value, $\langle n_e B_\parallel \rangle$. Information on $\overline{B}_\parallel$ can be extracted only if a reliable model for the distribution of n_e is available, which is often not the case. If, for example, the thermal gas has a low filling factor, any result concerning $\overline{B}_\parallel$ may not be representative.

2.3 *Small-Scale Field Structures*

Any unresolved field structures will lead to beam depolarization and thus to polarizations significantly below the theoretical limit of $P_0 \simeq 75\%$. This effect is independent of wavelength and can be used to estimate the spatial scale and strength of field irregularities using observations at short wavelengths, where Faraday effects are weak.

At longer wavelengths, varying field orientations along the line of sight give rise to dispersion in rotation measures (Faraday dispersion), which also leads to depolarization (Burn 1966). Faraday dispersion is expected to arise from small H II regions (of $\simeq 1$ pc in size) in the thin galactic disk (Ehle & Beck 1993) as well as from larger scale fluctuations ($\simeq$ 10–100 pc) in the diffuse ionized medium of the thick disk (e.g. Krause 1993, Neininger et al 1993). This effect makes the Faraday rotation angle no longer proportional to λ^2 because the effective Faraday depth decreases with increasing λ. It was recently discovered that at wavelengths greater than or equal to 10 cm, galaxies are generally *not* transparent to polarized radio waves (Sukumar & Allen 1991, Beck 1991, Horellou et al 1992). Even at $\lambda \simeq 6$ cm complete Faraday depolarization may occur in spiral arms or in the plane of edge-on galaxies.

To obtain full rotation measures, only observations in the Faraday-thin regime ($\lambda \leq 6$ cm) should be used (Vallée 1980, Beck 1993). Rotation measures between longer wavelengths are lower and are weighted to regions near to the observer. Variations in Faraday depth may also lead to a spatial variation of the observed RM, which complicates the interpretation of observations. On the other hand, Faraday depolarization allows the study of layers at different depths sampled at different wavelengths (EM Berkhuijsen et al, in preparation).

2.4 *Comparison with Optical Polarization Data*

Optical polarization observations have revealed spiral magnetic patterns in M51 (Scarrott et al 1987), NGC 1068 (Scarrott et al 1991), NGC 1808 (Scarrott et al 1993), and other galaxies (Scarrott et al 1990) (see also the review by Hough 1996). In the western half of M51, field orientations as derived from optical polarization disagree by up to about 60° from the spiral pattern as derived at several wavelengths in the radio continuum (Beck et al 1987). Optical polarization is contaminated by highly polarized light due to scattering at large angles. Polarization observations at far-infrared or submillimeter wavelengths are free from scattering effects and reveal the magnetic field structure in Galactic dust clouds (Davidson et al 1995) and near the Galactic Center (Hildebrand et al 1990, Hildebrand & Davidson 1994).

3. MAGNETIC FIELDS IN SPIRAL GALAXIES

3.1 *Field Strengths*

Mean equipartition strengths of the total field $\langle B^{3+\alpha_s}\rangle^{1/(3+\alpha_s)}$ (averaged over the volume of the visible radio disk) range from $\simeq 4$ μG in M33 (Buczilowski & Beck 1991) to $\simeq 12$ μG in NGC 6946 and NGC 1566 (Ehle & Beck 1993, Ehle et al 1996); they are proportional to surface brightness in the far-infrared range (Hummel et al 1988b) and to average gas density (S Niklas et al, in preparation). Hummel's (1986) sample of 88 Sbc galaxies has a mean minimum-energy field of $\simeq 8$ μG, using $K = 100$. Using the same value of K for the sample of 146 late-type galaxies by Fitt & Alexander (1993), one obtains a mean total minimum-energy field strength of 10 ± 4 μG. Extremal values found in normal galaxies can be up to 20 μG in spiral arms, as in NGC 6946 (Beck 1991) and NGC 1566 (Ehle et al 1996). In the mildly active galaxy M82, Klein et al (1988) found a field strength $\simeq 50$ μG.

The regular field strengths $\overline{B}_\perp$ as obtained from the intensity of polarized emission are typically a few μG. Such values are roughly consistent with regular field strengths $\overline{B}_\parallel$ as derived from Faraday rotation data, if we assume typical electron densities of a few 10^{-2} cm^{-3} (see e.g. Buczilowski & Beck 1991). Because polarized intensity and rotation measure depend differently on the filling factor of the field, the fact that $\overline{B}_\perp \simeq \overline{B}_\parallel$ implies that the filling factor is not very small. The ratio of regular to turbulent field strengths is typically $\simeq 0.5$ if observed with a spatial resolution of a few kpc (Buczilowski & Beck 1991).

In NGC 2276 the regular field strength reaches 10 μG (Hummel & Beck 1995), probably due to its interaction with the ambient intracluster gas. The total field is also unusually strong in that galaxy. Galaxies in clusters generally contain stronger fields (Gavazzi et al 1991, Niklas et al 1995).

3.2 *M31: The Nearly Perfect Magnetic Torus*

Radio observations of M31 reveal a $\simeq 20$ kpc-diameter torus with the regular magnetic field, as determined from a Faraday rotation model, aligned *uniformly* with respect to the circumference (Sofue & Takano 1981, Beck 1982, Beck et al 1989, Ruzmaikin et al 1990). A superposition of helical field loops (as expected from dynamo theory) can explain the asymmetric distribution of polarized emission with respect to the minor axis (Donner & Brandenburg 1990, Urbanik et al 1994). The mean equipartition field strength in the torus is $\simeq 7$ μG for the total field and $\simeq 4$ μG for the regular field.

M31 has a low star-formation rate globally and no grand-design spiral structure. Strong compression of magnetic fields by density waves seems to be absent. However, there are many small dust lanes and cloud complexes, which are traced by field lines. Comparison between the total radio intensity and the gas emission in CO and HI confirms a close coupling of the field to the gas clouds (Berkhuijsen et al 1993).

3.3 *Density-Wave Galaxies*

Radio polarization observations show that the regular magnetic fields follow approximately the optical spiral structure in M81 (Krause et al 1989b), M83 (Neininger et al 1991, Ehle 1995), NGC 1566 (Ehle et al 1996), and M51 (Neininger 1992; Neininger & Horellou 1996; EM Berkhuijsen et al, in preparation), but that the streamlines of the rotation models of the gas do not follow the optical spiral structure. The streamlines have strongly varying pitch angles (e.g. Otmianowska-Mazur & Chiba 1995) and are almost closed in the disk. However, some regions of M51 may be exceptional: Neininger (1992) has claimed that some field lines are carried along with streaming motions. The field lines in the central region of M83 are aligned with the bar.

Strong shocks should compress the magnetic field and lead to high degrees of polarization of 40–70% in the radio continuum (Beck 1982) at the inner edges of the spiral arms (see Section 8.4). Only in M51 are the strongest aligned fields indeed found at the positions of the prominent dust lanes on the inner edges of the optical spiral arms (Figure 1). This is best visible along the eastern arm where the aligned field even follows the dust lane crossing the optical arm. However, some regular fields extend far into the interarm regions. Furthermore, the 10–30% polarization at $\lambda 6$ cm is in contrast to the higher polarizations expected from shock alignment. Hence the radio data only tell us that the regular fields in M51 are somehow coupled to the cool gas as traced by dust lanes.

The aligned fields in M81 and NGC 1566 are strongest in interarm regions (Krause et al 1989b, Ehle et al 1996), whereas the total synchrotron intensity (tracing the total field) is highest in the optical spiral arms. Strongly aligned interarm fields have also been detected in the outer parts of M83, where the star formation rate is low (Allen & Sukumar 1990). High-resolution observations of M81 (Schoofs 1992; see Figure 2) confirmed that the regular fields extend across almost the entire interarm region, but are somewhat stronger near the inner edge of the prominent western spiral arm, where some dust clouds are visible. We stress that the distribution of magnetic pitch angles exhibits a weaker arm-interarm variation than that of the regular magnetic field strengths. Soida et al (1996) showed that strength and pitch angle of the regular fields in NGC 4254 reveal much less arm-interarm variations than expected from density-wave compression in its two major arms. They also showed that regular fields even exist in regions of chaotic optical pattern.

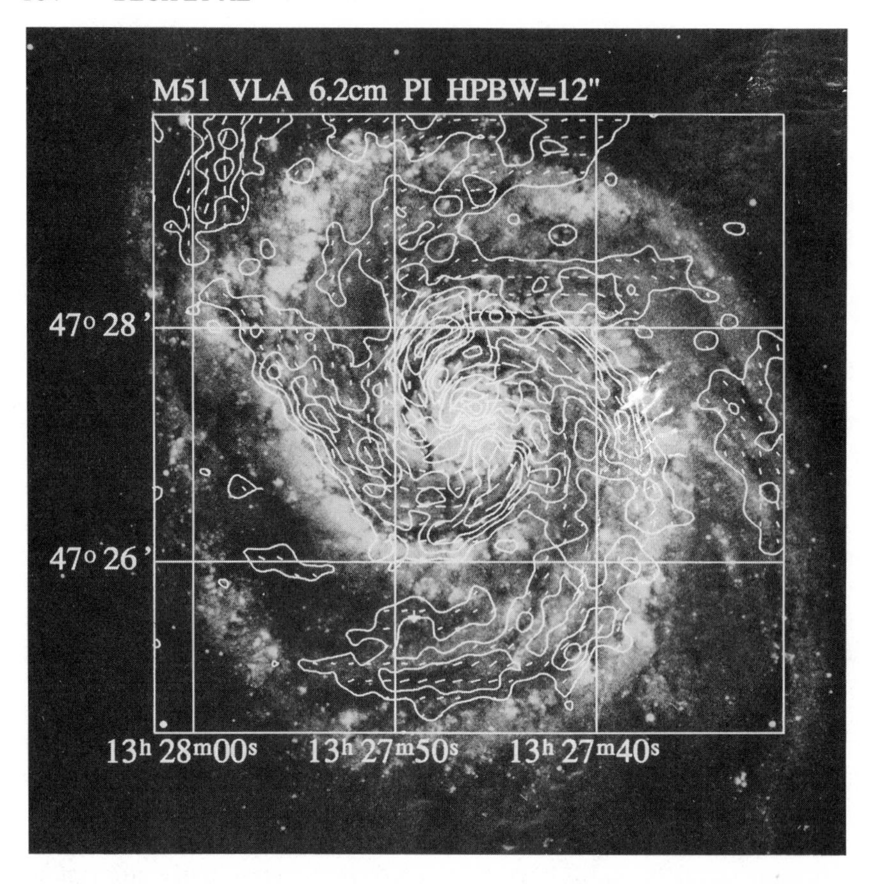

Figure 1 Polarized synchrotron intensity (*contours*) and magnetic field orientation of M51 (obtained by rotating the E-vectors by 90°), observed at λ6.2 cm with the VLA (12 arcsec synthesized beam). (From Neininger & Horellou 1996.)

3.4 *IC 342 and NGC 6946: Magnetic Spiral Arms*

These two galaxies exhibit high star-formation rates, but their spiral structure is less regular than that of M51. Long arms of polarized emission are present in IC 342 (Krause et al 1989a, Krause 1993).

Recent high-resolution observations of the similar galaxy NGC 6946 (Beck & Hoernes 1996; see Figure 3) revealed a surprisingly symmetric distribution of polarized emission with two major spiral features, in the north and in the south, located between optical spiral arms and running perfectly parallel to the adjacent optical arms over at least 12 kpc. This regular two-armed structure is much more symmetric than the distribution of total field, gas, and stars, which all show a quite irregular, multiarmed pattern. Two further, weaker, magnetic spiral arms are visible between the two main ones (Figure 3).

The main magnetic spiral arms in NGC 6946 do not fill all of the interarm regions, unlike the polarized emission in M81, but are only about 0.5–1 kpc wide. As they are also visible in total emission, both the regular and total magnetic fields are enhanced there. The strength of the (resolved) regular field varies between 3 and 13 μG along the arm. The peak values of polarized intensity and degree of polarization occur in the northern magnetic arm of NGC 6946, in a region between the optical arms, where the density of warm

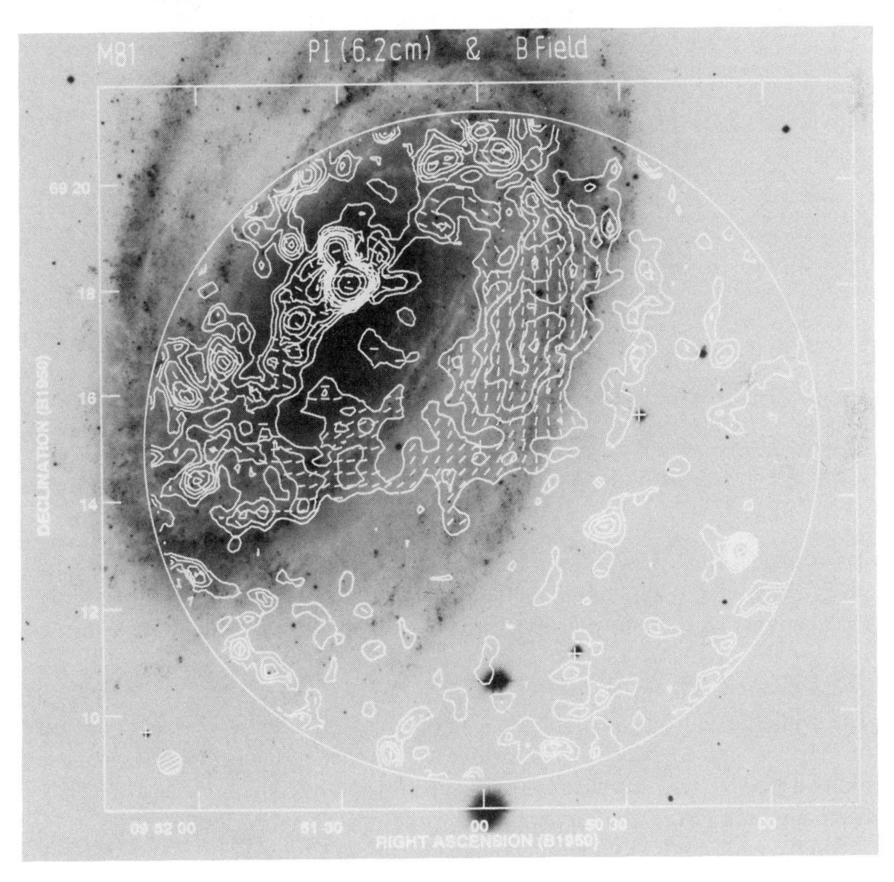

Figure 2 Polarized synchrotron intensity (*contours*) and magnetic field orientation in the southwestern part of M81 (obtained by rotating the E-vectors by 90°), observed at λ6.2 cm with the VLA (25 arcsec synthesized beam). The circle indicates the half-power diameter of the primary beam. (From Schoofs 1992.)

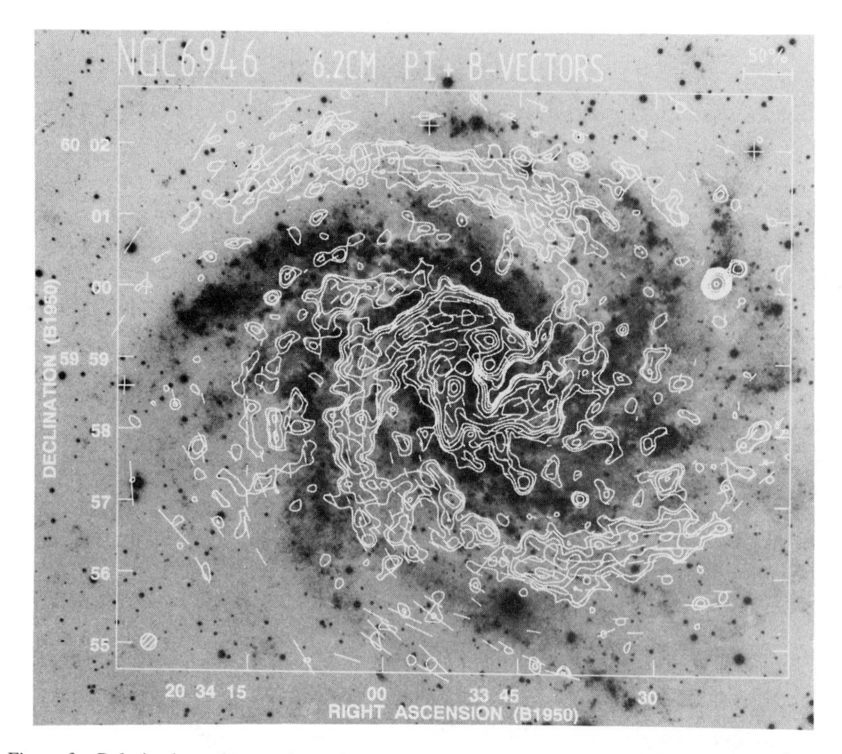

Figure 3 Polarized synchrotron intensity (*contours*) and magnetic field orientation of NGC 6946 (obtained by rotating E-vectors by 90°) observed at λ6.2 cm with the VLA (12.5 arcsec synthesized beam) and combined with extended emission observed with the Effelsberg 100-m telescope (2.5 arcmin resolution). The lengths of the vectors are proportional to the degree of polarization. (From Beck & Hoernes 1996.)

gas is exceptionally low. Subtraction of the diffuse, unpolarized background gives a degree of polarization of 30–65%, with the implication that the fields in the magnetic spiral arms must be almost totally aligned with the optical arms.

3.5 *ASS, BSS, or What?*

The available data on global magnetic structures in spiral galaxies are compiled in Tables 2 and 3. Most of the results were obtained using the RM analysis method (see Section 2.2); the more advanced ψ-analysis method has as yet only been applied to M51 (and in simplified form to M31, IC 342, and M81).

Singly-periodic RM variations indicative of ASS have been detected in the disks of M31 (Sofue & Takano 1981, Beck 1982) and IC 342 (Gräve & Beck 1988, Krause et al 1989a, Sokoloff et al 1992). In M31 Ruzmaikin et al (1990) found evidence for the presence of the $m = 1$ mode at lower amplitude in the

Table 2 Magnetic field structures of normal galaxies with low or moderate inclination as derived from synchrotron polarization data

Galaxy	Instrument[a] and wavelength	Field structure	References
Milky Way		ASS?	See Section 3.8
M31	E 11, 6 cm V 20, 6 cm	ASS (with weaker BSS)	Beck (1982), Ruzmaikin et al (1990)
M33	E 21, 18, 11, 6, 2.8 cm	Spiral (BSS?)	Buczilowski & Beck (1991)
M51	E 6, 2.8 cm V 20, 18, 6 cm W 21, 6 cm	MSS, magnetoionic halo	EM Berkhuijsen et al, in prep., Horellou et al (1992), Segalovitz et al (1976)
M81	E 6, 2.8 cm V 20, 6 cm	BSS (with weaker ASS)	Krause et al (1989b), Sokoloff et al (1992)
M83	E 6, 2.8 cm V 20, 6 cm A 13 cm	Spiral and ‖ bar	Neininger et al (1993), Sukumar & Allen (1989), Ehle (1995)
M101	E 6 cm	Spiral	Gräve et al (1990)
SMC	P 21, 12 cm	‖ main ridge	Haynes et al (1986)
LMC	P 21, 12, 6 cm	Loop south of 30 Dor	Klein et al (1993)
IC 342	E 11, 6 cm V 20, 6, 3.5 cm	ASS, magnetic spiral arms	Sokoloff et al (1992), Krause et al (1989a)
NGC 1566	A 20, 13, 6 cm	Spiral	Ehle et al (1996)
NGC 1672	A 20, 13, 6 cm	Spiral	M Ehle et al (in prep.)
NGC 2276	V 20, 6 cm	Spiral (BSS?)	Hummel & Beck (1995)
NGC 2903	E 6, 2.8 cm V 18, 20 cm	Spiral	R Beck et al (in prep.)
NGC 3627	E 2.8 cm	‖ dust lane	M Urbanik et al (in prep.)
NGC 4038/39	V 20, 6, 3.6 cm	‖ tidal arm	E Hummel & R Beck (in prep.)
NGC 4254	E 6, 2.8 cm	‖ compression region	Soida et al (1996)
NGC 4258	E 6, 2.8 cm W 49, 21 cm V 20, 6 cm	‖ anomalous arms	Hummel et al (1989)
NGC 4449	E 2.8 cm	‖ HII chains	U Klein et al (submitted)
NGC 5055	E 2.8 cm	‖ disk	M Soida et al (in prep.)
NGC 6946	E 6, 2.8 cm V 20, 18, 6, 3.5 cm	Spiral (MSS?), magnetic spiral arms	Ehle & Beck (1993), Beck & Hoernes (1996)

[a]Instruments: E = Effelsberg 100-m, A = Australia Telescope Compact Array, P = Parkes 64-m, W = Westerbork Synthesis Radio Telescope, V = Very Large Array.

Table 3 Magnetic field structures of (almost) edge-on galaxies

Galaxy	Instrument[a] and wavelength	Field structure	References
M82	V 6, 3.6	Radial	Reuter et al (1994)
NGC 253	P 6 cm V 20, 6 cm E 2.8 cm	 ∥ plane (ASS?)	Harnett et al (1990), Carilli et al (1992), Beck et al (1994b)
NGC 891	V 20, 6 cm E 2.8 cm	∥ & inclined to plane	Sukumar & Allen (1991), Dumke et al (1995)
NGC 1808	V 20, 6 cm	Extensions ⊥ plane	Dahlem et al (1990)
NGC 3628	V 20, 6 cm E 2.8 cm	 ∥ & inclined to plane	Reuter et al (1991), Dumke et al (1995)
NGC 4565	V 20, 6 cm E 2.8 cm	∥ plane	Sukumar & Allen (1991), Dumke et al (1995)
NGC 4631	E 6, 2.8 cm V 20, 6, 3.5 cm	⊥ plane (inner region), ∥ & inclined to plane (outer regions)	Hummel et al (1991a), Golla & Hummel (1994)
NGC 4945	P 6 cm	Extensions ⊥ plane	Harnett et al (1989),
NGC 5775	V 6 cm	∥ plane	Golla & Beck (1990)
NGC 5907	V 6 cm	∥ plane	M Dumke & M Krause (in prep.)
NGC 7331	E 2.8 cm V 20, 6 cm	Almost ∥ plane	Dumke et al (1995), E Hummel (unpubl.)
Circinus	A 13, 6 cm	⊥ northern plume	Elmouttie et al (1995)

[a]Instruments: E = Effelsberg 100-m, A = Australia Telescope Compact Array, P = Parkes 64-m, W = Westerbork Synthesis Radio Telescope, V = Very Large Array.

outer regions, superimposed on the dominating $m = 0$ mode. In NGC 6946 the phase of the azimuthal RM variation differs significantly from the value of the mean magnetic pitch angle (Ehle & Beck 1993). Recent high-resolution data for this galaxy (R Beck & P Hoernes, in preparation) indicate some correlation of RM with the optical spiral arms, suggesting local enhancements of RM due to thermal gas, rather than to field geometry. The magnetic spiral arms (where thermal gas density is low) seem to have RMs of opposite sign (R Beck & P Hoernes, in preparation), indicative of the $m = 0$ mode or, more realistically, a superposition of the $m = 0$ and the $m = 2$ mode with about equal amplitudes. In NGC 253, seen almost edge-on, the large-scale magnetic field has opposite directions on the "left" and "right" of the rotation axis of the inner disk. NGC 253 is thus another candidate for an ASS disk field (Beck et al 1994b).

The only clear candidate for a BSS symmetry is M81 (Krause et al 1989b, Sokoloff et al 1992). The analysis of Krause et al (1989b) indicated that the magnetic neutral lines are in the interarm space. In M33 the weak polarized emission leads to large uncertainties in RM, and a bisymmetric field can be claimed only with some caution (Buczilowski & Beck 1991). The same is true for NGC 2276 (Hummel & Beck 1995). The galaxies M33, M81, and NGC 2276 show signs of gravitational interaction, which can be important in producing nonaxisymmetric dynamo fields (see Sections 6.2 and 8.2). Thus these are all candidates for MSS status. Other claims of a dominating bisymmetric field (Sofue et al 1985, 1986) are of much lower significance.

The strongly interacting galaxy M51 is a special case where the pitfalls of data interpretation can be demonstrated. M51 was thought to contain a bisymmetric field (Tosa & Fujimoto 1978, Horellou et al 1990), but this was not confirmed by later Effelsberg and VLA measurements. At $\lambda \geq 10$ cm, Faraday depolarization is strong and the observed polarized emission originates in the upper disk or halo (Horellou et al 1992). By analyzing all available data in terms of the ψ angles, EM Berkhuijsen et al (in preparation) have found that the field in M51 can be described as MSS, with axisymmetric and bisymmetric components having about equal weights in the disk, together with a horizontal axisymmetric halo field with opposite direction.

The RM variation in M83 is doubly-periodic (Neininger et al 1993), but the phase is inconsistent with BSS symmetry. A future analysis of polarization angles including recent observations at $\lambda 13$ cm (Ehle 1995) might clarify the case. The RM pattern in M83 indicates a nonaxisymmetric distribution of gas or velocity field, an MSS field, or both.

3.6 *Magnetic Fields in Galactic Halos*

Vertical dust lanes are often seen in edge-on galaxies, which may indicate vertical magnetic field lines (Sofue 1987). Their initial detection via polarized radio emission in NGC 4631 by Hummel et al (1988a) prompted a systematic search in several nearby edge-on galaxies. Radio halos were detected also in NGC 253 (Carilli et al 1992) and in NGC 4666 (M Dahlem et al, in preparation). A survey of 181 edge-on galaxies observed with the Effelsberg and VLA radio telescopes (Hummel et al 1991b) disclosed no other cases with pronounced halos.

In contrast, NGC 891 (Hummel et al 1991a), NGC 3628 (Reuter et al 1991), NGC 5775 (Golla & Beck 1990), and most other edge-on galaxies (Hummel 1990) do not possess extended radio halos. These galaxies have thick disks with typical synchrotron scale heights of $\simeq 1$ kpc. In most of these galaxies the observed field orientations are approximately parallel to the disk (Dumke et al 1995; see also Table 3). The same is true for NGC 4945 (Harnett et al 1989) and NGC 1808 (Dahlem et al 1990), but there the polarized emission is restricted to two localized regions, one on each side of the plane. In the disks themselves, the polarized emission at $\lambda \geq 6$ cm is weak due to Faraday depolarization.

The other extremes are NGC 4565 (Sukumar & Allen 1991) and M31. The radio emission from any thick disk of M31 is not detectable and must be at least 200 times weaker than for NGC 891 (Berkhuijsen et al 1991). Either the low star-formation rates in M31 and NGC 4565 are below the threshold for the chimney-type outflows (Dahlem et al 1995) or the dynamo does not operate in the halos of these galaxies.

The increase of the degree of polarization with height above the disk of NGC 891 has been analyzed by Hummel et al (1991a). The data can be well modeled by Faraday depolarization in a thermal gas of scale height $\simeq 1$ kpc together with a turbulent magnetic field of scale height $\simeq 4$ kpc. The scale height of the thermal gas as derived from the radio data agrees well with that observed in Hα (Rand et al 1990, Dettmar 1990). The scale height of the turbulent halo field is consistent with equipartition between the field and cosmic-ray energy densities, where $z_b = 2z_{CR} = (3 + \alpha_s)z_{syn} \simeq 3.6$ kpc for a synchrotron scale height of $z_{syn} \simeq 0.9$ kpc and $\alpha_s \simeq 1.0$ (Hummel et al 1991a).

NGC 253 is the edge-on galaxy with the brightest and largest halo observed so far (Carilli et al 1992), extending to at least 9 kpc above the plane. It also has the brightest X-ray halo (Pietsch 1994), so that a strong outflow from the disk or the nucleus driven by the high star-formation rate seems probable. Gas outflow from the nucleus has indeed been found (Dickey et al 1992). Nevertheless, the regular magnetic field is predominantly parallel to the plane in the disk and in the halo (Beck et al 1994b; see Figure 4), possibly due to strong differential rotation even near the center.

NGC 4631 is another rare case of an extended radio halo, possibly driven by a strong galactic wind. The synchrotron scale height of $\simeq 2$ kpc is twice as large as for the bulk of edge-on galaxies (Hummel 1990). The magnetic field lines are roughly perpendicular to the inner disk, which is almost rigidly rotating (Hummel et al 1991a, Golla & Hummel 1994). In this respect, NGC 4631 is exceptional compared with most edge-on galaxies. NGC 4631 also shows signs of gravitational interaction. A few regions with field orientations parallel to the disk are visible in the (differentially rotating) outer disk.

A striking case of a strong galactic wind is M82, which has quasi-radial field lines (Reuter et al 1994). Even a field of $\simeq 50\ \mu$G strength (Klein et al 1988) cannot resist the flow with a velocity in excess of 1000 km s^{-1}.

Vertical magnetic fields may be a result of disk-halo interactions. The Parker instability produces alternating vertical magnetic fields. A vertical galactic wind of speed V_z could also drag the field from the disk. However, an azimuthal gradient of V_z is required to produce B_z from B_ϕ and a radial gradient of V_z to obtain B_z from B_r (see Section 7).

Rotation measures in galactic halos are important for revealing the direction of the field and thus its parity with respect to the midplane. Golla & Hummel (1994) could not find a clear RM pattern from their data of NGC 4631. Beck et al (1994b) determined rotation measures at a few positions in the lower halo of NGC 253 and found weak evidence for RMs of the same sign at $\simeq 5$ kpc above and below the plane, as expected for an even-parity mode.

In face-on galaxies "coronal holes" have been observed as regions of high rotation measure (with ensuing higher depolarization) with neither enhanced plasma density (Hα or X-ray emission) nor enhanced total field strength (total synchrotron emission). In these regions magnetic lines probably open into the halo. The RM maps of IC 342 (Krause et al 1989a) and NGC 6946 (Beck 1991) seem to show such phenomena. The rotation measures in NGC 6946, determined between λ18 cm and λ20 cm, are small and almost constant, except

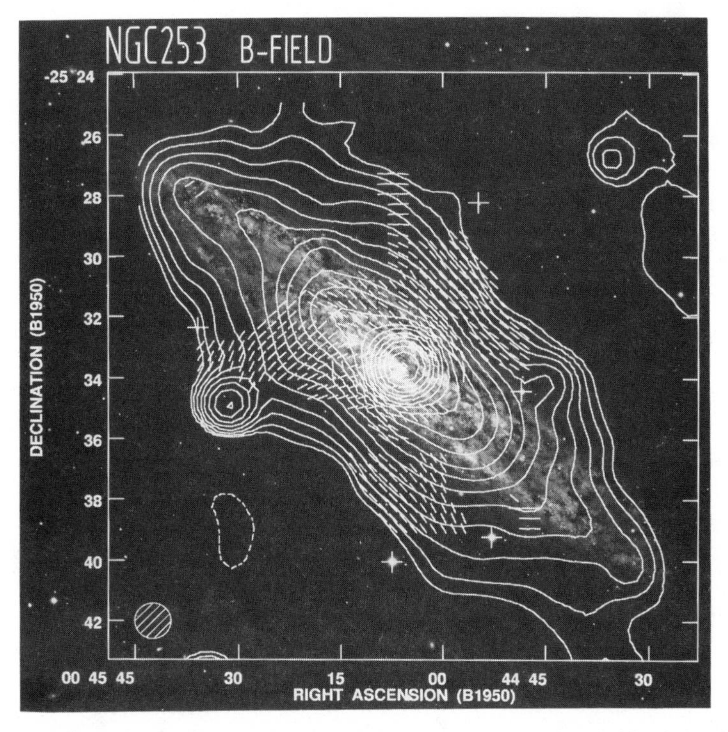

Figure 4 Total radio intensity (*contours*) and magnetic field orientation of NGC 253 (obtained by rotating E-vectors by 90°), observed at λ2.8 cm with the Effelsberg telescope (disk field) and at λ20 cm wavelength with the VLA (halo field). The resolution is 70 arcsec. (From Beck et al 1994b.)

in the southwest quadrant, where both high and low values occur in a region of $\simeq 10$ kpc in extent (Beck 1991). The spiral arms in the southwest quadrant of NGC 6946 are more diffuse and the X-ray emission is weaker (Schlegel 1994) than in the remainder of the galaxy. Thus galactic coronal holes may occur in regions of low star-forming activity.

3.7 *Magnetic Fields in High-Redshift Galaxies*

It is likely that spiral galaxies have possessed their large-scale magnetic fields at least 6×10^9 yr ago (corresponding to a redshift $z \simeq 0.5$) (Kronberg 1994, Perry 1994). The most convincing evidence is the detection of Faraday rotation attributed to a galaxy at $z = 0.395$ (Kronberg et al 1992). The inferred large-scale magnetic field strength is 1–4 μG and its direction reverses on a scale of $\simeq 3$ kpc. Kronberg et al (1992) argue for a bisymmetric magnetic structure, but this may equally well be an axisymmetric field with reversals (Poezd et al 1993).

Statistical studies of quasar samples (Kronberg & Perry 1982, Welter et al 1984, Perry et al 1993) indicate that excess Faraday rotation correlates with the presence of intervening absorbers. The size of the absorbers has been estimated as 45 kpc, with their global magnetic fields of 2–10 μG; these are probably galactic disks and/or halos. Wolfe et al (1992) and Oren & Wolfe (1995) have argued that damped Lyα systems [i.e. putative young galactic disks (Wolfe 1988, Wolfe et al 1993)] possess μG-strength global magnetic fields at $z \simeq 1$–2 when they are only 1–3 Gyr old. (However, statistical analyses of this kind are extremely difficult, in particular because of poor statistics, different selection effects, complications in isolating contributions of other intervenors such as our Galaxy, galaxy clusters, etc (Perry et al 1993, Perry 1994). The earliest time at which galaxies possess their large-scale magnetic fields still has to be established. Theoretical models of magnetic fields in young galaxies are discussed in Section 5.3.

A straightforward implication of these studies is a lower limit on the seed magnetic field required for galactic dynamos. If an $\Omega = 1$ cosmology is assumed, then this limit is 2×10^{-18} G (Kronberg et al 1992), or even possibly 10^{-9}–10^{-11} G if a tentative identification of excess RM in the quasar $1331 + 170$ with an absorber at $z = 1.775$ is confirmed (Perry 1994) (see also Section 5).

3.8 *The Milky Way*

Observations in the Milky Way offer a unique opportunity for studying interstellar magnetic fields in a detail unobtainable for even nearby external galaxies. However, the plethora of local detail, which obscures any grand-design features of the magnetic field in the Milky Way, still prevents a reliable picture from being obtained.

3.8.1 MAGNETIC fiELD IN THE SOLAR VICINITY The most reliable estimates concerning the large-scale magnetic field near the Sun are obtained from statistical analyses of Faraday rotation measures of nearby pulsars (within 2–3 kpc from the Sun) and high-latitude extragalactic radio sources, because larger samples involve lines of sight passing through remote regions in the Galaxy for which the inferred magnetic field configuration is less reliable (see Simard-Normandin & Kronberg 1980, Rand & Kulkarni 1989, and references therein). The regular field strength is $\simeq 2\ \mu$G, probably stronger within the arm.

The field is directed towards a galactic longitude of about 90° (see, e.g. Ruzmaikin et al 1978, Rand & Lyne 1994), with an accuracy of 10–20°. The scatter between different determinations makes it difficult to say whether it is aligned with the local spiral arm (pitch angle of about $-15°$) or not. A tentative upper limit on the magnetic pitch angle, $|p| \leq 15°$, implies that $|\overline{B}_r| \leq 0.3|\overline{B}_\phi|$.

The best agreement with observations is provided by models with the horizontal global magnetic field similarly directed above and below the midplane (S-type field) (Gardner et al 1969; Vallée & Kronberg 1973, 1975). Claims of an odd symmetry (Morris & Berge 1964; Andreassian 1980, 1982) probably result from contamination by strong local distortions in the magnetic field. A similar problem prevents the reliable detection of the vertical magnetic field $\overline{B}_z$ near the Sun: It is so weak that it cannot be separated from local magnetic inhomogeneities, $|\overline{B}_z| \ll |\overline{B}_r|, |\overline{B}_\phi|$.

Because the warm interstellar medium is the main contributor to the electron density in the diffuse ISM, rotation measures sample mainly this phase of the ISM. Heiles (1996) has argued that the warm interstellar medium occupies only $\simeq 20\%$ of the total volume in the Milky Way, so that the resulting $\overline{B}$ does not reflect the true volume-averaged field. This argument would apply also to external galaxies, where Faraday rotation is also used to study $\overline{B}$. However, the observed coherency of RM patterns over large regions in many nearby galaxies indicates that the inferred magnetic field is global rather than restricted to a small fraction of the volume (see also Section 3.1).

3.8.2 REVERSALS OF THE MAGNETIC fiELD AND ITS AZIMUTHAL STRUCTURE The property of the magnetic field in the Milky Way that distinguishes it from probably most other galaxies investigated up to now is the reversals of the regular field along the radius. The reversal closest to the Sun between the local (Orion) and the next arm to the center (Sagittarius) was first detected by Simard-Normandin & Kronberg (1979). The reversal is located in the interarm region at about 0.4–0.5 kpc inside the solar circle (see Rand & Lyne 1994).

There are some indications of more reversals at both smaller and larger galactocentric distances, but this evidence is much more controversial because distant spiral arms occupy smaller areas on the sky. Simard-Normandin &

Kronberg (1980) and Vallée (1983) argued that there is no reversal between the local and the next outer (Perseus) arms, whereas other authors found some evidence for an outer reversal (Agafonov et al 1988, Rand & Kulkarni 1989, Lyne & Smith 1989, Clegg et al 1992). Two additional reversals were claimed for the inner Galaxy by Sofue & Fujimoto (1983) and Han & Qiao (1994), but most analyses more conservatively imply only one more, at a galactocentric radius of 5.5 kpc (Vallée et al 1988, Vallée 1991, Rand & Lyne 1994). The controversy about the number of reversals is partly due to difficulties in the analysis of Faraday rotation measures. There are natural complications associated with strong local distortions of magnetic field, e.g. the North Polar Spur or the Gum Nebula. However, there are also many pitfalls in the statistical analyses. Many results rely on simple "naked-eye" fitting of the observational data (e.g. Simard-Normandin & Kronberg 1980, Sofue & Fujimoto 1983), which is especially dangerous when the global structure is investigated; others are based on nonrigorous applications of statistical tests (e.g. Han & Qiao 1994). Vallée (1996) discusses some of these problems. More rigorous studies imply an axisymmetric field with two reversals (Rand & Kulkarni 1989, Rand & Lyne 1994), although more cannot be excluded. The radial distribution of the magnetic field strength is shown in Figure 5 (see also Heiles 1996).

The available statistical analyses adopt either a bisymmetric structure of the global magnetic field (Simard-Normandin & Kronberg 1980, Sofue & Fujimoto 1983, Han & Qiao 1994) or a concentric-ring model in which magnetic field lines are directed exactly in the azimuthal direction. Comparison between these two models often shows that the latter provides a better fit to the data (e.g. Rand & Kulkarni 1989); however, the concentric-ring model is unrealistically simplistic. The regular magnetic field cannot have a zero pitch angle everywhere (see Section 8.3), even if it does near the Sun. The model is consistent neither with theoretical ideas about galactic magnetic fields nor with observations of external galaxies (Section 3.3). The pitch angle of the magnetic field should be a model parameter, possibly a function of position, obtained from fits to data rather than fixed to be zero (or any other value) beforehand. Another problem is that the magnetic field may really correspond to a superposition of different azimuthal modes, so that attempts at fitting a purely axisymmetric or bisymmetric model may lead to erroneous results.

The presence of reversals in the Milky Way is often interpreted as an unambiguous indication of the bisymmetric global structure of the magnetic field. As we discuss in Section 8.5, axisymmetric magnetic structures may also contain reversals, and mean-field dynamo models for the Milky Way favor an axisymmetric field structure.

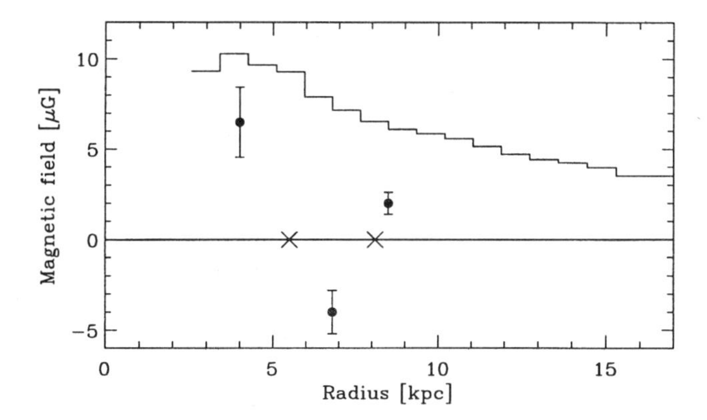

Figure 5 The strength of the large-scale magnetic field in the Milky Way (*full circles with error bars*) and positions of its reversals (*crosses*), as inferred from pulsar rotation measures (Rand & Lyne 1994). Note a gradual increase of $|\overline{B}|$ towards smaller radii (a positive $\overline{B}$ corresponds to the field direction towards the first and second Galactic quadrants). Error bars shown correspond to 30% uncertainty and are chosen tentatively to indicate a scatter of the available estimates at $r = 8.5$ kpc, the Galactic radius of the sun. The solid line shows the strength of the total magnetic field, averaged in azimuth as obtained by EM Berkhuijsen (in preparation) from the deconvolved surface brightness of synchrotron emission at 408 MHz (Beuermann et al 1985), assuming energy equipartition between magnetic field and cosmic rays; the accuracy of this estimate is probably $\simeq 30\%$.

Field reversals have rarely been observed in external galaxies, only in BSS candidates (see Table 2) and possibly in a galaxy at redshift 0.395 (Kronberg et al 1992; see Section 3.7). In some galaxies, the resolution of the observations is high enough to detect reversals if they were present: This is the case for M31 observed with a resolution of $\simeq 1$ kpc near the major axis (Beck 1982, Ruzmaikin et al 1990). In other galaxies the resolution of Faraday rotation data is lower (e.g. Krause et al 1989a, Buczilowski & Beck 1991, Ehle & Beck 1993) and reversals cannot be excluded. However, the number of reversals within the telescope beam cannot be large as this would average out any Faraday rotation.

Because the Sun is located fairly close to a reversal, the strength of the regular magnetic field at $r \simeq 8.5$ kpc is less likely to be a representative value for the bulk of spiral galaxies or even for the Milky Way itself. Values of order 4–6 μG seem to be more typical.

4. GALACTIC DYNAMO THEORY

We now discuss the mechanisms generating large-scale fields that have been presented in the previous section. We begin by considering first the small-scale magnetic fields.

4.1 *Random Magnetic Fields*

The interstellar medium is turbulent and thus any embedded magnetic field must have a random small-scale component. The presence of this component is crucial in all theories of large-scale dynamo action. There are several mechanisms that produce fluctuations in the interstellar magnetic fields: (*a*) tangling of the large-scale field by turbulence and from Parker and thermal instabilities, (*b*) compression of ambient magnetic fields by shock fronts associated with supernova remnants and stellar winds, and (*c*) self-generation of random magnetic fields by turbulence (small-scale dynamo). All of these mechanisms act together, and each imprints its own statistical properties onto the magnetic fields.

The available observational and theoretical knowledge of random magnetic fields and their maintenance in the ISM is rather poor. Instead, crude descriptions in terms of global quantities such as mean magnetic energy are usually applied. A widely used concept is that of equipartition between the magnetic and kinetic energy in the turbulence (Kraichnan 1965, Zweibel & McKee 1995), which implies that the rms random magnetic field strength is given by $b \simeq B_{\rm eq} \equiv (4\pi\rho v^2)^{1/2}$, with v the rms turbulent velocity and ρ the density. The equipartition value is significant in that the Lorentz force is expected to become comparable to the forces driving the turbulent flow as equipartition is approached. (This $B_{\rm eq}$ is not to be confused with the equipartition field strength in Sections 2 and 3, where equipartition refers to the estimated cosmic-ray energy density used to deduce the field strength from the synchrotron emission.) Interstellar turbulence is often treated as an ensemble of random Alfvén waves (McIvor 1977, Ruzmaikin & Shukurov 1982, McKee & Zweibel 1995) for which the equipartition holds exactly. Magnetic fluctuations are accompanied also by fluctuations in density (Armstrong et al 1995), so that other mechanisms, possibly nonpropagating fluctuations, must contribute to the interstellar turbulence (Higdon 1984). The random magnetic fields in the Milky Way are typically 4–6 μG (Ohno & Shibata 1993), close to $B_{\rm eq}$.

Another component of the random magnetic field, one that is associated with interstellar (super) bubbles, is observed in the Milky Way (Heiles 1989, Heiles et al 1993, Vallée 1993). The magnetic field in HI shells, detected via the Zeeman effect, seems to be concentrated in filaments with the magnetic pressure larger than the gas pressure. The field strength in magnetic bubbles around OB associations, as obtained from Faraday rotation measurements, follows the density dependence $b \propto \rho$ expected for a shocked medium.

The small-scale dynamo (Kazantsev 1968, Meneguzzi et al 1981) must be an important source of interstellar random magnetic fields (Sokoloff et al 1990). A distinctive feature of this component of the interstellar field, item (*b*) above, is that it is organized in intermittent magnetic ropes of small filling factor and lengths comparable to the correlation length of the turbulence (50–100 pc). The rms strength of the magnetic fluctuations generated by this mechanism is possibly close to the equipartition value, but the field within the filaments may be significantly higher (Belyanin et al 1993). For example, three-dimensional simulations of convective small-scale dynamo action at magnetic Reynolds numbers of about 1000 (Nordlund et al 1992) give $B_{\rm rms} = 0.4B_{\rm eq}$ and $B_{\rm max} = 3B_{\rm eq}$. We note that in the interstellar gas of elliptical galaxies a small-scale dynamo may be the only source of magnetic fields, generating random fields of μG strength and a few hundred parsecs in scale (Moss & Shukurov 1996).

4.2 *Large-Scale Fields*

The main challenge in the theory of galactic magnetism is to explain the origin and structure of the observed large-scale field. In Figure 6 we sketch different routes by which large-scale magnetic fields may arise. Large-scale flows (shear, compression) together with turbulence effects (swirling motions and inverse cascade—see below) can amplify weak seed magnetic fields (Section 5), thereby converting small-scale fields into large-scale fields. The amplifying effect of swirling motions on the large-scale field is described by the α-effect (Parker 1955, Steenbeck et al 1966, Moffatt 1978). Such motions also lead to an inverse cascade from the conservation properties of the magnetic helicity (Frisch et al 1975, Pouquet et al 1976) and from the cross-helicity effect (Yoshizawa & Yokoi 1993).

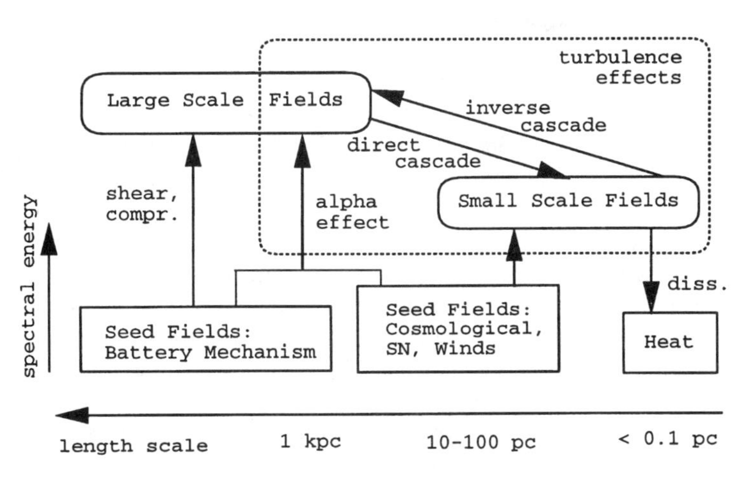

Figure 6 Sketch illustrating the various routes by which large-scale magnetic fields can arise. Turbulence effects (inverse cascade and α-effect) combined with shear and compression (differential rotation) amplify weak magnetic fields to produce strong large-scale fields.

These concepts were originally applied to stellar turbulence, where the existence of dynamos can almost be considered as an observational fact. It is not clear, however, how much galactic turbulence has in common with thermal turbulence in stars. Nevertheless, statistical properties of turbulence in molecular clouds seem to be remarkably similar to those determined from numerical simulations of ordinary compressible turbulence (Falgarone et al 1994).

There have been attempts to explain the large-scale magnetic field without invoking dynamo action. The turbulence must then be regarded as unimportant, and a large-scale seed magnetic field has to be amplified by large-scale shearing and compression alone. The inevitable eventual decay is assumed then to be considerably postponed (e.g. Kulsrud 1986). A model of this type has recently been proposed by Chiba & Lesch (1994), who consider fields that are maintained by an unspecified mechanism at large distances. Because these processes describe field amplification by shearing and compression alone, it is quite uncertain whether they can give fields of the strength and topology required at ages of about 10 Gyr; however, in conjunction with a dynamo, such motions might be important in certain galaxies.

4.3 *Treatment of Galactic Turbulence*

Several basic gas components are involved in galactic turbulence. The disk consists of warm gas, interspersed by cold clouds and hot bubbles. Hot bubbles result from local heating (e.g. OB associations and supernova and superbubble explosions) and eject hot gas into the halo (galactic fountains). These violent motions, in addition to stellar winds, help to drive the turbulence. Furthermore, random motions of molecular clouds may stir up the warm gas, because they are dynamically coupled by magnetic field lines. The Parker instability may also be a source of turbulence, or it may at least act as an agent causing the movement of flux tubes and thereby generate an α-effect (Parker 1992, Hanasz & Lesch 1993). In the model of Vázquez-Semadeni et al (1995), the turbulence is driven by gravity and density gradients that result from interstellar cooling and heating processes.

To understand the effect of these different gas components on the magnetic field we need to discuss the coupling of the magnetic field to those components. The magnetic fields in the hot component are rapidly ejected into the halo. They are then no longer directly important for magnetic processes in the disk, but are essential in the galactic halo. Clouds could be more important, because a large-scale field would be dragged with the gas into these clouds as they form, and the cloud motions would entangle the magnetic field lines (Beck 1991). This process is of only limited duration, because ambipolar diffusion (Mestel 1966) would decouple the clouds from the field on a timescale of 10^7 yr.

The outcome is that for most of the time the magnetic field remains attached to the diffuse ionized gas, and, to the extent that the field is associated with clouds, the effect of the clouds is to contribute to the turbulent dynamics of the magnetic field lines. Even if this is an important contributor to the chaotic driving of field lines (in addition to the turbulence mentioned above), it is reasonable to assume that the magnetic field in a galactic disk is on average linked to the warm, ionized medium and perhaps also to the warm neutral medium, both of which are in a turbulent state.

Dynamo action is well established from numerical turbulence simulations. In the absence of (rotational) velocity shear, the magnetic field is very intermittent (Meneguzzi et al 1981). In the presence of rotational shear, the resulting magnetic shear instability (e.g. Balbus & Hawley 1992) can lead to strong large-scale fields (Brandenburg et al 1995a). This mechanism yields coherent fields similar to those in ordinary $\alpha\Omega$-dynamos.

The classical α-effect quantifies the field-aligned electromotive force resulting from magnetic field lines twisted by the turbulence (cf simulations by Otmianowska-Mazur & Urbanik 1994). In the original picture the dynamics of these field lines is governed by external turbulent motions. Parker (1992) discussed a new, perhaps more appropriate, concept in which the motions result mostly from the dynamics of magnetic field lines themselves. The concept of an α-effect seems, however, sufficiently robust so that the form of the basic equations is always the same. In fact, the α-effect is only one of many effects relating the mean emf $\mathcal{E} \equiv \overline{\mathbf{u} \times \mathbf{b}}$ to the mean magnetic field and its derivatives. If the mean field is not too intermittent, we can expand

$$\mathcal{E}_i = \alpha_{ij}\overline{B}_j + \eta_{ijk}\frac{\partial \overline{B}_j}{\partial x_k} \tag{1}$$

(Krause & Rädler 1980), neglecting higher derivatives of $\overline{\mathbf{B}}$. This relation is used when solving the induction equation for the mean magnetic field,

$$\frac{\partial \overline{\mathbf{B}}}{\partial t} = \nabla \times (\overline{\mathbf{u}} \times \overline{\mathbf{B}} + \mathcal{E}). \tag{2}$$

The mean velocity $\overline{\mathbf{u}}$ comprises both the rotational velocity, as well as galactic winds and any other large-scale flows. This is where the observed rotation curves and other large-scale flow components of individual galaxies enter into the theory and models.

The α_{ij} and η_{ijk} tensors in (1) are anisotropic (Ferriére 1993, Kitchatinov et al 1994). Anisotropies can arise from stratification, rotation, shear, and magnetic fields. Stratification and rotation are most important, because without them there would be no $\alpha_{\phi\phi}$ component, which is needed to regenerate poloidal magnetic fields from $\overline{B}_\phi$. An important contribution to η_{ijk} comes from isotropic turbulent magnetic diffusion, $\epsilon_{ijk}\eta_t$, where η_t is the turbulent magnetic diffusivity. Explicit expressions in the framework of the first-order smoothing approximation (FOSA) were first derived by Steenbeck et al (1966) and Krause (1967), and more recently by Rüdiger & Kitchatinov (1993). They find expressions of the form

$$\alpha \approx -l^2 \Omega \cdot \nabla \ln(\rho v) F(\overline{\mathbf{B}}, \Omega), \qquad \eta_t \approx \frac{1}{3} v l G(\overline{\mathbf{B}}, \Omega), \tag{3}$$

where l is the correlation length of the turbulence α stands for $\alpha_{\phi\phi}$, and F and G are certain ("quenching") functions. The stratification of ρv is important, because it breaks the symmetry between upward and downward motions. If h is the scale height, a rough estimate gives

$$|\alpha| \sim \min(\Omega l^2/h, v), \tag{4}$$

which ensures that α does not exceed v (e.g. Zeldovich et al 1983).

The FOSA is valid either for small magnetic Reynolds numbers (which is irrelevant here) or in the limit of short correlation times (which is also not well satisfied in the ISM). However, although higher order terms may become important, they affect the results only quantitatively (Zeldovich et al 1988, Carvalho 1992). There are independent attempts to compute the transport coefficients resulting from evolving flux tubes (Hanasz & Lesch 1993) and from expanding supernovae and superbubbles rather than from turbulence (Ferriére 1993, Kaisig et al 1993). The resulting values of α and η_t are smaller than those expected from interstellar turbulence, suggesting that explosions are of lesser importance.

Turbulent diamagnetism (Zeldovich 1957) can be represented as a macroscopic velocity, $\overline{\mathbf{u}}_{\text{dia}} = -1/2\nabla\eta_t$ (Roberts & Soward 1975, Kitchatinov & Rüdiger 1992). It tends to expel magnetic fields from regions where η_t is large. This term can be considered as a contribution to the antisymmetric part of α_{ij} (Rädler 1969). Additional effects of this kind are magnetic buoyancy (Moss et al 1990) and topological pumping (Section 7.2).

4.4 *Basic Galactic Dynamo Models*

The simplest form of the mean field ($\alpha^2\Omega$) dynamo equation (2) that retains the basic physics (e.g. Parker 1979, Roberts & Soward 1992) is, in dimensionless form,

$$\frac{\partial \overline{\mathbf{B}}}{\partial t} = \nabla \times (R_\omega \overline{\mathbf{u}} \times \overline{\mathbf{B}} + R_\alpha F \alpha \overline{\mathbf{B}} - G\eta_t, \nabla \times \overline{\mathbf{B}}), \tag{5}$$

where $F(\overline{\mathbf{B}}, \Omega) = (1 + \overline{\mathbf{B}}^2/B_{\text{eq}}^2)^{-1}$ is the simplest form of "α-quenching" and $G(\overline{\mathbf{B}}, \Omega) = 1$. Distances and times are measured in units of h_* and h_*^2/η_{t*}, respectively, where $\overline{\mathbf{u}} = \boldsymbol{\Omega} \times \mathbf{r}$ and α and η_t are normalized by appropriately chosen characteristic values denoted by asterisks. Dimensionless numbers

$$R_\omega = h_*^2 \Omega_* / \eta_{t*}, \quad R_\alpha = h_* \alpha_* / \eta_{t*} \tag{6}$$

characterize the amplification of magnetic field by shearing of the mean velocity field and the α-effect, respectively. Using Equation (3), α and η_t can be expressed through observable parameters of the galaxies such as the rotation curve, rms velocity and scale, and the thickness of the ionized disk (a function of r). The quenching effects also require that the gas density is specified as a function of position. Equation (5) must be supplemented by boundary conditions. In models that treat the disk alone, these are usually vacuum boundary conditions in which one assumes that the turbulent magnetic diffusivity outside the disk is infinite. This proves to be a reasonable approximation to reality (Moss & Brandenburg 1992), as η_t varies by perhaps a factor of about 50 between the disk and the halo (see Brandenburg et al 1993, Poezd et al 1993). However, more advanced treatments employ the *embedded disk* model (Stepinski & Levy 1988). This includes a spherical galactic halo and appropriate boundary conditions are imposed at the surface of the halo, whereas the disk is modeled by appropriate distributions of $\overline{\mathbf{u}}$, α, and η_t. This concept has proved sufficiently adaptable to accommodate developing requirements, such as the inclusion of a flared disk, an α-effect extending into the halo (Section 7.1), and/or a galactic wind (Section 7.2).

Initial conditions for (5) are often chosen to correspond to a weak seed field. Exponentially growing solutions then arise, $\overline{\mathbf{B}} \propto \exp(\Gamma t)$, provided the *dynamo number* $D = R_\alpha R_\omega$ exceeds a certain value $D_{\rm crit} \approx 10$. Using Equation (3) one can show that $D \simeq 9(h_* \Omega_* / v)^2$. For $h_* \simeq 500$ pc, $\Omega_* \simeq 20$ km s^{-1} kpc^{-1}, and $v \simeq 10$ km s^{-1} we obtain $D \simeq 10$, so that the dynamo is expected to operate under typical galactic conditions. For $D \gg D_{\rm crit}$, the growth rate is estimated as $\Gamma \simeq C D^{1/2} \eta_t / h_*^2 \simeq C(\alpha_* \Omega_* h_*)^{1/2}$, with C a quantity of order unity depending on the galaxy model. A typical model gives $\Gamma^{-1} \simeq 5 \times 10^8$ yr; this is a lower estimate for the dynamo timescale. [We note, however, that the timescale for the magnetic shear instability is the inverse Oort A-value (Balbus & Hawley 1992), which is somewhat shorter (10^8 yr). This mechanism leads to dynamo action (Brandenburg et al 1995b) that would lower the effective value of Γ^{-1}.]

All classical dynamo models predict that the large-scale field in the outer parts of the disks in spiral galaxies has quadrupole (S0) symmetry, that is, both $\overline{B}_r$ and $\overline{B}_\phi$ are even in z, whereas $\overline{B}_z$ is odd (Parker 1971, Vainshtein & Ruzmaikin 1971). This mode is dominant in a disk (but not in a sphere). A dipole (A0) mode, with both $\overline{B}_r$ and $\overline{B}_\phi$ odd in z and $\overline{B}_z$ even, can be dominant near the axis of the disk. The large-scale field is amplified until α becomes significantly quenched, which occurs when $\overline{B}$ is of order $B_{\rm eq}$, typically a few μG.

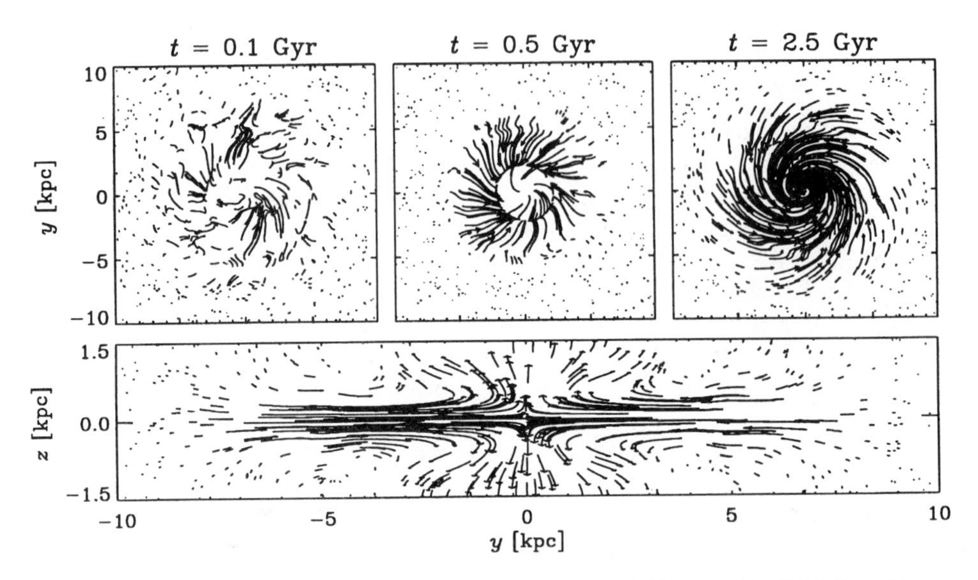

Figure 7 Face-on views showing the evolution of the magnetic field in a model of M83 (from KJ Donner & A Brandenburg, in preparation). The lower panel gives an edge-on view for $t = 8.1$ Gyr.

Field evolution is qualitatively different if the initial field is a random field with strength close to $B_{\rm eq}$. There is then no kinematic stage, because α-quenching is immediately important. The dynamo acts then to change the scale and spatial distribution of the field. An example of typical evolution of the magnetic field in a spiral galaxy as envisaged by the standard dynamo model is illustrated in Figure 7.

Over the past 5 to 10 years a large number of galactic dynamo models have been developed. The minimum ingredient of such models is a flat geometry. Such models were first computed in the 1970s, but computers can only now reach the regime applicable to the theory of asymptotically thin disks (Walker & Barenghi 1994 and references therein). Galactic models share the somewhat frustrating property that nonaxisymmetric solutions are always harder to excite than axisymmetric ones (Ruzmaikin et al 1988a, Brandenburg et al 1990, Moss & Brandenburg 1992). Not even the inclusion of anisotropies seems to change this conclusion (Meinel et al 1990). Stable nonaxisymmetric solutions have only been found if α and $\eta_{\rm t}$ vary azimuthally (Moss et al 1991, 1993a; Panesar & Nelson 1992). The inclusion of nonlinear effects demonstrated that mixed parity states can persist over rather long times, even comparable with galactic lifetimes (Moss & Tuominen 1990, Moss et al 1993a). When η-quenching is included ($G \neq 1$), linear calculations show that A0 and S1 modes may be more readily excited (Elstner et al 1996).

In most of these models α_{ij} and η_{ijk} were adopted using qualitative forms of (3) and (4), calibrated by observations. Significant conceptual progress has been made recently by deriving all these functions consistently from the same turbulence model, which includes stratification of density and turbulent velocity, derived from a condition of hydrostatic equilibrium (Schultz et al 1994, Elstner et al 1996). One should not forget, however, that such models still rely on important approximations and simplifications (e.g. FOSA and the lack of a reliable turbulence model).

4.5 *The Quenching Problem*

In recent years the feedback of the magnetic field on the turbulent diffusion and the α-effect has become a topic of major concern. Piddington (1970) was the first to suggest that for large magnetic Reynolds numbers the magnetic fluctuations would be strong enough to suppress turbulent diffusion. This idea was rejected by Parker (1973), who argued that the development of strong small-scale fields is limited by reconnection, so that they do not hinder turbulent mixing of field and fluid. In fact, without turbulent diffusion the galactic differential rotation would wind up the field so tightly that it would not resemble the magnetic field structure of any observed galaxy (Section 8.3).

The results of the two-dimensional numerical MHD experiment of Cattaneo & Vainshtein (1991) stimulated new interest in the problem of turbulent diffusion. They found that η_t is suppressed according to $\eta_t = vl/(1 + R_m\overline{\mathbf{B}}^2/B_{eq}^2)$, where $R_m = vL/\eta$ is the magnetic Reynolds number based on the microscopic diffusivity. Evidently, η_t would be significantly reduced when $\overline{B}$ is comparable to $R_m^{-1/2}B_{eq}$. In galaxies, $R_m = \mathcal{O}(10^{17})$, so η_t would essentially be zero. Even if we used a Reynolds number based on ambipolar diffusion, with $R_m^{AD} \gtrsim O(10^3)$, η_t would still be too small. This type of quenching is much stronger than the "traditional" quenching (Moffatt 1972), so something seems to be wrong (e.g. Field 1996).

In three dimensions the turbulent motions would continue to entangle the magnetic field in the direction perpendicular to $\overline{\mathbf{B}}$ (Krause & Rüdiger 1975, Parker 1992). This has now also been demonstrated numerically (Nordlund et al 1994) as well as analytically (Gruzinov & Diamond 1994). In other words, turbulent diffusion is really not significantly suppressed at field strengths somewhat below the equipartition value. The decay of sunspots is a good example of this (Krause & Rüdiger 1975).

Vainshtein & Cattaneo (1992) and Tao et al (1993) suggested that the α-effect might also be quenched dramatically, $\alpha = \alpha_{kin}/(1 + R_m\overline{\mathbf{B}}^2/B_{eq}^2)$, where α_{kin}

is the kinematic value of Equation (3). The analysis of Gruzinov & Diamond (1994) seems to support this result. On the other hand, the simulations of Tao et al (1993), as well as unpublished simulations by A Brandenburg, are reminiscent of an earlier result by Moffatt (1979), that the α-effect may fluctuate strongly and never converge to a finite value if R_{m} is large.

There is at present no conclusive resolution to this problem, but here are some possibilities: (*a*) The conventional α-effect might still work in reality, but the method used to estimate α from simulations is inappropriate (e.g. the boundary conditions preserve the magnetic flux, so the α-effect is forced to have zero effect on the average field; or the computational domain might be too small compared to the eddy size). (*b*) The conventional α-effect is really nonexistent, but instead some other mechanism (e.g. an inverse cascade mechanism, incoherent α-effect, or cross-helicity effect) generates large-scale fields in conjunction with shear. (*c*) An important contribution to α comes from the Parker instability: This mechanism would work especially for finite magnetic fields.

A somewhat different problem was raised by Kulsrud & Anderson (1992), who suggested that the growth of large-scale fields is suppressed by ambipolar diffusion at small scales. However, before we can draw any final conclusions, nonlinear effects need to be included. These can be important for two reasons: The inverse cascade process is inherently nonlinear, and nonlinear ambipolar diffusion can lead to sharp magnetic structures (Brandenburg & Zweibel 1995), which would facilitate fast reconnection and rapidly remove magnetic energy at small scales.

The problem raised by Vainshtein & Cattaneo (1992) is related to the assumption that most of the magnetic energy is at small scales, i.e. $\langle \mathbf{B}^2 \gg \rangle \mathbf{B}^2$. This, however, is only a result of linear theory and is not supported by observations (Section 3). A recent simulation by Brandenburg et al (1995a) is relevant in this context. Here a large-scale field is generated with $\langle \mathbf{B}^2 \rangle / \langle \mathbf{B}^2 \rangle \approx 0.5 \gg R_m^{-1/2} \approx 0.1$. The dynamo works even in the presence of ambipolar diffusion, which Kulsrud & Anderson (1992) thought to be effective in destroying large-scale dynamo action. Here, the incoherent α-effect is much larger than the coherent effect, but the estimated value of the dynamo number is nevertheless above the critical value, suggesting that conventional dynamo action might also be at work.

5. ORIGIN OF GALACTIC MAGNETIC FIELDS

5.1 *Cosmological Magnetic Fields*

Zeldovich (1965) noted that a Friedmannian cosmology admits a weak uniform magnetic field given as an initial condition at the Big Bang (see also Zeldovich & Novikov 1982, LeBlanc et al 1995). A hypothetical homogeneous magnetic field in the Universe has been never detected and only its upper limits are available. A uniform magnetic field $\overline{B} \gtrsim 10^{-7}$ G at the present day would lead to anisotropy in the expansion of the Universe, thereby affecting nucleosynthesis (e.g. Cheng et al 1994, Grasso & Rubinstein 1995). Analysis of Faraday rotation measures of extragalactic sources gives a stronger upper limit of 10^{-9}–10^{-10} G (Ruzmaikin & Sokoloff 1977). A magnetic field leads to transitions between left- and right-handed neutrinos (spin-flip) in the early Universe. Nucleosynthesis gives an upper limit to the abundance of right-handed neutrinos and thus yields the constraint

$$\overline{B}_{\text{proto}} \lesssim (1 - 30) \times 10^{-13}\ \text{G} \tag{7}$$

for the present-day uniform cosmological field (Sciama 1994).

Taking a cosmological magnetic field as a given initial condition at the Big Bang is rather unsatisfactory. Furthermore, it is not clear whether a homogeneous magnetic field can be incorporated into modern quantum cosmology, where it cannot be prescribed as an initial condition.

Several mechanisms of small-scale magnetic field generation by quantum effects in the early Universe have been proposed (Turner & Widrow 1988, Quashnock et al 1989, Vachaspati 1991, Ratra 1992). The resulting spatial scales of cosmological magnetic fields are very small and, even after cosmological expansion, they are negligible in comparison with protogalactic scales.

The strength and scale of the relic magnetic field can be estimated as follows. As magnetic diffusion smoothes the field, its scale at time t will be about $(\eta t)^{1/2}$, where η is the magnetic diffusivity, as the initial scale is much smaller. At the epoch of nucleosynthesis, the resulting scale is 10^4 cm, corresponding to a scale $l \approx 10^{-6}$ pc today. The same arguments as for Equation (7) give an upper limit on the magnetic field at nucleosynthesis of 10^{11} G. With allowance for a change in the equation of state at $t = t_* \approx 10^4$ yr, the frozen-in magnetic field at time t is diluted by cosmological expansion to $b(t_*/t)^{4/3}(1\ \text{min}/t_*)^{1/3}$. Since the protogalaxy includes $(L/l)^3$ correlation cells, the average field strength is smaller by a factor $(L/l)^{-3/2}$. This yields the following upper limit on the average magnetic field at the scale of the protogalaxy at the present time,

$$\overline{B}_{\text{proto}} \lesssim 2 \times 10^{-23}\ \text{G} \tag{8}$$

(see Enqvist et al 1993, 1995). Thus either the cosmological magnetic field is exactly homogeneous, and then the restriction (7) applies, or the field was produced in the early Universe, and then it must satisfy (8). We should note that the above estimates neglect ohmic losses. These constraints do not apply to magnetic fields generated at later stages of cosmological evolution. Battery mechanisms can contribute at more recent epochs, giving (Mishustin & Ruzmaikin 1971, see also Harrison 1970, Baierlein 1978),

$$\overline{B}_{\text{proto}} \lesssim 10^{-21}\ \text{G}. \tag{9}$$

5.2 *The Primordial Origin of Galactic Magnetic Fields*

We now assess the possibility that the large-scale magnetic field observed in galaxies is merely a result of the twisting of a cosmological magnetic field by galactic differential rotation (see e.g. Kulsrud 1986). Aiming at conservative estimates, we neglect any magnetic field dissipation. An isotropic contraction of the protogalaxy with a frozen-in magnetic field, from an intergalactic density $\rho_{\text{IG}} \approx 10^{-29}$ g cm^{-3} up to an interstellar density $\rho \approx 10^{-24}$ g cm^{-3}, results in amplification of the primordial magnetic field by a factor of 2×10^3. Differential rotation results in an amplification of the magnetic field in a young galaxy by the number of galactic rotations in 10^{10} yr, which is $N \sim 30$. Altogether, a conservative upper limit on the field in the galactic disk resulting from a primordial field is

$$\overline{B}_{\text{proto}} N (\rho/\rho_{\text{IG}})^{2/3} < 2 \times 10^{-7}\text{G}, \tag{10}$$

where the more favorable constraint (7) has been used. A primordial field wound up by differential rotation ultimately decays: In a region with closed streamlines (a galaxy in this case) this effect is known as *flux expulsion* (Moffatt 1978).

5.3 *The Dynamo Origin of Magnetic Field*

Any dynamo requires a seed field because Equation (5) is homogeneous in $\overline{B}$. There are two possibilities for the seed field: It can be essentially of cosmological origin or it can result from processes occurring in the ISM.

The large-scale dynamo timescale in a typical galaxy cannot be shorter than $\tau \sim 5 \times 10^8$ yr (see Section 4.4). A primordial field on a protogalactic scale then needs to be at least $O(10^{-18})$ G in order to be amplified to 10^{-6} G in 10^{10} yr (when the amplification by protogalaxy contraction is considered). With the estimates (8) and (9), we conclude that a cosmological magnetic field is not viable as a seed field for a galactic dynamo. Moreover, for the Milky Way and M31, the timescale is more like $\tau \approx 10^9$ yr, so that for these galaxies a primordial magnetic field needs to be at least about 2×10^{-14} G, assuming that τ has not varied significantly during galactic evolution.

A sufficiently strong seed field for the large-scale galactic dynamo can be generated by a small-scale dynamo. The scale height of the disk of a young galaxy is estimated as $h \simeq$ 100–500 pc (Briggs et al 1989) and the turbulent velocity as $v \simeq 10$ km s^{-1} (Turnshek et al 1989). Assuming that $l \simeq 100$–300 pc and $\rho \simeq 10^{-24}$ g cm^{-3}, we conclude that a random magnetic field $b \simeq (4\pi\rho v^2)^{1/2} \simeq 2$–$2.5\,\mu G$ of a scale 100–300 pc is generated by the fluctuation dynamo on a timescale of order $\tau_l \sim l/v \simeq 10^6$–$10^7$ yr. Because a galactic disk contains about $N_l = (h/l)(R/l)^2$ turbulent cells, the resulting mean field dynamo seed field is about $bN_l^{-1/2} \simeq 10^{-8}$ G. This is much larger than possible cosmological seed fields (8, 9), even if the field compression during galaxy formation is taken into account.

The resulting small-scale field is strong enough to produce, via a mean field galactic dynamo, a large-scale magnetic field of μG-strength in $\sim (1$–$2) \times 10^9$ yr (Beck et al 1994a). This means that even the presence of regular magnetic fields in galaxies with redshifts of $z \simeq 2$ or even $z \simeq 3.4$ (Wolfe et al 1992, White et al 1993) does not contradict the picture of generation and maintenance of large-scale fields by a mean-field dynamo mechanism. The possible role of the halo (Chiba & Lesch 1994) and radial motions (Camenzind & Lesch 1994) has also been investigated.

The fluctuation dynamo also needs a seed but, because of the very short fluctuation dynamo timescale, even the magnetic fields generated by the battery effects in stars (Biermann 1950, Mestel & Roxburgh 1962), and subsequently ejected into the ISM, or a cosmological field (Section 5.1) would suffice.

Thus, large-scale dynamo action in a galaxy is preceded by a small-scale dynamo that prepares the seed for the former. These may operate at different epochs. Small-scale dynamo action has been considered by Pudritz & Silk (1989) for the protogalaxy, by Zweibel (1988) during the post-recombination epoch, and before recombination by Tajima et al (1992).

A rather radical view of the role of the Galactic center in the origin of the global galactic magnetic field was proposed by Hoyle (1969), who suggested that the magnetic field observed in the solar vicinity had been ejected from the Galactic center. This idea was rejected because the required magnetic field in the nucleus is 10^9 G, and its energy exceeds the gravitational energy of a black hole with a mass of $10^8 M_\odot$. Nevertheless, Chakrabarti et al (1994) proposed a similar hypothesis, with the azimuthal field being amplified up to $\overline{B}_{\rm core} \simeq 3 \times 10^5$ G within $r_0 \simeq 3 \times 10^{11}$ cm of the center. A galactic wind is then supposed to carry this field to the outer parts of the Galaxy. However, this gives for the solar vicinity an extremely weak field of $\overline{B} \simeq (r_0/r_\odot)(h_0/h_\odot)\overline{B}_{\rm core} \simeq 6 \times 10^{-16}$ G, where $h_0 \simeq r_0$ and $r_\odot = 8.5$ kpc and $h_\odot = 500$ pc are the radius and half-thickness of the magnetoionic disk in the Solar vicinity. Chakrabarti et al obtained for $\overline{B}$ a value about 10^{10} times larger by overlooking a factor $h_0/h_\odot$.

6. EFFECTS OF THE DYNAMO ENVIRONMENT

6.1 *Starbursts*

Starburst galaxies are believed to contain regions of strongly enhanced star formation, particularly of massive stars. The rapid evolution of these stars, through phases with energetic stellar winds to supernovae, may possibly make the turbulence more energetic (for example, by increasing the fraction of hot gas and hence the mean sound speed), with several possible consequences for dynamo theory. Any increased turbulent pressure will inflate the disk, and the α-effect may be enhanced above the value appropriate to a quiescent galaxy. Both of these effects increase the dynamo number (Section 4.4). This enhancement may be preferentially concentrated in azimuth, perhaps lagging the spiral arms. Ko & Parker (1989) suggested that galactic dynamos may turn on and off in response to changing starburst activity. However, the timescale for starbursts is believed to be less than 10^8 yr, which is certainly no longer (and possibly considerably shorter) than a dynamo growth time. Thus it is hard to see how significant field growth can be caused by isolated starburst episodes; see also Vallée (1994). Nozakura (1993) presented a local model with several feedback loops, linking star formation via gravitational instability, dynamo action, and energy release into the ISM via supernovae. In some contrast to Ko & Parker, he concluded that there was only a limited parameter range in which strong star formation and dynamo action could coexist: Essentially star formation requires a high surface density of gas and/or a low sound speed, and so a thin disk, giving a smaller dynamo number. These are clearly matters requiring further attention. Moreover, in an active galaxy, fountain flows will be more frequent, enhancing the lifting of field from the disk into the halo—see Section 7.2.

6.2 *Galactic Encounters*

There is strong observational evidence that a number of spiral galaxies are interacting gravitationally with a neighbor. The clearest nearby example is M81, which is believed to have undergone a recent encounter with NGC 3077 (probably less than 10^9 yr ago). Because the orbit of NGC 3077 is approximately in the disk plane of M81, this system is particularly well suited to simulation, and Thomasson & Donner (1993) predict nonaxisymmetric velocities of order 10 $\mathrm{km\,s^{-1}}$ in the disk of M81. With $\eta_t \sim 10^{26}\,\mathrm{cm^2\,s^{-1}}$ and $L \sim 1$ kpc, the magnetic Reynolds number UL/η_t is then about 30, quite large enough to affect significantly the disk fields (Vallée 1986). Interestingly, M81 appears to have a strong bisymmetric field component. M33 also may have some bisymmetric field structure, and it is believed to be interacting with M31. Recently, at least weak evidence has been found for BSS in the interacting galaxy NGC 2276 (Hummel & Beck 1995) and for MSS in M51 (EM Berkhuijsen et al, in preparation).

If we consider a Fourier decomposition of $\overline{\mathbf{u}}$ and $\overline{\mathbf{B}}$ into parts $\overline{\mathbf{u}}_m$, $\overline{\mathbf{B}}_m$, corresponding to an azimuthal wave number m, then the induction term $\nabla \times (\overline{\mathbf{u}} \times \overline{\mathbf{B}})$ can give rise to a bisymmetric field component in two ways. If the dynamo basically generates an axisymmetric field $\overline{\mathbf{B}}_0$, then $\overline{\mathbf{u}}_1$ can generate a slaved $m = 1$ component $\overline{\mathbf{B}}_1$ from the $\overline{\mathbf{u}}_1 \times \overline{\mathbf{B}}_0$ interaction. Moss et al (1993b) investigated this possibility in a nonlinear model with a relatively thick disk, using a velocity field based on the Thomasson & Donner (1993) simulation. They found that a globally modest bisymmetric field component could be generated, concentrated to the outer part of the disk, where it may dominate. More subtly, the $\overline{\mathbf{u}}_2 \times \overline{\mathbf{B}}_1$ interaction (giving rise directly to $m = 1$ and $m = 3$ field components) may be such as to increase the linear growth rate of the bisymmetric field component compared to that of the axisymmetric component, so that in the nonlinear case a substantial bisymmetric field could survive. Moss (1995) showed that, in a simple linear model, the $m = 1$ field could then be excited at lower dynamo number than the $m = 0$ field, but a nonlinear investigation using a more realistic model is needed to clarify the importance of this mechanism. The remarks concerning the modal interactions apply, of course, whatever the mechanism providing the velocity field. In particular, it may be relevant that a $\overline{\mathbf{u}}_2 \times \overline{\mathbf{B}}_0$ interaction can give rise to a slaved $m = 2$ field component.

6.3 *Parametric Resonance with Spiral Arms*

A dynamo mechanism with selective amplification of BSS caused by swing excitation by the spiral arms has been proposed by Chiba & Tosa (1990). Unlike axisymmetric dynamo modes (which do not oscillate at realistic dynamo numbers), a bisymmetric magnetic field has the form of a dynamo wave, which propagates in the azimuthal direction as seen in an inertial frame. Because the spiral pattern modulates the dynamo efficiency, a parametric resonance between the spiral arms and the bisymmetric magnetic field might be expected. Applying the classical theory based on the Mathieu equation (see Landau & Lifshitz 1969), Chiba & Tosa argued that the $m = 1$ mode is amplified when its frequency ω_B is half that of the spiral pattern, ω_{SP}, and the growth rate of the $m = 1$ mode is increased proportionally to the increment of the dynamo number in a spiral arm. However, the classical theory of parametric resonance is valid only for simple, discrete, stable oscillatory systems and may not apply to a dynamo system (Schmitt & Rüdiger 1992).

Parametric resonance in a galactic dynamo, which is a *distributed* oscillatory system, was considered asymptotically in the thin-disk approximation by Kuzanyan & Sokoloff (1993). They showed that the resonant condition remains the same in terms of frequencies, but the resulting enhancement in the growth rate is much smaller than above and is proportional to the efficiency of the radial diffusive transport of the magnetic field, i.e. the aspect ratio of the disk h/R. Galactic parametric resonance has also been investigated numerically for a thin-disk model, keeping two explicit space directions, r and ϕ (Moss 1996). These results confirm that the effect is weaker than for a classical parametric resonance and, furthermore, demonstrate that the resonance remains efficient for a larger mismatch between $2\omega_B$ and ω_{SP} than implied by the Mathieu equation. Since the equality $2\omega_B = \omega_{SP}$ is not an intrinsic property of galaxies, this finding is very helpful for practical applications. Nevertheless, parametric resonance can be expected to occur at most in a fraction of galaxies, where these quasi-independent frequencies satisfy the appropriate condition.

Other attempts to enhance the effect involve dynamo solutions that oscillate even in the lowest approximation in h/R (Hanasz et al 1991, Hanasz & Chiba 1993), i.e. in the local dynamo equation. Such oscillatory solutions arise only for unrealistically large dynamo numbers, requiring a downward revision of the turbulent magnetic diffusivity by a factor of 10 (Hanasz & Lesch 1993).

A further type of parametric resonance that can occur only in a distributed system such as a galactic dynamo has been suggested by Mestel & Subramanian (1991) and Subramanian & Mestel (1993). They assume that the dynamo wave is comoving with a spiral arm and that the dynamo efficiency is larger inside the arm than in the interarm space. The resulting growth rate of the magnetic field, captured by the arms, is larger than on average over the disk; the resonance condition is thus $\omega_B = \omega_{SP}$. The resulting (regular) magnetic field is connected with the spiral arms rather than with the disk as a whole; in particular, significant vertical magnetic fields might be expected. It is not completely clear whether or not this mechanism favors the bisymmetric mode over the axisymmetric one. The predictions of these models deserve a careful confrontation with observations.

6.4 *Contrast Structures*

Suppose that the seed magnetic field in one part G_1 of a thin galactic disk has approximately the form of a growing eigensolution, while in another part G_2 the seed magnetic configuration is close to the same eigensolution, but with the opposite sign. After some time, advection and diffusion will bring these regions of oppositely directed magnetic fields into contact. The neutral surface at the boundary of these regions will move due to diffusion and advection, so the final stage of magnetic field evolution will be determined by magnetic field propagated, say, from the part G_1. The motion of the neutral magnetic surface is governed by the competition between advection and diffusion of field from G_1 towards G_2 and vice versa. Provided the nonlinear stage of magnetic field evolution begins before the field attains the form of the leading eigensolution, these two can balance each other. This balance is possible only if the neutral surface is at some special location in the galactic disk; then a long-lived magnetic structure appears (Belyanin et al 1994). This type of nonlinear solution of the dynamo equations is known as a contrast structure. The thickness of the transition region between G_1 and G_2 is approximately the disk thickness, and its lifetime can even be as long as the diffusion time along the disk, $R^2/\eta_t \sim 10^{11}$ yr. Inside the contrast structure, annihilation of the oppositely directed magnetic fields is balanced by generation and advection, similar to a solitonís behavior in the nonlinear wave equation.

Contrast structures in purely axisymmetric disks are expected to be most often axisymmetric, because they are not affected by differential rotation. In the Milky Way, such axisymmetric contrasting structures can survive until today, and they may be identified with the reversals discussed in Section 3.8.2 (Poezd et al 1993). Contrast structures supported by nonaxisymmetric velocity and density distributions might explain the dominance of BSS in some galaxies (Moss et al 1993b; D Moss, in preparation; A Bikov et al, in preparation).

6.5 *The Influence of Magnetic Fields on the Galactic Disk*

Early ideas that magnetic fields might universally give rise directly to spiral structure have now generally been abandoned, because large-scale fields would need to have strengths $\gtrsim 10\ \mu$G to cause the velocity perturbations of about 20 km s^{-1} associated with spiral arms (e.g. Binney & Tremaine 1987, p. 394). This can be compared with typical values of a few μG (Section 3.9). (Note that the above estimate is valid for a gas density appropriate to the Milky Way, and that for gas-rich galaxies, which tend to have larger fields, it would also be increased.) However, Nelson (1988) suggested, from study of a simplified, two-dimensional model, that magnetic fields might have a significant effect on gas dynamics at large galactocentric distances, where the gas density is lower.

Nevertheless, there may be more subtle effects. Magnetic pressure contributes significantly to the overall pressure balance in the ISM (e.g. Bowyer et al 1995), perhaps affecting the vertical distribution of the gas [scale height, etc (see Boulares & Cox 1990)]. This in turn can affect the dynamo efficiency, establishing a feedback loop (Dobler et al 1996). Magnetic fields, of both large and small scale, could affect the formation and motion of clouds, for example, by increasing their effective cross-section. More directly, magnetic fields are believed to mediate the star-formation process, inter alia helping to solve the "angular momentum problem" (see Mestel 1985). A locally stronger magnetic field may bias the initial mass function to more massive stars (e.g. Mestel 1989), which, with their more rapid and violent evolution, could result in a more energetic ISM and perhaps an enhanced α-effect, thus providing another feedback loop (Mestel & Subramanian 1991, see also the discussion by Nozakura 1993).

Even the relatively modest azimuthal magnetic torques might affect the centrifugal balance sufficiently to give a significant angular momentum transport. An investigation by Rüdiger et al (1993) suggests that in the case of fields of quadrupolar parity, a substantially subsonic gas inflow will result, with only a small effect on the dynamo field structure.

7. MAGNETIC FIELDS IN HALOS

From observations of external galaxies, magnetic fields are inferred in halos of spiral galaxies to distances of at least 5 kpc and maybe even 10 kpc from the disk plane, significantly beyond a synchrotron scale height (cf Section 3.6). Recently, dynamo models have directed some attention to out-of-disk fields. Here we address the two logical possibilities (while noting that they are not mutually exclusive): that such fields are generated in situ in the halo or that they are generated in the disk and then transported into the halo.

7.1 *In Situ Generation*

Interpretations of observations in the Milky Way suggest the presence of turbulent velocities of at least 50 km s^{-1} in galactic halos, compared to estimates of 10 km s^{-1} in disks. If we assume a length scale of order 0.5 kpc and that halo angular velocities are comparable with those in the disk, we get canonical estimates of $\alpha \sim 3$ km s^{-1} and $\eta_t \sim 5 \times 10^{27}$ cm^2 s^{-1}, to be compared with $\eta_t \sim 10^{26}$ cm^2 s^{-1} in the disk. [See, e.g. the discussion in Poezd et al (1993). Note that Schultz et al (1994) adopt halo turbulent velocities that are much smaller than those in the disk: This may be a direct consequence of their turbulence model with $\alpha \propto \partial \langle v^2 \rangle / \partial z$.] Taking $L \sim 10$ kpc gives standard dynamo numbers $R_\alpha = \alpha L/\eta_t \sim 2$ and $R_\omega = \Omega_0 L^2/\eta_t \sim 200$. These are large enough for a dynamo to be excited (Ruzmaikin et al 1988a, Section VIII.1; Kahn & Brett 1993). Note that such a dynamo would operate in a quasi-spherical volume, rather than a thin disk, that standard spherical $\alpha\Omega$ dynamos preferentially excite fields of dipolar (A0) topology, and that these are then often the only stable solutions of the full nonlinear equations. In contrast, S0 fields are usually preferred in thin disks. This situation immediately suggests the interesting possibility of simultaneous excitation of dynamo fields of opposite parity types in the two subsystems (halo and disk) (see Sokoloff & Shukurov 1990). A priori, the possible existence of magnetic structures asymmetric with respect to the midplane, of neutral sheets, and of other nonstandard phenomena cannot be dismissed, as has been shown in some detail by Brandenburg et al (1992). Growth times in the halo are substantially longer than in the disk, and the halo field may still be in a transient state after a Hubble time. Detailed integrations show that, starting from a seed field of mixed parity, the overall field is initially dominated by S0 topology and concentrated in the disk. This phase can persist for order a Hubble time, but the final configuration is usually of A0 type, and may even be oscillatory. Given the long-lived transient phase with mixed parity fields present, observers today may be presented not with the eventual stable configuration, but rather an intermediate state of quite arbitrary geometry. Note that magnetic fields in the disk and halo of M51 are oppositely directed (EM Berkhuijsen et al, in preparation): This argues for in situ generation. More satisfactory halo models will need better data than is currently available on the dependence of the angular velocity in the halo on z, but these results seem qualitatively robust. To summarize, in some circumstances, dynamo theory may not be able to make detailed predictions about field geometries in specific galaxies.

A largely unexplored possibility is that some sort of Ponomarenko ("screw") dynamo (e.g. Ruzmaikin et al 1988c) might operate in the halo, if large-scale quasi-radial outflows ("winds") are twisted into helical form by the galactic rotation. Such dynamos excite nonaxisymmetric fields. If we take a simple model investigated by Ruzmaikin et al and use their definitions, then a wind velocity of 100 km s^{-1} and a typical galactic angular velocity gives a magnetic Reynolds number R_{M} large enough for the dynamo to work. Naively, the minimum e-folding time would be about 10^9 yr, but this increases as $R_{\mathrm{M}}^{1/2}$ for larger R_{M}, because the screw dynamo is "slow." These estimates suggest that the mechanism might be of marginal importance in halos, but real galaxy velocity fields are likely to be less efficient dynamos than the idealized forms considered by Ruzmaikin et al. We note in passing that Spencer & Cram (1992) have discussed models of field amplification in which meridional flows ("winds") appear to play a central role. However, they solve the problem purely in the disk region; moreover, their solutions do not represent dynamo generation but rather local compression of field and hence the relevance to field generation processes in galaxies is unclear.

7.2 *Transport Out of the Disk*

Evidence for the existence of galactic winds, with speeds U of hundreds of kilometers per second, is seen in some galactic halos, notably NGC 4631 and M82 (Section 3.6), implying turbulent magnetic Reynolds numbers $R_{\mathrm{M}} = UL/\eta_{\mathrm{t}}$ of order 100. Strong field freezing will thus occur and, since the wind advection time (L/U) is much shorter than the dynamo growth time, the wind will markedly affect the near-disk fields. For halo magnetic fields that are strong enough for their energy density to be comparable with the kinetic energy density of the wind, the dynamical effect of the field on the wind needs also to be considered, as in the analogous stellar wind problem, although such studies are in their infancy (see, e.g. Breitschwerdt et al 1993). With typical values of $\overline{B} \sim 1\ \mu$G and $\rho \sim 10^{-27}$ g cm^{-3}, a kinematic treatment will be valid for winds of speed in excess of about 100 km s^{-1}. This outward advection of magnetic field may be partially offset near the disk by turbulent diamagnetism, which gives an effective velocity of field transport of a few kilometers per second towards the disk (if the diffusivity increases outwards), but for the larger wind velocities wind advection will dominate.

These problems were addressed in detail in the weak field approximation by Brandenburg et al (1993) and, with a rather different emphasis, by Elstner et al (1995). Brandenburg et al demonstrated that winds of plausible strength and geometry could drag out poloidal field lines almost radially into the halo and also move toroidal flux away from the disk. Moreover, by using realistic disk rotation curves for well-observed systems, and choosing appropriate (predominantly radial) wind velocity fields, solutions resembling the rather different halo fields of NGC 891 and NGC 4631, for example, can be generated without any careful "tuning." However, the halo field strengths are somewhat too low, and the field far from the disk makes too small an angle with the disk plane to provide a completely satisfactory model for NGC 4631.

However, a simple wind structure that is axisymmetric and varies smoothly with spherical polar angle θ may be inadequate; real galactic winds probably have considerably more structure, with streamers causing both azimuthal and latitudinal shear. Elstner et al (1995) presented a preliminary axisymmetric model (without azimuthal shear), with a wind velocity perpendicular to the disk and varying sinusoidally with distance from the rotation axis. They show that a short wavelength modulation (1.5 kpc) can markedly affect the field geometry and that odd parity "dipolar" fields may even be stable for some parameter values. Further work with a more realistic model is needed to elucidate the relation between such calculations and real galactic flows.

A priori, a quasi-radial or z-wise shearing flow is unlikely to produce a halo field that is predominantly parallel to the disk plane, although such fields are observed in some "edge-on" galaxies (e.g. NGC 253). A problem concerning mechanisms that advect field from the disk is that the gas dragging it into the halo belongs to the rarefied, hot phase of the ISM, where the field strength is typically about 0.1 μG (Kahn & Brett 1993), and so additional amplification outside the disk is necessary. Shearing by localized outflows can only amplify the vertical component. Brandenburg et al (1995b) pointed out that galactic fountain flows, especially in active starburst galaxies, may have the correct topology (upflows connected in horizontal cross section and isolated downdrafts) for a topological pumping mechanism to produce a strong mean horizontal field high in the halo. With realistic parameters, they showed that this mechanism might produce horizontal fields at a height of several kpc above the disk that were of comparable strength to those in the disk. As yet, this mechanism has not been included in a global dynamo calculation. Magnetic buoyancy in the disk may also play a role in moving field into the halo, but this mechanism has not yet been adequately quantified.

In general, an outflow that is symmetric both azimuthally and with respect to the disk plane will preserve in the halo any global parity or symmetry properties of the disk field. Clearly, if the outflow lacks such symmetries (as seems quite possible, a priori), then this connection between disk and advected halo fields will be lost.

8. MAGNETIC FIELD MODELS

Only the dynamo theory for galactic magnetic fields has been developed sufficiently to provide models of magnetic fields in particular galaxies that can be confronted with observations. Therefore, our discussion below is inevitably more detailed in the case of the dynamo theory. Wherever possible, we also mention inferences from the primordial field theory, ignoring the conceptual difficulties discussed in Section 5.

8.1 *The Parity*

It is generally believed that galactic magnetic fields have an even parity. As discussed in Section 3.8.1, the field parity near the Sun most plausibly is even. There is some evidence for an even symmetry of the regular magnetic field in the edge-on galaxy NGC 253 (Beck et al 1994b). In mildly inclined galaxies, Faraday rotation measures for even and odd fields of equal strengths would differ only by a factor of 2 (Krause et al 1989a), which makes it difficult to distinguish between the two configurations. All conventional dynamo models indicate that the quadrupole parity must be dominant in galactic disks.

A uniform primordial magnetic field trapped by a protogalaxy, with arbitrary inclination to the rotation axis, produces an S1 component from the action of the radial gradient of the angular velocity Ω on $\overline{B}_r$ (which is then even in z and nonaxisymmetric) and an A0 field from the action of $\partial\Omega/\partial z$ on $\overline{B}_z$ (which is odd and axisymmetric). Since $|\partial\Omega/\partial r| \gg |\partial\Omega/\partial z|$, at least during late stages of galactic evolution, the S1 field will become tightly wound and quickly decay because of reconnection. The resulting symmetry of a fossil field is then A0 or, possibly, a superposition of S1 and A0 configurations.

8.2 *Large-Scale Azimuthal Patterns*

Even the simplest asymptotic kinematic models of the mean-field dynamo in a thin disk have the promising property that only $m = 0$ modes are excited in those galaxies where the field is observed to be axisymmetric (M31 and IC 342), whereas the $m = 1$ mode is also excited (if it is not the fastest growing) in the galaxies with a dominant bisymmetric or mixed magnetic structure (e.g. M33, M51, and M81) (see Krasheninnikova et al 1990 and Ruzmaikin et al 1988a, b for reviews). The thinner the disk, the more readily the $m = 1$ mode can be maintained (Ruzmaikin et al 1988a Section VII.8; Moss & Brandenburg 1992). Weaker differential rotation is favorable for bisymmetric field generation. Even higher azimuthal modes might survive in galactic disks, e.g. the $m = 2$ mode (Starchenko & Shukurov 1989, Vallée 1992), which has a four-armed spiral pattern. An admixture of the $m = 2$ mode may arise as a distortion of an $m = 0$ field by a two-armed spiral pattern. An $m = 2$ mode superimposed on a $m = 0$ mode of similar amplitude would produce a pattern of the type possibly observed in NGC 6946 (Section 3.4).

The dominance of a bisymmetric field requires additional physical mechanisms to be invoked, as discussed in Sections 6.2 and 6.3; it seems, however, that these mechanisms are efficient only under certain conditions that can occur only in rare cases. Therefore, a general prediction of the galactic dynamo theory is that normally either axisymmetric magnetic structures (in the galaxies where only the $m = 0$ mode is excited) or a superposition of $m = 0$ and $m = 1$ modes (where both are maintained) should be found. The former situation is encountered in M31 and IC 342, whereas the latter is represented by M51. An admixture of even higher m-modes cannot be excluded, as possibly seen in NGC 6946. Only in those galaxies that provide a suitable environment for a fine tuning of the dynamo (Sections 6.2 and 6.3) should a dominant bisymmetric field be expected, as exemplified by M81. An important factor in maintaining BSS seems to be tidal interaction with a companion galaxy (Section 6.2).

In general, this picture is reasonably consistent with observations that most galactic fields do not have simple structures. Note that a superposition of even two or three azimuthal modes may give an appearance of a rather irregular large-scale magnetic field. So far, observations of only a few galaxies have been interpreted with allowance for such superpositions. We expect that new observations and analyses will extend the list of galaxies hosting mixed spiral structure.

A primordial magnetic field twisted by differential rotation is strongly dominated by the S1 or A0 modes (Section 3). An S0 field can arise only if it is assumed that the magnetic field had a very strong inhomogeneity across the protogalaxy (Sofue et al 1986), which appears to be a rather artificial requirement.

8.3 *Spiral Field Lines and Pitch Angles*

Plane-parallel magnetic fields with a dominant azimuthal component $\overline{B}_\phi$ prevail in spiral galaxies (see Section 3). This can be easily understood because differential rotation is strong in spiral galaxies (whether or not dynamos operate).

Dynamo theory predicts (Baryshnikova et al 1987), and observations of external galaxies show (Section 3.3), that the regular magnetic field must have the shape of a spiral, whether or not it is axisymmetric. Unlike spiral magnetic fields, a circular field produced within the galaxy (i.e. not supported by external currents) can not be maintained by any velocity field against turbulent magnetic diffusion. On average, the field must be a trailing spiral because differential rotation is important in producing $\overline{B}_\phi$ from $\overline{B}_r$. Of course, this does not preclude local deviations from a trailing spiral pattern, as observed, e.g. in M51 (Figure 1).

The pitch angle of the magnetic field p is a readily observable parameter sensitive to details of the mechanism of magnetic field generation. Hence the magnetic pitch angle is an important diagnostic tool for theories of galactic magnetic fields. Magnetic pitch angles in spiral galaxies are observed to lie in the range $p = -(10°–35°)$ (Figure 8). Galactic dynamo models even without spiral arms predict that p is close to these values (Krasheninnikova et al 1989, Donner & Brandenburg 1990, Elstner et al 1992, Panesar & Nelson 1992). A simple estimate for a kinematic dynamo in a thin axisymmetric disk gives (Krasheninnikova et al 1989)

$$p = \arctan(\overline{B}_r/\overline{B}_\phi) \simeq -(R_\alpha/R_\omega)^{1/2}, \tag{11}$$

and $p \simeq -20°$ under typical conditions. Note that $\overline{B}_r$ and $\overline{B}_\phi$ have opposite signs because of the action of differential rotation, and so p is negative (a trailing spiral). Asymptotic kinematic dynamo models using observed rotation curves have been applied to particular galaxies (see Ruzmaikin et al 1988a); the results agree fairly well with observations. Schultz et al (1994) discuss the dependence of the pitch angle on other parameters of turbulence.

It follows from Equations (3), (6), and (11) that $p \simeq -l/h$ (with l as the turbulent scale). The pitch angle $|p|$ thus decreases with r when $l =$ const, and h increases with r. This behavior is also typical of dynamos in a flat disk (Elstner et al 1992, Panesar & Nelson 1992) and is observed in spiral galaxies, as shown in Figure 8. The only exceptions are M81 and possibly also M33, both of which are candidates for bisymmetric magnetic structures due to interaction with companion galaxies (see Section 8.2).

As discussed in Section 3, magnetic pitch angles in spiral galaxies are surprisingly close to those of optical spiral arms, p_{SA}. Taken literally, Equation (11) implies that the equality $p \approx p_{SA}$ is a mere quantitative coincidence because the two depend on different physical parameters. Numerical simulations of the $\alpha\omega$-dynamo with spiral shock waves (Panesar & Nelson 1992) show that p is quite insensitive to the presence of the shocks. The interplay between the magnetic and spiral patterns is far from being completely understood (Section 8.4) and, possibly, there are deeper physical reasons for the observed correspondence of the pitch angles.

Concerning the primordial field theory, a straightforward idea is that the pitch angle of a magnetic field frozen into a differentially rotating disk is a decreasing function of time and, after N revolutions (with $N \simeq 30$ for the Solar vicinity in the Milky Way), we have $p \simeq -N^{-1}$ rad$\approx -2°$, so that $|p| \ll |p_{SA}|$. Furthermore, $|p|$ grows with r insofar as angular velocity decreases with r—a trend opposite to that observed.

We note that the ASS fields observed in the spiral galaxies M31 and IC 342 and the magnetic spiral arms in NGC 6946 are directed inwards. For the edge-on galaxy NGC 253, a similar conclusion follows if one assumes that the magnetic field is also aligned with the spiral arms. As the direction of a dynamo-generated field is determined by that of the initial field, this dominance, if it were to be confirmed by better statistics, might clarify the nature of the seed field. For example, it could indicate the importance of battery effects (relying on galactic rotation). Within the framework of the primordial field theory, such a dominance would imply a hardly plausible correlation between the directions of the intergalactic field and the sense of galactic rotation.

8.4 *Spiral Arms and Magnetic Fields*

A standard understanding of the interaction between spiral arms and large-scale magnetic fields is largely based on the idea that the spiral shock compresses the magnetic field and aligns it with the spiral arm (Roberts & Yuan 1970). This leads to a clear prediction that the regular magnetic field must be stronger at the inner edges of the arms and that there p is closer to p_{SA} than in the interarm space. This picture was believed to be supported by the observation that the

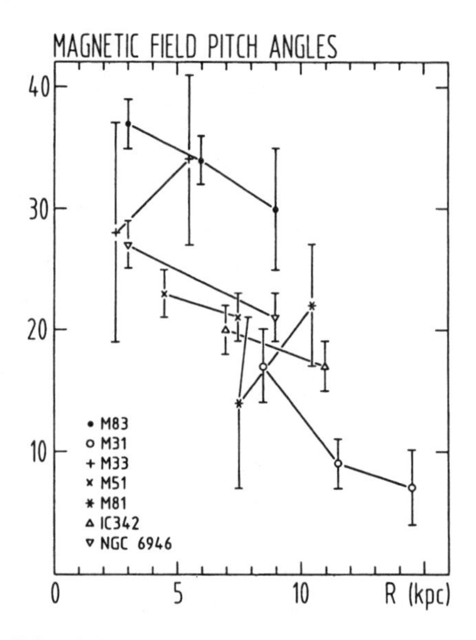

Figure 8 Observed radial variation of the magnetic pitch angle in the galaxy's plane averaged over azimuthal angle for several nearby spiral galaxies. (From Beck 1993.)

regular magnetic field in the Milky Way is enhanced within the local arm and that magnetic fields observed in nearby galaxies are well aligned with the spiral arms. It is, however, noteworthy that $p \neq p_{SA}$ near the Sun (Section 3.8), whereas the general alignment $p \approx p_{SA}$ can arise from dynamo action without any shock compression (Section 8.3).

However, recent observations of most nearby galaxies indicate that the regular magnetic fields are observed to be stronger between the arms, whereas the total field strength is stronger in the arms (Sections 3.3 and 3.4). The implication is straightforward: The action of the spiral pattern on galactic magnetic fields is not as direct and simple as passive compression (at least in these galaxies). (We note also that it is difficult to understand how the primordial theory, which gives only a passive role to the magnetic field, can explain its enhancement between the arms. Possibly, streaming motions induced by spiral arms could help, but this possibility has not been studied.)

The compression of magnetic field in spiral arms becomes much weaker if a large fraction of the interstellar medium is filled with hot gas, which prevents large-scale shocks from occurring. Star formation in spiral arms must then be triggered by, e.g. more frequent collisions of gas clouds (Roberts & Hausman 1984). The nearby spiral galaxies M51 and M81 exhibit strong density waves. In M51 prominent dust lanes, enhanced CO (García-Burillo et al 1993), and radio continuum emission at the inner edges of the optical spiral arms are indicators of narrow compression regions. In M81, however, the compression regions are much broader (Kaufman et al 1989) and can best be explained by the "cloudy" density-wave model of Roberts & Hausman (1984).

Beck (1991) has proposed a qualitative model to explain enhanced field tangling in the arms. He assumes that the field lines are trapped by gas clouds. As the clouds enter a spiral arm, they are decelerated, and their number density, collision rate, and turbulent velocity increase, which gives rise to field tangling and enhanced total field. However, the "magnetic arms" observed between the optical arms of NGC 6946 (Section 3.4) cannot be understood by this model and need a global mechanism such as the dynamo. How to include spiral arms adequately into the theory of galactic magnetic fields remains an important unresolved problem.

8.5 *Dynamo Models for Individual Galaxies*

The predictions of $\alpha^2\Omega$-dynamo models are roughly consistent with the large-scale field structures observed in spiral galaxies. In this section we discuss briefly a few individual galaxies for which detailed dynamo models have been developed and/or new problems have arisen.

Kinematic dynamo models for the Milky Way (see Ruzmaikin et al 1988a Sections VII.7 and VII.9) indicate that the axisymmetric mode is dominant, even though the $m = 1$ mode can also be maintained if the half-thickness of the ionized disk is within a narrow range (500–700 pc near the Sun, but these values are model dependent). In view of the uncertainty concerning the generation of bisymmetric fields in spiral galaxies, an ASS seems more likely but the presence of the BSS cannot be excluded; a superposition of the two modes (MSS) is also possible.

The presence of reversals (Section 3.8) is often considered as an indication of a bisymmetric global structure of magnetic field in the Milky Way. We again stress that this is not true. The possibility of such reversals in an axisymmetric spiral field was demonstrated in a dynamo model for the Milky Way by Poezd et al (1993). Even this simplified model exhibits a reasonable agreement with observations, yielding two or three reversals whose positions along the radius roughly agree with those observed. According to Poezd et al, the reversals represent transient nonlinear magnetic structures (cf Section 6.4).

Both dynamo theory and observations agree that the large-scale magnetic field in M31 is axisymmetric. A notable feature of this galaxy is that both the gas and the large-scale magnetic field are concentrated within a narrow ring of about 10 kpc radius (Section 3.2). The explanation provided by the dynamo models reviewed by Ruzmaikin et al (1988a, Section VII.7) relies on the rotation curve having a pronounced double-peaked shape (Deharveng & Pellet 1975). However, recent interpretations (with better allowance for radial motions) have resulted in a much less pronounced minimum in the rotation curve (Kent 1989, Braun 1991). Even though the new rotation curve has not yet been incorporated into dynamo models, it can be guessed that the kinematic dynamo modes will no longer show any concentration into a ring. Thus, the ring-like structure of magnetic field in M31 probably arises during the nonlinear stage of the dynamo and is associated with a similar distribution of the interstellar gas (Dame et al 1993).

M81 is the only nearby galaxy for which a dominant bisymmetric magnetic field is firmly indicated by observations (Section 3). Apart from kinematic asymptotic dynamo models (Krasheninnikova et al 1989, Starchenko & Shukurov 1989), a three-dimensional, nonlinear dynamo model has been developed for M81 based on the velocity field inferred from simulations of the interaction of this galaxy with its companion NGC 3077 (Moss et al 1993a). The interaction has been shown to result in a persistent bisymmetric structure. To reach a final conclusion about the nature of the magnetic field in M81, these numerical simulations must be extended to include better spatial resolution and a fully time-dependent representation of the velocity field. There is no minimum of polarized intensity observed near the probable location of the magnetic neutral line in M81 (Figure 2). Its absence probably indicates that the reversal in the BSS structure is rather abrupt, reminiscent of a contrast structure (see Section 6.2).

9. LAST WORDS

We have attempted to draw together various strands contributing to our current understanding of galactic magnetism. We feel that neither dynamo nor fossil theory is at present in a satisfactory state. Nonetheless, we believe that, while the problems with the primordial theory are quite fundamental, ways of resolving the difficulties of the dynamo theory exist, in principle at least. A primordial field may nevertheless be important; for example, it can provide a seed field for a dynamo (see Section 5).

We note the following. Axisymmetric spiral structures and more complicated field structures arise naturally from dynamo models. Pitch angles lie in the correct range. Dynamo models give generically plausible large-scale spatial field structures, which are in some cases quite realistic, and which readily allow the detailed modeling of specific galaxies. Finally, on general grounds, field strengths of order the equipartition value, as observed, seem explicable. These points support our view that a coherent explanation of galactic magnetism will only be achieved via the further development of some form of dynamo theory.

It is now possible to include realistic models of the ISM, including detailed data on the spatial distributions of turbulent velocity and scale, the vertical gradient in the overall galactic rotation, galactic fountains, etc, in dynamo models; however, this remains to be done. A detailed comparison of theory with observations is becoming increasingly, both because the theory is beginning to give results that are sufficiently generic, reliable, and detailed, and because observations have reached the stage where they can seriously constrain many aspects of the theory. Reliable and high-resolution information about the complex magnetic structures found in the disks and halos of spiral galaxies is needed, together with an improved theory of depolarization mechanisms.

ACKNOWLEDGMENTS

The authors acknowledge the hospitality of the Observatory of Helsinki University, where the work was initiated, and of Nordita (Copenhagen), where it was finished. We are grateful to EM Berkhuijsen, who provided the data used in Figure 5 prior to publication. Partial financial support from the NATO grants CRG921273 and CRG1530959 is acknowledged. AS thanks the Mathematics Department of Manchester University and the Max-Planck-Institut für Radioastronomie for their hospitality during his work on the paper. AS and DS acknowledge partial financial support from grants 93-02-3638 and 95-02-03724 from the Russian Foundation for Basic Research and MNP000/300 from the International Science Foundation.

Literature Cited

Agafonov GI, Ruzmaikin AA, Sokoloff DD. 1988. *Astron. Zh.* 65: 523–28

Allen RJ, Sukumar S. 1990. In *The Interstellar Medium in External Galaxies. Contributed Papers*, ed. DJ Hollenbach, HA Thronson, pp. 263–67. NASA Conf. Publ. 3084

Andreassian RR. 1980. *Astrofizika* 16:707–13

Andreassian RR. 1982. *Astrofizika* 18:255–62

Armstrong JW, Rickett BJ, Spangler SR. 1995. *Ap. J.* 443:209–21

Baierlein R. 1978. *MNRAS* 184:843–70

Balbus SA, Hawley JF. 1992. *Ap. J.* 392:662–66

Baryshnikova Y, Ruzmaikin A, Sokoloff DD, Shukurov A. 1987. *Astron. Astrophys.* 177:27–41

Beck R. 1982. *Astron. Astrophys.* 106:121–32

Beck R. 1991. *Astron. Astrophys.* 251:15–26

Beck R. 1993. In *Cosmic Dynamo*, ed. F Krause, K-H Rädler, G Rüdiger, pp. 283–97. Dordrecht: Kluwer

Beck R, Carilli CL, Holdaway MA, Klein U. 1994b. *Astron. Astrophys.* 292:409–24

Beck R, Hoernes P. 1996. *Nature* 379:47–49

Beck R, Klein U, Wielebinski R. 1987. *Astron. Astrophys.* 186:95–98

Beck R, Loiseau N, Hummel E, Berkhuisen EM, Gräve R, Wielebinski R. 1989. *Astron. Astrophys.* 222:58–68

Beck R, Poezd AD, Shukurov A, Sokoloff DD. 1994a. *Astron. Astrophys.* 289:94–100

Belyanin M, Sokoloff DD, Shukurov A. 1993. *Geophys. Astrophys. Fluid Dyn.* 68:237–61

Belyanin MP, Sokoloff DD, Shukurov AM. 1994. *Russ. J. Math. Phys.* 2:149–74

Berkhuijsen EM, Bajaja E, Beck R. 1993. *Astron. Astrophys.* 279:359–75

Berkhuijsen EM, Golla G, Beck R. 1991. In *The Interstellar Disk-Halo Connection in Galaxies. Proc. IAU Symp. 144*, ed. H Bloemen, pp. 233–36. Dordrecht: Kluwer

Beuermann K, Kanbach G, Berkhuijsen EM. 1985. *Astron. Astrophys.* 153:17–34

Biermann L. 1950. *Z. Naturforsch.* 5a:65–71

Binney J, Tremaine S. 1987. *Galactic Dynamics*. Princeton: Princeton Univ. Press

Bloemen JB, Strong AW, Blitz L, Cohen RS, Dame TM, et al. 1986. *Astron. Astrophys.* 154:25–41

Boulares A, Cox DP. 1990. *Ap. J.* 365:544–58

Bowyer S, Lieu R, Sidher SD, Lampton M, Knude J. 1995. *Nature* 375:212–14

Brandenburg A, Donner KJ, Moss D, Shukurov A, Sokoloff DD, Tuominen I. 1992. *Astron. Astrophys.* 259:453–61

Brandenburg A, Donner KJ, Moss D, Shukurov A, Sokoloff DD, Tuominen I. 1993. *Astron. Astrophys.* 271:36–50

Brandenburg A, Moss D, Shukurov A. 1995b. *MNRAS* 276:651–62

Brandenburg A, Nordlund Å, Stein RF, Torkelsson U. 1995a. *Ap. J.* 446:741–54

Brandenburg A, Tuominen I, Krause F. 1990. *Geophys. Astrophys. Fluid Dyn.* 50:95–112

Brandenburg A, Zweibel EG. 1995. *Ap. J.* 448:734–41

Braun R. 1991. *Ap. J.* 372:54–66

Breitschwerdt D, Völk HJ, Ptuskin V, Zirakashvili V. 1993. In *The Cosmic Dynamo*, ed. F Krause, K-H Rädler, G Rüdiger, pp. 415–19. Dordrecht: Kluwer

Briggs FH, Wolfe AM, Liszt HS, Davis MM, Turner KL. 1989. *Ap. J.* 341:650–57

Buczilowski UR, Beck R. 1991. *Astron. Astrophys.* 241:47–56

Burn BJ. 1966. *MNRAS* 133:67–83

Camenzind M, Lesch H. 1994. *Astron. Astrophys.* 284:411–23

Carilli CL, Holdaway MA, Ho PTP, DePree CG. 1992. *Ap. J. Lett.* 399:L59–62

Carvalho JC. 1992. *Astron. Astrophys.* 261:348–52

Cattaneo F, Vainshtein SI. 1991. *Ap. J. Lett.* 376:L21–24

Chakrabarti SK, Rosner R, Vainshtein SI. 1994. *Nature* 368:434–36

Cheng B, Schramm DN, Truran JW. 1994. *Phys. Rev. D* 49:5006–18

Chi X, Wolfendale AW. 1993. *Nature* 362:610–11

Chiba M, Lesch H. 1994. *Astron. Astrophys.* 284:731–48

Chiba M, Tosa M. 1990. *MNRAS* 244:714–26

Clegg AW, Cordes JM, Simonetti JH, Kulkarni SR. 1992. *Ap. J.* 386:143–57

Dahlem M, Aalto S, Klein U, Booth R, Mebold U, Wielebinski R, Lesch H. 1990. *Astron. Astrophys.* 240:237–46

Dahlem M, Lisenfeld U, Golla G. 1995. *Ap. J.* 444:119–28

Dame TM, Koper E, Israel FP, Thaddeus P. 1993. *Ap. J.* 418:730–42

Davidson JA, Schleuning D, Dotson JL, Dowell CD, Hildebrand RH. 1995. In *Airborne Astron. Symp. on the Galactic Ecosystem*, ed. MR Haas, JA Davidson, EF Erickson, pp. 225–34. San Francisco: Astron. Soc. Pac. Conf. Ser. 73

Deharveng JM, Pellet A. 1975. *Astron. Astrophys.* 38:15–28

Dettmar R-J. 1990. *Astron. Astrophys.* 232:L15–18

Dickey JM, Brinks E, Puche D. 1992. *Ap. J.* 385:501–11

Dobler W, Poezd A, Shukurov A. 1996. *Astron.*

Astrophys. In press
Donner KJ, Brandenburg A. 1990. *Astron. Astrophys.* 240:289–97
Dumke M, Krause M, Wielebinski R, Klein U. 1995. *Astron. Astrophys.* 302:691–703
Ehle M. 1995. Eigenschaften des diffusen interstelleren Mediums in den nahen Galaxien M51, M83, und NGC 1566. PhD Thesis. Univ. Bonn
Ehle M, Beck R. 1993. *Astron. Astrophys.* 273:45–64
Ehle M, Beck R, Haynes RF, Vogler A, Pietsch W, Elmouttie M, Ryder S. 1996. *Astron. Astrophys.* 306:73–85
Elmouttie M, Haynes RF, Jones KL, Ehle M, Beck R, Wielebinski R. 1995. *MNRAS* 275:L53–59
Elstner D, Golla G, Rüdiger G, Wielebinski R. 1995. *Astron. Astrophys.* 297:77–82
Elstner D, Meinel R, Beck R. 1992. *Astron. Astrophys. Suppl.* 94:587–600
Elstner D, Rüdiger G, Schultz M. 1996. *Astron. Astrophys.* 306:740–46
Enqvist K, Rez AI, Semikoz V. 1995. *Nucl. Phys. B* 436:49–64
Enqvist K, Semikoz V, Shukurov A, Sokoloff D. 1993. *Phys. Rev. D* 48:4557–61
Falgarone E, Lis DC, Phillips TG, Pouquet A, Porter DH, Woodward PR. 1994. *Ap. J.* 436:728–40
Ferrière K. 1993. *Ap. J.* 404:162–84
Field GB. 1996. In *The Physics of the Interstellar Medium and Intergalactic Medium*, ed. A Ferrara et al, pp. 1–2. *Astron. Soc. Pac. Conf. Ser.* Vol. 80
Fitt AJ, Alexander P. 1993. *MNRAS* 261:445–52
Frisch U, Pouquet A, Léorat J, Mazure A. 1975. *J. Fluid Mech.* 68:769–78
Garcia-Burillo S, Guélin M, Cernicharo J. 1993. *Astron. Astrophys.* 274:123–47
Gardner FF, Morris D, Whiteoak JB. 1969. *Aust. J. Phys.* 22:813–19
Gavazzi G, Boselli A, Kennicutt R. 1991. *Astron. J.* 101:1207–30
Golla G, Beck R. 1990. In *The Interstellar Disk-Halo Connection in Galaxies. Poster Proc. IAU 144*, ed. H Bloemen, pp. 47–48. Leiden Obs.
Golla G, Hummel E. 1994. *Astron. Astrophys.* 284:777–92
Grasso D, Rubinstein HR. 1995. *Astroparticle Phys.* 3:95–102
Gräve R, Beck R. 1988. *Astron. Astrophys.* 192:66–76
Gräve R, Klein U, Wielebinski R. 1990. *Astron. Astrophys.* 238:39–49
Gruzinov AV, Diamond PH. 1994. *Phys. Rev. Lett.* 72:1651–53
Han JL, Qiao GJ. 1994. *Astron. Astrophys.* 238:759–72
Hanasz M, Chiba M. 1993. *MNRAS* 266:545–68
Hanasz M, Lesch H. 1993. *Astron. Astrophys.* 278:561–68
Hanasz M, Lesch H, Krause M. 1991. *Astron. Astrophys.* 243:381–85
Harnett JI, Haynes RF, Klein U, Wielebinski R. 1989. *Astron. Astrophys.* 216:39–43
Harnett JI, Haynes RF, Wielebinski R, Klein U. 1990. *Proc. Astron. Soc. Aust.* 8:257–60
Harrison ER. 1970. *MNRAS* 147:279–86
Haynes RF, Klein U, Wielebinski R, Murray JD. 1986. *Astron. Astrophys.* 159:22–32
Heiles C. 1976. *Annu. Rev. Astron. Astrophys.* 14:1–22
Heiles C. 1989. *Ap. J.* 336:808–21
Heiles C. 1996. In *Polarimetry of the Interstellar Medium*, ed. W Roberge, D Whittet. San Francisco: Astron. Soc. Pac. In press
Heiles C, Goodman AA, McKee CF, Zweibel EG. 1993. In *Protostars and Planets III*, ed. EH Levy, JI Lunine, pp. 279–326. Tuscon: Univ. Ariz. Press
Higdon JC. 1984. *Ap. J.* 285:109–23
Hildebrand RH, Davidson JA. 1994. In *The Nuclei of Normal Galaxies*, ed. R Genzel, AI Harris, pp. 199–203. Dordrecht: Kluwer
Hildebrand RH, Gonatas DP, Platt SR, Wu XD, Davidson JA, et al. 1990. *Ap. J.* 362:114–19
Horellou C, Beck R, Berkhuijsen EM, Krause M, Klein U. 1992. *Astron. Astrophys.* 265:417–28
Horellou C, Beck R, Klein U. 1990. In *Galactic and Intergalactic Magnetic Fields*, ed. R Beck, PP Kronberg, R Wielebinski, pp. 211–12. Dordrecht: Kluwer
Hough JH. 1996. In *Polarimetry of the Interstellar Medium*, ed. W Roberge, D Whittet. San Francisco: Astron. Soc. Pac. In press
Hoyle F. 1969. *Nature* 223:936
Hummel E. 1986. *Astron. Astrophys.* 160:L4–L6
Hummel E. 1990. In *Windows on Galaxies*, ed. G Fabbiano, JS Gallagher, A Renzini, pp. 141–55. Dordrecht: Kluwer
Hummel E, Beck R. 1995. *Astron. Astrophys.* 303:691–704
Hummel E, Beck R, Dahlem M. 1991a. *Astron. Astrophys.* 248:23–29
Hummel E, Beck R, Dettmar R-J. 1991b. *Astron. Astrophys. Suppl.* 87:309–17
Hummel E, Davies RD, Wolstencroft RD, van der Hulst JM, Pedlar A. 1988b. *Astron. Astrophys.* 199:91–104
Hummel E, Krause M, Lesch H. 1989. *Astron. Astrophys.* 211:266–74
Hummel E, Lesch H, Wielebinski R, Schlickeiser R. 1988a. *Astron. Astrophys.* 197:L29–31
Kahn FD, Brett L. 1993. *MNRAS* 263:37–48
Kaisig M, Rüdiger G, Yorke HW. 1993. *Astron.*

Astrophys. 274:757–64

Kaufman M, Bash FN, Hine B, Rots AH, Elmegreen DM, Hodge PW. 1989. *Ap. J.* 345:674–96

Kazantsev AP. 1968. *Sov. Phys. JETP* 26:1031–34

Kent SM. 1989. *Astron. J.* 97:1614–21

Kitchatinov LL, Pipin VV, Rüdiger G. 1994. *Astron. Nachr.* 315:157–70

Kitchatinov LL, Rüdiger G. 1992. *Astron. Astrophys.* 260:494–98

Klein U, Haynes RF, Wielebinski R, Meinert D. 1993. *Astron. Astrophys.* 271:402–12

Klein U, Wielebinski R, Morsi HW. 1988. *Astron. Astrophys.* 190:41–46

Ko CM, Parker EN. 1989. *Ap. J.* 341:828–31

Kraichnan RH. 1965. *Phys. Fluids* 8:1385–86

Krasheninnikova Y, Ruzmaikin A, Sokoloff D, Shukurov A. 1989. *Astron. Astrophys.* 213:19–28

Krasheninnikova Y, Ruzmaikin A, Sokoloff D, Shukurov A. 1990. *Geophys. Astrophys. Fluid Dyn.* 50:131–46

Krause F. 1967. *Eine Lösung des Dynamoproblems auf der Grundlage einer linearen Theorie der magnetohydrodynamischen Turbulenz.* Univ. Jena: Habilitationsschrift

Krause F, Rädler K-H. 1980. *Mean-Field Electrodynamics and Dynamo Theory*. Berlin: Akademie-Verlag; Oxford: Pergamon

Krause F, Rüdiger G. 1975. *Sol. Phys.* 42:107–19

Krause M. 1990. In *Galactic and Intergalactic Magnetic Fields*, ed. R Beck, PP Kronberg, R Wielebinski, pp. 187–96. Dordrecht: Kluwer

Krause M. 1993. In *The Cosmic Dynamo*, ed. F Krause, K-H Rädler, G Rüdiger, pp. 305–10. Dordrecht: Kluwer

Krause M, Beck R, Hummel E. 1989b. *Astron. Astrophys.* 217:17–30

Krause M, Hummel E, Beck R. 1989a. *Astron. Astrophys.* 217:4–16

Kronberg PP. 1994. *Rep. Prog. Phys.* 57:325–82

Kronberg PP, Perry JJ. 1982. *Ap. J.* 263:518–32

Kronberg PP, Perry JJ, Zukowski ELH. 1992. *Ap. J.* 387:528–35

Kulsrud R. 1986. In *Plasma Astrophysics*, ed. TD Guyenne, LM Zeleny, pp. 531–37. Paris: ESA Publ. SP-251

Kulsrud RM, Anderson SW. 1992. *Ap. J.* 396:606–30

Kuzanyan KM, Sokoloff D. 1993. *Astrophys. Space Sci.* 208:245–52

Landau LD, Lifshitz EM. 1969. *Mechanics.* Oxford: Pergamon. 2nd Ed.

LeBlanc VG, Kerr D, Wainwright J. 1995. *Class. Quantum Grav.* 12:513–41

Longair MS. 1994. *High Energy Astrophysics*, Cambridge: Cambridge Univ. Press, 2:292–96

Lyne AG, Smith FG. 1989. *MNRAS* 237:533–41

McIvor I. 1977. *MNRAS* 178:85–100

McKee CF, Zweibel EG. 1995. *Ap. J.* 440:686–96

McKee CF, Zweibel EG, Goodman AA, Heiles C. 1993. In *Protostars and Planets III*, ed. EH Levy, JI Lunine, pp. 327–67. Tucson: Univ. Ariz. Press

Meinel R, Elstner D, Rüdiger G. 1990. *Astron. Astrophys.* 236:L33–35

Meneguzzi M, Frisch U, Pouquet A. 1981. *Phys. Rev. Lett.* 47:1060–64

Mestel L. 1966. MNRAS 133:265–84

Mestel L. 1985. *Physica Scripta* T11:53–58

Mestel L. 1989. In *Accretion Disks and Magnetic Fields in Astrophysics*, ed. G Belvedere, pp. 151–63. Dordrecht: Kluwer

Mestel L, Roxburgh IW. 1962. *Ap. J.* 136:615–26

Mestel L, Subramanian K. 1991. *MNRAS* 248:677–87

Mishustin IN, Ruzmaikin AA. 1971. *Sov. Phys. JETP* 61:441–44

Moffatt HK. 1972. *J. Fluid Mech.* 53:385–99

Moffatt HK. 1978. *Magnetic Field Generation in Electrically Conducting Fluids.* Cambridge: Cambridge Univ. Press

Moffatt HK. 1979. *Geophys. Astrophys. Fluid Dyn.* 14:147–66

Morris D, Berge GL. 1964. *Ap. J. Lett.* 139:1388–93

Moss D. 1995. *MNRAS* 275:191–94

Moss D. 1996. *Astron. Astrophys.* In press

Moss D, Brandenburg A. 1992. *Astron. Astrophys.* 256:371–74

Moss D, Brandenburg A, Donner KJ, Thomasson M. 1993a. *Ap. J.* 409:179–89

Moss D, Brandenburg A, Donner KJ, Thomasson M. 1993b. In *The Cosmic Dynamo*, ed. F Krause, K-H Rädler, G Rüdiger, pp. 339–43. Dordrecht: Kluwer

Moss D, Brandenburg A, Tuominen I. 1991. *Astron. Astrophys.* 247:576–79

Moss D, Shukurov AM. 1996. *MNRAS* 279:229–39

Moss D, Tuominen I. 1990. *Geophys. Astrophys. Fluid Dyn.* 50:113–20

Moss D, Tuominen I, Brandenburg A. 1990. *Astron. Astrophys.* 228:284–94

Neininger N. 1992. *Astron. Astrophys.* 263:30–36

Neininger N, Beck R, Sukumar S, Allen RJ. 1993. *Astron Astrophys.* 274:687–98

Neininger N, Horellou C. 1996. In *Polarimetry of the Interstellar Medium*, ed. W Roberge, D Whittet. San Francisco: Astron. Soc. Pac. In Press

Neininger N, Klein U, Beck R, Wielebinski R. 1991. *Nature* 352:781–82

Nelson AH. 1988. *MNRAS* 233:115–21

Niklas S, Klein U, Wielebinski R. 1995. *Astron. Astrophys.* 293:56–63
Nordlund Å, Brandenburg A, Jennings RL, Rieutord M, Ruokolainen J, et al. 1992. *Ap. J.* 392:647–52
Nordlund Å, Galsgaard K, Stein RF. 1994. In *Solar Surface Magnetic Fields*, ed. RJ Rutten, CJ Schrijver, pp. 471–98. Dordrecht: Kluwer
Nozakura T. 1993. *MNRAS* 260:861–74
Ohno H, Shibata S. 1993. *MNRAS* 262:953–62
Oren AL, Wolfe AM. 1995. *Ap. J.* 445:624–41
Otmianowska-Mazur K, Chiba M. 1995. *Astron. Astrophys.* 301:41–54
Otmianowska-Mazur K, Urbanik M. 1994. *Geophys. Astrophys. Fluid Dyn.* 75:61–75
Panesar JS, Nelson AH. 1992. *Astron. Astrophys.* 264:77–85
Parker EN. 1955. *Ap. J.* 122:293–314
Parker EN. 1971. *Ap. J.* 163:252–78
Parker EN. 1973. *Astrophys. Space Sci.* 22:279–91
Parker EN. 1979. *Cosmical Magnetic Fields.* Oxford: Clarendon
Parker EN. 1992. *Ap. J.* 401:137–45
Perry JJ. 1994. In *Cosmical Magnetism. Contributed Papers in Honour of Prof. L. Mestel*, ed. D Lynden-Bell, pp. 144–51. Cambridge: Inst. Astron.
Perry JJ, Watson AM, Kronberg PP. 1993. *Ap. J.* 406:407–19
Piddington JH. 1970. *Aust. J. Phys.* 23:731–50
Pietsch W. 1994. In *Panchromatic View of Galaxies*, ed. G Hensler, C Theis, JS Gallagher, pp. 137–154. Gif-sur-Yvette: Editions Frontières
Poezd A, Shukurov A, Sokoloff D. 1993. *MNRAS* 264:285–97
Pohl M. 1993. *Astron. Astrophys.* 279:L17–20
Pouquet A, Frisch U, Léorat J. 1976. *J. Fluid Mech.* 77:321–54
Pudritz RE, Silk J. 1989. *Ap. J.* 342:650–59
Quashnock JM, Loeb A, Sprengel DN. 1989. *Ap. J. Lett.* 344:L49–51
Rädler K-H. 1969. *Geod. Geophys. Veröff. II* 13:131–35
Rand RJ, Kulkarni SR. 1989. *Ap. J.* 343:760–72
Rand RJ, Kulkarni SR, Hester JJ. 1990. *Ap. J. Lett.* 352:L1–L4
Rand RJ, Lyne AG. 1994. *MNRAS* 268:497–505
Ratra B. 1992. *Ap. J.* 391:L1–L4
Reuter H-P, Klein U, Lesch H, Wielebinski R, Kronberg PP. 1994. *Astron. Astrophys.* 282:724–30; 293:287–88 (Erratum).
Reuter H-P, Krause M, Wielebinski R, Lesch H. 1991. *Astron. Astrophys.* 248:12–22
Roberge W, Whittet D. 1996. *Polarimetry of the Interstellar Medium.* San Francisco: Astron. Soc. Pac.
Roberts PH, Soward AM. 1975. *Astron. Nachr.* 296:49–64
Roberts PH, Soward AM. 1992. *Annu. Rev. Fluid Dyn.* 24:459–512
Roberts WW, Hausman MA. 1984. *Ap. J.* 277:744–67
Roberts WW, Yuan C. 1970. *Ap. J.* 161:877–902
Rüdiger G, Elstner D, Schultz M. 1993. *Astron. Astrophys.* 270:53–59
Rüdiger G, Kitchatinov LL. 1993. *Astron. Astrophys.* 269:581–88
Ruzmaikin AA, Shukurov AM. 1982. *Astrophys. Space Sci.* 82:397–407
Ruzmaikin AA, Shukurov AM, Sokoloff DD. 1988a. *Magnetic Fields of Galaxies.* Dordrecht: Kluwer
Ruzmaikin A, Shukurov AM, Sokoloff D. 1988b. *Nature* 336:341–47
Ruzmaikin AA, Sokoloff DD. 1977. *Astron. Astrophys.* 58:247–53
Ruzmaikin AA, Sokoloff DD. 1979. *Astron. Astrophys.* 78:1–6
Ruzmaikin AA, Sokoloff DD, Kovalenko AV. 1978. *Sov. Astron.* 22:395–401
Ruzmaikin A, Sokoloff D, Shukurov A. 1988c. *J. Fluid Mech.* 197:39–56
Ruzmaikin A, Sokoloff D, Shukurov A, Beck R. 1990. *Astron. Astrophys.* 230:284–92
Rybicki GB, Lightman AP. 1979. *Radiative Processes in Astrophysics.* New York: Wiley
Scarrott SM, Draper PW, Stockdale DP, Wolstencroft RD. 1993. *MNRAS* 249:L7–12
Scarrott SM, Rolph CD, Semple DP. 1990. In *Galactic and Intergalatic Magnetic Fields*, ed. R Beck, PP Kronberg, R Wielebinski, pp. 245–51. Dordrecht: Kluwer
Scarrott SM, Rolph CD, Walstencraft RD, Tadhunter CN. 1991. *MNRAS* 224:16–20
Scarrott SM, Ward-Thompson D, Warren-Smith RF. 1987. *MNRAS* 224:299–305
Schlegel EM. 1994. *Ap. J.* 434:523–35
Schmitt D, Rüdiger G. 1992. *Astron. Astrophys.* 264:319–26
Schoofs S, 1992. Hochaufgelöste VLA-Beobachtungen im Radiokontinuum zum Magnetfeld der Spiralgalaxie M81. Diploma Thesis, Univ. Bonn
Schultz M, Elstner D, Rüdiger G. 1994. *Astron. Astrophys.* 286:72–79
Sciama DW. 1994. In *Cosmical Magnetism. Contributed Papers in Honour of Prof. L Mestel*, ed. D Lynden-Bell, pp. 128–33. Cambridge: Inst. Astron.
Segalovitz A, Shane WW, de Bruyn AG. 1976. *Nature* 264:222–26
Simard-Normandin M, Kronberg PP. 1979. *Nature* 279:115–18
Simard-Normandin M. Kronberg PP. 1980. *Ap. J.* 242:74–94
Sofue Y. 1987. *Publ. Astron. Soc. Jpn.* 39:547–57
Sofue Y, Fujimoto M. 1993. *Ap. J.* 265:722–29

Sofue Y, Fujimoto M, Wielebinski R. 1986. *Annu. Rev. Astron. Astrophys.* 24:459–97
Sofue Y, Klein U, Beck R, Wielebinski R. 1985. *Astron. Astrophys.* 144:257–60
Sofue Y, Takano T. 1981. *Publ. Astron. Soc. Jpn.* 33:47–55
Soida M, Urbanik U, Beck R. 1996. *Astron. Astrophys.* In press
Sokoloff D, Ruzmaikin A, Shukurov A. 1990. In *Galactic and Intergalactic Magnetic Fields*, ed. R Beck, PP Kronberg, R Wielebinski, pp. 499–503. Dordrecht: Kluwer
Sokoloff DD, Shukurov A. 1990. *Nature* 347:51–53
Sokoloff D, Shukurov A, Krause M. 1992. *Astron. Astrophys.* 264:396–405
Spencer SJ, Cram LG. 1992. *Ap. J.* 400:484–501
Starchenko SV, Shukurov AM. 1989. *Astron. Astrophys.* 214:47–60
Steenbeck M, Krause F, Rädler K-H. 1966. *Z. Naturforsch.* 21a:369–76; see also transl. in PH Roberts, M Stix. 1971. *The Turbulent Dynamo*, Tech. Note 60. Boulder, Co: NCAR
Stepinski TF, Levy EH. 1988. *Ap. J.* 331:416–34
Subramanian K, Mestel L. 1993. *MNRAS* 265:649–54
Sukumar S, Allen RJ. 1989. *Nature* 340:537–39
Sukumar S, Allen RJ. 1991. *Ap. J.* 382:100–7
Tajima T, Cable S, Shibata K, Kulsrud R. 1992. *Ap. J.* 390:309–21
Tao L, Cattaneo F, Vainshtein SI. 1993. In *Solar and Planetary Dynamos*, ed. MRE Proctor, PC Matthews, AM Rucklidge, pp. 303–10. Cambridge: Cambridge Univ. Press
Thomasson M, Donner KJ. 1993. *Astron. Astrophys.* 272:153–60
Thomson RC, Nelson AH. 1980. *MNRAS* 191:863–70
Tinbergen J. 1996. *Astronomical Polarimetry.* Cambridge: Cambridge Univ. Press
Tosa M, Fujimoto M. 1978. *Publ. Astron. Soc. Jpn.* 30:315–25
Turner MS, Widrow LM. 1988. *Phys. Rev. D* 37:2743–54
Turnshek DA, Wolfe AM, Lanzetta KM, Briggs FH, Cohen RD, et al. 1989. *Ap. J.* 344:567–96
Urbanik M, Otmianowska-Mazur K, Beck R. 1994. *Astron. Astrophys.* 287:410–18
Vachaspati T. 1991. *Phys. Lett. B* 265:258–61
Vainshtein SI, Cattaneo F. 1992. *Ap. J.* 393:165–71
Vainshtein SI, Ruzmaikin AA. 1971. *Astron. Zh.* 48:902 9 (*Sov. Astron.* 16:365–67)
Vallée JP. 1980. *Astron. Astrophys.* 86:251–53
Vallée JP. 1983. *Astron. Astrophys.* 124:147–50
Vallée JP. 1986. *Astron. J.* 91:541–45
Vallée JP. 1991. *Ap. J.* 366:450–54
Vallée JP. 1992. *Astron. Astrophys.* 255:100–4
Vallée JP. 1993. *Ap. J.* 419:670–73
Vallée JP. 1994. *Ap. J.* 433:778–79
Vallée JP. 1996. *Astron. Astrophys.* In press
Vallée JP, Kronberg PP. 1973. *Nat. Phys. Sci.* 246:49–51
Vallée JP, Kronberg PP. 1975. *Astron. Astrophys.* 43:233–42
Vallée JP, Simard-Normandin M, Bignell RC. 1988. *Ap. J.* 331:321–24
Vázquez-Semadeni E, Passot T, Pouquet A. 1995. *Ap. J.* 441:702–25
Verschuur GL. 1979. *Fund. Cosmic Phys.* 5:113–91
Walker MR, Barenghi CF. 1994. *Geophys. Astrophys. Fluid Dyn.* 76:265–81
Welter GL, Perry JJ, Kronberg PP. 1984. *Ap. J.* 279:19–39
White RL, Kinney AL, Becker RH. 1993. *Ap. J.* 407:456–69
Wielebinski R, Krause F. 1993. *Astron. Astrophys. Rev.* 4:449–85
Wolfe AM. 1988. In *QSO Absorption Lines*, ed. JC Blades, DA Turnshek, CA Norman, pp. 297–317. Cambridge: Cambridge Univ. Press
Wolfe AM, Lanzetta KM, Oren AL. 1992. *Ap. J.* 388:17–22
Wolfe AM, Turnshek DA, Lanzetta KM, Lu L. 1993. *Ap. J.* 404:480–510
Yoshizawa A, Yokoi N. 1993. *Ap. J.* 407:540–48
Zeldovich YaB. 1957. *Sov. Phys. JETP* 4:460–62
Zeldovich YaB. 1965. *Sov. Phys. JETP* 48:986–88
Zeldovich YaB, Molchanov SA, Ruzmaikin AA, Sokoloff DD. 1988. *Sov. Sci. Rev. C Math. Phys.* 7:1–110
Zeldovich YaB, Novikov ID. 1982. *The Structure and Evolution of the Universe.* Chicago: Chicago Univ. Press
Zeldovich YaB, Ruzmaikin AA, Sokoloff DD. 1983. *Magnetic Fields in Astrophysics.* New York: Gordon & Breach
Zweibel EG. 1988. *Ap. J. Lett.* 329:L1–L4
Zweibel EG, McKee CF. 1995. *Ap. J.* 439:779–92

Annu. Rev. Astron. Astrophys. 1996. 34:207–40

THE FU ORIONIS PHENOMENON[1]

Lee Hartmann and Scott J. Kenyon

Harvard-Smithsonian Center for Astrophysics, 60 Garden Street, Cambridge, Massachusetts 01238

KEY WORDS: FU Ori objects, pre–main sequence evolution, accretion disks, outflows

ABSTRACT

We summarize the properties of FU Orionis variables, and show how accretion disk models simply explain many peculiarities of these objects. FU Ori systems demonstrate that disk accretion in early stellar evolution is highly episodic, varying from $\sim 10^{-7}\ M_{\odot}\ \mathrm{yr}^{-1}$ in the low (T Tauri) state to $10^{-4}\ M_{\odot}\ \mathrm{yr}^{-1}$ in the high (FU Ori) state. This variability in mass accretion is matched by a corresponding variability in mass ejection, with mass loss rates reaching $\sim 10^{-1}$ of the mass accretion rates in outburst. It appears that the FU Ori phenomenon is restricted to early phases of stellar evolution, probably with infall still occuring to the disk, which may help drive repetitive outbursts. Thermal instabilities are a promising way to produce FU Ori disk outbursts, although many uncertainties remain in the theory; triggering by interactions with companion stars on eccentric orbits may also play a role.

---

1. INTRODUCTION

The remarkable FU Orionis objects provide some of the clearest evidence for disk accretion during early stellar evolution. FU Ori outbursts occur when the mass accretion rate through the circumstellar disk of a young star increases by orders of magnitude. During outburst, the disk outshines the central star by factors of 100–1000, and a powerful wind emerges, which may have a significant impact on the surrounding interstellar medium.

In our present view of star formation, much of a typical low-mass star's mass must be accreted from the disk left over from the collapse of a rotating protostellar cloud (cf Shu, Adams & Lizano 1987). The large amounts of mass accreted in FU Ori outbursts reinforce the notion that disk accretion plays a major role in the formation of stars and not just their associated planetary systems. FU Ori outbursts also demonstrate that accretion rates through such disks can be highly time variable and unexpectedly large at times, with implications for disk physics and grain processing. Finally, the powerful winds of FU Ori objects have important implications for understanding the production of bipolar outflows and jets.

Figure 1 summarizes the current picture of a typical FU Ori object. A young, low-mass (T Tauri) star is surrounded by a disk normally accreting at $\sim 10^{-7}\ M_\odot$ yr^{-1}. This slow accretion is punctuated by occasional, brief FU Ori outbursts, in which the inner disk erupts, resulting in an accretion rate $\sim 10^{-4}\ M_\odot$ yr^{-1}. The disk becomes hot enough to radiate most of its energy at optical wavelengths, and it dumps as much as 0.01 $M_\odot$ onto the central star during the century-long

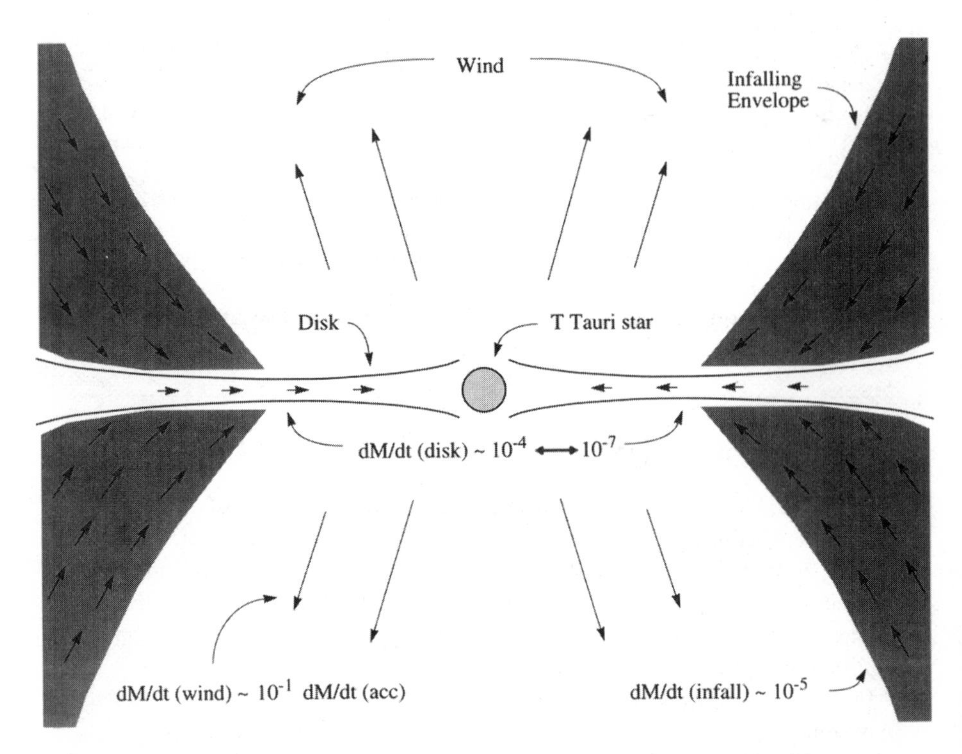

Figure 1 Schematic picture of FU Ori objects. FU Ori outbursts are caused by disk accretion increasing from $\sim 10^{-7}\ M_\odot$ yr^{-1} to $\sim 10^{-4}\ M_\odot$ yr^{-1}, adding $\sim 10^{-2}\ M_\odot$ to the central T Tauri star during the event. Mass is fed into the disk by the remanant collapsing protostellar envelope with an infall rate $\lesssim 10^{-5}\ M_\odot$ yr^{-1}; the disk ejects roughly 10% of the accreted material in a high-velocity wind.

outburst. High-velocity ($\gtrsim 300$ km s^{-1}) winds are generated during outbursts, with mass loss rates that are 10^{-1}–10^{-2} of the disk accretion rate. Indirect but suggestive arguments indicate that material from a protostellar envelope may still be falling onto the disk, providing the necessary material to replenish the disk between repetitive events of rapid disk accretion.

In Section 2 we outline the remarkable observational properties of FU Ori objects. We show how the basic properties of accretion disks provide simple explanations of these properties in Section 3. In Section 4, we discuss the progress that has been made in understanding the physical mechanisms over the past few years. Finally, in Section 5 we discuss how the powerful winds of FU Ori objects place interesting constraints on current theories of disk winds and jets.

Many recent investigations into the FU Orionis phenomenon, including those of the present authors, were stimulated by Herbig's (1977) seminal Russell Lecture. This paper remains an essential introduction to the observational aspects of this subject. Other reviews have been presented by Herbig (1966, 1989), Reipurth (1990), Hartmann, Kenyon & Hartigan (1993), Hartmann (1991), and Kenyon (1995a,b).

2. OBSERVATIONS OF FU ORI OBJECTS

2.1 *Overview*

The FU Ori objects are spatially and kinematically associated with star-forming regions (Herbig 1966, 1977; Hartmann & Kenyon 1985, 1987a,b; Hartmann et al 1989). Several other observational properties further suggest that FU Oris are quite young: All have reflection nebulae (cf Goodrich 1987, Figure 2); many are heavily extincted; and all have large infrared excess emission from circumstellar dust (see also Weintraub et al 1991).

FU Ori objects were originally identified as a class of young stars with large outbursts in optical light (Herbig 1966, 1977). Figure 3 illustrates light curves of the three best-studied objects. Although these light curves differ from one another, they all exhibit large increases in optical brightness of ~ 4 magnitudes or more and all these objects remain luminous for decades. The decay timescale for FU Ori is 50–100 yr.

The FU Ori systems also exhibit distinctive spectroscopic properties that set them apart from other pre–main sequence objects. At modest spectral resolution, all members of the class have optical spectral types of late F to G supergiants (effective temperatures of ~ 7200–6500 K and surface gravities of ~ 10 g cm^2 s^{-2}). (In contrast, other pre–main sequence objects have much higher surface gravities.) Broad absorption, blueshifted by several hundred

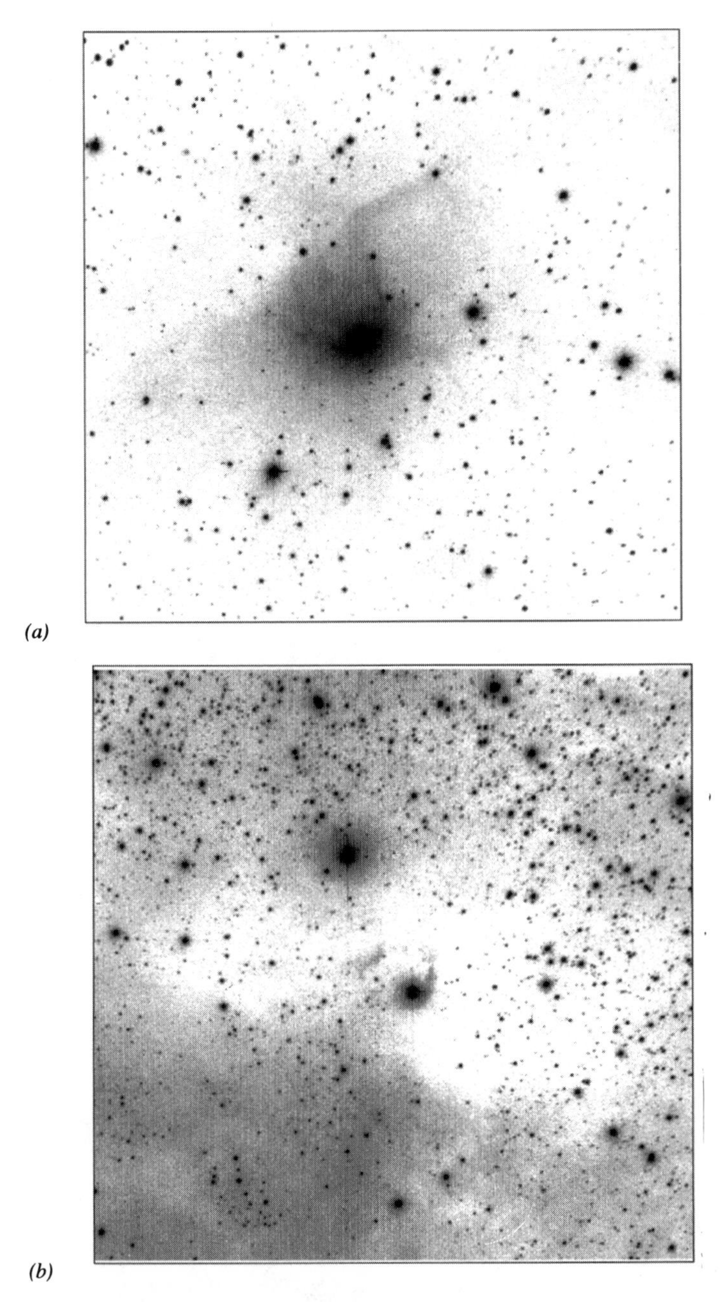

Figure 2 Optical (R) images of (*a*) FU Ori and (*b*) V1057 Cyg, illustrating reflection nebulae on scales ~ 0.1 pc. The images are approximately 5.6 arcmin on a side. (Courtesy C Briceño.)

kilometers per second, is typically observed in the Balmer lines, especially in Hα. The Na I resonance lines also show broad blueshifted absorption, sometimes in distinct velocity components or "shells." The emission component in the P Cygni Hα profile is often absent; when present, this emission extends to much smaller velocities redward than the blueshifted absorption. Infrared spectra of FU Oris show strong CO absorption at 2.2 μm and water vapor bands in the near-infrared ($\sim$ 1–2 μm) region. The near-infrared spectral characteristics are inconsistent with the optical spectra, if interpreted in terms of stellar photospheric emission; the infrared features are best matched with K–M giant-supergiant atmospheres (effective temperatures of $\sim$2000–3000 K). FGKM supergiants are rare in any case and are not commonly found in star formation regions; thus, optical and near-infrared spectra serve to identify FU Ori systems uniquely.

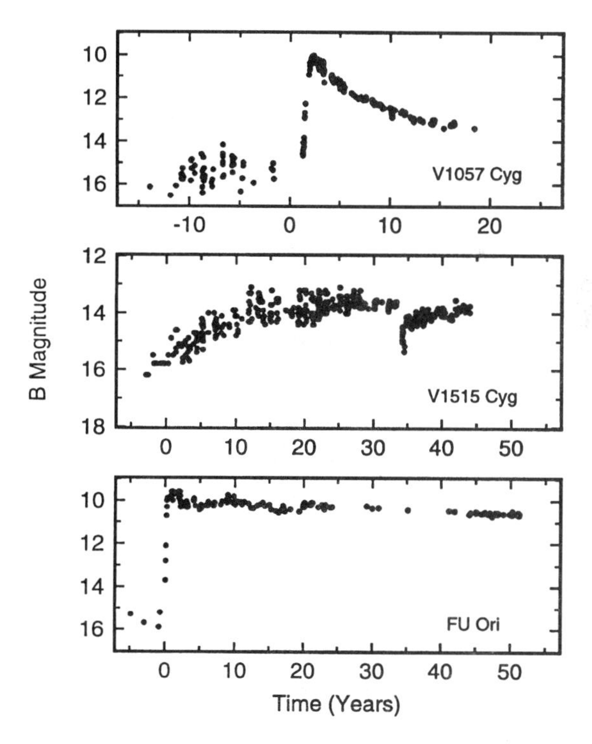

Figure 3 Optical (B) photometry of outbursts in three FU Ori objects. The FU Ori photometry is taken from Herbig (1977), Kolotilov & Petrov (1985), and Kenyon et al (1988); the V1057 Cyg photometric references are contained in Kenyon & Hartmann (1991); and the V1515 Cyg photometry is taken from Landolt (1975, 1977), Herbig (1977), Gottlieb & Liller (1978), Tsvetkova (1982), Kolotilov & Petrov (1983), and Kenyon et al (1991b).

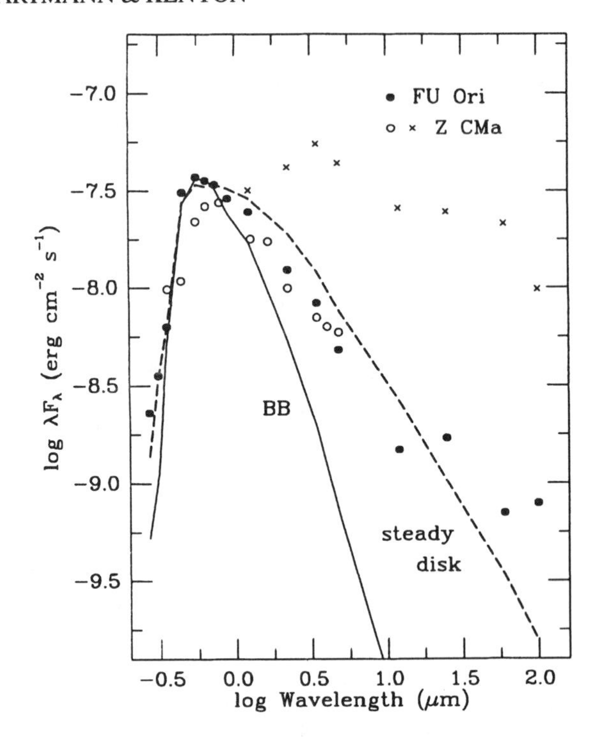

Figure 4 Dereddened spectral energy distributions (SEDs) of four FU Ori objects, compared with single temperature blackbody (stellar photosphere) distributions (*solid curve*) and steady disk models (*dashed curves*). The open circles denote the SED of the optical primary of the Z CMa binary (see text).

FU Ori objects all exhibit strong infrared excess emission (Figure 4). For several objects that are not heavily extincted, it is possible to reconstruct the intrinsic spectral energy distribution of the central object to a reasonable approximation. The results of this reconstruction show that, like many pre–main sequence, low-mass T Tauri stars, FU Ori objects have infrared excesses that can be modeled with simple steady accretion disks (Figure 4; Section 3). Other FU Ori objects have much more far-infrared excess emission than predicted by typical disk models, suggesting the presence of a surrounding circumstellar envelope of dust, which would not be surprising given the observed optical reflection nebulae. Unlike the T Tauri stars, there is no evidence of hot excess continuum in the near ultraviolet in FU Ori objects (cf Section 4.4).

For those FU Ori objects bright enough to be observed with high spectral resolution, further peculiarities are found. Many optical and infrared absorption lines, apparently photospheric in nature, are observed to be doubled, i.e. are split

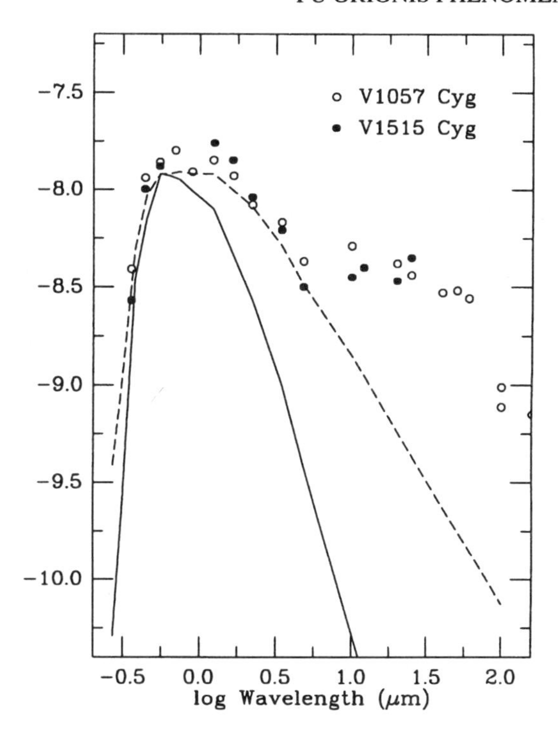

Figure 4 (*continued*)

into two components of roughly equal strength. FU Ori objects are differentially rotating objects; the velocity widths are smaller for lines in the infrared spectral region than for optical lines in FU Ori and V1057 Cyg; and in V1057 Cyg and Z CMa, careful analysis suggests that this variation of line width is a continuous function of wavelength through the optical spectral region (Section 3.1.4).

As discussed in Section 3, many of the peculiar features that set FU Ori objects apart from other pre–main sequence stars—the variation of spectral type with wavelength, the variation of rotation with wavelength, the doubled line profiles—can be explained by an accretion disk model in which the disk is much more luminous than the central star, dominating the observed emission at all wavelengths of observation.

2.2 *Definition(s) of the FU Ori Class; Statistics*

Membership in the FU Ori class depends upon whether observing an optical outburst is a necessary criterion. Several of the more recent candidates are heavily extincted, so that any prior outburst might easily have escaped detection.

The issue is further complicated by the nonuniformity of outburst light curves among the canonical members of the class and by the long durations of outburst, which suggests that other objects might have had outbursts before observing techniques permitted certain detection.

When discussing outburst statistics in this article, we restrict ourselves to considering only objects whose rise to maximum has been observed, while noting that this number is likely to be incomplete. In discussing the statistics of currently active objects, we rely on the unique spectroscopic properties of the class to include objects whose prior outbursts may have been missed. This point of view is motivated by our current understanding of the spectral peculiarities of FU Ori objects as due to extremely rapid disk accretion. Using a spectroscopic classification, we may select pre–main sequence stars that are rapidly accreting mass from their circumstellar disks—some of which may not have had a typical optical outburst. The distinction is not of great statistical significance, since the numbers of the class are small in any event (7 to 10) and inclusion or exclusion of spectroscopically selected objects does not change the size of the class by more than a factor of ~ 2.

The properties of the first three FU Ori objects listed in Table 1—FU Ori, V1057 Cyg, and V1515 Cyg—essentially define the class. In all, three outbursts have been observed (Figure 3), and all have the required spectroscopic properties of F–G supergiant optical spectral type and first-overtone CO absorption bands at 2 μm.

In the one FU Ori object in which a pre-outburst spectrum is available, V1057 Cyg (Herbig 1977), we know that the outburst was not caused by the removal of an obscuring screen of material because of the change in the spectral properties

Table 1 The FU Ori objects

Object	Outburst	t(Rise)	t(Decay)	d(kpc)	$L/L_\odot$	CO flow	Jet/HH	Ref.
FU Ori	1937	~ 1 yr	~ 100 yr	0.5	500	no	no	a,b,d
V1057 Cyg	1970	~ 1 yr	~ 10 yr	0.6	800–250	yes	no	a,b,d,q
V1515 Cyg	1950s	~ 20 yr	~ 30 yr	1.0	200	no	no	a,b,d
V1735 Cyg	$\sim$ 1957–65	< 8 yr	> 20 yr	0.9	>75	yes	no	c,d,q
V346 Nor	$\gtrsim 1984$	< 5 yr	> 5 yr	0.7	?	yes	yes	f,k
BBW 76	< 1930	?	~ 40 yr	1.7?	?	?	no	h,i,o
Z CMa	?	?	> 100 yr	1.1	600	yes	yes	d,e,m, n,p
L1551 IRS5	?	?	?	0.15	≥ 20	yes	yes	g,j,l
RNO 1B,C	?	?	?	0.8	?	yes?	no	r,s,t

References: a,b = Herbig 1977, 1989; c = Elias 1978; d = Levreault 1988; e = Covino et al 1984; f = Graham & Frogel 1985; g = Mundt et al 1985; h,i = Reipurth 1985, 1989b; j = Carr et al 1987; k = Cohen & Schwartz 1987; l = Stocke et al 1988; m = Hartmann et al 1989; n = Poetzel et al 1990; o = Eisloeffel et al 1990; p = T Ray 1990, private communication; q = Rodriguez et al 1990; r = Staude & Neckel 1991; s = Evans et al 1994; t = McMuldroch et al 1995.

of the object. The pre-outburst spectrum of V1057 Cyg is of low signal to noise, but Balmer, Ca II, Fe I, and Fe II emission lines characteristic of an accreting T Tauri star are observed. In outburst, the emission lines mostly disappeared (except for Hα), with an initial spectral type estimated as early A; the spectrum became later (cooler) as V1057 Cyg faded from maximum light.

Although V1735 Cyg and V346 Nor exhibited optical and near-infrared outbursts (Elias 1978, Graham & Frogel 1985), it is difficult to completely rule out dispersal of an obscuring dust cloud because of their large extinctions and the lack of pre-outburst spectroscopic observations. However, V346 Nor brightened so strongly in the near infrared—at least 2 mag at J (1.2 μm)—that the required change in optical extinction is extremely large, and an intrinsic outburst appears more plausible (Graham & Frogel 1985). No outburst has been detected in the southern object BBW76 (Reipurth 1985, 1989b; Eisloeffel et al 1990); however, over the past several years it has decayed in brightness with a timescale very similar to that of FU Ori, strongly suggesting that it has undergone an outburst previous to photographic monitoring (B Reipurth 1991, personal communication).

The situation for Z CMa is more complicated (see, e.g. Hessman et al 1991). Z CMa has historically undergone peculiar fluctuations of optical brightness that are unlike the canonical light curves shown in Figure 3 (Covino et al 1984). In addition, the first-overtone CO absorption at 2.2 μm in Z CMa is peculiarly weak, narrow, and slightly blueshifted compared with FU Ori and V1057 Cyg (Hartmann et al 1989). It now appears likely that these departures from typical FU Ori properties are due to radiation from a companion, non-FU Ori object. Z CMa is a close binary, with a projected separation of 0.1 arcsec ($\sim$ 100 AU) (Koresko et al 1991, Haas et al 1993). The optically brightest star (optical primary) is the FU Ori object, while the optical secondary is much brighter in the infrared and dominates the system luminosity. The peculiarities of the 2-μm CO absorption must therefore be attributed to the secondary, not to the FU Ori object. In addition, much of the optical variability is likely due to variations in brightness of the optical secondary seen in scattered light—an interpretation suggested by the high polarization of the Balmer emission lines (Whitney et al 1993). The simplest explanation of this polarization is that the secondary's strong emission lines and weak continuum emission are not completely absorbed by the dust cloud responsible for its very cool infrared color temperature; some optical radiation is scattered into our line of sight by the cloud, polarizing the emission lines more than the continuum, which represents a combination of (polarized) secondary and (unpolarized) primary emission. Variations in dust cloud geometry can produce large changes in the amount of scattered optical light from the secondary, which explains the observed irregular light curve.

Despite the lack of a demonstrable outburst and the difficulty in detecting the near-infrared CO features of the optical primary, we believe that Z CMa is so similar spectroscopically to FU Ori at optical wavelengths, and has such a similar luminosity, that it must be considered as a certain spectroscopic member of the class.

The last three objects in Table 1 are so heavily reddened that it is not possible to identify a historical optical outburst. L1551 IRS5 is a prototypical infall or Class I source in Taurus, with a well-developed bipolar molecular outflow. The optical spectrum can be observed only in scattered light, but it is typical of the FU Ori class (Stocke et al 1988); in addition, L1551 IRS5 exhibits CO absorption at 2 μm (Carr, Harvey & Lester 1987). The heavily extincted object RNO 1B brightened by several magnitudes at optical wavelengths (Staude & Neckel 1991) and has the optical spectrum of an FU Ori object, with lines showing evidence of profile doubling. A faint companion, RNO 1C, was detected at optical wavelengths; in the near infrared, 1B and 1C are of equal brightness. Both RNO 1B and 1C show the strong, rotationally broadened CO first-overtone absorption bands characteristic of FU Ori objects (Kenyon et al 1993). These observations provide strong evidence for a binary FU Ori system, separated by only $\sim$6000 AU in projection—a remarkable result, considering that the high accretion rates of FU Ori objects can only be sustained for periods of time much shorter than typical evolutionary timescales.

The P Cygni optical profiles of all these objects show that they have powerful winds at present. Whether this mass ejection produces detectable optical jets, Herbig-Haro objects, or sweeps up a substantial amount of interstellar matter to produce a CO outflow (Table 1), probably depends upon the amount of circumstellar matter nearby and the time history of the ejection. This suggestion is supported by the higher frequency of flows and jets among highly extincted objects.

Although there are relatively few members of the FU Ori class, the occurrence of outbursts is not infrequent compared with the stellar birthrate. If we assume that the mean star formation rate derived by Miller & Scalo (1979) represents an adequate average for the solar neighborhood, approximately one low-mass star should be formed in a 1-kpc cylinder centered on the Sun every hundred years. If we further assume a steady state and that only young stars undergo outbursts, the $\sim$5 FU Oris known to have erupted in this volume over the past 60 years imply that, if all (low-mass) stars have FU Ori outbursts, each star will undergo $\sim$10 outbursts (Hartmann & Kenyon 1985; see also Herbig 1977, 1989). We do not know whether all stars experience FU Ori outbursts, and it is hard to imagine that we have observed all such outbursts within 1 kpc over the past 50 years, especially for heavily reddened objects. Thus, despite the rarity of these objects, it is difficult to escape the conclusion that FU Ori outbursts are repetitive, occurring $\sim$10 or more times per object.

3. THE ACCRETION DISK MODEL

Observations place severe constraints on possible outburst mechanisms for FU Ori objects. The outburst must produce roughly 10^{45} to 10^{46} ergs over timescales of 10–100 yr, resulting in an object that resembles a rapidly rotating F–G supergiant at optical wavelengths and an M giant-supergiant in the near-infrared spectral region. The outburst mechanism must also repeat several times, on timescales of $\sim 10^5$ yr (to allow multiple outbursts during a lifetime $\lesssim 10^6$ yr).

Processes intrinsic to pre–main sequence stars have been proposed to explain FU Ori outbursts. Larson (1980) originally suggested that a rapidly rotating star might undergo instabilities and that the internal readjustment could deposit energy in outer layers to produce an outburst. Larson (1983) later modified this picture to propose that accretion of high–angular momentum material from a disk was responsible for depositing energy into the outer stellar layers. Stahler (1989) proposed that pre–main sequence stars of intermediate mass might exhibit a luminosity outburst during their evolution, but the frequency with which intermediate mass stars are formed is incompatible with FU Ori event statistics (see above). Any stellar outburst mechanism encounters insurmountable problems in explaining the infrared excess and differential rotation. The large infrared excesses require a very large emitting area compared with the optical-emitting area, in essence demanding a very extended object. Moreover, a star rotating at or near breakup velocities would exhibit faster rotation from the cooler equatorial regions, unlike the slower rotation observed.

In contrast to the stellar outburst model, accretion disk models naturally predict that the cooler, outermost regions should rotate more slowly. Moreover, other astrophysical accretion disks are observed to undergo outbursts, e.g. dwarf novae (Frank, King & Raine 1992 and references therein). At least one possible mechanism for dwarf novae outbursts (thermal instabilities in disks) may be directly applicable to FU Ori objects (Lin & Papaloizou 1985, Bell & Lin 1994). Based on these considerations, FU Oris were among the first young stellar objects to be interpreted in terms of disk accretion events (Lin & Papaloizou 1985); the first direct application of accretion disk theory to a young star was made for the FU Ori object V1057 Cygni (Hartmann & Kenyon 1985).

The fundamental basis for identifying FU Ori objects as accretion disks relies on the predicted spectrum of a Keplerian accretion disk. In the limit that accretion is steady, and if the disk is optically thick, the surface effective temperature T of the disk varies with cylindrical radius R as (Shakura & Sunyaev 1973)

$$T^4 = \frac{3GM_*\dot{M}}{8\pi\sigma R^3}[1 - (R_i/R)^{1/2}], \qquad (1)$$

where G is the gravitational constant, σ is the Stefan-Boltzmann constant, M_*

is the mass of the central star, $\dot{M}$ is the mass accretion rate through the disk, and R_i is the inner disk radius. The term in brackets results from the choice of inner boundary condition, namely that the flux of angular momentum at the inner disk radius is $\dot{J} = \dot{M}\Omega_K R_i^2$, where Ω_K is the Keplerian angular velocity at R_i (cf Pringle 1981). It is not clear whether this choice is appropriate for FU Ori objects (see Section 4.4), but we adopt this standard steady disk result for simplicity.

Equation (1) predicts a spectral energy distribution (SED) for a radially extended disk that is considerably broader than a single-temperature blackbody, with a long-wavelength flux distribution $\nu F_\nu = \lambda F_\lambda \propto \lambda^{-4/3}$ (Lynden-Bell & Pringle 1974). Moreover, the nature of the spectrum should vary with wavelength; observations at longer wavelengths will be sensitive to the outer, cooler disk regions. Thus, the spectral type or effective temperature, as indicated by absorption line features, should vary with wavelength, appearing increasingly cooler at longer wavelengths.

A disk in Keplerian rotation around a young star must be rapidly rotating. The rotational velocity broadening of spectral lines produced in the disk should decrease with increasing wavelength of observation, because slowly rotating outer disk regions dominate the spectrum at longer wavelengths. This variation of rotation with wavelength of observation provides the essential observational distinction between a disk and a rapidly rotating star. In addition, the line profiles predicted for a rotating disk are doubled, unlike the rotational broadening profile for a rotating spherical object. All of these features have been verified in the two best-studied FU Ori objects, as described below.

The use of steady (constant accretion rate) disk theory to model observations of outbursting objects may appear questionable at first sight. However, we are interested in modeling observations well after outburst, when the accretion rate is much more slowly-varying (Figure 3). Detailed time-dependent accretion disk calculations also suggest that, well after outburst, steady disks are a reasonable first approximation to the emission properties of FU Ori disks (Bell & Lin 1984, Bell et al 1995). We therefore use steady disk theory to test the disk hypothesis for FU Ori objects, which eliminates the need to introduce other parameters, such as the disk viscosity, time history of the outburst, etc.

3.1 *Tests of the Disk Hypothesis*

3.1.1 SPECTRAL ENERGY DISTRIBUTIONS Figure 4 shows the observed SEDs of four FU Ori objects, compared with the spectrum of a single temperature blackbody (i.e. a star) (*solid lines*) and the spectrum of a steady disk (*dashed lines*). It is evident that a blackbody cannot reproduce the emission of the FU Ori objects even limited to the optical and near-infrared regimes (Kenyon et al 1988). The steady disk model provides a reasonable approximation to the infrared excess emission of FU Ori (see also Thamm et al 1994). Although the overall spectrum of Z CMa departs strongly from the disk model (*crosses*), most of this excess is due to the infrared companion star; speckle interferometry shows that the optically dominant star intrinsically exhibits a spectrum very similar to that of FU Ori (*open circles*; Koresko et al 1991). The disk model also works reasonably well for the emission of V1057 Cyg and V1515 Cyg shortward of 10 μm; at longer wavelengths the excess emission probably arises from a circumstellar dust shell powered by light absorbed from the central regions (Section 3.3).

3.1.2 SPECTRAL TYPE VS WAVELENGTH The apparent temperature of emitting gas in a steady accretion disk varies with the wavelength of observation; at longer wavelengths, the outer cool regions dominate the emission. This feature enables disk models to account for the variation of spectral type with wavelength observed in FU Ori objects. Figure 5 shows that FU Ori has an optical spectrum of a $T \sim 7000$ K (F–G type) star, with its characteristic absorption lines, but the near-infrared spectrum appears much cooler, with strong water vapor and CO first-overtone absorption bands characterstic of much cooler stars (see also Sato et al 1992).

A quantitative test of the steady disk model can be made by assuming that Equation (1) represents the variation of effective temperature with radius across the disk, along with vertical hydrostatic and radiative equilibrium. This last assumption follows from a picture in which FU Ori disks are self-luminous and generate all their heat internally. With these assumptions, the spectrum of a disk model can be calculated by the typical stellar atmosphere radiative transfer methods, solved at each radius for the local effective temperature and surface gravity (Calvet et al 1991). The final spectrum is then the result of adding up the individual spectra of each annulus in the disk.

As shown in Figure 5, the spectrum of an optically thick, steady accretion disk model synthesized in this way (*dashed line*) can account for both the optical SED of FU Ori as well as the infrared water vapor and first-overtone CO absorption. The disk model needs no extra parameters to explain optical and infrared spectra simultaneously. The existence of strong absorption features in the FU Ori spectrum implies that the disk is very optically thick and that most of the energy release arises from the disk interior, which results in a temperature gradient that decreases vertically outward. External heating by the central star, which would tend to make the disk vertically isothermal, can be neglected because the star is much less luminous than the high accretion rate disk.

3.1.3 LINE PROFILES The line profiles produced by a simple disk model are qualitatively different than the parabolic shapes produced by a spherical rotating

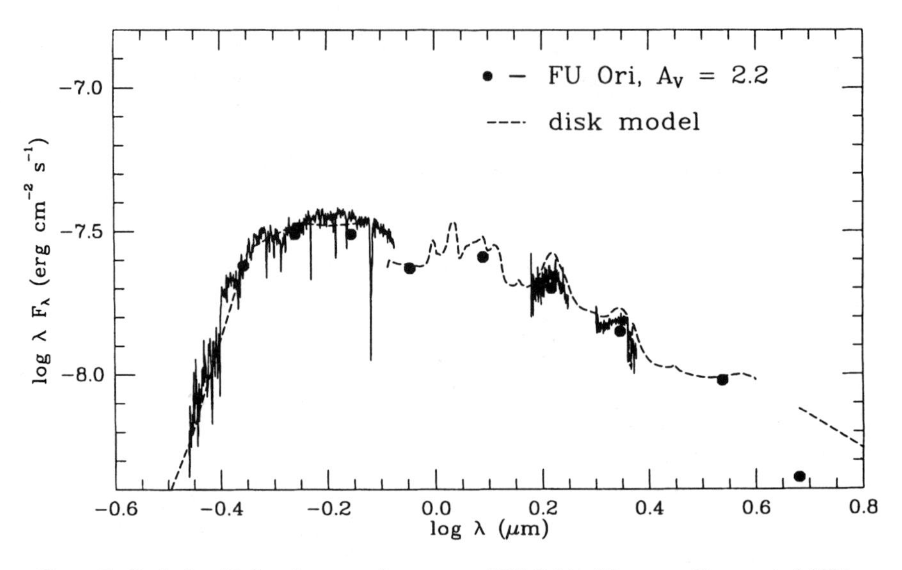

Figure 5 Optical and infrared spectrophotometry of FU Ori (*solid curves*: Kenyon et al 1991a, Mould et al 1978) and broad-band photometery (*filled circles*), compared with a steady accretion disk model (*dashed curve*) incorporating infrared water vapor and absorption bands (Calvet et al 1991). The disk model naturally explains both the presence of G-type spectral features in the optical spectrum as well as the deep water vapor absorption [the irregular continuum in the 1–2 μm spectrum that is characteristic of much cooler stars (see text)].

star. In the limit in which the rotational velocity is large compared with thermal or turbulent velocities, the regime appropriate to FU Ori objects, the line profiles produced by a rotating, flat, narrow ring as a function of velocity shift from line center, ΔV, are of the form (Hartmann & Kenyon 1985)

$$\phi(\Delta v) = [1 - (\Delta v/v_{\max})^2]^{-1/2}, \qquad -v_{\max} < \Delta v < v_{\max}, \tag{2}$$

where $v_{\max}$ is the rotational velocity of the ring corrected for its inclination to our line of sight. The final line profile observed at a given wavelength is the sum of profiles from several neighboring annuli with different rotational velocities, and this differential rotation smooths the profile indicated by Equation (2), resulting in a rounded double-peaked profile.

Figure 6 shows optical spectra of FU Ori objects that clearly demonstrate doubling of many absorption lines. A quantitative test of the disk model can also be made. A synthetic disk spectrum can be constructed as described in the previous subsection, by treating each disk annulus as a separate stellar atmosphere of appropriate effective temperature and gravity, but in addition convolving that annulus with the appropriate rotational velocity broadening profile (Equation 2). Because of the need to include the blanketing due to many lines, it is easier to accomplish this spectrum synthesis by using standard supergiant stars. The comparison between the spectra of disk models constructed in this way and the FU Ori spectra is remarkably good (Kenyon et al 1988).

Not all absorption lines in FU Ori are doubled, particularly those in the blue spectral region (Petrov & Herbig 1992). As discussed in Section 5, the observations suggest that the line doubling is masked by the expansion of the disk surface in an accelerating wind (Hartmann & Kenyon 1985, 1987a; Calvet et al 1993).

3.1.4 ROTATION VS WAVELENGTH The disk model predicts that the rotational velocity line broadening observed at infrared wavelengths should be very large in comparison with typical evolved stars, which are slowly rotating, but smaller than the optical line broadening, because the optical lines are produced in more rapidly rotating inner disk regions. In Figure 7, the uppermost spectrum is that of the M giant HR 867, which exhibits strong, sharp CO vibration-rotation absorption lines. Even though the spectra of the FU Ori objects V1057 Cyg and FU Ori are noisier, their larger velocity broadening is apparent, especially from the shape of the bandhead. The dotted line is a model disk spectrum, in which the rotational broadening is calculated by scaling the disk model rotation

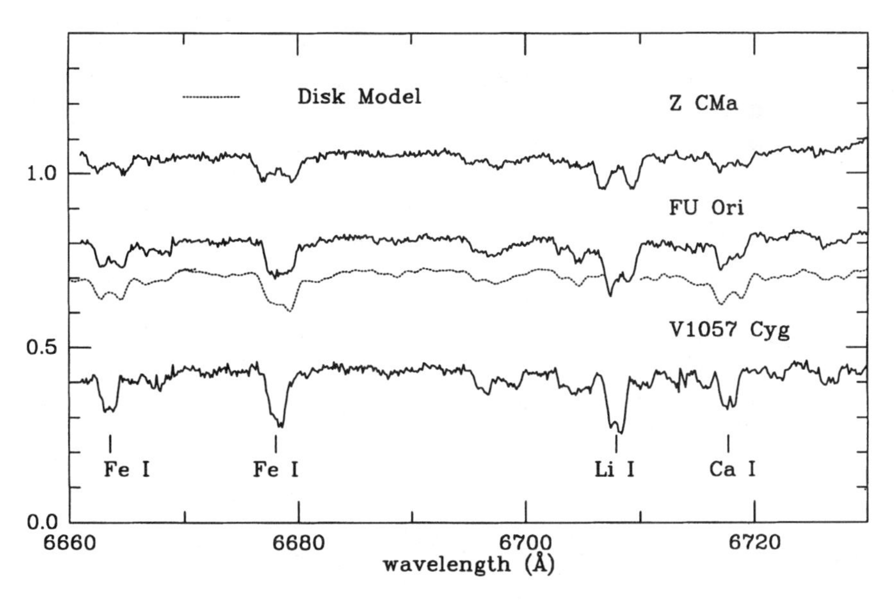

Figure 6 High-resolution optical spectra of three FU Ori objects, showing line profile doubling at different velocity widths. The dotted curve illustrates the spectrum synthesized for a disk model constructed for comparison with FU Ori (see text).

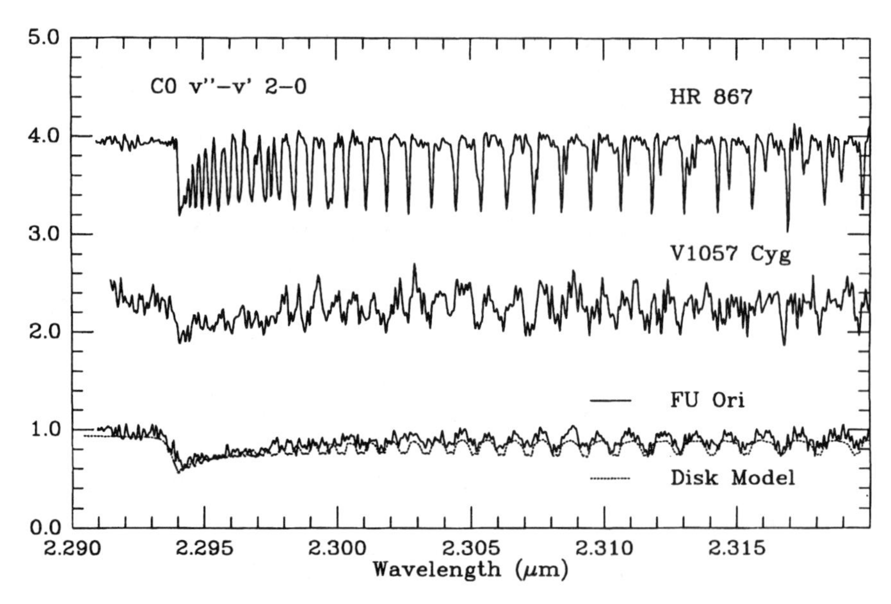

Figure 7 Near-infrared first-overtone ($v''-v' = 2-0$) CO absorption bands in two FU Ori objects. The CO lines are strongly rotationally broadened compared with the bands in the M giant HR 867; the amount of the broadening is consistent with the FU Ori disk model (*dashed curve*; see text).

to match the line broadening at optical wavelengths. The agreement between the theoretical prediction and the observations is fairly good, especially in reproducing the behavior of the bandhead.

To make a quantitative comparison with the optical line broadening, it proves convenient to adopt cross-correlation analysis to average over many lines. For rapidly rotating FU Ori objects, the cross-correlation peak width is a good measure of the rotational velocity broadening. Figure 8 compares optical and infrared cross-correlation peaks for FU Ori and V1057 Cyg. Note that in both objects the rotational broadening at 2.2 μm is $\sim 2/3$ of the rotational broadening observed at 0.6 μm, verifying the qualitative prediction of the disk model. To make a quantitative test of the Keplerian disk hypothesis, model disk spectra can be synthesized as described in the previous subsection and then correlated with the same template used for the observations to provide a direct comparison of cross-correlation peaks. Because the inclination and central mass are poorly known, we adjust the projected rotational velocity of the disk to match the observed optical line widths, and we then compare the predictions of this model to the observed infrared line broadening. The results (dashed lines) show that the model predictions agree quite well with the observations within uncertainties in the spectrum synthesis (see Kenyon et al 1988 for details).

The disk model predicts that the variation of rotation with wavelength should be continuous. Welty et al (1990) have also presented some evidence of such a continuous variation of rotational velocity with wavelength in the optical spectrum ($\sim$5000–8000 Å) of V1057 Cyg and Z CMa. Although the effect is subtle ($\sim$ 10%) and somewhat uncertain, it is consistent with the Keplerian disk model (Welty et al 1992). Optical differential rotation is not apparent in spectra of FU Ori, but this is probably due to the masking effects of the wind (Section 5).

3.2 *Disk and Central Star Properties*

Given the success of the steady accretion disk model in explaining the remarkable observational properties of FU Ori objects, we may then use the disk model to interpret the data in terms of physical properties of the star/disk system.

From optical and ultraviolet spectroscopic observations one can estimate the maximum temperature T_{max} of the disk, which constrains the parameter combination $M_* \dot{M} R_i^{-3}$ (Equation 1). The accretion luminosity through a standard steady disk is (Shakura & Sunyaev 1973)

$$L_{acc} = \frac{GM_* \dot{M}}{2R_i}. \tag{3}$$

The factor of two in the denominator results from the assumption that half the accretion luminosity is emitted in a narrow boundary layer between star and disk at R_i (Lynden-Bell & Pringle 1974). However, this assumption may not be applicable to FU Ori objects (Section 4.4); observations of FU Ori and Z CMa provide no evidence for this boundary layer emission (see below). Neglecting this problem, by combining Equations (1) and (2), one can derive an estimate of the inner disk radius R_i, which depends upon the generally unknown disk inclination i to the line of sight as $R_i \propto \cos^{1/2} i$. Inserting appropriate numbers, one finds estimates of $R_i \sim$ 4–5 $R_\odot$. Although these values are somewhat larger than typical radii of T Tauri stars ($\sim$2–3 $R_\odot$), the agreement is not bad considering the uncertainty in the boundary layer luminosity. In addition, it is conceivable that the central star expands somewhat under the onslaught of the rapid accretion of hot material (cf Section 4.5).

Equations (1) and (2) may be combined to eliminate R_i and constrain the product $M\dot{M}$ to values $\sim$ (2–8) $\times 10^{-5}$ $M_\odot^2$ yr^{-1}. To determine a mass accretion rate, one requires an estimate of the central mass. In principle, the rotational velocity measurements in the optical spectral region can be used to estimate the central mass. The low rotations of V1057 Cyg and V1515 Cyg imply very low masses, unless they are observed nearly pole-on, in which case the results are extremely sensitive to the assumed inclination. If we adopt an average $i = 60°$, then the estimated radii, masses, and mass accretion rates for FU Ori and Z CMa are 5.9 $R_\odot$, 0.35 $M_\odot$, and 1.9×10^{-4} $M_\odot$ yr^{-1}, and 6.5 $R_\odot$,

1.1 $M_\odot$, and 7.9×10^{-5} $M_\odot$ yr^{-1}, respectively. These estimated central star properties are reasonably consistent with typical values for low-mass pre–main sequence stars, although the radii are somewhat larger than typical of T Tauri stars (Kenyon et al 1988; see Section 4.4).

3.3 *Time Variability and Circumstellar Envelopes*

Although there is little spectral information available on the rise to maximum light in FU Ori objects, V1057 Cyg has faded substantially over the past twenty years, and its color evolution provides an important clue to physical conditions. After the outburst, the optical and near-infrared spectrum became redder as

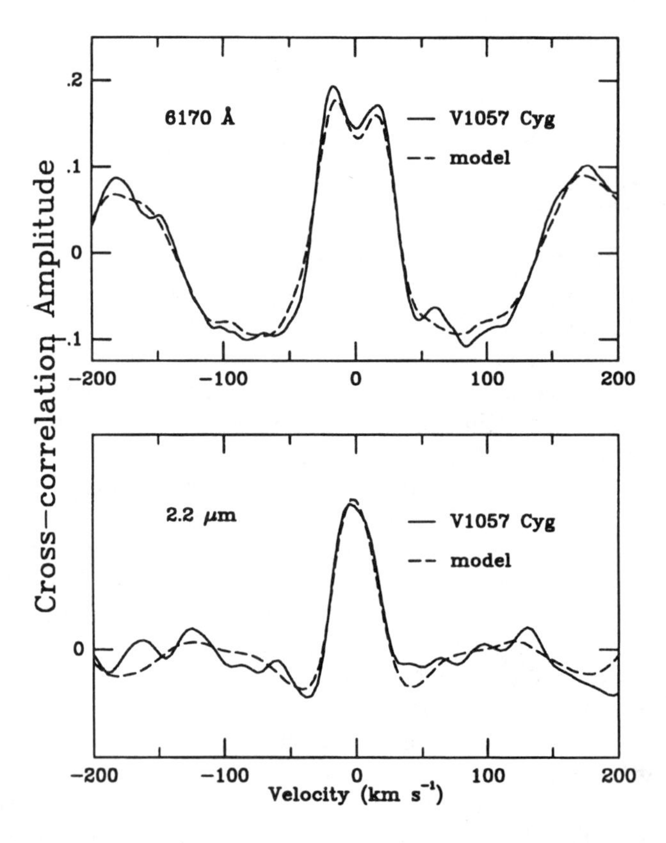

Figure 8 Differential rotation in FU Ori and V1057 Cyg. Cross-correlation peaks show that the intrinsic line widths are larger at optical wavelengths than observed in the $v'' - v' = 2 - 0$ CO absorption bands at 2.2 μm. Cross-correlation of synthetic disk model spectra with the same templates (*dashed curves*) demonstrate that the difference in optical and infrared line widths is consistent with the assumption of Keplerian rotation (see text). (From Kenyon et al 1988.)

V1057 Cyg faded, consistent with optical spectra that indicate that V1057 Cyg evolved from an A spectral type near maximum ($T_{eff} \sim 8000$ K; Herbig 1977) to a mid-G spectral type ($T_{eff} \sim 6500$ K) at the current epoch. A series of steady accretion disk models with fixed inner radii but decreasing mass accretion rates match the color evolution with decreasing accretion luminosity, fairly well from the optical to wavelengths of $\sim 5\ \mu$m, reproducing the observed decrease in amplitude of decay at increasing wavelengths (Kenyon & Hartmann 1991).

Disk models cannot explain the rapid decrease in brightness of V1057 Cyg at mid-infrared wavelengths $\lambda \gtrsim 10\ \mu$m. This mid-infrared fading is comparable to that observed at optical wavelengths, but is much faster than the slow decrease in brightness observed at $\lambda \sim 2$–$5\ \mu$m. The decay timescale of a few years is too fast for viscous disk evolution at the $\sim$ few AU distances appropriate for this

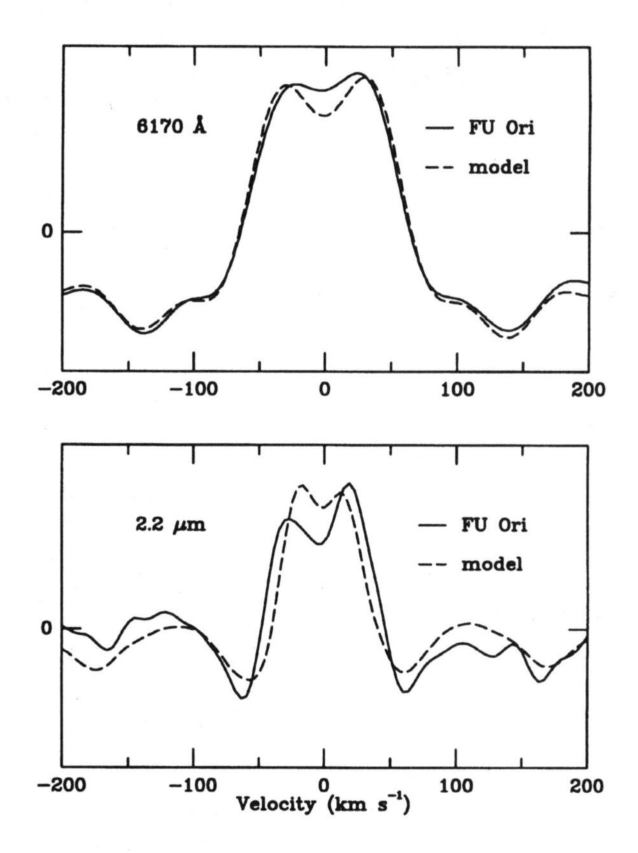

Figure 8 (*continued*)

emission in a disk model (Section 4.2). The break in the SED at 10 μm (Figure 4) similarly argues for another dust emission source at long wavelengths. The amount of flux emitted in long wavelength radiation is too large to be accounted for by a flat disk absorbing radiated accretion luminosity and reradiating this energy at longer wavelengths. However, a circumstellar envelope could easily subtend a sufficiently large solid angle from the inner disk regions to absorb the required fraction of the accretion luminosity, and would decrease in brightness as the inner accretion disk source faded, consistent with observations.

Kenyon & Hartmann (1991) showed that an opaque, dusty, infalling envelope, with the properties predicted for the protostellar infall phase, could explain the mid-IR observations of V1057 Cyg during its optical decline. Reproducing the mid-IR excess requires dusty material with $A_V \sim 50$ mag and a covering factor of $f \sim 1/2$ in solid angle, extending to an inner radius of 5–10 AU measured from the central star. The envelope must have a cavity or region of low extinction along our line of sight to the central object, to be consistent with the estimated extinction of the inner disk by $A_V \sim$ 4–5 mag. This picture is consistent with interpretations of the morphology of the scattered light nebulae of V1057 and V1515 Cyg (Figure 2; Goodrich 1987), which suggest that these objects are in fact observed through cavities in the surrounding material.

Further support for the idea that FU Ori objects may have dusty infalling protostellar envelopes comes from the identification of L1551 IRS5 as an FU Ori object (Mundt et al 1985, Carr et al 1987, Stocke et al 1988). Adams et al (1987), Butner et al (1991, 1994), and Kenyon et al (1993) have all modeled the SED of L1551 IRS5, with a dusty infalling envelope of $\dot{M} \sim$ a few $\times 10^{-6}$ $M_\odot$ yr^{-1}, similar to the result of Kenyon & Hartmann (1991) for V1057 Cyg, and comparable to expected infall rates for protostellar clouds (e.g. Terebey et al 1984). Not all FU Ori objects may have such dense envelopes, however; Adams, Lada & Shu (1987) suggested that FU Ori posesses a modest remnant envelope with $A_V \sim 2.5$ and a covering fraction $f \sim 0.1$.

If V1057 Cyg and V1515 Cyg are typical, and if FU Ori objects are oriented randomly in space, then roughly 50% of all FU Ori systems should be viewed through the envelope if the envelope covering factor is $\sim 1/2$ in most systems. These objects should have $L_{\rm IR} \gg L_{\rm opt}$, because nearly all optical radiation is absorbed. Five FU Ori objects—V1735 Cyg, V346 Nor, L1551 IRS5, RNO 1B, and RNO 1C—have SEDs that rise towards longer wavelengths, as expected for a source viewed through the envelope. Thus, it may be that the frequently large extinctions to spectroscopically identified FU Ori objects result in part from their own natal envelopes.

The picture of FU Ori objects as disks surrounded by infalling envelopes is attractive for its ability to explain multiple eruptions. If the central star accretes

$\sim 10^{-2}\ M_\odot$ from the disk during a typical event, infall from the envelope can replenish the disk in 10^3–10^4 yr for an infall rate of 1–10 $\times 10^{-6}\ M_\odot\ \mathrm{yr}^{-1}$.

4. OUTBURSTS

4.1 *Constraints on Outburst Mechanisms*

What causes the outbursts of disk accretion in FU Ori objects? To explain the fastest rise times of a year, the eruption must involve disk regions smaller than one AU [because disk evolution will occur on timescales much longer than an orbital period (Pringle 1981)]. The accretion of so much mass ($\sim 10^{-2}\ M_\odot$ in $\sim 10^2$ yr) is another important constraint. It is not easy to construct disk models satisfying these observations, although in recent years substantial progress has been made.

4.2 *Triggering of Outbursts*

One popular and recurring suggestion is that FU Ori outbursts of rapid disk accretion might result from the passage of a close companion star near the disk. This idea was first suggested by A Toomre (1985, personal communication), but the only calculation directly applied to this problem to date is that of Bonnell & Bastien (1992). As part of a series of studies of binary formation, Bonnell & Bastien found that disks forming around each member of a binary system were strongly affected when the pair plunge close to each other on an eccentric orbit. They found that the disks could be perturbed near periastron to high accretion rates that were comparable to that required for FU Ori systems; moreover, the evolution of one simulation suggested that disks might survive the collision process sufficiently well to undergo outbursts during three or four successive periastron passages.

This mechanism is attractive, since most stars are members of multiple systems and theoretical arguments suggest that at least some outbursts may be externally triggered (see Section 4.3). The discovery of RNO 1B and 1C as a spectroscopic FU Ori binary also lends some support to this suggestion. However, no apparent radial velocity shifts have been observed in the brightest, best-studied FU Ori objects, to a detection limit of a few kilometers per second (Herbig 1977, Hartmann & Kenyon 1987). The interaction must strongly perturb the disk at 1–10 AU to produce short rise times, because perturbations at larger distances would naturally show much longer (viscous) rise times; therefore, to explain the event statistics it would be necessary to assume that the orbital distribution among binaries was originally much more concentrated at small perihelion distances than presently observed in the field (e.g. Duquennoy & Mayor 1991) and that interactions, perhaps between circumstellar envelopes

(Clarke & Pringle 1991), caused evolution to the present-day distribution of binary orbits.

4.3 *The Thermal Instability Mechanism*

Accretion disks in binary systems are also known to undergo outbursts (Horne 1991). Thermal instability models have been constructed to explain outburst light curves for many interacting binaries, including low-mass X-ray binaries, cataclysmic binaries, and symbiotic stars (Mineshige & Osaki 1983; Smak 1984a,b; Meyer 1984; Duschl 1986a,b). These outbursts of disk accretion apparently are not due to changes in mass transfer into the disk, but due to an intrinsic instability. The most frequently considered situation is that of a thermal instability (cf Frank et al 1992), which causes the disk to pass from a low accretion rate state to a high accretion rate state (and vice versa).

The basic idea behind the thermal instability mechanism is as follows (see the discussions in Pringle 1981 and Frank et al 1992). In thermal equilibrium at a radial distance R in the disk, the viscous energy generation $F_{\rm vis}$ must be balanced by the radiative losses of the disk, $F_{\rm rad}$. Integrating the energy equations over the vertical structure of the disk at a fixed radial distance R, and adopting the usual "α" viscosity treatment (Shakura & Sunyaev 1973),

$$F_{\rm vis} = \frac{9}{4}\alpha\Omega\Sigma c_{\rm s}^2 \tag{4}$$

and

$$F_{\rm rad} = \sigma T_{\rm eff}^4. \tag{5}$$

Here Ω is the local Keplerian velocity at R, Σ is the total disk mass column density, $c_{\rm s}$ is the average internal disk sound speed, and $T_{\rm eff}$ is the local surface effective temperature at R. If we assume vertical radiative equilibrium and large optical depth, then

$$T^4 \approx \tau_r T_{\rm eff}^4, \tag{6}$$

where T is the disk internal temperature (near the midplane) and $\tau_r = \chi_r \Sigma$ is the Rosseland mean optical depth, with $\chi_r(T, P)$ as the Rosseland mean opacity for the appropriate internal temperature and pressure. This approximate radiative transfer equation accounts for the role of the disk opacity in trapping the internal heat produced by accretion.

Combining Equations (4) and (5) with (6), one finds

$$F_{\rm rad}/F_{\rm vis} \propto T^3 \chi_r(T, P)^{-1}\Sigma^{-2}\alpha^{-1} \tag{7}$$

(Kawazoe & Mineshige 1993, D'Alessio 1996). Now consider a disk annulus

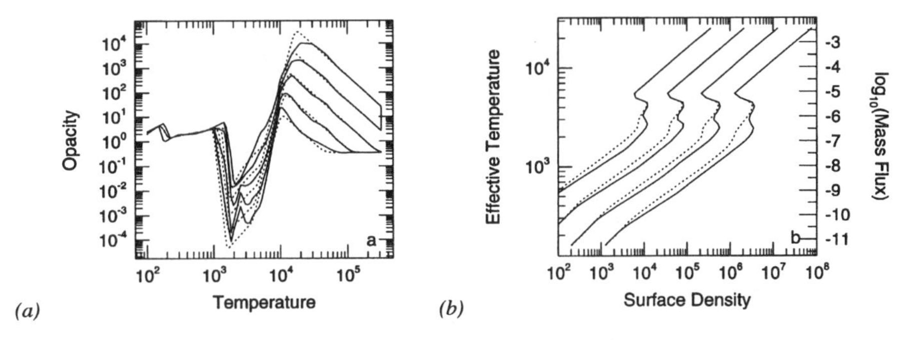

Figure 9 (*a*) Rosseland mean opacities (*solid curves*) as a function of temperature for densities ranging from 10^{-5} to 10^{-9} g cm^{-3}, in steps of a factor of 10. Dotted curves indicate a previous power-law fit. (*b*) "S curves" for accretion disks (see text) calculated using the opacities above. Four S curves are shown, each corresponding to a different value of the viscosity parameter α; from left to right, $\alpha = 10^{-1}, 10^{-2}, 10^{-3}$, and 10^{-4}. The dotted curves correspond to using the power-law opacities, illustrating the sensitivity of S curves to detailed opacity calculations. (From Bell & Lin 1994.)

initially in thermal equilibrium that is perturbed to a slightly higher internal temperature T. Because the thermal timescales for transferring energy are generally much shorter than the viscous timescales for transferring matter (Pringle 1981), it suffices to take $\Sigma \approx$ constant. Then Equation (7) indicates that the radiative cooling will exceed the viscous heating as long as the opacity does not increase more rapidly with temperature than T^3; the radiative cooling will therefore drive the disk back toward the thermal equilibrium. If, however, the opacity increases faster than T^3, heat is efficiently trapped within the disk and the surface cooling cannot keep pace with the increasing viscous heating. This leads to a thermal runaway, with the disk temperature increasing rapidly until the character of the opacity changes.

Figure 9*a* shows calculated Rosseland mean opacities for various gas densities as a function of temperature. Based on the above discussion, the steep dependence of the opacity on temperature between $\sim$3000 K and $\sim 10^4$ K makes it possible to have thermal instabilities in this temperature range (see also Hessman 1990 for a discussion of the possible role of dust grains). To explore the implications of this opacity dependence further, we consider the so-called S curves resulting from these assumptions (cf Frank et al 1992). The thermal equilibrium equations for a steady state accretion disk can be recast in terms of effective temperature (or mass accretion rates; cf Equation 1) and the surface density. Then the loci of thermal equilibrium in the T_{eff} vs Σ plane for a specific disk annulus form an S-shaped curve, with a "kink" where the opacity causes thermal instability.

Figure 9*b* shows a sample set of S curves, calculated for differing values of α at a fixed radial distance in an FU Ori disk model (Bell & Lin 1994). If we restrict our attention to a fixed value of α, and thus a single S curve, the regions above and to the left of the curve have faster cooling than heating; the regions below and to the right of the curve have faster heating than cooling. In the limit cycle theory of outbursts, a disk annulus begins in a low accretion rate state at low Σ, lying on the lower branch of an S curve. A perturbation that increases the central temperature tends to drive the disk annulus vertically upward into a region where cooling exceeds heating, so the disk is stable. However, as material piles up in the disk, moving the equilibrium point upwards along the curve, eventually the disk annulus reaches an inflection point in the S curve, at which point a positive temperature perturbation pushes the annulus vertically upward in the log $T_{\rm eff}$ vs log Σ plane, where the disk annulus is thermally unstable. The annulus must jump up to the upper branch of the S curve, where the effective temperature and mass accretion rate are much higher. Eventually, the enhanced accretion onto the central star causes disk material to drain away, dropping Σ to the point where the only stable solution is on the low accretion rate branch. The process can repeat itself, satisfying a key constraint of the outburst statistics, if there is a mass source that replenishes the material accreted during outburst.

Figure 9*b* shows one reason why thermal instability models are attractive for explaining FU Ori outbursts. Traveling up the lower branch of the S curve until the inflection point, an inner disk annulus is forced initially to jump up to a mass flux approaching $10^{-4}\ M_\odot\ {\rm yr}^{-1}$ and a peak effective temperature slightly below 10^4 K, in agreement with observations. As the figure shows, this general result depends little upon the value of α chosen (Bell & Lin 1994).

Another model prediction that is almost independent of α (for fixed α) is the mass accretion rate that would push the inner disk regions into outburst. At the low-temperature end of the unstable regime (Figure 9 *a*), the gas opacity is quite low, and it is difficult to make the disk enormously optically thick. Detailed disk calculations show that this results in an internal temperature that is not very much larger than the effective temperature. Thus, we can use the effective temperature as a useful guide to the disk regions that can become thermally unstable. Bell & Lin (1994) show that the critical effective temperature at the transition between the stable and unstable portions of the S curve is roughly ~ 2000 K (see Figure 9 *b*). If we impose a fixed mass accretion rate $\dot{M}_{\rm in}$ from the outer disk, we can then use this temperature constraint along with Equation (1) to derive an approximate outer limit radius for the region of thermal instability (Bell & Lin 1994),

$$R_{\rm max} \approx 20\ R_\odot \left(\frac{\dot{M}_{\rm in}}{3 \times 10^{-6} M_\odot\ {\rm yr}^{-1}} \right)^{1/3} \left(\frac{M_*}{M_\odot} \right)^{1/3} \left(\frac{T_{\rm eff}}{2000\ {\rm K}} \right)^{-4/3}. \qquad (8)$$

This result illustrates the important role that the input or background disk accretion rate plays in the thermal instability theory for protostellar disks.

Lin & Papaloizou (1985) first applied thermal instability models to FU Ori objects and generated optical outburst amplitudes similar to those observed. The eruptions were short-lived, however, and did not produce large infrared outbursts. The reason for this can be seen from Equation (8); at the background mass accretion rate of $\dot{M} = 10^{-7}\ M_\odot\ \mathrm{yr}^{-1}$, similar to the mass accretion rates of T Tauri stars, only the innermost disk regions became thermally unstable. This means that very little mass was involved in the eruption, and the decay times were too short. Clarke, Lin & Pringle (1990) managed to achieve an outburst with the same low background accretion rate by introducing a very large (50×) perturbation in the surface density. This perturbation had the effect of producing a much higher mass flux into the unstable regions, and so triggered an outburst lasting ~30 yr, which qualitatively resembled the decay of V1057 Cyg.

If FU Ori disks are acquiring mass from remnant infalling envelopes at $\sim 10^{-6}$–$10^{-5}\ M_\odot\ \mathrm{yr}^{-1}$ (Section 3.3), then the mass accretion rates of the outer disk would be driven closer to values needed to make the thermal instability operate more easily. Kawazoe & Mineshige (1993) first argued that FU Ori outbursts are produced by thermally unstable disks with high-mass input rates, and they noted the possible correspondence with infall rates. Kawazoe & Mineshige also argued that very low values of α were needed to produce sufficiently long recurrence times (and outburst durations). Bell & Lin (1994) and Bell et al (1995) calculated similar models in more detail and showed that many properties of the observed light curves can be reproduced with appropriate model parameters. Bell & Lin (1994) found that models simply fed at the appropriate accretion rate from the outer disk tended to become unstable at inner disk radii first; the instability propagates slowly outward to R_{max} (see also the discussion in Smak 1984b). Under these circumstances, Bell et al (1995) were able to reproduce the slow rise in luminosity of V1515 Cyg (Figure 3) but not the rapid rise times of FU Ori and V1057 Cyg. Bell et al (1995) were forced to return to the idea of external perturbations to obtain rapid rise times with an outside-in perturbation. However, with a higher background $\dot{M}_{\mathrm{in}}$, Bell et al were able to use much smaller perturbations than did Clarke et al (1990).

In summary, the thermal instability models are attractive for explaining FU Ori outbursts, because they predict high accretion states in reasonable accord with observation. Moreover, the importance of a high background accretion rate for the outer disk provides a natural explanation for why FU Oris are found preferentially in the earliest stages of stellar evolution; infall to the disk helps drive the instability. Quantitatively, it appears that the mass infall rates predicted theoretically and estimated observationally lie in the range needed to produce thermal instability over the large areas needed in the protostellar disk. Interactions with companion objects (Bonnell & Bastien 1992) also could be used to trigger thermal outbursts.

Unfortunately, many aspects of the theory are completely dependent upon the unknown value(s) of α, which determines the timescales for evolution. Relatively small values of α are needed to obtain sufficiently long evolutionary timescales and sufficiently large disk masses to explain FU Ori outbursts, and there is no independent justification for the choices made. Even worse, the calculations of Bell & Lin (1994) and Bell et al (1995) required a variable α, with values in the high accretion rate state about a factor of 10 higher than in the low accretion state to obtain a sufficiently large outburst. (Variable α also changes the S-curve shape in an ad hoc way.) Bell et al found that they needed $\alpha_{low} \sim 10^{-4}$, which implies a relatively massive disk; in fact, these models are barely gravitationally stable at their outer limits $\sim 1/2$ AU, and it is not clear whether such a massive disk is physically plausible.

One interesting consequence of the thermal instability picture is that, as infall ceases to the disk, and therefore $\dot{M}_{in}$ drops below $10^{-6}\ M_{\odot}\ yr^{-1}$ to approach typical T Tauri disk accretion rates of $\sim 10^{-7}\ M_{\odot}\ yr^{-1}$, one might expect to observe shorter and smaller amplitude outbursts, since only the innermost disk regions can become thermally unstable (Equation 8). Herbig (1977) drew attention to small optical outbursts observed in pre–main sequence stars that might constitute a separate class of "EXor" variables, and it is tempting to speculate that the EXor outbursts are also produced by thermal disk instabilities.

These considerations emphasize the importance of understanding disk viscosity and/or angular momentum transport mechanisms in more detail, without which it is difficult to make definite theoretical predictions. For example, if the disk viscosity is magnetic (Balbus & Hawley 1991a,b), the finite and variable conductivity predicted for protostellar disks indicates that the use of a constant α parameter for all parts of the disk is questionable. Gammie (1996) has suggested that "dead zones" of decreased accretion may occur where the disk ionization state is so low that magnetic viscosity is ineffective. Such dead zones might dam up material until other mechanisms permit accretion, providing a possible alternative explanation for FU Ori outbursts.

4.4 *The Boundary Layer Problem*

The standard disk accretion model has a boundary layer at its inner edge (Lynden-Bell & Pringle 1974). If the central star is slowly rotating, the kinetic energy that must be dissipated for disk material to come to rest on the stellar surface is roughly half of the total accretion luminosity. In typical disk models, the boundary layer is narrow, with a much smaller surface area than the star, and so the boundary layer is predicted to be both hot and luminous.

The standard boundary layer picture predicts large ultraviolet fluxes that are not observed in FU Ori objects (Kenyon et al 1989). *IUE* observations of FU Ori and Z CMa show absorption features characteristic of A–F supergiants instead of the expected $\sim 3 \times 10^4$ K boundary layer emission. The reddening-corrected UV fluxes in these two objects show no evidence for appreciable excess emission above that predicted by the disk model. The absence of strong emission lines also points to a lack of boundary layer emission, which, if hotter than 3×10^4 K and radiating half of the accretion luminosity, should produce enough EUV photons to ionize a substantial fraction of an FU Ori object's wind, which is inconsistent with observations.

The standard boundary layer model probably does not apply to T Tauri disks either, because the inner disks are disrupted by a stellar magnetic field, which channels the disk accretion into magnetospheric columns (Königl 1989). In the magnetosphere model, the hot ultraviolet continuum emission of T Tauri stars arises from the shocks at the base of the accretion column, where nearly free-falling gas collides with the star. FU Ori objects exhibit no evidence for this hot continuum, which might be explained by presuming that the dynamic pressure of the rapidly accreting disk is sufficient to crush the stellar magnetic fields back up against the outer stellar layers (Shu et al 1994). But then we are back to the original question: Why is no boundary layer emission apparent in FU Ori objects?

If the central star is rapidly rotating, then disk material does not need to lose much energy in a boundary layer to come to rest on the stellar surface. However, since T Tauri stars generally rotate slowly, such a case is unlikely to apply (Hartmann et al 1986, Bouvier et al 1986). It is also unlikely that the accreted material can spin up the central star sufficiently. In FU Ori, $\Delta M \lesssim 10^{-2}\ M_{\odot}$ has been accreted, and if the central star were initially completely convective with a moment of inertia of $0.2\,MR^2$ and a mass of $M \sim 0.5\ M_{\odot}$, the angular momentum $\Delta M \Omega_K R^2$ carried into the star cannot have spun it up to more than about 10% of the breakup angular velocity Ω_K (Kenyon et al 1989). [Suggestively, some symbiotic binaries, with mass accretion $\dot{M} \sim (2\text{–}4) \times 10^{-5}\ M_{\odot}\ \mathrm{yr}^{-1}$ onto low-mass main-sequence stars, similarly show evidence for decreasing importance of high-temperature emission in high disk accretion states (Kenyon et al 1991b; Mikolajewska & Kenyon 1992), and this cannot be due to spinup of the accreting star.]

The most promising solution for both FU Oris (and symbiotic stars) involves changes in the structure of the inner disk and boundary layer as $\dot{M}$ increases in an eruption. The width of the visible part of the boundary layer is comparable to the vertical scale height H_{bl}. Bath & Pringle (1982) first noted that the disk scale height can become a significant fraction of the stellar radius as $\dot{M}$ increases beyond 10^{-4} $M_\odot$ yr^{-1} in a disk surrounding a low-mass main sequence star. Lin & Papaloizou (1985) later found similar behavior in preliminary models for FU Ori eruptions, and Clarke et al (1989, 1990) confirmed these results with more detailed calculations. This behavior occurs because the scale height is set by the balance of thermal pressure and circular rotation, $H/R \sim c_s/v_\phi$, where c_s is the sound speed and v_ϕ is the orbital velocity. The sound speed increases with $\dot{M}$, whereas v_ϕ remains constant; H/R must then increase with $\dot{M}$.

None of these calculations explicitly treated the boundary layer along with the disk, nor could they determine if this region would also expand in an eruption. Popham et al (1993; see also Popham & Narayan 1991) modeled the entire accretion flow from the disk through the boundary layer and onto the central star and found steady-state solutions for both low and high $\dot{M}$ states. They found that as $\dot{M}$ increases to $\sim 10^{-4}$ $M_\odot$ yr^{-1} the dynamical boundary layer grows to 10–20% of a stellar radius and that the width of the thermal boundary layer also grows to the point that it becomes impossible to distinguish this region from the inner parts of the disk. Interestingly, at this point the disk is rotating at velocities well below Keplerian values, because the gas pressure is sufficiently large to help support material against gravity. The resulting high-temperature accretion could perturb the central star's radius considerably upward from values typical of T Tauri stars (see below).

4.5 *Outbursts and Stellar Evolution*

The statistics discussed in Section 2.2 suggest that, at present, about five to nine objects are accreting at $\sim 10^{-4}$ $M_\odot$ yr^{-1} within a roughly 1 kpc radius of the Sun. The birthrate estimates of Miller & Scalo (1979) suggest that within the same volume, approximately (5–10) $\times$ 10^{-3} $M_\odot$ yr^{-1} is being processed into new stars. Thus, the observations suggest that the typical low-mass star accretes $\sim$ 10% of its mass in "FU Ori" accretion (by which we mean high accretion rate states, not necessarily outbursts; cf Section 2.2). (Interestingly, typical parameters for T Tauri stars of $\dot{M} \sim 10^{-7}$ $M_\odot$ yr^{-1} over a lifetime of $\sim 10^6$ yr suggest a similar value for the mass accreted in the T Tauri phase.)

This estimate of the amount of mass accreted in FU Ori phases is an approximate lower limit if other as yet undiscovered objects exist. As techniques for observing heavily extincted objects improve, new candidates are being discovered (Staude & Neckel 1992, Strom & Strom 1993, Hanson & Conti 1995). Whether the source statistics are incomplete at the level of a factor ~ 5 needed to accrete most of the stellar mass in the FU Ori phase is not yet clear.

The FU Ori frequency needed to accrete all of a typical stellar mass can be estimated by assuming that FU Ori outbursts are confined to the protostellar infall phase. Then the fraction of time in the FU Ori state is simply the ratio of the infall rate to the FU Ori accretion rate: $f \sim \dot{M}_i/\dot{M}_{FU}$. FU Ori disk models require $\dot{M}_{FU} \sim 10^{-4}\ M_\odot\ yr^{-1}$, so $f \sim 0.05$ for a typical infall rate of $\dot{M}_i \sim 5 \times 10^{-6}\ M_\odot\ yr^{-1}$. Thus, a typical young, low-mass star must spend ~5% of the infall phase as an FU Ori object to accrete all of its mass in this high $\dot{M}$ state. The Taurus dark cloud contains ~20 embedded sources and one object (L1551 IRS5) in the FU Ori state, which agrees with the simple model prediction, but has no real statistical significance. Source statistics for embedded sources in other clouds are not as good as in Taurus, and the FU Ori population in these clouds is even less well known. However, both of these numbers can be improved with sensitive photometric and spectroscopic surveys of nearby molecular clouds to identify embedded sources and then determine which—if any—of these objects display the characteristic spectroscopic properties of FU Oris.

An independent reason for thinking that FU Ori outbursts may play a substantial role in the formation of stars comes from the apparent "luminosity problem" for protostar candidates in the Taurus molecular cloud (Kenyon et al 1990, 1993a, 1994). The luminosities of the protostellar or "Class I" sources in Taurus are roughly an order of magnitude smaller than would be predicted if the mass added by infalling envelopes is accreted steadily onto the central star. One way around this problem is to assume that material is piling up in the disk because the disk accretion rate is usually lower than the infall rate; then the bulk of the disk material would be accreted in brief, high-luminosity FU Ori events. To determine accretion rates onto central stars, and thus to test this general picture, requires near-infrared spectroscopy to set limits on stellar luminosities and disk infrared emission.

Because the state of the central star during an FU Ori outburst is very likely to be perturbed by the accretion of so much material over such a short period of time, it is important to understand just how much of the mass of a typical star is accreted rapidly. The calculations of Prialnik & Livio (1985) for accretion onto a convective (main-sequence) star suggest that, at FU Ori accretion rates, a convective pre–main sequence star may become radiative and expand substantially, particularly if the disk material retains enough thermal energy as it is incorporated into the star. Such expansion might be consistent, for example, with the current (though uncertain) estimates that inner disk radii are about a factor of two larger than the typical radii of T Tauri stars. Quite large amounts of thermal energy could be accreted into the star in the thick-disk models for the elimination of boundary layer emission (see above). The physical conditions produced by FU Ori accretion are substantially different than assumed in current calculations of the properties of very young stars (Stahler 1988), and this could result in stars appearing at different locations in the HR diagram at the end of infall than predicted by current theory.

5. FU ORI WINDS AND DISK ACCRETION

5.1 *Disk Winds*

FU Ori objects have very strong winds, as indicated by the very broad, deep P Cygni absorption seen in the Balmer lines and the low-ionization Na I resonance lines (Bastian & Mundt 1985). The estimated mass loss rate for the wind of FU Ori is $\dot{M}_w \sim 10^{-5}\ M_\odot\ yr^{-1}$ (Croswell et al 1987, Calvet et al 1993). For comparison, the corresponding mass loss rates for T Tauri stars are $\dot{M}_w \lesssim 10^{-8}$ $M_\odot\ yr^{-1}$ (Edwards et al 1993, Hartigan et al 1995), which demonstrates a direct relationship between mass accretion and mass ejection.

Because FU Ori objects have such strong winds, they exhibit spectroscopic signatures of mass loss that are not observable in T Tauri winds. Calvet et al (1993) showed that the wind of FU Ori itself is so strong that it can be detected in many photospheric absorption lines, which exhibit net blueshifts in absorption. From a detailed consideration of line profile shifts in FU Ori, Hartmann & Calvet (1995) were able to trace the accelerating wind back to the surface of the Keplerian disk. The basis of this analysis was an effect discovered by Petrov & Herbig (1992), who showed that the velocity shift of a spectral line depends directly on its strength; stronger lines show larger blueshifts. As Hartmann & Calvet demonstrated, this is exactly what one would predict for a differentially expanding disk wind; the outer regions of the wind show up only in the strongest lines, which therefore show the largest expansion velocities, whereas weak lines formed close to the disk photosphere exhibit nearly Keplerian rotation (Figure 6). The results support a magnetocentrifugally accelerated disk wind model (Blandford & Payne 1982, Pudritz & Norman 1983, Königl 1989), in which magnetic fields rotating with the disk acclerate the upper layers of the disk outward to high velocities.

5.2 *Mass Loss Energetics and Angular Momentum Transfer*

Mass loss rates for FU Ori winds are difficult to estimate because one requires the (unknown) velocity gradient to convert from column densities to densities, in addition to geometrical complexities. However, it seems fairly clear from a variety of methods (Croswell et al 1987, Calvet et al 1993, Hartmann & Calvet 1995) that the mass loss rates of FU Ori and Z CMa are $\sim 10^{-1}$ of the mass accretion rates; the same ratio of mass loss to mass accretion may be as much as an order of magnitude smaller in V1057 Cyg and V1515 Cyg (see also Rodriguez et al 1991, 1992).

These high mass loss rates suggest that FU Ori accretion disks are extraordinarily efficient in ejecting material, with somewhere between $\sim 1\%$ and 50% of the accretion energy being converted to kinetic energy of the outflowing material. However, this efficiency is not high enough for models in which magnetocentrifugally accelerated winds carry away all of the angular momentum of accretion (e.g. Königl 1989). In such models, most of the accretion energy is carried away by the wind, which means that most of the accretion energy is not radiated by the disk, as assumed in Section 3 to estimate accretion rates for FU Orionis objects. Since $L_{\rm wind}$ is not $\gg L_{\rm rad}$, it is clear that another mechanism of angular momentum transport must be operating in FU Ori disks.

However, as emphasized by Königl (1995), the winds of FU Ori objects can be magnetocentrifugally accelerated even if these winds do not carry away all of the angular momentum of accretion. Indeed, the observation of the rotating wind emerging from the disk supports the magnetic acceleration model. In the absence of meaningful thermal pressure in these low-temperature ($T \sim 6000$ K; Croswell et al 1987) winds, and the lack of a high-temperature source to make radiation pressure important, it is difficult to see how the winds of FU Oris can be accelerated without magnetic fields.

5.3 *Jets, Outflows, and Mass Accretion*

Because of the correlation between mass accretion and mass loss in FU Ori objects, FU Ori outbursts presumably must produce mass ejection events (e.g. Reipurth 1990). Multiple bow shocks in the Herbig-Haro objects HH 34, HH 47, and HH 111 suggest an elapsed time between ejections of 1000–2000 yr (Reipurth 1989a; Hartigan, Raymond & Meaburn 1990; Reipurth & Heathcote 1992; Reipurth, Raga & Heathcote 1992). Several sources with parsec-long jets also display quasi-periodic structures suggestive of multiple ejections on similar timescales (e.g. Bally & Devine 1994).

If, as suggested in Section 2, a typical low-mass star undergoes 10 FU Ori outbursts, and a typical infall phase (limiting the FU Ori lifetime) is $\sim 10^5$ yr, then the outbursts should occur every 10^4 yr. Although the resulting timescale is an order of magnitude larger than estimated from the jet observations, this prediction must be regarded as an upper limit because we probably have not detected all FU Ori outbursts, and if only a subset of stars undergo outbursts, the repetition rate for that subset must increase correspondingly. Thus at least some of the multiple jet ejections observed in young stellar objects could be due to FU Ori (or, perhaps, somewhat weaker) accretion events.

We next consider whether FU Ori mass loss can account for the momentum fluxes of $(Mv)_o = 1$–$100\ M_\odot$ km s^{-1} observed in bipolar molecular outflows from young stellar objects (Fukui 1989). If FU Ori's outburst lasts $\sim 10^2$ yr, then the mass-loss rates and ejection velocities noted in the previous section imply a total wind momentum $\sim 0.3\ M_\odot$ km s^{-1}; many such outbursts would be required to account for a typical molecular ouflow. Another way to look at this issue is to note that the observed flow momenta, if produced by mass ejection at ~ 200 km s^{-1}, would require total ejection masses of 0.05–0.5 $M_\odot$. In other words, a young stellar system must eject $\sim 10\%$ of its final mass to produce a typical molecular outflow. Even if we assume the FU Ori ratio of mass ejected to mass accreted of $\sim 10\%$, which implies a very efficient conversion of accretion energy to wind kinetic energy, one would still require total masses ~ 0.5–$5\ M_{\odot}$ to be accreted through the disk. Thus, apparently, if disk winds are to account for molecular outflows, a large fraction of the stellar mass must be accreted through the disk; if FU Ori outbursts are to account for outflows, much or most of a star's final mass must be accreted in FU Ori events. This possibility is still open (Section 4.5).

6. FUTURE DIRECTIONS

Although the interpretation of the FU Orionis phenomenon in terms of pre–main sequence disk accretion is secure, its general significance is less clear given the uncertainty in source statistics. This situation should improve in the near future, as new near-infrared spectrographs make it possible to survey spectral features of even heavily extincted objects in dense clouds of star formation (Greene et al 1994, Meyer 1994), and the strong, broad 2.2-μm CO first-overtone features of FU Ori objects should be easily detected. Spectroscopic surveys of nearby star-forming regions (e.g. Casali & Mathews 1992, Hanson & Conti 1995) almost certainly will yield much more definite statistics on FU Ori objects in the next few years, resulting in a much better understanding of FU Ori accretion in early stellar evolution.

ACKNOWLEDGMENTS

We are grateful to Robbins Bell for making available Figure 9, to Cesar Briceño for providing Figure 2, to Paola D'Alessio for material on thermal instabilities in disks, and to Nuria Calvet for helpful comments on the manuscript. This work has been supported in part by NASA grant NAGW-2306 and has made use of NASA's Astrophysics Data System Abstract Service.

Literature Cited

Adams FC, Lada C, Shu FH. 1987. *Ap. J.* 312:788–806
Balbus SA, Hawley JF. 1991a. *Ap. J.* 376:214–22
Balbus SA, Hawley JF. 1991b. *Ap. J.* 376:223–33
Bally J, Devine D. 1994. *Ap. J.* 428:L65–68
Bastian U, Mundt R. 1985. *Astron. Astrophys.* 144:57–63
Bath GT, Pringle JE. 1982. *MNRAS* 201:345–55
Bell KR, Lin DNC. 1994. *Ap. J.* 427:987–1004
Bell KR, Lin DNC, Hartmann LW, Kenyon SJ. 1995. *Ap. J.* 444:376–95
Blandford RD, Payne DG. 1982. *MNRAS*

199:883–903
Bonnell I, Bastien P. 1992. *Ap. J. Lett.* 401:L31–34
Bouvier J, Bertout C, Benz W, Mayor M. 1986. *Astron. Astrophys.* 165:110–19
Bouvier J, Cabrit S, Fernandez M, Martin EL, Mathews JM. 1993. *Astron. Astrophys.* 272:176–206
Butner HM, Evans NJ II, Lester DF, Levreault RM, Strom SE. 1991. *Ap. J.* 376:636–53
Butner HM, Natta A, Evans NJ II. 1994. *Ap. J.* 420:326–35
Calvet N, Hartmann L, Kenyon SJ. 1991. *Ap. J.* 383:752–56
Calvet N, Hartmann L, Kenyon SJ. 1993. *Ap. J.* 402:623–34
Carr JS, Harvey PM, Lester DF. 1987. *Ap. J. Lett.* 321:L71–74
Casali M, Mathews HE. 1992. *MNRAS* 258:399–403
Clarke CJ, Lin DNC, Papaloizou JCB. 1989. *MNRAS* 236:495–503
Clarke CJ, Lin DNC, Pringle JE. 1990. *MNRAS* 242:439–46
Clarke CJ, Pringle JE. 1991. *MNRAS* 249:588–95
Cohen M, Schwartz RD. 1987. *Ap. J.* 316:311–22
Covino E, Terranegra L, Vittone AA, Russo G. 1984. *Astron. J.* 89:1868–75
Croswell K, Hartmann L, Avrett EH. 1987. *Ap. J.* 312:227–42
D'Alessio P. 1996. PhD thesis. Universidad Nacional Autonoma de Mexico
Duquennoy A, Mayor M. 1991. *Astron. Astrophys.* 248:485–524
Duschl WJ. 1986a. *Astron. Astrophys.* 163:56–60
Duschl WJ. 1986b. *Astron. Astrophys.* 163:61–66
Edwards S, Ray T, Mundt R. 1993. In *Protostars and Planets III,* ed. EH Levy, JI Lunine, pp. 567–602. Tucson: Univ. Ariz. Press
Eisloeffel J, Hessman FV, Mundt R. 1990. *Astron. Astrophys.* 232:70–74
Elias JH. 1978. *Ap. J.* 223:859–75
Evans NJ II, Balkum S, Levreault RM, Hartmann L, Kenyon S. 1994. *Ap. J.* 424:793–99
Frank J, King A, Raine D. 1992. *Accretion Power in Astrophysics,* pp. 67–105. London: Cambridge Univ. Press. 2nd ed.
Fukui Y. 1989. In *Low Mass Star Formation and Pre–Main Sequence Objects,* ed. B Reipurth, pp. 95–118, *ESO Conf. and Workshop Proceedings No. 33.* Garching: ESO
Gammie CF. 1996. *Ap. J.* 457:355–62
Goodrich RW. 1987. *Publ. Astron. Soc. Pac.* 99:116–25
Gottlieb EW, Liller W. 1978. *Ap. J.* 225:488–95
Graham JA, Frogel JA. 1985. *Ap. J.* 289:331–41
Greene TP, Tokunaga AT, Carr JS. 1994. *Exp. Astron.* 3:309–12
Haas M, Christou JC, Zinnecker H, Ridgway ST, Leinert Ch. 1993. *Astron. Astrophys.* 269:282–90
Hanson MM, Conti PS. 1995. *Ap. J. Lett.* 448:L45–48
Hartigan P, Edwards S, Ghandour L. 1995. *Ap. J.* 452:736–68
Hartigan P, Raymond J, Meaburn J. 1990. *Ap. J.* 362:624–33
Hartmann L. 1991. In *Physics of Star Formation and Early Stellar Evolution. NATO Adv. Study Inst.,* ed. CJ Lada, ND Kylafis, pp. 623–48. Dordrecht: Kluwer
Hartmann L, Calvet N. 1995. *Astron. J.* 109:1846–55
Hartmann L, Hewett R, Stahler S, Mathieu RD. 1986. *Ap. J.* 309:275–93
Hartmann L, Kenyon SJ. 1985. *Ap. J.* 299:462–78
Hartmann L, Kenyon SJ. 1987a. *Ap. J.* 312:243–53
Hartmann L, Kenyon SJ. 1987b. *Ap. J.* 322:393–98
Hartmann L, Kenyon SJ, Hartigan P. 1993. In *Protostars and Planets III,* ed. EH Levy, JI Lunine, pp. 497–520. Tucson: Univ. Ariz. Press
Hartmann L, Kenyon SJ, Hewett R, Edwards S, Strom KM, et al. 1989. *Ap. J.* 338:1001–10
Herbig GH. 1966. *Vistas Astron.* 8:109–25
Herbig GH. 1977. *Ap. J.* 217:693–715
Herbig GH. 1989. In *ESO Workshop on Low-Mass Star Formation and Pre-Main Sequence Objects,* ed. B Reipurth, pp. 233–36. Garching: ESO
Hessman FV. 1990. *Astron. Astrophys.* 246:137–45
Hessman FV, Eisloeffel J, Mundt R, Hartmann L, Herbst W. 1991. *Ap. J.* 370:384–95
Horne K. 1991. In *Structure and Emission Properties of Accretion Disks,* ed. C Bertout, S Collin-Souffrin, JP Lasota, J Tran Thanh Van, pp. 3–18. Gif sur Yvette: Editions Frontieres
Kawazoe E, Mineshige S. 1993. *Publ. Astron. Soc. Jpn.* 45:715–25
Kenyon SJ. 1995a. *Astrophys. Space Sci.* 233:3–17
Kenyon SJ. 1995b. *Rev. Mex. Astron. Astrophys. (Conf. Ser.)* 1:237–45
Kenyon SJ, Calvet N, Hartmann L. 1993b. *Ap. J.* 414:676–94
Kenyon SJ, Gomez M, Marzke RO, Hartmann L. 1994. *Astron. J.* 108:251–61
Kenyon SJ, Hartmann L. 1987. *Ap. J.* 323:714–33
Kenyon SJ, Hartmann L. 1991. *Ap. J.* 383:664–73
Kenyon SJ, Hartmann L, Gomez M, Carr JS, Tokunaga A. 1993a. *Astron. J.* 105:1505–10
Kenyon SJ, Hartmann L, Hewett R. 1988. *Ap. J.* 325:231–51
Kenyon SJ, Hartmann L, Imhoff CL, Cassatella

A. 1989. *Ap. J.* 344:925–31
Kenyon SJ, Hartmann L, Kolotilov EA. 1991a. *Publ. Astron. Soc. Pac.* 103:1069–76
Kenyon SJ, Hartmann L, Strom KM, Strom SE. 1990. *Astron. J.* 99:869–87
Kenyon SJ, Oliverben NG, Mikolajewska J, Mikolajewski M, Stencel RE, et al. 1991b. *Astron. J.* 101:637–54
Kolotilov EA, Petrov PP. 1983. *Pis'ma Astron. Zh.* 9:171–76
Kolotilov EA, Petrov PP. 1985. *Pis'ma Astron. Zh.* 11:846–54
Königl A. 1989. *Ap. J.* 342:208–23
Königl A. 1995. *Rev. Max. Astron. Astrophys. (Conf. Ser.)* 1:275–83
Koresko CD, Beckwith SVW, Ghez AM, Matthews K, Neugebauer G. 1991. *Astron. J.* 102:2073–78
Landolt AU. 1975. *Publ. Astron. Soc. Pac.* 87:379–83
Landolt AU. 1977. *Publ. Astron. Soc. Pac.* 89:704–5
Larson RB. 1980. *MNRAS* 190:321–35
Larson RB. 1983. *Rev. Mex. Astron. Astrophys.* 7:219–27
Levreault RM. 1988. *Ap. J.* 330:897–910
Lin DNC, Papaloizou JCB. 1985. In *Protostars and Planets II,* ed. DC Black, MC Matthews, pp. 981–1072. Tucson: Univ. Ariz. Press
Lynden-Bell D, Pringle JE. 1974. *MNRAS* 168:603–37
McMuldroch S, Blake GA, Sargent AI. 1995. *Astron. J.* 110:354–65
Meyer F. 1984. *Astron. Astrophys.* 131:303–8
Meyer MR. 1994. *Bull. Am. Astron. Soc.* 185:91.04
Mikolajewska J, Kenyon SJ. 1992. *Astron. J.* 103:579–92
Miller GE, Scalo JM. 1979. *Ap. J. Suppl.* 41:513–47
Mineshige S, Osaki Y. 1983. *Publ. Astron. Soc. Jpn.* 35:377–96
Mould JR, Hall DNB, Ridgway ST, Hintzen P, Aaronson M. 1978. *Ap. J. Lett.* 222:L123–26
Mundt R, Stocke J, Strom SE, Strom KM, Anderson ER. 1985. *Ap. J. Lett.* 297:L41–45
Petrov PP, Herbig GH. 1992. *Ap. J.* 392:209–17
Poetzel R, Mundt R, Ray TP. 1990. *Astron. Astrophys. Lett.* 224:L13–16
Popham R, Narayan R. 1991. *Ap. J.* 370:604–14
Popham R, Narayan R, Hartmann L, Kenyon S. 1993. *Ap. J.* 415:L127–30
Prialnik D, Livio M. 1985. *MNRAS* 216:37–52
Pringle JE. 1981. *Annu. Rev. Astron. Astrophys.* 19:137–62
Pudritz RE, Norman CA. 1983. *Ap. J.* 274:677–97
Reipurth B. 1985. In *Proc. ESO-IRAM-Onsala Workshop on Sub-Millimeter Astronomy,* ed. PA Shaver, K Kjar, pp. 458–72. Garching: ESO
Reipurth B. 1989a. *Nature* 340:42–44
Reipurth B. 1989b. In *ESO Workshop on Low-Mass Star Formation and Pre-Main Sequence Objects,* ed. B Reipurth, pp. 247–80. Garching: ESO
Reipurth B. 1990. In *Flare Stars in Star Clusters. IAU Symp.*, ed. LV Mirzoyan, BR Petterson, MK Tsvetkov, 137:229–35. Dordrecht: Kluwer
Reipurth B, Heathcote S. 1992. *Astron. Astrophys.* 257:693–700
Reipurth B, Raga AC, Heathcote S. 1992. *Ap. J.* 392:145–58
Rodriguez LF, Hartmann LW. 1992. *Rev. Mex. Astron. Astrophys.* 24:135–38
Rodriguez LF, Hartmann LW, Chavira E. 1990. *Publ. Astron. Soc. Pac.* 102:1413–17
Sato S, Okeita K, Yamashita T, Mizutana K, Shiba H, et al. 1992. *Ap. J.* 398:273–77
Shakura NI, Sunyaev RA. 1973. *Astron. Astrophys.* 24:337–55
Shu FH, Adams FC, Lizano S. 1987. *Annu. Rev. Astron. Astrophys.* 25:23–81
Shu FH, Najita J, Ostriker E, Wilkin F, Ruden SP, Lizano S. 1994. *Ap. J.* 429:781–96
Smak J. 1984a. *Acta Astron.* 34:161–89
Smak J. 1984b. *Publ. Astron. Soc. Pac.* 96:5–18
Stahler SW. 1988. *Ap. J.* 332:804–25
Stahler SW. 1989. *Ap. J.* 347:950–58
Staude HJ, Neckel Th. 1991. *Astron. Astrophys. Lett.* 244:L13–16
Staude HJ, Neckel Th. 1992. *Ap. J.* 400:556–61
Stocke JT, Hartigan PM, Strom SE, Strom KM, Anderson ER, et al. 1988. *Ap. J.* 68:229–55
Strom KM, Strom SE. 1993. *Ap. J. Lett.* 412:63–66
Terebey S, Shu FH, Cassen P. 1984. *Ap. J.* 286:529–51
Thamm E, Steinacker J, Henning Th. 1994. *Astron. Astrophys.* 287:493–502
Tsvetkova KP. 1982. Photographic Photometry of V1515 Cygni. *Inform. Bull. of Variable Stars.* 2236:1–3
Weintraub DA, Sandell G, Duncan WD. 1991. *Ap. J.* 382:270–89
Welty AD, Strom SE, Edwards S, Kenyon SJ, Hartmann LW. 1992. *Ap. J.* 397:250–76
Welty AD, Strom SE, Strom KM, Hartmann LW, Kenyon SJ, et al. 1990. *Ap. J.* 349:328–34
Whitney BA, Clayton GC, Schulte-Ladbeck RE, Calvet N, Hartmann L, Kenyon SJ. 1993. *Ap. J.* 417:687–96

Annu. Rev. Astron. Astrophys. 1996. 34:241–77

CIRCUMSTELLAR PHOTOCHEMISTRY

A. E. Glassgold

Department of Physics, New York University, 4 Washington Place, New York, NY 10003-1113

KEY WORDS: circumstellar matter, AGB stars, molecules

ABSTRACT

The cooling flows or winds from evolved stars are ideal for the formation of molecules and dust. The main location of molecular synthesis is the outer circumstellar envelope, where UV radiation from the interstellar medium penetrates the envelope and, by photodissociating parent molecules, produces the high-energy radicals and ions that activate gas-phase neutral and ion-molecule chemistry. After introducing relevant observational results and theoretical ideas, the salient aspects of the photochemical model are described. The primary application is to the nearby C star, IRC + 10216, where 50 or more circumstellar molecules have been detected. Recent interferometer maps, with resolution approaching 1″, provide the means to verify the main ideas of the model and to indicate directions for its improvement.

1. INTRODUCTION

Circumstellar chemistry is a very broad subject because all stars have some circumstellar matter. Some of the most dramatic examples occur during the birth and death of stars, where mass gain and loss rates approach 10^{-5}–10^{-4} $M_{\odot}$ yr^{-1} for short periods of time. The present review is restricted to the circumstellar envelopes of evolved stars, with particular emphasis on very red stars on the asymptotic giant branch. This specialization is justified by the importance of this stage of stellar evolution and by restrictions on the length of this article. Circumstellar chemistry has not been reviewed previously in this series, except as part of Zuckerman's (1980) general introduction to the envelopes of late type stars.

0066-4146/96/0915-0241$08.00

The envelopes of asymptotic giant branch (AGB) stars occupy strategic intermediate ground between stellar and interstellar physics. For example, mass loss plays a decisive role in the final stages of stellar evolution. Indeed, the amount of mass lost by an intermediate-mass star on the AGB determines whether it explodes as a supernova or not. For the interstellar medium, AGB stars are the most important source of the seeds of interstellar dust. These envelopes also raise fundamental astrophysical problems, such as the origin of winds, the formation of dust, and the synthesis of complex molecules. With all of this astrophysical richness, it is fortunate that the envelopes of evolved stars are observable in considerable detail with infrared and millimeter-wave astronomy. Extensive efforts are currently underway to provide arcsec and subarcsec spatial resolution capabilities at both near-infrared and millimeter (and submillimeter) wavelengths and to extend observations into the 30–300 μm band inaccessible from the ground (Ishiguro & Welch 1994). Some of the most interesting applications of these new observational developments will be to circumstellar gas and dust.

Both the physics and chemistry of the winds are of paramount importance for interpreting observations of evolved stars. Because temperature and density decrease substantially with distance from a star, circumstellar material tends to undergo phase changes that favor molecules and solids over atoms and ions. Because the dynamical (flow) times are short, it is unlikely that these changes adiabatically follow the local gas temperature and pressure. The winds are also affected by "external agents" that drive circumstellar matter far from local thermodynamic equilibrium (LTE), e.g. UV radiation and high-energy particles from the star and the interstellar medium. Thus, a fundamental problem in circumstellar chemistry is to evaluate the degree to which AGB stars manifest "frozen-out" equilibrium chemistry or non-LTE chemistry. In either case, the character of circumstellar chemistry depends on the evolutionary stage of the star. For example, the UV radiation from a very hot planetary nebula star is far more destructive of molecules than the radiation from the spectrum of a cool AGB star.

Even with the restriction to AGB stars, circumstellar chemistry is complicated by the large range of physical conditions encountered. Three related subdisciplines are involved: the formation and evolution of dust, the catalysis of gas-phase molecules on dust grain surfaces, and "pure" gas-phase chemistry. In the dense inner regions of a circumstellar envelope, processes subsumed under all three categories are likely to occur. Even in the outer envelope, more than just gas-phase chemistry may be involved. Despite their importance in diverse branches of astronomy, both dust formation and catalysis are poorly understand, and we focus here on the better developed, gas-phase chemistry. Most of the observational progress of the past 10–15 years has involved spectroscopic studies at millimeter wavelengths of gas-phase molecules.

Although we emphasize gas-phase chemistry, we also discuss observations and physical processes that relate to dust. Indeed, an important reason to study the molecular properties of circumstellar envelopes is to obtain information about dust formation and catalysis, which tend to occur close to the star but are difficult to observe directly. Before evaluating the extent to which gas-phase chemistry can account for the observations of the outer envelope, we have to inquire into the reactions that occur in circumstellar envelopes. For this purpose, it is useful to identify three main classes of reactions: *ionic, neutral, and photo.* The key ionic processes are the ion-molecule reactions familiar from interstellar chemistry. Because the temperatures of the envelopes are low, the only efficient neutral reactions involve radicals. Although these have long been suspected to be astrophysically important, the necessary experimental basis for applying radical reactions to interstellar and circumstellar matter has been unavailable until recently, and it is now possible to critically compare the potential of ionic and neutral reactions for circumstellar chemistry.

We shall see that the chemical activity produced by both neutral and ionic reactions is generated by the interstellar radiation field that penetrates the envelope. We adopt the point of view that *circumstellar gas-phase chemistry is essentially photochemistry*. Accordingly, this review is organized as follows. An astrophysical overview is given in Section 2 to provide the general context for circumstellar chemistry. Relevant observations are discussed in Section 3, especially for the best studied envelope of the nearby C star, IRC + 10216. General chemical issues are considered in Section 4 and circumstellar chemistry in Section 5; Section 6 is a brief conclusion.

The scope of this review has been restricted to provide a detailed account of the most important chemical model of circumstellar envelopes. Many of the topics that might have been included in a longer article are well covered in recent conference proceedings. Among the most useful are: the carefully developed summary of general circumstellar chemistry by Omont (1991), the broad reviews of the observations and their interpretation by Olofsson (1994a,b), and a critical discussion of archetypical circumstellar envelopes by Huggins (1995). The evolutionary context is emphasized in Menessier & Omont (1990).

2. OVERVIEW

2.1 *Caricature of an AGB Star*

The astrophysical context for circumstellar chemistry is provided by the caricature of an AGB star shown in Figure 1. A thin cut has been made in the equatorial plane of the object. All such cuts look the same for a spherically symmetric envelope, which we assume is the case for purposes of orientation. Radial distances r have been marked by powers of 10. On a global scale, the lengths of general interest extend over 10 decades, from about 10^9 cm, the radius of the burnt-out core of the star, to 10^{19} cm, where the envelope is stopped by the interstellar medium. An angular distance scale can be introduced; on the assumption that the star is 200 pc distant; 1″ corresponds to $r = 3 \times 10^{15}$ cm.

Going outward, it is useful to recognize five regions: 1. the region just outside the stellar core, where nuclear burning occurs; 2. the pulsating envelope of the star, which includes the bulk of the stellar atmosphere; 3. the inner circumstellar envelope; 4. the main circumstellar envelope; and 5. the far outer envelope, which terminates in the interstellar medium. There is no sharp demarcation between these regions and, indeed, we are particularly interested in the connections between them. Furthermore, mass loss is intrinsically a time-dependent process that depends sensitively on the character of the star

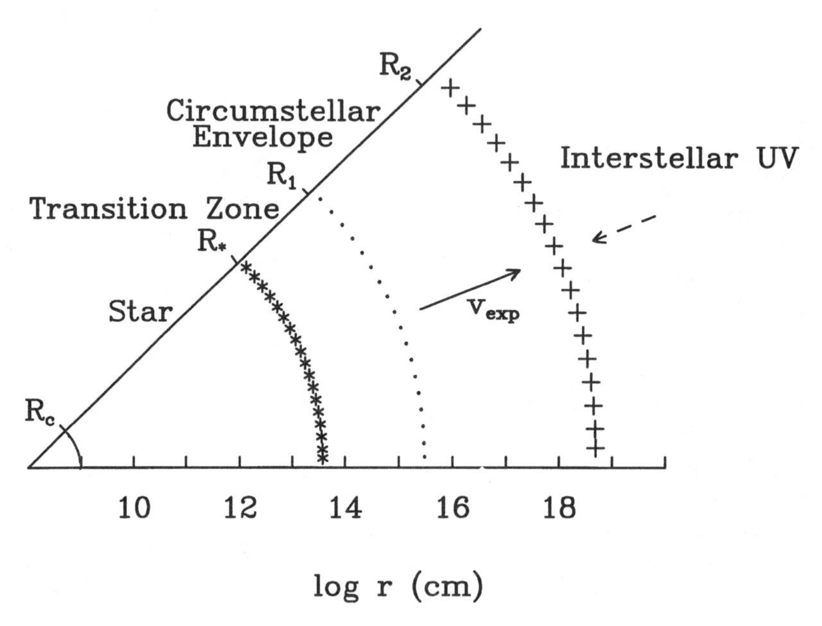

Figure 1 Caricature of an AGB Star. A slice of the star with the space divided into the star proper, circumstellar matter (an inner transition zone and the main circumstellar envelope), and the interstellar medium. Distances are shown on a logarithmic scale at the bottom. The inner and outer radii of the envelope, R_1 and R_2, are marked, as are the photospheric radius R_*, the stellar core radius R_c, and the expansion velocity v_{exp}. The angular distance of 1″ (not marked) is at $r = 3 \times 10^{15}$ cm, if the star is 200 pc away.

and its evolutionary stage, so that the properties of the five regions vary from envelope to envelope. By the *main* outer circumstellar envelope, we mean the region where the expanding wind has reached terminal speed and where most of the mass that will condense into solids has already done so. This is also the part of the system that can be studied effectively with infrared and radio astronomy.

2.2 *Properties of the Main Circumstellar Envelope*

To gain some perspective on the main circumstellar envelope, it helps to have in mind values for the inner and outer radii, even though these distances are not well defined. We conservatively take the inner radius of the main circumstellar envelope to be $R_1 = 3 \times 10^{15}$ cm, which is about $100R_*$ (since a typical photospheric radius for an AGB star is 3.75×10^{13} cm). For the nominal stellar distance $d = 200$ pc used in Figure 1, this corresponds to an angular distance from the star of $1''$. According to interferometric observations, the dust around Mira variables and C stars forms well within a distance of the order of R_1 (Danchi & Bester 1995). The outer radius of the envelope R_2 is determined by the lifetime or duration of the mass-loss phase under consideration. This is an important quantity that one would like to measure for individual envelopes, but good diagnostics of distant circumstellar material are hard to find. As discussed below, most easily observed molecules are destroyed by the interstellar radiation field in the middle of the envelope, between 10^{16} and 10^{17} cm. The continuum emission is weak because the dust temperature and emissivity become very small at large distances. In nearby sources, however, the extension of the CO emission provides a lower limit to the outer radius of the main envelope. CO is relatively abundant at large r because of strong self-shielding (Mamon et al 1988) so that, despite its small permanent dipole moment, it can be detected out to large distances. For example, CO has been been detected around IRC + 10216 out to 3–4$'$ (Huggins at al 1988), which gives $R_2 > 6 \times 10^{17}$ cm (for $d = 200$ pc). Models of the far-IR and sub-mm continuum emission of circumstellar envelopes suggest that the dust extends even further (Sopka et al 1985, Jura 1988). The *IRAS* 60-μm survey data support this expectation: About 15% of a sample or more than 500 AGB stars and planetary nebulae (PNe) have envelopes larger than 2$'$ (Young et al 1993).

Because the expansion velocities are constant in the main circumstellar envelope, radial position and time changes are linearly related, i.e. $dt = dr/v_{\rm exp}$. The terminal expansion speeds of the winds are typically of order 10 km s^{-1}, i.e. in the range 5–25 km s^{-1} (Olofsson 1994a). Thus the time Δt for a particle to travel $\Delta r = r_2 - r_1$ is

$$\Delta t = \Delta r/v_{\rm exp} = 317\,{\rm yr}\,(\Delta r_{16}/v_{\rm exp,6}), \tag{1}$$

where distances and velocities are measured in units of 10^{16} cm and 10^6 cm s^{-1}, respectively. Using the measured expansion velocity of IRC + 10216, $v_{\rm exp} = 14$ km s^{-1}, the observed size limit for CO implies a minimum duration of mass loss for this object of more than 1.4×10^4 yr.

The mass-loss rates inferred from observations are in the range 10^{-7}–10^{-4} $M_\odot$ yr^{-1} (Olofsson 1994a). If we continue to assume spherical symmetry, mass conservation leads to the gas density

$$n_{\rm H} = \frac{C_{\rm H}}{r^2}, \tag{2}$$

where

$$C_{\rm H} = 3 \times 10^{37}\ {\rm cm}^{-1} \left(\frac{\dot{M}_{-5}}{v_{\rm exp,6}} \right),$$

with the mass-loss rate $\dot{M}$ measured in units of 10^{-5} $M_\odot$ yr^{-1}. Even if spherical symmetry and uniform expansion do not apply in practice, this formula should still give a reasonable estimate of the average density. If we take the mass-loss rate of IRC + 10216 to be 3×10^{-5} $M_\odot$ yr $^{-1}$, Equation (2) gives $n = 6 \times 10^5\, r_{16}^{-2}$ cm^{-3}. With r_{16} varying from 0.3 (1″) to 300 (1000″), the density in the main circumstellar envelope of IRC + 10216 varies by a factor of 10^6, ranging from the densities observed in the densest molecular cloud cores to the those in diffuse interstellar clouds.

The kinetic temperature in the main circumstellar envelope probably varies much less than the density. According to the original theoretical considerations of Goldreich & Scoville (1976), the gas is heated mainly by the viscous drag exerted by the dust and cooled by line radiation. This general picture has been followed up in many contexts, usually on the asumption that the envelope is spherical. A recent comprehensive study for molecular envelopes by Groenwegen (1994) gives references to earlier work. It is usually possible to approximate the theoretical calculations by one or more power laws of the form

$$T(r) = T_o \left(\frac{R_o}{r} \right)^p, \tag{3}$$

where T_o is the temperature at a fiducial radius R_o and the exponent p ranges from 0.5 to 1. Groenwegen and other workers have also included the heating due to photoelectrons ejected from dust grains by the interstellar radiation field, which raises the temperature at large distances ($r > 30R_1$). The theory implies that the temperature varies inversely with mass-loss rate (Jura et al 1988, Kastner 1992). Most chemical models of IRC + 10216 use the results of Kwan & Linke (1982), which can be approximated by $R_o = 9 \times 10^{16}$ cm, $T_o = 12$ K, and $p = 0.72$ for $r < R_o$ and $p = 0.54$ for $r > R_o$, plus the limitation that $T > 10$

K. The main observational constraints used in the modeling are line shapes and spatial distributions of the integrated brightness temperature for the two or three lowest rotational transitions of the CO molecule (e.g. Huggins et al 1988, Groenwegen 1994). Although the more recent anaylses differ in detail from that of Kwan & Linke (1982), they indicate that the middle of the main envelope of IRC + 10216 is quite cool, e.g. $T \approx 25$ K at $r = 4.5 \times 10^{16}$ cm.

The main circumstellar envelope becomes optically thin at large distances (Figure 1), even to high-frequency radiation, if the mass loss has gone on for a sufficiently long time. Then Equation (2) applies, and the associated radial column density from ∞ to r, $N_H(r) = C_H/r$, helps measure the optical depth for radiation from the interstellar medium to penetrate to r (if we ignore various niceties of radiative transfer and assume $r \ll R_2$). On this basis, it is of interest to locate approximately the position where the interstellar radiation field acquires unit optical depth due to extinction by the dust. The wavelengths of primary concern for circumstellar chemistry are those responsible for molecular photodissociation, i.e. $900 < \lambda < 3000$ Å. On the assumption that the usual gas to dust ratio (ρ_g/ρ_d) is independent of position, dust and gas column densities are related by

$$N_d \left(\frac{\rho_g}{\rho_d}\right) \frac{m_g}{m_d} N_H,$$

where ρ_g and ρ_d are, respectively, the volume densities of the gas and the dust in the wind, and m_g and m_d are, respectively, the mean masses of gas and the dust particles. For the gas, $m_g \approx 1.3\, m_H = 2.2 \times 10^{-24}$ g, whereas for the dust, $m_d = \hat{\rho} V$, where $\hat{\rho}$ is the average internal density of the dust grains (about 2 and 3 g cm^{-3} for carbonaceous and silicaceous compositions, respectively) and V is their average volume. Expressing the specific extinction properties of the dust in terms of the average grain's extinction cross section divided by its volume, $(C(\lambda)/V)$[1], we obtain the familiar relation between optical depth and gas column density,

$$\tau_\lambda = \left(\frac{\rho_g}{\rho_d}\right) \frac{m_g}{\hat{\rho}} \left[\frac{C(\lambda)}{V}\right] N_H. \tag{4}$$

If we use $N_H = C_H/r$ with C_H from Equation (3), we can solve Equation (4) for r_1, the position where $\tau_\lambda = 1$:

$$r_1 = 9.7 \times 10^{16}\,\text{cm}\, \frac{\dot{M}_{-5}}{v_{\text{exp},6}} \left[\frac{C(\lambda)}{V}\right]_6, \tag{5}$$

[1]For spherical grains, $[C(\lambda)/V] = 3Q(\lambda, a)/4a$, where Q is the extinction coefficient (Spitzer 1978), so that for $a = 0.1\,\mu$m, $[C(\lambda)/V] = 7.5 \times 10^5 Q(\lambda, a)$ cm^{-1}.

with $C(\lambda)/V$ in units of 10^6 cm^{-1}. For the interstellar medium, $C(\lambda)/V = 1.6 \times 10^5$ cm^{-1} at visual wavelengths but $C(\lambda)/V = 7.4 \times 10^5$ cm^{-1} at 1000 Å (Draine & Lee 1984). If we use the calculations of Rouleau & Martin (1991) for amorphous carbon grains with 100-Å radii, the far-UV extinction is even larger, i.e. $C(\lambda)/V = 10^6$ cm^{-1} at 1000 Å.

Equation (5) implies that the dust in a thick circumstellar envelope ($\dot{M}_{-5} > 0.1$) shields the interior from interstellar far-UV radiation. The effective shielding radius also depends on the photodissociation cross section for each molecule. Abundant molecules with large cross sections can provide additional *self-shielding,* which may exceed that due to dust; H_2 and CO are the primary examples. In any case, Equation (5) suggests that molecules entering the main circumstellar envelope at R_1 will at first be protected from photodissociation but will eventually get destroyed at some radius $R_{\rm phdiss} \ll R_2$. In between, their dissociation products (radicals) may engage in chemical reactions, and some of these radicals and their progeny may be detectable by millimeter-wave radio astronomy.

Implicit in this first discussion of $R_{\rm phdiss}$ is the assumption that the duration of the mass-loss episode ($\tau_{\dot{M}}$) is much longer than the photodissociation time. In this case, we need to ensure that the outer radius of the circumstellar envelope R_2 is actually determined by $\tau_{\dot{M}}$ (i.e. by $R_2 = v_{\rm exp}\tau_{\dot{M}}$) and not by dynamical interaction with the interstellar medium. The termination of circumstellar shells has been discussed by Young et al (1993). The simplest situation involves the formation of a spherical shell and the deceleration of the wind by the sweeping up of interstellar gas. A characteristic time is that required to sweep up a mass equal to that of the envelope, which, for a spherical shell and no net motion relative to the interstellar medium, corresponds to a termination radius,

$$R_{\rm term} = \left(\frac{3C_{\rm H}}{n_{\rm o}}\right)^{1/2} = 9.5 \times 10^{18}\,{\rm cm}\left(\frac{\dot{M}_{-5}}{v_{\rm exp,6}}\frac{1}{n_{\rm o}}\right), \tag{6}$$

where $n_{\rm o}$ is the density of the interstellar medium. When the deceleration of the shell is included (Young et al 1993), $R_{\rm term}$ is increased, but the pressure of the interstellar medium (with thermal, turbulent, cosmic-ray, and magnetic components) has the opposite effect. Thus Equation (6) offers a reasonable first-order estimate of the dynamical size of an envelope determined by its interaction with the interstellar medium. As far as is known, this estimate is not in disagreement with any observations and, indeed, Young et al find a substantial number of large shells in the *IRAS* 60-μm data. The interstellar medium might induce more drastic effects on circumstellar envelopes if they are moving very fast or if the medium is dense ($n_o \gg 100$ cm^{-3}). Regarding the latter possibility, however, Kastner & Myers (1994) estimate that giant molecular clouds are unlikely to harbor many AGB stars. From the chemical point of view, the most important conclusion from Equation (6) is that $R_{\rm phdiss} \ll R_{\rm term}$ in most situations, so that *easily observable molecules are destroyed inside the circumstellar envelope.* [H_2 is expected to be destroyed close to the leading edge of the wind (Glassgold & Huggins 1983).]

2.3 *The Stellar Context*

Although we are mainly concerned with the outer circumstellar envelope, we should not lose sight of an important goal, which is to gain insights into the inner workings of AGB stars from observations of their outer envelopes. There are, of course, an immense number of red giants in the Milky Way, all losing mass at some level. However, most of them have modest mass-loss rates and imperceptible infrared excesses, and these can be studied in detail at optical and UV wavelengths. The highly reddened stars became widely known through the pioneering near-IR CIT/IRC (Neugebauer & Leighton 1969) and AFGL (Walker & Price 1975, Price & Walker 1976) surveys, which contain almost all of the main archetypes of current interest. Later the *IRAS* survey revealed an estimated 80,000 evolved stars and yielded a rich trove of data on broadband far-IR fluxes of many kinds of evolved stars (Habing 1990).

Variability appears to play a central role: Essentially all of the highly reddened stars are large-amplitude, long-period variables (e.g. Habing & Bloemmaert 1995). (Such stars have periods longer than 150 d, amplitudes greater than some minimum value, e.g. 2.5 mag in the visible or 0.5 mag at 2.2 μm, and luminosities greater than 3000 $L_\odot$.) Important examples are the optically visible Miras, including S stars (C/O $\approx$ 1) and C stars (C/O > 1), as well as M stars (C/O < 1). Indeed, all of the optical and infrared large-amplitude, long-period variables, including the OH/IR stars, have the properties expected for AGB stars. Habing & Bloemmaert (1995) have proposed classifying AGB stars according to period, going from optical Miras (P < 450 d), through infrared Miras (450 < P < 800 d), to OH/IR stars (P > 800 d). For circumstellar chemistry, we emphasize a related correlation, the general increase in mass-loss rate with period (e.g. Wood 1990). Because chemical activity increases with gas density, we can expect a correlation between chemical complexity and period. Because of their proximity, the sources in the CIT/IRC and AFGL catalogs, such as IRC + 10216, naturally play an important role in circumstellar chemistry.

Although we do not discuss stars on the red giant branch ($L < 2200\ L_\odot$), this does not mean that mass-loss from these stars is unimportant or that their envelopes are chemically inactive. This is also true for the M supergiants (usually $L > 5 \times 10^5\ L_\odot$), which we mention only in passing. The supergiants are particularly intriguing because their envelopes are the first circumstellar material encountered by Type II supernova blast waves. Huggins (1995) has summarized the properties of the archetype early-M supergiant, α Ori. Its circumstellar envelope is mainly atomic, as manifested by the detection of HI (Bowers & Knapp 1987, 21 cm), CI (Huggins et al 1994, 609 μm), OI (Haas & Glassgold 1993, 63 μm), and KI (e.g. Mauron 1990, 7700 Å)—all in different wavelength regions! Additional support for the atomic character of this envelope comes from the small abundance of CO (Huggins 1987, Huggins et al 1994) and of dust (Skinner & Whitmore 1987). Recent measurements of the 3-mm continuum and CO line emission for a large sample of O-rich stars have turned up objects with deficits in molecular gas, and a population of dusty red supergiants may be responsible (Josselin et al 1995). Some M supergiants have a few molecules (Bujarrabal et al 1994), unlike α Ori, where only CO has been detected. Finally, the M supergiants have been imaged in scattered resonance line radiation (Bernat & Lambert 1975, Mauron 1990), a technique now being extended to other kinds of circumstellar envelopes (Plez & Lambert 1994, Mauron & Guilain 1995), and of particular interest for the heavy elements.

3. OBSERVATIONS

The main gaseous constituents of the circumstellar envelopes of cool evolved stars are neutral or singly-ionized atoms and molecules. Although these systems offer transitions across the electromagnetic spectrum, the low temperature of the envelope favors *emission line spectroscopy* with the rotational transitions of molecules, the fine structure lines of atoms, and similar low-excitation lines. [Rotational transitions can be observed from low-lying, excited vibrational levels of complex molecules, as well as from the ground level (e.g. Ziurys & Turner 1986, Lucas & Guilloteau 1992).] *Absorption line spectroscopy* from the ground electronic level is also possible, but the wavelength coverage is limited by the extinction of the envelope. For example, the envelope of α Ori is sufficiently thin ($\approx$ 1 mag of visual extinction) that optical (Goldberg 1979, Querci 1986) and UV (e.g. Carpenter et al 1994) spectroscopy can be carried out. On the other hand, IRC + 10216 is so reddened ($\geq$ 100 mag) that observations of its spectrum in the red are only possible because scattered photospheric light escapes due to inhomogeneities and asymmetries in the envelope. Remarkably, Bakker et al (1994) have recently discovered C_2 and CN in the circumstellar envelope of IRC + 10216 in absorption near 8000 Å. We stress the radio (mainly mm-wave) and near-IR observations.

Our objective is not to review all potentially relevant observations for circumstellar chemistry, but to select some that, in conjunction with existing theory, raise basic questions about the chemistry. This selection process is a personal one, and the writer admits favoring in-depth observational studies of individual objects. The number of interesting objects that can be discussed at the present time is actually relatively small; indeed they involve only a few well-studied, nearby objects. Foremost among them is the nearby carbon star, IRC + 10216 (CW Leo). Its thick, dense envelope manifests a rich chemistry that rivals interstellar chemistry. Moreover, its proximity allows sensitive observations to be made on a sufficiently small spatial scale to obtain definitive conclusions on chemical processes. There are only a few other sources that can compare, so the focus on IRC + 10216 is well justified at this time.

3.1 *The Case of IRC + 10216*

A list of the species detected in the circumstellar envelope of IRC + 10216 as of late 1995 is shown in Table 1. Following the custom of an earlier review (Glassgold & Huggins 1986a), a rough abundance scale is indicated by the decade labels in the first column. The abundance of a species X is defined relative to the total number of H nuclei,

$$x(\mathrm{X}) = n(\mathrm{X})/n_{\mathrm{H}}; \tag{7}$$

$x(\mathrm{X})$ is close to $0.5f(\mathrm{X})$, the abundance $f(\mathrm{X})$ relative to molecular hydrogen (often favored by observers). Without going into the technical details (available in Omont 1991 and Olofsson 1994a), we note several problems in measuring absolute abundances. For example, circumstellar species have nonuniform spatial distributions as well as excitation properties, so both the abundance and excitation have to be determined as a function of position. This requires high spatial resolution measurements of at least two lines with significantly different excitation, followed by a detailed analysis with suitable radiative transfer models. It has not yet been possible to carry out this idealized procedure in many cases, and Table 1 consists of qualitative estimates made in different ways. Some abundances have been determined from IR absorption line spectroscopy of molecules located closer to the star than most of those studied at millimeter wavelengths, adding to the heterogeneity of the numbers. The uncertainties in the abundances in Table 1 range from 3–10, but relative values may be better than that.

An attempt has been made to segregate the species in Table 1 into families according to primitive, possibly arbitrary, chemical ideas:

Table 1 Species detected in IRC + 10216[1,2]

10^{-3}								
	CO							
10^{-4}								
			$C_2H_2^*$					
	C				HCN			
10^{-5}								
			C_2H		CN			
		CH_4^*	C_4H			SiS	CS	
10^{-6}					HC_3N			
			C_3^*		C_3N			
					HC_5N	SiC_2		AlCl
10^{-7}		SiH_4^*	C_6H		HNC	SiO		AlF
		NH_3	C_5^*		HC_7N	SiN		NaCN
			cC_3H_2		HC_9N	SiC		MgNC
10^{-8}			lC_3H	H_2C_4	$HC_{11}N$		C_2S	CP
			C_5H	HC_2N	CH_3CN		C_3S	NaCl
				H_2C_3				MgCN
10^{-9}			cC_3H			SiC_4	H_2S	KCl
	HCO^+							
10^{-10}								

[1] C_2 has been detected but its abundance has not been determined.
[2] The species marked with an asterisk have been detected only in the infrared.

column 2: CO, plus its protonated form HCO^+ and atomic C;

column 3: the few fully hydrogenated species;

column 4: simple hydrocarbons, starting with and probably derived from acetylene; almost all are linear chains and weakly hydrogenated;

column 5: carbene isomers and HC_2N;

column 6: HCN, its isomer HNC and radical CN, plus cyanopolyyne derivatives;

column 7: silicon species;

column 8: sulfur species not bound to silicon; and

column 9: heavy element molecules.

The species marked with an asterisk have been detected so far only in the near IR (e.g. Keady & Ridgway 1993, Hinkle 1994); CO, C_2H_2, C_2H, HCN, CN, NH_3, SiO, and CS have been detected at both near-IR and millimeter wavelengths.

Many isotopic variants have been detected, but this subject is not reviewed here; Olofsson (1994a) gives a brief summary.

With fifty or so molecular detections in IRC + 10216, it is easy to forget the remarkable effort by radio and infrared astronomers to obtain them over the past two decades. Many of the species are radicals whose spectra and structure were unknown beforehand. Their quantum mechanical characterization and astronomical detection had to be carried out in parallel and involved unprecedented collaboration between astronomers, molecular physicists, and quantum chemists. A recent illustration is the detection of MgNC in all three stable Mg isotopes (Guélin et al 1995), which involved laboratory spectra by Ziurys and Anderson, quantum calculations by Valiron, astronomical observations by Guélin and his associates, and nucleosynthetic calculations by Forestini.

Another fascinating aspect of molecular research on circumstellar envelopes has been the challenge to explain the detection of the cyanopolyynes in IRC + 10216: HC_3N (Morris et al 1975), HC_5N (Avery et al 1976), HC_7N (Kroto et al 1978), HC_9N (Broten et al 1978), and $HC_{11}N$ (Bell et al 1982). The idea that their synthesis is related to that of carbon chains and particles led to the discovery of the fullerenes (e.g. Kroto 1988). It is perhaps ironic that C_{60} has not been detected yet in IRC + 10216 (e.g. Hinkle & Bernath 1993).

One important aspect of Table 1 is the absence of atomic and molecular hydrogen, due to the fact that *detections* and not limits are listed. That pure hydrogen species have not yet been reported in IRC + 10216 reflects the difficulties of detecting them under cool or even warm conditions. The early theoretical and observational situation was reviewed by Glassgold & Huggins (1986a). Since then, Bowers & Knapp (1987) made a sensitive search for the 21-cm line with the VLA and obtained an upper limit to the H I mass: $M(\mathrm{H}) < 3\times10^{-4}\, M_\odot$ (for $d =$ 290 pc). This is only a factor of two larger than the mass of atomic hydrogen produced by photodissociation of H_2 and hydrocarbons by the interstellar UV radiation field, but much larger than the simplest but highly uncertain estimate of the H I frozen out from LTE in the upper atmosphere (Glassgold & Huggins 1983). The Bowers & Knapp H I limit is one of the most important measurements for the circumstellar chemistry of IRC + 10216. It corresponds to an abundance limit of $x(\mathrm{H}) < 2\times10^{-4}$, and it restricts the amount of H I in the extended atmosphere of the star. Thus it can provide a test of pulsation-driven shock models (e.g. Bowen 1988) and of chemical synthesis models for dust or dust precursors that depend sensitively on the H/H_2 ratio (e.g. Frenklach & Feigelson 1989). It also limits the role of H-initiated chemical reactions in the outer circumstellar envelope.

With so little atomic hydrogen, the hydrogen in such a reddened stellar envelope must be almost entirely molecular. Keady & Ridgway (1993) have considered the prospects of detecting H_2 using ground-based measurements of the pure-rotation and rotation-vibration lines of H_2 in the near IR. They attempted to detect the 2.122 μm $S(1)$ line and obtained the upper limit, $\dot{M}(H_2) < 4 \times 10^{-4}$ $M_\odot$ yr^{-1} (for a distance of 200 pc). This limit is consistent with, but not much larger than, the mass-loss rate determined from CO observations, $\dot{M}(H) \approx (2–4) \times 10^{-4}$ $M_\odot$ yr^{-1} (e.g. Huggins et al 1988). Again this is an important observation because it would measure the rate of loss of the most abundant species in the envelope and would help constrain the absolute abundance of CO.

Table 1 offers many other insights into the chemistry of IRC + 10216, some of which are:

1. The preponderance of C-bearing molecules indicates that the carbon to oxygen ratio is significantly greater than unity; all of the oxygen is taken up by CO with a small assist from SiO.

2. The abundance of oxygen, measured by CO, is approximately 1/2 Solar, consistent with visible C stars (Lambert et al 1986).

3. Molecular abundances are generally much higher than those estimated for interstellar clouds.

4. Simple, well-bound molecules are the most abundant, e.g. CO, C_2H_2, HCN, SiS, and CS.

5. Almost half of the observed species are radicals or isomers.

6. Polyatomic molecules with more than one heavy atom tend to form chains, rather than rings.

7. Only one ion has been detected, HCO^+.

8. There are many molecules with elements heavier than S and Si, including species not observed in the interstellar medium.

9. The abundances of molecules with atoms heavier than C, N, and O suggest that the depletion of heavy elements is less than in interstellar clouds.

Items 3 and 4 are consistent with a photospheric origin, but item 5 argues against the idea that the entire suite of molecules can be characterized by freeze-out from LTE (McCabe et al 1979). Items 5 and 6 have been challenges for circumstellar chemistry and are discussed in Section 5. The detection of a single ion is only partly the result of the lack of sensitivity or the unavailability of certain transitions from the ground, such as the 158 μm fine structure C II line. (Efforts will be made to detect this line in circumstellar envelopes with

ISO.) The very low abundance of HCO^+ indicates that the flux of high-energy particles ionizing H_2 must be quite small in IRC + 10216. Items 8 and 9 are relevant to the physics and chemistry of circumstellar dust.

The list in Table 1 supports another distinction, based on the *place of origin of the molecules*. In a circumstellar envelope, essentially all of the elements come from the star. Some species survive their escape from the stellar atmosphere and the traumatic events associated with wind generation and dust formation, whereas most of the others have to be synthesized in the outer envelope. The question of molecular origins can be addressed by interferometer maps at millimeter wavelengths. Even the largest single-dish telescopes have beams that are only commensurate with the photodissociation radii (cf Section 2). For example, the beam of the *IRAM* 30-m at 3 mm is about 25″, larger than almost all the molecular distributions in IRC + 10216, the most favorable case. Pioneering interferometer measurements by Bieging et al (1984) for the $J = 1 \rightarrow 0$ line HCN in IRC + 10216 at Hat Creek (*BIMA*), with an angular resolution of 15″, provided some of the earliest confirmation that circumstellar molecules are photodissociated by the interstellar radiation field. There has been considerable technical progress in the meantime, with four millimeter arrays (*BIMA*, *IRAM*, *NRO*, and *OVRO*) approaching 1″ angular resolution capability (Ishiguro & Welch 1994). Figure 2 shows some recent results for IRC + 10216 at 3 mm from the *IRAM* interferometer on the Plateau de Bure (M Guélin & R Lucas, private communication). The angular resolution ranges from 2″ for SiS to 4″ for MgNC. These maps confirm the gross distinction between molecules linked directly to the star (e.g. SiS) and molecules linked to outer envelope processes (SiC_2, MgNC, and C_4H). But there is much more in the maps, e.g. the SiC_2 is bright in the middle of the map as well as in a ring. Some of these features are discussed in Section 5.

3.2 *Other Stars*

Observational studies of the chemistry of other envelopes generally fall into two categories: archetypes for various stages of post-AGB evolution and surveys. Some well-studied objects are: α Ori (M supergiant); S Sct (C star with a detached envelope); CRL 2688, CRL 618, and OH231.8 + 4.2 (objects in transition between AGB and PN, often referred to as proto-PN); and NGC 7027 (PN). Excellent introductions to the observations of these objects have been provided by Olofsson (1994b) and Huggins (1995) and by Omont et al (1993) for O-rich circumstellar envelopes. Once the star leaves the AGB, the evolution is rapid and marked by dramatic changes in the star and the wind. Because dynamics (as well as radiation) drives the chemistry, all of the usual theoretical simplifications have to be abandoned, including the assumption of steady, spherically symmetric, homogeneous mass loss. A preliminary chemical study of transitional objects has been made (Howe et al 1992), but the extensive observations of proto-PN and PN have not yet been analyzed with a proper chemical-hydrodynamic theory.

Some recent molecular surveys of C stars are by Kastner et al (1993), Omont et al (1993a), and Olofsson et al (1993); Bieging & Latter (1994) and Sahai & Liechtl (1995) provide surveys of S stars; and stars of all types are surveyed by Loup et al (1993) and Bujarrabal et al (1994). The last work is the most extensive in terms of the number of species measured and the variety of sources. The survey approach helps avoid basing all conclusions about circumstellar chemistry on one or even a few objects, but it suffers from the disadvantage that most of the sources are more distant than IRC + 10216, for example, so that the information is bound to be less detailed. There is also the problem

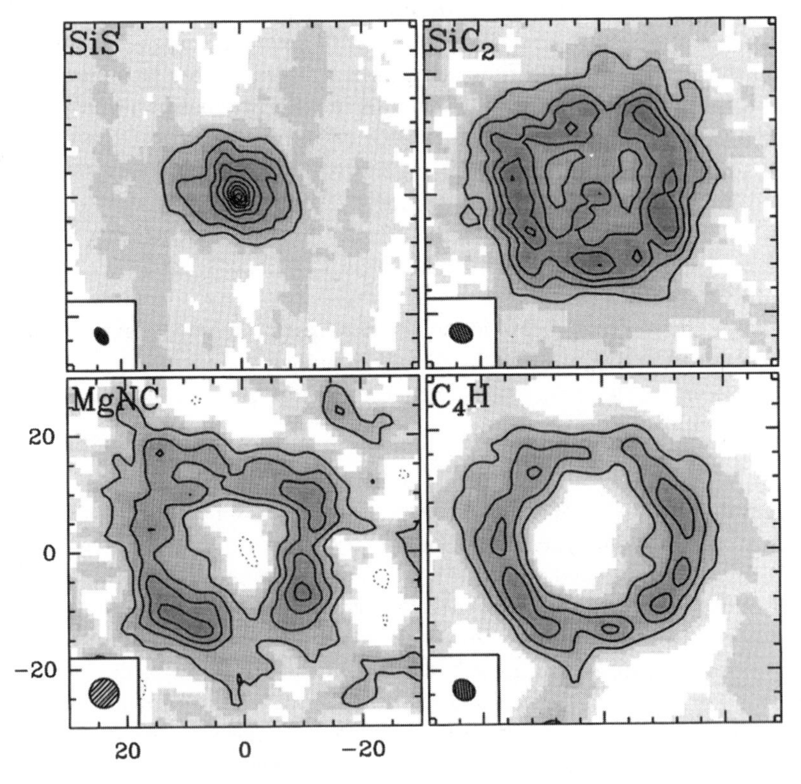

Figure 2 Interferometer maps of IRC + 10216 made with the *IRAM* Plateau de Bure interferometer at 3 mm (M Guélin & R Lucas, private communication). The angular scale is indicated in the lower left panel and is between 2 and 4 arcseconds.

of comparing objects with different dynamical and physical properties without adequate theory. Little work has been done so far to integrate the survey data from a unified chemical approach, but the attempt should be made because the results would help further our understanding of AGB and post-AGB stellar evolution.

4. CHEMISTRY

4.1 *The Role of LTE*

The constituents observed in a wind are determined in large part by the source, in this case the AGB star. Indeed, this simple fact provides the major motivation for studying circumstellar chemistry. If the photosphere is in chemical equilibrium, some signs of this or a closely related equilibrium state may be observable in the outer envelope. LTE will persist above the photosphere until the chemical and the dynamical timescales become comparable and the abundances "freeze out." The various species freeze out at different places because the chemical timescales depend sensitively on the abundances and on physical conditions. Non-LTE transformations can then occur while the wind material is transported into and through the outer envelope. Departures from LTE are also likely in the inner envelope due to shocks, dust formation, and wind generation. Thus a basic problem for circumstellar chemistry is to decide the extent to which the chemical abundances in AGB winds can be characterized by chemical equilibrium.

The LTE issue has practical consequences for calculating abundances. For LTE, all one needs is Gibbsian thermodynamics and the energies of the species (thermochemical properties). In contrast, any non-LTE theory, whether steady state or time dependent, requires the reaction rate coefficients, which are much less well known than the energies. However, one cannot decide that LTE does or does not apply without estimating timescales, and rate coefficients are again needed. Thus, in situations such as AGB winds with short timescales, there is no way to avoid actually identifying and characterizing the important reactions. This conclusion is supported by the observations of IRC + 10216 (Section 3.1), where a substantial fraction of the species is produced by non-LTE processes. Nonetheless, it is still necessary to have extensive calculations of LTE abundances in order to know what LTE means and the extent to which it applies.

Chemical equilibrium calculations have a long tradition in the theory of stellar atmospheres [the usual first reference is to Henry Norris Russell (1934)] and the influential early studies by Tsuji (1964, 1973) and Vardya (1966) contain references to older work. There is always a need for more complete and up-to-date calculations, particularly as our knowledge of the thermochemical properties improves. It is important to keep in mind that errors in the standard JANAF tables (Chase et al 1985) are often several kcal/mol; they are even larger for "exotic" species that are difficult to prepare in the laboratory. Note that 2.3 kcal/mol is equivalent to 0.1 eV or 1200 K, roughly the temperatures of interest. Chemical equilibrium calculations for hydrostatic stellar atmospheres are reviewed by Tsuji (1986), Vardya (1987), and Johnson (1994), the last for non-LTE effects in opacity calculations.

The application of LTE to circumstellar envelopes was pioneered by McCabe et al (1979) for the case of IRC + 10216. They introduced the freeze-out idea in this context and were able to account for about two thirds of the dozen or so species for which they had abundance estimates. The agreement for the others was poor, especially for CN and NH_3. [Later Lafont et al (1982) noted that many of the abundance estimates used by McCabe et al (1979) were inaccurate.] McCabe and associates suggested (incorrectly) that there might be no envelope chemistry at all, reasoning that the chemical timescales in *interstellar clouds* are much longer than the expansion timescale of the envelope. Lafont et al (1982) recalculated LTE abundances for both O-rich and C-rich envelopes, and they surveyed a broad range of non-LTE chemical processes likely to occur in circumstellar envelopes. They and, independently, Huggins & Glassgold (1982) ascribed the large abundances of CN and C_2H to the photodissociation of the parent molecules: HCN and C_2H_2.

Recent chemical equilibrium calculations (Tarafdar 1987, Sharp & Huebner 1990, Cherchneff & Barker 1992, Lodders & Fegley 1995) feature a greater variety of species (especially solids, e.g. Sharp & Huebner 1990, Lodders & Fegley 1995) and improved thermochemical information. Cherchneff & Barker (1992) emphasize large hydrocarbons and an atmospheric model based on Bowen's (1988) shock calculations. Lodders & Fegley (1995) study the effects of varying the pressure and the C/O ratio on the condensation temperatures in C-rich environments. They find that the most likely condensation sequence for C stars is TiC, C, and SiC (as a function of decreasing temperature), but C can occur first for sufficiently low pressures, e.g. $p < 3 \times 10^3$ dynes cm^{-2} and C/O ≥ 2. Figure 3 shows their heavy element phase diagrams, which are of considerable current interest for circumstellar chemistry. For the chosen pressure ($p = 10$ dynes cm^{-2}), many heavy elements remain gaseous down to $T \approx 1250$ K. Although atomic silicon disappears below about 1550 K, LTE predicts that only half goes into solid SiC; the remainder is mainly in gaseous SiS under these conditions. The partition of Mg compounds in Figure 3 is of interest in connection with the detection of MgNC and other Mg molecules in IRC + 10216 and the possible detection of solid MgS in proto-PN, discussed below in Section 5.4. It is important to keep in mind that the amount of MgS formed is limited by the abundance of S (less than Mg in the Sun) and that some of the S is taken up by CaS. Thus only 26% of the Mg is bound up in MgS at 1000 K in Figure 3, and Lodders (K Lodders, private comunication, 1995) estimates that, if all of the available S is used up at lower temperatures, only 40% would be in MgS, assuming that forsterite condensation is inhibited. Lodders & Fegley also calculated the abundances of trace elements for the "pre-Solar" SiC grains found in meteorites, believed to be derived from dust produced in carbon stars.

Following the critical discussion in Lafont et al (1982), Glassgold & Huggins (1986a), and Cherchneff & Barker (1992), it is clear that LTE should be applied with care because of numerous theoretical as well as observational uncertainties. The former include 1. the energies of the species, especially for radicals, isomers, or anything "exotic"; 2. the poorly known atmospheric profiles (run of pressure and temperature); 3. unknown photospheric abundances; 4. the effects of dust formation; and 5. the likely failure of LTE. If we refer to the rough abundances in Table 1 for IRC + 10216, a reasonable working hypothesis is that the most abundant molecules—CO, C_2H_2, HCN, SiS, and possibly CS—are described approximately by LTE conditions. However, these may not be the same as photospheric conditions because the molecules are mainly observed in regions where dust has already formed[2], i.e. depletion (e.g. SiS) or grain synthesis (e.g. CS) may already have occurred.

4.2 *Circumstellar Chemical Reactions*

Referring to the caricature in Figure 1, we identify three locales for circumstellar chemistry: 1. the photosphere, 2. a transition zone, and 3. the main circumstellar envelope. (These correspond to regions 2, 3, and 4 in the nomenclature of Section 2.) Region II, where the wind is generated and dust is formed, is the least well understood, and we can say little about it at this time. The hope is that detailed studies of the main circumstellar envelope will provide information on this transition zone and that high spatial resolution observations will eventually explicate this region. Within each region, a particular set of chemical reactions is important (Lafont et al 1982). Because of the low density of AGB atmospheres, three-body reactions play a minor role except near the photosphere. The important reactions in the main part of the envelope are familiar two-body reactions ($A + B \rightarrow C + D$) and one-body reactions involving a photon or high-energy particle, e.g. $h\nu + A \rightarrow C + D$. Because the densities of interest are similar to those in weakly ionized interstellar clouds, the chemical reaction base for circumstellar chemistry is much the same as for interstellar chemistry. One can then draw on a variety of sources for interstellar chemistry, such as the

[2]One exception is SiO masers, e.g. Danchi & Bester (1995), Greenhill et al (1995).

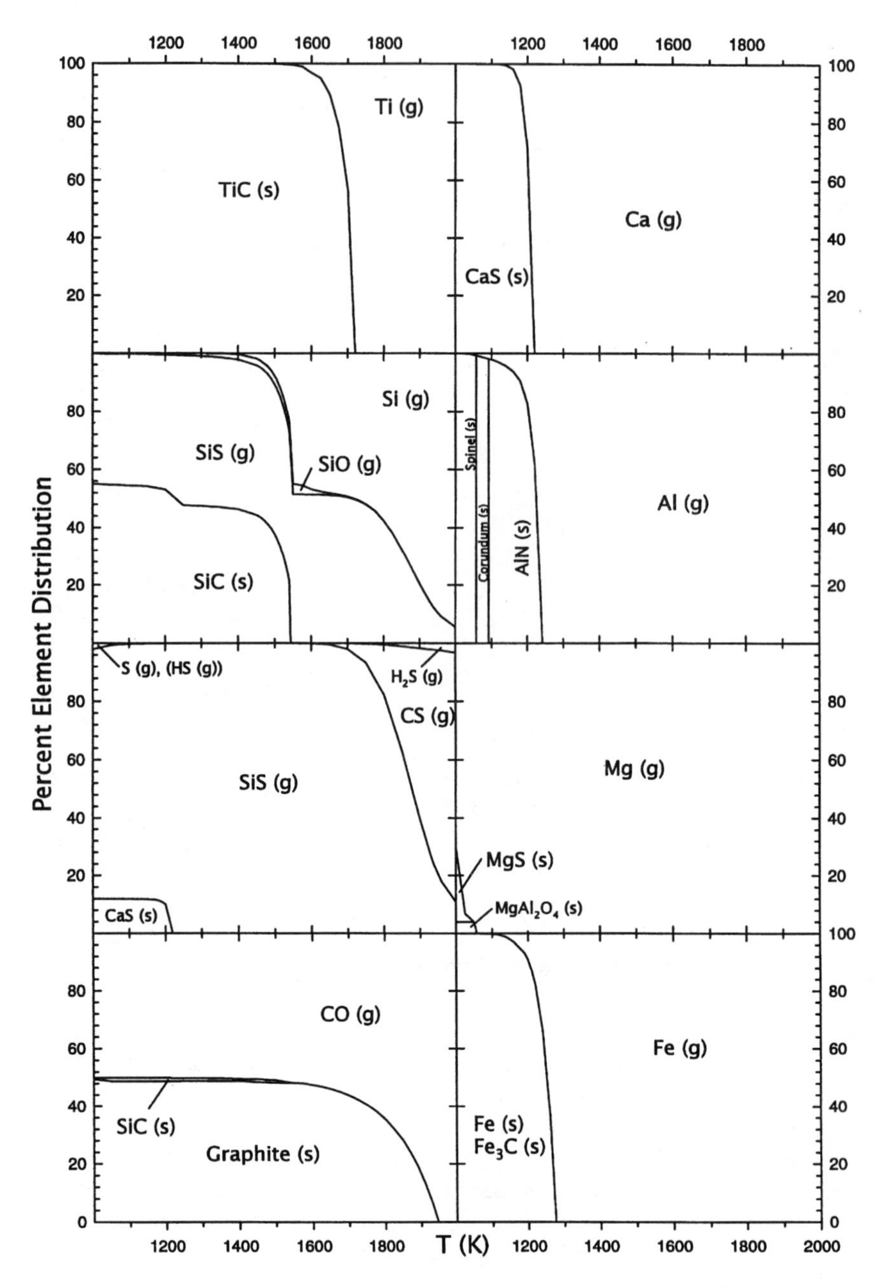

Figure 3 LTE phase diagrams for C/O = 2 showing only major compounds (K Lodders, private communication).

text by Duley & Williams (1984), conference proceedings on astrochemistry (e.g. Millar & Williams 1988, Singh 1992), and the UMIST reaction library (Millar et al 1991).

Although some interstellar chemical models employ thousands of reactions, only a subset are important in any particular situation. One important difference between circumstellar and interstellar chemistry is that the circumstellar timescales are much smaller, largely because the dynamical lifetimes of circumstellar envelopes are short ($< 50{,}000$ yr). Another difference is that the abundances of reactive species are larger, e.g. both the C_2H_2/CO and HCN/CO ratios are ~ 0.1. The relevant types of reactions have been discussed at length by Lafont et al (1982) and Glassgold & Huggins (1986a). A practical guide for circumstellar gas-phase chemistry is Appendix A of Cherchneff et al (1993). Here we give some examples of the most important classes of reactions:

- *Photodissociation* The reactions $h\nu + \mathrm{HCN} \rightarrow \mathrm{CN} + \mathrm{H}$ and $h\nu + \mathrm{C_2H_2} \rightarrow \mathrm{C_2H} + \mathrm{H}$ produce the most chemically active radicals in the outer envelope. A great deal is known about these two particular processes from experiments that measure the product radicals as well as the total absorption cross section. However, the photodissociation rates and branching ratios for most species are poorly known (van Dishoeck 1988).

- *Ionic reactions: dissociative recombination* For example, $\mathrm{e} + \mathrm{H_2CN^+} \rightarrow \mathrm{HCN} + \mathrm{H}$, $\mathrm{HNC} + \mathrm{H}$, and $\mathrm{CN} + \mathrm{H_2}$, are all important in determining the abundances of H_2CN^+ and the isomer HNC. Although there is good information on the total rate coefficient for dissociative recombination of molecular ions (Mitchell 1990), little information is available on branching ratios.

- *Ionic reactions: chemical synthesis* The reaction $\mathrm{C_2H_2^+} + \mathrm{HCN} \rightarrow \mathrm{H_2C_3N^+} + \mathrm{H}$, followed by dissociative recombination, produces the first cyanopolyyne HC_3N. The rate coefficient at room temperature is 1.3×10^{-10} cm^3 s^{-1} (Anicich 1993), but its efficiency is limited by the low abundance of $C_2H_2^+$ ($\sim 10^{-8}$).

- *Neutral reactions* For example, $\mathrm{CN} + \mathrm{C_2H_2} \rightarrow \mathrm{HC_3N} + \mathrm{H}$ is more effective in synthesizing HC_3N because CN is more abundant than $C_2H_2^+$ by at least two orders of magnitude. The rate coefficient approaches 5×10^{-10} cm^3 s^{-1} at the low temperatures characteristic of the outer circumstellar envelope (Sims et al 1993). The realization that ion-molecule reactions are ineffective for the synthesis of chain molecules in IRC + 10216 and that neutral radical reactions could do the job (Howe & Millar 1990) underscores the importance of laboratory measurements of astrophysically important reactions, especially at low temperatures (Rowe et al 1994).

- *Radiative association* $C + C_n \rightarrow C_{n+1} + h\nu$ is the neutral generalization of the reactions discussed by Freed et al (1982) with C^+. The efficiency is low if C is located far out in the envelope. Cherchneff & Glassgold (1993) also considered radiative association of weakly hydrogenated carbon chains, but the rate coefficients have not yet been calculated.

5. THE PHOTOCHEMICAL MODEL FOR CIRCUMSTELLAR ENVELOPES

The basic idea of the photochemical model for circumstellar envelopes is straightforward. A molecule X enters the main part of the envelope at $r = R_1$ with an abundance $x_1(X)$ determined by its photospheric value, but very likely altered in traversing the transition zone. The most important cases are well-bound molecules, similar to those predicted by LTE, but not necessarily with LTE abundances. The initial abundances $x_1(X)$ are best considered as *phenomenological parameters.* Because of the declining temperature and density, species X interacts weakly at first with other wind components including dust, but it eventually reaches a region where it experiences the effects of the interstellar radiation field. It is then photodissociated, and a radical is produced that may initiate chemical activity at large distances from the star. Thus the name *photochemical model* implies a two-step process in which photodissociation is followed by chemical reactions.

The radical produced in the first step is itself of observational interest, especially if it has a large abundance. This was discussed in the pioneering paper of Goldreich & Scoville (1976), who considered the photodissociation of H_2O as the source of the OH masers in O-rich circumstellar envelopes. Scalo & Slavsky (1980) were the first to focus on the second step, the initiation of O-rich circumstellar chemistry by OH. It was quickly realized that the photochemical model would have its greatest impact on the more chemically active, C-rich envelopes (Huggins & Glassgold 1982, Lafont et al 1982). A large part of the rich chemistry of IRC + 10216 is due to this two-step photochemical process. Many groups have contributed to this conclusion, and we refer to their work below.[3]

[3]When photochemical models were introduced for circumstellar chemistry, similar models were being developed for the surface regions of interstellar clouds (e.g. Tielens & Hollenbach 1985 and references therein). The application closest to circumstellar envelopes is the comas of comets (A'hearn & Festou 1990), where the photochemical model was introduced by Haser (1957).

5.1 *Timescales*

Timescales have been discussed extensively by Scalo & Slavsky (1980), Lafont et al (1982), Glassgold & Huggins (1986a), and Omont (1991). One use is to define chemical zones by the condition

$$\tau_{\rm dyn} = \tau_{\rm chem}, \tag{8}$$

where $\tau_{\rm dyn}$ and $\tau_{\rm chem}$ are chemical and dynamical timescales. The chemical timescale is, of course, specific to the reaction in question, and all timescales are functions of position. Thus Equation (8) has a solution $r_{\rm qu}$, called the "quench" radius following the usage in planetary science. For $r \ll r_{\rm qu}$, the reaction proceeds rapidly and for $r \gg r_{\rm qu}$, slowly. For a photo process with a local rate G (that includes shielding), $\tau_{\rm chem} \approx 1/G$; e.g. for HCN, this is $\sim 10^{10}$ s, even after several magnitudes of far-UV shielding. For a chemical reaction where species X_2 destroys species X_1 with rate coefficient k_{12}, the timescale for the destruction of X_1 is $\tau_{\rm chem} \approx 1/(k_{12} n x_2)$. For a typical density of $n \approx 2 \times 10^4$ cm^{-3} and a fast reaction ($k_{12} = 5 \times 10^{-10}$ cm^3 s^{-1}) with an abundant species ($x_2 = 10^{-5}$), this timescale is again $\sim 10^{10}$ s. It is customary to take the dynamical time to be the expansion time:

$$\tau_{\rm dyn} = \tau_{\rm exp} = \frac{r}{v_{\rm exp}} = 10^{10}\,{\rm s}\left(\frac{r_{16}}{v_{\rm exp,6}}\right). \tag{9}$$

The fact that all of these estimates are the same order of magnitude at intermediate distances from the star indicates that photochemistry can occur. For additional examples, see Scalo & Slavsky (1980) and Omont (1991).

5.2 *Technical Development*

In order to calculate abundances, one has to go beyond timescales and integrate the chemical rate equations for a model wind. For the case of a spherically symmetric flow, these are a set of coupled continuity equations of generic form,

$$v\frac{d}{dr}x({\rm Y}) = P({\rm Y}) - D({\rm Y})x({\rm Y}), \tag{10}$$

where P and D are the production and destruction rates of species Y. Because we use $dt = dr/v$ to convert time into spatial differences, we are assuming that the flow is prescribed as a function of distance, i.e. it has been derived independently of the chemistry. This is acceptable for the main part of a long-lived circumstellar envelope (constant $v = v_{\rm exp}$), but not for the transition zone, which is affected by the shocks in the pulsating atmosphere and by wind acceleration and dust formation. This equation would also be inappropriate for envelopes where the mass-loss rate is changing rapidly, in which case a full chemical-hydrodynamics approach is required. The terms P and D depend on the local physical conditions, and they are usually chosen along the lines of the Section 2, i.e. Equations (2) and (3). Otherwise, one must calculate the temperature simultaneously with the chemical kinetics equations.

The most important part of calculating the production and destruction terms in Equation (10) involves radiative rates, as indicated by the following discussion of some essential points of circumstellar chemistry:

- *Molecular photodissociation cross sections* Relatively few cross sections have been measured or calculated reliably, but good values are available for many abundant circumstellar molecules, e.g. H_2, CO, H_2O, C_2H_2, and HCN. The situation is not as promising for radicals and more complex molecules, where branching ratios are critical. Photodissociation rates for the interstellar medium have been reviewed by van Dishoeck (1988).

- *Optical properties of the dust* Circumstellar molecules are destroyed by UV radiation between 912 and 3000 Å. Unlike the situation in the infrared, little is known about the far-UV optical properties of circumstellar dust from observations. For heavily shrouded envelopes (large mass-loss rates), about the only direct method is a difficult one in which the envelope serves as a reflection nebula for the interstellar radiation field; see the discussion by Martin & Rogers (1987) of optical images of IRC + 10216. Something similar to the usual procedure used for measuring the UV properties of interstellar dust has been used for transparent envelopes (Snow et al 1987, α Sco; Buss & Snow 1988, supergiants with B-type companions; Buss et al 1989, two post-AGB stars). A longstanding hope is that the UV properties of circumstellar dust in thick envelopes like IRC + 10216 can be determined indirectly by applying the photochemical model to molecular maps (Jura 1983, 1994). In models for IRC + 10216, Cherchneff et al (1993) use the theoretical calculations of Rouleau & Martin (1991) for amorphous carbon, which fit the observed continuum fluxes from optical to millimeter wavelengths. Various uncertainties do not yet allow definitive conclusions, except that the interpretation of high spatial resolution observations requires chemical models with realistic dust models. Many molecules of interest, e.g. hydrocarbons, are destroyed at longer wavelengths ($\sim$ 1500 Å) than are species like H_2, CO, C, and CN (900–1100 Å), so that changes in the far-UV properties of dust between 1000 and 1500 Å are important.

- *Radiation field* The results of any photochemical model are sensitive to the radiation field, wheher interstellar or stellar. The effects of chromospheric UV radiation on the envelope α Ori have been considered by Clegg et al (1983) and Glassgold & Huggins (1986b). Local variations in the interstellar radiation field near IRC + 10216 have been discussed by Martin & Rogers

(1987) and Cherchneff et al (1993). Howe et al (1992) have considered how stellar radiation affects the chemistry of detached shells in the immediate post-AGB phase.

- *Shielding* The transfer of UV radiation through the envelope leads to a reduction in a photo-rate $G(r)$ at position r relative to the rate $G(r)_o$ in the absence of the envelope; $J(r) \equiv G(r)/G(r)_o$ is the *shielding factor.* The shielding is usually due to dust, but an abundant molecule with a large absorption cross section can also provide shielding. For example, H_2 and CO have large cross sections for absorbing UV radiation by line transitions to excited electronic levels, with finite probabilities for decay to the continuum of the ground level. This process was first identified by Stecher & Williams (1967) for interstellar H_2, and it was applied to circumstellar envelopes by Glassgold & Huggins (1983). H_2 can shield other species affected by radiation in the 900–1100 Å band, e.g. atomic C and CO. After laboratory studies confirmed that CO photodissociation proceeds in the same way (Letzelter et al 1987), the theory of CO line self-shielding was extended to circumstellar envelopes by Mamon et al (1988) and to interstellar clouds by Black & van Dishoeck (1988).

- *Wavelength grid* The above considerations indicate that the radiative rates depend on several wavelength-dependent quantities in the far UV: the radiation field, the dust optical properties, and the atomic and molecular cross sections. Thus an essential computational requirement of a photochemical code is a fine wavelength grid to calculate the photodissociation rates.

- *Radiation transfer* Most chemical calculations invoke approximations to the radiative transfer problem. The two most common are the Sobolev approximation for the lines and an exponential function for continuum absorption (Morris & Jura 1983),

$$J_{\rm cont}(r, \lambda) = \exp[-1.644\ \tau_{\rm cont}(r, \lambda)^{0.86}], \tag{11}$$

  where $\tau_{\rm cont}(r, \lambda)$ is the radial optical depth into r for continuum absorption at wavelength λ. For example, Cherchneff et al (1993) used Equation (11) for dust shielding, a straight exponential for molecular absorption (with no scattering) and a one-band approximation for the the many-line theory of CO self-shielding (Mamon et al 1988). A more consistent treatment of radiative transfer is needed to place the photochemical model on a sound basis.

In describing the calculation of photodissociation rates, we have touched on some of the limitations of existing versions of the photochemical model of circumstellar chemistry. These include the separation of the chemical from the thermal and the dynamical problems, the approximate treatments of radiation transfer, and the lack of good cross sections. Equally important are the assumptions of spherical symmetry and uniformity of the flow (not unrelated to the radiative transfer). In particular, the opacity of the envelope is particularly sensitive to the clumpiness of the wind.[4] With all of these reservations, it is best to regard the photochemical model as a phenomenological model and to use the results as a guide for discussing the meaning of the observations, rather than for obtaining a high level of quantitative agreement between theory and observations.

5.3 *Exemplary Results*

The development of the photochemical model proceeded in three stages. The basic idea was first expressed by Scalo & Slavsky (1980) for O-rich circumstellar envelopes and was then adapted to C-rich envelopes by Huggins & Glassgold (1982) and Lafont et al (1982). However, these papers did not go much beyond calculating the chain of photodissociation products derived from progenitor molecules. More complete, but mainly schematic, implementations of photochemistry were then presented by Millar and his Manchester collaborators (Nejad et al 1984; Nejad & Millar 1987, 1988) and by groups in New York and Grenoble (Glassgold et al 1986, 1987; Mamon et al 1987; Nercessian et al 1989). This work was largely based on the ion-molecule reaction chemistry developed for interstellar clouds. As more experimental information became available on neutral reactions, a third stage of circumstellar chemistry was started featuring these reactions (Howe & Millar 1990, Cherchneff et al 1993, Cherchneff & Glassgold 1993, Millar & Herbst 1994). We summarize the results of the photochemical model using this recent work and unpublished results.

Figure 4 shows the radial variation of the most abundant species in the photochemical chains arising from the photodissociation of C_2H_2 and HCN. The heavy solid lines are for the "1994 model" for IRC + 10216, so named to distinguish it from the very similar model of Cherchneff et al (1993), referred to as "CGM". The parameters distinguishing the two models are given in Table 2. The photodissociation of CO yields C and C^+, but at larger distances than that derived from C_2H_2 and HCN. This difference is due to the fact that the shielding of CO is due to the combined effects of dust and line self-shielding (Mamon et al 1988), whereas the shielding of C_2H_2 and HCN is mainly due to dust. The signature of the two sources of C and C^+ is the double peak in the atomic carbon abundance, which gets washed out, however, when more appropriate distributions for analyzing observations are used, e.g. the density. The main

[4]Bergmann et al (1993) have analyzed their CO measurements of S Sct with a clumpy model.

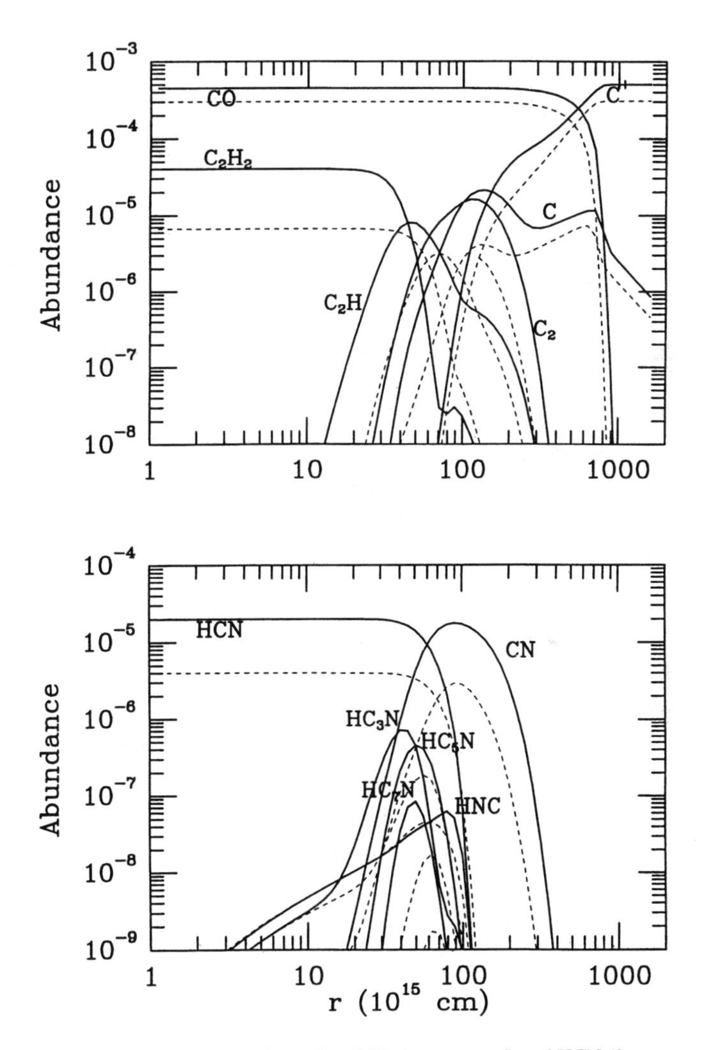

Figure 4 The basic photochemical chains for C_2H_2 (*upper panel*) and HCN (*lower panel*). Two models for IRC + 10216 are shown: the "1994" model (*heavy, solid lines*) and the "CGM" model (*light, dashed lines*); the parameters are given in Table 2.

Table 2 Model parameters for IRC +10216[1,2]

Parameter	CGM	1994
$\dot{M}_5$	3	2
$x_1(CO)$	3.0×10^{-4}	4.5×10^{-4}
$x_1(C_2H_2)$	6.7×10^{-6}	4.0×10^{-5}
$x_1(HCN)$	4.0×10^{-6}	2.0×10^{-5}

[1]Mass loss in $M_{\odot}$ yr^{-1}.
[2]Abundances relative to total H.

differences bewteen the CGM and the 1994 model are that (*a*) the abundances for CGM are significantly smaller and (*b*) the peaks for dissociation products occur farther out in the envelope. The latter effect is mainly due to the 50% larger mass-loss rate used by CGM and arises from the well-known shift in photochemical peaks with increased shielding (Huggins & Glassgold 1982, Netzer & Knapp 1987). The difference in abundance level reflects the long-standing discrepancy between initial progenitor abundances based on infrared and radio observations. The infrared measurements tend to sample the inner envelope, and the radio the outer envelope. The problem arose mainly in the context of C_2H: Early photochemical models produced too much C_2H, thus yielding low values for the initial C_2H_2. This problem has been gradually alleviated by finding new ways of processing C_2H into more complex species, especially via neutral reactions. In fact, the abundances in the 1994 model are close to those obtained by Keady & Ridgway (1993) in their analysis of infrared absorption measurements of molecules. It is important to bear in mind the warning of Section 3.1: Neither the radio nor the infrared abundances are accurate. A good guess is that a reasonable trial set of parameters lies somewhere between the two sets in Table 2, with the Keady-Ridgway recommendation favored, as in the 1994 model.

When the CGM model was published, the authors realized that the parameters used to illustrate the theory yielded photochemical shell radii that might be too large for IRC + 10216. For example, the typical shell radius measured with the *IRAM* interferometer (e.g. as shown in Figure 2) is 15″, which corresponds to 4.5×10^{16} cm, *if* the distance to this object is the customary $d = 200$ pc. The C_2H shell is about 50% larger for CGM parameters, whereas it comes out "right" in the 1994 model, when we ignore possible differences between shell radii for theoretical abundances and observed brightness. The main objective of models such as CGM and that of Millar & Herbst (1994) is to explore new chemical pathways and to understand how chemical abundances change when model parameters are varied. The need to use the models judiciously is illustrated by the recent observations of the 609-μm fine structure line of C I by Keene et al

(1993), who make direct comparisons with CGM. As Figure 4 (*upper panel*) shows, a simple parameter change (the 1994 model) removes most of the factor of 50 discrepancy cited by Keene et al. In addition to increasing the C I column density in the middle of the envelope, the 1994 model puts the C I where it is warmer, thus increasing the population of the upper level of the transition. Observations of C I can provide an important test of the photochemical model (Huggins & Glassgold 1982), and the suggestion by Keene et al that the far-UV opacity of the envelope may have been overestimated is likely on target.

The HCN photochain in Figure 4 shows that previous difficulties in accounting for the cyanopolyynes have been overcome by invoking enlarged pathways starting with neutral reactions of CN (Cherchneff & Glassgold 1993, Millar & Herbst 1994). Again, these results are not meant to be final fits, but they do suggest that, as more complete laboratory data become available on neutral reactions and photodissociation cross sections, a reasonable account of high spatial resolution observations may be possible. The basis for this approach is founded on the interferometric observations of species of the HCN photochain, such as those shown in Figure 5, which were made with the *BIMA* array with an angular resolution of 7–11″ (Bieging & Tafalla 1993, Dayal & Bieging 1995). We see that the brightness distribution of CN surrounds that of HCN, more or less as suggested in Figure 4, and a similar conclusion holds for C_3N and HC_3N. The CN distribution appears to be unusually large, as noted earlier (e.g. Huggins et al 1984, Truong-Bach et al 1987); however, it does not seem to be quite as large as predicted by theory. As discussed by Cherchneff et al (1993), this may be due to the use of the CN photodissociation rate recommended by van Dishoeck (1988). The larger value calculated by Lavendy et al (1987) cuts the CN distribution off more more sharply. A related factor is whether the shielding between 900–1000 Å has been overestimated, relevant also for C I (Keene et al 1993).

Figures 4 and 5 indicate that high spatial resolution interferometric observations of the HCN/CN and HC_3N/C_3N photodissociation pairs offer the possibility of making quantitative tests of the photochemical model, especially if done with several lines of different excitation and, for HCN and HC_3N, with the ^{13}C forms to reduce optical depth effects. Of course the theoretical predictions depend on several incompletely known quantites, e.g. the far-UV properties of dust, the photodissociation cross sections (especially for the product radicals), and the stellar distance, not to mention oversimplifications used in defining and calculating the model. Figure 6 compares the spatial distributions of the HC_3N and C_3N abundances for the 1994 model. The HC_3N peak is at 14″ if $d = 200$ pc, and the peaks are separated by about 5.5″. Bieging & Tafalla (1993) found 5″ for both the shift in the brightness peaks and the one for abundances based

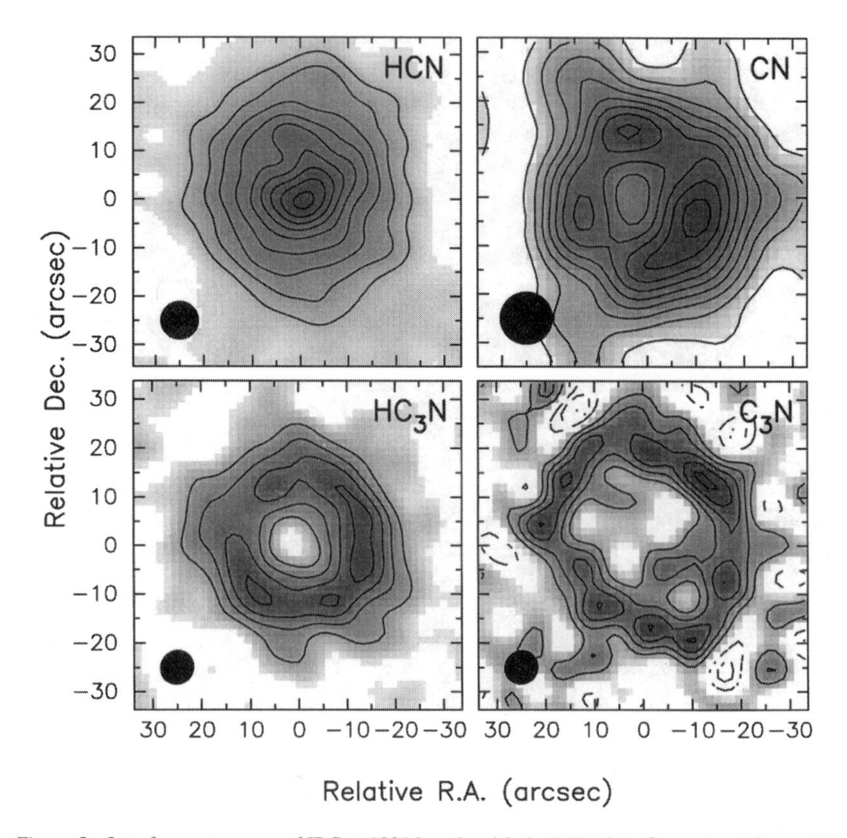

Figure 5 Interferometer maps of IRC + 10216 made with the *BIMA* interferometer at 3 mm (JH Bieging, private communication). The beam sizes are indicated in the lower left corner of each panel. The velocity width (km s^{-1}), peak brightness temperature (K), and contour interval (K) are: 1.1, 45, and 4.8 (HCN); 3.0, 7, and 0.7 (CN); 4.0, 9.3, and 2 (HC_3N); and 4.0, 2.1, and 0.4 (C_3N).

on a simplified excitation model. Considering the many uncertainties and the fact that the spatial resolution of the observations in Figure 5 is larger than 5″, it would be unwise to overemphasize the agreement, but the result is not discouraging.

Figure 6 also shows theoretical abundances for C_2H and C_4H. The C_2H peak is at 15″, and C_4H is shifted out by about 7″ to 22″. In contrast, the measured brightness C_4H distribution in Figure 2 peaks closer to 15″. The apparent failure of the theory in this case might well arise from bad guesses about the photodissociation of C_4H, for which there is essentially no firm information. It might also be part of a larger problem, however, that seems to be emerging

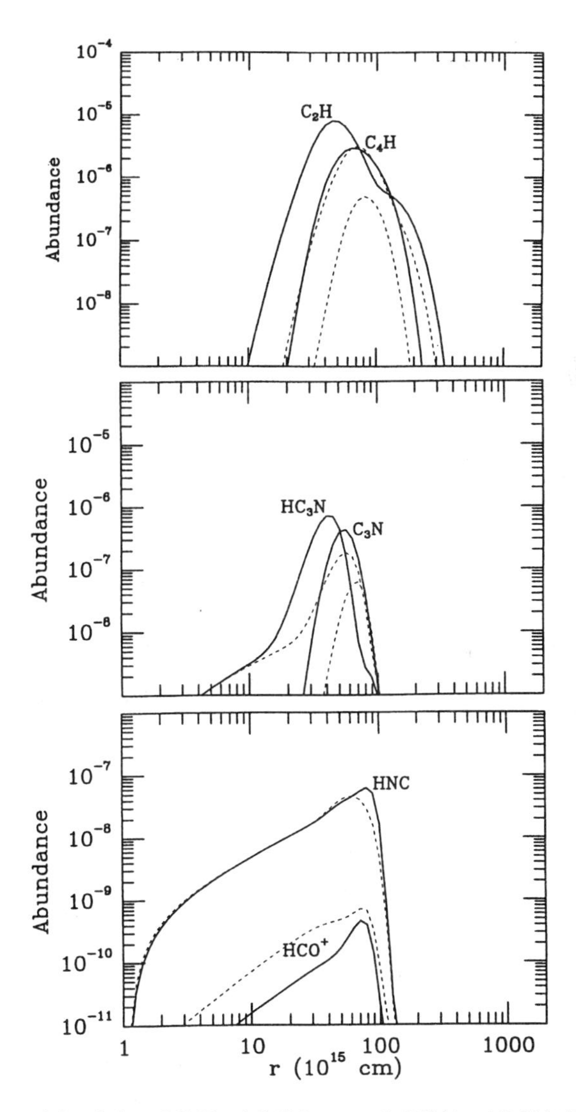

Figure 6 The spatial variation of C_2H and C_4H (*top panel*), HC_3N and C_3N (*middle panel*), and HCO^+ and HNC (*bottom panel*) for the models in Figure 4.

as the number of high spatial resolution interferometric maps increases. Lucas et al (1995) have mapped many species in IRC + 10216 and, for those with extended brightness distributions, there is a striking uniformity in shell size. Except for CN, which is clearly very extended, most of the shells have radii close to 15″. Even if the production of all these species is initiated by a small number of radicals, i.e. CN and C_2H, it is surprising that they would all have the same radius. Somewhat different synthetic pathways are involved and, at the very minimum, the molecules have different photodissociation cross sections. Guélin et al (1993) have offered an interesting alternative explanation, which combines photochemical and grain concepts. They suggest that the synthesis occurs on grain surfaces and that the molecules are released from the grains by the action of the interstellar radiation field. In this picture, all molecules appear at roughly the same location, determined by a penetration condition on the optical depth of far-UV radiation, like Equation (5). Even in this model, however, one might expect to find some variations associated, for example, with energetic barriers for detachment of radicals or with the wavelength variation of the dust extinction.

Figure 6 compares HCO^+ and HNC, species believed to be produced by ion-molecule reactions. As in interstellar clouds, HCO^+ is the direct descendant of H_3^+ produced by cosmic-ray or other high-energy ionization of H_2:

$$H_3^+ + CO \rightarrow HCO^+ + H_2. \tag{12}$$

HNC arises from the dissociative recombination of H_2CN^+, which is produced by a proton-transfer reaction like Equation (12),

$$MH^+ + HCN \rightarrow H_2CN^+ + M, \tag{13}$$

where MH^+ stands for H_3^+, N_2H^+, HCO^+, $C_2H_2^+$, and $C_2H_3^+$. Although single-dish measurements appeared to be in rough agreement with theoretical predictions (Glassgold et al 1987), recent interferometer maps show only a small amount of HNC emission interior to a very bright ring, similar to the MgNC map in Figure 2. The interior HNC comes from the recombination of H_2CN^+, produced by the cosmic-ray ions, H_3^+, N_2^+, and HCO^+. The interior abundance of H_2CN^+ produced by $C_2H_2^+$ and $C_2H_3^+$ is negligible; these ions are mainly produced by photoionization of C_2H_2 and not by cosmic rays. Thus a possible interpretation of the HNC observations is that the cosmic-ray ionization is suppressed inside IRC + 10216. This interpretation is consistent with the measurements of HCO^+ (Lucas & Guélin 1990), which indicate that the theory produces too much HCO^+.[5] A reduced cosmic-ray ionization rate could arise

[5]The calculations shown in all of the figures are based on a cosmic-ray ionization rate of $\zeta_{CR} = 5 \times 10^{-18}\ s^{-1}$, which is about 1/2 the local interstellar value.

from the sweeping up of the cosmic rays by the wind, as originally suggested by Parker (1963).

5.4 *Heavy Element Chemistry*

The discussion in Section 5.3 deals mainly with molecules containing C, N, O, and H. Although some consideration has been given to the photochemistry of S and Si compounds (Clegg et al 1983, Nejad & Millar 1988, Howe & Millar 1990, Glassgold & Mamon 1992, Omont et al 1993a), more detailed modeling would be worthwhile, especially for C-rich envelopes. The heavy molecules in the last column of Table 1 are of special interest, but little has been done on the chemistry of these systems. One naturally starts with LTE considerations, especially because the maps indicate that some species are concentrated in the center of IRC + 10216, e.g. SiS, CS, and NaCl. An important barrier to progress is the absence of reliable thermochemical information, not to mention rate coefficients. Turner (1991, 1995) has discussed many interesting possibilities for circumstellar heavy element chemistry.

An important observational result on heavy element chemistry is that MgNC is distributed in a shell in IRC + 10216 of about the same radius (15 ″) as most other shells except for CN. Chemical pathways for the synthesis of MgNC were first considered by Kawaguchi et al (1993), who also identified the species. In the absence of any obvious exothermic reaction, ionic or neutral, Kawaguchi et al proposed the radiative association,

$$Mg^+ + HCN \rightarrow MgNCH^+ + h\nu. \tag{14}$$

Photochemical model calculations support this idea and show that MgNC exists in a shell of about the right radius (W Liu & AE Glassgold, unpublished calculations). The basis for the shell distribution in this case is in the photoionization of Mg, which occurs in the same place in the outer envelope as most other photo processes. There is some question as to whether Equation (14) can give the right abundance level of MgNC. Again the problem is that we know so little about radiative association rates for heavy molecules. In the calculations just referred to, the rate coefficient used was of order of 10^{-10} cm^3 s^{-1} in the temperature range 20–30 K. Even so, nothing like the correct abundance level can be achieved unless Mg is *not* heavily depleted, as it is in the interstellar medium. This possibility is supported by the LTE calculations of Lodders & Fegley (1995) in Figure 3, which show that condensation into MgS is incomplete. MgS has long been considered as a potential component of circumstellar dust (Goebel & Moseley 1985, Cox 1993).[6] A tentative conclusion from the

[6]Recent *KAO* observations of the broad 30-μm feature in C-rich proto-PN (Omont et al 1995) appear to require incomplete condensation of Mg, if the observed feature is indeed due to MgS.

occurrence of heavy molecules in shells in IRC + 10216 is that depletion is not as severe in this envelope as it is in the interstellar mediium, even though one might expect a similar depletion pattern (Jura & Morris 1985).

6. CONCLUSION

The theoretical and observational results in Section 5 support the decisive role of photo processes in affecting the abundances and distribution of circumstellar molecules. Not only do photo processes limit their spatial extent, but photodissociation is the first step in the synthesis of the complex species observed to have shell distributions. Spectral imaging of circumstellar envelopes is now being done with an angular resolution of a few arcseconds—significantly smaller than the size of the molecular shells observed in IRC + 10216. Smaller scales will become accessible as improvements in existing interferometers are implemented and new ones built. The new instrumentation could be used to improve our understanding of the region interior to the shells and close to the source of the wind. It would also allow molecular shells to be imaged in envelopes that are far away or have small mass-loss rates.

Significant improvements in the modeling of circumstellar envelopes are required to do full justice to the high spatial resolution observations now being made with millimeter-wave interferometers. In addition to the limitations in current theoretical approaches discussed in Section 5.1, most chemical models do not include a calculation of the population of the molecular levels. As a result, there is a gap in comparing theoretical abundances and observed brightness distributions, only imperfectly bridged by the use of beam-averaged column densities and rough guesses or estimates of the excitation temperature (Omont 1991, Olofsson 1994e). Enlarging the scope of the chemical models only makes sense, however, if the interferometer maps are made in more than one line so that the excitation of the molecular lines can be determined as a function of position.

One of the most important results of the high spatial resolution maps obtained with the *IRAM* interferometer on Plateau de Bure is that the shells are indeed clumpy (Lucas et al 1995), as shown in Figure 2. The clump structure can be traced from one species to another, which suggests that they are density enhancements. Variations in density also imply variations in temperature and radiation field and thus changes in chemistry and excitation, not to mention the possibility of new hydrodynamic phenomena. The clumps present many theoretical and observational challenges, which have only begun to be faced (Olofsson 1994e). The advent of arcsecond spatial resolution means that the substructure of the clumps can be mapped in nearby objects like IRC + 10216. A particularly interesting aspect of the clumps already observed in IRC + 10216 is the suggestion in several species of multiple shells (e.g. Lucas et al 1995);

these may be related to those seen in optical images (Crabtree et al 1987). A rough estimate of the radial substructure of the shells in IRC + 10216 is several arcseconds, which corresponds to a distance of the order of 10^{16} cm and to a time of only 300 yr. Again, arcsecond resolution and better observations are called for.

In conclusion, it would appear that the photochemical model provides a good first-order description of the spatial distribution of molecular abundances observed in the outer circumstellar envelope of IRC + 10216. The model also promises to be useful as the starting point for analyzing similar, high-spatial observations of other envelopes.

ACKNOWLEDGMENTS

The author would like to express his appreciation to the Insitutut d'Astrophysique de Paris for hospitality during the summer of 1995 when this review was started; JH Bieging, K Lodders, R Lucas, M Guélin, GA Mamon for providing figures and other results prior to publication; many other colleagues for reprints (too many to quote individually); and NASA's Infrared and Radio branch and NSF's International Branch for support.

Literature Cited

A'hearn MF, Festou MC. 1990. In *Physics & Chemistry of Comets,* ed. WF Heubner, pp. 69–112. Heidelberg: Springer-Verlag
Amari S, Hoppe P, Zinner E, Lewis RS. 1995. *Meteoritics* 30:679–93
Anicich V. 1993. *Ap. J. Suppl.* 84:215
Avery LW, Broten NW, MacLeod JM, Oka T, Kroto HW. 1976. *Ap. J. Lett.* 205:L173–76
Bakker E, Lamers HJGLM, Waters LBFM, Schoenmaker T. 1994. *Astrophys. Space Sci.* 224:335–38
Bell MB, Kwok S, Feldman PA, Mathews HE. 1982. *Nature* 295:389–91
Bergmann P, Carlström U, Olofsson. 1993. *Astron. Astrophys.* 269:685–93
Bieging JH, Chapman B, Welch WJ. 1984. *Ap. J.* 285:656
Bieging JH, Latter WB. 1994. *Ap. J.* 422:765–82
Bieging JH, Tafalla M. 1993. *Astron. J.* 105:576–94
Black JH, van Dishoeck E. 1988. *Ap. J.* 334:771–802
Bowen GH. 1988. *Ap. J.* 329:299–317
Bowers PF, Knapp GR. 1987. *Ap. J.* 315:305
Broten NW, Oka T, Avery LW, MacLeod JM, Kroto HW. 1978. *Ap. J. Lett.* 233:L105–8
Bujarrabal V, Fuente A, Omont A. 1994. *Astron. Astrophys.* 285:247–71
Buss RH Jr, Lamers HJGLM, Snow TP Jr. 1989. *Ap. J.* 347:977–88
Buss RH Jr, Snow TP Jr. 1988. *Ap. J.* 335:331–71
Carpenter K, Robinson RD, Wahlgren GM, Linsky JL, Brown A. 1994. *Ap. J.* 428:329–44
Chase MW, Davies CA, Downes JR, Frurip DJ, McDonald RA, Syverad. 1985. *J. Phys. Chem. Ref. Data* 14:1
Cherchneff I, Barker JR. 1992. *Ap. J.* 394:703–16
Cherchneff I, Glassgold AE. 1993. *Ap. J. Lett.* 419:L41–44
Cherchneff I, Glassgold AE, Mamon GA. 1993. *Ap. J.* 410:188–201
Clegg RES, van Ijezendoorn LJ, Allamandola LJ. 1983. *MNRAS* 203:125–46

Cox P. 1993. In *Astronomical Infrared Spectroscopy: Future Observational Directions,* ed. S Kwok, *ASP Conf. Proc.* 41:163–70. San Francisco: Astron. Soc. Pac.
Crabtree DR, McLaren RA, Christian CA. 1987. In *Late Stages of Stellar Evolution,* ed. S Kwok, SR Pottasch, p. 227. Dordrecht: Reidel
Danchi WC, Bester M. 1995. *Astrophys. Space Sci.* 224:339–52
Dayal A, Bieging JH. 1995. *Ap. J.* 439:996–10006
Duley WW, Williams DA. 1984. *Interstellar Chemistry.* London: Orlando
Freed KF, Oka T, Sizuki H. 1982. *Ap. J.* 263:718–22
Frenklach M, Feigelson ED. 1989. *Ap. J.* 341:372–84
Glassgold AE, Huggins PJ. 1983. *MNRAS* 203:517–32
Glassgold AE, Huggins PJ. 1986a. In *The M-Type Stars,* ed. HR Johnson, FR Qureci, pp. 291–322. Washington, DC: NASA
Glassgold AE, Huggins PJ. 1986b. *Ap. J.* 306:605–17
Glassgold AE, Lucas R, Omont A. 1986. *Astron. Astrophys.* 157:35–48
Glassgold AE, Mamon GA. 1992. In *Chemistry & Spectroscopy of Interstellar Molecules,* ed. DK Bohme, E Herbst, N Kaifu, S Saito, pp. 261–66. Tokyo: Univ. Tokyo Press
Glassgold AE, Mamon GA, Omont A, Lucas R. 1987. *Astron. Astrophys.* 180:183–90
Goldberg L. 1979. *QJMNRAS* 20:361–82
Goldreich P, Scoville NZ. 1976. *Ap. J.* 205:144–54
Greenhill LJ, Colomer F, Moran JM, Backer DC, Danchi WC, Bester. 1995. *Ap. J.* In press
Groenwegen MAT. 1994. *Astron. Astrophys.* 290:531–43
Guélin M, Forestini M, Valiron P, Ziurys LM, Anderson AM, et al. 1995. *Astron. Astrophys.* 297:183–96
Haas MR, Glassgold AE. 1933. *Ap. J. Lett.* 410:L111–14
Habing HJ. 1990. In *From Miras to Planetary Nebula, Which Path for Stellar Evolution?,* ed. M Menessier, A Omont, pp. 16–41. Paris: Éditions Frontières
Habing HJ, Bloemmaert JADL. 1995. *Astron. Astrophys.* In press
Haser L. 1957. *Bull. Acad. R. Belgique, Classes Sci., Ser. 5* 12:233–41
Hinkle KH. 1994. In *Molecules in the Stellar Environment,* ed. UG Jørgenson, pp. 98–112. Heidelberg: Springer-Verlag
Hinkle KH, Bernath PF. 1993. In *Astronomical Infrared Spectroscopy: Future Observational Directions,* ed. S Kwok, *ASP Conf. Proc.* 41:125–26. San Francisco: Astron. Soc. Pac.
Howe DA, Millar TJ. 1990. *Astron. Astrophys.* 244:444–49
Howe DA, Millar TJ, Willimas DA. 1992. *Astron. Astrophys.* 255:217–26
Huggins PJ. 1987. *Ap. J.* 313:400–7
Huggins PJ. 1995. *Astrophys. Space Sci.* 224:281–92
Huggins PJ, Bachiller R, Cox P, Forveille T. 1994. *Ap. J. Lett.* 424:L127–30
Huggins PJ, Glassgold AE. 1982. *Ap. J.* 252:201–7
Huggins PJ, Glassgold AE, Morris M. 1984. *Ap. J.* 279:254–90
Huggins PJ, Olofsson H, Johansson LEB. 1988. *Ap. J.* 332:1009–18
Ishiguro M, Welch WJ. 1994. *Astronomy with Millimeter and Submillimeter Wave Interferometry.* San Francisco: Astron. Soc. Pac.
Johnson HR. 1994. In *Molecules in the Stellar Environment,* ed. UG Jørgenson, pp. 234–49. Heidelberg: Springer-Verlag
Josselin E, Loup C, Omont A, Barnbaum C, Forveille T, Nyman LA. 1995. *Astron. Astrophys.* In press
Jura M. 1983. *Ap. J.* 275:683–90
Jura M. 1988. In *Millimetre and Submillimetre Astronomy,* ed. RD Wolstencroft, WB Burton, pp. 189–206. Dordrecht: Kluwer
Jura M. 1994. *Ap. J.* 434:713–18
Jura M, Kahane C, Omont A. 1988. *Astron. Astrophys.* 201:80–88
Jura M, Morris M. 1985. *Ap. J.* 292:487–93
Kastner JH. 1994. *Ap. J.* 401:337–52
Kastner JH, Forveille T, Zuckerman B, Omont A. 1993. *Astron. Astrophys.* 275:163–86
Kastner JH, Myers PC. 1994. *Ap. J.* 421:605–14
Kawaguchi K, Kagi E, Hirano T, Taknao S, Saito S. 1993. *Ap. J. Lett.* 406:L39–42
Keady JJ, Ridgway ST. 1993. *Ap. J.* 406:199–214
Kroto H. 1988. *Science* 242:1139–45
Kroto HW, Kirby C, Walton DRM, Avery LW, Broten NW, et al. 1978. *Ap. J.* 219:L133–37
Kwan J, Linke RA. 1982. *Ap. J.* 254:587–93
Lafont S, Lucas R, Omont A. 1982. *Astron. Astrophys.* 106:201–13
Lambert DL, Gustafsson B, Erikssob K, Hinkle KH. 1986. *Ap. J. Suppl.* 62:373
Letzelter C, Eidelsberg M, Rostas F, Breton J, Thieblemont B. 1987. *Chem. Phys.*, 114:273
Lodders K, Fegley B Jr. 1995. *Meteoritics* 30:661–78
Loup C, Forveille T, Omont A, Paul J. 1993. *Astron. Astrophys. Suppl.* 99:291
Lucas R, Guélin M. 1990. In *Submillimeter Astronomy,* ed. GD Watt, AS Webster, pp. 97. Dordrecht: Kluwer
Lucas R, Guélin M, Kahane C, Audinos P, Cernicharo J. 1995. *Astrophys. Space Sci.* 224:293–96

Lucas R, Guilloteau S. 1992. *Astron. Astrophys.* 259:L23–26
Mamon GA, Glassgold AE, Omont A. 1987. *Ap. J.* 323:306–15
Mamon GA, Glassgold AE, Huggins PJ. 1988. *Ap. J.* 328:797–808
Martin PG, Rogers C. 1987. *Ap. J.* 322:374–92
Mauron N. 1990. *Astron. Astrophys.* 227:141–46
Mauron N, Guilain C. 1995. *Astron. Astrophys.* 298:869–78
McCabe EM, Connon Smith R, Clegg RES. 1979. *Nature* 281:263–66
Mennessier MO, Omont A. 1990. *From Miras to Planetary Nebulae: Which Path for Stellar Evolution?* Gif-sur-Yvette: Éditions Frontière
Millar TJ, Herbst E. 1994. *Astron. Astrophys.* 288:561–71
Millar TJ, Rawlings JMC, Bennett A, Brown PD, Charnley SB. 1991. *Astron. Astrophys. Suppl.* 87:585–619
Millar TJ, Williams DA. 1988. *Rate Coefficients in Astrochemistry.* Dordrecht: Kluwer
Mitchell JBA. 1990. *Rep. Prog. Phys.* 186:215–48
Morris M, Gilmore W, Palmer P, Turner BE, Zuckerman B. 1975. *Ap. J. Lett.* 199:L47–50
Morris M, Jura M. 1983. *Ap. J.* 264:546–53
Nejad LAM, Millar TJ. 1987. *Astron. Astrophys.* 183:279–86
Nejad LAM, Millar TJ. 1988. *Astron. Astrophys.* 230:79–86
Nejad LAM, Millar TJ, Freeman A. 1984. *Astron. Astrophys.* 134:129
Nercessian E, Guilloteau S, Omont A, Benayoun JJ. 1989. *Astron. Astrophys.* 210:225–35
Netzer N, Knapp GR. 1987. *Ap. J.* 323:734–48
Neugebauer G, Leighton RB. 1969. *Two-Micron Sky Survey. NASA SP-3047.* Washington, DC: NASA
Olofsson H. 1994a. In *Circumstellar Media in the Late Stages of Stellar Evolution,* ed. RES Clegg, IR Stevens, WPS Meikle. Cambridge: Cambridge Univ. Press
Olofsson H. 1994b. In *Molecules in the Stellar Environment,* ed. UG Jørgenson, pp. 113–33. Heidelberg: Springer-Verlag
Olofsson H, Eriksson K, Gustafsson B, Carlström U. 1993. *Ap. J. Suppl.* 89:267–304
Omont A. 1991. In *Chemistry in Space,* ed. JM Greenberg, V Pirronello, pp. 171–97. Dordrecht: Kluwer
Omont A, Loup C, ter Linetl-Hekkert, Habinh H, Sivagnanam P. 1993a. *Astron. Astrophys.* 267:515–48
Omont A, Lucas R, Morris M, Guilloteau S. 1993b. *Astron. Astrophys.* 267:490–514
Omont A, Moseley SH, Cox P, Glaccum W, Casey S, et al. 1995. *Ap. J.* In press
Parker EN. 1963. *Interplanetary Dynamical Processes.* New York: Interscience
Plez B, Lambert DL. 1994. *Ap. J.* 377:526–40
Price SD, Walker RG. 1976. *AFCRL Four Color Infrared Sky Survey. AFCRL TR-76-0208.* Hanscom AFB: Air Force Geophys. Lab.
Querci, M. 1986. In *The M-Type Stars,* ed. HR Johnson, FR Qureci, pp. 113–207. Washington, DC: NASA
Rouleau F, Martin PG. 1991. *Ap. J.* 377:526–40
Rowe B, Sims IR, Bocherel P, Smith IWM. 1994. In *Molecules & Grains in Space,* ed. I Nenner, pp. 445–61. New York: AIP
Russell HN. 1934. *Ap. J.* 79:317
Sahai R, Liechtl S. 1995. *Astron. Astrophys.* 293:198–207
Scalo JM, Slavsky DB. 1980. *Ap. J.* 239:L77–77
Sharp CM, Huebner WF. 1990. *Ap. J. Suppl.* 72:417–31
Sims IR, Queffelec JL, Travers D, Rowe BR. 1993. *Chem. Physics. Lett.* 211:461
Singh PD. 1992. *Astrochemistry of Cosmic Phenomena.* Dordrecht: Kluwer
Skinner CJ, Whitmore B. 1987. *MNRAS* 224:335
Snow TP Jr, Buss RH Jr, Gilra DP, Swings JP. 1987. *Ap. J.* 321:921–36
Sopka RJ, Hildebrand R, Jaffe DT, Gatley I, Roellig T, et al. 1985. *Ap. J.* 294:242–55
Stecher TP, Williams DA. 1967. *Ap. J. Lett.* 149:L29
Tarafdar SP. 1987. In *Astrochemistry,* ed. MS Vardya, SP Tarafdar, pp. 559–64. Dordrecht: Reidel
Tielens AGGM, Hollenbach D. 1985. *Ap. J.* 291:722–46
Tsuji T. 1964. *Ann. Tokyo. Astron. Obs.* 9:1
Tsuji T. 1973. *Astron. Astrophys.* 23:411–31
Tsuji T. 1986. *Annu. Rev. Astron. Astrophys.* 24:89–125
Turner BE. 1991. *Ap. J.* 376:573–98
Turner BE. 1995. *Astrophys. Space Sci.* 224:297–303
van Dishoeck E. 1988. In *Rate Coefficients in Astrophysics,* ed. TJ Millar, DA Williams, pp. 49–72. Dordrecht: Reidel
Vardya MS. 1966. *MNRAS* 134:347
Vardya MS. 1987. In *Astrochemistry,* ed. MS Vardya, SP Tarafdar, pp. 395–406. Dordrecht: Reidel
Walker RG, Price SD. 1975. *AFCRL Infrared Sky Survey. AFCRL TR-75-0373.* Hanscom AFB: Air Force Geophys. Lab.
Wood PR. 1990. In *From Miras to Planetary Nebulae, Which Path for Stellar Evolution?,* ed. M Menessier, A Omont, pp. 67–84. Paris: Éditions Frontières
Young K, Phillips TG, Knapp GR. 1993. *Ap. J.* 409:725–38
Ziurys L, Turner B. 1986. *Ap. J.* 300:L19–23xd
Zuckerman B. 1980. *Annu. Rev. Astron. Astrophys.* 18:263–88

Annu. Rev. Astron. Astrophys. 1996. 34:279–329

INTERSTELLAR ABUNDANCES FROM ABSORPTION-LINE OBSERVATIONS WITH THE *HUBBLE SPACE TELESCOPE*

Blair D. Savage

Department of Astronomy, University of Wisconsin, Madison, Wisconsin 53706

Kenneth R. Sembach

Center for Space Research, Massachusetts Institute of Technology, Cambridge, Massachusetts 02139

KEY WORDS: interstellar gas, interstellar dust, halo gas, ultraviolet spectra

ABSTRACT

The Goddard High-Resolution Spectrograph (GHRS) aboard the *Hubble Space Telescope* (*HST*) has yielded precision abundance results for a range of interstellar environments, including gas in the local medium, in the warm neutral medium, in cold diffuse clouds, and in distant halo clouds. Through GHRS studies, investigators have determined the abundances of elements such as C, N, O, Mg, Si, S, and Fe in individual interstellar clouds. These studies have provided new information about the composition of interstellar dust grains, the origin of the Galactic high-velocity cloud system, and the processes that transport gas between the disk and the halo. Precision measurements of the interstellar D to H ratio and of the abundances of r- and s-process elements have also provided fiducial reference values for cosmological and stellar evolutionary observations and theoretical models.

0066-4146/96/0915-0279$08.00

1. INTRODUCTION

The absorption lines of most atoms and molecules found in the interstellar medium (ISM) occur at ultraviolet (UV) wavelengths. Their direct detection through absorption-line spectroscopy requires instrumentation above the Earth's atmosphere. The first UV observations of interstellar absorption lines were obtained toward bright O and B stars with small spectrometers carried on sounding rockets (e.g. Morton & Spitzer 1966). These early studies were followed by the 1972 launch of the very successful Princeton telescope-spectrometer on the *Copernicus* satellite (Rogerson et al 1973; for a review of *Copernicus* results see Spitzer & Jenkins 1975). Throughout its 10-year lifetime, the *Copernicus* satellite provided fundamental information about elemental abundances in diffuse interstellar clouds (Cowie & Songaila 1986; Jenkins 1987). During the 1980s many interstellar programs were pursued with the *International Ultraviolet Explorer* (*IUE*) satellite (de Boer et al 1987). The lower spectral resolution of the *IUE* [full width at half maximum (FWHM) $\approx$25 vs 13 km s^{-1}] limited its ability to probe element abundances in the neutral gas as accurately as *Copernicus*, but the *IUE* was used effectively to study interstellar dust (Mathis 1987), atomic hydrogen (Shull & van Steenberg 1985; Diplas & Savage 1994), the highly ionized ISM (Sembach & Savage 1992), gas kinematics (Jenkins 1990), and nebular emissions (Dufour 1987; Köppen & Aller 1987). Very high-resolution far-UV spectra of bright stars have been obtained with the Interstellar Medium Absorption Profile Spectrograph (IMAPS) (Jenkins et al 1988). This instrument was first flown on a sounding rocket (Jenkins et al 1989; Joseph & Jenkins 1991) and was part of the recent ORFEUS-SPAS mission (Jenkins 1995). Precision high-resolution interstellar absorption-line spectroscopy was greatly enhanced with the launch of the Goddard High-Resolution Spectrograph (GHRS) aboard the *Hubble Space Telescope* (*HST*) in 1990. In this review we discuss many of the ISM abundance results obtained through GHRS absorption-line observations. We also discuss a few interstellar absorption-line results from the *HST* Faint Object Spectrograph (FOS). However, because of its relatively low resolution ($\sim$300 km s^{-1}), the FOS is more useful for studies of the emission from interstellar gas, which is not the subject of this review. The recent review of light elements and isotope ratios by Wilson & Rood (1994) complements the results discussed in this paper.

Measures of ISM elemental abundances provide important information about the physical conditions, chemical composition, and Galactic evolution of the gaseous material in the Milky Way. Heavy elements regulate gas temperatures through a variety of interstellar heating and cooling processes. The formation rates of molecules through gas-phase chemistry or on grain surfaces depend on the abundances of the reacting species. Many heavy elements have gas phase abundances that are less than the expected cosmic abundances because of varying levels of incorporation into interstellar dust grains. This phenomenon is referred to as depletion.

Gas-phase abundance measurements provide the most direct way by which to obtain information about elemental depletion in different Galactic environments and to gain insight into the composition of interstellar grains and the exchange of matter between the gaseous and solid forms through depletion and grain destruction processes (Jenkins 1987; Mathis 1990). Absolute interstellar abundances (gas + dust) serve as fundamental benchmarks for interpretations of abundances in galaxies and gas clouds in the distant Universe. Eventually, comparisons of these Milky Way abundances with those found in high-redshift quasar absorption-line systems will enable astronomers to study elemental abundance evolution over approximately 90% of the age of the Universe.

2. UV ABSORPTION-LINE DIAGNOSTICS

Ground-based optical absorption-line observations of interstellar gas are limited to a few molecules and a small number of ions from elements of relatively low cosmic abundance. Species observed from the ground include the atoms Li I, Na I, Ca I, Ca II, K I, Fe I, and Ti II and the molecules CH, CH^+, CN, C_2, and NH. In contrast, Table 1 lists the atomic and molecular species with resonance or low-excitation energy lines in the 1150 to 3200 Å region, as detected by the GHRS. Access to UV wavelengths allows the direct detection of absorption by such abundant atoms as C, N, O, Mg, Si, and Fe in a number of ionization states, including those found in cool neutral gas (C I, C II, N I, O I, etc) and in the hot ISM (C IV and N V). Adjacent ionization stages in the UV of the same element are useful for determining physical and ionization conditions in the gas because the ionic ratios for a given element do not depend on assumptions about relative elemental abundances. Examples of adjacent ions are C I-II, Mg I-II, Si I-II-III-IV, S I-II-III, and P I-II-III. UV observations also make possible studies of rare isotopes (i.e. D) and of elements of low cosmic abundance such as B, Ga, Ge, As, Se, Kr, Sn, Te, Tl, and Pb. An understanding of the abundances of these species may lead to information about primordial nucleosynthesis occurring in the Big Bang and about the enrichment of the interstellar gas with heavy elements created through both slow and rapid neutron capture processes.

The GHRS can perform sensitive searches for interstellar molecules that provide insights into interstellar chemical processes. Important molecules with lines in the accessible wavelength region include CH_2, CO, C_2, CO^+, N_2, CN^+, NO, NO^+, H_2O, OH, MgH^+, SiO, CS, and HCl. Molecules already detected are listed at the bottom of Table 1. Studies of abundant molecules such as CO are valuable in investigating interstellar isotopic abundances, the role of chemical

Table 1 Atoms and molecules with absorption lines detected in the ISM with the GHRS

Atoms[a] (1150 < λ < 3200 Å)	Z[b]	IP(eV)[c] (I to II)	IP(eV)[c] (II to III)	$\log(X/H)_m + 12$[d]
H I	1	13.60	...	12.00
D I	1	13.60	...	....
B II	5	8.30	25.15	2.88±0.04
C I, C I*, C I**, C II, C II*, C IV	6	11.26	24.38	8.55±0.05
N I, N V	7	14.53	29.60	7.97±0.07
O I, O I*	8	13.62	35.12	8.87±0.07
Mg I, Mg II	12	7.65	15.04	7.58±0.02
Al II, Al III	13	5.99	18.83	6.48±0.02
Si I, Si II, Si II*, Si III, Si IV	14	8.15	16.35	7.55±0.02
P I, P II, P III	15	10.49	19.73	5.57±0.04
S I, S II, S III	16	10.36	23.33	7.27±0.05
Cl I	17	12.97	23.81	5.27±0.06
Cr II	24	6.77	16.50	5.68±0.03
Mn II	25	7.44	15.64	5.53±0.04
Fe II	26	7.87	16.18	7.51±0.01
Co II	27	7.86	17.06	4.91±0.03
Ni II	28	7.64	18.17	6.25±0.02
Cu II	29	7.73	20.29	4.27±0.05
Zn II	30	9.39	17.96	4.65±0.02
Ga II	31	6.00	20.51	3.13±0.03
Ge II	32	7.90	15.93	3.63±0.04
As II	33	9.81	18.63	2.37±0.05
Se II	34	9.75	21.19	3.35±0.03
Kr I	36	14.00	24.36	3.23±0.07
Sn II	50	7.34	14.63	2.14±0.04
Tl II	81	6.11	20.43	0.82±0.04
Pb II	82	7.42	15.03	2.05±0.03

Molecules: $H_2(v = 3)$, OH, ^{12}CO, ^{13}CO, $C^{17}O$, $C^{18}O$, C_2, HCl

[a]The dominant ions found in neutral H regions are underlined. Because little ionizing radiation with $E > 13.6$ eV occurs in H I regions, the dominant ions are simply determined by whether the first ionization potential IP(I to II) is less than or greater than 13.6 eV. For Cl, Cl I sometimes is the dominant ion in regions containing H I and H_2 since chemical exchange reactions involving, H_2, establish the ionization equilibrium (Jura 1974, Jenkins et al 1986).

[b]Atomic number.

[c]First and second ionization potentials in eV from Moore (1970) are listed.

[d]The Solar System meteoritic abundances are from Anders & Grevesse (1989) except for C, N, and O, which are photospheric values from Grevesse & Noels (1993).

fractionation, and differences in photodestruction rates. Unfortunately, the electronic transitions for the most abundant interstellar molecule, H_2 (in the ground $v'' = 0$ vibration level), occur at wavelengths $\lambda < 1110$ Å, which are inaccessible to the GHRS[1]. As a result, information about this important molecule must be obtained from the *Copernicus* results (Spitzer & Jenkins 1975;

[1]The short-wavelength Digicon detector on the GHRS has a LiF window and can detect wavelengths as short as ~ 1070 Å with low efficiency. However, with the addition of the Corrective Optics Space Telescope Axial Replacement (COSTAR) (see next section) and two additional reflections from mirrors with MgF overcoats, the GHRS efficiency at these short wavelengths is very low.

Savage et al 1977). However, the GHRS can record absorption by H_2 from excited vibrational levels, and 2σ detections of the H_2 B-X (0-3) R(0) and R(1) lines at $\lambda\lambda$1274.535 and 1274.922 Å have been reported (Federman et al 1995).

3. THE GODDARD HIGH-RESOLUTION SPECTROGRAPH

The GHRS is the primary first-generation instrument aboard the *HST* for absorption-line studies of the Galactic interstellar gas at UV wavelengths from 1150 to 3200 Å (Brandt et al 1994; Heap et al 1995). The GHRS contains first-order diffraction gratings for low-resolution ($\lambda/\Delta\lambda \approx 2000$; $\Delta v = 150$ km s^{-1}) and intermediate-resolution ($\lambda/\Delta\lambda \approx 20{,}000$; $\Delta v = 15$ km s^{-1}) spectroscopy as well as an echelle grating (in combination with two cross-dispersers) for high-resolution ($\lambda/\Delta\lambda \approx 85{,}000$; $\Delta v = 3.5$ km s^{-1}) spectroscopy. A carousal rotates to bring the desired grating into the optical path and place the wavelength region of interest onto one of two 512-diode linear array photon-counting Digicon detectors.

The GHRS has both large ($2'' \times 2''$) and small ($0.25'' \times 0.25''$) entrance apertures for science observations. During its first two years of operation, spectra obtained with the large aperture had a degraded spectral resolution because of the spherical aberration present in the *HST* 2.4-m primary mirror. Spectra obtained in the small aperture achieved the full spectroscopic resolutions listed above but required integrations approximately three to four times longer than what would have been needed with a non-aberrated mirror because of light loss. The successful Space Shuttle repair and refurbishment mission in December 1993 fixed these problems with the addition of the Corrective Optics Space Telescope Axial Replacement (COSTAR), which provided corrective optics to the GHRS and other *HST* instruments. The post-COSTAR performance of the GHRS (Soderblom et al 1995) is close to the original GHRS design goals except for some loss of sensitivity at the shortest wavelengths.

Compared with previous instruments for UV interstellar studies, the GHRS offers higher spectral resolution, low-noise photon-counting detectors with modest multiplexing capability, and a relatively large aperture (2.4 m) of the *HST* primary mirror. The 3.5-km s^{-1} resolution of the echelle mode permits study of conditions in individual interstellar clouds. With poorer spectral resolution, severe blending can occur because interstellar H I clouds have mean velocity differences of $\approx$ 6 km s^{-1} (Spitzer 1978). Some interstellar clouds may remain unresolved even at the echelle resolutions offered by the GHRS; high-resolution observations from the ground reveal that velocity structure exists at the level of 1 km s^{-1} in some cold H I clouds (Wayte et al 1978; Blades et al

1980; Welty et al 1994).

The GHRS 512-channel Digicon detectors are capable of very high signal-to-noise (S/N) spectroscopy. The combination of high-resolution and high S/N spectroscopy makes possible searches for elements with low cosmic abundances as well as studies of weak interstellar features, which are important for accurate abundance measurements. By detecting faint objects well beyond the reach of the *IUE* satellite, the GHRS enables observers to probe gas through the entire halo of the Milky Way, to study conditions in dense interstellar regions with large extinctions, and to obtain absorption-line data for extragalactic systems.

Spectra obtained with the GHRS have stable and well-defined line spread functions, particularly when the small entrance aperture is used. The scattered light in the spectrograph is extremely small ($\ll 1\%$) in the first-order grating modes and is larger (≈ 3–10%) but accurately characterized in the echelle modes (Cardelli et al 1993a). The GHRS wavelength calibration is achieved by viewing a Pt lamp through the entire optical path of the spectrograph. Pt calibration spectra obtained at the same grating carrousel position as the observations result in a wavelength calibration accurate to ≈ 0.3 resolution element (or ≈ 5 and 1 km s^{-1} at intermediate and high resolution, respectively). Without the special calibration exposures, the wavelength calibration is approximately three times less accurate but can sometimes be improved through reference to terrestrial absorption lines. Although the windows and photocathodes of the Digicon detectors introduce fixed-pattern noise, this noise can be removed by obtaining multiple spectra with different detector alignments and solving for the noise pattern in the resulting data (Fitzpatrick & Spitzer 1994; Cardelli & Ebbets 1994). Through this process, Meyer et al (1994) and Lambert et al (1994) have obtained spectra with signal-to-noise ratios approaching 1200.

Figure 1 provides an example of GHRS high-resolution data from Spitzer & Fitzpatrick (1993) that illustrates the spectroscopic richness of the UV wavelength region covered by the GHRS. Interstellar absorption-line profiles are shown for the bright 09.5 Vp star HD 93521, which is situated in the halo 1.5 kpc from the Galactic plane. The complex multicomponent nature of the gas absorption makes it possible to measure the abundances and physical conditions in nine different absorbing structures situated in the Galactic disk and low halo.

4. TECHNIQUES

To obtain accurate column densities, either weak, unsaturated absorption lines or lines so strong they have developed radiation damping wings should be observed. In the weak-line limit, the line is said to lie on the linear portion of the curve of growth. The column density N and equivalent width of the line

W_λ are related through the following equations from Spitzer (1978):

$$W_\lambda = \int [1 - e^{-\tau(\lambda)}] d\lambda; \tag{1}$$

$$\tau(\lambda) = \frac{\pi e^2}{m_e c^2} f \lambda^2 N(\lambda); \tag{2}$$

$$N(\text{cm}^{-2}) = 1.13 \times 10^{17} \frac{W_\lambda(\text{mÅ})}{f\lambda^2(\text{Å})} \qquad (\tau(\lambda) \ll 1), \tag{3}$$

where λ is the rest wavelength of the line and f is its oscillator strength. In Equation 1, W_λ measures the amount of energy removed by the absorption and is independent of the instrumental resolution. When a line has well-developed damping wings, the following equation may be used to relate W_λ and N on the square root portion of the curve of growth:

$$N(\text{cm}^{-2}) = \frac{m_e c^3}{e^2 \lambda^4} \frac{W_\lambda^2}{f\gamma} = 1.07 \times 10^{33} \frac{W_\lambda^2(\text{mÅ})}{f\gamma\lambda^4(\text{Å})} \qquad (\tau(\lambda) \gg 1), \tag{4}$$

where γ is the radiation damping constant. In practice, N is generally estimated in this regime through a continuum reconstruction (Bohlin 1975; Diplas & Savage 1994) since the width of the line is usually much larger than the width of the instrumental spread function (i.e. the line is resolved). The H I Lyα line at 1215.67 Å is the most common example of a damped line in the GHRS wavelength range, but in a few cases some strong metal lines (e.g. Mg II $\lambda\lambda$2796, 2803) may also have damping wings (Sofia et al 1994).

The vast majority of interstellar lines normally observed in the GHRS wavelength range have values of $\tau(\lambda)$ that fall somewhere between the two limiting cases of equations 3 and 4. Column density estimates are commonly obtained from lines of intermediate strength through either (*a*) a curve-of-growth analysis in which the total equivalent widths of several lines of the same species are observed and their distribution in the log $W_\lambda/\lambda - \log Nf\lambda$ plane compared with a theoretical curve of growth for a single component subject to a Maxwellian velocity distribution in order to produce a value of N and a Doppler spread parameter b or (*b*) a simultaneous fitting of multiple absorption components having central velocities, widths, and column densities obtained by minimizing the intensity residuals between the observed and predicted absorption profiles for several lines of a species. Rather than describe these methods in detail, we refer the reader to Spitzer (1978). The common pitfalls frequently encountered in abundance studies have been described by Cowie & Songaila (1986) and Jenkins (1987). These problems include (but are not limited to) improper use of the standard curve of growth for strongly saturated lines, invalid application

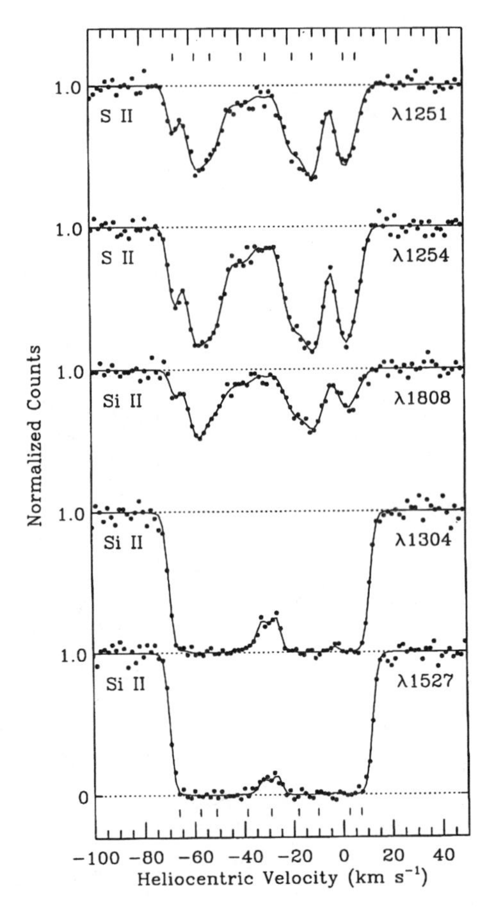

Figure 1 Normalized intensity vs heliocentric velocity for a suite of interstellar lines toward the halo star HD 93521. These GHRS high-resolution data (shown as solid points) reveal a rich velocity structure spanning nearly 90 km s^{-1}. Data of this quality make possible study of the abundances and physical conditions in individual clouds in the ISM along the sight line. The absorption lines shown cover a large range of line strengths, from relatively weak lines (Mg II λ1240 and Mn II λ2606) to very strong lines (Fe II λ2600 and Si II λ1526). Tick marks at the top and bottom of each panel indicate the velocities of the absorption components used to construct the theoretical profiles drawn with solid lines (from Spitzer & Fitzpatrick 1993).

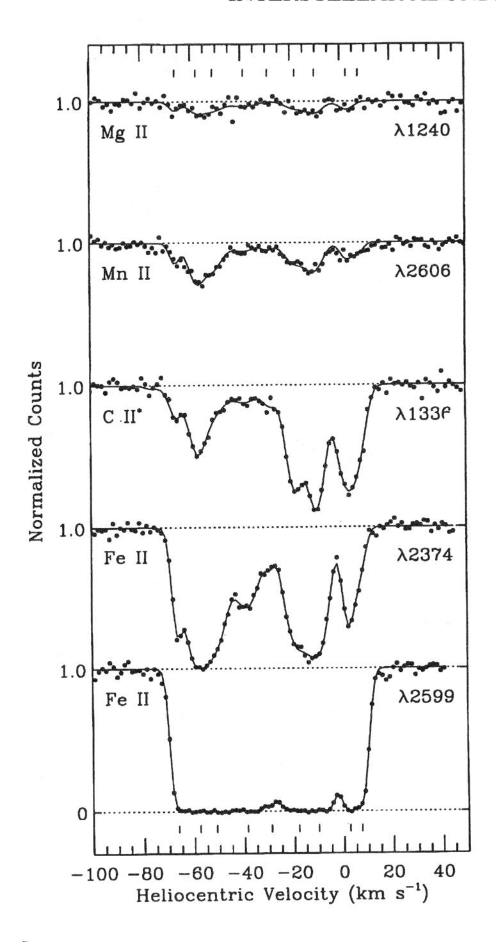

Figure 1 (*Continued*)

of curves of growth across species, inability to distinguish H I from H II region contributions, improper or insufficient treatment of errors, and uncertainties in f-values and cosmic reference abundances. Throughout this paper we reiterate some of these potential problems in the context of higher-resolution data available with the GHRS.

When the instrumental resolution is sufficiently high that an absorption profile (or portion thereof) is close to being resolved or the optical depth of the absorption is small, some of the problems caused by the assumption of a velocity distribution for the absorbing clouds can be circumvented by examining the profiles in terms of their apparent column density profiles (see Savage &

Sembach 1991). The apparent optical depth for the intensity of an observed line at a velocity v is given by

$$\tau_a(v) = -\ln[I_{obs}(v)/I_o(v)] = -\ln[e^{-\tau(v)} \otimes \phi_I(v)], \tag{5}$$

where I_o is the continuum intensity in the absence of absorption and $\phi_I(v)$ is the instrumental spread function. The apparent column density $N_a(v)$ is related to $\tau_a(v)$ through an equation similar to Equation 2:

$$N_a(v) = \frac{m_e c}{\pi e^2} \frac{\tau_a(v)}{f\lambda}$$
$$= 3.768 \times 10^{14} \frac{\tau_a(v)}{f\lambda(\text{Å})} \ [\text{atoms cm}^{-2}(\text{km s}^{-1})^{-1}]. \tag{6}$$

$N_a(v)$ is an apparent column density per unit velocity because its value depends on the resolution of the spectrograph and on the apparent shape of the line. The total apparent column density is $N_a = \int N_a(v)dv$. In the limit where the absorption line is weak ($\tau \ll 1$) or fully resolved [FWHM (line) > FWHM (ϕ_I)], the total apparent column density, N_a, and the true column density, N, are equal.

Many interstellar lines arising in H I regions are not fully resolved even at the highest resolution of the GHRS. In such cases, a comparison of the $N_a(v)$ profiles for two or more lines having different values of $f\lambda$ provides information about the amount and velocity of unresolved saturated structure within the lines. When unresolved saturated structure exists in the lines, the $N_a(v)$ profile of a stronger line underestimates the $N_a(v)$ profile of the weaker line at those velocities, and $N_a(v) < N(v)$. Those portions of the $N_a(v)$ profiles that agree provide valid instrumentally smeared versions of $N(v)$ and, in principle, of $N_a = N$ over those velocity ranges. For doublet lines having values of N_a agreeing to within 20%, the difference between the true column density and the value of N_a for the weaker member of the doublet is equal to the difference in the values of N_a for the doublet (i.e. $N = N_a + \Delta N_a$). Savage & Sembach (1991) have considered the analysis of apparent column density profiles for doublet lines and describe the use of this method in detail.

The main advantage of using the apparent column density method over traditional curve-of-growth techniques for analyses of intermediate-strength absorption lines is the conversion of the data into a form directly suitable for species-to-species comparisons as a function of velocity. No requisite assumption is made about the velocity distribution of the gas since the velocity information is retained in the analysis.

5. OSCILLATOR STRENGTHS

The accuracy of atomic oscillator strengths often limits the accuracy of interstellar abundance measurements. In a compilation important for interstellar studies, Morton (1991) surveyed the available atomic data and recommended a set of oscillator strengths for absorption lines with $\lambda > 912$ Å for a large number of abundant elements through $Z = 32$. Many observers use these oscillator strengths. In this section, we list some of the improvements made since Morton's compilation and extend the list to include some of the rarer elements not found in his tabulation.

Several other recent compilations of oscillator strengths have been performed. Fuhr & Wiese (1991) critically evaluated oscillator strengths for 63 elements through $Z = 83$ (bismuth). Their compilation includes 8300 spectral lines arising from various atomic levels. Verner et al (1994) assembled a list of absorption-line f-values from the ground level of atoms through $Z = 83$ for $\lambda > 228$ Å. Their work merges the compilation of Morton (1991) and that of Fuhr & Wiese (1991) with recent theoretical calculations from the Opacity Project (OP) (Seaton et al 1992), which has produced a complete set of accurate atomic data for permitted transitions involving all stages of ionization of abundant elements with $Z = 1$–14, 16, 18, 20, and 26. A comparison of these theoretical oscillator strengths with experimental results confirms the general reliability of the OP data (Seaton et al 1992; Mendoza 1992; Verner et al 1994). However, the OP f-value calculations assume good LS coupling, and in specific cases, configuration interaction can introduce large systematic errors.

Wavelengths and oscillator strengths for very heavy elements ($Z > 32$) with lines in the GHRS wavelength range likely detectable in absorption in diffuse clouds with $N(\mathrm{H}) > 10^{21}$ cm^{-2} can be found in Cardelli et al (1993b). Their list does not include absorption lines for which the expected line strength is less than ≈ 0.1 mÅ. Such a limit is probably reasonable since to date the weakest ISM lines detected with $> 3\sigma$ significance in very high signal-to-noise (S/N) spectra obtained with the GHRS have equivalent widths of $\approx 0.3 - 0.5$ mÅ (Cardelli et al 1993c; Federman et al 1995). Note that the Ge II λ1237.06 f-value listed in Morton (1991) and carried over to Verner et al (1994) contains a decimal error. The correct value should be 0.876. Since the compilation of Cardelli et al (1993b), Brage & Leckrone (1995) have reported new f-values for As II.

In Table 2 we list new or revised f-values important in determining abundances for dominant ions of abundant elements in interstellar H I regions. The footnotes to the table indicate the sources of the f-values; some are from new experimental measurements, others are from theory, and a few are based on GHRS ISM data.

An important effort in the pursuit of accurate oscillator strengths for ISM studies is the experimental program of Bergeson & Lawler (Refs. 2, 5, and 7 in Table 2), in which laser-induced fluorescence lifetime measurements of selectively excited upper levels are used to determine accurate transition lifetimes. These data, together with reliable branching ratios (when necessary), have yielded f-values with $\approx 10\%$ accuracy for Si II λ1808, Cr II $\lambda\lambda$2056, 2062, 2066; Zn II $\lambda\lambda$2026, 2062; and Fe II λ2249, 2260. These lines often are observed in abundance studies of gas in the Milky Way and in galaxies in the distant Universe.

Interstellar studies with the GHRS also have led to improved values of some atomic and molecular oscillator strengths. For example, new empirical oscillator strengths have been determined for lines of Si II (Spitzer & Fitzpatrick

Table 2 Oscillator strength update[a]

Ion	λ(vacuum) Å	f(Morton)	f(Revised)	Technique[b]	Source[c]
C II.........	2325.403	$4.48\text{x}10^{-8}$	$5.80\text{x}10^{-8}$	E	1
Mg II.......	1240.3947	$1.34\text{x}10^{-4}$	$6.25\text{x}10^{-4}$	I	4
	1239.9253	$2.68\text{x}10^{-4}$	$1.25\text{x}10^{-3}$	I	4
Si II........	2335.123	$3.72\text{x}10^{-6}$	$4.25\text{x}10^{-6}$	E	8
	1808.0126	$5.53\text{x}10^{-3}$	$2.18\text{x}10^{-3}$	E,T	2,11
	1304.3702	$1.47\text{x}10^{-1}$	$8.60\text{x}10^{-2}$	I,T	3,11
	1526.7006	$2.30\text{x}10^{-1}$	$1.10\text{x}10^{-1}$	I,T	3,11
Cl I.........	1088.0589	$1.59\text{x}10^{-2}$	$8.10\text{x}10^{-3}$	E	9
	1097.3692	$4.23\text{x}10^{-2}$	$8.80\text{x}10^{-3}$	E	9
	1347.2396	$1.19\text{x}10^{-1}$	$1.53\text{x}10^{-1}$	E	9
	1363.4476	$9.77\text{x}10^{-2}$	$5.50\text{x}10^{-2}$	E	9
Ar I.........	1048.2199	$2.44\text{x}10^{-1}$	$2.57\text{x}10^{-1}$	E	10
	1066.6599	$6.65\text{x}10^{-2}$	$6.40\text{x}10^{-2}$	T	10
Cr II........	2056.254	$1.40\text{x}10^{-1}$	$1.05\text{x}10^{-1}$	E	5
	2062.234	$1.05\text{x}10^{-1}$	$7.80\text{x}10^{-2}$	E	5
	2066.161	$6.98\text{x}10^{-2}$	$5.15\text{x}10^{-2}$	E	5
Fe II........	1608.4511	$6.19\text{x}10^{-2}$	$6.19\text{x}10^{-2}$	I,E	6,13
	1611.2005	$2.22\text{x}10^{-4}$	$1.02\text{x}10^{-3}$	I	6
	2249.8768	$2.51\text{x}10^{-3}$	$1.82\text{x}10^{-3}$	E	7,12
	2260.7805	$3.72\text{x}10^{-3}$	$2.44\text{x}10^{-3}$	E	7,12
	2374.4612	$2.82\text{x}10^{-2}$	$3.26\text{x}10^{-2}$	I,E	6,12
	2586.6500	$6.46\text{x}10^{-2}$	$6.84\text{x}10^{-2}$	I,E	6,12
Zn II........	2026.136	$5.15\text{x}10^{-1}$	$4.89\text{x}10^{-1}$	E	5
	2062.664	$2.53\text{x}10^{-1}$	$2.56\text{x}10^{-1}$	E	5

[a]We list important improvements to oscillator strengths made since the compilation of Morton (1991) for dominant ion lines of abundant elements found in H I regions. Results from the extensive Opacity Project (OP) data base (Seatonet al 1992) are not listed since they have been compiled by Verner et al (1994). In some cases, OP f-values are superiorto other experimental or theoretical values.

[b]The technique used to obtain the listed f-value is indicated. E and T indicate experimental and theoretical f-values, respectively. I indicates that the f-value is based on an interstellar absorption-line analysis method.

[c]The new f-values are from the following references: (1) Fang et al (1993) and branching ratios from Lennon et al(1985); (2) Bergeson & Lawler (1993b); (3) Spitzer & Fitzpatrick (1993); (4) Sofia et al (1994); (5) Bergeson & Lawler (1993a); (6) Cardelli & Savage (1995); (7) Bergeson et al (1994); (8) Calamai et al (1993); (9) Schectman et al(1993); (10) Federman et al (1992); (11) Dufton et al (1983, 1992); (12) Bergeson et al. (1996a); (13) Bergeson et al. (1996b).

1993), S I (Federman & Cardelli 1995), and Fe II (Cardelli & Savage 1995). Sofia et al (1994) compared column densities for C, N, O, and Mg derived from very strong (damped) lines and from weak lines with little saturated structure. In all cases, the strong-line f-values were well-known, thus permitting a check on the accuracy of the weak-line f-values. For O I (λ1302.2 vs λ1355.6), C II (λ1334.5 vs λ2325.4), and N I (λ1200.7 vs λ1160.9), the strong and weak lines yielded consistent column densities within the experimental errors of $\approx \pm 0.1$ dex. However, for Mg II (λ2803.5 vs $\lambda\lambda$1239.9, 1240.4), the strong lines gave a Mg II column density 4.7 times smaller than did the weak lines when using the Morton (1991) f-values. The new empirical f-values for Mg II $\lambda\lambda$1239.9, 1240.4 listed in Table 2 reflect this difference. Given the importance of Mg in astrophysical environments, these new f-values should be verified experimentally.

Morton & Noreau (1994) present an extensive compilation of wavelengths and oscillator strengths for 1500 electronic transitions of CO between 1000 and 1545 Å. A comparison of their results with the existing CO interstellar absorption-line literature provides consistency checks on f-values and information about CO line saturation corrections. Several of the small f-value transitions revealed through this investigation will be useful in studying the amount of CO in high-column density clouds.

6. THE GHRS ISM ABSORPTION-LINE DATA BASE

Many GHRS interstellar absorption-line observations have come from programs specifically designed for ISM science, whereas others have been by-products of the in-orbit scientific verification of the spectrograph. Table 1 contains a listing of ions observed with the GHRS. Many, but not all, of these ions have been detected in the interstellar clouds toward ζ Oph. Table 3 provides a summary of some of the sight lines for which extensive interstellar GHRS observations have been published. The list includes sight lines for which at least five elements have been studied and does not include objects such as β Pictoris, for which the absorption is primarily circumstellar. For each object the table lists the Galactic coordinates, distance, B-V color excess, total hydrogen column density $\equiv [N(\text{H I}) + 2N(\text{H}_2)]$, GHRS resolution mode (H = high; I = intermediate), elements studied, and primary references.

Few high-resolution observations have been made for objects fainter than $V \approx 9$, but substantial intermediate-resolution data are available. The 3C 273 sight line is the most extensively studied extragalactic direction at the intermediate resolution of the GHRS. Archival research has provided information about individual interstellar species along many sight lines (Cardelli 1994; Roth & Blades 1995; Sembach et al 1995b).

Table 3 The GHRS ISM absorption line database (studies of ≥ 5 elements)[a]

HD	Name	l (deg)	b (deg)	d (pc)	E(B-V) (mag)	log N(H) (cm^{-2})	Mode[b]	Elements studied	Ref.[d]
18100	....	217.9	-62.7	3100	0.02	20.14	I/H	Mg, Si, S, Mn, Cr, Fe	1
22586	....	264.2	-50.4	2000	0.06	20.35	I	O, Mg, Al, Si, S, Fe, Ni	2
24912	ξ Per	160.4	-13.1	540	0.32	21.29	I/H	Many elements - see Table 5	3
35149	23 Ori	199.2	-17.9	430	0.11	20.74	I	Cu, Ga, Ge, Kr, Sn	4
38666	μ Col	237.3	-27.1	1070	0.02	19.85	H	C, O, Mg, Al, Si, S, Cr, Mn, Fe, Ni, Zn	5
47839	15 Mon	202.9	+2.2	700	0.07	20.40	I	Cu, Ga, Ge, Kr, Sn	6
49798	....	253.7	-19.1	650	0.02	...	I	O, Mg, Al, Si, S, Fe	7
68273	γ^2 Vel	262.8	-7.7	450	0.04	19.74	H	C, Mg, Si, P, S, Mn, Fe	8
72089	....	263.2	-3.9	1700	...	...	I	O, Mg, Al, Si, S, Fe, Ni	9,11
72127[c]	....	262.6	−3.4	600	0.08	...	I	C, O, Mg, Si, P, S, Ge	10
93521	....	183.1	+62.2	1700	0.02	20.10	H	C, Mg, Si, S, Mn, Fe	12
116852	....	304.9	-16.1	4800	0.22	20.96	I/H	Mg, Si, P, S, Mn, Cr, Fe, Ni, Zn, Ge	13
120086	....	329.6	+57.5	1000	0.04	20.41	I	O, Mg, Al, Si, S, Fe, Ni	14
141637	1 Sco	346.1	+21.7	170	0.20	21.20	I	Cu, Ga, Ge, Kr, Sn, Pb	15
143018	π Sco	347.2	+20.2	170	0.08	20.75	I	Cu, Ga, Ge, Kr, Sn	16
149757	ζ Oph	6.3	+23.6	140	0.32	21.13	I/H	Many elements - see Table 5	17
149881	....	31.4	+36.2	2100	0.07	20.57	H	Mg, Si, S, Cr, Mn, Fe	18
154368	....	350.0	+3.2	800	0.82	21.62	I/H	C, O, Mg, Al, Si, P, S, Mn, Ni, Zn	22
167756	....	351.5	-12.3	4000	0.09	20.81	H	Mg, Si, Cr, Fe, Zn	19
212571	π Aqr	66.0	-44.7	315	0.23	20.56	I	Cu, Ga, Ge, Kr, Sn	20
....	3C 273	290.0	+64.4	......	.....	20.10	I	Mg, Si, S, Mn, Fe, Ni	21

[a]Sight lines for which extensive ISM data have been obtained and reported in the literature. For many sight lines not listed here, fewer than five interstellar species have been detected and studied in detail.
[b]I and H refer to the GHRS intermediate- and high-resolution modes, respectively.
[c]Binary system with a separation of 4.5". Absorption toward both stars has been studied.
[d]References: (1) Savage & Sembach (1994), Sembach & Savage (1996); (2) Jenkins & Wallerstein (1996); (3) Cardelli et al (1991b), Savage et al (1991), Smith et al (1991), Lambert et al (1995); (4) Hobbs et al (1993); (5) Sofia et al (1993); (6) Hobbs et al (1993); (7) Jenkins & Wallerstein (1996); (8) Fitzpatrick & Spitzer (1994); (9) Jenkins & Wallerstein (1995); (10) Wallerstein et al (1995a); (11) Jenkins & Wallerstein (1996); (12) Spitzer & Fitzpatrick (1992, 1993); (13) Sembach & Savage (1994, 1996); (14) Jenkins & Wallerstein (1996); (15) Hobbs et al (1993), Welty et al (1995); (16) Hobbs et al (1993), Lambert et al (1995); (17) Cardelli et al (1991a, 1993b, 1993c, 1994), Savage et at (1992), Federman et al (1993, 1994), Hobbs et al (1993), Cardelli (1994), Lambert et al (1994, 1995), Lyu et al (1994), Sembach et al (1994), Tripp et al (1994); (18) Spitzer & Fitzpatrick (1995); (19) Savage et al (1994), Cardelli et al (1995); (20) Hobbs et al (1993); (21) Savage et al (1993b); (22) Snow et al (1996).

7. NOTATION AND TERMINOLOGY

We have adopted a system of notation similar to that used in the stellar abundance literature. N(X) refers to the total column density (atoms cm^{-2}) of species X in a H I region. Generally, N(X) is closely approximated by either N(X I) or N(X II), depending whether the ionization potential of the neutral atom is greater or less than 13.6 eV. For hydrogen, $N(\mathrm{H}) = N(\mathrm{H\ I}) + 2N(\mathrm{H_2})$, where $N(\mathrm{H_2}) = \Sigma N(\mathrm{H_2})_J \approx N(\mathrm{H_2})_0 + N(\mathrm{H_2})_1$. Most of the molecular hydrogen is in the two lowest rotational levels ($J = 0$ and 1) in interstellar clouds in which $n(\mathrm{H_2})/n$(H I) is more than $\approx 1\%$ (Savage et al 1977). Therefore, the gas-phase abundance of species X with respect to hydrogen is $(X/H)_g = N(X)/N(H)$, where g refers to gas. The normalized gas-phase abundance with respect to cosmic abundances is $(X/H)_g/(X/H)_c$, where c refers to cosmic. We adopt the standard logarithmic notation system used in stellar astrophysics: $[X/H] \equiv \log (X/H)_g - \log(X/H)_c$.

In the ISM literature, the linear depletion $\delta(X)$ of species X is defined as $\delta(X) = (X/H)_g/(X/H)_c$, and the logarithmic depletion is $D(X) = [X/H]$. The linear and logarithmic depletions are simply the linear and logarithmic gas-phase abundances of a species referenced to cosmic or solar abundances. If $\delta(X) = 1.0$ or $D(X) = 0.00$, then element X has an interstellar gas-phase abundance equal to its cosmic abundance. An element is said to be depleted if $\delta(X) < 1.0$. For lightly depleted elements, $0.3 < \delta(X) < 1.0$, whereas highly depleted elements have $\delta(X)$ as small as 0.001. Highly depleted elements have large depletion factors, where the term depletion factor is defined as $1/\delta(X)$. One generally assumes that the amount of a given species missing from the gas is contained in the interstellar dust. Therefore, we also define a linear dust-phase abundance $(X/H)_d = (X/H)_c - (X/H)_g$ and a corresponding logarithmic dust-phase abundance $\log(X/H)_d = \log\{(X/H)_c - (X/H)_g\}$. For a highly depleted element [small $(X/H)_g$], the dust-phase abundance is essentially equal to the cosmic abundance $(X/H)_c$.

8. COSMIC REFERENCE ABUNDANCES

To understand the significance of interstellar gas-phase abundance measurements, we need measures that represent the total (gas + dust) abundances for young Population I matter. However, the required set of abundances does not exist, so solar system abundances commonly are adopted as the cosmic references. However, several other reference standards also could be used. Table 4 lists recent abundance measurements for H II regions, main-sequence B stars, and the Sun.

Temperature fluctuations in H II regions affect the abundances derived from collisionally excited emission lines (for reviews see Peimbert 1995 and Mathis 1995). The H II region abundances in Table 4 are averages for M8, M17, and Orion (Peimbert et al 1993) based on the assumption that strong temperature fluctuations occur within the nebulae, producing collisionally excited lines in hotter regions and recombination lines in cooler regions. The gas abundances derived from collisionally excited emission lines in the absence of temperature fluctuations are typically 0.25 dex lower than the values given in Table 4 (see Table 7 of Peimbert et al 1993). The most compelling argument for large temperature fluctuations in H II regions is that the abundances derived from collisionally excited emission lines and recombination lines for C and O can be brought into agreement (Peimbert 1995). However, Mathis (1995) concludes that nebular abundance measurements may contain large systematic errors, and he believes that present-day published nebular abundances are suspect even if they include the effects of temperature fluctuations. Similar concerns have been expressed by Kingdon & Ferland (1995), who have calculated nebular models including the effects of temperature fluctuations. As we discuss in Section 12, absorption-line measures of the interstellar oxygen abundance are difficult to interpret if either the H II region abundances or the solar abundances listed in Table 4 are the appropriate reference abundances of ISM studies.

The B-star abundances listed in Table 4 come from two sources: Gies & Lambert (1992) and Figure 2 of Kilian-Montenbruck et al (1994). Both studies focus primarily on main-sequence B stars in the solar neighborhood, and results are from both line-blanketed LTE model atmospheres and NLTE model atmospheres. These separate B-star results compare favorably for C, N, and O (difference < 0.15 dex), but there are large differences (> 0.2 dex) for Ne, Al, Si, S, and Fe. The B-star abundances listed in Table 4 generally span the range of values found for independent LTE model atmosphere calculations for B stars in the solar neighborhood (Fitzsimmons et al 1990) and for B stars in clusters within 4–5 kpc of the Sun (Rolleston et al 1993, 1994). However, inhomogeneities in the ISM may result in significantly different abundances for

Table 4 Reference abundance comparison

X	$\log(X/H)^a_{H\ II}$	$\log(X/H)^b_{B\ star}$		$\log(X/H)^c_\odot$	$\Delta(\langle B\ star \rangle - Sun)^d$
		GL	KM		
C	-3.40	-3.80	-3.73	-3.45	-0.31
N	-4.11	-4.19	-4.20	-4.03	-0.17
O	-3.23	-3.32	-3.44	-3.13	-0.25
Ne	-3.97	-4.03	-3.80	-3.91	-0.01:
Mg	...	...	-4.62	-4.42	-0.20
Al	...	-5.55	-5.81	-5.52	-0.16:
Si	...	-4.42	-4.80	-4.45	-0.16:
S	-4.69	-4.79	-5.03	-4.73	-0.18:
Fe	-5.41	-4.28	-4.60	-4.49	+0.05:

[a]H II region reference abundances are averages of values for M8, M17, and the Orion Nebula (Peimbert et al 1993) and assume substantial nebular temperature fluctuations. If there are no temperature variations, the reported H II region abundances decrease by typically 0.25 dex. For Fe, and possibly for C and O, the listed H II region gas-phase abundances may be smaller than the total abundance (gas + dust) if these elements are constituents of nebular dust.

[b]B-star reference abundances listed in the column labeled GL are from Gies & Lambert (1992), and those listed in the column labeled KM are from a least-squares fit to the measurements in Figure 2 of Kilian-Montenbruck et al (1994) for $R_g = 8.5$ kpc. Note the substantial disagreement between GL and KM (> 0.2 dex) for Ne, Al, Si, S, and Fe.

[c]Solar reference abundances are the meteoritic abundances given by Anders & Grevesse (1989) except for C, N, and O, which are the solar photospheric values from Grevesse & Noels (1993), who allow for the effect of recent revisions of the Fe abundance on gas and electron pressures in solar model atmospheres.

[d]We list the logarithmic differences between the average B-star abundance and the solar abundance. When the B-star abundances disagree by more than 0.20 dex, we attach a colon to the value.

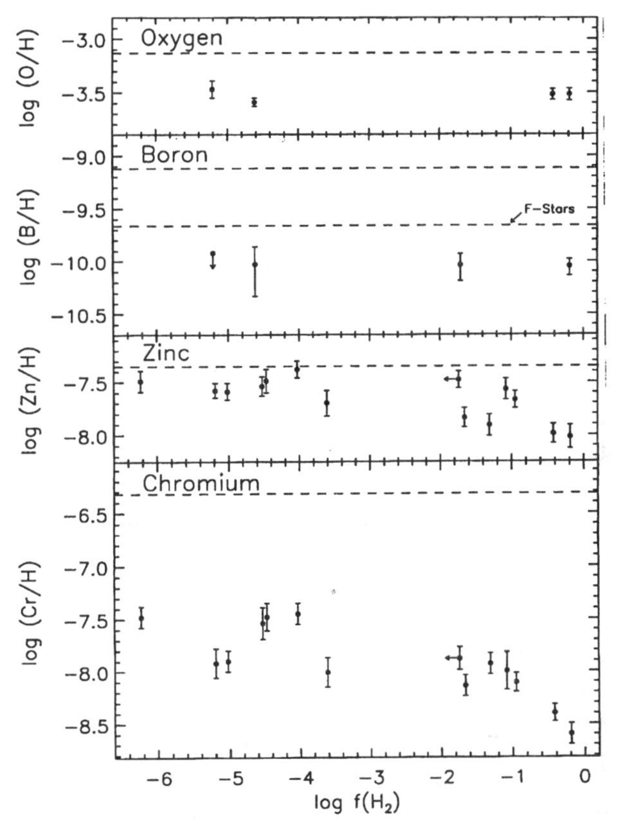

Figure 2 GHRS measurements of the logarithmic abundance of O, B, Zn, and Cr relative to H vs molecular hydrogen fraction $f(H_2) = 2N(H_2)/[N(H\,I) + 2N(H_2)]$. The dashed line immediately below the element name in each panel is the value of log(X/H) for the solar system (see Table 1). For boron, the lower dashed line indicates the reference abundance favored by Federman et al (1993). The two points having the highest molecular hydrogen fraction are ξ Per and ζ Oph. Data for this figure come from the following studies and references cited therein: O (Meyer et al 1994), B (Jura et al 1996) Zn and Cr (Roth & Blades 1995, Sembach et al 1995b).

individual clusters separated by less than a few kiloparsecs (see Lennon et al 1990; Rolleston et al 1994).

The solar abundances listed in Table 4 are the meteoritic abundances given by Anders & Grevesse (1989), with the exception of C, N, and O, which are the solar photospheric values from Grevesse & Noels (1993). These studies yield meteoritic and photospheric abundances that generally agree to within ≈ 0.04 dex and show that several long-standing discrepancies for key elements such as Fe have been resolved.

The last column of Table 4 lists the logarithmic differences between an average of the two B-star abundances and the solar abundances. When the two B-star abundance values differ by ≥ 0.20 dex, we attach a colon to the value of $\Delta(\langle\text{B star}\rangle - \text{Sun})$. With the exception of Ne and Fe, the average B-star abundances are lower than the solar abundances by approximately 0.2 dex. For C, N, and O, the differences range from -0.31 to -0.17 dex. A 4.6-Gyr star may have higher heavy-element abundances than the youngest stars in the solar neighborhood because of incomplete mixing of the Galactic gas, which results in abundance inhomogeneities (Gies & Lambert 1992). Alternatively, the local region of the Galaxy may have experienced a recent inflow of metal-poor material (Meyer et al 1994; Jura et al 1996).

In principle, X-ray astronomy could provide a reliable oxygen reference abundance. Schattenburg & Canizares (1986) reported an interstellar value of $N(\text{O}) = (2.78 \pm 0.55) \times 10^{18}$ cm^{-2} toward the Crab Nebula based on the detection of the oxygen K-shell edge. Because K-shell absorption records the presence of O in both the gas and solid phases, the value should reflect the total line-of-sight oxygen column density. These authors found that log(O/H) $= -3.08^{+0.11}_{-0.14}$, which is in agreement with the solar value. Unfortunately, because the total hydrogen column density toward the Crab Nebula is uncertain, this oxygen abundance may be subject to additional systematic errors.

Given the above considerations, we reference the ISM gas-phase abundances to the solar abundances listed in Table 1. We also discuss the effects of a 0.20-dex lowering of the abundances of all elements compared with the solar abundances in order to explore changes in the results if B-star abundances are used. This approach seems to be more practical than choosing a whole suite of new reference abundances for each of the many elements measured in B stars and in H II regions.

9. IONIZATION EFFECTS

For most elements, the singly ionized stage is dominant in the neutral ISM because the first ionization potential is below 13.6 eV and the second is above the H ionization threshold (see Table 1). Notable exceptions to this rule are N, O, Ar, and Kr, which have first ionization potentials ≥ 13.6 eV, and Ca II, which has a second ionization potential < 13.6 eV. Moreover, for some elements (e.g. Mg, Al, and Cl), dielectronic recombination or ion-molecule reactions may alter the balance of ion stages, depending on the conditions of the environment encountered (Jura 1974; York & Kinahan 1979).

In abundance studies of interstellar clouds, the total amount of an element is usually assumed to be equal to the amount present in the primary stage of ionization in H I regions. This assumption can lead to substantial abundance errors if a mix of H I and H II region gas is present along the sight line or within the cloud under study. One possible means to correct for the relative contributions of the two types of regions is to assume an equilibrium situation in which recombinations balance photonionization. The ratio of gas densities in adjoining ionization stages is then given by the familiar equation:

$$n(\mathrm{X}^{\mathrm{i}+1})/n(\mathrm{X}^{\mathrm{i}}) = \Gamma(\mathrm{X}^{\mathrm{i}})/[n_{\mathrm{e}}\alpha(T, \mathrm{X}^{\mathrm{i}+1})]. \tag{7}$$

If collisional ionization processes are not significant, the ionic ratio in this simplest case depends on the electron density n_{e}; on the photoionization rate $\Gamma(\mathrm{X}^{\mathrm{i}})$, which converts X^{i} into $\mathrm{X}^{\mathrm{i}+1}$; and on the temperature through the recombination coefficient $\alpha(T)$, which may contain both a radiative and a dielectronic component. Additional terms describing charge-exchange reactions between neutral and singly ionized atoms are not included in Equation 7 but may be important for the production or destruction of some species in H I regions and should not be ignored. Péquignot & Aldrovandi (1986) list H I region charge-exchange rates for various ion-neutral pairs as well as updated estimates for radiative recombination coefficients (Aldrovandi & Péquignot 1973; Gould 1978; see also Péquignot et al 1991). Photoionization rates are given by de Boer et al (1973), Witt & Johnson (1973), Gondhalekar & Wilson (1975), and Draine (1978; see also Reilman & Manson 1979). Dielectronic recombination coefficients for many ions can be found in Aldrovandi & Péquignot (1973, 1974, 1976) and Nussbaumer & Storey (1983, 1984, 1986).

Knowledge of several of the quantities in Equation 7 and measurements of $N(\mathrm{X}^{\mathrm{i}})$ and/or $N(\mathrm{X}^{\mathrm{i}+1})$ yield information about the remaining unknown quantities. The volume densities in Equation 7 are usually replaced by column densities, i.e. $n(\mathrm{X}^{\mathrm{i}})/n(\mathrm{X}^{\mathrm{i}+1}) \approx N(\mathrm{X}^{\mathrm{i}})/N(\mathrm{X}^{\mathrm{i}+1})$, although this substitution may not be valid in some cases and is best verified through high-resolution observations, which reveal the velocity structure of the absorption lines or apparent column density profiles.

Equation 7 also provides a means for determining H I region abundances when only trace ionization states of elements are observed. For example, Morton (1975) applied this technique successfully to derive abundance measurements for Li, Na, K, and Ca from optical measurements of the trace ions Li I, Na I, K I, and Ca II by using *Copernicus* satellite observations of dominant ion lines of other elements in the UV to determine n_{e} and T in the cool diffuse interstellar cloud toward ζ Oph. More recently, Federman et al (1993) used GHRS measurements of the weak S I lines and ionization balance to determine a S abundance for the cloud since the S II lines are strongly saturated.

Because temperatures and densities determined from dominant ion lines may differ from those appropriate for regions forming the trace ion lines, abundance estimates that rely on trace ions and on the assumption of ionization equilibrium often incur additional uncertainties. Relative depletion and ionization effects between trace and dominant ions may affect the results, as compositional differences and physical conditions change within individual clouds or from one cloud to the next since the local density scales as $n_{\rm H}^2$ for trace ions and as $n_{\rm H}$ for dominant ions (Jenkins 1987). Preferential incorporation of refractory elements into dust grains (see Sections 10, 12–13) and their subsequent release back into the gas phase under different conditions make comparisons of these elements with volatile elements particularly troublesome.

The assumption of ionization balance appears valid for the diffuse neutral interstellar cloud (the Local Cloud) surrounding the Sun. GHRS measurements of Mg I and Mg II absorption toward Sirius yield an electron density $n_{\rm e} = 0.19$–0.39 cm^{-3} and a temperature $T = 7600 \pm 3000$ K under the assumption of ionization balance defined in Equation 7 (Lallement et al 1994). These results are compatible with the electron density $n_{\rm e} = 0.22$–0.44 cm^{-3} required to explain the carbon component of the anomalous cosmic ray population in the Solar System (Frisch 1994) and with the temperatures $T = 7000 \pm 200$ K and $T = 6700 \pm 200$ K determined from GHRS line width measurements of D I and H I in the local ISM toward Capella and Procyon (Linsky et al 1993, 1995). The value of $\Gamma(\text{Mg I}) = 4.0 \times 10^{-11}$ s^{-1} (Frisch et al 1990) is rather well-known for the solar neighborhood since the 1200–1620-Å photons that create Mg II pass freely into and through H I regions.

Extreme Ultraviolet Explorer Satellite (EUVE) measurements of EUV (504–703 Å) radiation from ε CMa (Cassinelli et al 1995) imply that the star dominates the local stellar EUV radiation field. The measurements set a lower limit of 10–20% for the hydrogen ionization fraction in the local ISM due to stellar sources (Vallerga & Welsh 1995). An ionization fraction close to 15% for the very local ISM is compatible with the higher value of $\sim$50% found for the heliosphere if charge-exchange processes operate at the heliopause (Ripken & Fahr 1983; see also Clarke et al 1995). The *EUVE* result implies an ionization fraction for the local ISM of 70–80% for $n(\text{H}^{\circ}) = 0.1$ cm^{-3} (Frisch 1994) unless the interstellar cloud surrounding the Sun is subject to ionizing sources other than local hot stars, such as residual radiation from a recent ($t < 300{,}000$ years) supernova or from a conductive interface between the Local Cloud and the surrounding hot ($T \sim 10^6$ K) ISM of the Local Bubble (Slavin 1989; Vallerga & Welsh 1995). Recent GHRS measurements of the C IV and Si III-IV absorption toward ε CMa (Gry et al 1995) support the idea that a thermal conduction front at the boundary of the Local Cloud contributes to the ionization of the local interstellar gas in this direction.

In addition to the assumption of photon ionization balance, other methods can be used to determine temperatures and densities in various interstellar environments. For example, Jenkins & Wallerstein (1995) used the GHRS to study the C I fine-structure levels in recombining gas behind a shock front in the Vela SNR. The pressure of the gas is sufficiently high to produce observable fine-structure lines of neutral carbon. In this particular instance, the C I fine-structure levels are populated by collisions in proportion to their level degeneracies, resulting in estimates of $1000 < n_H < 2900$ cm^{-3} and $300 < T < 1000$ K in the absence of observable O I fine-structure lines. Smith et al (1991) also used the ratios of C I fine-structure lines to determine pressures in several of the diffuse cloud components toward ξ Per. They found a very high pressure for the strongest component at $v_{helio} = +6$ km s^{-1}, $\log(P/k) \geq 4.3$ for $T = 32$ K as well as a pressure which is a factor of ten lower in the component near $v_{helio} = +10$ km s^{-1}. Substantial changes occur in the abundances of atomic species between these two clouds as well (see Section 10). Collisional population of the fine-structure levels of C II and Si II also has been used to study conditions in the neutral and ionized gases in diffuse interstellar disk and halo clouds and in H II regions (e.g.) Spitzer & Fitzpatrick 1993, 1995; Fitzpatrick & Spitzer 1994).

High-resolution GHRS observations have been used successfully to identify narrow absorption components resulting from ionized gas tracers such as Al III, P III, S III, Si IV, and C IV in H II regions toward stars with diffuse H I clouds at nearby velocities (ζ Oph: Sembach et al 1994; γ^2 Vel: Fitzpatrick & Spitzer 1994). However, even if accurate column densities can be obtained for some H II region species, the interaction of stellar winds, depletion onto dust, and insufficient knowledge of the stellar flux distribution below 912 Å can seriously affect conclusions about H II region abundances derived from absorption-line data (see also Section 8). This last problem appears to be particularly acute in light of the recent *EUVE* measurement by Cassinelli et al (1995; see also Vallerga & Welsh 1995), which showed that the local ISM hydrogen ionization parameter $\Gamma(\text{H})_{LISM} = 1.1 \times 10^{-15}$ s^{-1} resulting from ε CMa alone is approximately six to seven times greater than previously estimated for the integrated value from all nearby stars combined (Bruhweiler & Cheng 1988).

H II regions outside the immediate vicinity of hot stars also may affect interstellar abundance determinations. Models of ionic ratios in partially ionized diffuse gases (Dömgorgen & Mathis 1994) are often necessary for studies of the ISM in the low Galactic halo since Hα background measurements suggest that the Milky Way ISM contains a diffuse, ionized gas component with a filling fraction of $\sim 20\%$ (Reynolds 1993 and references therein). Useful combinations of ions observable with the GHRS that yield ionization information about interstellar clouds include C I-II-IV, Mg I-II, Al II-III, P I-II-III, Si I-II-III-IV, and S-II-III. Sembach & Savage (1996) found that corrections of ≈0.15–0.20 dex are required in order to convert the abundances derived from standard H I region assumptions to total (H I + H II region) abundances if the diffuse halo clouds toward HD 116852 ($d = 4.8$ kpc; $l = 304.9°$; $b = -16.1°$) are partially ionized by the dilute radiation responsible for the diffuse Hα background.

Photons from O stars in the Galactic disk may be the primary source of ionization of the warm extended ($|z| \sim 1$ kpc) medium in the Milky Way (Miller & Cox 1993; Dömgorgen & Mathis 1994; Dove & Shull 1994). However, Spitzer & Fitzpatrick (1993) noted that the warm neutral and warm ionized gases seem to be well-mixed in the clouds toward the halo star HD 93521. They believe that the partial ionization in these clouds probably cannot be produced by starlight photoionization and suggest a number of alternate ionization processes, including collisional ionization from shocks, X-ray photoionization, and energetic charged-particle ionization.

The integrated flux of extragalactic background radiation from active galactic nuclei may strongly affect the ionization properties of more distant interstellar clouds (Bregman & Harrington 1986). Therefore, studies of abundances in the Magellanic Stream or distant outer Galaxy also must rely on photoionization models (Lu et al 1994a,b). In the distant high-velocity clouds toward Markarian 509, ionization of the interstellar gas is unusual; the clouds are seen in absorption only in C IV, not in NV or Si II (Sembach et al 1995a). This ionization structure is more typical of the high-ionization quasar metal-line absorption systems (see Sargent et al 1979; Steidel 1990) than it is of the quasar mixed-ionization systems, which resemble the absorption characteristics found in the outer Milky Way toward the Magellanic Clouds (Savage & Jeske 1981).

GHRS observations of C IV, N V, and Si IV toward stars in the low halo (Spitzer & Fitzpatrick 1992; Savage & Sembach 1994) and disk (Huang et al 1995) confirm that the high-ionization stages in these regions result primarily from collisional ionization. High-resolution GHRS observations of HD 167756 reveal the presence of C IV and N V at velocities at which no lower ions are detected. This suggests that the C IV and N V absorption occurs in a hot Galactic supershell (Savage et al 1994). Detections of the high ions toward HD 167756 and other stars reveal a second type of highly ionized gas that has lower ionization absorption and that probably traces gas near $T \sim 10^5$ K (see Spitzer 1990 and McKee 1993 for reviews of hot gas in the Galaxy). Rapid progress in theoretical models of conductive interfaces (Borkowski et al 1990) cooling flows (Shapiro & Benjamin 1993), turbulent mixing layers (Slavin et al 1993), and grain destruction in hot regions (Jones et al 1994) eventually may lead to a better understanding of the relationship of ionization and metal abundances in these types of environments.

10. ABUNDANCES IN DIFFUSE CLOUDS

The study of diffuse cloud abundances with the *HST* builds on the pioneering contributions of the *Copernicus* satellite, the use of which demonstrated that many elements in the ISM have lower gas-phase abundances than they do in the Solar System and that the amount of the underabundance (or depletion) depends on sight-line properties (see Jenkins 1987 for a review of the *Copernicus* results). For example, a number of studies (Savage & Bohlin 1979; Murray et al 1984; Harris & Bromage 1984; Harris & Mas Hesse 1986; Jenkins et al 1986; Crinklaw et al 1994) revealed that interstellar depletions correlate much better with average sight-line density $\langle n_{\mathrm{H}} \rangle = N(\mathrm{H})/d$ than with total hydrogen column density or total dust extinction. The *Copernicus* observations are not of high enough spectral resolution to help determine gas-phase abundances within individual clouds, but they provide accurate integrated sight-line column densities.

GHRS analyses of correlations with similar quantities that provide indirect information about the presence of dust, such as the molecular fraction of hydrogen $f(\mathrm{H_2}) = 2N(\mathrm{H_2})/[N(\mathrm{H\,I}) + 2N(\mathrm{H_2})]$, confirm that the depletions of some elements are linked to cloud properties. The relationship between dust and $f(\mathrm{H_2})$ is indirect because molecules such as H_2 in interstellar clouds appear to form on grain surfaces (Spitzer 1978), but these molecules are destroyed through processes such as photodissociation without significant dust destruction. In Figure 2 we plot GHRS measurements of O/H, B/H, Zn/H, and Cr/H for sight lines covering a large spread in $f(\mathrm{H_2})$. The dashed line in each panel indicates the solar value of the ratio. The points in the plot sample two regimes: an intercloud or diffuse cloud medium with little molecular material $[f(\mathrm{H_2}) < 0.01]$, and a cloudy medium with enough material to shelter molecules from photodissociation by UV starlight $[f(\mathrm{H_2}) > 0.01]$. Although the data points are few, the abundances of O and B are similar over the six-decade range spanned by the measurements, implying that these elements are not readily incorporated into dust in a density-dependent manner. In fact, Meyer et al (1994) postulated that the small deficiency (approximately a factor of two) of interstellar gas-phase O seen toward the four stars shown in Figure 2 could be due to dilution of the ISM by the infall of a metal-poor cloud in the solar neighborhood. This interpretation gains support from the B measurements if the reference abundance is $\log(\mathrm{B/H})_c = -9.66$, as found for two F stars (Lemke et al 1993; see also Bosegaard & Heacox 1978). Federman et al (1993) noted that this reference abundance also brings the interstellar B abundance for the ζ Oph sight line into agreement with the abundances of other elements having similar condensation temperatures.

The lack of a trend in the values of O/H and B/H in Figure 2 can be compared with the modest trend (~0.5 dex) for Zn/H and the large trend (~1.0 dex) for Cr/H between log $f(H_2) = -6$–0. The more pronounced correlations observed for Zn and Cr are consistent with $f(H_2)$ trends for other elements having modest to large depletions in dense interstellar environments [e.g. Mg (Cardelli 1994) and Fe (Savage & Bohlin 1979)] and with the ensemble of relationships between elemental depletion and sight-line density reviewed by Jenkins (1987). They also provide a basis for linking the observed behavior of gas-dust interactions in the Galaxy with those in quasar absorption-line systems, in which the sight-line density is not easily measured but $f(H_2)$ is known to be low (log $f(H_2) \lesssim -3$) (Foltz et al 1988; Levshakov et al 1992 and references therein).

The average sight-line density dependence of the elemental depletions in H I regions provided the impetus for a simple explanations for the variations in depletions or gas-phase abundances along different sight lines. Spitzer (1985) proposed that the depletion of an element along a sight line is equal to the average of contributions from three basic types of H I gases: a warm, low-density medium, which predominates at $\langle n_H \rangle \lesssim 0.2$ cm^{-3}; standard diffuse clouds, which contribute most strongly at $\langle n_H \rangle \approx 0.7$ cm^{-3}; and large cold clouds or complexes, the main constituents for sight lines with $\langle n_H \rangle \gtrsim 3$ cm^{-3}. Additional support for this suggestion comes from the work of Joseph (1988), who determined that, for a given level of overall depletion, the sight line-to-sight line variations of the depletions of individual elements are small. If the relative elemental abundances are established early in the lifetime of the grains in the different types of H I regions, then on average this idealized model probably describes well the general character of the neutral ISM.

Although often overused in the ISM literature as a typical interstellar sight line, the ζ Oph line of sight ($l = 6.3°, b = +23.4°, d = 140$ pc) presents an excellent example of the dependence of elemental depletions on cloud types. In Figure 3 we show high-resolution (FWHM = 3.5 km s^{-1}) GHRS spectra of various interstellar lines in the ζ Oph spectrum. The two main clouds (or groups of clouds) along the sight line at $v_{helio} = -27$ and -15 km s^{-1}, respectively, are easily separated in this echelle data and clearly have different ion-to-ion absorption strengths owing to differences in their physical conditions (n_H, T, etc). The cloud at -27 km s^{-1} is warm and has properties consistent with those of the warm neutral medium. The cloud(s) at -15 km s^{-1} contain(s) molecules and resembles a blend of cool diffuse clouds and a large cold cloud.

The left panel of Figure 3 contains a series of interstellar lines for elements that range from lightly depleted (N I, O I) to moderately depleted (Mg II, Mn II) to highly depleted (Fe II, Ni II, Cr II). In this progression, the warm cloud absorption strength increases relative to the cool cloud absorption strength

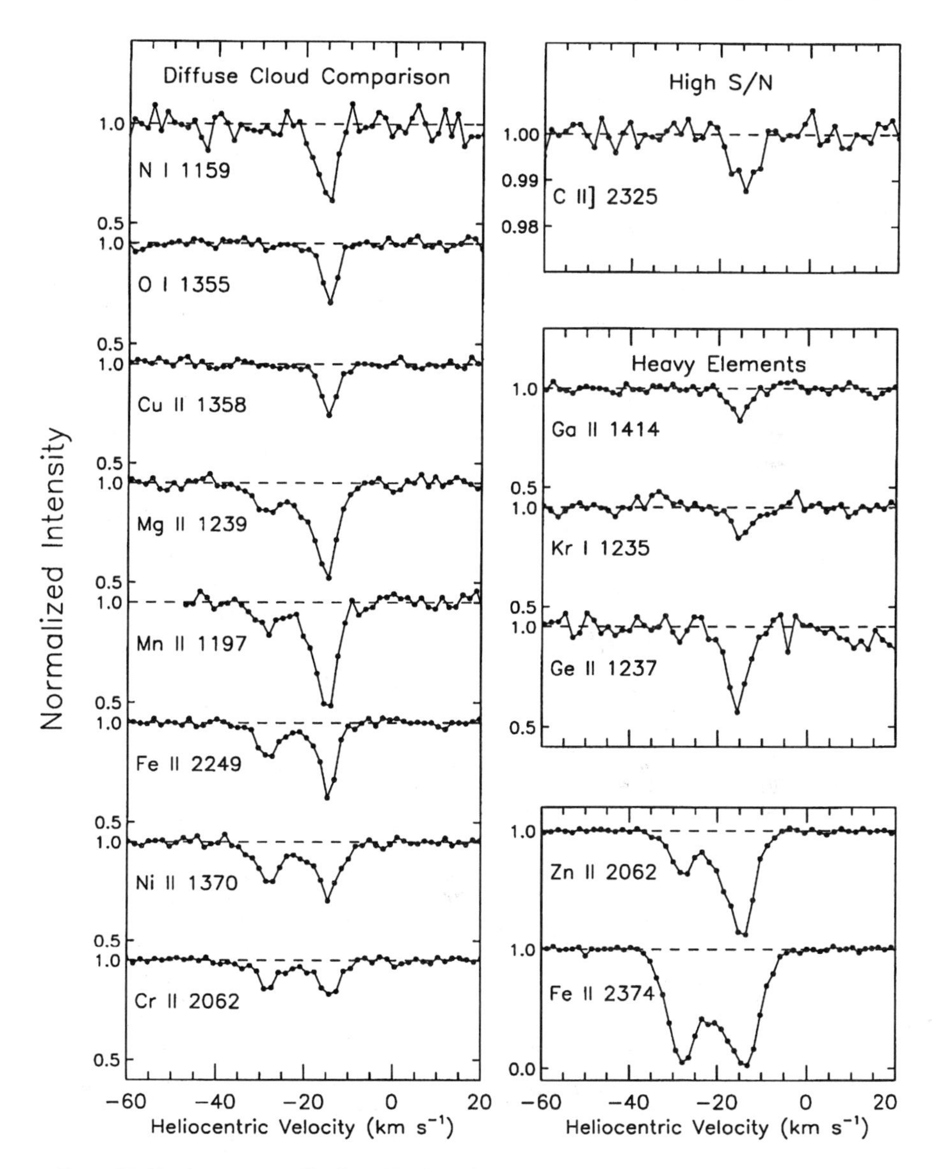

Figure 3 Continuum normalized profiles for selected interstellar lines in the direction of ζ Oph. The left panel shows a series of weak lines of elements that are lightly depleted (N I, O I, Cu II), moderately depleted (Mg II, Mn II), and highly depleted (Fe II, Ni II, Cr II). The bottom right panel shows stronger lines for a weakly depleted (Zn II) and a highly depleted (Fe II) element. The absorption near -27 km s^{-1} is tracing gas in a warm neutral cloud with $\log N(\mathrm{H}) = 19.74$ while the absorption near -15 km s^{-1} is tracing a cool diffuse cloud with $\log N(\mathrm{H}) = 21.12$ (see Table 5). The upper right panel shows a high S/N observation of the C II] line at 2325 Å. This weak line is valuable in determining reliable column densities of C in the ISM (see Cardelli et al 1993c). The middle right panel shows several examples of weak-line detections of heavy ($Z > 30$) elements

Table 5 Gas-phase abundances in the diffuse clouds toward ξ Persei and ζ Ophiuchi[a,b,c]

			[X/H] = log(X/H) − log(X/H)$_\odot$ (+1σ, −1σ)				
			ξ Persei	ζ Ophiuchi			
	log(X/H)$_\odot^{d}$	T_c^{e}	Cool	Cool	Warm	Ref. [f]	Note [g]
Li	-8.69	1225	...	-1.58 (0.05,0.05)	...	1	1
B	-9.12	650	...	-0.93 (0.07,0.09)	...	9	2
C	-3.45	75	-0.16 (0.11,0.14)	-0.41 (0.08,0.13)	...	7	3
N	-4.03	120	...	-0.07 (0.05,0.06)	...	5	
O	-3.13	180	-0.39 (0.05,0.06)	-0.39 (0.04,0.06)	0.00 (0.18,0.31)	5,7	
Na	-5.69	970	...	-0.95 (0.10,0.10)	...	1	1
Mg	-4.42	1340	-1.24 (0.04,0.04)	-1.55 (0.02,0.03)	-0.89 (0.05,0.05)	3,4,5	4
Si	-4.45	1311	> -1.26	-1.31 (0.03,0.03)	-0.53 (0.01,0.01)	4,5,10,13	5,6
P	-6.43	1151	-0.61 (0.02,0.03)	-0.50 (0.16,0.24)	-0.23 (0.02,0.02)	3,4,12	10
S	-4.73	648	> -1.05	+0.18 (0.17,0.30)	...	4,9	6,7
Cl	-6.73	863	...	0.00 (0.20,0.20)	...	1	
Ar	-5.44	25	...	-0.48 (0.16,0.16)	...	1	
K	-6.87	1000	...	-1.09 (0.24,0.25)	...	1	1
Ca	-5.66	1518	...	-3.73 (0.04,0.06)	...	1	1
Ti	-7.07	1549	-2.59 (0.05,0.05)	-3.02 (0.03,0.04)	-1.31 (0.06,0.08)	2,5	8
V	-7.98	1450	...	< -1.96	...	12	9
Cr	-6.32	1277	-2.08 (0.01,0.02)	-2.28 (0.03,0.03)	-1.07 (0.04,0.04)	3,4,5	
Mn	-6.47	1190	-1.32 (0.02,0.03)	-1.45 (0.03,0.03)	-0.90 (0.05,0.05)	3,4,5	
Fe	-4.49	1336	-2.09 (0.03,0.03)	-2.27 (0.02,0.03)	-1.25 (0.03,0.04)	3,4,5	
Co	-7.09	1351	...	-2.76 (0.10,0.12)	-1.54 (0.13,0.19)	9	
Ni	-5.75	1354	-2.46 (0.04,0.05)	-2.74 (0.02,0.02)	-1.51 (0.03,0.03)	3,4,5	
Cu	-7.73	1037	-1.53 (0.06,0.07)	-1.35 (0.03,0.02)	-0.82 (0.13,0.17)	3,4,5	
Zn	-7.35	660	-0.64 (0.02,0.02)	-0.67 (0.11,0.11)	-0.03 (0.02,0.01)	3,4,12	11
Ga	-8.87	918	...	-1.14 (0.06,0.06)	...	5	
Ge	-8.37	825	-0.67 (0.08,0.11)	-0.62 (0.04,0.04)	...	5,8	
As	-9.63	1157	...	-0.21 (0.09,0.09)	...	6	
Se	-8.65	684	...	+0.10 (0.23,0.23)	...	6	
Kr	-8.77	25	...	-0.26 (0.06,0.07)	...	5	
Sn	-9.86	720	...	+0.02 (0.09,0.09)	...	6	
Te	-9.76	680	...	< +0.77	...	6	
Tl	-11.18	450	...	+0.45 (0.12,0.12)	...	11	
Pb	-9.95	520	...	-0.71 (0.15,0.15)	...	11	

[a]Values of [X/H] = log(X/H) − log(X/H)$_\odot$ for elements in italics are from *Copernicus* or ground-based measurements. All others are from GHRS measurements.

[b]Log N(H) = 21.30 ± 0.17 for the cool diffuse clouds towards ξ Per. Log N(H) = 21.12 ± 0.10 for the cool diffuse clouds toward ζ Oph. Log N(H) = 19.74 for the warm diffuse cloud(s) toward ζ Oph. This value is derived by holding the O and Zn abundances in the cloud to within ≈10% of solar abundances (see Savage et al 1992) and is consistent with 21-cm measures of the sight line (Cappa de Nicolau & Poppel 1986). The uncertainty in N(H) for the component is probably ≈0.2–0.3 dex.

[c]All values of [X/H] have been converted into the system of atomic constants and reference abundances used throughout this paper. Errors (±1σ) are based on measurement errors N(X) only and do not include uncertainties in N(H), solar reference abundances, or atomic constants. Limits are 2σ estimates.

[d] Reference solar abundances, derived from meteoritic data, except for C, N, and O, which are derived from solar photosphere data (Grevesse & Noels 1993). The typical error (1σ) in these values is ≈ 0.04 dex (see Table 1).

[e]Condensation temperatures appropriate for the solar nebula with an initial gas pressure of 10^{-4} atm. The condensation temperature is the temperature at which 50% of the element has been removed from the gas phase. All values are from Wasson (1985 and references therein), except those for Pb and Tl (Grossman & Larimer 1974); C,N, O, Ar (Field 1974); and Kr (which we assume to be equal to the value for Ar). The noble gases Ar and Kr may be removed from the gas phase at temperatures between 450 and 700 K through solubility in magnetite (Lancet & Anders 1973; Grossman & Larimer 1974), which may help explain their modest subsolar abundances.

[f]ξ Per and ζ Oph abundance references: (1) Morton 1975 and references therein; (2) Stokes 1978; (3) Cardelli et al 1991b; (4) Savage et al 1991; (5) Savage et al 1992; (6) Cardelli et al 1993b; (7) Cardelli et al 1993c; (8) Cardelli et al 1991a; (9) Federman et al 1993; (10) Cardelli et al 1994; (11) Cardelli 1994; (12) New result derived for this article; (13) Sofia et al (1994).

because of the dependence of the elemental depletions on cloud conditions. For example, the O I absorption is hardly detected in the warm cloud compared to the cool cloud which is consistent with the H I column density being approximately ten times smaller in the warm cloud than in the cool cloud. In contrast, the Cr II absorption at the bottom of the panel has comparable strength in the two clouds. Cr II is lightly depleted in the warm cloud and heavily depleted in the cool cloud. A similar example between the strong lines of a lightly depleted element (Zn II) and a highly depleted element (Fe II) is shown in the lower right panel of Figure 3. Savage et al (1992) interpreted the depletion behavior and physical condition differences within the two clouds as direct evidence for Spitzer's (1985) model explaining the mean density dependence of interstellar depletions.

In Table 5 we list values of $[X/H] = \log(X/H) - \log(X/H)_{\odot}$ observed in the cool and warm diffuse clouds toward ζ Oph by the GHRS together with values observed in the cool diffuse clouds toward another well-studied object, ξ Per ($l = 160.4°, b = -13.1°, d = 540$ pc). We converted all gas-phase abundances into our preferred system of atomic constants (see Section 5) using the solar reference abundances listed in the second column of the table. The errors on [X/H] reflect measurement errors in N(X) only and do not account for uncertainties in oscillator strengths or reference abundances. Changes in [X/H] due to errors in N(H) are systematic across all elements. Errors for N(H) are given in footnote b of Table 5; the value of N(H) for the warm cloud toward ζ Oph is particularly uncertain. The condensation temperature listed in the third column of Table 5 is the temperature at which half of the initial amount of an element is removed from a gas of solar composition owing to the formation of solid matter under conditions close to thermal and chemical equilibrium. The values of T_c are appropriate for a solar nebula with an initial pressure of 10^{-4} atmospheres (Wasson 1985). For lower initial pressures, the values of T_c are larger but are generally within a few tens of degrees of the listed values. For some refractory elements, such as Fe, the differences may be as large as 150° if the initial pressures are as low as 10^{-6} atm (see Wai & Wasson 1977).

[g]Notes: (1) Abundance for ζ Oph cool cloud derived from trace ionization stage and ionization equilibrium considerations, with $n_e = 0.7$ cm^{-3} and $T = 56$ K; (2) Federman et al (1993) prefer log $(B/H)_{\odot} = -9.66$ based on the work of Lemke et al (1993); (3) [C/H] determined from weak C II λ2325 intersystem line; (4) The f-values of the Mg II λλ1239.9, 1240.4 lines still may contain substantial uncertainties; (5) [Si/H] for the ζ Oph cool cloud agrees with the value obtained from the weak Si II] λ2335 intersystem line; (6) Lower limit for ξ Per derived from profile integration of moderately saturated line; (7) ζ Oph result based on observations of SI and ionization balance considerations; (8) Result based on the optical line at 3384 Å (Stokes 1978); (9) Derived from a measured upper limit of 1.0 mÅ for the V II line at 2683.887 Å using an f-value of 0.1026; (10) The ζ Oph cool cloud P II value is derived from component fitting to the line at 1152.818 Å. Measurements of the weaker line at 1532.533 Å would enable stronger constraints to be placed on the P II abundance in the cloud; (11) The ζ Oph cool cloud Zn II value is derived from component fitting to the lines at 2036.136 and 2062.664 Å. This result improves on the previously published estimate of [Zn/H].

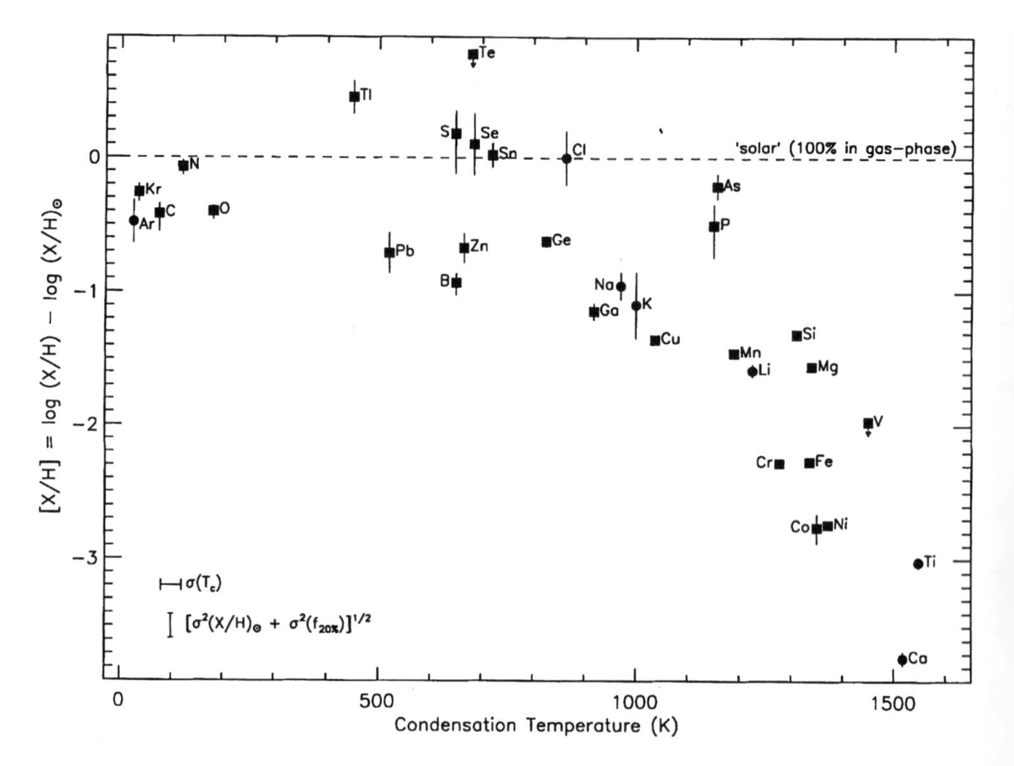

Figure 4 Gas-phase abundance, [X/H] = log(X/H) − log(X/H)$_{\odot}$, vs condensation temperature for the cool diffuse interstellar cloud toward ζ Oph. The data used to construct this plot are listed in Table 5. The condensation temperature is the temperature at which 50% of an element has been removed from the gas phase. GHRS data points referenced to solar abundances are shown as filled squares. *Copernicus* satellite and optical data points are indicated by filled circles. The error bars on all points represent measurement errors only. The data points for Kr and Ni have been shifted slightly in the horizontal direction for clarity. The 1σ errors in condensation temperature ($\pm$20 K) and solar reference abundances combined with f-value uncertainties ($\pm$0.04 dex) are shown in the lower left corner of the plot.

In Figure 4 we present the cool diffuse cloud abundance results for ζ Oph plotted in the familiar form of gas-phase abundance vs condensation temperature. The GHRS data (*filled squares*) are supplemented by *Copernicus* and ground-based observations (*filled circles*) for a few elements. In the cool cloud, C, N, O, S, Ar, Kr, and some heavy elements have depletion factors of less than three. P, Zn, and Ge have slightly larger depletion factors. Ca, Ti, V, Cr, Fe, Co, and Ni have depletion factors in excess of 100. This is the most complete set of elemental abundances available for any interstellar cloud. The depletion pattern exhibited by this cloud, in which elements with larger condensation temperatures generally have greater depletion factors as well, was first studied by Field (1974), who noted that the abundance deficiencies generally correlate with the temperatures derived for particles condensing out of the gas phase in cool stellar atmospheres. Since then it has become increasingly clear that the correlation of depletion with T_c also may depend on (or be lessened by) the growth of grains in molecular clouds and their subsequent stripping and destruction by shocks in the warm ISM. Time-scale considerations (Draine 1990) show that most grains experience at least one period of regrowth in the ISM during their lifetime and that the extreme depletion of various elements created in supernova explosions (such as Ca, Ti, Fe) can be explained only by elemental condensation within molecular clouds unless $> 99.9\%$ of the interstellar gas is cycled through a cool star atmosphere or through a surrounding nebula (Jenkins 1987). This realization is reinforced by the different depletions and properties of the two ζ Oph clouds. A more complete discussion of the origin and evolution of interstellar grains can be found in the monograph by Whittet (1992).

Figure 5 is a graphical comparison of the abundance results for the two ζ Oph diffuse clouds for those elements measured in both. For clarity of presentation, the ordering of the figure is one of decreasing abundance or increasing depletion factor. The differences in the gas-phase abundances within the two clouds are small (< 0.5 dex) for lightly depleted elements such as O and large (> 1 dex) for heavily depleted refractory elements such as Fe and Cr. The lower dashed line in the figure indicates the neutral hydrogen column density–weighted averages of the abundances within the ζ Oph clouds, which are dominated by the cool cloud values. The cool cloud values listed in Table 5 for the ξ Per sight line are close to the ζ Oph average values for many elements, indicating that the integrated sight-line values for the diffuse clouds along these two directions are likely similar as well. The cool clouds toward ξ Per show a velocity-dependent depletion effect that may result from the mixing of different interstellar gases. For example, between $v_{\text{helio}} = +5$ and $+20$ km s^{-1}, the Cr and Fe abundances relative to Zn and O increase by a factor of two (Cardelli et al 1991b; Savage et al 1991), suggesting that this absorption traces a mixture of cloud types.

In addition to providing the velocity resolution necessary to distinguish the absorptions in different interstellar clouds, the GHRS, through its ability to obtain high S/N data, can detect weak lines of cosmically abundant elements (in some cases even those with large depletion factors) or large f-value lines of elements with low cosmic abundances. Such studies have extended the basic interstellar detection list established by *Copernicus* to include elements as heavy as Pb ($Z = 82$; see Table 1). An example of the high S/N spectrum for the C II] λ2325 line toward ζ Oph from Cardelli et al (1993c) is shown in the top right panel of Figure 3. This weak line, which provides an example of the type of data necessary to detect weak intersystem lines of abundant elements (see also Meyer et al 1994), is particularly useful for establishing the C abundance in the interstellar medium because the resonance line of C II near 1334 Å is very strong and is almost always too saturated to yield a reliable carbon abundance. The absorption lines of rare elements also often have equivalent widths of less than 1 mÅ and must have high S/N ratios in order to be detected. Examples of weak-line detections for several heavy elements (Ga II, Ge II, Kr I) for the ζ Oph sight line are shown in the middle right panel of Figure 3. Heavy elements such as these (and others such as As, Sn, Te, Tl, and Pb) extend the study of interstellar chemical behavior into the fifth and sixth rows of the periodic table and provide information

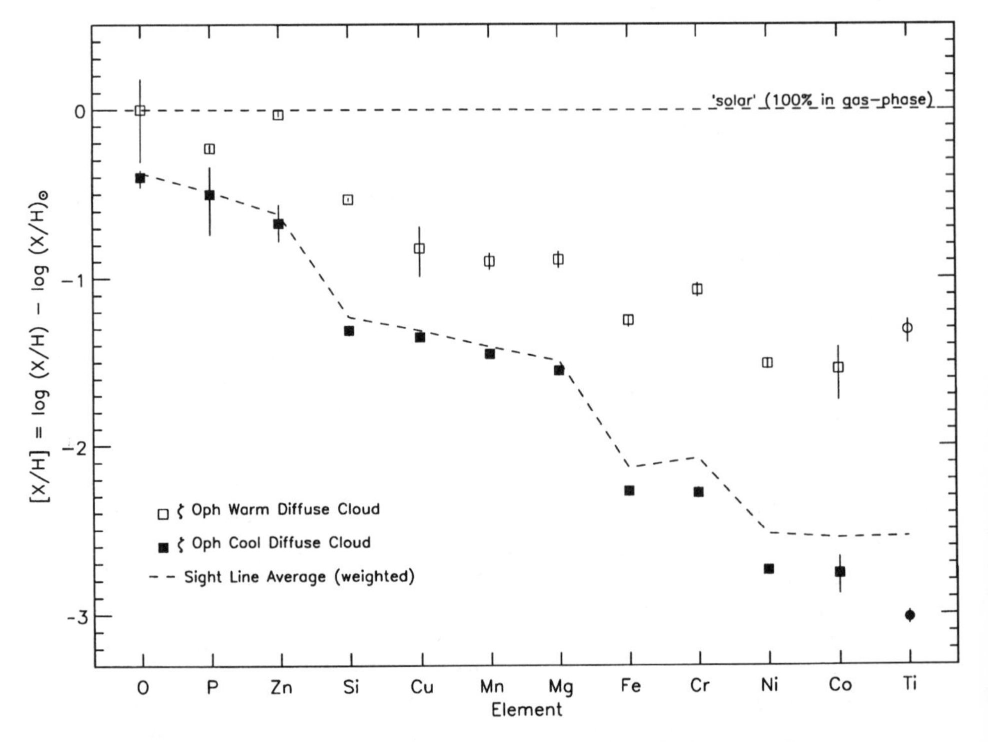

Figure 5 A comparison of the gas-phase abundances, $[X/H] = \log(X/H) - \log(X/H)_\odot$, in the cool and warm diffuse clouds toward ζ Oph at heliocentric velocities of -15 km s^{-1} and -27 km s^{-1}, respectively. The elements are arranged in order of decreasing gas-phase abundance (which is approximately one of increasing condensation temperature). The dashed line indicates the column density-weighted sight-line average abundances. There is a general progression in abundance differences as a function of elemental depletion (see text). The data used to construct this plot are listed in Table 5.

Table 6 Summary of diffuse cloud gas-phase abundances for cosmically abundant elements[a]

	$[X/H] = \log(X/H) - \log(X/H)_\odot$						
Cloud type[b]	Mg	Si	S	Mn	Cr	Fe	Ni
Halo	(<-0.28,-0.56)	(-0.09,-0.47)	(-0.23,+0.16)	(-0.47,-0.72)	(-0.38,-0.63)	(-0.58,-0.69)	(-0.77,-0.91)
Disk+ Halo	(-0.59,-0.62)	(-0.23,-0.28)	(+0.03)	(-0.66)	(-0.72,-0.88)	(-0.80,-1.04)	(-1.15)
Warm Disk	(-0.73,-0.90)	(-0.35,-0.51)	(-0.03,+0.14)	(-0.85,-0.99)	(-1.04,-1.15)	(-1.19,-1.24)	(-1.44,-1.48)
Cool Disk	(-1.24,-1.56)	(-1.31)	(~ 0.00)	(-1.32,-1.45)	(-2.08,-2.28)	(-2.09,-2.27)	(-2.46,-2.74)

[a]The values listed for each element represent the range of $[X/H] = \log(X/H) - \log(X/H)_\odot$ found by Sembach & Savage (1996) for each type of cloud. For some sight lines in the halo, disk + halo, and warm disk categories we assumed $[X/H] \approx [X/Zn] \approx [X/S]$ since no H I or H_2 estimates were available for the individual clouds studied. Zn and S are nearly undepleted in such environments. When only one value is listed, the element was measured for only one sight line.

[b]Halo cloud sight lines: HD 38666 ($v_{helio} = 41$); HD 93521 ($v_{helio} < -22$); HD 116852 ($v_{helio} = -29$ to -64 and -4 to -29); HD 149881, 3C 273. Disk + halo cloud sight lines: HD 18100, HD 167756. Warm disk cloud sight lines: HD 38666 ($v_{helio} = 23$), HD 93521 ($v_{helio} > -22$), HD 149757 ($v_{helio} = -27$). Cool disk cloud sight lines: HD 24912 ($5 < v_{helio} < 20$), HD 149757 ($v_{helio} = -15$).

about r- and s-process nucleosynthesis enrichment of the ISM or about rare elements not previously detected in interstellar clouds (Cardelli et al 1991a; Hobbs et al 1993; Cardelli 1994; Wallerstein et al 1995b; Welty et al 1995).

The GHRS has been used to study individual diffuse clouds along many different lines of sight through the Milky Way disk and halo. These sight lines exhibit various elemental mixes and different types of diffuse clouds. For example, the 1.7-kpc sight line to the low halo star HD 93521 intersects nine distinct warm interstellar clouds ($T \sim 6000$ K) and one cool interstellar cloud ($T \sim 500$ K) (Spitzer & Fitzpatrick 1993). Spectra for several interstellar lines in this direction are shown in Figure 1. The abundances within the warm diffuse clouds toward HD 93521 depend in part on the velocities of the clouds. The warm slow clouds, which have $|v_{helio}| \lesssim 10$ km s^{-1} and are located in the Galactic disk, have depletions very similar to those of the warm diffuse cloud toward ζ Oph (see Figure 5). The faster-moving warm clouds with $|v_{helio}| \gtrsim 35$ km s^{-1} are located in the low halo and have abundances closer to solar abundances and to those of other halo clouds than do the slow clouds. The velocity dependence of the depletions along the HD 93521 sight line is probably due to acceleration of the higher-velocity clouds by shocks that destroy a portion of the dust within the clouds (see Section 14 and Spitzer & Fitzpatrick 1993).

To provide a summary of the diffuse cloud depletions in different environments, we assembled the available GHRS data for nine sight lines in Table 6, where we list [X/H] for several abundant elements (e.g. Mg, Si, S, Mn, Cr, Fe, and Ni). Information for C, N, and O is provided in Section 12 for selected sight lines. The sight lines fall into four general classes divided into disk and halo regions. The two numbers for each element represent the observed range in values of [X/H]. For some of the warm clouds, we use values of [X/Zn] or

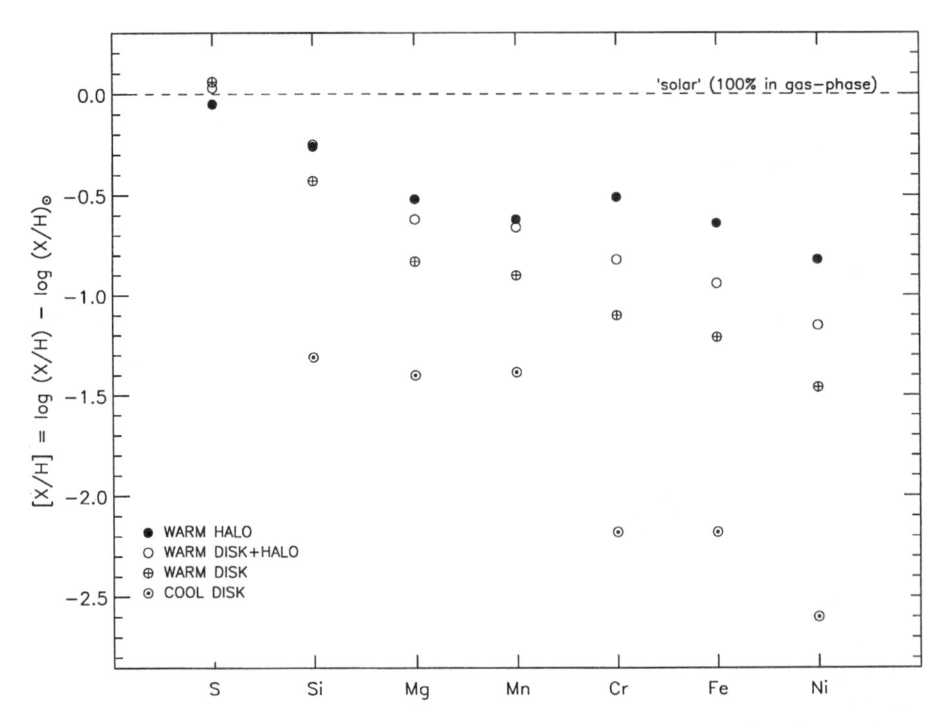

Figure 6 Gas-phase abundances, $[X/H] = \log(X/H) - \log(X/H)_{\odot}$, of 7 abundant elements for diffuse cloud sight lines in the Galactic disk and halo. The values of [X/H] for halo clouds have been derived from measures of [X/Zn] and the assumption that $[X/Zn] \sim [X/H]$. The data used to construct this figure are in Table 6. The cool disk data points are averages for the cool diffuse clouds toward ξ Per and ζ Oph (Table 5). Note the clear differences in the abundance patterns of the disk and halo gases. The upper envelope of abundances established by the halo cloud values indicates that the halo clouds contain a substantial amount of dust with core material that is difficult to destroy through the processes that inject gas and dust into the halo.

[X/S] from Sembach & Savage (1996) as measures of [X/H] since Zn and S have nearly solar abundances relative to H for these types of clouds (Harris & Mas Hesse 1986; Spitzer & Fitzpatrick 1992, 1995; Sembach et al 1995b; Roth & Blades 1995) and since accurate H I information was unavailable for most individual clouds studied. The elemental abundances of diffuse clouds in these regions exhibit systematic differences. Figure 6 shows the average gas-phase abundances for the halo and disk cloud sight lines listed in Table 6. Systematic differences of this type provide new insight into the physical processes that process dust and transport gas within the Galaxy. For the first time in interstellar studies, halo cloud abundances can be studied in as great detail as nearby disk clouds. We discuss Figure 6 and its implications further in Section 13.

11. ISOTOPIC AND MOLECULAR ABUNDANCES

11.1 *The Abundance of Deuterium*

Deuterium in interstellar gas can be atomic or chemically combined in molecules such as HD. Because fractionation affects the molecular abundance of D, the best way to obtain the interstellar ratio of D to H is through direct measures of atomic D I and H I. The larger mass of D compared with H shifts the UV Lyman series absorption of D I by -81.55 km s^{-1} from the Lyman series of H I. The *Copernicus* satellite provided the first reliable measures of the D-to-H ratio in the ISM and produced a large data base of D I and H I Lyman series observations (Rogerson & York 1973; York & Rogerson 1976). In a reanalysis of the *Copernicus* data and of measurements from the *IUE* satellite, McCullough (1992) estimated the mean value of the D-to-H ratio in the local interstellar medium to be $(D/H)_{LISM} = (1.5\pm0.2)\times10^{-5}$ and found no evidence of changes in the ratio from cloud to cloud. However, this result is not universally accepted (Laurent et al 1979; Bruston et al 1981), and higher-precision abundance measurements are required in order to reduce the errors.

Linsky et al (1993, 1995) used the GHRS to obtain an accurate measure of the D-to-H ratio toward Capella, a binary star system at a distance of 12.5 pc. They obtained high S/N observations of the Lyα lines of D I and H I at near opposite orbital quadratures, thus enabling accurate derivations of both the stellar Lyα line profiles and the interstellar H I and D I absorption lines. The full-profile analysis for the direction to Capella yields $(D/H)_{LISM} = 1.60 \times 10^{-5}$, with 1σ random errors of $\pm0.09 \times 10^{-5}$ and systematic errors of $(+0.05, -0.10) \times 10^{-5}$ (Linsky et al 1995). GHRS observations obtained for the sight line to Procyon ($d = 3.5$ pc) by the same investigators are consistent with the Capella result, but the shape of the intrinsic stellar Lyα profile prevents them from setting stronger constraints on the constancy of the D-to-H ratio in the LISM.

Because Galactic chemical evolution destroys D, the Capella sight-line result $(D/H)_{LISM} \geq 1.41 \times 10^{-5}$ sets a strong lower limit on the primordial D abundance. This limit, together with standard Big Bang nucleosynthesis calculations that assume a cosmological constant of zero (Walker et al 1991; Dar 1995), places a strong upper limit on the local baryon density of $\Omega_B h_{50}^2 \lesssim 0.125$, where Ω_B is the ratio of the local baryon density to the critical density and h_{50} is the Hubble constant in units of 50 km s^{-1} Mpc^{-1}. A smaller upper limit of $\Omega_B h_{50}^2$ is implied by the value of D/H $\approx (1.9\text{–}2.5) \times10^{-4}$ reported by Songaila et al (1994), which is based on the analysis of a Keck Observatory spectrum of a QSO absorption line system at $z = 3.32015$ toward Q0014 + 813. The D-to-H ratio must be observed in a number of astrophysical sites with a range of metallicities in order to determine its primordial value and to rule out the possibility that particular candidate D I lines are H I Lyα forest lines.

11.2 *The Molecular Fractionation of the Isotopes of C and O*

UV absorption-line studies of CO provide an opportunity to examine the isotopic fractionation of C and O associated with various molecule formation and destruction processes in diffuse interstellar clouds. For example, the fractionation of CO is quite severe in the diffuse cloud toward ζ Oph – $^{12}CO/^{13}CO \approx 167$, $C^{16}O/C^{18}O \approx 1550$, and $C^{16}O/C^{17}O > 5900$ (Lambert et al 1994). These molecular ratios are larger than atomic isotopic ratios of O for the Sun ($^{16}O/^{18}O \approx 500$ and $^{16}O/^{17}O \approx 2600$; Anders & Grevesse 1989) or of C for the ζ Oph sight line ($^{12}C/^{13}C \approx 66$). The $^{12}C/^{13}C$ ratio for ζ Oph is derived from optical measures of $^{12}CH^+/^{13}CH^+ \approx 66$ and from the assumption that isotopic fractionation does not affect CH^+ since the molecule is produced in warm gas and photodissociation of the two species is nonselective (Crane et al 1991; Hawkins et al 1993).

Two basic processes affect isotopic fractionation in the CO molecule: ion-molecule charge-exchange reactions and selective photodissociation. The ion-molecule charge-exchange reaction

$$^{13}C^+ + {}^{12}CO \leftrightarrow {}^{13}CO + {}^{12}C^+ + \Delta E \tag{8}$$

is exothermic and favors the production of ^{13}CO relative to ^{12}CO. However, photodissociation of ^{12}CO and ^{13}CO favors the selective destruction of ^{13}CO since self-shielding is more effective for the more abundant ^{12}CO molecule. The observed ratio $^{12}CO/^{13}CO \approx 167$ therefore provides information about the effectiveness of the two competing processes. Selective photodissociation appears to dominate the equilibrium ratio of the two species, which means that the standard models for the conditions in the well-studied diffuse cloud in front of ζ Oph must be revised (Lambert et al 1994). Ion-molecule charge-exchange reactions do not influence the equilibrium balance involving the different isotopes of O in the CO molecule since the dominant form of O in diffuse clouds is neutral. However, selective photodissociation leads to molecular fractionation, as is observed.

11.3 *Molecular Abundances*

Wavelength identifications and f-values for many molecules with electronic transitions in the GHRS spectral range can be found in Jenkins et al (1973) and van Dishoeck & Black (1986). Apart from the isotopically substituted molecules discussed above, to date the only new discoveries of molecular interstellar species with the GHRS are tentative detections of HCl and vibrationally excited H_2.

Federman et al (1995) reported a tentative 2.5σ detection of the HCl C-X (0-0) R(0) 1290.257 Å line toward ζ Oph. Column density estimates for Cl, Cl^+, H_2, and HCl provide a meaningful check on diffuse cloud chemical models describing the formation of HCl through reactions involving Cl^+ and H_2. The Federman et al model for the conditions in the neutral, diffuse cloud toward ζ Oph, which is based on a previous model by van Dishoeck & Black (1986), adequately predicts the observed column densities for Cl, H_2, and HCl. The inability of the model to predict the Cl^+ column density to within a factor of approximately three has led the investigators to suggest that the observed value of N(Cl^+) for the cloud might be contaminated by Cl^+ absorption from the H II region surrounding the star.

The *Copernicus* satellite revealed that H_2 is the most abundant molecule in interstellar space with typical fractional abundances $f(H_2) = 2N(H_2)/[N(\text{H I}) + 2N(H_2)]$ of ≈ 0.2 toward stars with B-V color excesses of approximately 0.3 magnitudes (Savage et al 1977). Spitzer et al (1974) studied in detail the rotational excitation of H_2 up to $J = 6$, but the search for vibrationally excited H_2 with *Copernicus* was unsuccessful (Frisch & Jura 1980). Detections of vibrationally excited H_2 created by fluorescent pumping by UV radiation would yield information about the rate of pumping and the mean intensity of UV radiation in diffuse clouds. Federman et al (1995) reported tentative 2σ detections of the H_2 B-X(0-3) R(0) and R(1) lines at 1274.537 and 1274.922 Å in the diffuse clouds toward ζ Oph. Confirmation of these detections would imply $H_2(v = 3)$ column densities a factor of two to three times smaller than the detailed model predictions of van Dishoeck & Black (1986) and a less intense radiation field than is usually assumed. Such detailed checks on the gas and radiation conditions in the ζ Oph clouds are important since this well-studied interstellar sight line provides many critical tests of theories of interstellar atomic and molecular processes.

12. IMPLICATIONS FOR THE COMPOSITION OF INTERSTELLAR DUST

Measures of the gas-phase abundances (g) of the elements in interstellar clouds provide indirect information about the composition of the interstellar dust (d), provided the total (gas + dust) or cosmic abundance (c) is known. The assumption

$$(\text{X/H})_d = (\text{X/H})_c - (\text{X/H})_g \qquad (9)$$

allows the absolute and relative numbers of atoms incorporated into grains found in various types of interstellar clouds to be determined. Studies of the most abundant elements (such as C, N, O, Si, S, Mg, and Fe) yield information about the primary grain constituents, whereas knowledge of the rarer elements provides insight into the processes by which different types of atoms are incorporated into and removed from interstellar grains. As discussed in Section 8, we adopt $(X/H)_c = (X/H)_\odot$, but we also study the implications of using cosmic reference abundances appropriate for the elements found in young main-sequence B stars by assuming $\log(X/H)_{B\ star} = \log(X/H)_\odot - 0.20$.

By measuring the gas-phase abundances in individual interstellar clouds with the GHRS, we can determine dust-phase abundances in clouds of different types and study the composition of dust with different histories of formation and destruction. In our discussions of dust composition, we assume that dust grains found in cool diffuse clouds consist of resilient grain cores covered with mantles. This assumption is consistent with many of the current theoretical ideas about dust formation, growth and destruction (Barlow & Silk 1977; Barlow 1978a,b; Joseph 1988; Mathis 1990) and is supported by the new GHRS observations (see Section 10 and below).

12.1 *Composition of Grain Cores and Mantles*

Dust-phase abundances measured for individual interstellar clouds yield information about the overall composition of dust (core + mantle) in the cloud. We summarize the dust-phase abundances of important elements in several types of interstellar clouds in Table 7, in which we list values of $10^6(X/H)_d$, the dust-phase atomic abundances. This quantity refers to the number of atoms of X in the dust compared to H in the gas. For example, $10^6(O/H)_d = 450$, implies a dust phase O abundance of 450 atoms compared to 10^6 H atoms in the gas.

The left half of the table lists dust composition values derived using solar abundances from Table 1 as the reference cosmic abundance, and the right half of the table lists dust composition values derived using B-star abundances as the cosmic reference standard. The dust-phase abundances given for the cool diffuse clouds toward ζ Oph and ξ Per provide information about the composition of grain cores and mantles, whereas the average results for the warm halo clouds likely represent measures of dust-phase abundances in resilient dust grain cores stripped of their mantles (see Figure 6). In this section we concentrate on results for the combined effects of mantles and cores and examine results for the cool clouds toward ζ Oph and ξ Per.

OXYGEN GHRS determinations of the abundance of oxygen improve on the earlier *Copernicus* survey results based on the weak O I] λ1355 line, which showed that O has a subsolar abundance in the ISM (York et al 1983). Sofia et al (1994) found that the weak O I] λ1355 and damped O I λ1302 lines yield similar column densities, indicating that the unexpectedly low interstellar gas-phase O abundances measured by *Copernicus* and the GHRS cannot be attributed to problems with the oscillator strength of the weak line. If the solar O abundance is used as the reference abundance, then ~450 O atoms per 10^6 H atoms must reside in the dust. The only other atoms with cosmic abundances adequate to combine chemically with this much O are H and possibly C. However, the interstellar H_2O ice feature at 3.1 μm is not seen in absorption along paths through diffuse interstellar clouds. For example, toward VI Cyg No. 12, the absence of the ice band (Sandford et al 1991) implies that less than 0.02% of O is in the form of ice. This problem of the "missing" O can be eased substantially, but not completely eliminated, if a B-star reference abundance of 0.20 dex less than the solar abundance is adopted. In this case, ~170 O atoms per 10^6 H atoms could be chemically incorporated into silicates and various oxides of Mg and Fe in the grains. If we adopt the full 0.25-dex difference between the B-star and solar O abundances given in Table 4, the problem of the missing O is resolved. The values of O/Si in the grains toward ζ Oph are 13 and 8 for solar and B-star reference abundances, respectively. If the grains were pure silicates, the expected value would be between three and four. Some of the O is evidently in the form of oxides (see Section 12.2). In the C1 carbonaceous chondrites, the value of O/Si is eight (Whittet 1984; Wasson & Kallemeyn 1988). Correcting for the O combined with H in the meteoric material, which presumably is in

Table 7 Dust-phase abundances[a]

X	10^6(X/H)	$10^6(X/H)_d$				10^6(X/H)	$10^6(X/H)_d$			
	Sun	ζ Oph (core + mantle)[b]	ξ Per	Halo (core)[c]	(mantle)[d]	B-Star	ζ Oph (core + mantle)[b]	ξ Per	Halo (core)[c]	(mantle)[d]
O	740	450	440	...	...	470	170	170	...	...
C	360	220	110	...	...	220	89	0.0	...	...
N	93	14	...	...	...	59	0.0	...	...	...
Mg	38	37	36	27	10	24	23	22	12	10
Si	36	34	<34	16	18	23	21	<20	2.9	18
Fe	32	32	32	25	7.0	20	20	20	13	7.0
S	19	0.0	<17	2.0	0.0	12	0.0	<10	0.0	0.0
Ni	1.8	1.8	1.8	1.5	0.3	1.1	1.1	1.1	0.87	0.23
Cr	0.48	0.48	0.48	0.33	0.15	0.30	0.30	0.30	0.15	0.15
Mn	0.34	0.33	0.32	0.26	0.07	0.21	0.20	0.20	0.13	0.07

[a]Dust-phase element abundances are listed in the units 10^6 $(X/H)_d$ for solar reference abundances (left side of table; see Table 1) and B-star reference abundances (right side of table; see Section 8). The reference is to the amount of H in the gas. Therefore, the value 10^6 $(X/H)_d$ = 360 listed for C means there are 360 C atoms in the dust for every 10^6 H atoms in the gas.

[b]Dust-phase abundances in grain cores + mantles are based on the abundances listed in Table 5 for the cool, diffuse clouds toward ζ Oph and ξ Per. The sources of the data are referenced in the footnotes to Table 5.

[c]Dust-phase abundances in grain cores are based on the average halo cloud abundances from Sembach & Savage (1996). The listed values of 10^6 $(X/H)_d$ refer mainly to atoms in the resilient cores of interstellar dust.

[d] Dust-phase abundances in grain mantles are derived by subtracting the listed dust core abundances from an average of the ζ Oph and ξ Per (core + mantle) abundances.

the form of H_2O, yields O/Si = 5 in the solid matter (see Fitzpatrick & Spitzer 1994). The abundances in primitive meteoric matter also appear to require the presence of O in the form of oxides.

CARBON Prior to the launch of the *HST*, reliable measures of the C abundance in diffuse interstellar clouds existed only for the sight line to δ Sco and were based on *Copernicus* satellite measurements of the very weak intersystem transition of C II] at 2325 Å (Hobbs et al 1982). The GHRS has provided measures of this weak line toward five stars: ζ Oph, ξ Per, ζ Per, λ Ori, and β^1 Sco (Cardelli et al 1996). The values of N(C II) obtained from the C II] λ2325 and damped C II λ1334 lines for ζ Oph and ξ Per agree to within 0.1 dex and confirm the reliability of the oscillator strength for the λ2325 line (Sofia et al 1994). The amount of C in the dust cores and mantles listed in Table 7 indicates a dust-phase abundance approximately a factor of two larger for ζ Oph than that for ξ Per. However, the gas-phase C abundances found for δ Sco, β^1 Sco, λ Ori, and ζ Per are similar to that found for ζ Oph, with an average result for these five stars of $10^6(C/H)_g \approx 130$. This average implies that $10^6(C/H)_d \approx 230$ and $10^6(C/H)_d \approx 90$ for Solar and B-star reference abundances, respectively.

Many theories of the composition of interstellar dust (see Mathis 1996) require an appreciable amount of C in order to account for the 2175-Å (bump) absorption feature [$10^6(C/H)_d \approx 50$] and for aspects of the continuous extinction [$10^6(C/H)_d \approx 150$]. The polycyclic aromatic hydrocarbons are expected to use $10^6(C/H)_d \approx 30$ (Tielens 1990). A total requirement of $10^6(C/H)_d \approx 230$ in the dust is compatible with the GHRS average result for the five stars assuming solar abundances. However, for the B-star reference abundances and for the ξ Per measurement, the amount of C in the dust is not sufficient to explain the continuous extinction.

NITROGEN The observed gas-phase N abundances for the sight line to ζ Oph imply that little or no N is incorporated into interstellar grains (see Table 7). This result is supported by the fact that weak and damped N I line measurements yield similar N I column densities (Sofia et al 1994). It is also consistent with the expectation that N is unimportant in the formation of dust cores because the large activation energy of N_2 prevents it from participating in gas-phase reactions in stellar atmospheres that lead to dust formation (Gail & Sedlmayr 1986). Evidently, the gas-phase accretion of N onto grains in diffuse clouds is also impeded.

Mg, Fe, Si, Ni, Cr, Mn These elements have high degrees of incorporation into dust. Their relative dust-phase abundances are close to their cosmic abundances.

SULFUR Sulfur is found mainly in the gas phase.

THE RARER ELEMENTS The rare elements that have more than 70% of their atoms in dust [i.e. high depletion factor or (X/H) < -0.5] in the cool diffuse cloud toward ζ Oph include Li, B, Na, K, Ca, Ti, V, Co, Cu, Zn, Ga, Ge, and Pb (see Figure 4 and Table 5). Those that have significant gas-phase abundances [i.e. low depletion factor or (X/H) > -0.5] include P, Cl, Ar, As, Se, Kr, Sn, and Tl. Although we have yet to understand the exact behavior of all these elements in terms of various accretion and destruction processes, some interesting behaviors have been revealed. In particular, the subsolar gas-phase abundance of -0.3 dex for the noble gases Kr (Cardelli et al 1991a) and possibly Ar (Morton 1975) implies either incorrect reference abundances or the presence of noble gases in interstellar grains. GHRS abundance measurements for Kr are particularly secure because 1. the absorption line (Kr I λ1235) is weak, 2. the f-value is well determined, 3. the first ionization potential of 14.00 eV is close to that of H I, and 4. similar gas-phase abundances [(Kr/H) ≈ -0.26 dex] are measured for five sight lines with a wide range of conditions (Cardelli 1994). The estimate of the solar value of Kr/H is somewhat uncertain since it is based on meteoritic and solar wind data (Anders & Grevesse 1989).

The modest depletions of Kr and of various other heavy elements (Sn, Tl) may provide vital clues about the nucleosynthetic enrichment of the interstellar gas once some of the aforementioned reference abundance issues have been resolved. These elements are created through both slow and rapid nucleosynthetic processes and could yield information about advanced stages of nucleosynthetic enrichment in the Galaxy. Other heavy elements also may be useful in such studies once the degree of variation in gas-phase abundances resulting from incorporation into dust is better understood.

12.2 *Composition of Grain Cores*

By studying cloud-to-cloud variations in gas-phase abundances, Sembach & Savage (1996) determined that many interstellar clouds in low-density environments have similar depleted gas-phase abundances. In particular, interstellar clouds at moderate distances from the Galactic plane ($0.5 \leq |z| \leq 1.5$ kpc) have such regular gas-phase abundances that the dust in these clouds appears to have been stripped to a resilient core. If this interpretation is correct, the composition of those grain cores follows directly from the dust-phase composition inferred for halo clouds. The dust-phase abundances for the average of many halo clouds observed by Spitzer & Fitzpatrick (1993, 1995) and by Sembach & Savage (1996) are listed in Table 7.

Study of the relative dust-phase abundances of Mg, Fe, and Si in grain cores is interesting since the existence of silicate grains in the ISM is well established because of the detection of the 9.7- and 18.5-μm SiO stretch-and-bend features in absorption along sight lines with large extinctions (Roche & Aitken 1985).

The observed strength of these IR features requires that nearly all the interstellar Si must be in silicate grains (Draine & Lee 1984). The fact that these same features are also seen in emission toward cool highly evolved stars rich in O (Pégourié & Papoular 1985) suggests that the primary source of silicate grains in interstellar space may be mass ejection from cool stars. The two primary forms of silicate grain cores expected to be produced in stellar atmospheres and to occur in the ISM are pyroxene [(Mg, Fe)SiO_3] and olivine [$(Mg, Fe)_2SiO_4$] (Ossenkopf et al 1992). For pure pyroxene grain cores, the expected ratio of (Mg + Fe)/Si is 1.0, whereas the expected value for olivine cores is 2.0. From Table 7 we see that the implied value of (Mg + Fe)/Si for the grain cores in halo cloud dust is (27 + 25)/16 = 3.3 for solar reference abundances and (12 + 13)/2.9 = 8.6 for B-star reference abundances. In both cases the observations imply that an appreciable fraction of the Mg and Fe in these dust cores must exist in chemical forms that do not involve Si. Obvious candidate substances are oxides of Mg and Fe, including MgO, Fe_2O_3, and Fe_3O_4 (Nuth & Hecht 1990; Fadeyev 1988), and pure Fe. However, interstellar shock models imply a higher destruction rate for pure Fe grains compared with oxide grains, leading Sembach & Savage (1996) to suggest that the extra Mg and Fe atoms are more likely found in oxides (see Section 14).

If the Mg II $\lambda\lambda$1239, 1240 f-values from Hibbert et al (1983) are used instead of those recommended by Sofia et al (1994), the derived values of (Mg + Fe)/Si are (0 + 25)/16 = 1.6 and (0 + 13)/2.9 = 4.5 for solar and B-star reference abundances, respectively. Here, the case for oxide grains is less secure, and the implied grain core composition is consistent with Fe-bearing silicate grains (Spitzer & Fitzpatrick 1993, 1995).

Information about C, N, and O in grain cores is not available through halo cloud studies, but Meyer et al (1994) have shown that the interstellar gas-phase abundance of O is relatively constant over five decades in $f(H_2) = 2N(H_2)/[N(\text{H I}) + 2N(H_2)]$ (see Figure 2). This result suggests that the amount of O depleted from dense diffuse clouds such as those toward ζ Oph and ξ Per is similar to that depleted from the lower-density warm ISM sampled along low-reddening sight lines. For the short path through the local warm cloud to Capella, Linsky et al (1995) obtained $10^6(\text{O/H})_g = 479$, which implies $10^6(\text{O/H})_d = 260$ and $10^6(\text{O/H})_d = 0.0$ for solar and B-star reference abundances, respectively.

12.3 *Composition of Grain Mantles*

If dust in clouds at large distances from the Galactic plane has the elemental composition of resilient grain cores, we can use that information to study the composition of grain mantles. The columns labeled mantle in Table 7 list the mantle composition obtained by subtracting the grain core composition from the average of the grain core + mantle entries for ζ Oph and ξ Per. For both solar and B-star reference abundances, (Mg + Fe)/Si = (10 + 7)/18 = 0.94. The relative abundances of Mg, Fe, and Si in the mantles resemble those found in pyroxene [(Mg, Fe) SiO_3], provided the Mg II f-values are accurate. Regardless of the reference abundance chosen or the absolute value of the Mg II oscillator strengths, the cores and mantles of grains appear to differ significantly in composition.

13. ABUNDANCES IN HALO GAS

The study of elemental abundances in interstellar clouds at large distances from the Galactic plane has progressed rapidly since the launch of the *HST*. We draw heavily upon the work of Spitzer & Fitzpatrick (1993, 1995) and Sembach & Savage (1996) in presenting the basic results relevant to the gas- and dust-phase abundances in clouds in the low Galactic halo. In Figure 6 we show the average gas-phase abundance results for seven elements (S, Si, Mg, Mn, Cr, Fe, and Ni) detected in interstellar clouds at $|z| > 300$ pc toward two or more distant halo stars. For comparison, we also show the warm neutral medium gas-phase abundances for clouds in the Galactic disk (*crossed circles*) and for halo clouds having absorption blended with disk gas absorption (*open circles*). Average values for the cool diffuse disk clouds toward ξ Per and ζ Oph are shown as dotted circles.

From Figure 6 we see a clear progression toward increasing gas-phase abundance of these species from the disk to the halo. The abundance pattern is consistent with a more severe destruction of dust in the halo clouds than in the disk clouds. This destructive processing may result from either more frequent or more severe shocking of the halo clouds compared with the disk clouds. The GHRS results confirm, in a quantitative way, previous optical and *IUE* studies showing that the vertical Galactic scale heights of refractory elements (Ca, Ti, Fe) are larger than those of H I (Morton & Blades 1986; Edgar & Savage 1989; Sembach & Danks 1994; Lipman & Pettini 1995). Furthermore, the spread in the gas-phase abundances for individual halo clouds sampled for $R_g \approx 7$–10 kpc using the GHRS is less than 0.1 dex for Mg, Mn, Fe, and Ni and less than 0.15 dex for Si, S, and Cr, suggesting that the grain cores are difficult to destroy and do not deviate strongly in composition over widely different paths into the halo.

That dust cores with similar properties can survive at large distances from the Galactic plane is somewhat surprising since the transfer of gas into the halo is likely accomplished through supernova explosions in the disk. Models predict that only a few shocks with $v_{sh} \sim 100$ km s^{-1} would be necessary to liberate enough material from the grains to produce the diffuse halo cloud abundances shown in Figure 6 (Sembach & Savage 1996). Some support for this interpretation comes from direct comparisons of the halo cloud abundances with those in strongly shocked environments, which indicate that the halo cloud abundances resemble those of intermediate- and high-velocity gas in the Vela supernova remnant (Jenkins & Wallerstein 1996). Still, the persistence of some types of grains in the ISM (such as the ultrasmall grains with radii < 10 Å) poses a problem for grain destruction and formation theories, partly because how material is cycled from one type of medium to another is unclear and partly because competing growth mechanisms cannot be distinguished from processing mechanisms without additional observations (see Draine 1990). Searches for the enrichment of Fe-peak elements (Fe, Ni) relative to α-process elements (Mg, Si) in halo gas by Type Ia supernovae at large distances from the Galactic plane suggest that the gas-transport processes that cycle material between the disk and the halo operate effectively on time scales short enough ($t \sim 10^7$–10^8 yr) to mask any vertical abundance gradients from S/N ejecta (Jenkins & Wallerstein 1996). These circulation processes include, but probably are not limited to, general turbulence in the disk, photolevitation of diffuse clouds, and movement of material by the flow of a Galactic fountain.

14. SHOCK DESTRUCTION OF GRAINS

It is generally accepted that shocks are responsible for most of the dust grain destruction in the ISM (Draine & Salpeter 1979a,b). Supernovae are likely the primary generators of these shocks in the disk and the halo of the Galaxy. A rich literature describes the effects of shocks on interstellar material and the consequences of such interactions (e.g. Bárlow 1978a,b; Shull & McKee 1979; Seab & Shull 1983; McKee et al 1987; Tielens et al 1994; Jones et al 1994; Vancura et al 1994). Recent reviews of the subject can be found in McKee (1989), Dwek & Arendt (1992), and Draine & McKee (1993).

The thermalization of supernova kinetic energy and the betatron acceleration of grains in the postshocked regions behind supernova blast waves lead to grain-grain collisions and sputtering (both thermal and nonthermal). These processes are most efficient in the warm neutral ISM, where the gas densities are moderate ($n_{\rm H} \sim 1$ cm^{-3}) and temperatures are high ($T \sim 10^4$ K) (see Draine & Salpeter 1979b; McKee et al 1987). Additional factors governing the destructiveness of shocks on grains include the porosity and mean molecular weight of the particles, the interstellar magnetic field strength, and partial grain vaporization (Jones et al 1994). Detailed models incorporating these (and many other) variables require high-quality data in order to test their validity and to place meaningful bounds on the range of possible postshock densities and temperatures. The halo cloud abundance measurements yield a ratio of (Mg + Fe)/Si $= 3.3 \pm 0.6$ for the chemical makeup of the grain cores (see Section 12), which is inconsistent with a pure silicate (olivine + pyroxene) composition for which (Mg + Fe)/Si = 1.0–2.0, and indicates that there is probably another carrier of Fe in the dust besides silicates. As discussed in Section 12, possible carriers include a population of pure Fe grains or various Fe oxides. Shock models can be used to predict that pure Fe grains are destroyed more rapidly behind fast shocks than grains with lower mass densities (McKee et al 1987). We therefore reasonably can conclude that oxides are carriers of Fe since shocks generated by supernovae and cloud-cloud collisions within the Galactic disk are two likely mechanisms that eject gas and dust into the low halo of the Milky Way.

The abundance pattern for resilient grain cores in diffuse halo clouds derived through GHRS absorption-line measurements provides a reference point for the effectiveness of grain destruction and evolution considerations in models used in attempts to predict the emergent spectrum of shocked gas in interstellar environments. The composition of the halo cloud grain core indicates that localized enrichment of the gas behind a shock is probably most pronounced when the grain cores are destroyed. Grain destruction models suggest that large shock velocities are necessary in order to disrupt the cores completely (see Jones et al 1994). Observational studies of gas in supernova remnants may help show how elements are liberated into the gas phase by shocks. An example of variations in line strength with velocity has been given by Jenkins & Wallerstein (1995, 1996) for the interstellar direction toward HD 72089, which lies behind the Vela SNR. They found that absorption strengths of elements readily incorporated into dust grain, such as Al and Fe, increase by at least an order of magnitude relative to elements found mainly in gas, such as O and S, over the velocity range of ≈ 120 km s^{-1} spanned by absorption in the remnant. In particular, in the $+90$ to $+120$ km s^{-1} absorption toward HD 72089, the ratio of Fe to S is nearly solar (i.e. [Fe/S] ≈ 0) (see Figure 2 of Jenkins & Wallerstein 1995). The dependence of elemental depletion on cloud velocity was first documented by Routly & Spitzer (1952) on the basis of observations that the ratio of Ca II to Na I generally increased with increasing cloud velocity, presumably as a result of shock destruction of dust grains or changes in the ionization of the cloud (Routly & Spitzer 1952; Spitzer 1978 and references therein). Vallerga et al (1993) gave an excellent example of the effect for Ca II and Na I. UV observations with *Copernicus* and the GHRS, such as those for HD 72089, have extended the study of the velocity dependence of depletions to dominant ion stages and have made possible exploration of the physical conditions in the gas that give rise to the depletion changes.

15. ABUNDANCES IN HIGH-VELOCITY CLOUDS

The Milky Way is enveloped in H I high-velocity clouds (HVCs), which are detected in 21-cm emission with $|v_{\rm LSR}| \geq 100$ km s^{-1} (for a review see Wakker

1991). The nature and origin(s) of HVCs are poorly understood, but absorption-line abundance measurements are providing important new information about them. Table 8 lists the known optical and UV detections of heavy-element absorption lines associated with HVCs.

Optical observations from the ground have led to detections of Ca II toward many HVCs. Uncertainties associated with ionization corrections and the depletion of Ca prevent observers from obtaining direct information about the total Ca abundance in the HVCs. However, the Ca II observations confirm that the detected HVCs do not consist of primordial (metal-deficient) gas. In addition, ground-based observations of Ca II are crucial for setting constraints (usually upper limits) on the distances to the HVCs.

The GHRS and the Faint Object Spectrograph (FOS) have yielded measures of N(Mg II)/N(H I) toward a number of HVCs (Table 8). Many of the values are lower limits because the degree of absorption-line saturation is unknown. For the -209 and -147 km s^{-1} HVCs toward Mark 205, the footnotes to Table 8 explain how we estimated [Mg/H] from the equivalent widths reported by Bowen et al (1995).

Detailed information on abundances in a HVC is available for the $+240$ km s^{-1} gas toward the Seyfert galaxy NGC 3783 ($l = 287.5°, b = +22.9°$). Lu et al (1994a) detected the HVC in the lines of S II, Si II, and possibly C I with the GHRS in intermediate-resolution mode (20 km s^{-1} FWHM). Sulfur is the most useful element for deriving a metallicity for this HVC because it is not readily depleted onto interstellar dust and because the observed lines of S II are not saturated. Furthermore, S II (IP = 23.3 eV) should be the dominant ionization stage of S in the HVC, which has $N(\text{H I}) = 1.21 \times 10^{20}$ cm^{-2}. Assuming the ionization corrections are not large, the observed measure of $N(\text{S II})/N(\text{H I}) = 2.8 \times 10^{-6}$ implies a gas-phase abundance relative to the Sun of $\delta(\text{S}) = 0.15 \pm 0.05$. For the same cloud, the Si II λ1260 line is saturated and yields only a lower limit $\delta(\text{Si}) \geq 0.006$. On the basis of these results, Lu et al (1994a) concluded that the HVC complex toward NGC 3783 is most likely associated with gas stripped from the Galaxy by an extragalactic object(s) such as the Magellanic Clouds.

The Seyfert galaxy Fairall 9 lies in the direction of the Magellanic Stream, which is an $\sim 180°$ narrow band of H I extending from the Magellanic Clouds to the south Galactic pole and beyond. In the direction to Fairall 9 ($l = 295.1°, b = -57.8°$), the 21-cm emission associated with the Magellanic Stream reveals two components, one near $v_{\text{LSR}} \approx +160$ km s^{-1}, with $N(\text{H I}) \approx 2 \times 10^{19}$ cm^{-2} and the other at $v_{\text{LSR}} \approx +200$ km s^{-1}, with $N(\text{H I}) \approx 6 \times 10^{19}$ cm^{-2} (Morras 1983). Lu et al (1994b) used the GHRS to study the abundances of S II and Si II in these two clouds. S II was not detected, and Si II produced saturated absorption.

Table 8 H I-HVC metal line detections

Object	l (deg)	b (deg)	HVC	v_{LSR} (km s^{-1})	Ion	[X/H]	Reference[h]
Mark 106	161.1	42.9	Complex A	-157	Ca II	...[a]	1
I Zw 18	160.5	44.8	Complex A	-165	O I	> -1.57[f]	2
PG 1351+640	112.0	52.0	Complex C	-154	Ca II	...[a]	9
PG 1259+592	120.6	58.1	Complex C	-127	Mg II	...[b]	3
Mark 205	125.4	41.7	125+41-209	-209	Mg II	-2.24(+0.31,-0.05)[c,e]	4
"	"	"	Complex C	-147	Mg II	-0.59(+1.09,-0.38)[c,e]	4
Mark 290	91.5	47.9	Complex C	-138	Ca II	...[a]	9
3C351	90.1	36.4	Complex C	-180	Mg II	...[b]	3
BD +38 2182	182.2	62.2	Complex M	-90	Si II	> -0.80[c]	5
"	"	"	Complex M	-96	Ca II	...[a]	13
PG 0043+039	120.2	-58.7	Mag. Stream	-348	Mg II	> -0.92[c]	3
PKS 2251+11	82.8	-41.9	Mag. Stream	-374	Mg II	> -1.23[c]	3
3C 454.3	86.1	-38.2	Mag. Stream	-397	Mg II	> -0.49[c]	3
Fairall 9	295.1	-57.8	Mag. Stream	+195	Ca II	...[a]	10
"	"	"	Mag. Stream	+160	Si II	> -0.70[c]	6
"	"	"	Mag. Stream	+200	Si II	> -1.15[c]	6
Q0637-752	286.4	-27.2	Outer Galaxy	+125	Mg II	...[g]	14
PKS 0837-12	237.2	17.4	242+17+106	+105	Ca II	...[a]	11
NGC 3783	287.5	23.0	287+22+240	+240	Ca II	...[a]	12
"	"	"	287+22+240	+240	S II	-0.82(+0.12,-0.18)	7
"	"	"	287+22+240	+240	Si II	> -2.22[c]	7
H1821+643	94.0	27.4	Outer Arm	+120	Si II	~ -1.0:[c,d]	8
"	"	"	Outer Arm	+120	Mg II	~ -1.0:[c,d]	8

[a]Abundances have not been derived from the Ca II observations since ionization corrections and Ca depletion into dust grains make any such measures uncertain.

[b]Detection is from FOS measurements. Strong blending with lower-velocity gas prevents reliable measure of absorption-line strength for the HVC alone.

[c]Result may be influenced by the effects of dust depletion.

[d]Value uncertain. A large correction for the effects of ionization by the extragalactic background is required (see Savage et al 1995).

[e]The numbers listed for the Mg abundances in the HVCs toward Mark 205 are calculated from the revised Mg II equivalent widths listed by Bowen et al (1995), from the curve-or-growth error method of Savage et al (1990), and from the values of *N*(H I) from Lockman & Savage (1995). The values of log *N*(Mg II) $\pm 1\sigma$obtained for the clouds at -209 and -147 km s^{-1} are 12.62 (+0.3, −0.05) and 13.14 (+1.09, −0.38), respectively. The errors for *N*(Mg II) associated with the cloud at -147 km s^{-1} are much larger than suggested by the formal profile fit results of Bowen & Blades (1993). The -147 km s^{-1} cloud also has a poorly determined value of *N*(H I).

[f]The uncertain O I equivalent width for the HVC toward I Zw18 obtained by Kunth et al (1994) has been used to estimate a lower limit for *N*(O I), assuming the observed absorption is produced on the linear part of the curve of growth. The listed limit for [O/H] follows from *N*(H I) $= 2.1 \times 10^{19}$ cm^{-2} (Kunth et al 1994). We do not list a result for Si II since the feature reported for the Si II λ1304 line has less than 2σ significance.

[g]Bowen et al (1995) detected Mg II absorption toward Q0637-752 in the velocity range from +100 to +150 km s^{-1}. These authors attribute the high-velocity gas to corotating high $|z|$ gas in the outer galaxy. Mg II and H I column densities are not reported. Note that gas at similar velocities is seen toward many stars in the Large Magellanic Cloud in the direction $l \approx 279°$, $b \approx -32°$ (Savage & de Boer 1981).

[h]References: (1) Schwarz et al 1995; (2) Kunth et al 1994; (3) Savage et al 1993a; (4) Bowen & Blades 1993, Bowen et al 1995; (5) Danly et al 1993; (6) Lu et al 1994b, upper limits also reported for S II; (7) Lu et al 1994a; (8) Savage et al 1995; (9) Wakker et al 1996; (10) Songaila 1981; (11) Robertson et al 1991; (12) West et al 1985; (13) Keenan et al 1995; (14) Bowen et al (1995).

Assuming the ionization corrections are not large, the observations show that $\delta(\mathrm{Si}) \geq 0.2$ and $\delta(\mathrm{S}) \leq 0.9$ for the + 160 km s^{-1} HVC and that $\delta(\mathrm{Si}) \geq 0.07$ and $\delta(\mathrm{S}) \leq 0.3$ for the + 200 km s^{-1} HVC. These limits are consistent with an origin for the Magellanic Stream closely tied to gas in the Magellanic Clouds.

UV absorption-line measurements of abundant species provide an estimate of the sky-covering factor of high-velocity gas with $|v_{\mathrm{LSR}}| > 100$ km s^{-1} containing metals. The first results based on FOS 300-km s^{-1} resolution spectra of quasars (Savage et al 1993a) revealed that 6 of 14 sight lines exhibited high-velocity Mg II absorption, excluding H1821 + 643, which has high-velocity absorption associated with the Galactic warp. Two of the seven extragalactic sight lines studied by Bowen et al (1995) have high-velocity absorption: Mark 205 and the + 100 to + 150 km s^{-1} absorption toward Q0637-752. When the two data sets are combined, 8 of 21 extragalactic sight lines exhibit high-velocity Mg II absorption. This observation suggests a high-velocity Mg II sky-covering factor of 38 ± 13%. Although this result is still subject to large errors owing to the small sample size, it is consistent with the large sky-covering factor of 37% for H I HVCs recently detected by Murphy et al (1995) through a sensitive HVC 21-cm emission line search. These authors found a large population of low–column density HVCs with $7 \times 10^{17} < N(\mathrm{H\,I}) < 2 \times 10^{18}$ cm^{-2} not detected in earlier surveys. Virtually all newly detected high-velocity, 21-cm emission is associated with previously known emission complexes, which evidently have extensive low–column density outer envelopes.

Highly ionized HVCs also have been discovered in spectra obtained by the GHRS. For example, they have been observed in the C IV doublet toward Mark 509 (Sembach et al 1995a) PKS 2155-304 (Bruhweiler et al 1993). In the direction of Mark 509, the C IV absorption extending from −170 to −340 km s^{-1} appears to be associated with H I-HVCs at similar velocities 1.5° from the line of sight. Sembach et al (1995a) favor the interpretation that this C IV-HVC is the low-density photoionized boundary of the H I-HVC. However, an origin in collisionally ionized gas with $T > 10^5$ K cannot be eliminated without further observations.

ACKNOWLEDGMENTS

The ISM abundance results discussed in this review would not have been obtained without the dedicated efforts of the many people responsible for the construction, launch, and operation of the *HST*, GHRS, and FOS. We thank them all. We appreciate comments on draft versions of our review by Steve Federman, Edward Jenkins, John Mathis, David Meyer, Todd Tripp, Alan Sandage, Lyman Spitzer, and Bart Wakker. We also thank Edward Fitzpatrick for providing an electronic version of Figure 1 and Jason Cardelli for providing the C II] data in Figure 3. KRS acknowledges support from a Hubble Fellowship from NASA through Grant number HF-1038.01-92A from the Space Telescope Science Institute, which is operated by AURA under NASA contract NAS5-26555. BDS appreciates support from NASA Grant NAG5-1852.

Literature Cited

Aldrovandi SMV, Péquignot D. 1973. *Astron. Astophys.* 25:137–40
Aldrovandi SMV, Péquignot D. 1974. *Rev. Bras. Fis.* 4:491
Aldrovandi SMV, Péquignot D. 1976. *Asron. Astrophys.* 47:321 (Erratum)
Anders E, Grevesse N. 1989. *Geochim. Cosmochim. Acta* 53:197–214
Barlow MJ. 1978a. *MNRAS* 183:367–95
Barlow MJ. 1978b. *MNRAS* 183:397–415
Barlow MJ, Silk J. 1977. *Ap. J.* 215:800–4
Bergeson SD, Lawler JE. 1993a. *Ap. J.* 408:382–88
Bergeson SD, Lawler JE. 1993b. *Ap. J. Lett.* 414:L137–40
Bergeson SD, Mullman KL, Lawler JE. 1994. *Ap. J. Lett.* 435:L157–59
Bergeson SD, Mullman KL, Wickliffe ME, Lawler JE. 1996a. *Ap. J. Lett.*, In press
Bergeson SC, Mullman KL, Lawler JE. 1996b. *Ap. J. Lett.* In press
Blades JC, Wynne-Jones I, Wayte RC. 1980. *MNRAS* 193:849–66
Bohlin RC. 1975. *Ap. J.* 200:402–14
Borkowski KJ, Balbus SA, Fristrom CC. 1990. *Ap. J.* 355:501–17
Bosegaard AM, Heacox WD. 1978. *Ap. J.* 226:888–96
Bowen D, Blades JC. 1993. *Ap. J. Lett.* 403:L55–58
Bowen D, Blades JC, Pettini M. 1995. *Ap. J.* 448:662–66
Brage T, Leckrone DJ. 1995. *J. Phys. B* 28:1201–10
Brandt JC, Heap SR, Beaver EA, Boggess A, Carpenter KG, et al. 1994. *Publ. Astron. Soc. Pac.* 106:890–908
Bregman JN, Harrington JP. 1986. *Ap. J.* 309:833–45
Bruhweiler FC, Boggess A, Norman DJ, Grady CA, Urry M, Kondo Y. 1993. *Ap. J.* 409:199–204
Bruhweiler FC, Cheng KP. 1988. *Ap. J.* 355:188–96
Bruston P, Audouze J, Vidal-Madjar A, Laurent C. 1981. *Ap. J.* 243:161–69
Calamai AG, Smith PL, Bergeson SD. 1993. *Ap. J. Lett.* 415:L59–62
Cappa de Nicolau CE, Poppel WGL. 1986. *Astron. Astrophys.* 164:274–99
Cardell JA. 1994. *Science* 265:209–13
Cardelli JA, Ebbets DC. 1994. In *Calibrating Hubble Space Telescope*, ed. JC Blades, SJ Osmer, pp. 322–31. Baltimore: Space Telesc. Sci. Inst.
Cardelli JA, Ebbets DC, Savage BD. 1993a. *Ap. J.* 413:401–15
Cardelli JA, Federman SR, Lambert DL, Theodosiou CE. 1993b. *Ap. J. Lett.* 416:L41–44
Cardelli JA, Mathis JS, Ebbets DC, Savage BD. 1993c. *Ap. J. Lett.* 402:L17–20
Cardelli JA, Meyer DM, Jura M, Savage BD. 1996. *Ap. J.* In press
Cardelli JA, Savage BD, Ebbets DC. 1991a. *Ap. J. Lett.* 383:L23–28
Cardelli JA, Savage BD. 1995. *Ap. J.* 452:275–85
Cardelli JA, Savage BD, Bruhweiler FC, Smith AM, Ebbets DC, et al. 1991b. *Ap. J. Lett.* 377:L57–60
Cardelli JA, Sembach KR, Savage BD. 1995. *Ap. J.* 440:241–53
Cardelli JA, Sofia UJ, Savage BD, Keenan FP, Dufton PL. 1994. *Ap. J. Lett.* 420:L29–32
Cassinelli JP, Cohen DH, MacFarlane JJ, Drew JE, Lynas-Gray AE, et al. 1995. *Ap. J.* 438:932–49
Clarke JT, Lallement R, Berteaux J-L, Quemerais R. 1995. *Ap. J.* 448:893–904
Crane P, Hegyi DJ, Lambert DL. 1991. *Ap. J.* 378:181–85
Cowie LL, Songaila A. 1986. *Annu. Rev. Astron. Astrophys.* 24:499–535
Crinklaw G, Federman SR, Joseph CL. 1994. *Ap. J.* 424:748–53
Danly L, Albert CE, Kuntz KD. 1993. *Ap. J. Lett.* 416:29–32
Dar A. 1995. *Ap. J.* 449:550–53

de Boer KS, Jura MA, Shull JM. 1987. See Kondo 1987, pp. 533–59
de Boer KS, Koppenaal K, Pottasch SR. 1973. *Astron. Astrophys.* 28:145–46
Diplas A, Savage BD. 1994. *Ap. J. Suppl.* 93:211–28
Dömgorgen H, Mathis JS. 1994. *Ap. J.* 428:647–53
Dove JB, Shull JM. 1994. *Ap. J.* 430:222–35
Draine BT. 1978. *Ap. J. Suppl.* 36:595–619
Draine BT. 1990. In *Evolution of the Interstellar Medium*, ed. L Blitz, pp. 193–205. San Francisco: Astron. Soc. Pac. Conf. Ser. Vol. 12
Draine BT, Lee HM. 1984. *Ap. J.* 285:89–108
Draine BT, McKee CF. 1993. *Annu. Rev. Astron. Astrophys.* 31:373–432
Draine BT, Salpeter EE. 1979a. *Ap. J.* 231:77–94
Draine BT, Salpeter EE. 1979b. *Ap. J.* 231:438–55
Dufour RJ. 1987. See Kondo 1987, pp. 577–87
Dufton PL, Hibbert A, Kingston AE, Tully JA. 1983. *MNRAS* 202:145–50
Dufton PL, Keennan FP, Hibbert A, Ojha PC, Stafford RP. 1992. *Ap. J.* 387:414–16
Dwek E, Arendt RG. 1992. *Annu. Rev. Astron. Astrophys.* 30:11–50
Edgar RJ, Savage BD. 1989. *Ap. J.* 340:762–74
Fang Z, Kwong VHS, Wang J, Parkinson WH. 1993. *Phys. Rev. A* 48:1114–22
Fadeyev Y. 1988. In *Atmospheric Diagnostics of Stellar Evolution*, ed. K Nomoto, pp. 174–80. Berlin: Springer-Verlag
Federman SR, Beideck DJ, Schectman RM, York DG. 1992. *Ap. J. Lett.* 401:367–70
Federman SR, Cardelli JA. 1995. *Ap. J.* 452:269–74
Federman SR, Cardelli JA, Sheffer Y, Lambert DL, Morton DC. 1994. *Ap. J. Lett.* 432:L139–42
Federman SR, Cardelli JA, van Dishoeck EF, Lambert DL, Black JH. 1995. *Ap. J.* 445:325–29
Federman SR, Sheffer Y, Lambert DL, Gilliland RL. 1993. *Ap. J. Lett.* 413:L51–54
Field G. 1974. *Ap. J.* 187:453–59
Fitzpatrick EL, Spitzer L. 1994. *Ap. J.* 427:232–58
Fitzsimmons A, Brown PJF, Dufton PL, Lennon DJ. 1990. *Astron. Astrophys.* 232:437–42
Foltz CB, Chaffee CB, Frederic H Jr, Black JH. 1988. *Ap. J.* 324:267–78
Frisch PC. 1994. *Science* 265:1423–27
Frisch PC, Jura MJ. 1980. *Ap. J.* 242:560–67
Frisch PC, Welty DE, York DG, Fowler JR. 1990. *Ap. J.* 357:514–23
Fuhr JR, Wiese WL. 1991. In *Atomic Transition Probabilities*, ed. DR Lide, pp. 10–128. *CRC Handbook of Chemistry and Physics*. Cleveland: CRC Press. 72nd ed.
Gail H, Sedlmayr E. 1986. *Astron. Astrophys.* 166:225–36
Gies DR, Lambert DL. 1992. *Ap. J.* 387:673–700
Gondhalekar PM, Wilson R. 1975. *Astron. Astrophys.* 38:329–33
Gould RJ. 1978. *Ap. J.* 219:250–61
Grevesse N, Noels A. 1993. In *Origin of the Elements*, ed. N Prantzos, E Vangioni-Flam, M Cassé, pp. 15–25. Cambridge: Cambridge Univ. Press
Gry C, Lemonon L, Vidal-Madjar A, Lemoine M, Ferlet R. 1995. *Astron. Astrophys.* 302:497–508
Grossman L, Larimer JW. 1974. *Rev. Geophys. Space Phys.* 12:71–101
Harris AW, Bromage GE. 1984. *MNRAS* 208:941–53
Harris AW, Mas Hesse JM. 1986. *MNRAS* 220:271–78
Hawkins I, Craig N, Meyer DM. 1993. *Ap. J.* 407:185–97
Heap SR, Brandt J, Randall CE, Carpenter KG, Leckrone DS, et al. 1995. *Pub. Astron. Soc. Pac.* 107:871–87
Hibbert A, Dufton PL, Murray MJ, York DG. 1983. *MNRAS* 205:535–41
Hobbs LM, York DG, Oegerle W. 1982. *Ap. J. Lett.* 252:L21–23
Hobbs LM, Welty DE, Morton DC, Spitzer L, York DG. 1993. *Ap. J.* 411:750–55
Huang J-S, Songaila A, Cowie LL, Jenkins EB. 1995. *Ap. J.* 450:163–78
Jenkins EB. 1987. In *Interstellar Processes*, ed. DJ Hollenbach, HA Thronson, pp. 533–59. Dordrecht: Reidel
Jenkins EB. 1990. In *Evolution in Astrophysics: IUE Astronomy in the Era of New Space Missions,* Vol. Sp-310, pp. 133–41. Noordwijk: ESA
Jenkins EB. 1995. In *Laboratory and Astronomical High Resolution Spectra*, ed. AJ Sauval, R Blomme, N Grevesse, pp. 453–58. San Francisco: Astron. Soc. Pac. Conf. Ser. Vol. 81
Jenkins EB, Drake FJ, Morton DC, Rogerson JB, Spitzer L, York DG. 1973. *Ap. J.* 181:L122–27
Jenkins EB, Joseph CL, Long D, Zucchino PM, et al. 1988. *Proc. SPIE* 932:213–29
Jenkins EB, Lees JF, van Dishoeck EF, Wilcots EM. 1989. *Ap. J.* 343:785–810
Jenkins EB, Savage BD, Spitzer L. 1986. *Ap. J.* 301:355–79
Jenkins EB, Wallerstein G. 1995. *Ap. J.* 440:227–40
Jenkins EB, Wallerstein G. 1996. *Ap. J.* In press
Jones AP, Tielens AGGM, McKee CF, Hollenbach DJ. 1994. *Ap. J.* 433:797–810

Joseph CL. 1988. *Ap. J.* 335:157–67
Joseph CL, Jenkins EB. 1991. *Ap. J.* 368:201–14
Jura M. 1974. *Ap. J.* 191:375–79
Jura M, Meyer DM, Hawkins I, Cardelli JA. 1996. *Ap. J.* 456:598–601
Keenan FP, Shaw CR, Bates B, Dufton PL, Kemp SN. 1995. *MNRAS* 272:599–604
Kilian-Montenbruck J, Gehren T, Nissen PE. 1994. *Astron. Astrophys.* 291:757–64
Kingdon JB, Ferland GJ. 1995. *Ap. J.* 450:691–704
Kondo Y, ed. 1987. *Exploring the Universe with the IUE Satellite*. Dordrecht: Reidel
Köppen J, Aller LH. 1987. See Kondo 1987, pp. 589–602
Kunth D, Lequeux J, Sargent WLW, Viallefond F. 1994. *Astron. Astrophys.* 282:709–16
Lallement R, Bertin P, Ferlet R, Vidal-Madjar A, Bertaux JL. 1994. *Astron. Astrophys.* 286:898–908
Lambert DL, Sheffer Y, Federman SR. 1995. *Ap. J.* 438:740–49
Lambert DL, Sheffer Y, Gilliland RL, Federman SR. 1994. *Ap. J.* 420:756–71
Lancet MS, Anders E. 1973. *Geochim. Cosmochim. Acta* 37:1371–88
Laurent C, Vidal-Madjar A, York DG. 1979. *Ap. J.* 229:923–41
Lemke M, Lambert DL, Edvardsson B. 1993. *Publ. Astron. Soc. Pac.* 105:468–75
Lennon DJ, Dufton PL, Fitzsimmons A, Gehren T, Nissen PE. 1990. *Astron. Astrophys.* 240:349–56
Lennon DJ, Dufton PL, Hibbert A, Kingston AE. 1985. *Ap. J.* 294:200–6
Levshakov S, Chaffee FH, Foltz CB, Black JH. 1992. *Astron. Astrophys.* 262:385–94
Linsky JL, Brown A, Gayley K, Diplas A, Savage BD, et al. 1993. *Ap. J.* 402:694–709
Linsky JL, Diplas A, Wood BE, Brown A, Ayres TR, Savage BD. 1995. *Ap. J.* 451:335–51
Lipman K, Pettini M. 1995. *Ap. J.* 442:628–37
Lockman FJ, Savage BD. 1995. *Ap. J. Suppl.* 97:1–47
Lu L, Savage BD, Sembach KR. 1994a. *Ap. J.* 426:563–76
Lu L, Savage BD, Sembach KR. 1994b. *Ap. J. Lett.* 437:L119–22
Lyu C-H, Smith AM, Bruhweiler FC. 1994. *Ap. J.* 426:254–68
Mathis JS. 1987. See Kondo 1987, pp. 517–29
Mathis JS. 1990. *Annu. Rev. Astron. Astrophys.* 28:37–70
Mathis JS. 1995. *Rev. Mex. Astron. Astrophys.* Serie de conf. 3:207–14
Mathis JS. 1996. In *Polarimetry of the Interstellar Medium*, ed. W Roberge, DCCM Whittet. San Francisco: Astron. Soc. Pac. In press
McCullough PR. 1992. *Ap. J.* 390:213–18
McKee CF. 1989. In *IAU Symp. 135: Interstellar Dust,*. ed. LJ Allamandola, AGGM Tielens, pp. 431–44. Dordrecht: Kluwer
McKee CF. 1993. In *Back to the Galaxy*, ed. S Holt, F Verter, pp. 499–513. New York: Am. Inst. Phys.
McKee CF, Hollenback DJ, Seab CG, Tielens AGGM. 1987. *Ap. J.* 318:674–701
Mendoza C. 1992. In *Atomic and Molecular Data for Space Astronomy: Needs, Analysis, and Availability, Lecture Notes in Physics*. ed. PL Smith, WL Wiese, pp. 85–119. Berlin: Springer-Verlag
Meyer DM, Jura MJ, Hawkins I, Cardelli JA. 1994. *Ap. J. Lett.* 437:L59–61
Miller WW, Cox DP. 1993. *Ap. J.* 417:579–94
Moore CE. 1970. *Ionization Potentials and Ionization Limits Derived from the Analysis of Optical Spectra. Rep. No. NSRDS-NBS34*. Washington, DC: US Dept. Commerce
Morras R. 1983. *Astron. J.* 88:62–66
Morton DC. 1975. *Ap. J.* 197:85–115
Morton DC. 1991. *Ap. J. Suppl.* 77:119–202
Morton DC, Blades JC. 1986. *MNRAS* 220:927–48
Morton DC, Noreau L. 1994. *Ap. J. Suppl.* 95(1):301–43
Morton DC, Spitzer L. 1966. *Ap. J.* 144:1–12
Murphy E, Lockman FJ, Savage BD. 1995. *Ap. J.* 447:642–45
Murray MJ, Dufton PL, Hibbert A, York DG. 1984. *Ap. J.* 282:481–84
Nussbaumer H, Storey PJ. 1983. *Astron. Astrophys.* 126:75–79
Nussbaumer H, Storey PJ. 1984. *Astron. Astrophys. Suppl.* 56:293–312
Nussbaumer H, Storey PJ. 1986. *Astron. Astrophys. Suppl.* 64:545–55
Nuth JA, Hecht JH. 1990. *Astrophys. Space Sci.* 163:79–94
Ossenkopf V, Henning Th, Mathis JS. 1992. *Astron. Astrophys.* 261:567–78
Pégourié B, Papoular R. 1985. *Astron. Astrophys.* 142:451–60
Peimbert M. 1995. In *Analysis of Emission Lines*, ed RE Williams. Cambridge: Cambridge Univ. Press. In press
Peimbert M, Torres-Peimbert S, Dufour RJ. 1993. *Ap. J.* 413:242–50
Péquignot D, Aldrovandi SMV. 1986. *Astron. Astrophys.* 161:169–76
Péquignot D, Boisson C, Pettijean P. 1991. *Astron. Astrophys.* 252:680–88
Reilman RF, Manson ST. 1979. *Ap. J. Suppl.* 40:815–80
Reynolds RJ. 1993. In *Back to the Galaxy*, ed. S Holt, F Verter, pp. 156–65. New York: Am. Inst. Phys.
Ripken HW, Fahr HJ. 1983. *Astron. Astrophys.* 122:181–92

Robertson JG, Morton DC, Schwarz UJ, van Woerden H, Murray D. 1991. *MNRAS* 248:508–14

Roche PF, Aitken DK. 1985. *MNRAS* 215:425–35

Rolleston WRJ, Brown PJF, Dufton PL, Fitzsimmons A. 1993. *Astron. Astrophys.* 270:107–16

Rolleston WRJ, Dufton PL, Fitzsimmons A. 1994. *Astron. Astrophys.* 284:72–81

Rogerson JB, York DG. 1973. *Ap. J. Lett.* 186:L95–98

Rogerson JB, York DG, Drake FJ, Jenkins EB, Morton DC, Spitzer L. 1973. *Ap. J. Lett.* 181:L110–15

Roth K, Blades JC. 1995. *Ap. J. Lett.* 445:L95–98

Routly PMcR, Spitzer L. 1952. *Ap. J.* 115:227–43

Sandford SA, Allamandola LJ, Tielens AGGM, Sellgren K, Tapia M, Pendleton Y. 1991. *Ap. J.* 371:607–20

Sargent WLW, Young PJ, Bocksenberg A, Carswell RF, Whelan JAJ. 1979. *Ap. J.* 230:49–67

Savage BD, Bohlin RC. 1979. *Ap. J.* 229:136–46

Savage BD, Bohlin RC, Drake JF, Budich W. 1977. *Ap. J.* 216:291–307

Savage BD, Cardelli JA, Bruhweiler FC, Smith AM, Ebbets DC, Sembach KR. 1991. *Ap. J. Lett.* 377:L53–56

Savage BD, Cardelli JA, Sofia UJ. 1992. *Ap. J.* 401:706–23

Savage BD, de Boer KS. 1981. *Ap. J.* 243:460–84

Savage BD, Edgar RJ, Diplas A. 1990. *Ap. J.* 361:107–15

Savage BD, Jeske NA. 1981. *Ap. J.* 244:768–76

Savage BD, Lu L, Bahcall JN, Bergeron J, Boksenberg A, et al. 1993a. *Ap. J.* 413:116–36

Savage BD, Lu L, Weymann RJ, Morris SL, Gilliland RL. 1993b. *Ap. J.* 404:124–43

Savage BD, Sembach KR. 1991. *Ap. J.* 379:245–59

Savage BD, Sembach KR. 1994. *Ap. J.* 434:145–61

Savage BD, Sembach KR, Cardelli JA. 1994. *Ap. J.* 420:183–96

Savage BD, Sembach KR, Lu L. 1995. *Ap. J.* 449:145–55

Schattenburg ML, Canizares CR. 1986. *Ap. J.* 301:759–71

Schectman RM, Federman SR, Beideck DJ, Ellis DG. 1993. *Ap. J.* 406:735–38

Schwarz UJ, Wakker BP, van Woerden H. 1995. *Astron. Astrophys.* 302:364–81

Seab CG, Shull MJ. 1983. *Ap. J.* 275:652–60

Seaton MJ, Zeippen CJ, Tully JA, et al. 1992. *Rev. Mex. Astron. Astrophys.* 23:19–43

Sembach KR, Danks AC. 1994. *Astron. Astrophys.* 289:539–58

Sembach KR, Savage BD. 1992. *Ap. J. Suppl.* 83:147–201

Sembach KR, Savage BD. 1994. *Ap. J.* 431:201–22

Sembach KR, Savage BD. 1996. *Ap. J.* 457:211–27

Sembach KR, Savage BD, Jenkins EB. 1994. *Ap. J.* 421:585–99

Sembach KR, Savage BD, Lu L, Murphy EM. 1995a. *Ap. J.* 451:616–23

Sembach KR, Steidel CC, Macke R, Meyer DM. 1995b. *Ap. J.* 445:L27–30

Shapiro PR, Benjamin RA. 1993. In *Star-Forming Galaxies and Their Interstellar Media*, ed. JJ Franco. Dordrecht: Reidel

Shull MJ, McKee CM. 1979. *Ap. J.* 227:131–49

Shull MJ, van Steenberg ME. 1985. *Ap. J.* 294:599–614

Slavin JD. 1989. *Ap. J.* 346:718–27

Slavin JD, Shull JM, Begelman MC. 1993. *Ap. J.* 407:83–99

Smith AM, Bruhweiler FC, Lambert DL, Savage BD, Cardelli JA, et al. 1991. *Ap. J. Lett.* 377:L61–64

Snow TP, Black JH, van Dishoeck EF, Burks G, Crutcher RM, Lutz BL, Hanson MM, & Shuping RY. 1996. *Ap. J.* In press

Soderblom DR, Gonnella A, Hulbert SJ, Leitherer C, Schultz A. 1995. *Instrument Handbook for the Goddard High Resolution Spectrograph*, Version 6.0. Baltimore: Space Telesc. Sci. Inst.

Sofia UJ, Cardelli JA, Savage BD. 1994. *Ap. J.* 430:650–66

Sofia UJ, Savage BD, Cardelli JA. 1993. *Ap. J.* 413:251–67

Songaila A. 1981. *Ap. J. Lett.* 243:L19–22

Songaila A, Cowie LL, Hogan CJ, Rugers M. 1994. *Nature* 368:599–604

Spitzer L. 1978. *Physical Processes in the Interstellar Medium.* New York: Wiley

Spitzer L. 1985. *Ap. J. Lett.* 290:L21–24

Spitzer L. 1990. *Annu. Rev. Astron. Astrophys.* 28:71–101

Spitzer L, Cochran WD, Hirshfeld A. 1974. *Ap. J. Suppl.* 28:373–89

Spitzer L, Jenkins EB. 1975. *Annu. Rev. Astron. Astrophys.* 13:133–64

Spitzer L, Fitzpatrick EL. 1992. *Ap. J. Lett.* 391:L41–44

Spitzer L, Fitzpatrick EL. 1993. *Ap. J.* 409:299–318

Spitzer L, Fitzpatrick EL. 1995. *Ap. J.* 445:196–210

Steidel CC. 1990. *Ap. J. Suppl.* 74:37–91

Stokes GM. 1978. *Ap. J. Suppl.* 36:115–41

Tripp T, Cardelli JA, Savage BD. 1994. *Astron. J.* 107:645–50

Tielens AGGM. 1990. In *Submillimeter and Millimeter Astronomy*, ed. G Watt, A Webster, pp. 13–17. Dordrecht: Kluwer
Tielens AGGM, McKee CF, Seab CG, Hollenbach DJ. 1994. *Ap. J.* 431:321–40
Vallerga JV, Welsh BY. 1995. *Ap. J.* 444:702–7
Vallerga JV, Vedded PW, Craig N, Welsh BY. 1993. *Ap. J.* 411:729–49
Vancura O, Raymond JC, Dwek E, Blair WP, Long KS, Foster S. 1994. *Ap. J.* 431:188–200
van Dishoeck EF, Black JH. 1986. *Ap. J. Suppl.* 62:109–45
Verner D, Barthel P, Tytler D. 1994. *Astron. Astrophys. Suppl.* 108:287–340
Wakker BP. 1991. In *IAU Symp. 144: The Interstellar Disk-Halo Connection in Galaxies*, ed. H Bloemen, pp. 27–40. Dordrecht: Kluwer
Wakker BP, van Woerden H, Schwarz UJ, Peletier RF, Douglas N. 1996. *Astron. Astrophys.* Submitted
Wallerstein G, Vanture AD, Jenkins EB. 1995a. *Ap. J.* 455:590–97
Wallerstein G, Vanture AD, Jenkins EB, Fuller GM. 1995b. *Ap. J.* 449:688–94
Walker TP, Steigman G, Schramm DN, Olive KA, Kang H-S. 1991. *Ap. J.* 376:51–69
Wai CM, Wasson JT. 1977. *Earth Planet. Sci. Lett.* 36:1–13
Wasson JT. 1985. *Meteorites: Their Record of the Early Solar System History.* New York: Freeman
Wasson JT, Kallemeyn GW. 1988. *Philos. Trans. R. Soc. London Ser. A* 325:535–44
Wayte RC, Wynne-Jones I, Blades JC. 1978. *MNRAS* 182:5P–10P
Welty DE, Hobbs LM, Kulkarni VP. 1994. *Ap. J.* 436:152–75
Welty DE, Hobbs LM, Lauroesch JT, Morton DC, York DG. 1995. *Ap. J. Lett.* 449:L135–38
West KA, Pettini M, Penston M, Blades JC, Morton DC. 1985. *MNRAS* 215:481–97
Whittet DCB. 1984. *MNRAS* 210:479–87
Whittet DCB. 1992. *Dust in the Galactic Environment.* New York: Inst. Phys.
Wilson TL, Rood RT. 1994. *Annu. Rev. Astron. Astrophys.* 32:191–226
Witt AN, Johnson MW. 1973. *Ap. J.* 181:363–68
York DG, Kinahan BF. 1979. *Ap. J.* 228:127–46
York DG, Rogerson JB. 1976. *Ap. J.* 203:378–85
York DG, Spitzer L, Bohlin RC, Hill J, Jenkins EB, et al. 1983. *Ap. J. Lett.* 265:L55–59

Annu. Rev. Astron. Astrophys. 1996. 34:331–81

GEMINGA: Its Phenomenology, Its Fraternity, and Its Physics

Giovanni F. Bignami

Istituto di Fisica Cosmica del CNR, Via Bassini 15, 20133 Milano, Italy and Dipartimento di Ingegneria Industriale, Universitá di Cassino, Cassino, Italy

Patrizia A. Caraveo

Istituto di Fisica Cosmica del CNR, Via Bassini 15, 20133 Milano, Italy

KEY WORDS: stars:neutron, pulsars:Geminga, radiation machanisms:non-thermal, thermal

> Le sujet s'accorde toujours avec le verbe sauf les occasions où le sujet ne s'accorde pas.
> Gustave Flaubert, Bouvard et Pecuchet, 1881

---

1. INTRODUCTION

After a twenty-year chase, Geminga, the first and brightest of the unidentified γ-ray sources, was recognized in 1993 as an isolated neutron star (INS), which rotates like a pulsar but is invisible in radio. When first observed, Geminga was in nuce the first INS not discovered through the radio channel and is the first astronomical object discovered through its γ-ray emission.

In part because of the long chase leading to its identification, Geminga is probably the best studied of the fraternity of multiwavelength INSs (MWINSs). MWINSs represent an interesting new class of astronomical objects numbering about two dozen. Their electromagnetic emission covers a panorama of mechanisms, which is unique for a single, compact astronomical body. As MWINS increase in age and slow down, or their nonthermal emission, dominating in the Crab pulsar case gives way to an intrinsic, slowly fading thermal emission in the ~ 20 radio pulsars recently observed with *ROSAT*. In other words, mechanisms dominated by rapidly rotating, huge dipole fields yield to Planckian emission of surfaces, slowly cooling from the initial collapse temperature. A

0066-4146/96/0915-0331$08.00

number of factors make it difficult to distinguish between the two classes of mechanisms. Such factors can be intrinsic and rather unpredictable, such as differential beaming at various wavelengths, or absolute, such as the distance of each neutron star from the Earth. Finally, not all MWINS have been searched for with the dogged pursuit that seems necessary to form a reasonably complete picture. Geminga has been the subject of such an effort and thus was discovered in exactly the canonical 20 years predicted by Trimble (1991) for this and other astronomical puzzles. The resulting panorama of observations was particularly rewarding owing to two fortunate circumstances: the "middle age" of Geminga, and its relative proximity to the Earth. The first enables nonthermal and thermal processes to coexist and makes them accessible in the sense that the former does not completely smother the latter (as in the case of the Crab pulsar), nor is it extinct, as in the majority of the faint *ROSAT* MWINSs. The second (and, it will be seen, other fortunate circumstances) makes the full range of electromagnetic radiation, from γ-ray to optical, visible with current instruments, although the most powerful telescopes must be used in each waveband.

In the end, the kind reader will see that Geminga follows the same strict rules as Flaubert's verb and subject: It is constantly in good accordance with its fraternity except in those cases when it is not. A surprising similarity is seen here between French "grammaire" and multiwavelength astronomy.

2. THE CHASE

2.1 *An Interesting γ-Ray Source (1972–1982)*

Once upon a time (circa 1973), NASA'S *SAS-2* γ-ray astronomy mission (Fichtel et al 1975) performed the first survey of the galactic disc using a digitized spark chamber, a γ-ray imaging instrument. In the anticenter region, well away from the Crab Nebula and its pulsar, an enhancement comprising little more than 100 photons became visible (see Figure 1). It was quite significant, however, and although its position was poorly defined, search for possible correlation with known objects was performed, with inconclusive results.

One reason for the inconclusiveness was that, in the very early days of γ-ray astronomy, the spark chamber of *SAS-2* was the first imaging detector used to look at the γ-ray sky. The concept of point spread function (PSF), in use today, was only intuitively understood and not quantitatively developed at the time. Thus, the spreading of photons around $\gamma\,195+5$, as the source was then called owing to its most probable position in galactic coordinates, was also thought to be compatible with emission from an extended object. This hypothesis made sense: apart from the Crab (Kniffen et al 1974) and, after a while, the Vela pulsars (Thompson et al 1975), point-like γ-ray sources were not known to

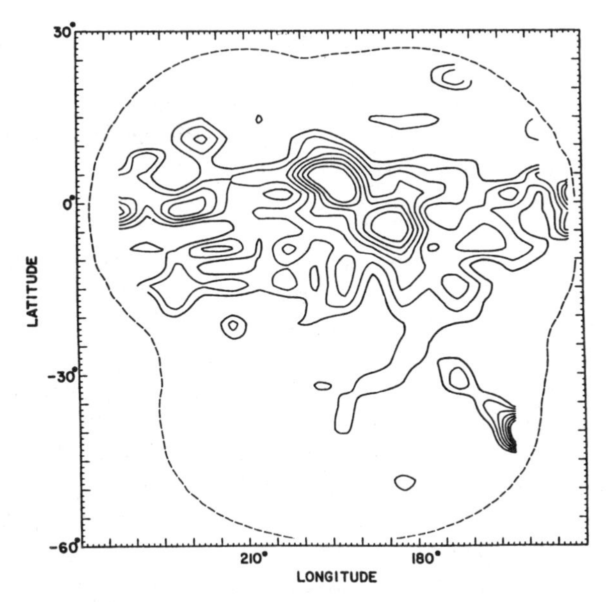

Figure 1 Contour map of the anticenter region (in galactic coordinates), as detected with *SAS-2* at energies above 35 MeV. The Crab pulsar is visible in the lower right and Geminga (at the time $\gamma 195 + 5$) in the upper left portion of the figure. The dashed line indicates the limit of *SAS-2* exposure in this region (from Thompson et al 1977).

exist in the sky. Instead, a well-understood mechanism for γ-ray production was present in diffuse regions (see e.g. Bignami & Fichtel 1974). As they did for the galactic disk as a whole, localized amounts of interstellar gas penetrated by cosmic rays (CR) offered a natural site for γ-ray production through π° decay (Hayakawa 1952, Stecker 1970). A spectacular example in this direction is the first strong and well-defined γ-ray enhancement in the Vela region of the sky (Thompson et al 1974), which coincided well with a bright, local supernova remnant (SNR). Energetically, the γ-ray emission could be ascribed to diffuse CR-matter interaction processes. A few months later, realizing that the second-fastest (at the time) pulsar in the Galaxy (PSR0833 − 45) was inside the SNR, Thompson et al (1975) performed photon timing, with immediate, spectacular results. The source was the pulsar, a point-like object that was recognized thanks to the radio timing signature applied to the γ-ray arrival times.

In the absence of a radio pulsar around $\gamma 195 + 5$, however, extended, gaseous objects were considered as possible origins of the γ rays. The papers announcing the discovery of $\gamma 195 + 5$ (Kniffen et al 1975, Fichtel et al 1975) make explicit mention of the IC443 SNR in the vicinity, or of similar enhancements in the extended radio continuum maps of the region. It thus came as no surprise when, in 1975, the announcement of the discovery of a satellite galaxy close to our own (Simonson 1975) in the anticenter region immediately led investigators to view this galaxy as the source counterpart (Bignami et al 1976, Cesarsky et al 1976), and created speculation as to its CR energy density. However, this idea soon evaporated, as did that of the existence of a nearby satellite galaxy altogether.

An interesting by-product nonetheless arose from the idea of an unseen companion of the Milky Way. A radio observation of that field was made in 1975–1976 by Bignami et al (1977) at 610 MHz from the Westerbork synthesis Radio Telescope (WSRT). A name for this radio field was needed, and thus the term Geminga was born to describe a gamma-ray source in the Gemini constellation. Vaguely inspired by the name Origem for a nearby feature between Orion and Gemini, the name Geminga was really driven by the possibility of a pun in Milanese dialect, in which "gh'è minga" means "it is not there" or "it does not exist." The name Geminga, pronounced with a hard G, as appropriate in Holland, stuck, propagating the noble Milanese argot the world over. Nothing significant came from the radio observations.

A more correct road had been taken, from the start, by Julien & Helmken (1978) and by Lamb & Worrall (1979), who tried to correlate the γ-ray data with the X-ray data then available. Julien & Helmken searched for weak *UHURU* sources near the newly discovered *COS-B* sources, whereas Lamb & Worrall concentrated on $\gamma 195 + 5$, using the sensitive, but not imaging, A-2 counter aboard *HEAO-1*. Both groups indeed found a possible coincidence with a weak source, probably the same one, although the *UHURU* source had approximately twice the count rate and was displaced by 1.8° with respect to the *HEAO-1* (A-2). These may well have been the first X-ray detections of Geminga.

The first pair of high-energy astronomy satellites comes into play in the long Geminga chase. With *SAS-2* γ rays and *HEAO-1* (A-2) and *UHURU* X rays, ideas were on the right track, but these sources were approaching the limit of the instruments sensitivities and positioning ability. In retrospect, *HEAO-1* (A-2) in particular, which preceded the imaging era of the X-ray sky, did not provide satisfactory source positioning. In addition, the poor positional accuracy of *SAS-2* (resulting from its limited statistics) rendered impossible any unambiguous identification. So, for a few years, fantasy roamed wild: from diffuse emission again (Abdulwahab & Morrison 1978) with ad hoc localized CR enhancements to compact object (Maraschi & Treves 1977b, Schlickeiser 1981) to the nearby star γ-Geminorum (Davies et al 1978).

The launch of *COS-B*, the first satellite of the European Space Agency (ESA), in August 1975 was the next milestone in the identification and understanding of Geminga. The mission was very similar in concept to *SAS-2*. The payload featured a digitized wire spark chamber of comparable surface and efficiency (Bignami et al 1975) plus a calorimeter that permitted crude spectral measurements ranging from several tens of MeV to nearly 1 GeV. This instrument in particular made it possible for *COS-B* to investigate the spectral characteristics of bright and hard sources such as Geminga. Of course, *COS-B* was first pointed toward the galactic anticenter. Within a month it produced statistics comparable to those obtained from the entire *SAS-2* mission. The anticenter γ-ray isophotes exhibited the typical "sunny-side up" two-egg structure with the Crab pulsar just below the galactic plane and Geminga nicely symmetrical above it (Figure 2). This observation was more than a confirmation; it was the beginning of a new phase in γ-ray astronomy dedicated to the study of unidentified gamma objects (UGOs), and Geminga seemed the logical starting point.

The too-short, eight-month life of *SAS-2* came to an end in 1973. The data from the global mission (Fichtel et al 1975) included some localized

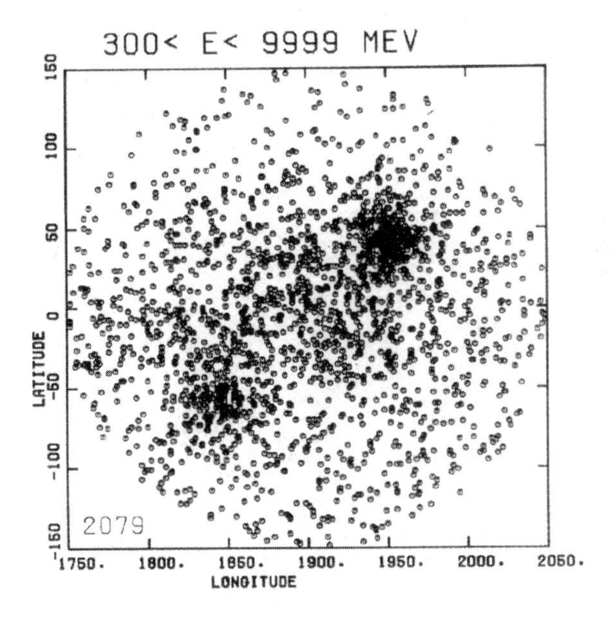

Figure 2 Event map of the anticenter, as observed with *COS-B*, integrating several one-month observing periods over the satellite's seven-year lifetime. Each γ-ray photon with $E > 300$ MeV recorded from the anticenter region is plotted. The Crab pulsar now appears at the lower left and Geminga to the upper right portion of the figure. Because Geminga spectrum is harder than that of the crab pulsar, Geminga appears brighter in this energy range. A total of 2079 photons are plotted on this map.

enhancements. What was to become Geminga was then defined as an excess region "too broad to be consistent with a point source." This definition was soon corrected by Thompson et al (1977) with the final *SAS-2* picture for the anticenter. The source $\gamma 195 + 5$ is now compatible with a point source and associated with a positional error box, albeit a discouraging $4.4^\circ \times 3^\circ$ in size. More importantly, this source seems to show a peculiar periodicity of ~ 59 sec, with an enormous period derivative of $\sim 2 \times 10^{-9}$ sec/sec. With a total of 121 photons, this periodicity is defined as "not statistically compelling, but worth investigating by future experiments." The implication for *COS-B* was all too clear, and on the basis of only marginally better statistics, the Caravane Collaboration (Masnou et al 1977) confirmed the effect. This result was seen as a major breakthrough: Two independent missions had identified an effect that linked Geminga to some kind of compact object, albeit one with a periodicity different from any observed to date in radio pulsars, for example. *COS-B*, with its greater sensitivity at high energies and ability to perform spectral measurements, also produced a good spectrum of Geminga that could be fitted with a power law with $\alpha = 1.8$ (Masnou et al 1981). This spectrum was harder than that of the Crab pulsar, with a break that made it appear to be π° induced. Coupled with the unusual periodicity, this finding presented an interesting challenge for investigators attempting to interpret this spectrum. Ideas ranged from a precessing (Maraschi & Treves 1977a) or accreting (Bisnovatyi-Kogan et al 1980) neutron star to a black hole (Maraschi & Treves 1977b; Giovannelli et al 1982).

Meanwhile, however, *COS-B* continued to function; statistics accumulated for Geminga; and the periodicity slowly became less significant, as reported by Masnou et al (1981), who also produced a final *COS-B* error box with an unprecedented (in γ-ray astronomy) radius of only 24 arcmins. The result was finalized in the second *COS-B* catalog of sources (Swanenburg et al 1981): No evidence of periodicity was mentioned; the given, absolute size of the error box (but slightly shifted in the sky) was the same; and a hard spectrum with a possible break at a few hundred MeV was discussed.

On the basis of the results in this catalog, a determined search was undertaken for counterparts of Geminga as well as of all other UGOs, which were here to stay (Bignami & Hermsen 1983). In light of the very preliminary (but lexically seminal) work of Bignami et al (1977), the radio wavelength was an obvious candidate for study: Mandolesi et al (1978), Mayer-Hasserlwander et al (1979), Ozel et al (1980), Seiradakis (1981), and Manchester & Taylor (1981) all searched the shrinking Geminga error box for radio sources, especially pulsars, in view of the reported/retracted periodicity and of the recognition of Crab and Vela as γ-ray pulsars. However, no positive result was obtained; upper limits varied depending on frequency, sensitivity, period range, etc. Sieber &

Schlickeiser (1982) took a different approach, using deep imaging of the error box at four different frequencies to search for new, peculiar continuum sources in accordance with an idea of Schlickeiser (1981). Many sources were found, of which one appeared more interesting than the others. This radio source was subsequently optically identified with a $z = 1.5$ quasar (Moffat et al 1983), giving rise to speculation that the object might in fact be a proton quasar with a spectacular (10^{48}–10^{49} ergs/sec) γ-ray luminosity.

2.2 *An Interesting X-Ray Source (1981–1985)*

The search at X-ray wavelengths (by pure luck proven a posteriori to be the right track to pursue) was initiated in part for general high-energy astronomy and in part because of the availability of a specific instrument, the *Einstein Observatory*. In retrospect, the first reason, although possibly justifiable for a Crab-like object, would not hold for Geminga, whose X-ray emission appears to be unrelated to its γ-ray emission and could easily have been undetectable. The use of the *Einstein Observatory*, on the other hand, was certainly a good idea. Originally NASA's *HEAO-2* mission, the *Einstein Observatory* was so named because it was launched in the centenary year of Einstein's birth. The *Einstein Observatory* was the first mission to produce images of the soft (.2- to 4-keV) X-ray sky, with source positioning at the arcmin level over the $\sim 1°$ field of view of the imaging proportional counter (IPC), which also provided reasonably accurate spectral information. Alternatively, the high-resolution imager (HRI), at the expense of a smaller field of view and little spectral information, could position sources at the 3–4 arcsec level, deemed sufficient for most optical identification work, even in crowded galactic fields. A program of coverage of *COS-B* error boxes with a mosaic of IPC pointing was started in 1979 (Caraveo 1983), with the aim of discovering and studying interesting, and possibly new, X-ray sources as candidate counterparts of γ-ray sources.

The results for Geminga are given in Bignami et al (1983). The data refer to two IPC observations and one HRI follow-up observation. As expected, in the IPC image (Figure 3) a number of sources, all new, were present in the region. To identify and understand the weak sources as either stars or extragalactic objects, probably serendipitous, was relatively easy. The brightest one, 1E0630 + 178, was probably the counterpart of the *HEAO-1* (A-2) and *UHURU* source marginally seen by Lamb & Worrall (1979) and Julien & Helmken (1978) (although this was not fully recognized at the time). The IPC revealed a very soft spectrum with little interstellar absorption, suggesting a distance not much in excess of 100 pc. More importantly, the accurate (3.5″ radius) HRI position did not produce an optical counterpart.

For a source of $\sim 10^{-12}$ erg/cm^2 per second in the unabsorbed region of the Galaxy to have no optical counterpart on the Palomar Observatory Sky

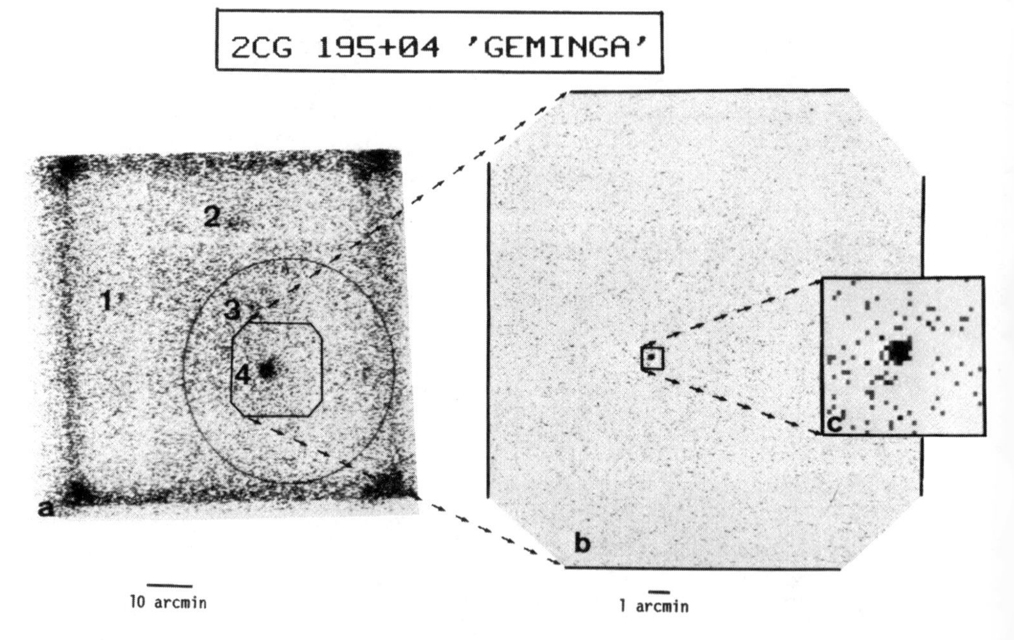

Figure 3 X-ray images of the Geminga field obtained with the *Einstein Observatory* [from Bignami et al (1983)]. (*a*) The final COS-B error circle is superimposed to the IPC data. Source N.4 is 1E0630 + 178. This source was observed with the HRI (*b*) and found to be point-like (*c*).

Survey (POSS) was unique. Coupled with the small distance, this observation suggested that an underluminous object with an L_x/L_{opt} ratio significantly higher than that expected for most known sources was the counterpart. Taking into account the special characteristics of the source, Bignami et al (1983) proposed it as the counterpart of Geminga, drawing analogies with the Vela pulsar: high $L\gamma/L_x$, high L_x/L_{opt}, and a relatively local nature. The main difference was the absence of a radio pulsar, which was especially significant now that a precise position was available. The standard periodicity search was performed on the ~ 800 IPC photons, with no significant result. [We now know that Geminga is a 237-msec pulsar, discovered in X rays. The *Einstein* data folding at the (now known) correct period give a puny $\sim 2\sigma$ peak, which was swept over by the original Fourier analysis and thus was not considered significant]. However, indirect evidence strongly indicated that 1E0630 + 178 was indeed the X-ray counterpart of Geminga.

Historically, the chase from this point on concentrates on this source and its optical identification. The chase is thus devoted to proving its association with Geminga, and no concrete alternatives are put forward.

2.3 *The Search for an Optical Counterpart (1984–1986)*

Almost 10 years after the actual source discovery, work on Geminga was about halfway finished thanks to the second pair of γ- and X-ray satellites, *COS-B* and the *Einstein Observatory*. There remained the all-important search for an optical/radio counterpart to 1E0630 + 178. The X and γ data pointed first and foremost to a neutron star, but no radio signal, pulsating or otherwise, came from the well-determined 1E0630 + 178 position. This result had been reported in the continuum not only by Caraveo et al (1984c) in their 6- and 20-cm VLA data and by Spölstra & Hermsen (1984) for the 21-cm WSRT but also by Fauci et al (1984) in their ad hoc Arecibo pulsar searches. In the optical, Caraveo et al (1984b) used data obtained from the *CFHT* at the end of the observing season (April 1983), going significantly deeper than the POSS results to uncover a candidate (dubbed G) at the boundary of the X-ray HRI box. At $m_v \sim 20.5$, the object was just below the POSS sensitivity and appeared a priori to be very plausible.

At this point, one of several blind alleys in the Geminga chase developed. In his search for G (or anything else) on the POSS plates using microdensitometry tracing, Bloemen (1984) instead found another object of comparable magnitude about 7″ away that was not present in the more sensitive CCD images of Caraveo et al (1984b). Led by the (very reasonable) assumption that Geminga could have a high proper motion if is it close (100 pc), Bloemen proposed that what he had observed on the POSS was in fact G, which would have moved 7″, a fully acceptable proper motion, in the $\sim$ 30 years that had elapsed between the POSS and the 1983 observations. In retrospect, now that we know that star G does not move and is not the counterpart of Geminga, we can conclude that Bloemen's object, which was never seen again, was probably a faint asteroid. At the small ecliptic latitude of Geminga, such faint, distant objects abound, leaving- not a track but a star-like image in a normal POSS exposure (for which the real G was indeed too faint). Within a few years, the nature of the Geminga optical counterpart had changed from that of a distant QSO to a that of solar system asteroid.

In the end, however, Bloemen's idea of proper motion proved prophetic. Meanwhile, deeper data of Sol et al (1985) proved the nonreality of a proper motion for star G. Halpern et al (1985) produced more convincing spectrophotometric work but could not confirm the identification of G, a standard field star, with 1E0630 + 178 and Geminga.

However, the work of Halpern et al (1985) yielded an increased value of L_X/L_{opt}, indicating that the true optical counterpart had to be searched for at a level fainter than G, where the error box remained unexplored. Meanwhile, additional X-ray observations of 1E0630 + 178 with ESA's *EXOSAT* X-ray observatory (Caraveo et al 1984a) confirmed its position (see Figure 4) and spectrum and yielded the accumulation, over several exposures, of nearly 1,000 additional X-ray photons.

Bignami et al (1984) ran into another blind alley when they reanalyzed the arrival times of all X-ray photons from the source using three *Einstein* and two *EXOSAT* observations. They were spurred on in part by an analysis of Zynskin & Mukanov (1983), who observed Geminga at very high energies (VHE) with the 59-second periodicity reported, but then retracted their conclusion. The strongest evidence came from the 800 IPC photons and the adjacent 91 HRI photons. Together they yielded a double-peaked light curve with a 50% modulation and a χ^2 of 4.72. Background data were completely flat, and an a posteriori scan did not reveal spurious effects in the data. Timing analysis of a previous, shorter IPC image and of two subsequent images taken with *EXOSAT* confirmed the presence of the pulsation, albeit at a lower statistical significance, probably owing to the smaller number of photons. Buccheri et al (1984) criticized the statistical significance of the result and the method used but did not reanalyze the data sets. Whatever the outcome of that particular analysis (never confirmed but also never disproved), the light curves reported in Bignami et al (1984) for the 59–60-second periodicity are of a statistical significance vastly superior to that of the light curve resulting from the folding at the (now known) period value, which can be precisely computed on the basis of the pulsar timing parameters.

The reported 59-second periodicity was all the more startling because the high value of its $\dot{P}$ implied an extremely short ($\sim$ 1500 years) system lifetime. On the one hand, this finding triggered the imaginative possibility of a connection with a supernova event observed in the region by the Chinese in 437 A.D. that led to speculation as to CR sources (Maddox 1984). On the other hand, theoreticians proposed a system, necessarily binary, that could produce the Geminga characteristics, such as a close pair of magnetized neutron stars (Nulsen & Fabian 1984), or a neutron star–black hole system (Bisnovatyi-Kogan 1985). Taking a sound observational approach, Kulkarni & Djorgovski (1986) failed to observe any 59-second pulsation from the optical candidate counterpart (the G object and the immediately adjacent region).

Although it caused significant conceptual difficulties at the interpretative level, the "59-second blind alley" did not stop optical investigations designed to identify 1E0630 + 178. Djorgovski & Kulkarni (1986) were the first to probe deep into the error box, as was required following exclusion of G as a candidate. Using CCD imaging from the 3-m Shane telescope at the Lick Observatory, these investigators confirmed the existence of the fainter object G′ (already suggested by Sol et al 1985) inside the HRI error box and reported a nearby, even fainter object, dubbed G″. As infernal bad luck would have it, however, this object turned out to be merely a sky fluctuation (to be expected when working close to the plate limit) that perturbed the positional measurement of G′, leading to a report of proper motion. When an even deeper image [taken in January 1984 but first published in Bignami et al (1987)] became available, the state of confusion in the optical counterpart chase reached its apogee (see e.g. Bignami 1987). The deep plate taken from CFHT, shown in Figure 4, did reveal for the first time the real (and only) G″ close to the plate limit, suggesting a strong proper motion of G″ [from its true position to that of the Djorgovski & Kulkarni (1986) fluctuation]. Bloemen's asteroid, the positional errors for G′, and deep sky fluctuations all pointed to proper motion for the three objects in the 1E0630 + 178 error box. Coupled with the report of 59-second

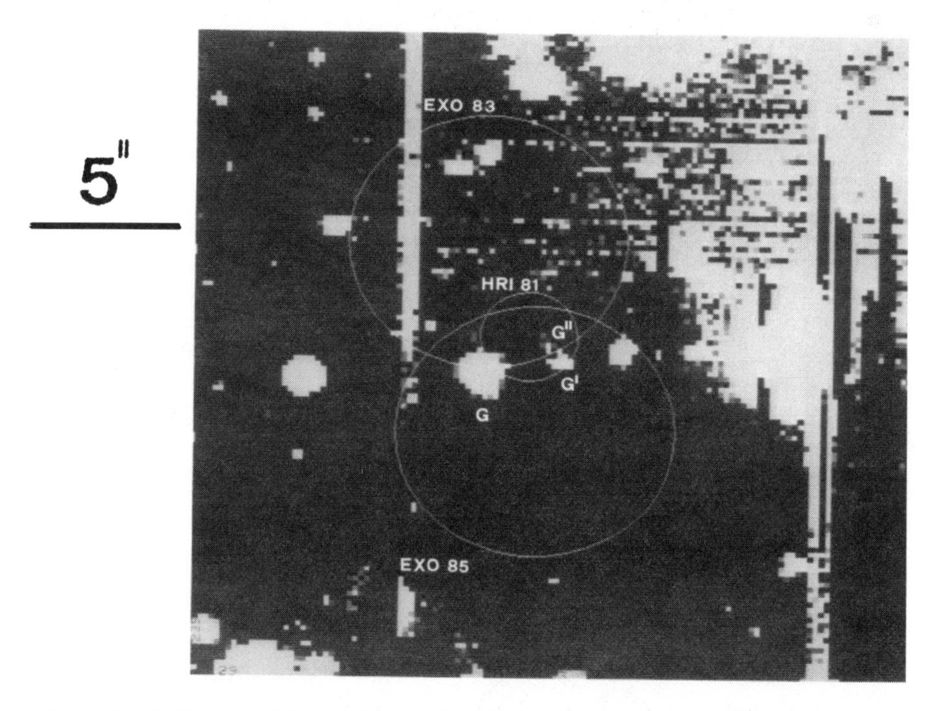

Figure 4 The first in-depth image of the Geminga field, obtained at *CFHT* in January 1984 (from Bignami et al 1987). Summing 12 exposures of 15 min each through an R filter brings the limiting magnitude to $m_1 \sim 26$. The *Einstein* HRI and two later *EXOSAT* error boxes (EXO 83 and 85), which all nicely overlap, are also shown. This is the first image to depict clearly the presence of two additional objects inside the HRI error box, which we subsequently refer to as G′ and G″ (north at top, east at left).

periodicity, which implied some form of binary system but whose accuracy was severely disputed, these findings presented investigators with a difficult puzzle. Romani & Trimble (1985) commented that all this made for more fun and for an interesting source.

2.4 *At Last, an Interesting Optical Object (1987–1988)*

Bignami et al (1987) published the deep 1984 *CHFT* plates, adding to them an exposure from January 1987 taken with the 3.6-m *ESO Telescope* on La Silla. Their work permitted confirmation of the existence of the *CHFT* G″ and recognition of the Djorgovski & Kulkarni (1986) G″ as a fluctuation. Of equal importance, for the first time a comparison between the R and V colors for the various objects was performed. Although results for G and G′ were unremarkable, the very last words of a note added in proof at the end of that paper indicated that such comparison "seems to suggest a bluer colour for G″."

The next breakthrough came quickly, again from Halpern's group (Halpern & Tytler 1988), who rigorously reanalyzed the *Einstein* X-ray data in order to assess the spectral shape of the emission and to find the best possible distance from its interstellar absorption. The soft X-ray emission was well-described by an approximately one million °K (or less) thermal spectrum, and the distance was likely between several hundred parsecs and 1 kpc. Both results were essentially correct but were still limited by the poor quality of the X-ray data and by the intrinsic difficulties of the method used. More important than these findings was the proposal of G″ as the optical counterpart because of its blue color. A detailed comparison of the g and r Gunn's colors of the objects in the region using observations from the 5-m *Hale Telescope* revealed that G″ stood out from the pack of field stars. Thus the Vela-like picture proposed by Bignami et al (1983) was revived owing to its γ-ray, X-ray, and optical energy distribution (about a factor of 1000 between each band, starting with the γ rays); neutron star nature, albeit still speculative; and vague similarity in their distances. In retrospect, the limits of such a comparison, i.e. no radio emission from Geminga (waved away as obviously due to different beaming in γ and radio) and, more importantly, no evidence for the fast periodicity of a neutron star, become evident. Moreover, although there was little doubt that the X-ray/optical emission from Geminga was thermal in nature, the Vela pulsar must have a nonthermal component, at least in the optical, since it obeys the Pacini law (Pacini 1971). Whatever the limits of such a comparison, however, they do not diminish the importance of Halpern & Tytler's (1988) singling out of G″, now by far the best candidate for the optical counterpart of Geminga because of its colors.

Intrigued by the report that G″ was blue, Bignami et al (1988) tried deep imaging in that color, again from the 3.6-m *ESO*. This measurement, which required painstaking calibrations, yielded a B magnitude for G″ of $\sim 26.5 \pm 0.5$, a value very close to the limit of ground-based astronomy at the time. These investigators concluded that G″ was perhaps green (Halpern & Tytler result through Gunn g filter) but certainly not blue, (even after instrumental limitations were accounted for). It was remarked that the colors of G″ were unusual, with what looked like an abnormal peak of emission in V or an equally abnormal depression of emission in B. G″ was nonetheless confirmed as the best candidate for the optical counterpart owing to its peculiar colors

2.5 *An X- and γ-ray Pulsar (1992–1993)*

Pending better (or more clever) optical observations, the chase in the early 1990s shifted into the hands of the third and fateful pair of high-energy satellites: the second in NASA's Great Observatories series, the Compton *Gamma-Ray Observatory (GRO)*, and the German/UK/US X-ray satellite (*Röntgen Satellite*) *ROSAT*. For *GRO*, launched in April 1991 and designed to take up where *COS-B* had left off, there were three objectives: to accumulate more γ-ray photons between a few tens of MeV to a few GeV from Geminga, to reduce the size of the original error box, and to measure the source γ-ray spectrum and time variability.

Apart from *GRO*, *ROSAT* (launched in 1990), was to contribute the most toward the resolution of the mystery of Geminga. With the majority of its observing time open to competition (mostly among the three collaborating nations), Geminga was a prime target, assigned in the end to Halpern and his collaborators. Following a series of observations with the *PSPC* in March 1991, the soft X-ray photons collected from the source numbered more than 7000, over three times the total amount collected prior to *ROSAT*. As Halpern & Holt (1992) remarked, "the soft X-ray band is the most promising one, because the count rate is high and the background is negligible...." A clear periodic signal with a period of 237 msec, typical of a young-to-middle age pulsar, was apparent from the *ROSAT* data. However, the first preliminary analysis showed the pulsed fraction to be small, starting at $\sim$ 24% between 70 and 200 eV, $\sim$ 15% above 1 KeV. [This result, coupled with the limited statistics, explains why the signal is not significant, even a posteriori, in the *Einstein* IPC or in the *EXOSAT* data (Mereghetti et al 1993).] The *EGRET* team was immediately notified. They tested the new periodicity on their Geminga data and found that the signal was indeed present (Bertsch et al 1992). At last, the unequivocal proof that 1E0630 + 178 was the same object, indeed the same rotating neutron star, as Geminga. Geminga was the first pulsar detected through its γ-ray emission as well as the first pulsar not seen in radio but rather discovered and recognized on the basis of its high-energy emission. In a sense, the outcome was what everyone expected, if for no other reason than the increasingly strong circumstantial evidence on 1E0630 + 178. But real proof has a different—and better—taste.

EGRET provided not only confirmation of the period value but also a first estimate of its derivative. This estimate is necessary in γ-ray astronomy, in which photon arrivals are so widely separated that the period derivative must be taken into account when folding their arrival times. The *EGRET* team reported a tentative value of $\sim 1.1(\pm 0.17) \times 10^{-14}$ sec/sec, but precision was limited by the relatively short observational time base ($<$ 3 months) then available. In contrast, with a seven-year life in orbit, *COS-B* certainly was not lacking in observational time base. Thus Bignami & Caraveo (1992) began searching the old *COS-B* data base, finding that "it was there all the time." This long time coverage allowed a refinement of the period derivative value ($\dot{P} = 1.099 \pm 0.001\ 10^{-14}$ sec/sec) and thus a better assessment of the theoretical values for the neutron star surface magnetic field ($\sim 1.5 \times 10^{12}$ G), age ($\sim 3.4 \times 10^5$ years), and rotational energy loss ($E \sim 3.2 \times 10^{34}$ ergs/sec). A later, independent analysis of the *COS-B* data by Hermsen et al (1992) largely confirmed the results of Bignami & Caraveo (1992). Mattox et al (1992) completed the timing analysis panorama, adding the photons detected by *SAS-2* in 1972–1973. The Geminga light curves obtained by the three γ-ray telescopes flown to date are compared in Figure 5, which shows the substantial improvement, both in quality and quantity, in γ-ray astronomy over the past 20 years.

A frequently asked question is: How could *COS-B* have missed it? The answer is simple: Without external knowledge of the source period and period derivative, which would reduce the number of trials performed, any Fourier (or folding, etc) blind periodicity scan requires so many trials that any result loses statistical significance. Therefore, the right P and $\dot{P}$ combination may well have been present in the repeated timing analysis of the *COS-B* data but overlooked in some large scan because they simply could not be considered significant.

Having come to terms with history and with the Geminga signal in the *COS-B* data, Bignami & Caraveo (1992) reached the same elementary conclusion as Bertsch et al (1992): Given a measure of the total rotational energy loss ($\dot{E} = I\Omega\,\dot{\Omega} = 3.2 \times 10^{34}$ ergs/sec) and given the observed flux in γ rays, an extreme upper limit to the object's distance can be set by requiring that all its mechanical slowing-down energy go into γ rays. This unlikely physical situation yields (barring beaming) an upper limit of $\sim$400 pc. Mindful of the next identification step, i.e. from X-ray to optical, Bignami & Caraveo (1992) remarked that, as is the case for Vela pulsar, a measurable proper motion could characterize the Geminga optical counterpart if it were a nearby neutron star.

2.6 *Proper Motion and Point of Origin (1993–1994)*

Typically, a neutron star with a transversal velocity in the sky of $\sim$ 100 km/sec at 100 pc would have a proper motion of 0.2″/year. Immediately in hand, however,

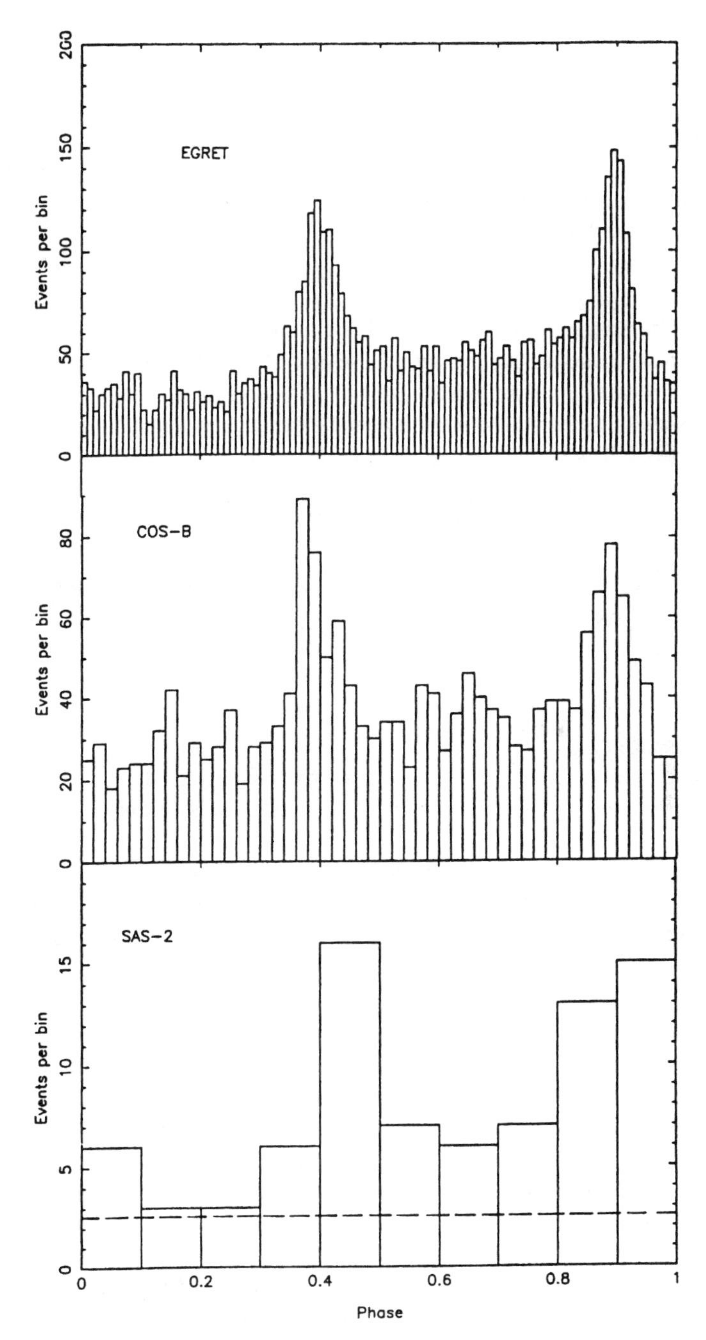

Figure 5 γ-ray light curves of the Geminga photon detected over the past 20 years by *EGRET*, *COS-B*, and *SAS-2* (Mattox et al 1992). The alignment in phase is arbitrary.

were two sets of images taken only three years apart (in 1984 and 1987) with two very different telescopes and detectors (*CFHT* and *ESO*) with pixel sizes of 0.4″ and 0.6″ respectively. The 1986 data of Halpern & Tytler were not readily available in raw form, and the *ESO* 1988 data had a marginal detection of only G″ in the blue. The first opportunity to observe Geminga came in autumn 1992, and on November 4 it was perfectly observed, in service mode, by A Smette, with the *ESO* 3.5-m *New Technology Telescope* coupled with the Super Seeing Imager (SUSI), capable of taking advantage of the very good seeing (0.6″–0.8″). The stack of 10 exposures of 15 min each was summed directly on the mountain by A Moneti and FTPed to Milan. It was relatively easy to compare this image (Figure 6) with those already in hand. Comparisons of the 1984 and 1992 data showed an obvious ($\sim 1.5''$) displacement of G″ to the NE. However, the clincher came later, when, on the screen, the 1987 *ESO* position was seen to fall between the 1984 and 1992 ones (Figure 6). Bignami et al (1993) reported a preliminary value for the total proper motion of $\sim 0.17''$/year, which was surprisingly close to expectations. The proper motion was taken as evidence that G″ was the optical counterpart of Geminga. This neutron star, recognized from its pulsations in X and γ rays, now had, in a small error box, a low-luminosity, high-velocity object with all the traits of a real neutron star. The association was compelling.

Given the proper motion magnitude and direction, as well as the object's age, the position of Geminga at the time of its birth is a point in the sky $\sim 16°$ SW of its present position. This position, however, is not uniquely defined since the proper motion reflects only the transverse velocity of the source while the radial component remains unknown. Uncertainties as to the source distance and its proper motion also make it difficult to trace the Geminga trajectory. Gehrels & Chen (1993) addressed this problem by propagating backwards the angular displacement of Bignami et al (1993) to obtain a starting position of $\alpha = 5^h 40^m \delta = 8° \; 24'$, or $l = 197°, b = -11.7°$. This would place the supernova event at an unknown (but not too large) distance, making it likely responsible for the so-called Local Bubble (Gehrels & Chen 1993).

A more general approach was taken by Frisch (1993), who analyzed the problem from a statistical point of view. The starting points of the source trajectories were computed for distances ranging from 150 to 400 pc and for radial velocities between $+400$ and -400 km/sec. The general Orion region was found to be a more likely place of origin than the Local Bubble. Smith et al (1993) went one step further and pointed out that the backward extrapolation of the Geminga position fell almost exactly on the λ Ori association at a distance of 400 ± 40 pc in the direction $l = 195.26°$ and $b = -11.62°$. This remarkable coincidence may point to a rather high radial velocity of the source.

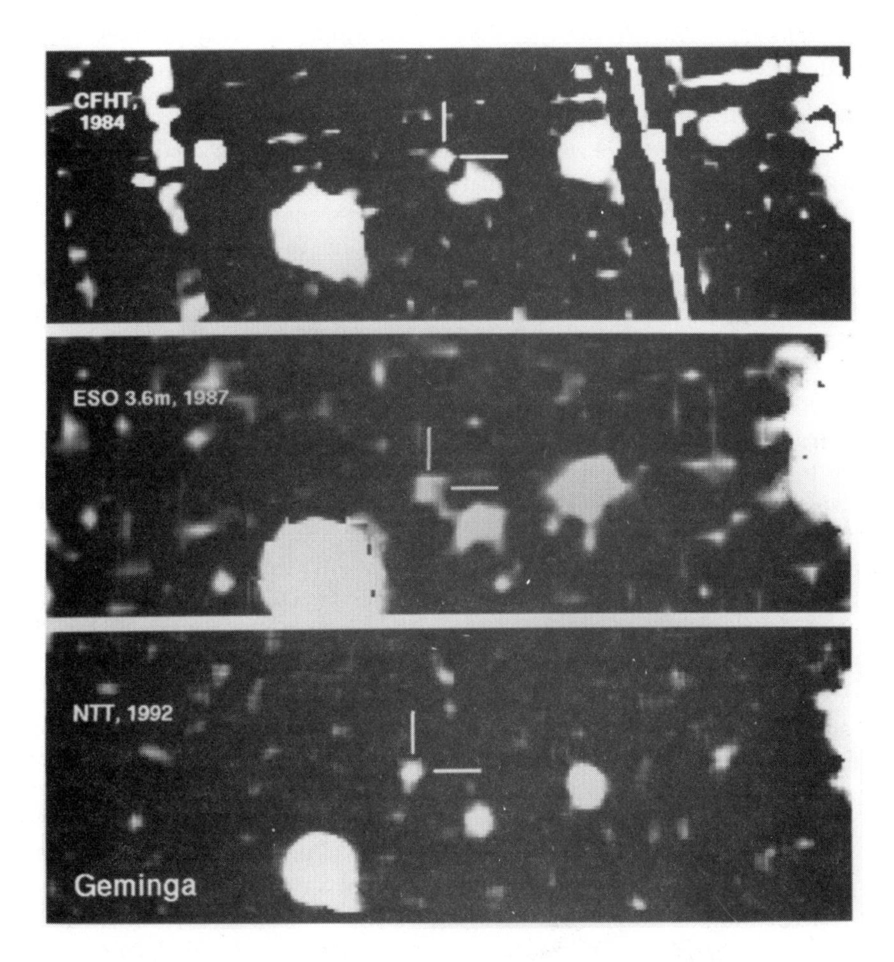

Figure 6 Comparison of optical images of the Geminga field taken in 1984 (*CFHT*), 1987 (*ESO* 3.6), and 1992 (*ESO* NTT equipped with SUSI). The displacement of G″ is evident (north at top, east at left).

Because such a starting position lies well outside the official boundaries of the Gemini constellation, its name could not have been given at birth. However, *ESO* Press Release #09/92 contains a reassuring note: The observations imply that Geminga is now moving toward the border between the Gemini and Lynx constellations, some 20° away to the NE and will cross it only in about half a million years; until then, the present name should be appropriate.

2.7 *Parallax and Distance (1994–1995)*

Meanwhile, however, collection of data on the proper motion continued. Their discovery of a fourth point in January 1994 enabled Mignani et al (1994) to refine the proper motion knowledge and present a complete picture. Their report in turn triggered a new observation campaign to measure the parallactic displacement of Geminga. The rationale behind this new effort is clear: The measured proper motion, coupled with our general knowledge of the average neutron star speed, points to a relatively nearby object. Distances of the order of a few hundred parsecs can be measured precisely through the annual parallactic displacement induced in the apparent position of the star by the rotation of the Earth around the Sun. Although the magnitude of displacement depends on the source distance, its direction, together with the season of maximum displacement, can be predicted a priori only on the basis of the source coordinates. Because of Geminga's position on the ecliptic plane, its parallactic displacement is almost all in Right Ascension, with the maximum parallactic factor occurring near the spring and fall equinoxes. Of course, any parallactic displacement would appear superimposed to the known proper motion of 170 mas/year. Figure 7 shows the expected annual path of a source with the proper motion at the position of Geminga at a distance of 100 pc. The ondulating path, induced by the hypothetical presence of the parallactic displacement is clearly asymmetrical. The March to September trajectory is longer than the September to March one because for the former, the parallax (actually twice its value) is added to the proper motion, whereas for the latter it is subtracted.

Figure 7 depicts quantitatively the magnitude of the effect. In order to measure parallactic displacements of < 10 mas, one should be able to determine relative positions within a few mas. For a source as faint as G″, this is impossible from the ground and becomes a task for the refurbished *Hubble Space Telescope (HST)*.

During Cycle 4 we used the Planetary Camera 2 on the *HST* to measure the parallactic displacement of G″. The instrument, which offers a good compromise between positional accuracy (pixel size) and dimension of the field of view, was pointed at Geminga on three occasions during 1994 and 1995 near the spring and fall equinoxes. Caraveo et al (1996) give details of the data analysis, including object centering correction for the instrumental geometrical

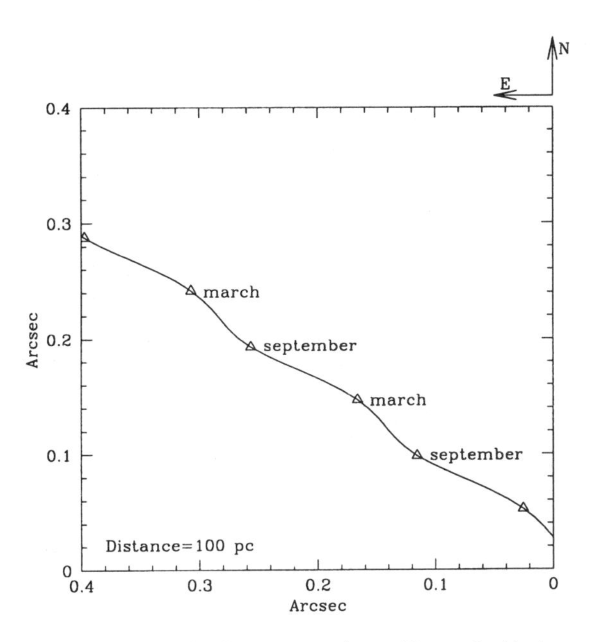

Figure 7 Expected ondulating path of a source at the position and with the proper motion of Geminga, assuming a distance of 100 pc.

distortion, alignment in Right Ascension and Declination, and superposition of the images.

In order to assess the parallactic displacement, a different movement of G″ is sought in the two semesters. In the absence of a measurable parallax, the two six-month trajectories should be identical. However, the March 1994 to September 1994 trajectory is larger than the September 1994 to March 1995 one. When all corrections, including systematics, are taken into account, a significant difference of $\sim 0.6 \pm 0.1$ pixel remains in the Right Ascension displacement. This difference is the expected signature of the parallactic effect. Figure 8 shows the real trajectory of Geminga, as determined from the three *WFPC2* observations. The parallactic displacement of $0.0064''$ (± 0.0017) corresponds to a distance of 157 ($+59$, -34) parsecs, the lower limit of the range allowed according to the X-ray data of Halpern & Ruderman (1993).

Figure 8 also provides the best available measure of the proper motion of G″ to date. Although 10 years of ground-based data yielded a proper motion of $\mu\alpha = 0.140 \pm 0.040$, $\mu\delta = 0.100 \pm 0.040$, one year of PC data yielded $\mu\alpha = 0.138 \pm 0.004$, $\mu\delta = 0.097 \pm 0.004$, which confirmed the ground-based results but reduced dramatically the error bars.

The measure of the Geminga distance from its annual parallax and its precise proper motion determination with the *HST* also have restricted the parameter space of any backward extrapolation toward its point of origin. The knowledge of the distance translates into a transverse velocity of 122 km/sec, while the more precise knowledge of the proper motion shrinks significantly the region of origin.

The star λ Ori remains a viable candidate from a positional point of view. Cuhna & Smith (1995) suggested that the expanding ring of gas surrounding the star could be due to a supernova explosion that occurred between 300,000 and 370,000 years ago. This time frames is in good agreement with the dynamical age of Geminga. However, the required radial velocity of -700 km/sec, although not unheard of in the pulsar family, is definitely on the high side. A positive radial velocity on the other hand, would have enabled Geminga to travel 250 pc in the opposite direction, starting e.g. from σ Sgr at 65 pc from

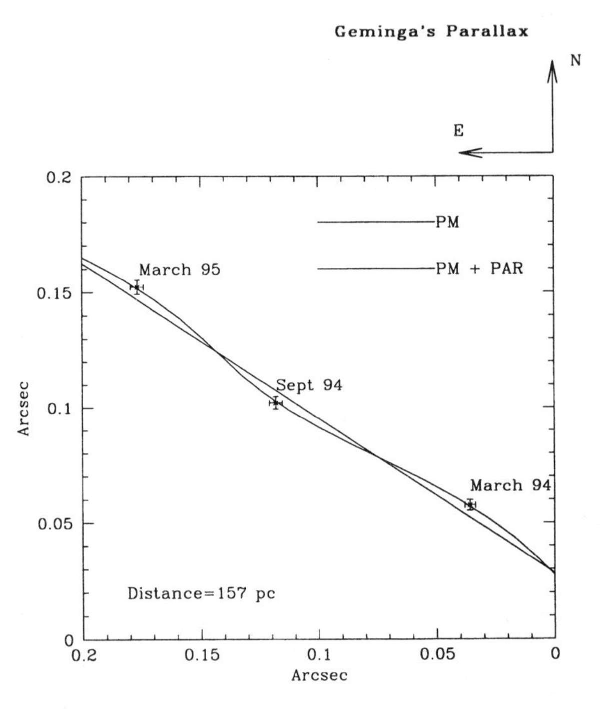

Figure 8 Actual path of Geminga, as inferred by the three measures of its apparent position at the time of its maximum displacements in March 1994, September 1994, and March 1995. The derived annual parallax is $\sim 0.064''$.

the Sun but at coordinates $l = 10^\circ$ $b = -12^\circ$, which are diametrically opposed to the current Geminga coordinates.

The distance inferred from the parallactic displacement was taken as additional evidence that G″ was indeed the counterpart of Geminga. However, a direct link between G″ and the X and γ rays emitting Geminga remained to be established.

2.8 *The Finishing Touch*

It is perhaps fitting that, in the end, the clincher comes once again from γ-ray astronomy: The proper motion of G″ is now apparent in γ-ray data timing. After a first attempt by Cheng et al (1993) using *COS-B* data, Mattox et al (1996a) succeeded in proving this hypothesis. With the known proper motion parameters and with the long *EGRET* observation time base now available for Geminga, its γ-ray light curve (phase diagram of photon arrival times) exhibits a marked improvement when the different positions of G″ as a function of time are used for the barycentric correction. Imposing the varying G″ positions as origins of the γ-ray photons at the epochs of the *EGRET* observations markedly sharpens the definition of the light curve. Thus, the γ rays are emitted from G″.

2.9 *Coda*

The chase for Geminga is over, but it may be of interest to reflect briefly on the role of chance during the pursuit. On the negative side one might list the inability to detect 237-msec pulsations in the *Einstein* data on an otherwise significant source. Were such a detection possible, Geminga would have been understood 10 years earlier from the *COS-B* data. The "59-second blind alley" also complicated the issue.

All in all, however, the good luck far outweighed the bad. For example, the proximity of the γ-ray source to the Crab pulsar gave Geminga considerable free exposure through the γ-ray telescopes used in the chase and beyond. Moreover, Geminga was in a low-background region, and the hard γ-ray spectrum of the source allowed its precise positioning, thus facilitating the search for counterparts.

In the search for the optical counterpart, there were some uncertainties and blind alleys, but the value of the V magnitude of G″, which is well above the extrapolation of the black-body emission seen in extreme ultraviolet (EUV) and soft X rays, was a stroke of luck. Had the V flux been present on such an extrapolation, it would have been undetectable, rendering impossible the proper motion and parallax studies.

The position of the moving G″ at the time of the first detection (1984) of the moving counterpart was very close to that of G′, which we now know to be, most probably, a background galaxy. The trajectory of the proper motion, however, is such that only a few years earlier G″ would have appeared projected over G′ and therefore would have been undetectable. Finally, measurement of the parallactic displacement, and hence of the source distance, was made easier by the position of the source, which causes the displacement to be almost entirely in the Right Ascension direction.

So, close to 20 years after the first γ-ray observation, we have a secure X-ray and optical identification. Together with γ rays, these channels appear to be the only means to obtain information about this object, considering the persistent radio silence. A mildly ironic coda can be appended here. During the 20-year chase, which relied primarily on photon-gathering efforts, an independent, powerful tool for astronomy, i.e. power timing analysis, was developed. Spurred by Arecibo's work on pulsars (especially millisecond ones), these techniques are now so powerful that Brazier & Kanbach (1996) and Mattox et al (1996b) used them successfully in an independent search for the Geminga period, in the *EGRET* data. What seemed impossible at the time of *COS-B* and at the beginning of *EGRET* is now reality: The pulsed γ-ray signal of Geminga can be recognized even without a priori knowledge of the period.

The avenue we and others pursued, however, leaves us with no regrets: First, such a technique was not available 20 years ago and probably could not have been applied to the sparse *COS-B* data anyway. Second, the extensive photon gathering was not in vain in that it yielded a remarkably complete phenomenology on Geminga.

3. THE PHENOMENOLOGY

The end of the chase gave us a new MWINS. This relatively young pulsar was the first discovered through its γ-ray emission and the first that did not emit in the radio band. We now review all the available Geminga data, both those emerging directly from the chase and those resulting from measurements performed after the identification had been made. This process will place Geminga in the context of other MWINSs, for which the available data, notably on X and γ rays, are briefly reviewed.

3.1 *Nonthermal Radiation: γ Rays*

The best data available to date in the γ-ray regime, the electromagnetic channel in which Geminga emits the vast majority of its energy, come from the *EGRET* experiment aboard *GRO* (Kanbach et al 1988). The most recent and accurate compilations of the *EGRET* data are those of Mayer-Hasselwander et al (1994) and Ramanamurthy et al (1995), who reported on 13 observing periods of about 2 weeks each for a total of approximately 6000 photons (recall that *COS-B* only detected ~ 1000 events in its entire 7-year mission). Of course, any global property of the object can be discussed only after any problem of variability between observations has been resolved. The *EGRET* γ-ray light curve for Geminga does not appear to vary in shape, as can be seen by comparing the individual data sets with the normalized template light curve. This stability allows for an accurate definition of phase intervals in the light curve over which a detailed spectral analysis can be performed and compared with the time-averaged spectra.

For Geminga, these time-averaged spectra allow for a good spectral index determination. The data are well-fit by a power law with a photon spectral index of -1.50 ± 0.08 in the interval between 30 MeV and 2 GeV (or of -1.47 in the interval between 100 MeV and 4 GeV). No change in slope close to the lower end of 30 MeV is reported, but a roll-off above 2–4 GeV is present at the high-energy end. Figure 9 shows different light curves for different energy intervals. Following Mayer-Hasselwander et al (1994), it is immediately apparent that: (*a*) peak 2 (phase 0.07–0.20) is significantly harder (-1.22 ± 0.05) than any other phase component; (*b*) peak 1 (phase 0.57–0.70) and the interpulse 1 (phase 0.70–0.07) have similar spectral indexes (-1.53 ± 0.05 and -1.47 ± 0.09, respectively); and (*c*) interpulse 2 (phase 0.20–0.57) appears to have a somewhat softer (-1.67 ± 0.23) spectral index, which is albeit very limited by the available statistics.

Significant spectral differences that seem to be present in the Geminga γ-ray phase diagram limit the value of time-average data, which benefit from the maximum of statistics. Moreover, analysis of the Geminga light curve leaves no room for unpulsed emission. Every γ-ray photon from Geminga lies in the phase interval of 0.07–0.70.

Related to the problem of spectral differences is that of time variability of emission or of spectral shape (in the end, observationally, the same thing). Grenier et al (1993, 1994) had already reported a strong indication on the limited but lengthy *COS-B* data base. The latest *EGRET* data (Ramanamurthy et al 1995) suggest some variability in the total Geminga flux that is possibly the reflex of the more significant variability reported for the spectral index(es) of the various phase components. A systematic difference, apparently quite significant, has been observed between the overall *COS-B* spectral index for Geminga (-1.84 ± 0.05) and that (-1.5 ± 0.08) obtained by *EGRET*. If confirmed, this finding would indicate a significant long-term (1975–1982, 1991–1994) time variability, but comparisons between two different missions are always difficult. In contrast, the Crab pulsar does not exhibit any difference in spectral shape between the *COS-B* and *EGRET* indexes.

In summary, although the high-energy emission from a neutron star is, not surprisingly, somewhat variable (possibly in some of its components), Geminga appears to be a rather stable pulsar whose timing history can be traced back

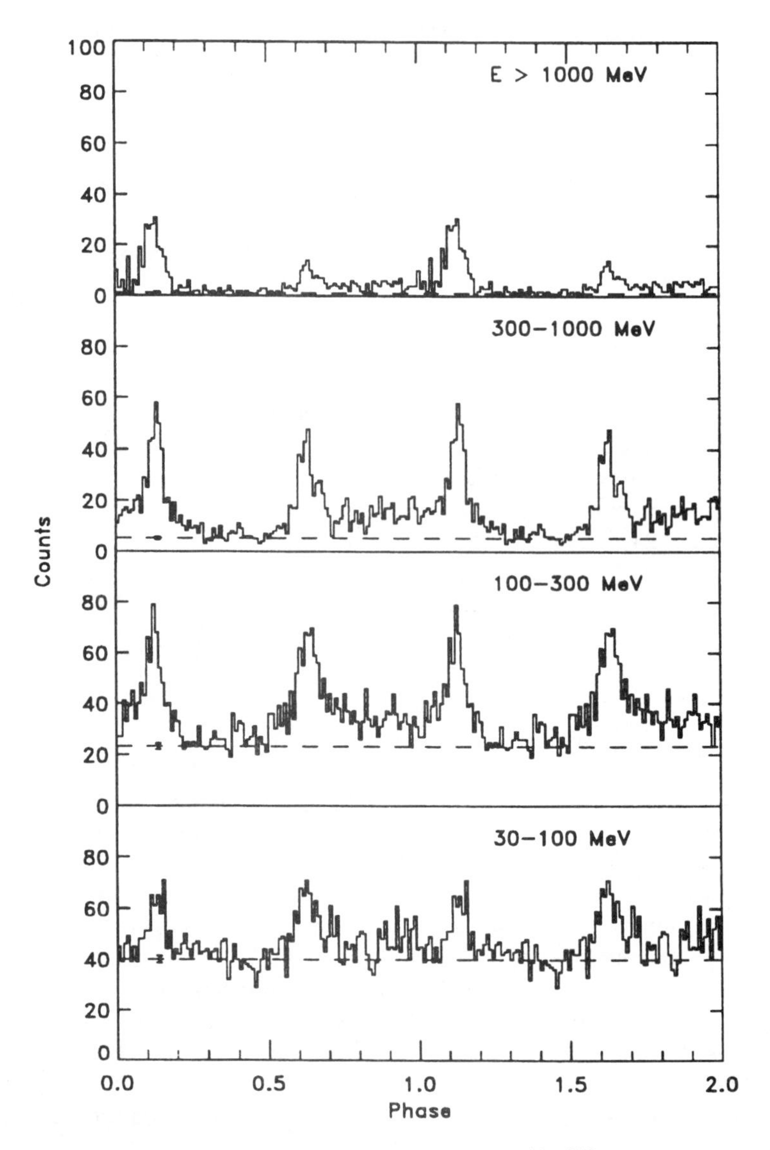

Figure 9 γ-ray light curves obtained using *EGRET* data grouped in different energy ranges. Different spectral shapes of the peak and intrapeak emissions are apparent (from Mayer-Hasselwander et al 1994). Two modulation cycles are shown.

more than 20 years, and entirely through its γ-ray observations. The *EGRET* data yielded the first (and best) timing information: P = 237.0974531 msec, $\dot{P}$ = 1.0976 10^{-14} sec/sec (T_o = May 24, 1991; Mayer-Hasselwander et al 1994). Following Bignami & Caraveo's (1992) initial attempt, Mattox et al (1992) developed a more complete timing history by reanalyzing the *SAS-2* data. The period and period derivative of the object appear remarkably constant over 20 years (see Figure 10); in fact, this pulsar does not seem to experience any glitches or period noise. Indeed, over the *GRO/EGRET* era, the timing solution could be carried out in phase, i.e. every pulsar revolution (about 4/sec) occurring over the several years of the mission could be counted. It took time, however, to recognize this stable character of the Geminga pulsar: The first timing solutions seemed to yield a measurable value of $\ddot{P}$. This value was in fact so high as to be quickly recognized as unphysical and in turn gave rise to the breaking index value $n = (\Omega\ \ddot{\Omega}/\dot{\Omega}^2)$ in the region of several tens (Alpar et al 1993; Bisnovatyi-Kogan & Postnov 1993). According to the braking index definition, $\dot{\Omega} = -k\Omega^n$; because $\dot{E} = I\Omega\ \dot{\Omega}$ and, for a dipole field, $\dot{E} \sim \Omega^{-2}$, the value expected for such a field is $n = 3$ [to wit the cases of PSR0531 + 21 (Crab), where $n = 2.51$ (Lyne et al 1988); PSR0540 − 69, where $n = 3.6$

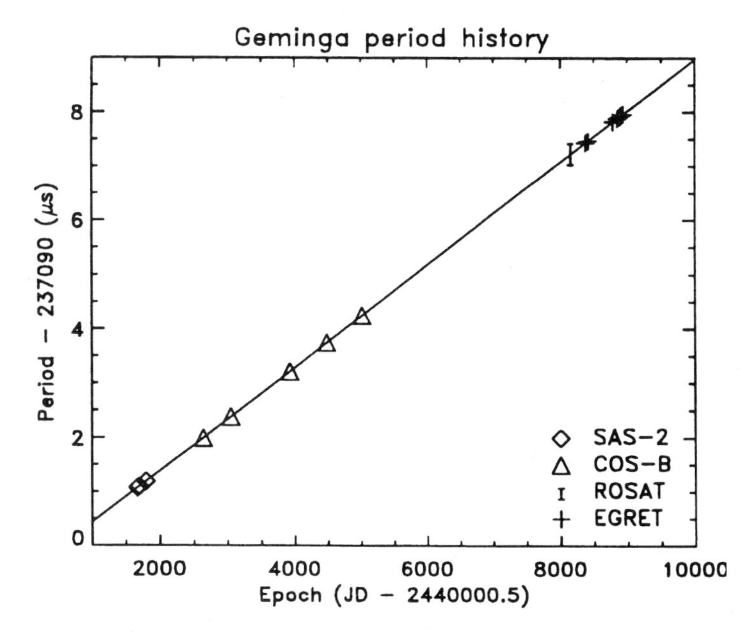

Figure 10 Evolution of Geminga's period, as measured by *SAS-2*, *COS-B*, *ROSAT*, and *EGRET* over more than 20 years (from Mayer-Hasselwander et al 1994).

(Middleditch et al 1987) or 2.04 (Guiffes et al 1992); and PSR1509 − 58, where $n = 2.837$ (Kaspi et al 1994)].

The original perturbations in the Geminga timing solutions were likely due to the failure to account for its proper motion. The question of the braking index value must remain open pending a meaningful measurement of the period second derivative. An analysis (Mattox et al 1996a) with $\ddot{P} = 0$ and including the proper motion yields a statistically better timing solution, resulting in a sharper light curve. As mentioned earlier, this finding independently suggests that G″ is Geminga.

The Geminga γ-ray data are the electromagnetic channel in which the majority of the energy flux is emitted ($\sim 3 \times 10^{-9}$ ergs/cm^2 per second between 30 MeV and 3 GeV). They contain nearly all the information on the nonthermal processes occurring in the vicinity of that neutron star and, in the absence of radio data, optical pulsed data, and long baseline X-ray data, nearly all the information on the timing of the pulsar. The Geminga results obtained by Akimov et al (1993) in their GAMMA-1 experiment overlap with part of the energy range covered by *EGRET*. Despite the failure of the telescope's spark chamber (which could have yielded an unprecedented angular resolution), the other counters on board were used to identify Geminga as a pulsating source in the interval of 400–4000 MeV and to measure qualitatively its spectral slope above 300 MeV. The data were obtained just prior to the launch of *GRO*, from November 1990 to February 1991. The results, with limited statistics, broadly agree with those of *EGRET*: a two-peaked light curve separated by .5 in phase, a timing law in agreement with the *COS-B/ROSAT/EGRET* law, and a spectrum above 300 MeV reportedly harder than that of *COS-B*. In sharp contrast to other MWINSs such as PSR0531 + 21 (the Crab pulsar) and PSR1509 − 58 but in line with the behavior of e.g. PSR0833 − 45 (the Vela pulsar), Geminga is difficult to detect in the vast energy range from a few keV to few tens of MeV. Pending hard X-ray observations by the upcoming *XTE*, what is currently available is an upper limit taken from the OSSE instrument (Schroeder et al 1995) and a preliminary measurement with the COMPTEL instrument (Hermsen et al 1994), both aboard the *GRO*.

For the VHE ($\sim$ TeV) γ-ray energy region, Vishwanath et al (1993), and Akerlof et al (1993) published conflicting reports, oscillating between marginal and null detections for both steady and pulsating emission. More work is needed at $\sim 10^{12}$ eV before a positive detection can be made.

3.2 *Thermal (?) Radiation: Soft X-rays*

1E0630 + 178 (Geminga) has been detected in soft X rays by three observatories: *Einstein* (1979–1981) *EXOSAT* (1983–1985), and *ROSAT* (1990–1991). (At the time of writing, more data from *ROSAT* are awaiting full analysis, and

ASCA observations are still unpublished). From here on we use the *ROSAT* data because of their superior accuracy, spectral range, and statistics. The *Einstein* HRI data (Bignami et al 1983) are nonetheless historically important because they enabled an accurate and absolute X-ray position of $\alpha_{(J\ 2000)} = 6$ h 33 m 54.11 s, $\delta = +17^\circ\ 46'\ 11''.61$ to be established for Geminga in 1981. The *ROSAT* HRI coordinates from 1991 are $\alpha_{(J\ 2000)} = 6$ h 33 m 54.36 s, $\delta = +17^\circ\ 46'\ 14''.6$ (Becker et al 1993). An uncertainty of approximately $\pm 3.5''$ and of $\pm 5''$ must be assigned to the first and second set of coordinates, respectively. However, difference between these coordinates ($< 6''$), although not statistically significant, is exactly in the direction expected given the proper motion of the source, which has moved $\sim 2''$ NE over ~ 10 years. The *Einstein* HRI nonetheless remains an important reference point for future X-ray observations.

The *ROSAT* PSPC data of Halpern & Holt (1992) & Halpern & Ruderman (1993) now include most of the Geminga X-ray results, with more than 7900 photons collected at a rate of $> .5$ photon/sec (to be compared with e.g. the *Einstein* IPC rate of $< .1$ photon/sec). These data show that the timing solution is in perfect agreement with the γ-ray low based on the much wider *EGRET* time span. More importantly, the amount and phase of the 237-msec modulation vary with the X-ray energy. Halpern & Ruderman (1993) observed both a soft and a hard component in the *ROSAT* PSPC data from Geminga that loosely comprise the $\sim$0.1–0.6-keV and $\geq$0.6-keV energy regions, respectively. The pulsed fractions of these components are fairly similar ($\sim$30 and $\sim$40%, respectively), but the shapes of their light curves differ considerably. The soft component is characterized by a rather broad dip, whereas the hard component has a narrow peak. Also, the position of the two features in absolute phase, i.e. with respect to the γ-ray data, is different. Figure 11 of Halpern & Ruderman (1993) shows a template outline of the Geminga pulses in the soft and hard X-ray region compared with the γ-ray region. Becker et al (1993) reported similar results, using the more limited *ROSAT* HRI data. The marked contrast of behavior in phase between the X rays (in their two components) and the γ rays is a key phenomenon against which any interpretation of the Geminga emission must be measured.

Halpern & Ruderman (1993) used the PSPC *ROSAT* data in their spectroscopic analysis and presented two possible solutions: First they suggest a two-temperature black-body fit, with $T_1 = (5.2 \pm 1.3)\ 10^5$ °K and $T_2 = (3 \pm 0.7) 10^6$ °K, $N_H \sim 1.5\ 10^{20}$ cm^{-2}. In this fit, the majority of the bolometric flux is due to the low-temperature curve, and the two components cross at $\sim$0.6 keV, close to the energy value at which the modulation in the light curve (discussed above) undergoes its change. Second, they propose two-component

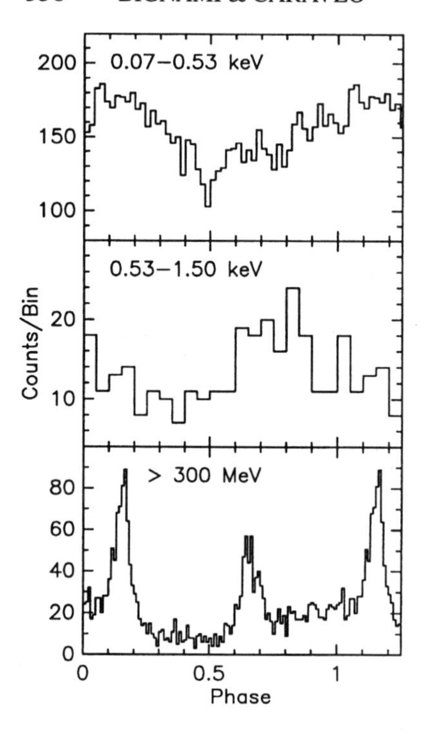

Figure 11 Comparison of the absolute phase of the X-ray light curves (Halpern & Ruderman 1993; J Halpern, private communication) with those of the γ-ray light curve. One and one half modulation cycles are shown.

fit with a black-body curve, with $T = 4.5 \pm 1.5\ 10^5$ °K, $N_H \sim 1.85\ 10^{20}$ cm^{-2} and a power law, with $\alpha = 1.47 \pm 0.3$.[1]

The goodness of both fits to the data is comparable. However, the second compound fit, with the power law extension, best fits the *ROSAT* data of PSR1055 − 52, a MWINS that closely resembles the Geminga phenomenology (Ögelman & Finley 1993). However, although the power law tail of PSR1055 − 52 could continue up to the high-energy γ rays detected by *EGRET* (Fierro et al 1993), Geminga is much brighter than PSR1055 − 52 in high-energy γ rays.

Both fits give similar values of the column density toward Geminga. Combined with our (rather poor) knowledge of the local interstellar medium, this value translates into a distance range for the source. Unfortunately, the constraints are not very stringent: Distances of 20–500 pc could be compatible with the X-ray data. However, taking into account the firm upper limit from the

[1]Recent ASCA data, unpublished at the time of writing, show a detection of the source, albeit at a weak level, up to at least 5 keV, with the GIS, gas proportional counter experiment. This is the first time that Geminga has been observed in X rays above 2 keV, and this detection strongly favors a power law shape for the hard component discovered with *ROSAT*.

high-energy γ-ray data and the optical measurements in V and B, Halpern & Ruderman (1993) concluded that "the best estimate of the distance to Geminga is 250 pc, with an allowed range of 150–400 pc." Any modification of the spectral shape or of the assumed instrument (PSPC) response matrix would change the best column density value and hence the distance estimate. In their exhaustive work, these authors recognized the potential importance of any circumstellar absorption near the neutron star, arising, for example, from the presence of an atmosphere. Until now, the topic of neutron star atmospheres has been a difficult one, with little observational evidence available.

Romani (1987) was the first to point out that neutron stars are likely to have an atmosphere and that this atmosphere would modify, perhaps extensively, the pure black-body spectrum supposed to emerge from the surface through a geometrically thin (but optically thick) atmosphere of plasma [see e.g. Meyer et al's (1994) treatment of Geminga]. The neutron star atmosphere consists of partially ionized H and He. Temperature and density increase with depth and with successive layers at different temperatures emitting (and absorbing) at different frequencies. Moreover, the centimeter-to-meter scale height of the atmosphere corresponds to the region in the immediate vicinity of the surface, where the magnetic field is highest ($B > 10^{12}$ G). This association further complicates the modeling of the emerging radiation (e.g. Shibanov et at 1992; Pavlov et al 1995), requiring one to tackle the physics of a highly magnetized plasma. In their detailed work, Meyer et al (1994) conclude that the atmosphere lowers the temperature value best fitting the soft *ROSAT* X rays to $T = 2\text{–}3\ 10^5$ °K. A harder, second component is still required.

The fits to the *ROSAT* data imposing an atmosphere also imply a smaller value for the distance than originally proposed by Halpern & Ruderman (1993). Now that the distance to Geminga is known through its parallactic displacement, this point is of academic interest. The argument now can be reversed and the distance to Geminga used to address the influence of an atmosphere on the emerging radiation.

3.3 *Thermal Radiation: Optical/UV/EUV*

We use the distance measurement to review the optical and EUV data and to place them, together with the softest X rays, in the context of global thermal emission from the neutron star surface (Bignami et al 1996).

The *EUVE* satellite with the 100-Å (Lexan) filter on the Deep Survey Instrument (DSI) and the Short Wavelength Spectrometer (SWS) was used to observe Geminga at the beginning of 1994. The wavelengths covered were ~70–90 Å for the DSI and 88–92 Å for the SWS, and the exposures yielded a good detection in the former (more than 2000 photons) and a marginal detection in the latter. The data also permitted an evaluation of the modulation of the signal with the pulsar's periodicity, resulting in a shallow ($\sim$20%) dip of $\sim$.1 extent in phase and located close to phase .6, with a profile similar to that found for low-energy X-ray data.

Having thus established the *EUVE* source identity with Geminga, an isothermal black-body emission can be modeled for a source with the observed (time-averaged) flux, the measured distance, and the standard dimensions of a neutron star. Figure 12 shows such an emission and includes the *EUVE* and *ROSAT* data. In particular, the large triangles and squares represent the *EUVE* DSI and SWS fluxes, respectively, and are superimposed to the total black-body fits (also shown). Two values for the neutron star radius (10 and 15 km) and two temperature/N_H combinations (2.8 10^5 °K and $N_H \sim 1.0\ 10^{20}$ cm^{-2}; -2.2 10^5 °K and $N_H \sim 5\ 10^{19}$ cm^{-2}) are shown. For comparison, the dotted line represents the Planckian that best fits the *ROSAT* data of Halpern & Ruderman (1993). This fit of the X-ray data disagrees somewhat with the *EUVE* data on the Wien side and around the maximum of the black-body curve. This discrepancy may be related to the modifications introduced by a neutron star atmosphere in the shape and apparent temperature of its black-body emission (see above discussion on the work of Meyer et al 1994). The temperature range suggested by the *EUVE* data is $\sim$ 2–3 10^5 °K, a result more in agreement with that of Meyer et al 1994 than with that of Halpern & Ruderman (1993). Nonetheless, for all the Planckians shown in Figure 12, the Rayleigh-Jeans side is in substantial agreement, with a modest spread in predicted flux depending on the (limited) parameter choice.

The optical range results for G″, shown in Figure 13, consist of: (*a*) an upper limit in the I filter (6000–8000 Å) obtained in February 1995 from the *ESO* NTT; (*b*) the R and V points (also ground based), which are fully compatible with those of Halpern & Tytler (1988); (*c*) the difficult B point of Bignami et al (1988), also from *ESO*; and (*d*) three *HST* points: a WFPC2 point through the F555 that is roughly equivalent to the V filter (in reality averaged over several WFPC2 observations for the measurement of the parallax), a WFPC2 upper limit through F675 that is roughly equivalent to R, and an FOC point through the F342W ($\lambda = 3400$ Å and $\Delta\lambda = 702$ Å) filter.

The I-to-FOC spectral region, with the colors now available, provides considerable information on Geminga. Taken as a whole, the cluster of ground-based and *HST* observations is largely compatible with the Rayleigh-Jeans side of the Planckians fitting the *EUVE* and *ROSAT* data. This coincidence, in absolute flux units, of independent measurements lends additional support to the identification of G″/Geminga and provides evidence for a thermal origin of the optical flux.

However, the I-to-FOC colors cannot all be fit by the Planckians shown in Figure 12. Although the FOC and B points as well the I upper limit are in agreement with the Rayleigh-Jeans slope and flux values, the R, V, and WFPC2 points are not. In fact, no single monotonic law can fit all the Geminga colors.

The simplest interpretation of the color distribution is the presence of a spectral feature on the thermal continuum. This feature could be either an emission peak (centered around the V color) or an absorption trough with a minimum around B. Although such a feature can be interpreted as due to the atmosphere of a magnetized neutron star, its existence first must be confirmed. For the B point in particular, the error ($m_b = 26.5 \pm .5$) reflects a measurement at the limit of the instrumental capability. In contrast, the *HST* magnitude

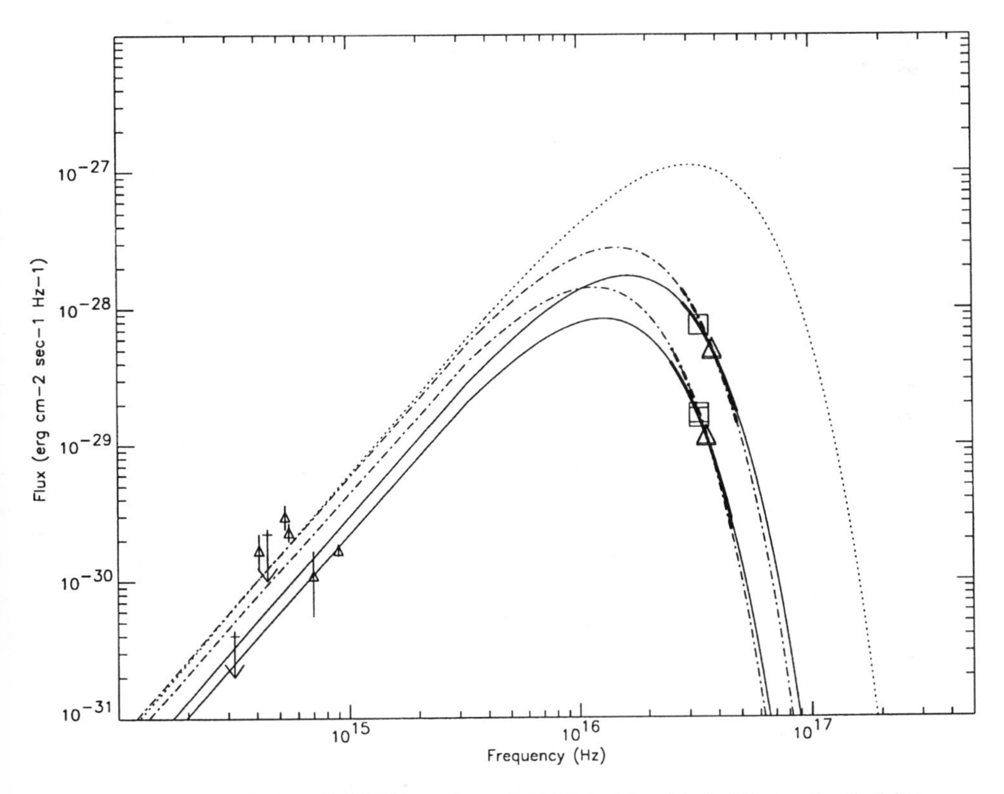

Figure 12 Compilation of *ROSAT (dotted curve)*, *EUVE (solid and dashed line)*, and optical data (*small triangles*) available for Geminga (Bignami et al 1996). The *EUVE* data are shown as large triangles and squares. Two examples are given: a solid curve indicating the flux of an object with a 10-km radius and a dashed curve indicating a neutron star with a 15-km radius. Only one black-body curve describes the eV-to-keV thermal emission from Geminga.

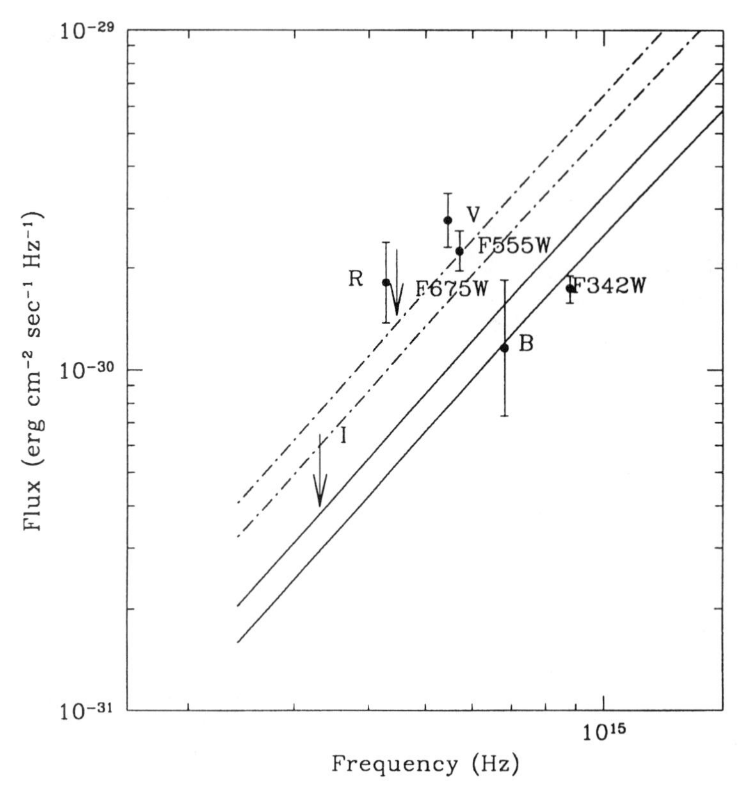

Figure 13 Zoom on optical points plotted along with the Rayleigh-Jeans extrapolation of the black-body curves best fitting the *EUVE* data. Data points are labeled according to the filters through which they were taken (Bignami et al 1996).

determinations are quite accurate, and the R and V points are the result of repeated ground-based measurements. The *HST FOC* data in the B and UV filters could be used to confirm the existence of this spectral feature while awaiting accurate multiband photometry with the VLT.

3.4 *Geminga and Its Fraternity*

Although unique owing to its lack of radio emission, Geminga does not otherwise differ from the hundreds of INSs observed to date.

The fraternity of MWINSs, however, is not large; it encompasses only 3% of the radio pulsar population. To date, $\sim$20 objects have been detected in soft X rays, of which six (or possibly eight) have been detected in the optical, four in the EUV, three in hard X rays, and six in high-energy γ rays (see Table 1).

The energy ranges used in Table 1 match the sensitivity of either the current or past generation of instruments. Optical refers to ground and *HST* data in the range of 1,000–10,000 Å; *EUV* refers to *EUVE* data; X_{soft} refers to *ROSAT* and *Einstein* data; X_{hard} refers to SIGMA as well as *BATSE/GRO* and *OSSE/GRO* data; γ_{soft} refers to *COMPTEL/GRO* data; and γ_{hard} refers to *EGRET/GRO* data.

To be included in Table 1, an INS must have been seen in at least one energy band in addition to the radio band. The objects have been ordered as a function of their characteristic age. P indicates that the identification is based on pulsation, whereas D stands for detection and reflects a good positional coincidence between the radio pulsar and a point source detected at other wavelengths. In the soft X-ray band, some of the pulsed emission is thermal in origin (indicated by a *T*). In the optical column, F130 signifies a detection with the *HST* FOC through the L130 filter, an open passband covering the complete FOC range (Pavlov et al 1996), while F342 implies detection through the FOC V-like filters. Geminga scores very well in Table 1, rivaling the number of covered spectral domains of the Crab, Vela, and PSR1509 − 58, three much younger pulsars with a much larger rotational energy loss.

The best tools to identify the preferential energy domain where NSs channel the bulk of their emission are multiwavelength efficiencies, i.e. ratios between the luminosity in a given energy range and the rotational energy loss of the NS. This is certainly of more general nature than the use of fluxes or luminosities (see Goldoni et al 1995). Such a procedure can be applied to Crab, Vela, 1509

Table 1 Multiwavelength detection of INSs

Pulsar	Radio	Optical	*EUV*	X_{soft}	X_{hard}	γ_{soft}	γ_{hard}
0531 + 21	P	P		P	P	P	P
0540 − 69	P	P		P			—
1509 − 58	P	D		P	P	P	—
0833 − 45	P	P		P,T	—	P	P
1706 − 44	P	—		D	—	—	P
1800 − 21	P			D,T			—
1823 − 13	P			D,T			—
2334 + 61	P			D,T			—
1951 + 32	P			P			P
0656 + 14	P	D,F130	D	P,T			?
Geminga	—	D,F342	P	P,T		—	P
1055 − 52	P	—		P,T	—	—	P
0355 + 54	P			D,T			—
1929 + 10	P	F130, F342	D	P,T			—
0823 + 26	P			D,T			—
0950 + 08	P	F130	D	D,T			—

– 58, and Geminga, i.e. the MWINSs with a sufficient number of detections. The results are shown in Figure 14 where, for each pulsar, the values of the efficiencies have been plotted as a function of energy.

The multiwavelength behavior of the four objects appears vastly different. Unfortunately, very little can be said about the optical emission, which is energetically irrelevant, and involving nonthermal mechanisms for Crab, Vela, and PSR1509 – 58 and a probable thermal nature for Geminga. Concentrating on the high-energy, bona fide non thermal radiation, it is possible to recognize a preferential energy output channel for the four objects. While the preferential channel for Geminga (open circles) is in the high-energy γ-ray domain with no detectable emission in hard X rays, in the Vela pulsar (filled squares) hard X-rays

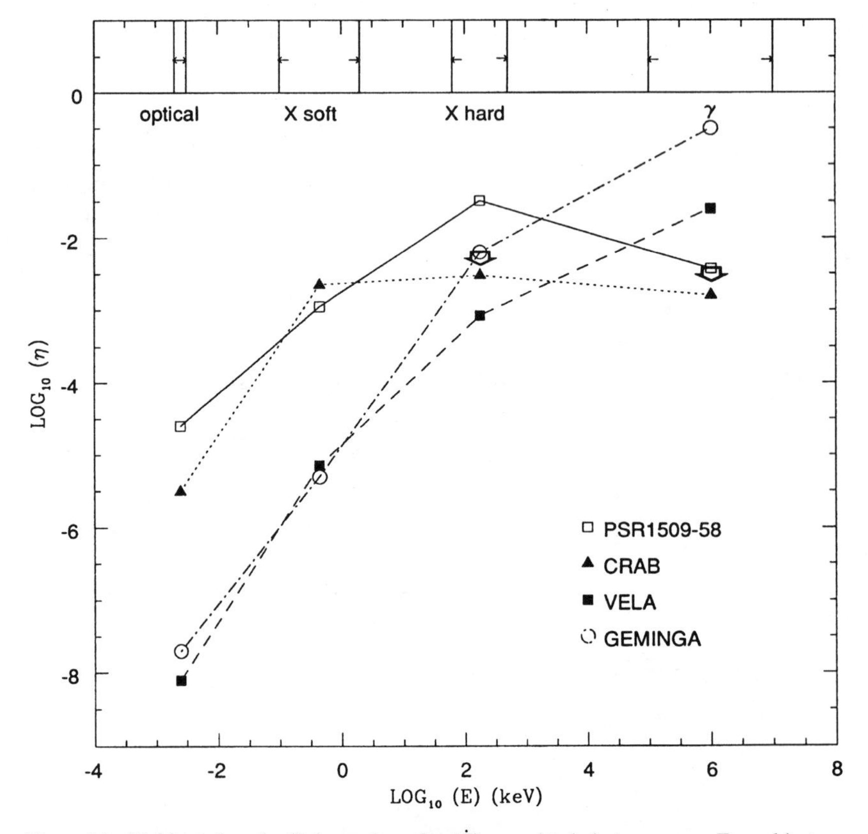

Figure 14 Multiwavelength efficiency ($\eta = L_W/\dot{E}$) vs emitted photon energy. Four objects are depicted: Geminga (*open circles*), the Vela pulsar (*filled squares*), the Crab (*filled triangles*), and PSR1509 – 58 (*open squares*). Variation on $\eta(E)$ are apparent.

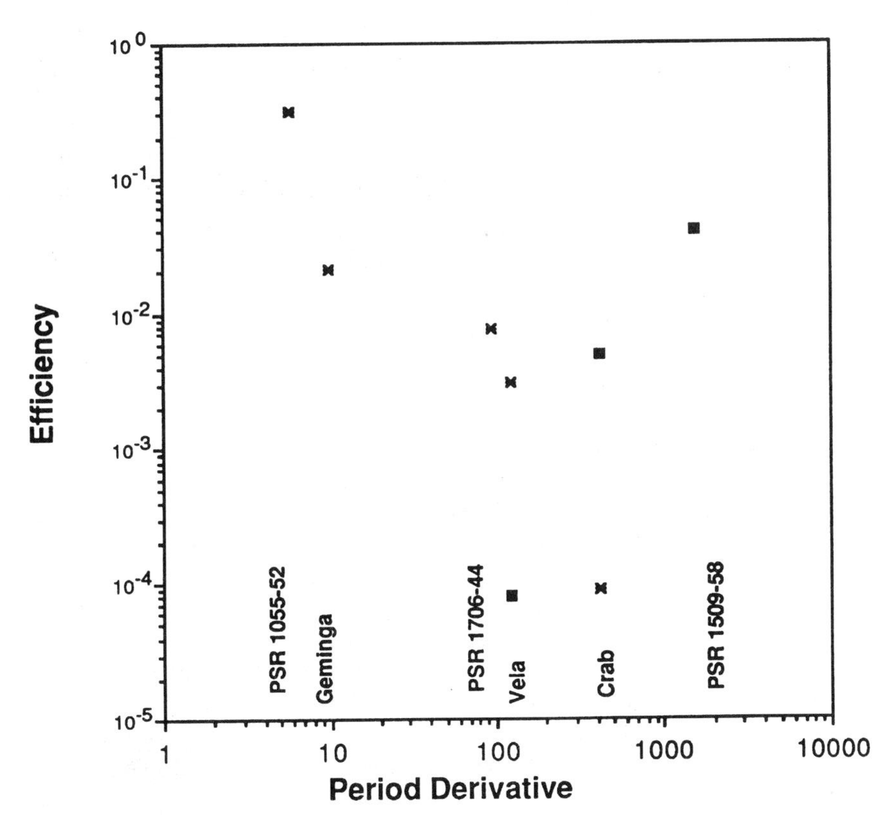

Figure 15 Hard X-ray (*filled squares*) and high-energy γ-ray (*) efficiencies for the same four objects plus PSR1055 − 52 and PSR1706 − 44 (Caraveo 1995). Opposing trends are seen for the X and γ rays.

start to emerge. The trend smoothly continues for the Crab (filled triangles), which counterbalances a markedly lower efficiency in γ rays with a higher efficiency in hard X rays, and ends up with PSR1509 − 58 (open squares), which channels the bulk of its energy output in hard X rays, with a negligible fraction in high-energy γ rays. This is in agreement with the spectral indexes measured by EGRET in the γ-ray range (Fierro et al 1993) whereby going from Geminga to PSR1509 − 58, the radiation emitted softens progressively.

This result is again confirmed by comparing the hard X-ray and high-energy γ-ray emissivities for all the objects listed in Table 1. This trend is shown in Figure 15 in which the efficiencies are plotted as a function of the pulsar period derivative. The choice of $\dot{P}$ is based on the results of Goldoni et al (1995) and Caraveo (1995), who identified $\dot{P}$ as the parameter effectively driving the high-energy behavior of the MWINSs detected to date.

This trend is as clear as it is undisputable: The hard X-ray efficiency increases at the expense of the high-energy γ-ray output, in close correlation with the value of the objects $\dot{P}$ and hence of their magnetic field. The higher the magnetic field, the higher the output in hard X rays and the lower the high-energy γ-ray throughput. The nondetections of PSR1509 − 58 by *EGRET* on one side and of Geminga by *SIGMA* on the other provide further confirmation of this intriguing anticorrelation.

4. THE PHYSICS

Geminga offers a good opportunity for the study of INS physics for two reasons. First, its middle age favors the presence, or rather the detectability, of both thermal and nonthermal processes occurring either near its surface or in its magnetosphere. In younger objects, such as the Crab pulsar (Becker & Aschenbach 1995), PSR0540 − 69, PSR1509 − 58, and the Vela pulsar (Ögelman 1995), any thermal emission is either overpowered or hopelessely mingled with magnetospheric nonthermal processes, which may even create extended nebulosity around the compact object, further clouding the issue.

Second, because the distance to Geminga is so small ($\sim$ 160 pc), multiwavelength observations have led to the phenomenology presented above. So far, only two other MWINSs have a similar combination of middle age and relatively small distance, namely PSR0656 + 14 and PSR1055 − 52 (the latter is probably further away), but none enjoy the same completeness of observation.

4.1 *A Hot, Cooling Neutron Star*

Following their formation in supernova (SN) event, INSs are rapidly cooled by their very intense neutrino emission, as observed for SN1987A, which yielded the first concrete evidence of INS formation in a SN event.

Neutrinos are created in processes in the core of the star and can be assimilated to β and inverse β decays. A good synopsis of the absolute and relative emissivities of the most important cases can be found in Page (1993). The various neutrino emission rates all depend strongly on the star's internal temperature, which ranges from T^6 to T^8. Photon cooling, on the other hand, goes with the fourth power of the star's surface temperature and, owing to internal heat transport, inevitably at a lower power of the internal temperature. Thus, photon emission will become the dominant process after a relatively short time, typically $\sim 10^5$ years.

For a canonical (but not yet measured) braking index of $n = 3$, the spin-down age of Geminga $\tau \sim 3.4\ 10^5$ years, at which photon cooling dominates. Note, here again, the importance of a solid determination of a braking index for Geminga, i.e. of the measurement of a finite second period derivative, or at least of a variation of the first. Let us assume that Geminga's age places the star in the photon cooling are and that neutrino cooling, either fast or slow, successfully explains its temperature upon entering such an era (Page 1994).

The next physical boundary condition on Geminga's apparent emission temperature is the presence of an atmosphere. As can be seen from the work of Shibanov et al (1992) and Meyer et al (1994), especially for Geminga, any H- (or He-) dominated atmosphere creates an excess of emission on the Wien side of the black-body spectrum that alters the standard black-body shape of the curve and lowers the value of the measured temperature. The effect of the gravitational redshift on the photons emitted close to the surface also should be taken into account. The value of such redshift, $z = (1\text{–}2\ \mathrm{GM}/Rc^2)^{-1/2} - 1$, ranges from 0.2 to 0.45, depending on the neutron star equation of state.

In the broad picture, Geminga is considered a cooling INS. To see the "surface" of a neutron star is rare. Until *ROSAT*, very little bona fide soft X-ray emission attributable to MWINSs was known. Geminga remained the best candidate from its surface, for showing thermal optical emission but the apparent temperature of its surface given by the preliminary X-ray data implied optical magnitudes much lower than those observed. Further analysis of Halpern & Ruderman's (1993) results revealed that the temperature of 5×10^5 °K from the *ROSAT* data really should be considered an upper limit. As discussed in Section 3, more data have become available in the optical and, more importantly, in the EUV; the X-ray calculation has been revised downward to account for any atmosphere; and an accurate distance measurement of ~ 160 pc has been obtained. As shown in Figure 12, a single black-body curve now provides a good overall fit to the Geminga data in the $\sim$eV to $\sim$KeV photon energy region, with a temperature of 2.2–2.8 10^5 °K.

The next-best candidate for a similar global view appears to be PSR0656 + 14, provided its optical counterpart is confirmed and measured in more detail.

4.2 *Evidence of a Magnetized Atmosphere*

The data of Figure 12 go beyond the broad picture, however. On the optical/UV side, they provide evidence of a spectral feature, either in absorption or, more likely, in emission. Such a feature may be associated with a magnetized atmospheric layer, the possible source of most of the radiation seen at optical frequencies. Each frequency corresponds to a different depth in the layer. What happens to the H (and He, and possible traces of metal) in such an atmosphere? Pavlov & Potekhin (1995) recently studied one effect, the so-called decentered

H atoms, which have binding energies in the optical/UV range. If the emission observed for Geminga can be interpreted as an absorption, these peculiar atoms may be responsible for this emission (see e.g. Ventura et al 1995). A more straightforward interpretation is possible for the I-to-FOC spectral region shown in Figure 13: At the high magnetic field close to the surface of Geminga, $\hbar w_{\mathrm{B}} \sim (\mathrm{P\dot{P}})^{1/2} \sim 1.55\ 10^{12}$ G, the electron cyclotron energy $\hbar\omega_{\mathrm{B}} = \hbar e\ \mathrm{B}/m_{\mathrm{e}}$ c is $\sim 11.6\ (\mathrm{B}/10^{12}\ \mathrm{G})$ keV (Pavlov et al 1995).

This energy is on the Wien side of the Planckian and may be unaccessible because at energies above a few keV, a probably nonthermal process dominates what is shown in Figure 13. In any case, Geminga has never been observed, even with *ASCA*, at X-ray energies above ~ 5 keV. On the other hand, the proton (ion) cyclotron energy is a factor of $m_{\mathrm{p}}/m_{\mathrm{e}} \bullet \mathrm{Z/A}$ lower. As such, it will fall in the 1–10 eV region, depending on the ion species considered, and on the exact B value at the height above the surface where the effect occurs. Additionally, the luminosity in the line should be corrected for the typically 20% effect due to the gravitational redshift $L_{\infty} = L(1+z)^{-2}$, again depending on the actual radial distance $R_{\infty} = R(1+z)$.

Given the impossibility of fitting the I-to-FOC data with any acceptable monotonic law, in particular the Rayleigh-Jeans law, the most logical explanation for the feature is that an ion cyclotron emission line is centered close to the V-band frequency. Alternatively, a (cyclotron) absorption trough centered around B may fit the data. The width of such a trough could be due to the Doppler effect, especially if $\hbar\omega_{\mathrm{B}} \sim \mathrm{kT}$. The possibility of absorption is less attractive than that of emission since the FOC and B points as well as the I upper limit seem to be in good absolute agreement with the Rayleigh-Jeans slope.

For Geminga, determination of the deformation foreseen (Meyer et al 1994; Shibanov et al 1995) on the Wien side of the Planckian due to the presence of an atmosphere awaits further EUV and soft X-ray observations. However, the spectral feature is in itself a strong indication of the presence of an atmospheric layer close to the surface of Geminga. This feature is the first observational evidence of an atmosphere in an INS.

4.3 *Geminga's Smoking Gun*

The X-ray light curve places important constraints on the geometry of the Geminga emission. The presence of just one broad pulse, as opposed to the two spiky pulses measured in high-energy γ rays, requires ad hoc assumptions about the magnetic field configuration of the highly inclined rotator. Halpern & Ruderman (1993) suggested that an out-of-center dipole was responsible for the sunspot polar cap configuration required by the shape of the X-ray light curve.

The picture is further complicated by the peculiar difference in the light curve shapes vs energy known as the Geminga effect, i.e. the observed decrease in the depth of modulation between the < 0.1 keV and the 0.2–0.6 keV regions, which is clearly visible in Figure 16. One would not expect to see such behavior in the case of a pure black-body emission. According to Page (1995) and Page et al (1995), if the thermal emission could be described using a pure black-body, the pulsed fraction would increase with increasing energy, contrary to observations. Halpern & Ruderman (1993) ascribed this behavior to the contamination of the harder tail, itself with a different modulation pattern, that dominates the spectrum above 0.6 keV, assuming that the harder component is due to starward accelerated particles.

In fact, heating due to backward particles is a controversial issue. In the context of the Harding et al (1993) polar cap model, the luminosity of this effect is evaluated to be $\sim 5 \times 10^{28}$ erg/sec. Halpern & Ruderman (1993) found that polar cap heating resulting from starward accelerated particles in the outer atmosphere produced a luminosity of 2.6 10^{32} erg/sec (and a temperature incompatible with the *ROSAT* hard component data, even if interpreted as thermal emission). Thus the first model falls short of the mark and the latter overshoots it, as can be determined from the absolute luminosity stemming from the distance measurement.

Moreover, in light of the *ASCA* data, which favor a power law, nonthermal interpretation for the emission in the > 0.6 keV region, the observed decrease in modulation likely results from a genuine variation inside the soft thermal component caused by the heated surface of the neutron star.

Geminga surface heating must result from heat transport from the star's interior. Because transport is in turn dominated by the crustal magnetic field, regions of the star in which the field is normal to the surface, thus facilitating heat transport, will be warmer than the regions in which heat transport across field lines parallel to the surface is more difficult.

Using the general relativistic effects of gravitational redshift and lensing, plus the effect of a magnetic atmosphere, one can model a patchwork of surface temperatures. Recall that in his plate tectonics theory, Ruderman (1991) discussed the association between such a patchwork and the marks left on the surface by any tectonic activity.

Thus Geminga has been caught with smoking-gun evidence. The Geminga effect cannot be explained using standard black-body emission; rather, it can be attributed to surface thermal anisotropies coupled to a magnetic atmosphere.

4.4 *Transition Region and Global Energetics*

Above ~ 0.6 keV, the spectrum of emission from Geminga changes shape, as does its light curve. In X rays, the object remains visible only up to < 5 keV,

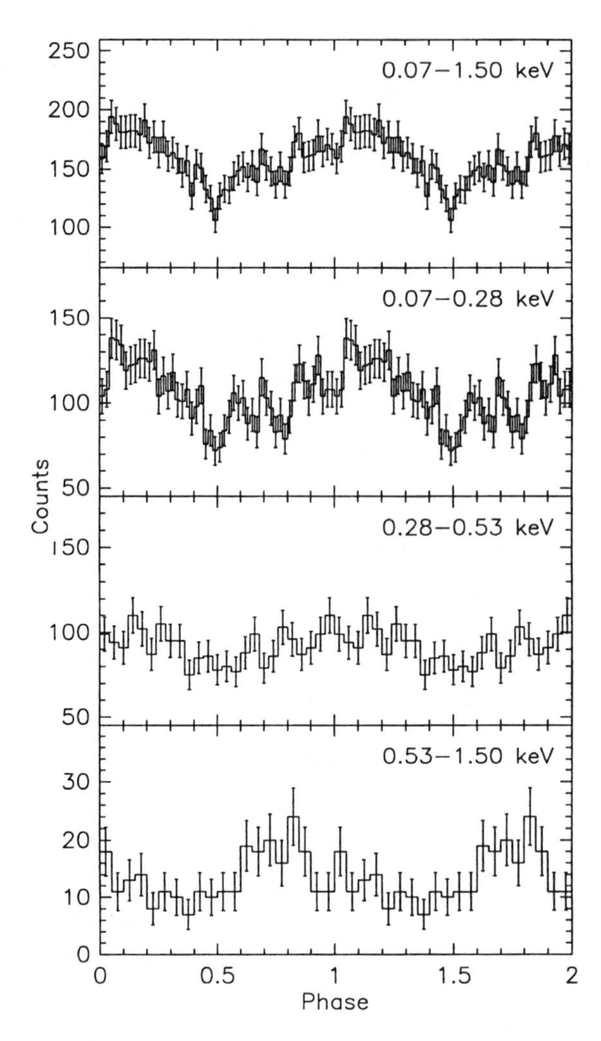

Figure 16 Geminga light curves (repeated for two cycles) as a function of the X-ray photon energy. The top panel shows the light curve for the total energy range of the *ROSAT* PSPC. In the three lower panels, these same data are divided into three energy intervals. A marked decrease of modulation occurs in the 0.28–0.53 keV energy range. This decrease has been dubbed the Geminga effect (from Halpern & Ruderman 1993).

but little is known about this spectral region. Its shape is hard and clearly nonthermal in origin owing to the shape of the power law observed using the *ASCA*, but its modulation pattern differs considerably from that of Geminga's nonthermal emission par excellence, the γ rays.

Interestingly, PSR0656 + 14 and PSR1055 − 52 also have a hard tail, at least on *ROSAT* X rays. Moreover, for the confirmed γ-ray emitter of this pair, PSR1055 − 52, the extrapolation of the power law end of such a tail seems to join the $\sim$ 100-MeV γ rays (Ögelman & Finley 1993).

In the case of Geminga, however, this daring extrapolation (more than approximately five orders of magnitude of photon energy) falls well below the observed γ-ray flux, as reflected in the much smaller amount of the energy-per-unit frequency carried by these nonthermal X rays compared with γ rays. Because for PSR1055 − 52 the modulation patterns of the two components are also quite different, it is tempting to attribute the successful extrapolation of this particular object to chance. For Geminga it seems advisable to separate the physics of this transition region from that of the γ rays. To date, no explanation has been put forward for the nonthermal nature of the $\sim$ keV emission because of the limited statistics and detection range.

As discussed in Section 3, Geminga's spectrum is unobservable at higher energies, and can be seen once again above tens of MeV using the *EGRET* γ rays (of which the possible detection at $\sim$ 10 MeV by *COMPTEL*, if confirmed, would be only a precursor).

Before discussing the physics behind γ-ray generation, we quantitatively examine the overall energy distribution in the emission of Geminga. The energetics of Geminga are actually very simple now that, thanks largely to the distance measurement, the eV to keV emission can be ascribed to a single Planckian with temperatures in the 2.2–2.5 10^5 °K region. The bolometric luminosity under such a Planckian is $\sim 1\ 10^{31}$ ergs/sec and is fully compatible with that of the surface of a neutron star with a 10–15 km radius seen at its measured distance of $\sim$ 160 pc.

The presence of an atmosphere does not alter significantly the energetics of Geminga's thermal spectrum but merely introduces distortions in its shape that are irrelevant on a global scale. The source of the surface heating is most probably transport to the surface of the internal heat. Energy deposit from starward accelerated particles hitting e.g. a polar cap region is unlikely (again on a global scale) for two reasons. First, the spot would be small ($\ll 10^{10}$ cm^2) but the evidence indicates an emission from an $\sim 10^{11}$–10^{12} cm^2 surface. Second, the energetics, which depend on nonthermal processes related to particle acceleration, is either too small (Harding et al 1993) to be relevant or too large to be correct (Halpern & Ruderman 1993). In any case, what was once considered a possible hot thermal component is now understood to be of different origin. Energetically, this hard X-ray component, covering from 0.6 keV to the end of the data at $\sim$5 keV, is small ($\sim$5 10^{29} erg/sec, or a few percent of the thermal component).

Finally, the γ rays of the observed *EGRET* spectrum contain $\sim$ 1.5 10^{34} ergs/sec in the simplest assumption of isotropic emission. This value is almost half of the overall spindown energy of $I\Omega\,\dot{\Omega} = 3.2\ 10^{34}$ ergs/sec. Given the shape of the γ-ray light curve, some beaming must be present, decreasing by a corresponding factor the necessary power.

4.5 *Why Not a Radio Pulsar?*

To date, the instruments aboard the *GRO* have been used to detect seven MWINSs (see Table 1 for a summary of the observational panorama). All but Geminga were already known radio pulsars, which begs the question, as to why is Geminga not a radio emitter as well.

As suggested by Bignani et al (1983), the simplest explanation seems to lie in the geometry. To be detectable in any energy band, the beam of radiation produced by any pulsar must intercept the observer's line of sight. Because different production mechanisms produce radiation in different energy domains, the beaming geometry can differ as well. Indeed, the compilation of the radio, optical X-, and γ-ray light curves of the *GRO* MWINS (Thompson et al 1994; DJ Thompson, private communication) shows that diversity is the rule for all objects except the Crab pulsar, which has similar light curves in the different energy ranges (see Figure 17). Moreover, because the beaming factor seems to be smaller in radio than at higher energy, it is more likely to miss the radio beam than the γ-ray one.

Unfortunately, this hypothesis is difficult to either prove or disprove. On the one hand, the difference in beam size cannot be extremely important since six of seven objects are detected in radio. On the other hand, one could argue that our sample of high energy–emitting MWINSs is far from complete: A fraction of the unidentified *EGRET* sources could indeed be radio-quiet MWINSs waiting to be discovered. This finding would change the ratio of radio-loud to radio-quiet MWINSs and render any geometrical explanation less ad hoc or, more appropriately, ad Gemingam. Until then, however, this simplistic argument cannot be considered fully satisfactory.

An alternate hypothesis has been put forward by Halpern & Ruderman (1993), who propose that the lack of radio detection of Geminga is due to a genuine lack of radio emission. In their orthogonal rotator model, radio emission is suppressed by the starward flow of particles accelerated in the outer magnetosphere. The large number of accelerated particles is close to the limit of self quenching owing to the considerable power required by γ rays. The radio

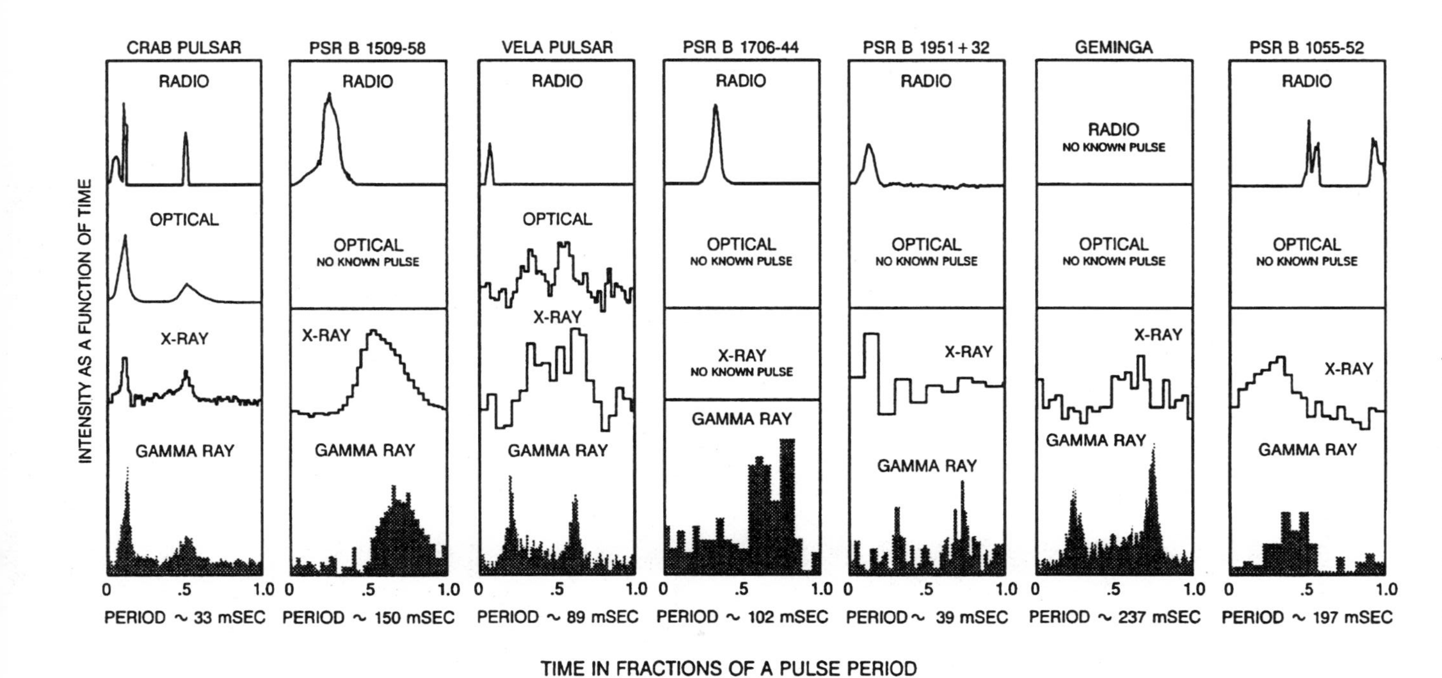

Figure 17 Compilation of the multiwavelength light curves of all the pulsars detected with the GRO (DJ Thompson, private communication). The γ-ray light curves refer to *EGRET* data except in the case of PSR1509 − 58, which was detected by *OSSE* and *COMPTEL* but not by *EGRET*.

pulsar mechanism is a delicate, coherent one with little power that occurs close to the star's surface. It is easy to see how such a mechanism could be disrupted or rendered unobservable by a process many orders of magnitude more energetic and possibly occurring above it. Until another bona fide Geminga is found, however, this explanation must be considered as valid as that given for different beam geometries.

4.6 *The γ Rays*

Gamma rays, i.e. photons in the interval from a few tens of MeV to a few GeV, are the channel through which most of the energy loss of the rotating neutron star occurs. The physics related to the production of γ rays is necessarily simple, at least at the basic level. In order to produce high-energy photons, the energy of the oblique rotator must be able to accelerate electrons and positrons, which then radiate photons. Thus, in the end, sufficiently large potential difference ΔV must be created and maintained.

Once produced, however, γ rays must be able to escape. They might lie in a region with so high a magnetic field that pair production against the B quanta creates absorption, subsequent reradiation, etc in a cascade process that depletes the primary photon energy of an amount dependent on the number of iterations in the cascade process.

The models for γ-ray production in a pulsar like Geminga are of two basic types. The so-called outer gap, originally developed by Ruderman & Sutherland (1975) and by Cheng et al (1986), was applied to Geminga by Halpern & Ruderman (1993).

In the outer gap model, the accelerator operates in the outer magnetosphere ($r \gg R \sim 10^6$ cm), close to the light cylinder (~ 800 km for Geminga). Here, the B value is reduced to $< 10^8$ G, alleviating the problem of transparency at least for those photons, up to a few GeV, detected by *EGRET*. If the acceleration takes place along the open field line bundle between the star surface (necessarily originating close to the pole) and the light cylinder, the maximum number of accelerated particles ($\dot{N}$) of charge e is

$$N = \frac{\Omega B_{\rm p} R^3}{ec} \simeq 8 \times 10^{31} \ \mathrm{sec}^{-1},$$

where $B_{\rm p}$ is the surface magnetic field (Halpern & Ruderman 1993). Compared with the total $\mathrm{I}\Omega\,\dot{\Omega}$ available power, the large $\mathrm{L}\gamma$ of Geminga ($1.5\ 10^{34}$ ergs/sec) yields a maximum potential drop along the field line between the surface and the light cylinder of

$$\Delta V_{\max} \simeq \frac{\Omega^2 B_{\rm p} R^3}{2c^2} \simeq \frac{\mathrm{I}\Omega\,\dot{\Omega}}{e\dot{N}} \simeq 2 \times 10^{14} \ \mathrm{volts},$$

which provides just enough power for Geminga's γ rays.

However, the predicted dependence of this outer gap model on the star's period is problematic. Ruderman & Cheng (1988), extrapolating from the Crab and Vela examples, predicted a turn off for γ-ray production at ~ 130 msec. In contrast, a periodicity of 237 ms was later detected by Geminga. This discrepency is resolved by taking into account the inclination angle between the star's dipole moment and its spin axis (Chen & Ruderman 1993; Halpern & Ruderman 1993). The value of such an angle at which the outer gap model is compatible with Geminga's γ-ray luminosity is $\sim 65°$. The beam geometry, necessarily reflected in the γ-ray light curve, calls for a highly inclined dipole. Both accelerators, one per pole, are visible for inclination values $> 70°$.

Although more work is needed in order to explain its detailed features, the outer gap model appears to be quite successful for Geminga's γ rays. The downward revision of the distance from the preferred 250 pc of Halpern & Ruderman (1993) to the observed ~ 160 pc, simply reduces the efficiency requirement.

Although in the classic outer gap model γ rays are produced via curvature radiation by particles accelerated by the potential drop along field lines, Sturmer & Dermer (1994) introduced variations on this theme. The acceleration mechanism was the same, but it was close to the polar caps and had a nearly aligned rotator. These authors claim they can reproduce the spectral shape and pulse profile of Geminga using magnetic Compton scattering of electrons on soft photons. This physics of production seems to allow greater freedom of choice of the emission geometry.

More detailed work on all γ-ray pulsars, including Geminga, was recently performed in the framework of the outer gap model by Chiang & Romani (1992, 1994) and by Romani & Yadigaroglu (1995). These authors were able to reproduce the shape of the pulse profiles as well as several geometric characteristics of the pulsars, including the Crab pulsar polarization pattern. They also successfully reproduced the spectral variability observed at different phases (see Section 3) simply by using various emission processes at different locations in the magnetosphere, which is mapped by the different light curve phase intervals. Moreover, they account for the radio to γ pulse offsets and, in the case of Geminga, explain the absence of a radio detection using different beaming geometries. Although this latter explanation may be somewhat lacking, on the whole Romani and collaborators do a good job of elucidating the physics of γ-ray emission from Geminga (and its fraternity) using the outer gap model.

In the polar cap model, originally proposed by Arons (1981) and by Daugherty & Harding (1982), acceleration of electrons occurs in response to the strong E field generated close to the polar cap regions near the surface of the star. Moving along field lines, the highest-energy particles radiate energetic photons, which in turn initiate a cascade process, starting with the production of one photon pair against the strong B field close to the neutron star surface. In this model, particles originally are extracted from the star surface, whereas in the outer gap model, they are formed primarily by pair production in the magnetosphere.

The polar cap model was applied to Geminga by Harding et al (1993) with some success, especially in the key parameter L_X/L_γ, observationally set at $\sim 1/1000$. However, the model underestimates the absolute values of such luminosities by about two orders of magnitude. In other words, for the work of Harding et al (1993) to be accurate, the distance to Geminga would have to be significantly smaller than that recently observed. Although of little importance for X rays (which we now know to be thermal in origin), this limitation is significant for γ rays.

Usov (1994) proposed a critical appraisal of the outer gap model for Geminga after comparison with the Vela pulsar. Difficulties are encountered [as mentioned in Halpern & Ruderman (1993)] for the ratio L_X/L_γ, predicted in this case to be about one order of magnitude higher than that observed, and possibly for the TeV γ-ray luminosity, again predicted to be higher than that observed.

In both models, photons interact via pair production to produce a more or less developed cascade process, depending on the depth in the magnetosphere of the first interaction. Lu et al (1994) performed a quantitative assessment of this absorption process in the context of the polar cap model. Here, photons generated close to the surface undergo pair production in the intense magnetic field and thus cascade (Hardee 1977), increasing in number but decreasing in energy at each step. We can then define a generation-order parameter describing the number of generations photons undergo before escaping as a function of the star's P and $\dot{P}$. The smaller such a parameter, the less degraded the outgoing photons and the harder the objects's spectrum. When applied to Geminga and its fraternity, this parameter is in surprisingly good agreement with observations, which explains in particular the correlation of spectral hardness with pulsar age, or with $\dot{P}$. Thus, the spectral shape of Geminga is naturally harder than that of the Crab pulsar, for example, but is softer than that of PSR1055 − 52.

5. THE FUTURE AND CONCLUSIONS

In order of increasing photon energy, several observations are needed to close the gap between phenomenology and physics as well as between reality and our understanding of it.

In the optical region, photon timing at the pulsar frequency is a must. A combination of a 4-m telescope, a *MAMA* detector, excellent seeing, and a good dose of luck and patience should reveal a light curve, assuming it has any significant modulation. Equally desirable would be a spectrum that could confirm the existing feature and search for new ones. It will be necessary to wait a few years for the VLT, although one wonders about possible developments with the Keck Telescope.

In the UV, the best possibility, now and in the foreseeable future, is offered by the FOC on the *HST*. Photometry in the 3000–2000 Å band is crucial to confirm the Rayleigh-Jeans slope of Figure 12.

In soft X rays, after *ROSAT* and *ASCA*, we must await the upcoming generation of high-throughput and high-resolution observatories: *XMM* (with EPIC and RGS; see e.g. Mason et al 1995) and *AXAF*. These will make possible not only better timing statistics but also time-resolved spectroscopy.

In X rays > 10 keV, after *GRO*, *XTE* will likely be the most valuable tool but in the search for spectral features in the 10–15 keV range, i.e. the electron correspondent of the ion cyclotron resonance possibly observed close to the V band.

In soft ($\sim$MeV) γ rays, any detection supporting the nonthermal modeling is desirable. However, we must await ESA's INTEGRAL mission, now just starting.

For hard (greater than approximately tens of MeV) γ rays, *EGRET* is probably still the best source of long-term timing information. If we could nail down a real $\ddot{P}$, we could in turn determine a braking index value with physical meaning. Unfortunately, no mission has been planned in this domain.

To increase the fraternity, more Gemingas awaiting discovery must be found, even if not all UGOs can be Gemingas (Helfand 1994).

Experience gained from Geminga can be put to use here. First, the 30 UGO error boxes of the second *EGRET* catalog (Thompson et al 1995) must be correlated with *ROSAT* soft X-ray sources showing the right spectrum and no obvious optical counterpart. To a list of such candidates one could then apply the new powerful software tools for timing analysis.

In general, more MWINSs should be found (Bignami 1996), probably by examining the *ROSAT* data. For example, if there are several thousand HRI serendipitous sources, which by now are essentially public domain, a small fraction of these will have no optical counterpart on the *POSS/ESO* surveys. These sources would comprise the candidate list for correlation with *EGRET* UGOs or with more in-depth studies, including timing. *ROSAT* has been used to detect $\sim$20 known (from radio) INSs and can access these objects up to $\sim$1 kpc. How many more radio-quiet MWINSs are present in such volume? Some probably already have been detected but not yet recognized. Brighter objects (i.e. hotter younger ones) such as $1E\ 1207-52$ (Mereghetti et al 1996, Bignami et al 1992) and the point sources in RCW103 (Becker 1995) and in Puppis A (Petre et al 1996), as indicated by their association with a SNR, can be seen from further away.

For each new candidate, it will be impossible to perform a Geminga-like chase, nor would such a chase likely be necessary given the currently available hardware and software tools (and lots of luck). Naturally, experience gained during the chase could come in handy, enabling us to take advantage of the few lucky breaks that even the most stubborn object must offer sooner or later. The potential reward is access to the physics of neutron stars through a greater variety of phenomena than that covered by radio, as evidenced by the basic progress made with only a handful of MWINSs.

For Geminga, progress in understanding its physics followed closely from its multiwavelength identification. From eV to keV, we now know that its emission is exactly that expected from a black body (with possible atmospheric modification) of a standard neutron star ($R \sim 10$ km) at the measured distance (~ 160 pc) and of the inferred age ($\tau = 3.4\ 10^5$ years) and temperature ($T = 2.2\ 10^5$ °K, following standard photon cooling). This observation in itself is certainly a successful finding in multiwavelength astronomy. A short, nonthermal tail ($0.6 < E < 5$ keV) has not yet been explained, owing mostly to data limitations, whereas the vast majority of the luminosity is in the nonthermal high-energy γ rays.

Important questions remain unanswered for both the thermal and nonthermal processes reviewed in Section 4. For example, surface thermal inhomogeneities and conductivity for the crust of a neutron star as well as the efficiency and time scale of the interior-to-surface heat transfer, remain to be determined, as does the geometry of high-energy electromagnetic processes in the magnetosphere: height above surface, absorption, etc.

However, questions as to the astrophysics of an object can be asked only after the object has been discovered. Wouldn't it be nice to be able to ask such questions about the counterpart of γ-ray bursts, for example? That we can ask such questions is a measure of the progress made, giving us confidence to forge ahead.

ACKNOWLEDGMENTS

It is a pleasure to thank G Lampis, S Mereghetti, and J Paul for careful reading of the manuscript. We are also grateful to P Goldoni, J Halpern, J Mattox, H Mayer-Hasselwander, R Mignani, and D Thompson for providing the originals of their figures.

Literature Cited

Abdulwahab M, Morrison P. 1978. *Ap. J.* 221:L33–36

Akerlof CW, Breslin AC, Cawley MF, Chantell M, Fegan DJ, et al. 1993. *Astron. Astrophys.* 274:L17–20

Akimov VV, Afanas'ev VG, Blokhintsev ID, Kalinkin LF, Leikov NG, et al. 1993. *Astron. Lett.* 19:229–30

Alpar MA, Ögelman H, Shaham J. 1993. *Astron. Astophys.* 273:L35–37

Alpar MA, Kiziloglu Ü, Van Paradijs J. eds. 1995. *Lives of Neutron Stars.* Dordrecht: Kluwer

Arons J. 1981. *Ap. J.* 248:1099–116

Becker W. 1995. Invited talk, Wurzburg Conf., Oct. 95

Becker W, Aschenbach B. 1995. See Alpar et al 1995, pp. 43–46

Becker W, Brazier KTS, Trümper J. 1993. *Astron. Astophys.* 273:421–424

Bertsch DL, Brazier KTS, Fichtel CE, Hartman RC, Hunter SD, et al. 1992. *Nature* 357:306–7

Bignami GF. 1996. *Science* 271:1372–73

Bignami GF. 1987. *In High Energy Phenomena Around Collapsed Object,* ed. F Pacini, NATO ASI Ser., pp.. 297–316. Dordrecht: Reidel

Bignami GF, Boella G, Burger JJ, Keirle P, Mayer-Hasselwander HA, et al. 1975. *Space Sci. Instrum.* 1:245–68

Bignami GF, Caraveo PA. 1992. *Nature* 257:287–287

Bignami GF, Caraveo PA, Lamb RC. 1983. *Ap. J.* 272:L9–12 (BCL)

Bignami GF, Caraveo PA, Mereghetti S. 1992. *Ap. J.* 389:L67–69

Bignami GF, Caraveo PA, Mereghetti S. 1993. *Nature* 361:704–6

Bignami GF, Caraveo PA, Mignani R, Edelstein J, Bowyer S. 1996. *Ap. J. Lett.* 456:L111–14

Bignami GF, Caraveo PA, Paul JA. 1984. *Nature* 310:464–69

Bignami GF, Caraveo PA, Paul JA. 1988. *Astron. Astrophys.* 202:L1–4

Bignami GF, Caraveo PA, Paul JA, Salotti L, Vigroux L. 1987. *Ap. J.* 319:358–61

Bignami GF, Fichtel CE. 1974. *Ap. J.* 189:L65–67

Bignami GF, Gavazzi G, Harten RH. 1977. *Astron. Astrophys.* 54:951–54

Bignami GF, Hermsen W. 1983. *Annu. Rev. Astron. Astrophys.* 21:67–108

Bignami GF, Maccacaro T, Paizis C. 1976. *Astron. Astrophys.* 51:319–21

Bisnovatyi-Kogan GS. 1985. *Nature* 315:555–57

Bisnovatyi-Kogan GS, Khlopov MYu, Chechetkin VM, Eramzhyan RA. 1980. *Sov. Astron.* 24(6):716–21

Bisnovatyi-Kogan GS, Postnov KA. 1993. *Nature* 366:663–65

Bloemen JBGM. 1984. *Astron. Astrophys.* 131:L7–10

Brazier K, Kanbach G. 1996. *Astron. Astrophys.* In press

Buccheri R, D'Amico N, Hermsen W, Sacco B. 1984. *Nature* 316:131–32

Caraveo PA. 1983. *Space Sci. Rev.* 36:207–21

Caraveo PA. 1995. *Adv. Space Res.* 15:(5)45–52

Caraveo PA, Bignami GF, Giommi P, Mereghetti S, Paul JA. 1984a. *Nature* 310:481–82

Caraveo PA, Bignami GF, Mignani R, Taff LG, 1996. *Ap. J. Lett.* 461:L1–4

Caraveo PA, Bignami GF, Vigroux L, Paul JA. 1984b. *Ap. J.* 276:L45–47

Caraveo PA, Bignami GF, Vigroux L, Paul JA, Lamb RC. 1984c. *Adv. Space Res.* 3(10–12):77–81

Cesarsky CJ, Casse M, Paul JA. 1976. *Astron. Astrophys.* 48:481–82

Chen K, Ruderman M. 1993. *Ap. J.* 402:264–70

Cheng KS, Ho C, Ruderman M, 1986. *Ap. J.* 300:500–21

Cheng LX, Li TP, Ma YQ, Sun XJ, Wu M. 1993. *Astron. Astrophys.* 277:L13–14

Chiang J, Romani RW. 1994. 436:754–761

Chiang J, Romani RW. 1992. 400:629–637

Cunha K, Smith VV. 1996. *Astron. Astrophys.* In press

Daugherty JK, Harding AK. 1982. *Ap. J.* 252:337–47

Davies RE, Fabian AC, Pringle JE. 1978. *Nature* 271:634–35

Djorgovski S, Kulkarni SR. 1986. *Astron J.* 91:90–97

Fauci F, Boriakoff V, Buccheri R. 1984. *Nuovo Cimento* 7C(6):597–603

Fichtel CE, Hartman RC, Kniffen DA, Thompson DJ, Bignami GF, et al. 1975. *Ap. J.* 198:163–82

Fierro JM, Bertsch DL, Brazier KT, Chaing J, D'Amico N, et al. 1993. *Ap. J. Lett.* 413:L27–30

Frish PC. 1993. *Nature* 364:395–96

Gehrels N, Chen W. 1993. *Nature* 361:706–8

Giovannelli F, Karakula S, Tkaczyk W. 1982. *Astron. Astrophys.* 197:376–77

Goldoni P, Musso C, Caraveo PA, Bignami GF. 1995. *Astron. Astrophys.* 298:535–43

Grenier IA, Bennett K, Buccheri R, Gros M, Henriksen RN, et al. 1994. *Ap. J. Suppl.* 90:813–16

Grenier IA, Hermsen W, Henriksen RN. 1993. *Astron. Astrophys.* 269:209–18
Guiffes C, Finley JP, Ögelman H. 1992. *Ap. J.* 394:581–85
Halpern JP, Grindlay JE, Tytler D. 1985. *Ap. J.* 296:190–96
Halpern JP, Holt SS. 1992. *Nature* 357:222–24
Halpern JP, Ruderman M. 1993. *Ap. J.* 415:286–97
Halpern JP, Tytler D. 1988. *Ap. J.* 330:201–17
Hardee PE. 1977. *Ap. J.* 216:873–80
Harding AD, Ozernoy LM, Usov VV. 1993. *MNRAS* 265:921–25
Hayakawa S. 1952. *Prog. Theor. Phys.* 8:571–80
Helfand DJ. 1994. *MNRAS* 267:490–500
Hermsen W, Kuiper L, Diehl R, Lichti G, Schönfelder W, et al. 1994. *Ap. J. Suppl.* 92:559–66
Hermsen W, Swanenburg BN, Buccheri R, Scarsi L, Sacco B, et al. 1992. *IAU Circ. No. 5541*
Julien PF, Helmken HF. 1978. *Nature* 272:699–701
Kanbach G, Bertsch DL, Favale A, Fichtel CE, Hartman RC, et al. 1988. *Space Sci. Rev.* 49:69–84
Kaspi VM, Manchester RN, Siegman B, Johnston S, Lyne AG. 1994. *Ap. J. Lett.* 422:L83–86
Kniffen DA, Hartman RC, Thompson DJ, Bignami GF, Fichtel CE, et al. 1974. *Nature* 251:397–98
Kniffen DA, Bignami GF, Fichtel CE, Hartman RC, Ogelman H, et al. 1975. *Int. Cosmic Ray Conf.* 1:100–5
Kulkarni SR, Djorgovski S. 1986. *Astron. J.* 91:98–106
Lamb RC, Worral DM. 1979. *Ap. J.* 231:L121–24
Lu T, Wei DM, Song LM. 1994. *Astron. Astrophys.* 290:815–17
Lyne AG, Pritchard RS, Smith FG. 1988 *MNRAS* 233:667–76
Maddox J. 1984. *Nature* 310:447
Manchester RN, Taylor JH. 1981. *Astron. J.* 86:1953–73
Mandolesi N, Morigi G, Sironi G. 1978. *Astron. Astrophys.* 67:L5–6
Maraschi L, Treves A. 1977a. *Astron. Astrophys.* 61:L11–13
Maraschi L, Treves A. 1977b. *Ap. J.* 218:L113–15
Masnou JL, Bennett K, Bignami GF, Buccheri R, Caraveo PA, et al. 1977. *Proc. 12th ESLAB Symp.* ESA SP 124:33–37
Masnou JL, Bennett K, Bignami GF, Bloemen JBGN, Buccheri R, et al. 1981. 17th ICRC Paris 1:177–80
Mason KO, Bignami G, Brinkman AC, Peacock A. 1995. *Adv. Space Res.* 16:41–50
Mattox JR, Bertsch DL, Fichtel CE, Hartman RC, Kniffen DA, Thompson DJ. 1992. *Ap. J.* 401:L23–26
Mattox JR, Halpern JP, Caraveo PA. 1996a. *Proc. 3rd Compton Symp. Astron. Astophys. Suppl.* In press
Mattox JR, Koh T, Lamb RC, Macomb DJ, Prince TA, Ray PS. 1996b. *Proc. 3rd Compton Symp. Astron. Astrophys. Suppl.* In press
Mayer-Hasselwander HA, Bertsch DL, Brazier KTS, Chaing J, Fichtel CE, et al. 1994. *Ap. J.* 421:276–83
Mayer-Hasselwander HA, Kanbach G, Sieber W. 1979. *Proc. 16th Int. Cosmic Ray Conf.* 1:206–9
Mereghetti S, Bignami GF, Caraveo PA. 1996. *Ap. J.* In press
Mereghetti S, Caraveo PA, Bignami GF. 1993. *Adv. Spce Res.* 13:343–46
Meyer ED, Pavlov GG, Meszaros P. 1994. *Ap. J.* 433:265–75
Middleditch J, Pennypacker CR, Burns MS. 1987. *Ap. J.* 315:142–48
Mignani R, Caraveo PA, Bignami GF. 1994. *The Messenger* 76:32–34
Moffat AFJ, Schlickeiser R, Shara MM, Sieber W, Tuffs R, Kühr H. 1983. *Ap. J.* 271:L45–48
Nulsen PEJ, Fabian AC. 1984. *Nature* 312:48–50
Ögelman H. 1995. See Alpar et al. 1995, pp. 101–20
Ögelman H, Finley JP. 1993. *Ap. J.* 413:L31–34
Ozel ME, Dickel RJ, Webber JC. 1980. *Nature* 285:645–47
Pacini F. 1971. *Ap. J. Lett.* 163:L17–19
Page D. 1993. Proc. of "First Symposium on Nuclear Physics in the Universe" IOP Publishing Ltd 151–62
Page D. 1994. *Ap. J.* 428:250–60
Page D. 1995. *Ap. J.* 442:273–85
Page D, Shibanov YuA, Zavlin VE. 1995. *Ap. J. Lett.* 451:L21–24
Pavlov GG, Shibanov YuA, Zavlin VE, Meyer RD. 1995. See Alpar et al 1995, pp. 71–90
Pavlov GG, Potekhin A. 1995. *Ap. J.* 450:883–95
Pavlov GG, Stringfellow GS, Cordova FA. 1996. *Ap. J.* In press
Petre R, Becker CM, Winkler PF. 1996. *Ap. J. Lett.* In press
Ramanamurthy PV, Bertsch DL, Fichtel CE, Kanbach G, Kniffen DA, et al. 1995a. *Ap. J.* 450:791–804
Romani RW. 1987. *Ap. J.* 313:718–26
Romani RW, Trimble V. 1985. *Nature* 318:230–31
Romani RW, Yadigaroglu IA. 1995. *Ap. J.*

438:314–21

Ruderman M. 1991. *Ap. J.* 382:576–86

Ruderman M, Cheng KS. 1988. *Ap. J.* 335:306–18

Ruderman MA, Sutherland PG. 1975. *Ap. J.* 196:51–72

Schlickeiser R. 1981. *Astron. Astrophys.* 94:57–60

Schroeder PC, Ulmer MP, Matz SM, Grabelsky DA, Rurcell WR, et al. 1995. *Ap. J.* 450:784–90

Seiradakis JH. 1981. *Astron. Astrophys.* 101:158

Shibanov YuA, Zavlin VE, Pavlov GG, Ventura J. 1992. *Astron. Astrophys.* 266:313–20

Shibanov YuA, Zavlin VE, Pavlov GG, Ventura J. 1995. See Alpar et al 1995, pp. 91–96

Sieber W, Schlickeiser R. 1982. *Astron. Astrophys.* 113:314–23

Simonson SC. 1975. *Ap. J. Lett.* 201:L103–8

Smith VV, Cunha K, Plez B. 1994. *Astron. Astrophys.* 281:L41–44

Sol H, Tarenghi M, Vanderriest C, Vigroux L, Lelievre G. 1985. *Astron. Astrophys.* 144:109–14

Spoelstra TATh, Hermsen W. 1984. *Astron. Astrophys.* 135:135–40

Stecker FW. 1970. *Astrophys. Space Sci.* 6:377–90

Sturmer SJ, Dermer CD. 1994. *Ap. J. Lett.* 420:L79–82

Swanenburg BN, Bennett K, Bignami GF, Buccheri R, Caraveo PA, et al. 1981. *Ap. J. Lett.* 243:L69–73

Thompson DJ, Bertsch DL, Dingus BL, Esposito JA, Etienne A, et al. 1995. *Ap. J. Suppl.* 101:259–86

Thompson DJ. 1994. "The Second Compton Symposium" ed. CE Fichtel, N Gohsels, JP Nouis. AIP Conference Proc. 304:57

Thompson DJ, Bignami GF, Fichtel CE, Kniffen DA. 1974. *Ap. J. Lett.* 190:L51–53

Thompson DJ, Fichtel CE, Hartman RC, Kniffen DA, Lamb RC. 1977. *Ap. J.* 213:252–62

Thompson DJ, Fichtel CE, Kniffen DA, Ögelman H. 1975. *Ap. J. Lett.* 200:L79–82

Trimble V. 1991. In *Gamma-Ray Bursts*, ed C Ho, RF Epstein, EE Fenimore, pp. 479–86. Cambridge Univ. Press

Usov VV. 1994. *Ap. J.* 216:873–80

Ventura J, Herod H, Kopidakis N. 1995. See Alpar et al 1995, pp. 97–100

Vishwanath PR, Sathyanarayana GP, Ramanamurthy PV, Bhat PN. 1993. *Astron. Astrophys.* 267:L5–7

Yadigaroglu IA, Romani RW. 1995. *Ap. J.* 449:211–15

Zynskin YuL, Mukanov DB. 1983. *Sov. Astron. Lett.* 9(2):117–18

Annu. Rev. Astron. Astrophys. 1996. 34:383–418

CHARGED DUST DYNAMICS IN THE SOLAR SYSTEM

Mihály Horányi

Laboratory for Atmospheric and Space Physics, University of Colorado, Boulder, Colorado 80309-0392

KEY WORDS: dust-plasma interactions, planetary rings, Moon

ABSTRACT

In most space environments, dust particles are exposed to plasmas and UV radiation and, consequently, carry electrostatic charges. Their motion is influenced by electric and magnetic fields in addition to gravity, drag, and radiation pressure. On the surface of the Moon, in planetary rings, or at comets, for example, electromagnetic forces can shape the spatial and size distribution of micron-sized charged dust particles. The dynamics of small charged dust particles can be surprisingly complex, leading to levitation, rapid transport, energization and ejection, capture, and the formation of new planetary rings.

This review briefly discusses the most important processes that determine the charge state of dust particles immersed in plasmas and the resulting dynamics on exposed dusty surfaces and in planetary magnetospheres.

1. INTRODUCTION

The study of dusty plasmas is an emerging new field that bridges traditionally separate subjects: celestial mechanics and plasma physics. Dust particles immersed in plasmas and UV radiation collect electrostatic charges and respond to electromagnetic forces in addition to all the other forces acting on uncharged grains. Simultaneously, they can alter their plasma environment. Dust particles in plasmas are unusual charge carriers. They are many orders of magnitude heavier then any other plasma particles, and they can have many orders of magnitude larger (negative or positive) time-dependent charges. Dust particles can

0066-4146/96/0915-0383$08.00

communicate nonelectromagnetic effects (gravity, drag, radiation pressure) to the plasma that can represent new free energy sources. Their presence can influence the collective plasma behavior, for example, by altering the wave modes and by triggering new instabilities. Dusty plasmas represent the most general form of space, laboratory, and industrial plasmas. Interplanetary space, comets, planetary rings, dusty surfaces in space, aerosols in the atmosphere, are all examples where electrons, ions, and dust particles coexist.

The present review cannot discuss all aspects of the rapidly expanding field of dusty plasmas. It is directed towards the recent or near future space observations related to the charging and dynamics of dust grains on exposed surfaces and in planetary magnetospheres. Several excellent reviews of other aspects of the physics of dusty plasmas are available in the literature (Grün et al 1984, Mendis et al 1984, Goertz 1989, De Angelis 1992, Hartquist et al 1992, Northrop 1992, Mendis & Rosenberg 1994). The large body of literature on the electromagnetic effects shaping the fine dust distribution around nonmagnetized bodies (e.g. comets, Mars) and in interplanetary space is also not covered here (see, e.g. Mendis & Horányi 1991, Juhász et al 1993, Leinert & Grün 1990).

In Section 2 we first discuss how a single isolated dust particle collects its electrostatic charge due to the most common charging currents in space plasmas: electron and ion bombardment and the production of secondary and photoelectrons. The charge on a small dust particle is of fundamental interest, for it allows for the coupling between the fields and particles environment to the dynamics of the dust grains. In Section 3 the interaction of dusty surfaces with plasmas is discussed. There is a renewed interest in the unresolved observational and theoretical problems of possible electrostatic dust transport on the lunar surface because of the planned observations by the *Rosetta* mission of a cometary nucleus at a large heliocentric distance, where the solar wind directly impinges on the surface, as occurs on the Moon. The dynamics of charged dust grains in planetary magnetospheres is discussed in Section 4. The inclusion of this topic is motivated by the recent *Ulysses* and *Galileo* observations at Jupiter, and the future *Cassini* mission to Saturn. In Section 5 we show that most space experiments comprising the "standard" payload of modern space missions are related to the study of the dynamics of charged dust particles.

Professor Hannes Alfvén (1908–1995) was one of the very first scientists to suggest that electromagnetic forces acting on small dust particles played an important role in the evolution of the entire Solar System. He recognized planetary rings, for example, as an important laboratory for studying processes that not only act today, but also shaped the evolution of the early Solar System billions of years ago (Alfvén 1954). The present review is dedicated to his memory.

2. CHARGING ISOLATED GRAINS IN PLASMAS

The evolution of the electrical charge Q of a dust grain in a plasma is described by the current balance equation:

$$\frac{dQ}{dt} = \sum_k J_k, \tag{1}$$

where J_k represent the charging currents. In most space plasmas electron and ion collection currents and secondary and photoelectron emission currents dominate. All of these currents are functions of the plasma properties (density, composition, energy distribution) and also the properties of the dust grain (size, velocity, composition, surface roughness). As a dust grain collects charges it changes the electrostatic potential distribution in its environment. If a grain initially collects more electrons than ions, the developing negative potential well around it will enhance the ion flux and lower the electron flux. The electrostatic charge on the grain that balances these fluxes is the equilibrium charge. The collisional mean free path of the plasma particles, λ, is generally larger than the characteristic size of the potential well, the Debye shielding distance λ_D; hence the distribution function of the ambient plasma can easily be related to the distribution function at the surface of the dust particle. Below we summarize the most commonly used expressions in space and astrophysical plasmas for the charging currents and give examples of the expected charge states of dust particles in various Solar System plasma environments.

2.1 *Electron and Ion Currents*

The flux of electrons and ions bombarding a dust grain with radius a, for the case $a \ll \lambda_D \ll \lambda$, using polar velocity space coordinates (v, θ, ψ), is

$$\begin{aligned} J_\alpha &= 4\pi a^2 \int_{v=v_\alpha^*}^{\infty} \int_{\theta=0}^{\pi/2} \int_{\psi=0}^{2\pi} v \cos\theta f_\alpha(v) v^2 \sin\theta d\theta d\psi dv \\ &= 4\pi a^2 \int_{E=\max(0,\pm e\phi)}^{\infty} \left[1 \pm \left(-\frac{e\phi}{E}\right)\right] \frac{dj_\alpha}{dE} dE, \end{aligned} \tag{2}$$

where α represents electrons or ions (possibly a number of different ions in a multicomponent plasma), ϕ is the surface potential of the grain, and v_α^* is to be chosen for each plasma species so that their energy, $E = (1/2)m_\alpha v_\alpha^{*2} - e\phi$, remains positive. Finally, $dj_\alpha/dE = (2\pi E/m_\alpha^2) f_\alpha(E)$ is the energy-differential particle flux, where m_α and $f_\alpha(E)$ are the mass and the distribution function of the ambient electrons or ions. For the most commonly used Maxwellian energy distribution corresponding to a temperature T_α, this particle flux is

$$\frac{dj_\alpha}{dE} = \frac{2\pi E}{m_\alpha^2} n_\alpha \left(\frac{m_\alpha}{2\pi k T_\alpha}\right)^{3/2} \exp\left(-\frac{E}{kT_\alpha}\right), \tag{3}$$

where n_α is the undisturbed electron or ion density. Substituting equation (3) into (2) results in

$$J_\alpha = J_{0\alpha} \times \begin{cases} \exp(-\chi_\alpha) & \text{if } \chi_\alpha \geq 0 \\ 1 - \chi_\alpha & \text{if } \chi_\alpha < 0, \end{cases} \tag{4}$$

where $\chi_\alpha = \mp e\phi / kT_\alpha$, $J_{0\alpha} = 4\pi a^2 n_\alpha (kT_\alpha / 2\pi m_\alpha)^{1/2}$, and the signs $-$ and $+$ correspond to electrons and ions, respectively.

In a case where the electron and ion thermal currents are the only charging currents, the equilibrium potential of a grain in a plasma with $T_e = T_i = T$ is (Spitzer 1941)

$$\phi_{\text{eq}} = -\beta \frac{kT}{e}, \tag{5}$$

where, for example, $\beta = 2.5$, 3.6, and 3.9 for H^+, O^+, and S^+ plasma, respectively. At this grain potential $J_e(\phi_{\text{eq}}) = J_i(\phi_{\text{eq}})$. For an isolated dust grain the surface potential ϕ can be easily related to the net charge via $\phi = Q/a$, which results in

$$Q_e \simeq 700 \phi_{\text{V}} a_\mu, \tag{6}$$

where Q_e is the number of extra/missing electrons, ϕ_{V} is the surface potential measured in volts, and a_μ is the radius of the grain in microns. In a 1 eV hydrogen plasma $\phi_{\text{eq}} = -2.5$ V, independent of a; thus a 1 μm radius particle will collect approximately 1800 extra electrons. Because of the statistical nature of the collection currents, charges fluctuate around their equilibrium value and the electron and ion fluxes balance only on average. The electrons and ions bombarding a grain that is in charge equilibrium might recombine on the surface and return to the plasma as neutral particles.

2.2 *Secondary Electron Currents*

Even at relatively low plasma temperatures, some of the bombarding particles can be energetic enough to ionize the material of the grain and produce secondary electrons. The escape flux, if any, of the secondary electrons represents a positive grain-charging current. The ratio of the emitted secondary electrons to incident ones is a function of the primary electron's energy and also the material and surface properties of the dust grains. It generally exhibits a maxima, δ_M, at an optimum incident energy, E_M, indicating that low-energy primary electrons will not produce secondaries and that energetic primaries penetrating the grain produce secondaries that cannot escape. The secondary electron yield is often approximated as (Sternglass 1954)

$$\delta(E) = 7.4 \delta_M (E/E_M) \exp[-2(E/E_M)^{1/2}], \tag{7}$$

and the energy distribution of the emitted electrons is usually described with a Maxwellian distribution with $kT_s \approx$ 1–5 eV. The secondary electron yield for small grains with sizes comparable to the range of the primary electrons can significantly differ from equation (7) (Chow et al 1993).

The flux of secondary electrons can be calculated by integrating the energy-differential flux of the primaries with the secondary electron yield. For positively charged dust particles, however, a secondary electron with insufficient energy will not escape, but fall back to the grain. The secondary electron flux is

$$J_{\text{sec}} = 4\pi a^2 \int_{E=-e\phi}^{\infty} \left(1 + \frac{e\phi}{E}\right) \frac{dj_{e,i}}{dE} \delta(E + e\phi) dE, \tag{8}$$

for $\phi < 0$ and

$$J_{\text{sec}} = 4\pi a^2 (1 - \chi_s) \exp(\chi_s) \int_{E=0}^{\infty} \left(1 + \frac{e\phi}{E}\right) \frac{dj_{e,i}}{dE} \delta(E + e\phi) dE, \tag{9}$$

for $\phi > 0$, where $\chi_s = -e\phi/kT_s$ and the terms in front of the integral on the right-hand side of Equation (9) represent the fraction of the Maxwell distributed secondary electrons with $E > e\phi$.

If we assume a Maxwellian distribution of the ambient plasma, the secondary electron flux can be calculated by substituting equations (3) and (7) into (8) and (9) (Meyer-Vernet 1982). For negatively charged grains ($\phi < 0$),

$$J_{\text{sec}} = 3.7\delta_M J_{0e} \exp(-\chi_e) F_5(E_M/4kT_e), \tag{10}$$

where

$$F_5(x) = x^2 \int_0^{\infty} u^5 e^{-(xu^2+u)} du. \tag{11}$$

For positively charged grains ($\phi > 0$),

$$J_{\text{sec}} = 3.7\delta_M J_{0e} (1 - \chi_s) \exp(\chi_s - \chi_e) F_{5,B}(E_M/4kT_e), \tag{12}$$

where

$$F_{5,B}(x) = x^2 \int_B^{\infty} u^5 \exp[-(xu^2 + u)] du$$

and

$$B = \left(\frac{-\chi_e}{E_M/4kT_e}\right)^{1/2}.$$

To find the equilibrium potential now, one has to solve for the current balance equation including the electron and proton thermal fluxes as well as the

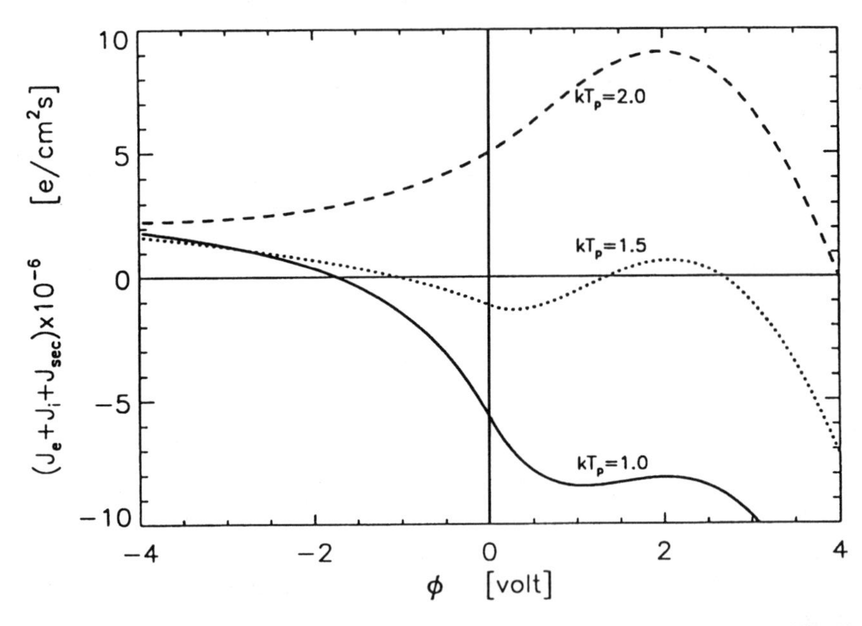

Figure 1 The total current-potential characteristics for an Al_2O_3 grain ($\delta_m = 9$, $E_M = 350$ eV, and $kT_s = 3$ eV) in a hydrogen plasma with $kT = 1$, 1.5, and 2 eV.

secondary electron flux. Figure 1 shows the characteristic total current versus surface potential for an Al_2O_3 grain in a hydrogen plasma. As a function of the plasma temperature the net current becomes zero at a negative or positive potential, or it may even have multiple roots. The roots where $dJ/d\phi < 0$ are stable; perturbations generate currents that drive the potential towards the equilibrium value. The middle solution of the multiple-root case is unstable. A small increase (decrease) in the ion current caused by a fluctuation, for example, results in a potential change that further enhances (lowers) it, and the potential evolves towards the positive (negative) stable equilibrium solution. However, grains with either one of the two stabile equilibrium potentials could possibly coexist in a plasma environment. The coexisting charge states in a thermal plasma with an additional monoenergetic beam of electrons were shown in laboratory experiments (Walch et al 1995, Horányi et al 1995).

2.3 *Photoelectron Currents*

Dust particles in space are often exposed to ultraviolet radiation—photons that are energetic enough to generate photoelectrons. The escaping flux of photoelectrons represents a positive current and is a function of the material properties of the dust particle. The photoelectron production due to the Sun's UV radiation is given by

$$J_\nu = \begin{cases} \pi a^2 e f & \text{if } \phi < 0 \\ \pi a^2 e f \exp(-e\phi/kT_\nu) & \text{if } \phi \geq 0, \end{cases} \tag{13}$$

where kT_ν ($\approx$ 1–3 eV) is the average energy of the generally assumed Maxwellian distribution of the photoelectrons, $f \approx 2.5 \times 10^{10}\kappa/d^2$ cm^{-2} s^{-1}, with the efficiency factor κ close to unity for conducting and close to 0.1 for dielectric materials, and d is the distance from the Sun measured in AU (Whipple 1981). The exponential factor for the $\phi > 0$ case means that photoelectrons with $E < e\phi$ will not escape but will fall back onto the grain.

2.4 *Ion Currents to a Moving Grain*

Frequently, the dust particles are not at rest in the plasma. For example, dust grains comprising planetary rings travel at approximately Keplerian speeds, immersed in a plasma that tends to corotate with the planet. In general the dust-to-plasma relative velocity is small compared to the electron thermal speeds, but it might become comparable to or even exceed the ion thermal velocities so that the ion flux becomes anisotropic. A Maxwellian distribution in the plasma frame translates into a drifting Maxwellian distribution in the frame fixed to the dust grain. In this case, for $\phi < 0$, the ion current becomes (Kanal 1962, Whipple 1981, Northrop & Birmingham 1995)

$$J_i = \frac{J_{oi}}{2}\left[\left(M^2 + \frac{1}{2} - \chi_i\right)\frac{\sqrt{\pi}}{M}\operatorname{erf}(M) + \exp(-M^2)\right], \tag{14}$$

where $M = w/(2kT_i/m_i)^{1/2}$ is the relative Mach number—the ratio of the dust-to-plasma relative velocity w over the ion thermal speed—and

$$\operatorname{erf}(x) = \frac{2}{\sqrt{\pi}}\int_0^x \exp(-y^2)dy$$

is the error function. For $\phi > 0$ the ion current is given as

$$\begin{aligned} J_i = \frac{J_{oi}}{2}\Bigg\{&\left(M^2 + \frac{1}{2} - \chi_i\right)\frac{\sqrt{\pi}}{M}[\operatorname{erf}(M + \sqrt{\chi_i}) + \operatorname{erf}(M - \sqrt{\chi_i})] \\ &+\left(\sqrt{\frac{\chi_i}{M}} + 1\right)\exp[-(M - \sqrt{\chi_i})^2] \\ &-\left(\sqrt{\frac{\chi_i}{M}} - 1\right)\exp[-(M + \sqrt{\chi_i})^2]\Bigg\}. \end{aligned} \tag{15}$$

In the limit of $M \to 0$ these expressions become identical to equation (4), and as $M \to \infty$ they become proportional to M and describe a sweep-up type ion collection.

In most astrophysical applications these (electron and ion thermal currents and secondary and photoelectron production) are the dominant charging processes. However, the plasma distributions are often non-Maxwellian (Rosenberg & Mendis 1992) and so the integrals in equations (2), (8), and (9) are often replaced by a summation over an appropriate bin structure describing the energy distribution of the plasma. Often grains traverse regions where the plasma parameters change rapidly and the grains do not accommodate these fast enough to assume their equilibrium potential.

Electrostatic charging can change the physical characteristics of the dust particles. For example, the electrostatic tension might overcome the tensile strength of a grain and result in the chipping off of protruding fingers or even full disruption (Opik 1956, Hill & Mendis 1980a). The growth of particles via nucleation or coagulation is generally handicapped between grains or droplets with like charges due to Coulomb repulsion. However, due to fluctuations in the plasma parameters or size- or composition-dependent secondary and photoelectron yields, small grains can end up oppositely charged from the bigger ones, resulting in greatly enhanced growth rates (Feuerbacher et al 1973, De 1979, Horányi & Goertz 1990, Chow et al 1993).

The presence of dust can alter the properties of any plasma environment acting as a source or sink for its density, momentum, and energy. In a low-density region, for example, photo and secondary electron production might become an important plasma source. Similarly, energetic enough plasma particles can sputter off atoms and molecules from dust grains, changing the density as well as the composition of the plasma environment (Johnson 1990).

In fact, if the presence of dust significantly alters the plasma density, then the charging equations for a single dust particle discussed above are no longer valid. In the case of significant plasma depletion ($Qn_{\mathrm{d}} \simeq en_{\mathrm{p}}$, where n_{d} is the density of the dust grains with a mono-dispersed size distribution for simplicity, Q is the charge calculated for a single grain, and n_{p} is the plasma density), the charge of a dust particle in a cloud of other dust particles becomes reduced (Goertz & Ip 1984; Whipple et al 1985; Havnes et al 1984, 1987, 1990). The theoretical predictions of this effect were also demonstrated in laboratory experiments (Xu et al 1993). Generally, when the characteristic Debye shielding distance in a plasma is comparable or smaller than the average intergrain distance ($\lambda_{\mathrm{D}} \leq n_{\mathrm{d}}^{-1/3}$), the dust particles can not be treated any longer as test particles, and the ensemble of electrons, ions, and the dust becomes a real dusty plasma that exhibits unusual wave modes and instabilities (Mendis & Rosenberg 1994). In this review we restrict our discussions to the dynamics of charged particles in various Solar System environments where collective effects are not important.

2.5 *Examples*

Our first example is an interplanetary dust particle at 1 AU from the Sun, exposed to solar UV radiation and also the solar wind plasma with $n \simeq 5\ \text{cm}^{-3}$ and $kT_e \simeq kT_i \simeq 10$ eV. The average solar wind speed $w = 400\ \text{km s}^{-1}$, which is supersonic compared to the proton thermal velocity ($M \sim 10$). Substituting equations (4), (12), (13), and (15) into (1) we can follow the charging history of a dust particle. Figure 2 shows the history of the currents for an initially uncharged micron-sized ($a_\mu = 1$) dielectric dust particle with photoelectron yield $\kappa = 0.1$ and secondary electron production coefficients $\delta_M = 1$, $E_M = 300$ eV, and $kT_s = 2.5$ eV. In this environment photoelectron production dominates charging. As the grain charges more and more positive, the photoelectron current drops because fewer and fewer of the produced electrons escape. Simultaneously the electron current increases. In about 200 s the sum of all the currents approaches zero and the dust grain reaches charge equilibrium. The time needed to reach it, unlike the equilibrium potential itself, does depend on the size of the dust particle. From equation (6), the equilibrium charge $Q_{\text{eq}} \sim a$, and all the currents are proportional to the surface area of a grain, $J \sim a^2$. The

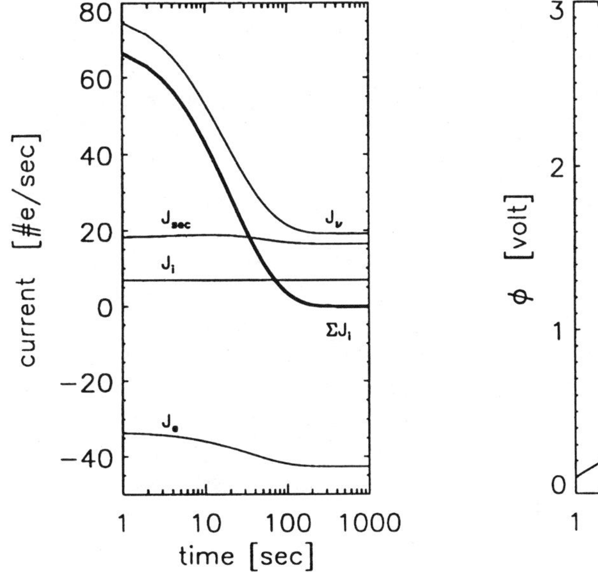

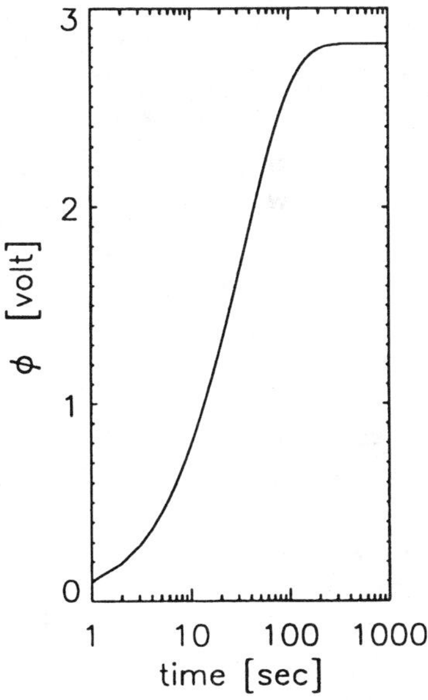

Figure 2 The history of the electron and ion currents, J_e, J_i; secondary and photoelectron currents J_{sec}, J_ν, and their sum $\sum J_k$ (*left panel*); and the electrostatic surface potential (*right panel*) of an initially uncharged micron-sized ($a = 1\ \mu$m) dielectric dust particle orbiting at 1 AU from the Sun.

characteristic charging time $\tau \sim Q_{eq}/\Sigma J \sim a^{-1}$. Bigger grains reach their equilibrium potential faster.

The magnetosphere of the Earth can be characterized as a high-energy, low-density plasma environment. For the average condition $n_e = n_{H^+} \approx 1$ cm^{-3}, $kT_{H^+} = 5$, and $kT_e = 2$ keV. The Earth does not have rings (yet!), but during solid rocket propellant burns a large number of submicron-sized Al_2O_3 spherules are dumped into the magnetosphere. The characteristic surface potential of these particles is on the order of +10 V, due to the dominating secondary electron fluxes (Horányi et al 1988).

The most prominent feature of the Jovian plasma environment is the plasma torus maintained by the strong production of oxygen and sulfur atoms from the volcanos on the moon Io (located at $r \approx 6$ R_J, where $R_J = 7.1 \times 10^4$ km is the radius of Jupiter). The plasma density peaks close to the orbit of Io with $n \approx 3000$ cm^{-3}, and the plasma temperature has a strong minima $kT_e = kT_i \approx 1$ eV at $r = 5$ R_J and rises sharply to $kT_e \approx 50$ and $kT_i \approx 80$ eV at $r = 8$ R_J (Bagenal 1995). The characteristic surface potential in this region varies in the range of $-30 \leq \phi \leq +3$ V. The change in the sign of ϕ is mainly due to changes in T_e, which controls the relative contribution of the secondary electron current (Figure 3*a*).

In the vicinity of the moon Enceladus (at $r \approx 4$ R_S, where $R_S = 6 \times 10^5$ km is the radius of Saturn) Saturn's plasma environment is characterized by a bi-Maxwellian electron distribution ("cold" and "hot" electron populations) with $n_e^{cold} \approx 100$ cm^{-3}, $kT_e^{cold} \approx 5$ eV, $n_e^{hot} \approx 0.2$ cm^{-3}, and $kT_e^{hot} = 100$ eV. The ion population consists of H^+ and O^+ with $n_H \approx 10$ cm^{-3}, $kT_H \approx 10$ eV, $n_O \approx 60$ cm^{-3}, and $kT_O \approx 50$ eV (Richardson & Sittler 1990). Here the expected surface potential of a dust particle is in the range of $-8 \leq \phi \leq -4$ V (Horányi et al 1992, Jurac et al 1995) (Figure 3*b*).

Generally, the distribution of astrophysical plasmas is not uniform. The plasma composition, density, and temperature might all exhibit spatial and temporal variations. Consequently, grains can have complicated charging histories. Their charge will not only depend on the instantaneous plasma environment, but also on the previous charge states. In all planetary magnetospheres, for example, dust grains will collect only a modest charge and the resulting electrodynamic forces acting on centimeter-sized or bigger grains are negligible compared to gravity. However, toward the lower end of the mass distribution ($a \leq$ few microns) the Lorentz force might become the most important perturbation and result in significant deviation from Kepler orbits. This size range of dust particles is frequently represented in planetary rings: Saturn's E-ring, the Jovian ring, and Neptune's arcs are examples where a significant portion of the optical depth is attributed to micron-sized grains. Electrodynamic perturbations

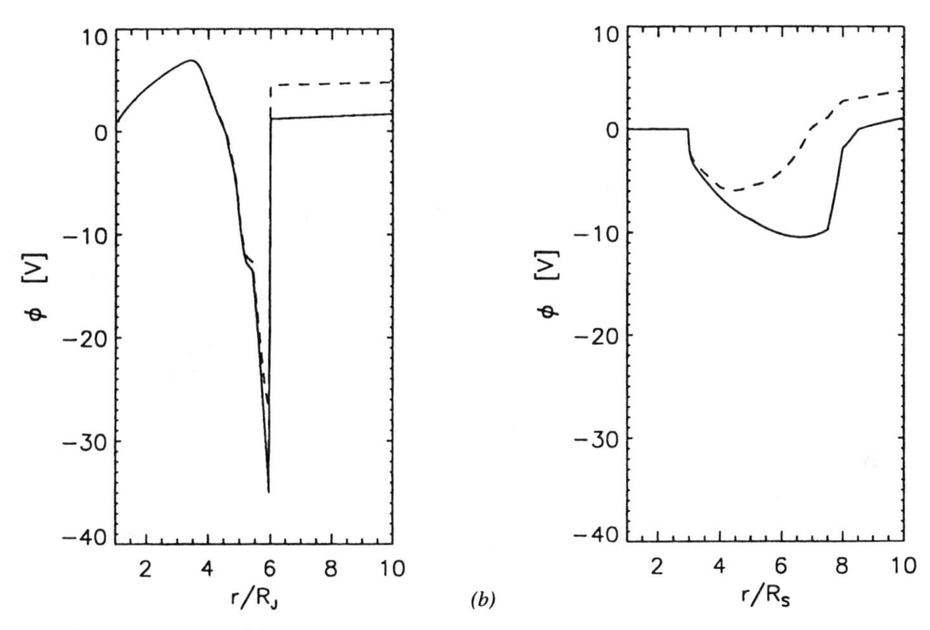

Figure 3 (*a*) The equilibrium surface potential of dielectric dust particles moving in an unperturbed circular Kepler orbit in the equatorial plane of Jupiter as function of distance from the planet. We assumed a secondary electron production parameter of $E_M = 500$; $\delta_m = 1$ (*continuous lines*) and 2 (*dashed lines*). The radial dependence of the equilibrium potential primarily reflects the variations in the plasma parameters. (*b*) Same as in (*a*), but for the case of Saturn.

often couple with other perturbations (oblateness, radiation pressure, plasma and neutral drag, etc) and can lead to unusual dynamics: transport, capture, and ejection.

3. PLASMA INTERACTIONS WITH DUSTY SURFACES

The surfaces of airless, nonmagnetized bodies in our Solar System (e.g. the Moon and asteroids, or comets at large heliocentric distances) are directly exposed to the solar wind plasma and UV radiation. The illuminated side of these objects is likely to have, on average, a few volts positive surface potential due to the dominating photoelectron flux. The night side, however, could be charged to $\simeq -1$ kV, which is the potential needed to balance the thermal currents of subsonic electrons and the supersonic flow of ions (Mendis et al 1981). We can expect large local deviations from the average surface potential caused by topography and/or compositional differences, resulting in strong local electric fields. These fields might be strong enough to overcome surface forces (adhesion and cohesion) and gravity for small charged dust particles, possibly resulting in electrostatic dust transport or levitated dust clouds.

The possibility of electrostatic dust transport on the Moon was envisioned by Gold (1955). He argued that the general appearance of large boulders—which were partially embedded in the soil, but which had their protruding parts virtually free from soil—was evidence for surface transportation of the lunar dust. The first observational evidence for electrostatic processes acting on the lunar surface was the horizon glow observed by the television systems on board the *Surveyor 5, 6,* and *7* spacecrafts. During each of these missions a line of light along the western lunar horizon was observed following the local sunset (Rennilson & Criswell 1973). The probable cause of this horizon glow is forward-scattered sunlight by a cloud of dust particles with radii $a \leq 10\ \mu$m and optical depth $\tau \sim 10^{-6}$ extending 10–30 cm in both vertical and horizontal direction along the camera's line of sight just above the sunlight/shadow boundaries in the terminator zone. The time evolution showed a monotonic decrease in brightness after sunset with a duration of 0.5–3 h (O'Keefe et al 1968). The detailed analysis of the luminance, temporal decay, and morphology prompted electrostatic levitation models (Criswell 1973, Rennilson & Criswell 1973). An independent set of observations probably related to dust levitation phenomena is the description of the visual observations of the *Apollo 17* crew during sunrise as seen from lunar orbit. They reported the appearance of bright streamers with fast temporal brightness changes (minutes as well as seconds) extending in excess of 100 km above the lunar surface. McCoy & Criswell (1974) argued for the existence of a significant population of lunar dust grains scattering the solar light. The rough estimates indicated that the scatterers are submicron-sized ($\simeq 0.1\ \mu$m) grains. These drawings were recently reanalyzed by Zook & McCoy (1991), who verified most of the earlier conclusions. This new study also yielded an estimate for the scale height of this "dusty-exosphere" of $H \simeq 10$ km, and the authors suggested that dust levitation could be observed using ground-based telescopes.

The most direct in situ observation was made by the Lunar Ejecta and Meteorites Experiment (LEAM) deployed by the *Apollo 17* astronauts in the Taurus-Littrow area of the Moon. This instrument was equipped with three omnidirectional sensor systems (east, west, and vertical directions), each gathering information on the impactor's velocity, energy, and angular distribution (Berg et al 1973). The original objective was the measurement of the impact parameters of cosmic dust on the lunar surface, but analyzing the early data soon revealed that the majority of the registered events were associated with lunar soil transport (Berg et al 1974). The data showed a 100-fold increase in the event rate registered by the east and the vertical sensors in conjunction with the passage of the sunrise terminator (Figure 4). The onset of this enhancement preceded the passage of the sunlit/shadow boundary by 3–40 h, but consistently vanished within 30 h following sunrise. The sunset terminator passage also triggered increased event rates, though much less consistently (Berg et al 1976). The emerging global picture suggests that micron-sized lunar dust particles tend to migrate away from the sunlit hemisphere of the Moon, but because the sunrise/sunset terminator fluxes are not equal, this process might result in large-scale organized dust transport.

Recently, several images taken of the lunar limb by the star-tracker camera of the *Clementine* spacecraft showed a faint glow along the lunar surface, stunningly similar to the sketches drawn by the *Apollo 17* astronauts (H Zook 1994, personal communication). The interpretation of these images is complicated by the presence of the scattered light from zodiacal dust particles; however, the spatial distribution and the size of the zodiacal dust particles are relatively well known, and their contribution can be subtracted by using images of various phase-angles. The preliminary analysis of the images indicates that *Clementine* also recorded the lunar horizon glow (H Zook 1995, personal communication).

Though the available theoretical models do not explain the observed phenomena (Singer & Walker 1962, Criswell 1973, De & Criswell 1977, Criswell & De 1977), they all indicate electrostatic processes. The simplest scenario to demonstrate the physics of dust levitation is the case of a flat surface that is

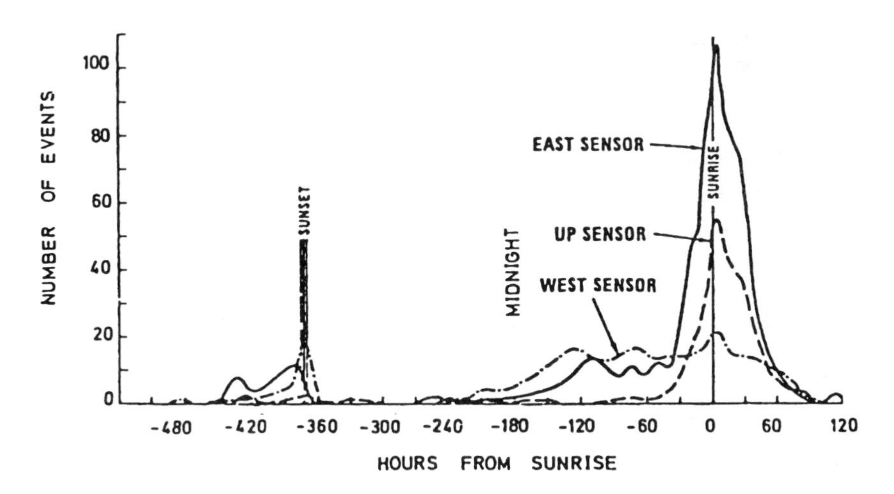

Figure 4 Number of dust impacts onto the Lunar Ejecta and Meteorite (LEAM) experiment sensors per 3-hr period, integrated over 22 lunar days as function of the local time. (From Berg et al 1976.)

exposed to plasma. An initially uncharged exposed surface will collect electrostatic charge from the plasma to reduce the electron and enhance the ion flux in order to achieve charge equilibrium. The potential distribution in the plasma sheath above the surface is the solution of Poisson's equation

$$\frac{d^2U}{dx^2} = 4\pi(n_e - n_i) = 4\pi n\left[\exp\left(\frac{eU}{kT}\right) - \exp\left(-\frac{eU}{kT}\right)\right]. \tag{16}$$

The solution of equation (16) is the one-dimensional Debye potential:

$$U(x) = U_o \exp\left(\frac{-x}{\lambda_D}\right), \tag{17}$$

where $\lambda_D = (kT/4\pi ne^2)^{0.5}$ is the Debye shielding distance and $U_o = -\beta kT/e$, according to equation (5). An example of the electron and ion densities as well as the plasma potential in the sheath is shown in Figure 5*a*.

A small dust grain, with radius a above the surface, also collects charges. At any distance from the surface its equilibrium potential ϕ is the solution to the transcendental equations resulting from Equation (4)

$$\frac{n_e}{n_i}\sqrt{\frac{m_i}{m_e}}\left(1 + \frac{e\phi}{kT}\right) = \exp\left(-\frac{e\phi}{kT}\right) \qquad \text{for} \quad \phi \geq 0$$

$$\frac{n_e}{n_i}\sqrt{\frac{m_i}{m_e}}\exp\left(\frac{e\phi}{kT}\right) = \left(1 - \frac{e\phi}{kT}\right) \qquad \text{for} \quad \phi < 0. \tag{18}$$

The electrostatic force on the grain is

$$F_E = -Q\nabla U = \frac{a\phi U}{\lambda_D}. \tag{19}$$

Figure 5*b* shows the grain potential and the resulting electrostatic force. The electrostatic force close to the surface ($x \leq 0.3\lambda_D$) points downward, since the dust particles and the surface are oppositely charged. At larger distances the force turns repulsive; it shows a maxima at a distance of $\sim \lambda_D$. This simple model demonstrates only the most basic elements of the physics of dust levitation. In reality, surfaces are three dimensional, the plasma environment is more complex, and the grains charge can be a function of time as well as location. Also, the presence of charged grains can significantly alter the potential distribution in the plasma sheath. As dust particles accumulate in a levitated layer, equation (16) has to be solved self-consistently with terms to account for the dust charge density (Nitter & Havnes 1992, Nitter et al 1994).

Clearly, more theoretical work is needed to understand the observations of electrostatic dust transport on the Moon. This is particularly important in view of current plans to renew lunar exploration, which will culminate with

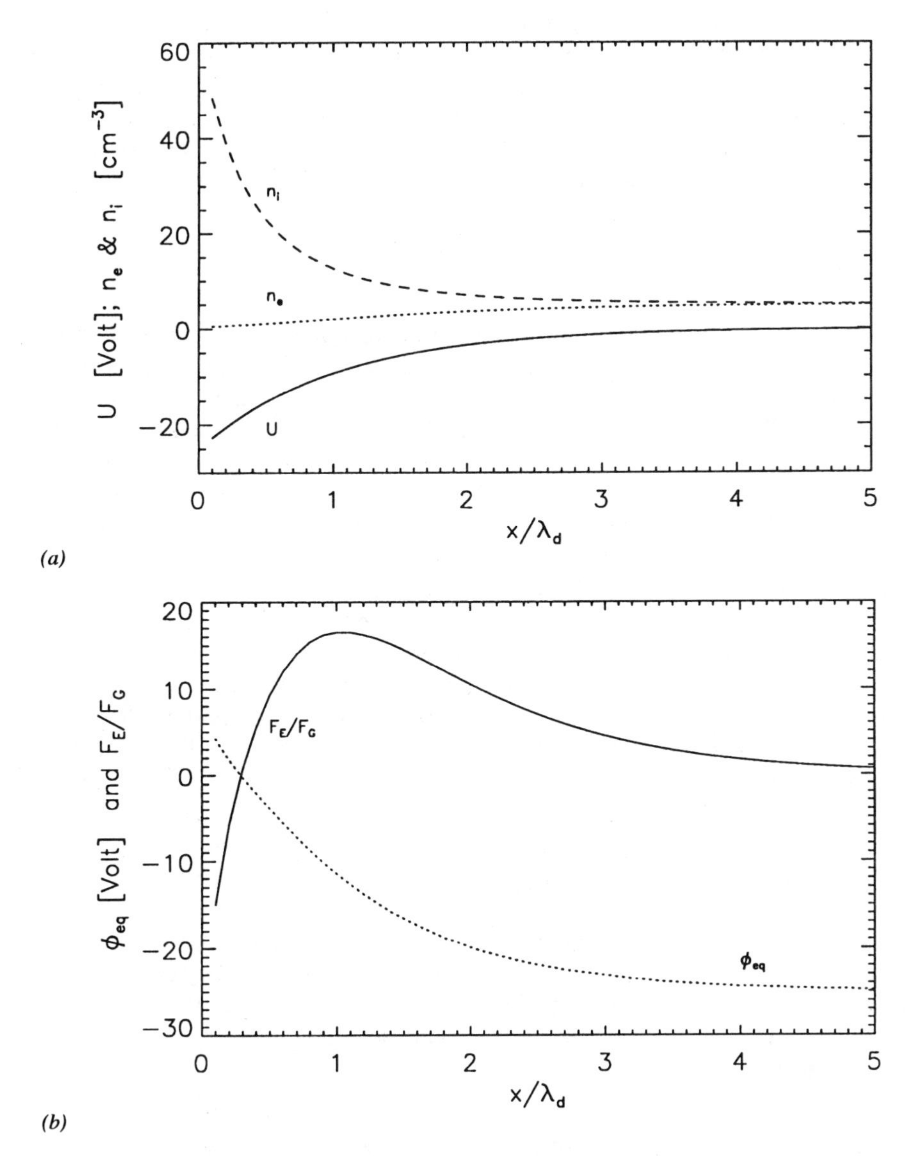

Figure 5 (*a*) The plasma potential U, the electron density n_e, and ion density n_i as functions of distance in the sheath above a flat surface exposed to H^+ plasma with $kT_e = kT_i = 10$ eV and density $n = 5$ cm^{-3}. The Debye length in this case is $\lambda_D \simeq 10^3$ cm. (*b*) The equilibrium surface potential ϕ_{eq} and the ratio of electrostatic to gravitational forces for a 1-μm-radius dust particle on the surface of the Moon.

the establishment of lunar optical observatories in the early part of the next century. Future plans for the *Rosetta* mission include the observations of a cometary nucleus at large heliocentric distances where the solar wind and its effects on the surface will be monitored. Finally, we point out that similar issues are now discussed in the semiconductor manufacturing industry. Dust particles collecting in the sheath above the silicon wafers in plasma etching devices can lead to significant production losses (Selwyn 1989).

4. DUST IN PLANETARY MAGNETOSPHERES

One of the most unexpected findings of the *Voyager* missions at Saturn was the periodically appearing, radially expanding dust clouds above the dense B-ring (Smith et al 1982). These "spokes" were perhaps the first to show that, contrary to general belief, gravity alone can not explain all of the complex dynamics in planetary rings. Successful models for the formation of the spokes recognized the importance of electromagnetic forces acting on small charged dust grains (Goertz & Morfill 1983, Morfill et al 1983, Hill & Mendis 1982, Shan & Goertz 1991).

There are many exciting phenomena associated with the interaction of magnetospheric fields and plasmas with the embedded dust grains. Lorentz resonances (Schaffer & Burns 1987), gyrophase-drifts due to compositional and/or plasma density and/or plasma temperature gradients (Northrop et al 1989), transport due to charge (Morfill et al 1980a,b,c,d) or magnetic field fluctuations (Consolmagno 1980), shadow resonance (Horányi & Burns 1991) and the coupling between radiation pressure and electrodynamic forces (Horányi et al 1992), for example, might all contribute to shaping the fine dust distribution in planetary rings. The ring-halo transition at Jupiter or the structure of Saturn's E-ring are recent examples in which the observed radial and vertical structure clearly demonstrates the effect of magnetospheric perturbations. The dust becomes an integral component of the magnetosphere since it acts as a source and sink of the plasma. The produced low-energy photo and secondary electrons or the sputtered-off ions might significantly alter the magnetospheric plasma distribution. Though many of these processes are now recognized, dusty planetary magnetospheres still hold surprises.

4.1 *Dynamics*

The equation of motion of a charged dust grain (of mass m and charge Q), as written in Gaussian units in an inertial coordinate system fixed to the planet's center, is

$$\ddot{\mathbf{r}} = \frac{Q}{m}\left(\frac{\dot{\mathbf{r}}}{c} \times \mathbf{B} + \mathbf{E}_c\right) - \frac{\mu}{r^3}\mathbf{r}, \tag{20}$$

where $\mathbf{r}$ is the grain's position vector, c is the speed of light, $\mathbf{B}$ is the magnetic field, and μ equals the gravitational constant times the planet's mass. For an infinite-conductivity magnetosphere that rigidly corotates with the planet with a rotation rate of Ω, $\mathbf{E_c} = (\mathbf{r} \times \Omega) \times \mathbf{B}/c$ is the corotational electric field. We have neglected the planet's oblateness, as well as the forces due to radiation pressure and the plasma and neutral drags.

To make a connection to the familiar Kepler problem, let us assume that the magnetic field is that of a simple dipole located at the center with its magnetic moment aligned with the rotation axis of the planet. This is a reasonable first approximation for the magnetic fields at Jupiter (J) and Saturn (S). Now, in the equatorial plane $B = B_o(R/r)^3$, where B_o is the magnetic field at the surface and R is the radius of the planet ($B_o^J = 4.2$ G, $B_o^S = 0.21$ G, and $R_J = 7.1 \times 10^4$ km, $R_S = 6 \times 10^4$ km). The magnetic field lines pierce the equatorial plane at right angles pointing antiparallel to Ω. The resulting corotating electric field points radially outward with an amplitude of $E_c = E_o(R/r)^2$ ($E_o^J = 5.8 \times 10^{-5}$ V m^{-1}, $E_o^S = 2.4 \times 10^{-6}$ V m^{-1}). Naturally, the force acting on dust particles associated with the corotational electric field also depends on their charge, $F_E = QE_c$. The charge in turn, as discussed above, is a function of the plasma environment, the material properties of the grain, the charging history, relative velocity, etc. Generally, the grain's charge can have a complicated history, but for the moment let us assume that it remains a constant.

If we assume an average density of $\rho = 1$ g cm^{-3} then the ratio of the electrostatic force to gravity is

$$F_E/F_G = \alpha\phi_v a_\mu^{-2} \tag{21}$$

($\alpha^J = 5.7 \times 10^{-3}$, $\alpha^S = 5.3 \times 10^{-4}$). In the typical range of $-50 \leq \phi_v \leq 10$, for particles with $a_\mu \gg 1$, electrostatic forces represent a perturbation only and the grains follow approximate Kepler orbits. In contrast, particles with $a_\mu \ll 1$ can be dominated by electromagnetic forces and gravity becomes a perturbation.

Let us return to Equation (20) and rewrite it in the equatorial plane of a planet using polar coordinates:

$$\ddot{r} = r\dot{\varphi}^2 + \frac{q}{r^2}(\dot{\varphi} - \Omega) - \frac{\mu}{r^2} \tag{22}$$

$$\ddot{\varphi} = -\frac{\dot{r}}{r}\left(\frac{q}{r^3} + 2\dot{\varphi}\right), \tag{23}$$

where we have introduced $q \equiv QB_oR^3/(mc)$, so that the combination $q/r^3 = \omega_g$ becomes the local gyrofrequency (the angular rate that dust particles circle about magnetic field lines).

On a circular equilibrium orbit, where the the sum of the radial components of all the forces is zero ($\ddot{\varphi} = \ddot{r} = \dot{r} = 0$ and $\dot{\varphi} = \text{constant} = \psi$), Equation (22) yields an algebraic equation for the angular velocity ψ,

$$\psi^2 + \omega_g \psi - \omega_g \Omega - \omega_k^2 = 0, \tag{24}$$

where $\omega_k = (\mu/r^3)^{1/2}$ is the Kepler angular velocity. For big particles, terms that contain ω_g can be dropped and we recover $\psi = \pm\omega_k$. For very small particles, terms that are not multiplied with ω_g are to be dropped and $\psi = \Omega$. Very small grains are picked up by the magnetic field and corotate with the planet.

Equations (22) and (23) can be integrated to yield constants of the motion

$$\mathcal{E} = \frac{1}{2}(\dot{r}^2 + r^2\dot{\varphi}^2) - \frac{\mu + q\Omega}{r} \tag{25}$$

$$J = r^2\dot{\varphi} - \frac{q}{r}. \tag{26}$$

For large particles ($q \to 0$) these constants become the Kepler energy and angular momentum. The Jacobi constant, $H = \mathcal{E} - \Omega J$, remains a constant even if Q changes with time (Northrop & Hill 1983, Schaffer & Burns 1994).

The right-hand side of Equation (22) can be written solely as a function of r using Equation (26) to replace $\dot{\varphi}$, and we can express $\ddot{r} = -(\partial U/\partial r)$, where the effective potential is

$$U(r) = -\frac{\mu + q\Omega}{r} + \frac{J^2}{2r^2} + \frac{qJ}{r^3} + \frac{q^2}{2r^4}. \tag{27}$$

The equilibrium orbit with a given J can be found from here by solving the equation $(\partial U/\partial r) = 0$. For any r, the initial conditions $\dot{r} = 0$ and $\dot{\varphi} = \psi$ [that is the solution to equation (24)] satisfy $(\partial U/\partial r) = 0$.

Small grains are constantly generated by micrometeoroid bombardment of the moons, or in the case of Jupiter, probably also by the volcanoes on Io. Generally, their escape velocity is small so that $\dot{\varphi}(t = 0) = \omega_k$. For the initial Kepler orbit, $J = r^2(\omega_k - \omega_g)$. Figure 6 shows $U(r)$ for particles originating from Io ($r_o = 5.9\ R_J$) at Jupiter for the two typical values (-30, $+3$) for ϕ_v (Horányi et al 1993a,b). Particles with negative surface potentials remain confined in the vicinity of r_o. However, grains in a certain size range with positive charges are not confined [$U(r)$ shows no minima]. What sets this size range? In the case of positively charged grains the force due to the corotational electric field points radially out, opposing gravity. The upper limit in size for ejection, $a_\mu^{\max}$, is set by the condition $F_E/F_G > 1$.

The lower limit in dust size for ejection results from the fact that very small grains behave like ions or electrons circling magnetic field lines. The radius

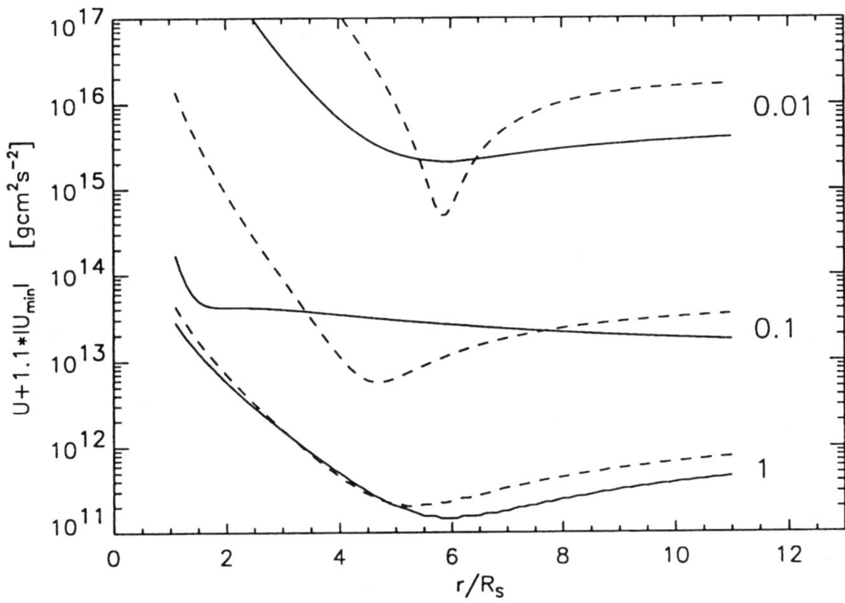

Figure 6 The effective potential for dust grains with $a_\mu = 0.01$, 0.1, and 1 started from Io on circular Kepler orbits with $\phi_V = -30$ (*dashed lines*) and +3 (*continuous lines*). To avoid the overlap of these curves, since only their shape is important, we have shifted them apart by plotting $U + 1.1\,|\text{minimum}\,(U)|$ instead of U itself.

of their trajectory is the gyroradius $r_g = |wmc/(QB)| = |w/\omega_g|$, where w is the relative velocity between the corotating magnetic fields and the particle. For Kepler initial conditions, $w = r(\Omega - \omega_k)$. The motion of these grains is well described by the guiding center approximation if the size of their orbit is smaller than the characteristic length scale for variations in the magnetic fields, $|r_g \nabla B/B| < 0.1$ ($|\nabla B/B| = 3/r$ in the equatorial plane of an aligned centered dipole). The upper limit of grain size satisfying this condition (i.e. the smallest grains that will be ejected) is

$$a_\mu^* = \left(\frac{10^{-3} B_o R^3 \phi}{4\pi r^2 \rho w c}\right)^{1/2}. \tag{28}$$

Grains in the range $a_\mu^* < a_\mu < a_\mu^{\max}$ will be ejected from the magnetosphere. As these positively charged grains move outward they gain energy from the corotational electric field

$$W = \int_{r_o}^{r_1} E Q dr = E_o R^2 Q \left(\frac{1}{r_o} - \frac{1}{r_1}\right), \tag{29}$$

where the upper limit of the integration, r_1, is the characteristic size of the magnetosphere. This mechanism was suggested to explain the recent *Ulysses* observations of small dust grains streaming away from Jupiter.

THE ULYSSES DUST STREAMS Let us first summarize the observations (Grün et al 1993). The mass of detected stream particles is $1.6 \times 10^{-16} < m < 1.1 \times 10^{-14}$ g and their velocity is $20 < v < 56$ km s^{-1}. If we assume an average density of 1 g cm^{-3}, the size range of these particles is $0.03 < a < 0.14$ μm. The streams were spaced approximately 500 R_J apart and they seem to be timed with a period of 28 ± 3 days (Figure 7*a*). These grains originate from within the Jovian system, suggesting the volcanos on Io (Horányi et al 1993a,b) or the dusty regions of the Jovian magnetosphere (Hamilton & Burns 1993) as a source of the escaping dust flux. Perhaps, Io offers the most natural way to explain the observed periodicities.

From equation (29), as positively charged particles escaping from Io move outward they gain energy from the corotating electric field

$$W = \int_{6R_J}^{50R_J} EQdr \simeq 0.2a_\mu \text{ erg},$$

where the lower limit of the integral is the location of Io and the upper limit is the approximate size of Jupiter's magnetosphere. If we ignore the initial gravitational binding energy, then the work done by the electric field is the kinetic energy of the ejected dust particle; their exit velocity at the outer boundary of the magnetosphere can be estimated as

$$v_{\text{exit}} \simeq \frac{3}{a_\mu} \text{km s}^{-1}. \tag{30}$$

From equations (21), (28), and (30), we anticipate that particles in the size range of $0.03 < a_\mu < 0.1$ to leave the Jovian magnetosphere with velocities in the range of $100 > v_{\text{exit}} > 30$ km s^{-1}. This simple size-velocity relationship is further complicated by the tilted nature of the magnetic field, but it holds within $\pm$ 10% in computer simulations, in agreement with the observations (Horányi et al 1993a,b). Figure 8 shows the "dusty ballerina skirt" comprised of small dust particles from Io streaming away from Jupiter.

The *Galileo* spacecraft is now in orbit around Jupiter. It carries a similar dust detector to the one on *Ulysses*. The dust impact rates prior to its arrival are shown in Figure 7*b*. These intermittent dust fluxes are very different from the *Ulysses* streams. The rates, in general, are higher, and there appears to be no clear timing between the streams. Although the approach geometry and hence the dust detector's view angle is very different between the two spacecrafts, it seems that *Galileo* approaches a very different dust environment around Jupiter

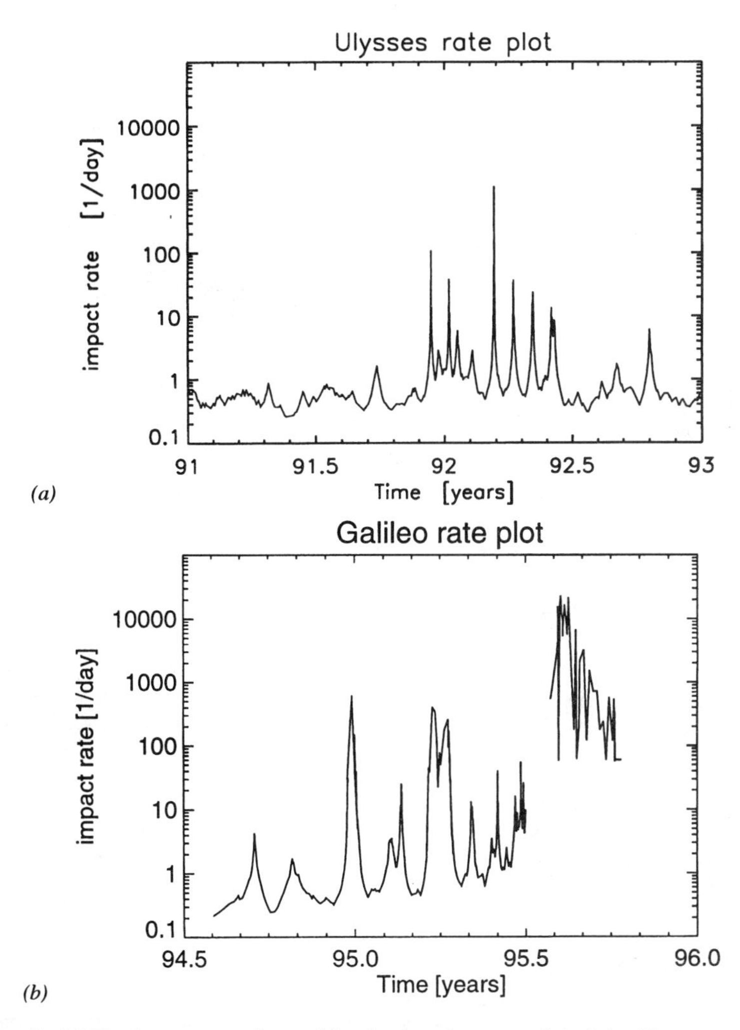

Figure 7 (*a*) The impact rates observed by the dust detector on board the *Ulysses* spacecraft during its encounter with Jupiter. (From Grün et al 1993.) *Ulysses*' relative velocity with respect to Jupiter during this period was $\simeq 14$ km s^{-1}; it traveled approximately a distance of 6300 R_J in a year. It flew by at a closest distance of $\simeq 6$ R_J to Jupiter on February 8, 1992. (*b*) The impact rates observed by the dust detector on board the *Galileo* spacecraft as it is approaching Jupiter. The gap in the data is due to the release of the atmospheric probe. *Galileo*'s relative velocity with respect to Jupiter is $\simeq 5$ km s^{-1}; it travels approximately a distance of 2200 R_J in a year. It arrived at Jupiter on December 7, 1995. (From the *Galileo* Dust Team, unpublished data.)

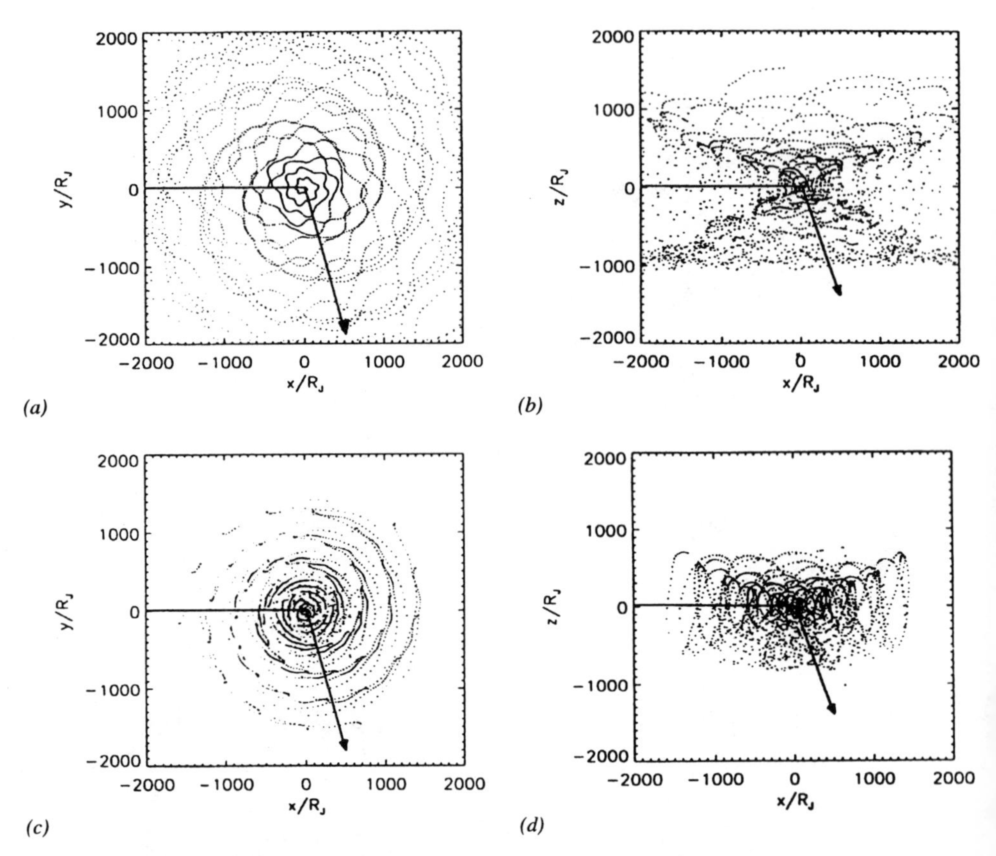

Figure 8 The scatter plot of Jupiter's "dusty ballerina skirt" comprised of 0.05-μm-size grains (*a, b*) and 0.1-μm-size grains (*c, d*) projected onto the ecliptic plane (Figures 8*a* and 8*c*) and to a plane perpendicular to the ecliptic (Figures 8*b* and 8*c*). The continuous lines represent the trajectory of *Ulysses* (the x axis is rotated in the ecliptic plane to point towards the inbound leg of the trajectory). One particle was started in every 15 minutes from Io, and the evolution of the entire growing population of dust grains was followed for 35 days. Each dot represents a dust grain in this snapshot. To a good approximation, these structures expand radially at a constant rate. (From Horányi et al 1993b.)

than *Ulysses* found only three years earlier. These changes might be related to the variable nature of the volcanic activity on Io. Several ground-based observers reported increased volcanic activity on Io in 1995 (Silverstone et al 1995; JR Spencer, personal communication). Alternatively, the changed dust environment could be due to comet Shoemaker-Levy-9. We return to this point below.

4.2 *First-Order Theory*

For micron-sized grains (which are often present in planetary rings) electromagnetic effects represent a potentially important perturbation. Let us further explore their motion, ignoring terms second order in q. For particles started at a radial position r_o on circular Kepler orbits with $J = r_o^2(\omega_k - \omega_g)$, the potential minima from equation (27) defines a reference orbit about which a particle oscillates. This reference orbit is located at

$$a = r_o(1 + \beta), \tag{31}$$

where $\beta = \omega_g(\Omega - \omega_k)\omega_k^{-2}$.

The displacement of the minimum energy orbit can be positive or negative depending on the sign of ω_g and the location, because $(\Omega - \omega_k)$ changes sign at $r = r_{syn}$ (r_{syn} is the position of the synchronous orbit, where $\Omega = \omega_k$). The amplitude of the displacement for $a_\mu = 1$ grains is given by

$$\beta = \beta_o \phi_v \left[1 - \left(\frac{r_{syn}}{r_o}\right)^{3/2}\right] \tag{32}$$

(where $\beta_o^J = 6.2 \times 10^{-2}$ and $\beta_o^S = 3.2 \times 10^{-3}$). Orbits close to the minimum energy reference orbit will harmonically oscillate about the bottom of the effective potential with an amplitude δ ($\ll a$), so

$$r(t) = a + \delta(t). \tag{33}$$

Substituting equation (33) into (22) and keeping only first-order terms yields

$$\ddot{\delta} = -\omega_r^2 \delta, \tag{34}$$

where the radial frequency is

$$\omega_r = \omega_k \left(1 + \frac{\beta}{2} + 2\frac{\omega_g}{\omega_k}\right) = \psi + 2\omega_g. \tag{35}$$

The frequency for the radial oscillations of grains with positive (negative) charges is slower (faster) than the angular rate of the circular reference orbit, $\omega_r > (<)\ \psi$. These orbits precess.

For perturbed motions, it is informative to follow the evolution of the osculating orbital elements, which provide a description of the fictitious Kepler orbit that the particle would assume if all perturbations were instantaneously turned off. To track the history of an uninclined orbit, for example, we can follow the evolution of the semimajor axis a_k, which describes the size of the orbit, the eccentricity e, which determines the orbit's shape, and the longitude of pericenter $\tilde{\omega}$, which fixes the orientation of the ellipse.

Because $\omega_r > (<) \,\psi$, the period of the radial oscillation is shorter (longer) than the period on the reference orbit, forcing the angular position of the pericenter to move forward (backward). From equation (35) the rate of change of the longitude of pericenter is

$$\dot{\tilde{\omega}} = \psi - \omega_r = -2\omega_g. \tag{36}$$

From here, the angular rate at which the pericenter of the orbit of a 1-μm grain with surface potential ϕ_v moves due to its electric charge is

$$\dot{\tilde{\omega}}_\phi \simeq -\gamma\phi_v\left(\frac{R}{a_k}\right)^3 \text{ deg/day}, \tag{37}$$

(where $\gamma_J \simeq 100$ and $\gamma_S \simeq 5$).

The planetary oblateness, J_2, is also known to cause precession of the longitude of the pericenter. The rate of change of the node due to oblateness (Danby 1988) is

$$\dot{\tilde{\omega}}_{J_2} = \frac{3}{2}\omega_k J_2\left(\frac{R}{a_k}\right)^2 \simeq \xi\left(\frac{R}{a_k}\right)^{3.5} \text{ deg/day} \tag{38}$$

(where $\xi_J = 65$ and $\xi_S = 50$). For charged grains the net pericenter motion ($\dot{\tilde{\omega}} = \dot{\tilde{\omega}}_\phi + \dot{\tilde{\omega}}_{J_2}$) can be positive, negative, or even zero depending on the particle's size, charge, and location in the magnetosphere. The possible close cancellation of the precession rates due to charging and oblateness was suggested to explain many of the observed features of Saturn's E-ring.

SATURN'S E-RING Let us first summarize the observations (Showalter et al 1991). This faint ring extends from 3 to 8 R_S with a sharp peak in its optical depth ($\tau \simeq 10^{-6}$) at the orbit of Enceladus at $r \simeq 4\ R_S$. The suggested particle size distribution to explain the wavelength dependence of the brightness in this region is surprisingly narrow, concentrated around $1 \pm 0.3\ \mu$m.

If the radiation pressure force in equation (20) is included, than this constant force pointing away from the Sun causes periodic changes in the particle's angular momentum and hence its orbital eccentricity. The maximum eccentricity that solar radiation pressure can induce (Horányi et al 1992) is

$$e_{\max} \approx \frac{3}{2}\left(\frac{hf}{\mu\dot{\tilde{\omega}}}\right), \tag{39}$$

where h and f are the specific (i.e. per unit mass) angular momentum and acceleration due to solar radiation pressure. Because charging slows precession caused solely by oblateness ($\dot{\tilde{\omega}} \rightarrow 0$), we anticipate large eccentricities according to Equation (39). In the plasma environment of Saturn's E-ring (Figure 3*b*) the characteristic surface potential of the dust particles is expected in the range of $-8 < \phi_V < -4$, resulting in a close cancellation of the precession of $a_\mu \simeq 1$ grains. If Enceladus is assumed to be the source of particles with a range of sizes, the magnetosphere selects particles with $a_\mu = 1$ and forces them to follow eccentric orbits (Figure 9). Smaller or bigger particles will stay close to the orbit of their source. The vertical structure of this ring is caused by the

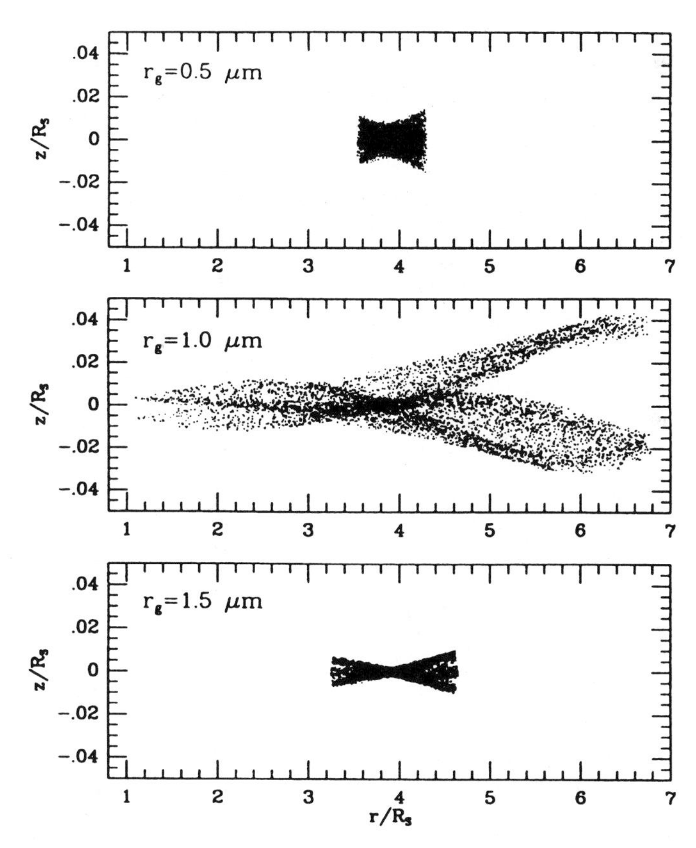

Figure 9 A scatter diagram of Saturn's simulated E-ring shown in the r, Z plane from randomly sampled orbits of 0.5 (*top*), 1 (*middle*), and 1.5 μm (*bottom*) size particles. (From Horányi et al 1992.)

component of the radiation pressure that is out of the ring plane (Horányi et al 1992).

Enceladus and all the other moons in this region are constantly bombarded by interplanetary dust fluxes. The impact-generated ejecta is a source of dust in the region. However, as discussed above, the ejected dust particles are forced to move on orbits with large eccentricities, leading to possible collisions with the moons. Due to the large expected eccentricities, dust particles strike the surface of the moons with large enough velocities to generate even more dust. The runaway dust production levels off at densities when grain-grain collisions become important. This self-sustained dust generation is suspected to be more important for the E-ring than the bombardment by interplanetary dust (Hamilton & Burns 1994).

We anticipate that further detailed information on the three-dimensional structure of this ring will be obtained since excellent opportunities to make ground-based observations exist in 1995 and 1996. The rings of Saturn will be seen edge-on from the Earth on three occasions, and once also from the Sun's perspective. In 2002 the *Cassini* spacecraft will arrive at Saturn and provide in situ observations of the E-ring during its many passes through this region.

4.3 *Magnetospheric Transport and Capture*

Charged dust grains in planetary magnetospheres can rapidly trade energy and angular momentum with the electric and magnetic fields because of their time-dependent electrostatic charges. Hence, the evolution of the spatial distribution of small dust particles is often shaped by the plasma environment. The energy change of a charged dust particle in one, approximately closed, orbital loop can be estimated as

$$\Delta E \simeq \oint Q\, \dot{\mathbf{r}} E_c \, dt, \tag{40}$$

and the angular momentum change as

$$\Delta \mathbf{J} \simeq \oint Q \mathbf{r} \times \left(\mathbf{E_c} + \frac{\mathbf{B}}{c} \times \dot{\mathbf{r}} \right) dt. \tag{41}$$

B does not appear in equation (40), since the Lorentz force does not do work.

Surprisingly rapid transport can result from the variable charge on a dust particle due to gradients in the plasma parameters (Northrop et al 1989), the velocity modulation of the ion current given in equations (14) and (15) (Burns & Schaffer 1989), the modulation of the photoelectron current given in equation (13) by the planet's shadow (Horányi & Burns 1991), or simply due to the stochastic nature of the charging processes (Morfill et al 1980a). A similar transport mechanism was recently suggested to explain the spatial structure of Jupiter's ring. Swift energy and angular momentum loss can also lead to dust particle capture in planetary magnetospheres, perhaps even the formation of new planetary rings.

THE JOVIAN RING Let us first summarize the observations (Showalter et al 1985, 1987). The Jovian ring system has a complex spatial structure. Its main ring is located at $1.71\ R_J \leq r \leq 1.81\ R_J$ with an optical depth $\tau \simeq 10^{-5}$. The main ring consists of "large bodies" in the centimeter to meter (perhaps km?) size range that are responsible for the charged particle absorption detected by *Pioneer 11* and also a large population of small dust particles in the $0.1 \leq a \leq 100\ \mu$m size range. The small dust grains and the macroscopic bodies contribute approximately equally to the optical depth. The thickness of the ring in back-scattered light (macroscopic bodies) is ≤ 30 km, and in forward-scattered light (small dust) it is ≤ 300 km. At its inner edge the main ring suddenly turns into a vertically extended, doughnut-shaped halo with a maximum scale height of 3000 km. The halo vanishes from the images at $\sim 1.4\ R_J$. The halo is comprised of small dust grains with a similar size distribution to the dust population of the main ring. The size distribution of the dust in the main ring and in the halo appears to follow a power-law size distribution $n(a)da \sim a^{\gamma}da$, with $\gamma = -2.5 \pm 0.5$. The third component of the ring system is a much fainter outward extension of the main ring that reaches near the orbit of the moon Amalthea at $2.54\ R_J$.

Our understanding of the spatial structure of the ring/halo region is mainly based on the processing technique where images are transformed into density contours of cross sectional slices (Showalter et al 1987). Figure 10*a* shows such an example. It was recognized early on that electromagnetic effects are likely to be responsible for maintaining the structure seen in Figure 10*a*. Common to all theoretical models is the assumption that unseen large parent bodies continuously supply the small dust particles via micro-meteoroid and/or Iogenic dust bombardment. The ejected dust grains become exposed to magnetospheric plasmas and collect electrostatic charges. Perhaps motivated by the discovery of Io's stunning volcanic activity and its energetic and dense plasma torus at 4–6 R_J, earlier theoretical models suggested that inward diffusing sulfur and oxygen ions dominate the plasma environment of the ring. The estimated equilibrium surface potential of small dust grains was generally believed to be -10 V. Plasma drag (Burns et al 1980, Morfill et al 1980a,b,c,d), fluctuations of the magnetic field (Consolmagno 1980, 1983) and the grain's electrostatic charge (Morfill et al 1980a,b,c,d), and resonance forcing by the higher order terms in the magnetic field (Burns et al 1985; Schaffer & Burns 1987, 1994), for example, were all suggested as causes of dust transport from the main ring.

Alternatively, the plasma environment of the ring could be dominated by photoelectrons escaping from Jupiter's ionosphere (Mendis & Axford 1974, Siscoe 1978, Luhmann & Walker 1980, Waite et al 1983, Nagy et al 1986). In

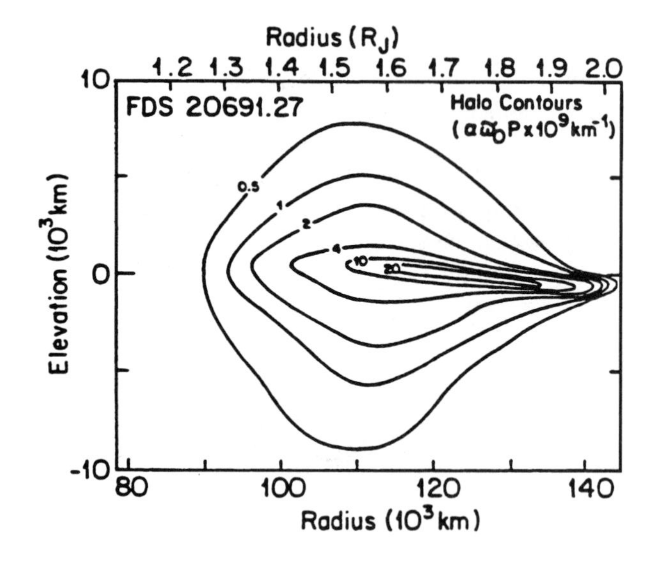

(a)

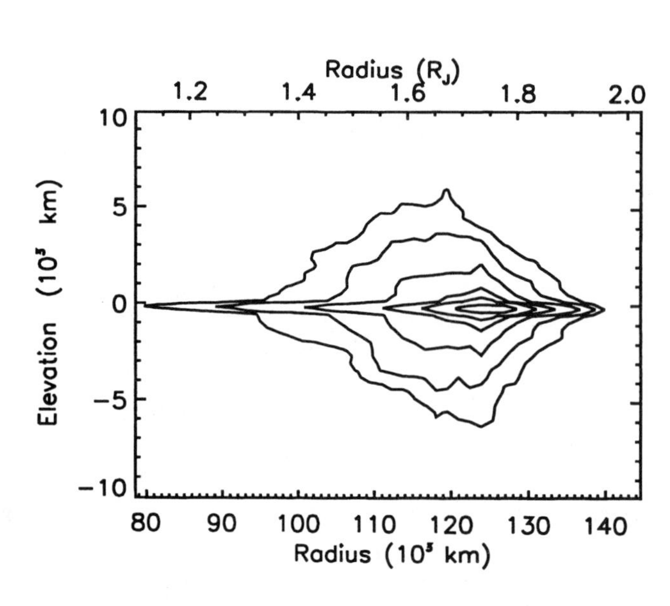

(b)

Figure 10 (*a*) The intensity contours of a halo slice reprojected into a rectangular coordinate frame. The original *Voyager* image (FDS 20691.27) was taken through a violet filter, $\lambda = 0.431\ \mu$m, at an average phase angle of $\theta = 175.8°$. (From Showalter et al 1987.) (*b*) The calculated contour plots of the normalized brightness distribution for dust particles with $\rho = 1$ g cm^{-3}. The dust production is assumed to follow a power law, $\dot{n}(a)da \sim a^{-5.5}da$; the grain's index of refraction is $m = 1.5 - i0.01$. The parameters to generate the intensity (λ, θ) were identical to those in (*a*). The contour levels are 0.0125, 0.05, 0.1, 0.25, and 0.5. (From Horányi & Cravens 1996.)

this case the expected plasma densities are much below the previous estimates. The characteristic energy of the escaping electrons is $\simeq$ 10 eV and their flux $F \simeq 6 \times 10^6 (r/R_J)^3$ cm^{-2} s^{-1} (Nagy et al 1986). The charging of the dust particles is dominated by photoelectron production from the grains with a small but important contribution from the ionospheric electron fluxes. Close to Jupiter ($r < 3\ R_J$) the equilibrium potential of the dust particles $\phi_{eq} > 0$ and, due to the decreasing contribution of the ionospheric electrons at larger distances, $\partial\phi_{eq}/\partial r > 0$ (Figure 3*a*). This radial dependence of ϕ_{eq} can be responsible for much of the observed structure of the Jovian ring (Horányi & Cravens 1996).

Newly released dust particles in the main ring will charge up positively and start oscillating towards and away from Jupiter owing to their slightly eccentric orbits. Because $\partial\phi_{eq}/\partial r > 0$, grains moving towards (away from) Jupiter tend to have larger (smaller) charges than the equilibrium charge at any r due to the charging time delay of these particles (Figure 11). These grains will

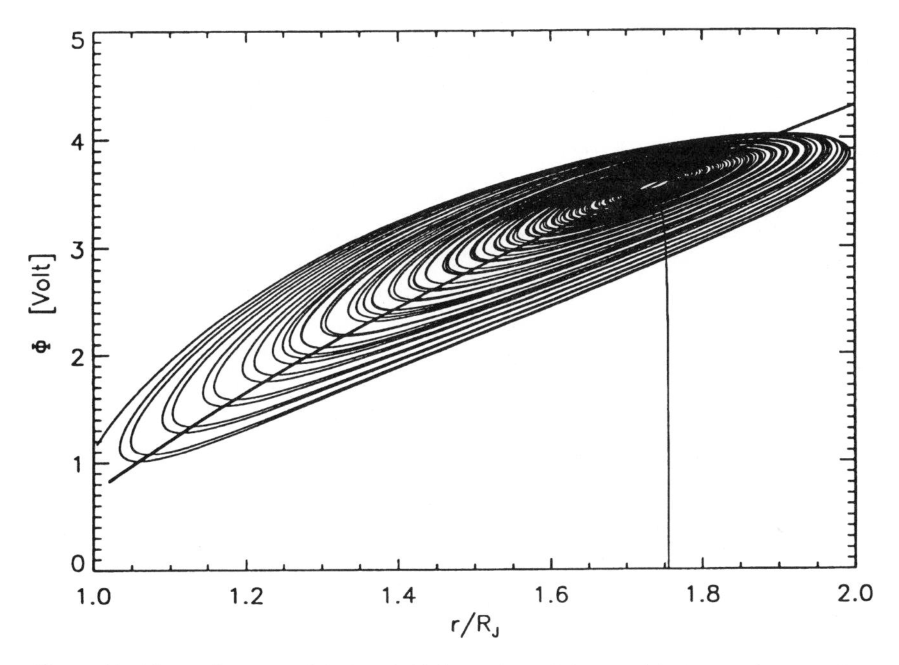

Figure 11 The surface potential of an initially uncharged dust particle ($a_\mu = 1$, $\rho = 1$ g cm^{-3}) (*thin line*) and the equilibrium potential (*thick line*) vs distance from Jupiter. The dust grain was started on a circular Kepler orbit with $a_K = 1.75$, and it circulates counter-clockwise on an expanding spiral. Moving inward the surface potential tends to stay above the equilibrium while moving outward it tends to stay below the equilibrium potential, resulting in swift energy and angular momentum loss (i.e. decreasing semimajor axis $\dot{a}_K < 0$ and an increasing orbital eccentricity $\dot{e} > 0$). The grain hits Jupiter when $a_K(1 - e) = 1$ (From Horányi & Cravens 1995.)

rapidly lose energy and angular momentum as described by equations (40) and (41). Consequently, they will follow orbits with decreasing semimajor axes and increasing eccentricities, crashing into Jupiter's atmosphere as soon as their perihelion distance $a_K(1-e) = 1$ (Figure 12). The lifetime of these grains is very short ($T \simeq 100a_\mu^3$ day). The tilted nature of the Jovian magnetic field results in forces and torques that are out of the ring plane. Particles moving around the equatorial plane will see a component of the magnetic field $\mathbf{B}_\parallel$ that is parallel to the equatorial plane and that is periodically changing in orientation. Particles looping around Jupiter on a prograde (counter-clockwise) trajectory experience a Lorentz force pushing them above (below) the ring plane as they cross the sectors where $\mathbf{B}_\parallel$ points toward (away from) Jupiter. This effect will force particles into orbits with increasing inclinations. The spatial structure seen in Figure 10*a* is approximately reproduced in Figure 10*b* by following a large number of particles generated via bombardment of the bigger boulders in the main ring and transported towards Jupiter's atmosphere following orbits with increasing eccentricities and inclinations. Perhaps a better match will be possible in the near future using *Galileo* images of this region. *Galileo* will not take in situ fields and particles measurements this close to Jupiter, but images showing the macroscopic sources, the detailed spatial structure, and the size distribution of the dust particles comprising this ring will reveal the plasma conditions.

A POSSIBLE NEW JOVIAN RING The comet Shoemaker-Levy-9 (SL9) produced a lot of dust while breaking apart during its close encounter with Jupiter in July 1992, and perhaps also on its final return path through the magnetosphere as its fragmented nuclei plunged into Jupiter's atmosphere in July 1994 (Figure 13). The orbital evolution of these grains strongly depends on their size; some of them can get captured to form a new ring about Jupiter (Horányi 1994). This process of magnetospheric capture was first discussed by Mendis & Axford (1974); later it was extended as a suggestion for the origin of the Jovian ring from captured small fragments of a hypothetical broken-up comet by Hill & Mendis (1979, 1980b).

Particles in the micron-size range from SL9 will be strongly perturbed by radiation pressure and electromagnetic forces and, unlike the bigger grains, they will avoid collision with Jupiter. Instead they will weave in and out of the magnetosphere. Particles with $a_\mu \sim 2$ can be shown to lose energy and angular momentum during their consecutive trips through the magnetosphere and settle into circular orbits, perhaps forming a new ring around Jupiter (Figure 14).

It is difficult to estimate how much mass will end up in this new ring. For a rough estimate, let us assume that between $0.1 < \eta < 1\%$ of the mass of the original 4-km-radius object ends up as dust in the range of $0.1\ \mu\text{m} < a < 1$ cm

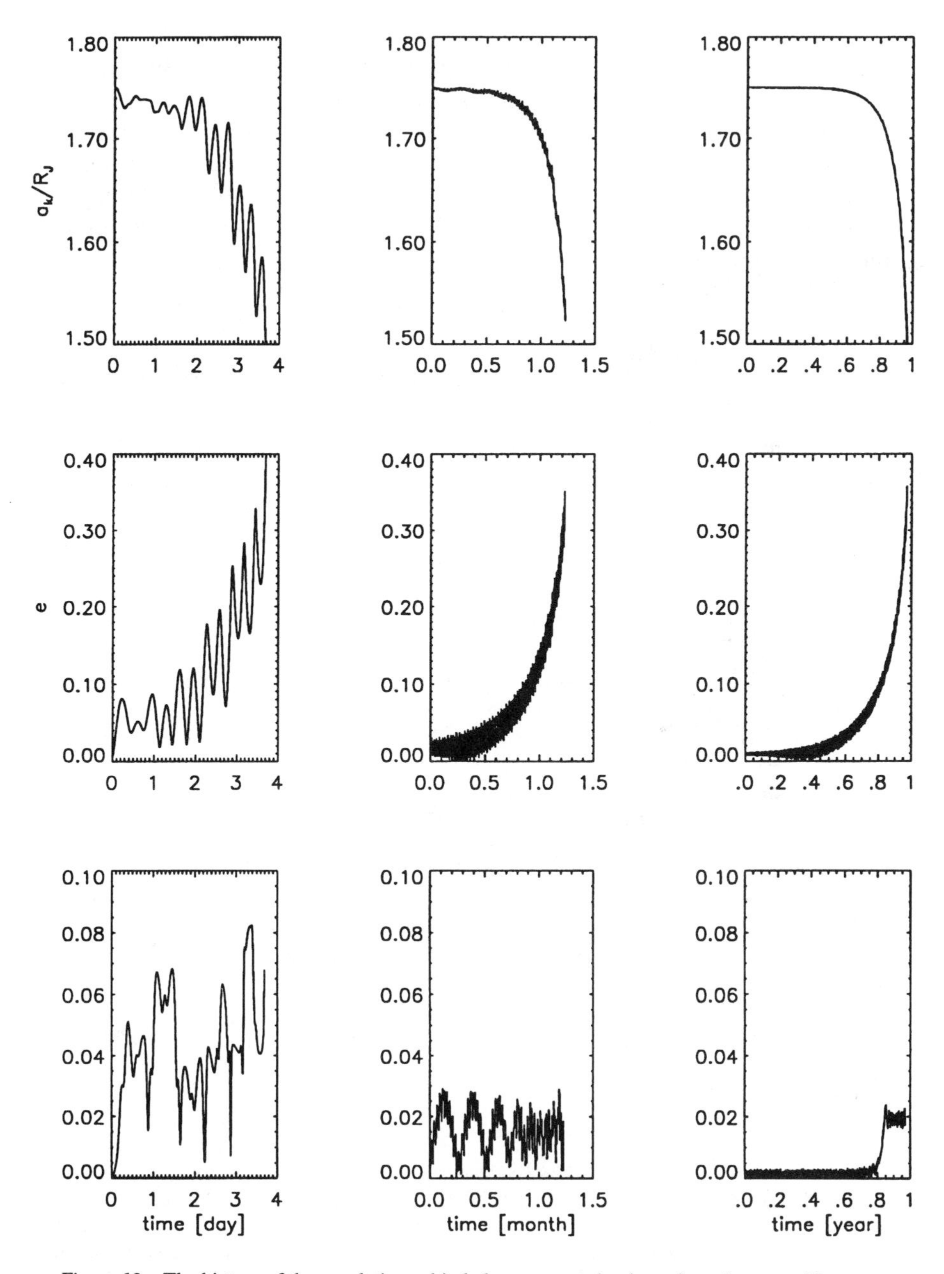

Figure 12 The history of the osculating orbital elements: semimajor axis a_{K} (*top panels*), eccentricity e (*middle panels*), and inclination i (*bottom panels*) for dust particles with $\rho = 1$ g cm^{-3} and $a_{\mu} = 0.5$ (*left column*), 1 (*middle column*), and 2 (*right column*).

and assume a power-law mass distribution $n(m)dm \sim m^{-p}dm$ with an index $2 < p < 2.5$, which is typical for cometary dust. With these assumptions, particles in the range of $1.5 < a < 2.5$ μm represent a fraction of the dust mass between $2 \times 10^{-4} > \xi > 5 \times 10^{-6}$, assuming $p = 2$ and 2.5, respectively. We estimate the mass of our new ring ($M = \eta\xi M_{\mathrm{SL9}}$, where the original mass $M_{\mathrm{SL9}} \sim 3 \times 10^{17}$ g) to be in the range of $2 \times 10^9 < M < 5 \times 10^{11}$ g. The upper limit would be comparable to the mass of the main Jovian ring. We estimate the average optical depth to be

$$\overline{\tau} = \frac{3M}{4\pi\rho a(R_2^2 - R_1^2)}, \tag{42}$$

where $R_1(= 4.5\ R_{\mathrm{J}})$ and $R_2(= 5.9\ R_{\mathrm{J}})$ are the inner and outer radial limits of

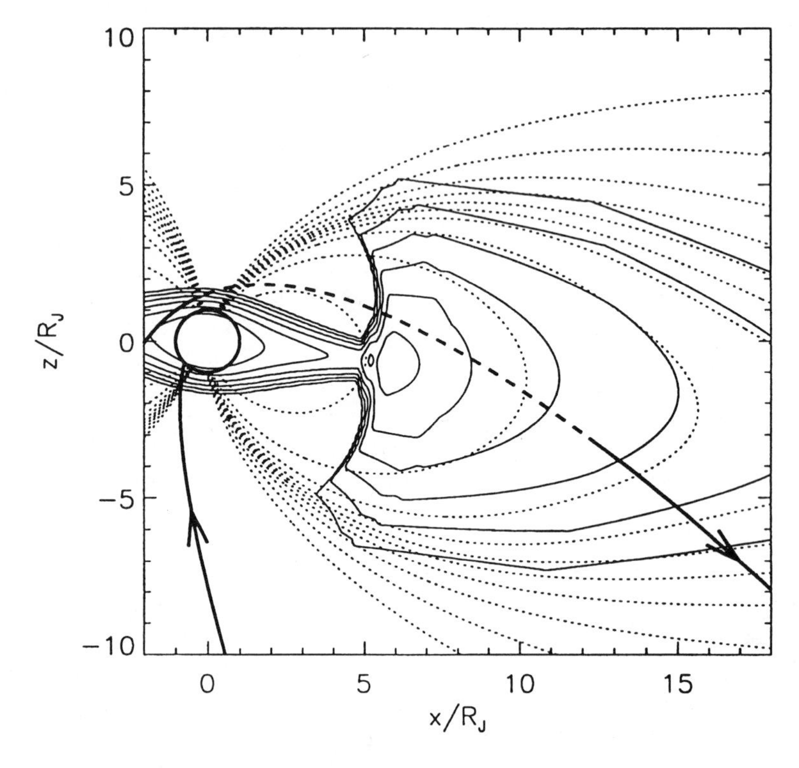

Figure 13 The trajectory of SL9 (*thick continuous and dashed line*) in a jovicentric coordinate system (z is along the rotational axis and x is in the equatorial plane of Jupiter pointing towards the Sun's position on July 20, 1994). Dust production was assumed to continue during the first 12 h following the break up, along the trajectory of SL9 (*dashed section*). The arrows show the direction of the motion. The snapshot of plasma density contours (*thin lines*) mark the values of $n_{\mathrm{p}} = 10^{-2}$, 10^{-1}, 10^{0}, 10^{1}, 10^{2}, 10^{3} cm^{-3} and the magnetic field lines shown (*dotted lines*) pierce the magnetic equator at $L = 4, 8, \ldots R_{\mathrm{J}}$ (From Horányi 1994.)

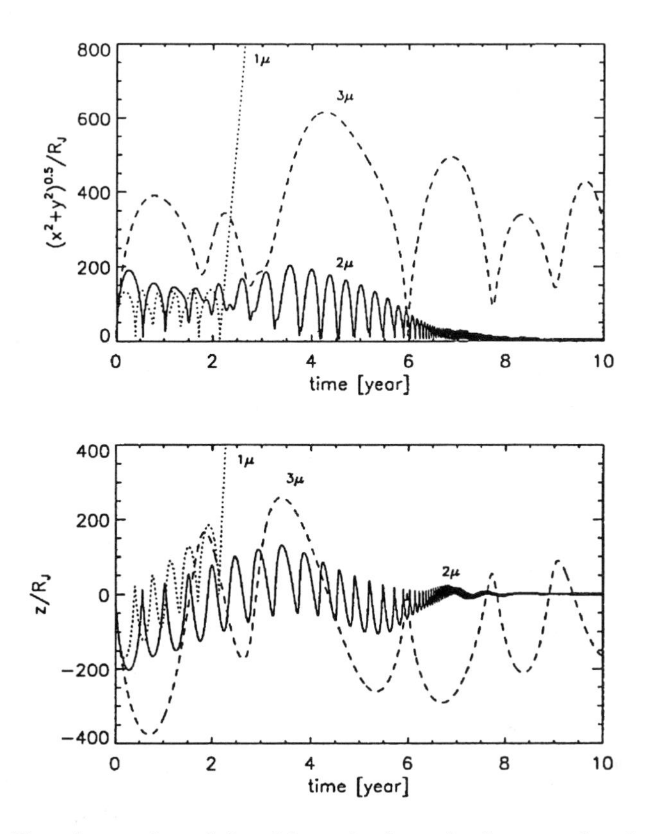

Figure 14 The trajectory of $a = 1$, 2, and 3 μm-size dust grains that were released at the break up of SL9 ($t = 0$). The top panel shows their distance in the equatorial plane of Jupiter; the bottom panel shows their position above/below that plane (From Horányi 1994.)

the new ring, respectively. Keeping $\rho = 1$ g cm^{-3} we expect the mean optical depth to lie in the range $3 \times 10^{-9} < \overline{\tau} < 8 \times 10^{-7}$. The upper limit would signal a new ring comparable in brightness to the main Jovian ring, whereas the lower limit puts our brightness estimate an order of magnitude below that of the faintest part of the Jovian ring system, the "gossamer ring."

The enhanced population of small dust grains detected by the *Galileo* spacecraft approaching Jupiter (Figure 7) is perhaps the precursor of the new ring. In 2002, when *Cassini* encounters Jupiter on its way to Saturn it might already observe the nascent ring descendant of SL9.

5. MORAL

In most Solar System plasma environments the dynamics of small charged particles is strongly influenced, if not dominated, by electromagnetic forces acting simultaneously with gravity, drag, and radiation pressure. Dust particles traversing various regimes adjust their electrostatic charges as dictated by the changing plasma conditions, and in fact they act as active electrostatic probes, continuously adjusting their surface potential towards the local equilibrium value. The fields and particles environment can uniquely shape the size and the spatial distribution of the dust grains. Studies of the motion of charged dust particles connect a number of observations that are often thought to be unrelated. Space missions, in general, are designed to make simultaneous in situ and remote observations with some combination of the following experiments:

- A dust detector provides in situ measurement of the mass and the velocity vector of the dust grains.
- A plasma detector provides the composition, density, and energy distribution of the plasma. These data are used to calculate the charging currents of the grains and to learn whether grains are in charge equilibrium or will have significant charge variations due to fluctuations and/or gradients in composition and/or density and/or temperature of the plasma.
- The plasma wave and radio science experiment on board *Voyager 2* detected broad-band noise passing through the ring planes at Saturn, Uranus, and Neptune. This noise is believed to be caused by small dust grains bombarding the body of the spacecraft. The few kilometer per second relative velocity between the spacecraft and the dust grains is sufficient to fully vaporize the impacting grains and in part ionize the produced gas. The expanding plasma cloud causes the detected noise. This phenomena led to the recognition that all giant planets are surrounded by vast tenuous sheets of small grains that could not have been discovered via imaging.
- A magnetometer provides the magnetic field measurements. These data are essential to calculate the trajectories of charged dust particles.
- An imaging experiment can supply images taken through filters at various phase angles to show the spatial and size distribution of the dust particles. Ultimately, the spatial distributions of the fine dust can be independently modeled based on the transport processes at work and compared with the images.

It is a unique and powerful consistency test if our models describing dust transport, based on particles and fields data, match the observations of the dust detectors and the images. However, without in situ data on particles and fields, images showing the spatial distribution of small dust grains can be used to infer the plasma conditions. The ongoing and planned future missions—*Ulysses* on polar orbit about the Sun, *Galileo* at Jupiter, *Cassini* at Saturn, *Rosetta* to a comet, or possible returns to the Moon—are all expected to find further examples on the complex dynamics of charged dust particles in our Solar System.

ACKNOWLEDGMENTS

It is my good fortune to have a number of outstanding collaborators: Drs JA Burns, E Grün, O Havnes, DA Mendis, G Morfill, T Northrop, and the late CK Goertz played important roles in my work. Support from the grants NASA NAGW-3355 and NSF 9322448 & 9526287 are gratefully acknowledged.

Literature Cited

Alfvén H. 1954. *On the Origin of the Solar System.* Oxford: Clavenden. 191 pp.
Bagenal F. 1995. *J. Geophys. Res.* 99:11,043
Berg OE, Richardson FF, Burton H. 1973. *NASA SP* 330:16
Berg OE, Richardson FF, Rhee JW, Auer S. 1974. *Geophys. Res. Lett.* 1:289
Berg OE, Wolf H, Rhee JW. 1976. *Lect. Notes Phys.* 48:233
Burns JA, Schaffer LE. 1989. *Nature* 337:340
Burns JA, Schaffer LE, Greenberg RJ, Showalter MR. 1985. *Nature* 316:115
Burns JA, Showalter MR, Cuzzi JN, Pollack JB. 1980. *Icarus* 44:339
Chow VW, Mendis DA, Rosenberg M. 1993. *J. Geophys. Res.* 98:19,065
Consolmagno GJ. 1980. *Nature* 285:557
Consolmagno GJ. 1983. *J. Geophys. Res.* 88:5607
Criswell DR. 1973. In *Photon and Particle Interaction in Space,* ed. RJL Grard, p. 545. Dordrecht: Reidel
Criswell DR, De BR. 1977. *J. Geophys. Res.* 82:1005
Danby JMA. 1988. *Fundamentals of Celestial Mechanics.* Richmond: Willmann-Bell 466 pp. 2nd ed.
De BR. 1979. *J. Chem. Phys.* 70:2046
De BR. Criswell DR. 1977. *J. Geophys. Res.* 82:999
De Angelis U. 1992. *Phys. Scr.* 45:465
Feuerbacher B, Willis RF, Fitton B. 1973. *Astrophys. J.* 181:102
Goertz CK. 1989. *Rev. Geophys.* 27:271
Goertz CK, Ip W-H. 1984. *Geophys. Res. Lett.* 11:349
Goertz CK, Morfill G. 1983. *Icarus* 53:219
Gold T. 1955. *MNRAS* 115:585
Grün E, Morfill GE, Mendis DA. 1984. In *Planetary Rings,* ed. R Greeneberg, A Brahic, p. 275. Tucson: Univ. Ariz. Press
Grün E, Morfill GE, Schwehm G, Johnson TV. 1980. *Icarus* 44:326
Grün E, Zook HA, Baguhl M, Balogh A, Bame SJ, Fechtig H, et al. 1993. *Nature* 362:428
Hamilton DP, Burns JA. 1993. *Nature* 364:695
Hamilton DP, Burns JA. 1994. *Science* 264:550
Hartquist TW, Havnes O, Morfill GE. 1992. *Fundam. Cosmic Phys.* 15:107
Havnes O, Aanesen TK, Melandsø F. 1990. *J. Geophys. Res.* 95:6581
Havnes O, Goertz CK, Morfill GE, Grün E, Ip W-H. 1987. *J. Geophys. Res.* 92:2281
Havnes O, Morfill GE, Goertz CK. 1984. *J. Geophys. Res.* 89:10,999
Hill JR, Mendis DA. 1979. *The Moon and the Planets* 21:3
Hill JR, Mendis DA. 1980a. *Can. J. Phys.* 59:897
Hill JR, Mendis DA. 1980b. *The Moon and the*

Planets 23:53
Hill JR, Mendis DA. 1982. *Geophys. Res. Lett.* 9:1069
Horányi M. 1994. *Geophys. Res. Lett.* 21:1039
Horányi M, Burns JA. 1991. *J. Geophys. Res.* 96:19,283
Horányi M, Burns JA, Hamilton D. 1992. *Icarus* 97:248
Horányi M, Cravens TE. 1996. *Nature*. In press
Horányi M, Goertz CK. 1990. *Ap. J.* 361:105
Horányi M, Houpis HLF, Mendis DA. 1988. *Astrophys. Space Sci.* 144:215
Horányi M, Morfill GE, Grün E. 1993a. *Nature* 363:1993
Horányi M, Morfill GE, Grün E. 1993b. *J. Geophys. Res.* 98:21,245
Horányi M, Robertson S, Walch B. 1995. *Geophys. Res. Lett.* 22:2079
Johnson RE. 1990. *Energetic Charged-Particle Interactions with Atmospheres and Surfaces.* New York: Springer-Verlag. 232 pp.
Juhász A, Tátrallyay M, Gévai G, Horányi M. 1993. *J. Geophys. Res.* 98:1205
Jurac S, Baragiola A, Johnson RE, Sittler EC Jr. 1995. *J. Geophys. Res.* 100:14,821
Kanal M. 1962. U. Mich. Sci. Rep. No. JS-5
Leinert C, Grün E. 1990. In *Physics of the Inner Heliosphere,* ed. R Schwenn, E Marsch, p. 207. New York: Springer-Verlag
Luhmann JG, Walker RJ. 1980. *Icarus* 44:361
McCoy JE, Criswell DR. 1974. *Proc. 5th Lunar Conf.* 3:2991
Mendis DA, Axford WI. 1974. *Rev. Earth Planet. Sci.* 2:419
Mendis DA, Hill JR, Houpis HLF, Whipple ECJ. 1981. *Ap. J.* 249:787
Mendis DA, Hill JR, Ip W-H, Goertz CK, Grün E. 1984. In *Saturn,* ed. T Gehrels, MS Matthews, p. 54. Tucson: Univ. Ariz. Press
Mendis DA, Horányi M. 1991. In *Cometary Plasma Processes,* ed. AJ Johnston, p. 17. *Geophys. Monogr. No. 6.* Washington DC: Am. Geophys. Union
Mendis DA, Rosenberg M. 1994. *Annu. Rev. Astron. Astrophys.* 32:419
Meyer-Vernet N. 1982. *Astron. Astrophys.* 105:98
Morfill GE, Grün E, Goertz CK, Johnson TV. 1983. *Icarus* 53:230
Morfill GE, Grün E, Johnson TV. 1980a. *Planet. Space Sci.* 28:1087
Morfill GE, Grün E, Johnson TV. 1980b. *Planet. Space Sci.* 28:1101
Morfill GE, Grün E, Johnson TV. 1980c. *Planet. Space Sci.* 28:1111
Morfill GE, Grün E, Johnson TV. 1980d. *Planet. Space Sci.* 28:1115
Nagy AF, Barakat AR, Schunk RW. 1986. *J. Geophys. Res.* 91:351
Nitter T, Aslaksen TK, Melandsø F, Havnes O. 1994. *IEEE Trans. Plasma Sci.* 22:159
Nitter T, Havnes O. 1992. *Earth, Moon and Planets* 56:7
Northrop TG. 1992. *Phys. Scr.* 45:475
Northrop TG, Birmingham TJ. 1995. *J. Geophys. Res.* In press
Northrop TG, Hill JR. 1983. *J. Geophys. Res.* 88:1
Northrop TG, Mendis DA, Schaffer L. 1989. *Icarus* 79:101
O'Keefe JA, Adams JB, Gault DB, Green J. 1968. *JPL Tech. Rep. 32-1262*
Opik EJ. 1956. *Irish Astron. J.* 4:84
Rennilson JJ, Criswell DR. 1973. *The Moon* 10:121
Richardson JD, Sittler EC. 1990. *J. Geophys. Res.* 95:12,019
Rosenberg M, Mendis DA. 1992. *J. Geophys. Res.* 97:14,773
Schaffer L, Burns JA. 1987. *J. Geophys. Res.* 92:2264
Schaffer L, Burns JA. 1994. *J. Geophys. Res.* 99:17,211
Selwyn GS. 1989. *J. Vac. Sci. Technol. A* 7:2758
Shan L-H, Goertz CK. 1991. *Ap. J.* 367:350
Showalter MR, Burns JA, Cuzzi JN, Pollack JB. 1985. *Nature* 316:526
Showalter MR, Burns JA, Cuzzi JN, Pollack JB. 1987. *Icarus* 69:458
Showalter MR, Cuzzi JN, Larson SM. 1991. *Icarus* 94:451
Silverstone MD, Becklin EE, Tsetsenkos KI. 1995. *Div. Planet. Sci. AAS 27th Annu. Meet.* 39.12-P (Abstr.)
Singer SF, Walker EH. 1962. *Icarus* 1:112
Siscoe GL. 1978. *J. Geophys. Res.* 83:2118
Smith BA, Soderblom LA, Beebe RF, Boyce J, Briggs GA, et al. 1982. *Science* 212:163
Spitzer L. 1941. *Ap. J.* 93:396
Sternglass EJ. 1954. *Sci. Pap. 1772,* Westinghouse Res. Lab., Pittsburgh
Waite JH Jr, Cravens TE, Kozyra J, Nagy AF, Atreya SK, Chen RH. 1983. *J. Geopys. Res.* 88:6143
Walch B, Horányi M, Robertson S. 1995. *Phys. Rev. Lett.* 75:838
Whipple EC Jr. 1981. *Rep. Prog. Phys.* 44:1197
Whipple EC Jr, Northrop TG, Mendis DA. 1985. *J. Geophys. Res.* 90:7405
Xu W, D'Angelo N, Merlino RL. 1993. *J. Geophys. Res.* 98:7843
Zook HA, McCoy E. 1991. *Geophys. Res. Lett.* 18:2117

Annu. Rev. Astron. Astrophys. 1996. 34:419–59

GRAVITATIONAL MICROLENSING IN THE LOCAL GROUP

Bohdan Paczyński

Princeton University Observatory, Princeton, New Jersey 08544-1001

KEY WORDS: brown dwarfs, dark matter, Milky Way galaxy, planetary systems, gravitational lensing

ABSTRACT

The status of searches for gravitational microlensing events of the stars in our galaxy and in other galaxies of the Local Group, the interpretation of the results, some theory, and prospects for the future are reviewed. The searches have already unveiled ~ 100 events, at least two of them caused by binaries, and have already proven to be useful for studies of the Galactic structure. The events detected so far are probably attributable to the effects of ordinary stars, and possibly to substellar brown dwarfs; however, a firm conclusion cannot be reached yet because the analysis published to date is based on a total of only 16 events. The current searches, soon to be upgraded, will probably allow determination of the mass function of stars and brown dwarfs in the next few years; these efforts will also provide good statistical information about binary systems, in particular their mass ratios. They may also reveal the nature of dark matter and allow us to detect planets and planetary mass objects.

1. INTRODUCTION

The topic of gravitational lensing has a long history, as described, for example, in the first book entirely devoted to the subject (Schneider et al 1992). The first known theoretical calculation of a light ray being bent by a massive object was published by Soldner (1801), who used Newtonian mechanics and determined that the deflection angle at the solar limb should be $0.''84$, half the value calculated with the general theory of relativity (Einstein 1911, 1916). The

0066-4146/96/0915-0419$08.00

first observational detection of this effect came soon afterwards (Dyson et al 1920). Zwicky (1937) pointed out that distant galaxies may act as gravitational lenses. Almost all essential formulae used today to analyze gravitational lensing were derived by Refsdal (1964). Walsh et al (1979) discovered the first case of a double image created by gravitational lensing of a distant source, the quasar 0957 + 561. Arclike images of extended sources—the galaxies—were first reliably reported by Lynds & Petrosian (1989). In these three cases the Sun, a galaxy, and a cluster of galaxies were acting as gravitational lenses. There are many recent review articles (Blandford & Narayan 1992, Refsdal & Surdej 1994, Roulet & Mollerach 1996) and international conferences (Moran et al 1989, Mellier et al 1990, Kayser et al 1992, Surdej et al 1993, Kochanek & Hewitt 1996) on the subject of gravitational lensing. Ansari (1995) has published a review of gravitational microlensing experiments.

The effect of double imaging of a distant source by a point mass located close to the line of sight and acting as a gravitational lens has been proposed many times over. Chang & Refsdal (1979) and Gott (1981) noted that even though a point mass in a halo of a distant galaxy would create an unresolvable double image of a background quasar, the time variation of the combined brightness of the two images could be observed. In this way the effect of nonluminous matter in a form of brown dwarfs or Jupiter-like objects could be detected. The term "microlensing" was proposed by Paczyński (1986a) to describe gravitational lensing that can be detected by measuring the intensity variation of a macro-image made of any number of unresolved micro-images.

Paczyński (1986b) suggested that a massive search of light variability among millions of stars in the Large Magellanic Cloud could be used to detect dark matter in the Galactic halo. Luckily, the technology needed for such a search became available soon afterwards, and the 1986 paper is credited with triggering the current microlensing searches: EROS (Aubourg et al 1993), MACHO (Alcock et al 1993), OGLE (Udalski et al 1992), and DUO (Alard 1996, Alard et al 1995a). Griest (1991) proposed that objects responsible for gravitational microlensing be called massive astrophysical compact halo objects (MACHO). The name became very popular, and it is commonly used to refer to all objects responsible for the observed microlensing events, no matter where they are located and what their mass may be.

It is not possible to discuss all theoretical and observational papers related to microlensing in the Local Group within the modest volume of this article. The selection of references at the end of this review is limited, and I apologize for all omissions and for the way I made the selection. Let us hope that somebody will write a more careful historical review before too long. Fortunately, there is a bibliography of over one thousand papers related to gravitational lensing that is available electronically.[1] The MACHO and the OGLE collaborations provide up-to-date information about their findings and a complete bibliography of their work on the World Wide Web (WWW) and by anonymous ftp. The photometry of OGLE microlensing events, finding charts, and a regularly updated OGLE status report, including more information about the "early warning system," can also be found on the Internet.[2]

In the following section a simple model of lensing by an isolated point mass is presented. This is all the theory one needs to understand most individual microlensing events. Section 3 presents a model for the spatial distribution and kinematics of lensing objects in order to provide some insight into problems in relating the observed time scales of microlensing events to the masses of lensing objects. Section 4 provides a glimpse into the diversity of light curves due to lensing by double objects, such as binary stars or planetary systems. Some of the special effects that make microlensing more complicated than originally envisioned are described in the Section 5. The most essential information about the current searches for microlensing events, and some of the results as well as problems with some results, are presented in Section 6. The last section is a rather personal outline of the prospects for the future of microlensing searches.

2. SINGLE POINT MASS LENS

Let us consider a single point mass M (a deflector) at a distance D_d from the observer and a point source S at a distance D_s from the observer as in Figure 1. Let there be two planes perpendicular to the line of sight, at the deflector and the source distances, respectively. The deflector has angular coordinates (x_m, y_m) in the sky, as seen by the observer. This projects into points M and M_s in the two planes, with the corresponding linear coordinates

[1]This bibliography has been compiled and is continuously updated by J Surdej & A Pospieszalska. It can currently be found on the World Wide Web at http://www.stsci.edu/ftp/stsci/library/grav_lens/grav_lens.html.

[2]This information is available from the host sirius.astrouw.edu.pl (148.81.8.1), using anonymous ftp service (directory = ogle, files = README, ogle.status, and early.warning). The file ogle.status contains the latest news and references to all OGLE related papers and PostScript files of some publications. These OGLE results are also available over the World Wide Web at http://www.astrouw.edu.pl (in Europe) and at http://www.astro.princeton.edu/~ogle/ (in North America).

Complete information about MACHO results is available at http://wwwmacho.mcmaster.ca/ (in North America), with a duplicate at http://wwwmacho.anu.edu.au/ (in Australia). For information about MACHO alerts see http://darkstar.astro.washington.edu/.

Information about EROS is at: http://www.lal.in2p3.fr/EROS/. I shall do my best to keep a guide to this information as part of the OGLE at http://www.astro.princeton.edu/~ogle.

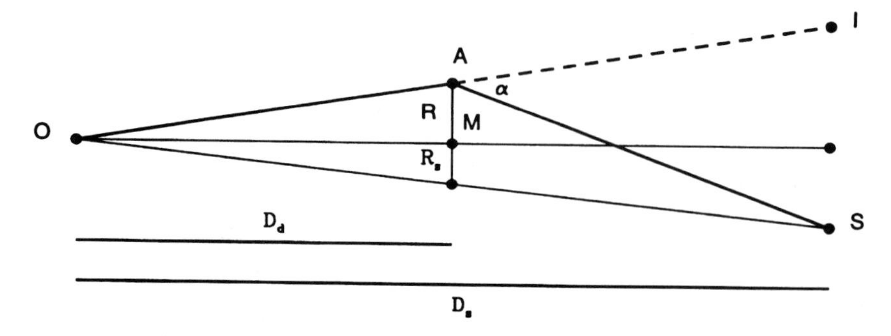

Figure 1 The geometry of gravitational lensing: The observer, the lensing mass, and the source are located at the points O, M, and S, respectively. Light rays are deflected near the lensing mass by the angle α, and the image of the source appears to be located at the point I, not at S. The distances from the observer to the lens (deflector) and to the source are indicated as $D_{\rm d}$ and $D_{\rm s}$, respectively.

$$X_{\rm M} = x_{\rm m} D_{\rm d}, \qquad Y_{\rm M} = y_{\rm m} D_{\rm d}, \tag{1a}$$

and

$$X_{\rm M_s} = x_{\rm m} D_{\rm s}, \qquad Y_{\rm M_s} = y_{\rm m} D_{\rm s}. \tag{1b}$$

Let the observer, located at point O, look at the sky in the direction with angular coordinates (x, y). The line of sight intersects the deflector plane at point A, which has coordinates

$$X_{\rm A} = x D_{\rm d}, \qquad Y_{\rm A} = y D_{\rm d}. \tag{2a}$$

If there were no deflection of light by the massive object, then the line of sight would intersect the source plane at point I with coordinates

$$X_{\rm I} = x D_{\rm s}, \qquad Y_{\rm I} = y D_{\rm s}. \tag{2b}$$

In fact, the light ray passes the deflector at a distance

$$R = \left[(X_{\rm A} - X_{\rm M})^2 + (Y_{\rm A} - Y_{\rm M})^2\right]^{1/2}. \tag{3}$$

As a consequence of general relativity the light ray is deflected by the angle

$$\alpha = 4GMRc^2, \tag{4}$$

with the two components

$$\alpha_x = \alpha \frac{X_{\rm A} - X_{\rm M}}{R}, \qquad \alpha_y = \alpha \frac{Y_{\rm A} - Y_{\rm M}}{R}. \tag{5}$$

The deflected light ray intersects the source plane at the point S with the coordinates

$$X_{\rm S} = X_{\rm I} - \alpha_x (D_{\rm s} - D_{\rm d}), \qquad X_{\rm S} = Y_{\rm I} - \alpha_y (D_{\rm s} - D_{\rm d}). \tag{6}$$

Passing through three points (O, A , and S), the light ray defines a plane. The lensing mass M and the point I are also located in the same plane, as shown in Figure 1. If there were no effect of the mass M on the light rays then the source of light would be seen at the point S, at the angular distance $R_{\rm s}/D_{\rm d}$ from the point M. However, because the light rays are deflected by M the image of the source appears not at S but at the point I, at the angular distance $R/D_{\rm d}$ from the point M. It is clear that the distance $(R + R_{\rm s})$ in the deflector plane is proportional to the distance between I and S in the source plane, and the latter can be calculated with Equations (6). Combining this result with the Equations (5) and (4) we obtain

$$\begin{aligned} R + R_{\rm s} &= [(X_{\rm S} - X_{\rm I})^2 + (Y_{\rm S} - Y_{\rm I})^2]^{1/2} \frac{D_{\rm d}}{D_{\rm s}} \\ &= \alpha (D_{\rm s} - D_{\rm d}) \frac{D_{\rm d}}{D_{\rm s}} = \frac{4GM}{Rc^2} \frac{(D_{\rm s} - D_{\rm d}) D_{\rm d}}{D_{\rm s}}. \end{aligned} \tag{7}$$

This equation may be written as

$$\frac{R_{\rm s}}{R_{\rm E}} = -\frac{R}{R_{\rm E}} + \frac{R_{\rm E}}{R}, \qquad R^2 + R_{\rm s} R - R_{\rm E}^2 = 0, \tag{8a}$$

where

$$R_{\rm E}^2 \equiv 2 R_{\rm g} D, \qquad R_{\rm g} \equiv \frac{2GM}{c^2}, \qquad D \equiv \frac{(D_{\rm s} - D_{\rm d}) D_{\rm d}}{D_{\rm s}}, \tag{8b}$$

and $R_{\rm E}$, $R_{\rm g}$, and D are called the linear Einstein ring radius of the lens, the gravitational radius of the mass M, and the effective lens distance, respectively.

Equation (8a) has two solutions:

$$R_{+,-} = 0.5[R_{\rm s} \pm (R_{\rm s}^2 + 4R_{\rm E}^2)^{1/2}]. \tag{9}$$

These two solutions correspond to the two images of the same source, located on the opposite sides of the point M, at the angular distances of $R_+/D_{\rm d}$ and $R_-/D_{\rm d}$, respectively. Figure 2 shows the appearance of a small circular source (in the absence of lensing) and its two distorted images in a telescope with very high resolving power. Since any lensing conserves surface brightness (cf Schneider et al 1992, Section 5.2), the ratio of the image to source intensity is given by the ratio of their areas. With the geometry shown in Figure 2, this ratio can be calculated as

$$A_{+,-} = \left| \frac{R_{+,-}}{R_{\rm s}} \frac{dR_{+,-}}{dR_{\rm s}} \right| = \frac{u^2 + 2}{2u(u^2 + 4)^{1/2}} \pm 0.5, \tag{10}$$

where $u \equiv R_s/R_E$ and it is assumed that the source is very small. The quantity A is called magnification or amplification. The term "magnification" will be used throughout this paper, for it better describes the lensing process. The total magnification of the two images can be calculated as

$$A = A_+ + A_- = \frac{u^2+2}{u(u^2+4)^{1/2}}. \tag{11}$$

It is interesting to note that this number is always larger than unity and that the difference in the magnification of the two images is constant:

$$A_+ - A_- = 1. \tag{12}$$

Let us consider a typical Galactic case, with a lens of $\sim 1\ M_\odot$ mass at a distance of a few kiloparsecs, and the source at a larger distance. The angular Einstein ring radius r_E can be calculated with Equations (8b):

$$r_E \equiv \frac{R_E}{D_d} = \left[\left(\frac{4GM}{c^2}\right)\left(\frac{D_s - D_d}{D_s D_d}\right)\right]^{1/2}$$
$$= 0.902 \text{ mas}\left(\frac{M}{M_\odot}\right)^{1/2}\left(\frac{10\text{ kpc}}{D_d}\right)^{1/2}\left(1 - \frac{D_d}{D_s}\right)^{1/2}. \tag{13}$$

With the image separation $\sim 2r_E$, i.e. of the order of a milliarcsecond (mas), we can only see the combined light intensity, rather than two separate images

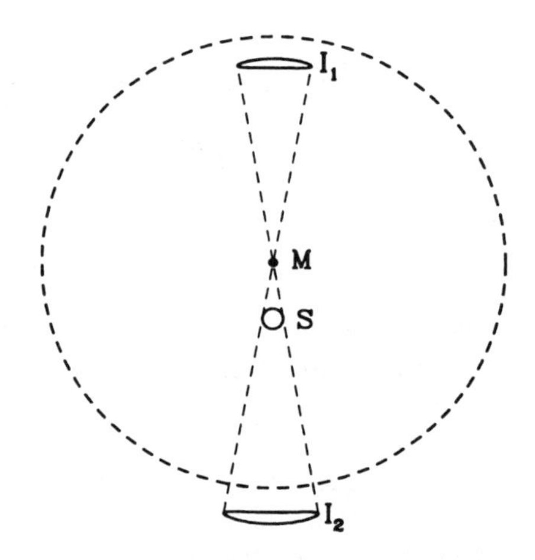

Figure 2 The geometry of gravitational lensing is shown. The lensing mass, the small circular source, and the two images are marked with M, S, I_1, and I_2, respectively. Of course, in the presence of mass M the source is not seen at S but only at I_1 and I_2. The Einstein ring is shown as a dashed circle. A typical radius of the circle is ~ 1 milliarcsecond for microlensing by stars in our galaxy.

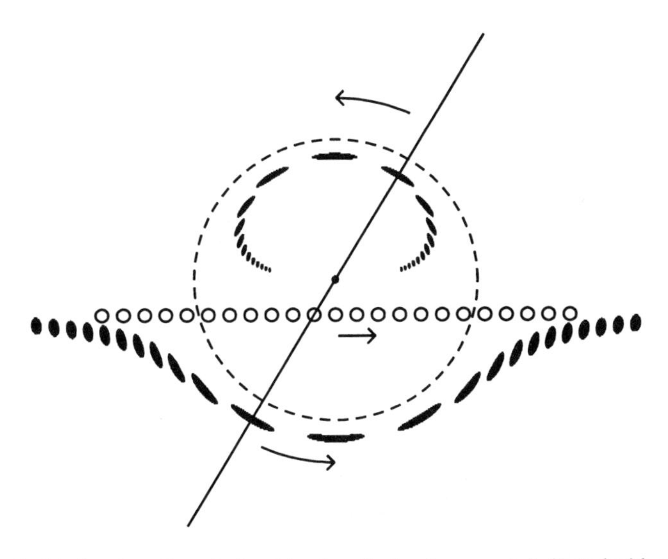

Figure 3 In this drawing of gravitational lensing, the lensing mass is indicated with a dot at the center of the Einstein ring, which is marked with a dashed line; the source positions are shown with a series of small open circles; and the locations and the shapes of the two images are shown with a series of dark ellipses. At any instant the two images, the source, and the lens are all on a single line, as shown in the figure for one particular instant.

because of the finite resolution of the optical telescopes. Fortunately, all objects in the Galaxy move, and we may expect a relative proper motion to be

$$\dot{r} = \frac{V}{D_{\rm d}} = 4.22 \text{ mas yr}^{-1} \left(\frac{V}{200 \text{ km s}^{-1}}\right)\left(\frac{10 \text{ kpc}}{D_{\rm d}}\right), \tag{14}$$

where V is the relative transverse velocity of the lens with respect to the source. Combining the last two equations we can calculate the characteristic time scale for a microlensing phenomenon as the time it takes the source to move with respect to the lens by one Einstein ring radius:

$$\begin{aligned} t_0 &\equiv \frac{r_{\rm E}}{\dot{r}} \\ &= 0.214 \text{ yr} \left(\frac{M}{M_\odot}\right)^{1/2} \left(\frac{D_{\rm d}}{10 \text{ kpc}}\right)^{1/2} \left(1 - \frac{D_{\rm d}}{D_{\rm s}}\right)^{1/2} \left(\frac{200 \text{ km s}^{-1}}{V}\right). \end{aligned} \tag{15}$$

This definition is almost universally accepted, with one major exception: The MACHO collaboration multiplies the value of t_0 as given by Equation (15) by 2.

While the lens moves with respect to the source, the two images change their position and brightness, as shown in Figure 3. When the source is close to the lens the images are highly elongated, and their proper motion is much higher than that of the source. Note that for each source position the two images, the source, and the lens are all located on a straight line that rotates around the lens. Unfortunately, none of this geometry can be observed directly for the stellar mass lenses, because the angular scale is of the order of a milliarcsecond. However, the total light variations can be observed, and the light curve can be calculated with Equation (11). The variable u can be determined as

$$u = \left[p^2 + \left(\frac{t - t_{\max}}{t_0} \right)^2 \right]^{1/2}, \tag{16}$$

where p is the dimensionless impact parameter (the smallest angular distance between the source and the lens measured in units of Einstein ring radius) and $t_{\max}$ is the time of the maximum magnification by the lens. Figures 4 and 5 show the geometry of lensing for six values of impact parameter and the corresponding time variability expressed in stellar magnitudes ($\Delta m \equiv 2.5 \log A$).

It is customary to define the cross section for gravitational microlensing to be equal to the area of the Einstein circle. The combined magnification of the two images of a source located within that circle is larger than

$$A \geq \frac{3}{5^{1/2}} = 1.3416, \qquad \Delta m \geq 0.3191 \text{ mag} \qquad (u \leq 1) \tag{17}$$

(cf Equation 11), i.e. it is easy to detect the intensity variation with reasonably accurate photometry.

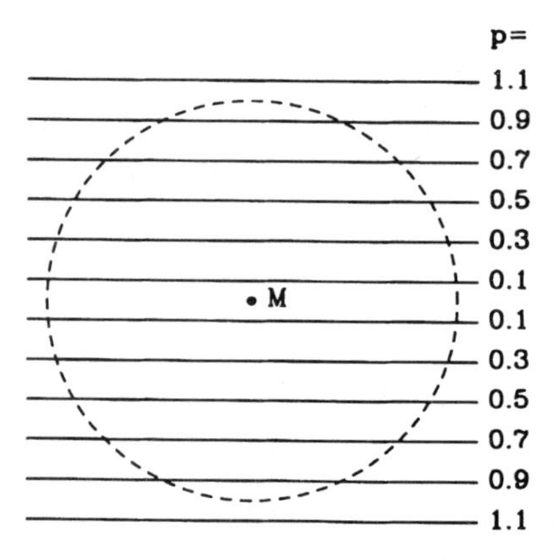

Figure 4 The geometry of gravitational lensing is shown. The lensing mass M is located at the center of the Einstein ring, which is marked with a dashed line. The 12 horizontal lines represent relative trajectories of the source, labeled with the value of dimensionless impact parameter p.

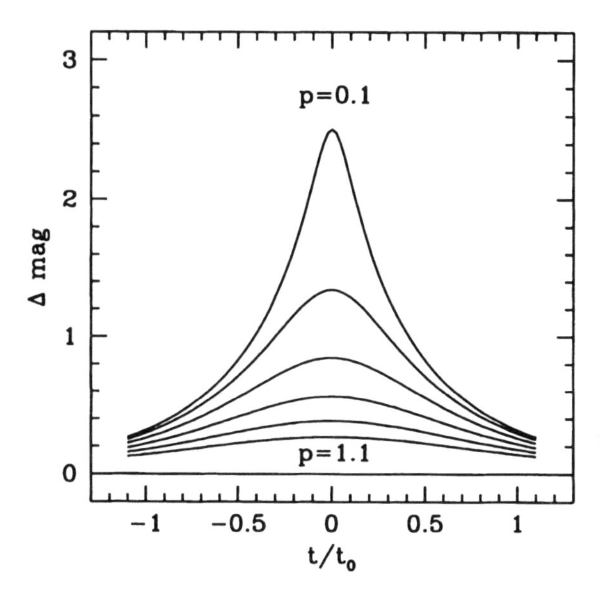

Figure 5 The variation of the magnification due to a point gravitational lensing is shown in stellar magnitudes as a function of time. The unit t_0 is defined as the time it takes the source to move a distance equal to the Einstein ring radius, $r_{\rm E}$. The six light curves correspond to the six values of the dimensionless impact parameter: $p = 0.1,\ 0.3,\ 0.5,\ 0.7,\ 0.9,\ 1.1$.

Suppose there are many lensing objects in the sky. The fraction of solid angle covered with their Einstein rings is called the optical depth to gravitational microlensing. Let all lensing objects have identical masses M. In a thin slab at a distance $D_{\rm d}$ and a thickness $\Delta D_{\rm d}$, there is, on average, one lens per surface area $\pi R_{\rm M}^2 = M/(\rho \Delta D_{\rm d})$, where ρ is the average mass density due to lenses in the volume $\pi R_{\rm M}^2 \Delta D_{\rm d}$. Each lens has a cross section $\pi R_{\rm E}^2$, where the Einstein ring radius $R_{\rm E}$ is determined from Equation (8b). The slab contribution to the optical depth is given as

$$\Delta\tau = \frac{\pi R_{\rm E}^2}{\pi R_{\rm M}^2} = \left[\frac{4\pi G\rho}{c^2}\,\frac{D_{\rm d}\,(D_{\rm s} - D_{\rm d})}{D_{\rm s}}\right] \Delta D_{\rm d}. \tag{18}$$

Then the total optical depth due to all lenses between the source and the observer can be calculated as

$$\begin{aligned}\tau &= \int_0^{D_{\rm s}} \frac{4\pi G\rho}{c^2}\,\frac{D_{\rm d}\,(D_{\rm s} - D_{\rm d})}{D_{\rm s}}\, dD_{\rm d} \\ &= \frac{4\pi G}{c^2} D_{\rm s}^2 \int_0^1 \rho(x)\, x(1-x)\, dx,\end{aligned} \tag{19}$$

where $x \equiv D_d/D_s$. Note that the optical depth τ depends on the total mass in all lenses, but it is independent of the masses of individual lenses, M. If the density of matter is constant, we have

$$\tau = \left(\frac{2\pi}{3}\right)\left(\frac{G\rho}{c^2}\right) D_s^2. \tag{20}$$

If the system of lenses is self-gravitating, then a crude but very simple estimate of the optical depth can be made. Let us suppose that the distance to the source D_s is approximately equal to the size of the whole system—a galaxy of lenses. The virial theorem provides a relation between the velocity dispersion V^2, the density ρ, and the size D_s:

$$\frac{GM_{tot}}{D_s} \approx \frac{G\rho D_s^3}{D_s} \approx V^2. \tag{21}$$

Combining Equations (20) and (21) we obtain

$$\tau \approx \frac{V^2}{c^2}. \tag{22}$$

A more accurate estimate of the optical depth can be obtained by evaluating the integral in Equation (19) for any distribution of mass density along the line of sight.

3. THE EVENT RATE AND THE LENS MASSES

We now proceed to the estimate of the number of microlensing events N that may be expected if n sources are monitored over a time interval Δt. We consider only those microlensing events that have peak magnification in excess of $3/(5)^{1/2}$, i.e. their dimensionless impact parameters are smaller than unity (cf Figures 4 and 5). We begin with the simplest case: All lensing objects have the same mass M, and all have the same three-dimensional velocity V. We also assume that the velocity vectors have a random but isotropic distribution, the source located at the distance D_s is stationary, and the number density of lensing objects is statistically uniform between the observer and the source.

The time scale of a microlensing event is given as

$$t_0 = \frac{r_E}{\dot{r}} = \frac{R_E}{V_t} = \frac{R_E}{V \sin i} \tag{23}$$

(cf Equation 15), where i is the angle between the velocity vector and the line of sight and $V_t = V \sin i$ is the transverse velocity of the lens. If all events had identical time scales, then the number of microlensing events expected in a time interval Δt would be given as

$$N = \frac{2}{\pi} n\tau \frac{\Delta t}{t_0} \tag{24}$$

for $t_0 = const$, where $2/\pi$ is the ratio of Einstein ring diameter to its area, in dimensionless units, and τ is the optical depth.

There is a broad distribution of event time scales because lenses (all with the same mass M and the same spatial velocity V) have transverse velocities in the range $0 \leq V_t \leq V$ and distances in the range $0 \leq D_d \leq D_s$. Straightforward but tedious algebra leads to the equation

$$N = \frac{3\pi}{16} n\tau \frac{\Delta t}{t_m} = \int_0^\infty N'(t_0) dt_0, \tag{25}$$

where

$$t_m \equiv \left(\frac{R_E}{V}\right)_{D_d = 0.5 D_s} = \left(\frac{GMD_s}{c^2}\right)^{1/2} \frac{1}{V} \tag{26}$$

is the time scale for a microlensing event due to a lens located halfway between the source and the observer and moving with the transverse velocity V. A detailed analysis is given by Mao & Paczyński (1996).

The probability distribution of event time scales is very broad. It can be shown to have power-law tails for very short and for very long time scales:

$$P(t_0 \leq t') = \frac{128}{45\pi^2} \left(\frac{t'}{t_m}\right)^3, \qquad \text{for} \quad t' \ll t_m, \tag{27}$$

$$P(t_0 \geq t') = \frac{128}{45\pi^2} \left(\frac{t_m}{t'}\right)^3, \qquad \text{for} \quad t' \gg t_m. \tag{28}$$

Note that the power-law tails in the distribution of event time scales are generic to almost all lens distributions ever proposed. The very short events are due to lenses that are either very close to the source or very close to the observer, whereas the very long events are caused by the lenses that move almost along the line of sight. Also note that only the first two moments of the distribution are finite; the third and higher moments diverge. Therefore, it is convenient to use a logarithmic probability distribution defined as

$$p(\log t_0) d \log t_0 = \left(\frac{\ln 10}{N}\right) t_0 N'(t_0) d \log t_0, \tag{29}$$

because all moments of this distribution are finite.

De Rújula et al (1991) proposed using the moment analysis to deduce the distribution of lens masses. We carry out such an analysis, but first we have to make our model more realistic. Since most astrophysical objects have a broad range of velocities, we adopt a three-dimensional Gaussian distribution:

$$p(V) \frac{dV}{V_{rms}} = 3 \left(\frac{6}{\pi}\right)^{1/2} \exp\left(-\frac{3V^2}{2V_{rms}^2}\right) \frac{V^2}{V_{rms}^2} \frac{dV}{V_{rms}},$$

$$\text{for} \quad 0 \leq V < \infty, \tag{30}$$

where the three-dimensional rms velocity $V_{\rm rms}$ is defined as

$$V_{\rm rms}^2 \equiv \int_0^\infty V^2 p(V) \frac{dV}{V_{\rm rms}}. \tag{31}$$

We also adopt a general power-law distribution of the number of lensing masses,

$$n(M)dM \sim M^\alpha dM, \qquad \text{for} \quad M_{\rm min} \le M \le M_{\rm max}, \tag{32}$$

or, in a logarithmic form,

$$n(\log M) d\log M = \left[\frac{(\alpha+1)\ln 10}{M_{\rm max}^{\alpha+1} - M_{\rm min}^{\alpha+1}} \right] M^{\alpha+1} d\log M, \qquad \text{for} \quad \alpha \ne -1, \tag{33}$$

$$n(\log M) d\log M = \left[\frac{\ln 10}{\ln(M_{\rm max}/M_{\rm min})} \right] d\log M, \qquad \text{for} \quad \alpha = -1. \tag{34}$$

The case $\alpha = -2$ corresponds to equal total mass per decade of lens masses, i.e. each decade has the same contribution to the optical depth. The case $\alpha = -1.5$ corresponds to an equal rate of microlensing events per decade of lens masses. If the mass function is very broad, i.e. $M_{\rm max}/M_{\rm min} \gg 1$, then the distribution of event rate will be flat in this case, with an equal number of events per logarithmic interval of t_0 and with power-law tails (cf Equations 27 and 28). If $\alpha = -1$ then there is an equal number of lenses per logarithmic interval of their masses.

A characteristic mass scale for the lenses can be related to the first moment of the t_0 distribution:

$$M_0 \equiv \frac{c^2 V_{\rm rms}^2}{G D_{\rm s}} t_{0,\rm av}^2, \tag{35}$$

where $\log t_{0,\rm av} \equiv \langle \log t_0 \rangle$ (cf Equations 26 and 30). The value of M_0 is indicative of the most common lens mass. However, for $\alpha \ll -1.5$ the mass spectrum is dominated by low-mass objects, the event time scales have a small effective range, and the events are mostly due to lenses with $M \approx M_{\rm min}$. In the other extreme case, with $\alpha \gg -1.5$, the microlensing events are dominated by lenses with $M \approx M_{\rm max}$.

For a given mass range the standard deviation of $\log t_0$ is the largest for $\alpha = -1.5$, and it can be estimated analytically. First, imagine that there is a unique relation between the lens mass M and the event time scale t_0. The exponent $\alpha = -1.5$ implies a uniform distribution of event time scales in $\log t_0$ over the range $\log t_{\rm min} \le \log t_0 \le \log t_{\rm max}$, with $\log(t_{\rm max}/t_{\rm min}) = 0.5 \log(M_{\rm max}/M_{\rm min}) = 0.5\Delta \log M$. This distribution would have a standard deviation $\sigma_{\log t_0} = (\Delta \log M)^2/48$. However, there is no unique relation between the lens mass and the corresponding time scale t_0 in our model. Even if all lenses had the same mass, the standard deviation would be $\sigma_0 = 0.268$, according to numerical calculations. Therefore, the following formula can be used to calculate the second moment:

$$\sigma_{\log t_0} = \left[\sigma_0^2 + \frac{(\Delta \log M)^2}{48}\right]^{1/2}, \quad \sigma_0 = 0.268, \quad \text{for} \quad \alpha = -1.5. \tag{36}$$

There is another serious problem affecting the relation between the observed event time scales and the lens masses. In general we do not know which model should be adopted for the spatial distribution of lenses and for their kinematics. For the purpose of our exercise we adopted a uniform space density and a uniform velocity distribution, all the way from the source to the observer. Unfortunately, there is no consensus yet on the distribution and the kinematics of the dominant lensing component towards the Galactic bulge and towards the Large Magellanic Cloud (LMC). To reach a consensus it will be necessary to determine observationally the variation of optical depth to microlensing with the galactic coordinates and to use this information to determine the spatial distribution of the dominant lens population. Next, a model of the galactic gravitational potential will have to be used to estimate the kinematics of the dominant lens population. In order to have confidence that the observed distribution of event time scales is not truncated by instrumental effects, the power-law tails, like those given with the Equations (27) and (28), will have to be well sampled. When all this work is done, a relation between the distribution of event time scales and the distribution of lens masses will be sound.

4. DOUBLE LENSES AND PLANETS

Let us consider now a large number of point-mass lenses, all located at the same distance D_d, in front of a point source at D_s. Let a lens i with mass M_i be located at (X_i, Y_i). A light ray crossing the deflector's plane at (X, Y) would pass at a distance R_i from the lens i:

$$R_i = \left[(X - X_i)^2 + (Y - Y_i)^2\right]^{1/2}, \tag{37}$$

(cf Equation 3). The contribution to the light-ray deflection by the mass i is given by the angle

$$\alpha_i = \frac{4GM_i}{R_i c^2}, \tag{38}$$

with the two components

$$\alpha_{x,i} = \alpha_i \frac{X - X_i}{R_i}, \qquad \alpha_{y,i} = \alpha_i \frac{Y - Y_i}{R_i} \tag{39}$$

(cf Equations 4 and 5). As a result of the combined deflection by all lensing masses the light ray will cross the source plane at (X_S, Y_S):

$$X_S = X\frac{D_s}{D_d} - \sum_i \alpha_{x,i}(D_s - D_d), \quad Y_S = Y\frac{D_s}{D_d} - \sum_i \alpha_{y,i}(D_s - D_d) \quad (40)$$

(cf Equation 6). Note that X, Y, X_i, Y_i, and R_i are all measured in the deflector plane, while X_s and Y_s are measured in the source plane.

Combining Equations (38–40) we obtain

$$x_s = x - \frac{(D_s - D_d)}{D_s D_d} \sum_i \frac{4GM_i}{c^2} \frac{(x - x_i)}{r_i^2}, \quad (41)$$

$$y_s = y - \frac{(D_s - D_d)}{D_s D_d} \sum_i \frac{4GM_i}{c^2} \frac{(y - y_i)}{r_i^2}, \quad (42)$$

where the angles are defined as

$$x_s = \frac{X_S}{D_s}, \qquad y_s = \frac{Y_S}{D_s}, \quad (43)$$

and

$$x = \frac{X}{D_d}, \quad y = \frac{Y}{D_d}, \quad x_i = \frac{X_i}{D_d}, \quad y_i = \frac{Y_i}{D_d}, \quad r_i = \frac{R_i}{D_d}. \quad (44)$$

With dimensionless masses m_i defined as

$$m_i = \frac{(D_s - D_d)}{D_s D_d} \frac{4GM_i}{c^2}, \quad (45)$$

Equations (41) and (42) may be written as

$$x_s = x - \sum_i \frac{m_i(x - x_i)}{r_i^2} \quad (46)$$

and

$$y_s = y - \sum_i \frac{m_i(y - y_i)}{r_i^2}. \quad (47)$$

All these are angles as seen by the observer.

The set of Equations (46–47) can be used to find all images created by a planar lens system made of any number of point masses (cf Witt 1993 and references therein). Here we restrict ourselves to a double-lens case, since binary stars are known to be very common (Abt 1983, Mao & Paczyński 1991). A very thorough analysis of double-star microlensing is provided by Schneider & Weiss (1986). A major new phenomenon not present in a single lens case is the formation of caustics caused by the lens astigmatism. When a source crosses a caustic, a new pair of images forms or disappears. A point source placed at a caustic is magnified by an infinite factor, while a source with a finite size is subject to a large, but finite magnification. A double lens is vastly more complicated than a single one.

Equations (46–47) applied to the binary case are

$$x_s = x - \frac{m_1(x - x_1)}{r_1^2} - \frac{m_2(x - x_2)}{r_2^2}, \tag{48}$$

$$y_s = y - \frac{m_1(y - y_1)}{r_1^2} - \frac{m_2(y - y_2)}{r_2^2}. \tag{49}$$

It is customary to adopt $m_1 + m_2 = 1$. This makes all the angles to be expressed in units of the Einstein ring radius for a lens with a unit mass. If the binary orbital motion is neglected (a static binary case) then, in addition to all standard parameters describing single point mass lensing, there are three new dimensionless parameters: the mass ratio (m_1/m_2), the binary separation in units of the Einstein ring radius, and the angle between the source trajectory and the line joining the two components of the lens. The diversity of possible light curves is staggering. Mao & Di Stefano (1995) have developed the first computer code that can not only generate theoretical light curves for a binary lens, but can also fit the best theoretical light curve to the actual data. It has been applied to determine the parameters of two events that were almost certainly caused by double lenses: OGLE #7 (Udalski et al 1994d) and DUO #2 (Alard et al 1995b).

A few examples of light curves generated by models of a double lens are shown in Figure 6, with the geometry of lensing indicated in Figure 7. The binary model is composed of two identical point masses (shown in Figure 7 with two large points), $M_1 = M_2 = 0.5\,M$, with the separation equal to the Einstein ring radius corresponding to M, the total binary mass. The complicated closed figure drawn with a solid line is the caustic. When a source crosses the caustic, two images appear or disappear somewhere on the critical curve, shown with a dashed line. The critical line is defined as the location of all points in the lens plane that are mapped by the lensing onto the caustic, which is located in the source plane. When the source is outside the caustic, three images are formed—one outside of the critical line and two inside—usually very close to one of the two point masses. When the source is inside the region surrounded by the caustic, two additional images are present—one inside and the other outside the region surrounded by the critical curve. In our example, the five sources, marked with small open circles, have radii equal to 0.05 of the Einstein ring radius. The five straight trajectories are also marked.

The mass ratio of a double lens may be very extreme if one of the two components is a planet. Mao & Paczyński (1991) proposed that microlensing searches may lead to the discovery of the first extra solar planetary system. That suggestion turned out to be incorrect as the first extra–solar system with three or even four planets has been discovered by other means (Wolszczan & Frail 1992, Wolszczan 1994). Still, it is possible that numerous planetary systems will be detected through their microlensing effects. Gould & Loeb (1992) and Bolatto & Falco (1994) refined the analysis by Mao & Paczyński (1991), but the real problem is setting up a practical detection system. Here we consider just one aspect of the problem, with more discussion to follow in Section 7.

Although a search for Jupiter-like objects might prove very interesting, the detection of Earth-like planets would be far more exciting. In fact, such planets have been discovered around the radio pulsar PSR B1257+12 by Wolszczan & Frail (1992). Until recently, the many searches for Jupiter-mass planets yielded negative results, leading to the conclusion that "... and the absence of detections is becoming statistically significant ..." (Black 1995). A recent discovery of a Jupiter-mass companion to a nearby star 51 Pegasi by Mayor & Queloz (1995)

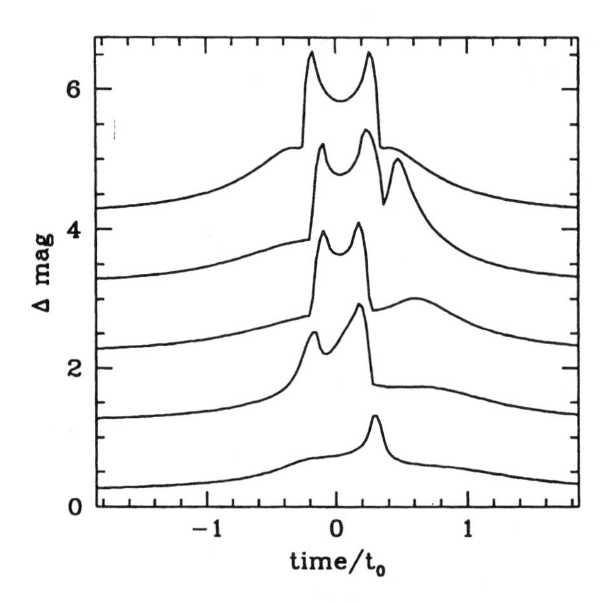

Figure 6 Five light curves are shown as examples of binary microlensing. The two components of the binary have identical masses and are separated by one Einstein ring radius. The corresponding source trajectories are shown in Figure 7. The top light curve shown here corresponds to the top trajectory in Figure 7. The sharp spikes are due to caustic crossings by the source. (The light curves are shifted by one magnitude for clarity of the display.)

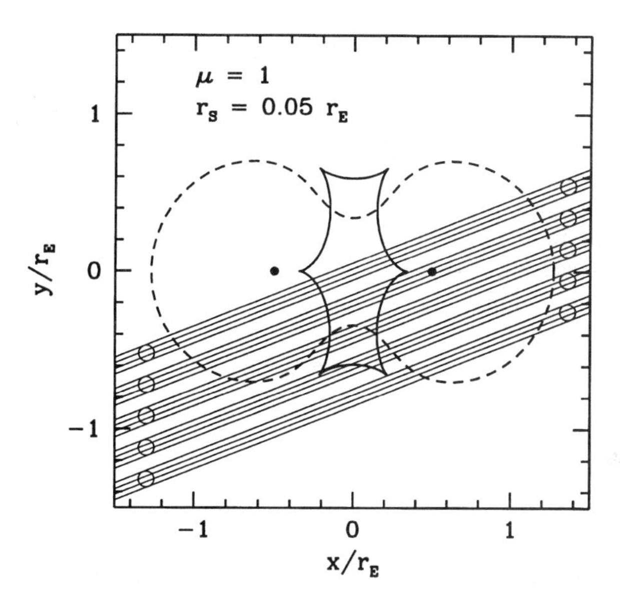

Figure 7 The geometry of gravitational microlensing responsible for the light curves presented in Figure 6 are shown. The two identical point masses, $M_1 = M_2 = 0.5\,M$, are indicated with two points separated by one Einstein ring radius, r_E. The closed figure drawn with a thick solid line is the caustic located in the source plane. The closed figure drawn with a thick dashed line is the critical curve. A source placed on a caustic creates an image located on the critical curve. Five identical sources are moving along the straight trajectories, as marked. All sources have radii equal to $r_s = 0.05\,r_E$, as shown with small open circles.

and the discovery by Marcy & Butler (1996) and by Butler & Marcy (1996) of similar companions to the stars 70 Vir and 47 UMa , respectively, demonstrate that such planets exist.

If all stars had Jupiters at a distance of a few astronomical units then a few percent of all microlensing events might show a measurable distortion of their light curves (Mao & Paczyński 1991, Gould & Loeb 1992, Bolatto & Falco 1994). If a small fraction of all stars have Jupiters at such distances then the fraction of microlensing events that may be disturbed by giant planets is correspondingly reduced. Note that the duration of the likely disturbance is of the order of ~ 1 day, i.e. very frequent observations are necessary to detect them. According to Butler & Marcy(1996), $\sim 5\%$ of all stars have super-Jupiter planets within 5 astronomical units.

The detection of Earth-like planets is beyond the range of traditional techniques: Radial velocity as well as astrometric measurements are not accurate enough, while pulsar timing is obviously not applicable to ordinary main sequence stars. It is interesting to check if Earth-like planets are within reach for gravitational lensing searches. Earth's mass is about $3 \times 10^{-6}\,M_\odot$, and a typical star is somewhat less massive than the Sun. Therefore, we consider now a very extreme mass ratio: $M_2/M_1 = 10^{-5}$.

An isolated planetary mass object would create an Einstein ring of its own, with the radius $r_{EP} = r_E \times (M_2/M_1)^{1/2}$. However, the same object placed close to a star, develops a complicated small-scale magnification pattern superimposed on the large-scale pattern generated by the star. It turns out that this effect is much more pronounced than a small disturbance which a planet may cause near the peak of the stellar magnification. An example of the planetary disturbances is shown in Figure 8, in which a very high magnification stellar microlensing event with impact parameter equal to zero is disturbed by eight Earth-like planets placed along the source trajectory. Needless to say, the disturbance caused by each planet is practically independent of the disturbances caused by all other planets.

Naturally, this is a very artificial arrangement, but it makes it possible to present a variety of microlensing effects of Earth-like planets in a single figure.

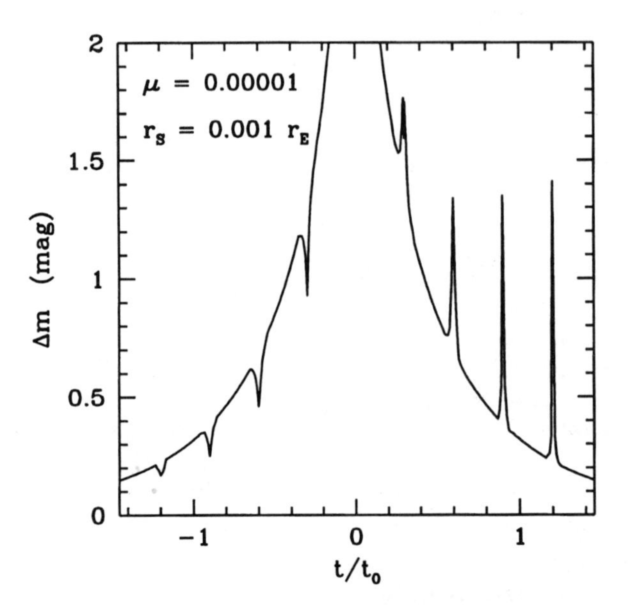

Figure 8 Variation of the magnification by a planetary system shown as a function of time. The system is made of a star and eight planets, each with the mass fraction $\mu = 10^{-5}$, all located along a straight line. The source with a radius $r_S = 10^{-3}\, r_E$ is moving along the line defined by the planets, with the impact parameter equal to zero. The planets are located at the distances from the star: $r_p/r_E = 0.57,\ 0.65,\ 0.74,\ 0.86,\ 1.16,\ 1.34,\ 1.55,\ 1.76$ in the lens plane, which corresponds to the disturbances in light variations at the times $t/t_0 = -1.2,\ -0.9,\ -0.6,\ -0.3,\ 0.3,\ 0.6,\ 0.9,\ 1.2$, as shown in the figure. Note that planetary disturbances create local light minima for $r_p/r_E < 1$ $(t/t_0 < 0)$ and local maxima for $r_p/r_E > 1$ $(t/t_0 > 0)$.

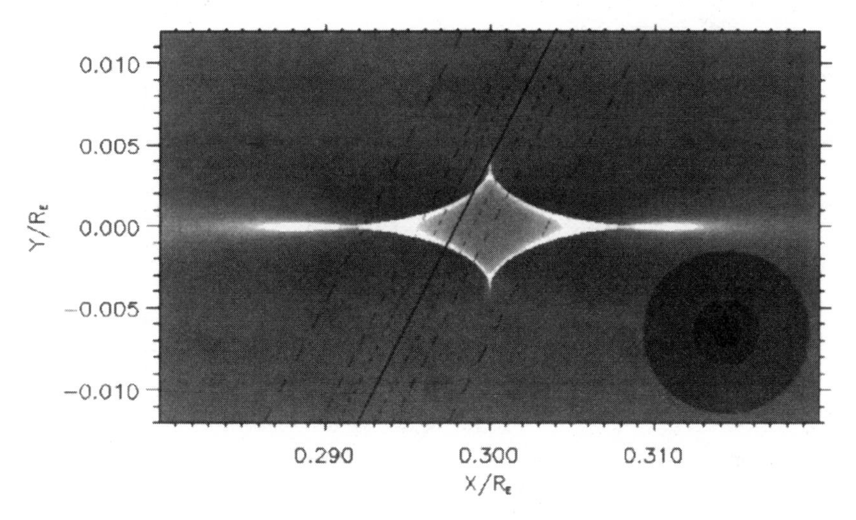

Figure 9 The illumination pattern in the source plane created by a planet with the mass fraction $\mu = 10^{-5}$. The planet is located at the distance $r_p/r_E = 1.16$ from the star in the lens plane, which corresponds to $r_p/r_E = 0.3$ in the source plane. The bright rims are due to the caustics. The centers of three circular sources are moving upwards along the solid straight line, with the dashed lines indicating the trajectories of the source edges. The sources have radii $r_S/r_E = 0.001,\ 0.002,\ 0.005$, and their sizes are shown in the lower right corner. The brightness variations caused by this planetary microlensing event are shown in Figure 10. Note that the light curve shown in Figure 8 corresponds to a source with the radius $r_S/r_E = 0.001$ moving along the X-axis. Also note that the area significantly disturbed by the planet is larger than $\pi\mu^{1/2}$, which is the microlensing cross section for an isolated planet.

The dimensionless time of a planetary disturbance t/t_0 is equal to the dimensionless position of the source r_s/r_E, and both are related to the dimensionless location of the planet with Equation (8a): $t/t_0 = r_s/r_E = r_p/r_E - r_E/r_p$. The planets located close to the star, at $r_p/r_E < 1$, create local minima in the microlensing light curve presented in Figure 8; they do this by reducing the brightness of the image corresponding to I_1 in Figure 2. The planets located farther away from the star, at $r_p/r_E > 1$, create local maxima (or double maxima) in the microlensing light curve presented in Figure 8; they do this by splitting the image corresponding to I_2 in Figure 2, and enhancing the combined brightness. If a planet is located close to the Einstein ring, i.e. if $r_p \approx r_E$, then it affects the peak of stellar microlensing light curve by disturbing one of the two images. While these disturbances are moderately large for Jupiter-like planets (Mao & Paczyński 1991), they turn out to be very small for Earth-like planets. A detailed description and the explanation of these phenomena is provided Bennett & Rhie (1996).

A close-up view of a single planetary event is presented in Figures 9 and 10. The two-dimentional magnification pattern is shown in Figure 9, together with

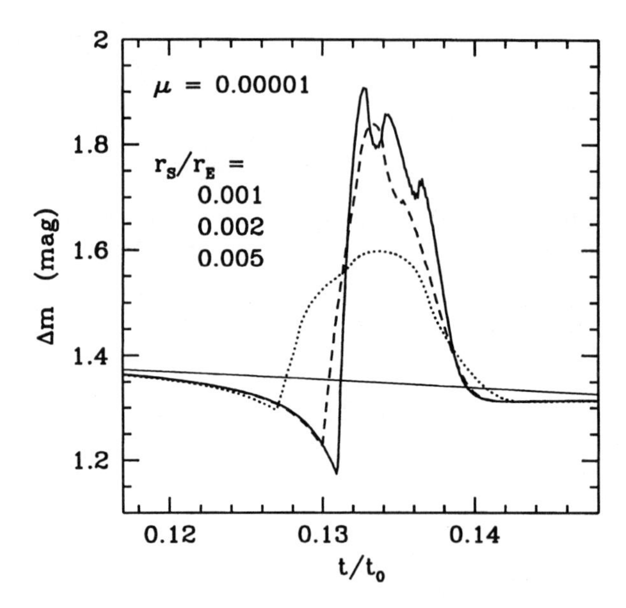

Figure 10 Variations of the magnification shown in stellar magnitudes as a function of time. The three light curves, shown with thick solid, dashed, and dotted lines, are caused by an Earth-like planet with the mass fraction $\mu = 10^{-5}$, and they correspond to the three sources with the radii $r_S/r_E = 0.001,\ 0.002,\ 0.005$, following trajectories shown in Figure 9. The thin, slowly descending solid line corresponds to the fragment of the stellar microlensing light curve in the absence of planetary disturbance. The full time interval shown in this figure corresponds to ~ 10 h, and $t_0 = 0$ corresponds to the peak of the stellar magnification light curve.

the trajectories of three sources. The corresponding magnification variations are shown in Figure 10. This event corresponds to the magnification disturbance at $t/t_0 = 0.3$ in Figure 8, but the source trajectory is different in the two cases: It was along the X-axis in Figure 8, but it is inclined to the X-axis at an angle $\sim 64°$ in Figures 9 and 10. The planetary event lasts longer in Figure 8 than in Figure 10 because the region of high magnification is strongly stretched along the X-axis, as shown in Figure 9. This stretching enhances the cross section for planetary microlensing. The diversity of possible light curves is very large, just as it is in the case of a binary lens (cf Figure 6). Figures 8, 9, and 10 presented in this section were prepared by Dr Joachim Wambsganss. I owe my insight into the diversity of planetry microlensing phenomena to Dr David Bennett and to Dr Jordi Miralda-Escudè.

5. VARIOUS COMPLICATIONS

It is often claimed that gravitational microlensing of stars is achromatic, that it does not repeat, and that a symmetric light curve is described by a single dimensionless quantity: the ratio of the impact parameter to the Einstein ring radius. Although it is true that the above description is a good approximation to the majority of microlensing events, it is well established that none of these claims is strictly correct.

Stars are not distributed randomly in the sky; they are well known to be clustered on many scales. Most stars are in binary systems, with the separations ranging from physical contact to 0.1 pc. Many stars are in multiple systems. Note that one of the brightest stars in the sky, Castor, is a sixtuple system. These statistics apply to the stars that are sources as well as to those that are lenses. The case of a double source is simple, as it generates a linear sum of two single-lens light curves (Griest & Hu 1992). Naturally, because the two stellar components may have different luminosities and colors, the composite light curve may well be chromatic. If the two source stars are well separated and the lens trajectory is along the line joining the two, we may have a perception of two microlensing events separated by a few months or a few years, but both acting on apparently the same star, because images of the two binary components are unresolved. This would have the appearance of a recurrent microlensing event and may affect a few percent of all events (Di Stefano & Mao 1996).

A much more complicated light curve is generated by a double lens as described in the previous subsection. At least two types of double-lensing events are expected. In the resonant lensing case, i.e. when the two-point lenses of similar mass are separated by approximately one Einstein ring radius, caustics are formed and dramatic light variations are expected, as in the case of OGLE #7 (Udalski et al 1994d) and DUO #2 (Alard et al 1995b).

Since microlensing events are very rare, all current searches are done in very crowded fields to measure as many stars as possible in a single CCD frame. This means that the detection limit is set not by the photon statistics but by overlapping of images of very numerous faint stars. A typical seeing disk is about one arcsecond across, whereas the cross section for microlensing is about one milliarcsecond (cf Equation 13). Therefore, an apparently single stellar image may typically be a blend of two or more stellar images separated by less then an arcsecond but by more than a milliarcsecond, i.e. only one of the two (or more) stars contributing to the blended image is subject to microlensing. Thus, when a theoretical light curve is fitted to the data it should always include a constant light term (Di Stefano & Esin 1995). The contribution from a constant light was found in the double lenses OGLE #7 (Udalski et al 1994d) and DUO #2 (Alard et al 1995b) and with at least one single lens OGLE #5 (S Mao, private communication). Because the various components to the apparently single stellar image may differ in color, a microlensing event of only one of them may appear as chromatic (Kamionkowski 1995). Of course, there should be a linear relation between the variable components in all color bands.

No star is truly a point source. The finite extent of a star affects a microlensing light curve when the impact parameter of a single lens is smaller than the source diameter, and also when the source crosses a caustic of a double lens. Accounting for the finite extent of a star complicates the light curve by adding one more adjustable parameter. This effect, when measured, may be used to calculate the relative proper motion of the lens-source system (Gould 1994a). It may also be used to study the distribution of light across the stellar disk, the limb darkening, and the spots.

So far we have assumed that the relative motion between a lens and a source is a straight line. However, the trajectory may be more complicated because of Earth's acceleration toward the Sun and, in the case of a double lens, because the two components orbit each other. This nonlinear motion of Earth and of binary components makes a single-lensing event asymmetric if its time scale is longer than a few months (Gould 1992). A complicated double-lens event thus becomes even more complicated.

6. RESULTS FROM CURRENT SEARCHES

6.1 *Highlights*

The most outstanding result of the current searches for microlensing events was the demonstration that they were successful. By the time this article is printed over 100 microlensing events will have been discovered, mostly by the MACHO collaboration, making this the most numerous class of gravitational lenses known to astronomers. Just a few years ago, when these projects had just begun, the skepticism about their success was almost universal, yet at least three teams, DUO, MACHO, and OGLE, now have numerous and believable candidate events. An example of a single microlensing event, OGLE #2, is shown in Figure 11 (following Udalski et al 1994a). The microlensing event took place in 1992. This object was in the overlap area of two separate fields, so it had a large number of measurements in the I band: 93, 187, and 94 in the observing seasons 1993, 1994, and 1995, respectively. The star was constant in luminosity during these three years, with average I band magnitudes of 19.07, 19.10, and 19.13, respectively; the standard deviations of individual measurements were 0.13, 0.10, and 0.09 magnitude, respectively (M Szymański, private communication).

The distribution of the OGLE events in the $(V - I) - V$ color-magnitude diagram of the stars seen through Baade's Window, i.e. at $(l, b) \approx (1, -4)$, is shown in Figure 12. Note that the events are scattered over a broad area of the diagram populated by Galactic bulge stars, in proportion to the local number density of stars multiplied by the local efficiency of event detection (Udalski et al 1994b). A very similar distribution for over 40 MACHO events detected toward the Galactic bulge was shown by Bennett et al (1995). It is interesting that three out of ~40 MACHO events appear to be at the location occupied by the bulge horizontal branch stars as well as the disk main sequence stars, near $(V - I) \approx 1.2$, $V \approx 17$. Spectroscopic analysis should be able to reveal whether the lensed stars are in the disk or in the bulge.

The most dramatic result is the estimate by the MACHO collaboration that the optical depth to microlensing through the Galactic halo is only $9^{+7}_{-5} \times 10^{-8}$ (based on three events) and can contribute no more than 20% of what would be needed to account for all dark matter in the halo (Alcock et al 1995a, Alcock et al 1996). The most surprising result is the OGLE discovery that the optical depth is as high as $3.3 \pm 1.2 \times 10^{-6}$ (based on nine events) toward the Galactic bulge (Udalski et al 1994b). This result was found independently by MACHO with four events (Alcock et al 1995c) and confirmed qualitatively by DUO (Alard 1996). It should be remembered that all quantitative analyses of the optical depth as published to date are based on only the 16 events mentioned above. The statistics improved with the recent preprint (Alcock et al 1995e), with the analysis of 45 MACHO events detected in the direction of the Galactic bulge.

Some nonstandard effects that had been first predicted theoretically were also discovered. These include very dramatic light curves caused by stellar sources crossing caustics created by double lenses (OGLE #7: Udalski et al 1994d, Bennett et al 1995; DUO #2: Alard et al 1995b; and probably one of

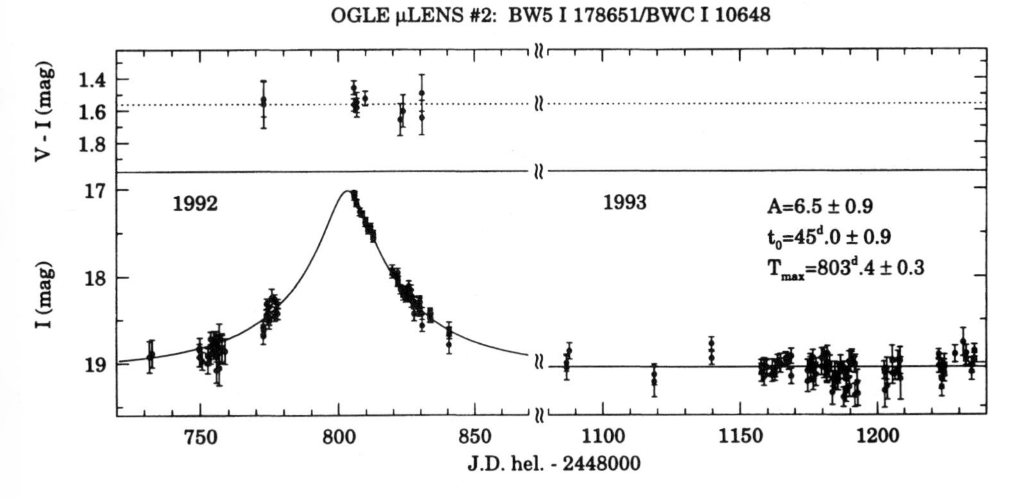

Figure 11 An example of the observed light curve due to a single point mass lensing: the OGLE lens candidate #2 (Udalski et al 1994a).

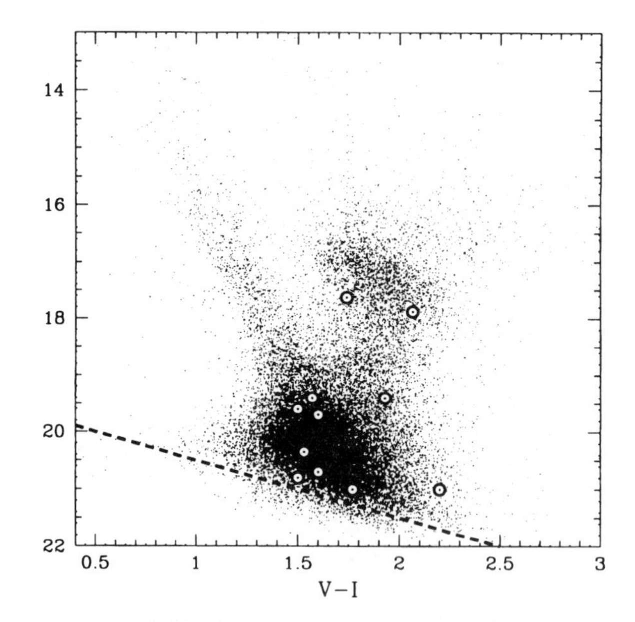

Figure 12 The color-magnitude diagram for stars in Baade's Window. The location of 11 OGLE lens candidates is shown with circles. The dashed line indicates the detection limit applied for the search (Udalski et al 1994b). The majority of stars are at the main sequence turn-off point of the Galactic bulge, near $V - I \approx 1.6$, $V \approx 20$. The bulge red clump stars are near $V - I \approx 1.9$, $V \approx 17$. The disk main sequence stars form a distinct band between $(V - I, V) \approx (1.5, 19)$ and $(V - I, V) \approx (1.0, 16)$ (Paczyński et al 1994 and references therein).

the MACHO Alert events: Pratt et al 1996), as predicted by Mao & Paczyński (1991); the parallactic effect of Earth's orbital motion (Alcock et al 1995d), as predicted by Gould (1992); the light curve distortion of a very high magnification event by a finite source size (C Alcock, private communication), as predicted by Gould (1994a), Nemiroff & Wickramasinghe (1994), Sahu (1994), Witt & Mao (1994), and Witt (1995); and the chromaticity of apparent microlensing caused by blending of many stellar sources, only one of them lensed (C Stubbs 1995, private communication), as predicted by Griest & Hu (1992).

The first theoretical papers with the estimates of optical depth towards the Galactic bulge (Griest et al 1991, Paczyński 1991) ignored the effect of microlensing by the Galactic bulge stars. That effect was noticed to be dominant by Kiraga & Paczyński (1994), who still ignored the fact that there is a bar in the inner region of our galaxy (de Vaucouleurs 1964, Blitz & Spergel 1991). Finally, the OGLE results forced upon us the reality of the bar (Udalski et al 1994b; Stanek et al 1994; Paczyński et al 1994; Zhao et al 1995, 1996). This rediscovery of the Galactic bar by the microlensing searchers, who were effectively ignorant of its existence, demonstrates that the microlensing searches are becoming a useful new tool for studies of the Galactic structure (Dwek et al 1995). We are witnessing a healthy interplay between theory and observations in this very young branch of astrophysics.

Figure 13 gives an example of a dramatic light curve of the first double microlensing event, OGLE #7, for 1992 and 1993 (following Udalski et al 1994d). This star was found to be constant in 1992, 1994, and 1995. The average magnitudes based on 32, 45, and 41 I-band measurements in these three observing seasons were 17.53, 17.52, and 17.54, respectively, with the variance of single measurements being 0.07, 0.04, and 0.03 magnitudes, respectively (M Szymański, private communication). The objects, were also found in the MACHO database (Bennett et al 1995), confirming the presence of the second caustic crossing event near JD 2449200 and demonstrating that the light variation was achromatic.

It is hard to decide which was the first detected microlensing event. If we take the time at which a microlensing event reached its maximum, then the first

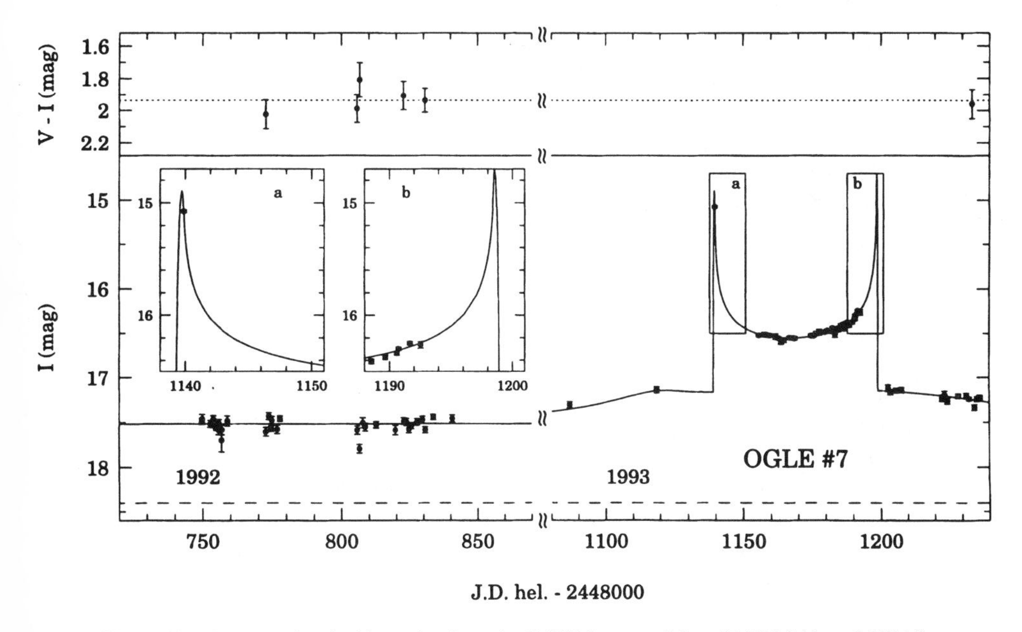

Figure 13 An example of a binary lensing: the OGLE lens candidate #7 (Udalski et al 1994d). The region of the two caustic crossings, (*a*) and (*b*), is shown enlarged in the two inserts. The MACHO collaboration has a few dozen additional data points in two bands, demonstrating that the light variations were achromatic; three MACHO data points cover the second caustic crossing (*b*) (Bennett et al 1995).

was OGLE #10, which peaked on June 29, 1992 (Udalski et al 1994b). It was followed by six other OGLE events that were observed that summer. However, these were not uncovered until the spring of 1994, when the automated computer searches finally caught up with the backlog of unprocessed data. The OGLE collaboration discovered its first event, OGLE #1, on September 22, 1993, but it peaked on June 15, 1993, almost a full year later than OGLE #10. The first event ever noticed by a human was MACHO #1—Will Sutherland saw it come out of a computer on September 12, 1993 (C Alcock, private communication). The first three papers officially announcing the first computer detections were published almost simultaneously by EROS (Aubourg et al 1993), by MACHO (Alcock et al 1993), and by OGLE (Udalski et al 1993).

Unfortunately, it is likely that the two events reported by EROS might have been due to intrinsic stellar variability rather than microlensing. EROS #1 is now known to be an emission line Be type star (Beaulieu et al 1995). The MACHO collaboration has identified a new class of variable stars, referred to as bumpers (Cook et al 1995, Alcock et al 1996). These are Be type stars, and it is possible that EROS #1 exhibited a bumper phenomenon. Recently, EROS #2 has been found to be an eclipsing binary, possibly with an accretion disk (Ansari et al 1995a). Stars with accretion disks are known to exhibit diverse light variability, and the EROS #2 event might have been due to disk activity. In any case, the fact that both EROS events were related to rare types of stars makes them highly suspect as candidates for gravitational microlensing.

About 100 gravitational microlensing events have been reported so far by the DUO, MACHO, and OGLE collaborations. The DUO collaboration reported the detection of 13 microlensing events toward the Galactic bulge, one of them double (Alard 1996b; Alard et al 1995a,b). The MACHO collaboration reported at various conferences a total of 8 events toward the LMC and "about" 60 events toward the Galactic bulge; many more events exist in the data currently being analyzed. The OGLE collaboration has detected 18 events toward the Galactic bulge, one of them double. In addition, various collaborations confirmed events detected by the others. I am not aware of a single case in which there would be a discrepancy among the data obtained by various collaborations, though there is plenty of difference of opinion about some aspects of the interpretation. Although quantitative analysis of these ~ 100 events lags behind their discovery, we expect that robust determinations of the optical depth toward the Galactic bulge and toward the LMC will be published in the very near future, as the data are already at hand.

A major new development in microlensing searches is on-line data processing. The OGLE collaboration implemented its Early Warning System (EWS), a full on-line data processing system, from the beginning of its third observing season, i.e. from April 1994 (Udalski et al 1994c). As a result, all eight OGLE events detected in 1994 and 1995 were announced over the Internet in real time. The MACHO collaboration, with its vastly higher data rate, implemented the "Alert System" with partial on-line data processing in the summer of 1994 (Stubbs et al 1995), and full on-line processing by the beginning of 1995 (Pratt et al 1996). The total number of events detected with the MACHO Alert System as of October 15, 1995, is about 40, and all of them were announced over the Internet in real time.

The first papers based on the follow-up of the real-time announcements of microlensing events have already been published: Szymański et al (1994) presented the first light curve of the first event announced by the MACHO Alert System. Benetti et al (1995) presented the first spectra for the ongoing microlensing event.

The implementation of PLANET (Probing Lensing Anomalies NETwork; Albrow et al 1996) was a major development in 1995. The aim of the project is to follow the announcements of real-time detection of microlensing events (currently implemented by the OGLE and MACHO collaborations) with frequent multicolor observations on four telescopes: the Perth Observatory 0.6-m telescope at Bickley, Australia; the 1.0-m telescope near Hobart in Tasmania; the South African Astronomical Observatory 1.0-m at Sutherland, South Africa; and the Dutch-ESO 0.9-m at La Silla, Chile.

6.2 *Data Analysis*

To translate the observed rate of microlensing events into quantitative information about the optical depth and the lens masses it is necessary to calibrate the detection system. This is fairly straightforward, at least in principle, as all data processing is done with computers using automated software. Naturally, for the calibration to be possible, every step of the detection process has to be done according to well-defined rules, following the same algorithm for the duration of the experiment. The calibration is done by introducing artificial microlensing events into the data stream and using the same algorithm to "detect" them as the one used for real detections. This process can be performed at two very different levels: 1. using pixels or 2. using star catalogs.

If the original data are collected with a CCD camera then from the beginning the data are stored in computer memory in a digital form. Every CCD image obtained by OGLE had 2048×2048 pixels for a total of 8 Mbytes of pixel data. Every MACHO exposure generated 8 such CCD frames, 4 in each of 2 color bands, for a total of 64 Mbytes. On a clear night, between 30 and 100 exposures may be taken. The OGLE collaboration was given ~ 70 nights on the Swope 1-m telescope at the Las Campanas Observatory in Chile (operated by the Carnegie Institution of Washington) for each of the four observing seasons (1992–1995) and generated ~ 20 Gbytes of data every year. The MACHO collaboration refurbished a dedicated 1.3-m telescope at the Mount Stromlo Observatory in Australia. Their observations began in mid-1992, but the routine operation started in January 1993. Currently, the MACHO collaboration collects ~ 800 Gbytes of data every year.

If the original data are obtained with photographic plates, as was done for the main part of the EROS project and the whole DUO project, then the plates have to be scanned and digitized before the data can be stored in computer memory. At typical scanning resolutions, a single Schmidt plate has $28{,}000 \times 28{,}000$ pixels for a total of ~ 1.6 Gbytes of data.

Once the data are available in a form of digital pixel images, a dedicated software routine is used to measure the location and the brightness of stellar images. The MACHO and OGLE collaborations modified DoPhot (Schechter et al 1993) to make it faster (Udalski et al 1992, Bennett et al 1993). The coordinates of stellar images were determined on good CCD frames, which were chosen to be the templates. All other frames were first shifted to coincide with the templates using a few bright stars, and the coordinates of all other stars were adopted from the template. Next, the brightnesses of all template stars were measured on all frames. Although this speeded up the data processing by a large factor, it restricted photometry to those stars that were found on the templates. Also, this procedure makes it impossible to detect proper motion. Using a modified DoPhot software, up to 2×10^5 stars could be measured on a CCD frame with $\sim 4 \times 10^6$ pixels, which corresponds to the effective number of ~ 20 pixels per star. The DUO software (Alard 1996b, Alard et al 1995a) measured not only the brightness but also the position of every stellar image on every digitized Schmidt plate, for a total of $\sim 1.4 \times 10^7$ stars per plate, i.e. ~ 56 pixels per star.

The results of stellar photometry are combined into a database of photometric measurements, which is then searched for microlensing events. Very stringent criteria have to be used for detection because the events are very rare. For example, one of the OGLE conditions was the requirement that at least five consecutive measurements of the candidate object had to be brighter by more than three standard deviations than the normal, well-measured brightness of the star (Udalski et al 1994b). This conservative approach was necessary to prevent from being swamped with fictitious "events," but it made the efficiency of the detection rather low. If N is the total number of photometric measurements of all stars, n the number of microlensing events detected in the database, and τ the optical depth to microlensing as estimated on the basis of these detections, then, using the published MACHO and OGLE results, one finds that $N\tau/n \approx 50$–100. This implies that the effective number of photometric measurements needed for a detection of a single event is 50–100.

The first catalog-level estimate of the efficiency of microlensing detections was published by the OGLE collaboration (Udalski et al 1994b). They introduced artificial microlensing events into the database of photometric measurements and followed exactly the same criteria for detection as the criteria used in the original search. The efficiency was approximately constant for events in the time scale range 30 days $< t_0 <$ 100 days, but it dropped rapidly toward shorter time scales: down by a factor 10 at $t_0 \approx 3$ days and a factor 100 at $t_0 \approx 1$ day. Very similar results were found by the MACHO collaboration with a much more thorough pixel-level calibration (Alcock et al 1995a). In that case artificial events were added as artificial stars to the images obtained with the CCD detector. The effect of blending of stellar images is therefore automatically taken into account. The pixel-level calibration is certainly the correct way to proceed, but it is much more time consuming than the catalog-level calibration. Fortunately, the results are the same to within $\sim 20\%$ because the two opposite effects nearly cancel out: Stellar blending means that there are really more stars that may be subject to microlensing, and this increases the number of events per stellar image; however, because the apparent brightening is reduced, some events are impossible to detect.

Provided that the procedure used for the detection of artificial events is exactly the same as the procedure used for the original search, estimates of the event rate and the optical depth are not influenced in a systematic way by the specific criteria. Naturally, if the criteria are too stringent then there are fewer detections (in the real data as well as in the simulations), and the random errors of the estimated values increase. If the criteria are too lax then in addition to genuine microlensing events a number of intrinsically variable stars, or even artifacts of the detector system, may enter the sample, introducing uncontrollable systematic errors. We have as yet no quantitative procedure to define the optimum criteria.

Because current detection systems are sensitive to events over only a limited range of time scales, estimates of the optical depth and/or the event rates can only provide the lower limits. As the duration of searches increases so does their sensitivity to ever longer events. A different observing procedure is needed to improve the sensitivity to very short events. Most searches done so far made only 1 or 2 photometric measurements per star per clear night. The only CCD search with up to 46 photometric measurements per star per clear night was done by the EROS collaboration, and the null result was reported by Aubourg et al (1995).

6.3 *Consensus and No Consensus*

There is currently a consensus about a number of important issues related to microlensing in our galaxy and near our galaxy. The most important is that the microlensing phenomenon has been detected. It is not possible to tell at this time what fraction of so-called candidate events is real and what fraction is due to poorly known types of stellar variability. My personal guess is that the contamination fraction is probably below 10%. My arguments in favor of microlensing as the dominant source of the newly detected variability are the following:

1. The observed light curves are achromatic; their shapes are well described by simple theoretical formulae.
2. The distribution of magnification factors is consistent with the theoretical expectations (Udalski et al 1994a), with some events magnified by a factor of up to ~ 50 (C Stubbs 1995, private communication).
3. The double-lensing events have been detected, as expected, and roughly at the expected rate.
4. The parallax effect has been detected, as expected.
5. The spectrum of the only event monitored spectroscopically is constant throughout the intensity variation (Benetti et al 1995).
6. The Galactic bar has been rediscovered through the enhanced optical depth.

There is also a consensus on some scientific issues:

1. The optical depth towards the Galactic bulge is large: $\sim 3 \times 10^{-6}$.
2. The optical depth towards the LMC is small: $\sim 10^{-7}$.
3. The huge databases generated with the searches can be used to perform very interesting additional science, including creation of color magnitude diagrams and analysis of many types of variable stars.

On some issues, no consensus has yet been reached:

1. What is the location of objects that dominate the lensing observed towards the Galactic bulge? Are these predominantly the Galactic bulge stars (Kiraga & Paczyński 1994; Paczyński et al 1994; Zhao et al 1995, 1996) or are the lenses mostly in the Galactic disk (Alcock et al 1995c)?
2. What is the dominant location of the objects responsible for the lensing observed towards the LMC? Are these in the Galactic disk, Galactic halo, the LMC halo, or in the LMC itself? Are they stellar-mass objects or are they substellar brown dwarfs (Sahu 1994, Alcock et al 1995a)?
3. What fraction of microlensing events is caused by double lenses?

A consensus on these issues is likely to emerge within a few years, but no doubt new controversies will develop as the volume of data increases.

6.4 *Serendipitous Results*

The microlensing searches worked because practical implementation of massive data acquisition and processing systems became possible. These searches generated huge databases of multiband photometry of millions of stars and led to the discovery of thousands of variable stars, most of them new. The OGLE Galactic bulge variable star catalogs published so far contain full data for 1656 pulsating, eclipsing, and other short-period variables, including their coordinates and finding charts (Udalski et al 1994e, 1995a,b). The other OGLE papers include data on RR Lyrae-type stars in the Sagittarius dwarf galaxy (Mateo et al 1995b) and Sculptor dwarf galaxies (Kałużny et al 1995a). The MACHO collaboration presented a period-luminosity diagram for ~ 1500 cepheids in the LMC and identified 45 double-mode pulsators among them (Alcock et al 1995b); it has a total of $\sim 90,000$ variables in its archive (Cook et al 1995). The EROS collaboration published data on 80 eclipsing binaries in the LMC (Grison et al 1995, Ansari et al 1995a). The DUO collaboration has data on $\sim 15,000$ Galactic bulge variables (Alard 1996a). These include ~ 1200 pulsating stars of the RR Lyrae-type ab in the Galactic bulge and ~ 300 such stars in the Sagittarius dwarf, considerably extending the known size of that galaxy and clearly demonstrating the usefulness of variable stars as tracers.

Perhaps the most important serendipitous result is the discovery by Kałużny et al (1995b) of the first detached eclipsing binaries at the main sequence turn-off point of the globular cluster Omega Centauri. The follow-up spectroscopic observations will determine, for the first time, the masses of stars at the globular cluster main sequence turn-off point and hence will provide a sound basis for the reliable age and helium content determinations. The point is that although the theoretical color-magnitude diagrams are affected by our lack of understanding of mixing-length theory, the mass-luminosity relation is insensitive to large changes of the mixing length (Paczyński 1984).

The same detached binaries will allow very accurate determination of distances to globular clusters, while the follow-up spectroscopic observations of the detached eclipsing binaries discovered by the EROS collaboration (Grison et al 1995) will provide a very accurate distance to the LMC (Paczyński 1996). We may expect that, in the near future, detached eclipsing binaries will provide accurate distances to all galaxies of the Local Group (Hilditch 1995). It should be pointed out that in order to discover a detached eclipsing binary one needs a few hundred photometric measurements. Only one out of a few thousand stars is of this type, with deep and narrow primary and secondary eclipses and with no anomalies in the light curve. Thus one needs $\sim 10^6$ photometric measurements to detect one good system. This is a lot of work, but it will lead to the determination of some of the most important numbers for cosmology: accurate ages of globular clusters, their helium content, and the Hubble constant.

Color-magnitude diagrams obtained in the standard (V,I) photometric system by the OGLE collaboration for the Galactic bulge region show evidence for the Galactic bar (Stanek et al 1994). They also indicate that the Galactic disk has a low number density of luminous (i.e. massive and young) as well as faint (i.e. low mass and old) main sequence stars in the inner ~ 4 kpc (Paczyński et al 1994), which seems to be consistent with the presence of the strong bar. The color-magnitude diagrams obtained for the recently discovered Sagittarius dwarf allowed Mateo et al (1995a) to determine the distance of 25.2 ± 2.8 kpc to this galaxy and to estimate its age and metallicity to be 10 Gyr and [Fe/H] $= -1.1 \pm 0.3$, respectively. Alard (1996) and Mateo et al (1996) found a large extension of this galaxy discovery of a large number of RR Lyrae variables, a by-product of massive photometry carried out by DUO and OGLE.

7. THE FUTURE OF MICROLENSING SEARCHES

7.1 *The Near Future*

The near future of microlensing searches is easy to predict: more of the same or, rather, very much more of the same. While the MACHO collaboration has streamlined its data processing and does it in real time, the EROS and OGLE collaborations are building in Chile their 1-m class telescopes to be dedicated to massive microlensing searches. Many other groups are either planning or developing new detection systems. In particular, a few groups intend to monitor M31 (Crotts 1992; Ansari et al 1995b; the latter program is called— AGAPE, for Andromeda Galaxy and Amplified Pixels Experiment).

It is not known how many stars can be monitored from the ground in the direction of the Galactic bulge and all members of the Local Group galaxies, but a fair estimate is $\sim 2 \times 10^8$, i.e. about one order of magnitude more than the current $\sim 2 \times 10^7$ of the MACHO collaboration. Note that stars towards the Galactic bulge and towards the LMC and SMC are bright, and even with a 1-meter class telescope the detection limit is set by the crowding of stellar images, not by the sky or the photon statistics. Therefore, the better the seeing, the more stars can be detected and accurately measured. It is easily imaginable that within a few years the detection rate will increase to ~ 300 events per year towards the Galactic bulge and ~ 10 events per year towards the LMC and SMC. With ~ 1000 events detected towards the bulge and ~ 30 towards the LMC and SMC, it will be easy to map the distribution.

The observed distribution of lensing events in the sky will soon reveal the spatial distribution of the lenses towards the Galactic bulge (cf Evans 1994,

Kiraga 1994). A rapid increase of the optical depth towards the center will indicate that the lenses are located predominantly in the bulge of our galaxy. A more or less uniform optical depth will point to the Galactic disk as the dominant site of the lensing objects. The task will be more difficult in the case of the Large Magellanic Cloud. The observed event rate is low, and it will take longer to accumulate good statistics. A rapid increase of the optical depth towards the center will indicate that the lenses are located predominantly in the bulge of the LMC. A more or less uniform optical depth would be more difficult to interpret, as such a result would be compatible with the lenses located either in the Galactic disk or in the Galactic halo. It may be necessary to measure the optical depth in as many directions as possible, towards all members of the Local Group of galaxies. This is a difficult task as these are far away, and hence their stars are very faint. It is likely that 2- to 4-meter class telescopes will be required for the observations (Crotts 1992, Colley 1995, Ansari et al 1995b).

It is interesting to note that the statistics of microlensing of the Galactic bulge red clump stars will allow the determination of the geometrical depth of the bulge/bar system (Stanek 1995). The lensed objects will be found predominantly among the stars located on the far side. Therefore, these stars will appear to be fainter, at least on average, than typical red clump stars.

Within a few years the range of event time scales between one hour and three years will likely be very well sampled. With the observed distribution of event time scale known over this whole range, or with stringent upper limits available for some part of the range, it will become meaningful to translate this distribution into the mass function of the lensing events. The necessary prerequisite will be the knowledge of the spatial distribution and kinematics of the lenses. It seems safe to expect that the spatial distribution will be revealed through the variation of the optical depth over the sky, while the kinematics will follow from the improved understanding of the Galactic structure. This analysis will not be good enough to know the mass of any particular lens to better than a factor 2 or 3, but it is likely that the overall range of masses will be known with a reasonable accuracy.

Many papers and preprints have been written about the possible distribution of mass within the dark halo, or within the Galactic disk, as well as on the possible nature of the lensing objects (including gas clouds and axion miniclusters). These are not reviewed here, as I consider them too speculative. A more direct analysis of the data will provide the answers to most interesting current questions in a matter of a few years.

The expansion of microlensing searches will certainly be followed by the expansion of the follow-up observations. There is a very natural division of labor in this area as well as in the area of supernovae searches. To discover very rare events one needs detectors with as many pixels as possible. Multiband coverage is not essential. The pixels are expensive and difficult to purchase, and it is practically impossible to fill the whole field of view of a modern 1-m class telescope with CCDs, because the field may be 1.5–5 degrees across. Given a limited number of pixels, we have to decide what is more efficient: Do we put them all in one plane, making the search area as large as affordable, or do we cover only 50% of the area in two bands? Every detection system is most sensitive in a particular band. For example, the OGLE system can record more stellar images in the I band than in the V band in a fixed exposure time. Also, when the moon is bright, the sky intensity in the I band is smaller than it is in the V band. Thin CCD chips are currently so expensive that extending the search to the B and U bands is not possible within a realistic budget. The implementation of real-time data analysis, coupled to the distribution of information over the Internet and the availability of follow-up systems like PLANET (Albrow et al 1996), makes the multiband search unnecessary.

In contrast, follow-up observations must be as multiband and as frequent as possible. Because only a relatively small number of pixels is needed for the follow-up, it should be possible to use thin CCD chips for this purpose. Frequent time coverage is most essential since many interesting phenomena happen on the time scale on which a star moves across its own diameter, which takes a few hours. Such source-resolving events may allow the measurement of the relative proper motion in the lens-source system (Gould 1994a), as well as enabling us to study the structure of stellar photospheres. They may also lead to the detection of planets around stars (Mao & Paczyński 1991).

A dramatic improvement in the data processing software may be expected when the current photometry of stellar images is supplemented with a search for variable point sources using digital image subtraction, also known as "CCD frame subtraction." The image subtraction is very elegant, and its principles are very simple, as described and implemented by Ciardullo et al (1990). Imagine that we have two images of the same area in the sky taken under identical seeing conditions and the same atmospheric extinction. Every point source has an image spread over many pixels according to the point spread function (PSF). When the two images are subtracted from each other there should be no residuals if nothing has changed. If some stars changed their brightness then there should be some residuals, positive or negative, with the profile defined by the PSF. These residuals can be measured, and thus the stellar variability can be detected. If we do not care about stars that remain constant, this is by far the most efficient way to proceed. Also, this method may allow the detection of variables that are too crowded to measure the nonvariable component of their light due to the constant contaminiating stars.

Unfortunately, it is difficult to implement the image subtraction technique because there are many practical problems, as described by Ciardullo et al (1990). Therefore, proof of its practical usefulness will likely be provided by the determination of light curves of periodical variables, such as pulsating or eclipsing stars. It is much easier to detect a periodic signal, and to confirm its reality, than to detect convincingly a nonperiodic and nonrepeating signal like a microlensing event. However, recently published efforts (Crotts 1992, Baillon et al 1993, Ansari et al 1995b, Gould 1996) seem to concentrate on the detection of nonperiodic signals.

7.2 *A More Distant Future*

Any discussion of the distant future is obviously speculative, but it is also entertaining. A rather obvious idea is to observe microlensing with space instruments. The natural reasons include stable weather, perfect seeing, and access to the UV and IR. In addition, there are special advantages for microlensing in putting an instrument at a distance of ~ 1 AU from the ground-based telescopes, since this would allow the detection of the microlensing parallax effect as pointed out by Refsdal (1966) and by Gould (1994b, 1995a, 1996). Any observations done at a single site provide only one quantity that has physical significance: the time scale t_0. Unfortunately, this time scale is a function of the lens mass, its distance, and its transverse velocity (cf Equation 15), none of which is known. However, the illumination pattern created by the lens varies substantially on a scale of a fraction of the Einstein ring radius or its projection onto the observer's plane. Hence, even a small space telescope placed at a solar orbit would reveal a different light curve of the lens and thereby provide some additional information about the lens properties. Unfortunately, the extra data obtained will not be sufficient to solve for all three unknowns. However, if the same event happens to resolve the source, then the relative proper motion of the lens-source system can be determined (Gould 1994a), and perhaps a complete solution could be obtained. In particular, the determination of the lens mass might be possible.

There are some problems with this idea. One of them is the high cost, likely to be of the order of 100 million dollars, as compared with 1 million for a typical ground-based microlensing search (though supposedly "dollars in space weigh less"). Also, even though the parallax effect alone can provide some constraint on the lens, it cannot provide a unique determination of the lens mass, with a possible exception of some special cases, such as the very high magnification events during which the source is resolved. Unfortunately, such events are very rare. Perhaps the single most dramatic impact of the space measurements would be a direct proof that a particular event is indeed caused by gravitational microlensing: No other phenomenon could be responsible for the difference in the observed light curves.

Many of the problems, including a decent statistical determination of the lens masses, will be solved in the near future with ground-based observations. When the optical depth to the Galactic bulge is mapped with $\sim 10^3$ lensing events, the spatial distribution of the lenses (Galactic disk versus Galactic bulge/bar) will be definitely understood. Because the kinematics of the disk and the bulge/bar stars can be directly observed, an adequate model for the lens statistics will be developed as a refinement of the approach outlined in the Section 3. Any particular lens is not likely to have its mass determined to better than a factor ~ 3, or so, but the average mass, and perhaps the mass range and the slope of the mass function, will be deduced. This task will be relatively easy for the lenses observed towards the Galactic bulge because of their high event rate. It will take much more time to obtain equally reliable information about the lenses observed at high galactic latitudes because of their very low rate. Yet, continuous monitoring of the stars in all Local Group galaxies will provide the basis for the determination of the lens distribution and masses within a decade or so. This is basically the task of detecting enough events to map the variation of the optical depth over the sky.

There are at least two cases in which the masses of individual lenses can be measured with no ambiguity. These are lenses belonging to globular clusters seen against the rich stellar background of the Galactic bulge, LMC, or SMC (Paczyński 1994) and the nearby high-proper-motion stars seen against the distant stars of the Milky Way, LMC, or SMC (Paczyński 1995a). In both cases the distances and proper motions of the lenses can be measured directly, or indirectly (the lenses in the globular clusters will be too dim to be seen), and the lens mass remains the only unknown quantity on the right-hand side of the Equation (15).

7.3 *The Limits of Ground-Based Searches and the Search for Planets*

The range of lens masses that can give rise to observable lensing events is very broad. At the low end the practical limit is imposed by the finite size of the sources. The amplification of a point source that is perfectly aligned with a lens is infinite (cf Equation 11 with $u = 0$). However, when a source with a finite angular radius r_s is perfectly aligned, its circular disk forms a ringlike image, which has its inner and outer radius, r_{in} and r_{out}, given with a slightly modified

Equation (9):

$$r_{\rm in} = \left[\left(r_{\rm E}^2 + 0.25 r_{\rm s}^2\right)^{1/2} - 0.5 r_{\rm s}\right],$$
$$r_{\rm out} = \left[\left(r_{\rm E}^2 + 0.25 r_{\rm s}^2\right)^{1/2} + 0.5 r_{\rm s}\right], \tag{50}$$

where all quantities are expressed as angles. Assuming, for simplicity, a uniform surface brightness of the source, we can calculate the maximum magnification as the ratio of the two areas:

$$A_{\rm max} = \frac{\pi r_{\rm out}^2 - \pi r_{\rm in}^2}{\pi r_{\rm s}^2} = \left[4\left(\frac{r_{\rm E}}{r_{\rm s}}\right)^2 + 1\right]^{1/2}. \tag{51}$$

For the event to be reasonably easy to detect, we choose $A_{\rm max} \geq 2^{1/2}$, which is equivalent to the condition $r_{\rm s}/r_{\rm E} \leq 2$.

An angular radius of a star is given as

$$r_{\rm s} = \frac{R_{\rm s}}{D_{\rm s}} = 2.3 \times 10^{-12}\ {\rm rad} \left(\frac{R_{\rm s}}{R_\odot}\right)\left(\frac{10\ {\rm kpc}}{D_{\rm s}}\right)$$
$$= 0.45\ \mu{\rm sec} \left(\frac{R_{\rm s}}{R_\odot}\right)\left(\frac{10\ {\rm kpc}}{D_{\rm s}}\right). \tag{52}$$

This may be combined with Equation (13) to obtain

$$\frac{r_{\rm s}}{r_{\rm E}} = 0.00050 \left(\frac{R_{\rm s}}{R_\odot}\right)\left(\frac{M_\odot}{M}\right)^{1/2}\left(\frac{10\ {\rm kpc}}{D_{\rm s}}\right)$$
$$\times \left(\frac{D_{\rm d}}{10\ {\rm kpc}}\right)^{1/2}\left(1 - \frac{D_{\rm d}}{D_{\rm s}}\right)^{-1/2}. \tag{53}$$

The smallest mass for which the Einstein ring radius is no less than half the source radius is given as

$$M \geq M_{\rm min} \approx 6 \times 10^{-8}\ M_\odot \left(\frac{R_{\rm s}}{R_\odot}\right)^2 \left(\frac{10\ {\rm kpc}}{D_{\rm s}}\right)\left(\frac{D_{\rm d}}{D_{\rm s} - D_{\rm d}}\right). \tag{54}$$

If the source is in the Galactic bulge, at $D_{\rm s} \approx 8$ kpc, and the lens is at $D_{\rm d} \approx 6$ kpc (Zhao et al 1995, 1996), then condition (54) becomes

$$M_{\rm min} \approx 2 \times 10^{-7}\ M_\odot \left(\frac{R_{\rm s}}{R_\odot}\right)^2 \qquad \text{(Galactic bulge)}. \tag{55}$$

If the source is in the LMC, at $D_{\rm s} \sim 55$ kpc, and the lens is in the Galactic halo, at $D_d \sim 10$ kpc, then condition (54) becomes

$$M_{\rm min} \approx 1.0 \times 10^{-7}\, M_\odot \left(\frac{R_s}{7\, R_\odot}\right)^2 \qquad \text{(LMC)}. \tag{56}$$

Note that the scaling factor for the source radius is 7 $R_\odot$ in Equation (56), whereas it is $R_\odot$ in Equation (55) to allow for the fact that the distance to the LMC is about seven times larger than to the Galactic center and that the source stars near the detection limit are correspondingly larger. The time scale of a microlensing event caused by lenses at the lower mass limit can be calculated by combining the Equations (15), (53), and (54) to obtain

$$t_0 \geq 30 \text{ min} \left(\frac{R_s}{R_\odot}\right)\left(\frac{D_d}{D_s}\right)\left(\frac{200 \text{ km s}^{-1}}{V}\right). \tag{57}$$

The estimate of $M_{\rm min}$ implies that it should be possible to extend the searches down to masses as small as that of our moon. These objects may be planets lensing together with their parent stars, as described in Section 4, or these may be planetary-mass objects roaming interstellar space, rather than orbiting any star. Clearly, the task of finding them will be very difficult unless they are very numerous. Let us make a modest assumption that every star has just one Earth-mass planet and that a typical mass of the lenses actually responsible for the events detected towards the bulge and the LMC is $\sim 0.3\, M_\odot$. This implies that the mass ratio is $\sim 10^{-5}$, and hence the optical depth to microlensing by those planets is $\sim 3 \times 10^{-6} \times 10^{-5} = 3 \times 10^{-11}$ towards the Galactic bulge and $\sim 10^{-7} \times 10^{-5} = 10^{-12}$ towards the LMC. These numbers may be somewhat larger as the cross section for planetary microlensing is enhanced by the presence of a nearby star (cf Figure 9). Clearly, searches aimed at finding such planets must be vastly more thorough than anything operating now.

Let us make a crude estimate of the most sensitive ground-based search for microlensing that could be done by simply making multiple copies of the existing systems. Let all stars that may be microlensed, perhaps 2×10^8 of them, be measured every few minutes to detect all events with a time scale $t_0 \geq 30$ minutes. Let there be ~ 3000 hours of clear observing per year, combining all ground-based sites. This translates into six events per year with $t_0 \sim 1$ hour if the optical depth is $\tau_{1\,\rm h} \approx 10^{-11}$. The good news is that there are enough stars in the sky to detect a moderately large number of planetary microlensing events with the 1-m class ground-based telescopes if a megaproject is carried out for a few years. The bad news is that the required data rate would be about 1000 times larger than the current MACHO rate. The price tag would certainly be very impressive, but probably not in excess of a single space mission.

This analysis demonstrates that with a rather immodest increase of current microlensing searches the dark objects down to asteroid-mass range could be detected even if their fractional contribution to the Galactic mass was as small as $\sim 10^{-5}$. It is also possible, at least in principle, to extend the search to masses as large as 10^5–10^6 $M_\odot$. The events due to lenses that massive have a duration of ~ 100 years, yet they may be detected, at least in principle, with two very different methods. First, a parallax effect due to Earth's orbital motion will introduce $\sim 1\%$ modulation in the light curve, with a period of one year (Gould 1992). Second, numerous stars that are normally below the detection threshold will be occasionally microlensed by a very large factor. According to Equation (11), the highly magnified source is within factor of 2 of its peak brightness for a time interval $t_{1/2} = t_0 3^{1/2}/A_{\max}$, which may be reasonably short for $A_{\max} \geq 100$ (Paczyński 1995b). Thus, either of the two methods may allow the extension of the microlensing searches to the range of supermassive MACHOs.

ACKNOWLEDGMENTS

It is a great pleasure to acknowledge the discussions with, and comments by, Drs. C Alcock, D Bennett, J Kalużny, A Kruszewski, M Kubiak, S Mao, J Miralda-Escudè, M Pratt, K Sahu, KZ Stanek, C Stubbs, M Szymański, A Udalski, and J Wambsganss. This work was supported by NSF grants AST-9216494 and AST-9313620.

Literature Cited

Abt HA. 1983. *Annu. Rev. Astron. Astrophys.* 21:343
Alard C. 1996a. *Ap. J. Lett.* 458:L17
Alard C. 1996b. See Kochanek & Hewitt 1996, p. 215
Alard C, Guibert J, Bienayme O, Valls-Gabaud D, Robin AC, et al. 1995a. *The Messenger* 80:31
Alard C, Mao S, Guibert J. 1995b. *Astron. Astrophys. Lett.* 300:L17
Albrow M, Martin R, Sahu K, Birch P, Menzies J, et al. 1996. See Kochanek & Hewitt 1996, p. 227
Alcock C, Akerlof CW, Allsman RA, Axelrod TS, Bennett DP, et al. 1993. *Nature* 365:621
Alcock C, Allsman RA, Alves D, Axelrod TS, Bennett DP, et al. 1995d. *Ap. J.* 454:L125
Alcock C, Allsman RA, Axelrod TS, Bennett DP, Cook KH, et al. 1995a. *Phys. Rev. Lett.* 74:2867
Alcock C, Allsman RA, Axelrod TS, Bennett DP, Cook KH, et al. 1995b. *Astron. J.* 109:1653
Alcock C, Allsman RA, Axelrod TS, Bennett DP, Cook KH, et al. 1995c. *Ap. J.* 445:133
Alcock C, Allsman RA, Axelrod TS, Bennett DP, Cook KH, et al. 1996. *Ap. J* 461:84
Alcock C, Allsman RA, Alves D, Axelrod TS, Bennett DP, Cook KH, et al. 1995e. Preprint: astro-phys/9512146
Ansari R. 1995. *Nucl. Phys. B (Proc. Suppl.)* 43:108
Ansari R, Auriére M, Baillon P, Bouquet A, Coupinot G, et al. 1995b. *Nucl. Phys. B (Proc. Suppl.)* 43:165
Ansari R, Cavalier F, Couchot F, Moniez M, Perdereau O, et al. 1995a. *Astron. Astrophys. Lett.* 299:L21
Aubourg E, Bareyre P, Bréhin S, Gros M, de Kat J, et al. 1995. *Astron. Astrophys.* 301:1

Aubourg E, Bareyre P, Bréhlin S, Gros M, Lachiéze-Rey M, et al. 1993. *Nature* 365:623
Baillon P, Bouquet A, Giraud-Héraud Y, Kaplan J. 1993. *Astron. Astrophys.* 277:1
Beaulieu JP, Ferlet R, Grison P, Vidal-Madjar A, Kneib JP, et al. 1995, *Astron. Astrophys.* 299:168
Benetti S, Pasquini L, West RM. 1995. *Astron. Astrophys. Lett.* 294:L37
Bennett DP, Akerlof C, Alcock C, Allsman R, Axelrod T, et al. 1993. *Ann. NY Acad. Sci.* 688:612
Bennett DP, Alcock C, Allsman R, Axelrod T, Cook KH, et al. 1995. *AIP Conf. Proc. 336. Dark Matter,* ed. SS Holt, DP Bennett, p. 77
Bennett DP, Rhie SH. 1996. Preprint: astro-ph/9603158
Black DC. 1995. *Annu. Rev. Astron. Astrophys.* 33:359
Blandford RD, Narayan R. 1992. *Annu. Rev. Astron. Astrophys.* 30:311
Blitz L, Spergel DN. 1991. *Ap. J.* 379:631
Bolatto AD, Falco EE. 1994. *Ap. J.* 436:112
Butler RP, Marcy GW. 1996. *Ap. J.* 457:93
Chang K, Refsdal S. 1979. *Nature* 282:561
Ciardullo R, Tamblyn P, Phillips AC. 1990. *Publ. Astron. Soc. Pac.* 102:1113
Colley WN. 1995. *Astron. J.* 109:440
Cook KH, Alcock C, Allsman R, Axelrod T, Freeman K, et al. 1995. In *Proc. IAU Colloq. 155, ASP Conf. Ser. 83, Astrophysical Applications of Stellar Pulsations,* ed. R Stobie, p. 221.
Crotts APS. 1992. *Ap. J. Lett.* 399:L43
De Rújula A, Jetzer Ph, Massó E. 1991. *MNRAS* 250:348
de Vaucouleurs G. 1964. *IAU Symp. 20. The Galaxy and the Magellanic Clouds,* ed. FJ Kerr, AW Rotgers, p. 195.
Di Stefano R, Esin AA. 1995. *Ap. J. Lett.* 448:L1
Di Stefano R, Mao S. 1996. *Ap. J.* 457:93
Dwek E, Arendt RG, Hauser MG, Kelsall T, Lisse CM, et al. 1995. *Ap. J.* 445:716
Dyson FW, Eddington AS, Davidson CR. 1920. *Mem. R. Astron. Soc.* 62:291
Einstein A. 1911. *Ann. Phys.* 35:898
Einstein A. 1916. *Ann. Phys.* 49:769
Evans NW. 1994. *Ap. J. Lett.* 437:L31
Gott RJ. 1981. *Ap. J.* 243:140
Gould A. 1992. *Ap. J.* 392:442
Gould A. 1994a. *Ap. J. Lett.* 421:L71
Gould A. 1994b. *Ap. J. Lett.* 421:L75
Gould A. 1995a. *Ap. J. Lett.* 441:L21
Gould A. 1995b. *Ap. J.* 455:44
Gould A. 1996. See Kochanek & Hewitt 1996, p. 365
Gould A, Loeb A. 1992. *Ap. J.* 396:104
Griest K. 1991. *Ap. J.* 366:412
Griest K, Alcock C, Axelrod TS, Bennett DP, Cook KH, et al. 1991. *Ap. J. Lett.* 372:L79
Griest K, Hu W. 1992. *Ap. J.* 397:362
Grison P, Beaulieu J-P, Pritchard JD, Tobin W, Ferlet R, et al. 1995. *Astron. Astrophys. Suppl.* 109:447
Hilditch RW. 1995. *Binaries in Clusters, ASP Conf. Ser.,* ed. G Milone, J-C Mermilliod. In press
Kalużny J, Kubiak M, Szymański M, Udalski A, Krzemiński W, et al. 1995a. *Astron. Astrophys. Suppl.* 112:407
Kalużny J, Kubiak M, Szymański M, Udalski A, Krzemiński W, et al. 1995b. *Binaries in Clusters, ASP Conf. Ser.,* ed. G Milone, J-C Mermilliod. In press
Kamionkowski M. 1995. *Ap. J.* 442:L9
Kayser R, Schramm T, Nieser L, eds. 1992. *Gravitational Lenses. Lect. Notes Phys. 406.* Berlin: Springer-Verlag
Kiraga M. 1994. *Acta Astron.* 44:241
Kiraga M, Paczyński B. 1994. *Ap. J.* 430:101
Kochanek CS, Hewitt JN, eds. 1996. *Astrophysical Applications of Gravitational Lensing. Proc. IAU Symp. 173.* Dordrecht: Kluwer
Lynds R, Petrosian V. 1989. *Ap. J.* 336:1
Mao S, Di Stefano R. 1995. *Ap. J.* 440:22
Mao S, Paczyński B. 1991. *Ap. J. Lett.* 374:L37
Mao S, Paczyński B. 1996. Preprint: astro-ph/9604002
Marcy GW, Butler RP. 1996. Preprint
Mateo M, Kubiak M, Szymański M, Kalużny J, Krzemiński W, Udalski A. 1995b. *Astron. J.* 110:1141
Mateo M, Mirabel N, Udalski A, Szymański M, Kalużny J, et al. 1996. *Ap. J. Lett.* 458:L13
Mateo M, Udalski A, Szymański M, Kalużny J, Kubiak M, Krzemiński W. 1995a. *Astron. J.* 109:588
Mayor M, Queloz D. 1995. *Nature* 378:355
Mellier Y, Fort B, Soucail G, eds. 1990. *Gravitational Lensing. Lect. Notes Phys. 360.* Berlin: Springer-Verlag
Moran JM, Hewitt JN, Lo KY, eds. 1989. *Gravitational Lenses. Lect. Notes Phys. 330.* Berlin: Springer-Verlag
Nemiroff RJ, Wickramasinghe WADT. 1994. *Ap. J. Lett.* 424:L21
Paczyński B. 1984. *Ap. J.* 284:670
Paczyński B. 1986a. *Ap. J.* 301:503
Paczyński B. 1986b. *Ap. J.* 304:1
Paczyński B. 1991. *Ap. J. Lett.* 371:L63
Paczyński B. 1994. *Acta Astron.* 44:235
Paczyński B. 1995a. *Acta Astron.* 45:345
Paczyński B. 1995b. *Acta Astron.* 45:349
Paczyński B. 1996. See Kochanek & Hewitt 1996, p. 199
Paczyński B, Stanek KZ, Udalski A, Szymanński M, Kalużny J, et al. 1994. *Ap. J. Lett.* 435:L113
Pratt MR, Alcock C, Allsman RA, Alves D, Axelrod TS, et al. 1996. See Kochanek & Hewitt

1996, p. 221
Refsdal S. 1964. *MNRAS* 128:295
Refsdal S. 1966. *MNRAS* 134:315
Refsdal S, Surdej J. 1994. *Rep. Prog. Phys.* 56:117
Roulet E. Molleroch S. 1996. Physics Reports, in press: astro-ph/9603119
Sahu K. 1994. *Nature* 370:275
Schechter PL, Mateo M, Saha A. 1993. *Publ. Astron. Soc. Pac.* 105:1342
Schneider P, Ehlers J, Falco EE. 1992. *Gravitational Lensing.* Berlin: Springer-Verlag
Schneider P, Weiss A. 1986. *Astron. Astrophys.* 164:237
Soldner J. 1801. *Berl. Astron. Jahrb. 1804,* p. 161
Stanek KZ. 1995. *Ap. J.* 441:L29
Stanek KZ, Mateo M, Udalski A, Szymański M, Kałużny J, Kubiak M. 1994. *Ap. J.* 429:L73
Stubbs C, Alcock C, Cook K, Allsman R, Axelrod T, et al. 1995. *Bull. Am. Astron. Soc.* 26:1337
Surdej J, Fraipont-Caro D, Gosset E, Refsdal S, Remy M, eds. 1993. *Gravitational Lenses in the Universe. Proc. 31st Liege Coll.* Liège: Univ. Liège
Szymański M, Udalski A, Kałużny J, Kubiak M, Krzemiński W, Mateo M. 1994. *Acta Astron.* 44:387
Udalski A, Kubiak M, Szymański M, Kałużny J, Mateo M, Krzemiński W. 1994e. *Acta Astron.* 44:317
Udalski A, Olech A, Szymański M, Kałużny J, Kubiak M, et al. 1995b. *Acta Astron.* 45:433
Udalski A, Szymański M, Kałużny J, Kubiak M, Krzemiński W, et al. 1993. *Acta Astron.* 43:289
Udalski A, Szymański M, Kałużny J, Kubiak M, Mateo M. 1992. *Acta Astron.* 42:253
Udalski A, Szymański M, Kałużny J, Kubiak M, Mateo M, et al. 1994c. *Acta Astron.* 44:227
Udalski A, Szymański M, Kałużny J, Kubiak M, Mateo M, Krzemiński W. 1994a. *Ap. J.* 426:L69
Udalski A, Szymański M, Kałużny J, Kubiak M, Mateo M, et al. 1995a. *Acta Astron.* 45:1
Udalski A, Szymański M, Mao S, Di Stefano R, Kałużny J, et al. 1994d. *Ap. J. Lett.* 436:L103
Udalski A, Szymański M, Stanek KZ, Kałużny J, Kubiak M, et al. 1994b. *Acta Astron.* 44:165
Walsh D, Carswell RF, Weymann RJ. 1979. *Nature* 279:381
Witt H. 1993. *Ap. J.* 403:530
Witt H. 1995. *Ap. J.* 449:42
Witt H, Mao S. 1994. *Ap. J.* 430:505
Wolszczan A. 1994. *Science* 264:538
Wolszczan A, Frail DA. 1992. *Nature* 355:145
Zhao HS, Spergel DN, Rich RM. 1995. *Ap. J. Lett.* 440:L13
Zhao HS, Spergel DN, Rich RM. 1996. *MNRAS.* In press
Zwicky F. 1937. *Phys. Rev.* 51:290

Annu. Rev. Astron. Astrophys. 1996. 34:461–510

THE AGE OF THE GALACTIC GLOBULAR CLUSTER SYSTEM

Don A. VandenBerg[1]

Department of Physics and Astronomy, University of Victoria, P.O. Box 3055, Victoria, British Columbia, Canada V8W 3P6

Michael Bolte

UCO/Lick Observatory, University of California, Santa Cruz, California 95064

Peter B. Stetson

Dominion Astrophysical Observatory, Herzberg Institute of Astrophysics, National Research Council of Canada, 5071 West Saanich Road, Victoria, British Columbia, Canada V8X 4M6

KEY WORDS: globular clusters, stellar structure, stellar evolution, ages, subdwarfs, RR Lyrae stars, chemical abundances, cosmology

ABSTRACT

A careful assessment of current uncertainties in stellar physics (opacities, nuclear reaction rates, equation of state effects, diffusion, rotation, and mass loss), in the chemistry of globular cluster (GC) stars, and in the cluster distance scale, suggests that the most metal-poor (presumably the oldest) of the Galaxy's GCs have ages near 15 Gyr. Ages below 12 Gyr or above 20 Gyr appear to be highly unlikely. If these $\approx 2\sigma$ limits are increased by ~ 1 Gyr to account for the formation time of the globulars, and if standard Friedmann cosmologies with the cosmological constant set to zero are assumed, then the GC constraint on the present age of the Universe ($t_0 \geq 13$ Gyr) implies that the Hubble constant $H_0 \leq 51$ km s^{-1} Mpc^{-1} if the density parameter $\Omega = 1$ or ≤ 62 km s^{-1} Mpc^{-1} if $\Omega = 0.3$.

[1]Killam Research Fellow

0066-4146/96/0915-0461$08.00

1. INTRODUCTION

As fossil relics dating from the formation of the Galaxy and as the oldest objects in the Universe for which reliable ages can be derived, the Galaxy's globular star clusters have been the subject of intensive investigation for more than four decades. Their age distribution and the trends that they define of age with metallicity, position in the Galaxy, and kinematic properties are direct tracers of the chronology of the first epoch of star formation in the Galactic halo. Whether the globular cluster (GC) system encompasses an age range of several billion years or whether the majority of the GCs are nearly coeval is still the subject of lively debate. In a companion review, Stetson, VandenBerg & Bolte (1996) summarize the many advances that have been made in the determination of *relative* GC ages and assess their implications for Galactic formation scenarios. *Absolute* cluster ages—which are the focus of the present study—provide a vital constraint on the age of the Universe and thereby on the cosmological models that are used to describe it. Globular clusters may well have been the first stellar systems to form in the Universe (Peebles & Dicke 1968), probably within approximately 10^9 yr after the Big Bang (see Sandage 1993c).

The current widespread interest in securing accurate globular cluster ages results from the dilemma that these ages pose for the presently preferred model in cosmology—a matter-dominated, Einstein-de Sitter universe. This model is characterized by the choice of $\Omega_{\mathrm{Total}} = 1$ (as required by most formulations of inflation theory), with the cosmological-constant term, Ω_Λ, taken to be zero, implying $\Omega_{\mathrm{Matter}} = 1$. In this case, the expansion age of the Universe is given by $t_0 = (2/3)H_0^{-1}$, which works out to 8.3 Gyr if the Hubble constant H_0 is taken to be 80 km $\mathrm{s}^{-1}\mathrm{Mpc}^{-1}$. Support for this particular value of H_0, or one within $\pm$10–15% of it, has been boosted by the detection and analysis of Cepheid variables in Virgo cluster galaxies using both the Canada-France-Hawaii Telescope (Pierce et al 1994) and the *Hubble Space Telescope* (Freedman et al 1994; Kennicutt, Freedman & Mould 1995). Moreover, very similar estimates have been favored in most recent reviews of H_0 determinations (e.g. Jacoby et al 1992; Huchra 1992; van den Bergh 1992, 1994).

These results notwithstanding, significant support persists for $H_0 < 65$ km s^{-1} Mpc^{-1} (e.g. Saha et al 1994, 1995; Birkinshaw & Hughes 1994; Hamuy et al 1995; Sandage et al 1996); consequently, such lower values cannot yet be ruled out. But, even if H_0 were as low as 55 km $\mathrm{s}^{-1}\mathrm{Mpc}^{-1}$, the implied age for the Universe from the standard cosmological model is only 12.2 Gyr, which is also inconsistent with the GC-based estimate of ~ 16 Gyr.[2] Thus the standard model would appear to fail the "age concordance" test, which is simply that the age of all things in the Universe must be smaller than the elapsed time since the Big Bang. Although there is increasing observational evidence for $\Omega_{\text{Matter}} \approx 0.3$ (Vogeley et al 1992, Carlberg et al 1996, Squires et al 1996), even for a low-density, $\Omega_\Lambda = 0$ Universe (for which $t_0 \approx H_0^{-1}$), it may not be possible to achieve compatibility with the GC age constraint if H_0 is as high as many people believe. This has lead to increasing speculation that the cosmological constant is nonzero (e.g. Efstathiou 1995); however, before stellar age estimates can be used to rule out any cosmologies, a reappraisal of the errors associated with GC age determinations is worthwhile. It is our intent to do just that.

In Section 2 we review the uncertainties in the stellar evolution models due to possible errors in the relevant input physics: nuclear reaction rates, opacities, nonideal gas law effects in the equation of state, and the treatment of convection. In this section we also discuss the effects of input physics that are not normally a component of standard models: rotation, diffusion, and main-sequence mass loss. The observed chemical abundance trends among GC giants are highlighted therein because they provide perhaps the strongest indication of inadequacies in the stellar models for very metal-poor stars. We then briefly consider the possible role of unconventional physics, describe some pertinent observational tests of stellar evolution theory, and briefly recall the very first estimates of GC ages.

As has been recognized for a number of years, the dominant error in the derivation of ages from the luminosity of main-sequence turnoff stars in GCs [which we designate as L^{TO}, $M_{\text{bol}}(\text{TO})$, or $M_V(\text{TO})$] is the uncertainty in the Population II distance scale (cf Renzini 1991). In Section 3 we discuss the issue of globular cluster distances. There appears to be a dichotomy developing, with a "long" distance scale based on nearby subdwarfs, the calibration of the horizontal branch (HB) in the LMC, and analyses of the pulsational properties of cluster RR Lyrae variables and a "short" distance scale based on Baade-Wesselink and statistical parallax studies of field RR Lyraes. The disagreement

[2]Since the 1970s there has been general agreement that the oldest of the Galaxy's GCs has an age somewhere between 14 and 19 Gyr, with 16 ± 3 Gyr being perhaps the most frequently mentioned estimate: See, for instance, Demarque & McClure (1977); Carney (1980); Sandage, Katem & Sandage (1981); VandenBerg (1983); Gratton (1985); Peterson (1987); Buonanno, Corsi & Fusi Pecci (1989); Lee, Demarque & Zinn (1990); Rood (1990); Iben (1991); Renzini (1991); Salaris, Chieffi & Straniero (1993); Sandage (1993c); Chaboyer (1995); Bolte & Hogan (1995); and Mazzitelli, D'Antona & Caloi (1995). These represent a small fraction of the published reviews and original investigations over this period that have reached basically the same conclusion.

in the implied luminosity of the HB from the two calibrations is $\gtrsim 0.25$ mag. We express some preference for the long distance scale, in which case the implied age for the metal-poor cluster M92 is $\sim$ 15 Gyr: For the short distance scale its age is increased to $\gtrsim$ 18 Gyr. A brief summary of the ramifications of such ages for cosmology is given in Section 4.

2. THE STELLAR EVOLUTION CLOCK

Iben & Renzini (1984) and Iben (1991) have written fine reviews of our understanding of the evolution of low-mass stars, and Renzini & Fusi Pecci (1988) have carried out an equally valuable analysis of the degree to which canonical stellar evolutionary sequences satisfy the constraints provided by GC color-magnitude diagrams (CMDs). These papers are well worth reading again: Much of what they contain (which is not repeated here) serves to bolster one's confidence in the adequacy and accuracy of computed stellar models. Although many aspects of the more evolved stages of stars remain problematic, the overall picture of stellar evolution is certainly correct. The main point to stress in this section is that the dependence of the turnoff luminosity, L^{TO}, on age—which constitutes the stellar evolution clock—appears to be an especially robust prediction. (The turnoff is defined to be the hottest point along an isochrone, marking the end of the main-sequence stage and the beginning of the subgiant phase.)

2.1 *Uncertainty in L^{TO} Due to Basic Stellar Physics Inputs*

Chaboyer (1995) has recently used the direct approach to ascertain the impact of changes to the basic input physics on GC ages, i.e. he has determined how derived ages would be affected if the nuclear reaction rates, opacities, etc were varied, in turn, by amounts equal to reasonable estimates of their probable errors. Consequently, we use a more indirect means to show, just as Chaboyer has concluded, that present uncertainties in these physical inputs can be expected to have only very minor effects on the ages that are obtained from turnoff luminosities. Our modus operandi reveals, in addition, some of the differences between modern evolutionary calculations and those carried out at earlier times.

2.1.1 NUCLEAR REACTIONS AND OPACITIES Figure 1 shows plots of the M_{bol} (TO) [$= 4.72 - 2.5\log(L^{\mathrm{TO}}/L_{\odot})$] versus age relationships that have been derived by a number of researchers over the past 25 years. All of the calculations are based on the assumptions that $Y = 0.20$ and $Z = 0.0001$ for the mass fraction abundances of helium and the metals, respectively. (Throughout our examination of absolute GC ages, we concentrate on the low-metallicity systems, which are likely to be the oldest.) The locus attributed to Iben (1971)

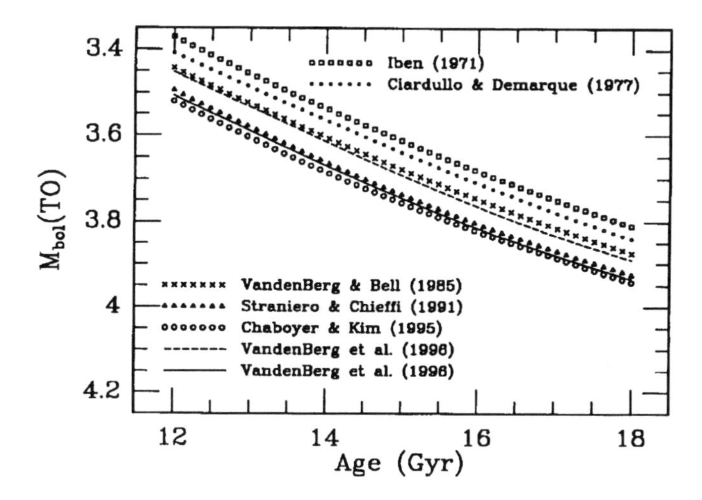

Figure 1 Turnoff luminosity vs age relations from the indicated investigations for the particular choice of $Y = 0.20$ and $Z = 0.0001$ for the mass-fraction abundances of helium and the heavier elements, respectively. The M_{bol}(TO) values were calculated on the assumption that the solar value is 4.72 mag. Small amounts of smoothing were applied in some instances. The differences between the various loci are discussed in the text.

is based on an analytic expression contained therein, which provides a good approximation to the computations by Iben & Rood (1970) and Simoda & Iben (1970). The latter assume pre-1966 nuclear reaction rates, for the most part, along with the Hubbard & Lampe (1969) set of conductive opacities and the Cox & Stewart (1970) radiative opacity data. Very similar input physics was used in the extensive grid of evolutionary tracks computed by Mengel et al (1979), which were the basis of the Ciardullo & Demarque (1977) isochrones. These were subsequently revised by Green, Demarque & King (1987) to make them better represent real stars. To be specific, the original Yale isochrones were shifted in effective temperature (T_{eff}) to compensate for the fact that Mengel et al tracks were computed for the choice of $\alpha_{\mathrm{MLT}} = 1.0$ instead of the more realistic value of 1.5. [The quantity α_{MLT} is an adjustable parameter in the mixing-length theory (MLT) of convection, which is commonly used in the construction of stellar evolutionary sequences.] Hence both sets of isochrones predict the same dependence of M_{bol}(TO) on age (the dotted curve in Figure 1).

VandenBerg & Bell (1985) adopted the updated nuclear reaction rates given by Harris et al (1983) and opacities derived from the Los Alamos Astrophysical Opacity Library (Huebner et al 1977). They noted that revisions to the nuclear physics had a $\sim 2\%$ effect on the calculated age-luminosity relation for a given evolutionary track. Because, at low Z, opacities are completely dominated by the free-free transitions of H and He (cf Schwarzschild 1958), the ramifications of improved determinations of the metal contribution are simply not significant. For instance, the use of the even more modern OPAL opacities (Rogers & Iglesias 1992), which is the main difference between the dashed curve in Figure 1 and the VandenBerg-Bell results, yields essentially the same relation between $M_{\rm bol}$(TO) and age. Indeed, the insensitivity of such relations to the particular generation of opacities assumed gives one considerable confidence in the predictions for, especially, the most metal-poor stars. But, even at higher Z, turnoff luminosity versus age relations are affected much less by improvements to the opacity than, say, the mass-luminosity relation for zero-age main-sequence stars (e.g. see Figure 4 in VandenBerg & Laskarides 1987). Although enhanced opacities will increase the main-sequence lifetime of a fixed mass, metal-rich star, they will also decrease the turnoff luminosity, such that nearly the same relationship between $L^{\rm TO}$ and age is obtained (Rood 1972, VandenBerg 1983). Fortunately, there is good reason for believing that current opacities are uncertain by no more than $\pm$ 10–20%, given that the OPAL data have led to the resolution of several longstanding discrepancies between the predictions of stellar models and actual observations (see the review by Rogers & Iglesias 1994).

2.1.2 EQUATION OF STATE The three lowermost curves in Figure 1 differ from the others in one important respect: They allow for Coulomb interactions in the equation of state. Proffitt (1993) was the first to show that this nonideal gas effect causes an $\approx$ 4% reduction in age at a given turnoff luminosity for stellar masses and chemical compositions appropriate to the globular clusters. This is close to the difference between the dashed and solid curves, which represent otherwise identical calculations except that the former ignores, and the latter includes, a Coulomb correction term in the free energy. Particularly noteworthy are the Chaboyer & Kim (1995) results: These authors used (in tabular form) the OPAL equation of state (Rogers 1994), which treats several other nonideal effects. They found a 6–7% reduction in age at a given $M_{\rm bol}$(TO), compared with the case when using the ideal gas law with radiation pressure and electron degeneracy assumed. [Their findings agree well with those of VandenBerg et al (1996) (see Figure 1), whose equation of state was set up to provide a good approximation of the more general OPAL code.]

Judging from the difference (in Figure 1) between the Iben (1971) and the Chaboyer & Kim (1995) results, there has been about a 15% reduction in the predicted age at a fixed $M_{\rm bol}$(TO) over the past 25 years (for the chemical composition that we have been considering). This reduction has resulted from steady refinements in the nuclear reaction rates, opacities, and equation of state during this time. These aspects of stellar physics are now believed to be sufficiently well understood that future developments in these areas are unlikely to affect predicted ages at more than the few percent level. This conclusion has also been reached by Chaboyer (1995), whose paper contains a useful table giving the fractional age errors as a function of the input physics (also see Renzini 1991).

2.1.3 CONVECTION THEORY A more serious concern may be the mixing-length theory of convection. Chaboyer's (1995) calculations show that, although the predicted age-luminosity relation for a given track is not greatly affected by changes in $\alpha_{\rm MLT}$, the age-color (or, equivalently, the luminosity-color) relation is altered in such a way as to shift significantly the luminosity of the hottest point on the track. This is quite an unexpected result. However, while previous studies (e.g. Demarque 1968, VandenBerg 1983) have shown that different assumptions about $\alpha_{\rm MLT}$ have profound implications for the temperature scale of an evolutionary track, without affecting its turnoff luminosity, it has apparently not been checked that the turnoffs of isochrones are similarly independent of $\alpha_{\rm MLT}$. In fact, Chaboyer has shown that this is not the case. This is perhaps not too surprising given that the temperature shift induced by a change in $\alpha_{\rm MLT}$ is a nonlinear function of mass and evolutionary state (see Figure 3 in VandenBerg 1983).

From a consideration of isochrones computed for values of $\alpha_{\rm MLT}$ in the range of 1.0 to 3.0, Chaboyer (1995) has surmised that uncertainties in how to treat convection lead to about a 10% uncertainty in GC ages as inferred from the turnoff luminosity. This is arguably a very generous error estimate given that there is no compelling evidence at the present time to suggest that $\alpha_{\rm MLT}$ differs by a large factor between stars of different mass or chemical composition or that it depends sensitively on evolutionary state. Rather, the present observational indications are that the value of $\alpha_{\rm MLT}$ needed to produce a realistic solar model is very similar to (possibly even the same as) that needed to explain the lower main-sequence slopes of young open clusters on the CMD (VandenBerg & Bridges 1984), to fit the CMD positions of the local Population II subdwarfs (see Section 3.1 in the present study; VandenBerg 1988), to match the properties of well-observed binaries whose components are in widely separated evolutionary phases (Andersen et al 1988, Fekel 1991), and to reproduce the effective temperatures of GC giant branches as determined by Frogel, Persson & Cohen (1981) from $V - K$ photometry (Straniero & Chieffi 1991, VandenBerg et al 1996). The last of these is potentially one of the most powerful constraints since the predicted position of the giant branch is highly dependent on the choice of $\alpha_{\rm MLT}$ and the comparison between theory and observation is largely independent of the GC distance scale (because the giant branch rises so vertically and its position varies only slightly with age). Frogel et al (1981) suggest that the uncertainty in their inferred temperatures is ± 90 K; even if the error were as large as ± 150 K, this could be accommodated by adopting a value of α_{MLT} that differs by as little as ± 0.3 (see VandenBerg 1983).

The value of α_{MLT} cannot be constrained any better than this, given the current observational uncertainties and the sensitivity of model temperatures to many other factors besides convection theory—notably the low-temperature opacities and the treatment of the model atmosphere boundary condition (see VandenBerg 1991). For this reason, the recent suggestion by Chieffi, Straniero & Salaris (1995) that α_{MLT} appears to be a weak function of metallicity is not convincing. Their fits to GC giant branches on the M_{bol}–log T_{eff} plane required a value of $\alpha_{\mathrm{MLT}} = 1.91 \pm 0.05$ for clusters of intermediate metal abundance ($Z \approx 0.001$), whereas the slightly smaller value, 1.75 ± 0.1, was needed for the globulars having $Z \approx 0.0001$. Obviously, such a small variation is well within the noise of its determination. One is instead impressed (once again) by the fact that stellar models are able to reproduce the observed properties of very different stars with little (or no) variation in α_{MLT}. It would be an astonishing result if α_{MLT} were constant, because there is no reason whatsoever why it should be; however, what variation there is in this parameter appears to be quite small.

The Chaboyer (1995) investigation does, however, raise the specter that a more realistic theory for convection than the MLT may have significant ramifications for GC ages. This possibility has been given considerable impetus by Mazzitelli, D'Antona & Caloi (1995), who have found that the predicted ages of the most metal-poor GCs are reduced by ~ 2 Gyr simply as a consequence of replacing the MLT by the Canuto & Mazzitelli (1991, 1992; hereafter CM) theory of turbulent convection. This theory, unlike the MLT, allows for a full spectrum of turbulent eddies, and it has essentially no free parameters: The mixing length is taken, at any point in the convective envelope, to be the geometrical depth from the upper boundary of the convection zone. [This choice for the scale length is claimed to be reasonable on the grounds of physical analogies (e.g. with the Earth's atmosphere) and its consistency with the physical scale length at which the superadiabatic zones inside stars grow and fade. Indeed, from the observed p-mode solar oscillation frequencies, Basu & Antia (1994) have found that envelope models based on the CM formalism provide a much closer match to the inferred structure of the Sun's convection zone than those constructed assuming the MLT. More recent developments (see Rosenthal et al 1995), however, suggest that this agreement may not necessarily imply such a clear-cut preference for the CM theory over the MLT.]

Mazzitelli et al (1995) find that T_{eff} effects alone lead to an apparent decrease in the turnoff luminosities of CM isochrones relative to those obtained using the MLT. (A representative example of their results is illustrated in Figure 2 for an assumed metallicity $Z = 10^{-4}$.) They verified that the temporal variation of luminosity and central hydrogen abundance along an evolutionary track is independent of how surface convection is treated (as it should be), concluding that it is the differences in morphology of the CM tracks that give rise to the decrease in M_{bol}(TO) in the corresponding isochrones, compared with MLT predictions. However, an age estimate based strictly on the turnoff luminosity will necessarily have a large uncertainty (in addition to those arising from, e.g. distance or chemical composition errors) because of the inherent difficulty in determining that point. By definition, an observed color-magnitude diagram is vertical at the turnoff; consequently, random photometric scatter or small systematic errors in the color calibration can easily cause the estimated magnitude of the bluest point to be in error by 0.1 mag (if not more)—thereby changing the derived age by at least 10%. But this uncertainty can be significantly reduced if, once the cluster distance is set using one or more standard candles (see Section 3), theoretical isochrones for the applicable chemical abundances are shifted horizontally (i.e. in color) by whatever amount is necessary to obtain a best-fit to the main-sequence photometry, and then the age is inferred from the coincidence of the predicted and observed subgiant-branch loci. The level of the subgiant branch (say, midway between the turnoff and the base of the RGB) is clearly a much better luminosity diagnostic than the turnoff point and, moreover, it is insensitive to the choice of convection theory (see Figure 2). (Granted, in the case of models that employ the MLT, large variations in α_{MLT}

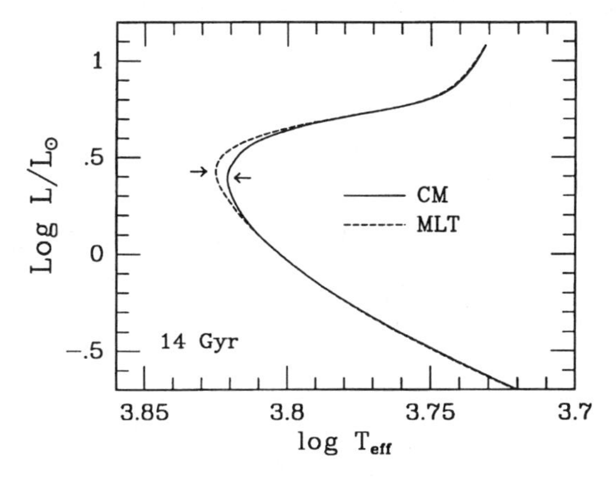

Figure 2 Comparison of Mazzitelli et al (1995) isochrones for $Z = 0.0001$, an age of 14 Gyr, and two different treatments of convection (see text). To obtain a superposition of the main-sequence loci, the MLT predictions were shifted by -0.0025 in $\log T_{eff}$. Arrows indicate the turnoff points.

would have some impact on the location/shape of the subgiant branch, but the value of this parameter appears to be fairly well constrained.) The treatment of convection need not, therefore, be a serious concern for the determination of GC ages.

2.2 *Uncertainty in L^{TO} Due to Additional Physics Usually Ignored*

There are (at least) three additional physical processes that can potentially influence the estimated ages of globular cluster stars: 1. atomic diffusion (or gravitational settling), 2. rotation, and 3. mass loss.

2.2.1 ATOMIC DIFFUSION Noerdlinger & Arigo (1980) were the first to construct models for low-mass, Population II stars in which helium was allowed to settle under the influence of gravity and thermal diffusion. They found that He diffusion tends to speed up a star's main-sequence evolution, with the result that the evolutionary tracks had slightly lower turnoff luminosities and effective temperatures compared with their nondiffusive counterparts. This translated into about a 22% reduction in the estimated ages of the globulars if the turnoff luminosity were used as the sole criterion for determining age. The follow-up study by Stringfellow et al (1983) added the interesting result that, as stars ascend the red-giant branch (RGB), the remixing of the outer layers by the deepening envelope convection erases much of the evidence of the settling of helium, and the tracks with and without diffusion gradually converge.

Nearly a decade later, Proffitt & Michaud (1991) computed a new set of diffusive models for metal-poor dwarfs using the improvements to the input physics that had occurred in the meantime—mainly to the diffusion coefficients (Paquette et al 1986). The turnoff luminosities of these models appeared to be significantly less affected by diffusion than the earlier calculations had predicted. And, in fact, the isochrones computed shortly thereafter by Proffitt & VandenBerg (1991) and by Chaboyer, Sarajedini & Demarque (1992) revealed that the age at a given L^{TO} is reduced by $\lesssim 10\%$ due to the gravitational settling of helium. [These investigations also suggested that the impact of diffusion on cluster ages would be appreciably less than this if the latter were obtained from a calibration of the magnitude difference between the horizontal branch and the turnoff. Somewhat reduced HB luminosities, compensating for $\gtrsim 1/2$ of the decrease in L^{TO}, is the expected consequence of differences in the envelope helium contents in the precursor red giants: Not all of the helium that had previously settled below the surface convection zone is dredged back up when the convection attains its deepest penetration on the lower RGB. Hence the envelopes of stars in more advanced evolutionary stages will be characterized by lower Y, which has the stated effect on HB luminosities (see, e.g. Sweigart & Gross 1976).]

However, atomic diffusion is not without its difficulties. As shown by, e.g. Michaud, Fontaine & Beaudet (1984) and Chaboyer & Demarque (1994), stellar models that allow for diffusion appear to be unable to explain the lithium abundance plateau (Spite & Spite 1982), which is the near constancy of Li abundance in halo stars having $T_{\rm eff} > 5500$ K (see, as well, Thorburn 1994). They also yield isochrones that are morphologically distinct from observed globular cluster CMDs (Proffitt & VandenBerg 1991). In contrast, the same investigations show that such data can be matched extremely well by standard, nondiffusive calculations. (The shapes of isochrones are altered by diffusion because it causes a rapid settling of helium in the very metal-poor stars, in particular, from the thin surface convection zones that they possess during their main-sequence phases. This leads to reduced turnoff temperatures by 200–300 K, whereas, as already mentioned, giant-branch effective temperatures remain relatively unaffected. One must always be wary of drawing strong conclusions from $T_{\rm eff}$/color comparisons, but it seems unlikely that current estimates of the temperatures of turnoff stars are uncertain by much more than ± 100 K.)

Why, then, is diffusion so problematic for Population II stars when it is not for, e.g. the Sun? Indeed, helioseismic data indicate a clear preference for solar models that include its effects (see Guzik & Cox 1992, 1993; Christensen-Dalsgaard, Proffitt & Thompson 1993). Because diffusion is such a fundamental physical process, which should occur in all stars, one can only conclude that something must be inhibiting its importance in metal-deficient stars. Suggested possibilities include turbulence (Proffitt & Michaud 1991), rotation (Chaboyer & Demarque 1994), and mass loss at the level of $\approx 10^{-12}\ M_\odot\ {\rm yr}^{-1}$ (Swenson 1995). Turbulent mixing below the surface convection zone will slow the rate at which the surface He abundance decreases, but as demonstated by Proffitt & Michaud (1991), it cannot eliminate the gravitational settling of helium without destroying more lithium than is consistent with the Spite plateau. The combined rotation-diffusion models of Chaboyer & Demarque (1994) are able to match the Li observations reasonably well, but they predict essentially the same evolutionary tracks on the H-R diagram as the pure diffusion calculations; consequently, the $T_{\rm eff}$ scale problems noted above would remain. Finally, Swenson's (1995) work has revealed that stellar models that treat diffusion can be made compatible with the Li data, and possibly even with globular cluster CMDs, if mass loss is assumed to occur at modest rates during main-sequence evolution.

Although stellar models that include these additional processes, which must operate to some extent in real stars, do not satisfy the observational constraints quite as well as one would hope, they do go a considerable distance towards overcoming the initial objections to diffusion. It is entirely possible that improved treatments of turbulence, rotation, and/or mass loss will reduce the remaining discrepancies concerning the surface properties of Population II stars, but they would not affect the shortening of main-sequence lifetimes due to He diffusing into the stellar cores (unless rapid core rotation could do so). Accordingly, in our view, there is really very little basis for ignoring the implications of diffusion for GC ages, which amount to less than a 10% reduction in age at a given turnoff luminosity. [This value may be slightly revised when models for GC stars become available that allow for the settling of heavier elements such as C and Fe, whose diffusion velocities are comparable with that of helium (Michaud et al 1984). Proffitt's (1994) latest solar models indicate that heavy-element settling causes only minor structural changes beyond those arising from He diffusion alone.]

2.2.2 ROTATION Rotation clearly has considerable potential in its own right to alter stellar ages (cf Law 1981), and given the abundance of direct and circumstantial evidence for rotation in GC stars, one might surmise that it has a significant impact on the ages of these objects. From the broadened lines evident in echelle spectra, Peterson (1985a,b) determined that the blue HB stars in a number of GCs rotate at significant rates (typical $v \sin i$ values of ~ 10–20 km s^{-1}). Moreover, she found that the mean rotation speeds were directly correlated with the ratio $B/(B+R)$, where B represents the number of HB stars to the blue of the instability strip and R denotes the number to the red. That is, the bluer a cluster's horizontal branch, the faster its stars rotate, on average. A qualitatively similar correlation exists between the apparent cluster ellipticity and HB type (Norris 1983): None of the most highly flattened globulars has a red horizontal branch. Although the ellipticity is presumably a reflection of the total cluster angular momentum, the Peterson data suggest that it may also be indicative of the amount of rotational angular momentum contained within individual member stars. Nonetheless, the Peterson observations do provide some support for those suggestions (e.g. by Fusi Pecci & Renzini 1975, Renzini 1977) that rotation could be a significant factor in determining the morphology of the horizontal branch. Importantly, they also show that at least some GC stars are able to retain substantial amounts of angular momentum in their interiors throughout their evolutionary histories.

Another possible signature of rotation is the observed spread in color/$T_{\rm eff}$ encompassed by a globular cluster's HB population. Such data seem to require that there is a large variation in mass among the core He-burning stars and that their mean mass is significantly smaller than that of stars presently near the turnoff (cf Rood 1973). Variable amounts of mass loss, driven (perhaps) by star-to-star differences in rotation rate, must therefore occur either during the giant-branch evolution and/or as a consequence of the helium flash event itself.

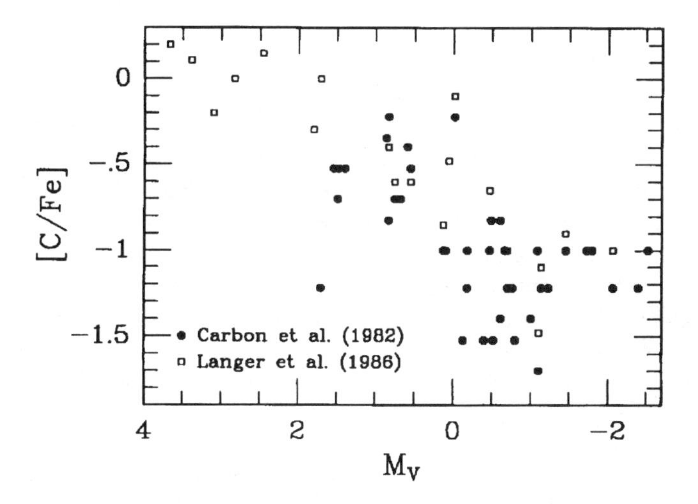

Figure 3 Carbon abundance as a function of M_V in M92. Only those data not flagged as being uncertain in the indicated studies, from which the observations were obtained, have been plotted.

The huge variations in the measured strengths of the CN, CH, and NH bands, and in the inferred or derived abundances of C, N, and O among bright GC giants (see the review by Kraft 1994 and references therein) provide further indirect evidence for the presence of rotation. These variations are not predicted by canonical evolutionary theory, but are plausibly explained in terms of circulation currents spawned by rotation (Sweigart & Mengel 1979, Smith & Tout 1992); or they may arise as the result of thermal instabilities in the H-burning shell (see Schwarzschild & Härm 1965; Von Rudloff, VandenBerg & Hartwick 1988). Especially compelling are those observations that show a dependence of molecular band strengths on giant-branch luminosity. For instance, as shown in Figure 3, the mean C abundance in M92 stars appears to decline continuously with advancing evolutionary state: Very similar trends have been observed in M15 (Trefzger et al 1983) and in NGC 6397 (Briley et al 1990). The data that have been plotted exhibit a spread of up to ~ 0.7 dex in [C/Fe] at a given M_V, which could be accounted for if some stars mix more than others due to differences in their angular velocities. ([C/Fe] represents the logarithm of the C/Fe number abundance ratio in an observed star minus the logarithm of the same quantity in the Sun, e.g. $[\mathrm{C/Fe}] = -1$ means that the measured carbon-to-iron ratio is one tenth of the solar value.) Low (or super-low) oxygen abundances are also being found in the same GCs—see Pilachowski (1988), Sneden et al (1991), and Bell, Briley & Norris (1992) regarding M92, M15 and M92, and NGC 6397, respectively—as well as in M13 (Brown, Wallerstein & Oke 1991;

Kraft et al 1993). The critical point is that nitrogen tends to be anticorrelated with C and O, often (though probably not always) to the extent that C+N+O is a constant (see Pilachowski 1989). This strongly suggests that the envelopes of bright giants in many globular clusters are somehow connected to the nuclear burning shell regions and are experiencing ongoing C→N and O→N processing. Indeed, ad hoc simulations that extend and maintain convective mixing down to the vicinity of the H-burning shell appear to be able to reproduce the observed abundance trends (see VandenBerg & Smith 1988).

The hypothesis that rotationally induced deep mixing is responsible for "anomalous" abundances in GC giants has become more credible during the past couple of years. Such mixing might also explain the correlations between, e.g. the strengths of sodium and aluminum lines with that of the CN band (Peterson 1980), which have been particularly difficult to fathom. It was generally supposed (cf Cottrell & Da Costa 1981) that such data indicated the existence of primordial abundance fluctuations, which is to say that the gas out of which the cluster stars formed was not well-mixed chemically. But a separate explanation for these anomalies may not be needed. Furthering the work of Denisenkov & Denisenkova (1990), who first explored the possibility that ^{22}Ne(p,γ)^{23}Na operated at the same temperatures as O→N burning, Langer, Hoffman & Sneden (1993) showed that deep mixing to this region of a star would naturally produce an N-Na correlation. In addition, they pointed out that the rate of ^{25}Mg(p,γ)^{26}Al was essentially the same as that of the aforementioned reaction and hence that, as long as ^{22}Ne and ^{25}Mg were present, the production of nitrogen by O→N cycling would be accompanied by the production of ^{23}Na and ^{26}Al. Because this occurred at a somewhat cooler temperature than the location of the H-burning shell, mixing into this region would not supply additional fuel into the hydrogen shell and would not, therefore, alter stellar evolution lifetimes.

Thus, a reasonably satisfactory explanation could be offered for the many observations revealing large overabundances of Na and Al, N-Na and N-Al correlations, and O-Na and O-Al anticorrelations (e.g. Cohen 1978; Norris & Freeman 1983; Paltoglou & Norris 1989; Pilachowski 1989; Lehnert, Bell & Cohen 1991; Drake, Smith & Suntzeff 1992; Kraft et al 1992). The main point of disagreement concerned the Langer et al (1993) prediction that the range in Al abundances should be Δ[Al/Fe]$\lesssim$ 0.3, whereas the observed variation can be as high as Δ[Al/Fe]$\sim$ 1.2 (also see Norris & Da Costa 1995). But with even deeper mixing, and with large initial abundances of the ^{25,26}Mg seed nuclei [the ^{26}Mg(p,γ)^{27}Al reaction can operate at a significant rate just below the O→N shell], these data can also be matched by proton-capture nucleosynthesis models (Langer & Hoffman 1995). These models predict that large aluminum enhancements should be accompanied by observable (~ 0.2 dex) depletions in Mg (initially mostly ^{24}Mg, which does not burn). These depletions may already have been detected: Smaller magnesium abundances appear to distinguish the super-oxygen-poor stars in M13 from those having higher oxygen abundances (MD Shetrone, private communication 1995).

An important consequence of the extra-deep mixing, according to Langer & Hoffman (1995), is that significant depletions of the envelope H abundance (or, equivalently, enhancements in the surface helium abundance) amounting to ~ 3–10% or more would likely occur. This would have some implications for the upper-RGB lifetimes of the affected stars, and it would influence their later evolution. As is well known (cf Rood 1973), higher envelope-helium contents make for hotter and somewhat brighter core He-burning stars. Langer & Hoffman suggest that this may help to explain why M13 has such a blue HB: The bluest stars could be characterized by higher Ys in their envelopes. (Alternatively, or in addition to this, these stars could have been subject to especially severe mass-loss rates.) Curiously, Moehler, Heber & de Boer (1995) find that high helium abundances seem to be necessary to explain the spectra of extremely blue HB stars in M15.

Not surprisingly, there are some concerns. For one, the meridional circulation mechanism should not work prior to the H-burning shell contacting the chemical composition discontinuity that is produced when the envelope convection attains its greatest penetration (near the base of the RGB): The significant mean molecular weight gradient between the energy-producing shell and that discontinuity (see Sweigart & Mengel 1979) should inhibit circulation (Tassoul & Tassoul 1984) until contact between the two is made. This contact should occur near $M_V = 0$ in very metal-poor stars (VandenBerg 1992), at which point there should be a brief hesitation in the rate of evolution up the giant branch as the H-burning shell adjusts to a higher hydrogen abundance (also see Iben 1968a, Sweigart & Gross 1978). In fact, Fusi Pecci et al (1990) claim to have detected the consequent bump in the RGB luminosity function at very close to this magnitude. Yet the progressive depletion of carbon in, e.g. M92 apparently begins fainter than $M_V = 3$ (see Figure 3), which would seem to be totally at odds with the theory and the Fusi Pecci et al observation.

This dilemma has yet to be resolved. On the one hand, the RGB bump is a very subtle feature in metal-deficient clusters; indeed, Fusi Pecci et al (1990) had to coadd the data for several of these objects in order to improve the signal-to-noise ratio. On the other hand, the measured C abundances are the least secure for the faintest stars, and perhaps these data are in need of significant revision. One point in favor of the abundance data, though, is that the subgiant and giant branch luminosity functions of [Fe/H] ~ -2 GCs show anomalous features and cannot be adequately matched by standard evolutionary predictions (see Bergbusch 1990, Stetson 1991, Bolte 1994). This point is discussed further in Section 2.5. Clearly, further and better observations are required.

Another concern is that deep mixing cannot be invoked to explain all of the abundance data. In M92, for instance, not all stars with low C abundances are nitrogen rich (Carbon et al 1982). Furthermore, despite the evidence for progressively lower $^{12}C/^{13}C$ ratios with increasing luminosity in some GCs (Smith & Suntzeff 1989), low carbon isotope ratios are often found in CN-normal stars, which have presumably not undergone substantial mixing (Bell, Briley & Smith 1990). Moreover, many clusters show bimodal distributions of CN-band strengths (Smith & Norris 1982; also see the review by Smith 1987), which persist right down to the main-sequence turnoffs in some globular clusters (e.g. 47 Tucanae; see Briley, Hesser & Bell 1991). The all-important point here is that the ratio of CN-strong to CN-weak stars does not appear to change with evolutionary state (Smith & Penny 1989, Smith & Norris 1993, Briley et al 1994). The existence of some level of primordial abundance variations would seem to be the inescapable conclusion.

Several general results bring home the complexity of globular cluster abundance work. First, as already implied, RGB mixing tends to become less severe as the cluster metallicity increases (Bell & Dickens 1980, Briley et al 1992). In metal-poor systems, CNO abundances seem to vary with M_V along much of the giant branch, whereas CN bimodalities with little or no dependence on evolutionary state appear to be characteristic of the more metal-rich globulars. Are we to conclude from this that the mean rotation rates of stars in GCs vary in some systematic way with [Fe/H]? If so, does this impact on derived ages and the age-metallicity relation that describes these objects? Second, even at the same metal abundance, different clusters show a considerable variation in their observed chemistry. Adding to the Suntzeff (1981) study of M3 and M13, which have essentially identical [Fe/H] values, Kraft et al (1992) report that they have been unable to find any super-low-oxygen stars in M3, whereas they comprise $\sim 15\%$ of the brightest giants in M13. Furthermore, although M13 stars tend to have low oxygen and high sodium abundances, those in M3 are O rich and Na poor. Do M13 stars rotate more rapidly than those in M3 and is this the reason why the two clusters also exhibit very different horizontal-branch morphologies? And finally, Population II field giants do not show the extreme abundance patterns seen in GC giants. No CN-strong stars are found in the field (Langer, Suntzeff & Kraft 1992), and many have noted the lack of field stars with high [Na/Fe] or [Al/Fe] (e.g. Brown & Wallerstein 1993, Norris & Da Costa 1995). Is it, then, a risky procedure to use field RR Lyraes to determine cluster distances? For that matter, how safe is it to use canonical horizontal-branch models to set the globular cluster distance scale given that the precursor RGB stage is problematic?

It is hard to deny the importance of rotation in GC stars and how it affects their evolution. However, precisely defining and quantifying the role that it plays are not easily accomplished in view of the relatively crude understanding that we presently have of turbulence, circulation, and angular momentum transport in rotating stars (see, e.g. Zahn 1992). Still, exciting progress is being made through such studies as those by Wasserburg, Boothroyd & Sackmann (1995) and Charbonnel (1995), who have been able to account for, among other things, the low $^{12}C/^{13}C$ ratios in bright giants by invoking meridional circulation. Further investigations along these lines are strongly encouraged.

But are the rotation rates of stars in clusters sufficient to significantly affect the relation between age and L^{TO} predicted by standard, nonrotating stellar models? The answer to this question is "probably not." Deliyannis, Demarque & Pinsonneault (1989) have computed a number of evolutionary sequences for low-mass, low-metallicity stars in which internal rotation is followed using the moderately sophisticated code described by Pinsonneault et al (1989). The transport of angular momentum due to rotationally induced instabilities, the angular momentum loss due to a magnetic wind, and the effects of rotation on the chemical abundance profiles are all calculated. In addition, the various free parameters in the theory have been constrained to satisfy the global properties of the Sun, including its present rotation rate and oblateness. Very encouraging is the fact that, as noted by Pinsonneault et al, the predicted rotation in the solar interior is in qualitative agreement with the estimates from oscillation data, especially at radii $> 0.6R_{\odot}$. The code has proven successful in modeling the observed surface Li abundances and rotation rates of stars in young open clusters (Pinsonneault, Kawaler & Demarque 1990) and in the halo (Pinsonneault, Deliyannis & Demarque 1992).

The calculations of Deliyannis et al (1989) for globular cluster parameters predict that rotation will not change the age at a given turnoff luminosity by more than 1%. Moreover, they suggest that, due to angular momentum redistribution and losses at the stellar surface, the angular momentum of the core is kept at a level that is insufficient to alter canonical estimates of the core mass at the helium flash. As a result, HB luminosities should not be affected nor should calibrations of the age dependence of the magnitude difference between the turnoff and the horizontal branch. At the same time, the models are expected to possess sufficient differential rotation with depth (according to Pinsonneault, Deliyannis & Demarque 1991), to be capable of matching the rotational velocity data that Peterson (1985a,b) has obtained for the HB stars in several GCs. (These inferences are based on the rotational characteristics of turnoff models: As far as we are aware, the tracks have not yet been extended past the lower RGB.) Worth repeating is the comment by Pinsonneault et al that "the thin surface convection zones of halo stars allow differential rotation with depth to begin much further out and to reach a greater contrast between central and surface rotation." This offers the reason why one might expect rotation (and mixing?) to become more important as the cluster metallicity decreases.

All of this represents a really superb theoretical effort and further progress is eagerly anticipated. In particular, it will be interesting to learn whether or not these models can explain the wealth of chemical abundance data previously described. Until those constraints are matched, we suspect that it is still within the realm of possibility that rotation has a bigger effect on turnoff ages than Deliyannis et al (1989) have estimated. However, extremely high rotation rates can be precluded simply because the turnoff stars in GCs follow very tight color-magnitude relationships—see, e.g. Stetson's (1993) review, wherein he reports that M92's photometric sequence is only 0.0078 mag thick in $B - V$ in the range $17.8 < V < 18.4$ (which is just above the turnoff). If the stars in clusters had high rotation rates, then much larger intrinsic spreads would be expected because of star-to-star differences in rotational velocity and the dependence of a star's photometry on the particular aspect being viewed (e.g. see Faulkner, Roxburgh & Strittmatter 1968). On the other hand, judging from the very simplistic models by Mengel & Gross (1976), who treat rotation in the spherically symmetric approximation, fairly large rotation rates are needed to cause significant departures from the evolutionary track that a nonrotating star of the same mass and chemical composition would follow. That is, the tightness of observed CMDs does not preclude rotation rates that are large enough to have small effects on the core mass at the helium flash or to change computed age versus turnoff luminosity relations at the few percent level (see the Mengel & Gross study). Note that rotation tends to increase the age at a given L^{TO}.

2.2.3 MASS LOSS A few years ago, Willson, Bowen & Struck-Marcell (1987) suggested that significant mass loss ($\gtrsim 10^{-9} M_{\odot}$ yr^{-1}) may occur in the region of the main sequence that overlaps with the extension of the Cepheid instability strip. This mass loss would be driven by pulsation as well as the rapid rotation normally possessed by the early-A to mid-F stars that occupy this region. They suggested, for instance, that the Sun's very low Li abundance (Steenbock & Holweger 1984) could be explained if it started out as as a 2 $M_{\odot}$ star and lost half its initial mass during the first 10^9 yr of its existence as it evolved through this critical zone on the H-R diagram. According to Willson et al, this mechanism might also account for blue stragglers, and it may even help to alleviate the apparent conflict between GC ages and the age of the Universe implied by high values of H_0 (should they prove to be correct).

Noting that the instability strip crosses the main sequence very near to where the so-called Li dip[3] occurs in Population I stars of type F, Schramm, Steigman & Dearborn (1990) considered whether or not the Willson et al (1987) hypothesis might also work here. They found that models that lose mass at $\gtrsim 7 \times 10^{-11} M_{\odot}$ yr^{-1} are able to match the shape of the Li dip in the Hyades quite well. However, the mass-loss rate had to be $< 1 \times 10^{-10} M_{\odot}$ yr^{-1} in order to avoid being in conflict with the observation that beryllium is not depleted (Boesgaard & Budge 1989). In their much more extensive study, Swenson & Faulkner (1992) agreed that mass-loss models are capable of matching the Li contents of Hyades F stars, but that very little leeway is allowed in the mass-loss rates, which must vary nonmonotonically with initial stellar mass in a very well-defined way, with little star-to-star deviation. [Note that there are alternative explanations for all or part of the Li-dip observations, including atomic diffusion (Michaud 1986) and rotationally induced mixing (e.g. Charbonnel & Vauclair 1992).]

Based on the existence of a few extremely metal-deficient stars with very low Li abundances, Dearborn, Schramm & Hobbs (1992) suggested that an analogous lithium dip might be present on the Population II main sequence. Further, they found that such data could be explained by the same mass-loss model as was used for the Hyades if mass-loss rates of $\sim 10^{-11} M_{\odot}$ yr^{-1} were assumed to apply within an instability strip lying in the range $6600 \leq T_{\rm eff} \leq 6900$. They commented that mass loss of this type would make GCs look ~ 1 Gyr older than they really are (also see Shi 1995). However, Molaro & Pasquini (1994) have detected lithium in a turnoff star of the $[Fe/H] \approx -2.1$ globular cluster NGC 6397, and, moreover, their measured Li abundance is the same as those of field halo stars (e.g. Thorburn 1994) to within the errors. This observation provides a very strong argument against the high mass-loss hypothesis, especially given that the field star data themselves preclude mass-loss rates $\gtrsim 2 \times 10^{-12} M_{\odot}$ yr^{-1} (Swenson 1995). (It also gives a reassuring indication of the similarity between cluster and field main-sequence stars at low Z.)

As discussed by Shi (1995), there is another way to test whether or not the turnoff stars in GCs are losing significant amounts of mass. If they are, then a step-like feature, reflecting the sudden onset of high mass-loss rates, should manifest itself in the luminosity function plane at the point where the CMD intersects the instability strip. Although there are some anomalous features in the observed luminosity functions for GCs, mainly for the most metal-deficient systems (see Section 2.5), they appear to be restricted to post-turnoff evolutionary phases. That is, no obvious bumps or steps are seen at turnoff luminosities.

[3]The Li dip refers to the striking variation of Li abundance with $T_{\rm eff}$ that Boesgaard & Tripicco (1986) discovered in the Hyades: Stars with $T_{\rm eff}$'s near 6600 K show severe Li depletions compared with those 300 K cooler or hotter.

All in all, the possibility that mass-loss rates are high enough to affect GC ages seems remote.

2.3 *Uncertainty in L^{TO} Due to Unconventional Physics*

In the late 1970s and early 1980s the possibility that the Gravitational constant G varied with time received considerable attention, due largely to the fact that this feature was common to three prominent cosmologies—Brans-Dicke (1961), Hoyle-Narlikar (1972a,b), and Dirac (1974). Some of the early tests of the implications of these theories for stellar evolution seemed to lead to satisfactory results (e.g. see Canuto & Lodenquai 1977, VandenBerg 1977, Maeder 1977). Even when potential difficulties, such as the apparent incompatibility of Dirac's theory with the observed characteristics of the microwave background, were pointed out (Steigman 1978), it was often possible to accommodate those objections by revising the theory (cf Canuto & Hsieh 1978). Using their flavor of gravitational theory, Canuto & Hsieh (1981) showed that it was possible for the estimated ages of GCs to decrease from 15 Gyr, under canonical assumptions, to < 10 Gyr, if G varied at a rate ($\dot{G}/G \approx -6 \times 10^{-11}$ yr^{-1}) that was consistent with observed limits at that time (see Van Flandern 1981).

However, those limits are now very much tighter. Taylor & Weisberg (1989) have determined that $\dot{G}/G = (1.2 \pm 1.3) \times 10^{-11}$ yr^{-1} from pulse time-of-arrival observations of the binary pulsar PSR1913+16 over the previous 14 years. Their data are completely consistent with Einstein's Theory of General Relativity. In addition, Müller et al (1991) obtain $\dot{G}/G = (0.01 \pm 1.04) \times 10^{-11}$ yr^{-1} from 20 years worth of lunar laser ranging data. These results essentially eliminate the possibility of temporal variations in G being a factor in the determination of GC ages.

More promising, perhaps, is the following idea: If nonbaryonic Weakly Interacting Massive Particles (or WIMPs) constitute the dark matter in the Universe, then they might be accreted by stars and affect their evolution (Steigman et al 1978, Press & Spergel 1985). Being massive, they would tend to collect in the cores of stars, and by virtue of being weakly interacting, they would provide an efficient means of central energy transport. If such particles resided in the Sun, for instance, they could lower the central temperature enough to enable a solution to the solar neutrino problem (Faulkner & Gilliland 1985, Spergel & Press 1985). This would require WIMP masses between approximately 2 and 7 GeV and interaction cross sections with nuclei within an order of magnitude (or so) of 10^{-35} cm^2 [see Dearborn, Griest & Raffelt (1991), who also discuss recent experimental limits on these properties]. Furthermore, according to Faulkner & Swenson (1988, 1993), the deduced turnoff ages of globular clusters would be $\sim$20% less than canonical estimates, if their member stars acquired sufficient numbers of WIMPs to isothermalize the innermost 10% of their masses.

The main testable prediction, as far as GCs are concerned, is that stars containing WIMPs will leave the main sequence somewhat sooner than canonical stellar models would predict, due to the isothermal core effect, and spend more time on the subgiant branch, because they have extra hydrogen to burn in the shell-narrowing phase. That is, an observed luminosity function should show an excess of subgiants and giants relative to the number of turnoff stars, if WIMP models are more realistic than standard calculations. Surprisingly, this is actually seen in the luminosity–function data for a number of GCs (see Stetson 1991, VandenBerg & Stetson 1991, Faulkner & Swenson 1993, Bolte 1994). However—and this poses a problem—these "anomalies" appear to be present in the observations of only the extremely metal-deficient clusters; i.e. the same ones that show strong evidence for progressive mixing along the RGB. As shown in Section 2.5, new observations for M5 appear to conform remarkably well to standard evolutionary predictions, as do the available luminosity–function data for 47 Tuc (see Bergbusch & VandenBerg 1992). One is tempted to think that something to do with the deep-mixing phenomenon, rather than WIMPs, is the more likely cause of the unexpected luminosity function features.

However, models have not yet been constructed for GC stars that incorporate the very detailed theory for the accretion (and evaporation) of WIMPs that has been developed by Gould (1990) and Gould & Raffelt (1990a,b). These models may predict something quite different from calculations that attempt to mimic the effects of WIMPs by imposing (albeit in a self-consistent way) an isothermal core structure on an otherwise normal stellar model. Thus further work is certainly warranted—even though there are other indications that the WIMP hypothesis faces an uphill battle. For instance, using the Gould/Gould-Raffelt theory, Turck-Chièze et al (1993) find that solar models containing WIMPs do not appear to satisfy helioseismic constraints as well as Standard Solar Models. Also, VandenBerg & Stetson (1991) have suggested that WIMPs would likely suppress the formation of convective cores in the $\approx 1.3 M_{\odot}$ turnoff stars in the old open cluster M67. If this happened, then the observed gap feature at $M_V \approx 3.5$ (see the recent CMD by Montgomery, Marschall & Janes 1993) would not be produced. (It is possible, of course, that the density of WIMPs is much greater in the halo of the Galaxy than in the disk and that the evolution of the Sun and M67 stars would be little affected.) In addition, WIMPs may (Renzini 1987) or may not (Spergel & Faulkner 1988) cause difficulties for our understanding of the horizontal-branch phase of low-mass stars (also see Dearborn et al 1990).

It would be premature to conclude that WIMPs, or the very similar "halons" that Finzi (1991, 1992) has proposed, or other dark matter candidates like axions (Peccei & Quinn 1977; Dearborn, Schramm & Steigman 1986; Isern, Hernanz

& Garcia-Berro 1992) do not affect stellar ages (if they exist). But neither can one give very serious consideration to the possibility that they do, at least at this time. There appears to be a number of difficulties for the WIMP hypothesis to overcome, and the other suggestions have simply not been adequately developed and tested to pose a serious challenge to standard stellar evolutionary theory.

2.4 *Uncertainty in L^{TO} Due to the Assumed Chemistry of Stars*

It has long been known that the predicted age of a star of a given mass depends on its initial helium and heavy-element abundances (e.g. Demarque 1967, Iben & Rood 1970). Even the special importance of the CNO elements for stellar ages was appreciated early on (e.g. Simoda & Iben 1968). This has driven a huge, ongoing effort by many observers to define the detailed run of chemical abundances in field and halo stars as accurately as possible. Thanks to that effort, we now know (for instance) that [C/Fe] and [N/Fe] ~ 0 over $0.3 \lesssim$ [Fe/H] $\lesssim -2$ and that the elements synthesized by α-capture processes (e.g. O, Ne, Mg, Si, etc) are enhanced, relative to iron, in metal-poor stars by a factor of 2–3 (see the comprehensive review by Wheeler, Sneden & Truran 1989). (In the standard notation, this corresponds to [α/Fe] $\approx$ 0.3–0.5, where α represents O or Ne or Mg, etc.) It is not yet definite that all of the so-called α-elements scale together as there is considerable scatter in the field star observations (some of it real): Thus, the precise shapes of the mean relations between the various [element/Fe] ratios as a function of [Fe/H] still have some degree of uncertainty. Also, whether or not field and GC dwarfs of the same iron content are chemically indistinguishable remains a matter of some concern. But the chemistry of stars appears to be largely under control.

High-resolution spectroscopy (e.g. Cohen 1979, Sneden et al 1991) and the tightness of observed CMDs (e.g. Stetson 1993; Folgheraiter, Penny & Griffiths 1993) have established that the dispersion in Fe abundances is very small in nearly all GCs (ω Cen and possibly M22 being exceptions). Moreover, the spectroscopic data now yield [Fe/H] values that are accurate to within $\approx \pm 0.2$ dex, if not better. According to the upper panel of Figure 4—which shows plots of the turnoff luminosity versus age relations that VandenBerg et al (1996) have computed for various choices of [Fe/H], [α/Fe], and Y—this implies an uncertainty in the age at a given $M_{\rm bol}$(TO) of about $\pm$ 1 Gyr ($\approx \pm 7\%$). Furthermore, since the α-element contents of stars in the extremely metal-deficient clusters like M92 appear to be within ± 0.15 dex of [α/Fe] $= 0.4$ (e.g. Sneden et al 1991; McWilliam, Geisler & Rich 1992), the corresponding age uncertainty is expected to be about $\pm 4\%$ (judging from Figure 4). This makes a total

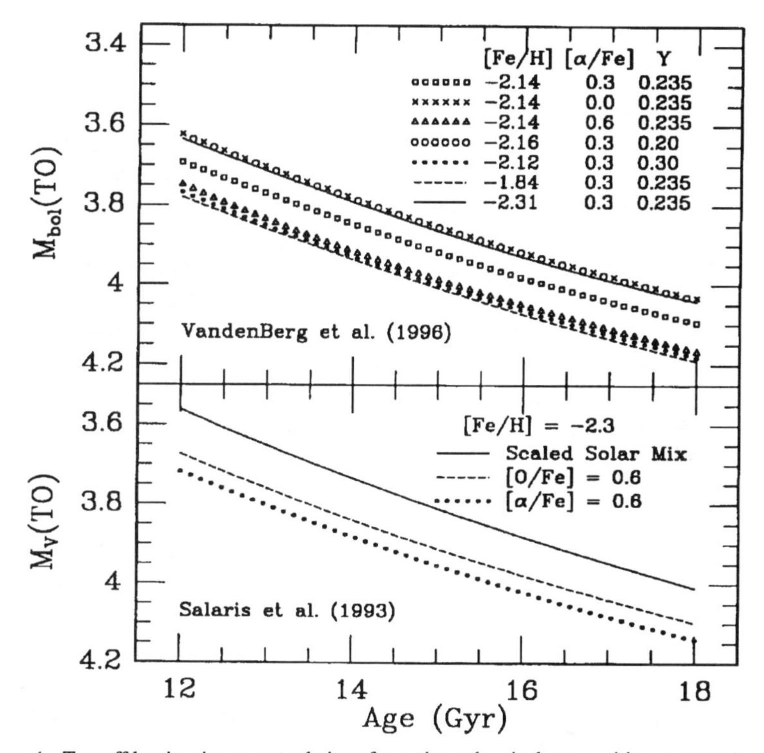

Figure 4 Turnoff luminosity vs age relations for various chemical composition parameters.

uncertainty of $\pm$ 11% in the turnoff ages due to current errors in heavy-element abundance determinations.[4]

Helium–abundance uncertainties could potentially affect age estimates at the few percent level (see the upper panel of Figure 4), but Y appears to be rather well determined, in spite of the fact that the methods used are indirect. [Spectral features due to helium can be detected in hot HB stars, but gravitational settling is known to be important in them (e.g. Heber et al 1986).] Foremost among

[4]At first sight, Figure 4 would appear to contradict the claim by Chieffi, Straniero & Salaris (1991) that enhancements in the α-elements do not lead to younger ages for the GCs (also see Bencivenni et al 1991). But, in fact, the reason why they obtained similar ages using either α-enhanced or scaled-solar abundance isochrones is that they set the distances to the globulars using theoretical horizontal-branch calculations, which predict that the HB luminosity should decrease as [α/Fe] increases. Only by an appropriate adjustment of the GC distance scale is it possible to reach the conclusion that ages are insensitive to [α/Fe]: The turnoff age-luminosity relations computed by Salaris, Chieffi & Straniero (1993) both for [α/Fe] = 0.0 and for [α/Fe] > 0.0 are very similar to those derived by VandenBerg et al (1996).

these techniques is the so-called R-method (Iben 1968b), which compares the ratio of the predicted HB and RGB lifetimes, t_{HB}/t_{RGB}, as a function of Y, with the observed number ratio of stars in these phases. Using mainly the calibration of Buzzoni et al (1983) (also see Caputo, Martinez Roger & Paez 1987), nearly all applications of the R-method (e.g. Buonanno, Corsi & Fusi Pecci 1985; Ferraro et al 1992, 1993) have yielded $Y = 0.23 \pm 0.02$. Discrepant results have been obtained for a few globulars, such as M68 (Walker 1994), for which the R-method implies $Y \sim 0.17$; however, in that particular case, the analogous ratio of the numbers of asymptotic-giant branch to RGB stars gives an estimate of the helium abundance that is within $1\,\sigma$ of $Y = 0.23$. (Why M68 has such an anomalous R value is presently unknown.)

Fits to the morphologies of observed HB populations (e.g. Dorman, VandenBerg & Laskarides 1989; Dorman, Lee & VandenBerg 1991) and to the red edges of the RR Lyrae instability strips in clusters (Bono et al 1995) reinforce the R-method results. Pulsation models have traditionally favored $Y \approx 0.30$, but due to the advent of the OPAL (Rogers & Iglesias 1992) and OP (Seaton et al 1994) opacities, lower values of Y can now be accommodated (Kovács et al 1992, Cox 1995). The adoption of $Y \approx 0.23$ in models for GC stars is further supported by the fact that this value is very close to that predicted by standard and inhomogeneous Big Bang nucleosynthesis calculations (see, e.g. Krauss & Romanelli 1990 and Mathews, Schramm & Meyer 1993, respectively), as well as empirical determinations of the pregalactic helium abundance (Pagel et al 1992; Izotov, Thuan & Lipovetsky 1994; Olive & Steigman 1995).

We conclude this section by emphasizing the importance of oxygen to stellar age determinations. Plotted in the lower panel of Figure 4 are the age versus turnoff luminosity relations that Salaris et al (1993) have derived for [Fe/H] $= -2.3$ and various assumptions about the element mix. This plot shows that most of the reduction in age at a given L^{TO} that results from an enhancement in the α-elements is due to oxygen. Getting the oxygen abundance right is, therefore, a much bigger concern than having precise abundances for most of the other heavy elements. This result is not unexpected given the large abundance of oxygen and its role as a catalyst in the CNO-cycle and as a major contributor to bound-free opacities in stellar interiors (see, e.g. VandenBerg 1992).

2.5 *Tests of Stellar Models*

The interior structures of low-mass, main-sequence stars are believed to be much simpler, and therefore (presumably) better understood, than those of their higher-mass ($M \gtrsim 1.15\,M_{\odot}$) counterparts because, in part, they do not contain convective cores and so are unaffected by the uncertainties (e.g. the extent of overshooting) associated with them. Perhaps the main evidence for possible inadequacies in the theory has been the longstanding failure of canonical models to reproduce the observed flux of neutrinos from the Sun, but solar oscillation studies have considerably diminished that concern. As Dziembowski et al (1994, 1995) have concluded, the inferred structure of the Sun from helioseismology is now so close to that predicted by the standard model, throughout its interior, that there is little room left for an astrophysical solution to the solar neutrino problem. Certainly, there are many examples in the scientific literature demonstrating how well current stellar evolutionary theory can match superb observational data. One of the nicest of these is the study of the Hyades by Swenson et al (1994). They obtained a self-consistent fit to the CMD, to the mass-luminosity relation defined by the cluster binaries, and to the Li abundances in the G stars, using opacities for the observed [Fe/H] value and without applying any ad hoc adjustments of any kind. Indeed, the *best-observed* binaries (e.g. AI Phe—see Andersen et al 1988) and the mass-luminosity relations derived from them appear to agree rather well with the predictions of standard models (cf Andersen 1991).

Considering the more evolved, post-turnoff phases, the main challenge to the theory would appear to be the observed chemical abundance variations among bright GC giants (already summarized in Section 2.2.2) and some anomalies in the luminosity function (LF) data for a few clusters (see below). These difficulties will probably be resolved once rotation is accurately treated (which, admittedly, is not easily done). Otherwise, as extensively reviewed by Renzini & Fusi Pecci (1988), there appears to be little basis for believing that the sorts of models that have been computed for the past 25 years or so are seriously in error. Although many discrepancies between theory and observation can be identified, it is much more likely that they are due to deficiencies in, for instance, the opacity or convection theory, than to a problem with the basic stellar structure equations themselves. But it may be the case that the LF anomalies have a different origin.

Relatively little work has been done on the luminosity functions of GCs, in spite of the fact that they provide a superior test of stellar models compared with the fitting of CMDs and despite some tantalizing results from early studies. For instance, Simoda & Kimura (1968) suggested that the LFs of M3 and M13 differed from one another—which might be an important clue (yet to be followed up) as to the cause of the differences in the HB morphologies of these common-[Fe/H] clusters. Also, making use of the fact that LFs provide one of the few ways to infer the helium abundances in GCs (see the recent study by Ratcliff 1987), Hartwick (1970) derived $Y \sim 0.35$ from such an analysis of M92. However, not until Bergbusch's (1990) study of the latter cluster was a possible inconsistency between an observed LF and theoretical predictions identified. Depending on how the synthetic and observed LFs were matched, the M92 data showed either a broad dip between $19 < V < 20$ or a bump near $V = 18.6$ that was not present in the models. Stetson's (1991) combined LF for M68, M92, and NGC 6397, based on new CCD observations of these clusters, confirmed the existence of these features. In addition, it revealed that, when theoretical LFs for the appropriate Y and Z were normalized to the turnoff data, the observed RGB had a significant excess of stars compared with the number predicted. These anomalies are not readily explained in terms of variations in any of the usual parameters, but they have turned out to be the anticipated signature of WIMPs (as already recounted in Section 2.3).

The lower panel in Figure 5 illustrates Bolte's (1994) LF for M30, and it too poses the same problems for the theory as the data for the other [Fe/H]

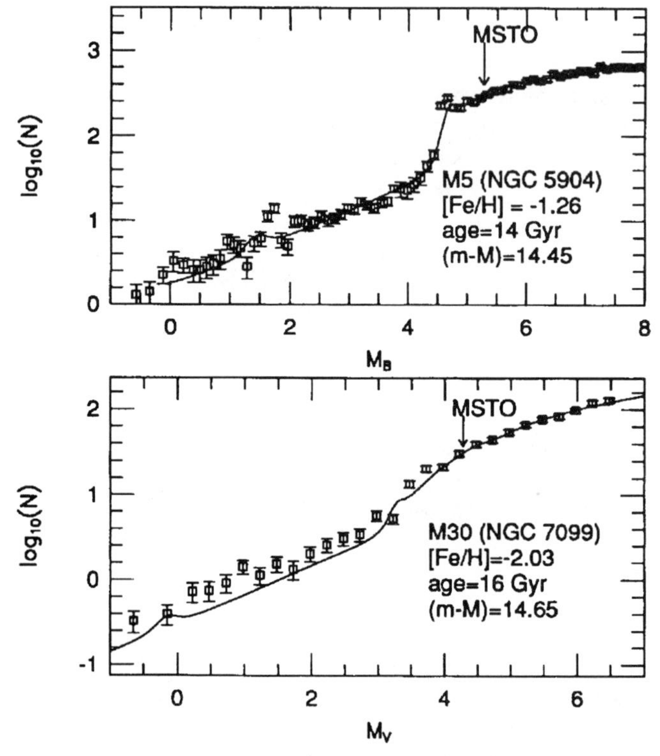

Figure 5 Comparisons of the observed luminosity functions for M5 (*upper panel*) and M30 (*lower panel*) with Bergbusch & VandenBerg (1992) isochrones, on the assumption of the indicated ages, [Fe/H] values, and distance moduli. The M5 data are from Sandquist et al (1996); the M30 results are as reported by Bolte (1994).

~ -2.1 GCs. But note that the M5 LF (in the *upper panel*) shows no such anomalies; indeed, it conforms remarkably well to the theoretical predictions. Similarly, Bergbusch & VandenBerg (1992) have not found any obvious difficulties in fitting the available luminosity function data for 47 Tuc (though their matching of the brighter to the fainter data is somewhat uncertain). And Stetson & VandenBerg's (1996) Canada-France-Hawaii Telescope photometry for a sample of $\sim 10^5$ stars in M13 shows no evidence of a subgiant bump either. Curiously, their very preliminary analysis suggests that the RGB in M13 may be underpopulated relative to the turnoff; i.e. the opposite to what is seen in the more metal-poor clusters.

There is clearly much to be learned from such LF studies, but from these first results, one has the impression that the anomalous subgiant bump is characteristic of only the extremely metal-poor clusters—and hence it can hardly be due to WIMPs, which should not show a preference for a particular [Fe/H] value. Is that feature somehow connected with the deep-mixing phenomenon? We do not know. The differences in the relative RGB-to-turnoff populations might be due to differences in helium abundance. Alternatively, it may be an indication of differences in core rotation. Using the simplest possible treatment of rotation (cf Mengel & Gross 1976), Larson, VandenBerg & De Propris (1995) have found that the number of giants relative to the number of turnoff stars is larger if the stars have significant internal rotation. Perhaps the main point to be made here is that, unless and until the LF data are satisfactorily explained, one should be wary of trusting the application of standard, nonrotating, unmixed-envelope models to those clusters (apparently the most metal-poor ones) whose luminosity functions cannot be reproduced by such models.

We conclude that, although there remain unexplained observations of evolved stars in globular clusters, the stellar models for GC stars at the main-sequence turnoff and probably to a few magnitudes down the presently observed main sequence are reliable. In particular, the agreement between predicted and observed mass-luminosity relations suggests that the theory is basically correct and essentially complete. The main outstanding issue for models (in the context of predicting the main-sequence lifetimes of low-mass stars) is the extent to which helium diffusion may reduce cluster age estimates. Our best estimate for the maximum reduction in ages due to this effect is $\lesssim 1$ Gyr. Although it could be postulated that some currently unconsidered physics will in the future reduce measured cluster ages, there currently does not appear to be any viable candidate physical processes, nor is there any clear motivation for seeking them other than to try to reduce GC ages.

2.6 *The First Estimates of Globular Cluster Ages*

It is instructive to look back to the first papers that were written on the subject of globular cluster ages. Using hand computation, Sandage & Schwarzschild (1952) produced the first evolutionary tracks for low-mass Population II stars to somewhat beyond central hydrogen exhaustion. From these calculations they inferred an age of 3.5×10^9 yr for the two globulars whose turnoffs had just been detected—M92 (Arp, Baum & Sandage 1953) and M3 (Sandage 1953). However, they had not modeled the earliest, low-luminosity phases, and when this was taken into account, the estimated age rose to 6.2×10^9 yr (Hoyle & Schwarzschild 1955). Essentially the same result (6.5 Gyr) was obtained by Haselgrove & Hoyle (1956), who were the first to use a digital computer to solve the stellar structure equations.

In the 1950s, it was generally supposed that the original matter in the Galaxy was pristine (i.e. "uncooked"); consequently, the stellar models that were computed at that time assumed $Y \approx 0.0$. It was not realized until somewhat later that it would be very difficult for conventional stellar nucleosynthesis to explain the increase from such low Y values to the observed high helium contents of Population I stars (see Hoyle & Tayler 1964), and it was later still that the microwave background was discovered (Penzias & Wilson 1965) and the notion that the Universe began as a singularity took hold. Big Bang nucleosynthesis calculations carried out shortly thereafter (e.g. Wagoner, Fowler & Hoyle 1967) predicted that the primordial helium abundance would be somewhere in the range $0.2 \leq Y \leq 0.3$.

However, even before these developments, Hoyle (1959) had computed an evolutionary track for $Y = 0.249$ and $Z = 0.001$ to explore the consequences of higher Y. Because that Y, Z combination is very close to what is assumed in present-day stellar models, we thought that it would be interesting to compare the track that Hoyle computed (as tabulated in his paper) with one for the same mass (1.163 $M_\odot$) and chemical composition using the latest version of the University of Victoria code (see VandenBerg et al 1996). That comparison is shown in Figure 6. Considering the primitive state of our understanding of stellar physics nearly 40 years ago—even the relative importance of the *pp*-chain versus the CNO-cycle was largely unknown—the agreement is remarkably good. The turnoff temperatures agree to within 240 K, the turnoff luminosities to within $\Delta(\log L/L_\odot) = 0.05$, and the turnoff ages to within $\approx 40\%$ (4.3 Gyr for Hoyle's model versus 2.7 Gyr for ours).

The point of this exercise is to show that the first stellar models computed for GC stars predicted a higher, not a lower, age at a fixed turnoff luminosity than do modern calculations. (The adoption of $Y \approx 0.0$ would tend to further increase that age.) Therefore, the low ages reported in those initial investigations

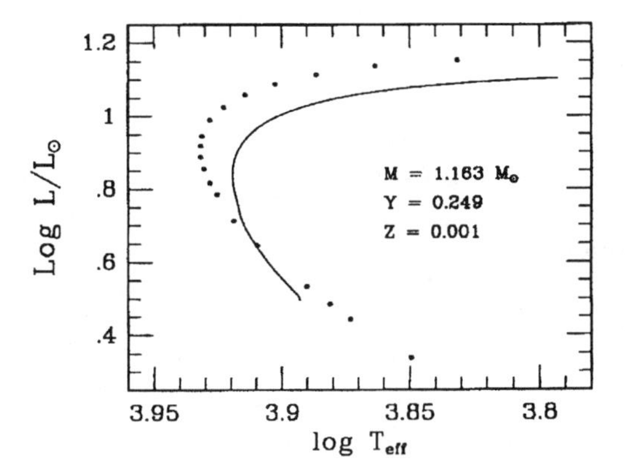

Figure 6 Comparison of an evolutionary track computed by Hoyle (1959) (*dotted curve*) with one for the same mass and chemical composition, as specified, but using the latest version of the University of Victoria code (*solid curve*).

must be attributed to something other than the evolutionary models that were used. In fact, they resulted from the then conventional assumption that the RR Lyrae variables, which were used to set the GC distance scale, had $M_V =$ 0.0. Only after such studies as that by Eggen & Sandage (1959), who used trigonometric parallax stars encompassing a range in [Fe/H] and the nearby Groombridge 1830 group of low-metallicity subdwarfs to do main-sequence fits, did it become accepted that the actual luminosities of the RR Lyraes must be near $M_V = 0.5$ [although there were earlier indications that this might be the case (e.g. Pavlovskaia 1953)]. Hoyle (1959) noted that this revision would imply an age for the Galaxy of greater than 10^{10} years. Thus, ages much more similar to current estimates would have resulted had the cluster distances been known more accurately. Even today, as we show in the next section, distance uncertainties continue to dominate over all other sources of error.

3. GLOBULAR CLUSTER DISTANCES

Given that the most reliable indicator of a globular cluster's age is its turnoff luminosity, the determination of precise distances to these systems is arguably the single most crucial observational input into the evaluation of accurate ages (see, e.g. Renzini 1991, Chaboyer 1995, Bolte & Hogan 1995). Nearly everything that we know about GC distances is based on two standard candles—namely, the nearby subdwarfs and the RR Lyrae variable stars. Thanks to the development of the *Hubble Space Telescope* (*HST*), we will soon be able to add white dwarfs to this very short list. These stars have the advantage of being essentially free of metallicity and convection complications (cf Fusi Pecci & Renzini 1979), and local white dwarf calibrators are much more numerous than subdwarfs. Although we can anticipate that the fitting to white dwarf cooling sequences will involve a number of difficulties (some unanticipated), it is encouraging that the first *HST* results, for M4 by Richer et al (1995), indicate a distance very similar to the one adopted by Richer & Fahlman (1984) on the assumption that $M_V(\mathrm{HB}) = 0.84$. These results lead to their determination of an age of 13–15 Gyr for this cluster. We also recognize the potential of direct astrometric methods (see Cudworth & Peterson 1988, Rees 1992) and the existence of a number of other approaches (e.g. using the RGB tip magnitude) to constrain cluster distances. We, however, restrict the present discussion to the two classical distance calibrators.

3.1 *Subdwarf-Based Distances*

The nearby subdwarfs—metal-poor stars with halo kinematics whose orbits have brought them near enough to the Sun for them to have measurable trignometric parallaxes—play two critical roles in the measurement of GC ages. First, with well-determined values of M_V, these objects provide a direct test of the model predictions for the position of the zero-age main-sequence as a function of [Fe/H] in the low-metallicity regime. Second, under the (testable) assumption that the subdwarfs are local versions of the unevolved main-sequence stars in globular clusters, they can be used to tie the cluster distances directly into the most reliable distance scale that exists in extra-Solar-system astronomy (that defined by trignometric parallaxes).

The recognition of the importance of the subdwarfs and of their relation to the RR Lyraes and the halo GCs is itself an interesting story (see the review by Sandage 1986). An important landmark was Sandage's (1970) identification of eight subdwarfs with sufficiently good π measures for them to be useful for deriving the distances to GCs. He also used them to calibrate the absolute magnitude of the horizontal branch at the position of the instability strip in M3, M15, and M92. Carney (1979) and Laird, Carney & Latham (1988) improved the [Fe/H] determinations of that sample. In the pre-CCD era of photometry, however, the subdwarfs were of limited usefulness for establishing the Population II distance scale because of fairly large random and (in retrospect) scale errors in the measurement of faint main-sequence cluster stars [see, e.g. Figure 4 in Fahlman, Richer & VandenBerg (1985) and Figure 30 in Stetson & Harris (1988)]. CMDs derived from CCD data, beginning in the mid-1980s, made the adoption of a subdwarf-based distance scale a much more viable alternative to purely HB-based distance estimates. With CCDs and 4-m telescopes, the main sequences of nearby clusters could be defined very accurately down to

$M_V \sim 10$ (e.g. see Figure 36 in Stetson & Harris 1988). In the CCD era, the limiting factors in the derivation of cluster distances via subdwarf fitting became the scatter in the Population II main-sequence fiducial defined by the subdwarfs and the lingering uncertainties in the reddening and color calibrations of the cluster data.

Table 1 contains our compilation of relevant data for all stars in the 1991 edition of the Yale Trigonometric Parallax Catalogue with $\sigma_\pi/\pi < 0.5$ and spectroscopic measures of [Fe/H] $\lesssim -1.3$. This list includes the original eight stars from Sandage (1970) minus HD 140283, which appears to be an evolved star (Magain 1989, Dahn 1994), plus an additional eight stars, which generally have large σ_π values. The tabulated σ_π values were taken from the Yale Catalogue; the apparent colors and magnitudes are from the compilation given in the Hipparcos Input Catalogue (Turon et al 1992). The absolute magnitudes were calculated from the usual equation: $M_V = V + 5 + 5\log(\pi)$. Because trignometric parallax measurements are subject to a Malmquist-like bias, arising from the coupling of the measuring errors with the steep slope of the true parallax distribution, there is a tendency for the observed parallaxes to be larger than their actual values. (This is true in the statistical sense for entire catalogues as well as for individual measurements.) The resultant so-called Lutz–Kelker (or L-K) corrections (Lutz & Kelker 1973) were determined to compensate for this effect. To be specific, we have applied the correction $\delta M_V = -5.43(\sigma/\pi)^2 - 25.51(\sigma/\pi)^4$, according to the formulation of Hanson (1979), who used the distribution of proper motions of objects in the parallax

Table 1 Subdwarfs with π and [Fe/H] determinations

ID	[Fe/H]	V	$B-V$	π ($''$)	σ_π ($''$)	M_V	$\sigma(M_V)$	M_V(L-K)	$(B-V)_{-2.14}$
HD 7808	−1.78	9.746	1.008	0.0663	0.0126	8.854	0.412	8.624	0.974
HD 19445	−2.08	8.053	0.475	0.0252	0.0052	5.060	0.448	4.783	0.471
HD 25329	−1.34	8.506	0.863	0.0548	0.0047	7.200	0.186	7.159	0.800
HD 64090	−1.73	8.309	0.621	0.0405	0.0023	6.346	0.123	6.328	0.591
HD 74000	−2.20	9.62	0.43	0.0155	0.0048	5.572	0.672	4.816	0.434
HD 84937	−2.12	8.324	0.421	0.0280	0.0064	5.560	0.496	5.206	0.420
HD 103095	−1.36	6.442	0.754	0.1127	0.0016	6.702	0.031	6.701	0.693
HD 134439	−1.4	9.066	0.770	0.0365	0.0025	6.877	0.149	6.851	0.714
HD 134440	−1.52	9.445	0.850	0.0365	0.0025	7.256	0.149	7.230	0.804
HD 149414	−1.39	9.597	0.736	0.0281	0.0035	6.841	0.270	6.750	0.679
HD 194598	−1.34	8.345	0.487	0.0194	0.0014	4.784	0.157	4.755	0.424
HD 201891	−1.42	7.370	0.508	0.0325	0.0027	4.929	0.180	4.891	0.462
HD 219617	−1.4	8.160	0.481	0.0280	0.0055	5.396	0.426	5.148	0.431
BD+66 268	−2.06	9.912	0.667	0.0216	0.0026	6.584	0.261	6.500	0.661
BD+11 4571	−3.6	11.170	1.060	0.0316	0.0047	8.668	0.323	8.536	1.080

catalogues to estimate the magnitudes of the L-K corrections. This expression for δM_V is strictly valid only for $\sigma_\pi/\pi < 0.33$. (The always-negative L-K corrections are added to the M_V values because the true luminosities are larger than the uncorrected estimates.)

The last column in Table 1 contains the predicted color that each star would have if its metallicity were [Fe/H] = −2.14 (chosen to illustrate the subdwarf-fitting procedure for the specific case of M92). At a fixed mass, main-sequence stars of different [Fe/H] will encompass a range in color and M_V; consequently, to define a fiducial for distance determinations by the main-sequence fitting technique using subdwarfs, it has become common practice to derive a mono-metallicity subdwarf sequence. This is obtained by correcting the color of each subdwarf, at its observed M_V, by the difference between the predicted colors of stars with the [Fe/H] of the subdwarf and that of the cluster itself. Thus, the model colors are used only differentially. Bi-cubic interpolation through a table of $B - V$ colors at different [Fe/H] and M_V values, generated from the Bergbusch & VandenBerg (1992) isochrones, was used to generate the color corrections: These take into account the dependence of radius on metallicity at fixed luminosity as well as purely atmospheric line blanketing effects.

Figure 7 shows how well 16 Gyr, [α/Fe] = 0.3 isochrones for [Fe/H] = −1.31, −1.71, and −2.14 (from VandenBerg et al 1996[5]) coincide with the positions of the local subdwarfs on the color-magnitude plane. We have plotted all of the stars in Table 1 (specifically, the data in the fourth, eighth, and ninth columns) whose metallicities fall within ± 0.15 dex of the isochrone [Fe/H] values. The agreement is about as good as one could hope for. Note, in particular, how well the models satisfy the constraint provided by the best of the subdwarfs (HD 103095, also called Groombridge 1830) and that the lower metal abundance subdwarfs tend to be displaced from those of higher Z in roughly the direction and amount suggested by the theory.

A main-sequence fit of M92 to the subdwarfs, using the data in the eighth, ninth, and tenth columns of Table 1 for those stars with $\sigma(M_V) < 0.3$ mag, is illustrated in Figure 8. When a foreground reddening of 0.02 mag (see Stetson & Harris 1988) is assumed, an apparent distance modulus of 14.65

[5]We make fairly extensive use of these calculations in this study, obviously because they are immediately at hand, but also because they represent the most up-to-date models presently available. In particular, they employ opacities for the adopted α/Fe number abundance ratios and are not based on the renormalization of scaled-solar-mix calculations, as has been advocated by Salaris et al (1993). Their procedure does appear to work well at low Z values, but not for $Z > 0.002$ (or so) according to VandenBerg et al (1996; also see Weiss, Peletier & Matteucci 1995): At high Z, the RGB location becomes insensitive to [α/Fe]. Importantly, as shown by VandenBerg (1992), Salaris et al, and the three lowermost curves in Figure 1 of this paper, virtually identical results are obtained when completely independent codes employing similar physics are used.

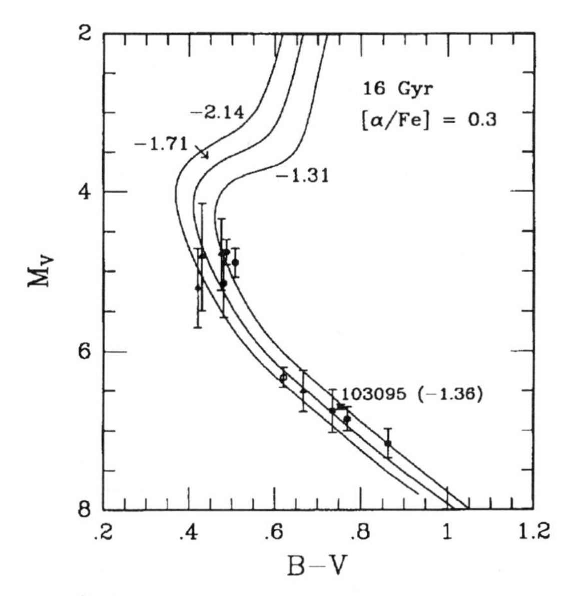

Figure 7 Comparison of the CMD locations of the nearby subdwarfs, whose properties are in the fourth, eighth, and ninth columns of Table 1, with VandenBerg et al (1996) isochrones. The closed circles, open circle, and closed triangles represent those subdwarfs whose tabulated [Fe/H] values are within ± 0.15 dex of those of the three isochrones; namely, −1.31, −1.71, and −2.14, respectively. All of the isochrones assume [α/Fe] = 0.3 and an age of 16 Gyr, though the latter choice is inconsequential.

mag is obtained. The VandenBerg et al (1996) isochrones, for the indicated parameters, have simply been overlayed on this figure, i.e. no color adjustments of any kind have been applied to them. [Their temperature and color scales are very close to those of the Bergbusch & VandenBerg (1992) calculations, which were used to produce the $B-V$ data in the last column of our table.] One has the impression that a small redward color shift should be applied to the isochrones at the fainter magnitudes, but what differences exist are clearly small.

An age of 15.5–16 Gyr is indicated from the observed location of the turnoff and subgiant branch relative to their theoretical counterparts. Allowing for helium diffusion would reduce this estimate to $\approx$ 15 Gyr (see Section 2.2.1), which should not be in error by more than $\pm$ 1.5 Gyr due to chemical composition uncertainties (see Section 2.4). According to Section 2.1.3, it is possible that deficiencies in convection theory could contribute a small age uncertainty, but other than this minor concern, remaining uncertainties in stellar physics should have little impact. Assuming no systematic error in the distance scale defined by the L-K corrected trigonometric parallax measures, the M92 distance modulus error is dominated by three terms [see Stetson & Harris (1988) for a more complete discussion of the errors associated with the subdwarf fit]. There is a goodness-of-fit term, which we approximate with the RMS vertical scatter (after correcting the colors to [Fe/H] = −2.14) of the subdwarf distribution around the distance-modulus-adjusted M92 main-sequence; a term for the uncertainty in the reddening towards M92,

$$\delta E(B-V) \times \frac{\partial M_V}{\partial (B-V)};$$

and a term for the uncertainty in the [Fe/H] value for M92 stars,

$$\delta[Fe/H] \times \frac{\partial M_V}{\partial (B-V)} \times \frac{\partial (B-V)}{\partial [Fe/H]}.$$

If we take $\delta E(B-V) \sim 0^m\!.02$ and $\delta[Fe/H] \sim 0.2$ dex, then these three terms added in quadrature give $\sigma(m-M) \sim 0^m\!.16$, which translates into an uncertainty in the age of ~2.0 Gyr (68% confidence interval). The δ[Fe/H]

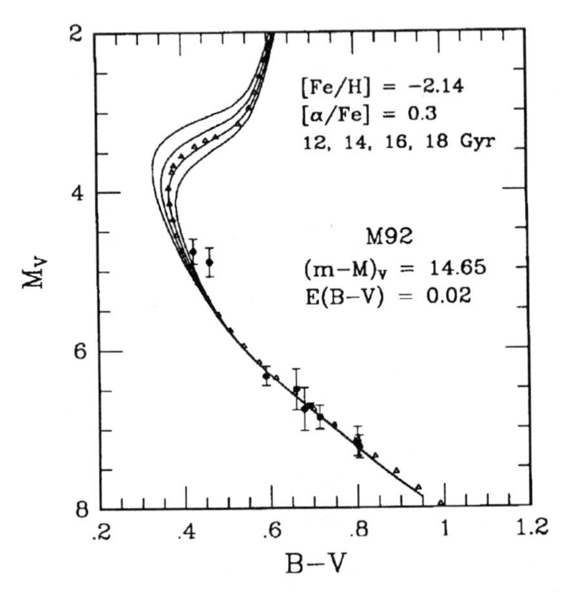

Figure 8 Main-sequence fit of the Stetson & Harris (1988) M92 main-sequence fiducial (*open triangles*) to the subdwarfs (*closed circles*), after the colors of the latter have been adjusted to compensate for differences between their [Fe/H] values and that of the cluster (see text). These revised colors are as given in the last column of Table 1. Only those data for which $\sigma(M_V) < 0.3$ mag have been plotted. VandenBerg et al (1996) isochrones for the indicated chemical composition and ages have been overlayed onto (not fitted to) the observations.

term enters the age uncertainty a second time because M_{bol}(TO) has an [Fe/H] dependence, and the formal *observational* uncertainty in the age that we derive for M92, *assigning no errors to the models and assuming the subdwarf distances have no systematic errors,* is 2.2 Gyr.

3.2 *Distances Based on RR Lyraes*

As noted in Section 2.6, the level of the horizontal branch at the color of the instability strip, which we refer to as M_V(HB) (although any bandpass can be used), has been used for $\sim$40 years to set the distance to a globular cluster. This value is defined to be the mean absolute magnitude of the cluster RR Lyrae stars after a proper averaging over each star's pulsational cycle. However, because horizontal-branch stars evolve to brighter magnitudes on their way to the asymptotic giant branch, there is an evolutionary width in the brightness of the HB (see Sandage 1990a); consequently, when comparing literature data for M_V(HB), care must be taken to ensure that the level of the HB is being compared for stars that have undergone the same amount of evolution.

It has long been suspected (cf Sandage 1958) that M_V(HB) is a function of [Fe/H]—considering equivalent evolutionary states—in the sense that more metal-poor HB stars are more luminous. As a result, a linear relation of the form $M_V(\mathrm{HB}) = c_0 + c_1$ [Fe/H] has generally been assumed, and a concerted effort has been made to try to determine the constants c_0 and c_1. The first of these constants is of critical importance for determining the age of the oldest GCs, while the second has a strong influence on the inferred age-metallicity relation that describes these systems (see, e.g. Sandage & Cacciari 1990, Walker 1992). Unfortunately, even the nearest RR Lyrae is too far away for a direct trignometric parallax distance; consequently, it has been necessary to use more indirect approaches to determine their M_V values. These include statistical parallaxes of field variables (e.g. Hawley et al 1986, 1996), Baade–Wesselink (B-W) analyses of field and cluster RR Lyraes (e.g. Liu & Janes 1990a,b; Storm, Carney & Latham 1994), main-sequence fits to GCs (Buonanno et al 1990, Bolte & Hogan 1995), and pulsation theory (Sandage, Katem & Sandage 1981; Sandage 1993b).

Figure 9 provides a graphical summary of the current status of this endeavor. The filled squares give the B-W results of Jones et al (1992), supplemented by an additional two stars from earlier work by Liu & Janes (1990a) that were not considered by the former. (Both investigations analyzed essentially the same sample of field RR Lyraes and both obtained very similar M_V values, generally agreeing to within 0.03 mag.) The solid curve gives the linear fit to these data adopted by Storm et al (1994); specifically, $M_V(\mathrm{HB}) = 1.02 + 0.16$ [Fe/H]. The slope of this relation is very similar to that of the dotted line, which represents the variation that Dorman (1993) computed from his zero-age main-sequence

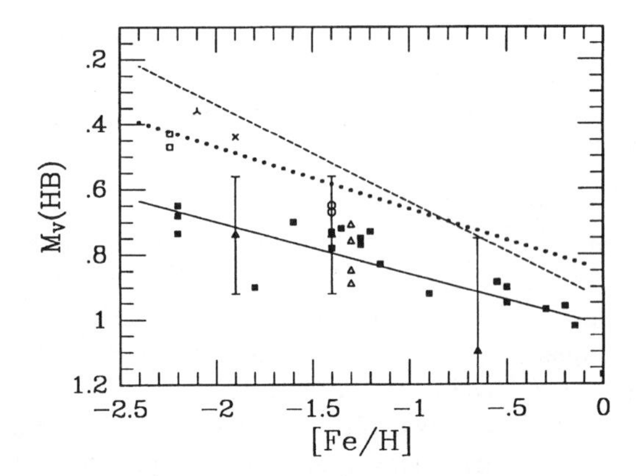

Figure 9 M_V(HB) vs [Fe/H] results from various sources. The closed squares represent the data for individual field RR Lyraes as obtained from the Baade-Wesselink analyses of Jones et al (1992) and Liu & Janes (1990a). The solid line gives the fit to these observations that was adopted by Storm et al (1994). Also based on the B-W method, the Liu & Janes (1990b) findings for four M4 variables are indicated by open triangles and the Storm et al results for RR Lyraes in M5 and M92 are denoted by open circles and open squares, respectively. The statistical parallax determinations by Hawley et al (1995) are represented by the closed triangles with attached error bars. The cross indicates Walker's (1992a) LMC estimate. The three-pointed star depicts the mean magnitude of M15 RR Lyraes as determined by Silbermann & Smith (1995). The dotted line gives the predicted ZAHB relation, as computed by Dorman (1993). The dashed line illustrates the relation between RR Lyrae magnitudes and [Fe/H] derived by Sandage (1993b) from his analysis of the Oosterhoff-Arp period-metallicity relation.

(ZAHB) models for scaled-solar abundances. It has $c_0 = 0.85$ and $c_1 = 0.19$, which are exceedingly close to the coefficients that Renzini (1991) determined from the ZAHBs in Sweigart, Renzini & Tornambè (1987).

Why the B-W M_V values are fainter, at a given [Fe/H], than the theoretically predicted values is hard to explain unless 1. the field RR Lyraes have a helium abundance that is significantly lower than the $Y \approx 0.23$ assumed in the models, 2. the application of the B-W method introduces a 0.2 mag zero-point error, or 3. the stellar interior computations are somehow deficient. The first option seems improbable in view of the R-method results and the present consensus that the primordial helium abundance was near $Y = 0.23$ (see Section 2.4). Concerning the last option, the only possibility that occurs to us is that the reduction in the envelope helium abundance due to diffusion has been underestimated (see Section 2.2.1). (Smaller core masses might also work, but there is no other reason to doubt our present understanding of the neutrino emission processes that largely determine the thermal structure of the core during RGB evolution.) Otherwise, any noncanonical process that might be going on, such as deep mixing or rapid core rotation, would tend to make the models brighter rather than fainter. A zero-point error could well be the most probable solution given that Carney, Storm & Jones (1992) themselves suspect that c_0 has a ± 0.15 mag uncertainty (also see Cohen 1992, Fernley 1994). This question awaits a satisfactory resolution.

Statistical parallaxes of field RR Lyrae stars also give a faint value for the HB brightness zero-point, but with a very large uncertainty. (This approach is subject to the assumption that the kinetmatic properties of the RR Lyrae population are, to first order, constant over the volume sampled.) Although the studies of Hawley et al (1986), Barnes & Hawley (1986), and Strugnell, Reid & Murray (1986) treated the existing data with a sophisticated set of analysis tools, there have remained some question marks regarding systematic errors in the proper motion lists available at the time and the use of a heterogeneous mix of [Fe/H] and radial velocity data. However, the recent reanalysis of the statistical parallax solution using a homogenous set of proper motion data based on an extragalactic coordinate system and new observations of [Fe/H] and radial velocity (Hawley et al 1995) has yielded essentially the same brightness zero-point as the earlier studies. (Note the location of the closed triangles with attached error bars in Figure 9.)

Beginning with the Cohen & Gordon (1987) investigation, there have been a number of attempts to carry out B-W analyses of cluster (as opposed to field) RR Lyraes. The open symbols in Figure 9 represent the results that Liu & Janes (1990b) obtained for M4 along with those derived for M5 and M92 by Storm et al (1994). (There has been a considerable evolution in the application of the B-W method over the years with the switch from BV to near-infrared photometry, the recognition that certain phases of the light curves give inconsistent results due to well-understood violations of assumptions, and the development of improved procedures for fitting the data. Thus only the latest determinations have been included in our figure.) Whereas the M4 RR Lyraes appear to be completely consistent with the field-star relation between M_V(HB) and [Fe/H], that is apparently not true of the two M92 pulsators, and possibly not of the M5 variables, although Storm et al (1994) suggest that the difference is not significant in the latter case.

Based on the two RR Lyraes studied, Storm et al (1994) derived $(m - M)_0 = 14.60 \pm 0.26$ for M92, i.e. effectively the same distance that Stetson & Harris (1988), Bolte & Hogan (1995), and we (in Section 3.1) obtained from the fitting of the nearby subdwarfs to the cluster main sequence. Storm et al expressed the concern that the two M92 variables might be highly evolved from the ZAHB, which would explain their displacement from the field-star relation, but there is additional evidence in support of the brighter luminosity scale. From CCD observations of 182 RR Lyraes in seven Large Magellanic Cloud clusters, Walker (1992) determined a mean M_V of 0.44 mag at [Fe/H] $= -1.9$ (the cross in Figure 9), assuming the Cepheid-based distance modulus of 18.5 (which should be accurate to within ± 0.1 mag). Curiously, Walker (1989) found the field RR Lyraes in the vicinity of the LMC cluster NGC 2257 to be 0.17 mag fainter, in the mean, than the cluster variables. Although this could simply be telling us that the cluster is closer than the average distance of the field stars, it could also be indicating a fundamental difference between the two stellar populations (a possible interpretation of the M92 results, as well). In addition, we note the determination of $M_V = 0.36 \pm 0.12$ (the three-pointed star in Figure 9) for the M15 variables from an analysis of their pulsational properties (Silbermann & Smith 1995).

Lastly, there is the Sandage (1993b) relation, $M_V(\mathrm{HB}) = 0.94 + 0.30\,[\mathrm{Fe/H}]$ (the dashed curve in Figure 9), which is based on his analysis of the Oosterhoff-Arp period versus metallicity correlation: Oosterhoff (1939, 1944) showed that GCs separate into two groups according to the mean periods of their respective RR Lyrae populations, while Arp (1955) discovered that the separation was one of cluster metal abundance. Based on several pieces of evidence, Sandage (1993a) concluded that the Oosterhoff-Arp effect is well described by $d \log P/d[\mathrm{Fe/H}] = -0.12 \pm 0.02$, where P is the mean period (in days) of the *ab*-type RR Lyraes. Then, using the fundamental pulsation equation, $P\sqrt{\bar{\rho}} = constant$, which can be turned into an equation in which P is given as a function of the pulsator's mass, luminosity, and effective temperature—namely,

$$\log P = 0.84 \log L/L_\odot - 0.68 \log \mathcal{M}/\mathcal{M}_\odot - 3.48 \log T_{\mathrm{eff}} + 11.502$$

(van Albada & Baker 1973)—he inferred that the relation between M_V(HB) and [Fe/H] must be steeply sloped, with the most metal-poor variables having rather bright magnitudes. This result made use of his deduction (in Sandage 1993b) that the instability strip is shifted towards cooler temperatures by $\Delta \log T_{\mathrm{eff}} = 0.012$ for each dex decrease in [Fe/H]. Thus, he contended that both a luminosity *and* a temperature shift must be taken into account to explain the observed period data.

It is unfortunately the case that the pulsation period depends sensitively on T_{eff} (see above), which is always very difficult to determine reliably. Prior to the Sandage (1993a,b) papers, a concerted effort had been made (see, e.g. Sandage 1982; Gratton, Ortolani & Tornambè 1986; Lee et al 1990; Sandage 1990b; Carney et al 1992; Catelan 1992, 1994; and references therein) to understand the so-called period-shift phenomenon, which is the term given to the dependence of pulsation period on metallicity at fixed amplitude, subsequently taken to be fixed T_{eff}. If canonical stellar models for $Y = 0.23$ (or so) are read at fixed T_{eff}, they are unable to produce a significant period shift between, for instance, M3 and M15 (whose observations have been central to this issue), *if* the RR Lyraes are near their respective ZAHB locations and *if* standard reddening values are assumed. Sweigart et al (1987) carried out an exhaustive examination of the models and of the relevant input physics and were unable to come up with a satisfactory explanation for the observations, unless helium is anticorrelated with metallicity (cf Sandage 1982). However, this would be completely contrary to current ideas about how chemical enrichment proceeds, as well as being in conflict with He abundance determinations from the R-method (e.g. Buonanno et al 1985) and (probably) from fits to the observed luminosity widths of cluster HB populations (e.g. Dorman et al 1989).

Hence, to maintain the canonical framework, suggestions were put forward that either the cluster reddening values that have generally been assumed are incorrect (e.g. Caputo 1988) or the RR Lyrae stars in the most metal-deficient clusters are highly evolved and are therefore much brighter than ZAHB stars (Lee, Demarque & Zinn 1990). But both of these alternatives seem indefensible. There is no doubt about M3 being essentially free of reddening, whereas the reddening of M15 has to be very close to 0.10 for the reason that this is required in order for the intrinsic colors of the turnoff stars in this cluster (see Durrell & Harris 1993) to be the same as those observed in M92 (Stetson & Harris 1988), which has the same age (VandenBerg, Bolte & Stetson 1990) and metallicity (Sneden et al 1991). The reddening of M92 is uncontroversial at $E(B-V) =$ 0.02 mag (see Stetson & Harris 1988). [Similar arguments have been put forward concerning M68, which belongs in the same metallicity group and which shows the same period shift relative to M3 as M15 (see Walker 1994).]

As Renzini & Fusi Pecci (1988), among others, have noted, the Lee et al (1990) explanation can hardly work in clusters that have very substantial RR Lyrae populations, such as M15. The variables should not constitute a big fraction of the total number of HB stars if the former are all in high-evolved states, where the evolutionary rates are particularly rapid. Lee (1991) has attempted to counter this argument by showing that the predicted period changes from his HB simulations agree well with those observed for the M15 RR Lyraes, though the uncertainties are large and his results are not entirely satisfactory because they fail to account for the existence of some stars whose periods are decreasing with time (see Silbermann & Smith 1995). According to the Lee et al hypothesis, all of the RR Lyrae stars should be evolving towards cooler temperatures and have periods that are increasing with time.

M68, however, poses an even greater challenge than M15, because it is much richer in RR Lyraes (see Table 2 by Carney et al 1992) and has many red HB stars (Walker 1994). There is little doubt that many of the RR Lyraes in this cluster are near the ZAHB. Furthermore, if one simply superimposes its CMD onto that for

M15 such that their respective turnoffs coincide, then one finds that their HBs also match (see McClure et al 1987); hence, the M15 variables are presumably also relatively near the ZAHB. If that is the case, which is by no means certain because even the color-magnitude data are not as secure as they should be (cf Figure 6 by Dorman et al 1991), then this difficulty for period-shift considerations would remain a problem for understanding the Oosterhoff-Arp effect. Even when a metallicity-dependent temperature shift of the instability strip is taken into account, it seems that canonical HB models can be reconciled with a steeply sloped period-metallicity relation only if the variables in the most metal-deficient clusters (such as M15) are significantly more evolved than those found in M3-like clusters, which are of intermediate metallicity (see Sandage 1993b).

Simon (1992), among others, has argued against such an evolutionary scenario, and if it were shown to be untenable, then canonical HB theory could well be called into question. In this regard, one cannot help but wonder whether there might be some connection with the inability of current models to account for either the chemical abundance trends along the RGBs of especially the most metal-poor GCs (see Section 2.2.2) or the luminosity functions of these same clusters (see Section 2.5). Certainly the HB of M15, in particular, has always been very hard to fathom (cf Crocker, Rood & O'Connell 1988). The main point to be emphasized here is that, although the precise slope of the relationship between $\log P$ and [Fe/H] is uncertain at the $\sim 20\%$ level, the Oosterhoff-Arp effect is beyond dispute and it must therefore be satisfactorily explained. A steeply sloped M_V(HB) versus [Fe/H] relation could well be the only way to accommodate the pulsation data.

Our understanding of the Population II distance scale is clearly less than satisfactory. The nearby subdwarfs appear to define a tight main-sequence locus that can be used to derive the distance to any globular cluster with accurate photometry of its main-sequence stars and a reliable reddening estimate. When applied to M92, this approach suggests that $M_V(\mathrm{HB}) \approx 0^m.40$ at the metal-poor end. This distance scale is consistent with the Galactic Cepheid scale as applied to the LMC and cluster RR Lyrae stars there. It is also consistent with the (fairly model-dependent) magnitudes derived for the most metal-poor cluster RR Lyraes, with B-W results for M92 (possibly), and with the luminosities inferred from the Oosterhoff-Arp period-metallicity relation. Taking the subdwarf distance for M92 as being free of systematic errors, we find an age for M92 of 15.8 ± 2 Gyr based on the VandenBerg et al (1996) models. (There is then an additional possible systematic error with the evolutionary calculations; in particular, we expect a reduction of ~ 1 Gyr if unhibited helium diffusion occurs). However, the unexplained discrepancy between this bright RR Lyrae magnitude zero-point and the fainter one derived via B-W and statistical parallax studies of field RR Lyrae stars leaves open the possibility that systematic errors remain in the distance scale. If the fainter scale turns out to be the correct one, then the age derived for M92 based on the same models mentioned above would be ~ 19 Gyr.

As a final remark, we point out that the CMDs of metal-rich GCs—like 47 Tuc, which is the most thoroughly studied of such systems (Hesser et al 1987)—appear to pose few difficulties for canonical stellar evolutionary theory. For instance, Bell (1992) has obtained a superb match to the entire CMD of 47 Tuc brighter than the turnoff (including the RGB, the HB, and the asymptotic giant branch), using stellar models for the observed [Fe/H] $= -0.8$ (Brown, Wallerstein & Oke 1990). His fits assumed a true distance modulus of $m - M =$ 13.33, which is identical to that recently derived by Montegriffo et al (1995) from their extensive photometry, very similar to those estimates contained in catalogues of cluster properties (cf Webbink 1985, Djorgovski 1993), and within 0.05 mag of that adopted by Hesser et al (1987), who derived an age of 13.5 Gyr. (This age should be reduced to perhaps 12 Gyr given that the models used by Hesser et al did not allow for He diffusion or Coulomb interactions in the equation of state.) Our supposition that the most metal-poor globular clusters are the oldest ones is almost certainly correct.

4. SUMMARY

The quest to determine accurate globular cluster ages and to ascertain when the first of these objects formed in the Galaxy (and how long that formation epoch lasted—see the companion review by Stetson et al 1996) is, without a doubt, one of the grand adventures in astronomy. It involves nearly all aspects of stellar astronomy and has profound importance for some of the biggest questions our species has ever asked: How did our Galaxy form? How old is the Universe? Is the Universe infinite, and will it exist forever? It has taken the effort of many researchers in many countries around the world to get to where we are now. Despite the enormous progress that has been made, the answers to such age-related questions remain elusive. Although the globular clusters are simple in many respects, being composed of low-mass stars of essentially the same age and initial chemical composition, our understanding of stellar evolution has not yet progressed far enough to be able to explain, in a fully self-consistent way and with sufficient precision, the entire wealth of information that we have garnered through the use of sophisticated observational techniques. This is particularly true for the later stages of evolution: Models for upper main-sequence and turnoff stars appear to meet the challenge of the observational tests so far devised.

As many others have found previously, our best estimate of the ages of the most metal-poor GCs, which are presumably the oldest, is 15^{-3}_{+5} Gyr (allowing for the full impact of helium diffusion, which was not treated in the models that were fitted to the M92 CMD). This figure could easily be off by 1–2 Gyr in either direction, but it would be very difficult (in our opinion) to reduce it to below 12 Gyr, or to increase it much above 20 Gyr. (These can probably be regarded as $\approx 2\sigma$ limits, though it is difficult to assign confidence intervals in this way because the errors in the models and in the various procedures used to obtain an age estimate are likely not Gaussian in sum total.) We favor an imbalance in the attached error bar for two reasons. First, the effects of He diffusion were allowed for in this estimate: Ignoring them would imply about a 7% increase in age. Second, we have opted for the distance scale defined by the local subdwarfs, which is within 0.1–0.2 mag of that implied by the calibration of RR Lyraes in the LMC (using the Cepheid-based distance to this system) and studies of the pulsational properties of cluster variable stars. The use of the distance scale based on B-W and statistical parallax measures of field RR Lyraes would also imply higher ages for the GCs. This estimate, which has remained essentially unchanged for (at least) the past 25 years despite steady refinements in both theory and observations during this period, should be regarded as quite a robust result by the cosmology community.

Although it is a common practice to simply add 1 Gyr to the best estimate of globular cluster ages to account for the formation time of these objects, there is potentially a fairly large range in the number that must be added to derive the age of the Universe. As shown in Figure 10 (for a more detailed analysis see Tayler 1986), the actual correction depends sensitively on the values of H_0, $\Omega_{\rm Matter}$, and the formation redshift of the GCs. The redshift at which galaxies like the Milky Way formed remains one of the most important open questions in observational cosmology. However, based on chemical-abundance measurements in absorption-line systems along the line of sight to distant quasars, it appears that gas in the Universe underwent significant enrichment between redshifts z of 3.5 to 2 (e.g. Lanzetta, Wolfe & Turnshek 1995, Wolfe et al 1995), and it therefore seems likely that the formation epoch of GCs was earlier than $z = 3.5$ (also see Sandage 1993c). Although there are theoretical reasons for believing that globular clusters formed before galaxies (Peebles & Dicke 1968), perhaps at redshifts as large as 10, the existence of field halo stars in the halo of the Milky Way that are significantly more metal-poor than GC stars may argue against this hypothesis. Still, with these limits on z, the age of the Universe is very likely $\lesssim 10^9$ yr (see Figure 10) older than the Galactic GC system.

Solutions to the field equations of General Relativity for isotropic, homogeneous universes are referred to as Friedmann, Friedmann-Lemaître, or Friedmann-Robertson-Walker models. These include, as a special case, the Einstein–de Sitter solution, in which $\Omega_{\rm Total} = 1$ (and the curvature of space is zero). Einstein–de Sitter universes are currently favored because $\Omega_{\rm Total} = 1$ appears to be a natural consequence of inflationary theory, which provides (*a*) a solution to the "horizon" problem posed by the smoothness of the cosmic microwave background on large scales, (*b*) a physical basis for the inhomogeneities that seeded galaxy formation, and (*c*) an explanation for the apparently very small amount of curvature in the Universe (the "flatness" problem). A choice motivated largely by elegance, and the application of Occam's razor, is the setting of the cosmological constant (Λ) to zero: The resultant matter-dominated Einstein–de Sitter model is arguably the standard model in cosmology today.

The solid curves in Figure 11 indicate loci of constant expansion age on the $\Omega_{\rm Matter}$ versus H_0 plane for Friedmann models with $\Lambda = 0$. Because we believe that a firm lower limit to GC ages is 12 Gyr (equal to our best estimate minus a generous error bar of 3 Gyr), the 12 Gyr curve should be shifted to somewhat

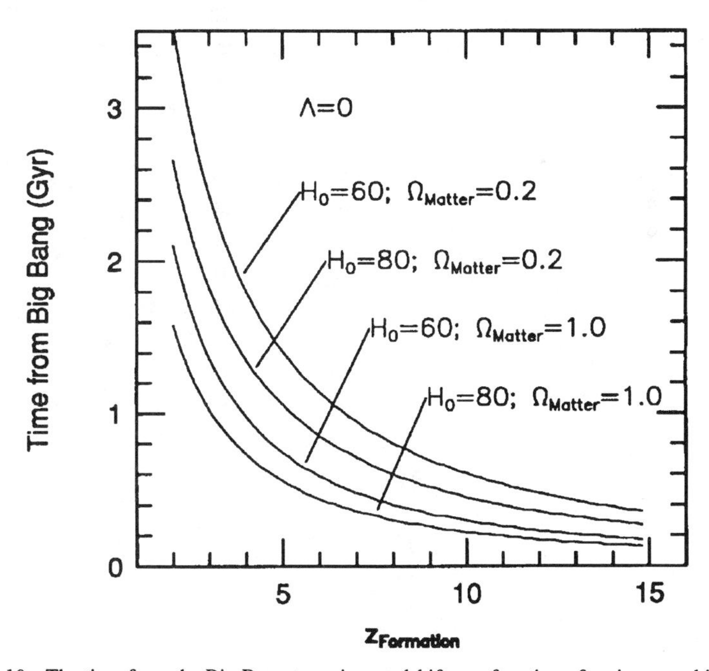

Figure 10 The time from the Big Bang to a given redshift as a function of various combinations of H_0 and $\Omega_{\rm Total} = \Omega_{\rm Matter}$ (i.e. the cosmological constant is assumed to be zero). Friedmann cosmological models are assumed.

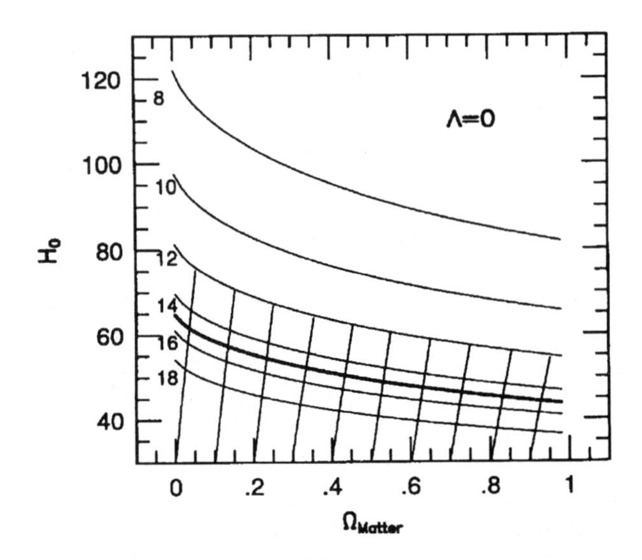

Figure 11 Expansion-age isochrones (for ages from 8 to 18 Gyr, as indicated) as a function of H_0 and $\Omega_{\rm Matter}$, assuming Friedman cosmological models. Given that globular clusters set a firm lower limit of 12 Gyr for the age of the Universe (see text), those combinations of H_0 and $\Omega_{\rm Matter}$ outside of the hatched area are precluded, unless the cosmological constant is nonzero. Our best estimate of GC ages is represented by the thick 15 Gyr locus.

lower H_0 values (at fixed $\Omega_{\rm Matter}$) to allow for the elapsed time between the Big Bang and GC formation (see Figure 10). But even as it stands, $\Omega_{\rm Matter} = 1$, $\Lambda = 0$ Einstein–de Sitter universes are rejected at the 95% confidence level for $H_0 = 65 \pm 10\%$ km s^{-1} Mpc^{-1}. Furthermore, if $H_0 \sim 80 \pm 8$ km s^{-1} Mpc^{-1} [see van den Bergh's (1995) summary of the *HST* H_0 Key Project results], then our age-based upper limit to H_0 is inconsistent at the $\sim 3\sigma$ level.

The two most widely discussed alternatives to the standard model to bring expansion ages into concordance with those derived for GCs are low-$\Omega_{\rm Matter}$, $\Lambda = 0$, spatially open universes or low-$\Omega_{\rm Matter}$, spatially flat universes that have a nonzero value of Λ. For the first case, if we assume a large value for the formation redshift, then there is no 1σ overlap between $H_0 = 80 \pm 8$ km s^{-1} Mpc^{-1} and our GC-age constraint on the Hubble constant: For $H_0 < 70$, the 1σ error bars do overlap. For the second case (see the excellent review on nonzero Λ models by Carroll, Press & Turner 1992), a positive value of Λ provides a term $[\Omega_\Lambda = \Lambda/(3H_0^2)]$ that can be added to $\Omega_{\rm Matter}$ to give a spatially flat Universe (and preserve inflation). For instance, for $\Omega_\Lambda = 0.8$, and assuming $\Omega_{\rm Matter} = 0.2$, the expansion age is 13.5 Gyr if $H_0 = 80$ km s^{-1} Mpc^{-1}. Although possibilities clearly exist for this alternative, there are already volume-z tests (e.g. the fraction of gravitationally lensed quasars; see Ostriker & Steinhardt 1995) that may exclude values for Ω_Λ as high as this. Also, because the effects of nonzero Λ change with time, a whole new set of fine-tuning problems may be introduced into cosmology. The implications of stellar ages $\sim$ 15 Gyr may, indeed, become profound in the next few years as the efforts to determine H_0 reduce the total (internal plus external) distance scale errors to $\lesssim 10\%$.

ACKNOWLEDGMENTS

We thank Márcio Catelan, Brian Chaboyer, Francesca D'Antona, Flavio Fusi Pecci, Bob Kraft, Charles Proffitt, Harvey Richer, Bob Rood, and Matt Shetrone for helpful information. We are especially grateful to Allan Sandage for his careful reading of the manuscript and for offering a number of helpful suggestions that have served to improve this paper. The tremendous support and encouragement from Jim Hesser and David Hartwick are also much appreciated. DAV acknowledges, with gratitude, the award of a Killam Research Fellowship from The Canada Council and the support of an operating grant from the Natural Sciences and Engineering Council of Canada.

Literature Cited

Andersen J. 1991. *Astron. Astrophys. Rev.* 3:91
Andersen J, Clausen JV, Gustafsson B, Nordström B, VandenBerg DA. 1988. *Astron. Astrophys.* 196:128
Arp HC. 1955. *Astron. J.* 60:317
Arp HC, Baum WA, Sandage A. 1953. *Astron. J.* 58:4
Barnes TG III, Hawley SL. 1986. *Ap. J. Lett.* 307:L9
Basu S, Antia HM. 1994. *J. Astrophys. Astron.* 15:143
Bell RA. 1992. *MNRAS* 257:423
Bell RA, Briley MM, Norris JE. 1992. *Astron. J.* 104:1127
Bell RA, Briley MM, Smith GH. 1990. *Astron. J.* 100:187
Bell RA, Dickens RJ. 1980. *Ap. J.* 242:657
Bencivenni D, Caputo F, Manteiga M, Quarta ML. 1991. *Ap. J.* 380:484
Bergbusch PA. 1990. *Astron. J.* 100:182
Bergbusch PA, VandenBerg DA. 1992. *Ap. J. Suppl.* 81:163
Birkinshaw M, Hughes JP. 1994. *Ap. J.* 420:33
Boesgaard AM, Budge KG. 1989. *Ap. J.* 338:875
Boesgaard AM, Tripicco MJ. 1986. *Ap. J. Lett.* 302:L49
Bolte M. 1994. *Ap. J.* 431:223
Bolte M, Hogan CJ. 1995. *Nature* 376:399
Bono G, Caputo F, Castellani V, Marconi M, Staiano L, Stellingwerf RF. 1995. *Ap. J.* 442:159
Brans C, Dicke RH. 1961. *Phys. Rev.* 124:925
Briley MM, Bell RA, Hoban S, Dickens RJ. 1990. *Ap. J.* 359:307
Briley MM, Hesser JE, Bell RA. 1991. *Ap. J.* 373:482
Briley MM, Hesser JE, Bell RA, Bolte M, Smith GH. 1994. *Astron. J.* 108:2183
Briley MM, Smith GH, Bell RA, Oke JB, Hesser JE. 1992. *Ap. J.* 387:612
Brown JA, Wallerstein G. 1993. *Astron. J.*

106:133
Brown JA, Wallerstein G, Oke JB. 1990. *Astron. J.* 100:1561
Brown JA, Wallerstein G, Oke JB. 1991. *Astron. J.* 101:1693
Buonanno R, Cacciari C, Corsi CE, Fusi Pecci F. 1990. *Astron. Astrophys.* 230:315
Buonanno R, Corsi CE, Fusi Pecci F. 1985. *Astron. Astrophys.* 145:97
Buonanno R, Corsi CE, Fusi Pecci F. 1989. *Astron. Astrophys.* 216:80
Buzzoni A, Fusi Pecci F, Buonanno R, Corsi CE. 1983. *Astron. Astrophys.* 128:94
Canuto VM, Hsieh S-H. 1978. *Ap. J.* 224:302
Canuto VM, Hsieh S-H. 1981. *Ap. J.* 248:801
Canuto VM, Lodenquai J. 1977. *Ap. J.* 211:342
Canuto VM, Mazzitelli I. 1991. *Ap. J.* 370:295
Canuto VM, Mazzitelli I. 1992. *Ap. J.* 389:724
Caputo F. 1988. *Astron. Astrophys.* 189:70
Caputo F, Martinez Roger C, Paez E. 1987. *Astron. Astrophys.* 183:228
Carbon DF, Langer GE, Butler D, Kraft RP, Suntzeff NB et al. 1982. *Ap. J. Suppl.* 49:207
Carlberg R, Yee HKC, Ellingson E, Abraham R, Gravel P, et al. 1996. *Ap. J.* 462:32
Carney BW. 1979. *Ap. J.* 233:211
Carney BW. 1980. *Ap. J. Suppl.* 42:481
Carney BW, Storm J, Jones RV. 1992. *Ap. J.* 386:663
Carroll SM, Press WH, Turner EL. 1992. *Annu. Rev. Astron. Astrophys.* 30:499
Catelan M. 1992. *Astron. Astrophys.* 261:457
Catelan M. 1994. *Astron. Astrophys.* 285:469
Chaboyer B. 1995. *Ap. J. Lett.* 444:L9
Chaboyer B, Demarque P. 1994. *Ap. J.* 433:510
Chaboyer B, Kim Y-C. 1995. *Ap. J.* 454:767
Chaboyer B, Sarajedini A, Demarque P. 1992. *Ap. J.* 394:515
Charbonnel C. 1995. *Ap. J. Lett.* 453:L41
Charbonnel C, Vauclair S. 1992. *Astron. Astrophys.* 265:55
Chieffi A, Straniero O, Salaris M. 1991. In *The Formation and Evolution of Star Clusters,* ed. K Janes, *Astron. Soc. Pac. Conf. Ser.* 13: 219
Chieffi A, Straniero O, Salaris M. 1995. *Ap. J. Lett.* 445:L39
Christensen-Dalsgaard J, Proffitt CR, Thompson MJ. 1993. *Ap. J. Lett.* 403:L75
Ciardullo RB, Demarque P. 1977. *Trans. Obs. Yale Univ.* 33
Cohen JG. 1978. *Ap. J.* 223:487
Cohen JG. 1979. *Ap. J.* 231:751
Cohen JG. 1992. *Ap. J.* 400:528
Cohen JG, Gordon GA. 1987. *Ap. J.* 318:215
Cottrell PL, Da Costa GS. 1981. *Ap. J. Lett.* 245:L79
Cox AN. 1995. In *Astrophysical Applications of Powerful New Databases,* ed. SJ Adelman, WL Wiese, *Astron. Soc. Pac. Conf. Ser.* 78:243
Cox AN, Stewart JN. 1970. *Ap. J. Suppl.* 19:261
Crooker DA, Rood RT, O'Connell RW. 1988. *Ap. J.* 332:236
Cudworth KM, Peterson R. 1988. In *The Extra-Galactic Distance Scale,* ed. S. van den Bergh, CJ Pritchet, *Astron. Soc. Pac. Conf. Ser.* 4:172
Dahn CC. 1994. In *Galactic and Solar System Optical Astrometry. Proc. R. Greenwich Obs. Workshop,* ed. LV Morrison, GF Gilmore, p. 55. Cambridge: Cambridge Univ. Press
Dearborn DSP, Griest K, Raffelt G. 1991. *Ap. J.* 368:626
Dearborn DSP, Raffelt G, Salati P, Silk J, Bouquet A. 1990. *Ap. J.* 354:568
Dearborn DSP, Schramm DN, Hobbs LM. 1992. *Ap. J. Lett.* 394:L61
Dearborn DSP, Schramm DN, Steigman G. 1986. *Phys. Rev. Lett.* 56:26
Deliyannis CP, Demarque P, Pinsonneault MH. 1989. *Ap. J. Lett.* 347:L73
Demarque P. 1967. *Ap. J.* 149:117
Demarque P. 1968. *Astron. J.* 73:669
Demarque P, McClure RD. 1977. In *Evolution of Galaxies and Stellar Populations,* ed. BM Tinsley, RB Larson, p. 199. New Haven: Yale Univ. Obs.
Denisenkov PA, Denisenkova SN. 1990. *Sov. Astron. Lett.* 16:275
Dirac PAM. 1974. *Proc. R. Soc. London Ser. A* 338:439
Djorgovski S. 1993. In *Structure and Dynamics of Globular Clusters,* ed. S Djorgovski, G Meylan, *Astron. Soc. Pac. Conf. Ser.* 50: 373
Dorman B. 1993. In *The Globular Cluster–Galaxy Connection,* ed. GH Smith, JP Brodie, *Astron. Soc. Pac. Conf. Ser.* 48:198
Dorman B, Lee Y-W, VandenBerg DA. 1991. *Ap. J.* 366:115
Dorman B, VandenBerg DA, Laskarides PG. 1989. *Ap. J.* 343:750
Drake JJ, Smith VV, Suntzeff NB. 1992. *Ap. J. Lett.* 395:L95
Durrell PR, Harris WE. 1993. *Astron. J.* 105:1420
Dziembowski WA, Goode PR, Pamyatnykh AA, Sienkiewicz R. 1994. *Ap. J.* 432:417
Dziembowski WA, Goode PR, Pamyatnykh AA, Sienkiewicz R. 1995. *Ap. J.* 445:509
Efstathiou G. 1995. *MNRAS* 274:L73
Eggen OJ, Sandage A. 1959. *MNRAS* 119:255
Fahlman GG, Richer HB, VandenBerg DA. 1985. *Ap. J. Suppl.* 58:225
Faulkner J, Gilliland RL. 1985. *Ap. J.* 299:994
Faulkner J, Roxburgh IW, Strittmatter PA. 1968. *Ap. J.* 151:203
Faulkner J, Swenson FJ. 1988. *Ap. J. Lett.* 329:L47

Faulkner J, Swenson FJ. 1993. *Ap. J.* 411:200
Fekel FC. 1991. *Astron. J.* 101:1489
Fernley J. 1994. *Astron. Astrophys.* 284:L16
Ferraro FR, Clementini G, Fusi Pecci F, Sortino R, Buonanno R. 1992. *MNRAS* 256:391
Ferraro FR, Clementini G, Fusi Pecci F, Vitiello E, Buonanno R. 1993. *MNRAS* 264:273
Finzi A. 1991. *Astron. Astrophys.* 247:261
Finzi A. 1992. *Astron. Astrophys.* 255:115
Folgheraiter EL, Penny AJ, Griffiths WK. 1993. *MNRAS* 264:991
Freedman WL, Madore BF, Mould JR, Hill R, Ferrarese L, et al. 1994. *Nature* 371:757
Frogel JA, Persson SE, Cohen JG. 1981. *Ap. J.* 246:842
Fusi Pecci F, Ferraro FR, Crocker DA, Rood RT, Buonanno R. 1990. *Astron. Astrophys.* 238:95
Fusi Pecci F, Renzini A. 1975. *Astron. Astrophys.* 39:413
Fusi Pecci F, Renzini A. 1979. In *Astronomical Uses of the Space Telescope*, ed. F Macchetto, F Pacini, M Tarenghi, p. 181. Geneva: Eur. South. Obs.
Gould A. 1990. *Ap. J.* 356:302
Gould A, Raffelt G. 1990a. *Ap. J.* 352:654
Gould A, Raffelt G. 1990b. *Ap. J.* 352:669
Gratton RG. 1985. *Astron. Astrophys.* 147:169
Gratton RG, Tornambè A, Ortolani S. 1986. *Astron. Astrophys.* 169:111
Green EM, Demarque P, King CR. 1987. *The Revised Yale Isochrones and Luminosity Functions.* New Haven: Yale Univ. Obs.
Guzik JA, Cox AN. 1992. *Ap. J.* 386:729
Guzik JA, Cox AN. 1993. *Ap. J.* 411:394
Hamuy M, Phillips MM, Maza J, Suntzeff NB, Schommer RA, Avilés R. 1995. *Astron. J.* 109:1
Hanson RB. 1979. *MNRAS* 186:875
Harris MJ, Fowler WA, Caughlan GR, Zimmerman BA. 1983. *Annu. Rev. Astron. Astrophys.* 21:165
Hartwick FDA. 1970. *Ap. J.* 161:845
Haselgrove CB, Hoyle F. 1956. *MNRAS* 116:527
Hawley SL, Jeffreys WH, Barnes TG III, Wan L. 1986. *Ap. J.* 302:626
Hawley SL, Hanley C, Layden AC, Hanson RB. 1996. In *The Formation of the Galactic Halo...Inside and Out*, ed. H Morrison, A Sarajedini, *Astron. Soc. Pac. Conf. Ser.* 92:188
Heber U, Kudritzki RP, Caloi V, Castellani V, Danziger J, Gilmozzi R. 1986. *Astron. Astrophys.* 162:171
Hesser JE, Harris WE, VandenBerg DA, Allwright JWB. Shott P, Stetson PB. 1987. *Publ. Astron. Soc. Pac.* 99:739
Hoyle F. 1959. *MNRAS* 119:124
Hoyle F, Narlikar JV. 1972a. *MNRAS* 155:305
Hoyle F, Narlikar JV. 1972b. *MNRAS* 155:323
Hoyle F, Schwarzschild M. 1955. *Ap. J. Suppl.* 2:1
Hoyle F, Tayler RJ. 1964. *Nature* 203:1108
Hubbard WB, Lampe M. 1969. *Ap. J. Suppl.* 18:297
Huchra JP. 1992. *Science* 256:321
Huebner WF, Merts AL, Magee NH, Argo MF. 1977. *Los Alamos Sci. Lab. Rep. No. LA-6760-M*
Iben I Jr. 1968a. *Ap. J.* 154:581
Iben I Jr. 1968b. *Nature* 220:143
Iben I Jr. 1971. *Publ. Astron. Soc. Pac.* 83:697
Iben I Jr. 1991. *Ap. J. Suppl.* 76:55
Iben I Jr, Renzini A. 1984. *Phys. Rep.* 105:329
Iben I Jr, Rood RT. 1970. *Ap. J.* 159:605
Isern J, Hernanz M, Garcia-Berro E. 1992. *Ap. J. Lett.* 392:L23
Izotov YI, Thuan TX, Lipovetsky VA. 1994. *Ap. J.* 435:647
Jacoby GH, Branch D, Ciardullo R, Davies RL, Harris WE, et al. 1992. *Publ. Astron. Soc. Pac.* 104:599
Jones RV, Carney BW, Storm J, Latham DW. 1992. *Ap. J.* 386:646
Kennicutt RC Jr, Freedman WL, Mould JR. 1995. *Astron. J.* 110:1476
Kovács G, Buchler JR, Marom A, Iglesias CA, Rogers FJ. 1992. *Astron. Astrophys.* 259:L46
Kraft RP. 1994. *Publ. Astron. Soc. Pac.* 106:553
Kraft RP, Sneden C, Langer GE, Prosser CF. 1992. *Astron. J.* 104:645
Kraft RP, Sneden C, Langer GE, Shetrone MD. 1993. *Astron. J.* 106:1490
Krauss LM, Romanelli P. 1990. *Ap. J.* 358:47
Laird JB, Carney BW, Latham DW. 1988. *Astron. J.* 95:1843
Langer GE, Hoffman RD. 1995. *Publ. Astron. Soc. Pac.* 107:1177
Langer GE, Hoffman RD, Sneden C. 1993. *Publ. Astron. Soc. Pac.* 105:301
Langer GE, Kraft RP, Carbon DF, Friel E, Oke JB. 1986. *Publ. Astron. Soc. Pac.* 98:473
Langer GE, Suntzeff NB, Kraft RP. 1992. *Publ. Astron. Soc. Pac.* 104:523
Lanzetta KM, Wolfe AM, Turnshek DA. 1995. *Ap. J.* 440:435
Larson A, VandenBerg DA, De Propris R. 1995. *Bull. Am. Astron. Soc.* 27:1431
Law W-Y. 1981. *Astron. Astrophys.* 102:178
Lee Y-W. 1991. *Ap. J.* 367:524
Lee Y-W, Demarque P, Zinn R. 1990. *Ap. J.* 350:155
Lehnert WD, Bell RA, Cohen JG. 1991. *Ap. J.* 367:514
Liu T, Janes KA. 1990a. *Ap. J.* 354:273
Liu T, Janes KA. 1990b. *Ap. J.* 360:561
Lutz TE, Kelker DH. 1973. *Publ. Astron. Soc. Pac.* 85:573
Maeder A. 1977. *Astron. Astrophys.* 56:359

Magain P. 1989. *Astron. Astrophys.* 209:211
Mathews GJ, Schramm DN, Meyer BS. 1993. *Ap. J.* 404:476
Mazzitelli I, D'Antona F, Caloi V. 1995. *Astron. Astrophys.* 302:382
McClure RD, VandenBerg DA, Bell RA, Hesser JE, Stetson PB. 1987. *Astron. J.* 93:1144
McWilliam A, Geisler D, Rich RM. 1992. *Publ. Astron. Soc. Pac.* 104:1193
Mengel JG, Gross PG. 1976. *Ap. Space Sci.* 41:407
Mengel JG, Sweigart AV, Demarque P, Gross PG. 1979. *Ap. J. Suppl.* 40:733
Michaud G. 1986. *Ap. J.* 302:650
Michaud G, Fontaine G, Beaudet G. 1984. *Ap. J.* 282:206
Moehler S, Heber U, de Boer KS. 1995. *Astron. Astrophys.* 294:65
Molaro P, Pasquini L. 1994. *Astron. Astrophys.* 281:L77
Montegriffo P, Ferraro FR, Fusi Pecci F, Origlia L. 1995. *MNRAS* 276:739
Montgomery KA, Marschall LA, Janes KA. 1993. *Astron. J.* 106:181
Müller J, Schneider M, Soffel M, Ruder H. 1991. *Ap. J. Lett.* 382:L101
Noerdlinger PD, Arigo RJ. 1980. *Ap. J. Lett.* 237:L15
Norris J. 1983. *Ap. J.* 272:245
Norris J, Da Costa GS. 1995. *Ap. J. Lett.* 441:L81
Norris J, Freeman KC. 1983. *Ap. J.* 266:130
Olive KA, Steigman G. 1995. *Ap. J. Suppl.* 97:49
Oosterhoff PT. 1939. *Observatory* 62:104
Oosterhoff PT. 1944. *Bull. Astron. Inst. Netherlands* 10:55
Ostriker JP, Steinhardt PJ. 1995. *Nature* 377:600
Pagel BEJ, Simonson EA, Terlevich RJ, Edmunds MG. 1992. *MNRAS* 255:325
Paltoglou G, Norris J. 1989. *Ap. J.* 336:185
Paquette C, Pelletier C, Fontaine G, Michaud G. 1986. *Ap. J. Suppl.* 61:177
Pavlovskaia P. 1953. *Variable Stars* 9:349
Peccei RD, Quinn H. 1977. *Phys. Rev. Lett.* 38:1440
Peebles PJE, Dicke RH. 1968. *Ap. J.* 154:891
Penzias AA, Wilson RW. 1965. *Ap. J.* 142:419
Peterson CJ. 1987. *Publ. Astron. Soc. Pac.* 99:1153
Peterson RC. 1980. *Ap. J. Lett.* 237:L87
Peterson RC. 1985a. *Ap. J.* 289:320
Peterson RC. 1985b. *Ap. J. Lett.* 294:L35
Pierce MJ, Welch DL, McClure RD, van den Bergh S, Racine R, Stetson PB. 1994. *Nature* 371:385
Pilachowski CA. 1988. *Ap. J. Lett.* 326:L57
Pilachowski CA. 1989. *The Abundances Spread Within Globular Clusters,* ed. G Cayrel de Strobel, M Spite, TL Evans, p. 1. Paris: Obs. Paris
Pinsonneault MH, Deliyannis CP, Demarque P. 1991. *Ap. J.* 367:239
Pinsonneault MH, Deliyannis CP, Demarque P. 1992. *Ap. J. Suppl.* 78:179
Pinsonneault MH, Kawaler SD, Demarque P. 1990. *Ap. J. Suppl.* 74:501
Pinsonnneault MH, Kawaler SD, Sofia S, Demarque P. 1989. *Ap. J.* 338:424
Press WH, Spergel DN. 1985. *Ap. J.* 296:679
Proffitt CR. 1993. In *Inside the Stars, IAU Colloq. 137,* ed. WW Weiss, A Baglin, *Astron. Soc. Pac. Conf. Ser.* 40:451
Proffitt CR. 1994. *Ap. J.* 425:849
Proffitt CR, Michaud G. 1991. *Ap. J.* 371:584
Proffitt CR, VandenBerg DA. 1991. *Ap. J. Suppl.* 77:473
Ratcliff SJ. 1987. *Ap. J.* 318:196
Rees RF Jr. 1992. *Astron. J.* 103:1573
Renzini A. 1977. In *Advanced Stages of Stellar Evolution,* ed. P Bouvier, A Maeder, p. 149. Geneva: Geneva Obs.
Renzini A. 1987. *Astron. Astrophys.* 171:121
Renzini A. 1991. In *Observational Tests of Cosmological Inflation,* ed. T Shanks, AJ Banday, RS Ellis, CS Frenk, AW Wolfendale, p. 131. Dordrecht: Kluwer
Renzini A, Fusi Pecci F. 1988. *Annu. Rev. Astron. Astrophys.* 26:199
Richer HB, Fahlman GG. 1984. *Ap. J.* 277:227
Richer HB, Fahlman GG, Ibata RA, Stetson PB, Bell RA, et al. 1995. *Ap. J. Lett.* 451:L17
Rogers FJ. 1994. In *The Equation of State in Astrophysics, IAU Colloq. 147,* ed. G Chabrier, E Schatzman, p. 16. Cambridge: Cambridge Univ. Press
Rogers FJ, Iglesias CA. 1992. *Ap. J. Suppl.* 79:507
Rogers FJ, Iglesias CA. 1994. *Science* 263:50
Rood RT. 1972. *Ap. J.* 177:681
Rood RT. 1973. *Ap. J.* 184:815
Rood RT. 1990. In *Astrophysical Ages and Dating Methods,* ed. E Vangioni-Flam, M Cassé, J Audouze, J Tran Thanh Van, p. 313. Gif sur Yvette: Editions Frontières
Rosenthal CS, Christiansen-Dalsgaard J, Houdek G, Monteiro MJPFG, Nordlund Å, Trampedach R. 1995. In *Helioseismology: Proc. 4th SOHO Workshop,* ed. V Domingo. In press
Saha A, Labhardt L, Schwengler H, Machetto FD, Panagia N, et al. 1994. *Ap. J.* 425:14
Saha A, Sandage A, Labhardt L, Schwengler H, Tammann GA, et al. 1995. *Ap. J.* 438:8
Salaris M, Chieffi A, Straniero O. 1993. *Ap. J.* 414:580
Sandage A. 1953. *Astron. J.* 58:61
Sandage A. 1958. In *Stellar Popluations,* ed. DJK O'Connell, p. 41. Specola Vaticana No. 5

Sandage A. 1970. *Ap. J.* 162:841
Sandage A. 1982. *Ap. J.* 252:553
Sandage A. 1986. *Annu. Rev. Astron. Astrophys.* 24:421
Sandage A. 1990a. *Ap. J.* 350:603
Sandage A. 1990b. *Ap. J.* 350:631
Sandage A. 1993a. *Astron. J.* 106:687
Sandage A. 1993b. *Astron. J.* 106:703
Sandage A. 1993c. *Astron. J.* 106:719
Sandage A, Cacciari C. 1990. *Ap. J.* 350:645
Sandage A, Katem B, Sandage M. 1981. *Ap. J. Suppl.* 46:41
Sandage A, Saha A, Tammann GA, Labhardt L, Panagia N, Macchetto FD. 1996. *Ap. J. Lett.* 460:L15
Sandage A, Schwarzschild M. 1952. *Ap. J.* 116:463
Sandquist E, Bolte M, Stetson PB. 1996. *Astron. J.*, In press
Schramm DN, Steigman G, Dearborn DSP. 1990. *Ap. J. Lett.* 302:L49
Schwarzschild M. 1958. *Structure and Evolution of the Stars.* Princeton: Princeton Univ. Press
Schwarzschild M, Härm R. 1965. *Ap. J.* 142:855
Seaton MJ, Yan Y, Mihalas D, Pradham AK. 1994. *MNRAS* 266:805
Shi X. 1995. *Ap. J.* 446:637
Silbermann NA, Smith HA. 1995. *Astron. J.* 110:704
Simoda M, Iben I Jr. 1968. *Ap. J.* 152:509
Simoda M, Iben I Jr. 1970. *Ap. J. Suppl.* 22:81
Simoda M, Kimura H. 1968. *Ap. J.* 151:133
Simon NR. 1992. *Ap. J.* 387:162
Smith GH. 1987. *Publ. Astron. Soc. Pac.* 99:67
Smith GH, Norris J. 1982. *Ap. J.* 254:149
Smith GH, Norris J. 1993. *Astron. J.* 105:173
Smith GH, Penny AJ. 1989. *Astron. J.* 97:1397
Smith GH, Tout CA. 1992. *MNRAS* 256:449
Smith VV, Suntzeff NB. 1989. *Astron. J.* 97:1699
Sneden C, Kraft RP, Prosser CF, Langer GE. 1991. *Astron. J.* 102:2001
Spergel DN, Faulkner J. 1988. *Ap. J. Lett.* 331:L21
Spergel DN, Press WH. 1985. *Ap. J.* 294:663
Spite F, Spite M. 1982. *Astron. Astrophys.* 115:357
Squires G, Kaiser N, Babul A, Fahlman GG, Woods D, et al. 1996. *Ap. J.* In press
Steenbock W, Holweger H. 1984. *Astron. Astrophys.* 130:319
Steigman G. 1978. *Ap. J.* 221:407
Steigman G, Sarazin CL, Quintana H, Faulkner J. 1978. *Astron. J.* 83:1050
Stetson PB. 1991. In *The Formation and Evolution of Star Clusters,* ed. K Janes, *Astron. Soc. Pac. Conf. Ser.* 13:88
Stetson PB. 1993. In *The Globular Cluster–Galaxy Connection,* ed. GH Smith, JP Brodie, *Astron. Soc. Pac. Conf. Ser.* 48:14
Stetson PB, Harris WE. 1988. *Astron. J.* 96:909
Stetson PB, VandenBerg DA. 1996. In preparation
Stetson PB, VandenBerg DA, Bolte M. 1996. *Publ. Astron. Soc. Pac.* 108:In press
Storm J, Carney BW, Latham DW. 1994. *Astron. Astrophys.* 290:443
Straniero O, Chieffi A. 1991. *Ap. J. Suppl.* 76:525
Stringfellow GS, Bodenheimer P, Noerdlinger PD, Arigo RJ. 1983. *Ap. J.* 264:228
Strugnell P, Reid N, Murray C. 1986. *MNRAS* 220:413
Suntzeff NB. 1981. *Ap. J. Suppl.* 47:1
Sweigart AV, Gross PG. 1976. *Ap. J. Suppl.* 32:367
Sweigart AV, Gross PG. 1978. *Ap. J. Suppl.* 36:405
Sweigart AV, Mengel JG. 1979. *Ap. J.* 229:624
Sweigart AV, Renzini A, Tornambè A. 1987. *Ap. J.* 312:762
Swenson FJ. 1995. *Ap. J. Lett.* 438:L87
Swenson FJ, Faulkner J. 1992. *Ap. J.* 395:654
Swenson FJ, Faulkner J, Rogers FJ, Iglesias CA. 1994. *Ap. J.* 425:286
Tassoul M, Tassoul J-L. 1984. *Ap. J.* 279:384
Tayler RJ. 1986. *Q. J. R. Astron. Soc.* 27:367
Taylor JH, Weisberg JM. 1989. *Ap. J.* 345:434
Thorburn JA. 1994. *Ap. J.* 421:318
Trefzger CF, Carbon DF, Langer GE, Suntzeff NB, Kraft RP. 1983. *Ap. J.* 266:144
Turck-Chièze S, Däppen W, Fossat E, Provost J, Schatzman E, Vignaud D. 1993. *Phys. Rep.* 230:57
Turon C, Arenou F, Evans DW, van Leeuwen F. 1992. *Astron. Astrophys.* 258:125
van Albada TS, Baker N. 1973. *Ap. J.* 185:477
VandenBerg DA. 1977. *MNRAS* 181:695
VandenBerg DA. 1983. *Ap. J. Suppl.* 51:29
VandenBerg DA. 1988. In *The Extragalactic Distance Scale,* ed. S van den Bergh, CJ Pritchet, *Astron. Soc. Pac. Conf. Ser.* 4:187
VandenBerg DA. 1991. In *The Formation and Evolution of Star Clusters,* ed. K Janes, *Astron. Soc. Pac. Conf. Ser.* 13:183
VandenBerg DA. 1992. *Ap. J.* 391:685
VandenBerg DA, Bell RA. 1985. *Ap. J. Suppl.* 58:561
VandenBerg DA, Bolte M, Stetson PB. 1990. *Astron. J.* 100:445
VandenBerg DA, Bridges TJ. 1984. *Ap. J.* 278:679
VandenBerg DA, Laskarides PG. 1987. *Ap. J. Suppl.* 64:103
VandenBerg DA, Smith GH. 1988. *Publ. Astron. Soc. Pac.* 100:314
VandenBerg DA, Stetson PB. 1991. In *Challenges to Theories of the Structure of Moderate-Mass Stars. Lect. Notes Phys.,*

ed. D Gough, J Toomre, 388:367. Berlin: Springer-Verlag

VandenBerg DA, Swenson FJ, Rogers FJ, Iglesias CA, Alexander DR. 1996. In preparation

van den Bergh S. 1992. *Publ. Astron. Soc. Pac.* 104:861

van den Bergh S. 1994. *Publ. Astron. Soc. Pac.* 106:1113

van den Bergh S. 1995. *Science* 270:1942

Van Flandern TC. 1981. *Ap. J.* 248:813

Vogeley MS, Park, C, Geller MJ, Huchra JP. 1992. *Ap. J. Lett.* 391:L5

Von Rudloff IR, VandenBerg DA, Hartwick FDA. 1988. *Ap. J.* 324:840

Wagoner R, Fowler WA, Hoyle F. 1967. *Ap. J.* 148:3

Walker AR. 1989. *Astron. J.* 98:2086

Walker AR. 1992. *Ap. J. Lett.* 390:L81

Walker AR. 1994. *Astron. J.* 108:555

Wasserburg GJ, Boothroyd AI, Sackmann I-J. 1995. *Ap. J. Lett.* 447:L37

Webbink RF. 1985. In *Dynamics of Star Clusters. IAU Symp.*, ed. J Goodman, P Hut, 113:541. Dordrecht: Reidel

Weiss A, Peletier RF, Matteucci F. 1995. *Astron. Astrophys.* 296:73

Wheeler JC, Sneden C, Truran JW Jr. 1989. *Annu. Rev. Astron. Astrophys.* 27:279

Willson LA, Bowen GH, Struck-Marcell C. 1987. *Comm. Astrophys.* 12:17

Wolfe AM, Fan X-M, Tytler D, Vogt SS, Keane MJ, Lanzetta KM. 1995. *Ap. J. Lett.* 435: L101

Zahn J-P. 1992. *Astron. Astrophys.* 265:115

Annu. Rev. Astron. Astrophys. 1996. 34:511–50

OLD AND INTERMEDIATE-AGE STELLAR POPULATIONS IN THE MAGELLANIC CLOUDS

Edward W. Olszewski

Steward Observatory, University of Arizona, Tucson, Arizona 85721-0065

Nicholas B. Suntzeff

Cerro Tololo Inter-American Observatory, National Optical Astronomy Observatories, Casilla 603, La Serena, Chile

Mario Mateo

Department of Astronomy, University of Michigan, 830 Dennison, Ann Arbor, Michigan 48109-1090

KEY WORDS: stellar populations, local group galaxies, photometry, galaxy formation

ABSTRACT

The Magellanic Clouds have galactocentric distances of 50 and 63 kiloparsecs, making it possible to probe the older populations of clusters and stars in some detail. Although it is clear that both galaxies contain an old population, it is not yet certain whether this population is coeval with the date of formation of the oldest globulars in the Milky Way. The kinematics of this old population in the Large Magellanic Cloud (LMC) are surprising; no component of this old population is currently measured to be part of a hot halo supported by velocity dispersion. Spectroscopy of field stars is beginning to show the existence of a small population of stars with abundances [Fe/H] less than −1.4. These stars will help to unravel the star-formation history when the next generation of telescopes are commissioned. Asymptotic giant branch stars, long-period variables, planetary nebulae, and horizontal-branch clump stars can be used to trace the extent and kinematics of the intermediate-age population. Deep color-magnitude diagrams can be used to derive the relative proportions of stars older than 1 Gyr.

0066-4146/96/0915-0511$08.00

The age distribution of populous clusters and the age-metallicity relation are used to compare the evolution of the two Magellanic Clouds to each other. The issue of where the LMC's metals originated is explored, as is the question of what triggers star formation in the Clouds.

1. INTRODUCTION

The notion of a "stellar population" is now so commonly used that we generally do not challenge ourselves to provide a specific definition for what really constitutes a population of stars. In the IAU Symposium 164, a range of possible definitions was offered, but perhaps the simplest is that a stellar population in a galaxy comprises the stars formed during a major event in the life of a galaxy (King 1995, Mould 1995). It has long been tacitly assumed, especially for the earliest stages of star formation in galaxies, that there exist some populations that are coeval for all disk galaxies, or perhaps even for all galaxies. What evidence do we really have that there was a substantial early episode of star formation in all galaxies? Are there characteristics of a young galaxy that determine the subsequent star-formation history or do external factors play a substantial role? To begin to answer such questions we can study distant galaxies or protogalaxies (e.g. Koo 1986, Djorgovski 1992), the formation of spheroids and disks (e.g. Schechter & Dressler 1987), or the special population of faint blue galaxies seen in faint galaxy number counts (Kron 1980, Tyson 1988, Colless et al 1993). Local galaxies can provide important details to galaxy evolution models, which makes the star-formation histories of the nearest galaxies a source of interest for cosmological questions. We are able to combine stellar photometry with spectroscopy to determine the fundamental population characteristics: the stellar luminosity function, spatial distribution, age, chemical abundance, and kinematics. The only external galaxies for which we can currently hope to make such detailed observations are the Magellanic Clouds and the Local Group dwarf spheroidal galaxies. This review focuses on the Magellanic Clouds.

Many of the characteristics that we attribute to a stellar population are strongly influenced by the properties of the Milky Way. For instance, we use the words "halo," "metal-poor," "high velocity," and "old" interchangeably to describe a certain population in our Galaxy, yet these words in the context of other galaxies are not necessarily interchangeable. As we will see, the "old" stars in the Large Magellanic Cloud (LMC) are not "halo" stars, and the "metal-poor" stars in the Small Magellanic Cloud (SMC) are not necessarily "old." We hope to show the reader that the Magellanic Clouds, with their relatively low metal abundances and large numbers of star clusters of all ages, can provide an instructive contrast to the notions of the stellar populations in our Galaxy. In addition, by studying the Magellanic Clouds, we may begin to better understand the characteristics of the stellar populations in a vigorously evolving system: systems that may have at some earlier epochs been similar to our Galaxy, but have subsequently evolved very differently.

In this review we concentrate on a quite limited set of questions that relate to these issues in the context of the Magellanic Clouds. What old populations are there in the Magellanic Clouds? How old is the oldest population in each galaxy and how do these populations compare to the Milky Way? Are the oldest populations in the Magellanic Clouds simply scaled versions of the halo of the Milky Way? What are the age-metallicity relations in each galaxy? What is the history of star formation from the earliest populations to the population characterizing the Clouds about 1 Gyr ago? Does the history of star formation as seen in the clusters reflect the history of star formation as seen in the field stars of the Magellanic Clouds?

We show that the available evidence suggests that the LMC really did lie relatively dormant for a substantial fraction of a Hubble time, while the SMC seems to have experienced a more constant star-formation rate. We know of no compelling observation that will tell us why the LMC resumed making stars several billion years ago. Although it is tempting to use the current dynamical models of Cloud–Milky Way interactions as a vehicle for understanding the LMC and SMC star-formation histories, these models are just beginning to predict the times of major episodes of star formation in the intermediate and old populations.

We approach these questions by summarizing the recently published observations of the field stars and of the rich luminous star clusters of the Magellanic Clouds. We also point out observations currently being made and observations that could and should be made in the future. It is our feeling that we are on the verge of a deep understanding of the general evolution of the Clouds, based on expected new data from the *Hubble Space Telescope* (*HST*), the southern 6.5–8-m class telescopes, proper motion surveys, wide-field optical and infrared surveys, and multifiber spectroscopy.

To make this review tractable, we have chosen to take Westerlund's (1990) review as our starting point. Unless it is important to our argument, we assume that Westerlund (1990) and other reviews and symposia (Feast 1995; Olszewski 1988, 1993, 1995; Da Costa 1993; many papers in Baschek et al 1993; many papers in Haynes & Milne 1991; Suntzeff 1992a,b; Mateo 1992; Graham 1988) can lead the reader to the earlier literature on the Magellanic Clouds.

2. THE OLDEST POPULATIONS OF THE MAGELLANIC CLOUDS

2.1 *Definition and Nomenclature*

The population that we are interested in describing here should more rightly be called the oldest observable population or the first substantial population in the Large and Small Magellanic Clouds (we use MC when referring to both Clouds). If there were a small tail of stars more metal poor than [Fe/H] $= -3.0$ in the MCs, therefore presumably older than the first major population, it would be impossible to detect it given currently available sample sizes, just as stars in the Milky Way halo with [Fe/H] < -3.0 have only been found by heroic efforts to search large numbers of Galactic halo stars (e.g. Beers et al 1992). We do not call the oldest major population in the MC "Population II" or the "halo population," as these words have connotations of abundance, age, and kinematics that may be inappropriate for the oldest MC populations. The word "halo" is used when these connotations are intended.

In this section we concentrate on describing the properties of the oldest population, including the mean age and kinematics. It is important to determine with certainty whether this "ancient" population exists in either Cloud or if the first dominant population is demonstrably younger in those galaxies than in the Milky Way. By an ancient population we mean a population coeval with the oldest globular clusters in the Milky Way, currently believed to be ~ 15 Gyr old. We acknowledge that the age scale may change in the future, but we believe that the age ranking of populations within the two Magellanic Clouds and relative to the Milky Way globulars will remain secure.

2.2 *The Old Cluster Population*

The Magellanic Clouds have a number of red clusters long suspected of being analogues to the Galactic globular clusters. A list of these old LMC clusters is given in Suntzeff et al (1992) and is repeated in Table 1 with a few revisions. Reticulum, NGC 1841, and NGC 1466 have at times been considered to be Milky Way globulars (e.g. Webbink 1985). All three have now been shown from color-magnitude diagrams (Gratton & Ortolani 1987, Walker 1992a for Reticulum; Walker 1990 for NGC 1841; Walker 1992b for NGC 1466) and velocities (Storm et al 1991, Suntzeff et al 1992, Olszewski et al 1991) to be LMC members. The few old clusters in the SMC are also listed in Table 1. The lower age cutoff in Table 1 is about 8 Gyr.

While the early papers of Arp (1958a, 1958b), Tifft (1962), and Hodge (1960) were among those that first showed that some of the redder Magellanic Cloud clusters were similar to Galactic globulars, direct evidence for the old ages of the red clusters comes only from deep color-magnitude diagrams (CMD),

Table 1 The oldest cluster in the Magellanic Clouds[a]

Cluster	[Fe/H]	M_V	n(RR)	Notes[b]
		LMC		
Hodge 11	−2.1	−7.0	0	1,2
NGC 1466	−1.8	−7.9	38	1,2
NGC 1754	−1.5	−7.1	—	1,2
NGC 1786	−1.9	−7.9	9	1,2
NGC 1835	−1.8	−9.2	35	1,2
NGC 1841	−2.1	−7.1	22	1
NGC 1898	−1.4	−7.4	—	1
NGC 1916	−2.1	−9.0	—	1,2
NGC 2005	−1.9	−7.5	—	1,2
NGC 2019	−1.8	−7.9	0	1,2
NGC 2210	−2.0	−8.1	12	1,2
NGC 2257	−1.8	−6.9	37	1
Reticulum	−1.7	−6.0	32	1
ESO 121-SC03	−0.9	−5.2	0	3
		SMC		
NGC 121	−1.3	−7.8	4	4
Lindsay 1	−1.3	−5.7	0	5
Kron 3	−1.3	−7.0	0	6

[a]Possible old clusters in the LMC: Hodge 7 (Bica et al 1991), SL244, NGC 1865, NGC 1928, NGC 1939 (Bica et al 1992). Walker (1989b) found no variables in the SMC clusters Kron 7, Kron 44, Lindsay 113, NGC 339, and NGC 416. Graham & Nemec (1984) found no variable in NGC 2121 and NGC 2155.
[b]1. Basic data given in Suntzeff et al (1992). 2. Confirmed to be metal poor by Dubath et al (1993). 3. Basic data in Mateo et al (1986). Cluster is 9 Gyr old and significantly younger than the other LMC clusters listed. 4. Spectral metallicity from Suntzeff et al (1989b). More variables may be in cluster center (Walker 1989b). Cluster is 12 Gyr old (Stryker et al 1985). 5. Younger than NGC 121. No variables (Walker 1989b). Age and metallicity given by Olszewski et al (1987, 1991). 6. Younger than NGC 121. No variables (Walker 1989b). Age and metallicity given by Rich et al (1984).

which can be measured from ground-based data for the clusters in uncrowded regions of the MCs. The expected main-sequence turnoff of the old population in the LMC will be at $V \sim 22.5$. The CMDs for the LMC clusters NGC 1466, NGC 1841, and Reticulum listed above, and for NGC 2257 (Testa et al 1995), all show main-sequence turnoffs roughly consistent with the ages of Galactic globular clusters, but the quality of the published color-magnitude diagrams is inadequate to say with certainty whether the LMC clusters are truly ancient or are a few Gyr younger. Two major programs have been undertaken in Cycle 5 of *HST* to observe the oldest LMC clusters listed in Table 1 to improve the age determinations.

Because all the LMC clusters known to be old based on CMD dating are also very metal poor, we have assumed all metal-poor LMC clusters to be old. The list of LMC clusters in Table 1 is therefore a compliation of all clusters with known old ages or low metallicities. In some cases, the old age is verified by the existence of RR Lyrae variables. A number of LMC clusters that could be considered old based on their red colors are excluded from Table 1 because their metallicity was found to be similar to the intermediate-age clusters (Olszewski et al 1991). No new metal-poor clusters have been found in recent spectroscopic surveys of MC clusters (Olszewski et al 1991).

In the SMC, NGC 121 (Stryker et al 1985) is the oldest cluster, but all three clusters listed in Table 1 are younger than the ancient Galactic globular clusters (Olszewski et al 1987, Rich et al 1984) based on ages from CMDs (see also Sarajedini et al 1995).

Some of the oldest MC clusters may no longer be in the Magellanic Clouds. Lin & Richer (1992) have argued, on the basis of cluster positions and velocities, that the Galactic globulars Pal 12 and Rup 106 have been captured from the LMC. This idea is actually not radical, for we know that the Sagittarius dwarf spheroidal is contributing four globular clusters to our Galaxy (see Da Costa & Armandroff 1995). van den Bergh (1994) has pointed out a possible problem with the Lin & Richer scenario. Pal 12 has a rather high metallicity to be an LMC cluster of its age, unless it is somehow an analogue to the unique LMC cluster ESO121-SC03. The possibility that Pal 12 came from the SMC should be examined.

It is probably true that no luminous old clusters ($M_V \lesssim -7$) remain to be found in the Clouds. The large clusters have had well-determined integrated *UBV* photometry for many years (van den Bergh 1981). Most have been or will be searched for RR Lyraes and observed with *HST* in the next few years. There may be old clusters among the generally less-luminous clusters cataloged by Olszewski et al (1988), Kontizas et al (1990), and Bica et al (1991, 1992). To find them efficiently, we will have to resort to indirect indicators of age such as low metallicity, integrated colors, and the presence of RR Lyrae variables.

2.2.1 RR LYRAES It is generally accepted that the existence of RR Lyrae variables or a blue horizontal branch (BHB) is prima facie evidence of the existence of an old population. Exactly how old is not well determined observationally or theoretically. RR Lyraes are low-mass helium core-burning stars. The thickness of the envelope determines whether such a star is redward of, blueward of, or in the instability strip. The envelope mass is a complicated function of age, metallicity, helium abundance, rotation, and mass loss; the latter is poorly understood (Lee et al 1994).

There is evidence that RR Lyraes are not all extremely old. Taam et al (1976) and Kraft (1977) have argued that the metal-rich RR Lyraes with thick-disk kinematics in the Milky Way may be younger than the ancient clusters, but the age difference is not known. How young can an RR Lyrae be? Clusters of known ages containing RR Lyraes provide a most direct way of estimating the youngest age for an RR Lyrae. We note, however, that it is difficult to age-date the metal-rich field RR Lyraes in either the Milky Way or the Magellanic Clouds. Although one can envision mechanisms that would make these stars very young, new data from HST (J Liebert, 1996, private communication) show that some metal-rich globular clusters do contain substantial blue horizontal-branch populations, thus allowing the possibility that metal-rich field RR Lyraes are old. Da Costa & Armandroff (1995) give the most recent list of Milky Way "young halo" globulars; many of these 21 objects have long been known (Sawyer Hogg 1973) to contain RR Lyraes. Ruprecht 106 is thought to be more than 3 Gyr younger than the typical globular (Buonanno et al 1993), yet it contains 12 RR Lyraes (Kaluzny et al 1995).

From Table 1, we see that the 12-Gyr old SMC cluster NGC 121 has four RR Lyraes whereas the 9-Gyr old cluster Lindsay 1 has none. This comparison is often cited as evidence that the youngest RR Lyraes are older than 10 Gyr (Olszewski et al 1987). However, L1 is 7 times fainter than NGC 121 (van den Bergh 1981). Scaling NGC 121's RR Lyrae population to L1, we expect less than one RR Lyrae. While the number of known RR Lyraes in the concentrated cluster NGC 121 is likely an underestimate, the expected number ot RR Lyraes in L1 will still be subject to small number statistics, and will not provide a good constraint on the age of the youngest-possible RR Lyraes. None of the other relatively luminous, older SMC clusters have RR Lyraes (Walker 1989b). For LMC clusters, it is unlikely that any of the luminous clusters in Table 1 are young enough to constrain the minimum age of the RR Lyraes. There are no known luminous LMC clusters in the age range 4–12 Gyr (see Section 3.4 below). It may be possible to constrain the age of the youngest RR Lyraes by seaching Milky Way old open clusters, the globulars in the Fornax dwarf spheroidal, and the luminous intermediate-color star clusters in M33.

In summary, the existence of RR Lyraes probably means that the population is old, but there is no strong observational evidence that can rule out their presence in a substantially younger population. The gravitational lens experiments such as MACHO should be very effective in searching for RR Lyraes in the less-luminous clusters in the MCs, which may overcome the low a priori chance of finding RR Lyraes in any individual low-luminosity cluster.

2.2.2 LOW METAL ABUNDANCES The metallicity of a cluster is often used to indicate old age (cf Andersen et al 1984) in analogy to our Galaxy where the metal-poor globular clusters are all old. For the Galaxy and the LMC, low metallicity is apparently a sufficient condition to indicate that a population is old. In the SMC, however, low abundance does not imply old age. The field giants near NGC 121 have ages of about 8 Gyr, with a mean metallicity [Fe/H] = −1.6 (Suntzeff et al 1986), and the 8-Gyr SMC cluster Kron 3 has a metallicity of [Fe/H] = −1.3 (Rich et al 1984).

The mean metal abundance of the LMC old clusters is [Fe/H] = −1.84, whereas that of the Galactic globulars outside of the solar circle is −1.70 (Suntzeff et al 1992). These similarities imply that fragments of long-destroyed galaxies not very different from the LMC could have assembled the halo (e.g. Searle & Zinn 1978). Van den Bergh (1993, 1994) has argued that objects like the current LMC could not have donated many clusters, but the differences between LMC clusters and Galactic halo clusters are subtle. The entire Galactic halo cannot simply be the debris from LMC-like protogalaxies, since inside the solar circle the Galaxy shows a metallicity gradient (Zinn 1985, Suntzeff et al 1991), implying (in part) a dissipational evolution involving chemical feedback from earlier generations of stars. Of course, a mechanism in which the largest density fluctuations in a protogalaxy were themselves sites of star and cluster formation would also be an explanation (Sandage 1990).

2.2.3 INTEGRATED COLORS A third indication of an old population can be found in the integrated colors or spectra. The broad-band integrated color of a star cluster changes continuously with age, although the color separation between very disparate ages is sometimes small (Searle et al 1980; Elson & Fall 1985, 1988; Bica et al 1991, 1992; Girardi et al 1995). Nevertheless, broad-band colors remain a useful indicator of clusters with potentially interesting properties. Bica and collaborators have obtained integrated *UBV* photometry for 624 LMC clusters and a number of SMC clusters (Bica et al 1986); this sample can be searched for possible old clusters. The LMC clusters Hodge 7 (Bica et al 1992) and NGC 2155 (Searle et al 1980) were both claimed to be ancient clusters. Our CMDs show that these clusters are not old, illustrating the potential problems of age classifications of lower luminosity clusters in crowded fields using broad-band photometry (see also Girardi & Bica 1993, Girardi et al 1995).

2.3 *Age of the Oldest Clusters*

Walker (1992b, Figure 9) and Da Costa (1993, Figure 2) have shown that LMC clusters have a redder HB morphology than expected for their metallicities. Zinn (1993) and van den Bergh (1993) have similarly shown the division of Galactic halo clusters into two major families in this metallicity–HB distribution plane. This division appears to be another manifestation of the "second parameter" effect in Galactic clusters and dSph galaxies. It is now common to associate this effect with an age difference of 2–3 Gyr (Lee et al 1994) between the two families of halo clusters. In our Galaxy, this effect is also accompanied by a kinematical signature (Rodgers & Paltoglou 1984, Chaboyer et al 1992, van den Bergh 1993, Zinn 1993, Da Costa & Armandroff 1995). If age is truly the second parameter determining the HB morphology, after metallicity, then we expect that the ancient LMC clusters will prove to be some 2 Gyr younger in the mean than old halo Milky Way clusters.

Suntzeff et al (1992) have used the data in Table 1 to argue that the old LMC cluster population is very similar to the Galactic cluster population outside the solar circle in terms of metallicity, cluster luminosity, number of RR Lyraes per unit cluster luminosity, and the ratio of luminous cluster mass to field star mass. The number of clusters in the LMC also is consistent with the luminosity difference between the LMC and the Galaxy. This similarity is somewhat surprising given that the dynamical mass of the Milky Way is 10 times that of the LMC, although the dynamical mass of the LMC is likely to be an underestimate (Suntzeff et al 1992).

If we scale the LMC to the SMC luminosity, we would expect the SMC to have about two old clusters: In Table 1, we find one old cluster with RR Lyrae and two substantially younger clusters. Little can be said of the general properties of the old clusters in the SMC due to small number statistics.

It is tempting to try to use the data in Table 1 (and the HB magnitudes given in Suntzeff et al 1992) to determine a M_V–[Fe/H] relationship for RR Lyraes. This relationship is an important and controversial one. We discuss it further in Section 2.5. The published cluster metallicities, however, have more scatter than can be tolerated for this problem—they are typically based on only one or two spectra or on the color of giant branches in the CMD. Even with better metallicities the relationship cannot be determined due to the small range in metallicities in these particular clusters and the unknown position of any cluster in front or behind the galaxy. It had been thought that the HB magnitude as a function of age was constant for stars older than a few Gyr (e.g. Olszewski et al 1987), allowing the metallicity effects to be derived by studying all LMC and SMC clusters older than a few Gyr. Recent results on the Carina dwarf galaxy (Smecker-Hane et al 1994; PB Stetson 1996, private communication) show, however, that 7- and 14-Gyr populations at [Fe/H] $= -2$ produce HB luminosities that differ by 0.25 mag, implying that age effects are as important as metallicity effects.

A more promising approach would be to search the large number of LMC RR Lyraes discovered by the MACHO project for large-amplitude, short-period RRab variables, which will tend to be metal rich. A few such stars are known in the MCs (Hazen & Nemec 1992), though they are clearly quite rare. Such a metal-rich sample, coupled with the much more numerous metal-poor RR Lyraes, should yield a definitive M_V–[Fe/H] relationship.

2.4 *Kinematics of the Oldest Clusters*

If the LMC contained a dynamically hot halo surrounding an inner disk, the velocity dispersion of this halo would be between $v_{circ}/\sqrt{2} = 56$ km s^{-1} and $v_{circ}/\sqrt{3} = 46$ km/s, where $v_{circ} = 79$ km s^{-1} is the circular velocity of the H I disk (Freeman et al 1983; see the derivation of equation 4-55 in Binney & Tremaine 1987). Unexpectedly, Freeman et al (1983) found that the oldest LMC clusters had a much smaller velocity dispersion than expected from these arguments. This very important result implies that the oldest clusters do not populate a kinematically hot halo supported by its velocity dispersion. In fact, the oldest LMC clusters form a rotating disk system. The velocities available to Freeman et al (1983) implied that the oldest LMC population, although in a disk, was not in the same disk as the H I gas.

Olszewski et al (1991) and Schommer et al (1992) have enlarged the sample of clusters and improved the observed cluster velocities. They confirmed that the oldest clusters were in a disk, and they eliminated the perceived difference between the old and young disks. They were able to derive a more precise kinematic solution, albeit with substantial errors due to the small number of old clusters. They found that the oldest clusters have a disk model solution in good agreement with the model fit to the H I gas, when velocities are corrected for the transverse motion of the LMC (Jones et al 1994) and when the oldest clusters superposed on the bar of the LMC are removed. This disk has $v/\sigma = 2$. This old-disk model has a significantly smaller circular velocity than does the H I disk (50 vs 79 km s^{-1}) and larger velocity dispersion (23 vs 10 km s^{-1}). In an adiabatic sense, the lower rotation speed and higher velocity dispersion are consistent in that the circular velocity of the old component lags the disk, and that some of that energy is now in the z component.

There are several interesting aspects to this set of "disk clusters" in the LMC defined by the oldest clusters. First, we return to the question discussed in Section 2.3: What is the age of this set of oldest clusters? If the oldest LMC clusters are indeed a few Gyr younger than an ancient population, then their disk kinematics could be explained merely by the fact that the clusters were created during disk formation, well after the initial population finished forming. In this case, where is this putative ancient population? The only other obvious old population is the population of RR Lyrae variables. But as shown by Suntzeff et al (1992), the number of field RR Lyraes in the LMC is consistent with the luminous mass in the old clusters, if we scale from the same populations in the outer Galactic halo. It therefore seems reasonable to associate the RR Lyraes and the oldest disk clusters with the same population. We feel the most natural explanation is that given by Freeman et al: The LMC does not have a prominent hot (supported by its velocity dispersion) halo. An obvious test would be to measure the velocity dispersion of the RR Lyraes or of the most metal-poor field stars.

A second aspect of the old cluster disk is that the disk is relatively large. The average distance from the bar of the LMC is 3.9 kpc for the total sample of old clusters and 6.1 kpc for those clusters away from the bar. This radius is two to four times the LMC disk scale length (Bothun & Thompson 1988).

Third, there are few clusters beyond 8 kpc from the bar of the LMC. Even though the velocities of these clusters formally give the same disk solution as that of the H I gas, it would be good to find other tracers to derive the rotation and the mass of the LMC out to the distances of the Reticulum cluster (11 kpc) or NGC 1841 (14 kpc). The best stellar tracer would presumably be young, because such objects could have a small velocity dispersion about a rotation solution, if no major recent perturbations of the LMC by interactions with the Milky Way or SMC exist. As pointed out above, the disk solutions for all major LMC components are similar to those of the H I gas. One can assume that the older components of the LMC continue to follow the same disk as the younger components, as the clusters imply, and use carbon stars (Demers et al 1993) or Cepheids or as-yet-uncataloged red giants to derive the rotation curve. As it stands, Figure 8 of Schommer et al (1992) is a good summary of current knowledge: The rotation curve is relatively flat from 2–8 degrees, with minimal information beyond 8 degrees (1 degree = 0.94 kpc). The mass implied by this rotation curve is $\sim 10^{10}\ M_{\odot}$. If the LMC were simply an exponential disk, the rotation curve should peak at 2.2 disk scale lengths, or 3.5 kpc. There is no evidence for a Keplerian falloff at this point, whose position would change outwardly slightly if a low-mass halo were added. If the rotation curve is flat out to the distance of NGC 1841, then substantial dark matter is needed.

Fourth, the cluster systems of the Magellanic Clouds, the Milky Way, M31, and M33 have been compared by Schommer et al (1992) and Schommer (1993) in terms of their spatial and kinematic distributions. The Milky Way and M31 both have disk globular cluster systems (Mayall 1946, Morgan 1958, Kinman 1959, Zinn 1985, Huchra et al 1991) that are metal rich and spatially concentrated to the inner parts of the cluster system. M33, with a wide range of cluster colors similar to the LMC and distinct from the Milky Way or M31, shows a trend of kinematics with color. The set of reddest M33 clusters (Schommer et al 1991) has no global rotation, a velocity dispersion of approximately the expected isothermal value of $\sim$ 70 km s^{-1}, and $v/\sigma = 0.6$. As we examine bluer M33 clusters (color presumably being related to age), rotation increases and line-of-sight velocity dispersion decreases. It is important to remember here that M33 is about twice as luminous as the LMC; it is hard to say if this mass difference is critical or incidental. It will be important to understand why some old cluster systems rotate and why some disk systems are inner-cluster system phenomena while others are outer-cluster system phenomena.

No global statements can be made about the kinematics of SMC clusters; no systematic survey of cluster abundances and velocities has been made. The dominant reason seems to be the small number of SMC clusters, coupled with the fact that most of the SMC clusters are in fairly crowded fields. A second reason is that it has always been tacitly assumed that the SMC kinematics will be quite complicated. It would be fascinating to know if the SMC has a halo that is even remotely spherical, or if it is unrecognizable because of its interactions.

2.5 *The Field RR Lyraes*

Although the Magellanic Clouds are known to be rich in variable stars, the published literature on field RR Lyraes is remarkably sparse. The Harvard Surveys were not deep enough to measure stars as faint as 19th magnitude. While there are 17 bright foreground RR Lyraes in the Hodge & Wright (1967) LMC, and 46 stars with periods less than 1^d in the Hodge & Wright (1977) SMC Atlas, most of these are foreground variables or possible bright members of the anomalous Cepheid population in the SMC.

The first major survey for RR Lyrae variables was done by Thackeray and collaborators (see Thackeray & Wesselink 1953), who found cluster variables in NGC 121 in the SMC and in NGC 1466 in the LMC and near NGC 1978 in the LMC. Graham (1975, 1977) initiated the first modern survey in the regions surrounding NGC 1783 in the LMC and NGC 121 in the SMC. Kinman et al (1991) summarize the major studies of field RR Lyraes in the LMC. As of 1991, there were 122 field RR Lyraes known, almost all in the two fields surrounding NGC 1783 and NGC 2210 in the inner regions of the LMC. In total, 4.3 square degrees have been thoroughly searched in the LMC, with most of this area in regions more than 5 kpc from the center of the LMC where the variable density is low. Kinman et al (1991) have fit a King model to the density distribution in the six LMC fields, and they estimate that the total number of LMC field RR Lyraes is $\sim 10{,}000$.

The MACHO group (Alcock et al 1993) is monitoring 41 square degrees of the LMC and 15 square degrees of the SMC for microlensing by objects in the halo of the Milky Way. A by-product of this survey is an extensive catalog of RR Lyrae variables in the areas surveyed. This catalog is probably complete except near the cores of the larger clusters. Alcock et al (1996) have published an initial review of their LMC RR Lyrae catalog, where they announce the discovery of 7902 LMC RR Lyraes in the innermost, densest regions (11 square degrees), based on light curve data with 250 individual observations per star over a 400-day observing cycle. Some basic properties of the LMC RR Lyrae population from this first reconnaissance of the data are: 1. The surface density of RR Lyraes in the central LMC regions is a factor of two higher than the Kinman et al (1991) estimate; 2. the mean period of the RRab variables is 0.583 days; 3. the mean metallicity based on the period-amplitude relationship and spectral observations is [Fe/H] = -1.7; 4. the period-amplitude relation for 500 randomly chosen LMC RR Lyraes is skewed to amplitudes lower than 1 magnitude, unlike the case for Milky Way RR Lyraes; and 5. about 1% of the variables may be second-overtone pulsators with $\langle P \rangle = 0.281^d$.

The Alcock et al results on the mean metallicities of the RR Lyraes are similar to previous metallicity measurments for field variables. Using image tube spectra, Butler et al (1982) derived [Fe/H] = -1.4. Hazen & Nemec (1992) and Walker (1989a), from the period-amplitude relation for the RR Lyraes in the clusters NGC 1783, NGC 2210, and NGC 2257 and their surrounding fields, measured mean abundances of [Fe/H] = -1.3, -1.8, and -1.8, respectively. The mean cluster RR Lyrae abundance from Table 1 (weighted by number of variables) is [Fe/H] = -1.8.

Both the Butler et al and Alcock et al metallicity studies are based on spectral data of rather poor quality by modern standards. With so many RR Lyraes with accurate periods and phases cataloged from the MACHO study, now is the time to consider a major new effort to study the abundance (and velocity) distribution of the RR Lyraes in the LMC, before the phase predictions are compromised by intrinsic period changes and period errors in the catalog.

Alcock et al (1996) provide some interpretation for the mean characteristics of the field RR Lyrae population. They argue that the excess of small-amplitude fundamental pulsators (RRb-type variables), the mean period of the RRab variables, and the "transition period" between the fundamental (RRab) and first overtone (RRc) pulsators imply that the HB stars are evolving from the red of the instability strip blueward. An underlying red horizontal branch in such a metal-poor population of variables may indicate that this population is not ancient but somewhat younger, something we have already seen in the HB characteristics of the LMC clusters. We note, however, that trends seen in mean properties of RR Lyraes in Galactic globular clusters are based on stars in a given cluster where all objects have the same age and metallicity. Interpreting the same mean properties in a population that almost certainly has a wide range in metallicity and possibly in age is problematical.

In the SMC, there have been two major surveys. Graham (1975) discovered 75 RR Lyrae stars in 1.3 square degrees surrounding NGC 121, and Smith et al (1992) surveyed 1.3 square degrees surrounding NGC 361, finding 22 definite and 20 probable RR Lyraes. The metallicity of the RR Lyraes is [Fe/H] = -1.6 from the period-amplitude relationship given by Smith et al and -1.8 from the spectra of three field stars (Butler et al 1982). Both estimates are quite uncertain.

Little work has been done on the kinematics of the RR Lyrae populations. The only quoted attempt to measure the velocity dispersion of the LMC RR Lyraes was made by Freeman a decade ago (Freeman 1996), who derived a dispersion of ~ 50 km s^{-1}. This result, if confirmed by modern data, is quite remarkable because, although it corresponds to the expected dispersion of a kinematical halo, it also disagrees with the LMC old cluster kinematics as discussed above. It should be noted that because RR Lyraes can have velocity amplitudes well in excess of 50 km s^{-1}, it is very important to have good phases in order to determine accurate corrections to the gamma velocities.

With accurate photometry, the mean magnitudes of the RR Lyraes can be used to derive the distance, tilt, and reddening of the LMC; to measure the relative distances between the Clouds; and to check the consistency of the RR Lyrae and the Cepheid distance scales. To date, however, there are only a few fields with relatively high-quality (error in magnitude < 0.05 mag) photometric zero points. In the LMC, three fields have adequate photometry: the NGC 2210 field (Hazen & Nemec 1992), the NGC 1466 field (Kinman et al 1991), and the NGC 2257 field (Walker 1989a). The mean B magnitudes, corrected for the presumed reddening, are 19.56, 19.55, and 19.29, respectively. In the SMC, the mean B magnitude for the NGC 121 field is 19.95, and for NGC 361 it is 19.91. Evidently the difference in mean RR Lyrae magnitude between the Clouds is ~ 0.45 (if we ignore the small difference in mean reddening to the Clouds), but accurate photometry in more fields across both Clouds are needed to reduce the uncertainty in this result to below 0.1 magnitudes. A final complication is that parts of the SMC are very extended along the line of sight (Mathewson et al 1986). Some of the fields discussed in Kinman et al (1991) have reasonably accurate relative photometry but need modern zero points.

In principle, the spread in magnitudes in a given field can be used to derive the line-of-sight depth in the galaxy, but two effects limit the usefulness of this statistic. The first effect is that Galactic globular clusters apparently have a natural dispersion in RR Lyrae luminosities, presumably due to evolution from the zero-age HB. For instance, 33 variables in M15 (Bingham et al 1984) give a dispersion of 0.15 mag, and 35 variables in M3 (Sandage 1981) give 0.07 mag, implying a natural dispersion in magnitude of about 0.1. The second effect is that, in a population with a range in abundances, there should be a natural range in luminosities due to the variation of M_V as a function of [Fe/H]. Although the precise level of this variation is controversial, it is roughly 0.2–0.3 mag per dex in [Fe/H] (Sandage & Cacciari 1990).

Do we resolve the depth in either of the Clouds? In the SMC, 34 RR Lyraes near NGC 121 (Graham 1975) have a dispersion of 0.12 mag in B, whereas near NGC 361, the data in Smith et al (1992) imply a dispersion of 0.19 mag for

17 stars with periods greater than 0.4 days. We may be marginally resolving a real depth in the SMC near NGC 361, but it is surprising that the dispersion is not much larger since NGC 361 lies close to the central body of the SMC where the Cepheid studies imply a line-of-sight depth of 15 to 20 kpc (Mathewson et al 1986, Caldwell & Coulson 1986). In the LMC we find the following dispersions: 0.16 mag for 55 stars in the NGC 1783 field (Graham 1977), 0.16 mag for 28 stars with periods longer than 0.4^d in the NGC 2210 field (Hazen & Nemec 1992), and 0.17 mag for 13 RRab variables near NGC 2257 (Nemec et al 1985, new zero points in Walker 1989a). Little can be said about the LMC depth with these data.

The distances to the Magellanic Clouds, the absolute magnitudes of the Population I Cepheid variables and the Population II RR Lyrae stars, the [Fe/H]-absolute magnitude relation for RR Lyrae stars, the ages of globular clusters, and the value of the Hubble Constant are all interrelated. It is beyond the scope of this review for us to try to provide best values for each of these parameters. We can, however, recapitulate the problems that occur when certain values are adopted. These problems will show that substantial fundamental work is still needed before local distance calibrators are on a firm footing.

If we assume that the LMC distance modulus $(m\text{-}M)_0 = 18.55$, based on the Cepheid distance scale (Feast and Walker 1987, Walker 1992c), the absolute magnitude of LMC and of SMC RR Lyraes follows. For the LMC field variables near NGC 2210, NGC 2257, and NGC 1866, we calculate the dereddened mean V to be $\langle V \rangle = 19.0$ (Walker 1989a, 1995; Hazen & Nemec 1992 with a transformation from B to V). The resultant absolute magnitude of $M_V = 0.45$ is identical to that found for the LMC cluster RR Lyraes (see Walker 1992c, 1993a).

The results for the SMC seem to be in mild discord with the LMC result. If we use the mean B magnitudes for the NGC 121 and NGC 361 fields given above, along with the mean B magnitude of 19.91 for the four variables in NGC 121 (Walker & Mack 1988), and deredden the NGC 361 field by E(B–V) = 0.06 (Smith et al 1992), and the NGC 121 field by 0.04, we find $\langle B \rangle = 19.75$. With an SMC Cepheid distance modulus of 18.8 (Feast and Walker 1987) and a B–V of 0.3 for the typical RR Lyrae, we derive an absolute V magnitude of ~ 0.65 for the SMC RR Lyraes. The NGC 121 field, at least, is in the portion of the SMC that has a small line-of-sight depth. Although no evidence for the depth of the SMC is seen in these small RR Lyrae samples, it is clear that the sample should be enlarged.

An interesting test of the relative photometry of the Cepheids and the RR Lyraes is to measure relative magnitudes of these types of stars in the same fields. One possible field is near the young, Cepheid-rich cluster NGC 1866, where Welch and Stetson (1993) and Walker (1995) have discovered a total of four field RR Lyraes.

Are the absolute V magnitudes of 0.45 or of 0.65 the expected values for metal-poor RR Lyraes? A number of authors have attempted to derive the metallicity-absolute magnitude relation for RR Lyraes. If we use the average relation given in Sandage and Cacciari (1990), and the relations of Carney et al (1992), and Layden et al (1996), we find $M_V = 0.7$ at $[Fe/H] = -1.7$. The LMC RR Lyraes would be brighter than this relation by 0.25 magnitudes, while the SMC RR Lyraes would be consistent with this Galactic calibration. Sandage (1993) has argued for a steeper slope in the relation, which makes the LMC RR Lyraes consistent with his calibration, and the SMC RR Lyraes inconsistent. Careful calibration of extant large samples of RR Lyraes in both Clouds is clearly of paramount importance, as is the discovery and photometry of metal-rich Magellanic Cloud RR Lyraes discussed above. We do note that an absolute magnitude of 0.45 for Galactic RR Lyraes does affect the calibration of globular cluster ages (Walker 1992c), and the value of the Hubble constant (van den Bergh 1996).

2.6 *The Old Long-Period Variables*

More than 1000 Long-Period Variables (LPVs) are known in the field of the LMC. A representative set of articles, with references to the other literature, are found in Bessell et al (1986), Hughes (1989, 1993), Hughes et al (1990, 1991), and Reid et al (1995). Most LPVs are luminous asymptotic giant branch (AGB) stars with $-7 < M_{\rm bol} < -4$ and come from a population younger than the ancient populations being considered here (Wood et al 1983). The small number of LPVs in Galactic globular clusters tend to have periods between $\sim 190^d$ and $\sim 230^d$, which places them at the lower end of the range of LPV periods. They are generally found in clusters with $[Fe/H] > -1$ (Frogel & Elias 1988). The globular cluster LPVs are generally up to a magnitude brighter than the core helium flash termination point of the first ascent red giant branch (RGB) $M_{\rm bol} \sim -3.5$, and therefore they are AGB stars close to the point of nuclear fuel exhaustion.

The Milky Way field short-period LPVs have halo kinematics, whereas the longer-period LPVs have disk kinematics (see the introduction to Bessell et al 1986). Under the assumption that the LMC LPVs would have similar properties, Bessell et al (1986) and Hughes et al (1990, 1991) define a class of old long-period variables (OLPVs) in the LMC. The theoretical pulsational masses for the OLPVs derived by Wood et al (1983) also support this assumption.

Hughes (1989) gives a large catalog of LPVs from which a sample of OLPVs were drawn. Hughes & Wood (1990) showed that these OLPV stars (defined as LPVs with periods of 150–225 days) in the LMC had absolute magnitudes consistent with those found in Galactic globulars, and Hughes et al (1991) measured velocities and the velocity distribution of the group. They find a trend of velocity dispersion with period (Figure 7 and Table 8 in Hughes et al 1991) that seems to vindicate their assumption that the LPV period is closely associated with the stellar age. The OLPVs have a dispersion of 35 km s^{-1}, which is significantly larger than that for any other population except the preliminary RR Lyrae result discussed above. This velocity dispersion is not corrected for the problem of observing these LPVs at random phase, but Hughes et al (1991) argue that the random-phase velocity errors are approximately the same as the measurement errors. This velocity dispersion implies that the OLPVs belong to a flattened spheroid and that the mean age derived from the pulsational masses is about 10 Gyr.

How certain are we that the OLPVs are truly an old population? The association with Galactic globulars and the halo-like kinematics of the field OLPVs certainly are consistent with these facts. Yet the OLPVs have some other curious properties. The association of the OLPVs with metal-rich clusters in the Galaxy would imply the Galactic OLPV population should have a kinematical signature that is more like a thick disk than a halo. In addition, almost all LPVs are either carbon (C) stars or M stars. Most of the OPLVs in the lists of Hughes et al (1991) are also C and M stars; many are late M stars. C stars are nonexistent in the Galactic globulars and are found in intermediate-age populations in the Clouds. M stars are found only in the metal-rich clusters ([Fe/H] $\gtrsim -1$). The OLPV population, therefore, seems signficantly more metal rich than the old clusters listed in Table 1, and if we are interpreting the presence of the C stars correctly, they are also signficantly younger. Finally, there is the disturbing fact that the Galactic globular cluster LPVs do not have the same period-luminosity relationship as the OPLVs of the LMC (Menzies & Whitelock 1991). These authors point out that unless both the Cepheid and RR Lyrae distances scales are in error, this difference implies that the OLPVs in the LMC are more massive than the variables in the Galactic clusters. Whether or not the OLPVs are as old as claimed by Hughes et al (1991), the fact that the velocity dispersion is *higher* than the old cluster population is provocative.

It would be very interesting to discover and study LPVs in Magellanic Cloud clusters to allow a direct age dating to verify the period-age assumptions and the pulsational masses. Frogel et al (1990) note that a few of their C and M stars found in a grism survey of clusters are clearly dusty LPVs. Periods are not known for these stars.

2.7 *The Old and Metal-Poor Field Stars*

There are two ways to find the ancient population of normal stars. The first is to derive abundances and velocities of red giants in a well-defined field to identify the most metal-poor stars. Simple photometry is inadequate because of Galactic foreground contamination. The second is to look for old and metal-poor main-sequence turnoffs and subgiant branches in the color-magnitude diagrams of the field.

There are two studies of complete samples of field giants in the Magellanic Clouds: Suntzeff et al (1986) for the field surrounding NGC 121 in the SMC and Olszewski (1993) for the NGC 2257 field in the LMC. The two RGB abundance distributions are quite different.

The NGC 2257 field is the same field used by Jones et al (1994) to derive the proper motion of the LMC as a whole. Many, but not all, foreground stars are therefore removed from the sample because they have large proper motions. This field is 8 kpc north of the LMC bar and would normally therefore be called a "pure halo" field. Its surface brightness is very low, $\sim$27 mag arcsec^{-2}, according to the surface brightness model of Bothun & Thompson (1988). The present-day neutral hydrogen density is less than 1.6×10^{20} cm^{-2}. Potential giants were picked from the CMD by superposing ridge lines of metal-poor and metal-rich Milky Way clusters, with care to ensure that no potentially metal-poor stars would be missed. Spectra were obtained for all such stars brighter than the red giant branch clump. Only 8 or 9 of the 36 resultant member giants are more metal poor than [Fe/H] ~ -1.3. The vast majority of the stars in this "halo" field have abundances centered on [Fe/H] $= -0.5$ (see Figure 3 in Olszewski 1993). This abundance distribution is not expected for a "Population II halo." It is not yet understood how such a field can be dominated by stars more metal rich than 47 Tuc. The velocity dispersion of the stars more metal poor than [Fe/H] $= -1.3$ is 29 km s^{-1} (23 km s^{-1} with one extreme-velocity star removed). The velocity dispersion for the more metal-rich stars is 16 km s^{-1}. These velocity dispersions are not obviously different from those of the clusters or OLPVs. Very large areas on the sky will need to be observed to derive the velocity dispersion of subsamples of the metal-poor population.

The NGC 121 field in the SMC is much more normal for a distant "halo" field. Astrometry was again used to cull foreground stars. Spectra were then obtained for 13 stars. The spectroscopy and photometry were then used to derive abundances for 31 stars in the field. The distribution of metallicities peaks at [Fe/H] $= -1.6$ (Suntzeff et al 1986, Figure 9) and is very similar to the abundance histogram of Galactic halo RR Lyraes and of Galactic halo globular clusters. There is no component similar to the dominant metal-rich component in the NGC 2257 field. The velocity dispersion is 24 km s^{-1}.

Old and metal-poor main-sequence turnoffs are an indication of the progenitors of the giants discussed above. Elson et al (1994) attempted to find this population with pre-Costar *HST* images of the young and crowded 30 Dor region. If their assumptions about the errors in the photometry are correct, they see a small excess of stars blueward of the dominant main-sequence population. They attribute these stars to a metal-poor population. We believe that this result is inconclusive. Elson et al (1994, Figure 6) show a color cut across the main sequence at $22 < V < 22.5$ in two 30 Dor fields approximately 30 arcmin apart. The color of the dominant population does not change, while the color of the putative metal-poor population changes by 0.08 mag. Fields known to contain BHB stars or strong subgiant branch populations will show the old metal-poor main sequence most clearly. This population is easily hidden. The clusters and RR Lyrae results suggest that the ancient population is no more than 10% of the luminous mass of these galaxies. Simple Monte Carlo simulations of synthetic color-magnitude diagrams show that the ancient population will be difficult to find because the lower main sequences of younger populations are superposed on the old turnoff.

The horizontal branch and red clump (RC) can be populated by populations with a wide range of ages (see Section 3.2). Gardiner & Hatzidimitriou (see references in Section 3.2) have made extensive studies of the color-magnitude diagrams of large areas of the SMC field using scans of SRC sky survey plates. A similar study is not currently published for the LMC, but surveys such as the MACHO project have already collected the appropriate photometry. Gardiner & Hatzidimitriou (1992) have argued that the majority of clump stars come from intermediate-age populations, though the bulk of the field populations near NGC 121 and L113 are claimed to be older than 10 Gyr. The RC stars closest to the instability strip can come from a population older than 10–12 Gyr (Hatzidimitriou 1991). Color cuts through the RC region of the CMD show this slightly bluer, fainter red HB population. More recent ages and abundances for the calibrating clusters (Armandroff et al 1992, Armandroff & Da Costa 1991, Buonanno et al 1990) lead to the conclusion that this population of red HB stars near the instability strip may be as much as several billion years younger than the oldest globulars. This subset of older clump stars is approximately 7% of the total number of clump stars. About 7% of the stars older than 2 Gyr are thus older than 10–12 Gyr. More importantly, there is little evidence for blue horizontal branch stars. Gardiner & Hatzidimitriou (1992) estimate that the number of BHB stars is an order of magnitude smaller than the number of old RHB stars. Given the low metal abundance of the SMC clusters and field it is more likely that BHB stars do not exist because ancient SMC stars do not exist. NGC 121 and its RR Lyraes are demonstrably younger than the oldest globulars. Therefore the SMC clusters do not contradict this reasoning. Important observational tests are needed to derive the metal abundance of the SMC RR Lyraes, to see if there exists significant populations with abundances more metal poor than [Fe/H] $= -1.6$. It is also important to derive the abundance distribution of the clump giants.

If one isolates a set of old clusters or field stars, then with the next generation of telescopes we can expect to make detailed analyses of elemental and isotopic ratios. The ratio most commonly discussed is [O/Fe], because oxygen is made in Type II supernovae, whose progenitors are very short lived, and iron is made in Type I supernovae, whose progenitors are very long lived. [Mg/Fe], [Ca/Fe], and [Si/Fe] should change in the same way as [O/Fe] and are easier to measure. As Gilmore & Wyse (1991) summarized, the ancient objects described in this review should be overabundant in oxygen if the initial star-formation burst were short, but could have a Solar ratio of [O/Fe] if these objects were formed near the end of a lengthy episode of star formation. If the clusters are not self polluted, we may be able to tell if the ancient clusters were formed before or after the ancient field.

2.8 *The CH Stars*

Another stellar type often associated with old populations are CH stars. They comprise a rather ill-defined class of carbon stars that are seen in dSph galaxies, the Galactic globular cluster ω Cen, and in the Galactic halo. Although CH stars do have a number of spectral peculiarities (strong CH bands, s-process enhancements, low metallicities, and bluer colors than N-type carbon stars), there is presently no spectroscopic way to identify a carbon star as a CH star unambiguosly. The existence of "CH-like" stars (Yamshita 1975), which have disk-like motions in the Galaxy but otherwise appear the same as CH stars, compounds the problem. The classification of a star as a CH star is therefore associated with its kinematics. While the kinematics coupled with spectral characteristics may allow the identification of a star as a CH star in our Galaxy, the unambigous identification of a star as a CH star in another galaxy cannot be made without other information, such as velocities or bolometric luminosities.

Hartwick & Cowley (1988) and Cowley & Hartwick (1991) have identified blue carbon stars with enhanced s-process features in the LMC. Infrared photometry of these stars by Suntzeff et al (1993) and Feast & Whitelock (1992) have shown these stars to be very luminous [$M_{\text{bol}} = -5.3$ in the mean]. They are probably carbon stars that are younger ($t < 1$ Gyr) than the dominant C-star population (Section 3.2 below) discovered with infrared plates (Blanco et al 1980). A deeper survey in the peripheries of the Clouds should reveal a number of true CH stars near the tip of the old RGB tip at $M_V = -3$, but no convincing population of these old carbon stars have been made to date.

3. THE INTERMEDIATE-AGE POPULATIONS OF THE MAGELLANIC CLOUDS

3.1 *Definitions*

In this section, we consider field stars and clusters with ages between approximately 1 Gyr and 10–12 Gyr (using the age scale adopted in Section 2.1) to represent the "intermediate-age" populations of the Clouds. For comparison,

Westerlund (1990) adopted the age range 0.2–7 Gyr for this purpose. Our definition of intermediate-age has the advantage of being empirically easy to apply. The upper age limit corresponds to the epoch of RR Lyr formation that was discussed in Section 2.2, while the younger age limit corresponds to the earliest appearance of the red giant branch. The precise ages of these limits remain uncertain because different treatments of convective overshoot significantly affect the ages of clusters younger than 1–2 Gyr (e.g. Mateo & Hodge 1987, Seggewiss & Richtler 1989, Bomans et al 1995). Moreover, the RGB-turnon age seems to exhibit an intrinsic spread of 250–500 Myr (Ferraro et al 1995).

3.2 *Evolved Field Stars*

3.2.1 ASYMPTOTIC GIANT BRANCH STARS For many years, "intermediate-age" in the Magellanic Clouds was taken to be nearly synonymous with the presence of luminous asymptotic giant branch stars. Numerous workers have identified and studied AGB stars in field and clusters of the MCs (e.g. Aaronson & Mould 1985, Frogel et al 1990, Mould 1992). Because the peak AGB luminosity increases steadily with decreasing age (due to the increase in core mass with increasing ZAMS mass), and because AGB stars are relatively easy to identify, the brightest AGB stars are useful as beacons for the intermediate-age populations throughout the Clouds and their environs. Infrared observations are required to determine AGB luminosities because these stars have low surface temperatures and correspondingly large optical bolometric corrections (e.g. Frogel et al 1990).

Magellanic Cloud clusters have played an important role in calibrating the luminosities of AGB stars in terms of age. Frogel et al (1990) provide the most recent calibration of AGB luminosities as a function of cluster age, but, as has been known for some time (Aaronson & Mould 1985, Hodge 1983), this correlation between AGB properties and age is very broad: For clusters of a given age, the observed peak AGB luminosities vary significantly in different objects. In practice, it becomes impossible to use the AGB to attain age resolution of better than a factor of two for populations older than about 1–2 Gyr. One important reason for this is that the AGB is a short-lived evolutionary phase (Iben & Renzini 1983). Thus, defining an AGB sequence or determining the peak AGB luminosity depends sensitively on the sample size. Metallicity differences between clusters at a given age and star-to-star variations in core mass and mass-loss rates also affect their AGB luminosities and lifetimes and contribute to the intrinsic scatter of AGB properties with age. The AGB is a blunt tool with which to constrain ages, but for many galaxies with only partially resolved stellar populations, it is often the only tool available.

While acknowledging these limitations, we can still use AGB stars to study intermediate-age populations in the MCs. Frogel & Blanco (1983, 1990) used the detailed features of AGB stars to estimate ages of the progenitor field populations in the Bar-West field of the LMC. They conclude that the luminosity and IR-color distribution of the AGB can be understood as two sequences. The bluer, more luminous AGB stars appear to be associated with stars as young as 10^8 yrs, while the redder, lower-luminosity AGB stars trace 1–5 Gyr populations (Frogel et al 1990, Feast & Whitelock 1994). Frogel & Blanco (1990) show that the young and intermediate-age populations are present in the ratio 1:4 in this inner LMC field. The results imply an episodic star-formation rate in the inner LMC coupled with a slow overall decline in star formation during the past few Gyr. These studies also illustrate the value of IR photometry as a way to identify AGB stars easily without the need for confirming optical spectroscopy.

Studies of carbon AGB stars—luminous, cool AGB stars with dredged-up carbon in their atmospheres—have recently been used to map the extent of intermediate-age populations in both MCs. Feast & Whitelock (1994) have analyzed the C stars identified by Demers et al (1993) to trace intermediate-age populations in the outer parts of the LMC, in the inter-cloud "bridge" connecting the two Clouds, and in the outer halo of the SMC. In all three cases, the mere presence of the C stars indicates the presence of a significant intermediate-age population in these regions.

The utility of AGB stars as intermediate-age population tracers is well illustrated by the Magellanic bridge. Demers and collaborators (Grondin et al 1990, 1992; Demers & Irwin 1991; Battinelli & Demers 1992; Demers et al 1991) have demonstrated the existence of young ($\lesssim$ 200 Myr) stars and loose clusters or associations in a stellar inter-cloud bridge connecting the two galaxies. This stellar bridge is closely associated with the H I bridge between the Clouds (Mathewson & Ford 1984, Westerlund 1990). The presence of luminous AGB stars in the bridge (Demers et al 1993) suggests that the young population—which may well have been born when the bridge itself formed (Demers et al 1991, Gardiner & Hatzidimitrou 1992)—may be accompanied by older stars that possibly predate the bridge. Whatever mechanism formed the bridge and its youngest stars was also probably responsible for stripping some intermediate-mass stars from one or both of the Magellanic Clouds (see Section 3.5).

3.2.2 ANOMALOUS CEPHEIDS Anomalous Cepheids (ACs) are believed to form during the late evolution of intermediate-mass metal-poor stars (Hirshfeld 1980). ACs have periods of $0.25–1.5^d$ and obey a period-luminosity relation (Nemec et al 1994). These stars have luminosities considerably higher than RR Lyrae stars, making ACs relatively easy to identify in external galaxies. The field of our Galaxy is not known to contain any ACs, but, even if present, they would be very difficult to identify due to confusion with RR Lyraes of similar periods (see

Mateo 1996a), and with type II Cepheids and evolved HB stars (Sandage et al 1994). Because the standard model for ACs requires them to be more massive than the turnoff masses of ancient clusters such as NGC 5466 (Hirshfeld 1980), it is generally believed that NEC 5466-V19 formed as the result of mass transfer in a now-merged binary progenitor (Zinn & King 1982, Nemec & Harris 1987).

In contrast, dwarf spheroidal galaxies (Mateo et al 1995, Nemec et al 1994) and the SMC (Smith & Stryker 1986, Smith et al 1992) appear to be rich in ACs. In the case of the SMC, this conclusion rests on observations of only four bona fide candidates. In dwarf galaxies—where complex star-formation histories appear to be the norm rather than the exception (e.g. Da Costa 1992, Smecker-Hane et al 1994, Mateo 1996a)—ACs are generally believed to result from the evolution of single, relatively massive metal-poor stars as described by Hirshfeld (1980; see also Smith et al 1992). Anomalous Cepheids therefore represent another potential tracer of the intermediate-age populations in the Clouds. One candidate AC has been reported for the LMC (Sebo & Wood 1995); however, these authors also note that the light curve of this star could be understood as that of a normal 0.51-day RR Lyrae star with a nonvariable but unresolved companion of comparable brightness. Thus it remains uncertain whether the LMC contains any ACs at all. Ongoing and planned large-scale MACHO and OGLE photometric surveys should soon provide comprehensive censuses of the ACs in both Clouds.

It is tempting to attribute the higher frequency of ACs in the SMC as evidence that this galaxy contains intermediate-age progenitors while the LMC does not. The details of how anomalous Cepheids relate to their progenitor population appear to be quite complex, with metallicity and age probably playing nearly equal roles (Smith & Stryker 1986, Mateo et al 1995). One can also confuse ACs with other types of pulsating, short-period variables found above the HB but which are generally evolved from old–not intermediate-age–progenitors (Sandage et al 1994). In dwarf spheroidal galaxies the frequency of ACs does *not* appear to correlate strongly with the age of the predominant field population (Mateo et al 1995). The LMC's deficiency of ACs may simply reflect the absence of suitably metal-poor intermediate-age progenitors rather than the complete absence of that population. New theoretical models of ACs applicable to Cloud metallicities and age ranges are badly needed to address this issue.

3.2.3 INTERMEDIATE-AGE LONG-PERIOD VARIABLES As described in Section 2.6, LPVs with periods less than about 250 days are generally classified as "old." Hughes et al (1991) have identified such stars with an old spheroidal-like population of the LMC. For LPVs with periods in the range 225^d–425^d, which are found in abundance in the LMC, the putative progenitors have main-sequence lifetimes of about 1–3 Gyr (Wood et al 1983, 1985; Hughes & Wood 1990).

This conclusion is consistent with their kinematics, which are disk-like and intermediate between the kinematical properties of young LMC components (e.g. H I clouds, very young clusters) and of older components (the ancient clusters, old LPVs). Hughes & Wood (1990) note that the relative frequency of old and intermediate-age LPVs suggests that the older population is considerably more abundant, by up to a highly uncertain factor of five.

We see below that very few field stars or clusters seemed to have formed 4–12 Gyr ago. The period distribution of LPVs shows no sign of a break corresponding to this hiatus in star formation, suggesting to us that the scatter in the mass-period relation for LPVs is comparable to the full mass range corresponding to intermediate-age stars. Alternatively, the progenitors of LPVs may have continued to form during an epoch when star-formation activity was otherwise very low in the LMC.

3.2.4 PLANETARY NEBULAE Planetary nebulae (PNe) are associated with the late stages of evolution in intermediate-mass stars. About 150 and 50 PNe are cataloged in the LMC and SMC, respectively (Meatheringham 1991, Vassiliadis et al 1992). In the late 1980s, Meatheringham and collaborators studied the spatial, kinematic, and physical properties of PNe in the LMC; Westerlund's (1990) review describes this work in detail.

How old are the PNe in the Magellanic Clouds? Using a simple orbital diffusion model based on heating by massive molecular clouds in the LMC, Meatheringham et al (1988) concluded that the mean age of the PNe sample is about 2–4 Gyr. This value is also consistent with stellar evolutionary models for PNe formation. It should be stressed that these models only very weakly constrain the PNe progenitor masses and therefore their ages. More recently, Vassiliadis et al (1992) confirmed the earlier kinematic solution and mean age for LMC PNe after adding to their sample radial velocities of 11 new nebulae located in the outskirts of the galaxy. Both the kinematic and stellar evolutionary ages of LMC PNe suggest that these objects belong to the galaxy's intermediate-age population. Given the large uncertainties in how the ages of PNe are determined in the Clouds, we cannot currently rule out that some of the PNe come from the ancient population.

In the SMC, Dopita et al (1985) measured a large velocity dispersion and found no clear evidence for rotation in a sample of 44 PNe. Interestingly, Hardy et al (1989) likewise found a kinematic solution very similar to that of a true spheroid or halo for C stars in the SMC. Unlike in the LMC, these two tracers of intermediate-age populations seem to be kinematically ancient. Whatever caused the complex structure and chaotic kinematics of the SMC as a whole are undoubtedly behind this unusual kinematic signature as well (e.g. Hatzidimitriou et al 1993); see section 3.5. If, as in the LMC, the PNe and C stars correspond to 2–6 Gyr populations, then we can conclude that whatever event induced the spheroidal-like kinematics in these objects probably did so within the past few Gyr. A lower limit of about 0.5–1 Gyr for the time when this event occurred is imposed by the fact that the H I in the SMC has approximately disk-like kinematics.

3.2.5 RED GIANT CLUMP STARS Virtually all intermediate-age stars experience a stable He core-burning phase. Among the oldest stars, this evolutionary phase corresponds to the horizontal branch seen in ancient star clusters (Sections 2.2 and 2.5). Stars in intermediate-age populations evolve into an analogous region located just blueward of the base of the red giant branch in the HR diagram. This evolutionary phase was first systematically studied in Galactic open clusters by Cannon (1970), and its location in the HR diagram is referred to as the red clump (RC). Data presented by Sarajedini et al (1995) nicely illustrate that the distinction between the RC and HB is fuzzy: As a population ages, the RC evolves smoothly into a red HB such as observed in 47 Tuc.

Unlike globular cluster horizontal branches (e.g. Lee & Demarque 1990), the structure and locations of red clumps do not change drastically over a very large range of age. Using photometry from MC and Galactic clusters with well-known ages, Hatzidimitrou (1991) noted that the (B-R) color separation of the RC and the RGB at the same luminosity varies systematically as a function of cluster age. This separation, d_{B-R}, varies systematically by about 0.15 mag as a population ages from 1 to 10 Gyr. This behavior is, to first order, independent of metallicity since RC and similar-luminosity RGB have similar surface temperatures and thus vary in color by similar amounts as line blanketing varies (though see Smecker-Hane et al 1994 for a counter-example of this simple evolutionary scenario observed in the Carina dwarf galaxy). For composite stellar populations it is nearly impossible to separate stars with different ages within this age range because the color spread of the RC itself is comparable to the full range of d_{B-R}. However, one can determine the mean age by measuring the color of the RC centroid.

Gardiner & Hatzidimitriou (1992) applied this technique to seven SMC fields, deriving mean ages for each. The individual age determinations scatter between 3 and11 Gyr, with a mean age of 7.6 ± 1.2 Gyr. We see below that the LMC seems to have formed very few stars at this time. These authors also claim the existence of an ancient population in their SMC fields on the basis of the color distribution of RC stars (see Section 2.7). Hatzidimitrou et al (1989, 1993), Gardiner & Hawkins (1991), and Gardiner & Hatzidimitrou (1992) have also used RC stars to determine the distribution of intermediate-age populations in the outer SMC, to study the line-of-sight depth of the SMC, and to measure the SMC kinematics as a function of depth.

Numerous photometric studies have identified a prominent RC population throughout the LMC (e.g. Hodge 1987, Brocato et al 1989, Bencivenni et al 1991, Bertelli et al 1992, Vallenari et al 1994a,c, Walker 1993b, Elson et al 1994, Reid & Freedman 1994, Gilmozzi et al 1994, Bhatia & Piotto 1994, Westerlund et al 1995) and the SMC (Brück 1980, Bolte 1987, Hilker et al 1995). In some cases, detailed photometric studies have subsequently revealed the intermediate-age main-sequence progenitors of these RC stars. The fact that the RC is visible in all fields suggests that the intermediate-age populations of the MCs are ubiquitous. With the possible exception of the inter-cloud bridge (Grondin et al 1992) and some fields located quite far from the MCs (e.g. near NGC 1841; Walker 1990), we know of no MC fields with adequate photometry that do not reveal RC stars. Such stars are even seen as a pervasive background in regions located close to active sites of present-day star formation (e.g. Elson et al 1994).

3.3 *Main-Sequence Stars*

Butcher (1977) identified a break in the slope of the main-sequence luminosity function of field stars in the LMC. This change in slope occurs about 1 mag brighter than the turnoff in globular clusters, leading him to conclude that the field was predominately composed of stars with ages 3–4 Gyr or younger. Later studies by Stryker (1983) and Hardy et al (1984) reached nearly identical conclusions for other fields in the LMC. In the SMC, Brück and collaborators (e.g. Brück 1980) and others (Hardy & Durand 1984, Bolte 1987) have also identified the presence of an intermediate-age population, but in this case with a mean age that seems to be significantly older than in the LMC.

More recently, studies using CCDs have begun to map out the detailed composition of the intermediate-age populations in the MCs. In order to adequately sample the old and intermediate-age stars, such data must reach to $V \sim 24$ to obtain good photometric precision at the level of the ancient turnoff, at $V \sim 22.5$ in the LMC and ~ 22.9 in the SMC. They should also cover a large area on the sky to adequately sample the rarest population present in a given field. Table 2 lists all MC fields for which deep photometry has been obtained that can constrain the frequency of populations as old as 10–15 Gyr using main-sequence stars. Linde et al (1995) provide a chart showing the locations of some of these fields in the LMC.

Bertelli et al (1992) analyzed deep CCD data in three LMC fields with the aim of deriving the star-formation histories at each location. They calculated a number of well-defined parameters—such as the ratio of red subgiants to main-sequence stars in a specified magnitude interval—and compared these with predictions from synthesized color-magnitude diagrams generated from the Padova stellar evolutionary models (e.g. Bertelli et al 1994). The indices were designed to be easy and robust to measure, to be relatively insensitive to photometric incompleteness, and yet to remain sensitive to various important physical parameters, such as the slope of the mass function (assumed to be constant with time), the mean metallicity, and the relative numbers of stars formed at different epochs. Remarkably, only a limited range of star-formation histories and metallicities could simultaneously account for the observed values of these parameters. Bertelli et al (1992) suggest that all three fields underwent a significant enhancement in the star-formation rate—by at least a factor of 10—some 3–5 Gyr ago. Though the precise age of this "burst" epoch depends on the models, the conclusion that the bulk of star formation commenced contemporaneously in all three fields is practically model independent.

Subsequent studies have broadly confirmed the Bertelli et al (1992) results, with some interesting differences. Westerlund et al (1995; see Linde et al 1995 for the input data) have studied two LMC fields, one each in the NW and SW regions of the galaxy. They find in both fields that the oldest major population corresponds to stars with ages of 1–3 Gyr. The ages of the youngest stars in the

Table 2 Deep main-sequence field-star studies in the Magellanic Clouds[†]

Field	α_{2000}	δ_{2000}	Galaxy	V^{a}_{max}	Area (arcmin^2)	N^{b}_{f}	N^{c}_{*}	Age range (Gyr)
B77	05 14	− 65 46	LMC	23.0	0.6 (2.5)[d]	1	120	<3–5
H84	05 09	− 68 53	LMC	21.3	72	1	18000	<3
DH84	01 08	− 72 34	SMC	21.4	33	3	~ 2000	0.1→3
S84	06 30	− 64 04	LMC	22.8	540	11	3700	1–6
A85	05 19	− 70 57	LMC	23.0	3.2	2	1150	0.1→1
B87	01 03	− 70 51	SMC	23.8	15	1	~800	0.2–8
H87	04 33	− 72 14	LMC	22.0	400	1	350	2–3
B92-N1783	04 58	− 65 58	LMC	23.5	15	1	2030	0.5–4.5
B92-N1866	05 13	− 65 26	LMC	23.5	15	1	2370	0.5–3.8
B92-N2115	05 58	− 65 28	LMC	23.8	15	1	1780	1.3–4
E94-F1	05 39	− 68 54	LMC	24.1	7.3	1	3000	0.7–5 + anc?[e]
E94-F2	05 35	− 69 10	LMC	24.1	7.3	1	4200	0.7–5 + anc?
W95-NW	05 03	− 65 52	LMC	22.9	15.5	4	2190	0.2–3
W95-SW	05 48	− 73 32	LMC	23.0	11.6	3	940	1–3

[a] V_{max} is the limiting V-band magnitude of the survey.
[b] N_f is the number of fields studied; in some cases different fields come from the same imaging data, but were measured and analyzed separately (e.g. S84, A85).
[c] N_* is the approximate number of stars observed; this could only be very roughly estimated in the case of HD84.
[d] Two field sizes are listed for B77; the smaller value refers to the field size in which stars as faint as $V \sim 23$ were measured.
[e] E94 claim detection of a possibly ancient field-star population, denoted "anc."
[†] This table lists only studies aimed principally to study intermediate-age MC field-star populations. Other field-star results can be found in the references listed in Section 3.3 of the text. Projects in progress are not listed here. Stryker (1984b) gives a summary of additional MC field-star studies. References: B77 = Butcher 1977; H84 = Hardy et al 1984; HD84 = Hardy & Durand 1984; S84 = Stryker 1984a; A85 = Ardeberg et al 1985; B87 = Bolte 1987; H87 = Hodge 1987; B92 = Bertelli et al 1992; E94 = Elson et al 1994; W95 = Westerlund et al 1995 (data for this study are described in Linde et al 1995).

two fields appear to differ significantly. In the NW field, stars as young as 0.3–0.8 Gyr are present, whereas the youngest population in the SW field appears to be slightly older than 1 Gyr. This is surprising because both are located beyond 5° from the LMC center, far from the well-known active star-forming sites in the LMC. Westerlund et al (1995) claim also to see evidence for a 7–10 Gyr population in their SW field, but the depth of their photometry and their sample size is probably only marginally adequate to constrain the presence of stars this old. This interesting result demands confirmation with deeper photometry. Vallenari et al (1994a,b) report preliminary results of a study patterned after that of Bertelli et al (1992) for several fields scattered throughout the LMC. They report that the age of the oldest significant population in one of the fields is 7–8 Gyr. Some of their data also seem to show evidence of distinct intermediate-age bursts in one of their fields, not seen in other fields.

No comparably deep photometric studies have been carried out in the SMC. The most comprehensive effort (Gardiner & Hatzidimitrou 1992) only probes main-sequence stars younger than about 2 Gyr. Gardiner & Hatzidimitrou conclude from these data that the SMC star-formation rate seems to be globally declining over the past 2 Gyr, with the obvious exceptions of active star-forming regions in the inner SMC and in the SMC "wing," which contain stars as young as 200 Myr. Deep CCD studies of the distribution of ages of field stars in the SMC are badly needed. Some care should be taken in selecting SMC fields. The very large line-of-sight depth of the galaxy (particularly on its eastern side; Hatzidimitrou et al 1989, Gardiner & Hawkins 1991) is troublesome. It would be very difficult to disentangle depth and age effects without independent kinematic data (Hatzidimitrou et al 1993). To avoid this problem, deep population studies should focus on the western portions of the SMC at present.

3.4 *Intermediate-Age Clusters*

3.4.1 THE AGE DISTRIBUTION It is now widely accepted that the distributions of cluster ages differ dramatically between the two Clouds (e.g. Olszewski 1993, Da Costa 1993, Feast 1995; see Figure 1). For example, we know of (almost) no LMC star clusters with precise age estimates obtained from deep, main-sequence photometry in the age range 4–12 Gyr. The one exception—ESO121SC-03—has an age of approximately 6–9 Gyr (Mateo et al 1986, Sarajedini et al 1995). In contrast, the SMC contains many clusters with ages between 4 and 12 Gyr: L 113 (age 4 Gyr; Mould et al 1984), K 3 (age 7 Gyr; Rich et al 1984), L 1 (age 9 Gyr; Olszewski et al 1987), NGC 121 (age 12 Gyr; Stryker et al 1985) all have ages that are either within or close to the LMC age gap. This suggests that ESO121SC-03 may have originally formed in the SMC and was swapped to the LMC (see also Lin & Richer 1992).

This dichotomy in cluster ages is not due to evolutionary fading of clusters. Hodge (1988a) has shown that the typical cluster destruction timescale is about 5–10 times longer in the LMC than in the Galaxy. In the LMC, there are numerous examples of 1–4 Gyr clusters that will remain prominent objects when they are older than 10 Gyr (Mateo 1993), including NGC 1978 (current age 2 Gyr; Bomans et al 1995), NGC 2155 (current age 3.5 Gyr; Mateo 1996b), and NGC 1831 (current age 1 Gyr; Vallenari et al 1992). There is also little doubt that the LMC age gap is real and not merely some statistitical fluke. Da Costa (1991) reported the preliminary results of a deep CCD survey designed to target clusters in the age gap exhibited by the LMC, selecting the clusters on the basis of their integrated colors. Of all the candidate clusters, none had ages in the 4–12 Gyr gap. Likewise, Mateo (1996b) reports the results of a study of clusters over a wide luminosity range in a small northern region of the LMC (see also Mateo 1988); none of these clusters are older than 4 Gyr. NGC 2155, noted in Section 2.2, is in fact one of the oldest intermediate-age clusters known. Olszewski et al (1991) note that the age gap is also a metallicity gap: The old clusters are metal poor; the younger ones are metal rich. This also strongly suggests a physical origin for the gap.

Girardi et al (1995) and Girardi & Bica (1993) have recently analyzed the integrated *UBV* photometric properties of a sample of over 624 newly observed LMC clusters (Bica et al 1992). Even though they span a range of over 10 Gyr in age, the intermediate-age and old clusters in this sample span a range of only 0.25 mag in $(B-V)$ and $(U-B)$ centered at approximately $(B-V) = 0.80$, $(U-B) = 0.90$. Clusters younger than 1 Gyr span a range of 0.8 and 1.1 mag in $(B-V)$ and $(U-B)$, respectively. The color degeneracy of intermediate-age and old MC clusters is also apparent in broad-band UV and IR colors (Cassatella et al 1987, Meurer et al 1990, Testa 1994). Broad-band photometry can lead to a distorted picture of the distribution of cluster ages (e.g. Elson & Fall 1988). This is not due to any sort of intrinsic error in the photometry or because of shortcomings in the evolutionary models used to interpret the photometry. Rather, the age resolution of broad-band colors is poorest precisely in the age range of greatest interest for intermediate-age and old MC populations (Girardi et al 1995). The only way to study the detailed differences in the distribution of ages of intermediate-age and old MC clusters is to obtain precise main-sequence stellar photometry.

Existing data clearly suggest that the LMC has formed clusters in at least two distinct bursts (4 Gyr and younger plus the ancient clusters), whereas the SMC has formed clusters more uniformly over the past 12 Gyr. Nevertheless, some additional work is desirable to strengthen and extend these conclusions. First, clusters are rarely selected for age determination on the basis of any clear

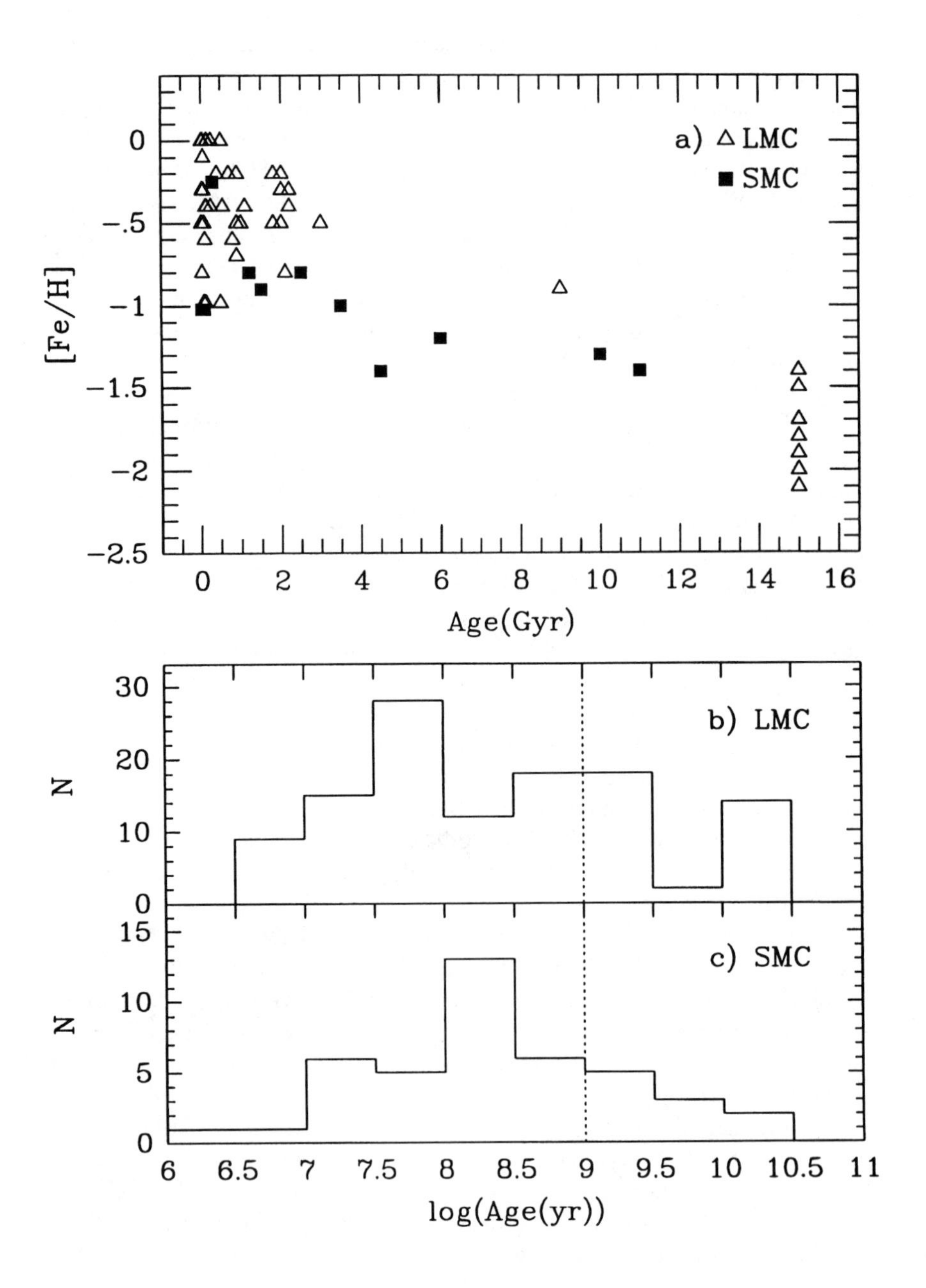

a) △ LMC
■ SMC
[Fe/H]
0
-.5
-1
-1.5
-2
-2.5
0
2
4
6
8
10
12
14
16
Age(Gyr)
b) LMC
N
30
20
10
0
c) SMC
N
15
10
5
0
6
6.5
7
7.5
8
8.5
9
9.5
10
10.5
11
log(Age(yr))

selection rules [one exception is the study of Da Costa (1992) mentioned above]. Before we can confidently rule out the existence of any 4–12 Gyr clusters in the LMC, reliable ages should be determined for a complete sample of clusters. The brute force approach to this problem—measuring main-sequence ages for all MC clusters—is still impractical at this time: The LMC contains about 4200 clusters and the SMC about 2000 (Hodge 1986, 1988b). Nonetheless, fewer than 2% of all MC clusters currently have reliable age determinations—a figure that can certainly be significantly improved upon. Some recent attempts at complete studies (e.g. Kubiak 1990a,b; Westerlund et al 1995; Mateo 1988, 1996b) have sampled only very small numbers of clusters and have been limited, so far, to the LMC. Such complete studies would also help to constrain the destruction rate of MC clusters and to obtain a better estimate of the absolute rate of cluster formation (cf Hodge 1988a).

A second difficulty in interpreting the age distribution of clusters is that in many cases one must combine age determinations based on very different assumptions about the relevant physical parameters (e.g. distance, reddening, metallicity) or based on comparisons with fundamentally different evolutionary models. For clusters younger than about 1–2 Gyr, modern models yield ages that can differ by a factor of up to two depending simply on how they treat convection at the core/envelope interface of intermediate-mass main-sequence stars (Mateo & Hodge 1987, Seggewiss & Richtler 1989). For the intermediate-age clusters discussed here, this effect is not too serious. But other effects are. Bomans et al (1995) have reanalyzed a number of CCD color-magnitude diagrams using a consistent set of evolutionary models and adopted astrophysical parameters. Although their new age determinations do not alter the basic conclusion above (no clusters are known older than about 2.5 Gyr in their reanalyzed LMC cluster sample) the ages of some of the clusters did change significantly from

←

Figure 1 (*a*) A plot of [Fe/H] vs age (in Gyr) for Magellanic Cloud clusters with precise metallicity and age determinations. The open triangles denote LMC clusters, while the filled squares denote SMC clusters. Results for 60 LMC and 11 SMC clusters are plotted. Following Da Costa (1992), we have plotted the age scale in *linear* units to emphasize the duration of the age gap in the LMC; no clusters seemed to have formed in that galaxy over $< 50\%$ of its lifetime. Note that the present-day metallicity of LMC clusters is considerably larger in the mean than for SMC clusters, whereas the metallicites were more similar 15 Gyr ago. (*b*) The age distribution histogram for LMC star clusters, now plotted as a function of log (Age). More clusters are represented in this panel than in panel (*a*) because of the considerable numbers of clusters with reasonable age estimates but only very crude metallicity determinations. (*c*) The same as panel (*b*), except for the SMC. In comparing panels (*b*) and (*c*), note that the intermediate-age and old populations reside to the right of the dotted lines. It is in this age range that the cluster age distributions show strikingly dissimilar behavior.

previously published values. Again, this does not reflect any sort of "error" in the initial studies, but rather the change in the age scale as one adopts different evolutionary models or astrophysical parameters. Until such time as all isochrones yield identical ages for a given cluster, it might be wise to establish a standard age scale based on a single set of models and parameters so that the relative ages of MC clusters can be compared with ease and confidence.

3.4.2 THE AGE-METALLICITY RELATION FOR STAR CLUSTERS Olszewski et al (1991) obtained abundances of approximately 80 LMC star clusters. This study represents by far the most extensive set of metallicities for intermediate-age clusters on a common scale, enlarging greatly the number of spectroscopic abundance determinations for these clusters (e.g. Cohen 1982, Cowley & Harwick 1982). Very few SMC clusters have spectroscopically determined abundances; their metallicities are generally derived from isochrone fits to the clusters' color-magnitude diagrams or from applications of empirical relations between the metallicity and the color of the RGB (e.g. Zinn & West 1984). Figure 1 illustrates the age-metallicity relations for clusters in both Magellanic Clouds. The data come from the compilations of Sagar & Panday (1989) and Seggewiss & Richtler (1989), supplemented by more recent results. We wish to highlight two important differences between the Clouds that are apparent from this Figure.

First, the mean LMC cluster metallicity jumps by a factor of about 40 during a time when virtually no clusters or—as we saw above—field stars were forming. Gilmore & Wyse (1991) and Köppen (1993) have shown that the chemical evolution of a galaxy characterized by a star-formation history punctuated by bursts can be very complex. In the case of the LMC, Gilmore & Wyse (1991) would argue that type I supernovae from the initial population steadily polluted the interstellar medium (ISM) during the long hiatus in star formation. When stars finally did begin to form again some 4 Gyr ago, the ISM was greatly enriched. One important implication of this idea—that O and other α elements should be enriched relative to Fe compared to their present day abundances—has already been described in Section 2.7.

But can this idea account for the very large observed increase in metallicity (Figure 1) during the age gap? A simple model in which only 2% of the LMC's dynamical mass formed in a burst more than 10 Gyr ago cannot account for this huge increase in metallicity. Supernova (type I and II) yields from Weaver & Woosley (1993) and Thielemann et al (1986) imply that the ancient LMC population could have only raised the mean iron abundance of the entire galaxy to $[Fe/H] = -1$ or less, far short of the value ~ -0.4 seen in 2–3 Gyr clusters. We do not seem to see sufficient numbers of field stars and clusters with ages between 4–12 Gyr to explain this discrepancy, unless the SN rates was much higher in the old MC populations than predicted by these models.

The second noteworthy feature of Figure 1 is that the two Magellanic Clouds have experienced quite distinct chemical enrichment histories (Da Costa 1992, Olszewski 1993, Feast 1995). This is clearly apparent in the shape of the age-metallicity relations, but also in the starting and ending values. Today the LMC is more metal rich than the SMC; 12 Gyr ago, these two galaxies may have had very similar abundances; it is not clear what the SMC abundance was 15 Gyr ago. In the case of the SMC, heavy-element enrichment (presumably from Type I SNe) may have lagged the star formation by a few Gyr. Currently, we see the metallicity rising as heavy elements are injected into the ISM. The lower mean abundance in the SMC simply reflects the fact that it has converted a smaller fraction of its gas to stars in accordance with a simple closed-box model of chemical evolution, and consistent with the present-day gas fractions of the two galaxies $(M_{\rm HI}/M_{\rm total}) \sim 0.35$ and 0.03 for the SMC and LMC, respectively.

Of course, all the arguments above assume that the MC clusters reliably track the chemical evolution of the LMC and SMC as a whole. Richtler (1993) summarizes the empirical and theoretical evidence for why this may not be the case.

3.4.3 KINEMATICS OF INTERMEDIATE-AGE CLUSTERS Freeman et al (1983) and Schommer et al (1992) obtained velocities for large numbers of intermediate-age clusters along with the results for ancient clusters described in Section 1. The intermediate-age clusters exhibit kinematics that are quite similar to those of young clusters and intermediate-age LPVs. Unlike the ancient clusters for which a disk rotation solution and small velocity dispersion was unexpected, the intermediate-age clusters appear to have kinematics that are "normal" for their ages. The orientation of the intermediate-age disk and the disk defined by the ancient clusters are statistically indistinguishable.

Feast (1995) has suggested that the commencement of the active star-formation epoch in the LMC some 4 Gyr ago may have coincided with the final collapse of the disk component of the LMC. The smooth transition of cluster kinematical properties is hard to understand in this scenario. Moreover, this idea does not offer any explanation why the old clusters also show disk kinematics: The LMC disk seems to have already been present when the oldest clusters formed.

3.5 *What Triggers Star Formation in the Magellanic Clouds?*

The Magellanic Clouds show extensive evidence of mutual interaction. The extreme depth of the SMC (Caldwell & Coulson 1986, Mathewson et al 1986, 1988, Gardiner & Hawkins 1991), the inter-cloud bridge, and the Magellanic Stream all testify that the Clouds have had a tumultuous recent interaction history. Lin & Richer (1992) and Muzzio (1988) argue that some MC clusters may have been lost to the Galactic halo on longer time scales. We have noted above (Section 3.4) that the unusual LMC cluster ESO121SC-03 may have formed originally in the SMC.

There is growing confidence that many of these global properties can be understood as originating in the tidal interactions of the Clouds with each other and with the Milky Way. Numerous workers (Murai & Fujimoto 1980, Lin & Lynden-Bell 1982, Shuter 1992, Gardiner et al 1994, Moore & Davis 1994, Heller and Rohlfs 1994, Lin et al 1995) have produced increasingly elaborate dynamical models that successfully account for many of the basic properties of the Magellanic Stream and the inter-cloud bridge and also properly account for the space velocity of the LMC (Kroupa et al 1994, Jones et al 1994). We have also seen in this review that the star-formation history of the Clouds are complex and distinct. Could tidal encounters between the MCs have provided the triggering mechanism needed to initiate this complex star-formation history? Interestingly, de Vaucouleurs & Freeman (1973) and Freeman (1984) pointed out that Magellanic systems tend to come in pairs, as if tidal interactions may be *necessary* to form galaxies similar to the Clouds.

All dynamical models of the LMC/SMC system agree that the two galaxies had a particularly close encounter within the past 100–500 Myr (Gardiner et al 1994, Moore & Davis 1994, Heller et al 1994, Lin et al 1995). For example, Gardiner et al (1994) note that the minimum separation of the Clouds was about 5 kpc compared to the 18 kpc separation we see today. Moreover, the best recent models also demand that in order to account for many of the present-day features of the Magellanic system and the space motions of the Clouds, the two galaxies must form a long-lived (> 10 Gyr) binary pair. Thus, a tidal trigger for the complex star-formation histories of the Clouds is plausible since they probably have tidally affected each other all their lives. One particular model of Gardiner et al (1994) shows that the LMC/SMC separation has had two particuarly close encounters during the past 15 Gyr: the recent encounter 200 Myr ago, and a similarly close mutual passage 4 Gyr ago. The latter time corresponds very closely to when the LMC recommenced forming stars and clusters after its long hibernation (see Sections 3.3 and 3.4; Figure 1).

Before we can conclude that the star-formation histories of the Clouds are driven by tidal triggers, we must address at least three basic problems. First, the *preferred* model of Gardiner et al (1994) does not show any close encounters prior to the one 200 Myr ago, and in fact the Clouds have maintained a nearly constant separation prior to that time. Second, we must somehow account for the fact that the LMC and SMC have such different star-formation histories. One possibility is that the tidal effects of the LMC on the SMC have been sufficient to maintain an enhanced star-formation rate in the smaller galaxy, while only very close encounters of the SMC can trigger star formation in the more massive LMC. Finally, we lack a solid physical model that can translate tidal interaction into star-formation activity.

Recent *HST* imaging of actively interacting galaxies has enabled the identification of vigorous star and cluster formation (Holtzman et al 1992, Whitmore & Schweizer 1995). These results show that encounters between massive galaxies can and do induce vigorous star and cluster formation. Can the much milder encounters between small galaxies such as the MCs also trigger activity? The answer is important not only to understand the Clouds, but also to explain the complex star-formation histories of other satellites in the outer halo and the evolution of faint blue galaxies seen in deep surveys.

4. WISH LIST

We conclude this paper by listing observations we hope will be made and papers we hope will be written on the old and intermediate-age populations in the Magellanic Clouds:

1. Deep, wide-field field star photometry in both Clouds.
2. Complete samples of ages and abundances of clusters in well-defined areas of the Clouds.
3. Precise ages of the oldest Cloud clusters.
4. Definitive evidence for or against a blue HB in the Cloud fields.
5. Red Clump magnitude distributions in the central and western portions of the SMC, keeping in mind that the northeast and east regions of the SMC have a large line-of-sight depth.
6. Magnitude and structure of the intermediate-age red clump and of the old red HB in the LMC.
7. Identification and spectroscopy of radially distant mass tracers in the LMC.
8. Detailed abundances, especially [α/Fe] for the ancient population and for the stars and clusters in the LMC burst that began 4 Gyr ago.
9. Metallicity distribution of clump giants.
10. Star-formation rates in both Clouds, based on deep field star photometry.

11. Abundances of a large sample of SMC clusters.
12. Identification of metal-rich RR Lyraes in the LMC and the abundance-luminosity relation for RR Lyraes.
13. An RR Lyrae survey in the SMC.
14. A firm photometric footing for the previous RR Lyrae work.
15. Age of the youngest RR Lyraes.
16. Resolution of the RR Lyrae and Cepheid distance scales in the Clouds.
17. Velocity dispersion of RR Lyrae stars and extremely metal poor giants in the LMC.
18. Catalogs of anomalous Cepheids in the Clouds.
19. Identification of long-period variables in the Cloud clusters.
20. Proper motions of the globular clusters thought to once be Cloud members, but are now Milky Way members.
21. Modeling of dynamical changes in Cloud clusters if they were removed from the Clouds and put into the Milky Way halo.
22. Continued detailed modeling of the Cloud/Galaxy interactions, with an eye to providing times of strongest interactions.
23. Search for tidal signatures in field star components of the Clouds and the Magellanic Stream.
24. Improved proper motions of the Clouds.

ACKNOWLEDGMENTS

EO was partly supported by the NSF through grants AST92-10830 and AST92-23967. NBS gratefully acknowledges the hospitality of the Dominion Astrophysical Observatory where part of this review was written. MM was partly supported by the NSF through grants AST91-18086 and AST92-23968.

Literature Cited

Aaronson M, Mould J. 1985. *Ap. J.* 288:551

Alcock C, Akerloff CW, Allsman RA, Axelrod TS, Bennett DP et al. 1993. *Nature* 365:621

Alcock C, Allsman RA, Axelrod TS, Bennett DP, Cook KH, et al. 1996. *Astron. J.* 111:1146

Andersen J, Blecha A, Walker MF. 1984. *MNRAS* 211:695

Ardeberg A, Linde P, Lyngå G, Lindgren H. 1985. *Astron. Astrophys.* 148:263

Armandroff TE, Da Costa GS. 1991. *Astron. J.* 101:1329

Armandroff TE, Da Costa GS, Zinn R. 1992. *Astron. J.* 104:164

Arp H. 1958a. *Astron. J.* 63:273

Arp H. 1958b. *Astron. J.* 63:487

Barbuy B, Renzini A, eds. 1992. *The Stellar Populations of Galaxies, IAU Symp. 149.* Dordrecht: Kluwer

Baschek B, Klare G, Lequeux J, eds. 1993. *New Aspects of Magellanic Cloud Research.* Berlin: Springer-Verlag

Battinelli P, Demers S. 1992. *Astron. J.* 104:1458

Beers TC, Preston GW, Shectman SA. 1992. *Astron. J.* 103:1987

Bencivenni D, Brocato E, Buonanno R, Castellani V. 1991. *Astron. J.* 102:137

Bertelli G, Bressan A, Chiosi C, Fagotto F, Nasi E. 1994. *Astron. Astrophys. Suppl.* 106:275

Bertelli G, Mateo M, Chiosi C, Bressan A. 1992. *Ap. J. Suppl.* 388:400

Bessell MS, Freeman KC, Wood PR. 1986. *Ap. J.* 310:710

Bhatia R, Piotto G. 1994. *Astron. Astrophys.* 283:424

Bica E, Claría JJ, Dottori H. 1992. *Astron. J.* 103:1859

Bica E, Claría JJ, Dottori H, Santos JFC, Piatti A. 1991. *Ap. J. Lett.* 381:L51

Bica E, Dottori H, Pastoriza MG. 1986. *Astron. Astrophys.* 156:261

Bingham EA, Cacciari C, Dickens, RJ, Fusi Pecci F. 1984. *MNRAS* 209:765

Binney J, Tremaine S. 1987. *Galactic Dynamics.* Princeton: Princeton Univ. Press

Blanco VM, McCarthy MF, SJ, Blanco BM. 1980. *Ap. J.* 242:938

Bolte M. 1987. *Ap. J.* 315:469

Bomans DJ, Vallenari A, de Boer KS. 1995. *Astron. Astrophys.* 298:427

Bothun GD, Thompson IB. 1988. *Astron. J.* 96:877

Brocato E, Buonanno R, Castellani V, Walker AR. 1989. *Ap. J. Suppl.* 71:25

Brück MT. 1980. *Astron. Astrophys.* 87:92

Buonanno R, Buscema G, Fusi Pecci F, Richer HB, Fahlman GG. 1990. *Astron. J.* 100:1811

Buonanno R, Corsi CE, Fusi Pecci F, Richer H, Fahlman GG. 1993. *Astron. J.* 105:184

Butcher HR. 1977. *Ap. J.* 216:372

Butler D, Demarque P, Smith HA. 1982. *Ap. J.* 257:592

Caldwell JAR, Coulson IM. 1986. *MNRAS* 218:223

Cannon RD. 1970. *MNRAS* 150:111

Carney BW, Storm J, Jones RV. 1992. *Ap. J.* 386:646

Cassatella A, Barbera J, Geyer EH. 1987. *Ap. J. Suppl.* 64:83

Chaboyer B, Sarajedeni A, Demarque P. 1992. *Ap. J.* 394:515

Cohen JG. 1982. *Ap. J.* 258:143

Colless M, Ellis RS, Broadhurst TJ, Taylor K, Peterson BA. 1993. *MNRAS* 261:19

Cowley AP, Hartwick FDA. 1982. *Ap. J.* 259:89

Cowley AP, Hartwick FDA. 1991. *Ap. J.* 373:80

Da Costa GS. 1991. See Haynes & Milne 1991, p. 183

Da Costa GS. 1992. See Barbuy & Renzini 1992, p. 181

Da Costa GS. 1993. See Smith & Brodie 1993, p. 363

Da Costa GS, Armandroff TE. 1995. *Astron. J.* 109:2533

De Vaucouleurs G, Freeman KC. 1973. *Vistas Astron.* 14:163

Demers S, Grondin L, Irwin MJ, Kunkel WE. 1991. *Astron. J.* 101:911

Demers S, Irwin MJ. 1991. *Astron. Astrophys. Suppl.* 91:171

Demers S, Irwin MJ, Kunkel WE. 1993. *MNRAS* 260:103

Djorgovski S. 1992. In *Cosmology and Large Scale Structures in the Universe, ASP Conf. Ser. 24*, ed. RR de Carvalho, p. 19. San Francisco: Astron. Soc. Pac.

Dopita MA, Ford HC, Lawrence C, Webster BL. 1985. *Ap. J.* 296:390

Dubath P, Mayor M, Meylan G. 1993. See Smith & Brodie 1993, p. 557

Elson RAW, Fall SM. 1985. *Ap. J.* 299:211

Elson RAW, Fall SM. 1988. *Astron. J.* 96:1383

Elson RAW, Forbes DA, Gilmore GF. 1994. *Publ. Astron. Soc. Pac.* 106:632

Feast MW. 1995. In *Stellar Populations, Proc. IAU Symp. 164,* ed. PC van der Kruit, G Gilmore, p. 153. Dordrecht: Kluwer

Feast MW, Walker AR. 1987. *Annu. Rev. Astron. Astrophys.* 25:345
Feast MW, Whitelock PA. 1992. *MNRAS* 259:6
Feast M, Whitelock P. 1994. *MNRAS* 269:737
Ferraro FR, Fusi Pecci F, Testa V, Greggio L, Corsi CE, et al. 1995. *MNRAS* 272:391
Freeman KC. 1984. In *Structure and Evolution of the Magellanic Clouds,* ed. S van den Bergh, K de Boer, p. 107. Dordrecht: Reidel
Freeman KC. 1996. In *Formation of the Galactic Halo...Inside and Out, ASP Conf. Ser.*, ed. HL Morrison, A Sarajedini, Vol. 92, p. 3. San Francisco: Astron. Soc. Pac.
Freeman KC, Illingworth G, Oemler A. 1983. *Ap. J.* 272:488
Frogel JA, Blanco VM. 1983. *Ap. J. Lett.* 274:L57
Frogel JA, Blanco VM. 1990. *Ap. J.* 365:168
Frogel JA, Elias JH. 1988. *Ap. J.* 324:823
Frogel JA, Mould J, Blanco VM. 1990. *Ap. J.* 352:96
Gardiner LT, Hatzidimitrou D. 1992. *MNRAS* 257:195
Gardiner LT, Hawkins MRS. 1991. *MNRAS* 251:174
Gardiner LT, Sawa T, Fujimoto M. 1994. *MNRAS* 266:567
Gilmore G, Wyse RFG. 1991. *Ap. J. Lett.* 367:L55
Gilmozzi R, Kinney EK, Ewald SP, Panagia N, Romaniello M. 1994. *Ap. J. Lett.* 435:L43
Girardi L, Bica E. 1993. *Astron. Astrophys.* 274:279
Girardi L, Chiosi C, Bertelli G, Bressan A. 1995. *Astron. Astrophys.* 298:87
Graham JA. 1975. *Publ. Astron. Soc. Pac.* 87:641
Graham JA. 1977. *Publ. Astron. Soc. Pac.* 89:425
Graham JA. 1988. In *Globular Cluster Systems in Galaxies, IAU Symp. 126,* ed. JE Grindlay, AGD Phillip, p. 151. Dordrecht: Kluwer
Graham JA, Nemec JM. 1984. In *Structure and Evolution of the Magellanic Clouds, IAU Symp. 108*, ed. S van den Bergh, KS de Boer, p. 37. Dordrecht: Reidel
Gratton RG, Ortolani S. 1987. *Astron. Astrophys. Suppl.* 71:131
Grondin L, Demers S, Kunkel WE. 1992. *Astron. J.* 103:1234
Grondin L, Demers S, Kunkel WE Irwin MJ. 1990. *Astron. J.* 100:663
Hardy E, Buonanno R, Corsi C, Janes K, Schommer R. 1984. *Ap. J.* 278:592
Hardy E, Durand D. 1984. *Ap. J.* 279:567
Hardy E, Suntzeff NB, Azzopardi M. 1989. *Ap. J.* 344:210
Hartwick FDA, Cowley AP. 1988. *Ap. J.* 334:135
Hatzidimitriou D. 1991. *MNRAS* 251:545
Hatzidimitriou D, Cannon RD, Hawkins MRS. 1993. *MNRAS* 261:873
Hatzidimitrou D, Hawkins MRS, Gyldenkerne K. 1989. *MNRAS* 241:645
Haynes R, Milne D, eds. 1991. *The Magellanic Clouds, IAU Symp. 148.* Dordrecht: Kluwer
Hazen ML, Nemec JM. 1992. *Astron. J.* 104:111
Heller P, Rohlfs K. 1994. *Astron. Astrophys.* 291:743
Hilker M, Richtler T, Gieren W. 1995. *Astron. Astrophys.* 294:648
Hirshfeld AW. 1980. *Ap. J.* 241:111
Hodge PW. 1960. *Ap. J.* 131:351
Hodge PW. 1983. *Ap. J.* 264:470
Hodge PW. 1986. *Publ. Astron. Soc. Pac.* 98:1113
Hodge PW. 1987. *Publ. Astron. Soc. Pac.* 99:730
Hodge PW. 1988a. *Publ. Astron. Soc. Pac.* 100:576
Hodge PW. 1988b. *Publ. Astron. Soc. Pac.* 100:1051
Hodge PW, Wright FW. 1967. *The Large Magellanic Cloud.* Washington, DC: Smithsonian Press
Hodge PW, Wright FW. 1977. *The Small Magellanic Cloud.* Seattle: Univ. Washington Press
Holtzman J, Faber SM, Shaya EJ, Lauer TR, Groth EJ et al. 1992. *Astron. J.* 103:691
Huchra J, Brodie J, Kent S. 1991. *Ap. J.* 370:495
Hughes SMG. 1989. *Astron. J.* 97:1634
Hughes SMG. 1993. In *New Perspectives on Stellar Pulsation and Pulsating Variable Stars,* ed. JM Nemec, JM Matthews, p. 192. Cambridge: Cambridge Univ. Press
Hughes SMG, Wood PR. 1990. *Astron. J.* 99:784
Hughes SMG, Wood PR, Reid N. 1990. *Publ. Astron. Soc. Aust.* 8:343
Hughes SMG, Wood PR, Reid N. 1991. *Astron. J.* 101:1304
Iben I Jr, Renzini A. 1983. *Annu. Rev. Astron. Astrophys.* 21:271
Jones BF, Klemola AR, Lin DNC. 1994. *Astron. J.* 107:1333
Kaluzny J, Krzeminski W, Mazur B. 1995. *Astron. J.* 110:2206
King IR. 1995. In *Stellar Populations, Proc. IAU Symp. 164,* ed. PC van der Kruit, G Gilmore, p. 337. Dordrecht: Kluwer
Kinman TD. 1959. *MNRAS* 119:538
Kinman TD, Stryker LL, Hesser JE, Graham JA, Walker AR, et al. 1991. *Publ. Astron. Soc. Pac.* 103:1279
Kontizas M, Morgan DH, Hatzidimitrou D, Kontizas E. 1990. *Astron. Astrophys. Suppl.* 84:527
Koo DC. 1986. In *Spectral Evolution of Galax-*

ies, ed. C Chiosi, A Renzini, p. 419. Dordrecht: Reidel
Köppen J. 1993. In *New Aspects of Magellanic Cloud Research,* ed. B Baschek, G Klare, J Lequeux, p. 372. Berlin: Springer-Verlag
Kraft RP. 1977. In *The Interaction of Variable Stars with Their Environment, IAU Colloq. 42,* ed. R Kippenhahn, J Rahe, W Strohmeir, p. 521. Bamberg: Remeis-Sternwarte
Kron, RG. 1980. *Ap. J. Suppl.* 43:305
Kroupa P, Röser S, Bastian U. 1994. *MNRAS* 266:412
Kubiak M. 1990a. *Acta Astron.* 40:349
Kubiak M. 1990b. *Acta Astron.* 50:355
Layden AC, Hawley SL, Hanson RB. 1995. *Astron. J.* in preparation
Lee Y, Demarque P. 1990. *Ap. J. Suppl.* 73:709
Lee Y-W, Demarque P, Zinn R. 1994. *Ap. J.* 423:248
Lin DNC, Jones BF, Klemola AR. 1995. *Ap. J.* 439:652
Lin DNC, Lynden-Bell D. 1982. *MNRAS* 198:707
Lin DNC, Richer HB. 1992. *Ap. J. Lett.* 388:L57
Linde P, Lyngå G, Westerlund BE. 1995. *Astron. Astrophys. Suppl.* 110:533
Mateo M. 1988. In *Globular Cluster Systems in Galaxies, IAU Symp. 126,* ed. JE Grindlay, AGD Phillip, p. 557. Dordrecht: Kluwer
Mateo M. 1992. See Barbuy & Renzini 1992, p. 147
Mateo M. 1993. See Smith & Brodie 1993, p. 387
Mateo M. 1996a. In *The Structure and Formation of the Galactic Halo,* ed. HL Morrison, A Sarajedini. San Francisco: *Astron. Soc. Pac.* 92:434
Mateo M. 1996b. in preparation
Mateo M, Fischer P, Krzeminski W. 1995. *Astron. J.* 110:2166
Mateo M, Hodge P. 1987. *Ap. J.* 320:626
Mateo M, Hodge P, Schommer RA. 1986. *Ap. J.* 311:113
Mathewson DS, Ford VL. 1984. In *Structure and Evolution of the Magellanic Clouds, IAL Symp. 108* ed. S van den Bergh, K de Boer, p. 125. Dordrecht: Reidel
Mathewson DS, Ford VL, Visvanathan N. 1986. *Ap. J.* 301:664
Mathewson DS, Ford VL, Visvanathan N. 1988. *Ap. J.* 333:617
Mayall NU. 1946. *Astron. J.* 104:290
Meatheringham SJ. 1991. See Haynes & Milne 1991, p. 89
Meatheringham SJ, Dopita MA, Ford HC, Webster BL. 1988. *Ap. J.* 327:651
Menzies JW, Whitelock PA. 1985. *MNRAS* 212:783
Meurer GR, Cacciari C, Freeman KC. 1990. *Astron. J.* 99:1124
Moore B, Davis M. 1994. *MNRAS* 270:209
Morgan WW. 1958. In *Stellar Populations,* ed. DJK O'Connell, SJ, p. 325. Vatican City: Vatican Obs.
Mould JR. 1992. See Barbuy & Renzini 1992, p. 181
Mould J. 1995. In *Stellar Populations, Proc. IAU Symp. 164,* ed. PC van der Kruit, G Gilmore, p. 349. Dordrecht: Kluwer
Mould JR, Da Costa GS, Crawford MD. 1984. *Ap. J.* 280:595
Murai T, Fujimoto M. 1980. *Publ. Astron. Soc. Jpn.* 32:581
Muzzio J. 1988. In *Globular Cluster Systems in Galaxies, IAU Symp. 126,* ed. JE Grindlay, AGD Phillip, p. 297. Dordrecht: Kluwer
Nemec JM, Harris HC. 1987. *Ap. J.* 316:172
Nemec JM, Hesser JE, Ugarte PP. 1985. *Ap. J. Suppl.* 57:287
Nemec JM, Nemec AFL, Lutz TE. 1994. *Astron. J.* 108:222
Olszewski EW. 1988. In *Globular Cluster Systems in Galaxies, IAU Symposium 126,* ed. JE Grindlay, AGD Phillip, p. 159. Dordrecht: Kluwer
Olszewski EW. 1993. See Smith & Brodie 1993, p. 351
Olszewski EW. 1995. In *Stellar Populations, Proc. IAU Symp. 164,* ed. PC van der Kruit, G Gilmore, p. 181. Dordrecht: Kluwer
Olszewski EW, Harris HC, Schommer RA, Canterna RW. 1988. *Astron. J.* 95:84
Olszewski EW, Schommer RA, Aaronson M. 1987. *Astron. J.* 93:565
Olszewski EW, Schommer RA, Suntzeff NB, Harris HC. 1991. *Astron. J.* 101:515
Reid N, Freedman W. 1994. *MNRAS* 267:821
Reid IN, Hughes SMG, Glass IS. 1995. *MNRAS* 275:331
Rich RM, Da Costa GS, Mould JR. 1984. *Ap. J.* 286:517
Richtler T. 1993. See Smith & Brodie 1993, p. 375
Rodgers AW, Paltoglou G. 1984. *Ap. J. Lett.* 283:L5
Sagar R, Panday AK. 1989. *Astron. Astrophys. Suppl.* 79:407
Sandage A. 1981. *Ap. J.* 248:161
Sandage A. 1990. *J.R. Astron. Soc. Can.* 84:70
Sandage AR, Cacciari C. 1990. *Ap. J.* 350:645
Sandage A, Diethelm R, Tammann GA. 1994. *Astron. Astrophys.* 283:111
Sarajedini A, Lee Y-W, Lee D-H. 1995. *Ap. J.* 450:712
Sawyer Hogg H. 1973. *Publ. DDO* 3:No. 6
Schechter PL, Dressler A. 1987. *Astron. J.* 94:563
Schommer RA. 1993. See Smith & Brodie 1993, p. 458

Schommer RA, Christian CA, Caldwell N, Bothun GD, Huchra J. 1991. *Astron. J.* 101:873
Schommer RA, Olszewski EW, Suntzeff NB, Harris HC. 1992. *Astron. J.* 103:447
Searle L, Wilkinson A, Bagnuolo WG. 1980. *Ap. J.* 239:803
Searle L, Zinn R. 1978. *Ap. J.* 225:357
Sebo KM, Wood PR. 1995. *Ap. J.* 449:164
Seggewiss W, Richtler T. 1989. In *Recent Developments in Magellanic Cloud Research,* ed. KS de Boer, F Spite, G Stasinska, p. 45. Paris: Obs. Paris
Shuter WLH. 1992. *Ap. J.* 386:101
Smecker-Hane TA, Stetson PB, Hesser JE, Lehnert MD. 1994. *Astron. J.* 108:507
Smith GH, Brodie JP, eds. 1993. *The Globular Cluster-Galaxy Connection, ASP Conf. Ser. 48.* San Francisco: Astron. Soc. Pac.
Smith HA, Silbermann NA, Baird SR, Graham JA. 1992. *Astron. J.* 104:1430
Smith HA, Stryker LL. 1986. *Astron. J.* 92:328
Storm J, Carney BW, Freedman WL, Madore BF. 1991. *Publ. Astron. Soc. Pac.* 103:661
Stryker LL. 1983. *Ap. J.* 266:82
Stryker LL. 1984a. *Ap. J. Suppl.* 55:127
Stryker LL. 1984b. In *Structure and Evolution of the Magellanic Clouds,* ed. S van den Bergh, K de Boer, p. 79. Dordrecht: Reidel
Stryker LL, Da Costa GS, Mould JR. 1985. *Ap. J.* 298:545
Suntzeff NB. 1992a. See Barbuy & Renzini, p. 23
Suntzeff NB. 1992b. In *Variable Stars and Galaxies,* ed. B Warner, Vol. 30, p. 161. San Francisco: Astron. Soc. Pac.
Suntzeff NB, Friel E, Klemola A, Kraft RP, Graham JA. 1986. *Astron. J.* 91:275
Suntzeff NB, Kinman TD, Kraft RP. 1991. *Ap. J.* 367:528
Suntzeff NB, Phillips MM, Elias JH, Cowley AP, Hartwick FDA, Bouchet P. 1993. *Publ. Astron. Soc. Pac.* 105:350
Suntzeff NB, Schommer RA, Olszewski EW, Walker AR. 1992. *Astron. J.* 104:1743
Taam RE, Kraft RP, Suntzeff NB. 1976. *Ap. J.* 207:201
Testa V. 1994. *Mem. Soc. Astron. Ital.* 65:761
Testa V, Ferraro FR, Brocato E, Castellani V. 1995. *MNRAS* 275:454
Thackeray AD, Wesselink AJ. 1953. *Nature* 171:693
Thielemann FK, Nomoto K, Yokoi K. 1986. *Astron. Astrophys.* 158:17
Tifft WG. 1962. *MNRAS* 125:199
Tyson JA. 1988. *Astron. J.* 96:1
Vallenari A, Chiosi C, Bertelli G, Meylan G, Ortolani S. 1992. *Astron. J.* 104:1100
Vallenari A, Aparicio A, Fagotto F, Chiosi C. 1994a. *Astron. Astrophys.* 284:424
Vallenari A, Bertelli G, Chiosi C. 1994b. *Mem. Soc. Astron. Ital.* 65:751
Vallenari A, Bertelli G, Chiosi C, Ortolani S. 1994c. *ESO Messenger* 76:30
Vallenari A, Chiosi C, Bertelli G, Meylan G, Ortolani S. 1992. *Astron. J.* 104:1100
van den Bergh S. 1981. *Astron. Astrophys. Suppl.* 46:79
van den Bergh S. 1993. *Ap. J.* 411:178
van den Bergh S. 1994. *Astron. J.* 108:2145
van den Bergh S. 1996. *Ap. J. Lett.* In press
Vassiliadis E, Metheringham SJ, Dopita MA. 1992. *Ap. J.* 394:489
Walker AR. 1989a. *Astron. J.* 98:2086
Walker AR. 1989b. *Publ. Astron. Soc. Pac.* 101:570
Walker AR. 1990. *Astron. J.* 100:1532
Walker AR. 1992a. *Astron. J.* 103:1166
Walker AR. 1992b. *Astron. J.* 104:1395
Walker AR. 1992c. *Ap. J. Lett.* 390:L81
Walker AR. 1993a. *Astron. J.* 105:527
Walker AR. 1993b. *Astron. J.* 106:999
Walker AR. 1995. *Astron. J.* 110:638
Walker AR, Mack P. 1988. *Astron. J.* 96:872
Weaver TA, Woosley SE. 1993. *Phys. Rep.* 227:65
Webbink RF. 1985. In *Dynamics of Star Clusters, IAU Symp. 113,* ed. J Goodman, P Hut, p. 541. Dordrecht: Reidel
Welch DL, Stetson PB. 1993. *Astron. J.* 105:1813
Westerlund BE. 1990. *Astron. Astrophys. Rev.* 2:29
Westerlund BE, Linde P, Lyngå G. 1995. *Astron. Astrophys.* 298:39
Whitmore BC, Schweizer F. 1995. *Astron. J.* 109:960
Wood PR, Bessell MS, Fox MW. 1983. *Ap. J.* 272:99
Wood PR, Bessell MS, Paltoglou G. 1985. *Ap. J.* 290:477
Yamashita Y. 1975. *Publ. Astron. Soc. Jpn.* 27:325
Zinn R. 1985. *Ap. J.* 293:424
Zinn R. 1993. See Smith & Brodie 1993, p. 38
Zinn R, King CR. 1982. *Ap. J.* 262:700
Zinn R, West MJ. 1984. *Ap. J. Suppl.* 55:45

Annu. Rev. Astron. Astrophys. 1996. 34:551–606

STELLAR PULSATIONS ACROSS THE HR DIAGRAM: Part II

Alfred Gautschy

Astronomisches Institut der Universität Basel, Venusstrasse 7, CH-4102 Binningen, Switzerland

Hideyuki Saio

Astronomical Institute, Tohoku University, 980 Sendai, Japan

KEY WORDS: stellar physics, stellar structure, stellar evolution, stellar variability

ABSTRACT

Stars over essentially the whole mass domain can become pulsationally unstable during various stages of their evolution. They will appear as variable stars with characteristics that are of much diagnostic value to astronomers. The analysis of such observations provides a challenging and unique approach to study aspects of the internal constitution and evolutionary status of these objects that are not accessible otherwise. This review touches on most classes of known pulsating variable stars and tries to elucidate connections to stellar physical aspects. To aid future investigations, we stress questions and problems that we believe are yet to be resolved satisfactorily.

1. INTRODUCTION

Cox (1975), in his introductory report at the 19th Liège conference, concluded that " ... overall, the [pulsation] theory and its application are in a fairly satisfactory state, except for a few disturbing problems. ..." Among these were the excitation mechanism for the β Cephei variables, the mass discrepancy for Cepheids, and the influence of physical mechanisms such as convection, rotation, and magnetic fields. Some of these problems remain with us to this day.

This article on stellar pulsations deals with the application of the theory presented in the first part of the review, published in Volume 33 of this series

0066-4146/96/0915-0551$08.00

(Gautschy & Saio 1995; GS95 in the following). We have tried to find a few distinct but conceptually connected paths across the Hertzsprung-Russell (HR) diagram along which the classes of pulsating variables can be linked to their evolutionary states. First, we deal with the pulsating stars on or near the main sequence. Next, we follow the evolution of low-mass stars and discuss the domains of pulsational instabilities that they encounter. Finally, we turn to massive stars and their variabilities and review the pulsation-based explanation attempts.

2. PULSATIONS CLOSE TO THE MAIN SEQUENCE

The traditional association with pulsating variables on or close to the main sequence has been confined either to β Cephei stars or pulsators at the intersection of the classical instability strip with the main sequence. This picture has had to be revised during the past decade. In the following subsections, we outline the present state of understanding of pulsating variables in the evolutionary phase of mainly core-hydrogen burning. Discussions of rapidly oscillating Ap (roAp) stars and more generally of the seismological aspects of solar-type stars are kept concise as there are numerous recent and comprehensive reviews available (Kurtz 1990, Gough & Toomre 1991, Matthews 1991, Shibahashi 1991, Brown & Gilliland 1994).

2.1 *Lower Main-Sequence Stars*

Our state of knowledge of pulsational instabilities in red dwarfs (main-sequence stars that are essentially fully convective and have masses below about 0.25 $M_{\odot}$) has not evolved far from what Cox (1974) described in his review. No new observational evidence indicating oscillations has emerged for low-mass stars on the main sequence. Gabriel (1969) performed quasi-adiabatic stability analyses on low-mass stellar models and found radial pulsation instabilities. The treatment of convection was, however, rather rudimentary. As there is no observational evidence yet for variability in such stars we might suspect that a fully nonadiabatic treatment and/or more sophisticated handling of convection might change the theoretical picture.

In the very low mass domain, the giant planet Jupiter is believed to oscillate nonradially (Deming et al 1989, Magalhães et al 1990, Mosser et al 1993). It is unclear whether this kind of behavior can also be expected in other low-mass objects.

In solar-type stars, observational data gathered to date does not prove the presence of oscillations unambiguously [see GONG92 (p. 599 below) for recent observational experiments]. The general theoretical belief is that solar-type stars should show oscillations with similar signatures as found in the sun itself.

Christensen-Dalsgaard & Frandsen (1983) estimated the amplitudes of such oscillations to be largest for stars with masses around 1.5 $M_{\odot}$. The photometric variability would remain in the μmag regime, and radial-velocity variations are expected to remain bounded by a few meters per second. Information from the many stochastically excited modes in such stars should provide constraints for the internal constitution similar to what low-ℓ modes do for the sun (Gough & Toomre 1991). A very promising approach to detecting solar-type oscillations was recently presented by Kjeldsen et al (1995). They observed changes of the equivalent widths of Balmer lines of η Boo with a low-dispersion spectrograph. The basic idea of their approach is that any solar-type oscillations cause temperature fluctuations that dominate over radial-velocity variability. The temperature fluctuations are monitored by the time variation of the equivalent width of selected spectral lines. Kjeldsen et al (1995) claimed to detect 13 distinct oscillation frequencies. The deduced frequency separations seem to agree reasonably well with theoretical expectations for a G0 IV star comparable with η Boo.

When solar-type oscillations are found in distant stars they will provide a new and accurate tool to determine fundamental stellar parameters. Basic asteroseismological aspects and their potential benefit for stellar astronomy were discussed recently in Brown & Gilliland (1994), Brown et al (1994), and in contributions in GONG92 and GONG94.

Near the intersection of the red edge of the classical instability strip with the main sequence—but usually at somewhat lower temperatures—a newly recognized class of variable stars is observed. Of the order of a dozen early F-type stars (Krisciunas & Handler 1995) exhibit photometric and radial-velocity variations with periods ranging from about 5 hours to roughly 2 days. The low-amplitude light variability of particular examples is described in Balona et al (1994a,b) (γ Dor), Krisciunas et al (1993) (9 Aur), and Lampens (1987) and Matthews (1990a) (HD 96008). The physical cause of the variability is unclear. The length of the periods of some of the F-type variables, those with rather high rotation velocities, could be compatible with rotationally induced phenomena such as starspots. Chromospheric activity is, however, generally restricted to spectral types later than about F7 (Radick et al 1983). Furthermore, the variable F-type stars themselves are not chromospherically active, and the persistence of the same periods over at least several years makes it difficult to attribute them to starspots. Multiperiodicity, as claimed for 9 Aur and γ Dor, suggests that indeed stellar oscillations are involved. The long periods observed can only be explained by high-order g-modes. Simple modeling shows that either the partial He II ionization zone or the rather weak Z-bump in the opacity data (cf GS95, Section 3.3.2) could potentially contribute to the excitation of such modes. But since very high overtones must be unstable, a large number of g-modes should be excited simultaneously if the classical κ-mechanism were the destabilizing agent. Observations, however, do not show the persistence of a large number of modes.

Close to the main sequence, within the boundaries of the classical instability strip, at masses of $\approx 2\ M_\odot$, lies the class of rapidly oscillating Ap (roAp) stars. Their light variation ranges from a few to roughly 50 mmag in the blue. The multiple periods are confined to between about 5 and 15 minutes. Many of the period multiplets with separations of a few μHz are believed to be rotational splittings; even when these components are removed, multiple periodicities often remain. The short periods of the observed pulsations must be due to high-order, low-degree p-modes. The oscillation amplitudes are modulated in accordance with the temporal variation of the stars' strong magnetic fields (some hundred to several thousand Gauss on the surface). The timescales of the modulation agree with the rotation periods of the stars. Mostly, roAp stars are slow rotators for their spectral type (rotation periods between 2 and 12 days); this is attributed to efficient braking due to their strong magnetic fields. The oscillations must be influenced by these strong fields. Phenomenologically, the basic pulsational behavior can be described successfully by the oblique pulsator model (Kurtz 1982). The pulsational axis is assumed to be aligned with the star's magnetic axis, which, by itself, has a nonvanishing obliquity relative to the rotation axis. For an external observer, such a geometrical setup leads to rotationally induced frequency splittings. For theoretical purposes, sophisticated perturbation schemes have been developed during the past few years. Using the relative amplitudes of frequency multiplets, they allow deductions about the alignments of the different axes on the star relative to the observer and estimates of averaged magnetic fields inside the star (Dziembowski & Goode 1985, 1986; Kurtz & Shibahashi 1986; Shibahashi & Takata 1993; Takata & Shibahashi 1995a). An alternative phenomenological model, which did not catch on so successfully, was the spotted pulsator model proposed by Mathys (1985). He assumed the pulsation axis to be aligned with the rotation axis. The oscillation patterns are hence always seen from the same aspect angle. In the spotted pulsator approach, the amplitude modulation results from the inhomogeneous distribution of the flux on the stellar surface, which is a function of the magnetic phase during a stellar rotation period.

The oblique pulsator model seems to more favorably account for the observed properties (Kurtz 1990). From a theoretical point of view, however, it is not clear how the oblique pulsation is maintained. The angular dependence of such an oblique pulsation of say $\ell = 1$ is expressed by a linear combination of spherical harmonics $Y_1^m(\theta, \phi)$ with $m = 0$ and $m = \pm 1$, where θ and ϕ are the spherical coordinates associated with the rotation axis. These components have frequencies $\sigma_m = \sigma_0 + m\Omega C_{n,1}$ in the corotating frame (cf GS95). To maintain the oblique pulsation, $C_{n,1}$ must vanish; otherwise the three components will have different frequencies in the corotating frame and the pulsation pattern would drift about the magnetic axis on a timescale of $1/C_{n,1}\Omega$. Nonmagnetic spherical stellar models predict $C_{n,1}$ to be of the order of 10^{-2} to 10^{-3} for high-order p-modes whereas observations, obtained from HR3881, lead to an upper limit of about 10^{-5} (Kurtz 1990). Dolez & Gough (1982) argued that it is difficult for the magnetic fields to suppress any drifting of pulsational patterns.

An important step towards identifying the instability mechanism is to map the roAp stars into a physically meaningful parameter space. Establishing a two-color diagram, for example, that clearly outlines and separates the instability region of roAp stars photometrically from stable stars has proven elusive (Matthews 1990b). The instability region of the roAp stars seems to at least partly overlap that of δ Sct variables (Section 2.2) on the HR diagram. The δ Sct pulsations are known to be driven by the partial He II ionization zone. In roAp stars, diffusion, which is believed to be responsible for the chemical anomalies (Michaud 1970, 1980), is thought to drain the He abundance in regions where He II ionization would otherwise be the driving mechanism. The excitation mechanism responsible for the oscillating Ap stars remains to be identified. Shibahashi (1983) and Cox (1984) contemplated magnetic overstable convection as a possible driving mechanism; in their picture, the restoring force results from the magnetic field, which resists being dragged along by convective motion. Matthews (1988), on the other hand, based on a local stability analysis, suggested that the κ-mechanism due to partial ionization of Si IV was the destabilizing agent. Vauclair & Dolez (1990) and Vauclair et al (1991) speculated that a very weak stellar wind (removing $\approx 10^{-14}$ $M_\odot$ year^{-1}) might transport enough He along magnetic field lines into the polar regions, where it settles and could excite oscillations. Detailed, self-consistent calculations for any of the above mechanisms are not available. Furthermore, observational evidence for prerequisites for the proposed mechanisms to work has not yet been collected. Not only does the excitation mechanism remain to be identified, but also, the physics of the mode selection remains unclear. Some of the roAp stars pulsate in multiple modes that are not consecutive radial orders. Many oscillation modes show very good long-term stability (at least over a decade), whereas others decay on the order of days.

As the oscillation frequencies are believed to be of high order, asymptotic mode analysis can be applied (see Tassoul 1980; Unno et al 1989, section 16). The often observed frequency differences (of the order of 20–80 μHz) between the multiple periods can be caused either by differences in radial order n of equal-ℓ modes or by alternating even and odd ℓ values. The first case allows one to measure the asymptotic quantity $\Delta\nu_0 \propto (M_*/R_*^3)^{1/2}$. The latter case provides a measure of $\Delta\nu_0/2$, however. Lines of equal $\Delta\nu_0$ on the HR diagram show that smaller values translate a star of known effective temperature to a later evolutionary phase and a higher mass (Shibahashi & Saio 1985, Gabriel et al 1985). A later evolutionary stage implies faster evolution as the stars would then be burning hydrogen already in a thick shell. Heller & Kawaler (1988) computed period changes that develop during and shortly after the main-sequence evolution. If roAp stars are indeed in their subgiant phase, evolutionary period changes should be detectable within a few years of monitoring. Martinez & Kurtz (1990) accumulated such data for HD 101065, a roAp star undergoing period changes, whose $dP/dt \equiv \dot{P}$ turns out to be about ten times larger than the maximum predicted by Heller & Kawaler (1988). Additionally, the observed sign of $\dot{P}$ is wrong compared with theory. Whether the observations are contaminated by binary motion is unclear. In any case, looking for evolutionary period changes promises to be a suitable method to solve the frequency-spacing dilemma.

Recently, Kurtz et al (1994) and Kurtz (1995) reported frequency variations in roAp stars on the timescale of $\sim$ 100 days. Such variations are considered now to be common among roAp stars. The signature is cyclic, and it is considered to be intrinsic to the pulsations and not of evolutionary origin. Kurtz (1995) attributed the frequency modulation to a changing magnetic field strength. Such a variation would modify the structure of the acoustic cavity and hence the magnitude of the eigenfrequencies. It is interesting that, based on our present understanding of magnetic fields in stars, magnetic cycles similar to the solar one are not expected for upper main-sequence stars.

2.2 *δ Scuti Stars*

Stars with masses of $1.5 \lesssim M/M_\odot \lesssim 2.5$ enter the lower instability strip (see Figure 1 of GS95) either in their core hydrogen-burning phase or when they evolve towards the base of the giant branch burning hydrogen in a shell. Such stars are commonly accepted as candidates for δ Sct-like oscillations. Very recently, some δ Sct variables were identified to be pre-main-sequence objects (Kurtz & Marang 1995). The pulsation periods of δ Sct variables range from about 0.02 to 0.25 days, indicating low-order radial or nonradial p-modes of low spherical degree. Some of the recent observational data hint at the presence of even g-modes (Breger et al 1995). We know of single/double mode δ Sct stars with large amplitudes (some tenths of a magnitude); the majority, however, are multiperiodic with small amplitudes (10^{-3}–10^{-2} mag). The lack of sufficiently long temporal baselines of observing runs prevented clear statements about the richness of the frequency spectrum in the past. Thanks to recent multisite campaigns we know now of multiply periodic δ Sct variables that show up to about a dozen oscillation modes (Belmonte et al 1993, Breger et al 1995). Results from Doppler imaging of rapidly rotating δ Sct stars indicate that high-degree modes ($12 \lesssim \ell \lesssim 16$) might also be excited (Kenelly et al 1992, Matthews 1993).

When considering the stars populating the lower instability strip, only a fraction ($\lesssim 50\%$) of them are observed to be photometrically variable, at least with amplitudes above the presently discernible limit of a few mmag. Because the number of known pulsators increases steeply towards low amplitudes (Baglin et al 1973, Breger 1979), we might suspect the putative stable stars to be pulsating with very low amplitudes. The presence of stable stars in the δ Sct variability domain, the different naming of allegedly differing subclasses of short-period pulsators, and the presence of both Population I and II objects in the lower instability strip make any survey rather confounding. Discussions of the classification issue from the observational viewpoint can be found in Eggen (1979), Breger (1979), Nemec & Mateo (1990), and references therein. For simplicity, we call variable stars with properties as mentioned above δ Sct variables—this is sufficient for the discussion of basic theoretical issues. On the observational side, examples of instructive reviews of δ Sct stars are those by Baglin et al (1973), Breger (1979), Wolff (1983), and Kurtz (1988) (who concentrated on those stars for which detailed frequency spectra were available).

The reviews by Petersen (1976) and Cox (1983) contain comprehensive accounts of early theoretical work on δ Sct variables, addressing in particular linear stability analyses and radial nonlinear simulations. Lee (1985b) reanalyzed the oscillatory stability of low-order nonradial modes of δ Sct-like stars in their subgiant phase. A generalization of Osaki's (1977) WKB treatment for the interior to a quasi-adiabatic treatment (Dziembowski 1977a, Lee 1985a) allowed the solution of the eigenvalue problem over the complete stellar models, including the condensed central parts of the star, which enforce short spatial wavelengths in the eigenfunctions. The very small local wavelength is caused by the large magnitude of the bump in the Brunt-Väisälä frequency in the deep interior (cf Figure 2 in GS95). This peak in the Brunt-Väisälä frequency close to the center lets the eigenmodes with δ Sct-like periods adopt dual character. They behave like p-modes in the envelope and turn locally into g-modes in the deep interior. If the evanescent zone (white area between the p- and g-mode propagation domains in the inlet of Figure 2 in GS95) between the two propagation regions is thin and the Brunt-Väisälä frequency in the inner regions is very large, then the deep interior constitutes a substantial energy sink. Dissipation in the deep interior will possibly stabilize p-modes that are driven in the He II ionization zone by the κ-mechanism. Nonetheless, Lee (1985b) found a large number of unstable modes in his subgiant model (see Figure 2 in Lee 1985b), a result very much like that of the earlier studies dealing with main-sequence-type stars (e.g. Dziembowski 1977a). The number of destabilized low-degree modes in these linear stability analyses is much higher than the number of modes presently observed. Dziembowski & Królikowska (1990) speculated that trapping of a fraction of the eigenmodes in the envelope could serve as a selection mechanism. Trapped modes with their reduced amplitudes in the deep interior and hence with reduced dissipation in that region would be favored to reach observable amplitudes. For low-degree modes, i.e. modes that are in principle photometrically detectable, the trapping efficiency was most pronounced for $\ell = 1$.

The amplitude-limiting process acting in δ Sct stars and other low-amplitude variables is thought to be different from the one that is relevant in large-amplitude variables like Cepheids or RR Lyrae stars. The observed frequency spectra of low-amplitude pulsators differ significantly from what simple linear pulsation theory predicts. Only a fraction of the linearly unstable modes reach an observable level. Besides the selection by trapping as mentioned above, nonlinear mode interactions between two or three modes, where one of them is linearly unstable, are proposed as efficient amplitude limiters (Dziembowski 1982). In a later investigation (Buchler & Goupil 1984), nonadiabaticity and nonlinearities in the growth rates were also accounted for by stepping up to second- and third-order nonlinearities. In case of δ Sct-like stars, the coupling of low-order, low-degree p-modes with pairs of g-modes limited amplitudes efficiently (Dziembowski & Królikowska 1985). Pairs of g-modes can be parametrically excited if the acoustic mode amplitudes exceed a threshold. This opens the possibility for a cascade-like spread of coupling of the now unstable g-modes with initially stable higher-order pairs of g-modes. For the case of three-mode interaction, the final amplitudes turn out to be too large compared with observation. By introducing rotation and associated frequency splitting, the chances for additional resonances, and hence for further reduction of the amplitudes, are increased (Dziembowksi et al 1988). Three-mode interactions result in steady amplitude solutions only in exceptional cases. More frequently, amplitude modulations in time are predicted (Moskalik 1985). Constraints for the theory of nonlinear mode coupling might hence come from observed long-term amplitude variations of oscillation modes in δ Sct stars.

The seeming coexistence of variable and stable stars in the same region of the HR diagram was attributed to the action of gravitational settling of helium and levitation of suitable metallic elements by radiation pressure in the stellar envelopes (Baglin 1972, 1976). This works only for slow pulsators since rapid rotation tends to destroy this effect. According to Kurtz (1989), diffusion and radiative levitation might indeed account for the mostly—but not exclusively—stable spectroscopically peculiar Am stars that are encountered in the lower instability strip. Physically, the disappearance or weakening of the He II ionization zone stabilizes the stars. Nonetheless, even for significantly He-depleted envelopes Cox et al (1979) found pulsational instabilities due to residual He II and some enhanced driving in the partial H I ionization zone close to the red edge of the classical instability strip. Not only can peculiar A and F stars be pulsationally stable in the δ Sct region, but also nonvariable stars with "normal" spectra were identified (Breger 1979). Because many of the stable candidates might be pulsating with low amplitudes, high-precision photometry needs to confirm the advocated ratio of stable to variable stars in the lower instability strip. The recent discovery of large-amplitude δ Scuti-like pulsation in a Am star casts doubt on the levitation mechanism being responsible for the abundance anomaly (Kurtz et al 1995).

For many years, it proved impossible to fit simultaneously the values observed for δ Sct and Cepheid variables to the theoretical Petersen diagrams(P_1/P_0 versus P_0) when standard stellar-evolution assumptions were invoked in the modeling. For such analyses, the periods of the radial fundamental mode P_0 and of the first overtone P_1 are usually calculated in the adiabatic approximation. Andreasen et al (1983) realized that the period ratios depend sensitively on the He content of the stellar envelopes. Indeed, noncanonically high values for He were suggested not only to accommodate unusually high period ratios (e.g. VZ Cnc, Cox et al 1984) but also to account for the theoretical, radial blue edge of δ Sct stars, which was considered too hot compared with observations. When Andreasen (1988) artificially increased the opacity at a few times 10^5 K he was able to reconcile observation and theory in the Petersen diagrams for double-mode δ Sct stars and for Cepheids. The new generation of opacity tables (OPAL and OP) fully support Andreasen's findings as was shown in preliminary calculations by Christensen-Dalsgaard (1993) (see Figure 1).

The location of the blue edge for radial instabilities in δ Sct-like models does not shift considerably when invoking the new opacities (Li & Stix 1994). The particular choice of the outer boundary in the computations seems, however, to influence the stability properties considerably. If the boundary conditions are formulated close to the photosphere, the eigenfunctions are sufficiently quenched in the H I/He I partial ionization zone to render it ineffective. When extending the eigensolutions further into the atmosphere, the H I/He I ionization zone is assigned more weight and it contributes to the driving so that the blue edge is shifted towards higher temperatures and fits the observations better.

The increase of the Brunt-Väisälä frequency (cf Figure 2 in GS95) that occurs in the central region when the star leaves the main sequence forces the

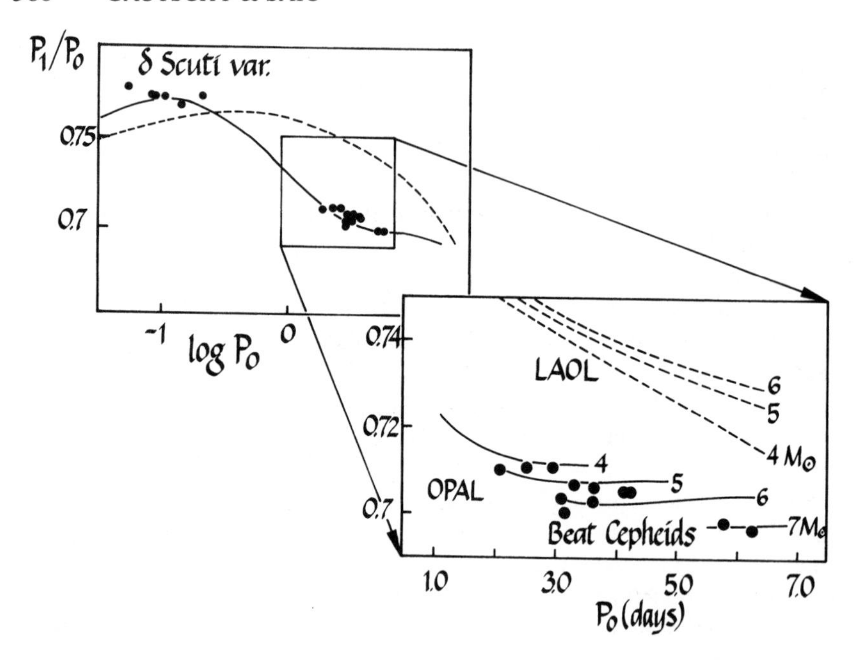

Figure 1 Petersen diagram for population I pulsators. The period ratios of the first to the fundamental mode are plotted against the logarithm of the fundamental mode in the upper diagram to accommodate δ Sct as well as beat Cepheids. The solid curve shows the convincing agreement resulting from calculations with new opacity data (OPAL). The broken curve displays the long-standing discrepancy due to the Los Alamos (LAOL) opacity data. The inset in the lower right allows a more detailed view of the beat Cepheid results, in particular the dependence of the period ratios on the stellar mass.

eigenfrequencies of some g-modes to penetrate the frequency domain of the p-modes. Assuming that those modes experiencing an avoided crossing (see GS95, Section 3.2.3) at some evolutionary phase—when a g-and a p-mode happen to be close to each other—are excited and also are observable, then we can potentially probe the very deep interior of such an evolved star. At the inner boundary of the radiative region, just outside the convective nuclear-burning core, the form of the Brunt-Väisälä frequency is governed by the detailed structure of the core/envelope (convection/radiation) transition. Dziembowski & Pamyatnykh (1991) proposed frequencies and irregularities in frequency separations of the core g-modes of δ Sct variables that proved useful for assessing their evolutionary state and unraveling properties of overshooting from the convective core.

Before structural and/or evolutionary aspects can be confidently constrained with δ Scuti variability, i.e. before asteroseismology is possible, the observed modes must be reliably identified. This is far from easy because the typical δ Scuti power spectrum is rather sparse. Presently, stellar evolution modeling is accompanying the mode identification to constrain the interpretations. Improvements can be expected from multicolor or simultaneous spectroscopic observations. Different degrees and orders of modes cause different signatures in color indices and in line variability that can contribute to positive identifications. Another promising approach is to search for δ Sct variables in clusters so that ensemble analyses can be performed. Common quantities that enter the modeling, such as age and maybe convection parameterization, can be determined for the whole ensemble of variables (cf Brown & Gilliland 1994).

Blue stragglers are stars whose positions in the HR diagram render them too young compared with the age of the agglomeration (often globular and old open clusters) with which they are associated. For a recent review of these objects see Stryker (1993) and Bailyn (1995) and references therein. Some of the blue stragglers happen to fall into the instability strip and exhibit δ Sct-like variability. Mateo (1993) reviewed the observational status of variable blue stragglers in old stellar systems (see also Nemec & Mateo 1990). The 24 stars Mateo (1993) mentioned are all attributed a single pulsation frequency, and they all seem to have rather large amplitudes, which is likely to be an observational bias. Gilliland et al (1991) discovered two multiperiodic blue stragglers in M67 from high-precision differential CCD photometry. The pulsational properties, in particular of multiperiodic blue stragglers, might prove important for constraining possible formation scenarios (Gilliland & Brown 1992). Should these mysterious stars be the result of coalesced binaries, then structure and chemical abundances can be expected to be peculiar compared with evolved single stars. Such abnormalities are likely to leave their traces in oscillaton periods and period spacings.

2.3 *Slowly Pulsating B Stars*

The slowly pulsating B (SPB) stars are multiperiodic variable mid-B-type stars (B3–B8) with periods between about 1 and 3 days. The prototype of this group, 53 Per (B4 IV, $v \sin i = 17$ km s^{-1}), was discovered as a line-profile variable star by Smith (1977). Its line-profile variations are well explained by velocity fields from low-ℓ nonradial pulsations on a rotating star (Smith & McCall 1978). The photometric variability of 53 Per was described in Percy & Lane (1977) and Africano (1977). Buta & Smith (1979) determined two periods (of about 2 days) for the light variation. Later, Waelkens & Rufener (1985) found a number of other mid-B-type stars showing small-amplitude light and color variations with periods between 1 and 3 days. In some of these stars line-profile variability could be detected (Waelkens 1987). Furthermore, Waelkens (1991) reported that all 7 of the then-known SPB stars were multiperiodic, and he confirmed *g*-mode pulsations as the cause of their variability. These findings suggested that variable mid-B stars, including 53 Per itself, form a distinct group of pulsators. The term "slowly pulsating B stars" was proposed by Waelkens (1991).

The SPB pulsations are recognized as *g*-modes because observed periods are longer than the expected period of the radial fundamental mode. Figure 1 displays the location of the SPB instability strip relative to the one of the β Cepheids and the blue part of the classical strip on the HR diagram. The low ℓ values were inferred from their photometric variability. Waelkens (1991) realized that the photometric amplitudes decrease towards longer wavelengths for his whole sample of stars, indicating that photospheric temperature modulation influences the light variation significantly.

Physically, the excitation of *g*-modes in SPB stars can be understood as an extension of the β Cephei instability (Gautschy & Saio 1993, Dziembowski et al 1993) (cf Section 2.4 below) towards longer periods. For main-sequence stars less massive than β Cepheids, high-order *g*-modes are excited. Because the modes populate the frequency domain rather densely, an increasing number of *g*-modes become excited simultaneously for lower masses. Hence, the multiperiodicity of the SPB stars (cf Smith et al 1984, Waelkens 1991) is in accordance with our theoretical understanding. Figure 2 shows the large number and the different radial orders of excited *g*-modes (enclosed in the dotted area of the inlet in the upper right) during the main-sequence evolution of a 5 $M_\odot$ star. The first turn-around of the evolutionary tracks of the appropriate stellar masses determines the width of the instability strip. During this evolutionary phase the hydrogen exhaustion in the stellar core induces a contraction of the central region, which then leads to a strong growth of the Brunt-Väisälä frequency. The large magnitude of the Brunt-Väisälä frequency is responsible for short spatial wavelengths—and enhanced dissipation—in the eigenfunctions of oscillation modes with periods relevant for SPB stars. The accompanying dissipation in the deep interior eventually overcomes the driving in the Z-bump region of the envelope (cf Figure 4 in GS95). Along the main sequence, the SPB instability strip terminates at around 2.5 $M_\odot$. The exact value depends somewhat on metallicity. At these low masses, the radial orders of the excitable *g*-modes become so high that strong radiative dissipation eventually overcomes the excitation by the κ-mechanism. The theoretical low-mass boundary for the *g*-mode instability agrees with the null result of a search for line-profile variations in late B (B8–B9.5) stars conducted by Baade (1989).

The low-ℓ, high radial order g-modes that are excitable in SPB stars let them assume essentially asymptotic modal properties (cf Section 3.2.2. in GS95). Deviations from the naively expected equal separations between adjacent periods of the g-modes of like degree were found in theoretical modeling (Dziembowski et al 1993). The periodic patterns in the period differences are induced by partial trapping of the modes around the outer edge of the nuclear-burning core. Only in variable white dwarfs are such signatures also encountered, which we can take advantage of for seismic analyses (see Section 3.2). The long periods of the SPB stars make it extremely challenging to obtain sufficiently accurate frequency data to perform seismological studies of value.

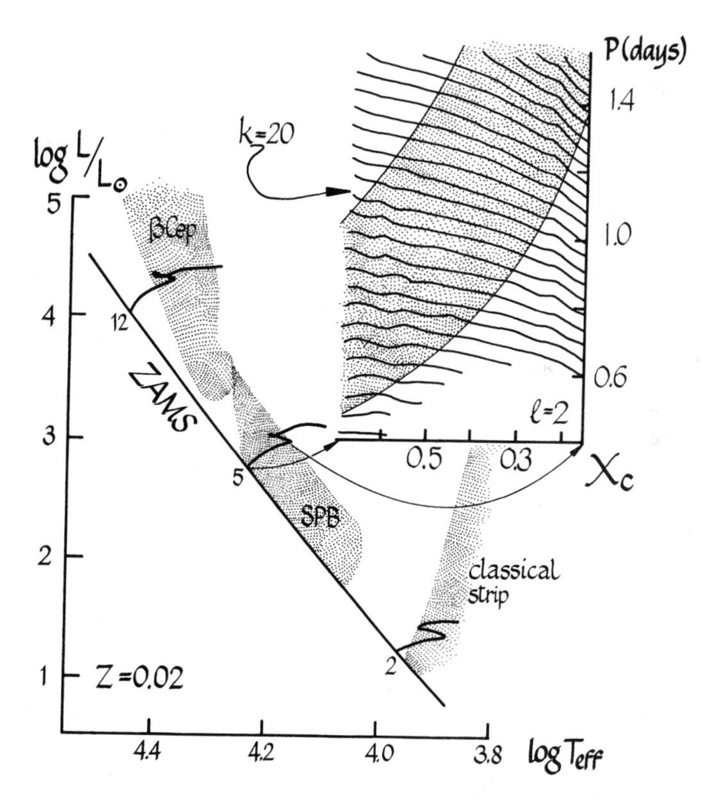

Figure 2 HR diagram of the upper main-sequence region. Evolutionary tracks of 2, 5, and 12 $M_{\odot}$ are shown leaving the ZAMS. Dotted areas mark pulsational instability regions. The finger at the upper left contains the β Cepheids. The area denoted by SPB contains the slowly pulsating B stars. On the lower right we show the blue part of the classical instability strip where δ Sct stars are found. The inset on the upper right shows the evolutionary variation of $\ell = 2$ g-mode periods of a 5 $M_{\odot}$ star. The temporal evolution is parameterized by the central hydrogen content X_C. The dotted strip indicates the location of the unstable modes.

All SPB stars seem to be slow rotators (Smith 1977, Waelkens 1987, Waelkens et al 1991). Since most SPB stars were discovered photometrically, the low rotation speeds can hardly be attributed to a selection effect. No SPB stars are found in the young open clusters NGC 3293 (Balona 1994) and NGC 4755 (Balona & Koen 1994) despite their containing many β Cephei stars. Balona & Koen (1994) suspected that rapid rotation in these stars suppresses the g-mode instability. The effect of rotation on the stability of high-order g-modes has not yet been tackled theoretically.

The presently known sample of SPB stars contains the star 53 Per and the 13 other stars listed by North & Paltani (1994). Sometimes, other members of the class of "53 Per variables"—defined spectroscopically as variable line-profile B stars by Smith (1980a)—are considered as members of the SPB group. However, a blind merging of the two groups, defined by differing aspects of their variability, might be inadequate.

2.4 *β Cephei Stars*

The β Cephei stars are a group of short-period ($\lesssim 0.3$ d) variables. Their oscillations are detectable in radial-velocity as well as in light variations. A list of the presently known β Cephei stars is given in Sterken & Jerzykiewicz (1993). The excitation of the pulsations of β Cephei stars and the slowly pulsating B stars was recently, after many years of futile attempts, found to be due to classical κ-mechanism driving in the Z-bump encountered in the new generation of opacity data (see Section 3.2.2 in GS95) (Kiriakidis et al 1992, Moskalik & Dziembowski 1992).

The κ-mechanism tends to excite pulsations with periods comparable to the thermal timescale of the excitation zone; this applies to radial as well as nonradial pulsations. In the region of β Cephei variables, oscillation modes with periods between ≈ 0.1 and 0.3 day are potentially excited. As effective temperature decreases, the depth of the driving zone increases and so do the periods of excited modes. On the main sequence, for models with 7–8 $M_\odot$ ($\log L/L_\odot \approx 3.5$) the length of the most favored period lies in the transition region between p-modes and g-modes (see Figure 2). Towards lower masses, long-period g-modes (with a dense frequency spectrum) are excited and observed in slowly pulsating B stars. A phenomenological distinction between the β Cephei stars and the SPB stars is introduced most naturally between 7 and 8 $M_\odot$ (the periods there are of the order of a few tenths of a day). Physics, though, does not enforce a formal subdivision of the two groups of variable stars.

The theoretically determined instability region (cf Figure 1) on the HR diagram for $Z = 0.02$ (Dziembowski & Pamyatnykh 1993), determined using a recent OPAL opacity release, comprises most of the observed β Cephei variables;

exceptions are some metal-rich stars. Waelkens et al (1991) found that the blue edge of the instability region is bluer (i.e. a more extended instability region on the HR diagram) for more metal-rich β Cephei stars. Another effect of the heavy-element dependence of the β Cephei pulsations is the lack of β Cephei variables in the Magellanic Clouds (Balona 1992, 1993). These facts are consistent with Z-bump driving. Theoretically, it was realized (see Moskalik 1995) that the detailed physical treatments of heavy elements in the opacity calculations, such as Ni, Cr, and Mn, influence the pulsational driving and hence the extent of the instability region, in particular in the low-mass region of the β Cephei stars.

The paucity of β Cephei variables at luminosities exceeding $\log L/L_{\odot} \approx 4.5$ is explained by the shifting of the instability region into the post-main-sequence phase (Dziembowski & Pamytnykh 1993). The evolutionary timescale is faster there; thus the probability of finding variable stars in this domain is much lower.

Pulsational instability regions deduced from linear theory and from nonrotating spherical star models do not yet coincide in all cases with observed domains of β Cephei variables. The β Cephei stars in the open cluster NGC 3293 (Balona & Engelbrecht 1981, Balona 1994) are confined to $4.4 < \log T_{\rm eff} < 4.44$, and all stars in this temperature range seem to pulsate. Comparing this instability domain with the theoretical result for $Z = 0.03$ obtained by Dziembowski & Pamyatnykh (1993) indicates that despite the observed high-$T_{\rm eff}$ boundary being roughly consistent with the theoretical blue edge, the observed low-$T_{\rm eff}$ boundary is much bluer than the theoretical one. Moreover, Balona & Koen (1994) have found a few constant stars within the β Cephei instability strip in NGC 4755.

The majority of β Cephei variables are multiperiodic (see e.g. Figures 22 and 23 in Sterken & Jerzykiewicz 1993). Some of the identified frequencies can be attributed to rotational m-splitting. These variables also show line-profile variations, which are mostly line broadening and narrowing (rather than traveling bumps and dips). Osaki (1971) showed that such line-profile variations are produced by the combination of rotation and nonradial pulsation with low $|m|$-values. Smith (1980b, 1983) and Campos & Smith (1980) compared theoretical line-profile variations with observed ones to determine pulsation modes and periods. These studies, performed by a trial-and-error method, were difficult because of the many parameters involved in the computation of the theoretical line profiles. Additionally, temperature variation, which was not accounted for, affects the line profiles of β Cephei stars markedly (Balona 1987, Cugier 1993). To overcome the difficulties, Balona (1986a,b, 1987, 1990a) proposed a moment method to identify pulsation modes. This approach is useful for nonradial pulsations with low $|m|$ in slowly rotating stars. Aerts et al (1992) and Mathias et al (1994) extended and applied the method to the β Cephei variables δ Ceti and α Lupi.

Pulsation modes may also be estimated by using the amplitudes of the variability in the visual and in the UV. This method makes use of the light variation resulting both from oscillatory variations in temperature (surface brightness) and from geometrical effects. The relative contributions vary with wavelength and with spherical degree of the oscillation mode (Dziembowski 1977b). Applying this method, Watson (1988) and Cugier & Boratyn (1992) concluded that large-amplitude, single-periodic stars pulsate in their fundamental mode. Recently, Cugier et al (1994) extended the method by including the amplitude of the radial-velocity variation.

Most β Cephei variables vary regularly; the amplitudes of a few of them change, however, drastically on short timescales. One example is Spica (α Vir), a double-lined spectroscopic binary with an orbital period of 4.01 day and a pulsation period of 0.1738 day discovered by Shobbrook et al (1969). The existence of the β Cephei-type variation was traced back to about 1890 (Smak 1970, Shobbrook et al 1972, Dukes 1974). The amplitude decreased in time and became undetectable by 1972 (Lomb 1978). In contrast, Smith (1985) observed traveling bumps and dips in the rotationally broadened Si III line profiles, indicating the existence of nonradial pulsations with high $|m|$. Balona (1985) suggested that the amplitude changes as the angle between the line of sight and the axis of nonradial pulsations (i.e. rotation axis) varies, due to the precession of the rotation axis around the orbital axis of the binary system. The period of precession is of the order of the period of apsidal motion, estimated to be $\approx$ 130–140 year (Shobbrook et al 1972, Dukes 1974). This hypothesis fails, however, if the light variations were caused by radial pulsations. Another example of a rapid change of the pulsation amplitude is the Be star 27 CMa, in which the β Cephei-type light variation grew from zero amplitude in less than two years (Balona & Rozowsky 1991). Since some Be stars are located in the domain of β Cephei variables, the existence of a β Cephei-type variability in a Be star is not surprising, but the cause for the rapid growth of the amplitude remains mysterious.

2.5 *Be-Type and Related Stars*

The Be stars are characterized by rapid rotation and by the occurrence of emission features in spectral lines. The emission components are attributed to a circumstellar disk (see Sletteback 1988 for a review). Two kinds of short periodic ($\lesssim$ day) variations are observed in Be stars (see e.g. Baade 1987 for a review): 1. high-order line-profile variations (e.g. ζ Oph (O9.5Vne), Walker et al 1979) in which bumps or dips are traversing rotationally broadened absorption lines from the blue to the red and 2. a photometric variability with periods of the order of a day (see Balona 1990a), which seems to be associated with low-order line profile (line asymmetry) variations (Baade 1982).

At least some of the high-order line-profile variations (LPVs) of Be stars are explained by nonradial pulsations of high azimuthal order $|m|$ (ζ Oph, Vogt & Penrod 1983; μ Cas, Baade 1984; γ Cas, Yang et al 1988; ζ Tau, Yang et al 1990). The line-profile variability of ζ Oph, the prototype of this group, was attributed to nonradial pulsations with $m = -8$ by Vogt & Penrod (1983), while Kambe et al (1990, 1993a) concluded that two modes ($m = -4$ with a period of 3.3 h and $m = -7$ with a period of 2.4 h) are simultaneously excited. In these investigations, nonradial pulsations were assumed to be sectoral ($\ell = |m|$) spheroidal modes. If this assumption is dropped, however, then the pulsation modes can no longer be identified from the LPVs alone (Osaki 1986a, Kambe & Osaki 1988).

The angular dependence of amplitude distribution of nonradial oscillation of a rotating star cannot be expressed by a single spherical harmonic. A spheroidal mode of given ℓ and m is accompanied by toroidal components of degree $\ell \pm 1$ and spheroidal components with $\ell \pm 2$. (Components with different m do not appear because axisymmetry is preserved in a rotating star.) The importance of these components is proportional to $|m|\Omega/\omega$ and $(\Omega/\omega)^2$, respectively, where ω is the oscillation frequency in the corotating frame and Ω stands for the angular frequency of rotation. Therefore, to fit theoretical LPVs to observed ones for a rapidly rotating star, theoretical modeling should account at least for the effect of toroidal components (Kambe & Osaki 1988). Aerts & Waelkens (1993) calculated LPVs including the effect of toroidal components; their results show that toroidal components induce additional bumps in the line profiles when Ω/ω exceeds 0.2. It is desirable, however, to use self-consistently derived relative amplitudes of toroidal components (which depend on $|m|\Omega/\omega$). Also, the influence of temperature variations during the pulsational cycle, which are not negligible in variable B stars (Balona 1987), should be included. For low-frequency g-modes, a somewhat simplified approach is possible (Berthomieu et al 1978). Lee & Saio (1990a,b,c) and Lee et al (1992) obtained theoretical line-profile variability due to low-frequency oscillations thus accounting in their analysis for temperature variations.

On the HR diagram, Be stars lie within the β Cephei or in the SPB instability regions, so it is reasonable to expect the high-$|m|$ modes in Be stars to be also excited by the Z-bump. Kambe et al (1993a) argued that the detected modes in ζ Oph are prograde modes with periods between 7 and 15 h in the corotating frame; they correspond to Q values[1] in the range 0.06–0.12. These periods are

[1] $Q \equiv P \cdot (\bar{\rho}/\bar{\rho}_\odot)^{1/2}$ where P is the period of the pulsation in days and $\bar{\rho}$ is the mean density of the star.

longer than those of β Cephei variables. Theoretically, the stability properties of corresponding oscillation modes in rapidly rotating stars are not yet clear (cf Lee & Baraffe 1995). But there is more to the LPVs in Be stars than only regular variability. In λ Eri, the high-order LPVs are irregular (Smith 1989, Kambe et al 1993b, Gies 1994). These variations are thought to be caused by stochastic small-scale activity above the photosphere (Smith 1989, Smith & Polidan 1993).

The other type of periodic variation, photometric variability, is ubiquitous in Be stars. Those Be stars exhibiting photometric variability are occasionally referred to as λ Eri stars (Balona 1990a). Numerous λ Eri stars were recently identified in open clusters of both the Small and Large Magellanic Clouds (SMC and LMC) (Balona 1992, 1993). The κ-mechanism associated with the Z-bump clearly cannot to be responsible for the light variations of these objects, even if their variability is attributable to oscillations.

The periodic light variations are strongly correlated with the rotation speed. The physical origin of the light variability, which is probably associated with low-order ($|m| \approx 2$) LPV, is not yet certain. Three possible mechanisms are proposed, two of which are rotational modulations caused by an inhomogeneous brightness distribution on the stellar surface and an asymmetric structure of the circumstellar envelope (Balona 1990a). The origin of these inhomogeneities was supposed to be connected with magnetic fields; there is, however, no clear observational evidence for the existence of strong magnetic fields in Be stars. The other mechanism involves a nonradial pulsation with frequency, in the frame corotating with the stellar surface, which is so small that observed periods are correlated with rotation. An excitation mechanism for such slow nonradial pulsations was proposed by Lee & Saio (1986, see also Lee 1988). They argued that oscillatory convection in the rotating convective core couples with a high-order g-modes in the envelope and so induces an overstable nonradial mode.

Because periodic light variations are common among Be stars, they are suspected to be related to the mechanism for the observed episodic mass loss. [Balona et al (1991) found that all rapid rotators in the open cluster NGC 3766 that showed low-order—i.e. small $|m|$—line-profile variations were in fact Be stars.] Assuming the periodic light variations to be induced by magnetic fields, Balona (1990b) proposed the episodic mass loss to be associated with giant flares on these stars. Alternatively, angular momentum transport by nonradial pulsations in rotating stars was proposed as an alternative mechanism for episodic mass loss (Ando 1986, Osaki 1986b, Lee & Saio 1993). In this picture, interior angular momentum is transported to the surface by the action of nonradial oscillations, eventually forcing the surface region to rotate supercritically, leading to a subsequent ejection.

2.6 *Very Massive Main-Sequence Stars*

The stability properties of very massive main-sequence stars were discussed in the past mainly in connection with the most massive stable stars expected to be encountered. Observational evidence exists for stars having masses exceeding 100 $M_\odot$ but probably below 200 $M_\odot$ (de Jager & Nieuwenhuijzen 1991). The numbers are estimated by comparing positions on the HR diagram with standard stellar evolution tracks, usually neglecting rotation, so presently stated values are likely to change in the future.

Ledoux (1941) investigated the possibility of nuclear burning to destabilize massive stars pulsationally through the action of the ϵ-mechanism (cf Section 3.3.1 in GS95). With very simple input physics he derived a mass of about 100 $M_\odot$ on the main sequence, above which stars pulsate due to the ϵ-mechanism of CNO burning. For most stars, the relative pulsation amplitudes derived from comparing the deep interior with the surface regions are small throughout the thermonuclear burning region so that the ϵ-mechanism is inefficient for pulsational destabilization. Above a critical mass, radiation pressure becomes large enough for the displacement to achieve sufficient amplitude also in the central region. The ϵ-mechanism can then contribute efficiently to the work integral and eventually overcome the radiative damping in the envelope. Hence, radiation pressure is an important side aspect for the action of the ϵ-mechanism. Cox (1974) reviewed attempts made before the mid-1970s to determine the upper mass limit on the main sequence due to ϵ-destabilization. This question turned out to be a delicate quantitative problem because the driving rates due to nuclear-burning terms are very low, usually of the order of $|\sigma_I/\sigma_R| = 10^{-6}$ (such expressions are obtained when the time dependence of the pulsation equations is parameterized by $\exp(i\sigma \cdot t)$, a ratio of 10^{-6} means that it takes $10^6/2\pi$ pulsation periods for the amplitude to grow by a factor of 2.72). Small nuclear driving has to compete with comparable radiative damping in the envelope. Hence, slight inaccuracies, e.g. in treating opacities and their derivatives, have serious consequences for the final outcome. Discussions in the literature of numerical results on this issue were therefore accordingly heated and involved in the past.

After some years of quiescence, the question of the upper mass limit received renewed attention after the advent of the new generation of opacity tables (OPAL and OP). Stothers (1992) concluded, on the basis of OPAL opacities, that an improved temperature dependence of the opacity leads to a higher central concentration in the stars, which reduces the efficiency of the ϵ-mechanism. The radial fundamental mode turned unstable for higher masses than in earlier studies (e.g. Ziebarth 1970), namely at 121 $M_\odot$ for $Z = 0.02$. In a more elaborate study, which included a larger mass range and which also involved analyzing a number of low-order overtones, Glatzel & Kiriakidis (1993a) arrived essentially at the opposite of Stothers' conclusions. The instability of the radial fundamental mode was computed to be of only minor importance because a much stronger instability (with $|\sigma_I/\sigma_R| \approx 0.1$) developed due to strange modes at lower masses. This mode-resonance instability appears at masses above about 60 $M_\odot$. Increasing the stellar mass, the instability extended from high frequencies towards the radial fundamental mode. However, even when considering only ϵ-destabilization, Glatzel & Kiriakidis (1993a) concluded that the critical mass was smaller than that determined from older studies.

The repeatedly recovered discrepancies in the magnitude of the critical mass determined by ϵ-destabilization may well be connected with details of the numerical treatment of the opacity data. Nonetheless, the existence of strange modes is undoubtedly a secure feature induced by the pronounced Z-bump and has an important impact on the stability of massive stars (see also Section 4). The growth of the strong strange-mode instabilities into their nonlinear regime has still to be thoroughly investigated. Glatzel & Kiriakidis (1993a) reported large mass-loss rates to have occurred in preliminary computations.

Whether ϵ-driven pulsational instabilities are crucial for terminating the massive end of the mass function remains unclear. Instabilities induced by nuclear burning are most efficient when the stars are essentially homogeneous, i.e. when they have settled on or close to the zero-age main sequence (ZAMS). The growth times of ϵ-induced instabilities are comparable with the main-sequence lifetimes and might hence be irrelevant for stellar evolution issues. The strange modes, on the other hand, show growth times of the order of the dynamical timescale of the stars; these envelope oscillations are much better candidates for shedding mass.

3. PULSATIONS IN EVOLVED LOWER-MASS STARS

First, lower-mass stars—understood as those climbing the asymptotic giant branch (AGB) and starting carbon burning in a degenerate state ($\lesssim 8\ M_\odot$ on the ZAMS)—are discussed with respect to the pulsational instabilities they encounter during their evolution. All stars originating from regions hotter than the classical instability strip on the ZAMS (i.e. $M_* \gtrsim 2\ M_\odot$) are prospective pulsation variables as they cross the HR diagram and pass through the instability strip on their way to the base of the giant branch. However, the evolutionary timescale is so short (therefore the notion of the Hertzsprung *gap*) that the probability of observing them therein is very low. The triple-mode pulsator AC And is presently considered as one of those rare objects in this transition phase (Fernie 1994). In the following, we restrict our attention to phases after the onset of central helium burning.

Combined photometric and radial-velocity observations of classical pulsators—such as Cepheids and RR Lyrae stars—are frequently used to derive radii and associated quantities by means of the Baade-Becker-Wesselink method. The basic principle underlying the approach is simple: From photometry and spectroscopy, radius ratios and radius differences are determined at suitably chosen phases during the pulsation cycle so that in principle absolute radii can be deduced. The detailed application to obtain accurate physical calibrations of pulsating stars is, however, neither very transparent nor easy. A number of assumptions must be introduced (e.g. on limb darkening, asymmetries of line profiles during pulsation, physical calibration of color indices, etc). The degree to which these assumptions are realized in stars restricts the level of accuracy achievable, and it may change from one type of pulsator to another. Gautschy (1987) and Moffett (1989) discussed the Baade-Becker-Wesselink method and its variants comprehensively; Moffett (1989) emphasized more recent developments.

3.1 *Through the He-Burning Stage*

When stars ascend the giant branch the more massive ones may reach the low-luminosity domain of red variables (see the upper right in Figure 1 in GS95) already during the first ascent, before the onset of central He burning. The general belief is, however, that the red variable stars are already in the asymptotic giant branch stage. During their first ascent stars spend only a short time in the topmost region of the giant branch. Hence, the probability is rather low of finding a red variable in that evolutionary stage.

3.1.1 RR LYRAE STARS[2] After the onset of degenerate central He burning, stars with masses between roughly 0.5 and 2.0 $M_\odot$ settle on or close to the horizontal branch (HB), but only stellar masses below about 0.75 $M_\odot$ are potential RR Lyrae pulsators during some phase of central helium burning. Accurate numbers depend on details of the assumed stellar physics (Dorman 1992a). The periods of this class of variable stars are around half a day. Stars appear as RR Lyrae variables either when they are close to the zero-age HB or else later upon their evolving to the blue or to the red (depending on the mass of the star). Observed HB stars are deduced to have had main-sequence masses above about 0.8 $M_\odot$ to reach the HB stage within a Hubble time.

The basic instability mechanism responsible for RR Lyrae pulsations is well understood on the basis of linear pulsation theory. Cox (1974, 1975) provided extensive reviews of the theoretical situation up to the early 1970s. Later,

[2]Three subclasses of RR Lyr pulsators, denoted by *a*, *b*, and *c*, exist. Types *a* and *b* have asymmetric light curves and are fundamental modes pulsators. The light curves of the first overtone type *c* RR Lyrae stars are essentially sinusoidal. For details see Cox (1974).

attention focused on nonlinear modeling (Stellingwerf 1975, 1982; Kovács & Buchler 1988a), the question of mode selection (Simon et al 1980, Buchler & Kovács 1986), and the redward extension of the instability strip including the role played by time-dependent convection. In an early attempt, Deupree (1977a,b) simulated two-dimensional convection and its interaction with pulsations of RR Lyr stars. He found convection to quench pulsations in low-$T_{\rm eff}$ models. Considerable efforts were invested in developing one-dimensional descriptions of time-dependent convection, which were coupled with hydrodynamical codes (Xiong 1981; Stellingwerf 1982, 1984a,b,c; Stellingwerf & Bono 1993; Gehmeyr 1992a,b, 1993). Such simulations find concurringly that the convective flux is enhanced during the compressed phases, diminishing the efficiency of the κ-mechanism, which couples to the radiative flux. Convection tends to reduce the amplitude of the pulsation; the actual amount depends on free parameters of the particular convection descriptions. In low-$T_{\rm eff}$ models with extended convection zones the variation of the convective efficiency during the pulsation cycle is advocated to cause a bump in the ascending branch of the light curve (Stellingwerf 1984c, Gehmeyr 1992b). At present it is unclear whether such a bump prevails in observed light curves of RR Lyrae stars. Accurate photometry of red RR Lyr stars should be useful to constrain the role played by convection theory in such pulsators.

RR Lyrae variables constitute easily identifiable and rather bright members of many globular clusters. As a consequence, RR Lyrae stars are used extensively as standard candles to determine distances to their host clusters. Based on the absolute magnitude of the RR Lyrae variables, ages of globular cluster systems are determined that bear important clues for galactic evolution and even for cosmology (Sandage 1982a,b or Sandage 1993a,b,c). To address questions of globular-cluster ages and distance scales the presently required accuracy of the physical calibration of RR Lyrae variables is very high. Delicate and complex issues of location and morphology of the horizontal branch (Buonanno et al 1989, Caputo et al 1989, Dorman 1992a, Lee et al 1994) and the range of acceptable evolutionary stages of RR Lyr variables (Lee et al 1990) are being investigated. On the pulsation-theoretical side, extensive numerical simulations (Kovács & Buchler 1988a, Simon 1989, Guzik & Cox 1993, Feuchtinger & Dorfi 1994) including the latest improvements in constitutional physics were performed. These attempts concentrate on the influence of envelope physics; modified input physics in the deep interior (such as opacity sources or elemental mixing processes) seems to affect mostly the structural properties (Dorman 1992b). Extensive grids of combined calculations for which pulsation properties are derived from stellar evolution models, both using the same input physics, do not exist. Such a procedure might at least eliminate the worry of how much disagreement between stellar evolution and pulsation theory is induced by inconsistent modeling procedures. Additional uncertainties are introduced by the lack of fully self-consistent temperature–color-index relations over the whole range relevant for RR Lyrae stars and by the choice of mean values of the periodically varying color indices (Sandage 1990a). Roughly speaking, the ultimate goal is eventually to reach a proper understanding, and therewith a correct quantification, of the Oosterhoff-period-shift (the mean period of RR Lyrae stars in Oosterhoff-type I clusters is 0.1 day shorter than that of Oosterhoff-type II clusters when determined by an ensemble mean over the periods) or of the Sandage-period-shift (the RR Lyrae periods decrease as the cluster metallicity increases, determined on a star-by-star basis at fixed $T_{\rm eff}$) (cf Sandage 1982a, 1990b; Bono et al 1994). This is done by arriving at acceptable RR Lyrae masses and luminosities over the whole parameter domain ([Fe/H], Y, [O/Fe],. . .) covered by globular clusters. A general agreement has not been reached, and the body of literature on that issue is considerable. The review of Rood (1990) pointed out the complexity associated with horizontal branch evolution, and he has sketched the large parameter space that must be dealt with correctly.

The elusive origin of the *Blazhko effect* is of much interest for stellar pulsation theory. This effect is a secular variation in form and amplitude of the fundamental-mode oscillation on a timescale between 20 and 100 days and is found in about 15–30% of the RR*ab* pulsators. A comprehensive review of the observational aspects was presented by Szeidl (1988). It seems, as also pointed out by Gloria (1990), referring to field RR Lyrae data from the 4th edition of the *General Catalogue of Variable Stars,* that the mean period of the Blazhko effect RR*ab* stars is shorter than the mean period of steadily pulsating RR*ab* variables. The theoretical meaning of this correlation, should it indeed not be an observational bias, is unclear. The Blazhko effect has been confirmed only for RR*ab* stars but not for RR*c* variables. One theoretical model—the oscillating oblique magnetic rotator (Cousens 1983)—attributed the Blazhko effect to a stellar magnetic field and its interaction with radial pulsation. The Blazhko period would then essentially be the stellar rotation period. At least for the star RR Lyrae itself a significant magnetic field that varies with the Blazhko period as well as with the pulsation period is observed (Babcock 1958, Romanov et al 1987). For other variables with a Blazhko effect, reliable data do not exist. The observed tertiary period of RR Lyrae, a modulation of the Blazhko effect itself, on a timescale of about four years, is preferably attributed to the magnetic cycle of the star. Very recently, Takata & Shibahashi (1995b) revisited and rederived the oscillating oblique magnetic rotator model. The method is very similar to the modeling of roAp stars, but in RR Lyrae stars the radial fundamental mode is self-excited and this mode picks up quadrupole components due to Lorentz and Coriolis forces. In contrast to Cousens (1983), Takata & Shibahashi (1995b) found a Blazhko amplitude that depends on the magnetic field strength. Moskalik (1985) attempted an explanation by looking for possible internal mode resonances in RR*ab* stellar models. His amplitude equations, including only lowest-order nonlinear couplings, revealed a resonance between the fundamental mode and the third overtone as the most likely way to produce a light variation similar to the Blazhko effect. The long-term evolution, however, could not be studied in his model. Since no clear discriminants emerge from either the theoretical or the observational side, none of the suggested explanations can yet be disqualified.

Despite RR Lyrae stars having usually low heavy-element abundances the new opacity generation (OP and OPAL) had some impact on the period ratios of double-mode RR Lyrae (RR*d*) stars. Kovács et al (1991) discussed the period ratios resulting from newly constructed pulsation models that were assumed to be appropriate for double-mode RR Lyrae variables in Oosterhoff I or II type clusters. Acceptable masses (compared with stellar evolution) could only be obtained when requiring $Z < 0.001$; within the parameter Z the distribution of the heavy elements was assumed to be solar. When allowing for nonsolar heavy-element ratios relative to Fe in the stellar material, an unexpected ambiguity in the results emerged due to competing effects from different chemical species (Kovács et al 1992).

In the Galactic field, only three RR*d* stars are presently known (Jerzykiewicz & Wenzel 1977, Clement et al 1991). Even in globular clusters, the RR*d* phenomenon was only recently appreciated (Sandage et al 1981). Szeidl (1988) estimated up to about 15% of RR Lyrae stars to be of RR*d* type. These variables are believed to be pulsating simultaneously in the fundamental and in the first overtone mode. Indeed, the colors or temperatures of RR*d* stars are confined to the transition region between the RR*c* and the RR*ab* instability domain. Observationally, the first overtone always has a higher amplitude than the fundamental mode. Kovács et al (1986) found the relative contributions of the fundamental and first overtone mode to have remained constant in the RR*d* stars of M15 over two decades. This result is not compatible with the simple mode-switching scenario of a star evolving accidentally through the transition region. Also, the observed number of RR*d*s appears to be too high for such a picture to apply. Bono & Stellingwerf (1993) pointed out that for their calculations the timescale for mode switching agrees better with predictions from stellar evolution. Nonetheless, the physical mechanism for long-term maintenance of double-mode pulsations is far from understood. Nonlinear pulsation calculations proved to have severe problems in simulating stably pulsating double-mode models (Stellingwerf 1974, Kovács et al 1992). Despite much effort, observed properties of double-mode RR*d* stars and double-mode Cepheids have not been reproduced satisfactory. For Cepheids, no stably pulsating double-mode models are known to exist. Some persistent double-mode pulsations have recently been constructed for RRd-type models (Stellingwerf & Bono 1993, Kovács & Buchler 1993). In most of these models, however, the amplitude of the fundamental mode is larger than the one of the first overtone—contradicting the observational evidence. The three-mode resonance $\sigma_F = \sigma_1 + \sigma_2$ seems to play an important role for persisting double-mode pulsations. The subscripts at the oscillation frequencies σ refer to the fundamental and the first two overtone modes. Kovács & Buchler (1993) noticed that the artificial viscosity required for the numerical codes to work had to be reduced below a certain threshold for the RR*d* models to reach the correct periods.

Observationally, the first overtone pulsators, the RR*c* variables, have light curves that are sinusoidal and very distinct from the asymmetric light curves of the fundamental-mode RR*ab* pulsators. This behavior is well reproduced by nonlinear simulations. The physical determinants of the light curve form have never been explained satisfactorily though. An attempt, based on one-zone models, was published by Stellingwerf et al (1987).

3.1.2 POPULATION II CEPHEIDS As the evolution after He core burning proceeds, stars with masses above approximately 0.51 $M_\odot$ evolve off the HB to approach and ascend the AGB. Either during the early evolution away from the HB or during shell flashes along the AGB, some stars can enter the instability strip again and appear as so-called population II or Type II Cepheids. The periods of population II Cepheids range from about 0.8 days—at the transition to RR Lyrae stars—to about 30 days, above which the stars in the classical instability strip are classified as RV Tau stars (see Section 3.1.3). For observational and theoretical reviews see Wallerstein & Cox (1984) and Gingold (1985), respectively. Linear pulsation calculations reveal that the combination of He II and H/He I ionization drives the pulsations. The fundamental radial or the first overtone mode seems to be excited (Nemec et al 1994). The instability strip of the metal-poor pulsators is considerably broader than that of population I variables at the same luminosity. In particular the red edge is shifted to lower temperatures. Wallerstein & Cox (1984), referring to Deupree & Hodson (1977), argue that this downward shift in temperature is connected with a reduced efficiency of convection. No recent, and in particular no systematic, studies are available in the literature that clarify what pulsation modes are potentially excited and where, in detail, the borders of the instability region of population II Cepheids lie on the HR diagram.

The short-period population II Cepheids or AHB1 (above horizontal branch; Diethelm 1990, Sandage et al 1994) are post-HB stars that pass through the instability strip during their evolution towards the AGB as they exhaust the helium in their cores. The period range of AHB1 stars is about 0.8–5 days. The observed properties and theoretical interpretations are thoroughly discussed in Sandage et al (1994). (Although these variables were sometimes called BL Her stars, the name is not suitable because the star BL Her itself is not metal deficient.) Since the timescale of evolution in the core helium exhaustion phase is much faster (of the order of 100 times) than during the core He burning phase, the number of AHB1 stars is much smaller than that of RR Lyrae stars. Because of the fast evolution, period changes for some of the AHB1 stars are actually observed (Wehlau & Bohlender 1992, Diethelm 1996). These data show that their periods are increasing on a timescale of 1–10 days / 10^6 years, in accordance with theoretical prediction. Below about $M = 0.51\ M_\odot$ (depending on the uncertainties of chemical composition) the stars do not evolve back to the AGB and hence do not cross the instability strip anymore; they instead turn towards high temperatures at some early phase of their post-HB evolution.

There is a group of population II variables called AC (anomalous Cepheids) that have periods similar to those of AHB1 stars. Their period-luminosity relation is, however, different from that of AHB1 variables, indicating that they are more massive than AHB1 stars (see e.g. Wallerstein & Cox 1984, Nemec et al 1994). The AC stars are thought to result from the coalescence of close binaries.

The period distribution of the population II Cepheids shows a gap between 5 and 10 days. The lack of stars in this period, and hence luminosity range, is thought to have evolutionary origin. Stars with short periods are leaving the HB to subsequently approach the AGB. Stars with long periods are either on a blueward excursion during late He-shell flashes or are already on their final departure from the AGB on their way towards the white-dwarf cooling region (Gingold 1976). The luminosity differences within the group of longer-period population II Cepheids (W Vir variables) can be attributed to a mass difference of the pulsators or to a different evolutionary status. Evolutionary calculations indicate that only below a critical remaining envelope mass do the He-shell flashing objects perform blueward loops. The number and distribution of periods of long-period population II Cepheids and their comparison with the estimated evolutionary timescales within the instability strip are not in satisfactory agreement (Gingold 1976, 1985). Therefore, a good statistical sample of the luminosity distribution of population II Cepheids should help considerably in clarifying the issue of their evolutionary status. Also, if the long-period population II pulsators have indeed terminated their AGB evolution and if they are not merely in an unstable He-shell burning episode, their periods should exclusively decrease in the long run as they evolve to higher temperatures.

3.1.3 RV TAU VARIABLES The longest-period W Vir stars seem to continuously change into what are classified as RV Tau variables, so they might in principle be considered as long-period population II Cepheids. Their pulsations are also driven by partial H and He ionization. The light curves show regularly alternating deep and shallow minima (double-wave form). Based on the luminosities ($\log L/L_{\odot} > 3$), the application of the coremass–luminosity relation, and the length of periods (50–150 d) larger masses are deduced for RV Tau variables than for the lower-luminosity W Vir stars. Jura (1986) saw indications in *IRAS* data of RV Tau stars leaving the AGB and evolving towards the white-dwarf domain; he inferred a very short duration—of the order of 500 years—of the pulsation phase. The double-wave light curves, which are a defining characteristic for RV Tau variables, are attributed to an internal resonance effect (similar to the Hertzsprung progression in Cepheids; see Section 3.1.5). Based on linear adiabatic theory, no satisfactory mode resonances could be identified by Takeuti & Petersen (1983). Fadeyev & Fokin (1985) reported a 2:1 resonance between the fundamental and the first overtone mode in their nonlinear modeling of RV Tau-like stars. Linear nonadiabatic pulsation calculations led Worrell (1987) to conclude that a single resonance is unlikely to be sufficient to understand the double-wave light curve of the RV Tau variables over the whole relevant $T_{\rm eff}$ and L range. The opposite conclusion was reached by Tuchman et al (1993) based on their linear nonadiabatic study.

In recent years, a number of nonlinear pulsation simulations have been performed to study the dynamical properties of population II pulsators. As the luminosity-to-mass ratio of the models was increased the oscillatory motions underwent transitions towards a low-dimensional chaotic behavior (Buchler & Kovács 1987, Aikawa 1987, Kovács & Buchler 1988b, Moskalik & Buchler 1990, Aikawa 1993). Period doublings, suggesting the so-called Feigenbaum sequence towards chaotic dynamics, occurred in a series of models with high luminosity-to-mass ratios upon reducing the effective temperatures of the equilibrium models. When increasing the L/M ratios, tangent bifurcations, leading to intermittency in the dynamics, were encountered. Based on such numerical studies, Kovács & Buchler (1988b) concluded that the RV Tau light curves represent early phases in a naturally occurring period-doubling bifurcation sequence for which the L/M ratio serves as a control parameter. In the simulations, the increasing degree of irregularity of the pulsations with rising luminosity (at constant mass) due to continuing period-doublings finds some observational correspondence.

The fully radiative pulsation models studied in the previously mentioned analyses, which adopt chaos-like properties, all approached low-dimensional chaotic attractors. A major deficiency of the modeling was the omission of time-dependent convection with its feedback on the pulsations. Whether chaotic dynamics still develops along the same routes and whether low-dimensional attractors will persist even in the presence of extensive convective envelopes remain unclear (Perdang 1991). To ensure the correct phase-space behavior, in particular for the long-term evolution, the quality of the numerical methods has to be very high; for critiques of numerical methods used for computing of chaotic phenomena see Miller (1991) and Yee et al (1991).

3.1.4 MIRA AND SEMIREGULAR VARIABLES The low-mass, long-period ($P \gtrsim$ 80 d) variables located at very low temperatures and luminosities above about $10^3\ L_\odot$ (cf Figure 1 in GS95) are known under a variety of names in the literature. These long-period AGB stars play an important and still controversial role in our understanding of strong mass loss, the formation of planetary nebulae, and the structure of the white-dwarf mass-function. From the point of view of pulsation theory we do not distinguish between the various observationally motivated classes that are conceivably superimposed on the same basic processes in these stars. The different behavior of the various families of cool variables can be caused by different masses, chemical compositions, or evolutionary stages along the AGB. For example, the observational distinction between Mira and semiregular (SR) variables is not necessarily a deep physical one. Often the distinction is based on the amplitude of the light variation. Large-amplitude variables are attributed to the class of Miras whereas low-amplitude pulsators are considered as SR variables. Both categories have, however, roughly the same regularity of their pulsational cycles (Whitelock 1990).

The driving mechanism of the pulsations is probably the combined action of partial H and He I ionization. Over a large fraction of the envelopes of these variables, energy transport by convection dominates and the timescale of convective overturn is of the same order as the pulsation cycle. Hence, any statements on the pulsational driving and on the extension of the instability region depend crucially on our understanding of the coupling of pulsation and convection and of the influence of convection on the equilibrium structure of the stars. Both aspects are not well comprehended at present. Balmforth et al (1990) showed that the inclusion of turbulent pressure in their models alters the equilibrium structure so that the excitation rates and also the periods of radial pulsation modes are considerably influenced. In pulsation calculations, not only should the perturbed convective energy flux be accounted for but also the perturbation of the turbulent pressure to fully describe the coupling of pulsation and convection in these envelopes. For their supposedly low-luminosity AGB model Balmforth et al (1990) obtained the first overtone as the dominantly destabilized pulsation mode; this agrees with the results of Fox & Wood (1982), who included the time dependence of the convective flux in a kind of flux retardation model. Only at luminosities leading to periods of about 320 days does the fundamental mode grow faster. Based on these calculations, most Mira variables were deduced to pulsate in the first overtone mode. Ostlie & Cox (1986) concluded from their linear pulsation calculations (in which turbulent pressure was included in some of the equilibrium models but the perturbation of the convective flux was neglected in the stability analyses) that Mira-type variability is consistent with fundamental mode pulsations. Using nonlinear initial-value simulations of Mira-type pulsations, Wood (1990a) concluded that, after artificially suppressing the growth of the fundamental mode in his models to force them into the first overtone, the velocity fields that built up were incompatible with observations. Hence, Miras were considered to pulsate in the fundamental mode. Tuchman (1991), applying his "acceleration analysis" to observed CO molecular lines in Mira variables, rejected the possibility of a fundamental-mode variability in his sample.

Spatial high-resolution observations of Mira (*o* Cet) itself (Haniff et al 1992) and of R Leo (Tuthill et al 1994), together with parallax estimates, allowed a direct estimate of the radius. Rather independent of the particular choice of the mass, a pulsation constant Q resulted that pointed to an excited first overtone. If, however, radii were derived from the excitation temperature of CO molecular lines the pulsation of *o* Cet was assigned to the fundamental mode (Hughes 1993). It must be kept in mind that the radii of Mira-type stars change by more than a factor of two when going from optically thin regions to the stellar photosphere. Thus, care has to be taken when radii derived by different techniques are compared.

Bessell et al (1989) attempted to reduce uncertainties in determining temperatures of Mira variables at different pulsation phases. Uncertain temperatures corrupt the accurate determination of Q values. They based their detailed radiative transfer calculations on density and temperature profiles obtained from nonlinear simulations.The resulting spectra are not yet fully satisfactory, possibly because detailed thermodynamic and radiative processes in the shock regions need to be dealt with in the hydrodynamic simulations. Furthermore, the quantifications of the pulsation constant Q rely on linear stability analyses (cf Wood 1995). To conclude, the debate on which mode is excited in Mira variables cannot presently be considered as settled (for a detailed recent discussion see Wood 1995).

Nonlinear pulsation simulations of Mira envelopes (incorporating with radiative energy transport only) show complicated multiple shock structures in their atmospheres (Bowen 1988, Wood 1979). Mostly, such calculations were not based on self-excited pulsations but on piston-driven motion of the stellar matter in the outermost layers. Although Feuchtinger et al (1993) presented promising preliminary results of high-quality numerics and coupled radiation-hydrodynamics, they still based their calculations on piston-driven pulsations in purely radiative model envelopes. Höfner et al (1995), using an extension of the same Viennese radiation-hydrodynamics code, presented interesting results of time-dependent dust formation in the atmospheres of long-period variables. They obtained quasi-periodic dust formation/destruction cycles and associated variable mass-loss rates even with static inner boundary conditions for certain abundance ratios of carbon and oxygen.

Observations indicate that the mass-loss rates of Mira variables are correlated with the pulsation period. Longer-period stars, which are also more luminous, tend to have higher mass-loss rates (Whitelock 1990 and references therein). The available simple nonlinear simulation models cannot reproduce the high values observed (Wood 1990b). Considerable improvements can be expected from properly dealing with convection in the envelopes of the long period variables and/or from the dynamical influence of grain formation and destruction (Höfner et al 1995). Pijpers & Habing (1989) estimated that dissipation of acoustic energy flux due to convection would be able to induce mass-loss rates between 10^{-7} and 10^{-4} $M_{\odot}$ year^{-1}.

A simple evolutionary scenario assumed Mira variables to evolve along the AGB towards higher luminosities and longer periods with an accompanying increase of mass loss until a critical luminosity was reached where the envelope would be shed. Thereafter, the Mira variables would be obscured by an optically thick envelope and become possibly observable as variable OH/IR sources that eventually evolve into a planetary-nebula system. Both observational and theoretical evidence speak against such a simple picture (Wood 1990a, Vassiliadis & Wood 1993, Whitelock et al 1994), however. Comparing the number densities of Mira stars with those of planetary nebulae or clump giants suggests that the Mira phase lasts for only about 5×10^4 years. For stars with masses of about 1 $M_{\odot}$ evolution on the AGB during a Mira lifetime does not result in a significant luminosity increase nor in any appreciable period change. The kinematic properties of Miras change with the length of their pulsation period, indicating that only small changes of the pulsation period occur during the lifetime of a Mira variable (Whitelock et al 1994). Additionally, Wood (1990a) argues that it is not possible for stars with masses below about 1.5 $M_{\odot}$ to ascend sufficiently high on the AGB to explain the very long periods (1000–2000 days) and the very low temperatures observed in OH/IR sources. Hence, the OH/IR variables are assumed to stem from a subgroup of stars that are more massive than what we see as shorter-period Miras. Under such circumstances the OH/IR sources were to be considered as late AGB stars rather than as post-AGB and pre–planetary nebulae objects. In terms of period, the variable AFGL objects, stars with strong

IR excesses, lie between the optically identified Mira variables and the radio-luminous OH/IR sources. The same applies with regard to the amount of mass loss and the degree of obscuration by circumstellar material (Jones et al 1990).

According to the interpretation of the Mira instability region mentioned above, the observed period-luminosity (PL) relation established for LMC Mira variables (Feast et al 1989 and references therein) would not represent an evolutionary sequence of stars but indicates a range of stellar masses occupying the Mira domain at different luminosities. Because red variables belong to the most luminous stars in a stellar system, the existence and robustness of a PL relation is of relevance for any kind of distance determination. Hence, the quality of the actual PL relation and estimates of the intrinsic scatter (either a stochastic one or one due to a hidden mass or color dependencies) are essential. Feast et al (1989) found that the Mira variables, in particular the O-rich Miras, obey well-determined period-luminosity-color relationships. This indicates that Mira variables occupy an instability strip of finite width on the HR diagram.

Because of their comparable kinematic properties the SR variables in the Galaxy are believed to emerge from the same populations as do the Mira variables (Jura & Kleinmann 1992). The periods of the SR variables are (at least in the sample of Jura & Kleinmann) usually shorter than those of Miras. Also, for the SR variables, the determination of the pulsation mode is controversial. Presently the short-period SR ($P \lesssim 150$ d) stars are assumed to be first or second overtone pulsators. For longer-period semiregular variables, on the other hand, pulsations in the fundamental mode are preferred.

Irregular variables show spectra with clear giant or supergiant characteristics. Jura & Kleinman (1992) referred also to the kinematical properties that led them to assume an affiliation of these stars with the same population as the long-period Mira variables with $300 < P < 400$ d. The evolutionary state of the irregulars seems, nevertheless, to be unclear in particular due to the uncertainties in assigning reliable luminosities to them. If the irregulars are situated at the low-luminosity end of the AGB then, in accordance with the picture of Wood & Cahn (1977), several higher overtone modes could be excited simultaneously, giving rise to a seemingly chaotic light variability. Another point of view is that the irregular variability is indeed a purely stochastic phenomenon (Perdang 1985). The latter suggestion is somewhat off the main line of thought as we do not know of any other class of stars being destabilized in this way. Careful monitoring of irregular variables over long time spans should enable the discrimination between the two models, at least if, according to the multimode picture, the number of simultaneously excited modes is small.

Some variables on the AGB with nonstrictly repeating light variations have undergone long-term temporal analyses to identify chaotic signatures in their light curves. [For an introduction into the concepts and the language of chaotic dynamics and its applications in astrophysics see Buchler et al (1985).] Blacher & Perdang (1988) analyzed a number of Mira variables by applying a "variance-function" approach. Cannizzo et al (1990) investigated long series of observations of the Mira variables *o* Cet, R Leo, and V Boo to reconstruct the underlying dynamical attractor. Kolláth (1990) studied 150 years of data for the RV Tau star R Sct. Based on a shorter temporal sequence of observations, Buchler et al (1995) claimed the identification of a four-dimensional embedding space for the quasi-regular variability of R Sct. In this case the amplitude modulations could be understood from nonlinear interaction of only two simultaneously excited pulsation modes. The irregular light variation of RU Cam—a W Vir star whose regular pulsations disappeared in the recent past—was analyzed by Kolláth & Szeidl (1993). Except for R Sct, none of these investigations unveiled compelling evidence for a low-dimensional chaotic attractor. Despite all the efforts we have no evidence that deterministic chaos occurs frequently in irregularly pulsating stars. Perdang (1991, 1993) argued that stars with convective layers (AGB stars have extensive ones) might not become chaotic with low-dimensional attractors. In these convective envelopes a large number of unstable degrees of freedom are unlocked by convective motion. It is not necessarily clear that a few outstanding modes would dominate the dynamics and therefore reduce the dimension of the attractor. It must also be remembered that transitions to chaos through tangent bifurcation or cascades of period doublings as found in simulations (Buchler & Kovács 1987) or in simple model systems (Buchler & Goupil 1988, Tanaka & Takeuti 1988, Takeuti 1990) are based on hydrodynamics incorporating only radiative energy transport.

3.1.5 CEPHEID PULSATIONS After the onset of nondegenerate core He burning, the luminosity of stars more massive than about 2.2 $M_\odot$ drops, and they very closely evolve down the track on which they ascended the giant branch the first time. Stars with masses between about 2.2 and 3 $M_\odot$ settle at lower luminosity to burn most of their central helium before they ascend the giant branch a second time (AGB evolution). Stars with $M \gtrsim 3\ M_\odot$ perform blueward loops during that evolutionary phase. Below roughly 5 $M_\odot$ (in the framework of standard stellar evolution without semiconvection and overshooting) the blue loops are not sufficiently pronounced to let the star enter the instability strip. Higher-mass stars cross the strip two or more times as they perform one or more loops, and they appear as Cepheids. The review of Cox (1975) provides an exhaustive account of the early theoretical developments in the field of Cepheid pulsations. The proceedings of a conference dedicated to two centuries of observed Cepheid variability (Madore 1985) contain important contributions to the advancement of our understanding to the mid-1980s. Becker (1985) and

Chiosi (1990) provide instructive accounts of evolutionary aspects and of the persisting uncertainties in connection with Cepheid pulsations. In that regard, the masses cited above should be taken as rough guidelines only; the particular values are likely to change depending on the various physical assumptions and on input data required by stellar evolution codes. The existence and also the topology of the blueward loops of massive stars during the core He burning phase and the double-shell burning stage are known to depend sensitively on subtle issues in the numerical treatment of stellar interiors (Lauterborn et al 1971, Höppner et al 1978, Becker 1985).

Cepheids are well known for the mass-discrepancy problem, which persisted for several decades. Cox (1980) reviewed the topic and elaborated on the various methods of mass determination applicable to Cepheids. The essential point in the mass-discrepancy controversy was that any kind of mass estimates inferred from stellar pulsation theory turned out to be systematically lower than the predictions from stellar evolution theory. Presently, the Cepheid mass discrepancy can be considered as essentially reconciled. The improvements were achieved with the help of the new generation of opacity data (OPAL and OP). Extensive pulsation calculations by Moskalik et al (1992) showed that the period ratios are considerably reduced when employing the new Rosseland opacity tables (see Figure 1). The Z-bump in the Rosseland opacity near 10^5 K alters the stellar structure such that the low-order pulsation frequencies shift differentially and lower the period ratios considerably at a fixed stellar mass. The microlensing searches of recent years provided, as side results, extensive data collections on variable stars. The MACHO consortium presented data of beat Cepheids in the LMC (Alcock et al 1995); some of them appear to be pulsating simultaneously in the first and second overtone. Christensen-Dalsgaard & Petersen (1995) reconciled the observed period ratios rather well with linear adiabatic computations based on the new opacities and otherwise standard stellar physics. The masses of these double-mode Cepheids with simultaneously excited fundamental and first overtone modes are presently attributed to values derived from linear pulsation theory, which are close to evolutionary masses. Masses of around 4 $M_\odot$ are now considered to be an appropriate low-mass domain for double-mode Cepheids; they might require some modifications in a stellar evolution treatment or a careful tuning of the X:Y:Z ratios to force them into the instability strip. New stellar evolution tracks based on OPAL opacity data (Schaller et al 1992) show that the Z-bump tends to diminish the sizes of the blueward loops. A 5 $M_\odot$ star with $Y = 0.3$ and $Z = 0.02$ does not—in contrast to calculations with old opacity data—enter the instability strip anymore. A reduction of the heavy-element abundances, for example (Schaerer et al 1993), produces sufficiently elongated blue loops for 5 $M_\odot$ stars to enable them to appear as short-period Cepheids.

The occurrence and the shift with period of the location of a secondary maximum (or bump) in the light curve of Cepheids in the period range from 4 to 20 days is known as the *Hertzsprung progression.* The phase of the bump within the pulsational cycle changes monotonically as a function of period and hence as a function of mass. The observed bump location can be used to infer the mass of the pulsating star. Since the early days of nonlinear modeling of stellar pulsations (Christy 1968) it was clear that the mass deduced for a given bump phase did not agree with the standard mass-luminosity relation. Again, the results based on the latest opacity data, reported by Moskalik et al (1992), indicate that stellar-evolution masses are now in much better accord with the bump masses. The bump is attributed to the accidental 2:1 ratio of the period of the second overtone to that of the fundamental mode (Simon & Schmidt 1976). Stars with periods below about 20 days pass through this 2:1 resonance (Kovács & Buchler 1989). Because a period ratio is also involved in the mass determination of bump Cepheids, it is understandable that the modified opacities affected them in the same way as the beat Cepheid masses.

Despite providing a resolution to the Cepheid mass problem the new opacity data introduced new and unexpected complications. The period ratios admitted by the pulsation models, much like the ones for the RR Lyrae pulsators, appear to show a noticeable dependence on the particular contributions of heavy-element abundances to the Z abundance parameter. An easy and accurate mass determination for pulsators seems no longer possible without taking additional care in determining the chemical composition of the object.

The pulsational instability of Cepheids is well explained by linear pulsation theory; the nonlinear behavior, however, is not so comprehensively understood. Fernie (1990) analyzed observations and found that despite large amplitudes being predominantly associated with longer periods, large- and small-amplitude pulsators mix over essentially the whole extent of the instability strip. In particular, the case of α UMi is presently of interest. The amplitude of Polaris diminished exponentially during this century, and it was expected to stop pulsating or at least drop below the detection limit of some millimagnitudes by 1994. In contrast to this prediction, Krockenberger et al (1995) reported continuing pulsations at low but detectable amplitude. Fernie et al (1993) established that α UMi is not on the verge to cross the red edge of the instability strip. It has cooler neighbors in the instability strip that pulsate at significantly higher amplitudes. Other Cepheids are known that are unusual either in terms of their amplitude (γ Cyg, Butler 1992) or in terms of the modal behavior (HR 7308, Burki et al 1986). We currently lack any systematic studies addressing the long-term evolution of nonlinear limit cycles that would help us understand such stars. Direct numerical integration of the initial-value problem might be of little use. The amplitude-equation ansatz (Buchler & Goupil 1984) together with qualitative methods from dynamical-system theory could serve as starting points.

Short-period Cepheids (with $P \lesssim 7$ d) having low-amplitude sinusoidal light curves are classified as s-Cepheids. Those s-Cepheids with periods smaller than about 3 days are believed to be first overtone pulsators, mainly from their sinusoidal light curves (cf review by Simon 1990). This assignment is consistent with the P-L relation of LMC Cepheids obtained by the MACHO project (Alcock et al 1995), which shows that most of the single-mode pulsators with $P \lesssim 2.5$ d are first overtone pulsators. For some longer-period s-type Cepheids that still show symmetric light curves, the mode assignment remains unclear. We must remember that we do not understand the mechanisms shaping the light curve. Despite the analyses of simple model systems (Stellingwerf et al 1987) it is conceivable that the inspection of the light curve alone is insufficient to unambiguously identify the prevailing pulsation mode.

Simon & Lee (1981) introduced analyses combining phases and amplitude ratios of various terms of Fourier-decomposed light and velocity curves of pulsating stars. This allowed them to quantify geometrical properties of stars' temporal light and velocity variations. The method found broad application in quantitatively describing the properties of simulated and observed stellar pulsations. The ultimate hope was, and still is, that directly derivable numbers from observed pulsations let us infer stellar-physical quantities. Simon (1988) wrote a review, with many important references, addressing applications and the level of understanding of the content of the different Fourier components. Simulations of nonlinear pulsations serving as a basis for physically calibrating the Fourier components are few since the necessary numerical quality is difficult to achieve.

3.2 *Late Evolutionary Phases and Degenerate Stars*

Roughly speaking, the chemical evolution of low-mass stars ends with helium core burning. Such stars evolve up the AGB with recurrent He-shell flashes and leave it when the envelope mass drops below a critical level. The stars evolve rather rapidly at roughly constant luminosity to high temperatures. After the remaining nuclear shell source extinguishes, the stars settle on the cooling track of the white dwarfs. During this whole evolutionary phase a number of opportunities exist for these stars to become pulsationally unstable.

As mentioned previously, low-mass stars leaving the AGB might be observable as RV Tau stars. Another group of pulsating stars, low-mass F- and G-type supergiants (also known as UU Her stars) are considered to be post-AGB stars.

Their semiregular low-amplitude variability has a timescale between 40 and 70 days. The high galactic latitudes and the strong infrared excesses in 89 Her and HD161796, typical members of the UU Her class, (Parthasarathy & Pottasch 1986, Likkel et al 1987) indicate that they are pre–planetary nebulae objects rather than population I, massive supergiants as some analyses suggested. Fernie & Sasselov (1989) studied the long-term behavior of UU Her stars and found their period and color changes to be one to two orders of magnitude smaller than what is expected if these variable stars had already left the AGB. Not only the evolutionary state but also the theoretical pulsation properties of their variability remain incompletely understood. Some of the variable F- and G-type supergiants are hotter than the blue edge of the classical instability strip. Pulsation calculations confirm the presence of pulsational instabilities at high temperatures for sufficiently large luminosity-to-mass ratios (Aikawa 1993, Gautschy 1993, Zalewski 1992, 1993). Seemingly, strange modes (cf GS95 Section 3.4) must be involved to explain the pulsational instabilities. These effects make a proper discussion of the pulsation physics of the strongly nonadiabatic envelopes of UU Her objects cumbersome. Nonlinear simulations of Aikawa (1993) and Zalewski (1993) indicated the presence of chaotic dynamics in certain $T_{\rm eff}$ and luminosity domains that could account for the observed irregular behavior. The light variations derived from nonlinear simulations were low (of the order of a few hundredths of a mag) and decrease towards high effective temperatures. This is not necessarily in agreement with observations. But again, the nonlinear modeling was restricted to purely radiative envelopes so that the relevance of these results still needs to be confirmed.

The peculiar variable star FG Sge, which was identified rather early as a post-AGB object [see Whitney (1978) for a historical account and references], is another example demonstrating that such stars can exhibit unusual pulsational behavior, at least in terms of the region of their pulsational instability. FG Sge is observed to have crossed the HR diagram from log $T_{\rm eff} \approx 4.7$ to 3.65 (van Genderen 1994) within about a century. Pulsations have been detected as far back as 1934 when the effective temperature was around 2×10^4 K (van Genderen & Gautschy 1995). These observed facts indicate a very broad instability domain. Application of the OPAL/OP data shows that the presence of the Z-bump strongly affects the pulsation modes and enhances their instability. Hence, the combination of the Z-bump, He II- and He I/H-ionization can drive pulsational instabilities essentially over the whole temperature range maximally coverable on the HR diagram by such objects if the L/M ratio is sufficiently high (Gautschy 1993, Zalewski 1993, van Genderen & Gautschy 1995).

When a post-AGB star reaches a surface temperature of around 3×10^4 K, it emits enough high-energy photons to efficiently photoionize the remaining circumstellar material and to let it appear as a planetary nebula. Observations of cool central stars show that some of them are variable (Méndez et al 1986, Bond & Ciardullo 1989, Hutton & Méndez 1993, Wlodarcyk & Zola 1990). Timescales are of the order of hours. The photometric amplitudes can reach a few hundredths of a magnitude. The variable cool central stars all show P Cygni-type line profiles, indicating strong stellar winds. Pulsation calculations (Gautschy 1993, Zalewski 1993) showed that cool central stars of planetary nebulae can be pulsationally unstable over a very broad effective-temperature range extending essentially from the AGB up to at least log $T_{\rm eff} \approx 4.9$ depending on the heavy-element abundances and the L/M ratio. The trend indicates that the higher the L/M ratio ($\gtrsim 10^4\ L_\odot/M_\odot$) the more unstable are the stars. Therefore, we expect the low-mass branch (below about 0.60 $M_\odot$) of the central-star mass function to be pulsationally stable and more massive ones to be pulsationally unstable. Observationally, the available data are insufficient as yet to support or disprove such predictions. Additionally, we do not know how pulsational instabilities with large growth rates, as found in linear stability analyses, behave in the nonlinear regime. Since several radial modes were simultaneously excited over a broad temperature range, the final observable pulsational pattern in such stars could be rather complicated.

In the following we turn to the oscillations found in pre–white dwarfs and in the different families of white-dwarf stars. This has been a very active field of research in recent years, with particular emphasis on using white dwarf pulsations as seismological laboratories. Recent comprehensive review articles include those by Winget (1988a), who also provides some historical perspective, Winget (1988b), Kawaler & Hansen (1989), Kawaler (1990), and Brown & Gilliland (1994). For a review of the general physical properties of white dwarfs, see Koester & Chanmugam (1990) and references therein. Tassoul et al (1990) provide a wealth of information on evolutionary models of hydrogen- and helium-rich white dwarfs and their characteristics in pulsation analyses.

3.2.1 VARIABLE PG1159 STARS In the region of the HR diagram where the post-AGB tracks bend towards the white dwarfs' cooling sequences, at the knee (see Figure 1 of GS95), are the so-called PG1159 stars. These stars have very high effective temperatures ($7 \lesssim T_{\rm eff}/(10^4{\rm K}) \lesssim 17$), and they show spectroscopically strong deficiency of H but pronounced C and He features and the presence of O (see Werner 1992 and Dreizler et al 1995 for recent reviews). Due to the very high temperatures, determining the hydrogen content is extremely difficult; only relatively poor upper limits can be guessed, such as $\approx 10\%$ by number for PG1159$-$035 (Werner 1995). Some of the PG1159 stars are known to be central stars of planetary nebulae.

A fraction of the PG1159 class shows photometric light variations with periods ranging from about 7 to about 30 min. Such oscillation modes are attributed to low-ℓ g-modes of high radial order. Variable PG1159 stars surrounded by planetary nebulae are sometimes referred to as variable planetary nebulae nuclei (PNNV), and those without signs of planetary nebulae as DOV or GW Vir stars. The pulsation periods of PNNVs are a factor of 2 or 3 longer than those of DOVs, indicating that the PNNVs have larger radii than the DOVs. From the evolutionary point of view it is plausible to assign the DOV stars to the vicinity or even to the early phases of the white dwarf cooling tracks. The PNNVs, in contrast, are still evolving towards higher temperatures at essentially constant luminosity and have not reached yet the knee. Bond et al (1993) list 9 PNNVs. Among them, RXJ2117.1+3412 can be regarded as a transition object between the PNNV and the DOV phase because the star has the shortest period among the PNNVs [ranging from about 11–22 min (Vauclair et al 1993)] and it has an extended low-surface-brightness planetary nebula. Long-term monitoring of the secular changes of periods of these hot pulsators should allow us to more precisely pin down their evolutionary status.

High-quality oscillation mode spectra were recently obtained from the WET consortium for the two DOV stars, GW Vir (PG1159−035) (Winget et al 1991) and PG2131+66 (Kawaler et al 1995). The results point towards $\ell = 1$ high-order g-modes being the dominant ones. From the periods and the period spacings, stellar masses of $\approx 0.6\ M_{\odot}$ were derived. From the departures from equidistant spacing between the periods, the depth of the composition transition (to a pure CO core) was estimated to be $\approx 3 \times 10^{-3}\ M_{\odot}$ for GW Vir (Kawaler & Bradley 1994) and PG2131 + 006 (Kawaler et al 1995). Furthermore, Winget et al (1991) obtained a rate of period change of $\dot{P} \approx -2.5 \times 10^{-11}$ for GW Vir, which corresponds to an evolutionary timescale of the order of 10^6 year.

The region on the HR diagram where variable PG1159 stars reside was determined by Werner et al (1995). However, constant and variable stars are intermixed in the instability region, indicating that luminosity and effective temperature are insufficient to characterize variable PG1159 stars. Pairs of spectroscopically identical stars are known in which one is variable and the other stable (Werner 1993).

The excitation mechanism of the PG1159 oscillations is thought to be partial ionization of the K-shell of C and/or O (Starrfield et al 1984, 1985). To obtain overstable pulsation modes, chemically homogeneous envelopes with sufficiently high C and O abundances must be invoked. The location and extent of the instability regions for stars evolving around the knee on the HR diagram depend, however, on the particular admixture of He in the CO-rich envelopes (Stanghellini et al 1991).

Vauclair (1990) proposed an alternative driving mechanism. His model focused on the levitation of chemical species in the strong radiation field of the PG1159 stars. Assuming equilibrium conditions, Vauclair (1990) found a strong nitrogen enhancement at a depth of the envelope where some of the g-modes achieve sufficiently high amplitudes and might be destabilized by the κ-mechanism by partial ionization of nitrogen. Unfortunately, however, the predicted surface composition contradicts the composition of GW Vir obtained by a detailed spectroscopic analysis by Werner et al (1991).

Kawaler (1988) investigated the effect of a H-burning shell on destabilizing H-rich nuclei of planetary nebulae. By accounting for temporal phase shifts between elemental abundances participating in the CNO cycle and the temperature perturbation, he found that pulsations with periods between 70 and 200 seconds were destabilized by the action of the ϵ-mechanism. For H-deficient central stars, Kawaler et al (1986) studied the effect of the ϵ-mechanism in a He-burning shell. Again, unstable g-modes were encountered. The low-ℓ g-modes had periods that are, however, about a factor of 3 to 4 shorter than those observed in DOV stars. Even under the favorable conditions for nuclear-driven instabilities as encountered in PNNV there is as yet no indication that nature permits such pulsations. Since the growth rates of modes excited by nuclear-burning shells are extremely small, the amplitudes of these modes may not grow sufficiently to be detected.

3.2.2 VARIABLE DB WHITE DWARFS (DBV) The eight presently known variable DB-type white dwarfs are all confined to a narrow effective-temperature range between about 21,500 and 24,000 K. The location of five of these variables on the HR diagram is shown in Figure 1 of GS95; the temperature determinations are those by Thejll et al (1991). A luminosity of $\log L/L_{\odot} = -1.3$ was assumed for all of them; only that of GD 358 is based on the estimate of Winget et al (1994). For DB stars, as for other white dwarf families, the quantification of physical parameters is still controversial. The effective temperatures obtained by Thejll et al (1991) are significantly lower than those by Liebert et al (1986) for example. The uncertainties in $T_{\rm eff}$, which exceed 1000 K, handicap the pulsational analyses of the driving mechanism by making it difficult to establish the borders of the instability strip.

The observed periods all fall into the interval between about 140 and 1000 seconds. Most DBV power spectra are complicated with many frequencies arranged in well discernible groups in frequency space. The power spectra are not always stable in time, although in part this may be due to undersampling of the data. Even in the WET multisite campaign this problem remained for PG1115 (Clemens et al 1993). Winget (1988b) considered the power spectrum of PG1351 to be the only resolved one; it is also the simplest one. Recently, a WET campaign on GD358 (Winget et al 1994) provided more than 180 significant peaks in its impressive power spectrum. From the observed triplet structures (indicating $\ell = 1$ modes) of the different radial orders, estimates of the amount of differential rotation were made. The period spacings between consecutive radial orders indicate a mass of 0.61 $M_{\odot}$. The different behavior of prograde and retrograde modes led to the postulation of a possible kG magnetic field in GD 358.

The pulsational driving of the DB variability is attributed to partial second He ionization. Driving of g-modes seems to occur, and to lead to results consistent with observations, for helium-rich surface layers ranging from 10^{-8} to 10^{-2} M_* (Bradley & Winget 1994). Helium-layer masses below about 10^{-6} M_* (Pelletier et al 1986) are, however, expected to lead to mixing that would transform DB white dwarfs into stars with carbon-enhanced surface layers. The important point is that the mass of the He-rich layers does not seem to be crucial for discriminating between stability and instability. Its thickness is pivotal, however, in determining the size of the mode trapping cycles and hence in the selection of eventually observable modes (Bradley et al 1993). Results from multisite studies of GD 358 and PG1115 indicate rather low-mass He blankets of $\approx 10^{-6} M_*$ and $\lesssim 10^{-4}$ M_*, respectively, on these two DBV stars (Clemens et al 1993, Winget et al 1994). In theoretical modeling the convective efficiency plays an important role in fixing the exact location of the edges of the instability strip. The blue boundaries are usually adjusted to concur with the hottest known DBV stars by adapting the convective efficiency in the model envelopes. With the presently adopted dialects of the mixing-length formalism for convection, unusually efficient convection zones are called for to reach agreement with observational data (see, for example, Bradley & Winget 1994 for a discussion).

3.2.3 VARIABLE DA WHITE DWARFS (DAV) At still lower effective temperatures along the white-dwarf cooling sequences are the variable hydrogen-rich white dwarf (DAV or ZZ Cet) stars. Spectroscopy of the atmospheres of DA white dwarfs shows a black-body continuum with superposed H-absorption lines; helium and metals are essentially absent (Koester & Chanmugam 1990). It is believed that most, if not all DA white dwarfs, will oscillate when they enter the appropriate temperature interval. There are some indications that some stable DA stars exist inside the ZZ Cet instability domain (Kepler & Nelan 1993). The exact temperature range of the instability domain is a matter of debate. As mentioned before, also for the DAV stars, spectroscopic determination of physical parameters is complex. According to the latest studies (Bergeron et al 1995) the blue edge of the ZZ Ceti variables is believed to lie between 12,500 K and 13,700 K. Their paper contains a comprehensive discussion of the involved difficulties.

The low-amplitude ($\lesssim 0.2$ mag) photometric variability of the ZZ Ceti stars is usually multiperiodic with periods ranging from about 100 to more than 1000 s. Some of the closely spaced periods are considered to be multipletsplittings due to rotation. Others must be sets of excited modes belonging to different radial orders. The power spectra of DAV stars have considerably fewer features than are encountered in DBV and DOV stars. Hence, they are less well suited for detailed seismological analyses. Due to the relatively low number of simultaneously excited oscillation modes, Clemens (1993) tried to identify similarities between the single ZZ Ceti stars to find a typical mean oscillation spectrum for a whole sample. One important conclusion from this approach was the inference of thick, i.e. $\approx 10^{-4}\ M_*$, hydrogen surface layers, a conclusion derived by assigning $\ell = 1$ to the dominant modes. This identification is supported, at least in G117−B15A, by multicolor observations by Robinson et al (1995). For some time the large-amplitude DAV stars were considered to be promising candidates to study nonlinear mode coupling and mode switching through the observed changes in their power spectra on short timescales. Kepler (1984) and O'Donoghue (1986) challenged the mode-switching interpretation by attributing the variable power to undersampling of the data. Theoretical modeling and discussions of ZZ Ceti power spectra were recently attempted by Brickhill (1992a,b).

If the mass of the surface hydrogen layer is large enough ($\gtrsim 10^{-14}\ M_*$), partial hydrogen ionization can in principle excite low-degree g-mode oscillations of DAV stars (Dolez & Vauclair 1981, Winget et al 1982). The major source of uncertainty in theoretically destabilizing DA white dwarfs is the coincidence of the driving region with the bottom of the convection zone, which is caused by the ionization of hydrogen. In the equilibrium models for DAV stars, below about 12,000 K, only a small fraction of the total energy passing through the convection zone is transported by photon diffusion. Hence, convection is dominating locally. Despite the convective turn-over timescale being shorter than the observed oscillation periods of ZZ Ceti stars, the nonradial stability analyses are usually carried out neglecting the perturbation of the convective flux.

It is generally agreed on that the theoretical location of the blue edge for the ZZ Ceti oscillations depends sensitively on the parameterization of convective efficiency in the equilibrium stellar model (e.g. Cox et al 1987, Bradley & Winget 1994). Only unusually extended convection zones (compared with other stellar applications) seem to lead to blue edges that agree with present observational data. Indeed, two-dimensional convection simulations on white dwarfs by Ludwig et al (1994) resulted in much shallower convection zones than those postulated on pulsational grounds. The thicknesses were, though, compatible with usual MLT convection zones and a mixing length of about 1.5 pressure scale-heights. The figures in Ludwig et al (1994) show that a constant mixing-length approach cannot adequately describe the stratification of the surface regions. Although a reliable time-dependent convection theory is necessary to exactly solve the convection-oscillation coupling, convection in ZZ Ceti stars seems to be tractable because the convective turn-over timescale is much shorter than the relevant oscillation periods. Brickhill (1990) discussed the effect of turbulent pressure and turbulent stress on the oscillatory motion. He concluded that the horizontal motion of oscillation should be nearly independent of depth in the convection zones. Applying this result to the stability analysis for ZZ Ceti stars, Brickhill (1991) showed that the convective perturbation tends to drive oscillations and that it is the most important agent for exciting g-modes in ZZ Ceti stars rather than the κ-mechanism. Brickhill's theory can be checked by comparing it with the observed range of effective temperatures of ZZ Ceti stars and with the observed periods. In reaching conclusions care is advised as the $T_{\rm eff}$ calibrations for DAV stars by spectroscopic analyses are still uncertain by the order of 500 K or more (Koester & Allard 1993, Bergeron et al 1995).

The question of the thickness of the superficial H-rich layers on DAV stars and their influence on the instability was extensively discussed in the 1980s. In contrast to long prevailing results admitting only thin hydrogen layers ($M_{\rm H}/M_* \lesssim 10^{-8}$, Winget et al 1982), Bradley & Winget (1994), and Fontaine et al (1994) joined—for different reasons—the conclusion of Cox et al (1987) that nonradial g-mode instabilities of DA white dwarfs can occur for hydrogen layers at least as massive as $10^{-4}\ M_*$. Instabilities of g-modes are presently considered rather insensitive to the hydrogen mass floating on the surface. For GD 165 (Bergeron et al 1993), observations pointed indeed to a thick hydrogen layer, exceeding possibly $10^{-4}\ M_*$, depending on the assumption about the spherical degree of the observed oscillation modes. Based on mean power spectra properties, Clemens (1993) also favored thick hydrogen layers. The thickness of the hydrogen layer controls the trapping properties, and hence it is believed to be responsible for selecting the eventually observable modes. An extensive study of the trapping properties in stratified DAV stars was undertaken by Brassard et al (1992).

The often very stable oscillation modes of white-dwarf variables are considered to be accurate clocks for measuring their evolutionary timescales. Attempts are being made to deduce cooling rates of these degenerate pulsators from the secular variation of the periods that were monitored over more than ten years (O'Donoghue & Warner 1987, Kepler et al 1990, Kepler 1993). Upper bounds determined for $\dot{P}/P$ as yet do not contradict the standard theory of cooling white dwarfs. For the DOV star PG1159−035, the observed negative $\dot{P}$ has at first sight the wrong sign if it is assumed to be located on the cooling sequence (Winget et al 1991). Detailed modeling of DOV stars suggested, though, that the way a particular mode is trapped influences the sign of its $\dot{P}$; oscillation modes confined in the outermost layers of stars having already reached their cooling track can well exhibit period decreases (Kawaler & Bradley 1994).

Radial mode instabilities in white dwarfs were studied for both DAV stars [see Cox (1974) for early references and e.g. Cox et al (1980), Saio et al (1983)] and for DBV stars (Kawaler 1993). The theoretical studies revealed instabilities for both classes of white dwarfs. Observationally, no signature of radial modes, with expected periods below a few seconds (Robinson 1984, Kawaler et al 1994), has been detected so far. It is unclear at present if the lack of observable radial modes points to theoretical inadequacies or to very low amplitudes that are below the present detection threshold.

4. PULSATIONS IN EVOLVED VERY MASSIVE STARS

Clear evidence was collected during the Geneva-photometry monitoring campaigns that stars of luminosity class I show low-level photometric variability in essentially all spectral types between O and K (Grenon 1993). In the following, we address some aspects of massive-star stability theory that might provide explanations in terms of pulsational instabilities. In particular, we consider stars more massive than about 30 $M_{\odot}$ as they evolve off the zero-age main sequence. When such stars return towards the main-sequence region after their red-supergiant phase they may have—due to significant mass loss—a considerably higher L/M ratio than when they left the ZAMS. Strong winds and hence large mass-loss rates are also considered essential for massive stars to evolve into Wolf-Rayet stars. A self-consistent physical picture for the evolution towards the Wolf-Rayet stage has not yet emerged, however. A number of difficult fluid-dynamical problems, such as semiconvection, convective overshooting, mass loss, and rotation with accompanying instabilities influence the evolution of massive stars crucially, and most of them defy a satisfactory treatment at present. For recent numerical studies concerning the evolution of massive stars we refer to e.g. Maeder & Meynet (1987), Schaller et al (1992), or Langer et al (1994) and the literature cited therein.

4.1 *Luminous Blue Variables (LBVs)*

By LBVs we designate all those variable massive stars with luminosities exceeding some 10^4 $L_{\odot}$ having envelopes with considerable amounts of hydrogen and which are located to the red of the ZAMS on the HR diagram. Hence, we discuss within the same framework all variable stars occupying the uppermost part of Figure 1 of GS95; the data for these variables were adopted from van Genderen

(1989). Our definition is somewhat wider than what is typically encountered in the literature (see e.g. Humphreys 1989, Humphreys & Davidson 1994). In our picture of LBV stars, the α Cyg-like supergiants with luminosities around $10^4\ L_\odot$ and masses of the order of 10 $M_\odot$ (Lucy 1976) as well as intermediate variable supergiants are also included. The very luminous variables discovered in nearby galaxies (in particular in M31 and M33) were called Hubble-Sandage variables in the past (Viotti 1992). Based on the characteristics of their variability there are, however, good reasons to assume that they can be attributed to what is called LBV in the Galaxy and in the Magellanic clouds.

The LBVs belong to the most luminous stars that exist and hence are of considerable interest for extragalactic applications. Light and radial velocity vary on timescales from weeks to possibly centuries. The long-term variations, with brightness changes of several magnitudes, are usually referred to as outbursts, which may or may not repeat (see references in van Genderen 1989 or Wolf 1992). Interestingly, the light variability is essentially a reflection of variable bolometric correction only. The bolometric luminosity remains roughly constant during an outburst so that the stars move horizontally across the HR diagram (see Figure 7 in Wolf 1992). Short-term variability, which is also observed, occurring on the timescale of weeks to months and having small amplitudes, shows recurring patterns that might indicate a pulsational origin (cf de Jager 1980, Sterken 1989). The observational data base is, despite considerable efforts from the observers' side, rather meager for conclusive analyses. Probably the best temporal coverage exists for HD 160529 (Sterken et al 1991). For this star, which is not unlike other LBVs, a characteristic timescale of 57 days was derived. This number is compatible with periods of radial modes for a star with the parameters suggested by Sterken et al (1991). It is presently unknown if the short-term variations are also occurring at constant bolometric luminosity. Short-term, small-scale variability is not considered to be an outstanding characteristic behavior for LBV stars as understood by Humphreys (1989) since this is also found for lower luminosity supergiants that do not show eruptions, such as e.g. R 71 or S Dor. For the discussion of the pulsation hypothesis as the origin of the short-term variability such a distinction is possibly not vital.

Recently, Glatzel & Kiriakidis (1993b) and Kiriakidis et al (1993) reported on extensive linear, radial stability analyses of massive stellar-evolution models. They followed stars between 20 and 200 $M_\odot$ through the hydrogen-burning phase. A sketch of the situation is shown in the upper left part of Figure 3. The first study (Glatzel & Kiriakidis 1993b), which was based on the old Los Alamos opacity data, showed radial instabilities when the stars evolved away from the ZAMS. An impressive enhancement of the strength of instability was

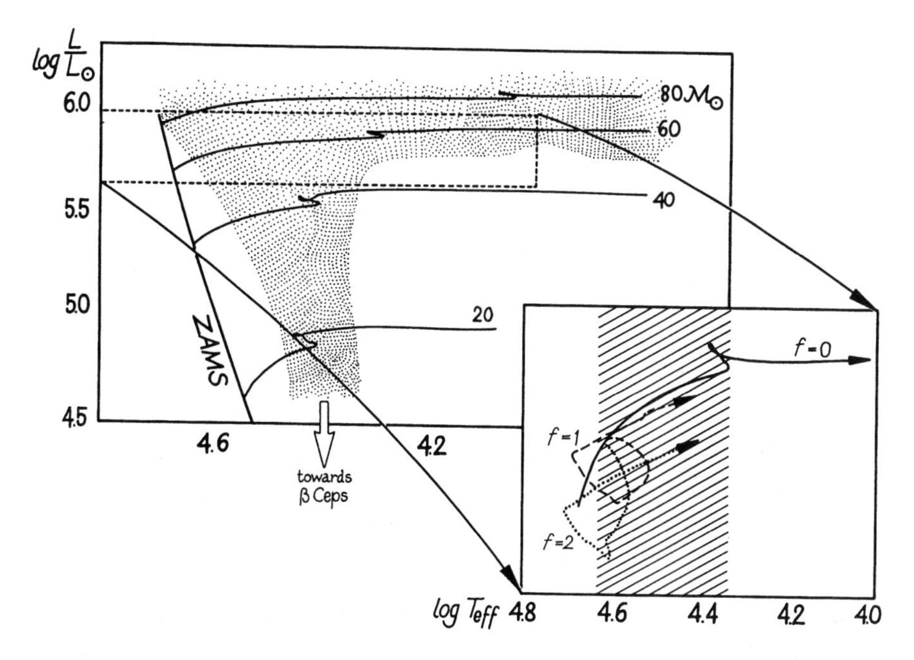

Figure 3 Luminous part of the HR diagram. Evolutionary paths of massive-star evolution according to Kiriakidis et al (1993). The dotted region outlines the extent of the radial instabilities encountered. The inset at the lower right, adapted from Langer et al (1994), shows the influence of mass loss on the evolutionary paths. The mass-loss rates were scaled with the growth rates of the most unstable mode from linear stability computations assuming different efficiencies f over the hatched interval.

later achieved with the use of OPAL opacity data (Kiriakidis et al 1993). The instability domain is sketched by the dotted area in Figure 3.

The extent of the instability regions depends on the amount of heavy elements. Even for low Z (= 0.004), pulsational instability was encountered at masses above 65 $M_\odot$ in regions of the HR diagram that are compatible with observed LBVs. For higher heavy-element abundances, the instability region extended to higher $T_{\rm eff}$ as well as to lower luminosities and seemed to confine itself to the S-bend phase of evolution and eventually towards the instability domain of the β Cepheids (see Figure 3).

The radial-mode instabilities in massive stars can achieve growth rates that are several magnitudes larger than those found in classical pulsators. This led to the suspicion that these pulsations become violent enough to induce mass loss. Preliminary nonlinear simulations (Kiriakidis 1992) resulted in very rapidly growing pulsations indeed. They appear to be a mixture of the different unstable modes found in the linear analyses. The eventual nonlinear motion led to considerable mass loss; stable limit cycles were not often found. Whether the pulsations occurring in the stellar models of Kiriakidis et al (1993) can be causally connected with the eruptive phases of the LBVs remains to be seen. At least part of the short-term behavior can be attributed to pulsational instabilities.

Again, as in other high L/M stars such as helium stars, the unstable modes in the massive stars of Kiriakidis et al (1993) were found to be strange modes. They must be attributed to the strong influence of the He II ionization zone and the Z-bump on the acoustic cavity of the stellar envelopes, giving rise to a rich unstable oscillation spectrum.

In a first attempt, Langer et al (1994) used results from the linear pulsation calculations of Kiriakidis et al (1993) to parameterize the mass loss in their stellar evolution calculations. The mass-loss rate within the hatched area in the lower-right inset of Figure 3 was assumed to be proportional to the growth rate of the most unstable pulsation mode. The influence of different scaling factors f of this mass-loss rate on the evolutionary tracks is displayed. Langer et al (1994) performed such calculations in their attempt to devise an evolutionary scenario to interweave consistently the different spectroscopic subgroups of very massive stars.

Model envelopes considered appropriate for α Cyg-like objects showed that the essential prerequisite for pulsational instabilities to develop is a sufficiently high L/M ratio (Gautschy 1992). Hence, the short-term variability of low-luminosity (compared with typical LBVs), intermediate-type supergiants might be understandable in basically the same framework as the LBVs. The quantitative results of Gautschy (1992) are probably outdated, due to the old Los Alamos data used to construct the envelopes. The existence of the Z-bump in the new opacity data reduces the L/M ratio for pulsational instabilities to occur in very luminous stars.

4.2 *Wolf-Rayet Stars*

Wolf-Rayet (WR) stars—either as type WN or WC—occupy the area on the HR diagram between $4.5 \lesssim \log L/L_\odot \lesssim 6$ and at $\log T_{\rm eff} \gtrsim 4.6$ (Langer et al 1994). The WR stars, whose masses are estimated to be higher than about 4 $M_\odot$, probably originate from stars more than 40 $M_\odot$ on the main sequence after experiencing heavy mass loss during their early evolution. Photometrically, as well as spectroscopically, some WR stars show variability timescales of several hours (e.g. Vreux 1986, Koenigsberger & Auer 1987, Gosset et al 1989, van Genderen et al 1990). Clear periodicities have not been established, and the physical origin of the variability is not known. The large mass-loss rates associated with WR stars and the estimates that the momentum in the wind frequently exceeds the momentum contained in the radiation field led rather early to the conjecture that pulsations could be involved. Only recently, Lucy & Abbott

(1993) showed that multiple scattering in expanding WR envelopes can, under suitable conditions, transfer sufficient momentum into the wind for it to achieve magnitudes comparable with observational estimates. Besides pulsations of the WR envelopes, hydrodynamic instabilities in the dense winds are considered to be the source of light and radial-velocity variations. Matthews & Beech (1987) argued that pulsations might not be observable at all spectroscopically due to the long geometrical paths of photons through the extensive expanding atmospheres, which would smear out variations on shor timescales.

For many years unstable pulsation modes were sought, in particular ones driven by the ϵ-mechanism. The rather compact structure of WR stars assures considerable relative amplitudes of the pulsational displacement in the innermost regions of the stars, facilitating the efficiency of nuclear burning on driving pulsational instabilities (Noels & Gabriel 1981, Maeder 1985, Cox & Cahn 1988). In sufficiently massive stars the radial fundamental mode is destabilized by the ϵ-mechanism during He core burning. All periods of unstable modes remained below one hour. The very low growth rates (of the order of the evolutionary timescale of the stars) could hardly be responsible for strong pulsation-driven mass loss. To investigate the longer-period domain the stability of g-modes, which can be trapped in shell-burning regions, was studied. These modes appeared particularly appropriate for explaining the WN phases during massive-star evolution. The quasi-adiabatic analyses of Noels & Scuflaire (1986) and Scuflaire & Noels (1986) revealed very brief evolutionary phases during which the g_1-mode at low and intermediate spherical degree ($\ell < 10$) became weakly unstable. The resulting periods at fractions of a day compared, though, rather favorably with the observed timescales. Cox & Cahn (1988), in their fully nonadiabatic analysis, could not verify the Noels-Scuflaire results. The origin of the discrepancy remains unresolved.

In view of the rather dramatic instabilities found recently by Glatzel et al (1993) the ϵ-driven instability loses much of its appeal. Glatzel and collaborators performed stability analyses on homogeneous helium main-sequence star models. Their models were chosen to be appropriate approximations for WC-type WR stars. Besides the well-known instability of the radial fundamental mode—setting in above about 15 $M_{\odot}$—due to ϵ-driving, they encountered strongly unstable modes crossing the regular acoustic mode spectrum. The growth times of these unstable modes eventually reached values of only a few dynamical timescales. Such violently unstable modes can be imagined as promising candidates to at least *initiate* mass loss. Presently available are only linear analyses on static background models. It would be highly desirable to follow the nonlinear evolution of these unstable modes, of which several can occur simultaneously. It is not yet clear how a possible onset of mass loss affects the pulsational instability, i.e. if the oscillation modes are stabilized and how strongly, due to the presence of a velocity field. In any case, from the point of view of relevant instabilities the situation in the WR domain is comparable to the one of LBV stars: The strange modes that were missed in earlier studies were identified as the dominating pulsation modes.

5. THE FUTURE

On the observational side, the CCD-photometry experiment of Gilliland & Brown (1992) proved that many smaller university observatories located in mediocre climatic environments could be revitalized to perform profitable variable star work if the data obtained are analyzed with appropriate care. Either space-borne experiments or collaborative campaigns around the globe to monitor variable stars and provide very long time series of observations will allow detailed analysis of multiperiodic oscillators. In particular, such approaches will enable the detection of closely spaced frequencies in the power spectra and suppress the cumbersome sidelobe-effects due to monitoring gaps.

The increasing spectral resolution and stability of spectrographs are going to provide radial-velocity data of stars with the attempted resolution below 1 m sec^{-1} in the near future. Such accuracy will allow solar-type oscillations to be discovered in distant stars. The monitoring of equivalent-width variations (Kjeldsen et al 1995) might turn out to be a competitive alternative to searching for solar-type oscillations in stars requiring off-the-shelf spectrographs only.

Large homogeneous observational data sets of survey projects (such as the microlensing projects MACHO, EROS, OGLE) provide an important basis for statistical studies with pulsating stars. First results from different projects are available (Cook et al 1995, Beaulieu et al 1995, Udalski et al 1994). The number densities of pulsating variables at their different locations throughout the HR diagram will help to test and improve, when deduced from statistically meaningful samples, our understanding of the underlying stellar evolution.

The points mentioned above contribute to establishing stellar pulsations as a reliable tool to study a variety of aspects of the internal constitution of stars. Pulsation theory is now entering the era of a detail-rich quantitative theory. Despite its glorious history several important and exciting topics within stellar pulsation theory still need to be developed or even correctly formulated. Pulsation-convection interaction, hydromagnetic waves, and pulsation-rotation coupling are examples of basic fluid-dynamical processes that still need much effort before their effects on stellar structure and pulsation are understood quantitatively.

Nonlinear pulsation simulations including time-dependent convection and detailed radiation transport need further development. In particular, in cases when mass loss is expected to set in, reliable nonlinear solutions are required to get even a glimpse of the final state of the pulsating system. The long-term nonlinear behavior of pulsating stars is far from clear. And results from decade-long monitoring efforts prove that even "simple" pulsating stars do not necessary have very stable limit cycles. As for nonradial pulsations, no numerical methods are known to exist that would allow, with sufficient spatial resolution, unstable modes to be followed into their nonlinear regimes.

However, phenomena will certainly be discovered that will provide new insights into these issues.

ACKNOWLEDGMENTS

AG is indebted to the Swiss National Science Foundation for the financial support through a PROFIL2 fellowship. We are grateful to NH Baker's critically improving the language in an early version of the manuscript.

CORRIGENDUM TO PART I

Equation (18) should read

$$-\sigma^2\xi_r - \frac{1}{r^4\rho}\frac{\mathrm{d}}{\mathrm{d}r}\left(\Gamma_1 p r^4 \frac{\mathrm{d}\xi_r}{\mathrm{d}r}\right) - \frac{1}{\rho r}\left\{\frac{\mathrm{d}}{\mathrm{d}r}[(3\Gamma_1 - 4)p]\right\}\xi_r = 0.$$

Literature Cited

The following abbreviations of conference proceedings are used in the text and in the reference list:

Bo90: *Confrontation between Stellar Pulsation and Evolution, ASP Conf. Ser.* Vol. 11, ed. C Cacciari, G Clementini. San Francisco: Astron. Soc. Pac. (1990)

Buda88: *Multimode Stellar Pulsations,* ed. G Kovács, L Szabados, B Szeidl. Budapest: Kultura (1988)

CT95: *Astrophysical Applications of Stellar Pulsation, ASP Conf. Ser.* Vol. 82, IAU Coll. 155, ed. RS Stobie, PA Withelock. San Francisco: Astron. Soc. Pac. (1995)

GONG92: *Seismic Investigation of the Sun and the Stars, ASP Conf. Ser.* Vol. 42, ed. TM Brown. San Francisco: Astron. Soc. Pac. (1993)

GONG94: *Helio- and Asteroseismology, ASP Conf. Ser.* Vol. 76, ed. RK Ulrich, EJ Rhodes Jr, W Däppen. San Francisco: Astron. Soc. Pac. (1995)

ITS92: *Inside the Stars, ASP Conf. Ser.* Vol. 40, IAU Coll 137, ed. WW Weiss, A Baglin. San Francisco: Astron. Soc. Pac. (1993)

PSSS: *Progress of Seismology of the Sun and Stars,* ed. Y Osaki, H Shibahashi, *Lecture Notes Phys.* 367. New York: Springer-Verlag (1990)

Vic92: *New Perspectives on Stellar Pulsation and Pulsating Variable Stars,* ed. JM Nemec, JM Matthews. Cambridge: Cambridge Univ.

Press (1993)
Wd88: *White Dwarfs,* IAU Coll. 114, ed. G Wegner, *Lecture Notes Phys.* 328. New York: Springer-Verlag (1989)
Wd92: *White Dwarfs: Advances in Observation and Theory,* ed. MA Barstow, NATO ASI Ser. C, Vol. 403. Dordrecht: Kluwer (1993)
*****

Aerts C, De Pauw MD, Waelkens C. 1992. *Astron. Astrophys.* 266:294–306
Aerts C, Waelkens C. 1993. *Astron. Astrophys.* 273:135–46
Africano J. 1977. *Inf. Bull. Variable Stars* 1301
Aikawa T. 1987. *Astrophys. Space Sci.* 139:81–93
Aikawa T. 1993. *MNRAS* 262:893–900
Alcock C, Allsman RA, Axelrod TS, Bennett DP, Cook KH, et al. 1995. *Astron. J.* 109:1653–62
Ando H. 1986. *Astron. Astrophys.* 163:97–104
Andreasen GK. 1988. *Astron. Astrophys.* 201:72–79
Andreasen GK, Hejlesen PM, Petersen JO. 1983. *Astron. Astrophys.* 121:241–49
Baade D. 1982. *Astron. Astrophys.* 105:65–75
Baade D. 1984. *Astron. Astrophys.* 135:101–6
Baade D. 1987. In *Physics of Be Stars,* ed. A Slettebak, TP Snow, pp. 361–83. Cambridge: Cambridge Univ. Press
Baade D. 1989. *Astron. Astrophys.* 222:200–4
Babcock HW 1958. *Ap. J. Suppl.* 3:141–210
Baglin A. 1972. *Astron. Astrophys.* 19:45–50
Baglin A. 1976. In *Multiple Periodic Variable Stars,* IAU Coll. 29, ASSL Vol. 29, pp. 223–46, ed. WS Fitch. Dordrecht: Reidel
Baglin A, Breger M, Chevalier C, Hauck B, le Contel J-M, et al. 1973. *Astron. Astrophys.* 23:221–40
Bailyn CD. 1995. *Rev. Astron. Astrophys.* 33:133–62
Balmforth NJ, Gough DO, Merryfield WJ. 1990. In *From Miras to Planetary Nebulae: Which Path for Stellar Evolution?* ed. MO Mennessier, A Omont, pp. 85–87. Gif-sur-Yvette: Editions Frontières
Balona LA. 1985. *MNRAS* 217:L17–21
Balona LA. 1986a. *MNRAS* 219:111–29
Balona LA. 1986b. *MNRAS* 220:647–56
Balona LA. 1987. *MNRAS* 224:41–52
Balona LA. 1990a. In PSSS, pp. 443–48
Balona LA. 1990b. *MNRAS* 245:92–100
Balona LA. 1992. *MNRAS* 256:425–36
Balona LA. 1993. *MNRAS* 260:795–802
Balona LA. 1994. *MNRAS* 267:1060–70
Balona LA, Engelbrecht CA. 1981. *MNRAS* 212:889–97
Balona LA, Hearnshaw JB, Koen C, Collier A, Machi I, et al. 1994b. *MNRAS* 267:103–10
Balona LA, Koen C. 1994. *MNRAS* 267:1071–80
Balona LA, Krisciunas K, Cousins AWJ. 1994a. *MNRAS* 270:905–13
Balona LA, Rozowsky J. 1991. *MNRAS* 251:L66–68
Balona LA, Sterken C, Manfroid J. 1991. *MNRAS* 252:93–101
Beaulieu JP, Grison P, Tobin W, Pritchard JD, Ferlet R, et al. 1995. *Astron. Astrophys.* 303:137–54
Becker SA. 1985. In *Cepheids: Theory and Observations,* ed. BF Madore, pp. 104–25. Cambridge: Cambridge Univ. Press
Belmonte JA, Roca Cortés T, Vidal I, Schmider FX, Michel E, et al. 1993. In ITS92, pp. 739–41
Bergeron P, Fontaine G, Brassard P, Lamontagne R, Wesemael F, et al. 1993. *Astron. J.* 106:1987–99
Bergeron P, Wesemael F, Lamontagne R, Fontaine G, Saffer RA, Allard NF. 1995. *Ap. J.* 449:258–79
Berthomieu G, Gonczi G, Graff Ph, Provost J, Rocca A. 1978. *Astron. Astrophys.* 70:597–606
Bessell MS, Brett JM, Scholz M, Wood PR. 1989. *Astron. Astrophys.* 213:209–25
Blacher S, Perdang J. 1988. In Buda88, pp. 283–99
Bond HE, Ciardullo R. 1989. In Wd88, pp. 473–76
Bond HE, Ciardullo R, Kawaler SD. 1993. *Acta Astron.* 43:425–30
Bono G, Caputo F, Stellingwerf RF. 1994. *Ap. J.* 423:294–304
Bono G, Stellingwerf RF. 1993. In Vic92, pp. 275–276
Bowen GH. 1988. *Ap. J.* 329:299–317
Bradley PA, Winget DE. 1994. *Ap. J.* 421:236–44
Bradley PA, Winget DE, Wood MA. 1993. *Ap. J.* 406:661–73
Brassard P, Fontaine G, Wesemael F, Hansen CJ. 1992. *Ap. J. Suppl.* 80:369–401
Breger M. 1979. *Publ. Astron. Soc. Pac.* 91:5–26
Breger M, Handler G, Nather RE, Winget DE, Kleinman SJ, et al. 1995. *Astron. Astrophys.* 297:473–82
Brickhill AJ. 1990. *MNRAS* 246:510–17
Brickhill AJ. 1991. *MNRAS* 251:673–80
Brickhill AJ. 1992a. *MNRAS* 259:519–28
Brickhill AJ. 1992b. *MNRAS* 259:529–35
Brown TM, Christensen-Dalsgaard J, Weibel-Mihalas B, Gilliland RL. 1994. *Ap. J.* 427:1013–34
Brown TM, Gilliland RL. 1994. *Annu. Rev. Astron. Astrophys.* 32:37–82
Buchler JR, Goupil M-J. 1984. *Ap. J.* 279:394–400
Buchler JR, Goupil M-J. 1988. *Astron. Astro-*

phys. 190:137–47
Buchler JR, Kovács G. 1986. *Ap. J.* 308:661–68
Buchler JR, Kovács G. 1987. *Ap. J. Lett.* 320:L57–62
Buchler JR, Perdang JM, Spiegel EA. 1985. *Chaos in Astrophysics,* NATO ASI Ser. C, Vol. 161. Dordrecht: Reidel
Buchler JR, Serre T, Kolláth Z, Mattei Z. 1995. *Phys. Rev. Lett.* 73:842–45
Buonanno R, Corsi CE, Fusi Pecci F. 1989. *Astron. Astrophys.* 216:80–108
Burki G, Schmidt EG, Arellano Ferro A, Fernie JD, Sasselov D, et al. 1986. *Astron. Astrophys.* 168:139–46
Buta RJ, Smith MA. 1979. *Ap. J.* 232:213–35
Butler RP. 1992. *Ap. J. Lett.* 394:L25–27
Campos AJ, Smith MA. 1980. *Ap. J.* 238:250–65
Cannizzo JK, Goodings DA, Mattei JA. 1990. *Ap. J.* 357:235–42
Caputo F, Castellani V, Tornambé A. 1989. *Astron. Astrophys.* 222:121–24
Chiosi C. 1990. In Bo90, pp. 158–92
Christensen-Dalsgaard J. 1993. In ITS92, pp. 483–96
Christensen-Dalsgaard J, Frandsen S. 1983. *Sol. Phys.* 82:469–86
Christensen-Dalsgaard J, Petersen JO. 1995. *Astron. Astrophys.* 299:L17–20
Christy RF. 1968. *Q. J. R. Astron. Soc.* 9:13–39
Clemens JC. 1993. *Baltic Astron.* 2:407–34
Clemens JC, Barstow MA, Nather RE, Winget DE, Bradley PA, et al. 1993. In Wd92, pp. 515–21
Clement CM, Kinman TD, Suntzeff NB. 1991. *Ap. J.* 372:273–80
Cook KH, Alcock C, Allsman RA, Axelrod TS, Freeman KC, Peterson BA, et al. 1995. In CT95, pp. 221–31
Cousens A. 1983. *MNRAS* 203:1171–82
Cox AN. 1980. *Annu. Rev. Astron. Astrophys.* 18:15–41
Cox AN. 1983. In *Astrophysical Processes in Upper Main Sequence Stars,* Saas-Fee Course No.13, pp. 82–100
Cox AN, Cahn JH. 1988. *Ap. J.* 326:804–12
Cox AN, Hodson SW, Starrfield SG. 1980. In *Nonradial and Nonlinear Stellar Pulsation,* ed. HA Hill, WA Dziembowski, *Lecture Notes Phys.* 125:458–66. New York: Springer-Verlag
Cox AN, King DS, Hodson SW. 1979. *Ap. J.* 231:798–807
Cox AN, McNamara BJ, Ryan W. 1984. *Ap. J.* 284:250–56
Cox AN, Starrfield SG, Kidman RB, Pesnell WD. 1987. *Ap. J.* 317:303–24
Cox JP. 1974. *Rep. Prog. Phys.* 37:563–698
Cox JP. 1975. *Mem. Soc. R. Sci. Liège,* Coll. 8, 6e Ser. 8:129–59
Cox JP. 1984. *Ap. J.* 280:220–27
Cugier H. 1993. *Acta Astron.* 43:27–38
Cugier H, Boratyn DA. 1992. *Acta Astron.* 42:191–209
Cugier H, Dziembowski WA, Pamyatnykh AA. 1994. *Astron. Astrophys.* 291:143–54
de Jager C. 1980. *The Brightest Stars, Geophys. Astrophys. Monogr.,* Vol. 19. Dordrecht: Reidel
de Jager C, Nieuwenhuijzen H. 1991. *Instabilities in Evolved Super- and Hypergiants.* Amsterdam: R. Netherlands Acad. Arts Sci.
Deming D, Mumma MJ, Espenak F, Jennings DE, Kostink T, et al. 1989. *Ap. J.* 343:456–67
Deupree RG. 1977a. *Ap. J.* 211:509–26
Deupree RG. 1977b. *Ap. J.* 214:502–9
Deupree RG, Hodson SW. 1977. *Ap. J.* 218:654–58
Diethelm R. 1990. *Astron. Astrophys.* 239:186–92
Diethelm R. 1996. *Astron. Astrophys.* 307:803–6
Dolez N, Gough DO. 1982. In *Pulsations in Classical and Cataclysmic Variable Stars,* ed. JP Cox, CJ Hansen, pp. 248–56. Boulder: JILA
Dolez N, Vauclair G. 1981. *Astron. Astrophys.* 102:375–85
Dorman B. 1992a. *Ap. J. Suppl.* 80:701–24
Dorman B. 1992b. *Ap. J. Suppl.* 81:221–50
Dreizler S, Werner K, Heber U. 1995. In *White Dwarfs, Proc. 9th European Workshop on White Dwarfs,* ed. D Koester, K Werner, *Lecture Notes Phys.* 443:160–70. New York: Springer-Verlag
Dukes RJ Jr. 1974. *Ap. J.* 192:81–91
Dziembowski WA. 1977a. *Acta Astron.* 27:95–126
Dziembowski WA. 1977b. *Acta Astron.* 27:203–11
Dziembowski WA. 1982. *Acta Astron.* 32:147–71
Dziembowski WA, Goode PR. 1985. *Ap. J. Lett.* 296:L27–30
Dziembowski WA, Goode PR. 1986. In *Seismology of the Sun and Distant Stars,* ed. DO Gough, NATO ASI C, Vol. 169, pp. 441–51. Dordrecht: Reidel
Dziembowski WA, Królikowska M. 1985. *Acta Astron.* 35:5–28
Dziembowski WA, Królikowska M. 1990. *Acta Astron.* 40:19–26
Dziembowski WA, Królikowska M, Kosovichev A. 1988. *Acta Astron.* 38:61–75
Dziembowski WA, Moskalik P, Pamyatnykh AA. 1993. *MNRAS* 265:588–600
Dziembowski WA, Pamyatnykh AA. 1991. *Astron. Astrophys.* 248:L11–14
Dziembowski WA, Pamyatnykh AA. 1993. *MN-*

RAS 262:204–12
Eggen OJ. 1979. *Ap. J. Suppl.* 41:413–34
Fadeyev YuA, Fokin AB. 1985. *Astrophys. Space Sci.* 111:355–74
Feast MW, Glass IS, Whitelock PA, Catchpole RM. 1989. *MNRAS* 241:375–92
Fernie JD. 1990. *Ap. J.* 354:295–301
Fernie JD. 1994. *MNRAS* 271:L19–20
Fernie JD, Kamper KW, Seager S. 1993. *Ap. J.* 416:820–24
Fernie JD, Sasselov DD. 1989. *Publ. Astron. Soc. Pac.* 101:513–15
Feuchtinger MU, Dorfi EA, Höfner S. 1993. *Astron. Astrophys.* 273:513–23
Feuchtinger MU, Dorfi EA. 1994. *Astron. Astrophys.* 291:209–25
Fontaine G, Brassard P, Wesemael F, Tassoul M. 1994. *Ap. J. Lett.* 428:L61–64
Fox MW, Wood PR. 1982. *Ap. J.* 259:198–212
Gabriel M. 1969. In *Low-Luminosity Stars,* ed. SS Kumar, pp. 267–77. New York: Gordon & Breach
Gabriel M, Noels A, Scuflaire R, Mathys G. 1985. *Astron. Astrophys.* 143:206–8
Gautschy A. 1987. *Vistas Astron.* 30:197–241
Gautschy A. 1992. *MNRAS* 259:82–88
Gautschy A. 1993. *MNRAS* 265:340–46
Gautschy A, Saio H. 1993. *MNRAS* 262:213–19
Gautschy A, Saio H. 1995. *Annu. Rev. Astron. Astrophys.* 33:75–113 (GS95)
Gehmeyr M. 1992a. *Ap. J.* 399:265–71
Gehmeyr M. 1992b. *Ap. J.* 399:272–83
Gehmeyr M. 1993. *Ap. J.* 412:341–50
Gies D. 1994. In *Pulsation, Rotation and Mass Loss in Early-Type Stars, IAU Symp. 162,* ed. LA Balona, H Herichs, JM Le Contel, pp. 89–99. Dordrecht: Kluwer
Gilliland RL, Brown TM. 1992. *Astron. J.* 103:1945–54
Gilliland RL, Brown TM, Duncan DK, Suntzeff NB, Lockwood GW, et al. 1991. *Astron. J.* 101:541–61
Gingold RA. 1976. *Ap. J.* 204:116–30
Gingold RA. 1985. *Mem. Soc. Astron. Ital.* 56:169–91
Glatzel W, Kiriakidis M. 1993a. *MNRAS* 262:85–92
Glatzel W, Kiriakidis M. 1993b. *MNRAS* 263:375–84
Glatzel W, Kiriakidis M, Fricke KJ. 1993. *MNRAS* 262:L7–11
Gloria KA. 1990. *Publ. Astron. Soc. Pac.* 102:338–43
Gosset E, Vreux J-M, Manfroid J, Sterken C, Walker EN, Haefner R. 1989. *MNRAS* 238:97–113
Gough DO, Toomre J. 1991. *Annu. Rev. Astron. Astrophys.* 29:627–85
Grenon M. 1993. In ITS92, pp. 693–707
Guzik JA, Cox AN. 1993. *Astrophys. Space Sci.* 210:307–9
Haniff CA, Gehz AM, Gorham PW, Kulkarni SR, Matthews K, Neugebauer G. 1992. *Astron. J.* 103:1662–67
Heller CH, Kawaler SD. 1988. *Ap. J. Lett.* 329:L43–46
Höfner S, Feuchtinger M, Dorfi EA. 1995. *Astron. Astrophys.* 297:815–27
Höppner W, Kähler H, Roth ML, Weigert A. 1978. *Astron. Astrophys.* 63:391–99
Hughes SMG. 1993. In Vic92, pp. 192–200
Humphreys RM 1989. In *Physics of Luminous Blue Variables,* ed. K Davidson, AFJ Moffat, HJGLM. Lamers, ASSL Vol. 157, pp. 3–14. Dordrecht: Kluwer
Humphreys RM, Davidson K. 1994. *Publ. Astron. Soc. Pac.* 106:1025–51
Hutton RG, Méndez RH. 1993. *Astron. Astrophys.* 267:L8–10
Jerzykiewicz M, Wenzel W. 1977. *Acta Astron.* 27:35–50
Jones TJ, Bryja CO, Gehrz RD, Harrison TE, Johnson JJ, et al. 1990. *Ap. J. Suppl.* 74:785–817
Jura M. 1986. *Ap. J.* 309:732–36
Jura M, Kleinmann SG. 1992. *Ap. J. Suppl.* 83:329–49
Kambe E, Ando H, Hirata R. 1990. *Publ. Astron. Soc. Jpn.* 42:687–710
Kambe E, Ando H, Hirata R. 1993a. *Astron. Astrophys.* 273:435–50
Kambe E, Ando H, Hirata R, Walker GAH, Kennelly EJ, Matthews JM. 1993b. *Publ. Astron. Soc. Pac.* 105:1222–31
Kambe E, Osaki Y. 1988. *Publ. Astron. Soc. Jpn.* 40:313–29
Kawaler SD. 1988. *Ap. J.* 334:220–28
Kawaler SD. 1990. In Bo90, pp. 494–512
Kawaler SD. 1993. *Ap. J.* 404:294–304
Kawaler SD, Bond HE, Sherbert LA, Watson TK. 1994. *Astron. J.* 107:298–305
Kawaler SD, Bradley PA. 1994. *Ap. J.* 427:415–28
Kawaler SD, Hansen CJ. 1989. In Wd88, pp. 97–108
Kawaler SD, O'Brien MS, Clemens JC, Nather RE, Winget DE, et al. 1995. *Ap. J.* 450:350–63
Kawaler SD, Winget DE, Hansen CJ, Iben I Jr. 1986. *Ap. J. Lett.* 306:L41–44
Kenelly EJ, Walker GAH, Merryfield WJ. 1992. *Ap. J. Lett.* 400:L71–74
Kepler SO. 1984. *Ap. J.* 278:754–60
Kepler SO. 1993. *Baltic Astron.* 2:444
Kepler SO, Nelan EP. 1993. *Astron. J.* 105:608–13
Kepler SO, Vauclair G, Dolez N, Chevreton M, Barstow MA, et al. 1990. *Ap. J.* 357:204–7
Kiriakidis M. 1992. *Stabilität und Pulsationen*

von massereichen Sternen. PhD thesis. Univ. Göttingen
Kiriakidis M, El Eid MF, Glatzel W. 1992. *MNRAS* 255:L1–5
Kiriakidis M, Fricke KJ, Glatzel W. 1993. *MNRAS* 264:50–62
Kjeldsen H, Bedding TR, Viskum M, Frandsen S. 1995. AJ 109:1313–19
Koenigsberger G, Auer LH. 1987. *Publ. Astron. Soc. Pac.* 99:1080–83
Koester D, Allard N. 1993. In Wd92, pp. 237–43
Koester D, Chanmugam G. 1990. *Rep. Prog. Phys.* 53:837–915
Kolláth Z. 1990. *MNRAS* 247:377–86
Kolláth Z, Szeidl B. 1993. *Astron. Astrophys.* 277:62–68
Kovács G, Buchler JR. 1988a. *Ap. J.* 324:1026–41
Kovács G, Buchler JR. 1988b. *Ap. J.* 334:971–94
Kovács G, Buchler JR. 1989. *Ap. J.* 346:898–905
Kovács G, Buchler JR. 1993. *Ap. J.* 404:765–72
Kovács G, Buchler JR, Marom A. 1991. *Astron. Astrophys.* 252:L27–30
Kovács G, Buchler JR, Marom A, Iglesias CA, Rogers FJ 1992. *Astron. Astrophys.* 259:L46–48
Kovács G, Shlosman I, Buchler JR. 1986. *Ap. J.* 307:593–608
Krisciunas K, Aspin C, Geballe TR, Akazawa H, Claver CF, et al. 1993. *MNRAS* 263:781–88
Krisciunas K, Handler G. 1995. *Inf. Bull. Variable Stars* 4195
Krockenberger M, Noyes RW, Sasselov DD. 1995. *Bull. Astron. Soc. Am.* 26:1366
Kurtz DW. 982. *MNRAS* 200:807–59
Kurtz DW. 1988. In Buda88, pp. 95–106
Kurtz DW. 1989. *MNRAS* 238:1077–84
Kurtz DW. 1990. *Annu. Rev. Astron. Astrophys.* 28:607–55
Kurtz DW. 1995. In GONG94, pp. 606–617
Kurtz DW, Shibahashi H. 1986. *MNRAS* 223:557–79
Kurtz DW, Marang F. 1995. *MNRAS* 276:191–98
Kurtz DW, Martinez P, van Wyk F, Marang F, Roberts G. 1994. *MNRAS* 268:641–53
Kurtz DW, Garrison RF, Koen C, Hofmann GF, Viranna NB. 1995. *MNRAS* 276:199–205
Lampens P. 1987. *Astron. Astrophys.* 172:173–78
Langer N, Hamann W-R, Lennon M, Najarro F, Pauldrach AWA, Puls J. 1994. *Astron. Astrophys.* 290:819–33
Lauterborn D, Refsdal S, Weigert A. 1971. *Astron. Astrophys.* 10:97–117
Ledoux P. 1941. *Ap. J.* 94:537–48
Lee U. 1985a. *Publ. Astron. Soc. Jpn.* 37:261–77
Lee U. 1985b. *Publ. Astron. Soc. Jpn.* 37:279–91
Lee U. 1988. *MNRAS* 232:711–24
Lee U, Baraffe I. 1995. *Astron. Astrophys.* 301:419–32
Lee U, Jeffery CS, Saio H. 1992. *MNRAS* 254:185–91
Lee U, Saio H. 1986. *MNRAS* 221:365–76
Lee U, Saio H. 1990a. *Ap. J.* 349:570–79
Lee U, Saio H. 1990b. *Ap. J. Lett.* 359:L29–32
Lee U, Saio H. 1990c. *Ap. J.* 360:590–603
Lee U, Saio H. 1993. *MNRAS* 261:415–24
Lee Y-W, Demarque P, Zinn R. 1990. *Ap. J.* 350:155–72
Lee Y-W, Demarque P, Zinn R. 1994. *Ap. J.* 423:248–65
Li Y, Stix M. 1994. *Astron. Astrophys.* 286:815–23
Liebert J, Wesemael F, Hansen CJ, Fontaine G, Shipman HL, et al. 1986. *Ap. J.* 309:241–52
Likkel L, Omont A, Morris M, Forveille T. 1987. *Astron. Astrophys.* 173:L11–14
Lomb NR. 1978. *MNRAS* 185:325–33
Lucy LB, Abbott DC. 1993. *Ap. J.* 405:738–46
Ludwig H-G, Jordan S, Steffen M. 1994. *Astron. Astrophys.* 284:105–17
Madore BF. 1985. *Cepheids: Theory and Observations*. Cambridge: Cambridge Univ. Press
Maeder A. 1985. *Astron. Astrophys.* 147:300–8
Maeder A, Meynet G. 1987. *Astron. Astrophys.* 182:243–63
Magalhães JA, Weir AL, Conrath BJ, Gierasch PJ, Leroy SS. 1990. *Icarus* 88:39–72
Martinez P, Kurtz DW. 1990. *MNRAS* 242:636–52
Mateo M. 1993. In *Blue Stragglers*, ed. RE Saffer, *ASP Conf. Ser.* Vol. 53, pp. 74–96
Mathias P, Aerts C, Pauw MD, Gillet D, Waelkens C. 1994. *Astron. Astrophys.* 283:813–26
Mathys G. 1985. *Astron. Astrophys.* 151:315–21
Matthews JM. 1988. *MNRAS* 235:L7–11
Matthews JM. 1990a. *Astron. Astrophys.* 229:452–56
Matthews JM. 1990b. In PSSS, pp. 385–91
Matthews JM. 1991. *Publ. Astron. Soc. Pac.* 103:5–19
Matthews JM. 1993. In GONG92, pp. 303–16
Matthews JM, Beech M. 1987. *Ap. J. Lett.* 313:L25–29
Méndez RH, Fortez JC, López RH. 1986. *Rev. Mex. Astron. Astrof.* 13:119–29
Michaud G. 1970. *Ap. J.* 160:641–58
Michaud G. 1980. *Astron. J.* 85:589–98
Miller RH. 1991. *J. Comput. Phys.* 93:469–76
Moffett TJ. 1989. In *The Use of Pulsating Stars*

in Fundamental Problems of Astronomy, IAU Coll. 111, ed. EG Schmidt, pp. 191–204. Cambridge: Cambridge Univ. Press
Moskalik P. 1985. *Acta Astron.* 35:229–54
Moskalik P. 1995. In CT95, pp. 44–55
Moskalik P, Buchler JR. 1990. *Ap. J.* 355:590–601
Moskalik P, Buchler JR., Marom A. 1992. *Ap. J.* 385:685–93
Moskalik P, Dziembowski WA. 1992. *Astron. Astrophys.* 256:L5–8
Mosser B, Mékarnia D., Maillard JP, Gay J, Gautier D, Delache P. 1993. *Astron. Astrophys.* 267:604–22
Nemec JM, Mateo M. 1990. In Bo90, pp. 64–85
Nemec JM, Nemec AFL, Lutz TE. 1994. *Astron. J.* 108:222–46
Noels A, Gabriel M. 1981. *Astron. Astrophys.* 101:215–22
Noels A, Scuflaire R. 1986. *Astron. Astrophys.* 161:125–29
North P, Paltani S. 1994. *Astron. Astrophys.* 288:155–64
O'Donoghue D. 1986. *MNRAS* 220:L19–22
O'Donoghue D, Warner B. 1987. *MNRAS* 228:949–55
Osaki Y. 1971. *Publ. Astron. Soc. Jpn.* 23:485–502
Osaki Y. 1977. *Publ. Astron. Soc. Jpn.* 29:235–48
Osaki Y. 1986a. In *Seismology of the Sun and the Distant Stars,* ed. DO Gough, pp. 453–63. Dordrecht: Reidel
Osaki Y. 1986b. *Publ. Astron. Soc. Pac.* 98:30–32
Ostlie DA, Cox AN. 1986. *Ap. J.* 311:864–72
Parthasarathy M, Pottasch SR. 1986. *Astron. Astrophys.* 154:L16–19
Pelletier G, Fontaine G, Wesemael F, Michaud G, Wegner G. 1986. *Ap. J.* 307:242–52
Percy JR, Lane MJ. 1977. *Astron. J.* 82:353–59
Perdang J. 1985. *Physicalia* 7:239–303
Perdang J. 1991. In *ESO Workshop on Rapid Variability of OB-Stars: Nature and Diagnostic Value,* ed. D Baade, pp. 349–61. Garching: ESO
Perdang J. 1993. In *Cellular Automata: Prospects in Astrophysical Applications,* ed. J Perdang, A Lejeune, pp. 342–68. Singapore: World Scientific
Petersen JO. 1976. In *Multiple Periodic Variable Stars, IAU Coll. 29,* ed. WS Fitch, pp. 195–222. Dordrecht: Reidel
Pijpers FP, Habing HJ. 1989. *Astron. Astrophys.* 215:334–46
Radick RR, Lockwood GW, Thomson DT, Warnock III A, Hartmann LW, et al. 1983. *Publ. Astron. Soc. Pac.* 95:621–34
Robinson EL. 1984. *Astron. J.* 89:1732–34
Robinson EL, Mailloux TM, Zhang E, Koester D, Stiening RF, et al. 1995. *Ap. J.* 438:908–16
Romanov YuS, Udovichenko SN, Frolov MS. 1987. *Sov. Astron. Lett.* 13:29–31
Rood R. 1990. In Bo90, pp. 11–21
Saio H, Winget DE, Robinson EL. 1983. *Ap. J.* 265:982–95
Sandage A. 1982a. *Ap. J.* 252:553–73
Sandage A. 1982b. *Ap. J.* 252:574–81
Sandage A. 1990a. *Ap. J.* 350:603–30
Sandage A. 1990b. *Ap. J.* 350:631–44
Sandage A. 1993a. *Astron. J.* 106:687–702
Sandage A. 1993b. *Astron. J.* 106:703–18
Sandage A. 1993c. *Astron. J.* 106:719–25
Sandage A, Diethelm R, Tammann GA. 1994. *Astron. Astrophys.* 283:111–20
Sandage A, Katem B, Sandage M. 1981. *Ap. J. Suppl.* 46:41–74
Schaerer D, Meynet D, Maeder A, Schaller G. 1993. *Astron. Astrophys Suppl.* 98:523–27
Schaller G, Schaerer D, Meynet G, Maeder A. 1992. *Astron. Astrophys Suppl.* 96:269–331
Scuflaire R, Noels A. 1986. *Astron. Astrophys.* 169:185–88
Shibahashi H. 1983. *Ap. J.* 275:L5–9
Shibahashi H. 1991. In *Challenges to Theories of the Structure of Moderate-Mass Stars,* ed. D Gough, J Toomre, Lecture Notes Phys. 388:393–410. New York: Springer-Verlag
Shibahashi H, Saio H. 1985. *Publ. Astron. Soc. Jpn.* 37:245–59
Shibahashi H, Takata M. 1993. *Publ. Astron. Soc. Pac.* 45:617–41
Shobbrook RR, Herbison-Evans D, Johnston ID, Lomb NR. 1969. *MNRAS* 145:131–40
Shobbrook RR, Lomb NR, Herbison-Evans D. 1972. *MNRAS* 156:165–80
Simon NR. 1988. In *Pulsation and Mass Loss in Stars,* ed. R Stalio, LA Willson, pp. 27–50. ASSL. Vol. 148, Dordrecht: Kluwer
Simon NR. 1989. *Ap. J. Lett.* 343:L17–20
Simon NR. 1990. In Bo90, pp. 193–208
Simon NR, Cox AN, Hodson SW. 1980. *Ap. J.* 237:550–57
Simon NR, Lee AS. 1981. *Ap. J.* 248:291–97
Simon NR, Schmidt EG. 1976. *Ap. J.* 205: 162–64
Slettebak A. 1988. *Publ. Astron. Soc. Pac.* 100:770–84
Smak J. 1970. *Acta Astron.* 20:75–91
Smith MA. 1977. *Ap. J.* 215:574–83
Smith MA. 1980a. In *Nonradial and Nonlinear Stellar Pulsation,* ed. HA Hill, WA Dziembowski, *Lecture Notes Phys.* 125:60–75 New York: Springer-Verlag
Smith MA. 1980b. *Ap. J.* 240:149–60
Smith MA. 1983. *Ap. J.* 265:338–53
Smith MA. 1985. *Ap. J.* 297:206–23
Smith MA. 1989. *Ap. J. Suppl.* 71:357–86
Smith MA, Fitch WS, Africano JL, Goodrich

BD, Halbedel W, et al. 1984. *Ap. J.* 282:226–35
Smith MA, McCall ML. 1978. *Ap. J.* 223:221–33
Smith MA, Polidan RS. 1993. *Ap. J.* 408:323–36
Stanghellini L, Cox AN, Starrfield S. 1991. *Ap. J.* 383:766–78
Starrfield S, Cox AN, Kidman RB, Pesnell WD. 1984. *Ap. J.* 281:800–10
Starrfield S, Cox AN, Kidman RB, Pesnell WD. 1985. *Ap. J. Lett.* 293:L23–27
Stellingwerf RF. 1974. *Ap. J.* 192:139–44
Stellingwerf RF. 1975. *Ap. J.* 195:441–66
Stellingwerf RF. 1982. *Ap. J.* 262:330–38
Stellingwerf RF. 1984a. *Ap. J.* 277:322–26
Stellingwerf RF. 1984b. *Ap. J.* 277:327–32
Stellingwerf RF. 1984c. *Ap. J.* 284:712–18
Stellingwerf RF, Bono G. 1993. In Vic92, pp. 252–60
Stellingwerf RF, Gautschy A, Dickens RJ. 1987. *Ap. J. Lett.* 313:L75–79
Sterken C. 1989. In *Physics of Luminous Blue Variables,* ed. K Davidson, AFJ Moffat, HJGLM Lamers, ASSL Vol. 157, pp. 59–66. Dordrecht: Kluwer
Sterken C, Gosset E, Jüttner A, Stahl O, Wolf B, Axer M. 1991. *Astron. Astrophys.* 247:383–92
Sterken C, Jerzykiewicz M. 1993. *Space Sci. Rev.* 62:95–171
Stothers RB. 1992. *Ap. J.* 392:706–9
Stryker LL. 1993. *Publ. Astron. Soc. Pac.* 105:1081–100
Szeidl B. 1988. In Buda88, pp. 45–65
Takata M, Shibahashi H. 1995a. *Publ. Astron. Soc. Jpn.* 47:219–31
Takata M, Shibahashi H. 1995b. Preprint
Takeuti M. 1990. In *The Numerical Modelling of Nonlinear Stellar Pulsations,* ed. JR Buchler, NATO ASI Series C, Vol 302, pp. 121–41. Dordrecht: Reidel
Takeuti M, Petersen JO. 1983. *Astron. Astrophys.* 117:352–56
Tanaka Y, Takeuti M. 1988. *Astrophys. Space Sci.* 148:229–37
Tassoul M. 1980. *Ap. J. Suppl.* 43:469–90
Tassoul M, Fontaine G, Winget DE. 1990. *Ap. J. Suppl.* 72:335–86
Thejll P, Vennes S, Shipman HL. 1991. *Ap. J.* 370:355–369
Tuchman Y. 1991. *Ap. J.* 383:779–83
Tuchman Y, Lèbre A, Mennessier MO, Yarri A. 1993. *Astron. Astrophys.* 271:501–7
Tuthill PG, Haniff CA, Baldwin JE, Feast MW. 1994. *MNRAS* 266:745–51
Udalski A, Kubiak M, Szymański M, Kałużny J, Mateo M, Krzemiński W. 1994. *Acta Astron.* 44:317–86
Unno W, Osaki Y, Ando H, Saio H, Shibahashi H, 1989. *Nonradial Oscillations of Stars.* Tokyo: Univ. Tokyo Press. 2nd ed.
van Genderen AM. 1989. *Astron. Astrophys.* 208:135–40
van Genderen AM. 1994. *Astron. Astrophys.* 284:465–76
van Genderen AM, Gautschy A. 1995. *Astron. Astrophys.* 294:453–68
van Genderen AM, van der Hucht KA., Larsen I. 1990. *Astron. Astrophys.* 229:123–32
Vassiliadis E, Wood PR. 1993. *Ap. J.* 413:641–57
Vauclair G. 1990. In PSSS, pp. 437–442
Vauclair G, Belmonte JA, Pfeiffer B, Chevreton M, Dolez N, et al. 1993. *Astron. Astrophys.* 267:L35–38
Vauclair S, Dolez N. 1990. In PSSS, pp. 399–403
Vauclair S, Dolez N, Gough DO. 1991. *Astron. Astrophys.* 252:618–24
Viotti R. 1992. In *Variable Star Research: An International Perspective,* ed. JR Percy, JA Mattei, C Sterken, pp. 194–204. Cambridge: Cambridge Univ. Press
Vogt SS, Penrod GD. 1983. *Ap. J.* 275:661–82
Vreux J-M. 1986. *Publ. Astron. Soc. Pac.* 97: 274–279
Waelkens C. 1987. In *Stellar Pulsation,* ed. AN Cox, WM Sparks, SG Starrfield. In *Lecture Notes Phys.* 274:75–78. New York: Springer-Verlag
Waelkens C. 1991. *Astron. Astrophys.* 246:453–68
Waelkens C, Rufener F. 1985. *Astron. Astrophys.* 152:6–14
Waelkens C, Van den Abeele K, Van Winckel H. 1991. *Astron. Astrophys.* 251:69–74
Walker GAH, Yang S, Fahlman GG. 1979. *Ap. J.* 233:199–204
Wallerstein G, Cox AN. 1984. *Publ. Astron. Soc. Pac.* 96:677–91
Watson RD. 1988. *Astrophys. Space Sci.* 140:255–90
Wehlau A, Bohlender D. 1982. *Astron. J.* 87:780–91
Werner K. 1992. In *The Atmospheres of Early-Type Stars,* ed. U Heber, CS Jeffery, *Lecture Notes Phys.* 401:273–87. New York: Springer-Verlag
Werner K. 1993. In Wd92, pp. 67–75
Werner K. 1995. *Astron. Astrophys.* In press
Werner K, Heber U, Hunger K. 1991. *Astron. Astrophys.* 244:437–61
Werner K, Rauch T, Dreizler S, Heber U. 1995. In CT95, pp. 96–97
Whitelock PA. 1990. In Bo90, pp. 365–378
Whitelock PA, Menzies J, Feast M, Marang F, Carter B, et al. 1994. *MNRAS* 267:711–42
Whitney CA. 1978. *Ap. J.* 220:245–50

Winget DE. 1988a. In *Advances in Helio- and Asteroseismology, IAU Symp. 123,* ed. J Christensen-Dalsgaard, S Frandsen, pp. 305–24. Dordrecht: Reidel
Winget DE. 1988b. In Buda88, pp. 181–97
Winget DE, Nather RE, Clemens JC, Provencal JL, Kleinman S, et al. 1991. *Ap. J.* 378:326–46
Winget DE, Nather RE, Clemens JC, Provencal JL, Kleinman S, et al. 1994. *Ap. J.* 430:839–49
Winget DE, Van Horn HM, Tassoul M, Hansen CJ, Fontaine G, Carroll BW. 1982. *Ap. J. Lett.* 252:L65–68
Wlodarczyk K, Zola S. 1990. In Bo90, pp. 586–88
Wolf B. 1992. In *Reviews in Modern Astronomy,* ed. G Klare, 5:1–15. Berlin: Springer-Verlag
Wolff SC. 1983. In *The A-type Stars, NASA SP-463,* pp. 93-111. Washington DC: NASA
Wood PR. 1979. *Ap. J.* 227:220–31
Wood PR. 1990a. In *From Miras to Planetary Nebulae: Which Path for Stellar Evolution?* ed. MO Mennessier, A Omont, pp. 67–83. Gif-sur-Yvette:Editions Frontières
Wood PR. 1990b. In Bo90, pp. 355–64
Wood PR. 1995. In CT95, pp. 127–38
Wood PR, Cahn JH. 1977. *Ap. J.* 211:499–508
Worrell JK. 1987. In *Stellar Pulsation,* ed. AN Cox, WM Sparks, SG Starrfield, *Lecture Notes Phys.* 274:289–92. New York: Springer-Verlag
Xiong D-R. 1981. *Acta Astron. Sinica* 22:350–56
Yang S, Walker GAH, Hill GM, Harmanec P. 1990. *Ap. J. Suppl.* 74:595–608
Yee HC, Sweby PK, Griffiths DF. 1991. *J. Comput. Phys.* 97:249–310
Zalewski J. 1992. *Publ. Astron. Soc. Jpn.* 44:27–43
Zalewski J. 1993. *Acta Astron.* 43:431–40
Ziebarth K. 1970. *Ap. J.* 162:947–62

Annu. Rev. Astron. Astrophys. 1996. 34:607–44

X-RAY NOVAE

Y. Tanaka

Max-Planck Institut für Extraterrestrische Physik, D-85748 Garching, Germany and Institute of Space and Astronautical Science, Yoshinodai, Sagamihara, Kanagawa 229, Japan

N. Shibazaki

Department of Physics, Rikkyo University, Nishi-Ikebukuro, Toshima-ku, Tokyo 171, Japan

KEY WORDS: X-ray binaries, mass accretion, neutron stars, black holes

---

1. INTRODUCTION

In our galaxy there exist about 200 X-ray binaries containing a neutron star or a black hole (van Paradijs 1995). About one third of them are not persistently visible and are detected as transient sources. The transient sources are divided into two different classes: one belonging to the high-mass X-ray binaries (HMXBs) and the other belonging to the low-mass X-ray binaries (LMXBs). The HMXBs are systems with an O or B star companion, while the LMXBs have primarily K or M dwarf companions.

Most of the HMXB transients are Be-binary pulsars. They exhibit outbursts with regular intervals that are considered to be periodic encounters of a neutron star in an eccentric orbit with a stellar wind zone around an early-type star. Because their X-ray spectra are significantly harder than those of LMXBs, they are called "hard X-ray transients."

LMXB transients are characterized by episodic X-ray outbursts. During an outburst, they show a soft X-ray spectrum that is characteristic of the persistent luminous LMXBs and do not exhibit regular pulsations. These are generally called "soft X-ray transients," or "X-ray novae." X-ray outbursts are accompanied by optical outbursts. The optical outbursts are considered to be due

0066-4146/96/0915-0607$08.00

to reprocessing of the X rays in the accretion disk, which indicates that the companion has a low intrinsic luminosity and thus a low mass. Since the X-ray and optical properties of the soft X-ray transients during outbursts are so similar to those of the persistent LMXBs, the soft X-ray transients are most likely the same class of object, i.e. a close binary system composed of a Roche-lobe-filling low-mass star and either a weakly magnetized neutron star or a black hole. Soft X-ray transients belong to this class of object. Many soft X-ray transients exhibit recurrent outbursts with intervals ranging from several months to tens of years. Perhaps all of them are recurrent.

Because all LMXBs are variable, the distinction of X-ray novae from persistent sources undergoing a large intensity increase is sometimes ambiguous. We employ the following criteria for X-ray novae for the convenience of this review:

1. The X-ray flux increases by more than two orders of magnitude within several days.

   In reality, the pre-outburst flux levels were often below the detection limit of the instruments. In the light of the recent results, a more appropriate criterion would be an increase of X-ray intensity from below 10^{33} erg s^{-1} to well above 10^{37} erg s^{-1} in the range 1–10 keV, whenever a distance estimate is available. Hereafter, the quoted X-ray luminosity values are those in the 1–10 keV range, unless otherwise mentioned. Because of the uncertainties in the distance of many of these objects, the estimated luminosities may be uncertain by a factor of 2 or sometimes even more.

2. The flux declines on time scales of several tens of days to more than one hundred days, and it eventually returns to the pre-outburst level.

   The light curves are various: Some show a rather monotonic decline, but in many cases the light curves are complex. In some cases, the sources are "turned on" and remain persistently visible for a year or longer after an outburst.

3. In recurrent transients, the duration of an outburst is shorter than the "quiescent" (pre- or post-outburst) period: The duty ratio over a long time span is smaller than unity.

4. There is no fixed periodicity of recurrence.

An outburst of X-ray novae is caused by a sudden dramatic increase of mass accretion onto the compact object. During an outburst, the optical counterpart also brightens significantly (> 5 mag) with a spectrum characteristic of an X-ray illuminated disk. Radio outbursts are often observed during X-ray outbursts. In some cases, relativistic jets such as seen from active galactic nuclei are found, though these are much smaller in scale and power. This evidence shows that an eruptive mass ejection occurs as a result of sudden mass accretion.

In this review, we first summarize the observed results of X-ray novae in Section 2. Following the classification of X-ray novae in Section 2.1, the properties of X-ray outbursts and the optical counterparts are briefly described in Section 2.2, and the associated radio outbursts are outlined in Section 2.3.

The long quiescent period of X-ray novae also allows detailed optical measurements of the secondary star. This is not feasible for persistent sources in which the optical flux is dominated by that from the accretion disk. Optical photometry and spectroscopy allow measurements of the type of the secondary star, the orbital parameters, and in favorable cases the mass functions (see Section 2.2).

The remarkable increase of sensitivity obtained with focusing X-ray telescopes has enabled detailed X-ray studies of these systems in the quiescent state to be made, especially using *ROSAT* and *ASCA*. The results of these observations are summarized in Section 2.4. X-ray emission during the quiescent period provides diagnostic information on the condition of lowest-rate mass accretion to the compact objects, which is crucial for understanding the outburst mechanism. In addition, because the accretion rate in X-ray novae varies over several orders of magnitude, these objects are extremely useful for studying the physics of mass accretion and the relationship of the X-ray properties to the mass accretion rate.

In Section 3, the following major questions are discussed in the light of the recent observational results:

1. What conditions make LMXBs persistent or transient?

2. What are the mechanisms of X-ray nova outbursts?

3. How many still "sleeping" LMXBs are there in the Galaxy?

The first question is addressed in Section 3.1, based on various observed characteristics of X-ray novae. The optical observations during quiescence also give an essential clue to the understanding of this problem. The mass accretion during the quiescent period is a new subject as discussed in Section 3.2, and has an important relation to the outburst mechanism.

The mechanisms of X-ray nova outbursts are discussed in Section 3.3. It is often noted that X-ray novae are similar to dwarf novae in several respects (e.g. van Paradijs & Verbunt 1984, Priedhorsky & Holt 1987). Mechanisms of X-ray novae outbursts have so far been discussed primarily based on two different kinds of models that were originally proposed to explain dwarf novae outbursts: disk instability models and mass transfer instability models.

From the current statistics of X-ray novae, we infer that still many more LMXBs await discovery, in particular those containing a black hole, which are in quiescence at present. We estimate the total number of them in Section 3.4. This issue relates to the problems of the origin of LMXB systems and of their birth rate.

For previous reviews related to X-ray novae, see White et al (1984), Priedhorsky & Holt (1987), Cowley (1992), van Paradijs & McClintock (1995), and Tanaka & Lewin (1995). For a general review of the properties of X-ray binaries, see e.g. White et al (1995).

2. OBSERVATIONAL RESULTS

2.1 *Classification of Soft X-Ray Transients*

In the recent catalog of X-ray binaries by van Paradijs (1995), which contains 124 LMXBs, 41 are categorized as transient sources and the remaining 83 as persistent sources. About 30 transients show a maximum flux greater than 100 μJy in 2–10 keV, which is roughly equal to ~ 0.1 the Crab Nebula intensity (the Crab Nebula is sometimes used for a flux standard: The Crab Nebula intensity = 1 Crab). These transients are listed in Table 1 along with more recently discovered transients observed by the *Compton Gamma-Ray Observatory* (*CGRO*) and *GRANAT*. Note, however, that Table 1 includes those for which sufficient outburst data are not available. Some of them are simply new detections of previously uncatalogued sources and may not fulfill the criteria for X-ray novae given in Section 1. Of these 33 bright transients, 18 belong to the class of black-hole LMXBs, i.e. LMXBs for which the compact object is a black hole. Remarkably, all the black-hole LMXBs currently known are transient sources. The three known persistent black-hole X-ray binaries are all high-mass systems; two are in the Large Magellanic Cloud (LMC X-1, LMC X-3), leaving only one (Cyg X-1) in our own galaxy. In contrast, amongst the LMXBs of which the compact object is a neutron star (neutron-star LMXBs), only 1/7 are transients. Although the above classification is not entirely secure, this contrast is robust and quite striking.

We briefly summarize below how LMXBs are classified into black-hole binaries and neutron-star binaries.

2.1.1 X-RAY BURST: A FIRM SIGNATURE OF A NEUTRON-STAR LMXB Matter accumulated on the surface of a neutron star in a LMXB undergoes a thermonuclear flash under certain conditions and generates an X-ray burst. To be exact, there are two distinctly different types of X-ray burst [designated Type I and Type II by Hoffman et al (1978)]. Type I burst are the ones due to thermonuclear flashes and exhibit gradual softening of the blackbody spectrum (cooling) during the burst decay, whereas Type II bursts are those caused by spasmodic accretion and do not show cooling with time. Thus, if a Type I burst is found from a source, it is an unequivocal signature that the compact object is a neutron star. Since Type II bursts are observed only from 1730–335 (see Section 2.2.1) and possibly Cir X-1, we hereafter call Type I bursts simply X-ray bursts.

Burst sources comprise a large fraction ($\sim$40%) of LMXBs, but not all the neutron-star LMXBs are known to produce X-ray bursts. This fraction is uncertain because bursts are not always frequent and some bursts could have been missed. Also, very luminous ($\sim 10^{38}$ erg s^{-1}) neutron-star LMXBs do not produce bursts. Among X-ray novae, X-ray bursts have been seen from about ten sources (see Table 1). They are definitely neutron-star LMXBs. For details of X-ray bursts, see e.g. a recent review by Lewin et al (1995).

2.1.2 MASS FUNCTION A precise optical measurement of the radial velocity curve of the secondary star provides a determination of the mass function, which, in turn, gives an absolute minimum mass of the compact primary. A more accurate estimate of the mass of the primary can be obtained with additional information on the secondary mass and the orbital inclination.

According to the current theory, a neutron star more massive than 3 $M_\odot$ cannot be stable and will collapse into a black hole. Therefore, a lower limit to the mass exceeding 3 $M_\odot$ has been considered to be reliable evidence for a black hole. The relativistic effects that are unique to black holes have not been confirmed as yet.

So far nine sources including three high-mass systems are known to have a compact primary with a mass $> 3\ M_\odot$. They are listed in Table 2. All six low-mass systems are X-ray novae.

2.1.3 X-RAY SPECTRUM X-ray emission is powered by mass accretion. If the compact object of a LMXB is a neutron star, X rays are considered to be emitted from two regions: an accretion disk and an optically thick neutron star envelope. As shown in Figure 1*a*, the observed spectra of luminous ($\geq 10^{37}$ erg s^{-1}) neutron-star LMXBs consist of two components, a blackbody component, which is most probably the emission from the neutron star envelope (an optically thick boundary layer), and a softer component, most probably from the accretion disk (Mitsuda et al 1984, White et al 1988). These two components can be separately determined observationally. The soft component is represented by a multicolored blackbody spectrum expected from an optically thick accretion disk (Mitsuda et al 1984, Tanaka 1992b). The characteristic temperature of the soft component is $\sim$1.5 keV (color temperature) when the X-ray luminosity L_X is $\sim 10^{38}$ erg s^{-1} and decreases as luminosity goes down. However, the above properties hold for $L_X < 10^{37}$ erg s^{-1}. At $L_X < 10^{37}$ erg s^{-1}, the X-ray

Table 1 List of X-ray novae

a. Black-hole LMXBs

Source[a] name	Outburst year	Spectrum[b]	Distance (kpc)	Optical[c] counterpart	Companion type	Orbital period (hour)	Ref.[d]
J0422+32 N Per	'92	PL	2	V518 Per	M0–5, M2V	5.09	1
0620–003 N Mon	'17, '75	U+PL	0.9	V616 Mon	K5V	7.75	1
1009–45 N Vel	'93	U+PL		[5]			4
1124–684 N Mus	'91	U+PL	3		K0–4V	10.4	1
1354–645 N	'67(?), '87	U+PL		BW Cir			2,3
1524–617 N	'74	U+PL	4.4 [6]	KY TrA			2,3
1543–475 N	'71, '83, '92	U+PL	4		A2V		2,3
1630–472 RN	(~ 600d)	U+PL					2,3
J1655–40 N Sco	'94	U+PL	3.2		F3–6	62.7	1
1659–487 RN (GX339–4)	(~ 460d)	U+PL	4?	V821 Ara		14.83	2,3
1705–250 N Oph	'77	U+PL		V2107 Oph			2,3
1716–249 N Oph (J1719–24)	'93	PL		[8]			7
1730–312 N	'94	U+PL					9
1741–322 N	'77	U+PL					2,3
1846+031 N	'85	U+PL					2,3
1915+105 N Aql	'92	U+PL	12.5	IR star			1
2000+251 N Vul	'88	U+PL	2	QZ Vul	early K	8.26	1
2023+338 N Cyg	'38, '56, '79(?),'89	PL	3.5	V404 Cyg	K0IV	155.4	1

[a]N: Nova outburst, RN: Recurrent nova outbursts (recurrent period).
[b]PL: Power law, U+PL: Ultrasoft + power law.
[c]IR: infrared star.
[d]References:
1. see Section 2, and references therein
2. van Paradijs 1995, and references therein
3. Tanaka & Lewin 1995, and references therein
4. Lapshov et al 1993, 1994; Sunyaev et al 1994
5. Della Valle & Benetti 1993
6. Barret et al 1995
7. Ballet et al 1993; Harmon et al 1993; Sunyaev et al 1994
8. Della Valle et al 1993; Della Valle et al 1994
9. Borozdin et al 1995

spectral characteristics of the sources change and the spectra become power law in form (see Section 2.1.4).

In contrast, nearly 20 sources commonly show a spectral shape characterized by a soft component and a hard power-law tail as shown in Figure 1*b*, which is distinctly different from the spectra of neutron-star LMXBs. The blackbody component seen in the neutron-star LMXBs is conspicuously absent. The soft components of these sources are also represented by a multicolored blackbody spectrum, from an optically thick accretion disk but they have a significantly lower characteristic temperature ($\leq$ 1.2 keV) than those observed from luminous neutron-star LMXBs and hence the spectra are called "ultrasoft." The

b. Neutron-star LMXBs

Source[a] name	Outburst year	Remark[b]	Distance (kpc)	Optical[c] counterpart	Companion type	Orbital period (hour)	Ref.[d]
0748–676		B		UY Vol		3.82	2,3
1455–314 N (Cen X-4)	'69,'79	B	1.2	V822 Cen	K5–7	15.10	1
1516–569 RN (Cir X-1)		B	9	BR Cir		398	2,3
1608–522 RN		B	3.6	IR			1
1658–298		B		V2134 Oph		7.11	2,3
1730–335 RN (Rapid Burster)		B	10	in Lil 1			1
1730–220 N	'72						2
1731–260		B					2
1735–28							2
1742–289		B(?)			K dwarf?		2
1744–361		LM or HM?					2
1745–203				in NGC 6440			2
1803–245		LM or HM?					2
1908+005 RN (Aql X-1)		B	2.5	V1333 Aql	G8–K0V	19.0	1
1947+300		LM or HM?					4

[a]N: Nova outburst, RN: Recurrent nova outbursts
[b]B: Burster, LM or HM?: Uncertain whether it is a low-mass system or a high-mass system.
[c]IR: infrared star.
[d]References: 1. see Section 2, and references therein
2. van Paradijs 1995, and references therein
3. White, Nagase & Parmar 1995, and references therein
4. Borozdin et al 1990

power-law component changes its intensity irregularly and is unrelated to the intensity of the soft component. The photon index remains essentially constant at around ~2.5 during the intensity changes (Tanaka 1992b).

White et al (1984) first suggested that this ultrasoft spectrum is a possible black-hole signature. In fact, seven of the nine reliable black-hole sources with large mass lower limits (Table 2) show this characteristic spectral shape (see Figure 2 of Tanaka 1994 and Section 2.2.2 below). The absence of a blackbody component is consistent with the compact object being a black hole, since no solid surface is present in a black hole. Also, X-ray bursts have never been observed from these sources at any luminosity level. The characteristic temperature is expected to be lower for an accretion disk around a black hole, since it scales as $M^{-1/2}$ for a given accretion rate with the mass M of the compact

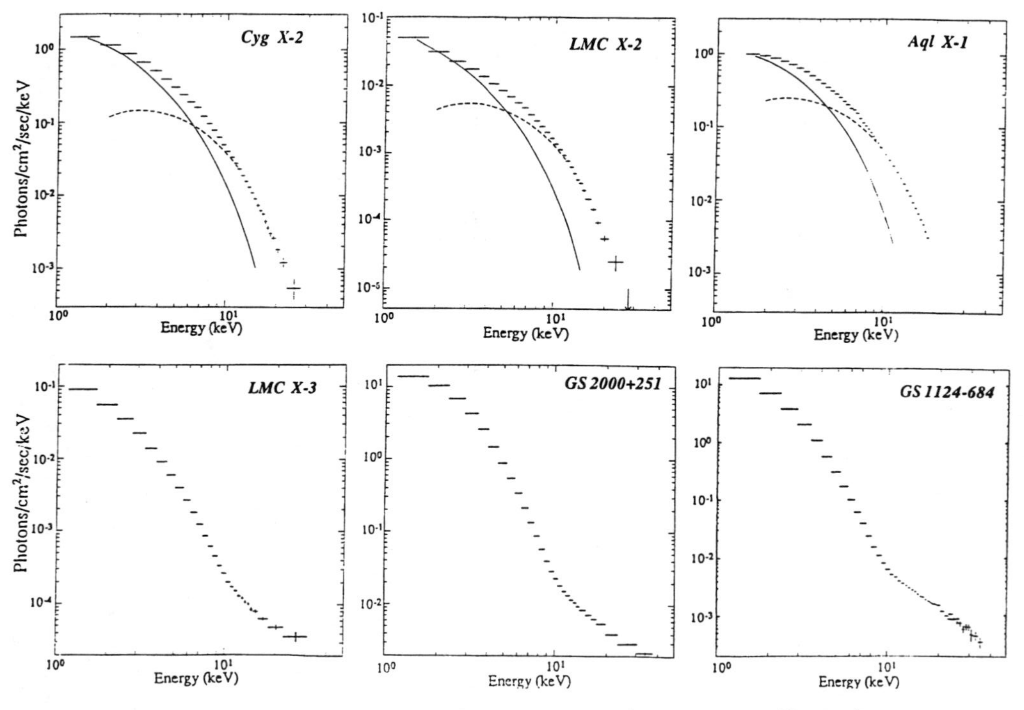

Figure 1 Examples of the energy spectra of X-ray binaries at high luminosities ($L_X > 10^{37}$ erg s^{-1}). (*a*) The spectra of the neutron-star LMXBs, which are composed of a soft component expected from an optically thick accretion disk (*solid curve*) and a blackbody component from the neutron star envelope (*dashed curve*) (Tanaka 1994). Cyg X-2 and LMC X-2 are persistent sources, whereas Aql X-1 is a transient source. (*b*) The spectra of the black-hole binaries (Tanaka 1994) established with the mass lower limit of the compact object exceeding 3 $M_\odot$ (see text). The spectra are dominated by an ultrasoft component expected from the accretion disk (softer than the corresponding soft component in the neutron-star LMXBs) and are accompanied by a hard tail. LMC X-3 is a persistent HMXB, whereas GS 2000+251 and GS/GRS 1124−684 are low-mass transients.

Table 2 Black-hole binaries established from the mass functions

Source name		Spectrum[a]	Companion[b]	$F(M)$ $(M_\odot)$	BH mass $(M_\odot)$	Ref.[c]
Cyg X-1	persistent	US+PL	HM O 9.7 Iab	0.241 ± 0.013	$\sim 16(>7)$	1
LMC X-3	persistent	US+PL	HM B 3 V	2.3 ± 0.3	>7	2
LMC X-1	persistent	US+PL	HM O 7–9 III	0.14 ± 0.06	$\sim 6(?)$	3
J0422+32	Nova Per	PL	LM M 2 V	1.21 ± 0.06	>3.2	4
0620−003	Nova Mon	US+PL	LM K 5V	3.18 ± 0.16	>7.3	5
1124−684	Nova Mus	US+OL	LM K 0–4 V	3.1 ± 0.4	~ 6	6
J1655−40	Nova Sco	US+PL	LM F 3–6	3.16 ± 0.15	4–5.4	7
2000+251	Nova Vul	US+PL	LM early K	4.97 ± 0.10	6–7.5	8
2023+338	Nova Cyg	PL	LM K 0 IV	6.26 ± 0.31	8–15.5	9

[a]US+PL: ultrasoft + power-law, PL: power law.
[b]HM: high-mass system, LM: low-mass system.
[c]References:
1. Gies & Bolton 1982
2. Cowley et al 1983
3. Hutchings et al 1987
4. Filippenko et al 1995a
5. McClintock & Remillard 1986
6. McClintock, Bailyn & Remaillard 1992
7. Bailyn et al 1995b
8. Filippenko et al 1995b
9. Casares et al 1992

object (Mitsuda et al 1984). For these reasons, an ultrasoft spectrum plus a hard power-law tail can be considered as a probable signature of a black-hole binary. For more discussion, see Tanaka & Lewin (1995).

2.1.4 SOFT-HARD TRANSITION: NO BLACK-HOLE SIGNATURE Two of the nine reliable black-hole binaries, GS 2023+338 and GRO J0422+32, showed an approximate single power-law spectrum at all luminosity levels observed (with no ultrasoft component; see Section 2.2.2), indicating the presence of black-hole binaries of another spectral type.

For some time, a hard power-law spectrum associated with flickering (rapid intensity fluctuations) or soft-hard (high-low) transitions such as that seen in Cyg X-1 has been considered to be a possible black-hole signature. However, this is not the case. Studies of X-ray novae clearly show that, regardless of whether the compact object is a neutron star or a black hole, the X-ray spectrum changes from a soft state at high luminosities (as described above) to a hard power-law state with a photon index of 1.5–2.0 at low luminosities (see White et al 1995; Tanaka 1989, 1994; Tanaka & Lewin 1995). Transitions between these two spectral states have been observed from several X-ray novae as they undergo luminosity changes; e.g. Aql X-1 and 1608−522 (neutron-star systems) and GS/GRS 1124−684 and GX 339–4 (black-hole systems), as shown

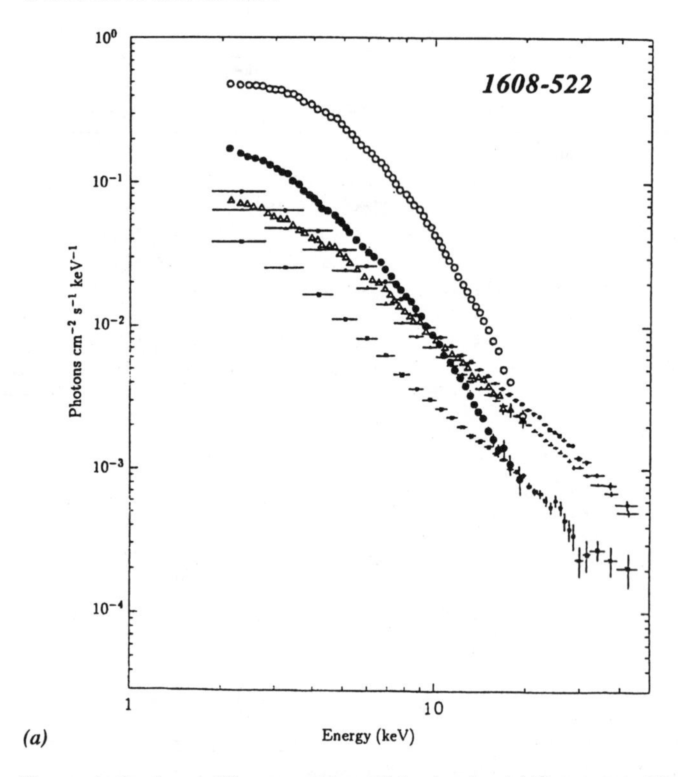

Figure 2 Changes in the shape of X-ray spectrum with luminosity. (*a*) The spectra of the neutron-star LMXB 1068−522 (Mitsuda et al 1989, Yoshida et al 1993). (*b*) The spectra of the black-hole LMXB GS/GRS 1124−684 (Ebisawa et al 1994), in which the *Ginga* observation dates are indicated. These spectra have been corrected for the detector response (Tanaka 1992b, 1994).

in Figure 2. Note that the power-law photon index of the black-hole systems becomes smaller (harder) when the ultrasoft component disappears (Tanaka 1992a). This transition seems to occur around $L_X \sim 10^{37}$ erg s^{-1} or lower. The change in the spectrum is most probably due to an accretion-dependent change in the structure of the accretion disks in these sources. Flickering (rapid intensity fluctuations on various time scales down to $\ll 1$ s) is also one of the common properties in the power-law state at low luminosities (Yoshida et al 1993). The power law extends to 100 keV and beyond, and is most probably formed by Comptonization of soft photons (see White et al 1995, Nakamura et al 1989).

Thus, a power-law spectrum alone is insufficient to classify a source. Since all neutron-star LMXBs observed show soft thermal spectra when L_X is well

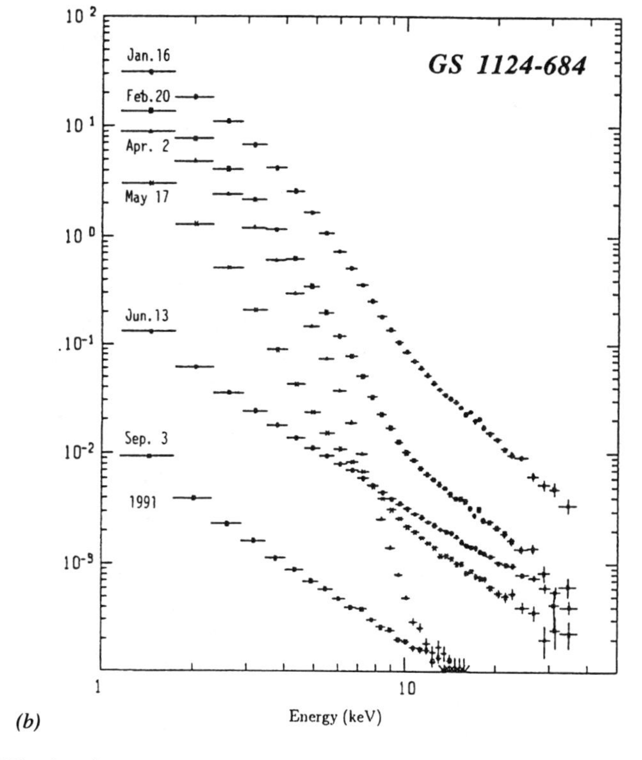

(b)

Figure 2 (*Continued*)

above 10^{37} erg s^{-1}, we may consider a source to be a black-hole binary only if it shows a single power-law spectrum when $L_X \gg 10^{37}$ erg s^{-1}. However, this classification is less secure than cases where the spectrum consists of an ultrasoft component and a power-law tail.

2.2 *Outbursts*

2.2.1 NEUTRON-STAR X-RAY NOVAE We summarize the characterisitics of X-ray nova outbursts of the systems in which the compact object is known to be a neutron star (neutron-star X-ray novae). Our discussion is based primarily on the results of the detailed studies of Cen X-4, Aql X-1, 1608−522, and 1730−335. All of these have been studied at various phases of activity over a wide range of luminosity.

Cen X-4 Two outbursts of Cen X-4 have been observed—the first in July 1969 (Evans et al 1970) and the second in May 1979 (Kaluzienski et al 1980). In the 1969 outburst, the X-ray intensity reached a maximum of $\sim 10^{38}$ erg s^{-1}

(for $d = 1.2$ kpc) in a few days followed by a decline that became steeper after about two months. The second outburst was smaller: a peak flux of $\sim 1/5$ of the first and a duration of about a month. The light curve of the second outburst showed a complex multipeaked structure for about 10 days, followed by a relatively smooth decay. The X-ray spectrum was typical of a luminous neutron-star LMXB (Kitamura et al 1971, Matsuoka et al 1980). An X-ray burst was detected during the decay of the 1979 outburst (Matsuoka et al 1980). The total energy emitted during an outburst was $\sim 3 \times 10^{44}$ erg for the 1969 outburst and $\sim 3 \times 10^{43}$ erg for the 1979 outburst (van Paradijs & Verbunt 1984).

The optical counterpart (V822 Cen), which had brightened from $V = 18.3$ to $V = 12.8$, was discovered by Canizares et al (1980). It was identified as a late K5–7V star (van Paradijs et al 1980, Cowley et al 1988, Chevalier et al 1989, McClintock & Remillard 1990) with a binary period of 15.1 h (Chevalier et al 1989). The mass function of 0.20 $M_\odot$ was obtained by McClintock & Remillard (1990). Liller (1979) noted the absence of earlier optical outbursts brighter than 16 mag since 1900. The distance is estimated to be ~ 1.2 kpc (Matsuoka et al 1980, Chevalier et al 1989).

A radio outburst was also observed (Hjellming et al 1988; see Section 2.3.).

Aql X-1 Aql X-1 undergoes major outbursts roughly once a year (Kaluzienski et al 1977a, Priedhorsky & Holt 1987). Minor outbursts (subflares) are also seen between major outbursts. The rise time is typically ~ 5 days. The light curve is complex and is often multi-peaked; it decays with a typical *e*-folding time of a month (though it is not an exponential decay). For the large outbursts in June 1978 (Charles et al 1980) and March 1987 (Kitamoto et al 1993), the peak luminosity was $\sim 2 \times 10^{37}$ erg s^{-1} (for $d = 2.5$ kpc). The total energy per outburst varied from 1 to 5 $\times 10^{43}$ erg (van Paradijs & Verbunt 1984). X-ray bursts were also detected (Koyama et al 1981, Czerny et al 1987).

The X-ray spectrum is typical of a luminous neutron-star LMXB (a soft component plus a blackbody component) during outburst, but it was a single power law when $L_X \sim 10^{35}$ erg s^{-1} (Czerny et al 1987, Tanaka 1994). There is an indication in the result of Czerny et al (1987) that this spectral change occurred around 4×10^{36} erg s^{-1}.

Priedhorsky & Terrel (1984) reported an underlying recurrence periodicity of 122–125 days based on 10 years of *Vela 5B* data from 1969, although only some cycles showed an outburst. However, the data of a longer span exclude the 125-day periodicity (Kitamoto et al 1993). Although the occurrence of outbursts appears quasiperiodic, there is no underlying clock in the system.

A positive correlation between the peak intensity (also the total energy emitted in an outburst) and the time since the preceding outburst was first pointed out by White et al (1984) (at 99.7% confidence; Priedhorsky & Holt 1987). This correlation was confirmed with increased samples including smaller outbursts (Kitamoto et al 1993). Although the scatter of data points is large, the data are consistent with a linear relation, suggesting that it releases in one outburst all the potential energy accumulated since the last. This result favors a disk instability model (see Section 3.3.1).

The optical counterpart (V1333 Aql) was discovered by Thorstensen et al (1978). The spectrum during X-ray quiescence was of a main-sequence G7-K3 star, near K0V. In a large outburst in June 1978, it brightened from $V \sim$ 19 in quiescence to 14.8 (Charles et al 1980). Chevalier & Ilovaisky (1991) discovered a probable binary period of 18.97 h. The distance was estimated to be between 1.7 and 4.0 kpc by Thorstensen et al (1978) and 1–2 kpc by Margon et al (1978). The peak luminosities of X-ray bursts (Koyama et al 1981, Czerny et al 1987) set an upper limit of $\sim$4 kpc. The most probable distance is $\sim$2.5 kpc (Thorstensen et al 1978).

A radio outburst was observed in an X-ray outburst (Hjellming et al 1990; see Section 2.3).

1608−522 Many outbursts of 1608−522 have been observed (see e.g. Lochner & Roussel-Dupré 1994 and references therein). During major outbursts, the peak luminosity goes up to $\sim 4 \times 10^{37}$ erg s^{-1} (for $d = 3.6$ kpc) (e.g. Nakamura et al 1989). The light curves of the outbursts are various (see Lochner & Roussel-Dupré 1994). In several outbursts, the X-ray intensity rose in a few days followed by a decay with an *e*-folding time of $\sim$ 10 days, but some outbursts lasted for more than 100 days (see Figure 1 of White et al 1984). Some outbursts showed a slow rise ($>$ 10 days). Outbursts occur randomly with intervals ranging from several months to years. The total energy per outburst varies in the range 10^{43}–10^{44} erg. No significant correlation between the outburst energy and the waiting time was observed (Lochner & Roussel-Dupré 1994).

1608−522 is exceptional in the following aspects: 1. It occasionally remains persistent. According to the *Vela 5B* 10-year history, the source stayed on at the $\sim 2 \times 10^{36}$ erg s^{-1} level for more than four years from around September 1971 through early 1976 (Lochner & Roussel-Dupré 1994). It was probably on between April 1983 and June 1984 as well (Nakamura et al 1989). 2. Some outbursts occurred even when the source was persistently on (see figures in Lochner & Roussel-Dupré 1994).

The evolution of the X-ray spectrum was studied throughout the decay of an outburst (Mitsuda et al 1989). As the luminosity decreased to 10^{37} erg s^{-1}, the thermal spectrum quickly hardened and became a single power-law for lower luminosities (Mitsuda et al 1989, Yoshida et al 1993), as shown in Figure 2*a*. 1608−522 is a well-known burster, generating frequent X-ray bursts (see Lewin et al 1995 and references therein). The distance was estimated to be 3.6 kpc from the bursts that show a luminosity saturation presumably at the Eddington limit (Nakamura et al 1989).

The optical counterpart was identified by Grindlay & Liller (1978) with a reddened, low-mass star that brightened by more than 2 I-magnitudes during a flare in July 1977. No further details of this star are known.

1730−335: The Rapid Burster 1730−335, discovered by Lewin et al (Lewin 1976), is a unique transient source that exhibits a train of rapidly repetitive bursts when active, hence it has been called the Rapid Burster. These bursts are due to spasmodic mass accretion (Type II bursts). The Rapid Burster also produces bursts due to thermonuclear flashes (Type I bursts) (Hoffman et al 1978). The source is located in a globular cluster (Liller 1977) at a distance of ~ 10 kpc. Recurrences of activity at a rate of once to twice a year have been observed, but sometimes no activity was detected for a few years. An active period (outburst) lasts about a month. The time-averaged luminosity and the total energy emitted vary from one active period to another. The total energy emitted in one outburst is rougly in the range (0.3–1) $\times 10^{44}$ erg.

The X-ray spectrum is best described with a blackbody, unlike the spectra of other neutron star LMXBs. For more details, see Lewin et al (1995) and references therein.

2.2.2 BLACK-HOLE X-RAY NOVAE In this section, we deal with X-ray nova outbursts of the systems in which the compact object is a black hole. Both reliable cases with mass lower limits $> 3\ M_\odot$ and probable ones based on the spectral characteristics are included.

The X-ray light curve of the outburst is different from source to source. However, the outbursts of four X-ray novae, A 0620−003, GS 2000+251, GS/GRS 1124−684, and GRO J0422+32, exhibit striking similarities: a fast rise followed by a relatively smooth exponential decay with similar time constants and the presence of a secondary increase. Some of these light curves are shown in Figure 3.

Other black-hole X-ray novae display much more complex light curves: multiple peaks and/or irregular decay (see White et al 1984, Harmon et al 1994). Also, there are at least two sources that show an unusually slow rise: $\sim$ a month for GX 339-4 and $\sim$ 3 months for GRS 1915+105 (Harmon et al 1994).

Brief summaries of individual outbursts that are relatively well studied are given below.

0620−003 (Nova Mon 1975) The outburst of A 0620−003 was detected on August 3, 1975, with *Ariel V* (Elvis et al 1975). After a small precursory peak, the X-ray luminosity reached a maximum of 1.3×10^{38} erg s^{-1} (for $d = 0.9$ kpc).

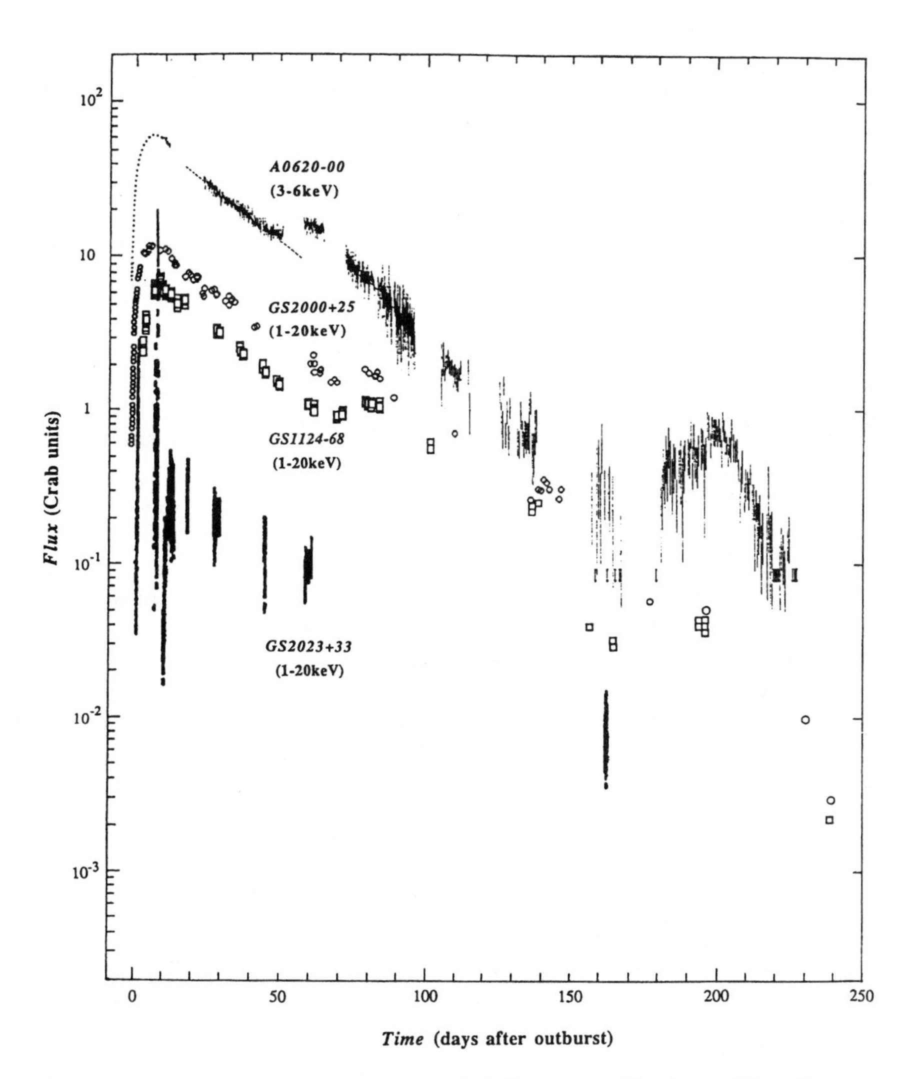

Figure 3 X-ray light curves of four bright black-hole X-ray novae. The observed X-ray fluxes are shown in units of the Crab Nebula intensity in the energy band separately indicated for each source. The dotted curve for A 0620−003 is from Elvis et al (1975), and the dots with vertical error bars are from Kaluzienski et al (1977b). The GS 2000+251 data (*open circles*) are from Tsunemi et al (1989) and Takizawa (1991). The GS 2023+338 data (*thick vertical bars*, which indicate actual large flux excursions) are from Tanaka (1992a). The GS/GRS 1124−684 data (*open squares*) are from Kitamoto et al (1992) and Ebisawa et al (1994).

The time to rise from 10% to 90% of the peak flux was ~ 4 days. As seen in Figure 3, the decay of the count rate (3–6 keV) was approximately exponential, with a decay constant $\tau \sim 24$ days (Kaluzienski et al 1977b). (Note that count rate decreases faster than the bolometric luminosity because of progressive softening of the spectrum; see below.) About ~ 50 days after onset, the flux increased again and reached a secondary peak (a factor of ~ 2 higher than the extrapolation of the initial decay) around 60 days; it then decreased at about the same time constant as before, and a third broad maximum was observed around 200 days (Kaluzienski et al 1977b). The count rate then declined steeply. The total energy emitted in the outburst is estimated to be $\sim 4 \times 10^{44}$ erg.

The X-ray spectrum was approximately a power law (with a photon index of ~ 1.6) in the first few days until the flux (2–18 keV) rose to $\sim 1/10$ the maximum; it then changed to an ultrasoft spectrum accompanied by a hard tail (Ricketts et al 1975). The spectrum steadily softened through the decay until at least 80 days after the outburst peak (Matilsky et al 1976).

The optical counterpart (V616 Mon) brightened to $V \sim 12$ (Boley et al 1976) from $V \sim 18.3$ in quiescence (Murdin et al 1980). The spectrum is characteristic of a K5V dwarf (Oke 1977, Whelan et al 1976, Murdin et al 1980), and its orbital period is 7.75 h (McClintock et al 1983). The distance is estimated to be ~ 0.9 kpc (Oke 1977). McClintock & Remillard (1986) determined the mass function to be $3.18 \pm 0.16\, M_\odot$ and estimated the mass of the compact object as $> 7.3\, M_\odot$. The optical record shows that a previous outburst occurred in 1917 (Eachus et al 1976).

A bright radio outburst was observed following the X-ray outburst (Davis et al 1975, Owen et al 1976, Hjellming et al 1988; see Section 2.3).

2000+251 (Nova Vul 1988) An outburst of GS 2000+251 was detected on April 23, 1988, with *Ginga* (sensitive in the range 1–40 keV) (Makino 1988, Tsunemi et al 1989). The flux increased from 10% to 90% of maximum in ~ 3 days and reached a maximum luminosity of $\sim 1.5 \times 10^{38}$ erg s^{-1} (for $d = 2$ kpc). The count rate decayed exponentially with $\tau \sim 30$ days (see the light curve in Figure 3). However, the decay of the bolometric luminosity was slower, with $\tau \sim 40$ days, due to steady softening (see below). A secondary peak with an increase of $\sim 50\%$ appeared around 80 days after the onset; this was followed by a decay with the same τ. Although there is no obvious tertiary increase, a flattening of decay is noted around 200 days, followed by a much steeper decay than before. The total energy emitted in the outburst was $\sim 5 \times 10^{44}$ erg.

The X-ray spectrum was ultrasoft, having a hard power-law tail with a photon index varying in the range 2.1–2.5 (Tanaka 1989). The ultrasoft component steadily softened during decay. The power-law component changed its flux widely and was irregularly independent of the ultrasoft flux (Tanaka 1989). This power-law component extended to ~ 300 keV (Sunyaev et al 1988; Döbereiner et al 1989, 1994). After 230 days, the spectrum became a single power law with a photon index of ~ 1.7. This spectral change occurred at an exceptionally low luminosity level of the order 10^{35} erg s^{-1}.

The optical counterpart was identified with a star (QZ Vul) that brightened to $V \sim 16.4$ from a reddened $R = 21$ mag state in quiescence (Charles et al 1988, Okamura & Noguchi 1988, Wagner et al 1988, Tsunemi et al 1989). This is likely to be an early K dwarf at a distance of 2–3 kpc, with a probable orbital period of 8.26 h (Callanan & Charles 1991; Chevalier & Ilovaisky 1990, 1993). Recently, Filippenko et al (1995b) determined the mass function to be $4.97 \pm 0.10\ M_{\odot}$, with a plausible mass of the compact object between 5.9 and 7.5 $M_{\odot}$.

Radio emission was detected in the observation during the decay (Hjellming et al 1988; see Section 2.3).

2023+338 (Nova Cyg 1989) GS 2023+338 was discovered on May 21, 1989, with *Ginga* (Makino 1989, Kitamoto et al 1989). The outburst of this source was unlike those of any other X-ray nova. The X-ray activities during the first 10 days and at later times were qualitatively different.

At first the flux increased from 0.1 Crab (the Crab Nebula Flux) to 4 Crab (1–6 keV) within a day and then declined quickly in a few days. Nine days later on May 30, a dramatic outburst occurred (see Figure 3.6 of Tanaka & Lewin 1995). The flux rose to a maximum of ~ 21 Crab and then showed repeated rapid changes between this level and a minimum of ~ 0.1 Crab for ~ 3 h. The luminosity at the maximum was very high, $\sim 2 \times 10^{39}$ erg s^{-1} in 1–40 keV (for $d = 3.5$ kpc), and the luminosity seems to saturate at this level, which is possibly the Eddington limit (Tanaka 1989).

In the first 10 days, not only the flux fluctuated by more than two orders of magnitude, but the spectrum also changed dramatically from time to time (Tanaka 1989, in't Zand et al 1992; see Figure 3.7 of Tanaka & Lewin 1995). Although the spectral features are complex, the main cause of spectral changes (consequently resulting in flux changes) was the occurrence of large changes in absorption. The absorption column on the line of sight fluctuated rapidly and sometimes exceeded 10^{24} H atoms cm^{-2}, suggesting that the outer accretion disk was suddenly flooded by a large amount of cool matter. From time to time, the source became essentially free from low-energy absorption. The intrinsic (unabsorbed) spectrum was approximately a single power law with a photon index of ~ 1.5 (Tanaka 1989). An ultrasoft component was never present. The power-law component was detectable up to ~ 300 keV and showed a high-energy cutoff (an exponential steepening of the slope) (Aref'ev et al 1989, Sunyaev et al 1991, Döbereiner et al 1994). There is a clear indication that the cutoff shifts to lower energies as the luminosity increases, probably due to Compton cooling (Inoue 1994).

After the first 10 days, the violent changes ceased and an approximately exponential decay started from a level of ~0.5 Crab with $\tau \sim 40$ days (see the light curve in Figure 3). The spectrum was a single power law with a photon index of ~1.7. Absorption was much reduced and changed little. At about the same time, strong flickering started and continued throughout the decay. The total energy emitted in the outburst was roughly 6×10^{43} erg.

The optical counterpart was identified with V404 Cyg (Hurst & Mobberley 1989, Marsden 1989, Wagner et al 1989). Previous outbursts were recorded optically in 1938 (Wachmann 1948), 1956, and possibly 1979 (Richter 1989). V404 Cyg brightened to $V \sim 11.6$ near X-ray maximum (Buie & Bond 1989, Wagner et al 1991, Leibowitz et al 1991). The increase in brightness was ~7.5 mag in V. The secondary star is either a K0 IV star or a stripped giant (Wagner et al 1992a, King 1993) and is probably a subgiant (Casares et al 1993). Distance estimates are 2–3 kpc (Charles et al 1989, Casares et al 1991, Gotthelf et al 1992), > 3 kpc (Han & Hjellming 1990), and ~3.5 kpc (Wagner et al 1992a). Casares et al (1992) obtained an orbital period of 6.47 days and a mass function of $6.26 \pm 0.31\ M_\odot$, which is the largest among known black-hole systems. The probable mass of the compact primary is in the range 8–15.5 $M_\odot$ (Casares et al 1992), with a firm lower limit of 7 $M_\odot$ (Casares et al 1993).

A strong radio outburst was also observed (Hjellming et al 1989; Han & Hjellming 1992a, see Section 2.3).

1124−684 (Nova Mus 1991) An outburst of GS/GRS 1124−684 was detected on January 8, 1991, with *Ginga* (Makino 1991, Kitamoto et al 1992) and independently with the *GRANAT WATCH* (sensitive in the range 8–60 keV) (Lund & Brandt 1991, Brandt et al 1992). With a rise time from 10 to 90% of the peak in 4–5 days, it reached a maximum of $\sim 2.5 \times 10^{38}$ erg s^{-1} (for $d = 3$ kpc). The count rate (1–20 keV) decayed exponentially with $\tau \sim 30$ days (see the light curve in Figure 3), while the decay of the bolometric luminosity was slower with $\tau \sim 40$ days (Ebisawa et al 1994). A secondary increase by a factor of ~2 occurred with a rise time of ~10 days and peaked around 80 days after the onset. The following decay was slightly slower than before. A tertiary peak was observed around 200 days, after which the flux decreased rapidly. These characteristics of the X-ray light curve are very similar to those observed in A 0620–003. The total energy emitted in the outburst was $\sim 10^{45}$ erg.

The X-ray spectrum was hard in the first few days, dominated by a power law (Brandt et al 1992, Ebisawa et al 1994). It then became ultrasoft with a hard power-law tail (Figure 2*b*) and remained so until mid-May (Ebisawa et al 1994). The soft component steadily softened during decay. The power-law component with a photon index of 2.5–2.7 extended to ~ 500 keV (Goldwurm et al 1992, Sunyaev et al 1992). The flux of the power-law component varied widely and was independent of the ultrasoft flux (see Figure 2*b*), which is similar to the case of GS 2000+251. After mid-June, the spectrum was a single power law with a photon index of ~ 1.6, distinctly harder than before (Ebisawa et al 1994). Flickering started at the same time. This change occurred when the luminosity had decreased to $(1 \sim 2) \times 10^{37}$ erg s^{-1}. A transient 480-keV line was observed during the outburst with the *GRANAT SIGMA*; this was interpreted as an electron-positron annihilation line (Goldwurm et al 1992, Sunyaev et al 1992).

The optical counterpart was identified with a star that brightened to $V \sim 13.5$ (Della Valle et al 1991) from $V \sim 20.4$ in quiescence. This star is of spectral type K0–4 and has an orbital period of 10.4 h (Remillard et al 1992). McClintock et al (1992) determined the mass function to be $3.1 \pm 0.4\ M_{\odot}$ and estimated the minimum mass of the primary to be $3.75 \pm 0.43\ M_{\odot}$. The optical spectrum in quiescence is strikingly similar to that of V616 Mon (A 0620–003) (Remillard et al 1992). The distance is estimated to be between 1 and 5 kpc—probably ~ 3 kpc (West 1991).

A strong radio outburst was observed during the X-ray outburst (Kesteven & Turtle 1991; see Section 2.3).

J0422+32 (Nova Per 1992) The outburst of GRO J0422+32 was discovered by the *CGRO BATSE* instrument (which is sensitive in the range 20–300 keV) on August 5, 1992 (Paciesas et al 1992). The X-ray spectrum is a power law with a photon index of ~ 1.5, extending from 2 keV to 500 keV, modified by an exponential cutoff with a characteristic energy around 100 keV (Sunyaev et al 1993, Döbereiner et al 1994, Harmon et al 1994). The source was detected up to 2 MeV during the outburst peak (van Dijk et al 1995). The conspicuous absence of an ultrasoft component is similar to the case in GS 2023+338 (Sunyaev et al 1993). The peak luminosity is estimated to be $\sim 3 \times 10^{37}$ erg s^{-1} in 2–300 keV (for $d = 2$ kpc) by extrapolation of the *Mir-Kvant* result to the outburst peak (Sunyaev et al 1993).

J0422+32 reached its maximum flux in 6 days (Harmon et al 1994) (the rise time from 10% to 90% of the peak was ~ 4 days). The decay was approximately exponential with $\tau \sim 40$ days (Harmon et al 1994) to 44 days (Vikhlinin et al 1995). These values are similar to those of the decay in bolometric luminosity for both GS 2000+251 and GS/GRS 1124−684 (~ 40 days), and probably also for A 0620−003 if the gradual softening during decay is taken into account. A distinct secondary increase began at ~ 125 days and peaked at ~ 140 days after the onset of the primary outburst; this was followed by a decay with the same τ as before (Harmon et al 1994). The secondary peak flux was ~ 3 times that extrapolated from the primary decay. The decay apparently steepened after ~ 230 days, which is similar to the last decay phase of A 0620−003, GS 2000+251, and GS/GRS 1124−684. However, a tertiary peak such as observed in A 0620−003 and GS/GRS 1124−684 did not occur. The total energy emitted in the outburst is estimated to be $\sim 1.2 \times 10^{44}$ erg.

The optical counterpart (V518 Per) is an M star; M0–5 (Martin et al 1995), $\sim$ M0V (Chevalier et al 1995, Orosz et al 1995) or M2V (Filippenko et al 1995a). It brightened by ~ 7 mag to $V \sim 13.2$ (Castro-Tirado et al 1992a, 1993; Wagner et al 1992b). No previous outburst was found from the search of optical archive data (Shao 1992). The distance to the source is > 1 kpc (Callanan et al 1995) and probably ~ 2 kpc (Chevalier et al 1995). A periodicity is found from photometry and spectroscopy at 5.09 h, which is probably the binary period (Kato et al 1993, Chevalier et al 1995, Orosz et al 1995, Callanan et al 1995, Casares et al 1995). The mass function was determined to be $0.9 \pm 0.4\ M_{\odot}$ by Oroz et al (1995) and more recently $1.21 \pm 0.06\ M_{\odot}$ by Filippenko et al (1995a). Although this does not preclude the possibility of the compact object being a neutron star, its mass may exceed $3\ M_{\odot}$ depending on the secondary mass and the inclination angle (Orosz et al 1995). In fact, Filippenko et al (1995a) estimate the mass of the primary to be $3.57 \pm 0.34\ M_{\odot}$ for a normal M2 secondary.

The initial optical outburst lasted about 250 days. Two smaller optical outbursts occurred later (5.5 mag in V) (see Chevalier & Ilovaisky 1995, Callanan et al 1995, and references therein). Separations of the major optical outbursts (including the one coincident with the secondary X-ray increase) from the initial one are found to be integer multiples of ~ 120 days (Chevalier & Ilovaisky 1995), suggesting echos (delayed mass transfer triggered by X-ray illumination) (Augusteijn et al 1993). Note, however, that the third X-ray/optical increase was missing.

Radio emission was also detected (Han & Hjellming 1992b).

1915+105 (Nova Aql 1992) GRS 1915+105 was discovered by the *GRANAT WATCH* (Castro-Tirado et al 1992b, 1994) in August 1992. The source had begun to be visible in May 1992 with the *CGRO BATSE* (Harmon et al 1994). The rise was unusually slow at ~ 3 months. At a distance of 12.5 kpc (Mirabel & Rodriguez 1994), the peak luminosity was close to 10^{39} erg s^{-1} (Sazanov et al 1994). Multiple flaring events, each lasting 10–20 days, were observed over two years, with recurrent periods of several months and a peak flux of ~ 1 Crab in 8–20 keV (Sazanov et al 1994). The hard X-ray light curve observed with the *CGRO BATSE* also shows violent flaring activity (Harmon et al 1994). The source still remains active as of August 1995.

The spectrum of 1915+105 is ultrasoft and is accompanied by a power-law tail. An intense ultrasoft component was observed with *Mir-Kvant*

(Alexandrovich et al 1994a) and also with *ASCA* (T Kotani 1994, private communication). The power-law component extends to $\sim$ 230 keV with a photon index of $\sim$2.5 (Sazanov et al 1994) and shows an exponential cutoff above several tens of keV (Harmon et al 1994).

Mirabel et al (1994) found a variable star in the near infrared at the position of the radio source. The reddening ($A_V \sim 30$ mag) and the absorption column of $\sim 5 \times 10^{22}$ H atoms cm^{-2} obtained with *ROSAT* (Greiner et al 1994) suggest a distance > 8 kpc. Mirabel & Rodriguez (1994) estimate that $d = 12.5 \pm 1.5$ kpc.

Radio jets revealing superluminal motion were also discovered by Mirabel & Rodriguez (1994): See Section 2.3.

J1655−40 (Nova Sco 1994) GRO J1655−40 was discovered in outburst on July 27, 1994, with the *CGRO BATSE* (20–300 keV) (Zhang et al 1994). The hard X-ray light curve is complex, showing four outbursts in the first five months (Harmon et al 1995). Each outburst is characterized by a fast rise (typically $\sim$ 1 day) and an abrupt fall. The source is still bright as of August 1995 with repeating episodic hard X-ray outbursts.

The spectrum consists of an ultrasoft component and a power-law component. The ultrasoft component was observed with *Mir-Kvant* in September 1994 and February 1995; flux values differed by a factor of $\sim$3 (Alexandrovich et al 1994b, 1995). The recent *ASCA* observation in the range 0.5–10 keV in August 1995 also revealed an intense ultrasoft component with a luminosity of $\sim 4 \times 10^{37}$ erg s^{-1} (for d = 3 kpc) accompanied by a power law (Y Ueda et al 1995, private communication). However, the long-term behavior of the ultrasoft component is not known. The power-law component is highly variable in intensity and extends to at least 600 keV, with a slope varying between 2.5 and 3.1 (Harmon et al 1995). The maximum luminosity of outburst is probably as high as 10^{38} erg s^{-1} in the range 1–300 keV.

The optical counterpart was discovered by Bailyn et al (1995a). The observation on August 10, 1994, yielded V = 14.4 mag. It had brightened by 3 mag. The estimated distance is around 3 kpc (McKay & Kesteven 1994, Tingay et al 1995, Bailyn et al 1995a, Greiner et al 1995). Hjellming & Rupen (1995) obtained $d = 3.2 \pm 0.2$ kpc from an analysis of the radio jets. The recent spectroscopic observation shows that the secondary star is an F3–6 star with a binary period of 62.7 h and that the mass function is 3.16 ± 0.15 $M_\odot$ (Bailyn et al 1995b). Photometric observations show primary and secondary eclipses, indicating a large inclination of the system (see Bailyn et al 1995b). The radio-jet axis is at an inclination angle of 85° to the line of sight (Hjellming & Rupen 1995). Given such a large system inclination, the primary mass will be between 4.0 and 5.4 $M_\odot$ for a secondary mass between 0.4 and 1.5 $M_\odot$ (Bailyn et al 1995b).

Radio jets showing superluminal motion were also discovered (Tingay et al 1995, Hjellming & Rupen 1995; see Section 2.3).

2.3 *Radio Outbursts*

Radio outbursts associated with the X-ray outbursts were discovered from Cen X-4, Aql X-1, A 0620−003, GS 2000+251, GS 2023+338, GS/GRS 1124−684 (see references in Sections 2.1 and 2.2), and more recently from GRS 1915+105 and GRO J1655−40 (see below). Most probably, every X-ray outburst is accompanied by a radio outburst. In particular, relativistic radio jets showing superluminal motion have been resolved from the last two sources, as described below.

The radio outbursts in X-ray novae are interpreted to be ejections of "synchrotron bubbles" that contain relativistic electrons and magnetic fields. Sudden commencement of a large mass inflow in an accretion disk would cause a dynamical event that produces an outward-moving shock in which relativistic electrons are accelerated (Dickel et al 1989). A spherically expanding synchrotron bubble model fits the observed multiwavelength radio data well (Hjellming & Han 1995). This model may also apply for the individual plasma bubbles ejected in radio jets.

Hjellming & Rupen (1995) note that the observed characteristics of the radio flares in GRO J1655−40 (see below) are similar to those in A 0620−003, GS 2000+251, GS 2023+338, GS/GRS 1124−684, Cen X-4, and Aql X-1; they suggest that these may have been collimated jet events as well.

For a review, see Hjellming & Han (1995) and references therein.

1915+105 Following the detection of strong radio outbursts (Rodriguez & Mirabel 1993, Gerard et al 1994), Mirabel & Rodriguez (1994) discovered superluminal motion of a pair of radio blobs associated with GRS 1915+105. This is the first Galactic superluminal source. The standard model of relativistic jets yields a velocity of the radio blobs of $(0.92 \pm 0.08)c$ for an estimated distance of 12.5 ± 1.5 kpc (Mirabel & Rodriguez 1994). Mirabel & Rodriguez (1994) noted that the large kinetic energy of the radio blobs would require a minimum rate of 10^{41} erg s^{-1} to accelerate the twin ejecta—much larger than the maximum X-ray luminosity of the source ($\sim 10^{39}$ erg s^{-1}; Sazanov et al 1994). This enormous energy generation rate, far exceeding the Eddington limit, cannot be explained in terms of accretion energy. One possibility is that the ejecta are electron-positron plasmas. Further study is required.

J1655−40 A strong radio outburst was detected during the decline of the initial X-ray outburst (Campbell-Wilson & Hunstead 1994). Radio outbursts were observed following each X-ray outburst (Hjellming & Rupen 1995, Harmon et al 1995).

Tingay et al (1995) discovered superluminal motion. The detailed structure of the jets was resolved by Hjellming & Rupen (1995). They show that the jets move at $(0.92 \pm 0.02)c$ almost perpendicular to the line of sight ($i \cong 85°$). Backward extrapolation of the motion of the subcomponents shows that ejection of radio-emitting plasma occurred several days to two weeks after the onset of each hard X-ray outburst (Hjellming & Rupen 1995). Tingay et al (1995) suggest that the formation and ejection of the radio component was suppressed while the inner accretion disk is radiation-pressure dominant and geometrically thick. Note, however, that any relationship between the radio outbursts and the soft X-ray component of the source (Section 2.2.2) is not known because no soft X-ray (< 10 keV) data around the radio events are available.

2.4 *Properties During Quiescence*

Owing to a dramatic increase in sensitivity in recent years, X rays from several X-ray novae in the quiescent state have been positively detected at luminosity levels from below 10^{31} to 10^{33} erg s^{-1}. The results of X-ray observations of X-ray novae in quiescence, along with the 3σ upper limits, are listed in Table 3. Several individual sources are briefly described below: three neutron-star LMXBs (Cen X-4, Aql X-1, and 1068–522) and two black-hole LMXBs (A 0620–003 and GS 2023+338). As we discuss in Section 3, the X-ray and optical properties of X-ray novae in quiescence are important not only for understanding mass accretion at extremely low rates but also for clarifying the mechanism of X-ray nova outbursts.

Cen X-4 Van Paradijs et al (1987) detected X rays from Cen X-4 with the *Einstein Observatory* and *EXOSAT* at $(4–8)\times10^{32}$ erg s^{-1} in 0.5–4.5 keV (for $d = 1.2$ kpc). This was the first X-ray detection of X-ray novae in quiesence. The source was recently detected with *ASCA* at a luminosity of $\sim 2.4 \times 10^{32}$ erg s^{-1} in 0.5–10 keV (Asai et al 1995). The spectrum is very soft, but is accompanied by a significant hard tail with a photon index of ~ 1.9. If the soft component is fitted with a blackbody spectrum, the temperature kT is approximately 0.2 keV and the emitting area is ~ 10 km^2.

Aql X-1 Aql X-1 was observed with *ROSAT* five times at various luminosity levels from $\sim 4 \times 10^{36}$ erg s^{-1} down to 4.4×10^{32} erg s^{-1} (for $d = 2.5$ kpc) in 0.4–2.4 keV (Verbunt et al 1994). When $L_X \geq 10^{35}$ erg s^{-1}, the spectrum is consistent with a hard power law (Czerny et al 1987, Tanaka 1994). However, it changes to a very soft spectrum when $L_X < 10^{33}$ erg s^{-1}. The best-fit blackbody temperature is $kT \cong 0.3$ keV, which gives an emitting area of ~ 1 km^2 (Verbunt et al 1994).

Table 3 X-ray luminosity and spectrum during quiescence

Source name	Class[a]	Observed range (keV)	Luminosity[b] (10^{32} erg/s)	kT (for b.b.) (keV)	Ref.[c]
Cen X-4	ns	0.5–4.5	4–8		1
		0.5–10	2.4	0.2	2
				+ hard tail (2–10 keV)	
Aql X-1	ns	0.2–2.4	4.4	0.3	3
1608–522	ns	0.5–10	6	0.3	2
0620–003	bh	0.2–2.4	0.06	0.16	4
2023+338	bh	0.2–2.4	80	0.2	3,5
2000+251	bh	0.2–2.4	< 0.1		2
		0.5–10	< 0.3 (Crab)		6
			< 0.1 (b.b)		
1124–684	bh	0.5–10	< 0.1 (Crab)		6
GX339–4	bh	0.5–10	< 0.3 (Crab)		6
			< 0.1 (b.b.)		
1705–250	bh	0.2–2.4	< 5 (d = 3 kps assumed)		3

[a]ns: neutron-star LMXB, bh: black-hole LMXB.
[b]b.b.: for a blackbody spectrum, Crab: for a Crob-like spectrum.
[c]References: 1. van Paradijs et al 1987
2. Asai et al 1995
3. Verbunt et al 1994
4. McClintock et al 1995
5. Wagner et al 1994
6. ASCA result, unpublished

1608−522 1608−522 was detected with *ASCA* at a luminosity of $\sim 6 \times 10^{32}$ erg s^{-1} (for $d = 3.6$ kpc) in 0.5–10 keV (Asai et al 1995). The spectrum is very soft: For a blackbody fit, $kT \cong 0.3$ keV, and the emitting area is ~ 10 km^2. No hard tail such as seen in Cen X-4 is present.

0620−003 McClintock et al (1995) detected A 0620−003 with *ROSAT* at a luminosity of 6×10^{30} erg s^{-1} (for $d = 0.9$ kpc): the lowest positive detection among X-ray novae in quiescence. A previous upper limit for A 0620–003 was 10^{32} erg s^{-1} (Long et al 1981). The observed spectrum was markedly soft: If fitted with a blackbody, $kT \cong 0.16$ keV, and the emitting area is ~ 1 km^2. Although a K dwarf can emit X rays in excess of 10^{30} erg s^{-1} (Verbunt 1995), most of the observed X rays are considered to come from the accretion disk. If the accretion disk is optically thick, the mass transfer rate to the black hole is extraordinarily small: $\dot{M} \sim 2 \times 10^{-15}$ $M_\odot$ y^{-1} or $\sim 10^{11}$ g s^{-1} (McClintock et al 1995).

2023+338 Mineshige et al (1992) reported detection of GS 2023+338 with *Ginga* at a luminosity of $\sim 3 \times 10^{34}$ erg s^{-1} (for $d = 3.5$ kpc) in 1.2–10 keV after the source had returned to its quiescent optical brightness. More recently,

Verbunt et al (1994) (see also Verbunt 1995) detected the source in the *ROSAT All-Sky Survey*. The count rate corresponds to a luminosity of $\sim 8 \times 10^{33}$ erg s^{-1} in the 0.4–2.4 keV band, if $N_{\rm H} = 1.5 \times 10^{22}$ cm^{-2} (Wagner et al 1994). Wagner et al (1994) also detected the source with *ROSAT* at about the same luminosity. The intensity showed marked variations: a factor of ~ 2 variability in ~ 0.5 h and a factor of ~ 10 decrease over an interval of < 0.5 day. The minimum luminosity observed ($\sim 8 \times 10^{33}$ erg s^{-1}) corresponds to a mass accretion rate into the black hole of $\sim 10^{14}$ g s^{-1} for an optically thick disk. The spectrum was very soft: The best-fit blackbody temperature is $kT \cong 0.21$ keV.

The above results show that the X-ray spectra commonly become very soft when $L_{\rm X}$ goes down below 10^{34} erg s^{-1}, in contrast to the hard power-law (photon index of 1.5–2.0) spectra when $L_{\rm X} = 10^{35} - 10^{37}$ erg s^{-1}. If the spectrum is fitted with a blackbody, the blackbody temperature kT is around $0.2 \sim 0.3$ keV. Whether this spectral change is a sudden transition or gradual is as yet unclear. Note, however, that the spectral shape in quiescence has not been too well constrained: A blackbody spectrum, a thin thermal (bremsstrahlung) spectrum, and a steep power law are equally acceptable for all cases. It is important to note that there is no characteristic difference in the X-ray spectrum in quiescence between the neutron-star systems and the black-hole systems (see the discussion in Section 3.2).

3. DISCUSSION

3.1 *Persistent vs Transient*

As mentioned in Section 2.1, there are many LMXBs that are persistently visible, whereas a number of LMXBs are not persistent and undergo transient outbursts. What causes this distinct difference?

Table 4 lists the estimates of time-averaged luminosity between outbursts: the total energy emitted in an outburst divided by the interval from the last. In those cases where no record of the previous outburst is available, the estimates are only qualitative. Interestingly, the time-averaged luminosities are all in the range from 10^{35} to 10^{36} erg s^{-1}. Two sources with short (~ 1 y) recurrence periods, 1608−522 and 1730−335, are on the high side of the range. This translates into a range of mass accretion rates of 10^{15}–10^{16} g s^{-1} (for a conversion factor $L_{\rm X}/\dot{M} \sim 10^{20}$ erg g^{-1}), which appears to be lower than that for the persistent LMXBs (although this needs confirmation). From this, White et al (1984) suggested that mass accretion becomes unstable below a critical rate around 3×10^{16} g s^{-1}.

Indeed, the observed facts (Section 2) seem to indicate that stable mass accretion onto the compact objects cannot be maintained below an accretion rate

Table 4 Time-averaged luminosity of outbursts

Source name	Outburst year (or recur. period)	Total energy (10^{44} erg/s)	Interval[a] (year)	Ave. luminosity (10^{35} erg/s)	Ave. accretion rate (10^{15} g/s)	Ref.[b]
Cen X-4	1969	3	70*	1.4	0.7	
	1979	0.3	10	1	0.5	2
Aql X-1	~1yr	0.1–0.5		3	1.5	2
1608–522	0.5–3yr	0.1–1		16–40	8–20	3
1730–335	0.5–3yr	0.3–1	1*	10~30	5~15	1
0620-003	1917					
	1975	4	58	2	2	2
200+251	1988	5	50*	3	3	1
2023+338	1956,1979?	0.6	if 10	2	2	1
	1989		if 33	0.6	0.6	
1124–684	1991	10	50*	7	7	1
J0422+32	1992	1.2	50*	0.8	0.8	1

[a]*: assumed interval.
[b]References: 1. see Section 2, and references therein
2. White et al 1984
3. Lochner & Rousel-Dupre 1994.

of around 10^{16} g s^{-1}. Four black-hole LMXBs, A 0620–003, GS 2000+251, GS/GRS 1124−684, and GRO J0422+32, that display similar exponential decays all show an abrupt steep fall in the last phase of the decay (see Figure 3). This steepening begins at around similar luminosity levels for all of them: $\sim 0.5 \times 10^{36}$ erg s^{-1} for GRO J0422+32, $\sim 2 \times 10^{36}$ erg s^{-1} for GS/GRS 1124–684, and $\sim 10^{36}$ erg s^{-1} for A 0620–003 and GS 2000+251. The light curve of the 1969 outburst of Cen X-4 also shows a steepening (Evans et al 1970), but at a somewhat higher luminosity ($\sim 4 \times 10^{36}$ erg s^{-1}). Another neutron-star LMXB, 1608–522, sometimes stays on persistently at a level of $\sim 2 \times 10^{36}$ erg s^{-1}, but it gets turned off below this luminosity. These results can be interpreted to show that the accretion flow into the inner part of the disk is choked below a luminosity of $\sim 10^{36}$ erg s^{-1}, or for an accretion rate $\dot{M} < 10^{16}$ g s^{-1}, and the source can no longer remain persistently X-ray luminous (see discussions in the next section).

On the other hand, the optical observations invariably show that mass transfer from the secondary star continues even during the quiescent periods of X-ray novae. The optical spectrum during quiescence clearly reveals the presence of emission from the accretion disk characterized by a blue continuum and the hydrogen Balmer emission lines (van Paradijs et al 1980), contributing $\sim 30\%$ in the V-band for Cen X-4 (Chevalier et al 1989) to 30–60% for GRO J0422+32 (Filippenko et al 1995a) and $\sim 55\%$ for A 0620–003 (McClintock et al 1995). The optical disk luminosities of these sources are $\sim 10^{32}$ erg s^{-1}. The continued mass transfer during quiescence supports the idea that the secondary star fills its Roche lobe. At such low X-ray luminosities as observed in quiescence, optical emission due to reprocessing of X rays is implausible. The optical emission may come from either the outer accretion disk heated by viscous dissipation or a hot spot in the stream-disk collision region (McClintock et al 1995).

From the optical disk luminosity of A 0620–003, McClintock et al (1995) estimate the mass transfer rate onto the outer disk $\dot{M}_{\rm d}$ to be $\sim 6 \times 10^{15}$ gs^{-1}, employing the empirical relations derived for dwarf novae. This rate is only slightly lower than the critical value inferred above for the stable accretion onto the compact object. Similar $\dot{M}_{\rm d}$ values in quiescence are inferred from the optical luminosity for Cen X-4 and GRO J1655−40 (see above) and for other X-ray novae in quiescence that show similar optical emission from the disk. Remarkably, this optically inferred mass transfer rate is the same order of magnitude as the long-term average of $\dot{M}$ (see Table 4). This is consistent with the idea that a large fraction of the transferred matter is stored in the disk during a quiescent interval. The potential energy of the accumulated matter can account for the total energy of a next outburst; this lends support to a disk instability for outbursts (see Section 3.3).

Conversion of the observed X-ray luminosity to the mass accretion rate depends on the assumption of the disk structure. If the inner accretion disk is optically thick, all the gravitational energy released is radiated away. In this case, the estimated values of $\dot{M}$ through the inner disk during quiescence range from 10^{11} (for A 0620+003; McClintock et al 1995) to 10^{13} g s^{-1} or even less, which are orders of magnitude smaller than the optically inferred rates $\dot{M}$d onto the outer disk. This would imply that very little of the transferred mass goes onto the compact object and provides strong evidence for steady mass accumulation in the disk during quiescence. However, the assumption of an optically thick disk may not be valid. If the disk is optically thin, estimating the accretion rate from the X-ray luminocity is not straightforward. Nonetheless, as discussed in the next section, the observed results still support mass accumulation in the outer disk.

3.2 *Mass Accretion During Quiescence*

The mass accretion during X-ray quiescence is a whole new subject. The origin of the observed soft X rays during quiescence is still quite uncertain. If the observed X rays are indeed blackbody radiation, the area of the emitting region is $\sim$ 1–10 km^2 for all except GS 2023+338 (which measures $\sim$400 km^2). This excludes an optically thick inner accretion disk for the possible site of the soft X-ray emission, because the inferred areas are much too small for the size of the inner accretion disk. Moreover, the observed blackbody temperature ($\sim$0.2–0.3 keV) is too high for the inner disk at $\dot{M} \sim 10^{11}$–10^{13}

g s^{-1}. According to the standard accretion disk model (Shakura & Sunyaev 1973), a disk will become optically thin below a certain accretion rate, which depends on the viscosity parameter α. The above result may imply that the inner accretion disk is no longer optically thick.

Narayan et al (1995) have introduced a stable disk model for low accretion rates that consists of an optically thick outer disk and an optically thin advection-dominated inner disk. Most of the thermal energy released by viscous dissipation is advected into the compact object, and only a small fraction ($\sim 10^{-4}$–10^{-3}) of the energy is radiated from the disk. Therefore, in a black-hole system, the X-ray luminosity will be far lower than that when the disk is optically thick. This can account for a very low X-ray luminosity against a relatively large $\dot{M}$ (e.g. $\sim 6 \times 10^{15}$ g s^{-1} for A 0620−003; McClintock et al 1995). This model can reproduce the observed spectra of the quiescent black-hole systems in the optical, UV, and X-ray bands. In this model, the soft X rays are mainly Comptonized synchrotron photons in the inner hot disk. Note that this model predicts a substantial flux of hard X rays and soft γ rays due to Comptonization.

The situation is very different in a neutron-star system, depending on whether or not the accretion flow reaches the neutron star. As long as the flow reaches the neutron star, the advected energy is eventually released on the surface and all goes into blackbody radiation. In other words, the conversion factor $L_X/\dot{M}$ is essentially the same for both an optically thick disk and an optically thin advection-dominated disk. Hence, for whichever disk, accretion at the optically inferred rate will give rise to an X-ray luminosity that is orders of magnitude higher than observed: e.g. for $\dot{M} = 10^{15}$ gs^{-1}, $L_X \sim 10^{35}$ erg s^{-1} with $kT \geq 0.3 f^{-1/4}$ keV, where f is the fraction of the X-ray emitting area of the neutron star surface. (The spectrum is not necessarily a blackbody spectrum, but may well be modified by Comptonization.) Thus, in this case, the advection-dominated disk is inadequate at least for the neutron-star systems, and the mass accretion rate through the inner disk is indeed low and cannot exceed the value inferred from the observed X-ray luminosity, i.e. $\sim 10^{12}$–10^{13} g s^{-1}.

However, there is a possibility that the accretion flow stops at some distance from the neutron star surface. Accretion onto a neutron star can occur only if the gravitational attraction is stronger than the centrifugal force exerted by the magnetosphere (the "propeller effect"; see Illarionov & Sunyaev 1975). This situation requires that the rotation speed of the magnetospheric boundary (the Alfvén surface) is slower than the local Keplerian velocity, which sets a minimum rate $\dot{M}_{min}$ to allow continued accretion onto the neutron star as

$$\dot{M}_{\rm min} \simeq 2 \times 10^{16} \left(\frac{B}{10^9\ \rm G}\right)^2 \left(\frac{P}{10^{-2}\ \rm s}\right)^{-7/3} \ \rm g\ s^{-1},$$

for a 1.4 $M_\odot$ neutron star of 10^6 cm radius with the surface magnetic field B and a spin period P. Therefore, for such a neutron star as those of millisecond radio pulsars, accretion onto the surface will be prevented below a rate of the order of 10^{16} g s^{-1}.

Suppose accretion is blocked ($\dot{M} < \dot{M}_{\rm min}$) at the Alfvén radius, which is given approximately by

$$r_{\rm A} \sim 10^7 \left(\frac{B}{10^9 {\rm G}}\right)^{4/7} \left(\frac{\dot{M}}{10^{16}\,{\rm gs}^{-1}}\right)^{-2/7} {\rm cm}$$

$$\simeq 10^7 \left(\frac{P}{10^{-2} s}\right)^{2/3} \left(\frac{\dot{M}}{\dot{M}_{\rm min}}\right)^{-2/7} {\rm cm}$$

If the accretion disk is advection dominated and the soft X rays come from the hot inner disk (Narayan et al 1995), the X-ray luminosity of the neutron-star systems will be systematically lower than those of the black-hole systems for a given accretion rate. Because the disk terminates far away ($\geq 10^7$ cm unless $P \ll 10^{-2}$ s^{-1}) from the neutron star, the available gravitational energy will be an order of magnitude less. This may also cause a difference in the spectral shape. However, the observed results for both systems are strikingly similar to each other and do not show such systematic differences.

Although the hypothesis of an advection-dominated inner disk cannot be ruled out, such a disk with a relatively high mass-flow rate ($\sim 10^{15}$–10^{16} g s^{-1} as optically inferred) is not entirely consistent with the observed similarities between the neutron-star systems and the black-hole systems. The main origin of the soft X rays is probably not an advection dominated disk. On the other hand, any other structure of the inner disk will give a much higher X-ray luminosity for this accretion rate than observed. Thus, regardless of the disk structure, consistency with the observed results seems to require that the accretion rate through the inner disk is much lower than the optically inferred rate, lending support for mass accumulation in the outer disk.

The remarkable similarities in the X-ray and optical properties, particularly the ratio of the X-ray to optical disk luminosities, between the neutron-star systems and the black-hole systems in quiescence seem to argue for a common origin of the observed soft X rays for both systems. One possible origin of the soft X rays is the inner disk which is not entirely optically thick but is a gray body, i.e. the emissivity is less than that of a blackbody disk. This would resolve the problem arising in the blackbody calculation, where the emitting area obtained comes out too small. Verbunt et al (1994) interpreted the observed small emitting area of Aql X-1 to be either the magnetic polar caps or a boundary layer. However, because of the similarities between the neutron-star systems and the black-hole systems, X rays from the neutron star would comprise at most a fraction of those from the disk. Alternatively, the soft X rays might originate in the outer disk. A simple estimation from the gravitational potential shows that to obtain $L_X \sim 10^{32}$ erg s^{-1} for an optically inferred $\dot{M}$ of $\sim 10^{15}$–10^{16} g s^{-1}, the inner radius of the disk must be $\leq 10^{10}$ cm for a neutron-star system. (This inner radius scales with mass of the compact object.) This disk should not be optically thick, otherwise the temperature is far too low to account for the observed result. Both these possibilities remain speculative and require detailed examination.

At present, whether or not accretion onto neutron stars continues during X-ray quiescence is still an open question. As discussed below, this has an important relation to the subject of millisecond pulsars. The recycled pulsar scenario postulates that the neutron-star LMXBs are the progenitors of millisecond radio pulsars: The weakly magnetized neutron stars have been spun up by the accretion torque during the X-ray binary lifetime (for a review, see e.g. Bhattacharya 1995). Yet, there has been no direct proof, such as X-ray pulsation, that the neutron stars in LMXBs are rotating at millisecond periods. If indeed accretion onto a neutron star is confirmed, it provides a strong limit on the rotation period of the neutron star and its magnetic field (e.g. Stella et al 1994, Verbunt et al 1994). The observed X-ray luminosity gives an upper limit to $\dot{M}$ onto the neutron star: $\dot{M} < 10^{12}$–10^{13} g s^{-1}. Assuming that the soft X rays of Aql X-1 come from the neutron star, Verbunt et al (1994) state that it is not spinning at a rate of millisecond pulsars, even if $B \sim 10^8$ G, or alternatively that the magnetic fields are even weaker (see the equation above for $\dot{M}_{\rm min}$). Thus, the confirmation of soft X-ray emission from the neutron star would lead to a conclusion that at least transient LMXBs are not suitable progenitors of recycled millisecond radio pulsars.

The absence of a detected radio millisecond pulsar from Cen X-4 during X-ray quiescence is also consistent with slow rotation or low magnetic fields of the neutron star in this system (Kulkarni et al 1992). If a millisecond-rotator with $B \sim 10^8$–10^9 G is present, the mass accretion rate for the obtained X-ray luminosity upper limit at that time (5×10^{32} erg s^{-1}) should be low enough to allow the neutron star to function as a radio pulsar (Shaham & Tavani 1991). Yet, the absence of a millisecond pulsar is not entirely definitive: Some propagation effects may exist that quench the radio pulsation. More radio observations of the neutron-star LMXBs in quiescence are important.

3.3 *Outburst Mechanism*

Two competing models have been discussed to explain outbursts of X-ray novae: 1. disk instability models (Osaki 1974; Meyer & Meyer-Hofmeister 1981; Cannizzo et al 1982, 1985; Cannizzo 1993; Lin & Taam 1984; Huang & Wheeler 1989; Mineshige & Wheeler 1989) and 2. mass transfer instability models (Osaki 1985; Hameury et al 1986, 1987, 1988, 1990). The recent observations, especially those in quiescence (see Sections 2.4 and 3.1), set important constraints on these models and also on the mass accretion process.

3.3.1 DISK INSTABILITY MODELS Thermal instability triggers the outburst in the disk instability models. This instability arises from a very steep temperature dependence of the opacity in a partially ionized accretion disk.

An accretion disk is stable both in a cool, neutral state and in a hot, fully ionized state. However, when the mass inflow rate onto the outer part of a disk is within a critical range, the disk undergoes a thermal limit cycle. In a quiescent state, the accretion disk is in the cool state. As matter accumulates in the disk, both the surface density and temperature increase. When the surface density reaches an upper critical value, a thermal instability sets in. The disk jumps to the hot state, which gives rise to a high accretion rate, causing rapid infall of matter onto the compact object and hence an X-ray outburst. When the surface density drops below a lower critical value, the disk returns to the cool state.

The disk instability models can fairly well explain the observed properties of X-ray nova outbursts: the rise, decay, and recurrence times of outburst. The decay time of an outburst is determined by the diffusion time of matter in the hot state, whereas the recurrence period is determined by the diffusion time in the cool state. Since the diffusion time is proportional to the mass of the compact object, systematically longer decay and recurrence times are expected for black-hole systems than for neutron star systems (Mineshige & Kusunose 1993). Note that, under standard assumptions, the light curve expected in the disk instability models is a power-law function of time, if the total angular momentum is conserved (Lyubarskii & Shakura 1987). However, Mineshige et al (1993) and Cannizzo (1994) argue that mass and angular momentum are rapidly transferred outward from the inner hot region to the outer cool region, due to an abrupt change in temperature and hence kinematic viscosity across the thermal transition front. This effect yields an exponential decay of the disk luminosity.

The secondary and tertiary maxima seen in the light curves of some black-hole X-ray nova outbursts (see Figure 3) may also hold important clues. The decay times are similar before and after the secondary maximum. This implies a sudden supply of extra mass into the disk. Possible mechanisms proposed for the sudden mass supply are: 1. evaporation of matter by X rays near the inner Lagrangean point (L_1) when the disk and ionized clouds become incapable of blocking X rays from the central region (Chen et al 1993, Mineshige 1994) and 2. a mass transfer instability caused by hard X-ray heating of the subphotospheric layers of the secondary during an outburst (Chen et al 1993). Mineshige (1994) points out that an abrupt heating of the outer portion of the disk by X rays is also required to explain the fast rise of the secondary maximum.

There is evidence supporting a disk thermal instability as the mechanism of X-ray nova outbursts. As described in Section 3.1, the optical observations indicate continued mass transfer into the outer disk during quiescence at a rate of 10^{15}–10^{16} g s^{-1}. This value is similar to that derived from the time-averaged outburst luminosities (Table 4). Hence, it is consistent with the idea that most of the transferred mass remains in the disk during quiescence and the total mass accumulated since the last outburst is consumed in an outburst. The low X-ray luminosities observed during quiescence are interpreted to show that only a small fraction of the transferred mass goes onto the compact object. (See the discussion in Section 3.2.)

However, some problems with the disk instability models have been pointed out. In order to explain recurrence times as long as $\sim$50 y, the viscosity parameter α in quiescence must be extremely small: $\alpha \leq 10^{-4}$ (Hameury et al 1986, 1993; Mineshige & Wheeler 1989; Mineshige & Kusunose 1993). However, α is expected to be $\sim$0.01 for those systems with recurrence times of the order of 1 year. This large difference in α between the systems whose other properties are similar to each other is difficult to explain. Furthermore, for such cases as A 0620-003, the small α-value ($\sim 10^{-4}$) inferred from a long recurrence time is not compatible with the $\alpha \sim 0.01$ required to reproduce the observed amplitude and duration of the outburst (Mineshige & Kusunose 1993).

3.3.2 MASS TRANSFER INSTABILITY MODELS In mass transfer instability models, the subphotospheric layers of a companion star are considered to be heated up by relatively hard (> 7 keV) X rays from a compact object. These layers slowly expand and ultimately bring the atmosphere into an unstable regime, which leads to a sudden mass transfer. An accretion disk builds up and thickens as the mass transfer rate from the secondary increases. Consequently, the mass inflow from the outer region of the disk onto the compact object is suddenly enhanced, giving rise to an outburst. The mass loss instabiltiy will cease, when the L_1 region is shielded by the accretion disk, and the outburst ends when all of the matter in the disk has been accreted onto the compact object. Once the system settles back into a quiescent state, it will take a long time for X rays to heat and expand the atmosphere of the secondary until it reaches the unstable condition again (next outburst).

The calculated results show that the mass transfer instability models can also account for the main characteristics of outbursts such as the light curve and the recurrence time (Hameury et al 1986, 1988, 1990). However, recent observational results impose a severe problem on the mass transfer instability models. In order for a mass transfer instability to trigger an outburst, the hard X-ray flux at the L_1 point must exceed the intrinsic stellar flux: $L_X(> 7\ \mathrm{keV}) > 2.5 \times 10^{34}(M_c/M_\odot)^2\ \mathrm{erg\ s^{-1}}$, where M_c is the mass of the companion (Mineshige et al 1992). However, as described in Section 2.4, the X-ray spectra during quiescence are all very soft, with a blackbody temperature $kT \sim 0.2$–0.3 keV, and the X-ray luminosities are mostly in the range 10^{31}–10^{33} erg s^{-1}. Thus, the hard X-ray flux appears far insufficient to induce a mass transfer instability.

As discussed above, the observed results seem to be more in support of a disk instability than a mass transfer instability for the mechanism of triggering an outburst. Nonetheless, a mass transfer instability may still be a viable mechanism for the secondary (and tertiary) maxima during decay. Also, many outbursts exhibit complex multipeaked light curves (see Section 2.2). Some X-ray novae (e.g. GRS 1915+105, GRO J1655−40) remain active for years, repeating outbursts. These complex activities may well be due to the secondary mass transfer from the companion as a consequence of X-ray heating. A further interesting case is the outbursts of 1608–522. Some of the outbursts of this source occur even when it is persistently on (see Section 2.2.1). This aspect is difficult to explain in terms of a disk instability.

3.4 *How Many LMXBs are Still in Quiescence?*

In the past 20 years, about 30 low-mass transients have been discovered. The discovery rate has increased substantially since late 1980s owing to a continued sky watch with *Ginga*, *GRANAT*, and *CGRO*. The average rate of detection of X-ray nova outbursts is approximately 2 y^{-1} in the past five years. There must be many more low-mass binaries that are currently in quiescence but that eventually undergo an outburst. An attempt to estimate the total number of them was made previously by Tanaka (1992a).

In estimating the total number of such low-mass binaries, the largest unknown is the average recurrence period for all X-ray novae. The recurrence period is largely different from source to source, distributed in a very wide range from ~ 1 y to more than 50 y. There is no apparent difference between the neutron-star systems and the black-hole systems. In fact, most X-ray novae were observed only once. From the available data, an average period shorter than 10 y is unlikely. Here we assume it to be 10–50 y.

For most X-ray novae observed, the estimated distances are less than 5 kpc. Even if distance estimates are not available, the assumption that the observed peak luminosity is $\sim 10^{38}$ erg s^{-1} gives distances < 5 kpc. There are only a few detected X-ray novae that are more distant (e.g. GRS 1915+105, 1730−335).

This is obviously due to a detection bias toward bright outbursts. Considering the fact that 90% of the observed low-mass transients are within a galactic longitude range of $\pm 80°$ (showing that almost all the low-mass transients lie within ~ 8 kpc of the Galactic center), the effective coverage of the Galactic plane would be $\sim 10\%$.

With the above considerations, the total number of LMXBs in quiescence is estimated to be ~ 200 for a modest average recurrence period of 10 y and ~ 1000 for an average period of 50 y. Although these estimates are crude, it is certain that the currently known LMXBs are but a minor fraction of the whole LMXB systems. In addition, it is remarkable that most of the bright X-ray novae observed thus far are black-hole systems. Although current statistics are still insufficient, it seems probable that the number of (quiescent) black-hole LMXBs is at least as many as or even greater than that of (quiescent+persistent) neutron-star LMXBs.

There is another striking difference between neutron-star LMXBs and black-hole LMXBs. As mentioned in Section 2.1, ~ 80 neutron-star LMXBs are known to be persistent, whereas not a single persistent black-hole LMXB has yet to be found. The reason for this distinct difference remains unclear. According to the discussion in Section 3.1, this difference implies that the doner stars of the black-hole low-mass systems are somehow unable to achieve a mass transfer rate $> 10^{16}$ g s^{-1} required to make them persistent sources. Yet, the optically inferred mass transfer rates of the observed black-hole LMXBs are only slightly lower than this critical rate. There is no noticeable difference in the types of secondary stars or in the orbital period distribution between the black-hole LMXBs and the neutron-star LMXBs. Although the conditions that determine the accretion rate are not yet clear, the subtle difference in the accretion rate may be a consequence of possible differences in the evolutionary state between the two classes of low-mass binary system.

The above results may provide important clues to the problems of formation and evolution of LMXB systems (for a review, see Verbunt & van den Heuvel 1995).

ACKNOWLEDGMENTS

The authors thank F Verbunt and R Mushotzky for carefully reading the manuscript and for valuable comments. They also thank S Mineshige for useful discussions and R Sunyaev for providing them with the *Mir/Kvant* and *GRANAT* results. This work has been done while YT stayed at Max-Planck Institut für Extraterrestrische Physik as an Alexander von Humboldt research fellow. YT is grateful to the staff of the Institut, in particular to J Trümper, for their kind hospitality and to the Alexander von Humboldt Foundation for support.

Literature Cited

Alexandrovich N, Borozdin K, Evremov V, Sunyaev R. 1994b. *IAU Circ. No. 6087*
Alexandrovich N, Borozdin K, Sunyaev R. 1994a. *IAU Circ. No. 6080*
Alexandrovich N, Borozdin K, Sunyaev R. 1995. *IAU Circ. No. 6143*
Arefév V, Borozdin K, Churazov E, Efremov V, Gilfanov M. 1989. *Proc. 23rd ESLAB Symp. Two-Topics X-Ray Astronomy, Bologna, Italy,* ESA SP-296, p. 255
Asai K, Dotani T, Mitsuda K, Hoshi R, Vaughan B, et al. 1996. *Publ. Astron. Soc. Jpn.* 48:257
Augusteijn T, Kuulkers E, Shaham J. 1993. *Astron. Astrophys.* 279:L13
Bailyn CD, Orosz JA, Girard TM, Jogee S, Della Valle M, et al. 1995a. *Nature* 374:701
Bailyn CD, Orosz JA, McClintock J, Remillard R. 1995b. *IAU Circ. No. 6173, Nature* 378:157
Ballet J, Denis M, Gilfanov M, Sunyev R. 1993. *IAU Circ. No. 5874*
Barret D, Motch C, Pietsch W, Voges W. 1995. *Astron. Astrophys.* 296:459
Bhattacharya D. 1995. See Lewin et al 1995, p. 233
Boley F, Wolfson R, Bradt H, Doxsey R, Gernigan G, et al. 1976. *Ap. J. Lett.* 203:L13
Borozdin K, Gilfanov M, Sunyaev R, Churazov E, Loznikov V, et al. 1990. *Sov. Astron. Lett.* 16(5):345
Borozdin KN, Aleksandrovich NL, Aref'ev, VA, Sunyaev RA, Skinner GK. 1995. *Astron. Lett.* 21:212
Brandt S, Castro-Tirado AJ, Lund N, Dremin V, Lapshov I, et al. 1992. *Astron. Astrophys.* 254:L39
Buie MW, Bond HE. 1989. *IAU Circ. No. 4786*
Callanan PJ, Charles PA. 1991. *MNRAS* 249:573
Callanan PJ, Garcia MR, McClintock JE, Zhao P, Remillard RA, et al. 1995. *Ap. J.* 441:786
Campbell-Wilson D, Hunstead R. 1994. *IAU Circ. No. 6052, 6055*
Canizares CR, McClintock JE, Grindlay JE. 1980. *Ap. J. Lett.* 236:L55
Cannizzo JK. 1993. *Accretion Disks in Compact Stellar Systems,* ed. JC Wheeler, p. 6. Singapore: World Sci.
Cannizzo JK. 1994. *Ap. J.* 435:389
Cannizzo JK, Wheeler JC, Gosh P. 1982. *Pulsations in Classical and Cataclysmic Variable Stars,,* ed. JP Cox, p. 13. Boulder: Univ. Colo./Natl. Bur. Stand.
Cannizzo JK, Wheeler JC, Gosh P. 1985. *Cataclysmic Variables and Low-Mass X-Ray Binaries,* ed. DQ Lamb, J Patterson, p. 307. Dordrecht: Reidel
Casares J, Charles PA, Jones DHP, Rutten RGM, Callanan PJ. 1991. *MNRAS* 250:712
Casares J, Charles PA, Naylor T. 1992. *Nature* 355:614
Casares J, Charles PA, Naylor T, Pavlenko EP. 1993. *MNRAS* 265:834
Casares J, Marsh TR, Charles PA, Martin AC, Martin EL, et al. 1995. *MNRAS* 274:565
Castro-Tirado AJ, Brandt SA, Lund N. 1992a. *IAU Circ. No. 5587*
Castro-Tirado AJ, Brandt SA, Lund N. 1992b. *IAU Circ. No. 5590*
Castro-Tirado AJ, Brandt SA, Lund N, Lapshov I, Sunyaev RA, et al. 1994. *Ap. J. Suppl.* 92:469
Castro-Tirado AJ, Pavlenko EP, Salyapikov AA, Brandt S, Lund N, et al. 1993. *Astron. Astrophys.* 276:L37
Charles P, Hassall B, Machin G, Smale A, Allington-Smith J. 1988. *IAU Circ. No. 4609*
Charles PA, Casares J, Jones DHP, Broadhurst T, Callanan PJ, et al. 1989. *Proc. 23rd ESLAB Symp. Two—Topics in X-Ray Astronomy, Bologna, Italy,* ESA SP-296, p. 103
Charles PA, Thorstensen JR, Bowyer S, Clark GW, Li FK, et al. 1980. *Ap. J.* 237:154
Chen W, Livio M, Gehrels N. 1993. *Ap. J. Lett.* 408:L5
Chevalier C, Ilovaisky SA. 1990. *Astron. Astrophys.* 238:163
Chevalier C, Ilovaisky SA. 1991. *Astron. Astrophys.* 251:L11
Chevalier C, Ilovaisky SA. 1993. *Astron. Astrophys.* 269:301
Chevalier C, Ilovaisky SA. 1995. *Astron. Astrophys.* 297:103
Chevalier C, Ilovaisky SA, van Paradijs J, Pedersen H, van der Klis M. 1989. *Astron. Astrophys.* 210:114
Cowley AP. 1992. *Annu. Rev. Astron. Astrophys.* 30:287
Cowley AP, Crampton D, Hutchings JB, Remillard R, Penfold JE. 1983. *Ap. J.* 272:118

Cowley AP, Hutchings JP, Schmidke PC, Hartwick FDA, Crampton D, et al. 1988. *Astron. J.* 95:1231
Czerny M, Czerny B, Grindlay JE. 1987. *Ap. J.* 312:122
Davis RJ, Edwards MR, Morison I, Spencer RE. 1975. *Nature* 257:659
Della Valle M, Benetti S. 1993. *IAU Circ. No. 5890*
Della Valle M, Jarvis BJ, West RM. 1991. *Nature* 353:50
Della Valle M, Mirabel IF, Cordier B. 1993. *IAU Circ. No. 5876*
Della Valle M, Mirabel IF, Rodriguez LF. 1994. *Astron. Astrophys.* 290:803
Dickel JR, Eilek JE, Jones EM, Reynolds SP. 1989. *Ap. J. Suppl.* 70:497
Döbereiner S, Englhauser J, Pietsch W, Reppin C, Trümper J, et al. 1989. *Proc. 23rd ESLAB Symp. Two—Topics X-Ray Astronomy, Bologna,* ESA SP-296, p. 387
Döbereiner S, Maisack M, Englhauser J, Pietsch W, Reppin C, et al. 1994. *Astron. Astrophys.* 287:105
Eachus L, Wright E, Liller W. 1976. *Ap. J. Lett.* 203:L17
Ebisawa K, Ogawa M, Aoki T, Dotani T, Takizawa M, et al. 1994. *Publ. Astron. Soc. Jpn.* 46:375
Elvis M, Page CG, Pounds KA, Ricketts MJ, Turner MJL. 1975. *Nature* 257:656
Evans WD, Belian RD, Conner JP. 1970. *Ap. J. Lett.* 159:L57
Filippenko AV, Matheson T, Ho LC. 1995a. *Ap. J.* 455:614
Filippenko AV, Matheson T, Barth AJ. 1995b. *Ap. J. Lett.* 455:L139
Gerard E, Rodriguez LF, Mirabel IF. 1994. *IAU Circ. No. 5958*
Gies DR, Bolton CT. 1982. *Ap. J.* 260:240
Goldwurm A, Ballet J, Cordier B, Paul J, Bouchet L, et al. 1992. *Ap. J. Lett.* 389:L79
Gotthelf E, Halpern JP, Patterson J, Rich RM. 1992. *Astron. J.* 103:219
Greiner J, Predehl P, Pohl M. 1995. *Astron. Astrophys.* 297:L67
Greiner J, Snowden S, Harmon BA, Kouveliotou C, Paciesas W. 1994. *The Second Compton Symposium,* ed. CE Fichtel, N Gehrels, JP Norris, p. 260. New York: AIP
Grindlay JE, Liller W. 1978. *Ap. J. Lett.* 220:L127
Hameury JM, King AR, Lasota JP. 1986. *Astron. Astrophys.* 162:71
Hameury JM, King AR, Lasota JP. 1987. *Astron. Astrophys.* 171:140
Hameury JM, King AR, Lasota JP. 1988. *Astron. Astrophys.* 192:187
Hameury JM, King AR, Lasota JP. 1990. *Ap. J.* 353:585
Hameury JM, King AR, Lasota JP. 1993. *Accretion Disks in Compact Stellar Systems,* ed. JC Wheeler, p. 360. Singapore: World Sci.
Han X-H, Hjellming RM. 1990. *Proc. 11th North American Workshop CVs and LMXBs,* ed. CW Mauche, p. 25. Cambridge: Cambridge Univ. Press
Han X-H, Hjellming RM. 1992a. *Ap. J.* 400:304
Han X-H, Hjellming RM. 1992b. *IAU Circ. No. 5593*
Harmon BA, Wilson CA, Zhang SN, Paciesas WS, Fishman GJ, et al. 1995. *Nature* 374:703
Harmon BA, Zhang SN, Paciesas WS, Fishman GJ. 1993. *IAU Circ. No. 5874*
Harmon BA, Zhang SN, Wilson CA, Rubin BC, Fishman GJ. 1994. *The Second Compton Symposium,* ed. CE Fichtel, G Gehrels, JP Norris, p. 210. New York: AIP
Hjellming RM, Calovini T, Han X-H, Córdova FA. 1988. *Ap. J. Lett.* 335:L75
Hjellming RM, Han X-H. 1995. See Lewin et al 1995, p. 308
Hjellming RM, Han X-H, Córdova FA. 1989. *IAU Circ. No. 4790*
Hjellming RM, Han X-H, Roussel-Dupré D. 1990. *IAU Circ. No. 5112*
Hjellming RM, Rupen MP. 1995. *Nature* 375:464
Hoffman JA, Marshall HL, Lewin WHG. 1978. *Nature* 271:630
Huang M, Wheeler JC. 1989. *Ap. J.* 343:229
Hurst GM, Mobberley M. 1989. *IAU Circ. No. 4783*
Hutchings JB, Crampton D, Cowley AP, Bianchi L, Thompson IB. 1987. *Astron. J.* 94:340
Illarionov AF, Sunyaev RA. 1975. *Astron. Astrophys.* 39:185
Inoue H. 1994. *Multi-Wavelength Continuum Emission of AGN, Proc. IAU Symp. 159, Geneva,* ed. TJ-L Courvoisier, A Blecha, p. 73. Dordrecht: Kluwer
in't Zand JJM, Pan HC, Bleeker JAM, Skinner GK, Gilfanov MR, et al. 1992. *Astron. Astrophys.* 266:283
Kaluzienski LJ, Holt SS, Boldt EA, Serlemitsos PJ. 1977a. *Nature* 265:606
Kaluzienski LJ, Holt SS, Boldt EA, Serlemitsos PJ. 1977b. *Ap. J.* 212:203
Kaluzienski LJ, Holt SS, Swank JH. 1980. *Ap. J.* 241:779
Kato T, Mineshige S, Hirata R. 1993. *IAU Circ. NO. 5704*
Kato T, Mineshige S, Hirata R. 1995. *Publ. Astron. Soc. Jpn.* 47:31
Kesteven MJ, Turtle AJ. 1991. *IAU Circ. No. 5181*
King AR. 1993. *MNRAS* 260:L5
Kitamoto S, Tsunemi H, Miyamoto S, Hayashida K. 1992. *Ap. J.* 394:609

Kitamoto S, Tsunemi H, Miyamoto S, Roussel-Dupré D. 1993. *Ap. J.* 403:315
Kitamoto S, Tsunemi H, Miyamoto S, Yamashita K, Mizobuchi S, et al. 1989. *Nature* 342:518
Kitamura T, Nakagawa M, Takagishi M, Matsuoka M, Miyamoto S, et al. 1971. *Nature* 229:31
Koyama K, Inoue H, Makishima K, Matsuoka M, Murakami T, et al. 1981. *Ap. J. Lett.* 247:L27
Kulkarni SR, Navarro J, Vasisht G, Tanaka Y, Nagase F. 1992. *X-Ray Binaries and Recycled Pulsars,* ed. EPJ van den Heuvel, SA Rappaport, p. 99. Dordrecht: Kluwer
Lapshov I, Sazanov S, Sunyaev R. 1993. *IAU Circ. No. 5864*
Lapshov IYu, Sazanov SYu, Sunyaev RA, Brandt S, Castr-Tirado A, et al. 1994. *Astron. Lett.* 20:205
Leibowitz EM, Ney A, Drissen L, Grandchamps A, Moffat AFJ. 1991. *MNRAS* 250:385
Lewin WHG. 1976. *IAU Circ. No. 2922*
Lewin WHG, van Paradijs J, Taam RE. 1995. See Lewin et al 1995, p. 175
Lewin WHG, van Paradijs J, van den Heuvel EPJ, eds. 1995. *X-Ray Binaries.* Cambridge: Cambridge Univ. Press
Liller W. 1977. *Ap. J.* 213:L21
Liller W. 1979. *IAU Circ. No. 3366*
Lin DNC, Taam RE. 1984. See Woosley 1984, p. 83
Lochner JC, Roussel-Dupré D. 1994. *Ap. J.* 435:840
Long KS, Helfand DS, Grabelsky DA. 1981. *Ap. J.* 248:925
Lund N, Brandt S. 1991. *IAU Circ. No. 5161*
Lyubarskii YuE, Shakura NI. 1987. *Sov. Astron. Lett.* 13(5):386
Makino F. 1988. *IAU Circ. No. 4587, 4600*
Makino F. 1989. *IAU Circ. No. 4782, 4786*
Makino F. 1991. *IAU Circ. No. 5161*
Margon B, Katz JI, Petro LD. 1978. *Nature* 271:633
Marsden BG. 1989. *IAU Circ. No. 4783*
Martin AC, Charles PA, Wagner RM, Casares J, Henden AA, et al. 1995. *MNRAS* 274:559
Matilsky T, Bradt HV, Buff J, Clark GW, Jernigan JG, et al. 1976. *Ap. J. Lett.* 210:L127
Matsuoka M, Inoue H, Koyama K, Makishima K, Murakami T, et al. 1980. *Ap. J. Lett.* 240:L137
McClintock JE, Bailyn CD, Remillard RA. 1992. *IAU Circ. No. 5499*
McClintock JE, Horne K, Remillard RA. 1995. *Ap. J.* 442:358
McClintock JE, Petro LD, Remillard RA, Ricker GR. 1983. *Ap. J. Lett.* 266:L27
McClintock JE, Remillard RA. 1986. *Ap. J.* 308:110
McClintock JE, Remillard RA. 1990. *Ap. J.* 350:386
McKay D, Kesteven M. 1994. *IAU Circ. No. 6062*
Meyer F, Meyer-Hofmeister E. 1981. *Astron. Astrophys.* 104:L10
Mineshige S. 1994. *Ap. J. Lett.* 431:L99
Mineshige S, Ebisawa K, Takizawa M, Tanaka Y, Hayashida K, et al. 1992. *Publ. Astron. Soc. Jpn.* 44:117
Mineshige S, Kusunose M. 1993. *Accretion Disks in Compact Stellar Systems,* ed. JC Wheeler, p. 370. Singapore: World Sci.
Mineshige S, Wheeler JC. 1989. *Ap. J.* 343:241
Mineshige S, Yamasaki T, Ishizaka C. 1993. *Publ. Astron. Soc. Jpn.* 45:707
Mirabel IF, Duc PA, Rodriguez LF, Teyssier R, Paul J, et al. 1994. *Astron. Astrophys.* 282: L17
Mirabel IF, Rodríguez LF. 1994. *Nature* 371:46
Mitsuda K, Inoue H, Koyama K, Makishima K, Matsuoka M, et al. 1984. *Publ. Astron. Soc. Jpn.* 36:741
Mitsuda K, Inoue H, Nakamura N, Tanaka Y. 1989. *Publ. Astron. Soc. Jpn.* 41:97
Murdin P, Allen DA, Morton DC, Whelan JAJ, Thomas RM. 1980. *MNRAS* 192:709
Nakamura N, Dotani T, Inoue H, Mitsuda K, Tanaka Y, et al. 1989. *Publ. Astron. Soc. Jpn.* 41:617
Narayan R, McClintock JE, Yi I. 1996. *Ap. J.* 457:821
Okamura S, Noguchi T. 1988. *IAU Circ. No. 4589*
Oke JB. 1977. *Ap. J.* 217:181
Oroz JA, Bailyn CD. 1995. *Ap. J. Lett.* 446:L59
Osaki Y. 1974. *Publ. Astron. Soc. Jpn.* 26:429
Osaki Y. 1985. *Astron. Astrophys.* 144:369
Owen F, Balonek T, Dickey J, Terzian Y, Gottesman S. 1976. *Ap. J. Lett.* 203:L15
Paciesas WS, Briggs MS, Harmon BA, Wilson RB, Finger MH. 1992. *IAU Circ. No. 5580*
Priedhorsky WC, Holt SS. 1987. *Space Sci. Rev.* 45:291
Priedhorsky WC, Terrel J. 1984. *Ap. J.* 280:661
Remillard RA, McClintock JE, Bailyn CD. 1992. *Ap. J. Lett.* 399:L145
Richter GA. 1989. *IBVS,* pt. 3362:1
Ricketts MJ, Pounds KA, Turner MJL. 1975. *Nature* 257:657
Rodriguez LF, Mirabel IF. 1993. *IAU Circ. No. 5900*
Sazanov SYu, Sunyaev RA, Lapshov IYu, Lund N, Brandt S, et al. 1994. *Astron. Lett.* 20:787
Shaham J, Tavani M. 1991. *Ap. J.* 377:588
Shakura NI, Sunyaev RA. 1973. *Astron. Astrophys.* 24:337
Shao Y. 1992. *IAU Circ. No. 5606*
Stella L, Campana S, Colpi M, Mereghetti S, Tavani M. 1994. *Ap. J. Lett.* 423:L47

Sunyaev R, Churazov E, Gilfanov M, Dyachkov A, Khavenson N, et al. 1992. *Ap. J. Lett.* 389:L75
Sunyaev RA, Borozdin KN, Aleksandrovich NL, Aref'ev VA, Kaniovskii AS, et al. 1994. *Astron. Lett.* 20:777
Sunyaev RA, Kaniovskii AS, Efremov VV, Aref'ev VA, Borozdin KN, et al. 1991. *Sov. Astron. Lett.* 17(2):123
Sunyaev RA, Kaniovsky AS, Borozdin KN Efremov VV, Aref'ev VA, et al. 1993. *Astron. Astrophys.* 280:L1
Sunyaev RA, Lapshov IYu, Grebenev SA, Efremov VV, Kaniovskii AS, et al. 1988. *Sov. Astron. Lett.* 14(5):327
Takizawa M. 1991. MSc. thesis. Univ. Tokyo
Tanaka Y. 1989. *Proc. 23rd ESLAB Symp. Two-Topics X-Ray Astronomy, Bologna, Italy,* ESA SP-296, p. 3
Tanaka Y. 1992a. *X-Ray Binaries and Recycled Pulsars,* ed. EPJ van den Heuvel, SA Rappaport, p. 37. Dordrecht: Kluwer
Tanaka Y. 1992b. *Ginga Memorial Symp., ISAS,* ed. F Makino, F Nagase, p. 19
Tanaka Y. 1994. *New Horizon of X-Ray Astronomy,* ed. F Makino, T Ohashi, p. 37. Tokyo: Universal Acad.
Tanaka Y, Lewin WHG. 1995. See Lewin et al 1995, p. 126
Thorstensen J, Charles P, Bowyer S. 1978. *Ap. J. Lett.* 220:L131
Tingay SJ, Jauncey DL, Preston RA, Reynolds JE, Meier DL, et al. 1995. *Nature* 374:141
Tsunemi H, Kitamoto S, Okamura S, Roussel-Dupré D. 1989. *Ap. J. Lett.* 337:L81
van Dijk R, Bennett K, Collmar W, Diehl R, Hermsen W, et al. 1995. *Astron. Astrophys.* 296:L33
van Paradijs J. 1995. See Lewin et al 1995, p. 536
van Paradijs J, McClintock JE. 1995. See Lewin et al 1995, p. 58
van Paradijs J, Verbunt F. 1984. See Woosley 1984, p. 49
van Paradijs J, Verbunt F, Shafer RA, Arnaud KA. 1987. *Astron. Astrophys.* 182:47
van Paradijs J, Verbunt F, van der Linden T, Pedersen H, Wamsteker W. 1980. *Ap. J. Lett.* 241:L161
Verbunt F. 1995. *Compact Stars in Binaries, IAU Symp. 165,* ed. E Kuulkers, EPJ van den Heuvel. Dordrecht: Kluwer. p. 333
Verbunt F, Belloni T, Johnston HM, van der Klis M, Lewin WHG. 1994. *Astron. Astrophys.* 285:903
Verbunt F, van den Heuvel EPJ. 1995. See Lewin et al 1995, p. 457
Vikhlinin A, Churazov E, Gilfanov M, Sunyaev R, Finoguenov A, et al. 1995. *Ap. J.* 441:779
Wachmann AA. 1948. *Ergänzungshefte zu Astronomische Nachrichten* 11(5):E42
Wagner RM, Bertram R, Starrfield SG, Howell SB, Kreidl TJ, et al. 1991. *Ap. J.* 378:293
Wagner RM, Bertram R, Starrfield SG, Shrader CR. 1992b. *IAU Circ. No. 5589*
Wagner RM, Henden AA, Bertram R. 1988. *IAU Circ. No. 4600*
Wagner RM, Kreidl TJ, Howell SB, Starrfield SG. 1992a. *Ap. J. Lett.* 401:L97
Wagner RM, Starrfield S, Cassatella A. 1989. *IAU Circ. No. 4783*
Wagner RM, Starrfield SG, Hjellming RM, Howell SB, Kreidl TJ. 1994. *Ap. J. Lett.* 429:L25
West RM. 1991. *Proc. Workshop Nova Muscae, Lyngby,* ed. S Brandt, p. 143
Whelan JAJ, Ward MJ, Allen DA, Danziger IJ, Fosbury RAE, et al. 1976. *MNRAS* 180:657
White NE, Kaluzienski LJ, Swank JL. 1984. See Woosley 1984, p. 31
White NE, Nagase F, Parmar AN. 1995. See Lewin et al 1995, p. 1
White NE, Stella L, Parmar AN. 1988. *Ap. J.* 324:363
Woosley S, ed. 1984. *High-Energy Transients in Astrophysics.* New York: AIP
Yoshida K, Mitsuda K, Ebisawa K, Ueda Y, Fujimoto R, et al. 1993. *Publ. Astron. Soc. Jpn.* 45:605
Zhang SN, Wilson CA, Harmon BA, Fishman GJ, Wilson RB et al. 1994. *IAU Circ. No. 6046*

Annu. Rev. Astron. Astrophys. 1996. 34:645–701

THE GALACTIC CENTER ENVIRONMENT

Mark Morris

University of California at Los Angeles, Los Angeles, California 90095-1562

Eugene Serabyn

California Institute of Technology, Pasadena, California 91125

KEY WORDS: Galactic bar, magnetic fields, SgrA*, circumnuclear disk, star formation, molecular clouds, black holes, X rays

ABSTRACT

The central half kiloparsec region of our Galaxy harbors a variety of phenomena unique to the central environment. This review discusses the observed structure and activity of the interstellar medium in this region in terms of its inevitable inflow toward the center of the Galactic gravitational potential well. A number of dissipative processes lead to a strong concentration of gas into a "Central Molecular Zone" of about 200-pc radius, in which the molecular medium is characterized by large densities, large velocity dispersions, high temperatures, and apparently strong magnetic fields. The physical state of the gas and the resultant star formation processes occurring in this environment are therefore quite unlike those occurring in the large-scale disk. Gas not consumed by star formation either enters a hot X ray–emitting halo and is lost as a thermally driven galactic wind or continues moving inward, probably discontinuously, through the domain of the few parsec-sized circumnuclear disks and eventually into the central parsec. There, the central radio source SgrA* currently accepts only a tiny fraction of the inflowing material, likely as a result of a limit cycle wherein the continual inflow of matter provokes star formation, which in turn can temporarily halt the inflow via mass-outflow winds.

1. INTRODUCTION

In its most inclusive sense, the Galactic center encompasses a wide variety of phenomena occurring on stellar to galactic scales. Because of this, and because

0066-4146/96/0915-0645$08.00

research in this area has advanced rapidly for almost three decades, volumes could be written to summarize our current state of knowledge of this region. Indeed, four conference volumes dedicated largely to this subject can now be consulted (Riegler & Blandford 1982, Backer 1987, Morris 1989, Genzel & Harris 1994), and the pages of this review have already seen three major summaries on Galactic center research: those of Oort (1977), Brown & Liszt (1984), and Genzel & Townes (1987). In related contributions, Frogel (1988) discusses the stellar content of the Galactic bulge and Reid (1993) summarizes research on the distance to the Galactic center; here, we adopt his suggested distance of 8.0 ± 0.5 kpc. The most recent reviews on the Galactic center are those by Blitz et al (1993) and Genzel et al (1994).

Given this plethora of both information and publications, we have chosen to concentrate this review on the interstellar environment and, in particular, on the causes and ramifications of radial mass flow toward the center. This encompasses the nature of the central molecular layer, the hot medium surrounding it, and the strong magnetic field in which they are found (Section 2); the central stellar bar and its effect on the gas (Section 3); star formation (Section 4); the nature and fate of the circumnuclear gas disk (Section 5); and the central nuclear radio source, Sagittarius A* (Section 6). In the final section we address the possibility of recurrent, highly energetic activity in the Galactic nucleus. Timely topics notably missing from this review include the compact sources of high-energy radiation near the Galactic center, large-scale jets, stellar populations and distributions within the Galactic bulge, stellar kinematics, collisions and mass segregation in the central stellar core, and determinations of the radial mass distribution beyond the central few parsecs. Recent information on these subjects can be found in previous reviews and in Sofue et al (1989), Sofue (1994), Lee (1994, 1995), and Blitz & Teuben (1996).

2. CONTENTS OF THE INTERSTELLAR MEDIUM

2.1 *The Arena*

The emissivity of our Galaxy's core reflects a unique environment. Line emission from both molecular CO and atomic C peak sharply in our Galaxy's central few hundred parsecs, and a somewhat smaller region is quite bright in radio and infrared continuum emission as well (Figure 1; Altenhoff et al 1978, Mezger & Pauls 1979, Odenwald & Fazio 1984, Dame et al 1987, Handa et al 1987, Cox & Laureijs 1989, Bennett et al 1994). This compact and luminous nuclear region, hereafter designated the central molecular zone or CMZ (to distinguish this largely molecular region from the more extensive "H I nuclear disk" in which it is ensconced), produces $\sim$ 5–10% of our Galaxy's infrared and

Lyman continuum luminosities and accounts for roughly 10% of our Galaxy's molecular gas content (Mezger 1978, Mezger & Pauls 1979, Hauser et al 1984, Scoville & Sanders 1987, Cox & Laureijs 1989, Güsten 1989, Scoville & Good 1989, Bloemen et al 1990, Wright et al 1991, Bennett et al 1994). The Galactic center is obscured at optical and UV wavelengths by a line-of-sight interstellar extinction of roughly 30 visual magnitudes, but becomes accessible again at energies above $\sim$ 1 keV, where X-ray line emission again peaks strongly on the Galactic center (Koyama et al 1989, Yamauchi & Koyama 1993). In contrast, the CMZ does not stand out in the continuum at high energies (Blitz et al 1985, Bloemen et al 1986, Cook et al 1991, Yamauchi & Koyama 1993, Diehl 1994, Goldwurm et al 1994). The focus of this review is on processes occurring in this bright central region of our Galaxy. In this section we outline the unique properties of the interstellar medium (ISM) in the CMZ and examine the links between its different components.

2.2 *The Central Molecular Zone*

The structure of our Galaxy's CMZ is best delineated by high-resolution observations of the CO molecule (Bania 1977; Liszt & Burton 1978; Heiligman 1987; Bally et al 1987, 1988; Oka et al 1996). At galactocentric radii in excess of a few hundred parsecs, the emissivity of CO is low (Scoville & Sanders 1987), reflecting an average molecular surface density $\sim 5\ M_\odot\ \mathrm{pc}^{-2}$ in the H I nuclear disk (Burton 1992, Burton & Liszt 1993, Boyce & Cohen 1994, Liszt 1996, and references therein), a mixed molecular/atomic (and markedly tilted) layer that occupies the region inside of our Galaxy's "4-kpc molecular ring." However, at a radius of roughly 200 pc, a transition occurs to a largely molecular, high-density ($n \gtrsim 10^4\ \mathrm{cm}^{-3}$), high volume filling factor ($f \gtrsim 0.1$) medium containing 5–10 $\times\ 10^7\ M_\odot$ of gas (Armstrong & Barrett 1985; Walmsley et al 1986; Bally et al 1987, 1988; Güsten 1989; Stark et al 1989; Tsuboi et al 1989). Such densities are usually found only in molecular cloud cores, with small net filling factor, but in the CMZ, stable clouds require such densities to withstand tidal shearing (Güsten & Downes 1980). However, although high densities are required, the high surface density (several hundred $M_\odot\ \mathrm{pc}^{-2}$) and total mass content are not, suggesting mass concentration due to inflow from larger radii.

The clouds in the CMZ also show significantly elevated temperatures (30–200 K, typically $\sim$ 70 K; Güsten et al 1981, Morris et al 1983, Armstrong & Barrett 1985, Mauersberger et al 1986, Hüttemeister et al 1993). Such gas temperatures are high in comparison both to outer Galaxy clouds and to coextensive dust, calling for a direct and widespread gas heating mechanism. Highly supersonic internal velocity dispersions ($\sim$ 15–50 km s^{-1}), comparable to the intercloud velocity dispersion, are also the rule. The elevated temperatures and linewidths may be linked, as dissipation of turbulent energy is a prime candidate for direct

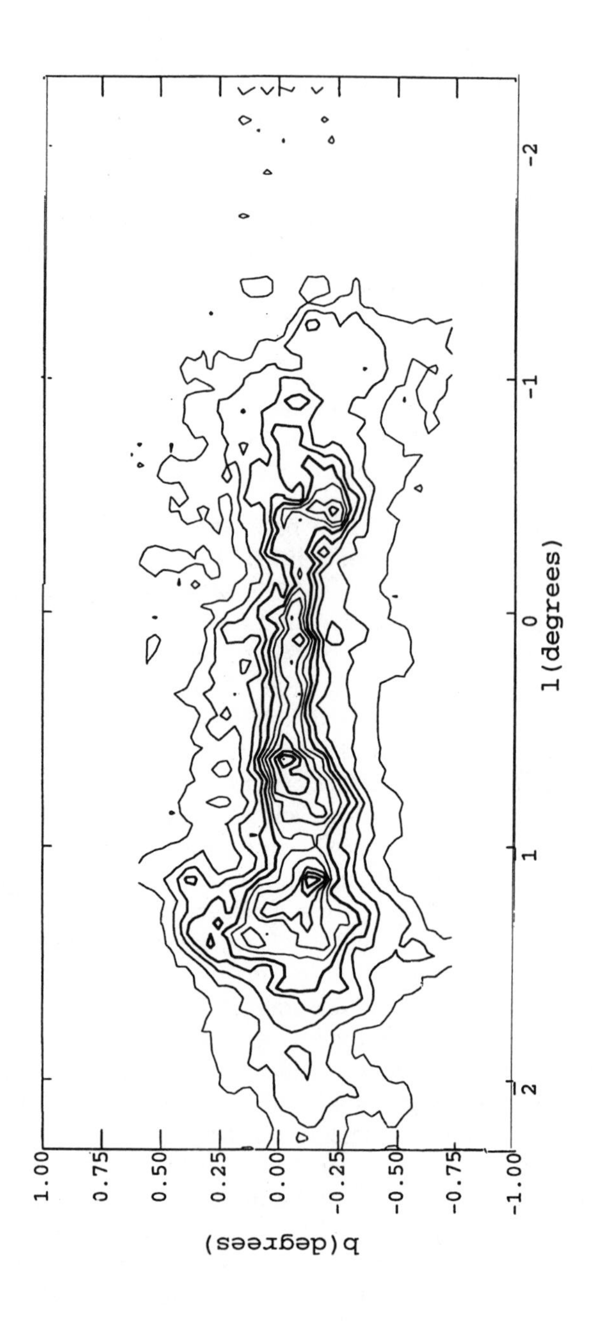
-2
-1
0
1
2
l(degrees)
1.00
0.75
0.50
0.25
0.00
-0.25
-0.50
-0.75
-1.00
b(degrees)

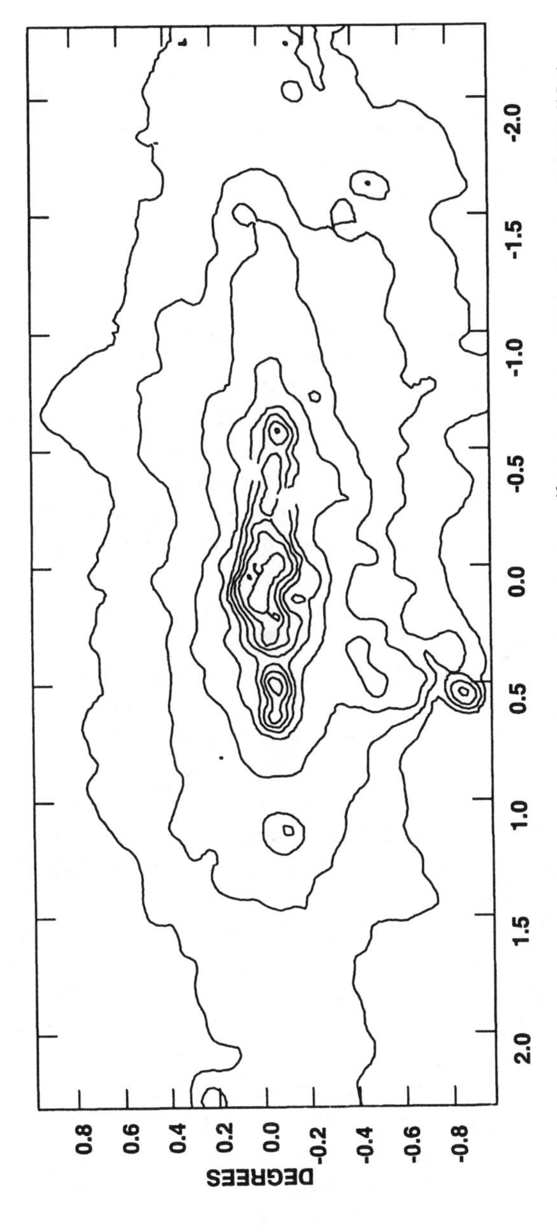

Figure 1 Our Galaxy's central molecular zone. (*Top*) integrated intensity of ^{12}CO, $J = 1\text{–}0$ emission, from the AT&T Bell Labs survey (Uchida et al 1996a). (*Bottom*) 60-μm emission map, showing the distribution of relatively warm dust, as measured by *IRAS*.

heating of the gas (Wilson et al 1982). This remains true if the cloud velocity dispersions are due to magnetosonic disturbances (Section 2.5). Although energy extraction from the tidal field has been proposed as the origin of the large linewidths (Fleck 1980), Das & Jog (1995) conclude that this mechanism is efficient only for a central bar potential, in which tidal fields vary markedly along elongated orbits. Viscous magnetic heating resulting from the drift of ions and grains through the neutral gas (ambipolar diffusion) also warrants strong consideration in this highly magnetized environment (Roberge et al 1995).

The distribution and kinematics of the gas in the CMZ are inconsistent with both axial symmetry and uniform circular rotation (Bania 1977, Liszt & Burton 1978, Morris et al 1983, Heiligman 1987, Bally et al 1988, Jackson et al 1996). The distribution is such that roughly three quarters of the dense molecular gas is located at positive longitudes, three quarters is at positive velocities, and large radial and vertical motions are present in a significant fraction (30%) of the gas (Bally et al 1988). On the basis of molecular kinematics, it is possible to divide the CMZ gas into two components. The first is a high-velocity (130–200 km s^{-1}), quasi-continuous ring structure surrounding the nucleus and having a radius ($\sim$ 180 pc) suggesting a location at the outer boundary of the CMZ. The kinematics of this ring (hereafter referred to as the "180-pc molecular ring") and its tilt relative to the Galactic plane are continuous with those of the exterior H I nuclear disk, indicating that this structure likely marks the location of an H I/H_2 transition (Binney et al 1991).

Inside of this "boundary" lies the second, mass-dominant molecular component: a lower-velocity (typically $\lesssim$ 100 km s^{-1}) population of dense and massive molecular clouds. This cloud population (which includes the familiar clouds associated with the Sgr A–Sgr E H II regions) lies quite close to the true Galactic plane and is referred to as the Galactic center "disk population" (Bally et al 1988, Heiligman 1987). This disk population contains filament-like clouds with coherent velocity gradients over scales of 30–100 pc, suggestive of dust lanes and tidally stretched arcs or arms of gas (Stark & Bania 1986; Serabyn & Güsten 1987; Bally et al 1988; Sofue 1995a,b). Several of these extended clouds contain a large fraction of the total molecular mass (Bally et al 1988, Güsten 1989). Sgr B2, the most massive molecular cloud in the Galaxy, contains about 5% of the gaseous mass present in the CMZ, while a handful of large clouds contributes up to one third of the CMZ's CO line flux. Because of our vantage point in the disk of the Galaxy, the true geometric arrangement of these clouds and structures remains ambiguous, but a framework for interpretation appears possible in the context of gas flows in a barred potential (Section 3).

One of these massive molecular clouds, the 50 km s^{-1} cloud, is adjacent to the Galactic center in projection. It is associated with the bright, central

Sgr A complex (Ekers et al 1983, Yusef-Zadeh & Morris 1987a, Pedlar et al 1989), which itself consists of three components: (*a*) the Sagittarius A West H II region and which is centered more or less on the central stellar cluster; (*b*) the somewhat larger-scale nonthermal shell source Sgr A East, which is offset from the center, but which encloses Sgr A West in projection (Section 5 provides further detail); and (*c*) the compact, central radio source Sgr A* (Section 6).

The distributions of both ionized gas and diffuse infrared emission in the CMZ (Altenhoff et al 1978, Odenwald & Fazio 1984, Handa et al 1987, Cox & Laureijs 1989) are more symmetric than the CO distribution (Figure 1), suggesting that heating sources are more evenly distributed than are the discrete clouds. In particular, the dust temperature drops regularly as $r^{-0.3}$ from its central peak to its asymptotic value of 23 K near a radius of 200 pc. The Lyman continuum production rate and far-infrared luminosity of the CMZ are $\sim 1\text{–}3 \times 10^{52}\ \mathrm{s}^{-1}$ and $10^9\ L_\odot$, respectively, and infrared excesses of roughly 10 and 30 are deduced for individual H II regions and the extended emission, respectively (Mezger & Pauls 1979, Reich et al 1987, Cox & Laureijs 1989). The former is typical of Galactic H II regions, while the latter requires a later population of stars. Star formation with an underabundance of O stars had been considered (Odenwald & Fazio 1984, Lis & Carlstrom 1994), but the abundant population of K and M giants in the Galactic center suffices to heat the dust in the extended region to the observed levels (Cox & Laureijs 1989) without adjusting the O/B ratio. A soft interstellar radiation field also receives support from the underabundance of C^+ emission from the CMZ (Nakagawa et al 1996).

2.3 *The Hot Component*

Extended X-ray emission centered on the Galactic nucleus has also been detected, with a size ($1.8^\circ \times 1.0^\circ$) roughly half that of the CMZ. This region stands out most clearly in the intense 6.7-keV Kα transition of He-like Fe (Koyama et al 1989, Yamauchi et al 1990), although it is also seen in the continuum (Kawai et al 1988, Sunyaev et al 1993, Markevitch et al 1993) and in the 6.4-keV Kα transition of neutral Fe (Koyama 1996, Koyama et al 1996). The continuum emission drops rapidly with frequency, and this central region is not seen at energies above 35 keV (Goldwurm et al 1994). Skinner (1989) provides a complete review of earlier X-ray observations; here we concentrate on recent results from the *ASCA, Ginga,* and *GRANAT* spacecraft.

The detection of the 6.7-keV line from He-like Fe clearly reveals the existence of a high-temperature plasma in the central hundred or so parsecs. Estimates for its temperature vary, from 10–15 keV (Koyama et al 1989, Yamauchi et al 1990, Nottingham et al 1993) down to 1–3 keV (Markevitch et al 1993). The latter estimate is lower because of the finding that the X-ray emission changes character for energies above $\sim$ 10 keV; the lower energy emission shows a roughly elliptical distribution, but the higher energy emission essentially mimics the CMZ's flattened distribution. This, and the image of the neutral Fe Kα transition, which shows a clear correlation with the dense clouds (Koyama 1996, Koyama et al 1996), strongly suggest that scattering of X-rays by the high-column-density medium in the CMZ is of importance, both in terms of the 6.4-keV line and of the continuum above 10 keV. X-ray data thus also provide information on the cold molecular medium, and by inference, on the discrete X-ray sources that illuminate it (Sunyaev et al 1993, Koyama et al 1996).

Although estimates vary, the hot plasma has a temperature of some 10^7–10^8 K. The upper end of this range is quite high, more akin to the temperatures found for intergalactic gas bound to clusters of galaxies than to the temperatures found for individual galactic nuclei or supernova remnants (Holt & McCray 1982, Pietsch 1994, Koyama et al 1996). At temperatures near 10^8 K, the gas would not be bound to the Galaxy at all (Yamauchi et al 1990, Sunyaev et al 1993), and a wind would expand from the center at a few thousand kilometers per second, implying a lifetime of $\sim 10^5$ yr. The energy and mass requirements would then be severe, with the plasma containing some 10^{53} ergs and $\sim 3000\,M_\odot$ (Yamauchi et al 1990, Koyama et al 1996), at an average density of 0.3–0.4 cm^{-3}. To account for the extreme inferred temperatures, these authors propose an energetic explosion in the Galactic center at a past epoch. However, the necessary prior luminosity, 10^{41}–10^{42} erg s^{-1}, dramatically exceeds anything observed today (Section 6). On the other hand, the lower temperatures implied by removal of the high-energy scattering component from the temperature determination yield plasma temperatures more in line with supernova remnant temperatures, but the emission measure then requires a rather large number of supernovae ($\sim$ 1000). It has thus been difficult to settle on a single scenario for this plasma that does not involve past releases of energy in the Galactic center far exceeding the current rate. Perhaps the simplest explanation would be a past starburst, in which the ensuing supernovae generate the high-temperature plasma.

Scattering of high-energy photons also provides a tracer for past activity through the source-scatterer time delay; from the observed distribution of high-energy X rays, Sunyaev et al (1993) conclude that the X-ray luminosity of the central $\sim$ 100 pc has not exceeded the current luminosity of $\sim 10^{37}$ erg s^{-1} by more than a factor of 10–100 in the past several hundred years, whereas scattering in the Kα line of Fe$^\circ$ leads Koyama et al (1996) to infer that a single source must have brightened to $> 10^{39}$ erg s^{-1} in roughly this same interval. The 1.8-MeV line of ^{26}Al, which traces past massive star formation on a few

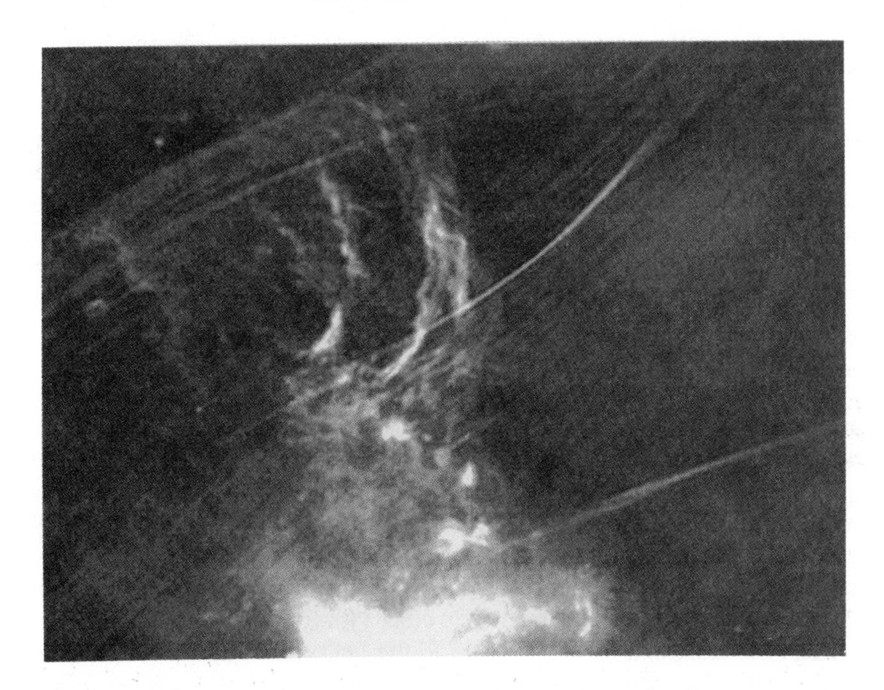

Figure 2 Radiograph of 20-cm emission arising from the Northern and Southern Threads, made with the VLA (Echevarria & Morris 1996). The vertical scale is 16.5 arcminutes (~ 40 pc). The Sgr A complex is visible at bottom center. The radio Arc (*upper left*) appears weak because it is beyond the telescope's primary beam.

million–year time scale, shows a peak near, but offset from, the Galactic Center (Diehl et al 1995), and so it may be relevant in tracing past activity in the central region also. However, a coherent picture of the recent past of our Galactic nucleus has not yet emerged.

2.4 *The Galactic Center Magnetosphere*

The magnetic field within the CMZ probably plays a significant role in the physical interactions occurring there. Its presence is revealed in several ways, the most striking of which is the observation of seven or eight systems of nonthermal radio filaments (NTFs) within ~150 pc of the Galactic center (Yusef-Zadeh 1989; Morris 1990, 1994, 1996). These polarized structures are tens of parsecs long and only a fraction of a parsec wide. They may occur in isolation (in which case they have been called threads; Morris & Yusef-Zadeh 1985) or in bundles, such as those comprising the linear portion of the prominent Radio Arc (Yusef-Zadeh et al 1984). Morphologically, the NTFs are strikingly uniform in brightness and curvature (Figure 2) and therefore quite different from the meandering, contorted filamentary structures one finds in supernova remnants and emission-line nebulae. The morphology of the NTFs suggests that they reflect the local magnetic field direction, and radio polarization evidence supports that notion. Tsuboi et al (1986) corrected the observed polarization angles within the NTFs of the Radio Arc for Faraday rotation and deduced that the intrinsic field is parallel to the filaments, a result confirmed by Reich (1994) and Tsuboi et al (1995) at high frequencies (32 and 45 GHz, respectively), where Faraday rotation is small.

A clue to the strength of the magnetic field in the NTFs is provided by the near absence of deformation or bending along their lengths. Every NTF that has been sufficiently well studied has been found to be associated with, and is probably interacting with, at least one molecular cloud (detailed below). However, the magnetic filaments are not subject to large distortions at the interaction sites, in spite of the large velocity dispersion within Galactic center molecular clouds, and in spite of the likelihood that, given the large intercloud velocity dispersion at the Galactic center, most clouds have a typical velocity of at least a few tens of kilometers per second with respect to the ambient magnetic field. By equating the apparent turbulent pressures within clouds (or the ram pressure associated with presumed cloud motion relative to the field) to the magnetic pressure, as a minimum condition on the strength of the magnetic field, Yusef-Zadeh & Morris (1987b,c; 1988) have determined that the magnetic field within the NTFs has milligauss strength. [The one known exception is G359.1-0.2, the Snake filament (Gray et al 1991, 1995), which is endowed with a few "kinks," suggesting that turbulence in its environment has partially overcome the rigidity of the field. However, this filament is the one with the greatest projected distance from the Galactic nucleus (125 pc), so it may imply a declining field strength within filaments at larger distances.] This "rigidity" method of constraining the magnetic field strength in NTFs needs to be recast if the clouds consist of a number of independently moving clumps (cf Serabyn & Morris 1994), because the field lines can then be deflected around and between the clumps as the ensemble of clumps moves through the field. A modified picture of this sort does not dramatically change the constraint on the magnetic field strength; the time scale to transport clump-induced distortions of the interclump field lines out of the cloud by Alfvén waves, $2 \times 10^4\ B_{\rm mG}^{-1}\ (n/10\ {\rm cm}^{-3})^{0.5}\ (D/10\ {\rm pc})$ yr for cloud size D and interclump density n, must be less than a clump crossing time (typically 2×10^4 years), which still implies $\gtrsim$ mG field strengths.

The fact that all of the known NTFs are perpendicular to the Galactic plane to within about 20° (Morris & Yusef-Zadeh 1985, Anantharamaiah et al 1991) suggests that these structures trace an ubiquitous dipole magnetic field occurring on a scale comparable to that of the CMZ. In this view, the observed filaments are the magnetic flux tubes that happen to be illuminated by the local injection of relativistic particles. A few alternative hypotheses have been considered, however. One—that the site of each NTF or NTF bundle represents a locally generated field enhancement—encounters difficulties because of the strength of the magnetic field in the NTFs. If the milligauss fields are present only in the NTFs, then a strong confinement mechanism operating along their full ~30-pc length is needed to prevent them from expanding at the Alfvén speed and thus disappearing on a time scale of $\sim 10^3$ yr. This is much faster than is needed to establish the currents necessary to generate the structure in the first place (comparable to a dynamical crossing time of $\gtrsim$ a few times 10^4 yr) and is shorter than the energy-loss time scale for relativistic electrons in the filaments (Anantharamaiah et al 1991, Sofue et al 1992). Another alternative is that the NTFs lie upon the surface of a cylindrical wall of compressed magnetic field surrounding the Galactic center (Uchida et al 1985, Heyvaerts et al 1988) and that they are again illuminated by some local mechanism for the generation of relativistic particles. Such a geometry could result from expansive gas motions from the center (Umemura et al 1988), if the gas momentum were sufficiently large, although the expansive events would have to be continuous, quite recent, or even in progress (such as would be manifested by the hypothetical, expanding molecular shell described in Section 3.1). This geometry cannot be ruled out, although it would imply a remarkable pattern of currents that might be difficult to maintain.

A twisted poloidal magnetic field geometry has been considered for the Galactic center, on the assumption that this would result from flux-freezing of field lines into the differentially rotating disk gas (Uchida et al 1985, Sofue et al 1987, Shibata & Uchida 1987). However, the evidence for such a geometry is based on rotation measures, which are difficult to localize to the Galactic center and which only probe the presumably relatively weak, line-of-sight magnetic field component. The more direct indicators of the dominant field geometry—the NTFs—pass through the gas layer in the Galactic disk without showing any appreciable large-scale twist or distortion, other than perhaps a slow divergence. A large-scale twist might be hidden by a fortuitous projection of one or two filaments, but it is unlikely that all seven or eight known NTF systems would be similarly projected. The apparent immunity of the NTFs to the inertia of the disk gas could be understood if the fields within molecular clouds are only weakly coupled to the external field (i.e. separated by magnetopauses).

2.4.1 MID AND FAR-INFRARED POLARIZATION The second most informative probe of the magnetic field near the Galactic center has been the polarization of thermal infrared emission from dust. Various mechanisms have been proposed for orienting the emitting dust grains in such a way that their spin axes are preferentially aligned along the magnetic field (Hildebrand 1988, Lazarian 1994, Roberge et al 1995). All potentially viable mechanisms imply the same relationship—orthogonality—between the polarization position angle and the projected magnetic field direction. A map of the distribution of polarization position angles is thus readily transformed into a map of the projection of the magnetic field onto the plane of the sky, weighted by the volume emissivity of the dust along the line of sight. The dust emissivity is maximized in dense clouds or at warm cloud surfaces, so in contrast to the NTFs, which depict the field geometry in the intercloud medium, far-IR polarization tends to probe the magnetic field within dense clouds. The gross uncertainties about the specifics of the grain alignment mechanisms make quantitative links between the field strength and the fractional polarization difficult, although the magnitude of field direction fluctuations can be indicative of the field strength (Morris et al 1992, 1996a,b).

Polarization measurements were first applied to the Galactic center by Aitken et al (1986, 1991), who, observing at 10 μm, found that the magnetic field within the northern arm of Sgr A West (cf Section 5) is parallel to that arm, a result that would arise naturally from the action of shear within this orbiting gas stream. Shear is expected to affect much of the gas present within the strong tidal field of the Galactic center, so this result may have widespread applicability. Strong far-IR polarization has now been mapped in four clouds near the Galactic center, as reviewed by Davidson (1996). In essentially all of the clouds, the inferred magnetic field direction is predominantly parallel to the Galactic plane (Sgr B2 is a complex mix of absorption and emission and is still ambiguous in this regard), which stands in marked contrast to the perpendicular field inferred for the intercloud medium. Figure 3 shows the 60-μm polarization vectors toward G0.18-0.04 (Morris et al 1996b), superimposed upon a 6-cm radiograph (Yusef-Zadeh & Morris 1987b) and the underlying cloud (Serabyn & Güsten 1991). The implied field direction within the cloud is parallel to the ridge of molecular and thermal radio emission, which is largely parallel to the Galactic plane. The implication that the magnetic field follows the ridge of the molecular cloud in this and in the nearby arched filament cloud suggests that the field within clouds is shaped by the stresses that shape the clouds themselves (Serabyn & Güsten 1987; Morris et al 1992, 1995). Furthermore, the general orthogonality of fields inside and outside of clouds suggests that the two are relatively independent, consistent with the apparent immunity of NTFs to the inertia of disk gas.

The shaping of the fields within clouds by the stresses which shape clouds themselves appears to apply to the circumnuclear disk (CND) as well (Section 5; Werner et al 1988; Hildebrand et al 1990, 1993). With the exception of a

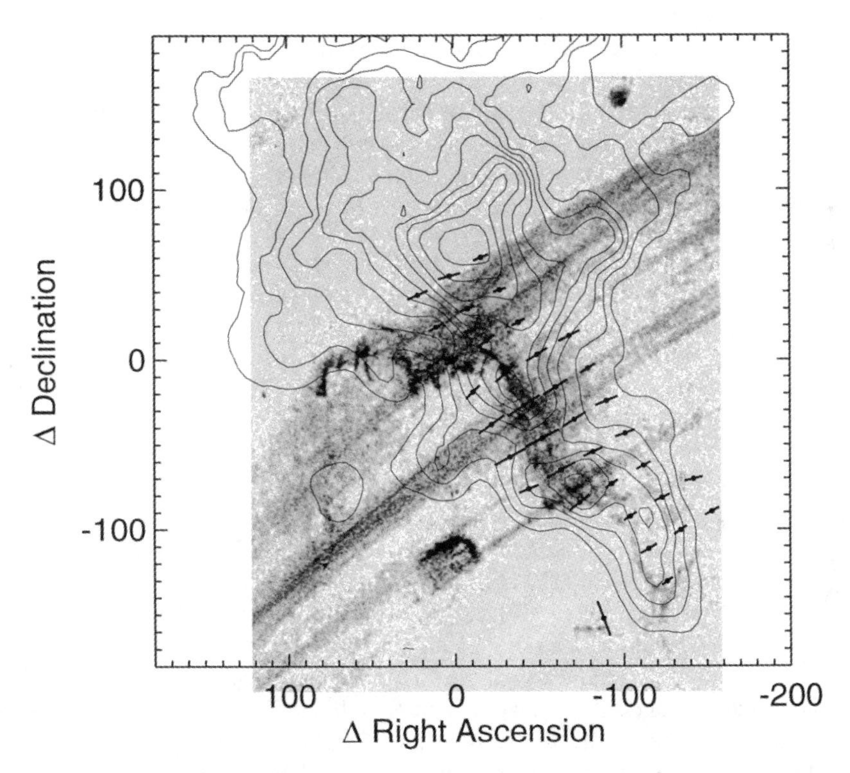

Figure 3 60-μm polarization measurements across the H II region G0.18-0.04 (from Morris et al 1996b), superimposed upon a λ6-cm continuum radiograph (from Yusef-Zadeh & Morris 1987b). The orientations of the line segments represent the position angles of the electric vectors; their length is proportional to the percent polarization. A horizontal segment representing 4% polarization is shown at bottom left. The position angle of the Galactic plane is $\sim 30^\circ$, so the polarization vectors are all close to being perpendicular to the Galactic plane. The superimposed contours show the intensity of CS $J = 3$–2 emission, from Serabyn & Güsten (1991).

presumably infalling stream of dust and gas coinciding with the Northern Arm of Sgr A West, all of the far-IR emission from dust in the CND is polarized in a manner consistent with a dominant magnetic field component parallel to the plane of the disk. Self-similar models for a poloidal field well outside the disk (i.e. perpendicular to the disk plane), which has been deformed and sheared within the CND by a combination of differential rotation and radial infall, were proposed by Wardle & Königl (1990). Variations of these models have generally been rather successful in accounting for the polarimetric mapping results on this source (Hildebrand et al 1993). Another model based on numerical magnetohydrodynamic calculations carried out by Meglicki et al (1994) is also able to reproduce the gross geometric and magnetic characteristics of the CND, although this model disk is undergoing rapid evolution, with inward radial motions ($\sim$50 km s^{-1}) well in excess of the observational limits ($\sim$20 km s^{-1}, cf Section 5).

2.4.2 THE ZEEMAN EFFECT Although the Zeeman effect has been used to probe the magnetic field in Galactic center clouds directly, the results are mixed. The large velocity breadth of the lines from essentially all Galactic center clouds have made Zeeman measurements difficult for all but the strongest fields. Using H I absorption, Schwarz & Lasenby (1990) derived a modest line-of-sight field, $B_{\|} = 0.5$ mG, toward the northern side of the CND. With 1667-MHz OH absorption, Killeen et al (1992) measured $B_{\|} = -2.0$ mG, averaged separately over broad portions of the northern and southern sides of the CND, with considerable spatial variation in $B_{\|}$ on scales of 5 to 10$''$. Most recently, Plante et al (1995) used H I absorption to detect fields ranging up to about -3 mG on the northern side of the CND. They attribute the strongest fields to a feature interpreted as a stream of infalling gas rather than a part of the disk. These results, all obtained with the VLA, and the negative H I Zeeman results of Marshall et al (1995), who obtained $B_{\|} < 0.5$ mG in 45×45 arcsec^2 regions of the CND, suggest that the average line-of-sight component of the field within the CND is $\sim$ 1 mG, but the likelihood that $B_{\|}$ changes considerably and even undergoes reversals along the line of sight (for example, in the model of Wardle & Königl 1990) implies that the Zeeman signals are diminished by projection effects.

Elsewhere in the Galactic center, Zeeman measurements have yielded negative results. Uchida & Güsten (1995) observed OH absorption in a number of clouds with a relatively large beam; they obtained typical upper limits to $B_{\|}$ of a few tenths of a milliGauss. Although their observations are subject to dilution of the Zeeman signal by averaging over spatial variations within the beam, as well as along the line of sight, these results raise questions about the conclusion that milligauss fields pervade the Galactic center region, unless either the line-of-sight component is small everywhere (i.e. a field that is everywhere within $\sim 10^\circ$ from the vertical to the Galactic plane) or the field has strong spatial variations.

2.4.3 FARADAY ROTATION MEASURES Faraday rotation measures (RMs) provide yet another probe of the field surrounding nonthermally emitting Galactic center radio structures. Vast areas of the Galactic center display polarized emission (Sofue et al 1987, Haynes et al 1992), so in principle the magnetic field within a large volume of the intervening medium can be studied, although this probe has not yet been extensively utilized. The RMs can be quite large at some sites within the Galactic center. Using 6-cm radio observations, Yusef-Zadeh & Morris (1987c, 1988) report RMs up to 5500 radians m^{-2} toward the Radio Arc, with somewhat smaller values (up to 1450 radians m^{-2}) at higher Galactic latitudes (see also Tsuboi et al 1995). In a 2-cm study of the Radio Arc (Inoue et al 1989), complex structure was seen in the foreground Faraday screen, even to the extent that depolarizing filaments were identified, but these have no obvious counterpart in total intensity images. When this kind of study is performed more thoroughly and quantitatively, with independent measures of electron density, it will provide much-needed information on the field strength and geometry of magnetic structures.

2.4.4 THE PRODUCTION OF RADIO FILAMENTS A number of hypotheses about the origin of the Galactic center NTFs have been offered (Chudnovsky et al 1986, Heyvaerts et al 1988, Benford 1988, Morris & Yusef-Zadeh 1989, Serabyn & Güsten 1991, Lesch & Reich 1992, Rosso & Pelletier 1993, Serabyn & Morris 1994), the merits and drawbacks of which are summarized by Morris (1996). Here, we describe only the hypothesis that, in our judgement, conforms best to the observational evidence. The key to this hypothesis is that all sufficiently well-studied NTFs appear to be interacting with molecular structures. It was originally suggested following examination of molecular CS emission from the cloud underlying the G0.18-0.04 H II region (Serabyn & Güsten 1991, Serabyn & Morris 1994; see also Morris 1995, Serabyn 1995). Yusef-Zadeh & Morris (1987b) had pointed out that most of the nonthermal filaments of the Galactic Center Radio Arc undergo abrupt discontinuities where they encounter the ionized ridge of G0.18-0.04, and Serabyn & Güsten (1991) demonstrated that this H II region lies at the surface of a relatively massive molecular cloud. With a partial interferometric view of the molecular cloud, we found that dense clumps are present at a number of the locations where the NTFs undergo brightness discontinuities or small deflections (Serabyn & Morris 1994). This led us to propose that the NTFs in the Radio Arc, and perhaps all Galactic center threads, originate at the surfaces of such dense molecular clumps. The hypothesis can be stated as follows: When certain conditions are satisfied by a molecular cloud moving through a more diffuse ambient medium, magnetic field line reconnection taking place at leading clump surfaces leads to the acceleration of charged particles, and the resultant relativistic particles then stream away from their point of origin along the field lines. The flux lines to which these particles are attached are then illuminated by synchrotron emission.

The hypothesized conditions for particle acceleration are: (*a*) The cloud surface must be ionized, presumably by a fortuitously placed local source of ionizing radiation. The ionized gas provides a source of free electrons for the acceleration mechanism to act upon, and the turbulence associated with the ionization front ensures that the cloud and intercloud magnetic fields are sufficiently mixed for reconnection to be efficient. (*b*) The cloud surface at which the acceleration takes place must be moving at a relatively large velocity into the ambient intercloud medium, so as to force the magnetic fields in the two media into contact. (*c*) The orientations of the fields in the cloud and intercloud media must be quite different for rapid energy extraction, a condition that has very recently been verified in the case of G0.18-0.04, where far-IR polarization measurements show that the internal cloud field is perpendicular to the linear radio filaments with which the cloud is interacting (Figure 3). The same may be true for a substantial fraction of all Galactic center clouds, if the trends from far-IR polarimetry are any guide. However, the conditions listed here are collectively stringent enough to explain why it is that not every molecular clump in the Galactic center magnetosphere has an associated NTF.

Conversely, it is possible that these conditions are satisfied by every filament or thread in the Galactic center. The NTFs associated with Sgr C arise (in projection) at the interface between that bright H II region and its associated molecular cloud (Liszt & Spiker 1995); the filament G359.5+0.18, located to the north of Sgr C, is apparently interacting with at least one, and possibly two, molecular clouds (Bally et al 1989, Staguhn et al 1996); the Northern thread (G0.08+0.15; Morris & Yusef-Zadeh 1985) undergoes an abrupt intensity discontinuity where it is superimposed upon the thermal arched filament H II region and associated cloud (Figure 2); and finally, an H II region/molecular cloud complex is located at the northern tip of the Snake (G359.1-0.2; Gray et al 1995, Uchida et al 1996b). However, in none of these cases is a cause-and-effect relationship between an NTF and a molecular cloud/H II region as convincing as the Radio Arc case because high-resolution observations are lacking.

2.4.5 ORIGIN OF THE STRONG POLOIDAL FIELD

If the Galactic center magnetosphere is pervasive on a scale of $\sim$ 100 pc, then its total energy content is $\sim 10^{54} \langle B(\mathrm{mG}) \rangle$ ergs, comparable to the energy content of the hot, X-ray emitting gas and only a few times larger than the kinetic energy associated with noncircular motions in the 180-pc molecular ring (Section 3). The maintenance of such a strong, dynamically important field requires a rather powerful ring current circulating about the center, presumably somewhere within, or at the boundary of, the CMZ. The maintenance of such a current, and the confinement of the field, might be accomplished by the Lorentz forces accompanying the quasi-steady radial inflow or outflow (or both) of gas through a vertical magnetic field (e.g. Lesch et al 1989), although the physical elements of such a dynamo mechanism still require elaboration.

One hypothesis for the vertical field at the Galactic center, which is appealing both for its simplicity and its seeming inevitability, is that of Sofue & Fujimoto (1987). Noting that Galactic evolution is characterized by the inexorable radial inflow of matter, these authors hypothesized that the vertical component of the early Galactic field is dragged inwards by the gas accreting to the Galactic center region over the lifetime of the galaxy; its rate of outward radial diffusion with respect to the gas is small compared to the rate of inward gas flow, thereby concentrating the vertical field at the nucleus. In contrast, the component of the field parallel to the disk can be transported vertically out of the Galaxy by diffusion or by the Parker instability on a time scale that is short compared to a Hubble time, so that the azimuthal field represents an equilibrium between the flux lost by vertical transport and that regenerated by a Galactic dynamo.

According to the Sofue & Fujimoto hypothesis, the strength of the vertical field concentrated at the nucleus reflects the strength of the primordial magnetic field. However, it also depends on the spectrum of spatial fluctuations, since the only fluctuations that could survive to form a uniform field when concentrated at the nucleus are those present initially on Galactic scales or larger (Morris 1994). Smaller-scale field fluctuations would have suffered reconnection and annihilation in the Galactic center mix. These processes, along with the conversion of gravitational potential energy of the inflowing gas, may have been important sources of heating for Galactic center gas throughout Galactic history.

2.5 *Pressure Balance*

The physical conditions present in the CMZ are consistent with thermal, nonthermal, and magnetic pressures several orders of magnitude higher than those present in the large-scale Galactic disk. For molecular densities of 10^4 cm^{-3} at the inferred temperatures (Section 2.2), the pressure $P_{\rm thermal} \sim 10^{-10}$ erg cm^{-3}. In the hot plasma, $P_{\rm hot} \sim 4 \times 10^{-10}$ erg cm^{-3}, close enough to consider pressure balance (Spergel & Blitz 1992). However, turbulent pressures in the clouds greatly exceed this value, approaching $P_{\rm turb} \sim 10^{-8}$ erg cm^{-3}. For a field strength of 0.1–1 mG, the magnetic pressure is comparably large: $P_{\rm mag} \sim 4 \times 10^{-10}$ to 4×10^{-8} erg cm^{-3}. Internal cloud turbulence is therefore likely to be linked to magnetic pressures. In particular, the observed velocity dispersions in Galactic center clouds likely reflect hydromagnetic waves (Arons & Max 1975), which could be generated by the angular momentum loss processes affecting Galactic center clouds (described in Section 3), including shocks, cloud collisions, and the irregular viscous interactions suffered by clouds moving through a magnetized, low-density medium.

3. THE CENTRAL BAR; GAS DYNAMICS NEAR THE GALACTIC CENTER

The presence of a substantial stellar bar in our Galaxy, first invoked to account for the noncircular motions of H I near the Galactic center (de Vaucouleurs 1964, Peters 1975), is now well established by a variety of methods, including techniques based on photometry (Blitz & Spergel 1991b, Sellwood 1993, Weiland et al 1994, Dwek et al 1995), kinematic studies of gas (Liszt & Burton 1980, Mulder & Liem 1986, Binney et al 1991, Wada et al 1994) and stars (Zhao et al 1994, Blum 1995), and counts of luminous stars (Nakada et al 1991; Weinberg 1992a,b; Whitelock & Catchpole 1992; Stanek et al 1994). The picture that has emerged is as follows: The bar is quite pronounced, with axial ratios of about 3:1:1 [though Binney et al (1991) and Wada et al (1994) prefer 1.5:1 for the axial ratio in the Galactic plane]; it extends at least to the corotation radius of about 2.4 kpc; it has a total mass of 1–3 $\times 10^{10} M_{\odot}$; and its long axis is oriented at a modest angle with respect to our line of sight (15 to 45° toward positive longitude, depending on the model).

The excess of microlensing events toward the Galactic bulge provides yet another indication that the Galactic bulge is strongly barred and that the bar is somewhat along our line of sight (Paczyński et al 1994, Han & Gould 1995, Zhao et al 1995). Leaving aside the details of these determinations, we turn to the issue of the interstellar material that coexists with the stellar bar. A strong bar is an essential element for understanding gas dynamics near the Galactic center and especially for appreciating the potential importance of bar-induced migration of interstellar gas toward the Galactic nucleus.

3.1 *Response of Interstellar Gas to a Bar Potential*

The response of orbiting gas to such a pronounced bar is quite strong and leads both to large deviations from circular motion and to strong shocks. Thus, when studying gas dynamics, it is essential to properly account for the $m = 2$ (quadrupole) deviation from axial symmetry represented by the bar (where the gravitational potential depends on azimuthal angle, ϕ, via the factor $e^{im\phi}$). The velocity field of the gas provides one of the best ways of probing the shape of the Galactic gravitational potential. The primary data to be explained by any bar model comprise the distribution of molecular emission (or absorption) throughout the longitude-latitude-velocity (l, b, v) cube, and the most remarkable feature of this distribution in our Galaxy is the 180-pc molecular ring (e.g. Bania 1977, Bally et al 1987), seen as a parallelogram, or ellipse, in projections of this cube onto the l-v plane (Figure 4). Models for the response of orbiting gas to a symmetric $m = 2$ bar potential (described more fully below) have met with considerable success in accounting qualitatively for the gas kinematics measured in both H I and CO surveys (Binney et al 1991), notably including this parallelogram, and it is now becoming the standard paradigm for Galactic center kinematics, although a quantitatively faithful model is still being sought (e.g. Binney 1994, Jenkins & Binney 1994). Most importantly, existing models readily account for the abundance of gas near the Galactic center having "forbidden" radial velocities (i.e. with sign opposite to that of Galactic rotation).

3.1.1 ORBITS OF GAS CLOUDS IN THE PRESENCE OF A STRONG BAR Gas moving in response to a bar potential tends to settle into closed, elongated orbits because cloud collisions and consequent energy dissipation act to enforce conformity. Angular momentum loss by the orbiting gas resulting from the processes described more fully in Section 3.2 causes it to drift inward along a family of nested, closed orbits. When the gas is orbiting at radii between that of corotation and the inner Lindblad resonance (ILR) of the bar pattern, it moves on the so-called X_1 orbits (Contopoulos & Mertzanides 1977), which are elongated oval orbits that have their major axes aligned with the bar (Figure 5).

As the ILR is approached, however, there is an innermost stable X_1 orbit inside of which these orbits become self-intersecting or cusped. Any angular momentum loss by gas in the innermost stable X_1 orbit leads to orbit crossings and shocks. The shocks, in turn, imply a further, abrupt loss of angular momentum, which causes the gas to fall inward until it settles onto a new family of closed, elongated orbits lying deeper in the potential well: the X_2 orbits. These oval orbits have their long axes oriented perpendicular to the bar. At their apocenter, the outermost X_2 orbits are thought to graze the pericenter of the innermost X_1 orbits. Thus, angular momentum loss by gas in the innermost X_1 orbit, for whatever reason, be it via viscous transport of angular momentum outward or by orbit crossings near the cusps, would lead it to collide with gas in the X_2 orbits, presumably creating a spray, which leads to a shock at the far side of the innermost X_1 orbit (Binney et al 1991, Athanassoula 1992, Jenkins & Binney 1994, Gerhard 1996).

Binney et al (1991) hypothesize that the 180-pc molecular ring corresponds to the innermost stable X_1 orbit. The small radial extent of this feature is responsible for the narrowness of the trace of emission defining the parallelogram and for its well-defined, sharp vertices. The shocks along the inside edge of the X_1 orbits are held responsible for compressing the gas into molecular form, which explains why predominantly molecular gas occupies only the

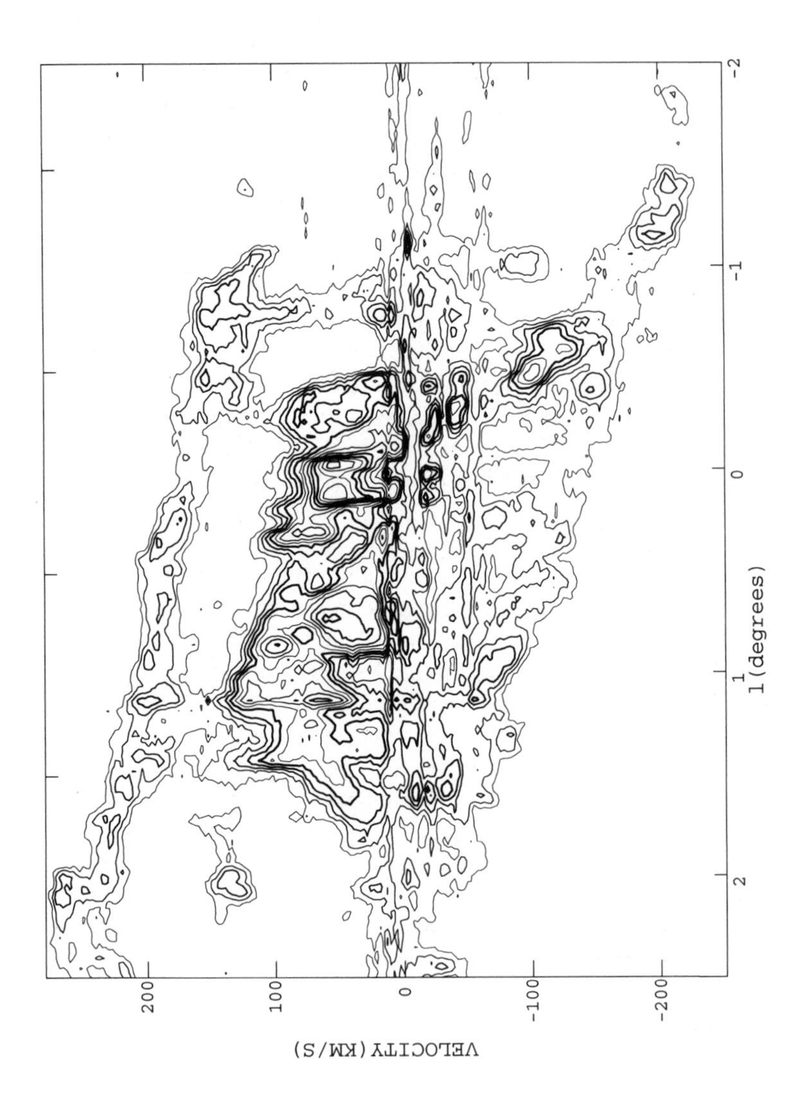

Figure 4 ^{12}CO $J = 1$–0 emission contours in a position-velocity plane in which position extends along the line from $l, b = 2.5°, -0.333°$ to $-2.0°, 0.1333°$, from the AT&T Bell Labs survey (Uchida et al 1996a). This line, tilted by $\sim 6°$ with respect to the Galactic plane, was chosen to reflect the tilt of the plane containing the 180-pc molecular ring (Uchida et al 1994a). Contour units are 1.4, 2.8, 4.4, 6.2, 8.2, 10.4, . . . K.

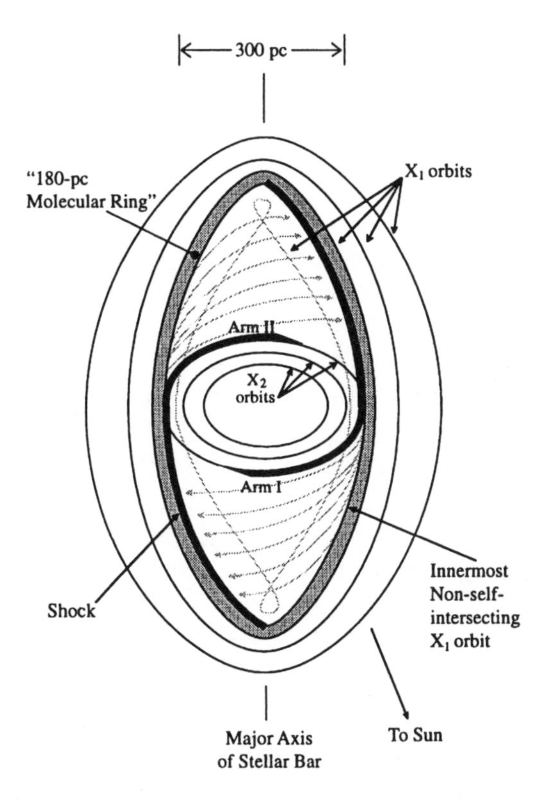

Figure 5 Schematic diagram illustrating the shapes and relative orientations of the X_1 and X_2 orbits and the locations of the shocks resulting from the interaction between the innermost X_1 orbit and the outermost X_2 orbit. The arms in the CMZ hypothesized by Sofue (1995a) are also shown.

innermost X_1 orbits. Indeed, gas located outside the 180-pc molecular ring is evident in H I data (Burton & Liszt 1978), and the dynamics of this H I nuclear disk can be modeled in terms of a hierarchy of X_1 bar orbits (Liszt & Burton 1980, Binney et al 1991). One potential problem with this interpretation of the parallelogram is the velocities of the vertices (Uchida 1993, Binney 1994): Although the velocity of the negative-longitude, intermediate-velocity vertex is near zero in the models, measurements show it to be large ($\sim$150 km s^{-1}) and in the forbidden quadrant (Figure 4).

Recall that the 180-pc molecular ring surrounds a complex, asymmetric concentration of disk population molecular clouds. Binney et al (1991) propose that these clouds are distributed on X_2 orbits. An attempt to identify some spatial order in this complex distribution was recently made by Sofue (1995a;

see also Scoville et al 1974), who defines two "arms" of molecular material, seen as continuous streams in the (l, b, v) data cube of ^{13}CO emission [Figure 3 of Liszt (1992) nicely shows the intersection of these streams with the $b = -3'$ plane] as the repository of most of the molecular mass interior to the 180-pc ring. These hypothetical arms have small pitch angles and are roughly symmetrically placed about the Galactic center at radii of $\sim$ 120 pc. Their proposed morphology is similar to the innermost, spiral density enhancements resulting from shocks in a numerical model of gas response to a bar (Gerhard 1996, Athanassoula 1992). It is also similar to the molecular distribution seen in several nearby galaxies (Kenney et al 1992, Turner et al 1993), so the molecular arms may have a natural explanation in the context of bar-induced kinematics. These shocks result from the transition from X_1 to X_2 orbits, as discussed above, and they should presumably extend continuously out to the 180-pc molecular ring, although l-v plots show no clear continuity between the arms and the ring.

3.1.2 THE EMR VERSUS THE BAR Perceived in early molecular data as an ellipse in the l-v plane, the 180-pc molecular ring was originally interpreted as a radially expanding molecular ring surrounding the Galactic center (Scoville 1972, Kaifu et al 1972) and was subsequently referred to as the EMR. It was hypothesized to be the result of an explosive event at the center $\sim 10^6$ years ago, which gave a radial momentum impulse to $\sim 10^7 M_\odot$ of nearby gas (Bania 1977). This hypothesis requires a rather extreme energy ($> 10^{55}$ ergs when dissipation, work against the gravitational potential, and the nonspherical geometry of the ring are taken into account; Sanders 1989, Saito 1990), although such energies can, in principle, be generated by episodic accretion events at the Galactic center (e.g. Bottema & Sanders 1986). Another challenge for the EMR hypothesis has been the dearth of evidence that non-EMR clouds in the CMZ have been affected by the passage of this putative flash flood of radially moving material. However, Uchida et al (1994a,b) recently pointed out that a molecular structure associated with the far-infrared source AFGL 5376 (at $l, b = 359.5, +0.43$) is apparently the site of a large-scale (100-pc) shock oriented perpendicular to the Galactic plane. They argue that it and another vertical complex at $l = 1.^\circ 2$ could be the signatures of the interaction of the EMR with the ambient medium. The projection of these shock structures out of the Galactic plane is attributable in part to the tilt ($\sim 20^\circ$) of the EMR with respect to the Galactic plane. Uchida et al (1994a) also cautioned against abandoning the EMR hypothesis by noting that, when CO emission within $0.^\circ 15$ of the Galactic plane is excluded from the average over latitudes, the resultant l-v plot reveals an elliptical envelope, as forseen by the original EMR hypotheses.

It would be curious indeed if an energetic explosion at the Galactic center left behind no kinematic signature outside of the single plane defined by the

EMR. In another examination of the latitudinal structure of the EMR, Sofue (1995b) presents evidence that the EMR is really an expanding molecular shell (EMS). Using the ^{13}CO survey data of Bally et al (1987), Sofue finds that the EMS is an oblate spheroid that is pinched around its equatorial waist because the expansion has suffered more impedance there by the higher density in the midplane of the CMZ. He also concludes that the EMS has gaps in it along radial rays that intersect the most prominent interstellar structures in the CMZ, as would be expected if these structures were ponderous and thick enough to block the passage of the EMS. Although this interpretation of the data is not completely unequivocal, it definitely warrants further investigation.

Which of the two competing hypotheses for the 180-pc molecular ring is more correct is not yet clear. Whereas the kinematic response of gas to the Galactic bar should inevitably produce a signature in the (l, b, v) cube looking roughly like the data, episodic explosions at the nucleus might also be both in evidence and inevitable (cf Section 7). Thus, although it appears that the dominant characteristics of the gas kinematics are naturally determined by the bar potential, the final analysis of Galactic center gas dynamics may well require an occasional injection of radial momentum.

3.2 *The Inward Transport of Gas to the Nucleus*

The mass of CMZ material said to be on X_2 orbits constitutes 85–90% of the total molecular mass in the Galactic center arena, so the residence time on these orbits is clearly much longer than in the innermost X_1 orbit. Nevertheless, this reservoir of gas can only be temporary. Angular momentum loss by orbiting disk gas is inevitable, in the face of the many processes acting near the Galactic center. Clouds on X_2 orbits circulate with velocities greater than the pattern speed of the bar (estimated at 19 km s^{-1} kpc^{-1} by Wada et al 1994); they thereby lose angular momentum to the stellar bar by gravitational torques. These clouds are also subject to tidal friction from stars in the bulge (Stark et al 1991), and the most massive of them are doomed to spiral into the Galactic center on time scales of a few times 10^8 years. The magnetic field is another contributor to angular momentum loss, particularly for the less massive clouds; if clouds continually move through a strong (mG), pervasive vertical magnetic field in the inner 100 pc or so, with a relative velocity equal to a substantial fraction of the cloud's orbital velocity, then the magnetic viscous force would cause angular momentum loss on a time scale of $\sim 10^8$ years (Morris 1996).

In addition to 1. bar-induced torques, 2. dynamical friction, and 3. magnetic viscosity, several other angular momentum loss processes contribute to the inexorable inward transport of matter: 4. shocks associated with the X_1-X_2 transition; 5. viscous drag in the differentially rotating Galactic disk, including that resulting from cloud-cloud collisions (von Linden et al 1993, Biermann et al 1993); and 6. dilution of the gaseous disk's specific angular momentum by stellar mass loss material raining down out of the slowly rotating Galactic bulge (Jenkins & Binney 1994). At present, assessing the relative importance of these mechanisms is rather difficult, but several of them are individually quite important, so gas brought into this arena is destined to migrate into the center on time scales much smaller than a Hubble time.

The rate of mass flow through the ILR, $\dot{M}_{\rm ILR}$, can be estimated using the mass of molecular gas in the 180-pc ring, $8 \times 10^6 M_\odot$ (Bania 1977, Sofue 1995a), and, following Gerhard (1992), noting that the gas cannot stay on this innermost (cusped) X_1 orbit for more than about one orbital period (2×10^7 yr). Thus, $\dot{M}_{\rm ILR} =\sim 0.4 M_\odot$ yr^{-1}. This estimate, which is larger than that of Gerhard because of the larger mass assumed for the 180-pc ring, is quite uncertain, and we regard a range of $\dot{M}_{\rm ILR} \approx 0.1$–$1\ M_\odot$ yr^{-1} as more appropriate. This range can be compared to the $0.07\ M_\odot$ yr^{-1} estimated by Jenkins & Binney (1994) to be the rate at which the bulge stars within 2 kpc shed mass. The actual contribution from bulge stars is likely to be larger than this, since this estimate is based on the assumption that the bulge is similar to an elliptical galaxy, whereas many bulge stars are younger than this would imply (Lindqvist et al 1992, Rich 1993, Rieke 1993). Of course, if it is to contribute to the mass budget of the Galactic center gas reservoir, the matter shed by bulge stars does indeed have to migrate down to the Galactic plane without being lifted off by a bulge wind. In addition, some fraction of the material moving inward at the ILR may come from further out in the disk, as a result of some of the same angular momentum loss processes, although the torque exerted by the bar outside the corotating region has the opposite sign and may partially counteract the other loss mechanisms. In any case, there appears to be no fundamental problem with finding enough material to maintain the gas inflow at its present rate.

The ratio of the total gas mass inside the 180-pc molecular ring, 4–9 $\times$ $10^7 M_\odot$, to $\dot{M}_{\rm ILR}$ provides an estimate of 0.4–1×10^9 yr for the mean residence time for gas in X_2 orbits, if we assume a steady state inflow. This range of time scales is similar to that for dynamical friction to extract angular momentum from massive clouds. The gas moves inward from ~ 150 pc at a mean rate of 0.2–2 km s^{-1}, which is comparable to the estimate of 0.3 km s^{-1} for inward radial motion near the solar circle in spiral structure models (Lacey & Fall 1985) and to the azimuthally averaged inflow velocities found in the bar models of Athanassoula (1992). This inwardly migrating gas meets one of three fates: star formation (0.3–0.6 $M_\odot$ yr^{-1}; Güsten 1989, Section 4), a thermally driven Galactic wind or fountain (0.03–0.1 $M_\odot$ yr^{-1}; Section 2.3), and accretion into and through the domain of the much smaller-scale circumnuclear disk (0.03–

0.05 $M_\odot$ yr^{-1}; Section 5). Although none of these routes is negligible, star formation evidently dominates.

In principle, the CMZ could undergo a global gravitational instability, creating a "bar within a bar," leading to yet further angular momentum loss and thus to a greatly enhanced inflow rate, as well as to a starburst (Shlosman et al 1989). However, the mass fraction of the gas inside the ILR, $\sim$ 5–10%, is apparently at or below the limiting value for such an instability. We therefore presume that the list of angular momentum loss processes given above is complete and that the inward flow of gas on 100-pc scales is relatively steady, although the possibility that the Galactic center has undergone substantial convulsions in the past must be kept in mind.

3.3 *Evidence for an $m = 1$ Wave*

In addition to the $m = 2$ bar mode, evidence exists for an $m = 1$ wave with substantial amplitude near the Galactic center. Such a wave would reveal itself as a displacement of the central stellar cluster from the centroid of the bar, of the bar from the center of mass of the Galaxy, and/or as a global tilt of the inner disk of the Galaxy; it would have implications for the stellar velocity field as well. The pronounced longitudinal asymmetry of molecular line emission both from the Galactic center disk population (cf Section 2.2) and from the molecular parallelogram (Section 3.1, Uchida et al 1994a, Blitz 1994) is a long-standing clue that the gas may be responding to an asymmetry in the Galactic potential. A pure $m = 2$ bar would show some displacement from $l = 0°$ because of projection effects, but the predicted offset is considerably less than that observed. Blitz (1994) finds that more distant gas—H I at radii out to 750 pc—shares this rotation center offset from the assumed Galactic center: Sgr A* and its surrounding stellar cluster.

A second clue for an $m = 1$ wave in our Galaxy is the velocity displacement with respect to the Local Standard of Rest (LSR) of 30 or 40 km s^{-1} of the appropriate velocity centroids of gas in the 180-pc molecular ring and the H I nuclear disk from the 0 km s^{-1} expected in an axisymmetric galaxy (Blitz 1994). A portion of this offset (15 km s^{-1}) can perhaps be attributed to motion of the LSR in response to a large-scale triaxial potential (e.g. Blitz & Spergel 1991a), but the remainder suggests that the entire central gas layer, and perhaps the bulge as well, are in motion, presumably oscillatory, with respect to the Galactic disk. This is reminiscent of observations of lopsided galaxies, i.e. barred spirals in which one spiral arm is much longer than the other (Baldwin et al 1980). In these galaxies, the velocity profiles near the center are invariably found to be highly anomalous, and attempts to pinpoint the rotation center always find it at a large distance from the center of the bar. Curiously, the gas in the circumnuclear disk, lying within several parsecs of the dynamical center of the Galaxy, shows no LSR velocity displacement to within about 20 km s^{-1} (Section 5), so the central stellar core and its immediate environment appear to be kinematically independent of the gas motions on scales of a few hundred parsecs.

Theoretically, $m = 1$ asymmetries in galaxies may arise in response to gravitational interaction with a passing companion, wherein the time scale for subsequent relaxation is different for the bulge and disk components. However, interaction with another galaxy is not required. Tagger & Athanassoula (1990), for example, point out that an $m = 1$ mode, or wave, can emerge in a galactic disk via nonlinear coupling to $m = 2$ spiral waves. From a different perspective, Miller & Smith (1992) find in numerical experiments that a test particle initially located at the Galactic center is subject to overstable oscillations about the galaxy's mass centroid. This result would imply that a mass concentration such as a galaxy's central stellar cluster should typically appear to be offset from the galaxy center as defined by nearby isophotes. These authors summarize the literature on off-center and lopsided galaxies that can perhaps be understood in this light.

Another form of an $m = 1$ wave at the Galactic center is the pronounced tilt of the plane defined by gas in X_1 orbits with respect to the large-scale Galactic plane (Liszt & Burton 1980; Uchida et al 1996a). A tilt of the presumably triaxial bulge has been suggested (Blitz & Spergel 1991a, Izumiura et al 1995), but *COBE* near-IR maps rule out a large tilt (Dwek et al 1995). The tilt of the gas layer, considered as a coherent, propagating warp, can perhaps be understood as a response to the central triaxial bar, even if the major axis of the bar is not itself tilted with respect to the Galactic plane (Binney 1978, Sparke 1984). Long-lived warps can also be fed by coupling to $m = 2$ spiral density waves (Masset & Tagger 1995, 1996). For our Galactic center, it remains to be seen whether the tilt of the gas layer and the longitudinal asymmetries near the Galactic center bear any relationship to each other. The understanding of both of these effects is still in its earliest stages.

4. STAR FORMATION

The initial conditions for star formation in the Galactic center environment differ dramatically from those found elsewhere in the Galaxy: The temperature, pressure, velocity dispersion, and estimated magnetic field strength were all much larger in the CMZ than in the Galactic disk, in some cases by several orders of magnitude (cf Section 2). Furthermore, potentially collapsing gas clouds are subject to the unusually strong tidal field near the nucleus, capable of overcoming a cloud's self-gravity for cloud densities $< 10^7$ cm^{-3} $(1.6 \text{ pc}/r)^{1.8}$, where r is the galactocentric distance (Güsten & Downes 1980). Consequently, self-gravity can initiate collapse only in the densest clouds. The conditions prevailing within the CMZ thus imply an extremely large Jeans mass ($\sim 10^5\ M_\odot$) and the inhibition of star formation via slow, quasi-static contraction of cloud cores. Nevertheless, star formation may proceed in this manner in the occasional exceptional cloud. The unusually dense and massive Sgr B2 cloud, located $\sim$ 100 pc in projection from the nucleus, is furiously forming stars (e.g. Mehringer et al 1993, Gaume et al 1995), albeit very probably with an initial mass function (IMF) quite different from that of the Galactic disk. Even in this cloud, however, the possibility has been raised that the star formation was provoked by a cloud collision (Hasagawa et al 1994).

In spite of all of the impediments to star formation within the CMZ, stars are clearly forming there, at many locations. It seems likely that much of the star formation is induced by events external to the clouds, notably by shocks associated with cloud collisions, supernova remnants, and perhaps violent gas outflows from the nucleus. This mode of star formation is likely to lead to an IMF skewed toward relatively massive stars. The lower mass cutoff may be elevated as well (Morris 1993 and references therein). Also, the enhanced metallicity of gas in the CMZ, which can be up to twice the solar value (Lester et al 1981, 1987; Lacy et al 1989; Wannier 1989; Shields & Ferland 1994), implies a correspondingly large opacity per gram. This affects the IMF by prolonging protostellar collapse, thus allowing more time for matter to accrete.

The global star formation rate in the CMZ, Φ, has been estimated at 0.3–0.6 $M_\odot\ yr^{-1}$ by Güsten (1989), who assumed a relatively normal IMF and used the global production rate of Lyman continuum photons derived from the radio continuum flux (Section 2). This estimate is sensitive to the IMF, however; if massive stars are favored, relative to the Galactic disk, then the global rate of star formation should be correspondingly reduced. Güsten (1989) gives a lower limit of 0.05 $M_\odot\ yr^{-1}$ from the luminosity of the discrete far-IR sources measured by Odenwald & Fazio (1984), but here again a normal IMF is assumed. A proper accounting of the mass budget of interstellar gas in the Galactic center will clearly require new information on the IMF, as well as an improved census of the sites of star formation.

Attempts to understand the spatial distribution of star formation near the Galactic center, and whether, for example, it reflects large-scale shocks in the CMZ, are still in their earliest stages. Infrared imaging has revealed several new star formation sites (Moneti et al 1992, 1994). Another promising approach has been to survey H_2O masers in *IRAS* sources (Taylor et al 1993, Levine 1995). After the masers associated with evolved stars were differentiated from those associated with sites of star formation, about a dozen sites of star formation in the Galactic center have been identified. This ongoing survey should eventually reveal large-scale star formation patterns.

Main sequence stars have been difficult to detect near the Galactic center because of confusion with the more luminous giants, so the IMF has not yet been directly probed. However, access to the upper main sequence is becoming possible with near-infrared observations. Three spectacular clusters of young stars are now known to be located near the Galactic center, each of which contains a sizeable number of exceptionally luminous ($10^{6\pm0.5}$ $L_\odot$) stars: 1. the well-known cluster occupying the central parsec of the Galaxy, centered roughly on Sgr A*; 2. the Quintuplet cluster, or AFGL 2004, located near the G0.15-0.05 H II region and the NTFs in the Galactic Center Arc (Nagata et al 1990, Okuda et al 1990, Glass et al 1990, Cotera et al 1995, Figer et al 1995); and 3. G0.121+0.017, or Object 17, located near the thermal arched filaments of the Radio Arc (Figure 6; Cotera et al 1994, 1996; Nagata et al 1995; Serabyn & Shupe 1996). All of these clusters have a population of massive and luminous emission-line stars, including Wolf-Rayet stars (late-type WN and WC stars, notably WN9/Ofpe and WC9) and blue supergiants (luminous blue variables, B[e] stars). The near-IR emission-line spectra of these stars indicate that they have high-velocity winds with speeds of 500 to 1000 km s^{-1} and that many of these stars are helium rich, consistent with their presumed post–main sequence status.

4.1 *Star Formation in the Central Parsec*

The evidence for recent massive star formation within the central parsec has grown over the years (Rieke & Lebofsky 1982, Lacy et al 1982, Allen 1987, Rieke & Rieke 1989, Allen et al 1990) and is now rather widely accepted. The cluster of emission-line stars centered on the core of the central stellar cluster has been the most thoroughly studied, and in many respects it is the most remarkable (Allen 1994; Genzel et al 1994; Blum et al 1995a,b; Eckart et al 1995; Krabbe et al 1995; Libonate et al 1995; Tamblyn et al 1996). The presence of at least two dozen He I/H I emission-line stars and four M or K supergiants ascribed to this cluster has been interpreted by Krabbe et al (1995) in terms of a modest burst of star formation between 3 and 7 million years ago. The young stars created in this burst are intermingled with the older population of the central stellar core, of which only the giant stars have so far been observed. The young stars dominate the luminosity of the central parsec, but the total mass of the stars formed in the burst is only $\sim 10^4$ $M_\odot$, far less than the mass of old stars in the central parsec of the Galaxy's central cluster ($\sim 10^6$ $M_\odot$). The newly formed stars have thus likely not yet equilibrated with their elder brethren. This scenario places the newly formed cluster in a relatively brief, windy phase, which should subside in a few million years. The picture is complicated somewhat by the presence of about 10 medium-luminosity, late-type stars in the central 8 arcseconds—objects which appear to be intermediate-mass AGB stars, signaling that at least one other star formation event took place there within the past $\sim 10^8$ years.

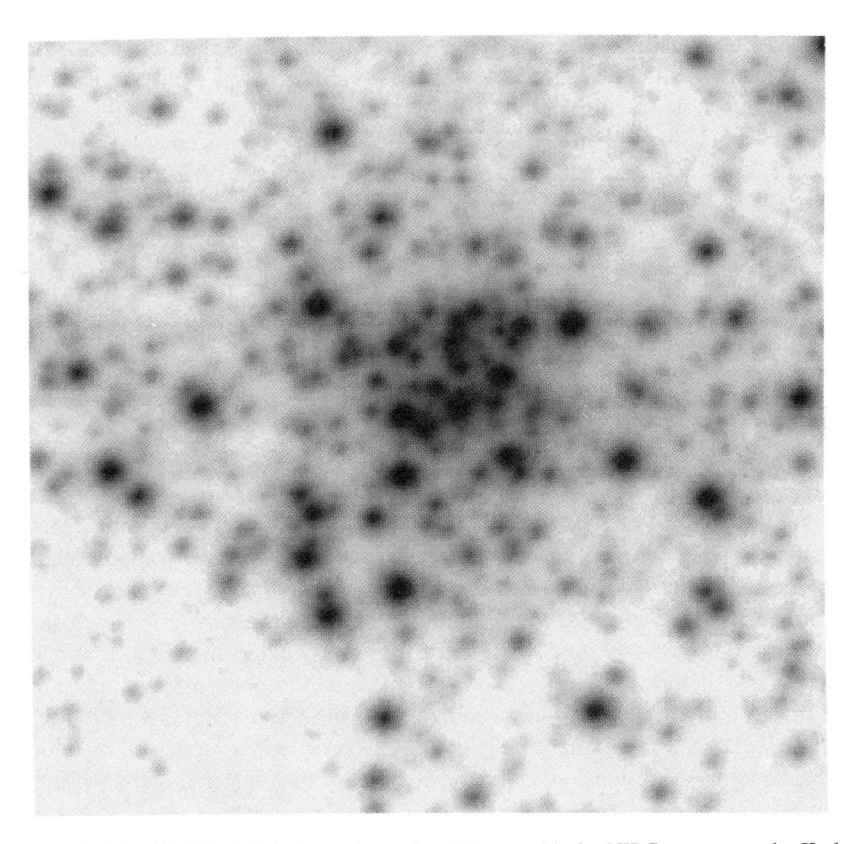

Figure 6 The G0.121+0.017 cluster, imaged at 2.2 μm with the NIRC camera on the Keck Telescope (Serabyn & Shupe 1996).

The implication that the central parsec is the site of repetitive bursts of star formation is perhaps not surprising, given the inevitability of radial inflow of gaseous matter (Section 3), although the factors inhibiting star formation—tidal forces, cloud turbulence, magnetic fields—are probably nowhere more extreme than in the central parsec. The star formation events taking place in the deep gravitational potential well of the Galactic nucleus must be unusually violent events unique to this environment.

4.2 *The Other Major, Young Clusters*

The possibility that the luminous, blue, objects which generate strong winds in the central parsec are something more exotic than young stars (Morris 1993, Eckart et al 1993) was rendered unlikely by the finding of the same kinds of stars in two other clusters located well outside the Galaxy's central stellar cluster: the Quintuplet and G0.121+0.017. The luminous post–main sequence stars in these clusters, again identified by their emission-line spectra in the near infrared, constitute a large fraction of the brightest cluster members (Figer 1995, Cotera 1995), especially in G0.121+0.017 (Figure 6), where a dozen stars show He I and H I emission lines (Cotera et al 1996). If these are WN stars, as Cotera et al suggest, then this one cluster contains 14% of all known Galactic WN stars. The brightest members of the Quintuplet cluster—the original five—are featureless in spectra measured so far, so their nature is unclear; they may be cocoon-like protostellar objects (Okuda et al 1990) or dust-enshrouded WC9 stars (DF Figer, personal communication). The bright stars dominating the Quintuplet and G0.121+0.017 are likely to be quite massive (50–100 $M_{\odot}$); thus, both of these clusters have masses of at least several thousand $M_{\odot}$, and possibly much larger if their IMF is not highly unusual.

The ages of these two remarkable clusters are comparable to that of the young cluster occupying the central parsec. A possible scenario for their formation is that the burst of energy accompanying the formation of stars in the central parsec strongly shocked nearby, massive clouds, inducing a relatively catastrophic gravitational collapse over a region much more widespread than the central parsec. In contrast, the star formation that gave rise to these clusters may resemble what is now going on in Sgr B2, where the star formation is presumably determined by local events. The two known, major, noncentral clusters are likely responsible for ionizing the surfaces of molecular clouds in their immediate environments (Figer 1995, Cotera 1995). However, the cloud near the Quintuplet does not share the velocity of the cluster (Figer 1995); the parent clouds of these clusters may no longer be in evidence.

There are several other sites of star formation near the Galactic center (e.g. Moneti et al 1992, 1994, Lis et al 1994), but no young clusters having the status of the Quintuplet or G0.121+0.017 are known. A survey carried out at 2 μm by Figer (1995) over a 60 $\times$ 30-pc region near the Galactic center revealed no new examples of clusters of emission-line stars, so if there are further instances of such clusters, they are either highly extincted or are located further from the center, in projection, than the known clusters.

One particularly interesting string of H II regions known as G-0.02-0.07, or the Sgr A East H II regions, lies quite close (10 pc) to the Galactic center in projection (Ekers et al 1983, Goss et al 1985). The H II regions, embedded in a ridge within the 50 km s^{-1} cloud (Mezger et al 1989, Ho et al 1991, Serabyn et al 1992b), are neatly aligned adjacent to, and along the edge of, the Sgr A East nonthermal shell source, giving the strong impression that the expandion of the shell has provoked the formation of the stars that ionize the H II regions.

However, there seems to be a mismatch between the expected expansion time of Sgr A East and the substantially longer time required for stars to form in response to the shock from the expanding shell and then evolve to the main sequence, as these stars have apparently done. Therefore, the origin of this cluster is not yet understood. Furthermore, although the cluster appears to contain at least one evolved emission-line star (Cotera et al 1994), star formation may still be adding to it, as evidenced by the rather high densities inferred for the molecular ridge (Serabyn et al 1992b), where an H_2O maser is located (Yusef-Zadeh & Mehringer 1995).

4.3 *The Environmental Effects of Stellar Winds*

The powerful winds emanating from the massive, post–main sequence stars at the Galactic center have a strong effect upon their surroundings, especially when they act collectively, as in the three major clusters described above. The cluster within the central parsec is the only one where the effect of the winds has been demonstrated, and there the result is profound. The 1.5-pc-radius cavity inside the circumnuclear disk (cf Section 5) may be largely swept clear by the cumulative mass outflow winds (Gatley et al 1984, 1986), aside from a few streamers of gas that appear to have enough inertia to maintain their integrity as they move through this region. Consequently, the matter currently accreting onto the central object, Sgr A*, may be almost entirely dominated by the material in the wind (Section 6). Other effects of the wind on gas at small galactocentric radii are described in Section 5.

The most spectacular effect of the Galactic center wind is the ablation of the envelope of the red supergiant, IRS7. The envelope of this well-studied star (e.g. Sellgren et al 1987) is not only ionized by hot stars near the center (Serabyn 1984, Rieke & Rieke 1989, Yusef-Zadeh et al 1989); it also has an extended, cometary "tail" of ionized gas pointing away from the sources of the wind and showing a pronounced velocity gradient (Yusef-Zadeh & Morris 1991, Serabyn et al 1991). This apparently is a case of colliding winds: the supergiant has its own radiation-pressure-driven wind (although this modest wind is quickly overcome by the ram pressure of the Galactic wind), and a bow shock is evident on the side of the star facing the wind source (Yusef-Zadeh & Melia 1992). The hope that the shape and size of this bow shock can allow one to derive the momentum in the wind is diminished by the conclusion of Dyson & Hartquist (1994) that the ram pressure of the Galactic wind plays no role in determining the size of the large-scale bow shock; the long tail implies that the stellar envelope is clumpy and permeable, so the wind is decelerated primarily in bowshocks around the individual clumps.

The same phenomenon should be happening at some level to the atmospheres of all red giants and supergiants in the central parsec, although no further cometary stellar wind tails have yet been observed. Coming years should see more examples of this phenomenon, as sensitive telescopes with improved spatial resolution are brought to bear.

5. CIRCUMNUCLEAR MATERIAL—DISK OR DEBRIS?

Our Galaxy's innermost molecular feature is a relatively compact (< 7 pc maximum radius) and dense disk- or torus-like structure that orbits about the center (Genzel et al 1994 and references therein). Although normally referred to as the circumnuclear disk, this nomenclature may actually hide a multitude of sins (Section 5.2). The molecular medium in general extends only to within about 1.5 pc of the center; its inner edge presents a rather sharply defined ionized boundary layer basking in the radiation from the encircled stellar cluster. Much of the filamentary Sgr A West H II region (Figure 7; Killeen & Lo 1989, Lacy et al 1991, Roberts & Goss 1993, Yusef-Zadeh & Wardle 1993), located in the innermost 1.5 pc radius region, can then be attributed to photoionization of the molecular medium's inward-facing surfaces. In order for centrally originating photons to propagate out to the CND's inner boundary, the wind-evacuated central cavity must be comparatively transparent (Becklin et al 1982). However, gas kinematics indicate that several ionized filaments and clumps are found in closer proximity to the center (Serabyn & Lacy 1985, Serabyn et al 1988, Serabyn 1989, Lacy et al 1991, Herbst et al 1993b, Lacy 1994), as is neutral and molecular gas (Davidson et al 1992, Marr et al 1992, Jackson et al 1993, Pauls et al 1993, Yusef-Zadeh et al 1993, Zhao et al 1995, Telesco et al 1996). Both are likely to be the result of dense gas plunging into the central cavity from the CND or beyond. However, the location of some of the molecular material remains controversial (Liszt & Burton 1993, Marshall & Lasenby 1994b). In this section, we discuss recent developments bearing upon this general scenario and also address evidence regarding its shortcomings. As this is a well-studied topic, we limit ourselves to a quick summary.

5.1 *The Standard Model*

The established picture of a clumpy, centrally illuminated torus or disk of largely molecular gas provides a good first-order description of the excitation, distribution, and kinematics of the gas making up the CND. Although a one-to-one identification of ionized structures with individual molecular clumps is far from complete, a large and increasing fraction of the ionized gas in the central few parsecs can indeed be accounted for in this manner (Telesco et al 1996). Thus, the intricate distribution of ionized gas evident in the Ne^+ 12.8-μm image of Sgr A West in Figure 7 (Lacy et al 1991) can likely be attributed largely to photoillumination of the clumpy molecular medium surrounding the center by

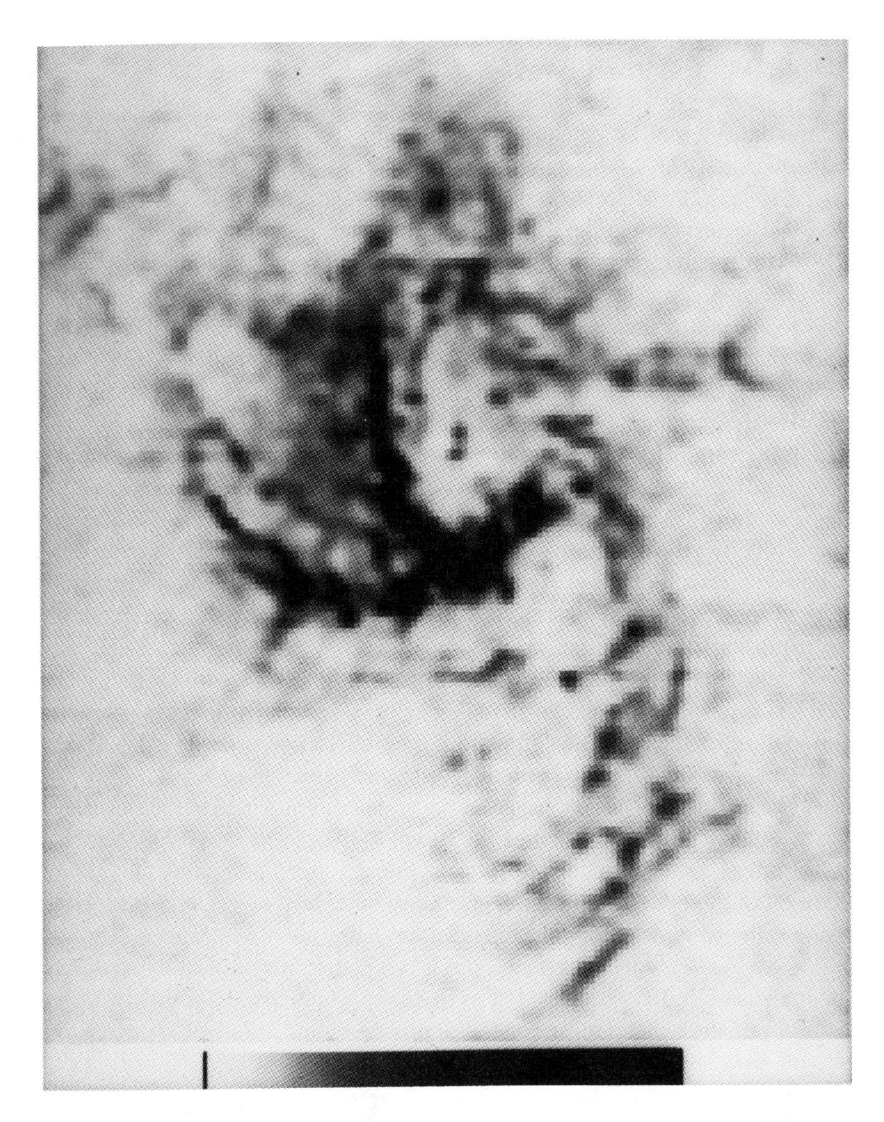

Figure 7 12.8-μm Ne^+ line emission from the Sgr A West H II region (Lacy et al 1991). The nomenclature for the ionized filaments is as follows: The Western Arc is the diffuse, slightly curved filament crossing the frame from top to bottom to the right (west) of center; the Northern and Eastern Arms are the well-defined filaments extending from the vicinity of the center of the image to the north and east, respectively; and the Bar is the bright, clumpy east-west structure near the center of the image.

stars in the central cluster. Local heating by individual stars is also observed (Smith et al 1990, Gezari 1992).

Copious measurements have by now probed the molecular medium lurking behind the outermost filaments in Figure 7 (most recently, by Sutton et al 1990, Jackson et al 1993, Marr et al 1993, Marshall & Lasenby 1994a, and Serabyn et al 1994). The HCN 1–0 map of the inner part of the CND, along with the 90-μm dust continuum emission, is shown in Figure 8. The gas near the CND's inner rim is both hot (a few hundred K) and dense (10^4–10^7 cm^{-3}), with line emission consistent with a dense photodissociation region (Genzel et al 1994). The column abundances and excitation states of the ionized and neutral gas components, as well as of the dust component, provide constraints on the incident radiation field's effective temperature and total luminosity (Section 6). The molecular medium is extremely clumpy (with clumps most likely taking the form of tidally sheared streamers), with low filling factor, thereby allowing for penetration of radiation well beyond the CND's innermost edge. Large linewidths ($\gtrsim$ 40 km s^{-1}) at the CND's inner edge (Marshall & Lasenby 1994a) likely reflect a large interclump velocity dispersion, the magnitude of which suggests frequent interclump collisions and shocks. Toward the outer edge of the CND, the linewidths decrease (Serabyn et al 1994), probably due to decreasing clump overlap.

Circular rotation provides a good first-order fit to extant kinematic data on the CND. Most kinematic models yield low values for radially directed velocities, although other interpretations remain possible. The differences in interpretation appear to stem largely from the method of selection of a disk major axis orientation: A major-axis selected to coincide with maximal velocities necessarily yields low radial velocities ($\lesssim$ 20 km s^{-1}; Jackson et al 1993, Marshall & Lasenby 1994a, Serabyn et al 1994), while an orientation based on the morphology of the intensity distribution (or any other criterion) can yield a larger radial velocity component (up to 50 km s^{-1}; Gatley et al 1986, Gatley & Merrill 1994; see Fig. 8). A common major-axis orientation may not apply at all radii (or for all molecular species), but even so, speculations about disk warp are likely premature. The most general approach would fit all disk parameters, but few data sets have been complete enough (beyond the CND's inner rim) to warrant such a treatment.

At smaller radii, gas kinematic modelling of the dense, ionized gas filaments in the Sgr A West HII region is possible. The ionized gas shows velocities as high as $\pm$300 km s^{-1}, with pronounced velocity gradients in the central 10 arcsec. For several of the inner filaments, eccentric orbits in the field of a point mass reproduce the gas velocities well (Serabyn and Lacy 1985, Serabyn et al 1988, Serabyn 1989, Herbst et al 1993a), under the assumption of gas flow

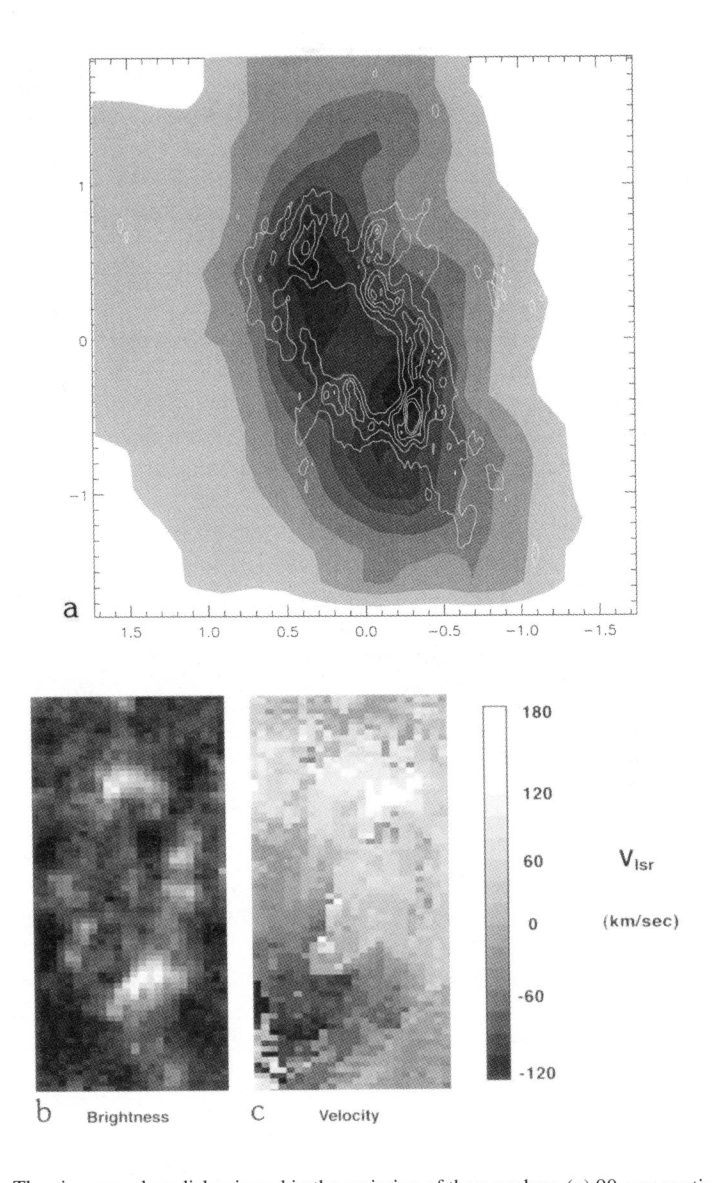

Figure 8 The circumnuclear disk, viewed in the emission of three probes: (*a*) 90-μm continuum emission (Davidson et al 1992), and HCN $J = 1$–0 emission (Wright et al 1987), (*b*) 2-μm molecular hydrogen emission (Gatley & Merrill 1995; I Gatley, personal communication). Panel *c* shows the velocity field of the H_2 emission, indicating that the maximum velocities do not lie at the ends of the apparent major axis of the elliptical emission trace and that the nodes in the velocity distribution do not lie along the apparent minor axis.

along the filaments. However, flow primarily along the filaments need not be taking place. The alternatives include tidally stretched cloud models (Quinn and Sussman 1985, Lacy et al 1991, Zhao et al 1995), which call for similarly elongated orbits with large radial motions, and the contrasting mini-spiral model of Lacy et al (1991), which allows an interpretation of the data (along a unified Western Arc-Northern Arm feature; see Fig. 7 for nomenclature) in terms of circular orbits. However, growing evidence indicates distinct natures for the Western Arc and the Northern Arm, the former being the inner photoionized boundary of the CND in near circular rotation, and the latter the inner photoionized boundary of a neutral gas feature presumably falling towards the center (the northern intruder; Davidson et al 1992, Jackson et al 1993, Telesco et al 1996). As even the eastern arm and a portion of the innermost bar feature (Fig. 7) have velocity patterns consistent with orbital trajectories transiting the central 0.5 pc (Serabyn 1989), it is difficult to avoid the conclusion that the kinematics of the innermost filaments call for an infall component.

However, while large radial motions are indicated for the ionized filaments interior to the CND (and the underlying molecular material, such as the northern intruder), the CND exhibits relatively small radial motions in most models. Thus, net mass inflow from the CND is probably not the result of small-scale viscosity in a homogeneous disk, but is perhaps more accurately described either in terms of sporadic infall of individual clumps resulting from dissipative inter-clump collisions, or as tongues of gas splitting off the inside edge of the CND as a result of an instability at this interface. Recent estimates for the inflow rate for the CND gas and the material in the central cavity are $\sim$ 3–5 $\times 10^{-2}\ M_{\odot}\ \mathrm{yr}^{-1}$ (Güsten et al 1987, Jackson et al 1993).

The inner rim of the CND also exhibits emission from vibrationally excited H_2, which may result from a strong shock caused by the impinging central wind (Gatley et al 1984, 1986; Fig. 8), although UV fluorescence has not been completely excluded (Pak et al 1996). This wind, arising from stars in the central stellar cluster (§4), probably helps to explain the abrupt density discontinuity at the CND's inner edge, where a quasi-steady equilibrium may be established between the pressures of the wind and the turbulent disk. The impact of the wind upon the disk may also contribute to the large linewidths at the disk's inner edge, and it has been held responsible for creating the radio continuum "streamers" that originate at the interface (Yusef-Zadeh & Morris 1987a, Yusef-Zadeh et al 1995). However, we note that the sharp inner edge of the CND might also be the result of a discontinuity in the radial gradient of the Galactic mass distribution at that point (Duschl 1988), a density wave in the field of a central point mass (Lacy et al 1991), a hydromagnetic instability (Fridman et al 1994), or as we discuss in the next section, the possible short-lived nature of the CND. The wind also appears to affect the distribution of the ionized filaments in the central-most few arcseconds, causing compressed and possibly rippled surfaces on the infalling gas streams (Yusef-Zadeh & Wardle 1993), and it may be responsible for the formation of the mini-cavity in the gas distribution within the bar (Morris & Yusef-Zadeh 1987, Yusef-Zadeh et al 1989, 1990, Wardle & Yusef-Zadeh 1992, Eckart et al 1992, Lutz et al 1993, 1994, Yusef-Zadeh 1994).

5.2 *A Symmetric Disk?*

The major objection to the simple model of the CND described above is that the material beyond the CND's inner edge is distributed highly asymmetrically about the center (Serabyn et al 1986, 1996b; Fukui & Churchwell 1987, Sutton et al 1990, Marshall & Lasenby 1994a): at negative Galactic longitudes the disk extends up to 7 pc from the center, whereas at positive longitudes the disk's radial extent is much smaller, $\lesssim 3$ pc. In galactic latitude, it is the ionized gas which is asymmetrically distributed, with a bright ionized filament (the Western Arc) outlining the positive latitude side of the CND (Fig. 7), but only a weak counterpart in the southeastern quadrant. Although this may result from selective shielding by interior features, the molecular gas distribution along the CND's inner rim is also not entirely symmetric (Jackson et al 1993, Marr et al 1993). In addition, molecular gas extends asymmetrically into the CND's central cavity from the northeast. Kinematically, the velocities are also not quite azimuthally symmetric, although it is not yet clear whether this is the result of non-circular streamlines or merely clumpiness (i.e., regions of "missing" gas). Finally, largely ignored and still unaccounted for is the positional coincidence between the southwestern rim of the nonthermal Sgr A East radio shell source and the thermal Western Arc filament in the Sgr A West HII region (Ekers et al 1983, Yusef-Zadeh & Morris 1987a, Pedlar et al 1989).

Symmetric models thus probably oversimplify matters. As symmetry implies longevity, asymmetry suggests time dependence, i.e. a short-lived disk. The asymmetric distribution of the CND beyond its inner rim suggests that the current appearance of the CND may result either from the gravitational capture and tidal disruption of a passing molecular cloud or from an energetic disruption of a stable disk. In either case, the relative symmetry of the CND's inner rim is a natural result of the shorter rotational time scales there.

The origin of either of these alternatives may lie in the expanding Sgr A East shell: The atomic carbon map presented in Figure 9 shows a very high degree of anticorrelation between the southern lobe of the CND and the ring of neutral carbon that evidently encloses the Sgr A East radio shell (Serabyn et al 1996b)—the carbon ring fades in intensity precisely at the location of the southern lobe of the CND. Thus, either the expansion of the Sgr A East shell has compressed

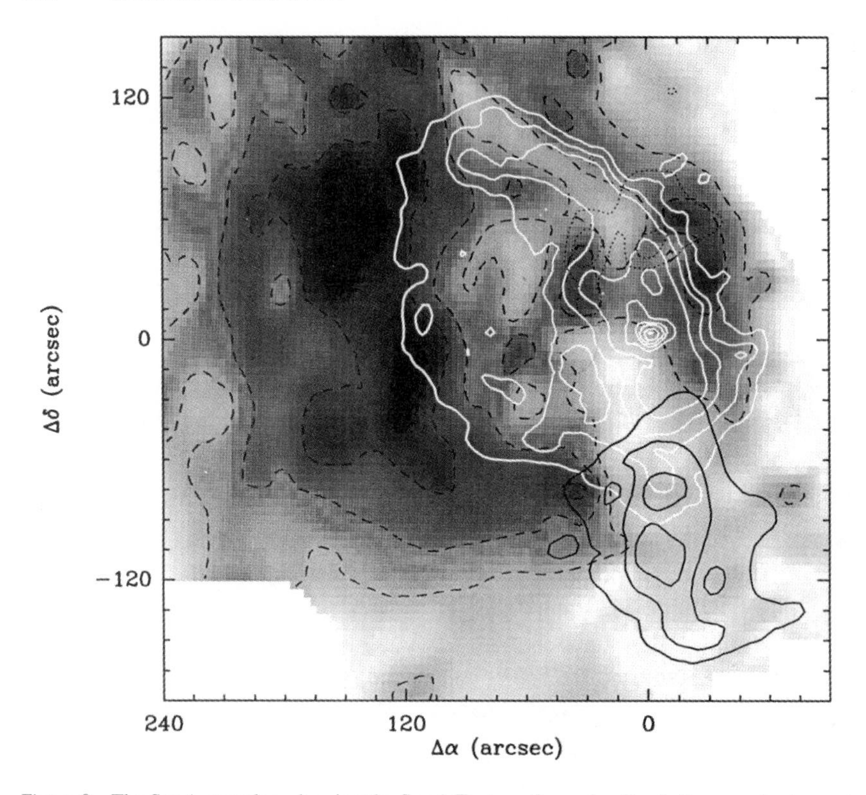

Figure 9 The Sgr A complex, showing the Sgr A East nonthermal radio shell source (*outermost white contours*), the Sgr A West thermal H II region (*fourth white contour and above*), the compressed 50 km s^{-1} molecular cloud layers surrounding the expanding radio shell (*gray scale* and *dashed contours*), and the circumnuclear disk (*solid black* and *dotted contours*). The (0,0) position marks the location of Sgr A*. All maps except the radio continuum map derive from observations of the 492-GHz fine-structure transition of CI (Serabyn et al 1996b). The velocity intervals are—50 km s^{-1} cloud: 30 to 80 km s^{-1}; blueshifted CND gas (*solid black contours*): −130 to −70 km s^{-1}; redshifted CND gas (*dotted contours*): 80 to 130 km s^{-1}.

and swept up molecular gas, pushing it toward the center (Mezger et al 1989), where it is now settling into a disklike configuration (in which case the notion of a wind-evacuated central cavity may be unnecessary), or the expansion has disrupted a preexisting and originally more symmetric disk. Alternatively, the geometric relationship could be radiatively highlighted; i.e. the central cluster could be illuminating the inner edge of the Sgr A East shell except where it is shielded by the nearer CND material, but even in this case the expansion of the radio shell must be considered in the context of mechanical interaction with the CND. Although the most likely scenario is not manifest at this point (except that

Sgr A East is largely behind Sgr A West; Yusef-Zadeh & Morris 1987a, Pedlar et al 1989), what does emerge clearly is that the expanding Sgr A East shell, and more specifically the molecular material that the shell is compressing (Mezger et al 1989; Genzel et al 1990; Zylka et al 1990; Ho et al 1991; Poglitsch et al 1991; Serabyn et al 1992b, 1996b; Ho 1994), is indeed intimately associated with the CND. As a result of this interaction, the long-lived nature of the CND must be seriously questioned.

6. THE COMPACT CENTRAL OBJECT

In common with several categories of external galaxies (Owen et al 1980), our own Galaxy harbors a compact radio source at its center (Lo 1989, 1994; Backer 1994, 1996). Referred to as Sgr A* to differentiate it from the extended Sgr A complex in which it is ensconced, our Galaxy's central radio source shares a unique set of characteristics with its siblings: central location, compact size, high brightness temperature, and a relatively flat radio spectrum. However, its luminosity falls near the low end of the observed range for compact nuclear radio sources, almost certainly the result of selection, as proximity makes our own nuclear source comparatively easy to detect. Thus, the standard argument calling for a massive central black hole on the basis of a prodigious energy output from a small volume is inapplicable in the case of our own Galactic nucleus. Indeed, independent of wavelength, the flux from the vicinity of Sgr A* is quite low compared both to observed active galactic nuclei and to theoretical expectations for the radiative output from an accreting compact object.

6.1 *Source Characteristics*

Although Sgr A* is frequently referred to as a point source, VLBI observations at centimeter wavelengths routinely resolve its emission, showing an apparent source size very closely proportional to λ^2, for wavelength λ. An axial size ratio of 0.5 is typically observed (Alberdi et al 1993, Lo et al 1993), with a major-axis position angle roughly east-west, and compilations of existing measurements yield a major-axis source size in microarcseconds of $\theta_{\mu as} = 14\,\lambda^2_{mm}$ (Jauncey et al 1989, Marcaide et al 1992, Backer 1994, Yusef-Zadeh et al 1994). This behavior is consistent with anisotropic scattering by the intervening interstellar medium, and because nearby maser sources show similar characteristics, the highly magnetized medium in the central few hundred parsecs is a prime candidate for the scattering agent (Frail et al 1994, Yusef-Zadeh et al 1994). Unfortunately, the resultant image broadening prevents direct observation of source structure in Sgr A* at cm wavelengths (and also masks any possible intrinsic size variation with λ).

VLBI observations have recently been extended to millimeter wavelengths, yielding tantalizing suggestions of intrinsic source structure (Krichbaum et al 1993, 1994) but as yet no confirmed evidence for a source size exceeding that set by interstellar scattering (Backer et al 1993, Rogers et al 1994). The most stringent observational limit gives an intrinsic source diameter $< 130\ \mu$as, or 1.1 AU, at $\lambda = 3.5$ mm (Rogers et al 1994) and an intrinsic brightness temperature $> 1.4 \times 10^{10}$ K, implicating synchrotron emission. Since the inverse Compton limit and the absence of short-time-scale scintillations at near-millimeter wavelengths both imply a source diameter $\gtrsim 10\ \mu$as, or 0.1 AU (Kellermann & Pauliny-Toth 1981, Gwinn et al 1991, Serabyn et al 1992a, Zylka et al 1992), the intrinsic source size at near-millimeter wavelengths is evidently constrained to within an order of magnitude. This scale size is quite interesting from a theoretical perspective. The smallest stable circular orbit about a massive black hole (with a diameter of 6 Schwarzschild radii, or 0.12 M_6 AU, where M_6 is the black hole mass in units of $10^6\ M_\odot$) falls near the lower end of the range, for masses suggested by kinematic data ($M_6 \sim$ 2–3; Section 6.3). VLBI observations are thus beginning to probe a very telling regime, although it is likely that the shortest millimeter wavelengths, and perhaps even submillimeter wavelengths, will be needed to overcome the scattering handicap and reveal the intrinsic source structure.

The spectrum of Sgr A* is also key to understanding its nature, and the radio source has now been detected at frequencies up to 670 GHz (Zylka et al 1992, 1995; Dent et al 1993). Most observations at frequencies below 100 GHz show a slowly rising approximately power-law spectrum (Lo 1989), but as moderate variability is established at cm and mm wavelengths (Zhao et al 1989, 1992), care is required in deriving the spectral slope. In particular, because Sgr A* shows short-time-scale ($\lesssim$ month-long), weak ($\sim$ a factor of 2) outbursts at cm and mm wavelengths several times per year, with concommitant spectral changes (Wright & Backer 1993), a concatenation of numerous flux density measurements taken over the years could be misleading. To avoid these concerns, we instead show in Figure 10 nearly simultaneous measurements of the radio/millimeter spectrum of Sgr A* taken inside a span of 12 days in March 1993 with the VLA, OVRO, and CSO-JCMT interferometers (Serabyn et al 1996a). These simultaneous data cover nearly two decades in frequency, from 5 to 354 GHz, and the measurements at frequencies below 100 GHz are best fit by a $\nu^{0.25}$ power law (Figure 10), consistent with earlier spectra. At frequencies above 200 GHz, an excess of 1–2 Jy above the extrapolated power law (up to a factor of 2) is evident, and the higher frequency simultaneous data points are in good agreement with prior interferometric and bolometric measurements (Zylka & Mezger 1988; Zylka et al 1992, 1995; Serabyn et al 1992a; Dent et al 1993).

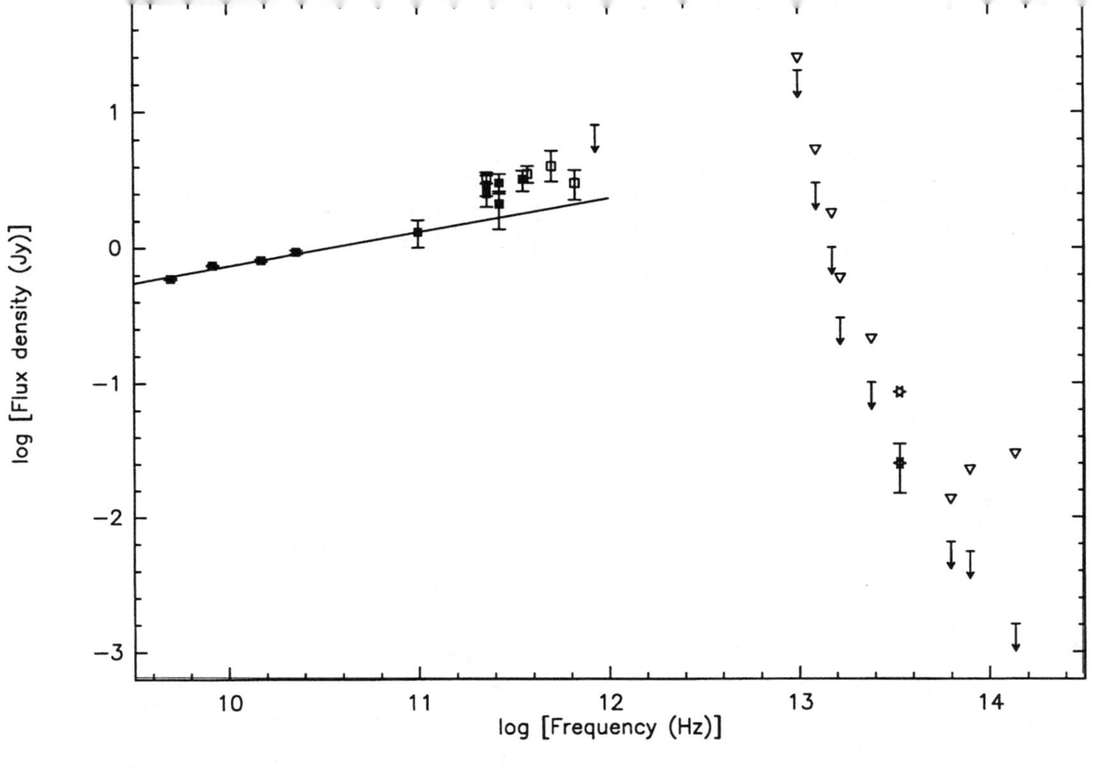

Figure 10 The radio/infrared spectrum of Sgr A*. The solid squares show the near-simultaneous measurements of Serabyn et al (1996a), the empty squares the submillimeter data of Zylka et al (1995), the downward-pointing arrows the measured upper limits to the source flux at shorter wavelengths, the inverted triangles the corresponding extinction-corrected limits, and the asterisks the measured 8.7-μ flux (25 mJy; Stolovy et al 1996) and its extinction-corrected value. The measured upper limits are: 8 Jy at 350 μ and 3 Jy at 24.3 μ (Serabyn et al 1996a), 20 Jy at 30 μ (Telesco et al 1996), 1 Jy at 20 μ (Gezari et al 1994), 0.3 Jy at 18.1 μ and 0.1 Jy at 12.4 μ (Gezari et al 1992), 6.5 mJy at 4.8 μ and 5.5 mJy at 3.8 μ (Herbst et al 1993b), and 1.6 mJy at 2.2 μ (Eckart et al 1995). The upper limits are corrected using the extinction curve of Mathis (1990) with $A_V = 30$. The conservative 1 μ limit of 75 μJy (Liu et al 1993) is too low to plot on this scale.

At frequencies beyond those measured interferometrically, the shape of the spectrum remains unsettled. Single-dish measurements between 300 and 670 GHz yield flux densities of roughly 3–4 Jy for Sgr A* across this range, although an initial report of a 1.5 Jy upper limit at 670 GHz (Dent et al 1993) conflicts. Large error bars leave spectral slopes indeterminate at present, and the extant database is insufficient to definitively address the question of source variability at submillimeter wavelengths. Nonetheless, the fact that Sgr A*'s flux density does not exceed the 4 Jy level even for frequencies approaching 700 GHz suggests that the meager near-millimeter excess is due either to a slight rise in an undulating spectrum, as is seen in external nuclear radio sources at lower frequencies (Owen et al 1980), or perhaps to a final hump just before turnover (Falcke 1996).

Because of its positive spectral index at radio frequencies, the luminosity of Sgr A* is dominated by its highest frequency emission, making submillimeter and shorter wavelength observations critical for a determination of this quantity. The power-law spectrum to 100 GHz has a total luminosity of $\approx 10^{34}$ erg s^{-1}, and inclusion of the emission to 700 GHz increases the total luminosity by roughly an order of magnitude, to 1.4×10^{35} erg s^{-1}. At just under $40\, L_{\odot}$, this is well below the Eddington limit of even a solar mass object ($3.4 \times 10^4\, L_{\odot}$).

Searches have also been carried out for counterparts to Sgr A* at many shorter wavelengths, with success only recently becoming apparent. Several far- and mid-infrared upper limits (Figure 10), as well as the mid-IR detection of a source toward Sgr A* (Stolovy et al 1996), constrain the high-frequency rolloff of Sgr A*'s synchrotron emission. It is clear from the extinction-corrected data at wavelengths shortward of 20 μm (Figure 10) that the synchrotron spectrum must cut off between 1 and 10 THz, with a falloff at least as steep as ν^{-1}. The maximum synchrotron luminosity then cannot exceed several hundred $L_{\odot}$. We expect that measurements of the submillimeter and mid-IR spectra will soon be dramatically improved by the new generation of high-resolution cameras.

In the near infrared, where angular resolution and sensitivity have increased markedly in the past decade, no definitive counterpart to Sgr A* has yet been identified, owing primarily to confusion by surrounding stellar sources. Increased pointing accuracy has eliminated several candidate detections, and most recently, Eckart et al (1995) have found a compact cluster of half a dozen near-IR sources located within a few tenths of an arcsecond of the position of Sgr A* (Figure 11), possibly representing a central cusp in the stellar distribution. While intriguing in its own right, this result makes the identification of any nonstellar near-IR source that may be associated with Sgr A* itself problematic. At the same time, the extremely low limits ($m_K \gtrsim 14$, Figure 10) to the

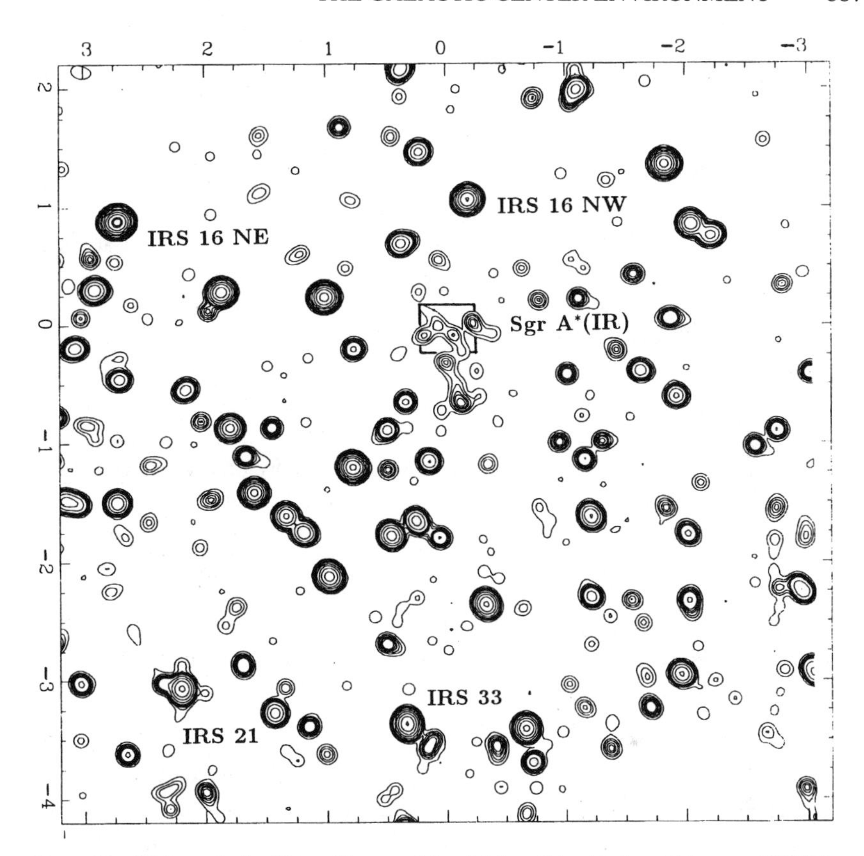

Figure 11 High-resolution K-band image of the vicinity of Sgr A*, from Eckart et al (1995).

possible near-IR flux attributable to Sgr A* severely constrain accretion disk models (Section 6.2).

At visual and UV wavelengths, interstellar extinction obscures the Galactic center completely, but constraints on the radiation field at these wavelengths can nevertheless be derived from the excitation of the ISM in and surrounding the central parsec (Becklin et al 1982, Lacy et al 1982, Serabyn et al 1985, Davidson et al 1992, Maloney et al 1992, Genzel et al 1994, Shields & Ferland 1994, Zylka et al 1995, Telesco et al 1996). The result for a single-component cloud model is a radiation temperature of about 35,000 K, an ionizing photon flux of 2×10^{50} s^{-1}, and a total luminosity of $\approx 2 \times 10^7 \, L_\odot$. However, because stellar sources apparently dominate the production of Lyman continuum photons in the central parsec (Rieke et al 1989; Krabbe et al 1991, 1994; Tamblyn & Rieke 1993), the optical/UV flux in the central parsec arising from Sgr A* itself is likely only a small fraction of the total ($\lesssim 10^6 L_\odot$) and is hence difficult to determine reliably.

At an energy of roughly 1 keV, it again becomes possible to search for emission from the direction of Sgr A*. A 5-arcmin resolution *GRANAT* ART-P map at 4–20 keV shows emission from the general direction of the extended Sgr A source (Syunyaev et al 1991, Pavlinsky et al 1994), but high- resolution 2–10 keV *ASCA* data (Koyama et al 1996) show the main emission to be associated not with Sgr A* but with the Sgr A East shell source. The 2–10 keV flux from Sgr A* itself is then limited by the *ASCA* observations to $\sim 10^{35}$ erg s^{-1} (Koyama et al 1996), which is an order of magnitude below the large-beam *GRANAT* estimate and more consistent with the 1.5×10^{35} erg s^{-1} seen below 4 keV by Watson et al (1981). The lower *ASCA* flux is also more consistent with data in the softer 1.2–2.5 keV *ROSAT* band (Predehl & Trümper 1994), where a source coincident with Sgr A* was also found. Comparison of the *ROSAT* and *GRANAT* data had earlier suggested the need for an anomalously high extinction at long X-ray wavelengths, but this need is removed by the lower *ASCA* flux actually attributable to the compact source. Of course, source variability may affect this discussion. At higher energies (35–150 keV), no source was detected toward Sgr A* by the Sigma/*GRANAT* telescope (Goldwurm et al 1994). Thus, the total X-ray flux that arises in a small enough region to be attributable to Sgr A* itself is no more than a meager 2.5×10^{35} erg s^{-1}.

6.2 *Source Models*

The preceding section establishes that Sgr A* is a rather weak source across the electromagnetic spectrum, with a luminosity of at most a few hundred $L_\odot$ at radio and X-ray wavelengths. The intermediate-frequency situation is not as clear, but allowable near-IR/optical/UV luminosities have decreased markedly over the past decade (by a factor of 100 at K band), ruling out luminosities for Sgr A* in excess of $10^6 L_\odot$. Because of the low radio and X-ray luminosities definitively attributable to Sgr A*, accretion onto a stellar source cannot be excluded solely on energetic grounds. However, the radio spectrum of Sgr A* is much flatter than that of a pulsar, too stable for a mass-transfer binary, and much more energetic than typical examples of either (Lo 1989). Furthermore, as proper motion studies (Backer 1996) establish a lower limit to the mass of Sgr A* of at least 100 $M_\odot$, and possibly as high as 1000 $M_\odot$ (depending on whether one assumes kinetic equilibrium with low- or high-mass stars, respectively), stellar sources are effectively ruled out. Thus, models advanced recently to account for the emission from this unique radio source are all variants on the general scenario of accretion onto a massive compact object.

Proposed models differ widely in their assumptions, including the size of the central mass, the accretion rate and geometry, the radiative mechanisms and efficiency, the source opacity, and the disposition of infalling material. Central masses of 500–3 $\times$ 10^6 $M_\odot$ appear in different models (Falcke et al 1993; Ozernoy 1993, 1994; Melia 1994a,b; Falcke 1996; Narayan et al 1995), and accretion rates vary even more widely, from 10^{-10} to 10^{-4} $M_\odot$ yr^{-1}, the high end of which is coincidentally the maximum possible average accretion rate over a Hubble time. Emission geometries range from spherical to disk-like to jet-like, with the curious result that the same radio spectrum can apparently be fitted (to some level of accuracy) with either a jet-like outflow or an accretion inflow. The radio spectrum as measured to date thus may be less than definitive, and indeed, models running the gamut from optically thick to thin have been proposed (de Bruyn 1976, Reynolds & McKee 1980, Duschl & Lesch 1994). Higher frequency continuum flux measurements and VLBI imaging will no doubt aid in discriminating between models, since a definitive measure of source geometry, size, and cutoff frequency would provide tangible constraints.

Potentially one of the most telling probes of SgrA* may well be its spectrum beyond the radio (i.e. synchrotron) regime. It is too soon to tell whether mid-IR emission might arise from the tail of the synchrotron spectrum (which is at least as steep as ν^{-1}), a dust component, inverse-Compton scattered photons, or from an accretion disk photosphere. With only upper limits available from the near infrared through the UV, definitive model fitting is unfortunately not yet possible at these wavelengths either. It has, however, long been clear that standard accretion disk photospheres are in conflict with stringent near-IR/UV source constraints (Rieke & Lebofsky 1982, Lacy et al 1982, Rieke et al 1989). The only potential means of evading the near-IR constraints, for a classical thin disk, is to assume a disk orientation very close to edge-on, certainly a distinct possibility (we are, after all, located in the plane of the Galaxy), but the UV constraint arising from the excitation of the surrounding medium (after removal of the stellar UV contribution) is relatively independent of disk orientation. Furthermore, general-relativistic light bending tends to isotropize emergent radiation, rendering even an edge-on disk visible (Falcke & Heinrich 1994, Hollywood & Melia 1995); hence, disk inclination may not be a panacea after all. Thus, near-IR/UV limits remain extremely restrictive. Finally, high-resolution X-ray data (Watson et al 1981, Koyama et al 1996) also provide very tight luminosity constraints.

How do existing models fare? Four schools of thought have recently emerged in rather distinct corners of parameter space, and we briefly outline the salient points. The first two models are related; both are based on the idea that a massive black hole would capture and accrete a sizable fraction of the observed stellar mass-outflow wind in its vicinity (Ozernoy 1989), a scenario that has the advantage of properly accounting for the specific environment of Sgr A*.

Expanding on this idea, Ozernoy (1993, 1994) and Mastichiadis & Ozernoy (1994) claim that the observed radio and X-ray luminosities of Sgr A* determine upper limits to its mass of $\sim 500\ M_\odot$ and $\sim 6000\ M_\odot$, respectively. Because the Bondi-Hoyle accretion rate is proportional to the square of the central mass, such a low mass results in very slow accretion ($10^{-10}\ M_\odot\ \mathrm{yr}^{-1}$), and so little near-IR radiation is produced. However, the low mass also introduces a small length scale, implying optically thick synchrotron emission at high frequencies, and so a further emission component not currently in the model is needed to account for the flat low-frequency spectrum observed (Falcke 1996). Beyond this, a detailed prediction for the source spectrum would certainly aid in evaluation.

A similar Bondi-Hoyle accretion scenario is considered by Melia (1994a,b), but diametrically opposite conclusions are reached: A central mass of 1–3 $\times 10^6\ M_\odot$ is called for, and an extremely high accretion rate ($1.6 \times 10^{-4}\ M_\odot\ \mathrm{yr}^{-1}$) emerges, necessitating an edge-on disk unless a general-relativistic treatment is applied (Hollywood & Melia 1995). However, although the low-frequency radio spectrum (below 100 GHz) can be fitted, the turnover frequency predicted tends to be too low, no excess emission at submillimeter wavelengths is predicted, and the most recent near-IR/UV and X-ray constraints to the flux from Sgr A* are likely violated.

Of course, the near-IR and UV luminosity problems are obviated if the accreting material radiates with low efficiency. This can occur if electrons and ions are decoupled in temperature in the inner parts of the disk (Rees et al 1982, Narayan et al 1995), a situation that can arise because the viscously heated heavy ions cannot easily transfer energy to the much lighter electrons via two-body collisions. Thus, the more efficient radiators do not have ready access to the thermal energy, and the energy is advected with the ions' flow (Abramowicz et al 1995, Narayan & Yi 1995). To remove angular momentum on a time scale short compared to the ion-electron equilibration time scale, a reasonably high viscosity is also needed, implying radial infall rates higher than in a standard thin disk (10^{-7}–$10^{-5}\ M_\odot\ \mathrm{yr}^{-1}$ in Narayan et al 1995). Accreting material is then able to fall through the Schwarzschild radius before radiating away its thermal energy content, yielding a "dim disk" in the near-IR/optical/UV range. The resulting spectrum (Narayan et al 1995) consists of a series of roughly equal energy humps, due to (in increasing frequency) synchrotron, inverse-Compton, and thermal bremsstrahlung emission. A central mass of rougly $10^6\ M_\odot$ provides a good match to the submillimeter luminosity, but not to the radio spectrum (without the addition of an ad hoc component), and there may also be difficulties in fitting the X-ray and submillimeter fluxes in a single model, especially when the data of Koyama et al (1995) are considered. However, because plasma instabilities may serve to couple the ions and electrons much more effectively than single-particle collisions, this scenario is based on an untested assumption.

Finally, in analogy with external galaxies, a core-jet model has been suggested (Blandford & Königl 1979, Falcke et al 1993, Falcke & Heinrich 1994, Falcke 1996), in which the observed radio emission arises not in the accreting material, but in an ensuing jet-like outflow. This model thus has the specific prediction that at high enough angular resolution, a jet should be seen, and it is also the only model that predicts a slight excess of emission at submillimeter wavelengths. The latter results from the emission from the jet nozzle, the smallest scale structure (Falcke 1996). The predicted accretion rate is $\sim 10^{-7}\, M_{\odot}$, a mid-range value, and is thought to be set by long viscous time scales in a remnant steady-state accretion disk. Because the accretion rate is governed not by the capture rate, but by viscosity, the central mass is not well constrained. Possible drawbacks to this model are the large discrepancy between the Bondi capture rate and the low steady-state accretion rate required, the fact that no jet has yet been conclusively detected, and the lack of a prediction for the X-ray emission.

While several of these models show promise, the contrast between them also serves to highlight the continuing scarcity of definitive source-structure information. However, with continually improving observational capabilities throughout the electromagnetic spectrum, this situation will no doubt soon improve.

6.3 *The Central Mass*

Clearly one of the most vital of Sgr A*'s unknown parameters is its mass, which cannot be determined uniquely from models of the emerging radiation. Nonetheless, mounting observational evidence over the past decade consistently shows that line-of-sight velocities increase toward the center (inside the central 0.5 pc or so), for both the gas and the stars (Lacy et al 1980, 1982, 1991; Serabyn & Lacy 1985; Serabyn et al 1988; McGinn et al 1989; Sellgren et al 1990; Herbst et al 1993a; Krabbe et al 1995; Haller et al 1996). The conversion of these findings to an enclosed mass distribution is fraught with uncertainties, for both the stellar and gaseous components, but different techniques yield remarkably good agreement both for the size ($2–3 \times 10^{6}\, M_{\odot}$) of the central dark mass concentration (observationally constrained to lie within the central few tenths of a parsec) and for the distributed stellar mass, $\sim 3 \times 10^{6}\, r_{\rm pc}\, M_{\odot}$, just beyond the core radius of the central cluster (cf Rieke & Rieke 1988, McGinn et al 1989, Lindqvist et al 1992, Serabyn et al 1994, Haller et al 1996). Although the radial size constraint on the dark mass concentration cannot be used to tie the mass excess directly to Sgr A*, new imaging techniques now being employed are pushing observations ever closer to the center, where the discrimination will no doubt improve. In particular, the possible stellar cusp around Sgr A* may soon provide important proper motion data (Eckart et al 1995).

7. THE EPISODIC NATURE OF ACTIVITY AT THE GALACTIC CENTER

The inexorable inward flow of matter in our Galaxy makes nuclear activity inevitable; if the accretion rate onto a presumed central black hole of mass M_6 were only $0.02\,M_6(0.1/\epsilon)\ M_\odot\ \mathrm{yr}^{-1}$, where ϵ is the efficiency with which the accretion energy is radiated away (often estimated at ~ 0.1), then the nucleus would emit at the Eddington rate, with an energy output of Seyfert proportion, $\gtrsim 10^{43}$ ergs s^{-1}. Given the mass flow budget outlined in Sections 3.2 and 5.1, a time-averaged mass flow rate into the central parsec equal to this value is plausible. However, the current accretion rate and luminosity of Sgr A* are many orders of magnitude smaller than this (Section 6). It therefore appears that the Galactic center may now be in a lull between brief accretion events. The massive young stars clustered within the inner parsec are indicative of a substantial mass accumulation event within the past 10^7 yr, while the probably inward-moving circumnuclear disk likely heralds the succeeding accretion epoch, possibly within the coming 10^5–10^6 yr.

The implied limit cycle is one that begins with the relatively sudden ($\lesssim 10^4$ yr) appearance of gas within the gravitational potential well defined by the core of the central stellar cluster (0.2–1.2 pc; Rieke & Lebofsky 1987, Allen 1994, Rieke & Rieke 1994, Eckart et al 1995). This would occur, for example, if the inward radial velocity of the CND were to be as high as 50 km s^{-1} during its last stages of inward migration, or if the CND underwent a strong disk instability during its descent, causing unusually rapid transfer of angular momentum outwards. Alternatively, dissipative intercloud or interclump collisions might be responsible for sending material rapidly inwards. The gas arriving in the inner few tenths of a parsec has no exit other than to accrete onto a compact object (either a supermassive black hole or one of a collection of smaller ones; Morris 1993, Lee 1995) or to form stars. However, as discussed in Section 4, star formation is inhibited until either the density grows above a few times 10^7 cm^{-3} or the gas is subjected to a violent shock. Such a shock may be provided by fast-moving cloud collisions in the small central volume or by the energy emerging from an accretion disk around the supermassive black hole. The combination of accretion energy, in the form of radially propagating winds and radiation, coupled with the energy released in the formation of dozens, if not hundreds, of massive stars, reverses the infall of gas and begins a period of winds (initially from the massive O stars, and later from the post–main sequence WR-type stars), aided considerably by radiation pressure on dust. This would be the moment at which an expanding molecular ring, if such exists in our Galaxy, could be launched. However, gas clouds initially in the near vicinity of the nucleus can only be pushed to a relatively large radius under the most extreme circumstances; for a 0.5-pc diameter cloud of density $10^5 n_5$ cm^{-3}, for example, the ratio of the outward radiative to the inward gravitational accelerations at galactocentric distance r_{pc}, even assuming a Seyfert-like central luminosity of 10^{43} ergs, is only $\sim 0.1 n_5^{-1} r_{pc}^{-1}$. The ratio of the ram pressure acceleration of that same cloud caused by a steady wind carrying $\dot{M}$ $M_\odot$ yr^{-1} at a velocity of 1000 v_3 km s^{-1} to the gravitational acceleration is $4\dot{M}\, v_3 n_5^{-1} r_{pc}^{-1}$. Thus, an exceptionally large mass outflow rate would have to be sustained for an extended time period ($\sim 10^4 (\dot{M}\, v_3)^{-1} n_5 r_{pc}$ yr) to give a significant radial impulse to even a modest cloud starting near the center. Alternatively, in place of a steady wind, this mass outflow could be accomplished with a succession of at least 10^2–10^3 supernovae occurring within the central parsec on a similar time scale. If and when these conditions are met, the radially moving gas would mix with orbiting gas, raising its specific angular momentum and preventing it from simply falling back in on a free-fall time scale.

At this stage, the Galactic nucleus would resemble that of an emission-line galaxy, a starburst galaxy, or even an AGN, depending on how much mass had accumulated before the reversal took place and on whether the central accretion disk generates more energy than the luminous, newborn stars. Also, those stars would resemble the kinds of stellar objects that have been considered as the source of activity in emission-line galaxies [e.g. "warmers" (Terlevich & Melnick 1985) or the stars in Wolf-Rayet galaxies (Conti 1991)].

As these massive stars evolve on a time scale of a few million years, their population declines as some of them become supernovae, and the total stellar luminosity decreases. The luminosity of the accretion disk also declines, as it exhausts the supply of gas with which it was impulsively endowed during the relatively brief accumulation phase. The surrounding gas transfers its angular momentum outwards by the processes described in Section 3.4 and migrates back in, against the resistance of the winds from the central stars, for as long as those stars stay in a windy stage of evolution. In broad terms, this is the state in which we now apparently find the Galactic center. However, the gas is not smoothly distributed, so the timing of the recurrent accretion episodes and the manner in which they occur can only be characterized stochastically. In a picture proposed by Sanders (1981), activity at the Galactic center is initiated when a cloud passes close enough to be accreted. We adopt that picture here, noting that a cloud that is "captured" near the Galactic nucleus is likely to first form a sheared, lumpy, and asymmetric circumnuclear disk much like the observed CND. The task before us is to unravel the final stages of this recurrent process by understanding how rapidly the CND is now moving inward and then how the gas will behave as it descends upon the denizens of the center.

Interstellar gas is not the only possible source of matter for the accretion disk around the presumed supermassive black hole at the center, since stars whose orbits carry them within the central object's Roche limit are susceptible to tidal disruption (Hills 1975, Ozernoy 1979, Lacy et al 1982, Rees 1988, and references therein). Some fraction of the resulting debris settles into a disk that can have a large accretion rate and a consequent luminosity approaching the Eddington luminosity. Phinney (1989) estimates that, in the Galactic center, this happens about once every 10^4 yr, with the accompanying luminosity impulse lasting 10–100 yr. This settling may affect the X-ray appearance of the Galactic center by producing, for example, a large flux that can be scattered in the 6.4-keV iron line (Koyama et al 1996), thereby giving rise to an iron-line glow persisting for hundreds of years after the initial, brief, production event. Another potential consequence of stellar disruption arises directly from the ejecta created during the tidal disruption event. Khokhlov & Melia (1996) note that half the stellar mass is asymmetrically ejected with an energy well in excess of that of a supernova, and so they propose that the energetic and asymmetric Sgr A East shell source was created in this manner. This stellar snacking will not typically have much effect on the limit cycle in which Galactic center gas is engaged; however, if it is timed to occur within the $\lesssim 10^4$-year interval during which gas is descending at a large rate upon the nucleus, then the pulse of radiation (and perhaps matter, in the form of a jet or a wind) can play a pivotal role in inducing star formation within the central parsec and controlling the rate at which gas accumulates onto the central black hole.

Finally, we remark that a profound shift of emphasis has taken place in Galactic Center research since the time of Oort's (1977) review. That review stressed outflows and violent ejections from the Galactic nucleus, whereas we have traced the thread of inflow and accretion. The recognition of the bar-like nature of the central potential has dramatically reduced the need for explosive origin models for observed kinematic structures. Still, winds and jets remain an important part of the story of Galactic center activity (e.g. Sofue 1994), although the ultimate driver for that activity is the far larger amount of mass flowing toward the Galactic center.

ACKNOWLEDGMENTS

We thank our many colleagues who avidly contributed with discussions, unpublished data, preparation of figures, and/or readings of the preliminary manuscript:

R Blandford, J Carlstrom, JA Davidson, D Dowell, I Gatley, RH Hildebrand, JH Lacy, DC Lis, N Mastrodemos, S Phinney, DL Shupe, S Stolovy, K Uchida, and MCH Wright. We also appreciate little Nicholas Serabyn's unrelenting interest in the flavor of the manuscript.

Literature Cited

Abramowicz MA, Chen X, Kato S, Lasota J-P, Regev O. 1995. *Ap. J. Lett.* 438:L37–39

Aitken DK, Briggs GP, Roche PF, Bailey JA, Hough JH. 1986. *MNRAS* 218:363–84

Aitken DK, Smith CH, Gezari D, McCaughrean M, Roche PF. 1991. *Ap. J.* 380:419–28

Alberdi A, Lara L, Marcaide JM, Elosegui P, Shapiro II, et al. 1993. *Astron. Astrophys.* 277:L1–4

Allen DA. 1987. See Backer 1987, pp. 1–7

Allen DA. 1994. See Genzel & Harris 1994, pp. 293–304

Allen DA, Hyland AR, Hillier DJ. 1990. *MNRAS* 244:706–13

Altenhoff WJ, Downes D, Pauls T, Schraml J. 1978. *Astron. Astrophys. Suppl.* 35:23–54

Anantharamaiah KR, Pedlar A, Ekers RD, Goss WM. 1991. *MNRAS* 249:262–81

Armstrong JT, Barrett AH. 1985. *Ap. J. Suppl.* 57:535–70

Arons J, Max CE. 1975. *Ap. J. Lett.* 196:L77–81

Athanassoula E. 1992. *MNRAS* 259:345–64

Backer DC, ed. 1987. *AIP Conf. Proc. No. 155, The Galactic Center.* New York: AIP. 204 pp.

Backer DC. 1994. See Genzel & Harris 1994, pp. 403–10

Backer DC. 1996. See Blitz & Teuben 1996. p. 193

Backer DC, Zensus JA, Kellerman KI, Reid M, Moran JM, Lo KY. 1993. *Science* 262:1414–16

Baldwin JE, Lynden-Bell D, Sancisi R. 1980. *MNRAS* 193:313–19

Bally J, Stark AA, Wilson RW, Henkel C. 1987. *Ap. J. Suppl.* 65:13–82

Bally J, Stark AA, Wilson RW, Henkel C. 1988. *Ap. J.* 324:223–47

Bally J, Yusef-Zadeh F, Hollis JM. 1989. See Morris 1989, pp. 189–93

Bania TM. 1977. *Ap. J.* 216:381–403

Becklin EE, Gatley I, Werner MW. 1982. *Ap. J.* 258:135–42

Benford G. 1988. *Ap. J.* 333:735–42

Bennett CL, Fixsen DJ, Hinshaw G, Mather JC, Moseley SH, et al. 1994. *Ap. J.* 434:587–98

Biermann PL, Duschl WJ, von Linden S. 1993. *Astron. Astrophys.* 275:153–57

Binney J. 1978. *MNRAS* 183:779–97

Binney J. 1994. See Genzel & Harris 1994, pp. 75–86

Binney J, Gerhard OE, Stark AA, Bally J, Uchida KI. 1991. *MNRAS* 252:210–18

Blandford RD, Königl A. 1979. *Ap. J.* 232:34–48

Blitz L. 1994. In *Physics of the Gaseous and Stellar Disks of the Galaxy,* ed. I King, pp 1–13. San Francisco: ASP

Blitz L, Teuben PJ ed. 1996. *IAU Symp. No. 169, Unsolved Problems of the Milky Way.* Dordrecht: Kluwer. In press

Blitz L, Binney J, Lo KY, Bally J, Ho PTP. 1993. *Nature* 361:417–24

Blitz L, Bloemen JBGM, Hermsen W, Bania TM. 1985. *Astron. Astrophys.* 143:267–73

Blitz L, Spergel DN. 1991a. *Ap. J.* 370:205–24

Blitz L, Spergel DN. 1991b. *Ap. J.* 379:631–38

Bloemen JBGM, Deul ER, Thaddeus P. 1990. *Astron. Astrophys.* 233:437–55

Bloemen JBGM, Strong AW, Blitz L, Cohen RS, Dame TM, et al. 1986. *Astron. Astrophys.* 154:25–41

Blum RD. 1995. *Ap. J. Lett.* 444:L89–92

Blum RD, DePoy DL, Sellgren K. 1995a. *Ap. J.* 441:603–16

Blum RD, Sellgren K, DePoy DL. 1995b. *Ap. J. Lett.* 440:L17–20

Bottema R, Sanders RH. 1986. *Astron. Astrophys.* 158:297–304

Boyce PJ, Cohen RJ. 1994. *Astron. Astrophys. Suppl.* 107:563–647

Brown RL, Liszt HS. 1984. *Annu. Rev. Astron. Astrophys.* 22:223–65

Burton WB. 1992. In *The Galactic Interstellar Medium,* ed. D Pfenniger, P Bartholdi, pp.

1–155. Berlin: Springer-Verlag. 400 pp.
Burton WB, Liszt HS. 1978. *Ap. J.* 228:815–42
Burton WB, Liszt HS. 1993. *Astron. Astrophys.* 274:765–74
Chudnovsky EM, Field GB, Spergel DN, Vilenkin A. 1986. *Phys. Rev. D* 34:944–50
Contopoulos G, Mertzanides C. 1977. *Astron. Astrophys.* 61:477–85
Conti PS. 1991. *Ap. J.* 377:115–25
Cook WR, Grunsfeld JM, Heindl WA, Palmer DM, Prince TA, et al. 1991. *Ap. J. Lett.* 372:L75–78
Cotera AS. 1995. *Stellar ionization of the thermal emission regions in the Galactic center.* PhD thesis. Stanford Univ. 165 pp.
Cotera AS, Erickson EF, Allen DA, Colgan SWJ, Simpson JP, Burton MG. 1994. See Genzel & Harris 1994, pp. 217–21
Cotera AS, Erickson EF, Allen DA, Colgan SWJ, Simpson JP, Burton MG. 1995. In *Airborne Astronomy Symp. on the Galactic Ecosystem: From Gas to Stars to Dust,* ed. MR Haas, JA Davidson, EF Erickson, 73:511–14. San Francisco: ASP. 737 pp.
Cotera AS, Erickson EF, Colgan SWJ, Simpson JP, Allen DA, Burton MG. 1996. *Ap. J.* In press.
Cox P, Laureijs R. 1989. See Morris 1989, pp. 121–28
Dame TM, Ungerechts H, Cohen RS, De Geus EJ, Grenier IA, et al. 1987. *Ap. J.* 322:706–20
Das M, Jog CJ. 1995. *Ap. J.* 451:167–75
Davidson JA. 1996. In *Polarimetry of the Interstellar Medium,* ed. WG Roberge, DCB Whittet. San Francisco: ASP. In press
Davidson JA, Werner MW, Wu X, Lester DF, Harvey PM, et al. 1992. *Ap. J.* 387:189–211
de Bruyn AG. 1976. *Astron. Astrophys.* 52:439–47
de Vaucouleurs G. 1964. In *IAU-URSI Symp. 20: The Galaxy and the Magellanic Clouds,* ed. FJ Kerr, AW Rodgers, pp. 88–91, 195–199. Canberra: Aust. Acad. Sci.
Dent WRF, Matthews HE, Wade R, Duncan WD. 1993. *Ap. J.* 410:650–62
Diehl R. 1994. See Genzel & Harris 1994, pp. 3–12
Diehl R, Dupraz C, Bennett K, Bloemen H, Hermsen W, et al. 1995. *Astron. Astrophys.* 298:445–60
Duschl WJ. 1988. *MNRAS* 240:219–23
Duschl WJ, Lesch H. 1994. *Astron. Astrophys.* 286:431–6
Dwek E, Arendt RG, Hauser MG, Kelsall T, Lisse CM, et al. 1995. *Ap. J.* 445:716–30
Dyson JE, Hartquist TW. 1994. *MNRAS* 269:447–50
Echevarria L, Morris M. 1996. In preparation
Eckart A, Genzel R, Hofmann R, Sams BJ, Tacconi-Garman LE. 1993. *Ap. J. Lett.* 407:L77–80
Eckart A, Genzel R, Hofmann R, Sams BJ, Tacconi-Garman LE. 1995. *Ap. J. Lett.* 445:L23–26
Eckart A, Genzel R, Krabbe A, Hofmann R, van der Werf PP, Drapatz S. 1992. *Nature* 355:526–28
Ekers RD, van Gorkom JH, Schwarz UJ, Goss WM. 1983. *Astron. Astrophys.* 122:143–50
Falcke H. 1996. See Blitz & Teuben 1996. p. 169
Falcke H, Heinrich OM. 1994. *Astron. Astrophys.* 292:430–38
Falcke H, Mannheim K, Biermann PL. 1993. *Astron. Astrophys.* 278:L1–4
Figer DF. 1995. *A search for emission-line stars near the Galactic center.* PhD thesis. Univ. Calif., Los Angeles. 214 pp.
Figer DF, McLean I, Morris M. 1995. *Ap. J. Lett.* 447:L29–32
Fleck RC. 1980. *Ap. J.* 242:1019–22
Frail DA, Diamond PJ, Cordes JM, Van Langevelde HJ. 1994. *Ap. J. Lett.* 427:L43–6
Fridman AM, Khoruzhii OV, Lyachovich VV, Ozernoy L, Blitz L. 1994. In *Physics of the Gaseous and Stellar Disks of the Galaxy,* ed. I King, pp. 285–303. San Francisco: ASP
Frogel JA. 1988. *Annu. Rev. Astron. Astrophys.* 26:51–92
Fukui Y, Churchwell E. 1987. See Backer 1987, pp. 110–13
Gatley I, Jones TJ, Hyland AR, Beattie DH, Lee TJ. 1984. *MNRAS* 210:565–75
Gatley I, Jones TJ, Hyland AR, Wade R, Geballe TR, Krisciunas TL. 1986. *MNRAS* 222:299–306
Gatley I, Merrill M. 1994. In *Infrared Astronomy with Arrays,* ed. IS McLean, pp. 551–60. Dordrecht: Kluwer
Gaume RA, Claussen MJ, De Pree CG, Goss WM, Mehringer DM. 1995. *Ap. J.* 449:663–73
Genzel R, Harris AI, eds. 1994. *Nuclei of Normal Galaxies: Lessons from the Galactic Center.* NATO ASI Ser. C. Dordrecht: Kluwer. 499 pp.
Genzel R, Hollenbach D, Townes CH. 1994. *Rep. Prog. Phys.* 57:417–79
Genzel R, Stacey GJ, Harris AI, Townes CH, Geis N, et al. 1990. *Ap. J.* 356:160–73
Genzel R, Townes CH. 1987. *Annu. Rev. Astron. Astrophys.* 25:377–423
Gezari D. 1992. In *The Center, Bulge, and Disk of the Milky Way,* ed. L Blitz, pp. 23–46. Dordrecht: Kluwer
Gezari D, Ozernoy L, Varosi F, McCreight C, Joyce R. 1994. See Genzel & Harris 1994, pp. 427–30
Gerhard OE. 1992. *Rev. Mod. Astron.* 5:174–87
Gerhard OE. 1996. See Blitz & Teuben 1996. p.

79
Glass IS, Moneti A, Moorwood AFM. 1990. *MNRAS* 242:55–58p
Goldwurm A, Cordier B, Paul J, Ballet J, Bouchet L, et al. 1994. *Nature* 371:589–62
Goss WM, Schwarz UJ, van Gorkom JH, Ekers RD. 1985. *MNRAS* 215:69–73p
Gray AD, Cram LE, Ekers RD, Goss WM. 1991. *Nature* 353:237–39
Gray AD, Nicholls J, Ekers RD, Cram LE. 1995. *Ap. J.* 448:164–78
Güsten R. 1989. See Morris 1989, pp. 89–106
Güsten R, Downes D. 1980. *Astron. Astrophys.* 87:6–19
Güsten R, Downes D. 1981. *Astron. Astrophys.* 99:27–30
Güsten R, Genzel R, Wright MCH, Jaffe DT, Stutzki J, Harris AI. 1987. *Ap. J.* 318:124–38
Güsten R, Walmsley CM, Pauls T. 1981. *Astron. Astrophys.* 103:197–206
Gwinn CR, Danen RM, Middleditch J, Ozernoy LM, Tran TK. 1991. *Ap. J. Lett.* 381:L43–46
Haller JW, Rieke MJ, Rieke GH, Tamblyn P, Close L, Melia F. 1996. *Ap. J.* 456:194–205
Han C, Gould A. 1995. *Ap. J.* 449:521–26
Handa T, Sofue Y, Nakai N, Hirabayashi H, Inoue M. 1987. *Publ. Astron. Soc. Jpn.* 39:709–53
Hasegawa T, Sato F, Whiteoak JB, Miyawaki R. 1994. *Ap. J. Lett.* 429:L77–80
Hauser MG, Silverberg RF, Stier MT, Kelsall T, Gezari DY, et al. 1984. *Ap. J.* 285:74–88
Haynes RF, Stewart RT, Gray AD, Reich W, Reich P, Mebold U. 1992. *Astron. Astrophys.* 264:500–12
Heiligman GM. 1987. *Ap. J.* 314:747–65
Herbst TM, Beckwith SVW, Forrest WJ, Pipher JL. 1993a. *Astron. J.* 105:956–70
Herbst TM, Beckwith SVW, Shure M. 1993b. *Ap. J. Lett.* 411:L21–24
Heyvaerts J, Norman C, Pudritz RE. 1988. *Ap. J.* 330:718–36
Hildebrand R. 1988. *Q. J. R. Astron. Soc.* 29:327–51
Hildebrand RH, Davidson J, Dotson J, Figer DF, Novak G, et al. 1993. *Ap. J.* 417:565–71
Hildebrand RH, Gonatas DP, Platt SR, Wu XD, Davidson JA, et al. 1990. *Ap. J.* 362:114–19
Hills JG. 1975. *Nature* 254:295–98
Ho PTP. 1994. See Genzel & Harris 1994, pp. 149–60
Ho PTP, Ho LC, Szczepanski JC, Jackson JM, Armstrong JT, Barrett AH. 1991. *Nature* 350:309–12
Hollywood JM, Melia F. 1995. *Ap. J. Lett.* 443:L17–20
Holt SS, McCray R. 1982. *Annu. Rev. Astron. Astrophys.* 20:323–65
Hüttemeister S, Wilson TL, Bania TM, Martin-Pintado J. 1993. *Astron. Astrophys.* 280:255–67
Inoue M, Fomalont E, Tsuboi M, Yusef-Zadeh F, Morris M, et al. 1989. See Morris 1989, pp. 269–74
Izumiura H, Deguchi S, Hashimoto O, Nakada Y, Onaka T, et al. 1995. *Ap. J.* 453:837–63
Jackson JM, Geis N, Genzel R, Harris AI, Madden S, et al. 1993. *Ap. J.* 402:173–84
Jackson JM, Heyer MH, Paglione TAD, Bolatto AD. 1996. *Ap. J. Lett.* 456:L91–95
Jauncey DL, Tzioumis AK, Preston RA, Meier DL, Batchelor R. 1989. *Astron. J.* 98:44–48
Jenkins A, Binney J. 1994. *MNRAS* 270:703–19
Kaifu N, Kato T, Iguchi T. 1972. *Nature Phys. Sci.* 238:105–7
Kawai N, Fenimore EE, Middleditch J, Cruddace RG, Fritz GG, et al. 1988. *Ap. J.* 330:130–41
Kellermann KI, Pauliny-Toth IIK. 1981. *Annu. Rev. Astron. Astrophys.* 19:373–410
Kenney JDP, Wilson CD, Scoville NZ, Devereux NA, Young JS. 1992. *Ap. J. Lett.* 395:L79–82
Khokhlov A, Melia F. 1996. *Ap. J. Lett.* 457:L61–64
Killeen NEB, Lo KY. 1989. See Morris 1989, pp. 453–55
Killeen NEB, Lo KY, Crutcher R. 1992. *Ap. J.* 385:585–603
Koyama K. 1996. See Blitz & Teuben 1996. p. 287
Koyama K, Awaki H, Kunieda H, Takano S, Tawara Y, et al. 1989. *Nature* 339:603–05
Koyama K, Maeda Y, Sonobe T, Takeshima T, Tanaka Y, Yamauchi S. 1996. *Publ. Astron. Soc. Jpn.* In press
Krabbe A, Genzel R, Drapatz S, Rotaciuc V. 1991. *Ap. J. Lett.* 382:L19–22
Krabbe A, Genzel R, Eckart A, Najarro F, Lutz D, et al. 1995. *Ap. J. Lett.* 447:L95–99
Krichbaum TP, Schalinski CJ, Witzel A, Standke KJ, Graham DA, et al. 1994. See Genzel & Harris 1994, pp. 411–14
Krichbaum TP, Zensus JA, Witzel A, Mezger PG, Standke KJ, et al. 1993. *Astron. Astrophys.* 274:L37–40
Lacey CG, Fall SM. 1985. *Ap. J.* 290:154–70
Lacy JH. 1994. See Genzel & Harris 1994, pp. 165–74
Lacy JH, Achtermann JM, Bruce DE. 1989. See Morris 1989, pp. 523–24
Lacy JH, Achtermann JM, Serabyn E. 1991. *Ap. J. Lett.* 380:L71–74
Lacy JH, Townes CH, Geballe TR, Hollenbach DJ. 1980. *Ap. J.* 241:132–46
Lacy JH, Townes CH, Hollenbach DJ. 1982. *Ap. J.* 262:120–34
Lazarian A. 1994. *MNRAS* 268:713–23
Lee HM. 1994. See Genzel & Harris 1994, pp. 335–42

Lee HM. 1995. *MNRAS* 272:605–17
Lesch H, Crusius A, Schlickeiser R, Wielebinski R. 1989. *Astron. Astrophys.* 217:99–107
Lesch H, Reich W. 1992. *Astron. Astrophys.* 264:493–99
Lester DF, Bregman JD, Witteborn FC, Rank DM, Dinerstein HL. 1981. *Ap. J.* 248:524–27
Lester DF, Dinerstein HL, Werner MW, Watson DM, Genzel R, Storey JWV. 1987. *Ap. J.* 320:573–85
Levine D. 1995. *A survey of 22-GHz water masers within 2.25° of the Galactic center.* PhD thesis. Univ. Calif., Los Angeles. 290 pp.
Libonate S, Pipher JL, Forrest WJ, Ashby MLN. 1995. *Ap. J.* 439:202–23
Lindqvist M, Habing HJ, Winnberg A. 1992. *Astron. Astrophys.* 259:118–27
Lis DC, Carlstrom JE. 1994. *Ap. J.* 424:189–99
Lis DC, Menten KM, Serabyn E, Zylka R. 1994. *Ap. J. Lett.* 423:L39–42
Liszt HS. 1992. *Ap. J. Suppl.* 82:495–503
Liszt HS. 1996. See Blitz & Teuben 1996. p. 297
Liszt HS, Burton WB. 1978. *Ap. J.* 226:790–816
Liszt HS, Burton WB. 1980. *Ap. J.* 236:779–97
Liszt HS, Burton WB. 1993. *Ap. J. Lett.* 407:L25–28
Liszt HS, Spiker RW. 1995. *Ap. J. Suppl.* 98:259–70
Liu T, Becklin EE, Henry JP, Simons D. 1993. *Astron. J.* 106:1484–89
Lo KY. 1989. See Morris 1989, pp. 527–34
Lo KY. 1994. See Genzel & Harris 1994, pp. 395–402
Lo KY, Backer DC, Kellermann KI, Reid M, Zhao JH, et al. 1993. *Nature* 362:38–40
Lo KY, Claussen MJ. 1983. *Nature* 306:647–51
Lutz D, Krabbe A, Genzel R. 1993. *Ap. J.* 418:244–54
Lutz D, Krabbe A, Genzel R, Blietz M, Drapatz S, van der Werf PP. 1994. See Genzel & Harris 1994, pp. 373–76
Maloney PR, Hollenbach DJ, Townes CH. 1992. *Ap. J.* 401:559–73
Marcaide JM, Alberdi A, Bartel N, Clark TA, Corey BE, et al. 1992. *Astron. Astrophys.* 258:295–301
Markevitch M, Sunyaev RA, Pavlinsky M. 1993. *Nature* 364:40–42
Marr JM, Rudolph AL, Pauls TA, Wright MC, Backer DC. 1992. *Ap. J. Lett.* 400:L29–32
Marr JM, Wright MC, Backer DC. 1993. *Ap. J.* 411:667–73
Marshall J, Lasenby AN. 1994a. See Genzel & Harris 1994, pp. 175–77
Marshall J, Lasenby AN. 1994b. *MNRAS* 269:619–25
Marshall J, Lasenby AN, Yusef-Zadeh F. 1995. *MNRAS* 274:519–22
Masset F, Tagger M. 1995. *Astron. Astrophys.* In press
Masset F, Tagger M. 1996. *Astron. Astrophys.* Submitted
Mastichiadis A, Ozernoy LM. 1994. *Ap. J.* 426:599–603
Mathis JS. 1990. *Annu. Rev. Astron. Astrophys.* 28:37–70
Mauersberger R, Henkel C, Wilson TL, Walmsley CM. 1986. *Astron. Astrophys.* 162:199–210
McGinn MT, Sellgren K, Becklin EE, Hall DNB. 1989. *Ap. J.* 338:824–40
Meglicki Z, Wickramasinghe D, Dewar RL. 1994. *MNRAS* 272:717–29
Mehringer DM, Palmer P, Goss WM, Yusef-Zadeh F. 1993. *Ap. J.* 412:684–95
Melia F. 1994a. *Ap. J.* 426:577–85
Melia F. 1994b. See Genzel & Harris 1994, pp. 441–48
Mezger PG. 1978. *Astron. Astrophys.* 70:565–74
Mezger PG, Pauls T. 1979. In *IAU Symp. No. 84, The Large-scale Characteristics of the Galaxy,* ed. WB Burton, pp. 357–66. Dordrecht: Reidel
Mezger PG, Zylka R, Salter CJ, Wink JE, Chini R, et al. 1989. *Astron. Astrophys.* 209:337–48
Miller RH, Smith BF. 1992. *Ap. J.* 393:508–15
Moneti A, Glass I, Moorwood A. 1992. *MNRAS* 258:705–14
Moneti A, Glass I, Moorwood A. 1994. *MNRAS* 268:194–202
Morris M, ed. 1989. *IAU Symp. No. 136, The Center of the Galaxy.* Dordrecht: Kluwer. 661 pp.
Morris M. 1990. In *IAU Symp. No. 140, Galactic and Extragalactic Magnetic Fields,* ed. R Beck, P Kronberg, R Wielebinski, pp. 361–67. Dordrecht: Kluwer
Morris M. 1993. *Ap. J.* 408:496–506
Morris M. 1994. See Genzel & Harris 1994, pp. 185–98
Morris M. 1996. See Blitz & Teuben 1996. p. 247
Morris M, Davidson JA, Werner MW. 1995. In *Airborne Astronomy Symposium on the Galactic Ecosystem,* ed. MR Haas, JA Davidson, EF Erickson, ASP Conf. Ser. 73:477–88. San Francisco: ASP
Morris M, Davidson JA, Werner MW, Dotson J, Figer DF, et al. 1992. *Ap. J. Lett.* 399:L63–66
Morris M, Davidson JA, Werner MW, Hildebrand RH, Dotson J, et al. 1996a,b. In preparation
Morris M, Polish N, Zuckerman B, Kaifu N. 1983. *Astron. J.* 88:1228–35

Morris M, Yusef-Zadeh F. 1985. *Astron. J.* 90:2511–13
Morris M, Yusef-Zadeh F. 1987. See Backer 1987, pp. 127–32
Morris M, Yusef-Zadeh F. 1989. *Ap. J.* 343:703–12
Mulder WA, Liem BT. 1986. *Astron. Astrophys.* 157:148–58
Nagata T, Woodward CE, Shure M, Kobayashi N. 1995. *Astron. J.* 109:1676–81
Nagata T, Woodward CE, Shure M, Pipher JL, Okuda H. 1990. *Ap. J.* 351:83–88
Nakada Y, Deguchi S, Hashimoto O, Izumiura H, Onaka T, et al. 1991. *Nature* 353:140–41
Nakagawa T, Doi Y, Yui YY, Okuda H, Mochizuki K, et al. 1996. *Ap. J. Lett.* In press
Narayan R, Yi I. 1995. *Ap. J.* 452:710–35
Narayan R, Yi I, Mahadevan R. 1995. *Nature* 374:623–25
Nottingham MR, Skinner GK, Willmore AP, Borozdin KN, Churazov E, Sunyaev RA. 1993. *Astron. Astrophys. Suppl.* 97:165–67
Odenwald SF, Fazio GG. 1984. *Ap. J.* 283:601–14
Oka T, Hasegawa T, Handa T, Hayashi M, Sakamoto S. 1996. *Ap. J.* 460:334–342
Okuda H, Shibai H, Nakagawa T, Matsuhara H, Kobayashi Y, Kaifu N, Nagata T, Gatley I, Geballe TR. 1990. *Ap. J.* 351:89–97
Oort J. 1977. *Annu. Rev. Astron. Astrophys.*, 15:295–362
Owen FN, Spangler SR, Cotton WD. 1980. *Astron. J.* 85:351–62
Ozernoi LM. 1979. In *Large-Scale Characteristics of the Galaxy,* ed. WB Burton, pp. 395–400. Dordrecht: Reidel
Ozernoy LM. 1989. See Morris 1989, pp. 555–66
Ozernoy LM. 1993. In *Back to the Galaxy,* ed. SS Holt, F Verter, pp. 69–72. New York: AIP
Ozernoy LM. 1994. See Genzel & Harris 1994, pp. 431–40
Paczyński B, Stanek KZ, Udalski A, Szymański M, Kaluzny J, et al. 1994. *Ap. J. Lett.* 435:L113–16
Pak S, Jaffe DT, Keller LD. 1996. *Ap. J. Lett.* 457:L43–46
Pauls T, Johnston KJ, Wilson TL, Marr JM, Rudolph A. 1993. *Ap. J. Lett.* 403:L13–16
Pavlinsky MN, Grebenev SA, Sunyaev RA. 1994. *Ap. J.* 425:110–21
Pedlar A, Anantharamaiah KR, Ekers RD, Goss WM, van Gorkom JH, et al. 1989. *Ap. J.* 342:769–84
Peters WL. 1975. *Ap. J.* 195:617–29
Phinney ES. 1989. See Morris 1989, pp. 543–53
Pietsch W. 1994. See Genzel & Harris 1994, pp. 13–19
Plante RL, Lo KY, Crutcher RM. 1995. *Ap. J. Lett.* 445:L113–16
Poglitsch A, Stacey GJ, Geis N, Haggerty M, Jackson J, et al. 1991. *Ap. J. Lett.* 374:L33–36
Predehl P, Trümper J. 1994. *Astron. Astrophys.* 290:L29–32
Quinn PJ, Sussman GJ. 1985. *Ap. J.* 288:377–84
Rees MJ. 1988. *Nature* 333:523–28
Rees MJ, Begelman MC, Blandford RD, Phinney ES. 1982. *Nature* 295:17–21
Reich W. 1994. See Genzel & Harris 1994, pp 55–62
Reich W, Sofue Y, Fürst E. 1987. *Publ. Astron. Soc. Jpn.* 39:573–87
Reid M. 1993. *Annu. Rev. Astron. Astrophys.* 31:345–72
Reynolds SP, McKee CF. 1980. *Ap. J.* 239:893–97
Rich RM. 1993. In *IAU Symp. No. 153. Galactic Bulges,* ed. HJ Habing, T DeJonghe, pp. 19. Dordrecht: Kluwer
Riegler GR, Blandford RD, eds. 1982. *AIP Conf. Proc. No. 83, The Galactic Center.* New York: AIP. 216 pp.
Rieke GH, Lebofsky MJ. 1982. See Riegler & Blandford 1982, pp. 194–203
Rieke GH, Lebofsky MJ. 1985. *Ap. J.* 288:618–21
Rieke GH, Lebofsky MJ. 1987. See Backer 1987, pp. 91–94
Rieke GH, Rieke MJ. 1988. *Ap. J. Lett.* 330:L33–37
Rieke GH, Rieke MJ. 1989. *Ap. J. Lett.* 344:L5–8
Rieke GH, Rieke MJ. 1994. See Genzel & Harris 1994, pp. 283-91
Rieke GH, Rieke MJ, Paul AE. 1989. *Ap. J.* 336:752–61
Rieke MJ. 1993. In *Back to the Galaxy,* ed. SS Holt, F Verter, pp. 37–42. New York: AIP
Roberge WG, Hanany S, Messinger DW. 1995. *Ap. J.* 453:238–55
Roberts DA, Goss WM. 1993. *Ap. J. Suppl.* 86:113–52
Rogers AEE, Doeleman S, Wright MCH, Bower GC, Backer DC, et al. 1994. *Ap. J. Lett.* 434:L59–62
Rosso F, Pelletier G. 1993. *Astron. Astrophys.* 270:416–25
Saito M. 1990. *Publ. Astron. Soc. Jpn.* 42:19–38
Sanders RH. 1981. *Nature* 294:427–29
Sanders RH. 1989. See Morris 1989, pp. 77–87
Schwarz UJ, Lasenby J. 1990. In *IAU Symp. No. 170, Galactic and Intergalactic Magnetic Fields,* ed. R Beck, PP Kronberg, R Wielebinski, pp. 383–84, Dordrecht: Kluwer
Scoville NZ. 1972. *Ap. J. Lett.* 175:L127–32
Scoville NZ, Solomon PM, Jefferts KB. 1974. *Ap. J. Lett.* 187:L63–66
Scoville NZ, Good JC. 1989. *Ap. J.* 339:149–62
Scoville NZ, Sanders DB. 1987. In *Interstellar*

Processes, ed. DJ Hollenbach, HA Thronson, pp. 21–50. Dordrecht: Reidel
Sellgren K, Hall DN, Kleinmann SG, Scoville NZ. 1987. *Ap. J.* 317:881–91
Sellgren K, McGinn MT, Becklin EE, Hall DNB. 1990. *Ap. J.* 359:112–20
Sellwood JA. 1993. In *Back to the Galaxy,* ed. SS Holt, F Verter, pp. 133–36. New York: AIP
Serabyn E. 1984. PhD dissertation. Univ. Calif., Berkeley
Serabyn E. 1989. See Morris 1989, pp. 437–41
Serabyn E. 1996. See Blitz & Teuben 1996. p. 263
Serabyn E, Carlstrom JE, Scoville NZ. 1992a. *Ap. J. Lett.* 401:L87–88
Serabyn E, Carlstrom JE, Scoville NZ, Lis DC, Lacy JH. 1996a. In prep.
Serabyn E, Güsten R. 1987. *Astron. Astrophys* 184:133–43
Serabyn E, Güsten R. 1991. *Astron. Astrophys* 242:376–87
Serabyn E, Güsten R, Walmsley CM, Wink JE, Zylka R. 1986. *Astron. Astrophys* 169:85–94
Serabyn E, Keene J, Lis DC, Phillips TG. 1994 *Ap. J. Lett.* 424:L95–98
Serabyn E, Keene J, Lis DC, Phillips TG. 1996b. In prep.
Serabyn E, Lacy JH. 1985. *Ap. J.* 293:445–58
Serabyn E, Lacy JH, Achtermann JM. 1991. *Ap. J.* 378:557–64
Serabyn E, Lacy JH, Achtermann JM. 1992b. *Ap. J.* 395:166–73
Serabyn E, Lacy JH, Townes CH, Bharat R. 1988. *Ap. J.* 326:171–85
Serabyn E, Morris M. 1994. *Ap. J. Lett.* 424:L91–94
Serabyn E, Shupe D. 1996. In prep.
Shibata K, Uchida Y. 1987. *Publ. Astron. Soc. Jpn.* 39:559–71
Shields JC, Ferland GJ. 1984. *Ap. J.* 430:236–51
Shlosman I, Frank J, Begelman MC. 1989. *Nature* 338:45–47
Sjouwerman LO, van Langevelde HJ. 1996. *Ap. J. Lett.* 461:L41–44
Skinner GK. 1989. See Morris 1989, pp. 567–80
Smith CH, Aitken DK, Roche PF. 1990. *MNRAS* 246:1–9
Sofue Y. 1994. See Genzel & Harris 1994, pp. 43–54
Sofue Y. 1995a. *Publ. Astron. Soc. Jpn.* 47:527–49
Sofue Y. 1995b. *Publ. Astron. Soc. Jpn.* 47:551–59
Sofue Y, Fujimoto M. 1987. *Publ. Astron. Soc. Jpn.* 39:843–48
Sofue Y, Murata Y, Reich W. 1992. *Publ. Astron. Soc. Jpn.* 44:367–72
Sofue Y, Reich W, Inoue M, Seiradakis JH. 1987. *Publ. Astron. Soc. Jpn.* 39:95–107
Sofue Y, Reich W, Reich P. 1989. *Ap. J. Lett.* 341:L47–49
Sparke LS. 1984. *MNRAS* 211:911–26
Spergel DN, Blitz L. 1992. *Nature* 357:665–67
Staguhn J, Stutzki J, Yusef-Zadeh F, Uchida KI. 1996. In preparation
Stanek KZ, Mateo M, Udalski A, Szymański J, Kaluzny J, Kubiak M. 1994. *Ap. J. Lett.* 429:L73–6
Stark AA, Bally J, Wilson RW, Pound MW. 1989. See Morris 1989, pp. 129–33
Stark AA, Bania TM. 1986. *Ap. J. Lett.* 306:L17–20
Stark AA, Gerhard OE, Binney J, Bally J. 1991. *MNRAS* 248:14–17p
Stolovy S, Hayward T, Herter T. 1996. In prep.
Sunyaev R, Markevitch M, Pavlinsky M. 1993. *Ap. J.* 407:606–10
Sutton EC, Danchi WC, Jaminet PA, Masson CR. 1990. *Ap. J.* 348:503–14
Syunyaev R, Pavlinskii M, Gil'fanov M, Churazzov E, Grebenev S, et al. 1991. *Sov. Astron. Lett.* 17:42–45
Tagger, M, Athanassoula, E. 1990. In *Dynamics of Galaxies and Their Molecular Cloud Distributions,* ed. F Combes, F Casoli, pp. 105–07. Dordrecht: Kluwer
Tamblyn P, Rieke GH. 1993. *Ap. J.* 414:573–79
Tamblyn P, Rieke GH, Hanson MM, Close LM, McCarthy DW Jr, Rieke MJ. 1996. *Ap. J.* 456:206–16
Taylor GB, Morris M, Schulman E. 1993. *Astron. J.* 106:1978–86
Telesco CM, Davidson JA, Werner MW. 1996. *Ap. J.* 456:541–56
Terlevich R, Melnick J. 1985. *MNRAS* 213:841–56
Tsuboi M, Handa T, Inoue M, Inatani J, Ukita N. 1989. See Morris 1989, pp. 135–40
Tsuboi M, Inoue M, Handa T, Tabara H, Kato T, et al. 1986. *Astron. J.* 92:818–24
Tsuboi M, Kawabata T, Kasuga T, Handa T, Kato T. 1995. *Publ. Astron. Soc. Jpn.* 47:829–36
Turner JL, Hurt RL, Hudson DY. 1993. *Ap. J. Lett.* 413:L19–22
Uchida KI. 1993. *The manifestations of energetic activity in the Galactic center.* PhD thesis. Univ. Calif., Los Angeles. 258 pp.
Uchida KI, Güsten R. 1995. *Astron. Astrophys.* 298:473–81
Uchida KI, Morris M, Bally J. 1994a. See Genzel & Harris 1994, pp. 99–103
Uchida KI, Morris M, Bally J. 1996a. In preparation
Uchida KI, Morris M, Güsten R, Serabyn E. 1996b. *Ap. J.* 462:768–76
Uchida KI, Morris M, Serabyn E, Bally J. 1994b. *Ap. J.* 421:505–16

Uchida Y, Shibata K, Sofue Y. 1985. *Nature* 317:699–701
Umemura S, Iki K, Shibata K, Sofue Y. 1988. *Publ. Astron. Soc. Jpn.* 40:25–46
von Linden S, Duschl WJ, Biermann PL. 1993. *Astron. Astrophys.* 269:169–74
Wada K, Taniguchi Y, Habe A, Hasegawa T. 1994. *Ap. J. Lett.* 437:L123–25
Walmsley CM, Güsten R, Angerhofer P, Churchwell E, Mundy L. 1986. *Astron. Astrophys.* 155:129–36
Wannier PG. 1989. See Morris 1989, pp. 107–19
Wardle M, Königl A. 1990. *Ap. J.* 362:120–34
Wardle M, Yusef-Zadeh F. 1992. *Nature* 357:308–10
Watson MG, Willingale R, Grindlay JE, Hertz P. 1981. *Ap. J.* 250:142–54
Weiland JL. et al. 1994. *Ap. J. Lett.* 425:L81–84
Weinberg MD. 1992a. *Ap. J.* 384:81–94
Weinberg MD. 1992b. *Ap. J. Lett.* 392:L67–69
Werner MW, Davidson JA, Morris M, Novak G, Platt SR, Hildebrand RH. 1988. *Ap. J.* 333:729–34
Whitelock P, Catchpole R. 1992. In *The Center, Bulge, and Disk of the Milky Way,* ed. L Blitz, pp. 103–10. Dordrecht: Kluwer
Wilson TL, Ruf K, Walmsley CM, Martin RN, Pauls TA, Batrla W. 1982. *Astron. Astrophys.* 115:185–89
Wright EL, Mather JC, Bennet CL, Cheng ES, Shafer RA, et al. 1991. *Ap. J.* 381:200–9
Wright MCH, Genzel R, Güsten R, Jaffe, DT. 1987. See Backer 1987, pp. 133–37
Wright MCH, Backer DC. 1993. *Ap. J.* 417:560–64
Yamauchi S, Kawada M, Koyama K, Kunieda H, Tawara Y, Hatsukade I. 1990. *Ap. J.* 365:532–38
Yamauchi S, Koyama K. 1993. *Ap. J.* 404:620–24
Yusef-Zadeh F. 1989. See Morris 1989, pp. 243–63
Yusef-Zadeh F. 1994. See Genzel & Harris 1994, pp. 355–72
Yusef-Zadeh F, Cotton W, Wardle M, Melia F, Roberts DA. 1994. *Ap. J.* 434:L63–6
Yusef-Zadeh F, Lasenby A, Marshall J. 1993. *Ap. J. Lett.* 410:L27–30
Yusef-Zadeh F, Mehringer DM. 1995. *Ap. J. Lett.* 452:L37–40
Yusef-Zadeh F, Melia F. 1992. *Ap. J. Lett.* 385:L41–44
Yusef-Zadeh F, Morris M. 1987a. *Ap. J.* 320:545–61
Yusef-Zadeh F, Morris M. 1987b. *Astron. J.* 94:1178–84
Yusef-Zadeh F, Morris M. 1987c. *Ap. J.* 322:721–28
Yusef-Zadeh F, Morris M. 1988. *Ap. J.* 329:729–38
Yusef-Zadeh F, Morris M. 1991. *Ap. J. Lett.* 371:L59–62
Yusef-Zadeh F, Morris M, Chance D 1984. *Nature* 310:557–61
Yusef-Zadeh F, Morris M, Ekers RD. 1989. See Morris 1989, pp. 443–51
Yusef-Zadeh F, Morris M, Ekers RD. 1990. *Nature* 348:45–47
Yusef-Zadeh F, Wardle M. 1993. *Ap. J.* 405:584–90
Yusef-Zadeh F, Zhao JH, Goss WM. 1995. *Ap. J.* 442:646–52
Zhao J-H, Ekers RD, Goss WM, Lo KY. 1989. See Morris, pp. 535–41
Zhao J-H, Goss WM, Ho PTP. 1995. *Ap. J.* 450:122–36
Zhao J-H, Goss WM, Lo KY, Ekers RD. 1992. In *Relationships between Active Galatic Nuclei and Starburst Galaxies,* ed. AV Filipenko, pp. 295–99. San Francisco: ASP
Zhao H-S, Spergel DN, Rich RM. 1994. *Astron. J.* 108:2154–63
Zylka R, Mezger PG. 1988. *Astron. Astrophys.* 190:L25–28
Zylka R, Mezger PG, Lesch H. 1992. *Astron. Astrophys.* 261:119–29
Zylka R, Mezger PG, Ward-Thompson D, Duschl WJ, Lesch H. 1995. *Astron. Astrophys.* 297:83–97
Zylka R, Mezger PG, Wink JE. 1990. *Astron. Astrophys.* 234:133–46

Annu. Rev. Astron. Astrophys. 1996. 34:703–47

THEORY OF ACCRETION DISKS II: Application to Observed Systems

D.N.C. Lin

Board of Studies in Astronomy and Astrophysics, University of California, Santa Cruz, California 95064

J.C.B. Papaloizou

School of Mathematical Sciences, Queen Mary & Westfield College, London E1 4NS, United Kingdom

KEY WORDS: star formation, interacting binary stars, AGNs, interstellar medium

ABSTRACT

Accretion disks are important for many astrophysical phenomena, including galactic nuclei, interacting binary stars, and young stellar objects. The central issue in the theory of accretion disks is to identify the dominant mechanisms that regulate angular momentum transfer and mass flow in a variety of contexts. In the first part of this review, we described some recent advances in the study of the physical processes that may be present in accretion disks. Concurrent with these theoretical developments, the arrival of high-resolution astronomical instruments has led to explosive progress on the observational side. In many cases, the study of accretion disks has evolved from their inferred presence based on circumstantial evidence to direct imaging and detailed spectral analyses. Here, we summarize the theoretical interpretation of these data. We review the constraints that may be imposed on the efficiency and nature of angular momentum transfer processes in a variety of astrophysical contexts.

1. INTRODUCTION

Accretion of gas and dust plays an important role not only in the formation of planets, stars, and galaxies but also in the evolution of binary stars and active galactic nuclei. In most cases, the scale lengths associated with the reservoirs from which matter is accreted are much larger than those of the central objects

0066-4146/96/0915-0703$08.00

onto which the accretion occurs. Thus, even a relatively small specific angular momentum associated with the material to be accreted naturally leads to the formation of a centrifugally supported disk.

Energy production occurs if material can lose angular momentum by transporting it outwards so that it can sink lower in the gravitational potential of the central object. In this way the accretion process provides a power source for astrophysical objects. Clearly, an important issue in the theory of accretion disks is the understanding of processes that lead to outward angular momentum transport and therefore enable mass accretion to occur.

In the first part of this review (Papaloizou & Lin 1995a), subsequently referred to as Paper I, we described recent work on physical processes that might lead to effective angular momentum transport in accretion disks. In Section 3 of that paper we outlined recent developments in the study of hydromagnetic winds, which could result in a relatively small amount of matter carrying away most of the angular momentum. In Sections 4, 6, 7, and 8 of Paper I we described ongoing attempts to isolate instabilities that may lead to turbulence and, where possible, reviewed estimates for the likely magnitude of the resulting effective viscosity parameter α. In Section 5 of the first paper we considered angular momentum transport induced by waves excited by external perturbers.

The advent of high-resolution astronomical instruments in the past few years has meant that the direct imaging of accretion disks has now become possible, where previously their existence had to be inferred. In the second part of this review, we begin with a brief summary of the relevant observational evidence for accretion disks around active galactic nuclei, interacting binary stars, and young stellar objects. We primarily focus our discussions on the most compelling observational support for accretion disks and the ranges of physical parameters, such as accretion rate, disk radii, and evolutionary timescale in each of these contexts. These quantitative data are important in relation to disk stability and evolution and, with regard to the construction of models to account for the origin of planetary systems, the formation and evolution of binary stars, and the evolution of stellar systems in the center of galaxies.

In astronomy and astrophysics, observational data not only provide incentives and guidance to theoretical developments but also serve as constraints that can be used to delineate the dominant physical processes involved. For a more comprehensive summary of the observational properties than we can give here, we refer readers to other recent reviews (cf Osterbrock 1993, Horne 1993, Robinson et al 1993, Beckwith & Sargent 1993b, Hartmann et al 1993, Mathieu 1994, Sargent 1995).

In this second part of the review, we also consider instabilities, such as thermal, viscous, and pulsational instabilities that may occur in disks once an angular momentum transport mechanism is in place. Our main purpose here is to discuss how the action of these instabilities, together with a proposed angular momentum transport mechanism, may regulate the flow in an accretion disk and how observation of unsteady disk flow may be used to constrain the theory of accretion disks in the context of active galactic nuclei (AGN), cataclysmic variables (CVs), and young stellar objects (YSOs). Finally, we discuss recent investigations on the interaction between disks and orbiting satellites.

2. OBSERVATIONAL PROPERTIES OF ACCRETION DISKS

2.1 *Active Galactic Nuclei*

A few percent of all galaxies exhibit intense nuclear activity with luminosities reaching 10^{47} erg s^{-1} (Schmidt & Green 1983, Osterbrock 1989). If AGN represent a generic phase in galactic evolution, their abundance relative to spiral galaxies would suggest that they may persist for $\sim 10^8$ yr (Woltjer 1959). During this phase the amount of energy radiated is comparable to the rest-mass energy of the residual gas observed in typical disk galaxies today (Sanders et al 1988).

2.1.1 DISK EXISTENCE The current paradigm for the energy source of AGN is based on gas accretion onto massive black holes (MBHs) (Lynden-Bell 1969). Rapid fluctuations in AGN light curves indicate that most of the energy is generated on the scale (r_e) of a few light hours (Nandra et al 1990). Fuel for the MBH may originate from the inflow of gas supplied by gas-star interactions (Petrosian 1982, Dahari 1985, Lin et al 1988, Hernquist 1989), the collisions between stars (Frank 1979, Murphy et al 1991), or the debris of stars tidally disrupted by the MBH (Hills 1975). In all these scenarios, the infall of gas is initiated well beyond r_e. If it carries even a small fraction of the specific angular momentum characteristic of the host galaxy, dynamical infall is expected to be halted by the formation of a centrifugally supported disk with dimension much larger than r_e.

The images and spectra obtained with the *Hubble Space Telescope* (*HST*) provide direct evidence for a rotating gaseous disk, on the scale of ~ 20 pc, around the nucleus of M87 (Ford et al 1994). The existence of accretion disks has also been inferred from the radiation characteristics of AGN. For example, the observed excess in the blue and UV continuum (with $\lambda \sim 1$–4×10^3 Å) has been interpreted as radiation emerging from disks on the scale $\sim 10^{-2}$ to 1 pc (Shields 1978, Malkan & Sargent 1982, Sanders et al 1989).

The correlation between fluctuations in the optical and UV continuum (Clavel et al 1991, Krolik et al 1991, Peterson et al 1991) suggests that disks (on the scale of $\sim 10^{-2}$ pc) may be the sites for reprocessing of radiation incident upon them rather than emission of energy dissipated within them (Collin-Souffrin 1991, Rokaki et al 1993, Rokaki 1994, Ulrich 1994). The existence of an obscuring torus or disk structure has also been inferred from spectral polarization properties (Miller et al 1991). Although the double-peaked emission-line profile in Arp 102B (Chen & Halpern 1989) is consistent with that expected for a rotationally supported disk (Mathews 1982), it is not common among AGN.

The most definitive evidence for an accretion disk is provided by the discovery of megamasers (Claussen et al 1984), which appear to be located in a molecular torus (Antonucci & Miller 1985) around the nucleus of the mildly active galaxy, NGC 4258 (Claussen & Lo 1986). Confirmation of Keplerian rotation was obtained from radio interferometer (Nakai et al 1993) and VLBI observations (Greenhill et al 1995). Based on the correlation between the spatial locations and radial velocities of the masers, Moran et al (1995) deduced that the masers are located at 0.13–0.26 pc around a MBH with a mass $3.5 \times 10^7\ M_\odot$ (Watson & Wallin 1994, Maoz 1995) (see Figure 1). Similar megamaser sources are also found in the nuclei of other galaxies, though the disk properties are less well determined (Koekemoer et al 1995).

2.1.2 TIMESCALES For the most luminous ($\sim 10^{13}\ L_\odot$) sources, a minimum mass accretion rate $\dot{M} \sim 1\ M_\odot\ \mathrm{yr}^{-1}$ is inferred based on the assumption that all the rest-mass energy of the accreted matter is converted into radiation. Larger values of $\dot{M}$ are required for modest or low conversion efficiency. If the emergent luminosity does not exceed the Eddington limit (Sun & Malkan 1989) such large accretion rates are only possible if the MBH has a mass $M_{\mathrm{bh}} > 10^8$–$10^9\ M_\odot$. This mass is comparable to the dynamical mass required to account for the rotation velocity of the gas at the center of M87 (Ford et al 1994) and the profile of the broad emission lines shaped by the effect of Doppler broadening (Netzer 1990). Additional support for this mass estimate comes from the large-amplitude variations in the X-ray luminosity, which occur on timescales comparable to the light travel time ($\sim 10^3$ s) across the Schwarzschild radius (r_s) of a $\sim 10^8\ M_\odot$ MBH (Yaqoob & Warwick 1991). For less active galactic nuclei such as NGC 4258 and the Galactic center, the estimated mass is typically an order of magnitude smaller (Genzel et al 1994).

If the MBH acquires most of its mass through disk accretion, the persistence timescale (τ_{p}) for the disk would be constrained by the MBH growth timescale. For a black hole accreting at the Eddington limit and converting all the rest-mass energy to radiation, $\tau_{\mathrm{p}} = \bar{\kappa}_{\mathrm{e}} c/(4\pi G)$, where $\bar{\kappa}_{\mathrm{e}}$ is the opacity due to scattering by free electrons. The magnitude of τ_{p} is independent of the black hole mass and is $\sim 10^8$ yr. This timescale is comparable to that inferred from the statistical arguments.

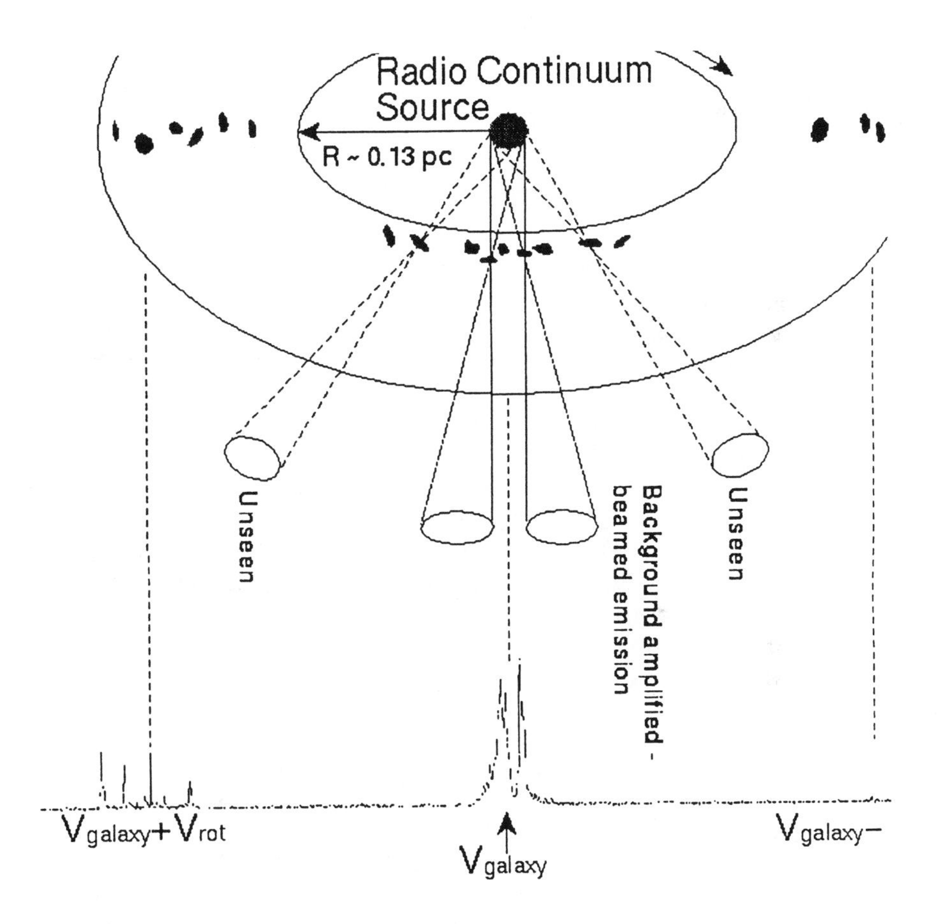

Figure 1 A schematic illustration of megamasers in NGC 4258 (See Greenhill et al 1995).

In AGN with extended (up to a few Mpc) radio jets (such as 3C236) (Barthel et al 1985), τ_p is constrained by the requirement that the central engines of the collimated outflows must be maintained for at least the light travel time along the jets ($\sim 10^7$ yr). Evidence for disks being the best candidates for the central engine of jets (Begelman et al 1984) includes the alignment between the radio jets with both the spin axis of a dusty torus in the central region of NGC 4261 (Jaffe et al 1993) and the orientation of polarization produced in the disks (Antonucci et al 1993).

Another important timescale in any accretion disk is the evolutionary timescale (τ_e) for gas to diffuse inward from the outer region of the disk. This timescale must be longer than the dynamical timescale ($\tau_d \sim 10^3$ yr associated with $10^8\ M_\odot$, within a length scale ~ 1 pc) and shorter than τ_p and the Hubble timescale (~ 2 Gyr) for the most distant ($z \sim 5$) QSOs.

The correlation between interacting galaxies and intense AGN activity (MacKenty 1989) suggests that the tidal disturbance produced by a companion may induce fresh fuel from a host galaxy to be brought to the vicinity of a central black hole on the galaxy-encounter timescale of $\tau_x \sim 10^8$ yr (Hernquist 1989), although the comparable time for a density wave to move from the outer to inner parts may be more relevant. Because the disk is much smaller than the host galaxy, we expect $\tau_e < \tau_x$, unless the angular momentum transport mechanism is very inefficient. Finally, along the jets of 4C29.47, regularly spaced knots have been interpreted as evidence for disk flow, which varies on the evolutionary timescale $> 10^7$ yr (Shields & Lin 1988), although they could also be due to jet precession (Condon & Mitchell 1984) or MHD instabilities in the jet flow.

2.1.3 PHYSICAL PROPERTIES The fiducial disk mass ($M_d \sim \dot{M}\tau_e$) is typically a significant fraction of M_{bh}. Although a mass constraint may be obtained for the gas associated with the broad emission lines (Kinney et al 1991), there is no consensus on whether this gas is mostly contained in the disk or is the dominant contributor to the bulk properties of the disk (Netzer 1990). The excitation of the broad emission lines and the emergence of the UV continuum indicate a temperature $T \sim 10^4$ K over a substantial region of the disk. The sound speed ($c_s \sim 10$ km s^{-1}) is about a few percent of the Keplerian speed (v_k), giving an aspect ratio between the thickness *(H)* and radius *(R)*, $H/R \sim 10^{-2}$. If the density and temperature of the disk are comparable to that associated with the broad line clouds, disks with $M_d \sim M_{bh}$ would become optically thin beyond 10 pc. At the distance of M87, this scale is marginally resolvable by *HST.*

In NGC 4258, the absence of significant deviation from Keplerian rotation implies $M_d < 4 \times 10^6\ M_\odot$ (Maoz 1995). The observed upper limit on the geometric aspect ratio, based on the unresolved disk thickness, is $H/r < 0.0025$, and the correspondent disk temperature is $< 10^3$ K (Moran et al 1995). The latter upper limit implies the disk would be gravitationally unstable if $M_d > 10^5 M_\odot$. The observed X-ray luminosity of NGC 4258 is 4×10^{40} erg s^{-1} (Makishima et al 1994). If this luminosity is powered by accretion onto the MBH with 1% efficiency, $\dot{M} \simeq 7 \times 10^{-5} M_\odot$ yr^{-1}. The characteristic evolution timescale would be $\tau_e \sim M_d/\dot{M} \sim 10^9$ yr, which corresponds to the dimensionless viscosity parameter $\alpha \sim 0.1$.

In most quasars, a large fraction of the total luminosity is contained in the 1–100 μm IR continuum (Sanders et al 1989). This is likely to be reprocessed optical, UV, or X-ray radiation incident on the disk at a distance $> 10^2$ pc. At such large radii, disks eventually become optically thin to continuum radiation, even when they have sufficient mass to become gravitationally unstable. In order to reproduce the observed power index of the spectral energy distribution (SED), the disk must either flare or warp (Phinney 1989). Warps may arise naturally because the symmetry axes of AGN are not generally aligned with the rotation axes of the host galaxies (Tohline & Osterbrock 1982).

2.2 *Interacting Binary Stars*

The first generally accepted identification of gaseous accretion disks was in cataclysmic variables, such as classical novae (CN), nova-like (NL) systems, and dwarf novae (DN), and these disks have been the most thoroughly analyzed (cf Warner 1995).

These systems contain semi-detached binaries consisting of a main sequence star and a white dwarf. Mass is transferred from the main sequence star in a stream, which, in the initial absence of a disk, swings around the white dwarf, strikes itself, and forms a ring (Kraft 1962, Prendergast & Burbidge 1968). If it is viscous, the ring can transport angular momentum outward, spreading in the process to form a disk (Bath et al 1974).

2.2.1 MAPPING OF ACCRETION DISKS When a disk is present, intense energy dissipation at the location where the mass transferring stream now impacts it produces a hot spot, which is conspicuous in the light curves of many CVs (Smak 1969, Warner 1976). Images of accretion disks in systems (such as UX UMa) have been reconstructed using eclipse mapping analysis (see Rutten et al 1993 and Figure 2). The typical disk size is a significant fraction of a solar radius.

Evidence for Keplerian rotation is found in the double-peaked profiles of Balmer emission lines (Robinson 1976) and the spectral evolution through eclipses (Young et al 1981a,b). Analogous disks occur in low-mass X-ray binary stars in which the accreting objects are neutron stars or black hole candidates (van Paradijs 1983, Inoue 1993, Marsh et al 1994).

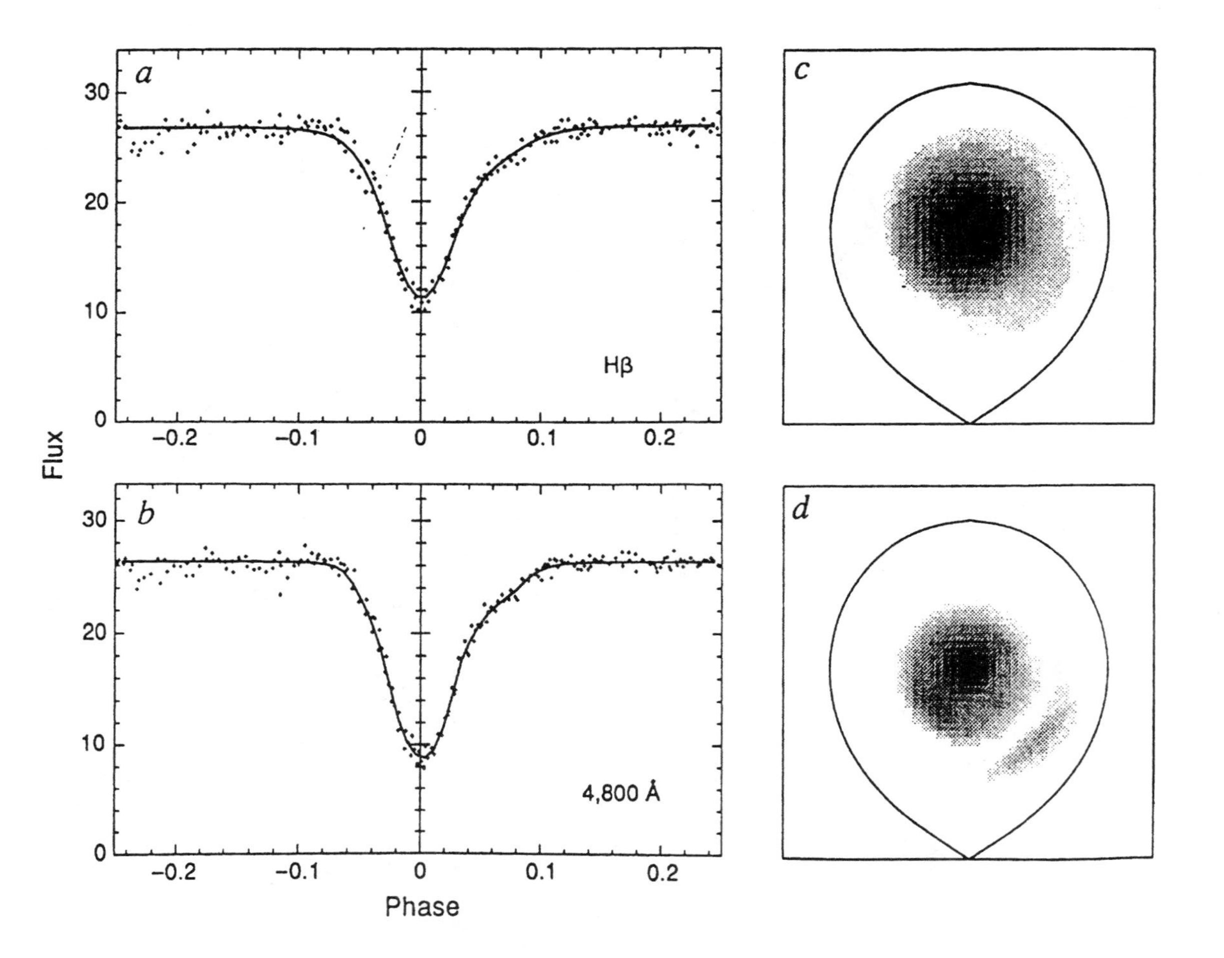

Figure 2 An eclipse mapping of UX UMa (See Rutten et al 1993).

In CN and NL systems, the SED of the UV and optical continuum is well approximated by steady state disk models (Haug 1987, La Dous 1989). In several sources, the radial temperature distribution deduced from the eclipse mapping method is also consistent with that expected from a steady state disk (Horne & Cook 1985), although there are some NL systems with flatter SEDs, which indicate an excess flux from the outer regions of the disk (Rutten et al 1992b). The dominant presence of Balmer absorption lines indicates that the disk gas is opaque and highly ionized in disks in CN and NL systems.

2.2.2 STABILITY AND EVOLUTION In contrast to CN and NL systems, DN undergo outbursts with a characteristic interval of a few weeks to months. At the beginning of these outbursts, the visual magnitude increases by 3–5 magnitudes within a few hours (Wade & Ward 1985). At its maximum light, the SED of DN is similar to that of CN. Thereafter, the visual magnitude declines on a timescale of a few days (identified with τ_e). During the decline, the SED gradually flattens from that expected for a stationary disk (Hassell et al 1983) and the Balmer lines evolve from absorption to emission (Hessman et al 1984). When quiescence is reestablished, the disk spectrum is dominated by emission lines (Williams 1983) and the disk appears to be optically thin, isothermal, and mostly neutral (Wood 1990).

2.2.3 PHYSICAL PROPERTIES The mass transfer rate ($\dot{M}$) in the disk is inferred from 1. the SED (Wade 1984), 2. disk luminosity in the few cases where the parallax is known (Patterson 1984), or 3. eclipse mapping (Horne 1993). These estimates generally indicate that $\dot{M} \sim 10^{-8}\, M_\odot\ \mathrm{yr}^{-1}$ in CN and NL systems (Rutten et al 1992b). But in DN, $\dot{M} \sim 10^{-10}\, M_\odot\ \mathrm{yr}^{-1}$ during quiescence, and it is $> 10^{-9}\, M_\odot\ \mathrm{yr}^{-1}$ during outbursts (Rutten et al 1992a). Some kind of instability in the disk may be responsible for producing such variations in the mass transfer rate (Osaki 1974).

In some CN and NL systems, such as HR Del and TT Ari, mass outflow is inferred from the P Cyg profiles of C IV and other UV lines (Krautter et al 1981). A similar inference is made from the *IUE* data from DN such as OY Car (Naylor et al 1988). The mass outflow rate is typically two orders of magnitude smaller than the estimated accretion rate (Mauche & Raymond 1987, Drew 1987). The large ($\sim$ 5000 km s^{-1}) velocities inferred from the line profiles indicate that the outflow originates near the white dwarf (Cordova & Mason 1982).

Eclipse mapping indicates that the outer boundary of the disk extends to about half of the Roche lobe radius during quiescence (Marsh et al 1990). The disk size appears to double during DN outbursts (Smak 1984b). Most DN emit hard X-rays during quiescence, and some emit soft X-rays during outburst (Cordova & Mason 1983). This radiation most likely originates from the boundary layer region where the disk joins the accreting white dwarf (Pringle 1977, Pringle & Savonije 1979, Kley 1989).

In CN and NL systems where the accretion flow is steady, it is difficult to estimate the disk mass. However, in DN, the amount of mass flushed during typical outburst events is $M_d \sim \dot{M}\tau_e \sim 10^{-11}\ M_\odot$, although higher values have been claimed for some systems (such as Z Cha and U Gem) (Anderson 1988). If this mass of gas exists in a disk with characteristic size $\sim 3 \times 10^{10}$ cm during quiescence, then the corresponding surface density would be $\Sigma \sim 5$ gcm^{-2}. For a temperature $T < 7$–8×10^3 K, the disk becomes optically thin. These physical conditions are consistent with those required to produce strong emission lines (Williams 1980, Williams & Ferguson 1982).

Perhaps the most promising candidate for providing the angular momentum transfer mechanism during the outburst is MHD turbulence (see Paper I). Although magnetic activity (Galeev et al 1979, Rozyczka et al 1996) has not been directly observed in disks, it has been inferred (Horne 1994) from the apparent correlation between the emission line surface brightness and the orbital rotation frequency (Horne & Saar 1991). This correlation is in close agreement with that found for the Calcium H and K lines in stars and in the quiet regions on the solar surface (Schrijver 1987, 1992).

2.3 *Young Stellar Objects*

The hypothesis that the planets in our Solar System were formed in a flattened gaseous disk was proposed (Laplace 1796, Cameron 1978) to account for the origin of their present dynamical properties. Recent observations (Goodman et al 1993) indicate that typical star-forming dense cores in dark molecular clouds have specific angular momenta $> 10^{21}$ cm^2 s^{-1}. When these clouds undergo gravitational collapse, an order of magnitude reduction from their present dimension (~ 0.1 pc) would lead to the formation of a rotationally supported disk analogous to the primordial solar nebula (Tereby et al 1984, Bodenheimer 1995).

2.3.1 DIRECT IMAGING The coronagraphic image of β Pic provided the most direct evidence that relic disks of dust particles do exist around main sequence stars (Smith & Terrile 1984). The progenitors of these particle disks are likely to be gaseous disks around young stellar objects similar to the primordial solar nebula. In the *HST* images of the Orion nebula, a flattened structure of ionized gas has been detected (O'Dell et al 1993, O'Dell & Wen 1994). A dark disk-shaped structure is seen as a silhouette (Figure 3) against the bright background of the Orion nebula in some *HST* images (McCaughrean & O'Dell 1996). The size of such structures is larger than that of the Solar System by at least an order of magnitude.

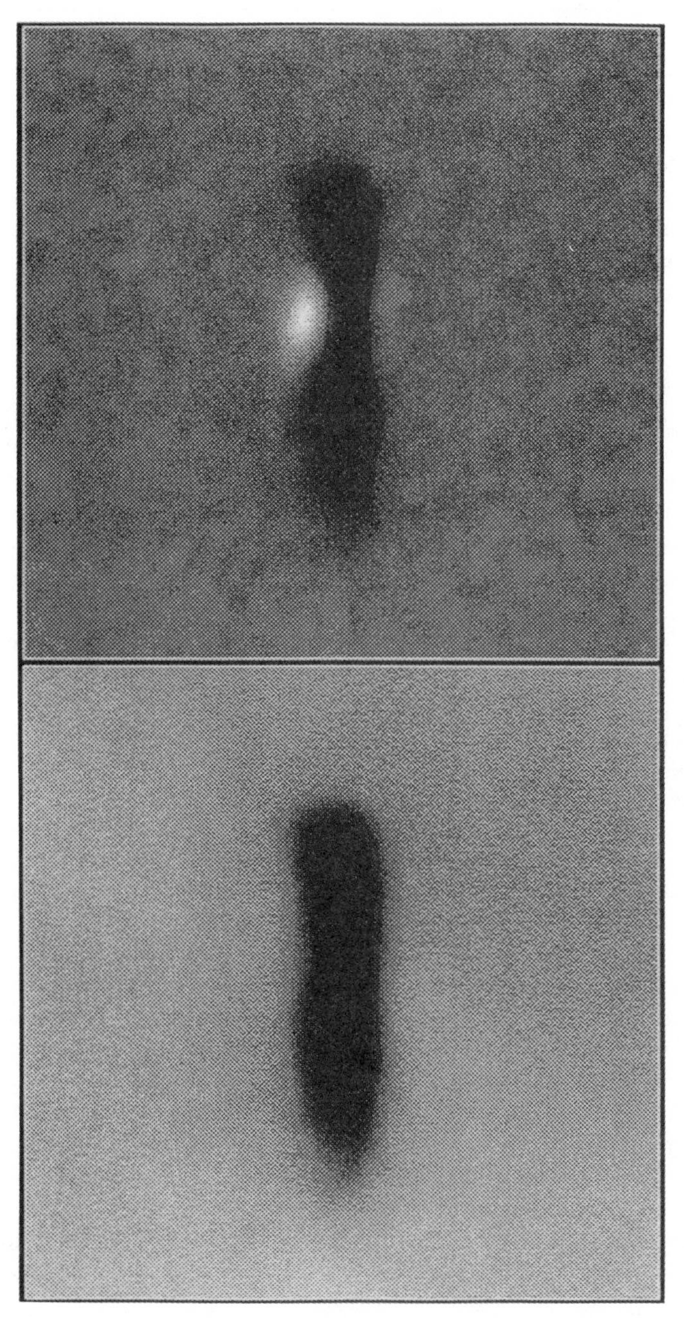

Figure 3 *HST* image of a protostellar disk in the Orion nebula (See McCaughrean & O'Dell 1996).

The presence of protostellar disks is also clearly implied from the *HST* image of the central star of HH 30 (Figure 4). The central star is embedded in a nearly edge-on dusty disk, and the optical emission is due to dust scattering by the atmosphere above a flaring disk (Burrows et al 1996). A nearly face-on disk is inferred from the scatter light around GM Aug (KR Stapelfeldt, private communication) and the ^{13}CO map (Koerner et al 1993).

Perhaps the best observed gaseous disk is that around HL Tau. The ^{13}CO map of HL Tau revealed an elongated image extending to 2000 AU (Sargent & Beckwith 1987, 1991). The velocity field on a scale of $\sim 10^3$ AU is compatible with that of a slowly rotating and mainly collapsing flattened structure in which the mass inflow rate is $\dot{M}_{\rm i} \sim 5 \times 10^{-6} M_\odot$ yr^{-1} (Hayashi et al 1993). The specific angular momentum carried by the infalling gas is adequate to induce the formation of a centrifugally supported disk with a size comparable to that of the primordial solar nebula (Lin et al 1994). Submillimeter and millimeter interferometer maps resolve the inner disk at $\sim$ 50–100 AU (Lay et al 1994, Sargent & Koerner 1995, Mundy et al 1996). At optical wavelengths, the disk is optically thick and does not appear in the *HST* image; however, it truncates a C-shaped reflection nebulosity to a radius consistent with that inferred from radio observations (Stapelfeldt et al 1995). Based on the inferred luminosity and its position in the H-R diagram (D'Antona & Mazzitelli 1994), the age of HL Tau is estimated to be $\sim 10^5$ yr. This age is consistent with that deduced from the accretion timescale $\tau_* \sim M_{\rm dyn}/\dot{M}_{\rm i} \sim 10^5$ yr (Hartmann et al 1994a, Lin et al 1994).

2.3.2 PERSISTENCE TIMESCALES With rapid advancement in IR instrumentation, the presence of protostellar disks has been inferred from the spectroscopic data. Among YSOs with an IR excess, many have flat SEDs (flat spectrum sources: FSS) (Rucinski 1985), which are most likely due to reprocessed radiation coming from a flattened infalling envelope (Kenyon et al 1993).

However, most YSOs with infrared (IR) excess have relatively steep SEDs; these are normally referred to as classical T Tauri (CTT) stars (Beckwith et al 1990). The existence of disks on scales of astronomical units is inferred from the power-law SED in the IR over more than 2 orders of magnitude in wavelength. The power-law index in these systems is consistent with that expected from either a steady state viscous accretion disk with $\dot{M} \sim 10^{-7} M_\odot$ yr^{-1} or the reprocessing of radiation emitted near the accreting central star by a geometrically thin opaque disk (Adams et al 1987). The latter possibility is consistent with the detection of a significant amount ($> 3\%$) of polarization in CTT stars (Bastien & Menard 1990, Menard & Bastien 1992). The colors and magnitudes of the CTT stars are best fitted by pre–main sequence stars with

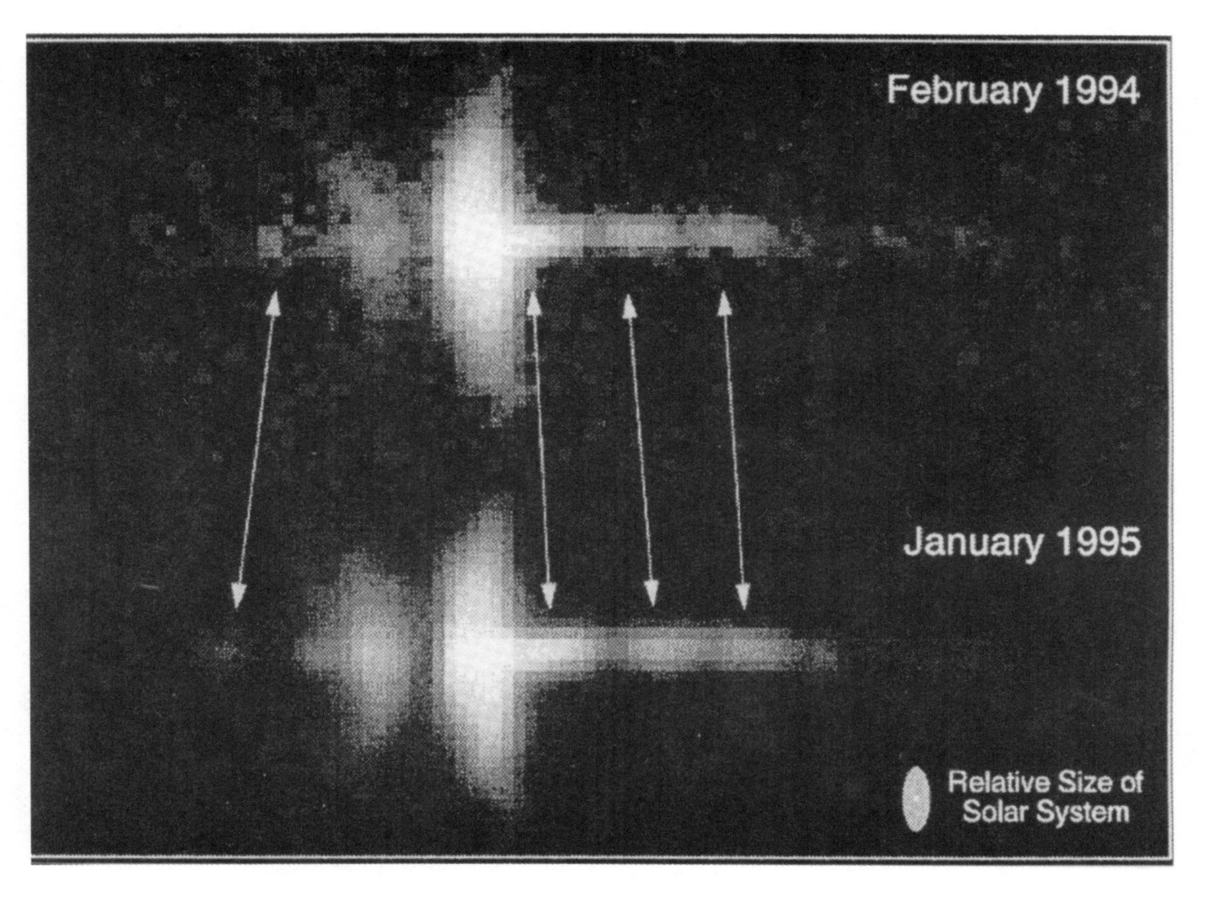

Figure 4 *HST* image of a protostellar disk and outflow around HH 30 (See Burrows et al 1996).

ages $\tau_* \sim 10^6$ yr (Strom et al 1993). The ratio between the estimated age of the FSS ($\sim 10^5$ yr) to that of the CTT stars is comparable to the ratio between the relative abundance of these two populations.

A recent survey based on the *HST* images indicates that between 25 and 75% of YSOs in the Orion nebula have disks (Prosser et al 1994, McCaughrean & Stauffer 1994). However, in the Taurus-Auriga complex, more than half the observed T Tauri stars have little or no IR excess; they are referred to as weak-line (or naked) T Tauri (WTT) stars.

In many cases, the central stars in these systems have colors and magnitudes consistent with a typical age of $\tau_* \sim 10^7$ yr (Walter et al 1988), although some are younger. A similar age may be inferred from the depletion of CO around YSOs of a few 10^6 years of age (Zuckerman et al 1995) and the apparent correlation between the age of young star clusters and the relative abundance of WTT stars (Lada et al 1993). The absence of IR excess around these stars has been interpreted as an indication that either the dust grains have settled and coagulated or gas has been ejected from the protostellar disks (Skinner et al 1991). The *IRAS* survey indicates the presence of dust around more than 30% of the main sequence A stars, which are typically $\sim 10^8$ years in age (Aumann et al 1984, Backman & Gillett 1987, Walker & Wolstencroft 1988). An alternative suggestion is that, for WTT stars, the disk matter has been expelled by a magnetic torque that becomes large during phases of a dynamo cycle when the stellar magnetic field is large (Clarke et al 1995).

2.3.3 PHYSICAL PROPERTIES The IR excess indicates that the disks have low temperatures, with the consequence that the gas is essentially neutral (Beckwith et al 1990). A temperature distribution with a range 10^3–10 K has been inferred from the intermediate ($\sim 10^{-1}$–10^2 AU) regions of the disk (Adams et al 1988). A similar distribution may be inferred from the equilibrium grain condensation temperature in the planet-forming regions (Lewis 1974).

For the solar nebula, a minimum mass ($\sim 10^{-2} M_\odot$) is obtained by augmenting the present mass contained in the planets to make up a solar composition (Cameron 1962, Weidenschilling 1977). In protostellar disks, a break in the SED at $\sim 100\ \mu$m signifies that the disk is optically thin- to long-wavelength photons. Disk masses estimated from the observed flux at long wavelengths are typically a significant fraction of the central stellar mass (M_*) in many systems (Adams et al 1990), but there are large uncertainties in such estimates because the dust opacity at long wavelengths is poorly known (Pollack et al 1994, Ossenkopf & Henning 1994, Henning et al 1995). Similar masses are also obtained for the outer regions of the disk from images at wavelengths of ^{13}CO, $C^{18}O$, and other molecular lines (Sargent & Beckwith 1991, Ohashi et al 1991), although the assumption that the disk is optically thin at these wavelengths may be questionable in some cases (Beckwith & Sargent 1993a).

2.3.4 STABILITY AND OUTBURSTS Although the inner regions of protostellar disks cannot be resolved, the mass accumulated there can be estimated in YSOs with variable $\dot{M}$. Some of the youngest (FSS) systems, such as FU Ori (FU), undergo outbursts with light curves remarkably similar to those of DN (Herbig 1977). Maximum light is reached on a timescale of months to years. At maximum light, the SED resembles that expected for a steady accretion disk and the estimated mass accretion rate onto the central star is $\sim 10^{-4} M_{\odot}$ yr^{-1} (Kenyon et al 1988).

In some FUs (such as FU Ori), maximum light is maintained, whereas in others (such as V 1057 Cyg) the blue and visual magnitude declines on the timescale of decades to centuries. During the decline, there is very little change in the IR flux so that the SED flattens (Kenyon & Hartmann 1991). This spectral evolution indicates that $\dot{M}$ decreases as mass is depleted in the inner regions ($<$ 1 AU), with $\dot{M}$ being maintained at a steady rate comparable to $\sim \dot{M}_{\rm i}$ in the outer regions of the disk (Bell et al 1995).

The total amount of mass flushed during an outburst event is $\sim 10^{-2} M_{\odot}$. Although there has not been sufficient time to observe the reestablishment of quiescence after any observed outbursts, circumstantial evidence suggests that FUs become flat spectrum sources, e.g. HL Tau (Lin et al 1994). If the disk is supplied by infall at a rate $\dot{M}_{\rm i}$ comparable to that inferred for HL Tau, the replenishment timescale ($\tau_{\rm r}$) would be several 10^3 yr. The ratio of outburst duration to $\tau_{\rm r}$ is comparable to the relative abundance ($\sim$ 10%) of FUs to FSS.

2.3.5 STAR-DISK BOUNDARY REGION Evidence for disks reaching the close vicinity of the stellar surface is provided by the UV excess observed in the spectra of some T Tauri stars (Bertout et al 1988) and the double-peaked line profile seen in the CO bandhead emission of WL 16 (Carr et al 1993, Najita et al 1996). However, in many cases there is no excess in the UV or near IR despite a considerable amount of excess in the far IR (Skrutskie et al 1990). In such cases, the inner disk boundary may be either optically thin (Basri & Bertout 1989) or at some distance from the stellar surface because of the presence of a magnetic field (Königl 1991, Calvet & Hartmann 1992, Hartmann et al 1994b).

Strong winds and bipolar outflows similar to the collimated outflows in AGN are found to be well correlated with the inferred presence of inner disks from the IR excess (Cabrit et al 1990) and UV veiling (Hartmann 1994). The mass-loss rate is typically a significant fraction (~ 0.1) of the estimated accretion rate in the disk (Croswell et al 1987, Calvet et al 1992). In the resolved cases, such as L1551 IRS5 (Snell et al 1980), the outflow is along the spin axis of the disk (Moriarty-Schieven & Snell 1988) and speeds up to the escape velocity of the central star ($\sim 10^7$ cm s^{-1}).

In Fu Ori, many neutral unblended metal lines are observed to have double-peak profiles. The velocity shift from the line center in the red wing of these lines decreases with the line depth, whereas that in the blue wing remains unchanged (Petrov & Herbig 1992). Calvet et al (1993) suggest that lines with increasing oscillator strength are produced at increasing distances above the photosphere. The apparent correlation between the line strength and the profile is an indication that, near the photosphere, the flow is mostly Keplerian rotation, whereas further up in the atmosphere it becomes predominantly an outflow (Hartmann 1994).

One possible driving mechanism for the outflow is hydromagnetic wind (Section 3 in Paper I). The field strength needed is $\sim 10^1$–10^3 Gauss (Königl 1994). Such a field may originate from the surface of a rotating central star (Najita & Shu 1994, Shu et al 1994, Ostriker & Shu 1995). It is also possible that this field is advected inwards by the accreting matter (Lubow et al 1994). However, at the present time, there is no direct observational evidence for such a strong field throughout the disk. Advection of all the local disk angular momentum is possible if the Alfvén radius is significantly larger than the radius of the location where the wind is launched. Nonetheless, the issue of causal relationship between the wind and accretion flow remains controversial (Königl & Ruden 1993).

Beyond a few stellar radii, the gas density is sufficiently high to shield the disk from cosmic rays (Umebayashi & Nakano 1988). Although the degree of ionization near the midplane is negligibly small, within a thin surface layer of the disk it may be sufficiently high for the gas to be coupled to the field (Gammie 1996). For a disk wind to be the primary mechanism for angular momentum removal (Pudritz & Norman 1983, Königl 1994), the outflow rate would have to be sufficiently large to reprocess the emerging radiation. Based on the IR excess, Safier (1993) estimates an outflow of up to $\sim 10^{-8}\ M_{\odot}\ \mathrm{yr}^{-2}$ in some CTTs; but for most CTTs and FU, the IR excess is inadequate to allow the presence of a disk wind with sufficiently high mass-loss rate (Hartmann 1994).

2.4 *A Brief Summary of Observational Data*

At the present time, direct evidence for accretion disks is irrefutable in CVs, convincing in YSOs, and remarkable in AGN. The dynamical range spanned by disk radii in CVs, YSOs, and AGN are 2, 6, and 7 orders of magnitude, respectively. In all these cases, gas cannot flow onto the accreting object unless it is able to reduce its specific angular momentum by up to ~ 3 orders of magnitude. Furthermore, in these systems, the disk evolves on the local mass inflow timescale, ranging from 10^3 to $10^6\ \Omega^{-1}$, where Ω is the angular velocity near the outer edge of the active region.

The physical conditions are vastly different in these three contexts. Within disks around YSOs (or AGN), a diverse range of densities and temperatures occur. In CV disks, the ionization fraction undergoes a transition during an outburst cycle. It is possible that a variety of mechanisms may be important according to the physical circumstances. Thus, many possible avenues should be explored before we can identify the most important angular momentum transport mechanisms operating in each context.

3. GLOBAL EVOLUTION OF AN ACCRETION DISK

In Paper I, we considered instabilities intrinsic to laminar Keplerian flow and the mechanisms that can provide efficient angular momentum transfer in the form of an anomalous viscosity. We now consider the utilization of theories of disk evolution, taken together with models of the disk atmosphere for the extraction of useful information from observational data.

When an opaque disk is in a steady state, the emerging SED has a characteristic wavelength dependence in which the monochromatic flux, $F_\lambda \propto \lambda^{-7/3}$, with λ in this context representing the wavelength. The only dependence of F_λ on the disk parameters is through a linear dependence on $\dot{M}$. Unless the disk mass (or Σ) can be deduced explicitly, very little information on the dominant angular momentum transfer mechanisms can be extracted from the observed properties of a stationary disk. However, in a nonstationary disk where Σ evolves with time, F_λ is also expected to change. In Section 2, we indicated that these changes are observed in many systems on timescales of hours to years.

In order to describe the global long-term evolution of a thin disk, it is desirable to work with equations for which behavior such as small-scale turbulence—which is expected to occur on characteristically short timescales $\sim \Omega^{-1}$ and on short lengthscales $\sim H$—is averaged out. In most accretion disks the azimuthal motion is hypersonic such that the disk semi-thickness $H \simeq c_s/\Omega \ll R$, where c_s is the sound speed (see Section 2). In this limit, the disk flow is approximately Keplerian, i.e. $\Omega \simeq (GM_*/r^3)^{1/2}$, where $M*$ is the central mass and r is the distance from it.

In the analysis of thin-disk evolution, it is adequate to adopt the approximation in which the properties of the gas are vertically averaged such that for any quantity Q,

$$\langle Q \rangle = \frac{\int_{-\infty}^{\infty} Q\rho dz}{\int_{-\infty}^{\infty} \rho dz},$$

where ρ is the density. For isolated disks it is also customary to assume the disk to be globally axisymmetric. In cylindrical coordinates (r, φ, z) based on the central mass, the structure and evolution of the disk are governed by a single diffusion equation for the surface density Σ, which takes the form (see Lüst 1952, Lynden-Bell & Pringle 1974, Lin & Papaloizou 1985)

$$\frac{\partial \Sigma}{\partial t} - \frac{1}{r}\frac{\partial}{\partial r}\left[3r^{1/2}\frac{\partial}{\partial r}\left(\Sigma\langle\nu\rangle r^{1/2}\right) - \frac{2S_\Sigma J}{\Omega} - \frac{2\Sigma\Lambda}{\Omega}\right] - S_\Sigma = 0, \tag{1}$$

where ν is the kinematic viscosity (Equation 5, Paper I) and the angled brackets denote vertical averaging.

A source of mass per unit area S_Σ may exist. In a CV this contribution might be due to the mass transfer stream. If S_Σ is negative it could represent the effects of mass loss through a wind. The symbol J represents the excess specific angular momentum carried by the source/sink material, and we have allowed for a local injection rate of angular momentum per unit mass Λ into the disk due to external torques or dissipation of wave action.

Equation (1) contains three possible contributions to the angular momentum transport: 1. local viscous transport, 2. advective loss, and 3. external perturbation. Physical processes that could contribute to each of these were reviewed in Part I.

Turbulent eddy viscosity (which could be of hydromagnetic origin) is usually assumed to be the dominant contributor to the kinematic viscosity coefficient ν, which is often parameterized through the α prescription (Shakura & Sunyaev 1973) in which $\nu = \alpha c_s^2/\Omega$, where α is a dimensionless constant that must be less than, but might be comparable to, unity.

In a disk with unsteady mass transfer, Equation (1) indicates that the rate of evolution of the disk is determined by the efficiency of angular momentum transport. Thus, detailed analyses of unsteady flows in accretion disks can provide useful constraints on the magnitude of the viscosity.

4. UNSTEADY FLOWS IN ACCRETION DISKS

Unsteady flows in the disk arise if the mass supply rate to the disk varies. In the context of AGN and YSOs, the mass infall rate varies on a timescale of $\sim 10^6$–10^8 yr. Much more rapid variations in $\dot{M}$ can occur if there are instabilities associated with disk evolution in the presence of the anomalous viscosity. Some of these instabilities may be a manifestation of inconsistent theoretical prescriptions (Pringle 1981), but others may be realizable in nature. We focus our discussion on the observable aspects of these instabilities.

The best examples of phenomena related to variable mass transfer (and consequently variable energy production) in accretion disks are DN and FU outbursts (see Section 2). There is also evidence for similar outbursts in X-ray transient sources (Bradt & McClintock 1983, Lin & Taam 1984, Mineshige & Kusunose 1993) and AGN (Lin & Shields 1986, Abramowicz et al 1988).

4.1 *Thermal Instability*

In an approximately steady accretion disk, the energy radiation rate is sufficient to remove the orbital kinetic energy on the inflow or accretion timescale. The thermal energy content per unit mass ($\simeq c_s^2$) is smaller (by a factor $H^2/r^2 \ll 1$) than the orbital kinetic energy per unit mass. Accordingly, the local thermal timescale is smaller than the inflow timescale by the same factor. This inequality suggests that gas in an accretion disk will try to adjust to a local thermal equilibrium in which the thermal energy generation through dissipation is balanced by radiative losses. The mass transfer process would then occur under conditions of slowly adjusting local thermal equilibrium.

In a thin disk, the most important radiative losses occur through the vertical component of the radiative flux. Ignoring the other components, external heat sources, and time dependence, one obtains a local thermal equilibrium in which the rate of viscous dissipation per unit volume

$$D_\nu = \frac{\partial F_z}{\partial z}, \tag{2}$$

where F_z is the vertical flux. Integrating (2) vertically through the disk gives the total dissipation per unit area under conditions of thermal equilibrium (counting both the upper and lower surface of the disk) as

$$\int_{-\infty}^{\infty} D_\nu dz = 2\sigma T_{\text{eff}}^4, \tag{3}$$

where T_{eff} is the effective temperature and σ here denotes Stefan's constant.

The specification of the form of the dissipation as a function of local state variables (such as ρ and the temperature T) provides a relation between T_{eff} and these state variables. For purely anomalous viscous dissipation,

$$\int_{-\infty}^{\infty} D_\nu dz = \langle \nu \rangle \Sigma r^2 \left(\frac{d\Omega}{dr}\right)^2, \tag{4}$$

where the local surface density $\Sigma(z)$ is obtained through

$$\frac{\partial \Sigma}{\partial z} = \rho. \tag{5}$$

For vertical transport of energy by radiation,

$$F_z = -\frac{4acT^3}{3\bar{\kappa}\rho}\,\frac{\partial T}{\partial z}, \tag{6}$$

where $\bar{\kappa}$ here denotes the opacity, c is the speed of light, and a is the Stefan-Boltzmann radiation constant. In addition, the condition for vertical hydrostatic equilibrium implies

$$\frac{\partial P}{\partial z} = -\rho\Omega^2 z. \tag{7}$$

Equations (2), (6), (7), and (5) constitute a set of four basic equations for the vertical structure of the disk, at any radial location, akin to those of stellar structure. For a star, if the mass is specified, the radius and effective temperature are determined, provided it is chemically homogeneous and the opacity and energy production rate are specified as functions of the state variables. The entirely analogous situation for accretion disks is that if the surface density is specified, the total vertical extent and effective temperature are determined—but note that there may be a multiplicity of solutions in some cases (see below).

In this way, local thermal equilibria are completely determined once a surface density is given. For simple power-law forms of $\bar{\kappa}$ and D_ν, the solutions have simple homology relations, which are useful for analyzing their properties (see, for example, Shakura & Sunayev 1973, Lin & Papaloizou 1980, Faulkner et al 1983). Furthermore, in a steady state, the mass transfer rate through a viscous disk is $\dot{M} = 3\pi\langle\nu\rangle\Sigma$, which is also determined as a function of Σ.

The above description of a disk achieving a state of steady accretion under conditions of local thermal equilibrium can only be valid if the thermal equilibria are stable. Thermal instability occurs if, for example, we make a small local increase in temperature, at constant surface density, that causes an adjustment that results in the disk being less able to radiate. Such a perturbation would result in the disk heating up locally in an unstable manner (Pringle et al 1973, Shakura & Sunyaev 1976).

It is possible to decide which equilibria are thermally unstable by considering the behavior of $\langle\nu\rangle\Sigma$, at a particular radius, calculated from the equilibrium models, as a function of Σ. *This function defines a series of equilibria as Σ is varied.* At a point of change of stability, two series must merge, or, equivalently, there is a point of bifurcation. At such a point

$$\frac{\partial(\langle\nu\rangle\Sigma)}{\partial\Sigma} = \infty. \tag{8}$$

If the series of equilibria constitute an S curve when viewed in the $(\langle\nu\rangle\Sigma, \Sigma)$ plane, then the points of infinite gradient, where the S bends round, constitute bifurcation points at which there is a change of stability. If the upper and lower branches of the S curve define an equilibrium series for $\langle\nu\rangle\Sigma$ as a monotonically increasing function of Σ, then these branches are expected to be stable with the intermediate section joining the upper and lower branch being unstable.

For steady viscous disks in local thermal equilibrium satisfying (4), where the series of equilibria in the $(\langle\nu\rangle\Sigma,\ \Sigma)$ plane form such an S curve, the above condition for thermal instability is equivalent to

$$\frac{\partial(\langle\nu\rangle\Sigma)}{\partial\Sigma} < 0. \tag{9}$$

This kind of thermal instability occurs for disk models undergoing partial ionization. At any radius, there is a range of $\langle\nu\rangle\Sigma$ or $T_{\rm eff}$, defined by the portion of the S curve with negative slope, for which stable thermal equilibrium is not possible.

4.2 *Viscous and Pulsational Overstabilities*

Thermal instability represents only one kind of linear instability that is possible in a viscous disk. Linear stability analyses also indicate the existence of instabilities known as diffusive or viscous instabilities. These instabilities occur on a long timescale during which thermal equilibrium is maintained (Pringle 1981). Disk evolution is then governed by Equation (1), where $\langle\nu\rangle\Sigma$ is given as a function of Σ at any particular location by the thermal equilibrium models.

The viscous/diffusive instability criterion, which is identical to the criterion for thermal instability in the case of gaseous disks in a state of partial ionization, is

$$\frac{\partial(\langle\nu\rangle\Sigma)}{\partial\Sigma} < 0. \tag{10}$$

When this criterion is satisfied, (1) gives instability because small surface density perturbations evolve with an effective negative diffusion coefficient (Lightman & Eardley 1974).

When the disk is thermally unstable, either because H^- opacity dominates or because radiation pressure and electron scattering opacity dominate (see discussions in Section 5.1), the disk is also viscously unstable. However, in these cases the disk is never able to achieve the unstable local thermal equilibria as is required for the viscous instability to occur. Accordingly, this instability is unlikely to have any physical significance in these cases.

To obtain a physically realizable viscous instability, one requires, for example, that the equilibrium plot of $\langle\nu\rangle\Sigma$ against Σ give a unique equilibrium series with negative slope but without bifurcation points. This equilibrium can then be thermally stable, at a surface density where the viscous instability criterion is satisfied. In the context of planetary rings, where the α prescription is not relevant, the equilibrium series might be of this form , where it is possible for the rings to be viscously unstable but thermally stable (Lin & Bodenheimer 1981, Ward 1981, Lukkari 1981).

Although the viscous instability modifies the efficiency of angular momentum and mass transfer, it does not lead to unsteady flows unless possibly the unstable range of Σ has no limit. In general, the saturation of the viscous instability leads to a bimodal distribution in Σ such that unstable values of Σ are eliminated. Stable regions of high surface density are separated from stable regions of low surface density with the integrated viscous stress ($\propto \langle \nu \rangle \Sigma$) being continuous across the interfaces as observed in Saturn's rings (Cuzzi et al 1984).

There is also a possible mode of pulsational overstability. In this case one looks for instabilities in which oscillations on the orbital timescale grow in amplitude because of the effects of viscosity. In the analysis of pulsational overstability, the time dependence of all physical quantities must be evaluated with all the basic equations (Kato 1978, Blumenthal et al 1984, Papaloizou & Stanley 1986). Because evolution may occur on short timescales, the assumption of a turbulent viscosity, derived after taking a time average, may not be valid. Consequently, the theoretical foundation of these instabilities is on less certain ground than the others.

Neglecting this complication, pulsational overstability is possible because, in some temperature regions, the efficiency of turbulent angular momentum transport can increase upon local compression. (This occurs, for vertically averaged models, when $\partial \ln\langle \nu \rangle / \partial \ln \Sigma \gg 1$, when the disk is expected to be thermally and viscously stable.) In this case, if a compressional normal mode of oscillation is established, the thermal energy generation arising from viscous dissipation of the shear flow increases during compression and can result in amplification of the oscillation. This effect is entirely analogous to the way nuclear energy generation can cause stellar pulsations to become overstable (Kato 1978).

Applying a local analysis to an α model, Blumenthal et al (1984) found that CV disks are pulsationally overstable for $\alpha > 0.06$. They also suggested that the quasi-periodic oscillations in CVs (Robinson & Nather 1979) and galactic black hole candidates (Chen & Taam 1995) may be due to the pulsational overstability (also see Okuda et al 1992). One-dimensional global nonlinear analysis showed that only small oscillations near the outer and inner disk boundaries may grow into a coherent response on a viscous timescale (Papaloizou & Stanley 1986). Similar analyses have been applied to relativistic disks in the context of AGN (Honma et al 1993). In the proximity of a massive black hole, this overstability may lead to short-period (days to months) oscillations similar to the p-modes and g-modes of helioseismology (Nowak & Wagoner 1993). These oscillations have also been found in two-dimensional (r-z), axisymmetric, nonlinear calculations, which also show that they are primarily radial and confined to the outer regions of the disk for modest values of α (Kley et al 1993). For small α, they are damped.

Nonaxisymmetric thin-disk analyses show that pulsational overstability may also induce normal-mode oscillations in narrow, eccentric planetary rings (Papaloizou & Lin 1988) such as those observed around Uranus (Elliot & Nicholson 1984).

In all these cases, the oscillations may generally be suppressed if the viscous stress tensor is anisotropic, with a larger viscosity coefficient applying to the meridional components of the equation of motion.

5. DISK EVOLUTION IN DN AND FU ORIONIS SYSTEMS

The best candidates for systems undergoing thermal instability are DN during normal outbursts (Meyer & Meyer-Hofmeister 1981, Smak 1982, Cannizzo et al 1982, Faulkner et al 1983). These ordinary outbursts are to be distinguished from the super outbursts seen in SU UMa systems, which may be related to variable mass transfer from the secondary star (Papaloizou & Pringle 1979).

5.1 *Thermal Instability Criteria*

In a DN outburst cycle, the observed $T_{\rm eff}$ is characteristically several thousand degrees during quiescence (see Section 2). Although it is important to distinguish between $T_{\rm eff}$ and the midplane temperature ($T_{\rm c}$), they do not differ by a large factor when the disk is optically thin. (Even in outburst, $T_{\rm c} \sim \tau^{1/4} T_{\rm eff}$ is only a few times larger than $T_{\rm c}$.) Similar temperatures are also found in the inner regions of protostellar disks around YSOs and in the outer regions of AGN disks where the blue bump in the continuum originates (also see Section 2).

At such temperatures, the gas is partially ionized such that the continuum opacity is primarily due to H^-, and it increases steeply with temperature. In this case, a small increase in T leads to a reduction in the radiative flux as a result of the increase in opacity. Because the rate of energy dissipation does not decrease with T, the more efficient trapping of the heat generated leads to a further increase in temperature and thermal runaway. Equilibrium models for this temperature range therefore satisfy the instability criterion in Eq (9) and are thermally unstable.

The high-temperature, low-density region around AGN (and neutron stars), where electron scattering opacity and radiation pressure dominate, may also be thermally unstable (Pringle et al 1973, Shakura & Sunyaev 1976). However, this situation is marginal since the opacity is independent of T and the existence of thermal instability depends on the temperature sensitivity of the viscous heating, which itself depends on the precise viscosity prescription (Sakimoto & Coroniti 1981, Taam & Lin 1984, Abramowicz et al 1988, Chen & Taam 1994).

In the low-T range where grain and molecular opacity dominates, and in the higher-T range where bound-bound, bound-free, and free-free processes dominate the opacity (appropriate for DN), the disk would be thermally stable provided the rate of energy dissipation does not increase very rapidly with the temperature (as is the case for the α prescription).

Thus, for equilibrium models spanning the range $2 \times 10^3 < T_{\rm eff} < 10^6$ for surface densities appropriate for DN, a characteristic S curve describes the $\dot{M}$ $(\Sigma\langle\nu\rangle)$ dependence on Σ. This equilibrium solution is composed of two separate, stable branches at high and low Σ (corresponding to high and low temperatures) for which $\partial(\langle\nu\rangle\Sigma)/\partial\Sigma > 0$. These branches are joined by an unstable branch, for which $\partial(\langle\nu\rangle\Sigma)/\partial\Sigma < 0$, at intermediate temperatures and surface densities on which the disk material is undergoing partial ionization.

For a particular radial location, the upper stable branch exists for $\dot{M}$ exceeding a critical value $\dot{M}_B$, where $\Sigma = \Sigma_B$, while the stable lower branch exists only for $\dot{M}$ below a critical value $\dot{M}_A$, where $\Sigma = \Sigma_A$. For values of $\dot{M}$ such that $\dot{M}_A < \dot{M} < \dot{M}_B$, there is no steady state solution in stable thermal equilibrium. Under the circumstance that gas is supplied to a particular location at a rate $(\dot{M}_{\rm i})$ in this intermediate regime, the disk parameters can only oscillate between equilibria on the upper and lower branches (Bath & Pringle 1982). For $\Sigma < \Sigma_A$ initially, the surface density gradually increases because the local mass diffusion rate $(\dot{M})$ is $< \dot{M}_{\rm i}$.

The disk parameters adjust to quasi-equilibrium states along the lower branch on a long time scale until $\Sigma = \Sigma_A$. Further increases in Σ force an upward transition to the upper branch (with a several-fold increase in the midplane temperature $T_{\rm c}$) on the thermal timescale. On the upper branch, $T_{\rm eff}$ is also much larger than that on the lower branch. To maintain an equilibrium, the viscous dissipation rate must also increase. Since the ultimate energy source is the release of gravitational energy through viscous dissipation (Lynden-Bell & Pringle 1974), an upward transition in the energy dissipation rate requires $\dot{M}$ to increase beyond $\dot{M}_{\rm i}$ so that Σ decreases. As gas is depleted in the disk on a viscous timescale, Σ reduces to Σ_B. Further decreases in Σ force a downward transition to the lower branch, which is colder and less viscous.

The actual values of Σ_A and Σ_B depend on the functional dependence of the effective viscosity on other disk properties. For example, in the ad hoc α prescription, viscosity is an increasing function of T as required for the S curve equilibrium solution. In this case (Lin & Papaloizou 1985),

$$\Sigma_A = \alpha_{\rm l}^{-19/47}\Omega^{-29/47}{\rm g\ cm}^{-2}, \tag{11}$$

$$\Sigma_B = 0.2\alpha_{\rm h}^{-26/45}\Omega^{-13/18}{\rm g\ cm}^{-2}. \tag{12}$$

The corresponding mass transfer rates are

$$\dot{M}_A = 10^{11} \alpha_l^{32/83} \Omega^{-138/83} \mathrm{g\ s^{-1}}, \tag{13}$$

$$\dot{M}_B = 4 \times 10^{12} \alpha_h^{3/10} \Omega^{-7/4} \mathrm{g\ s^{-1}}. \tag{14}$$

Because the dominant angular momentum transfer mechanism may depend on T and the degree of ionization, the values of α on the lower (α_l) and upper (α_h) stable branch need not be identical.

Both $\dot{M}_A$ and $\dot{M}_B$ are functions of Ω and therefore r. For example, near the white dwarf in a cataclysmic variable, the disk becomes partially ionized and thermally unstable even in systems where $\dot{M}_i$ is very small ($< \dot{M}_A \sim 10^{-12}\ M_\odot$ yr^{-1}), but near their outer boundaries, CV disks remain fully ionized and stable only in systems with relatively large $\dot{M}_i$ ($> \dot{M}_B \sim 10^{-8}\ M_\odot$ yr^{-1}). Since both $\dot{M}_A$ and $\dot{M}_B$ are increasing functions of r, the entire disk would be fully ionized and thermally stable if $\dot{M}_i > 10^{-8}\ M_\odot$ yr^{-1}, as in the disks of CN and NL systems (Smak 1984a). However, in DN systems, the mass transfer rate between the interacting binary stars is typically $\sim 10^{-9}\ M_\odot$ yr^{-1} (see Section 2), which is $< \dot{M}_B$ near the outer boundary of the disk. Consequently, these disks are unstable and may undergo repeated outbursts.

For protostellar disks in YSOs, we consider a scale of about 1 AU, where for $\alpha > 10^{-6}$, $\dot{M}_A$ is of order $10^{-6}\ M_\odot$ yr^{-1}. Outbursts are not expected to be observed in protostellar disks around typical T Tauri stars because $\dot{M}_i$ is much smaller than this value (Bertout et al 1988). Then any outbursts must be confined to a very small radial extent, so they are likely to be unobservable. It is also possible that any potentially unstable regions are absent if the central star has a magnetic field.

However, members of the FU subclass of YSOs have flat SEDs beyond $\sim 10\ \mu$m. If the flat SEDs are due to radiation reprocessing by infalling envelopes, then the inferred infall rate onto the disk would be $\dot{M}_i \sim 10^{-6}$–$10^{-5}\ M_\odot$ yr^{-1} (Kenyon & Hartmann 1991). This makes it possible for the inner region of protostellar disks to become thermally unstable. However, for $\alpha > \sim 10^{-3}$, $\dot{M}_A$ increases with r to values $> 10^{-5}\ M_\odot$ yr^{-1} beyond about 1 AU. Thus, transitions triggered by thermal instability, or any other mechanism associated with partial ionization, must be confined to the inner regions.

5.2 *Disk Evolution During an Outburst Cycle*

The above stability criteria are based on local analyses applicable to rings at single radial locations. In order to construct models for DN and FU outbursts, it is necessary to examine the possibility that a transition may be triggered at one location and propagate over extended regions of the disk with a relatively small time lag. This propagation is caused by the change in the viscous stress that occurs as a consequence of a transition.

In a thin disk, the redistribution of matter is governed by Equation (1). When an anomalous local viscous stress, such as may be provided by, for example, MHD turbulence, dominates the angular momentum transfer process, Equation (1) reduces to

$$\frac{\partial \Sigma}{\partial t} - \frac{3}{r}\frac{\partial}{\partial r}\left[r^{1/2}\frac{\partial}{\partial r}\left(\Sigma\langle\nu\rangle r^{1/2}\right)\right] = 0. \quad (15)$$

Along a series of stable thermal equilibrium models, $\langle\nu\rangle$ is solely determined as a function of Σ. Along such a series, the evolution of the disk surface density may be obtained from Equation (15) alone. However, when a point of bifurcation is reached, there is a transition between two different stable branches during which local thermal equilibrium is no longer maintained. Based on the fact that the transition occurs on a timescale shorter than the viscous evolution timescale, changes in the viscous stress during transition have been determined using various relaxation prescriptions (Meyer & Meyer-Hofmeister 1984, Smak 1984b, Mineshige & Osaki 1985, Cannizzo et al 1986, Pringle et al 1986).

These prescriptions neglect the nonlocal contributions to the energy transport, such as radiative and advective heat flow in the radial direction, as well as the ionization (or recombination) of the disk gas. These processes have strong effects in regulating the propagation of the transition front in the radial direction (Papaloizou et al 1983). In principle, the evolution of the vertical structure should be analyzed simultaneously with the adjustment to a new thermal equilibrium and the radial redistribution of matter (Kley & Lin 1996). Hydrostatic equilibrium in the vertical direction, however, is established on the local dynamical timescale, which is a fraction ($\sim \alpha$) of the thermal timescale and is much smaller than the viscous diffusion timescale (by a factor $\sim \alpha H^2/r^2$). Thus, it is customary to adopt the assumption of hydrostatic equilibrium in order that the disk structure may be vertically averaged and expressed as a function of r and t alone.

If the disk is idealized as a slab of constant semi-thickness H, in which the temperature $T \equiv T_c \neq T_{\rm eff}$ and density $\rho \equiv \Sigma/(2H)$ are independent of height, Equation (3) of Paper I gives for a viscous disk:

$$C_V\left[\frac{DT_c}{Dt} - (\Gamma_3 - 1)\frac{T_c}{\Sigma}\frac{D\Sigma}{Dt}\right] = \langle\nu\rangle r^2\left(\frac{d\Omega}{dr}\right)^2 - \frac{2F_s}{\Sigma} - \frac{2H}{r\Sigma}\frac{\partial(rF_r)}{\partial r}, \quad (16)$$

where $D/DT = \partial/\partial t + V_r\partial/\partial r$ is the advective derivative, V_r is the mean radial velocity, and F_r and $F_s = \sigma T_{\rm eff}^4$ are the radial and surface heat flux, respectively.

It is necessary to express $T_{\rm eff}$ as a function of $T_{\rm c}$ and Σ. In most temperature ranges, the Eddington approximation is adequate ($T_{\rm eff}^4 = T_{\rm c}^4/\tau$) (Shakura & Sunyaev 1973, Lin & Papaloizou 1980). However, just below 10^4 K, when H^- is the dominant contributor to the opacity, the vertical thermal structure is determined by the surface boundary condition that $\bar{\kappa}(\rho, T_{\rm eff})\Sigma \sim 1$ (Faulkner et al 1983). Nevertheless, hydrostatic equilibrium in the vertical direction gives the approximation for the pressure $P(\rho, T_{\rm c}) = \rho\Omega^2 H^2$, so that ρ and therefore $T_{\rm eff}$ can be expressed as functions of $T_{\rm c}$ and Σ.

The validity of these approximations can be verified by the numerical solutions of Equations (2–7) for thermal equilibrium in which the energy generated at any given radius is radiated locally. For an α model, the above prescriptions for $T_{\rm eff}(T_{\rm c}, \Sigma)$ reproduce the $(\langle\nu\rangle\Sigma, \Sigma)$ S curve obtained by numerical integration (Meyer & Meyer-Hofmester 1981, Faulkner et al 1983). Similar checks can be made for ranges of $T_{\rm c}$ and Σ corresponding to models appropriate for the transition phase between stable branches and for moderately thick regions where the first two terms on the right-hand side of Equation (16) do not cancel each other. Under the assumption that the contribution to the radiative and advective heat transport in the radial direction is concentrated near the midplane, the numerical integrations of the vertical structure equations (Bell & Lin 1994) lead to results that agree well with those inferred from the above $T_{\rm eff}(T_{\rm c}, \Sigma)$ prescription. The global evolution of a thermally unstable disk can then be examined with Equations (15) and (16).

In both CVs and YSOs, matter is fed to the disk through its outer regions. During the initial buildup of the disk, $\dot{M}$ in the disk is smaller than the mean infall flux ($\dot{M}_{\rm i}$), so that the newly arriving matter accumulates in the disk and Σ increases along the stable lower branch despite some mass diffusion. After the disk becomes thermally unstable and undergoes an upward transition at some location (which may be in either the inner or the outer parts of the disk depending on the circumstances), that location reaches a new thermal equilibrium in which both the viscous stress and T are substantially increased. Thereafter, the heated region evolves as a localized, hot, highly viscous annulus. The inner edge of this hot region spreads inward, promoting upward transitions at successively smaller radii; similarly, the outer edge spreads outward (Papaloizou et al 1983) and propagating transition fronts are established (Meyer 1984, 1986).

The propagation speed of the transition front can be estimated by noting that disk material changes state on the thermal timescale, which is $\sim C_V \mathcal{R}^{-1}(\alpha\Omega)^{-1}$ (Pringle 1981), where the ratio of heat capacity per unit mass to the gas constant is $C_V \mathcal{R}^{-1} \sim 1$ except when the gas is undergoing ionization, when it becomes > 10. During this time, viscous diffusion can occur over a length scale $\sim (C_V\langle\nu\rangle)^{1/2}(\mathcal{R}\alpha\Omega)^{-1/2}$. A characteristic speed U_r is obtained by dividing this length scale by the thermal timescale, resulting in

$$U_r \sim \langle\nu\rangle C_V^{-1/2}\mathcal{R}^{1/2}/H \sim \alpha C_V^{-1/2}\mathcal{R}^{1/2}c_s. \tag{17}$$

This value is the natural speed at which a transition occurring at one location can move to adjacent regions (Papaloizou & Pringle 1985b). It is important to note that the front speed is significantly reduced and the front width is significantly increased when the gas is undergoing ionization and $C_V\mathcal{R}^{-1} \sim 10$.

An alternative method of analysis is to examine solutions of (16) and (15) (or other comparable equations) in the neighborhood of some point $r = r_0$, by setting $r = r_0 + x$, letting $\partial/\partial r \rightarrow \partial/\partial x$, and then neglecting x where it appears explicitly. One can then find propagating solutions where variables are functions of $x - U_r t$ only, with U_r being given by a multiple of the above characteristic velocity (see Lin et al 1985 for an example).

As the heated region expands, the luminosity of the disk increases. Eventually, the light curves reach their asymptotic limit when the transition wave has either reached the outer edge of the disk or stalled. In CVs, the physical dimension of disks is limited by the Roche lobe of the binary ($\sim 0.5\ R_\odot$). The transition front can propagate throughout the entire disk on a timescale $\sim \alpha^{-1} \times 10^4$ s. The observed rise timescale (a few hours) in systems such as SS Cyg implies $\alpha_h \sim 0.1$ in the upper state during CV outbursts (Cannizzo & Kenyon 1987).

The observed rise timescale for YSO disks is typically a year to several decades. Although YSO disks are very extended ($> 10^{16}$ cm), the propagation of the front stalls at radii (~ 1 AU) well interior to the disk edge (Clarke et al 1989, 1990) because $\dot{M}_A$ is a rapidly increasing function of r. Values of $\alpha_h \sim 10^{-2}$ are compatible with the observed rise timescale for FU Ori outbursts (Bell et al 1995).

After front propagation is completed, a new thermal equilibrium is established in which the gas is fully ionized. If the mass supply rate to the disk $\dot{M}_i$ is larger than $\dot{M}_B$ everywhere in the disk, this stable equilibrium would be maintained. However, if $\dot{M}_i < \dot{M}_B$, somewhere, mass is diffused faster than supplied when the disk is fully ionized. Both Σ and T must decrease until the disk gas becomes partially recombined and unstable to a downward transition. The theoretical value of $\dot{M}_B \sim$ a few $\times 10^{-9}\ M_\odot\ \mathrm{yr}^{-1}$ for the outer regions of CV disks. In CN and NL, the observationally inferred $\dot{M}_i > \dot{M}_B$ in the outer parts of the disk, and the disk permanently remains in a fully ionized hot state (see Section 2).

However, in DN, $\dot{M}_i(\sim 10^{-9}\ M_\odot\ \mathrm{yr}^{-1})$ (Patterson 1984) is less than both $\dot{M}_B$ in the outer regions of the disk and the observationally inferred $\dot{M}(> 10^{-9}\ M_\odot\ \mathrm{yr}^{-1})$ during outbursts in systems such as Z Cha (Horne & Cook 1985) and OY Car (Rutten et al 1992a). Because matter in the disk is not being replenished at a sufficiently rapid rate, each outburst is followed by a decline to quiescence in which the disk has $\dot{M} < \dot{M}_{\rm i}$.

In protostellar disks, the observationally inferred $\dot{M}(\sim 10^{-4}$–$10^{-5}\ M_{\odot}\ {\rm yr}^{-1})$ during FU outbursts (Hartmann et al 1993) is much larger than typical values of $\dot{M}_{\rm i}(\sim 10^{-6}\ M_{\odot}\ {\rm yr}^{-1})$ (Lin et al 1994). Thus, these outbursts are followed by a decline, although in some cases (e.g. FU Ori) the decay proceeds very slowly. Theoretically, the FU outbursts would not persist indefinitely if $\dot{M}_B > \dot{M}_{\rm i}$. The value of $\dot{M}_B$ increases with r and is $\gg 10^{-5}\ M_{\odot}\ {\rm yr}^{-1}$ for $r > 1$ AU. Based on the observed luminosity and the estimated core mass, the mass infall rate is constrained to be $\sim 10^{-6}$–$10^{-5}\ M_{\odot}\ {\rm yr}^{-1}$ for these embedded protostars (Adams et al 1987). A smaller $\dot{M}_{\rm i}$ is inferred for the FU systems because they have less extinction.

During the decline, the depletion of disk material occurs on a viscous diffusion timescale $\sim r^2/\langle\nu\rangle$ that is longer than the rise timescale by a factor r/H, as indicated by the observed light curves of both DN and FUs. The disk temperature decreases with the gradual reduction in Σ until ionized hydrogen atoms begin to recombine and H^- opacity once more dominates. Thereafter, the thermally unstable regions undergo a downward transition in temperature, opacity, and viscous stress. This transition always occurs first at the outer edge of the fully ionized hot regions. In the region interior to the transition front, a quasi-steady state is established in which $\dot{M}$ is essentially independent of r—although it does decrease gradually. Consequently, the SED of the continuum radiation decreases self-similarly, as is observed during the initial stages of the decline from DN and FU outbursts.

When the transition reaches the inner boundary of the disk, the original quiescent configuration is restored. The observed evolution of the Balmer lines from absorption to emission lines indicates that the entire disk of a dwarf nova and the inner regions of an FU disk become optically thin during the decline. In this limit, turbulence may decay because it is possible for the thermal energy to be radiatively lost at a rate faster than it can be generated through viscous dissipation. The flattening of the SED of the continuum radiation at the advanced stages of the decline suggests that mass diffusion is no longer steady. As mass begins to accumulate, the cycle repeats itself.

5.3 *Confrontation Between Theories and Observations*

The thermal instability scenario provides an attractive model for the repeated occurrence, the relative strengths, and the absolute magnitudes of DN and FU outbursts (Cannizzo 1992). Unsteady disk flow associated with these outbursts provides a powerful diagnostic for evaluating the nature and efficiency of angular momentum transport in astrophysical accretion disks. The observed spectral and SED evolution indicates that during the outbursts the disk is highly ionized, opaque, and bright. In contrast, during quiescence it is mostly neutral, optically thin, and dim. This correlation indicates that the efficiency of angular momentum must be an increasing function of temperature. The conventional α prescription, although ad hoc, is consistent with this correlation and therefore may be used as a fiducial formula for the effective viscosity.

Nevertheless, there are remaining inconsistencies between observational data and preliminary models. For example, in DN systems, a potential problem has been raised by the observed 1-day delayed increase of the UV continuum during the rise of VW Hydri (Schwarzenberg-Czerny et al 1985). The first attempts to model this observation (Pringle et al 1986, Cannizzo & Kenyon 1987) are based on the numerical solution of Equation (15) with a relaxation prescription governing the transition of T_c and $\langle \nu \rangle$ during thermally unstable phases (Smak 1984c, Cannizzo et al 1986). In these calculations, the SED of emerging radiation is computed from an extrapolation of stellar atmosphere models for disks (Wade 1984), rather than the black body law. The results of these calculations indicate that an upward transition leads to a rapid increase in $T_{\rm eff}$ and a nearly simultaneous rise in both UV and optical flux. This inconsistency with observation has been used as evidence against thermal instability as the mechanism regulating DN outbursts.

However, these treatments do not account for the great changes in the ionization fraction of the gas and the associated internal energy during upward/downward transitions. Also, advective transport in the radial direction is ignored (see Section 5.2). The required energy input for ionization is about one order of magnitude larger than that needed to increase the kinetic energy of the gas particles during an upward transition. This is reflected in the fact that $C_V \mathcal{R}^{-1} \sim 10$ as the gas undergoes ionization (Lin et al 1985). Mineshige (1988) finds that the incorporation of the ionization contribution, together with a recalculation of the detailed form of the S curve, causes the increase in $T_{\rm eff}$ to be slowed at $T_c \sim 10^4$ K as the gas becomes fully ionized, which consequently enables the observed UV delay to be reproduced. Note that the release of "latent heat" during recombination will also lengthen the decay timescale in the light curve.

Another problem with the thermal instability scenario for DN is that the theoretical light curves generated with a uniform α prescription provide either no large coherent outbursts (Smak 1984c) or outbursts that are separated by very short quiescent intervals (Papaloizou et al 1983). In these models, the range of $\Sigma (= \Sigma_A - \Sigma_B)$ for which equilibria can be thermally unstable is small. Consequently, after each downward transition, very little replenishment is needed to trigger the onset of a subsequent upward transition. The outburst interval can be increased by either decreasing the magnitude of α through the cycle, making it smaller in the lower state (Cannizzo et al 1988, Livio & Spruit 1991), or by making α an increasing function of T_c (or equivalently H/r), so that it is larger during the upper fully ionized state (Smak 1984c, Cannizzo et al 1986).

Modifications of this type (Mineshige 1986) enable the disk to be mostly flushed during the outburst cycle so that a substantial replenishment is needed to increase Σ enough to trigger a subsequent upward transition. One suggestion (Meyer & Meyer-Hofmeister 1983, Vishniac et al 1990) for explaining a T_c-dependent α is that a magnetic dynamo may become more vigorous and efficient in thicker disks. It is also possible that an increasing ionization fraction may lead to conditions more favorable for dynamo mechanisms dependent on MHD turbulence arising from the Balbus-Hawley instability (Tout & Pringle 1992).

The magnitudes of α_h and α_l can be constrained by the observed light curves in DN. During quiescence, Equation (11) implies that upward transitions would occur in the outer disk regions ($R \sim 10^{10}$ cm) when $\Sigma > \Sigma_A \sim 10^2$ g cm^{-2}, if $\alpha_l < 10^{-2}$. But, the disk mass accreted by the white dwarf during outbursts is $\sim 10^{-11}\ M_{\odot}$. If this mass is a significant fraction of the disk gas accumulated during quiescence, the inferred Σ would be ~ 50 g cm^{-2}. The outer regions of the disk are unstable to upward transition at this surface density only if $\alpha_l \sim 0.05$.

Similarly, if $\alpha_h < 0.1$, Equation (12) would imply $\Sigma_B > 10$ g cm^{-2} in the outer regions. Then the disk would remain opaque throughout the downward transition and a new low state equilibrium would be established at $T_e < 3 \times 10^3$ K (Cannizzo et al 1982). However, strong emission lines emerge and dominate the spectrum during the decline (see Section 2). Detailed analysis of the equivalent width and decrement ratio of Balmer lines (Williams 1980, Williams & Ferguson 1982) suggests that, in quiescence, the disks become optically thin with $T_c \sim 6$–8×10^3 K and $\Sigma \sim$ a few g cm^{-2}. Such a state is only attainable with values of $\alpha_h \sim 0.1$–1).

The value of α may also be obtained more directly from observations. By fitting the brightness temperature distribution. Mineshige & Wood (1989) estimate $\alpha_l \sim 0.02(r/10^{10}\,\mathrm{cm})^{0.4}$ for Z Cha during quiescence. However, their atmosphere model has problems in reproducing the SED of the continuum for both the optically thick and thin regions of the disk (Wade 1988). An alternative method to extract the value of α_l is to utilize the values of T_c and Σ obtained from detailed modeling of the equivalent width, decrement, and profile of Balmer emission lines during quiescence.

Based on the assumption that the disk is optically thin and the observed disk flux is generated from local viscous dissipation, Cheng & Lin (1992) find $\alpha_l \sim 0.5$. Smaller values of α would not only reduce the equivalent width but also enhance Stark broadening to the extent that the double-peak profile (due to Doppler broadening) would be smeared out (Lin et al 1988). These values are consistent with those extrapolated from the evolution of the Balmer line profiles during eclipses (Rutten et al 1992a,b). Note that these values of α_1, deduced for the outer disk region, are larger than those required to reproduce the observed DN light curves.

This discrepency may be due to the additional dissipation of tidally excited density waves in the outer regions of the disk (Savonije et al 1994). When a viscous disk dissipates energy, an angular momentum flux F_H is transported to the outer regions where tidal dissipation takes place at a rate $\sim F_H\Omega$ (Papaloizou & Pringle 1977). As this occurs, angular momentum is transported to the companion. In a steady situation, this dissipation rate is equal to $\dot{M}r^2\Omega^2$, evaluated at the tidal radius. For an annulus of width Δr, the ratio of this tidal dissipation rate to the internal viscous dissipation rate is $2r/(3\Delta r)$. Thus tidal dissipation might be significant for an annulus that is accumulating near the tidal radius at the outside edge of the disk during quiescence. Furthermore, it is possible that the tidal dissipation takes place in strong shocks within several disk thicknesses of the disk edge (Lin & Papaloizou 1993). Note too that tidal dissipation in addition to that expected in a steady state may occur if tidal effects are causing the disk to contract from its expanded extent during outburst. Under these conditions, tidal dissipation can have a strong influence on the outer regions.

For modest values of $\alpha_1 (> 10^{-2})$, viscous diffusion during quiescence would cause a sufficiently large inward diffusion of mass to make the inner region become unstable first. This happens because a given T is achieved at smaller Σ in the inner region of the disk owing to the stronger gravity there.

The onset location of thermal instability in the disk may be inferred from the spectral evolution of the Balmer lines, which alternates between absorption during outbursts to emission during quiescence. These transitions are consistent with the thermal instability models in which the disk is opaque during the high state but optically thin in the low state (Cheng & Lin 1989, Shaviv & Wehrse 1989). During the rise in SS Cyg, the transition of Balmer lines from emission to absorption first occurs in the wings. Since the double-peaked profile indicates the lines are Doppler broadened, the line wings originate from the inner parts of the disk. The width of the emission component decreases as the absorption feature in the wings expands (Horne et al 1990). A symmetric spectral evolution is also observed during the decline (Clarke et al 1984, Hessman et al 1984). Detailed spectral analyses (Cheng & Lin 1992) indicate that outbursts may indeed be initiated at the central regions of the disk. Nevertheless, there are DN (such as VW Hydri) where the outbursts may have originated in the outer regions of the disk. In all cases, the decline is initiated at the outer regions of the disk.

Information on the spectral evolution of FU systems is much more scarce than that for DN. A pre-outburst spectrum is only available in one FU system (V1057 Cyg), and it is also characterized by strong emission lines (Herbig 1977). Candidates for FU in quiescence, i.e. flat spectrum sources (Lin et al 1994), also have strong emission lines that may originate from the optically thin inner regions of the disk (Bell & Lin 1994), similar to classical T Tauri stars (Bertout et al 1988).

Analogous spectroscopic data for the rise of FU outbursts is not available, but during the decline the spectra are characterized by absorption lines similar to the spectra of DN at a similar stage. In contrast to the spectral evolution of DN, the width of absorption lines continues to decrease well after the maximum light (Herbig 1989). This evolutionary trend may indicate that the outward propagation of the ionization wave has not yet come to a stall even after maximum light is attained (Bell et al 1995).

Thermal instability models of the observed light curves of FUs require $\alpha_l \sim 10^{-3}$ (Bell et al 1995) for sufficient matter to be accumulated during the $\sim 10^3$ yr intervals between outbursts. During the outbursts, the power-law SED indicates that, near the stellar surface, the disk is opaque and the gas near the midplane is fully ionized. However, the decade-long persistence of high states requires $\alpha_h \sim 10^{-2}$. These low values of α are in contrast to those inferred from DN outbursts and may be inconsistent with what is expected from fully developed MHD turbulence.

6. COMPANION-DISK INTERACTION

The presence of orbiting companions in the proximity of accretion disks is a common feature in CVs and YSOs. It has also been sugested that they could play an important role in AGN. Tidal perturbation by a companion can excite density waves, which carry a flux of angular momentum in the disk. The exchange of angular momentum can regulate both disk evolution and the orbit of the companions. The detailed physics associated with companion-disk interaction has been reviewed recently (Goldreich & Tremaine 1982, Lin & Papaloizou 1993). We briefly discuss its role in different astronomical contexts.

6.1 *Star-Disk Interaction in AGN*

Around the nucleus of M87, a cusp in the stellar density is observed (Lauer et al 1992). This extends into the region where a rotating gaseous disk is found. Accordingly, the direct passage of field stars in inclined orbits through the disk has been considered. This is not expected to induce a very strong tidal torque (Ostriker 1983), but, analogous to the situation in planetary rings (Shu 1984), corrugation waves are excited at the vertical resonances of the perturbing stars

(Artymowicz 1994, Ostriker 1994). Dissipation of the waves causes the stellar orbital energy and inclination to gradually decrease (Syer et al 1991).

When the disk viscosity is large, waves are dissipated near their point of origin. For small viscosity, warping modes are excited in the form of radially propagating waves (Papaloizou & Lin 1995b). Amplification of the oscillation amplitude in low surface density regions may increase the effective area that is exposed to the central quasar for radiation reprocessing.

Stars with relatively low initial inclination essentially corotate with the disk and are expected to be rapidly trapped by the disk. Artymowicz et al (1993) argue that the trapped stars are likely to evolve into upper main sequence stars through gas accretion in the dense environment of the disk. Because an outward radial pressure gradient results in a sub-Keplerian azimuthal speed of the disk gas, a drag on the orbits of the trapped stars is expected (Weidenschilling 1977). In addition, the trapped stars interact tidally with the disk. Resonant tidal interaction (Ward 1986, Korycansky & Pollack 1993) may result in radial migration in the disk (Ward & Hourigan 1989). The efficiency of this feedback process depends on the details of the disk response (Papaloizou & Lin 1984, Lin & Papaloizou 1986b, Takeuchi et al 1996). In combination, these two drag effects lead to a minimum starting radius ($\sim 10^{-2}$ pc) beyond which the trapped stars will evolve into type II supernovae before drifting into the central black hole.

Although supernovae explosions of these stars do not disrupt the disk flow, they do introduce a local stirring, which leads to an efficient outward transfer of angular momentum (Rozyczka et al 1995). The disk is also contaminated with heavy element ejecta. The concurrent contamination by trapped stars and the flushing of disk gas into the central massive black hole determine the asymptotic disk metallicity. Spectral models of the observed relative emission line intensities of Fe and the CNO group of elements indicate that the heavy element abundance of quasars (some with very large redshifts) is comparable to or larger than the solar value (Davidson 1977, Rees et al 1989, Netzer et al 1992). It is possible that the observed high metallicity could be generated globally during a brief phase of rapid star formation during the early epoch of galaxy evolution (Hamann & Ferland 1992). However, local contamination by trapped stars in the disk has the advantage of high efficiency in generating the metallicity indicated by the AGN emission lines without the need of prolific production of intermediate- and low-mass stars in the host galaxy.

The distribution of Galactic pulsars indicates that typical neutron-star remnants have a typical recoil velocity of a few hundred kilometers per second (Narayan & Ostriker 1990, Lyne & Lorimer 1994). The subsequent motion in the gravitational potential of the MBH would lead to a sufficiently small orbital inclination for the remnants to be retrapped. Upon settling back into the disk, it is possible that the compact remnants resume their rapid accretion despite the Eddington-limit barrier (Chevalier 1993) and become stellar mass black holes (SBHs) on timescales short compared with the active phase of AGN. When SBHs have acquired a mass $\sim (H/r)^3 M_{\mathrm{MBH}} (\sim 10^2 M_\odot)$, they may be able to tidally truncate the disk in the vicinity of their orbit (Lin & Papaloizou 1993). The formation of gaps in the disk prevents any further growth of the SBH.

If present, SBHs also exchange orbital angular momentum with the disk gas near their orbital radius through tidally induced density waves (cf Section 5 in Paper I). Embedded in the background of many low-mass field stars, the SBH orbits decay by dynamical friction (Binney & Tremaine 1987). The angular momentum removed from the disk is passed to the field stars in the galactic nucleus, which in turn undergo two-body relaxation with more distant stars. A self-consistent estimate indicates that at any given time, $\sim 10^4$ SBHs may be embedded in the disk, and their collective contribution could drive the disk to evolve on a timescale of $\sim 10^8$ yr, with a corresponding $\alpha \sim 1$ (Lin et al 1994).

The companion-disk interaction may also play an important role in gravitationally unstable regions of AGN disks (Paczyǹski 1978, Kozlowski et al 1979). Gravitational instability occurs in a disk when the ratio of the mass contained in the disk to that contained in the central object becomes larger than the ratio of the sound speed to the Keplerian speed, or if the Toomre parameter $Q < 1$. In most AGN disks, this condition is easily satisfied, particularly in the outer regions. Gravitational instability can then occur when the disk mass is small, with the characteristic wavelength of the most rapidly growing modes being much shorter than the radius. Gravitational instability may lead to fragmentation (Burkert & Bodenheimer 1993) or the formation of a system of clouds (Miyama et al 1984, Shlosman et al 1990) in regions with $Q \ll 1$. Subsequent gravitational scattering of the clouds against each other leads to some angular momentum transport (Fukunaga 1984, Fukunaga & Tosa 1989, Jog & Ostriker 1988, Gammie et al 1991), but in an analogous manner to what occurs in planetary rings (Goldreich & Tremaine 1978), these interactions cause an increase in the cloud velocity dispersion until physical collisions become important. Collisions may lead to either the disruption or coagulation of clouds. The consequences of these physical processes remain to be fully explored.

Finally, disks in AGN may be under the gravity of two or more MBHs interacting in galactic nuclei (Begelman et al 1980). The existence of MBH-binaries (and multiple-BH systems) follows naturally from two widely accepted assumptions: 1. that many galaxies contain central MBHs and 2. that large galaxies are built hierarchically through mergers of smaller galaxies. A precise understanding of the dynamical evolution of such binaries is needed to answer many pressing questions in extragalactic astronomy concerning the behavior of jets in AGN, the formation of double-nuclei galaxies, the formation of cores in galactic nuclei, and the viability of MBHs as a source of halo dark matter.

Although there are several analyses of dynamical interactions between MBH-binaries with a stellar background (Mikkola & Valtonen 1992, Makino et al 1993, Quinlan et al 1995), the role of companion-disk interaction has not yet been discussed widely in this case.

6.2 *Tidal Perturbations in CV Disks*

Tidal interaction induces disturbances in the disk. When these disturbances are dissipated, angular momentum is effectively transferred from the disk to the binary orbit (Section 6). A significant fraction of the tidal disturbance may be dissipated in the outer regions of the disk. During quiescence when the disks are not in a steady state (Section 2) this process could contribute to a large fraction of the total energy output from them and provide a natural explanation for the observed flat temperature distribution in the disks of Z Cha and OY Car (Wood et al 1986, 1989) and the strength of the emission lines in SS Cyg (Cheng & Lin 1992).

Tidal interaction also produces density waves that propagate toward the center of the disk. During the outbursts, the disk is relatively hot, viscous, opaque, and thermally stratified. Compressional density waves propagating through such a medium tend to be refracted and to propagate vertically into the tenuous atmosphere of the disk. When these waves become nonlinear, shock dissipation causes the conversion of the energy carried by the waves into thermal energy of the disk gas. The dissipation of energy above the midplane reduces the thermal stratification and the efficiency of the refraction process. During their propagation in the radial direction, the density waves steepen into shocks and deposit angular momentum in the disk gas. However, it is unlikely that the dissipation of these waves could produce an effective α sufficiently large to be significant during the outbursts, so that the importance of tides is limited to the outer regions of the disk.

We note that if the process leading to angular momentum transport is MHD turbulence, then expulsion of toroidal field from the midplane regions through a process such as magnetic buoyancy (Tout & Pringle 1992, Rozyczka et al 1995) may also lead to significant dissipation in the upper regions in the interior part of the disk.

When heated to above $T \sim 10^4$ K, the disk atmosphere may become thermally unstable and heat up (King & Shaviv 1984, Murray & Lin 1991). A temperature inversion and disk corona may be produced. Such a temperature inversion is consistent (Cheng & Lin 1992) with 1. the absence of any Balmer discontinuity, which is predicted from the extrapolation of stellar atmosphere models (Wade 1984); 2. the shallow profile of the absorption wings of Balmer emission lines (Hessman et al 1984); and 3. the coexistence of strong He I lines and Balmer emission lines with flat decrements (Williams 1980). The effects of energy dissipation in the upper regions of the disk may be sufficient to establish the observed flat temperature distribution in NL disks (Rutten et al 1992b), although details remain to be worked out.

Tidal interaction may also play a very important role in regulating the disk structure in the SU Ursae Majoris subclass of DNs. These systems have short orbital periods and have mass-losing companions of masses usually less than one third that of the star at the center of the disk. This class of stars undergoes super as well as normal outbursts (Warner 1995). The super outbursts occur more regularly and have larger amplitudes than regular outbursts. During the super outbursts, the light curves contain super humps, which have periods a few percent longer than the binary period. These super humps have been interpreted to be due to slowly precessing elliptical rings near the outer edge of the disk (Krzeminski & Vogt 1985). One possible mechanism for exciting such a precession is through resonant tidal interaction between the companion and the disk (Whitehurst 1988, Hirose & Osaki 1990). Numerical simulations based on Lagrangian (Whitehurst & King 1991) and smooth particle hydrodynamics (Lubow 1992) show that in binary systems with such an extreme mass ratio, eccentric flow pattern in the outer regions of the disk may be excited by a parametric 3:1 resonance. However, this excitation mechanism is weak and its occurrence may depend on the special form of the tidal potential. A simulation with a finite difference scheme indicates that no eccentric prograde disk evolves when the full potential is used (Heemskerk 1994). It remains possible that an eccentric flow pattern may be excited by the impact of a stream of newly transferred gas in the outer regions of the disk (Lubow 1994).

6.3 *Protoplanets in YSO disks*

The discovery of GW Ori (Mathieu et al 1991), GG Tau (Dutrey et al 1994, Roddier et al 1995), and T Tau (Dyck et al 1982, Ghez et al 1991) indicates that binary companions are common in or outside protostellar disks. Analogous to the situation occurring in CVs, the disk structure is strongly perturbed by the tidal field of the companion, especially in proximity to the orbit (Jensen et al 1994), and gap formation may occur. For unresolved sources, this structural alteration can lead to modification of the SED (Mathieu 1994). Detailed analysis of binary YSOs is important not only for probing disk structure but also for the investigation of the origin of binary stars (Artymowicz et al 1991, Artymowicz & Lubow 1994). For comprehensive discussions on the observational properties and theoretical analysis of binary star-disk interaction in YSOs, we refer readers to other recent reviews (Mathieu 1994, Bodenheimer 1995).

Protostellar disks also interact with any protoplanets embedded within them. According to current theories, protoplanetary formation proceeds with the initial condensation of grains (Weidenschilling 1977). Cohesive collisions between these grains then lead to the emergence of planetesimals (Aarseth et al 1992), which coagulate to produce protoplanetary cores (Safronov 1972, Wetherill 1980). When these cores acquire a sufficiently large mass (typically a few times the mass of the Earth), they accrete gas in the vicinity of their orbits and become protoplanets (Bodenheimer & Pollack 1986). Protoplanets interact tidally with the disk during their growth, generating density waves near their Lindblad resonances (Goldreich & Tremaine 1980). The waves exterior (interior) to a protoplanet carry an excess (deficit) of angular momentum compared to the local disk material. The dissipation of the waves induces the disk material to gain (lose) angular momentum and to move radially outward (inward) and thus away from the protoplanet's orbit (Section 5 of Paper I).

The response to the tidal perturbation of a low-mass protoplanet is dominated by the pressure of the gas. However, for protoplanetary mass $> 50\, M_*/\mathcal{R}$, where $\mathcal{R}$ is the Reynolds number, the angular momentum flux due to the tidal torque exceeds that due to the viscous stress. Provided that the mass is also large enough that its tidal or Roche radius $> H$, a gap would form so that further growth of the protoplanetary mass is quenched. These conditions imply that in order for Jupiter to acquire its present mass in the solar nebula, $H \sim 0.1\, r$ and $\alpha \sim 10^{-2}$–10^{-3} (Lin & Papaloizou 1980, 1986a, 1993).

The above inferred value for H is comparable to that observed in YSOs. The magnitude of α may also be constrained from the diffusion timescale of the disk. The observed SED of CTTs has a break in the power index at $\sim 100\, \mu$m, which is normally interpreted as a transition from optically thick to optically thin disk structure (Beckwith et al 1990). Whether the source of IR excess is due to reprocessing or viscous dissipation, the size of the opaque disk in a CTT is ~ 30 AU. Since the age of the central star in CTTs is typically a few million years or less, the disk evolution timescale is unlikely to be much longer. The corresponding value of α is $\sim 10^{-3}$ in most regions of the disk, consistent with the above estimate for Jupiter as well as that expected for convective turbulence (Section 6 of Paper I). After the formation of several protoplanets (each of mass M_t), the dissipation of their collective disturbance in the disk may also provide $\alpha \sim (M_t/50M_*)(R/H)^2 \sim 10^{-3}$.

In an unperturbed disk, any radial variation of disk parameters causes an imbalance in the magnitude of the angular momentum fluxes produced on opposite sides of its orbit (Ward 1988). This imbalance could induce a protoplanet to migrate radially at such a rate that there is not enough time for the local depletion of the gas (Hourigan & Ward 1984), with the result that gap formation is inhibited (Ward & Hourigan 1989). However, radial profiles in the disk are affected by the tidal torques. The Σ profile adjusts depending on where the density waves are dissipated. For a modest effective viscosity, density waves are dissipated near where they are launched close to the protoplanet's orbit. In such a situation, the disk adjusts to a quasi-equilibrium, in which the angular momentum flux becomes approximately constant, on a characteristic timescale $(\alpha\Omega)^{-1}$. The orbit then evolves on the much longer global viscous timescale (Goldreich & Tremaine 1980, Lin & Papaloizou 1986b). The tidal torque decreases with the local depletion of gas, and a narrow gap opens (Artymowicz & Lubow 1994).

If the effective viscosity is small, density waves can propagate to large distances, and then the resonant tidal torque is not affected by the back reaction of the disk. Nevertheless, a protoplanet with a sufficiently large mass can open a wide gap (Takeuchi et al 1996).

The width of such gaps may be inferred from SEDs (Mathieu et al 1991). The improved spacial resolution of millimeter interferometers should make it possible to image protoplanets inside gaps embedded in protostellar disks in the near future. Because the formation of the protogiant planets requires the initial buildup of rocky cores with a few Earth masses, protogiant planets are most likely to be found in protostellar disks that are depleted in heavy elements. Such disks may be around WTTs, and a search for them should be made with H_2 molecule absorption lines.

Note that a protoplanet can undergo radial migration on a viscous timescale because of tidal effects, even though the torques are not strong enough to open a gap. The recent discovery of a protoplanet within 0.05 AU of 51 Peg (Mayor & Queloz 1995) may be due to this type of orbital migration (Lin et al 1996). If the disk is not totally truncated, the disk-companion interaction through corotation resonances leads to a reduction in orbital eccentricity. In contrast, gap formation would lead to the clearing of gas near corotation resonances. Disk-companion interaction through Lindblad resonances leads to an increase in the orbital eccentricity. For a sufficiently large eccentricity, the protoplanet may undergo excursion outside the gap and therefore continue to grow.

7. SUMMARY

The above discussion clearly demonstrates that a variety of mechanisms may be important in regulating angular momentum flow in accretion disks. There are also many generic features in the structure and evolution of disks in AGN, CVs, and YSOs. In the past decade, major advances in both theory and observation were made possible through an interdisciplinary approach, which will continue to be fruitful in future investigations.

ACKNOWLEDGMENTS

This work is supported in part by NASA through grants NAGW-3599, NAGW-3408, NAGW-4967, and by NSF through grant AST 93-15578. We also benefitted from a NASA astrophysics program that supports a joint Center for Star Formation Studies at NASA/Ames Research Center and the Universities of California at Berkeley and Santa Cruz.

Literature Cited

Aarseth SJ, Lin DNC, Palmer PL. 1992. *Ap. J.* 403:351–76
Abramowicz MA, Czerny B, Lasota JP, Szuszkiewicz E. 1988. *Ap. J.* 332:646–58
Adams FC, Emerson JP, Fuller GA. 1990. *Ap. J.* 357:606–20
Adams FC, Lada CJ, Shu FH. 1987. *Ap. J.* 312:788–806
Adams FC, Lada CJ, Shu FH. 1988. *Ap. J.* 326:865–83
Anderson N. 1988. *Ap. J.* 325:266–81
Antonucci R, Kinney AL, Hurt T. 1993. *Ap. J.* 414:506–9
Antonucci RRJ, Miller JS. 1985. *Ap. J.* 297:621–32
Artymowicz P. 1994. *Ap. J.* 423:581–99
Artymowicz P, Clarke CJ, Lubow SH, Pringle JE. 1991. *Ap. J. Lett.* 370:L35–38
Artymowicz P, Lin DNC, Wampler EJ. 1993. *Ap. J.* 409:592–603
Artymowicz P, Lubow S. 1994. *Ap. J.* 419:155–65
Aumann HH, Gillett FC, Beichman CA, de Jong T, Houck JR, et al. 1984. *Ap. J. Lett.* 278:L23–27
Backman DE, Gillett FC. 1987. *Cool Stars, Stellar Systems and the Sun,* ed. JC Linsky, RE Stencel, pp. 340–50. Berlin: Springer-Verlag
Barthel PD, Schilizzi RT, Miley GK, Jägers WJ, Strom RG. 1985. *Astron. Astrophys.* 148:243–53
Basri G, Bertout C. 1989. *Ap. J.* 341:340–58
Bastien P, Menard F. 1990. *Ap. J.* 364:232–41
Bath GT, Evans DW, Papaloizou J, Pringle JE. 1974. *MNRAS* 169:447–70
Bath GT, Pringle JE. 1982. *MNRAS* 199:267–80
Beckwith SVW, Sargent AI. 1993a. *Ap. J.* 402:280–91
Beckwith SVW, Sargent AI. 1993b. See Levy & Lunine 1993, pp. 521–42
Beckwith SVW, Sargent AI, Chini RS, Güsten R. 1990. *Astron. J.* 99:924–45
Begelman MC, Blandford RD, Rees MJ. 1980. *Nature* 287:307–9
Begelman MC, Blandford RD, Rees MJ. 1984. *Rev. Mod. Phys.* 56:255–351
Bell KR, Lin DNC. 1994. *Ap. J.* 427:987–1004
Bell KR, Lin DNC, Hartmann LW, Kenyon SJ. 1995. *Ap. J.* 444:376–95
Bertout C, Basri G, Bouvier J. 1988. *Ap. J.* 330:350–73
Binney J, Tremaine SD. 1987. *Galactic Dynamics.* Princeton: Princeton Univ. Press
Blumenthal G, Yang LT, Lin DNC. 1984. *Ap. J.* 287:774–84
Bodenheimer PH. 1995. *Annu. Rev. Astron. Astrophys.* 33:199–238
Bodenheimer PH, Pollack E. 1986. *Icarus* 67:391–408
Bradt HVD, McClintock JE. 1983. *Annu. Rev. Astron. Astrophys.* 21:13–66
Burkert A, Bodenheimer PH. 1993. *MNRAS* 264:798–806
Burrows CJ, Stapelfeldt K, Watson AM, Krist J, Cabrit S, Edwards S, Strom SE, Strom KM. 1990. *Ap. J.* 354:687–700
Calvet N, Hartmann L. 1992. *Ap. J.* 386:239–47
Calvet N, Hartmann L, Hewett R. 1992. *Ap. J.*

386:229–38
Calvet N, Hartmann L, Kenyon SJ. 1993. *Ap. J.* 402:623–34
Cameron AGW. 1962. *Icarus* 1:13–69
Cameron AGW. 1978. *Moon Planet* 18:5–40
Cannizzo JK. 1992. *Ap. J.* 385:94–107
Cannizzo JK, Ghosh P, Wheeler JC. 1982. *Ap. J. Lett.* 260:L83–86
Cannizzo JK, Kenyon SJ. 1987. *Ap. J.* 320:319–32
Cannizzo JK, Shafter AW, Wheeler JC. 1988. *Ap. J.* 333:227–35
Cannizzo JK, Wheeler JC, Polidan RS. 1986. *Ap. J.* 301:634–40
Carr JS, Tokunaga AT, Najita J, Shu FH, Glassgold AE. 1993. *Ap. J. Lett.* 411:L37–40
Chen K, Halpern JP. 1989. *Ap. J.* 344:115–24
Chen X, Taam RE. 1994. *Ap. J.* 431:732–41
Chen X, Taam RE. 1995. *Ap. J.* 441:354–60
Cheng FH, Lin DNC. 1989. *Ap. J.* 337:432–65
Cheng FH, Lin DNC. 1992. *Ap. J.* 389:714–23
Chevalier RA. 1993. *Ap. J. Lett.* 411:L33–36
Clarke CJ, Armitage PJ, Smith KW, Pringle JE. 1995. *MNRAS* 273:639–42
Clarke CJ, Lin DNC, Papaloizou JCB. 1989. *MNRAS* 236:495–503
Clarke CJ, Lin DNC, Pringle JE. 1990. *MNRAS* 242:439–46
Clarke JT, Chapel D, Bowyer S. 1984. *Ap. J.* 287:845–55
Claussen MJ, Heiligman GM, Lo KY. 1984. *Nature* 310:298–300
Claussen MJ, Lo KY. 1986. *Ap. J.* 308:592–99
Clavel J, Reichert GA, Alloin D, Crenshaw DM, Kriss G, et al. 1991. *Ap. J.* 366:64–81
Collin-Souffrin S. 1991. *Astron. Astrophys.* 249:344–50
Condon JJ, Mitchell KJ. 1984. *Ap. J.* 276:472–75
Cordova FA, Mason KO. 1982. *Ap. J.* 260:716–21
Cordova FA, Mason KO. 1983. *Accretion Driven Stellar X-ray Sources,* ed. WHG Lewin, EPJ van den Heuvel, pp. 147–87. Cambridge: Cambridge Univ. Press
Croswell K, Hartmann L, Avrett EH. 1987. *Ap. J.* 312:227–42
Cuzzi JN, et al. 1984. See Greenberg & Brahic 1984, pp. 73–199
Dahari O. 1985. *Ap. J. Suppl.* 57:643–64
D'Antona F, Mazzitelli I. 1994. *Ap. J. Suppl.* 90:467–500
Davidson K. 1977. *Ap. J.* 218:20–32
Drew JE. 1987. *MNRAS* 224:595–632
Duschl WJ, Frank J, Meyer F, Meyer-Hofmeister E, Tscharnuter WM, eds. 1994. *Theory of Accretion Disks—2.* Dordrecht: Kluwer
Dutrey A, Guilloteau S, Simon M. 1994. *Astron. Astrophys.* 286:149–59
Dyck HM, Simon T, Zuckerman B. 1982. *Ap. J. Lett.* 255:L103–6
Elliot JL, Nicholson PD. 1984. See Greenberg & Brahic 1984, pp. 25–72
Faulkner J, Lin DNC, Papaloizou JCB. 1983. *MNRAS* 205:359–77
Ford HC, Harms RJ, Tsvetanov ZI, Hartig GF, Dressel LL, et al. 1994. *Ap. J. Lett.* 435:L27–30
Frank J. 1979. *MNRAS* 187:883–904
Fukunaga M. 1984. *Publ. Astron. Soc. Jpn.* 36:433–56
Fukunaga M, Tosa M. 1989. *Publ. Astron. Soc. Jpn.* 41:975–90
Galeev A, Rosner R, Vaiana GS. 1979. *Ap. J.* 229:318–26
Gammie CF. 1996. *Ap. J.* 457:355–62
Gammie CF, Ostriker JP, Jog CJ. 1991. *Ap. J.* 378:565–75
Genzel R, Hollenbach D, Townes CH. 1994. *Rep. Prog. Phys.* 57:417–79
Ghez AM, Neugebauer G, Gorham PW, Haniff CA, Kulkarni SR, et al. 1991. *Astron. J.* 102:2066–72
Goldreich P, Tremaine S. 1978. *Icarus* 34:240–53
Goldreich P, Tremaine S. 1980. *Ap. J.* 241:425–41
Goldreich P, Tremaine S. 1982. *Annu. Rev. Astron. Astrophys.* 20:249–83
Goodman AA, Benson PJ, Fuller GA, Myers PC. 1993. *Ap. J.* 406:528–47
Greenberg R, Brahic A, eds. 1984. *Planetary Rings.* Tucson: Univ. Ariz. Press
Greenhill LJ, Henkel C, Becker R, Wilson TL, Wouterloot JGA. 1995. *Astron. Astrophys.* 304:21–33
Hamann F, Ferland G. 1992. *Ap. J. Lett.* 391:L53–57
Hartmann L. 1994. See Duschl et al 1994, pp. 19–33
Hartmann L, Boss A, Calvet N, Whitney B. 1994a. *Ap. J. Lett.* 430:L49–52
Hartmann L, Hewett R, Calvet N. 1994b. *Ap. J.* 426:669–87
Hartmann L, Kenyon S, Hartigan P. 1993. See Levy & Lunine 1993, pp. 497–520
Hassall BJM, Pringle JE, Schwarzenberg-Czerny A, Wade RA, Whelan JAJ, Hill PW. 1983. *MNRAS* 203:865–85
Haug K. 1987. *Astrophys. Space Sci.* 130:91–102
Hayashi M, Ohashi N, Miyama SM. 1993. *Ap. J. Lett.* 418:L71–74
Heemskerk MHM. 1994. *Astron. Astrophys.* 288:807–18
Henning T, Begemann B, Mutschke H, Dorschner J. 1995. *Astron. Astrophys. Suppl.* 112:143–49
Herbig GH. 1977. *Ap. J.* 217:693–715

Herbig GH. 1989. *ESO Workshop on Low Mass Star Formation and Pre-Main Sequence Objects,* ed. B Reipurth, pp. 233–46. Garching: ESO
Hernquist L. 1989. *Nature* 340:687–91
Hessman FV, Robinson EL, Nather RE, Zhang E-H. 1984. *Ap. J.* 286:747–59
Hills JG. 1975. *Nature* 254:295–98
Hirose M, Osaki Y. 1990. *Publ. Astron. Soc. Jpn.* 42:135–63
Honma F, Matsumoto R, Kato S. 1994. See Duschl et al 1994, pp. 265–70
Horne K. 1993. See Wheeler 1993, pp. 117–47
Horne K. 1994. See Duschl et al 1994, pp. 77–91
Horne K, Cook MC. 1985. *MNRAS* 214:307–17
Horne K, LaDous CA, Shafter AW. 1990. *Accretion Powered Compact Binaries,* ed. CW Mauche, pp. 109–12. Cambridge: Cambridge Univ. Press
Horne K, Saar SH. 1991. *Ap. J. Lett.* 374:L55–58
Hourigan K, Ward WR. 1984. *Icarus* 60:29–39
Inoue H. 1993. See Wheeler 1993, pp. 303–59
Jaffe W, Ford HC, Ferrarese L, van den Bosch F, O'Connell RW. 1993. *Nature* 364:213–15
Jensen ELN, Mathieu RD, Fuller GA. 1994. *Ap. J. Lett.* 429:L29–32
Jog CJ, Ostriker JP. 1988. *Ap. J.* 328:404–26
Kato S. 1978. *MNRAS* 185:629–42
Kenyon SJ, Calvet N, Hartmann L. 1993. *Ap. J.* 414:676–94
Kenyon SJ, Hartmann L. 1991. *Ap. J.* 383:664–73
Kenyon SJ, Hartmann L, Hewett R. 1988. *Ap. J.* 325:231–51
King AR, Shaviv G. 1984. *Nature* 308:519–21
Kinney AL, Bohlin RC, Blades JC, York DG. 1991. *Ap. J. Suppl.* 75:645–717
Kley W. 1989. *Astron. Astrophys.* 222:141–49
Kley W, Lin DNC. 1996. *Ap. J.* 461:933–50
Kley W, Papaloizou J, Lin DNC. 1993. *Ap. J.* 409:739–47
Koekemoer AM, Henkel C, Greenhill LJ, Dey A, et al. 1995. *Nature* 378:697–99
Koerner DW, Sargent AI. 1995. *Ap. J.* 109: 2138–45
Koerner DW, Sargent AI, Beckwith SVW. 1993. *Icarus* 106:2–10
Königl A. 1991. *Ap. J. Lett.* 370:L39–43
Königl A. 1994. See Duschl et al 1994, pp. 53–67
Königl A, Ruden SP. 1993. See Levy & Lunine 1993, pp. 641–88
Korycansky DG, Pollack JB. 1993. *Icarus* 102:150–65
Kozlowski M, Witta PJ, Paczynski B. 1979. *Acta Astron.* 29:157–76
Kraft RP. 1962. *Ap. J.* 135:408–23
Krautter J, Klare G, Wolf B, Duerbeck HW, Rahe J, et al. 1981. *Astron. Astrophys.* 102:337–46
Krolik JH, Horne K, Kallman TR, Malkan MA, Edelson RA, Kriss GA. 1991. *Ap. J.* 371:541–62
Krzeminski W, Vogt N. 1985. *Astron. Astrophys.* 144:124–32
Lada CJ, Young ET, Greene TP. 1993. *Ap. J.* 408:471–83
La Dous C. 1989. *Astron. Astrophys.* 221:131–55
Laplace PS de. 1796. *Exposition du Système du monde.* Paris
Lauer TR, Faber SM, Lynds CR, Baum WA, Ewald SP, et al. 1992. *Astron. J.* 103:703–10
Lay OP, Carlstrom J, Hills RJ, Phillips TG. 1994. *Ap. J. Lett.* 434:L75–78
Levy GH, Lunine JI, eds. 1993. *Protostars and Planets III.* Tucson: Univ. Ariz. Press
Lewis JS. 1974. *Science* 186:440–43
Lightman A, Eardley D. 1974. *Ap. J. Lett.* 187:L1–3
Lin DNC, Artymowicz P, Wampler EJ. 1994. See Duschl et al 1994, pp. 235–46
Lin DNC, Bodenheimer PH. 1981. *Ap. J. Lett.* 248:L83–86
Lin DNC, Bodenheimer PH, Richardson D. 1996. *Nature.* 380:606–07
Lin DNC, Hayashi M, Bell KR, Ohashi N. 1994. *Ap. J.* 435:821–28
Lin DNC, Papaloizou JCB. 1980. *MNRAS* 191:37–48
Lin DNC, Papaloizou JCB. 1985. In *Protostars and Planets II,* ed. DC Black, MS Mathews pp. 981–1072. Tucson: Univ. Ariz. Press
Lin DNC, Papaloizou JCB. 1986a. *Ap. J.* 307:395–409
Lin DNC, Papaloizou JCB. 1986b. *Ap. J.* 309:846–57
Lin DNC, Papaloizou JCB. 1993. See Levy & Lunine 1993, pp. 749–835
Lin DNC, Papaloizou JCB, Faulkner J. 1985. *MNRAS* 212:105–50
Lin DNC, Pringle JE, Rees MJ. 1988. *Ap. J.* 328:103–10
Lin DNC, Shields GA. 1986. *Ap. J.* 305:28–34
Lin DNC, Taam RE. 1984. *High Energy Transients in Astrophysics,* ed. SE Woosley, pp. 85–102. New York: AIP
Lin DNC, Williams RE, Stover R. 1988. *Ap. J.* 327:234–48
Livio M, Spruit HC. 1991. *Astron. Astrophys.* 252:189–92
Lubow SH. 1992. *Ap. J.* 401:317–24
Lubow SH. 1994. *Ap. J.* 432:224–27
Lubow SH, Papaloizou JCB, Pringle JE. 1994. *MNRAS* 267:235–40
Lukkari J. 1981. *Nature* 292:433–35
Lüst R. 1952. *Z. Naturforsch. Teil A* 7:87–98
Lynden-Bell D. 1969. *Nature* 223:690–94

Lynden-Bell D, Pringle JE. 1974. *MNRAS* 168:603–37
Lyne AG, Lorimer DR. 1994. *Nature* 369:127–29
MacKenty JW. 1989. *Ap. J.* 343:125–34
Makino J, Fukushige T, Okumura SK, Ebisuzaki T. 1993. *Publ. Astron. Soc. Jpn.* 45:303–10
Makishima K, Fujimoto R, Ishisaki Y, Kii T, Loewenstein M, et al. 1994. *Publ. Astron. Soc. Jpn.* 46:L77–80
Malkan MA, Sargent WLW. 1982. *Ap. J.* 254:22–37
Maoz E. 1995. *Ap. J. Lett.* 447:L91–94
Marsh TR, Horne K, Schlegel EM, Honeycutt RK, Kaitchuck RH, et al. 1990. *Ap. J.* 364:637–46
Marsh TR, Robinson EL, Wood JH. 1994. *MNRAS* 266:137–54
Mathews WG. 1982. *Ap. J.* 258:425–33
Mathieu RD. 1994. *Annu. Rev. Astron. Astrophys.* 32:465–530
Mathieu RD, Adams FC, Latham DW. 1991. *Astron. J.* 101:2184–98
Mauche CW, Raymond JC. 1987. *Ap. J.* 323:690–713
Mayor F, Queloz D. 1995. *Nature* 378:355–59
McCaughrean MJ, O'Dell CR. 1996. *Ap. J.* In press
McCaughrean MJ, Stauffer JR. 1994. *Astron. J.* 108:1382–97, 1513–18
Menard F, Bastien P. 1992. *Astron. J.* 103:564–72
Meyer F. 1984. *Astron. Astrophys.* 131:303–8
Meyer F. 1986. *MNRAS* 218:P7–11
Meyer F, Duschl WJ, Frank J, Meyer-Hofmeister E, eds. 1989. *Theory of Accretion Disks.* Dordrecht: Kluwer
Meyer F, Meyer-Hofmeister E. 1981. *Astron. Astrophys.* 104:L10–12
Meyer F, Meyer-Hofmeister E. 1983. *Astron. Astrophys.* 121:29–34
Meyer F, Meyer-Hofmeister E. 1984. *Astron. Astrophys.* 132:143–50
Mikkola S, Valtonen MJ. 1992. *MNRAS* 253:115–20
Miller JS, Goodrich RW, Mathews WG. 1991. *Ap. J.* 378:47–64
Mineshige S. 1986. *Publ. Astron. Soc. Jpn.* 38:831–52
Mineshige S. 1988. *Astron. Astrophys.* 190:72–78
Mineshige S, Kusunose M. 1993. See Wheeler 1993, pp. 370–421
Mineshige S, Osaki Y. 1985. *Publ. Astron. Soc. Jpn.* 37:1–18
Mineshige S, Wood JH. 1989. *MNRAS* 241:259–80
Miyama S, Hayashi C, Narita S. 1984. *Ap. J.* 279:621–32
Moran J, Greenhill L, Herrnstein J, Diamond P, Miyoshi M, et al. 1995. *Proc. Natl. Acad. Sci. USA.* In press
Moriarty-Schieven GH, Snell RL. 1988. *Ap. J.* 332:364–78
Mundy L, Looney L, Ericson W, Grossman A, Welch WJ, et al. 1996. *Ap. J. Lett.* In press
Murphy BW, Cohn HN, Durisen RH. 1991. *Ap. J.* 370:60–77
Murray SD, Lin DNC. 1992. *Ap. J.* 384:177–84
Najita J, Carr JS, Glassgold AE, Shu FH, Tokunaga AT. 1996. *Ap. J.* In press
Najita JR, Shu FH. 1994. *Ap. J.* 429:808–25
Nakai N, Inoue M, Miyoshi M. 1993. *Nature* 361:45–47
Nandra K, Pounds KA, Stewart GC. 1990. *MNRAS* 242:660–68
Narayan R, Ostriker JP. 1990. *Ap. J.* 352:222–46
Naylor T, Bath GT, Charles PA, Hassall BIM, Sonneborn G, et al. 1988. *MNRAS* 231:237–55
Netzer H. 1990. *Active Galactic Nuclei,* ed. RD Blandford, H Netzer, L Woltjer, pp. 57–160. Berlin: Springer-Verlag
Netzer H, Laor A, Gondhalekar PM. 1992. *MNRAS* 254:15–20
Nowak MA, Wagnoner RV. 1993. *Ap. J.* 418:187–201
O'Dell CR, Wen Z. 1994. *Ap. J.* 436:194–202
O'Dell CR, Wen Z, Hu X. 1993. *Ap. J.* 410:696–700
Ohashi N, Kawabe R, Hayashi M, Ishiguro M. 1991. *Astron. J.* 102:2054–65
Okuda T, Ono K, Tabata M, Mineshige S. 1992. *MNRAS* 254:427–34
Osaki Y. 1974. *Publ. Astron. Soc. Jpn.* 26:429–36
Ossenkopf V, Henning T. 1994. *Astron. Astrophys.* 291:943–59
Osterbrock DE. 1989. *Astrophysics of Gaseous Nebulae and Active Galactic Nuclei.* Mill Valley: Univ. Sci. Bks.
Osterbrock DE. 1993. *Ap. J.* 404:551–62
Ostriker EC. 1994. *Ap. J.* 424:292–318
Ostriker EC, Shu FH. 1995. *Ap. J.* 447:813–28
Ostriker JP. 1983. *Ap. J.* 273:99–104
Paczyński B. 1978. *Acta Astron.* 28:91–109
Papaloizou JCB, Faulkner J, Lin DNC. 1983. *MNRAS* 205:487–513
Papaloizou JCB, Lin DNC. 1984. *Ap. J.* 285:818–34
Papaloizou JCB, Lin DNC. 1988. *Ap. J.* 331:838–60
Papaloizou JCB, Lin DNC. 1995a. *Annu. Rev. Astron. Astrophys.* 33:505–40
Papaloizou JCB, Lin DNC. 1995b. *Ap. J.* 438:841–51
Papaloizou JCB, Pringle JE. 1977. *MNRAS* 181:441–54
Papaloizou JCB, Pringle JE. 1979. *MNRAS*

189:293–97
Papaloizou JCB, Pringle JE. 1985. *MNRAS* 217:387–90
Papaloizou JCB, Stanley Q. 1986. *MNRAS* 220:593–610
Patterson J. 1984. *Ap. J. Suppl.* 54:443–93
Peterson BM, Balonek TJ, Barker ES, Bechtold J, Bertram R, et al. 1991. *Ap. J.* 368:119–37
Petrosian AR. 1982. *Astrofizika* 18:548–62 (Engl. transl. *Astrophysics* 18:312–21)
Petrov PP, Herbig GH. 1992. *Ap. J.* 392:209
Phinney ES. 1989. See Meyer et al 1989, pp. 457–70
Pollack JB, Hollenbach D, Beckwith S, Simonelli DP, Roush T, Fong W. 1994. *Ap. J.* 421:615–39
Prendergast KH, Burbidge GR. 1968. *Ap. J. Lett.* 151:L83–88
Pringle JE. 1977. *MNRAS* 178:195–202
Pringle JE. 1981. *Annu. Rev. Astron. Astrophys.* 19:137–62
Pringle JE, Rees MJ, Pacholczyk AG. 1973. *Astron. Astrophys.* 29:179–84
Pringle JE, Savonije GJ. 1979. *MNRAS* 187:777–83
Pringle JE, Verbunt F, Wade RA. 1986. *MNRAS* 221:169–94
Prosser CF, Stauffer JR, Hartmann L, Soderblom DR, Jones BF, et al. 1994. *Ap. J.* 421:517–41
Pudritz RE, Norman CA. 1983. *Ap. J.* 274:677–97
Quinlan GD, Hernquist L, Sigurdsson S. 1995.
Rees MJ, Netzer H, Ferland GJ. 1989. *Ap. J.* 347:640–55
Robinson EL. 1976. *Annu. Rev. Astron. Astrophys.* 14:119–42
Robinson EL, Marsh TR, Smak JI. 1993. See Wheeler 1993, pp. 75–116
Robinson EL, Nather RE. 1979. *Ap. J.* 39:461–80 (Suppl.)
Roddier F, Roddier C, Graves JE, Northcott MJ. 1995. *Ap. J.* 443:249–60
Rokaki E. 1994. *Reverberation Mapping of the Broad-Line Regions in Active Galactic Nuclei. ASP Conf. Ser.* 69, ed. PM Gondhalekar, K Horne, BM Peterson, pp. 257–64. San Francisco: Astron. Soc. Pac.
Rokaki E, Collin-Souffrin S, Magnan C. 1993. *Astron. Astrophys.* 272:8–24
Rozyczka M, Bodenheimer PH, Lin DNC. 1995. *MNRAS* 276:597–606
Rozyczka M, Bodenheimer PH, Lin DNC. 1996. *Ap. J.* 459:371–83
Rucinski SM. 1985. *Astron. J.* 90:2321–30
Rutten RGM, Dhillon V, Horne K, Kuulkers E, van Paradijs J. 1993. *Nature* 362:518–20
Rutten RGM, Kuulkers E, Vogt N, van Paradijs J. 1992a. *Astron. Astrophys.* 265:159–67
Rutten RGM, van Paradijs J, Tinbergen J. 1992b. *Astron. Astrophys.* 260:213–26
Safier PN. 1993. *Ap. J.* 408:115–47
Safronov VS. 1972. *Evolution of the Protoplanetary Cloud and Formation of the Earth and Planets.* Moscow: Nauka
Sakimoto PJ, Coroniti FV. 1981. *Ap. J.* 247:19–31
Sanders DB, Phinney ES, Neugebauer G, Soifer BT, Mathews K. 1989. *Ap. J.* 347:29–51
Sanders DB, Scoville NZ, Sargent AI, Soifer BT. 1988. *Ap. J. Lett.* 324:L55–58
Sargent AI. 1995. *Disks and Outflows Around Young Stars,* ed. SVW Beckwith, A Natta, J Staude. Berlin: Springer-Verlag. In press
Sargent AI, Beckwith S. 1987. *Ap. J.* 323:294–305
Sargent AI, Beckwith S. 1991. *Ap. J. Lett.* 382:L31–35
Savonije GJ, Papaloizou JCB, Lin DNC. 1994. *MNRAS* 268:13–28
Schmidt M, Green RF. 1983. *Ap. J.* 269:352–74
Schrijver CJ. 1987. *Astron. Astrophys.* 172:111–23
Schrijver CJ. 1992. *Astron. Astrophys.* 258:507–20
Schwarzenberg-Czerny A, Ward M, Hanes DA, Jones DHP, Pringle JE, et al. 1985. *MNRAS* 212:645–55
Shakura NI, Sunyaev RA. 1973. *Astron. Astrophys.* 24:337–55
Shakura NI, Sunyaev RA. 1976. *MNRAS* 175:613–32
Shaviv G, Wehrse R. 1989. See Meyer et al 1989, pp. 419–44
Shields GA. 1978. *Nature* 272:706–8
Shields GA, Lin DNC. 1988. *Supermassive Black Holes,* ed. M Kafatos, pp. 296–99. Cambridge: Cambridge Univ. Press
Shlosman I, Begelman MC, Frank J. 1990. *Nature* 345:679–86
Shu FH. 1984. See Greenberg & Brahic 1984, pp. 513–61
Shu FH, Najita J, Ruden SP, Lizano S. 1994. *Ap. J.* 429:797–807
Skinner SL, Brown A, Walter FM. 1991. *Astron. J.* 102:1742–48
Skrutskie MF, Dutkevitch D, Strom SE, Edwards S, Strom KM, et al. 1990. *Astron. J.* 99:1187–95
Smak J. 1969. *Acta Astron.* 19:155–64
Smak J. 1982. *Acta Astron.* 32:199–211
Smak J. 1984a. *Publ. Astron. Soc. Pac.* 96:5–18
Smak J. 1984b. *Acta Astron.* 34:93–96
Smak J. 1984c. *Acta Astron.* 34:161–89
Smith BA, Terrile RJ. 1984. *Science* 126:1421–24
Snell RL, Loren RB, Plambeck RL. 1980. *Ap. J. Lett.* 239:L17–22

Stapelfeldt KR, Burrows CJ, Krist JE, Trauger JT, et al. 1995. *Ap. J.* 449:888–93
Strom SE, Edwards SE, Skrutskie MF. 1993. See Levy & Lunine 1993, pp. 837–66
Sun WH, Malkan MA. 1989. *Ap. J.* 346:68–100
Syer D, Clarke CJ, Rees MJ. 1991. *MNRAS* 250:505–12
Taam RE, Lin DNC. 1984. *Ap. J.* 287:761–68
Takeuchi T, Miyama S, Lin DNC. 1996. *Ap. J.* In press
Tereby S, Shu FH, Cassen P. 1984. *Ap. J.* 286:529–51
Tohline JE, Osterbrock DE. 1982. *Ap. J. Lett.* 252:L49–52
Tout CA, Pringle JE. 1992. *MNRAS* 259:604–12
Ulrich M-H. 1994. See Duschl et al 1994, pp. 253–59
Umebayshi T, Nakano T. 1988. *Prog. Theor. Phys. Suppl.* 96:151–60
van Paradijs J. 1983. *Accretion Driven Stellar X-ray Sources,* ed. WHG Lewin, EPJ van den Heuvel, pp. 189–260. Cambridge: Cambridge Univ. Press
Vishniac ET, Jin L, Diamond P. 1990. *Ap. J.* 365:648–59
Wade RA. 1984. *MNRAS* 208:381–98
Wade RA. 1988. *Ap. J.* 335:394–405
Wade RA, Ward MJ. 1985. *Interacting Binary Stars,* ed. JE Pringle, RE Wade, pp. 129–76. Cambridge: Cambridge Univ. Press
Walker HJ, Wolstencroft RD. 1988. *Publ. Astron. Soc. Pac.* 100:1509–21
Walter FM, Brown A, Mathieu RD, Myers PC. 1988. *Astron. J.* 96:297–325
Ward WR. 1981. *Geophys. Res. Lett.* 8:641–43
Ward WR. 1986. *Icarus* 67:164–80
Ward WR. 1988. *Icarus* 73:330–48
Ward WR, Hourigan K. 1989. *Ap. J.* 347:490–95
Warner B. 1976. *Structure and Evolution of Close Binary Systems,* ed. PP Eggleton, S Mitton, JAJ Whelan, pp. 80–140. Dordrecht: Reidel
Warner B. 1995. *Cataclysmic Variable Stars.* Cambridge: Cambridge Univ. Press
Watson WD, Wallin BK. 1994. *Ap. J. Lett.* 432:L35–38
Weidenschilling S. 1977. *Astrophys. Space Sci.* 51:153–58
Wetherill G. 1980. *Annu. Rev. Astron. Astrophys.* 18:77–113
Wheeler JC, ed. 1993. *Accretion Disks in Compact Stellar Systems.* Singapore: World Sci.
Whitehurst R. 1988. *MNRAS* 232:35–51
Whitehurst R, King A. 1991. *MNRAS* 249:25–35
Williams G. 1983. *Ap. J. Suppl.* 53:523–51
Williams RE. 1980. *Ap. J.* 235:939–44
Williams RE, Ferguson DH. 1982. *Ap. J.* 257:672–85
Woltjer L. 1959. *Ap. J.* 130:38–44
Wood JH. 1990. *MNRAS* 243:219–30
Wood JH, Horne K, Berriman G, Wade RA. 1989. *Ap. J.* 341:974–96
Wood JH, Horne K, Berriman G, Wade RA, O'Donoghue D, Warner B. 1986. *MNRAS* 219:629–55
Yaqoob T, Warwick RS. 1991. *MNRAS* 248:773–86
Young P, Schneider DP, Shectman SA. 1981a. *Ap. J.* 244:259–68
Young P, Schneider DP, Shectman SA. 1981b. *Ap. J.* 245:1035–42
Zuckerman B, Forveille T, Kastner JH. 1995. *Nature* 373:494–96

Annu. Rev. Astron. Astrophys. 1996. 34:749–92

LUMINOUS INFRARED GALAXIES

D. B. Sanders
Institute for Astronomy, University of Hawaii, 2680 Woodlawn Drive, Honolulu, HI 96822

I. F. Mirabel
Service d'Astrophysique, Centre d'Etudes de Saclay, 91191 Gif sur Yvette, France

KEY WORDS: luminosity function, starbursts, active galactic nuclei, molecular gas, dust emission

ABSTRACT

At luminosities above $10^{11}\ L_{\odot}$, infrared galaxies become the dominant population of extragalactic objects in the local Universe ($z \lesssim 0.3$), being more numerous than optically selected starburst and Seyfert galaxies and quasi-stellar objects at comparable bolometric luminosity. The trigger for the intense infrared emission appears to be the strong interaction/merger of molecular gas-rich spirals, and the bulk of the infrared luminosity for all but the most luminous objects is due to dust heating from an intense starburst within giant molecular clouds. At the highest luminosities ($L_{\rm ir} > 10^{12}\ L_{\odot}$), nearly all objects appear to be advanced mergers powered by a mixture of circumnuclear starburst and active galactic nucleus energy sources, both of which are fueled by an enormous concentration of molecular gas that has been funneled into the merger nucleus. These ultraluminous infrared galaxies may represent an important stage in the formation of quasi-stellar objects and powerful radio galaxies. They may also represent a primary stage in the formation of elliptical galaxy cores, the formation of globular clusters, and the metal enrichment of the intergalactic medium.

---

1. INTRODUCTION

One of the most important discoveries from extragalactic observations at mid- and far-infrared wavelengths has been the identification of a class of "infrared galaxies", objects that emit more energy in the infrared ($\sim$ 5–500 μm) than at all other wavelengths combined. The first all-sky survey at far-infrared wavelengths carried out in 1983 by the *Infrared Astronomical Satellite* (*IRAS*)

0066-4146/96/0915-0749$08.00

resulted in the detection of tens of thousands of galaxies, the vast majority of which were too faint to have been included in previous optical catalogs. It is now clear that part of the reason for the large number of detections is the fact that the majority of the most luminous galaxies in the Universe emit the bulk of their energy in the far-infrared. Previous assumptions, based primarily on optical observations, about the relative distributions of different types of luminous galaxies—e.g. starbursts, Seyferts, and quasi-stellar objects (QSOs)—need to be revised.

The bulk of the luminosity produced in galaxies bolometrically more luminous than $\sim 4\ L^*$ (i.e. $L_{\rm bol} \gtrsim 10^{11}\ L_\odot$) appears to be produced in objects that are heavily obscured by dust. Although luminous infrared galaxies (hearafter LIGs: $L_{\rm ir} > 10^{11}\ L_\odot$) are relatively rare objects, reasonable assumptions about the lifetime of the infrared phase suggest that a substantial fraction of all galaxies with $L_{\rm B} > 10^{10}\ L_\odot$ pass through such a stage of intense infrared emission (Soifer et al 1987b).

Substantial progress has been made in cataloging infrared galaxies detected in the *IRAS* database, allowing for a good determination of the luminosity function over a wide range of infrared luminosity ($L_{\rm ir} \sim 10^7$–$10^{13}\ L_\odot$). A brief review of *IRAS* galaxy redshift surveys and a comparison of the infrared galaxy luminosity function with other classes of extragalactic objects is given in Section 3. Section 4 reviews published multiwavelength data for complete samples of the brightest infrared sources and selected samples of the most luminous infrared objects. Section 5 discusses the origin and evolution of LIGs and presents detailed data for several well-studied objects. Several important phenomena associated with LIGs that may significantly impact other areas of extragalactic research are reviewed in Section 6, and Section 7 briefly discusses how theoretical simulations are being used to more accurately interpret the observed morphology and kinematics of both the gas and dust in LIGs.

This review is the first to focus almost exclusively on the properties of luminous infrared galaxies. Pre-*IRAS* reviews of extragalactic infrared observations by Neugebauer et al (1971) and Rieke & Lebofsky (1979) include discussions of luminous infrared emission from optically selected objects. Soifer et al (1987a) present a broad overview of the extragalactic sky as seen by *IRAS*, including a detailed discussion of the infrared galaxy luminosity function and some discussion of the properties of a few selected LIGs, while the review by Telesco (1988) provides an excellent complimentary summary focusing on the properties of nearby, lower luminosity infrared galaxies. Recent reviews by Young & Scoville (1991) on the molecular gas properties of galaxies and by Barnes & Hernquist (1992) on theoretical models of interacting galaxies cover topics that are particularly relevant to the study of LIGs.

2. BACKGROUND

2.1 *Pre-IRAS*

The fact that some galaxies emit as much energy in the infrared as at optical wavelengths was established with the first mid-infrared observations of extragalactic sources (Low & Kleinmann 1968; Kleinmann & Low 1970a,b). Observations of both optical- and radio-selected objects at wavelengths of 2–25 μm uncovered several objects—including luminous starbursts, Seyferts, and QSOs—with "similar infrared continua" that appeared to emit most of their luminosity in the far-infrared. More accurate photometry of a larger number of sources (Rieke & Low 1972) provided further evidence for dominant infrared emission from Seyfert galaxies and the nuclei of relatively normal spiral galaxies and also singled out several "ultrahigh" infrared luminous galaxies whose extrapolated luminosity at far-infrared wavelengths rivaled the bolometric luminosity of QSOs. A tight correlation between the 21-cm radio continuum and 10-μm infrared fluxes was established for "Seyfert and related galaxies", although the relevance of this correlation for determining the nature of the dominant energy source in these objects was not discussed.

The first critical evidence that the infrared emission from Seyferts was not direct synchrotron radiation was provided by monitoring of the 10-μm flux from the archetypal Seyfert 2 galaxy NGC 1068, which failed to show evidence for variability (Stein et al 1974), plus measurements that showed the infrared source to be extended at 10 μm (Becklin et al 1973). The infrared spectrum appeared to be better explained by models of thermal reradiation from dust (e.g. Rees et al 1969, Burbidge & Stein 1970). More extensive mid-infrared photometry of larger samples of Markarian Seyferts and starbursts (Rieke & Low 1975, Neugebauer et al 1976), Seyfert galaxies (Rieke 1978), and bright spirals (Rieke & Lebofsky 1978, Lebofsky & Rieke 1979), plus far-infrared (30–300 μm) observations of nearby bright galaxies (Harper & Low 1973, Telesco & Harper 1980) showed that "infrared excess" was indeed a common property of extragalactic objects, and that the shape of the infrared continuum in most of these sources, with the possible exception of Seyfert 1 galaxies and QSOs, could best be understood in terms of thermal emission from dust. Although star formation seemed to be the most obvious explanation for the dominant energy source in normal galaxies and starbursts, a dust-enshrouded active galactic nucleus (AGN) remained a plausible model for Seyferts and QSOs.

A class of objects that would prove to be particularly relevant to LIGs were those objects in catalogs of interacting and peculiar galaxies (e.g. Vorontsov-Velyaminov 1959, Arp 1966, Zwicky & Zwicky 1971). The classic papers by Toomre & Toomre (1972), and Larson & Tinsley (1978) called attention to the role of interactions in triggering extreme nuclear activity, as well as more widespread starbursts[1]. Condon & Dressel (1978) and Hummel (1980) found that the 21-cm radio continuum in the nuclei of interacting galaxies was enhanced (by factors of 2–3) compared to isolated spirals. Condon et al (1982) later interpreted the radio continuum morphology of a class of "bright radio spiral galaxies" as evidence for powerful nuclear starbursts, the majority of which seemed to be triggered by galaxy interactions. Heckman (1983), following a suggestion by Fosbury & Wall (1979) that systems identified as "ongoing mergers" by Toomre (1977) might be exceptionally radio-loud, found that these and similar systems identified from the *Arp atlas* (Arp 1996) were ~ 8 times more likely to be radio-loud than single spirals with the same total optical luminosity, although it was not clear whether this enhanced radio activity was due to an AGN or a starburst.

Extremely strong mid-infrared and radio continuum emission in the interacting galaxy system Arp 299 (NGC 3690/IC 694) (Gherz et al 1983) was interpreted as evidence for "super starbursts" involving several regions, each forming $\gtrsim 10^9\ M_\odot$ of stars in bursts and lasting $\sim 10^8$ years (although the most luminous infrared source, associated with the nucleus of IC 694, appeared to be powered by an AGN). Surveys of interacting galaxies in the mid-infrared (Joseph et al 1984a, Lonsdale et al 1984, Cutri & McAlary 1985) revealed an enhancement of infrared emission in interacting systems (typically by factors of 2–3) compared to isolated galaxies. Joseph & Wright (1985) identified a subset of advanced mergers in the Arp atlas with extremely strong mid-infrared emission that they described as "ultraluminous" infrared galaxies; they argued that super starbursts may occur in the evolution of most mergers.

2.2 *Early IRAS Results*

IRAS was the first telescope with sufficient sensitivity to detect large numbers of extragalactic sources at mid- and far-infrared wavelengths (Neugebauer et al 1984). *IRAS* surveyed $\sim 96\%$ of the sky, producing an initial *IRAS Point Source Catalog* (1988; hearafter PSC) with a completeness limit of ~ 0.5 Jy at 12 μm, 25 μm, and 60 μm, and ~ 1.5 Jy at 100 μm. It contained $\sim 20{,}000$ galaxies, the majority of which had not been previously cataloged. Table 1 lists the definitions that have generally been adopted as standards for computing the broad-band infrared properties of *IRAS* galaxies.

Although some previously cataloged objects would prove to have extreme infrared properties, the vast majority were more modest infrared emitters as typified by the results reported by de Jong et al (1984) for galaxies in the Shapley-Ames catalog. In a sample of 165 SA galaxies, *IRAS* detected nearly

[1]We adopt here the definition of a starburst given by Larson & Tinsley (1978) as a "burst" in the star formation rate of duration $\sim 10^7$–10^8 years, involving up to 5% of the total stellar mass.

Table 1 Abbreviations and definitions[a]

$F_{\rm fir}$	1.26×10^{-14} $\{2.58\ f_{60} + f_{100}\}$ [Wm^{-2}]
$L_{\rm fir}$	L(40–500 μm) $= 4\pi\ D_{\rm L}^2\ {\rm C}\ F_{\rm fir}$ [$L_\odot$]
$F_{\rm ir}$	1.8×10^{-14} $\{13.48\ f_{12} + 5.16 f_{25} + 2.58 f_{60} + f_{100}\}$ [Wm^{-2}]
$L_{\rm ir}$	L(8–1000 μm) $= 4\pi\ {\rm D}_{\rm L}^2\ F_{\rm ir}$ [$L_\odot$]
$L_{\rm ir}/L_{\rm B}$	$F_{\rm ir}/\nu f_\nu$ (0.44 μm)
LIG	Luminous Infrared Galaxy, $L_{\rm ir} > 10^{11} L_\odot$
ULIG	UltraLuminous Infrared Galaxy, $L_{\rm ir} > 10^{12} L_\odot$
HyLIG	HyperLuminous Infrared Galaxy, $L_{\rm ir} > 10^{13} L_\odot$

[a]Throughout this review we adopt $H_0 = 75$ km s^{-1} Mpc^{-1}, $q_0 = 0$. A luminosity quoted at a specific wavelength refers to $\nu L_\nu(\lambda)$ and is given in units of solar bolometric luminosity (3.83×10^{33} ergs s^{-1}). The quantities f_{12}, f_{25}, f_{60}, and f_{100} are the *IRAS* flux densities in Jy at 12, 25, 60, and 100 μm. The broad-band far-infrared luminosity, $L_{\rm fir}$, is computed using the prescription given in Appendix B of *Cataloged Galaxies and Quasars Observed in the IRAS Survey* (1985). The scale factor C (typically in the range 1.4–1.8) is the correction factor required to account principally for the extrapolated flux longward of the *IRAS* 100 μm filter. $D_{\rm L}$ is the luminosity distance. $L_{\rm fir}$ has mostly been replaced by the quantity $L_{\rm ir}$, which better represents the total mid- and far-infrared luminosity. $L_{\rm ir}$ is computed by fitting a single temperature dust emissivity model ($\epsilon \propto \nu^{-1}$) to the flux in all four *IRAS* bands and should be accurate to $\pm$ 5% for dust temperatures in the range 25–65 K (Perault 1987).

all late-type spirals (Sb–Sd) and Irr–Am galaxies, approximately half of the early type, S0–Sa, galaxies and *none* of the ellipticals. For those galaxies detected, $L_{\rm ir}/L_{\rm B} = 0.1$–5, with a mean value of ~ 0.4. The few objects with $L_{\rm ir}/L_{\rm B} > 2$ were typically SBs or irregulars. Objects with higher $L_{\rm ir}/L_{\rm B}$ ratios tended to have warmer f_{60}/f_{100} colors. The classic starburst galaxies M82 and NGC 253 had $L_{\rm ir}/L_{\rm B}$ ratios of 3 and 5, and $L_{\rm ir} = 10^{10.3}$ and $10^{10.8}$ $L_\odot$ respectively. No objects were found with $L_{\rm ir} > 10^{11}$ $L_\odot$.

The more extreme infrared properties of infrared-selected samples are typified by objects in the *IRAS* minisurvey (Rowan-Robinson et al 1984). For a complete flux-limited sample of 86 infrared-selected galaxies from the minisurvey, Soifer et al (1984a) found that virtually all had $L_{\rm ir} > 10^{10}$ $L_\odot$ and ratios $L_{\rm ir}/L_{\rm B} = 1$–50, with the fraction of interacting galaxies being as high as one fourth. More intriguing were the 9 "unidentified" sources ($L_{\rm ir}/L_{\rm B} > 50$) which had no obvious optical counterparts in galaxy catalogs and often no visible counterpart on the Palomar Sky Survey plates (Houck et al 1984). Initial cross-correlation of larger *IRAS* source lists with galaxy catalogs had produced only one or two objects with similar extreme ratios, most notably the ULIG Arp 220 (Soifer et al 1984b) and NGC 6240 (Wright et al 1984, Joseph et al 1984b). Ground-based observations of the unidentified minisurvey objects quickly led to the discovery of faint galaxies, typically at redshifts 0.1–0.2 (Aaronson & Olszewski 1984, Houck et al 1985, Antonucci & Olszewski 1985, Allen et al 1985, Iyengar & Verma 1984), implying that these objects also had "ultrahigh" infrared luminosities, typically $L_{\rm ir} \gtrsim 10^{12}$ $L_\odot$, and $L_{\rm ir}/L_{\rm B} = 30$–400. None of these objects showed obvious evidence for an active nucleus.

IRAS surveys of optically selected Seyfert galaxies (Miley et al 1985) and QSOs (Neugebauer et al 1985, 1986) showed that active galaxies could be strong far-infrared emitters; most optically selected AGNs had ratios L_{ir}/L_B in the range 0.2 to 1.0 with higher values in only a small number of objects. However, the full range of infrared excess exhibited by active galaxies is indeed much larger (e.g. Fairclough 1986). de Grijp et al (1985) found that searches based on "warm" ($f_{25}/f_{60} \gtrsim 0.3$) colors could be useful for discovering new infrared-luminous active galaxies in the *IRAS* database. This technique appeared to have been motivated by the shape of the infrared spectrum of the Seyfert 2 galaxy NGC 1068 (Telesco & Harper 1980) and the discovery of a similar "warm" 25-μm component in the broad-line, infrared-luminous radio galaxy 3C 390.3 (Miley et al 1984). Early statistics suggested that the true space density of AGNs could be a factor of two larger than previously assumed with the majority of the new infrared selected objects being a mixture of LINERS and Seyfert 2s.

3. REDSHIFT SURVEYS

A more complete description of the properties of infrared galaxies became possible only after the determination of redshifts for relatively large unbiased samples of infrared selected objects. Table 2 lists the major published redshift catalogs for *IRAS* galaxies [Saunders et al (1990) provides a good reference for *IRAS* galaxy surveys prior to 1990].

3.1 *Luminosity Functions*

A comparison of the luminosity function of infrared bright galaxies with other classes of extragalactic objects is shown in Figure 1. At luminosities below 10^{11} $L_\odot$, *IRAS* observations confirm that the majority of optically selected objects are relatively weak far-infrared emitters (Bothun et al 1989, Knapp et al 1989, Devereux & Young 1991, Isobe & Feigelson 1992). Surveys of Markarian galaxies (Deutsch & Willner 1986, Mazzarella & Balzano 1986, Mazzarella et al 1991, Bicay et al 1995) confirm that both Markarian starbursts and Seyferts have properties (e.g. f_{60}/f_{100} and L_{ir}/L_B ratios) closer to infrared selected samples as does the subclass of optically selected interacting galaxies (e.g. Bushouse 1987, Kennicutt et al 1987, Bushouse et al 1988, Sulentic 1989); however relatively few objects in optically selected samples are found with $L_{ir} > 10^{11.5}$ $L_\odot$.

The high luminosity tail of the infrared galaxy luminosity function is clearly in excess of what is expected from the Schechter function. A better description (e.g. Soifer et al 1987b) is a double power law with slope $\lesssim -1$ at low luminosity, changing to a slope of $\sim$$-2.35$ at $L_{bol} \gtrsim 10^{10.3}$ $L_\odot$. For $L_{bol} = 10^{11} - 10^{12}$ $L_\odot$, LIGs are as numerous as Markarian Seyferts and $\sim$3 times more numerous than Markarian starbursts. Ultraluminous infrared galaxies (hereafter ULIGs: $L_{ir} > 10^{12}$ $L_\odot$) appear to be $\sim$ 2 times more numerous than optically selected QSOs, the only other previously known population of objects with comparable bolometric luminosities.

Table 2 *IRAS* galaxy redshift surveys

Name	Flux limit(s)	Area	Sources[a]	Reference[b]
		all sky[c]		
RBGS	$f_{60} \geq 5.24$ Jy	$\|b\| > 5°$	602 P	Sanders et al 1996a
1.2 Jy Survey	$f_{60} \geq 1.2$ Jy	$\|b\| > 10°$	5321 P	Fisher et al 1995
1 Jy ULIGs	$f_{60} \geq 1.0$ Jy	$\|b\| > 30°$	115 F	Kim & Sanders 1996
QDOT	$f_{60} \geq 0.59$ Jy	$\|b\| > 10°$	2387 F	Lawrence et al 1996
12 μm Survey	$f_{12} \geq 0.22$ Jy	$\|b\| > 25°$	893 F	Rush et al 1993
		small area		
2 Jy Survey	$f_{60} \geq 2.0$ Jy	1072 deg^2	70 P	Smith et al 1987
Bootes Void	$f_{60} \geq 0.75$ Jy	1423 deg^2	379 P	Strauss & Huchra 1988
KOS-KOSS	$f_{60} \geq 0.6$ Jy	142 deg^2	63 P	Vader & Simon 1987b
NGW	$f_{60} \geq 0.5$ Jy	844 deg^2	389 P	Lawrence et al 1986
FSS-z	$f_{60} \gtrsim 0.2$ Jy	1310 deg^2	~3600 F	Oliver et al 1996
Pointed Obs	$f_{60} \geq 0.15$ Jy	18 deg^2	66 A	Lonsdale & Hacking 1989
NEPR	$f_{60} \geq 0.05$ Jy	6 deg^2	76 D	Ashby et al 1996
		color selected		
AGN Candidates	$1 > f_{25}/f_{60} > 0.27$	$\|b\| > 20°$	563 P	de Grijp et al 1992
WEO	$3 > f_{25}/f_{60} > 0.25$	$\|b\| > 30°$	187 P	Low et al 1988
Tepid FIRGs	$f_{25}/f_{60} < 0.27$, $f_{60}/f_{100} > 0.78$		53 P	Armus et al 1989
60 μm Peakers	$1 > f_{25}/f_{60} > 0.25$, $f_{60}/f_{100} > 1$	$\|b\| > 10°$	51 P	Vader et al 1993

[a]*IRAS* Catalogs: (P) PSC (1988), (F) FSC (Moshir et al 1992), (A) Pointed Observations (Young et al 1986), (D) Deep Survey (Hacking & Houck 1987).
[b]References for earlier versions of surveys:
RBGS—Soifer et al 1986, 1987b, 1989 (BGS); Sanders et al 1995 (BGS-Part II),
1.2 Jy Survey—Strauss et al 1992 (1.936 Jy Survey),
QDOT—Lawrence et al 1989 (QCD),
12 μm Survey—Spinoglio & Malkan 1989,
Tepid FIRGs—Heckman et al 1987.
[c]1 Jy ULIGs—$|b| > 30°$ and $\delta \geq -40°$,
QDOT—1 in 6 random source selection.

Although LIGs comprise the dominant population of extragalactic objects at $L_{\rm bol} > 10^{11}\ L_{\odot}$, they are still relatively rare. For example, Figure 1 suggests that only one object with $L_{\rm ir} > 10^{12}\ L_{\odot}$ will be found out to a redshift of ~ 0.033, and indeed, Arp 220 ($z = 0.018$) is the only ULIG within this volume. The total infrared luminosity from LIGs in the *IRAS* Bright Galaxy Survey (BGS) is only $\sim 6\%$ of the infrared emission in the local Universe (Soifer & Neugebauer 1991).

Comparison of the space density of ULIGs in the 1-Jy Survey with "local" ULIGs from the BGS provides some evidence for possible strong evolution in the luminosity function at the highest infrared luminosities. Assuming pure density evolution of the form $\Phi(z) \propto (1 + z)^n$, Kim (1995) found $n \sim 7 \pm 3$ for the complete 1-Jy sample of ULIGs, the uncertainty being influenced primarily by the small range of redshift ($z_{\rm max} = 0.27$) and the apparent effects of local large scale structure: Nearly all of the evidence for strong evolution

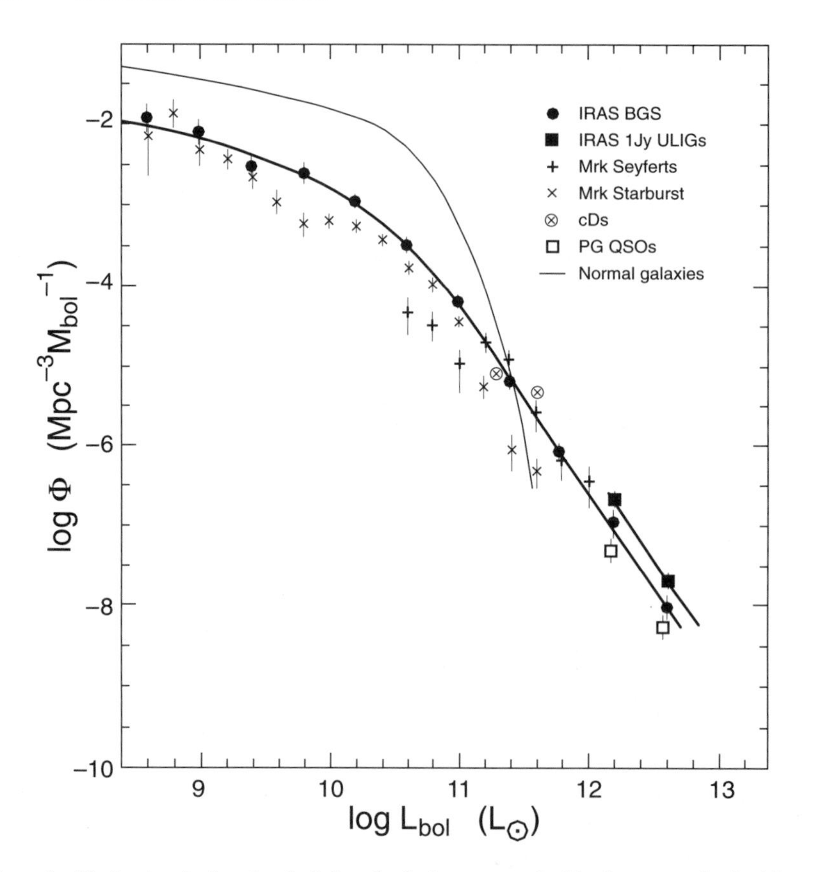

Figure 1 The luminosity function for infrared galaxies compared with other extragalactic objects. References: *IRAS* RBGS (Sanders et al 1996a), *IRAS* 1-Jy Survey of ULIGs (Kim 1995), Palomar-Green QSOs (Schmidt & Green 1983), Markarian starbursts and Seyfert galaxies (Huchra 1977), and normal galaxies (Schechter 1976). Determination of the bolometric luminosity for the optically selected samples was as described in Soifer et al (1986), except for the adoption of a more accurate bolometric correction for QSOs of $11.8 \times \nu L_{\nu}(0.43\ \mu m)$ (Elvis et al 1994).

comes from ULIGs at flux levels $f_{60} = 1$–2 Jy corresponding to sources at $z \gtrsim 0.13$. No evidence for evolution is found for the subsample of 2 Jy ULIGs, i.e. $n = 3.8 \pm 3$ (Kim & Sanders 1996). These results appear to be consistent with previous debates in the literature which find $n \sim 5.6$–7 for redshift surveys with flux limits $f_{60} \sim 0.5$ Jy (Saunders et al 1990, Oliver et al 1995) but only $n \sim 3$–4 for surveys with flux limits $f_{60} \gtrsim 1.5$ Jy (Fisher et al 1992), and with analyses of *IRAS* extragalactic source counts (Hacking et al 1987, Lonsdale & Hacking 1989, Lonsdale et al 1990, Gregorich et al 1995) that show evidence for strong evolution only at relatively low flux levels ($f_{60} \lesssim 1$ Jy). More definitive tests of whether the luminosity function for ULIGs indeed evolves strongly, and how this may compare, for example, with the strong evolution seen for the most luminous QSOs (e.g. Schmidt & Green 1983), will need to wait for future more sensitive infrared surveys.

3.2 *Spectral Energy Distributions*

The infrared properties for the complete *IRAS* BGS have been summarized and combined with optical data to determine the relative luminosity output from galaxies in the local Universe at wavelengths ~ 0.1–$1000\ \mu$m (Soifer & Neugebauer 1991). Figure 2 uses data from Sanders et al (1996a,b) and Kim (1995) to illustrate how the shape of the mean spectral energy distribution (SED) varies for galaxies with increasing total infrared luminosity. Systematic variations are observed in the mean infrared colors; the ratio f_{60}/f_{100} increases while

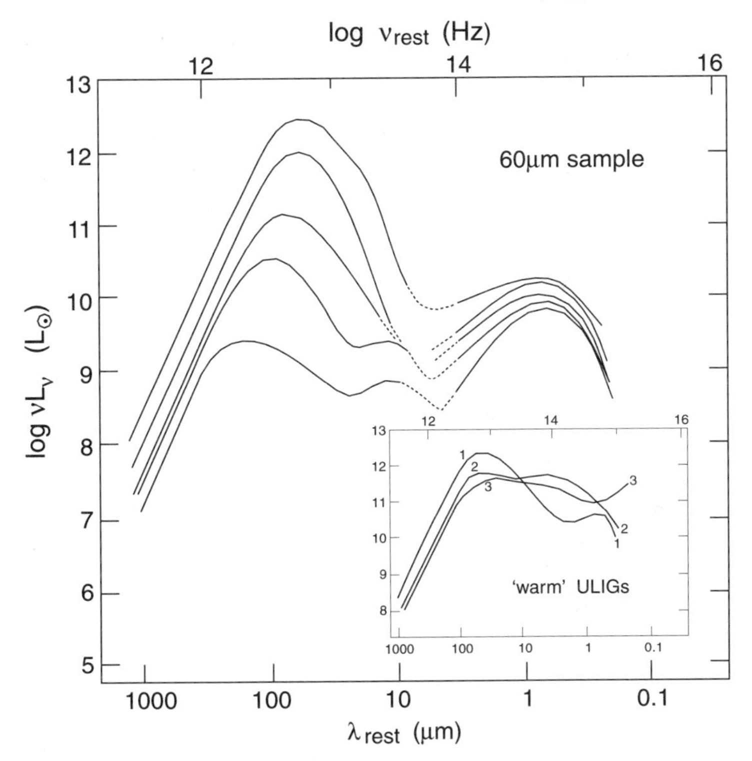

Figure 2 Variation of the mean SEDs (from submillimeter to UV wavelengths) with increasing L_{ir} for a 60 μm sample of infrared galaxies. (*Insert*) Examples of the subset ($\sim$ 15%) of ULIGs with "warm" infrared color ($f_{25}/f_{60} > 0.3$). Data for the three objects (1—the powerful Wolf-Rayet galaxy IRAS 01002–2238, 2—the "infrared QSO" IRAS 07598+6508, 3—the optically selected QSO I Zw 1) are from Sanders et al (1988b).

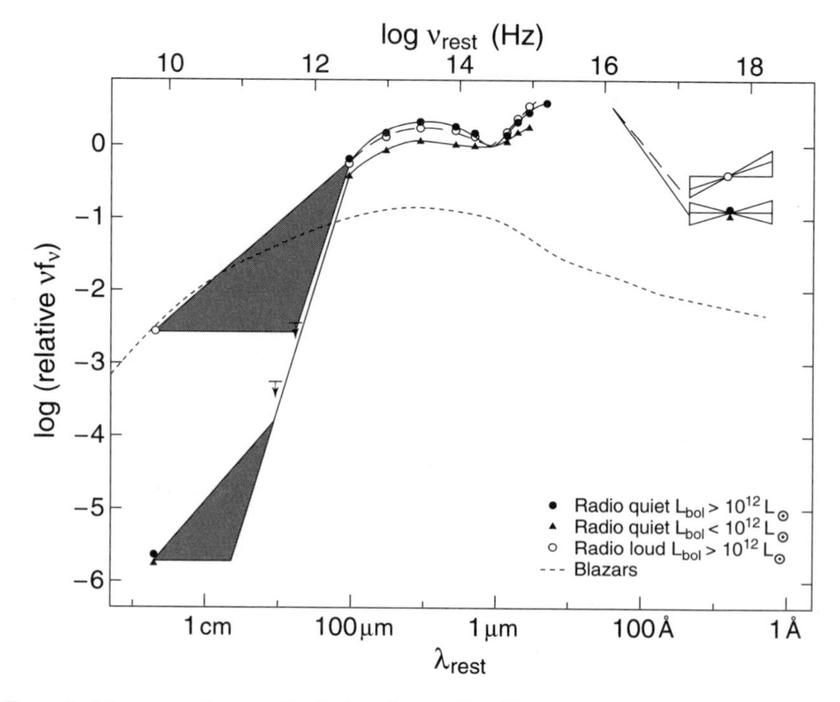

Figure 3 Mean spectral energy distributions from radio to X-ray wavelengths of optically selected radio-loud and radio-quiet QSOs (Sanders et al 1989a), and Blazars (Impey & Neugebauer 1988).

f_{12}/f_{25} decreases with increasing infrared luminosity. Figure 2 also illustrates that the observed range of over 3 orders of magnitude in L_{ir} for infrared-selected galaxies is accompanied by less than a factor of 3–4 change in the optical luminosity.

Various models of the infrared emission (e.g. Helou 1986, Rowan-Robinson 1986, Rowan-Robinson & Efstathiou 1993) have suggested that in lower luminosity "normal" galaxies the secondary peak in the mid-infrared is due to emission from small dust grains near hot stars, while the stronger peak at $\lambda \gtrsim$ 100–200 μm represents emission dominated by dust from infrared cirrus ($T_D \lesssim$ 20 K) heated substantially by the older stellar population. In more infrared luminous galaxies a "starburst" component emerges ($T_D \sim$ 30–60 K) with a peak closer to 60 μm, plus, in Seyfert galaxies, an even warmer component ($T_D \sim$ 150–250 K) peaking near 25 μm, presumably representing warm dust directly heated by the AGN.

Sanders et al (1988b) showed that a small but significant fraction of ULIGs, those with "warm" ($f_{25}/f_{60} > 0.3$) infrared colors, have SEDs with mid-infrared emission (~5–40 μm) over an order of magnitude stronger than the larger fraction of "cooler" ULIGs. These warm galaxies (Figure 2 insert), which appear to span a wide range of classes of extragalactic objects including powerful radio galaxies (PRGs: $L_{408MHz} \gtrsim 10^{25}$W Hz^{-1}) and optically selected QSOs, have been used as evidence for an evolutionary connection between ULIGs and QSOs (e.g. Sanders et al 1988a,b). This connection is strengthened by *IRAS* data for QSOs (Figure 3), which shows that the mean SED of optically selected QSOs is dominated by thermal emission from an infrared/submillimeter bump ($\sim$ 1–300 μm) in addition to the "big blue bump" ($\sim$ 0.05–1 μm); the former is typically 30% as strong as the latter and is presumably thermal emission from dust in an extended circumnuclear disk surrounding the active nucleus.

4. PROPERTIES OF LUMINOUS INFRARED GALAXIES

Substantial multiwavelength observations now exist for large samples of *IRAS* galaxies; the most extensive and highest spatial resolution observations are of objects in the BGS. Data for LIGs are summarized below by wavelength region. Wherever possible, emphasis is placed on trying to understand the variation of the properties of LIGs as a function of infrared luminosity.

4.1 *Optical and Near-Infrared Imaging*

Early optical imaging studies of *IRAS* galaxies are largely split into two groups: morphological classifications of relatively bright infrared sources that had previously been cataloged in optical surveys and observations of the most luminous infrared sources or sources selected for their extreme infrared properties (typically high infrared-to-blue or extreme color temperature). The former captured, almost exclusively, relatively nearby objects with $L_{ir} < 10^{11} L_{\odot}$ and can be summarized by the results from Rieke & Lebofsky (1986), who found that nearly all E's and most S0s have $L_{ir} < 10^9 L_{\odot}$, with most spirals having higher infrared luminosities; for $L_{ir} = 10^{10}$–$10^{11} L_{\odot}$, nearly all of the sources were Sb or Sc galaxies. It also appeared that $\sim$ 12–25% of LIGs were peculiar or interacting systems (e.g. Soifer et al 1984a). A much higher proportion of interacting and disturbed systems was reported from samples selected on the basis of extreme infrared properties. The most luminous sources ($L_{ir} \gtrsim 3 \times 10^{12} L_{\odot}$) in the *IRAS* database (Kleinmann & Keel 1987; Sanders et al 1987, 1988b; Hutchings & Neff 1987; Vader & Simon 1987a) were universally classified as strong interactions/mergers. Sources with high L_{ir}/L_B ratios (e.g. van den Broek 1990; Klaas & Elsässer 1991, 1993), or warm colors, either f_{60}/f_{100} or f_{25}/f_{60} colors (e.g. Armus et al 1987, 1990; Sanders et al 1988b; Heisler & Vader 1994), were predominantly ($\gtrsim$ 70%) strongly interacting/peculiar systems, with the remaining objects often being amorphous or elliptical-like in appearance.

More recent data for the complete BGS now shows that the fraction of objects that are interacting/merger systems appears to increase systematically with

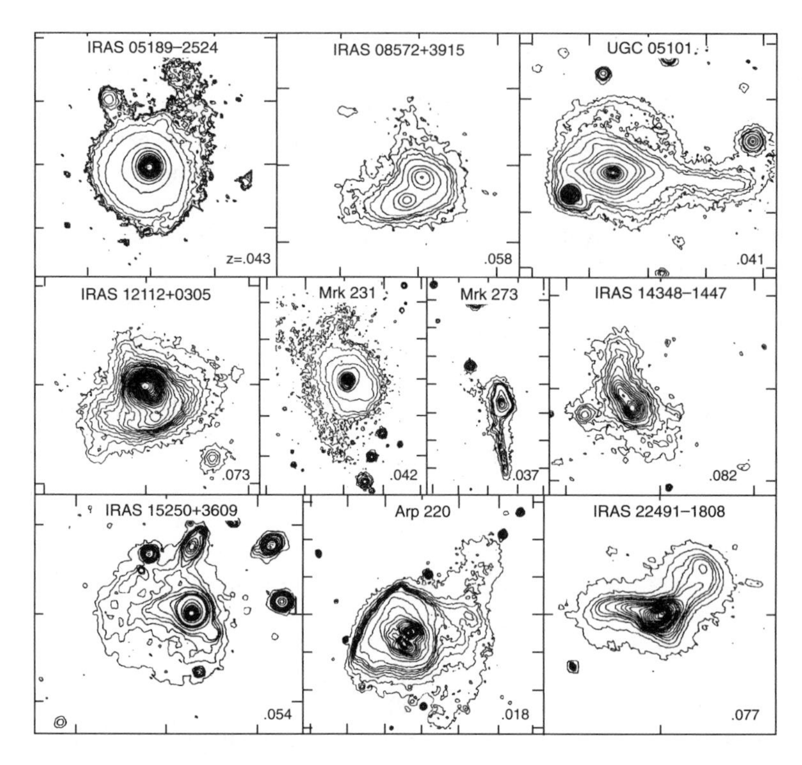

Figure 4 Optical (r-band) CCD images of the complete sample of ten ULIGs from the original BGS (Sanders et al 1988a). Tick marks are at 20″ intervals.

increasing infrared luminosity. Images of objects in the BGS (Sanders et al 1988a, 1996a; Melnick & Mirabel 1990) show that the fraction of strongly interacting/merger systems increases from $\sim$10% at $\log(L_{ir}/L_{\odot}) = 10.5$–11 to $\sim$100% at $\log(L_{ir}/L_{\odot}) > 12$. Figure 4 shows images of the complete sample of 10 ULIGs in the original BGS (Sanders et al 1988a). Other studies of ULIGs have generally reached a similar conclusion that $\gtrsim$ 95% are merger systems (Kim 1995, Murphy et al 1996, Clements et al 1996; although see Lawrence et al 1989, Leech et al 1994).

The improved angular resolution and lower optical depth in the near-infrared as compared to the optical has proved to be particularly useful for disentangling nuclear morphology not seen in the optical data (e.g. Carico et al 1990, Graham et al 1990, Eales et al 1990). The mean and range of projected nuclear separations for ULIGs in the BGS are $\sim$ 2 kpc and < 0.3 to 10 kpc respectively (Sanders 1992). More recent K-band imaging of larger samples of ULIGs (Murphy et al 1996, Kim 1995) generally confirms these results, although a few systems appear to have nuclear separations as large as 20–40 kpc.

Despite their extreme infrared luminosities, photometry of ULIGs confirms that most are only moderately luminous in the optical and near-infrared. For the 10 ULIGs in the original BGS the median blue absolute magnitude is $\bar{M}_{\rm B} = -20.7$ (Jensen et al 1996) [compared to the mean value $\langle M_{\rm B} \rangle = -20.2$ reported by Armus et al (1990) for their more distant sample of "Arp 220-like" objects], $\bar{M}_{\rm r} \sim -21.6$ (Murphy et al 1996; corrected by + 0.44 mag by J Surace, private communication), and $\bar{M}_{\rm K'} = -25.2$ (Jensen et al 1996). Compared to an L^* galaxy[2] the median total luminosities for ULIGs are $\sim 2.5\,L^*_{\rm B}$, $\sim 2.7\,L^*_{\rm r}$, and $\sim 2.5\,L^*_{\rm K'}$, where the range around the median (excluding Seyfert 1 objects) is -0.9 to $+1.6$ mag in $M_{\rm r}$ and -1.0 to $+2.2$ mag in $M_{\rm K'}$. Most ULIGs contain compact nuclei, with typically one quarter of the total K'-band luminosity originating in the inner $1''$ radius, except for the few Seyfert 1 galaxies (e.g. Mrk 231) where the pointlike nuclear source can be as much as a factor of ~ 5 times stronger than the surrounding galaxy. The typical host galaxies of ULIGs, therefore, appear to be $\sim 2L^*$ at K'.

4.2 *Optical and Near-Infrared Spectroscopy*

Although extensive optical redshift surveys have been carried out to identify *IRAS* galaxies, much of these data either have spectral resolution that is too low or are too limited in wavelength coverage to be of use for anything more than simple redshift determinations. Higher-resolution observations (typically $\Delta\lambda =$ 3–5 Å for $\lambda \sim 3800$–8000 Å) are now available for most of the *IRAS* galaxies in the imaging studies discussed above. Elston et al (1985) classified the majority of *IRAS* minisurvey objects as "starbursts taking place in dusty galaxies". On the other hand, $\sim 50\%$ of the objects with "warm" colors, f_{25}/f_{60}, from the sample of deGrijp et al (1985) were classified as Seyferts, with an additional $\sim 20\%$ classified as LINERS (Osterbrock & De Robertis 1985). Observations of "warm" objects at fainter flux levels in the *IRAS* database produced the first two infrared selected QSOs (Beichman et al 1986, Vader & Simon 1987a) plus several of the most intrinsically luminous infrared sources currently known, all of which are classified as Seyfert 2 in direct emission (e.g. Kleinman & Keel 1987, Hill et al 1987, Frogel et al 1989, Cutri et al 1994), but have been shown to contain hidden broad-line regions in polarized light (Hines 1991, Hines & Wills 1993, Hines et al 1995). Vader et al (1993) also report that $\sim 60\%$ of their "warm" sample of "60 μm Peakers" are Seyferts.

Heckman et al (1987) and Armus et al (1989, 1990) have used long-slit spectroscopy (4500–8000 Å) and narrow-band (Hα + [NII]) imaging to show that Arp 220-like "tepid" LIGs appear to contain huge ($\gtrsim 10$ kpc), powerful emission-line nebulae often characterized by spectacular loops and bubbles,

[2]For an L^* galaxy $M^*_{\rm B} = -19.7$ mag (Schechter 1976), $M^*_{\rm r} = -20.5$ mag assuming a typical B–$r = 0.75$, and $M^*_{\rm K'} = -24.2$ mag (Mobasher et al 1993).

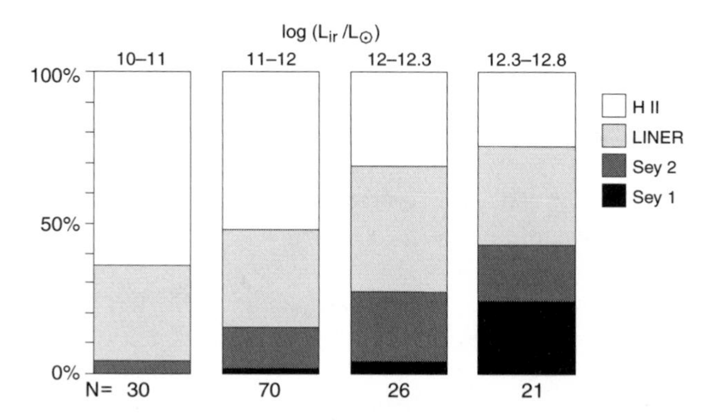

Figure 5 The optical spectral classification of infrared galaxies versus infrared luminosity (Kim 1995).

which they interpret as a starburst-driven superwind (see also Section 6.4). The visible spectrum of these objects appears to be dominated by young stars, with $\sim$ 20% showing evidence for a substantial intermediate-age population (few $\times\ 10^8$ years) and another $\sim$ 20% showing strong Wolf-Rayet lines, indicating a very young starburst ($\lesssim 10^7$ years). The total Hα + [NII] luminosity (corrected for extinction) is typically a factor of $\sim$ 30 larger than that for isolated spiral galaxies (Kennicutt & Kent 1983). As a class, these objects are similar to LINERs; about half are intermediate between LINERs and low-excitation H II regions. Steep Balmer decrements imply that they are being viewed through substantial amounts of obscuring dust. These data have been used to suggest that the "Arp 220–phase" may be characterized by an ongoing powerful circumnuclear starburst that may be rapidly clearing out the obscuring gas and dust in the inner few kiloparsecs of these objects.

More recently, the fraction of LIGs of different spectral type has been determined from an analysis of long-slit spectroscopy of complete samples of objects using several diagnostic emission-line ratios (e.g. Veilleux & Osterbrock 1987). Figure 5 shows the results from an analysis of a complete subsample of objects in the BGS (Kim et al 1995, Veilleux et al 1995), supplemented by a larger sample of ULIGs (Kim et al 1996). The percentage of Seyfert galaxies increases systematically from $\sim$ 4% at log $(L_{\rm ir}/L_\odot) = 10$–11, to $\sim$ 45% at log$(L_{\rm ir}/L_\odot) >$ 12.3, whereas the percentage of LINERs remains relatively constant ($\sim$ 33%) at all infrared luminosities log $(L_{\rm ir}/L_\odot) > 10$.

The amount of published near-infrared spectroscopy for *IRAS* galaxies is small compared to the available optical data. Rieke et al (1985) presented a detailed analysis of relatively low-dispersion infrared data at *K* and *L* bands for NGC 6240 and Arp 220, claiming that the former could be powered entirely by a very luminous starburst whereas as much as half of the luminosity in

Arp 220 appeared to be due to a Seyfert nucleus. Most of the early low-dispersion data for LIGs have been superseded by spectra covering the *K*-band window at resolving powers $\Delta\lambda/\lambda \sim$ 300–800 using infrared CCD arrays. Goldader (1995) and Goldader et al (1995) find that LIGs with Seyfert-like optical classifications also show evidence for dominant non thermal emission in the *K*-band, and most if not all ULIGs with Seyfert 2 optical spectra show evidence for Seyfert 1 linewidths in Paβ (Goodrich et al 1994) or Paα (Veilleux et al 1996); however, for "cool" LIGs, no new broad-line regions are discovered in the near-infrared that were not already seen at optical wavelengths. Perhaps the most intriguing new result is the systematic low value of the Brγ/H_2S(1) line ratio in ULIGs as compared with lower luminosity objects, a result that suggests that the dominant luminosity source in ULIGs is still highly obscured even at near-infrared wavelengths (Goldader et al 1995). A more detailed analysis of Arp 220 using several infrared lines (Armus et al 1995a,b) also suggests that as much as 80–90% of the total luminosity could be powered by an obscured AGN.

For the few hyperluminous infrared galaxies (HyLIGs: $L_{\rm ir} > 10^{13}\ L_{\odot}$) that have been discovered in the *IRAS* database, all are at $z \gtrsim 1$, so that near-infrared spectra provide the only means for observing several of the most prominent diagnostic lines (e.g. Hα and Hβ). All of these objects have rest-frame optical emission-line ratios characteristic of Seyferts [e.g. IRAS 15307+3252 (Soifer et al 1995, Evans et al 1996a); IRAS 10214+4724 (Soifer et al 1992, 1995; Elston et al 1994)].

4.3 *Mid- and Far-Infrared (Post–IRAS) Observations*

Ground-based observations in the mid-infrared, and far-infrared measurements with the *Kuiper Airborne Observatory* (*KAO*) have been carried out in an attempt to set meaningful constraints on the size of the infrared emitting region in a few LIGs. Becklin & Wynn-Williams (1987) reported that the 20-μm size of Arp 220 was smaller than 1.5″ (500 pc), and they estimated a visual extinction of at least 50 mag based on the depth of a deep silicate absorption feature at 10 μm. Dudley & Wynn-Williams (1996) use the depth of the silicate absorption feature to estimate 10-μm sizes of only a few parsecs for Arp 220 and the warm ULIG IRAS 08572+3915. Matthews et al (1987) used slit scans to show that the 10-μm source in Mrk 231 was smaller than 1″ (800 pc). Miles et al (1996) have obtained 10-μm maps for 10 LIGs and find that a large fraction ($\sim$ 65–100%) of the 10-μm emission in ULIGs and warm LIGs originates in an unresolved ($\lesssim$ 0.6″) core.

Observations with the *KAO* using drift scans at 50–100 μm (Joy et al 1986, 1989; Lester et al 1987) have shown that the emission regions in a few sources (e.g. Arp 220, Arp 299, NGC 1068) contain dominant compact ($\lesssim$ 10″) components at these wavelengths; however, larger telescopes or interferometers are clearly needed before more meaningful constraints can be set on the size of the far-infrared sources responsible for the bulk of the far-infrared luminosity in LIGs.

4.4 *Submillimeter Continuum*

Ground-based measurements in the submillimeter continuum ($\sim$350–860 μm) have been obtained for a few of the brightest LIGs. Emerson et al (1984) found that the far-infrared/submillimeter emission from Arp 220 could be fit by a single temperature dust model with $T_{\rm dust} \sim 60$ K, and they derived an optical depth of ~ 1 at 180 μm (!) for an assumed source size of 4″. More recently, Rigopoulu et al (1996a) have interpreted their submillimeter measurements for all ULIGs in the BGS as being consistent with thermal dust emission.

Submillimeter observations of LIGs have also been reported by Eales et al (1989) and Clements et al (1993), with the general result that the far infrared/submillimeter continuum in all of the objects can be reasonably fit by a single temperature dust model (assuming a ν^{-2} emissivity law) with dust temperatures of 30–50 K. There is no obvious evidence for large amounts of cooler dust, although large quantities of sufficiently cool dust (i.e. $T_{\rm dust} \lesssim 20$ K) cannot be ruled out (e.g. Devereux & Young 1991).

4.5 *Radio Continuum*

A "tight correlation" between the flux in the infrared and the radio continuum has been found in several studies of normal, starburst, and Seyfert galaxies (van der Kruit 1971, Rieke & Low 1972, Dickey & Salpeter 1984, Helou et al 1985, Sanders & Mirabel 1985, Wunderlich et al 1987). Figure 6*a* shows that the logarithmic ratio of far-infrared and radio continuum flux densities, $q = \log \{[F_{\rm fir}/(3.75 \times 10^{12}\ {\rm Hz})]/[f_\nu(1.49\ {\rm GHz})]\}$, is relatively constant, $q \sim 2.35$, for most LIGs in the BGS, with a rather small dispersion at a given $L_{\rm ir}$ ($\sigma \sim 0.2$). This relationship appears to hold for sources covering several orders of magnitude in $L_{\rm ir}$, from quiescent disk-like spirals like M31 to the powerful nuclear sources in ULIGs like Arp 220, although there is evidence that the mean changes slightly at both low and high infrared luminosities. At lower infrared luminosities, $\langle q \rangle$ appears to increase slightly as galaxies transit from heating dominated by young stars to heating by an old stellar population (Condon et al 1991a, Xu et al 1994), whereas at the highest infrared luminosities $\langle q \rangle$ increases, apparently due mainly to optical depth effects in the radio (Condon et al 1991b). More recently, Bicay et al (1995) have shown that $\langle q \rangle$ for Markarian starbursts is enhanced by a factor of ~ 3 relative to Markarian Seyferts, but the reason for this increase is not immediately clear.

Most radio galaxies and radio-loud QSOs have q-values much lower than 2.35 (typically by factors of $\sim$ 2–4 in the log), usually due to the presence of strong, very-compact radio cores combined with extended radio lobes/jets that are apparently decoupled from the infrared emission. However, evidence exists

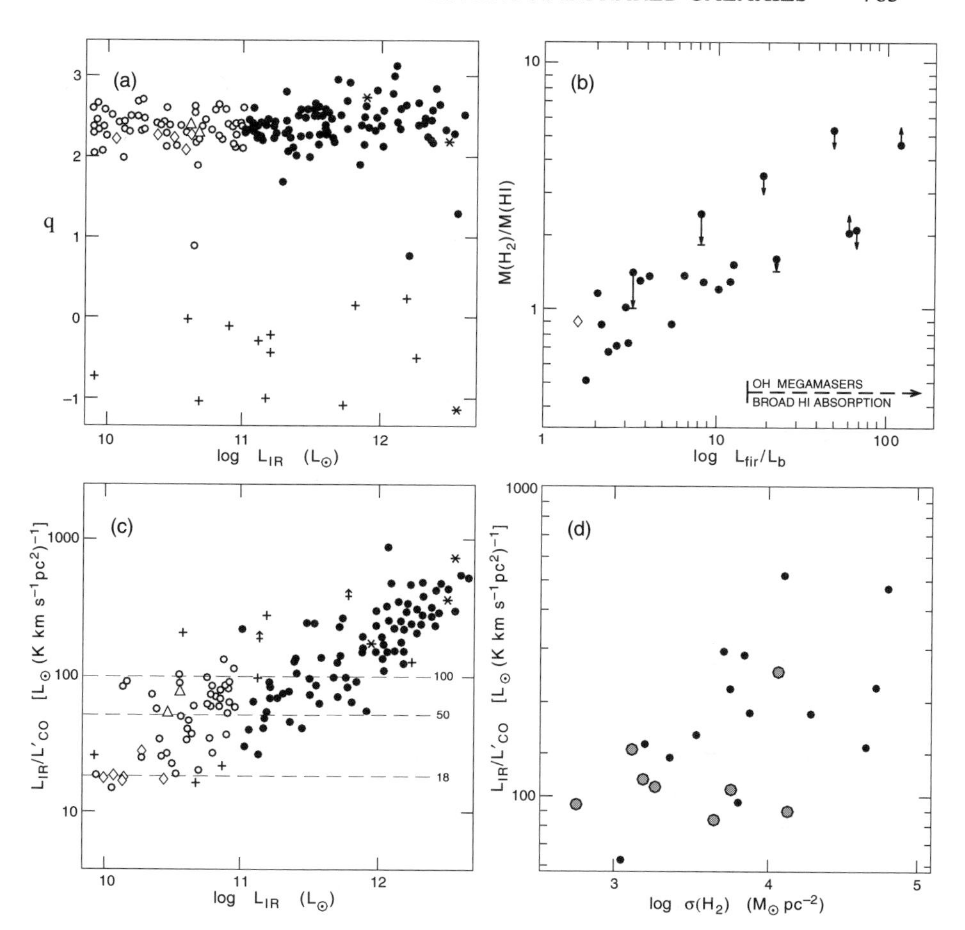

Figure 6 (*a*) q versus L_{ir}. Solid and open circles represent LIGs and lower luminosity BGS galaxies respectively. Open triangles refer to the nearby starburst galaxies M82 and NGC 253, and open diamonds represent the Milky Way and three "normal" nearby spiral galaxies (IC 342, NGC 6946, NGC 891). Asterisks refer to optically selected QSOs that have been detected in CO (I Zw1, Mrk 1014, 3C 48), and "+" signs represent PRGs detected by *IRAS*. (*b*) The ratio of molecular (H_2) to atomic (H I) gas versus the infrared-to-blue luminosity ratio in *IRAS* BGS galaxies (Mirabel & Sanders 1989). The arrows are due to lower and upper limits in the measurement of the H I fluxes. Most galaxies with $L_{ir}/L_B \gtrsim 20$ exhibit H I absorption and OH megamaser emission with velocity extents of several hundred km s^{-1}. (*c*) L_{ir}/L'_{CO} versus L_{ir}. The dashed lines represent mean values for nearby "normal" spirals ($L_{ir}/L'_{CO} \sim 18$), nearby starburst galaxies ($L_{ir}/L'_{CO} \sim 50$), and the most extreme star-forming GMC cores in the Milky Way ($L_{ir}/L'_{CO} \sim 100$). (*d*) Correlation of the central concentration of molecular gas with the L_{ir}/L'_{CO} ratio for LIGs in the *IRAS* BGS (Scoville et al 1991, Bryant 1996). Small black and larger grey circles represent objects where the spatial resolution was sufficient to resolve circumnuclear regions of < 1 kpc and 1–2 kpc diameter respectively.

that radio galaxies still show a correlation in their far-infrared and radio fluxes, but with a value $\langle q \rangle \sim -0.65$ (e.g. Golombek et al 1988, Knapp et al 1990, Impey & Gregorini 1993).

The "radio-infrared correlation" has been used in an attempt to overcome the poor angular resolution of far-infrared instruments. High-resolution radio surveys of LIGs in the BGS were made at 1.49 GHz (Condon et al 1990) and at 8.44 GHz (Condon et al 1991b). The VLA maps show that nearly all galaxies with $L_{\rm fir} \leq 10^{11}\ L_{\odot}$ are dominated by extended, diffuse radio emission, whereas most ULIGs are dominated by compact, sub-arcsec radio sources. Condon et al (1991b) concluded that most LIGs in the BGS—with the exception of the Seyfert 1 galaxy Mrk 231, which is dominated by a variable ultracompact radio source ($\leq$1 pc)—can be modeled by ultraluminous nuclear starbursts (see also Crawford et al 1996). These starburst regions would be so dense that they are optically thick even to free-free absorption at $\nu = 1.49$ GHz and to dust absorption at $\lambda \leq 25\ \mu$m ! If this is true, then X rays, infrared, and radio waves may not be able to probe the dense cores of ULIGs; the far-infrared luminosity is then at best a good calorimeter. Due to Compton scattering in such dense clouds, even a compact source of hard X rays will be hidden from the observer !

As a further probe of the size of the radio sources in LIGs, Lonsdale et al (1993) carried out a sensitive VLBI survey of 31 objects and found that typically ~12% of the radio flux arises in cores only 5–150 milliarcsec in size (which rules out a single supernova interpretation of the compact radio cores). These compact VLBI cores are comparable in power to the total radio power of typical Seyfert galaxies (Ulvestad & Wilson 1989) and radio-quiet QSOs (Kellermann et al 1989).

4.6 *Gas Content*

Single-dish observations of millimeter-wave emission from the rotational transitions of CO ($\theta_{\rm FWHM} \sim 20$–$60''$) and the 21-cm line of H I ($\theta_{\rm FWHM} \sim 3$–$10'$) now exist for the majority of objects in the BGS. These data have shown that the total neutral gas content, in particular the total mass of molecular gas, appears to play a critical role in the genesis of LIGs. More recently, interferometer maps of a few dozen LIGs and lower luminosity infrared-selected objects have provided dramatic pictures of the redistribution of the H I and H_2 gas that occurs during interactions and mergers.

4.6.1 ATOMIC GAS (H I) The first observations of LIGs at centimeter wavelengths revealed somewhat perplexing properties. At $\lambda = 21$ cm, the most luminous infrared galaxies showed very broad H I absorption lines, indicating rotation plus large amounts of unusually turbulent neutral gas (Mirabel 1982). The H I profiles typically show absorption features with widths of a few hundred to up to 1000 km s^{-1}, with total column densities $\gtrsim 10^{21-22}$ atoms cm^{-2}. VLA observations of the H I absorption had suggested that most of the absorbing H I is located in the inner few hundred parsecs of these galaxies (e.g. Baan et al. 1987), in front of the nuclear radio continuum sources. The total masses of H I in a complete sample of galaxies with $L_{\rm fir} \geq 2 \times 10^{10}\ L_\odot$ are in the range of 5×10^8 to $3 \times 10^{10}\ M_\odot$, with only a weak correlation between M(H I) and $L_{\rm fir}$ (Mirabel & Sanders 1988).

Mirabel & Sanders (1988) carried out a statistical analysis of the difference between the radial velocities of the H I 21-cm line absorptions and the optical redshifts. Although the discrepancy between the radio and optical velocities is usually smaller than the velocity width of the H I absorptions, there is a clear trend for the radio redshifts to be greater than the optical redshifts. Among 18 galaxies with H I absorption, 15 were found with $V_{\rm H\ I_{abs}} > V_{\rm opt}$ and only 3 with $V_{\rm H\ I_{abs}} \leq V_{\rm opt}$. The mean value of $V_{\rm H\ I_{abs}} - V_{\rm opt}$ is 90 km s^{-1}. From VLA observations Dickey (1986) found a similar trend, from which he estimated an accretion rate of $\sim$1 $M_\odot$ year^{-1} into the nuclear regions.

The optical redshifts may be affected by systematic errors and there are some caveats to the interpretation of the statistical discrepancy between optical and radio velocities as due entirely to infall of H I. The optical redshifts are often determined from emission lines in low dispersion spectra, which due to large scale superwinds (see Section 6.4) are usually asymmetric with extended blue wings. In this context, some of the statistical discrepancy between the radio and optical redshifts could be due to optical line-emitting gas mixed with dust that is moving radially, probably outward. However, from a careful analysis of the available data, Martin et al (1991) concluded that most of the H I seen in absorption in is indeed infalling toward the central source.

4.6.2 MOLECULAR GAS (H_2) Substantial information on the total molecular gas content of LIGs has been obtained from single-dish observations of millimeter-wave CO emission for large samples of *IRAS* galaxies. An important discovery has been that *all* LIGs appear to be extremely rich in molecular gas. Early CO observations of infrared selected galaxies (most with $L_{\rm ir} = 10^{10} - 10^{11}\ L_\odot$) found a rough correlation between CO and far-infrared luminosity (Young et al 1984, 1986a,b; Sanders & Mirabel 1985). Assuming a constant conversion factor between CO luminosity and H_2 mass, $M(H_2)/L'_{\rm CO} = 4.6[M_\odot({\rm K\,km\,s^{-1}pc^2})^{-1}]$ (e.g. Scoville & Sanders 1987), total H_2 masses were in the range $\sim$ 1–30 $\times 10^9\ M_\odot$, or approximately 0.7 to 20 times the molecular gas content of the Milky Way. Mirabel & Sanders (1989) found that the ratio of total H_2 to H I mass is typically >1 with some evidence that $M_T({\rm H_2})/M_T({\rm H\ I})$ increases with increasing $L_{\rm ir}/L_{\rm B}$ (Figure 6*b*). Multitransition CO measurements (e.g. Sanders et al 1990, Radford et al 1991, Braine et al 1993, Devereux et al 1994, Rigopoulu et al 1996b) and detection of strong emission from dense gas tracers such as HCN (Solomon et al 1992a), CS, and HCO^+ (see reviews by

Mauersberger & Henkel 1993, Radford 1994, Gao 1996) indicate that the mean molecular gas temperatures and densities in the central regions of LIGs are hot, $T_{kin} = 60$–90 K, and dense, $n(H_2) \sim 10^5 - 10^7$ cm^{-3}, similar to the conditions in massive Galactic giant molecular cloud (GMC) cores.

Significant improvements in detector performance in the late 1980s resulted in a dramatic increase in the number of extragalactic infrared sources detected in CO. The first objects detected at $z \gtrsim 0.03$ were ULIGs from the BGS; these included the first detection of CO emission from a Seyfert 1 galaxy (Mrk 231: Sanders et al 1987) and two of the most intrinsically luminous CO sources currently known (VII Zw 31: Sage & Solomon 1987; IRAS 14348–1447: Sanders et al 1988d). Detections of objects at $z > 0.1$ soon followed, including the first CO detections of UV-excess QSOs (Mrk 1014: Sanders et al 1988c; I Zw 1: Barvainis et al 1989), an infrared selected QSO (IRAS 07598+6508: Sanders et al 1989b), a PRG (4C 12.50: Mirabel et al 1989), and a radio-loud QSO (3C 48: Scoville et al 1993).

Figure 6*c* includes data from several single-dish CO(1→0) surveys of *IRAS* galaxies (Sanders et al 1991, Mirabel et al 1990, Tinney et al 1990, Downes et al 1993, Mazzarella et al 1993, Young et al 1995, Elfhag et al 1996, Solomon et al 1996, Evans 1996) illustrating both the general trend of increasing L_{ir}/L'_{CO} ratio with increasing L_{ir} and the fact that this ratio can vary by nearly a factor of 30 at a given L_{ir}. Assuming $M(H_2) = 4.6\, L'_{CO}$, Figure 6*c* shows that the *total* molecular gas mass in ULIGs—typically $M(H_2) \gtrsim 10^{10}\, M_{\odot}$—is on average more infrared luminous than any of the most active star-forming Galactic GMC cores [which have diameters typically of $\sim$2–5 pc and $M(H_2) = 10^3 - 10^4\, M_{\odot}$].

Millimeter-wave interferometer measurements of CO emission have been obtained for approximately two dozen LIGs (e.g. Scoville et al 1991; Okumura et al 1991; Bryant 1996; Yun & Hibbard, private communication). Figure 6*d* shows that for these objects, nearly all of which are advanced mergers, $\sim$ 40–100% of the total CO luminosity, or $M(H_2) = 1$–$3 \times 10^{10}\, M_{\odot}$ assuming the standard Milky Way conversion factor, is contained within the central $r <$ 0.5–1 kpc. The mean surface density in the central 1 kpc regions of ULIGs is typically in the range $\langle\sigma(H_2)\rangle = 1.5$–$7 \times 10^4\, M_{\odot}$ pc^{-2}, although the H_2 masses in these extreme regions may be overestimated by a factor of 2–3 (Downes et al 1993, Solomon et al 1996). Even allowing for such a decrease in the conversion factor, these values are $\sim$ 50–250 times larger than the mean gas surface density in the central 1 kpc of the Milky Way and would seem to imply enormous optical depths ($A_V \sim$ 200–1000 mag) along an average line of sight toward the nucleus of these objects. Such high values are consistent with the implied high optical depths in the nuclear regions of "cool" ULIGs at *K*-band (see Section 4.2) and optical depths near unity at $\lambda \sim 200\ \mu$m implied by the far-infrared/submillimeter measurements for Arp 220 (see Section 4.4).

4.6.3 OH MEGAMASERS The high-density molecular gas in the central regions of LIGs is the site of the most luminous cosmic maser sources known. The amplified main OH lines at 1667 and 1665 MHz correspond to transitions between the hyperfine splitting of a Λ-doubling level at λ18 cm. The isotropic luminosities in the OH lines can be as strong as $\sim 10^4\ L_\odot$, almost a million times that of the most luminous OH maser sources in the Galaxy, hence the name megamaser. Since the first detection of OH maser emission from the ULIG Arp 220 (Baan et al 1982), it became evident that this type of emission could be detected out to redshifts of $z \sim 0.5$, and therefore, that it could be used to probe the circumnuclear high-density interstellar gas in distant infrared galaxies (Baan 1985, Stavely-Smith et al 1987, Martin et al 1991, Kazès & Baan 1991). About 50 OH megamasers have now been identified (Baan 1993). The OH megamaser spectra usually exhibit broad linewidths and extended velocity wings that could be due to the rapid rotation of circumnuclear molecular disks (e.g. Montgomery & Cohen 1992), large-scale outflow motions (Mirabel & Sanders 1987, Baan 1989), and/or distinct components arising in the interactive system (e.g. Baan et al 1992).

The isotropic OH 1667MHz luminosity is proportional to $(L_{\rm ir})^2$ (Martin et al 1988, Baan 1989). This has been interpreted by Baan & Haschick (1984) as low-gain amplification of the nuclear continuum radio source by intervening OH that is being pumped by far-infrared radiation from dust. Inversion of the OH population requires a source of steep-spectrum emission in the far-infrared such as that expected from warm dust emission. Usually an OH megamaser galaxy requires a strong nuclear radio continuum source, a column density of gas along the line of sight $> 10^{22}$ cm^{-2}, a ratio $f_{60}/f_{100} \geq 0.7$, and $L_{\rm ir} > 10^{11}$ $L_\odot$ (Mirabel & Sanders 1987).

The location of megamasers in the $L_{\rm ir}$ versus $M({\rm H_2})$ plane shows that they occur in objects with the largest $L_{\rm fir}/M({\rm H_2})$ ratios (Mirabel & Sanders 1989). At present we consider it an open question whether the origin of the far-infrared luminosity in megamasers is entirely due to star formation. A considerable fraction of the energy could ultimately come from a compact nonthermal source at the dynamical center of these galaxies. Future high-resolution VLBI observations of the OH emission may be a way to answer this question.

There is increasing evidence that OH megamasers can eventually be used to successfully probe the molecular gas kinematics and the dynamic masses at the very centers of starburst galaxies and AGNs. Furthermore, given the observed $L_{\rm OH} \propto L_{\rm ir}^2$ relationship in megamaser galaxies, the spatial distribution of the far-infrared emitting region can be inferred with unprecedented angular resolution. Single-dish observations have in the past been interpreted as showing evidence for spatial extents of a few hundred parsecs in the OH emission from nearby systems (Baan 1993). However, MERLIN observations of III Zw 35 have shown

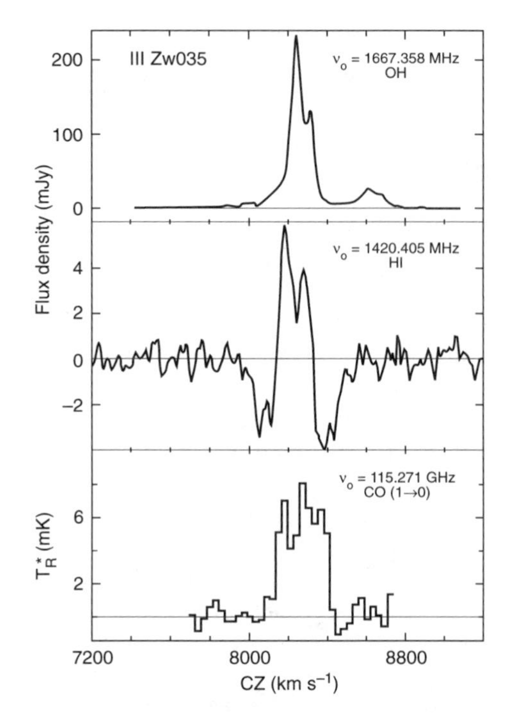

Figure 7 OH megamaser emission, H I 21-cm line emission, and CO(1→0) line emission from the high L_{ir}/L_B object III Zw 35 (Mirabel & Sanders 1987).

that most of the OH emission arises from a region only 100 pc × 60 pc in size, which is more compact than the radio continuum source (Montgomery & Cohen 1992). The velocity of the OH emission appears to trace out the characteristic signature of a rotating disk, which would imply a dynamical mass of $\sim 10^9\ M_\odot$ inside a region of ~15 pc in radius.

It is interesting that one of the most compact megamasers mapped so far with MERLIN, III Zw 35, is also among those megamasers with the largest implied infrared pumping efficiencies (Mirabel & Sanders 1987). Figure 7 shows OH emission from III Zw 35 up to 450 km s^{-1} below the systemic velocity, which raises the interesting possibility of extreme circular velocities in a compact molecular torus. Similar high velocity wings, perhaps due to rapidly rotating disks, have been observed in Mrk 231 and Mrk 273 (Stavely-Smith et al 1987). In addition, recent VLBI observations of Arp 220 strongly suggest that the OH peak emission originates in a structure $\lesssim 1$ pc across and that nearly all of the OH emission comes from a region ≤ 10 pc in size (Lonsdale et al 1994). This would imply that the maser is physically 10–100 times smaller than previously thought (Baan 1993) and that much of the far-infrared emission from Arp 220 arises in a very small region, possibly a torus surrounding a quasar nucleus. These new results suggest that OH megamasers may eventually be used to probe the central engines in AGNs, much as the H_2O megamasers have been used in NGC 4258 (Miyoshi et al 1995).

4.7 *High-Energy Observations*

A strong correlation has been found between the X-ray luminosity in the *Einstein* 0.2–4.5 keV energy band and the far-infrared and blue luminosities for 51 galaxies in the BGS (David et al 1992). The best fit for the X-ray luminosity, $L_X = 9.9 \times 10^{-5}\, L_B + 9.3 \times 10^{-5} L_{fir}$, has two components, one proportional to L_B due to the old population (Type I supernovae and low mass X-ray binaries) and a second component proportional to L_{fir} due to young objects (Type II supernovae, O stars, high-mass binaries). LIGs with large L_{ir}/L_B ratios are dominated by the second component and can have a significant excess of X-ray flux relative to the mean (up to a factor of 10 in the case of Arp 220). It has been proposed that LIGs could make a significant contribution (up to 40%) to the X-ray background (Griffiths & Padovani 1990).

The cross-correlation between the *IRAS* PSC and the *ROSAT* All-Sky Survey (0.1–2.4 keV) allowed the identification of 244 *IRAS* galaxies that appeared to be positionally coincident with *ROSAT* X-ray sources (Boller et al 1992). This sample is dominated by galaxies with active nuclei. An unexpected result was the discovery of a dozen spiral galaxies with X-ray luminosities up to 10^{43} erg s^{-1}, well above those found by the *Einstein* satellite. These objects have steep spectra with typical photon indexes of 2.5–3.6, which would make them preferentially detected by *ROSAT* because of its sensitivity in the soft X rays. Optical spectroscopy revealed active nuclei in half of these objects (Boller et al 1993). In the galaxy IRAS 15564+6359 a compact component can be disentangled from that of any extended component, since in the 0.1–2.4 keV energy range flux variations with a doubling time of 1500 sec (corresponding to a source size of $\lesssim 2 \times 10^{-5}$ pc) have been discovered (Boller et al 1994). The non-detection of optical AGN activity in this galaxy is intriguing.

ROSAT observations of Arp 220 reveal extended emission with $T_{kin} \sim 10^7$ K (Heckman et al 1996), presumably due to gas ejected from the active nuclear region (see also Section 6.4). The morphology is similar to, but much more powerful than, the extended X-ray components observed in the classic starburst galaxies NGC 253 and M82 (Fabbiano 1988) and the edge-on spiral NGC 3628 (Fabbiano et al 1990). *ROSAT* observations of NGC 4038/39 ("The Antennae") have shown that about 55% of the X-ray emission can be identified with the two galaxy nuclei plus a large ring of H II regions; the remainder appears to be diffuse emission from gas at $\sim 4 \times 10^6$ K that extends up to 30 kpc away from the merging disks (Read et al 1995).

To test whether it is possible to use X rays to search for obscured AGN in ULIGs, as suggested by Rieke (1988), it is instructive to consider the case of NGC 1068. Spectropolarimetry of NGC 1068 by Antonucci & Miller (1985) revealed a Seyfert 1 nucleus hidden in a dense torus of dust and gas. Millimeter-wave observations of HCN imply $\sim 1.6 \times 10^8\ M_{\odot}$ of molecular gas within a region $\lesssim 34$ pc in radius (Tacconi et al 1994). Assuming a spherical distribution implies a column density $\gtrsim 5 \times 10^{24}$ atoms cm^{-2}, which—given the expected high metalicity—would produce an almost total absorption of X rays by Compton scattering. In fact, *ASCA* observations revealed that the direct component in NGC 1068 is totally absorbed even at energies of 20 keV and that the observed spectrum below 20 keV is light scattered from the AGN by electrons over the absorption torus (Ueno et al 1994). Because the central gas densities in ULIGs are likely to be even greater than in NGC 1068, hard X rays most likely cannot be used to probe the central engines in these galaxies.

Gamma rays from extragalactic systems have so far mostly been detected from AGNs, either as quasi-isotropic emission from unobscured Seyferts or as beamed emission from highly variable Blazars. As expected, NGC 1068 was not detected at > 20 keV (Dermer & Gehrels 1995). Observations of the nearby lower-luminosity starburst galaxies NGC 253 and M82, over the energy range 0.05–10 MeV, yelded a 4σ detection of NGC 253 in continuum emission up to 165 keV, with an estimated luminosity of $\sim 10^{40}$ erg s^{-1}. No significant flux at high energies was detected from M82 (Bhattacharya et al 1994).

5. ORIGIN AND EVOLUTION OF LUMINOUS INFRARED GALAXIES

The data presented in Section 4 are sufficient to present a fairly detailed picture of the morphology of LIGs and to investigate how their properties change as a function of infrared luminosity. Table 3 summarizes some of the main results using observations of the complete sample of LIGs in the BGS supplemented by data for a subsample of less luminous BGS objects and data for a larger sample of ULIGs from the survey of Kim (1995).

5.1 *Strong Interactions and Mergers*

It now seems clear that strong interactions and mergers of molecular gas-rich spirals are the trigger for producing the most luminous infrared galaxies. The images of ULIGs from the BGS (see Figure 4) show clearly that maximum infrared luminosity is produced close to the time when the two nuclei actually merge. At $L_{\mathrm{ir}} < 10^{11}\ L_{\odot}$ the vast majority of infrared galaxies appear to be single, gas-rich spirals whose infrared luminosity can be accounted for largely by star formation. Over the range $L_{\mathrm{ir}} = 10^{11} - 10^{12}\ L_{\odot}$ there is a dramatic increase in the frequency of strongly interacting systems that are extremely rich

Table 3 *IRAS* galaxy properties versus L_{ir}

		10.5–10.99	11.0–11.49	11.5–11.99	12.0–12.50
			$\log(L_{ir}/L_{\odot})$		
No. of objects[a]		50	50	30	40
Morphology	merger	12%	32%	66%	95%
	close pair	21%	36%	14%	0%
	single (?)	67%	32%	20%	5%
Separation[b]	[kpc]	36.	27.	6.4	1.2
Opt Spectra	Seyfert 1 or 2	7%	10%	17%	34%
	LINER	28%	32%	34%	38%
	H II	65%	58%	49%	28%
L_{ir}/L_B[c]		1	5	13	25
L_{ir}/L'_{CO}[c]	$[L_{\odot}(\text{K km s}^{-1}\text{pc}^2)^{-1}]$	37	78	122	230

[a]Objects in the *IRAS* BGS plus additional ULIGs from Kim & Sanders (1996).
[b]Mean projected separation of nuclei for mergers and close pairs only.
[c]Mean values.

in molecular gas; at the low end of this range the luminosity appears to be dominated by starbursts with Seyferts becoming increasingly important at higher luminosities. Those objects that reach the highest infrared luminosities, $L_{ir} > 10^{12}\ L_{\odot}$, contain exceptionally large central concentrations of molecular gas; because of heavy dust obscuration it is hard to distinguish the relative roles of starburst and AGN activity, although the conditions are clearly optimal for fueling both enormous nuclear starbursts as well as building and/or fueling an AGN.

The enormous build-up of molecular gas in the centers of the most luminous infrared objects plays a fundamental role in LIGs and is best illustrated by showing data for several relatively nearby well-studied objects. The ultimate fate of these mergers, once their infrared excess subsides, is not completely clear. Examples are also shown below for a small but important subclass of objects which show both strong optical and infrared emission, and which plausibly represent a transition stage in the evolution of LIGs into optically selected AGN (e.g. Sanders et al 1988b). Finally we mention a few objects initially identified as HyLIGs that illustrate why caution needs to be taken when identifying objects at higher-*z*.

5.2 *Case Studies*

5.2.1 MERGERS OF MOLECULAR GAS–RICH SPIRALS *NGC 4038/39 (Arp 244 = VV 245 = "The Antennae")* This classic, nearby "early merger" system is composed of what appear to be two overlapping, distorted, late-type spiral disks (Figure 8*a*). The total infrared luminosity is the minimum required by our definition of LIG. The *K*-band image shows two nuclei ~ 15 kpc apart and what appears to be a large ring of bright H II regions in the northern disk. CO interferometer observations (Stanford et al 1990) show that $\sim 60\%$

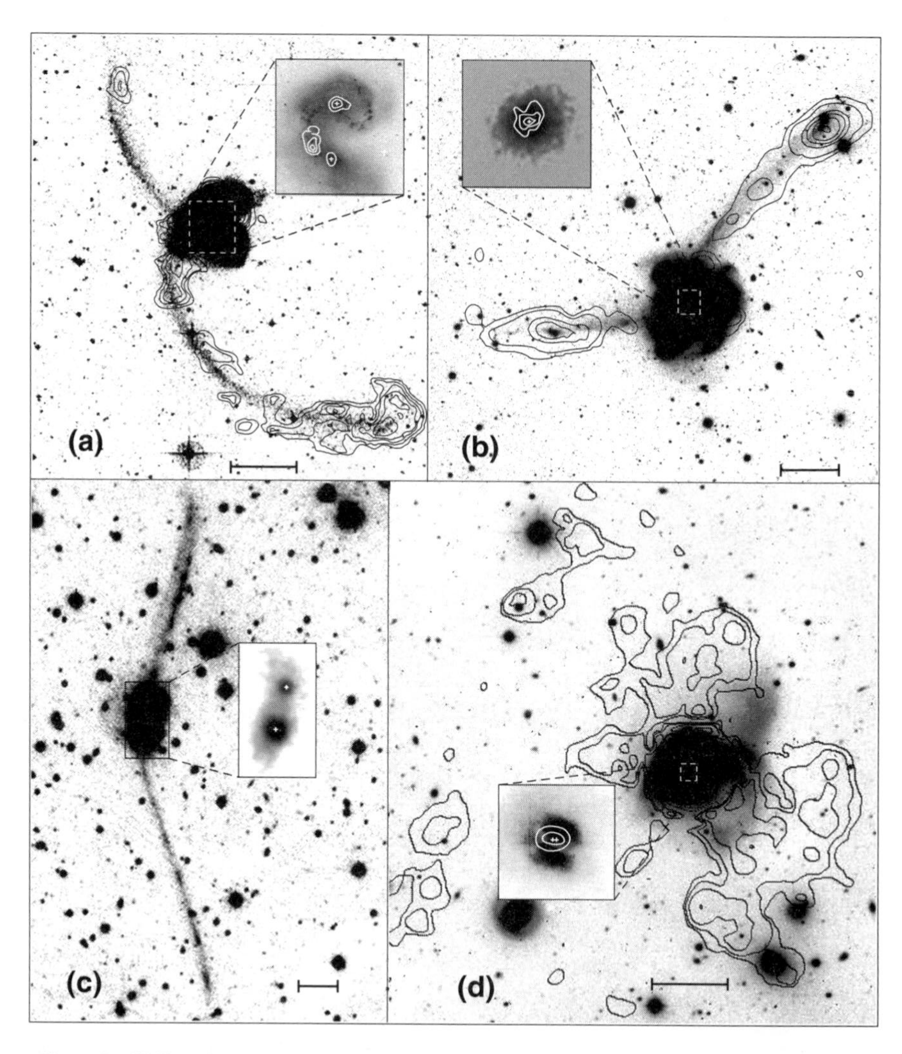

Figure 8 Well-studied mergers: (*a*) NGC 4038/39 (Arp 244 = "The Antennae"); (*b*) NGC 7252 (Arp 226 = "Atoms for Peace"); (*c*) IRAS 19254–7245 ("The Super Antennae"); (*d*) IC 4553/54 (Arp 220). Contours of H I 21-cm line column density (*black*) are superimposed on deep optical (*r*-band) images. Inserts show a more detailed view in the *K*-band (2.2 μm) of the nuclear regions of NGC 4038/39, NGC 7252, and IRAS 19254–7245, and in the *r*-band (0.65 μm) of Arp 220. White contours represent the CO(1$\rightarrow$0) line integrated intensity as measured by the OVRO millimeter-wave interferometer. No H I or CO interferometer data are available for the southern hemisphere object IRAS 19254–7245. The scale bar represents 20 kpc.

of the $3.9 \times 10^9\ M_\odot$ of molecular gas in the system (Sanders & Mirabel 1985, Young et al 1995) is concentrated in the two nuclei and in a large off-nuclear complex where the two disks strongly overlap. No CO has been detected in the extended tails. In contrast, about 70% of the total H I mass, $M(\mathrm{H\,I}) \sim 10^9\ M_\odot$, is associated with the long (total extent of ~ 100 kpc) tidal tails (van der Hulst et al, in preparation). About one quarter of the total H I mass is located at the tip of the southern "antenna". In this early merger the total ratio $M(\mathrm{H_2})/M(\mathrm{H\,I})$ is ~ 4.

NGC 7252 (Arp 226 = "Atoms for Peace") NGC 7252 has been considered the prototype of a late-stage merger (Schweizer 1978). This object has $L_{\mathrm{ir}} \sim 5 \times 10^{10}\ L_\odot$ and therefore is currently not a LIG. The *K*-band image shows a single nucleus inside what has been described as a relaxed elliptical body. The optical image shows relatively large symmetric tails, which rotate in opposite directions and have a total extent of ~ 130 kpc (Figure 8*b*). NGC 7252 has been described as a decaying merger of two massive gas-rich late-type spirals caught in the act of forming a single elliptical galaxy (Schweizer 1978, Casoli et al 1991, Fritze-von Alvensleben & Gerhard 1994). However, it is not yet clear whether this system has passed through a LIG phase. Single-dish (Dupraz et al 1990) and interferometer (Wang et al 1992) CO observations reveal a nuclear molecular gas disk at $r < 1.5$ kpc, which contains nearly 75% of the total $\mathrm{H_2}$ mass of $4.7 \times 10^9\ M_\odot$. Hibbard et al (1994) and Hibbard & van Gorkom (1996) detect a slightly smaller total mass of H I ($3.6 \times 10^9\ M_\odot$), but *all* of the H I is located in the tidal tails. Until most of the H I and $\mathrm{H_2}$ is either consumed, expelled, or ionized, it would appear that NGC 7252 still has too much cold gas to resemble most present-day ellipticals.

IRAS 19254–7245 ("The Super Antennae") IRAS 19254–7245 is a remarkable ULIG in which two distinct rotating merging disks can still be identified (Mirabel et al 1991). The colossal tidal tails have a total extent of ~ 350 kpc (Figure 8*c*). Among the 20 ULIGs in the *IRAS* BGS this object has the largest projected nuclear separation (~ 10 kpc; Melnick & Mirabel 1990). If the *IRAS* luminosity has the same spatial distribution as the observed 10-μm luminosity, then $\gtrsim 80\%$ of the total luminosity, $L_{\mathrm{ir}} = 1.1 \times 10^{12}\ L_\odot$, originates in the southern, heavily obscured Seyfert nucleus. The total molecular gas content of the system is $M(\mathrm{H_2}) \sim 3.0 \times 10^{10}\ M_\odot$ (Mirabel et al 1988).

IC 4553/54 (Arp 220) At a distance of 77 Mpc, Arp 220 is the nearest ULIG (by a factor of ~ 2). In contrast to "The Super Antennae", it shows two relatively short and wide tails (Figure 8*d*). The nuclear region is completely obscured in the optical, but two distinct nuclei separated by $0.8''$ (~ 300 pc) are detected at *K*-band (Graham et al 1990). The radial brightness profile at 2.2 μm is closely approximated by a $r^{-1/4}$ de Vaucouleurs' profile (Wright et al 1990, Kim 1995). The total mass of molecular gas is $M(\mathrm{H_2}) \sim 2 \times 10^{10}\ M_\odot$, with $\sim 2/3$ of it inside a projected radius of ~ 300 pc. Compact OH megamaser emission

($\leq$ 10 pc in diameter) has recently been discovered at the center of Arp 220 (Londsdale et al 1994). H I is detected in absorption against the nuclear radio continuum source (Mirabel 1982). In emission, all of the $\sim 2.3 \times 10^9 M_\odot$ of H I is located outside the main body (Hibbard & Yun 1996). Although as luminous as "The Super Antennae", the shorter tidal tails, smaller nuclear separation, and more relaxed central body of Arp 220 suggest that it is a more advanced merger. Because of the strong H I absorption it is not possible to estimate the total mass of H I in the merger disks; however, most of the cold gas there is likely to be in molecular form, and the overall ratio $M(H_2)/M(\text{H I})$ is probably close to 10.

5.2.2 ENCOUNTERS IN CLUSTERS Most LIGs detected by *IRAS* appear to involve strong interactions/mergers of molecular gas-rich spirals where the pairs are either isolated or part of small groups. In these interactions/mergers the relative mean velocity of the individual members is typically $\lesssim 200$ km s^{-1}. These conditions are not expected to be typical of cluster environments where the relative velocities are often much higher, and many galaxies may be either gas-poor spirals (from ram-pressure striping) or already transformed into ellipticals (perhaps due to past mergers). Although multiple high-speed encounters in clusters may produce some LIGs (e.g. Moore et al 1996), the relatively small number of *IRAS* LIGs found in clusters suggests that lower-speed mergers in pairs or loose groups may be a more efficient way of enhancing infrared activity. As examples of the types of strong interactions involving relatively large ($\gtrsim L^*$) galaxies in low-z clusters, we discuss here observations of an elliptical–spiral encounter (Arp 105) and a relatively high-speed lenticular–spiral encounter (NGC 5291A/B).

Arp 105 (Figure 9*a*) consists of an infrared luminous ($L_{\text{ir}} \sim 10^{11} L_\odot$) spiral galaxy (NGC 3561A) being torn apart by a massive elliptical (NGC 3561B) in the X-ray rich cluster Abell 1185. Duc & Mirabel (1994) find that the elliptical has already accreted gas-rich objects and may be the precursor of a cD galaxy. At the tip of one of the colossal tidal tails that emanate from the spiral are a compact dwarf galaxy and an irregular galaxy of Magellanic type. The H I and CO interferometer observations by Duc (1995) show that the collision has caused a marked spatial separation of the cold interstellar gas: Whereas all of the $\sim 10^{10} M_\odot$ of molecular gas (H_2) is found within the central ~ 6 kpc radius of the spiral, a similar mass of atomic gas (H I) is found outside that region, most of it far from the spiral galaxy near the tips of the tidal tails. H I clouds infalling at high velocities toward the nucleus of the elliptical are detected in absorption against a central compact radio source.

Figure 9*b* shows an optical image of NGC 5291A/B (Duc & Mirabel 1996, in preparation) and H I column density map (Malphrus et al 1996, in preparation). The nuclei of the lenticular galaxy NGC 5291A and the spiral NGC 5291B (the "Sea shell" galaxy) are at a projected separation of 12 kpc and are moving with a

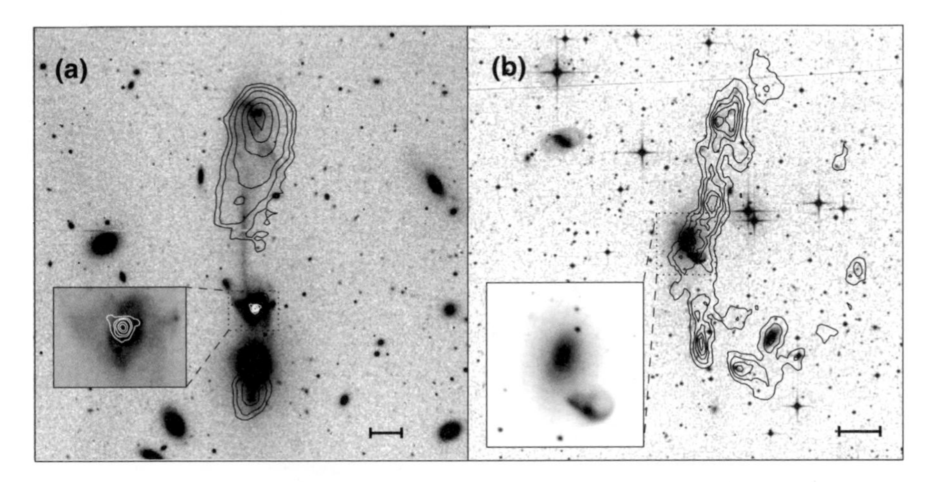

Figure 9 Encounters in clusters: (*a*) NGC 3561A/B (Arp 105); (*b*) NGC 5291A/B ("Sea shell"). Contours of H I 21-cm line column density (*black*) are superimposed on deep optical (*r*-band) images. Inserts show a more detailed view in *r*-band of the spiral galaxy NGC 3561A (Duc & Mirabel 1994), and of the interacting pair NGC 5291A/B. White contours represent the CO(1→0) line integrated intensity as measured by the IRAM millimeter-wave interferometer. CO emission has not been detected in NGC 5291A/B. The scale bar represents 20 kpc.

relative radial speed of $\sim 450\ \mathrm{km\ s^{-1}}$ (Longmore et al 1979). Figure 9*b* reveals a multitude of intergalactic H II regions superimposed on an arc-like distribution of H I, which has a diameter of ~ 200 kpc and a total H I mass of $\sim 10^{11}\ M_\odot$ (Simpson et al, private communication). However, optical spectroscopy of the spiral that is being torn apart shows no signs of recent star formation, which is consistent with its apparently low infrared luminosity ($L_{\mathrm{ir}} < 10^9\ L_\odot$); this source was not detected by *IRAS*.

5.2.3 MOLECULAR GAS-RICH MERGERS IN QSOs AND PRGs Infrared and millimeter-wave observations of QSOs and PRGs show that a significant fraction of these objects have values of L_{ir} and $M(\mathrm{H_2})$ that overlap values found for infrared-selected ULIGs. These data have been used as evidence for an evolutionary connection between QSOs, PRGs, and ULIGs (e.g. Sanders et al 1988b, Mirabel et al 1989). Figure 10 shows high-resolution optical images of the optically-selected QSO Mrk 1014, the radio-selected PRG Pks 1345+12, and the radio-selected QSO 3C 48, three objects that exhibit several properties in common with ULIGs.

Figure 10*a* shows the UV-excess QSO Mrk 1014 at a redshift of 0.163. The deep CCD image shows two tidal tails extending from a large distorted disk with a total diameter of ~ 90 kpc (e.g. Mackenty & Stockton 1984). This warm

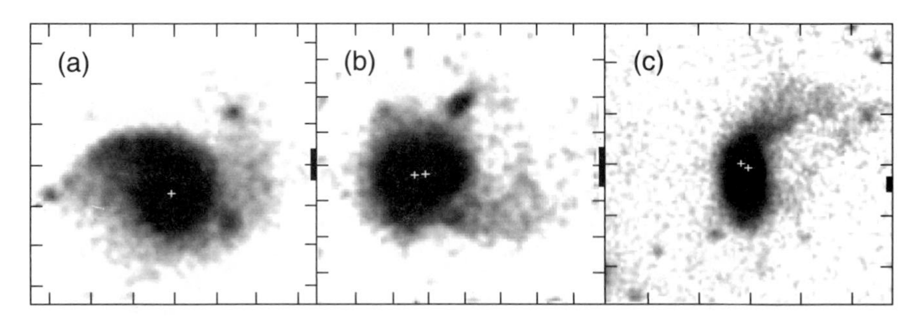

Figure 10 Optical images of QSOs and PRGs with strong infrared emission: (*a*) Mrk 1014 (PG 0157+001), (*b*) 4C 12.50 (Pks 1345+12), (*c*) 3C 48. The "+" sign indicates the position of putative optical nuclei. Tick marks are at 5″ intervals, and the scale bar represents 10 kpc.

infrared object has $L_{ir} \sim 3 \times 10^{12}\ L_\odot$ and $M(H_2) \sim 4 \times 10^{10}\ M_\odot$ (Sanders et al 1988c).

Figure 10*b* shows a deep optical image of the PRG 4C 12.50. This object exhibits a double nucleus embedded in a distorted disk ~ 85 kpc in diameter (see also Heckman et al 1986). *IRAS* observations imply $L_{ir} \sim 1.6 \times 10^{12}\ L_\odot$. CO emission (Mirabel et al 1989) and H I absorption (Mirabel 1989) with widths of ~ 950 km s^{-1} have been detected from this object. The CO data implies an extremely large mass of molecular gas, $M(H_2) \sim 6.5 \times 10^{10}\ M_\odot$. Mirabel (1989) proposed that objects like 4C 12.50 (which is classified as a compact steep-spectrum radio source) are relatively young PRGs that have evolved from progenitors like Arp 220, where a powerful compact source of nonthermal radio emission is still confined by a large nuclear concentration of gas and dust.

Figure 10*c* shows a recent deep optical image of 3C 48, the first radio source to be identified as a QSO (Matthews & Sandage 1963) and the second QSO to have its redshift determined (Greenstein & Matthews 1963). 3C 48 appears to have a double nucleus and at least one prominent tidal tail (Stockton & Ridgway 1991). Neugebauer et al (1985) used the strength and shape of the far-infrared continuum, and Stein (1995) used the f_{60}/f_{25} ratio versus the radio spectral index at 4.8 GHz to argue that the dominant luminosity source in this classic QSO is star formation in the host galaxy rather than the central AGN. *IRAS* observations of 3C 48 imply $L_{ir} \sim 5 \times 10^{12}\ L_\odot$, and the recent detection of CO emission by Scoville et al (1993) implies a molecular gas mass $M(H_2) \sim 4 \times 10^{10}\ M_\odot$ (see also Evans et al 1996b).

5.2.4 GRAVITATIONAL LENSING IN HIGH–Z OBJECTS The identification of the *IRAS* Faint Source 10214+4724 with an emission-line galaxy at a redshift of 2.286 (Rowan-Robinson et al 1991) has attracted considerable interest. The apparent enormous infrared luminosity, $L_{ir} \sim 2 \times 10^{14}\ L_\odot$, the detection of $\gtrsim$

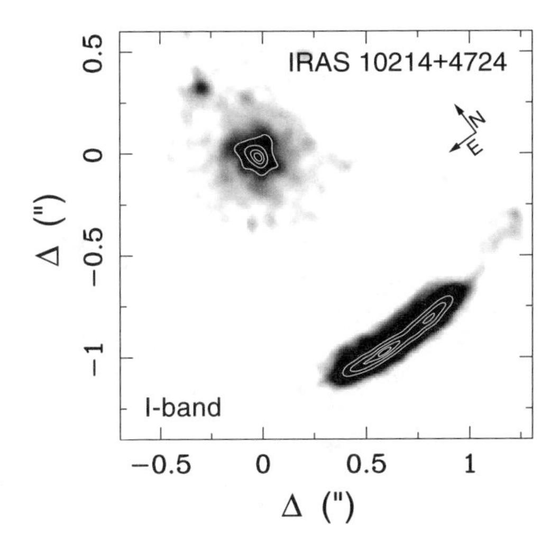

Figure 11 *HST* Planetary Camera (F814W) image of the gravitational lens object IRAS 10214+4724 (Eisenhart et al 1996).

10^{11} $M_{\odot}$ of molecular gas (Brown & Vanden Bout 1991; Solomon et al 1992b,c; Radford et al 1996), and the detection of strong submillimeter continuum emission from dust (Clements et al 1992, Downes et al 1992) led many observers to propose that this object was a prime candidate for a "primeval galaxy". However, IRAS 10214+4724 exhibits at least one unusual property for an object with such extreme luminosity and mass; the CO line width of ~ 250 km s^{-1} is small for such a massive object.

The idea that IRAS 10214+4724 might be gravitationally lensed by a foreground galaxy or group of galaxies (Elston et al 1994, Trentham 1995, Broadhurst & Lehár 1995) gained support from sub-arcsecond *K*-band images obtained with the Keck telescope (Matthews et al 1994, Graham & Liu 1995), and now appears certain from the image obtained with the *Hubble Space Telescope* (*HST*) (Figure 11). The geometry of the lensed arc suggests an amplification factor of ~ 30 for the infrared emission (Einsenhart et al 1996). CO interferometer maps indicate that the molecular gas may be somewhat more extended than the infrared emission (Scoville et al 1995, Downes et al 1995), and Downes et al (1995) suggest that the CO may be amplified by a factor $\lesssim 10$. Thus, the intrinsic infrared luminosity and molecular gas mass of this object may only be $L_{ir} \sim 7 \times 10^{12}$ $L_{\odot}$ and $M(H_2) \sim 2 \times 10^{10}$ $M_{\odot}$, respectively, suggesting that IRAS 10214+4724 is not a unique infrared selected object but is better described as a high-redshift analogue of ULIGs found in the local Universe.

At present, the only other *IRAS* object at $z > 2$ that has been detected in CO is the Cloverleaf QSO at $z \sim 2.5$, which is also a lensed object (Barvainis et al

1994). After correcting for amplification, the Cloverleaf appears to have a similar infrared and submillimeter spectrum as for IRAS 10214+4724 (Barvainis et al 1995), again suggesting that this object is yet another high-redshift analog of local ULIGs. However, several objects at $z > 4$ not detected by *IRAS* have recently been shown to be strong far-infrared (rest-frame) emitters (Omont et al 1996a), and one of these sources—the radio-quiet QSO BR1202–0725 at $z = 4.69$ (Irwin et al 1991)—was recently detected in CO (Ohta et al 1996, Omont et al 1996b) implying $M(H_2) \sim 6 \times 10^{10}\, M_{\odot}$. This object has been shown to have a submillimeter spectrum remarkably similar to IRAS 10214+4724 (McMahon et al 1994, Isaak et al 1994). An *HST* image (Hu et al 1996) shows no obvious evidence that this object is lensed, but it may be premature to rule out this possibility.

If one extrapolates the local luminosity function of ULIGs assuming relatively strong evolution (e.g. Kim & Sanders 1996), then the probability that a ULIG detected at $z \gtrsim 2$ is lensed by at least a factor of 10 can be as high as $\sim 30\%$ (Trentham 1995, Broadhurst & Lehár 1995). Future studies of LIGs at high redshifts will almost certainly require high-resolution imaging to test whether these objects are lensed.

6. ASSOCIATED PHENOMENA

6.1 *Formation of Ellipticals*

In disk-disk collisions of galaxies, dynamical friction and subsequent relaxation may produce a mass distribution similar to that in classic elliptical galaxies. From the relative numbers of mergers and ellipticals in the New General Catalogue Toomre (1977) estimated that a large fraction of ellipticals could be formed via merging. The first direct observational evidence for the transition from a disk-disk merger toward an elliptical was presented in the optical study of NGC 7252 by Schweizer (1982). The brightness distribution over most of the main body of this galaxy is closely approximated by a de Vacouleurs ($r^{-1/4}$) profile. However, NGC 7252 still contains large amounts of interstellar gas and exhibits a pair of prominent tidal tails (see Figure 9*b*); neither property is typical of ellipticals.

Near-infrared images are less effected by dust extinction and also provide a better probe of the older stellar population, which contains most of the disk mass and therefore determines the gravitational potential. *K*-band images of six mergers by Wright et al (1990) showed that the infrared radial brightness profiles for two LIGs—Arp 220 and NGC 2623—follow an $r^{-1/4}$ law over most of the observable disks. Among eight merger remnants, Stanford & Bushouse (1991) found *K*-band brightness profiles for four objects that were well fitted by an $r^{-1/4}$ law over most of the observable disks. Kim (1995) finds a similar proportion ($\sim 50\%$) of ULIGs whose *K*-band profiles are well fit by a $r^{-1/4}$ law.

More recently, Kormendy & Sanders (1992) have proposed that ULIGs are elliptical galaxies forming by merger-induced dissipative collapse. The extremely large central gas densities ($\sim 10^2$–$10^3 M_\odot$ pc^{-3}) observed in many nearby ULIGs (see Section 4.6.2) and the large stellar velocity dispersions found in the nuclei of Arp 220 and NGC 6240 (Doyon et al 1994) are comparable to the stellar densities and velocity dispersions respectively, in the central compact cores of ellipticals.

Despite the *K*-band and CO evidence that LIGs may be forming ellipticals, we still need to account for two important additional properties of ellipticals: 1. the large population of globular clusters in the extended halos of elliptical galaxies, which cannot be accounted for by the sum of globulars in two preexisting spirals (van den Bergh 1990), and 2. the need to remove the large amounts of cold gas and dust present in infrared-luminous mergers in order to approximate the relative gas-poor properties of ellipticals. These two issues are addressed below.

6.2 *Formation of Star Clusters*

Populations of bright blue pointlike objects have recently been discovered with *HST* in the galaxy at the center of the Perseus cluster, NGC 1275 (Holtzman et al 1992), and in the disk-disk mergers NGC 7252 (Whitmore et al 1993), "The Antennae" (Whitmore & Schweizer 1995), and Arp 299 (Meurer et al 1995, Vacca 1996). These objects have $M_V \sim -11$ to -16, and it has been proposed that they are young clusters that are, or may evolve into, globular clusters. If this hypothesis is true, then the number of globular clusters would indeed increase during the merger of gas-rich spirals, thus weakening one of the main arguments against ellipticals being formed through disk mergers.

The interpretation of these star clusters as protoglobulars has been questioned by van den Bergh (1995a,b), who argues that mergers simply increase the rate of normal star and cluster formation and do not promote a specific population of very massive clusters that will evolve into globulars. In fact, the luminosity function of the blue clusters in The Antennae has a power-law shape as do open clusters in the Milky Way, rather than the Gaussian shape of the luminosity function of globular clusters (Whitmore & Schweizer 1995, van den Bergh 1995b; but see Meurer 1995). Furthermore, the light radii of the young clusters in The Antennae seem to be larger than in typical globular clusters by a factor of about three. Future observations of these objects with corrected *HST* optics may provide better images that can prove if these blue clusters indeed have the star densities typical of globulars.

6.3 *Formation of Dwarf Galaxies*

Collisions between giant disk galaxies may trigger the formation of dwarf galaxies. This idea, which was first proposed by Zwicky (1956) and later by Schweizer (1978), has received recent observational support (Mirabel et al

1991, 1992; Elmegreen et al 1993; Duc & Mirabel 1994). Renewed interest in this phenomenon arose from the inspection of the optical images of ULIGs, which frequently exhibit patches of optically emitting material along the tidal tails (see Figures 8*a,b,c*). These objects appear to become bluer near the tips of the tails at the position of massive clouds of H I. These condensations have a wide range of absolute magnitudes, $M_V \sim -14$ to -19.2, and H I masses, $M(\text{H I}) \sim 5 \times 10^8$ to $6 \times 10^9\, M_\odot$. Mirabel et al (1995) have shown that objects resembling irregular dwarfs, blue compacts, and irregulars of Magellanic type are formed in the tails. These small galaxies of tidal origin are likely to become detached systems, namely, isolated dwarf galaxies. Because the matter out of which they are formed has been removed from the outer parts of giant disk galaxies, the tidal dwarfs we observe forming today have a metallicity of about one third solar (Duc 1995).

It is interesting that in these recycled galaxies of tidal origin there is—as in globular clusters—no compelling evidence for dark matter (Mirabel et al 1995). The true fraction of dwarf galaxies that may have been formed by processes similar to the tidal interactions we observe today between giant spiral galaxies will require more extensive observations of interacting systems. A recent step forward is the statistical finding that perhaps as much as one half of the dwarf population in groups is the product of interactions among the parent galaxies (Hunsberger et al 1996).

6.4 *Enrichment of the Intergalactic Medium*

X-Ray and optical evidence for galactic superwinds has been presented for both ULIGs and PRGs (Heckman et al 1990, 1996; Veilleux et al 1995). In these objects, it has been proposed that the combined kinetic energy from supernovae and stellar winds from powerful nuclear starbursts drive large-scale outflows that shock, heat, and accelerate the circumnuclear ambient gas. The morphology, kinematics, and physical properties of the optically emitting gas tend to support this model. Continuum-subtracted narrow-band images show emission-line nebulosity extending over tens of kiloparsecs; a good example of this phenomenon is the enormous Hα bubble found in Arp 220 (Armus et al 1987). Further kinematic signatures of outflows along a galaxy's minor axis are provided by observed double emission-line profiles with line splittings of 200–600 km s^{-1}; a spectacular example is the $\gtrsim 1500$ km s^{-1} splitting found in the nuclear superbubble of NGC 3079 (Veilleux et al 1994). Additional evidence for mass outflows in LIGs includes the statistical difference found between the H I 21-cm line and optical emission line redshifts. This difference is likely due to outflow motions of the optical line-emitting gas (Mirabel & Sanders 1988).

Galactic superwinds may play an important role in the metal enrichment of the intergalactic medium. Heckman et al (1990) calculate a mass loss rate, $dM/dt = 4\ (L_{\text{fir},11} M_\odot\ \text{year}^{-1})$, where $L_{\text{fir},11}$ is the far-infrared luminosity in units of 10^{11} $L_\odot$, implying that a galaxy like Arp 220 can be expected to inject $\sim 5 \times 10^8$ $M_\odot$ of metals and $\sim 10^{58}$ ergs over an estimated lifetime of $\sim 10^7$ years. Assuming no cosmological evolution and using a luminosity function for *IRAS* galaxies similar to that given in Figure 1, Heckman et al (1990) derived a total mass-injection rate of $\sim 2 \times 10^7$ ($M_\odot$ Mpc^{-3}), 25% of which is in metals. If cosmological evolution is included (see Section 3.1), then the injected mass and energy can increase by an order of magnitude.

The superwinds observed in nearby ULIGs may provide local examples that can be used to understand the X-ray iron abundance of $\sim 1/3$ solar in the intracluster medium (Arnaud et al 1992). Recent *ASCA* observations of large amounts of silicon and oxygen in four nearby clusters indicate that type II supernovae are responsible for the metal enrichment (Loewenstein & Mushotzky 1996). In turn, the metals found in nearby clusters may represent the fossil records of ancient starbursts (Elbaz et al 1992, Terlevich & Boyle 1993). More specifically, the iron-elliptical correlation observed in clusters may suggest that giant ellipticals indeed went through an ULIG-superwind phase.

On the question of the impact that these galactic superwinds may have on the interstellar gas content of the host galaxy, it is important to know the fraction of the total interstellar gas that these winds can entrain and expel into the intergalactic medium. Heckman et al (1990) suggest that the maximum amount of ambient material ejected by the wind is $M_{\rm ej,max} = 10^{10}\ (E_{\rm wind}/10^{58})\ M_\odot$, which implies that when the entrained material has velocities above the escape velocity of the host galaxy, the entire interstellar medium could be blown away. Since the observed velocities of the line-emitting gas in LIGs are close to the escape velocities, it appears possible that a large fraction of the interstellar gas may be driven out of the host galaxy, and that in some cases the entire neutral intersteller medium may have been lost, as appears to be the case in most giant ellipticals.

7. THEORETICAL MODELS

Recent theoretical models have been quite successful in explaining several important features that appear to be common to most gas-rich mergers. For example, theorists reacted rapidly to the observational discovery of high concentrations of cold gas in the central regions of some mergers and developed new dynamic models with gas+stars that have provided much needed insight into how such gas concentrations are produced (e.g. Barnes & Hernquist 1992, Barnes 1995). During collisions, the gas readily losses angular momentum due to dynamical friction, decouples from the stars, and inflows rapidly toward the merger nuclei. The typical half-mass radius of the gas is only 2.5% that of the stars, which scaled to Milky Way units implies $\sim 5 \times 10^9$ $M_\odot$ of neutral gas within a radius of ~ 140 pc and a gas density of $\sim 10^3 M_\odot$ pc^{-3} (Barnes 1995).

Although all gas-rich disk-disk mergers appear to end up with enormous gas densities at their centers, independent of the specific orbital parameters, the rate of gas inflow may depend on the structure of the progenitor galaxies, specifically, on the size of a dense central bulge (Mihos & Hernquist 1994). Galaxies without bulges develop bars that produce a more steady inflow that may last $\sim 1.5 \times 10^8$ years. By contrast, in objects with large bulges the disk tends to be stabilized against strong inflow until just before the galaxies finally coalesce, when strong dissipation finally drives the contained gas to the center in $\lesssim 5 \times 10^7$ years. If the level of infrared activity depends on the *rate* of gas infall into what is to be the maximum central gas density configuration, or if some other mechanism such as a binary black hole is required to create an extreme nuclear starburst (e.g. Taniguchi & Wada 1996), then it may be the case that not all mergers of gas-rich spirals will produce ULIGs.

Noninteracting spirals typically have large quantities of H I beyond their optical disks. For instance, in the Milky Way most of the H I mass is beyond the solar circle, whereas most of the molecular gas is at galactocentric radii $< 0.7\ R_{\odot}$. Computer models predict that a large fraction of the H I gas in the outer regions of the pre-encounter disks will be pulled out to large radii in the form of tidal tails, out of which dwarf irregular galaxies can be formed (Barnes & Hernquist 1992, Elmegreen 1993). Elmegreen et al (1993) have proposed that the heating of the interstellar medium due to the encounter increases the Jeans' mass and that the H I gas that leads the matter launched into intergalactic space will form self-gravitationally bound cloud complexes, which may collapse and appear as detached irregular dwarf galaxies. This model forms large H I clouds with masses as large as $10^9\ M_{\odot}$ at the end of the tidal tails opposite the companion (Mirabel et al 1992, Duc & Mirabel 1994). In the Barnes & Hernquist (1992) model, however, the clumps form from collapse of the stellar population, with mass fluctuations on all scales, producing also "failed dwarfs" of only old stars due to the fact that the potential wells are too shallow to capture enough H I to form new stars. This result is consistent with the wide range of *B–V* colors of the condensations along the tidal tails of some ULIGs (Mirabel et al 1991).

8. SUMMARY

Our knowledge of luminous galaxies has increased dramatically as a result of the *IRAS* all-sky survey. During the past decade, ground-based redshift surveys have succeeded in identifying thousands of LIGs, including several hundred ULIGs in the local Universe ($z \lesssim 0.3$), and detailed multiwavelength studies, particularly of the brightest objects, have provided substantial information on what types of galaxies emit these large infrared luminosities. It seems clear that strong interactions/mergers of gas-rich spirals are responsible for the great majority of LIGs and that enormous starbursts must be involved in generating a substantial fraction of the infrared luminosity, at least in the early phases of the interaction.

Theoretical models have added substantially to our understanding of the merger process, in particular by showing how it is possible to build up enormous concentrations of gas in the centers of the merger remnant. Interferometer measurements at millimeter wavelengths show that the most luminous infrared objects often contain as much as 10^{10} $M_{\odot}$ of gas within 0.5-kpc radius of the merger nucleus! Such enormous gas concentrations are an ideal breeding ground for a variety of powerful phenomena, including powerful starbursts with their accompanying superwinds that may add substantially to the metal enrichment of the intergalactic medium, the formation of massive star clusters, and most likely the building and/or fueling of an AGN. LIGs very likely represent an important link between starburst galaxies and the AGN phenomena exhibited by QSOs and PRGs. They also likely represent transition objects spanning the gap between merging spirals and emerging ellipticals.

Future infrared space missions and more sensitive ground-based submillimeter surveys will succeed in identifying more distant LIGs, thus allowing a test of whether the infrared luminosity function evolves as steeply as that of QSOs and whether LIGs were more numerous in clusters during the epoch when it is presumed that most of the ellipticals were formed from mergers of spirals.

However, detailed studies of objects already identified from the *IRAS* survey will probably play the largest role in our ability to determine the nature of the dominant energy sources in LIGs and to understand more precisely what parameters lead to the origin and ultimate fate of these objects. The following research areas could prove particularly productive:

1. Detailed multiwavelength studies of those LIGs already identified in the *IRAS* BGS that do not show obvious evidence of current or past interactions (e.g. IRAS 10173+0828, VII Zw 031). Do these "exceptions" to the merger hypothesis simply represent the end-stage of the merger after most of the obvious tidal features have disappeared, or do they represent an alternative way of producing LIGs ?

2. Identification of more "transition objects"—those galaxies that are currently LIGs but that also show broad Sy 1 emission lines or powerful radio cores/jets that are characteristics of QSOs and PRGs respectively. These objects currently represent $\sim 15\%$ of LIGs in the *IRAS* BGS and many more should be discovered in follow-up studies of sources in existing catalogs of fainter objects.

3. A systematic search for possible fossil remnants of the ULIG merger phase (e.g. molecular gas, faint tidal features, etc.) in the host galaxies of QSOs and PRGs.

4. Mid- and far-infrared spectroscopy with the *Infrared Space Observatory* (*ISO*) to better study the nature of the deeply embedded energy sources in LIGs.

5. More elaborate theoretical models that include a better treatment of star formation, stellar winds, and supernovae explosions to better follow the evolution and fate of the central gas concentration.

6. VLBI observations of extragalactic OH megamasers—ULIGs host a class of megamasers that may prove to be a powerful dynamical probe that can be used to test for the presence of supermassive black holes.

Finally, it should eventually be possible to better discriminate between AGN versus starbursts in ULIGs by direct measurement of the size of the emitting region at mid- and far-infrared wavelengths. This could happen in the next few years as ground-based submillimeter interferometers come into operation and later with proposed airborne and space-based platforms such as *SOFIA* and *FIRST*.

ACKNOWLEDGMENTS

It is a pleasure to thank the many people who sent us preprints and reprints of their work on luminous infrared galaxies. We also thank L Armus, M Arnaud, W Baan, J Ballet, J Barnes, E Brinks, R Cohen, L Cowie, P Duc, A Evans, T Heckman, E Hu, R Joseph, J Kormendy, J Mazzarella, R Norris, T Soifer, A Stockton, N Trentham, W Vacca, and S Veilleux for helpful discussions and comments while writing the manuscript. We are grateful to those people who generously made their H I data available to us for use in constructing overlays for the figures: E Brinks (Figure 9*a*), C Simpson and S Gottesman (Figure 9*b*), J Hibbard (Figures 8*b* and 8*d*), J van der Hulst (Figure 8*a*); to P Bryant for permission to use his new millimeter-wave interferometer CO data in Figure 6*d*; to J Wink for permission to use the CO map in Figure 9*a*; to A Stockton for permission to reproduce optical images obtained with the CFHT and the UH 2.2-m telescope (Figure 10); and to P Eisenhart for permission to use the *HST* image reproduced in Figure 11. We would particularly like to express our appreciation to Pierre-Alain Duc for reducing most of the grey-scale images displayed in Figures 8 and 9, to Karen Teramura for preparing final versions of all of the figures, and to John Kormendy for his *TeX* macro, which was used to prepare preprints of this article. DBS was supported in part by NASA grants NAG5–1741 and NAGW–3938 and would like to acknowledge the hospitality of the Service d'Astrophysique, Saclay, during extended visits while this article was being written.

Literature Cited

Aaronson M, Olszewski EW. 1984. *Nature* 309:414
Allen DA, Roche PF, Norris RP. 1985. *MNRAS* 213:67P
Antonucci RRJ, Miller JS. 1985. *Ap. J.* 297:621
Antonucci RRJ, Olszewski EW. 1985. *Astron. J.* 90:2203
Arp H. 1966. *Ap. J. Suppl.* 14:1
Armus L, Heckman TM, Miley GK. 1987. *Astron. J.* 94:831
Armus L, Heckman TM, Miley GK. 1989. *Ap. J.* 347:727
Armus L, Heckman TM, Miley GK. 1990. *Ap. J.* 364:471
Armus L, Neugebauer G, Soifer BT, Matthews K. 1995a. *Astron. J.* 110:2610
Armus L, Shupe DL, Matthews K, Soifer BT, Neugebauer G 1995b. *Ap. J.* 440:200
Arnaud M, Rothenflug R, Boulade O, Vigroux L, Vagnioni-Flam E. 1992. *Astron. Astrophys.* 254:49
Ashby MLN, Hacking PB, Houck JR, Soifer BT, Weisstein EW. 1996. *Ap. J.* 456:428
Baan WA. 1985. *Nature* 315:26
Baan WA. 1989. *Ap. J.* 338:804
Baan WA. 1993. In *Astrophysical Masers*, ed. AW Clegg, GE Nedoluha, p. 73. New York: Springer Verlag
Baan WA, Haschick AD. 1984. *Ap. J.* 279:541
Baan WA, Rhoads J, Fisher K, Altschuler DR, Haschick A. 1992. *Ap. J. Lett.* 396:99
Baan WA, van Gorkom J, Schmelz JT, Mirabel IF. 1987. *Ap. J.* 313:102
Baan WA, Wood PAD, Haschick AD. 1982. *Ap. J. Lett.* 260:49
Barnes JE. 1995. In *The Formation of Galaxies*, ed. C Muñoz-Tuñón, F Sanchez, p. 399. Cambridge: Cambridge University Press
Barnes JE, Hernquist L. 1992. *Annu. Rev. Astron. Astrophys.* 30:705
Barvainis R, Alloin D, Antonucci R. 1989. *Ap. J. Lett.* 337:69
Barvainis R, Tacconi L, Antonucci R, Alloin D, Coleman P. 1994. *Nature* 371:586
Barvainis R, Antonucci R, Hurt T, Coleman P, Reuter HP. 1995. *Ap. J. Lett.* 451:9
Becklin EE, Matthews K, Neugebauer G, Wynn-Williams GC 1973. *Ap. J. Lett.* 186:69
Becklin EE, Wynn-Williams CG. 1987. In *Star Formation in Galaxies*, ed. CJ Lonsdale-Persson, p. 643. Washington DC: US Govt. Print. Off.
Beichman CA, Soifer BT, Helou G, Chester TJ, Neugebauer G, et al. 1986. *Ap. J. Lett.* 308:1
Bhattacharya D, The L-S, Kurfess JD, Clayton DD, Gehrels N, Leising MD, et al. 1994. *Ap. J.* 437:173
Bicay MD, Kojoian G, Seal J, Dickinson DF, Malkan MA. 1995. *Ap. J. Suppl.* 98:369
Boller T, Fink H, Schaeidt S. 1994. *Astron. Astrophys.* 291:403
Boller T, Meurs EJA, Brinkmann W, Fink H, Zimmermann U, Adorf HM. 1992. *Astron. Astrophys.* 261:57
Boller T, Meurs EJA, Dennefeld M, Fink H. 1993. *Astrophys. Space Sci.* 205:43
Bothun GD, Lonsdale CJ, Rice WL. 1989. *Ap. J.* 341:129
Braine J, Combes F, Casoli F, Dupraz C, Gérin M, et al. 1993. *Astron. Astrophys. Suppl.* 97:887
Broadhurst T, Lehár J. 1995. *Ap. J. Lett.* 450:41
Brown RL, Vanden Bout PA. 1991. *Astron. J.* 102:1956
Bryant P. 1996. *High resolution observations of the molecular gas in luminous infrared galaxies.* PhD thesis, Calif. Inst. Tech., Pasadena
Burbidge GR, Stein WA. 1970. *Ap. J.* 160:573
Bushouse HA. 1987. *Ap. J.* 320:49
Bushouse HA, Lamb SA, Werner MW. 1988. *Ap. J.* 335:74
Carico DP, Graham JR, Matthews K, Wilson TD, Soifer BT, Neugebauer G, Sanders DB. 1990. *Ap. J. Lett.* 349:39
Casoli F, Dupraz C, Combes F, Kazès I. 1991. *Astron. Astrophys.* 251:1
Cataloged Galaxies and Quasars Observed in the IRAS Survey. 1985. Prepared by CJ Lonsdale, G Helou, JC Good, W Rice, Pasadena:JPL
Clements DL, Rowan-Robinson M, Lawrence A, Broadhurst T, McMahon R. 1992. *MNRAS* 256:35p
Clements DL, Sutherland WJ, McMahon RG, Saunders W. 1996. *MNRAS* 279:477
Clements DL, van der Werf PP, Krabbe A, Blietz M, Genzel R, Ward MJ. 1993. *MNRAS* 262:L23
Condon JJ, Anderson ML, Helou G. 1991a. *Ap. J.* 376:95

Condon JJ, Dressel LL. 1978. *Ap. J.* 221:456
Condon JJ, Condon MA, Gisler G, Puschell. JJ 1982. *Ap. J.* 252:102
Condon JJ, Helou G, Sanders DB, Soifer BT. 1990. *Ap. J. Suppl.* 73:359
Condon JJ, Huang Z-P, Yin QF, Thuan TX. 1991b. *Ap. J.* 378:65
Crawford T, Marr J, Partridge B, Strauss MA. 1996. *Ap. J.* 460:225
Cutri RM, Huchra JP, Low FJ, Brown RB, Vanden Bout PA. 1994. *Ap. J. Lett.* 424:65
Cutri RM, McAlary CW. 1985. *Ap. J.* 296:90
David LP, Jones C, Forman W. 1992. *Ap. J.* 388:82
de Grijp MHK, Keel WC, Miley GK, Goudfrooij P, Lub J. 1992. *Astron. Astrophys. Suppl.* 96:389
de Grijp MHK, Miley GK, Lub J, de Jong T. 1985. *Nature* 314:240
de Jong T, Clegg PE, Soifer BT, Rowan-Robinson M, Habing HJ, et al. 1984. *Ap. J. Lett.* 278:67
Dermer CD, Gehrels N. 1995. *Ap. J.* 447:103
Deutsch LK, Willner SP. 1986. *Ap. J. Lett.* 306:11
Devereux N, Taniguchi Y, Sanders DB, Nakai N, Young JS. 1994. *Astron. J.* 107:2006
Devereux NA, Young JS. 1991. *Ap. J.* 371:515
Dickey JM. 1986. *Ap. J.* 300:190
Dickey JM, Salpeter EE. 1984. *Ap. J.* 284:461
Downes D, Radford SJE, Greve A, Thum C, Solomon PM, Wink JE. 1992. *Ap. J. Lett.* 398:25
Downes D, Solomon PM, Radford SJE. 1993. *Ap. J. Lett.* 414:13
Downes D, Solomon PM, Radford SJE. 1995. *Ap. J. Lett.* 453:65
Doyon R, Wells M, Wright GS, Joseph RD, Nadeau D, James PA. 1994. *Ap. J. Lett.* 437:23
Duc PA. 1995. *Genèse de galaxies naines dans les systèmes en interaction.* PhD thesis, Univ. Paris
Duc PA, Mirabel IF. 1994. *Astron. Astrophys.* 289:83
Dudley CC, Wynn-Williams CG. 1996. *Ap. J.* submitted
Dupraz C, Casoli F, Combes F, Kazès I. 1990. *Astron. Astrophys.* 228:L5
Eales SA, Becklin EE, Hodapp KW, Simons DA, Wynn-Williams CG. 1990. *Ap. J.* 365:478
Eales SA, Wynn-Williams CG, Duncan WD. 1989. *Ap. J.* 339:859
Eisenhardt PR, Armus L, Hogg DW, Soifer BT, Neugebauer G, Werner MW. 1996. *Ap. J.* 461:72
Elbaz D, Arnaud M, Cassé M, Mirabel IF, Prantzos N, Vangioni-Flam E. 1992. *Astron. Astrophys.* 265:L9
Elfhag T, Booth RS, Hoglund B, Johansson LEB, Sandqvist A. 1996. *Astron. Astrophys. Suppl.* 115:439
Elmegreen BG. 1993. *Ap. J.* 411:170
Elmegreen BG, Kaufman M, Thomasson M. 1993. *Ap. J.* 412:90
Elston R, Cornell ME, Lebofsky MJ. 1985. *Ap. J.* 296:106
Elston R, McCarthy PJ, Eisenhardt P, Dickinson M., Spinrad H, et al. 1994. *Astron. J.* 107:910
Elvis M, Wilkes BJ, McDowell JC, Green RF, Bechtold J, et al. 1994 *Ap. J. Suppl.* 95:1
Emerson JP, Clegg PE, Gee G, Cunningham CT, Griffin MJ, et al. 1984. *Nature* 311:237
Evans AS. 1996. *The molecular gas content and ionization processes in distant powerful radio galaxies and hyperluminous infrared galaxies.* PhD thesis, Univ. Hawaii
Evans AS, Sanders DB, Cutri R, Radford SJE, Solomon P, et al. 1996a. *Ap. J.* submitted
Evans AS, Sanders DB, Mazzarella JM, Solomon PM, Downes D, et al. 1996b. *Ap. J.* 457:658
Fabbiano G. 1988. *Ap. J.* 330:672
Fabbiano G, Heckman T, Keel WC. 1990. *Ap. J.* 355:442
Fairclough JH. 1986. *MNRAS* 219:1p
Fisher KB, Huchra JP, Strauss MA, Davis M, Yahil A, Schlegel D. 1995. *Ap. J. Suppl.* 100:69
Fisher KB, Strauss MA, Davis M, Yahil A, Huchra JP. 1992. *Ap. J.* 389:188
Fosbury RAE, Wall JV. 1979. *MNRAS* 189:79
Fritze-von Alvensleben V, Gerhard OE. 1994. *Astron. Astrophys.* 285:775
Frogel JA, Gillett FC, Terndrup DM, Vader JP. 1989. *Ap. J.* 343:672
Gao Y. 1996. *Dense molecular gas in galaxies and the evolution of luminous infrared galaxies.* PhD thesis, SUNY at Stony Brook
Gherz RD, Sramek RA, Weedman DW. 1983. *Ap. J.* 267:551
Goldader J. 1995. *Near-infrared spectroscopy of luminous infrared galaxies.* PhD thesis, Univ. Hawaii
Goldader J, Joseph RD, Doyon R, Sanders DB. 1995. *Ap. J.* 444:97
Golombek D., Miley GK, Neugebauer G. 1988. *Astron. J.* 95:26
Goodrich RW, Veilleux S, Hill GJ. 1994. *Ap. J.* 422:521
Graham JR, Carico DP, Matthews K, Neugebauer G, Soifer BT, Wilson TD. 1990. *Ap. J. Lett.* 354:5
Graham JR, Liu MC. 1995. *Ap. J. Lett.* 449:29
Greenstein JL, Matthews TA. 1963. *Nature* 197:1041
Gregorich DT, Neugebauer G, Soifer BT, Gunn JE, Herter TL. 1995. *Astron. J.* 110:259
Griffiths RE, Padovani P. 1990. *Ap. J.* 360:483

Hacking PB, Condon JJ, Houck JR. 1987. *Ap. J. Lett.* 316:15

Hacking PB, Houck JR. 1987. *Ap. J. Suppl.* 63:311

Harper DA, Low FJ. 1973. *Ap. J. Lett.* 182:89

Heckman TM. 1983. *Ap. J.* 268:628

Heckman TM, Dahlem M, Eales SA, Fabbiano G, Weaver K. 1996. *Ap. J.* 457:616

Heckman TM, Armus L, Miley GK. 1987. *Astron. J.* 93:276

Heckman TM, Smith EP, Baum SA, van Breugel WJM, Miley GK, et al. 1986. *Ap. J.* 311:526

Heckman TM, Armus L, Miley GK. 1990. *Ap. J. Suppl.* 74:833

Heisler CA, Vader JP. 1994. *Astron. J.* 107:35

Helou G. 1986. *Ap. J. Lett.* 311:33

Helou G, Soifer BT, Rowan-Robinson M. 1985. *Ap. J. Lett.* 298:7

Hibbard JE, Guhathakurta P, van Gorkom JH, Schweizer F. 1994. *Astron. J.* 107:67

Hibbard JE, van Gorkom JH. 1996. *Astron. J.* 111:655

Hibbard JE, Yun MS. 1996. In *Cold Gas at High Redshift*, ed. M Bremer, H Rottgering, P van der Werf, C Carilli, p. 47. Dordrecht:Kluwer

Hill GJ, Wynn-Williams CG, Becklin EE. 1987. *Ap. J. Lett.* 316:11

Hines DC. 1991. *Ap. J. Lett.* 374:9

Hines DC, Schmidt GD, Smith PS, Cutri RM, Low FJ. 1995. *Ap. J. Lett.* 450:1

Hines DC, Wills BJ. 1993. *Ap. J.* 415:82

Holtzman JA, Faber SM, Shaya EJ, Lauer TR, Groth EJ, et al. 1992. *Astron. J.* 103:691

Houck JR, Schneider DP, Danielson GE, Beichman CA, Lonsdale CJ, et al. 1985. *Ap. J. Lett.* 290:5

Houck JR, Soifer BT, Neugebauer G, Beichman CA, Aumann HH, et al. 1984. *Ap. J. Lett.* 278:63

Hu EM, McMahon RG, Egami E. 1996. *Ap. J. Lett.* 459:53

Huchra G. 1977. *Ap. J. Suppl.* 35:171

Hummel E. 1980. *Astron. Astrophys.* 89:L1

Hunsberger SD, Charlton JC, Zaritsky D. 1996. *Ap. J.* 462:50

Hutchings JB, Neff SG. 1987. *Astron. J.* 93:14

Impey C, Gregorini L. 1993. *Astron. J.* 105:853

Impey C, Neugebauer G. 1988. *Astron. J.* 95:307

IRAS Point Source Catalog, ver. 2. 1988. Washington:GPO (PSC)

Irwin M, McMahon RG, Hazard C. 1991. In *The Space Distribution of Quasars*, ed. D Crampton, p. 117. San Francisco: ASP

Isaak KG, McMahon RG, Hills RE, Withington S. 1994. *MNRAS* 269:L28

Isobe T, Feigelson ED. 1992. *Ap. J. Suppl.* 79:197

Iyengar KUK, Verma RP. 1984. *Astron. Astrophys.* 139:64

Jensen JB, Sanders DB, Wynn-Williams CG. 1996. *Ap. J.* submitted

Joseph RD, Meikle WPS, Robertson NA, Wright GS. 1984a. *MNRAS* 209:111

Joseph RD, Wright GS. 1985. *MNRAS* 214:87

Joseph RD, Wright GS, Wade R. 1984b. *Nature* 311:132

Joy M, Lester DF, Harvey PM, Frueh M. 1986. *Ap. J.* 307:110

Joy M, Lester, DF, Harvey PM, Telesco C, Decher R, et al. 1989. *Ap. J.* 339:100

Kazès I, Baan WA. 1991. *Astron. Astrophys.* 248:L15

Kellermann KI, Sramek R, Schmidt M, Shaffer DB, Green R. 1989. *Astron. J.* 98:1195

Kennicutt RC, Keel WC, van der Hulst JM, Hummel E, Roettiger KA. 1987. *Astron. J.* 93:1011

Kennicutt RC, Kent SM. 1983. *Astron. J.* 88:1094

Kim DC. 1995. *The IRAS 1 jy survey of ultraluminous infrared galaxies*. PhD thesis, Univ. Hawaii

Kim DC, Sanders DB. 1996. *Ap. J.* submitted

Kim DC, Sanders DB, Veilleux S, Mazzarella JM, Soifer BT. 1995. *Ap. J. Suppl.* 98:129

Kim DC, Veilleux S, Sanders DB. 1996. *Ap. J.* submitted

Klaas U, Elsasser H. 1991. *Astron. Astrophys. Suppl.* 90:33

Klaas U, Elsasser H. 1993. *Astron. Astrophys.* 280:76

Kleinmann DE, Low FJ. 1970a. *Ap. J. Lett.* 159:165

Kleinmann DE, Low FJ. 1970b. *Ap. J. Lett.* 161:203

Kleinmann SG, Keel WC. 1987. In *Star Formation in Galaxies*, ed. CJ Lonsdale-Persson, p. 559. Washington DC: US Govt. Print. Off.

Knapp GR, Bies WE, van Gorkom JH. 1990. *Astron. J.* 99:476

Knapp GR, Guhathakurta P, Kim DW, Jura M. 1989. *Ap. J. Suppl.* 70:329

Kormendy J, Sanders DB. 1992. *Ap. J. Lett.* 390:53

Larson RB, Tinsley BM. 1978. *Ap. J.* 219:46

Lawrence A, Rowan-Robinson M, Leech K, Jones DHP, Wall JV. 1989. *MNRAS* 240:329

Lawrence A, Rowan-Robinson M, Saunders W, Parry IR, Xiaoyang X, et al. 1996. *MNRAS* In press

Lawrence A, Walker D, Rowan-Robinson M, Leech KJ, Penston MV. 1986. *MNRAS* 219:687

Lebofsky MJ, Rieke GH. 1979. *Ap. J.* 229:111

Leech KJ, Rowan-Robinson M, Lawrence A, Hughes JD. 1994. *MNRAS* 267:253

Lester DF, Joy M, Harvey PM, Ellis HB, Parmar PS. 1987. *Ap. J.* 321:755

Loewenstein M, Mushotzky RF. 1996. *Ap. J.*

466:695
Longmore AJ, Hawarden TG, Cannon RD, Allen DA, Mebold U, et al. 1979. *MNRAS* 188:285
Lonsdale CJ, Diamond PJ, Smith HE, Lonsdale CJ. 1994. *Nature* 370:117
Lonsdale CJ, Hacking P. 1989. *Ap. J.* 339:712
Lonsdale CJ, Hacking P, Conrow TP, Rowan-Robinson M. 1990. *Ap. J.* 358:60
Lonsdale CJ, Persson SE, Matthews K. 1984. *Ap. J.* 287:95
Lonsdale CJ, Smith HE, Lonsdale CJ. 1993 *Ap. J. Lett.* 405:9
Low FJ, Huchra JP, Kleinmann SG, Cutri RM. 1988. *Ap. J. Lett.* 327:41
Low FJ, Kleinmann DE. 1968. *Astron. J.* 73:868
MacKenty JW, Stockton A. 1984. *Ap. J.* 283:64
Martin JM, Bottinelli L, Dennefeld M, Gouguenheim L, Le Squeren AM. 1988. *Astron. Astrophys.* 201:L13
Martin JM, Bottinelli L, Dennefeld M, Gouguenheim L, Le Squeren A-M. 1991. In *Dynamics of Galaxies and their Molecular Cloud Distributions*, ed. F Combes, F Casoli, p. 447. Dordrecht:Reidel
Matthews K, Neugebauer G, McGill J, Soifer BT. 1987. *Astron. J.* 94:297
Matthews K, Soifer BT, Nelson J, Boesgaard H, Graham JR, et al. 1994. *Ap. J. Lett.* 420:13
Matthews TA, Sandage AR. 1963. *Ap. J.* 138:30
Mauersberger R, Henkel C. 1993. *Astron. Gesellschaft. Rev. Mod. Astron.* 6:69
Mazzarella JM, Graham JR, Sanders DB, Djorgovski G. 1993. *Ap. J.* 409:170
Mazzarella JM, Balzano VA. 1986. *Ap. J. Suppl.* 62:751
Mazzarella JM, Bothun GD, Boroson T. 1991. *Astron. J.* 101:2034
McMahon RG, Omont A, Bergeron J, Kreysa E, Haslam CGT. 1994. *Nature* 267:L9
Melnick J, Mirabel IF. 1990. *Astron. Astrophys.* 231:L19
Meurer GR. 1995. *Nature* 375:742
Meurer GR, Heckman TM, Leitherer C, Kinney A, Robert C, Garnett DR. 1995. *Astron. J.* 110:2665
Mihos JC, Hernquist L. 1994. *Ap. J. Lett.* 431:9
Miles JW, Houck JR, Hayward TL, Ashby MLN. 1996. *Ap. J.* 465:191
Miley G, Neugebauer G, Clegg PE, Harris S, Rowan-Robinson M, et al. 1984. *Ap. J. Lett.* 278:79
Miley GK, Neugebauer G, Soifer BT. 1985. *Ap. J. Lett.* 293:11
Mirabel IF. 1982. *Ap. J.* 260:75
Mirabel IF. 1989. *Ap. J. Lett.* 340:13
Mirabel IF, Booth RS, Garay G, Johansson LEB, Sanders DB. 1990. *Astron. Astrophys.* 236:327
Mirabel IF, Dottori H, Lutz D. 1992. *Astron. Astrophys.* 256:L19
Mirabel IF, Duc PA, Dottori H. 1995. In *Dwarf Galaxies*, ed. G Meylan, P Prugniel, p. 371. Garching bei Munchen:ESO
Mirabel IF, Kazès I, Sanders DB. 1988. *Ap. J. Lett.* 324:59
Mirabel IF, Lutz D, Maza J. 1991. *Astron. Astrophys.* 243:367
Mirabel IF, Sanders DB. 1987. *Ap. J.* 322:688
Mirabel IF, Sanders DB. 1988. *Ap. J.* 335:104
Mirabel IF, Sanders DB. 1989. *Ap. J. Lett.* 340:53
Mirabel IF, Sanders DB, Kazès I. 1989. *Ap. J. Lett.* 340:9
Miyoshi M, Moran J, Herrnstein J, Greenhill L, Nakai N, et al. 1995. *Nature* 373:127
Mobasher B, Sharples RM, Ellis RS. 1993. *MNRAS* 263:560
Montgomery AS, Cohen RJ. 1992. *MNRAS* 254:23p
Moore B, Katz N, Lake G, Dressler A, Oemler A. 1996. *Nature* 379:613
Moshir M, Kopan G, Conrow J, McCallon H, Hacking P, et al. 1992. *Explanatory Supplement to the IRAS Faint Source Survey, Version 2*, JPL D-10015 8/92, Pasadena: JPL
Murphy TW, Armus L, Matthews K, Soifer BT, Mazzarella JM, et al. 1996. *Astron. J.* 111:1025
Neugebauer G, Becklin EE, Hyland AR. 1971. *Annu. Rev. Astron. Astrophys.* 10:67
Neugebauer G, Becklin EE, Oke JB, Searle L. 1976. *Ap. J.* 205:29
Neugebauer G, Habing HJ, van Duinen R, Aumann HH, Baud B, et al. 1984 *Ap. J. Lett.* 278:1
Neugebauer G, Soifer BT, Miley GK. 1985. *Ap. J. Lett.* 295:27
Neugebauer G, Soifer BT, Miley GK, Clegg PE. 1986. *Ap. J.* 308:815
Ohta K, Yamada T, Nakanishi K, Kohno K, Akiyama M, Kawabe R. 1996. *Nature* 382:426
Okumura SK, Kawabe R, Ishiguro M, Kasuga T, Morita KI, Ishizuki S. 1991. In *Dynamics of Galaxies and Their Molecular Cloud Distributions*, ed. F Combes, F Casoli, p. 425. Dordrecht:Reidel
Oliver S, Broadhurst T, Rowan-Robinson M, Saunders W, Lawrence A, et al. 1995. In *Wide-Field Spectroscopy and the Distant Universe*, ed. SJ Maddox, A Aragon-Salamanca, p. 264. Singapore:World Scientific
Oliver S, Rowan-Robinson M, Broadhurst TJ, McMahon RG, Saunders W, et al. 1996. *MNRAS* 280:673
Omont A, McMahon RG, Cox P, Kreysa E, Bergeron J, et al. 1996a. *Astron. Astrophys.* In press

Omont A, Petitjean P, Guilloteau S, McMahon RG, Solomon PM, Pécontal E. 1996b. *Nature* 382:428

Osterbrock DE, DeRobertis MM. 1985. *Publ. Astron. Soc. Pac.* 97:1129

Perault M. 1987. *Structure et evolution des nuages moleculaires.* PhD thesis, Univ. Paris

Radford SJE. 1994. In *The Cold Universe*, ed. T Montmerle, CJ Lada, IF Mirabel, J Tran Thanh Van, p. 369. Gif-sur-Yvette:Ed Frontieres

Radford SJE, Downes D, Solomon PM, Barrett J, Sage LJ. 1996. *Astron. J.* 111:1021

Radford SJE, Solomon PM, Downes D. 1991. *Ap. J. Lett.* 368:15

Read AM, Ponman TJ, Wolstecroft RD. 1995. *MNRAS* 277:397

Rees MJ, Silk JI, Werner MW, Wickramasinghe NC. 1969. *Nature* 223:37

Rieke GH. 1978. *Ap. J.* 226:550

Rieke GH. 1988. *Ap. J. Lett.* 331:5

Rieke GH, Cutri R, Black JH, Kailey WF, McAlary CW, Lebofsky MJ, Elston R. 1985. *Ap. J.* 290:116

Rieke GH, Lebofsky MJ. 1978. *Ap. J. Lett.* 220:37

Rieke GH, Lebofsky MJ. 1979. *Annu. Rev. Astron. Astrophys.* 17:477

Rieke GH, Lebofsky MJ. 1986. *Ap. J.* 304:326

Rieke GH, Low FJ. 1972. *Ap. J. Lett.* 176:95

Rieke GH, Low FJ. 1975. *Ap. J. Lett.* 200:67

Rigopoulu D, Lawrence A, Rowan-Robinson M. 1996a. *MNRAS* 278:1049

Rigopoulu D, Lawrence A, White GJ, Rowan-Robinson M, Church SE. 1996b. *Astron. Astrophys.* 305:747

Rowan-Robinson M. 1986. *MNRAS* 219:737

Rowan-Robinson M, Broadhurst T, Lawrence A, McMahon RG, Lonsdale CJ, et al. 1991. *Nature* 351:719

Rowan-Robinson M, Clegg PE, Beichman CA, Neugebauer G, Soifer BT, et al. 1984. *Ap. J. Lett.* 278:7

Rowan-Robinson M, Efstathiou A. 1993. *MNRAS* 263:675

Rush B, Malkan MA, Spinoglio L. 1993. *Ap. J. Suppl.* 89:1

Sage LJ, Solomon PM. 1987. *Ap. J. Lett.* 321:103

Sanders DB. 1992. In *Relationships Between Active Galactic Nuclei and Starburst Galaxies*, ed. A Filippenko, p. 303. San Francisco:PASP

Sanders DB, Egami E, Lipari S, Mirabel IF, Soifer BT. 1995. *Astron. J.* 110:1993

Sanders DB, Mazzarella JM, Jensen J, Wynn-Williams CG, Hodapp KW. 1996b. *Ap. J. Suppl.* submitted

Sanders DB, Mirabel IF. 1985. *Ap. J. Lett.* 298:31

Sanders DB, Phinney ES, Neugebauer G, Soifer BT, Matthews K. 1989a. *Ap. J.* 347:29

Sanders DB, Sargent AI, Scoville NZ, Phillips TG. 1990. In *Submillimeter Astronomy*, ed. GD Watt, A Webster, p. 213. Dordrecht:Kluwer

Sanders DB, Scoville NZ, Soifer BT. 1988c. *Ap. J. Lett.* 335:1

Sanders DB, Scoville NZ, Soifer BT. 1988d. *Science* 239:625

Sanders DB, Scoville NZ, Soifer BT. 1991. *Ap. J.* 370:158

Sanders DB, Scoville NZ, Zensus A, Soifer BT, Wilson TL, et al. 1989b. *Astron. Astrophys.* 213:L5

Sanders DB, Soifer BT, Elias JH, Madore BF, Matthews K, et al. 1988a. *Ap. J.* 325:74

Sanders DB, Soifer BT, Elias JH, Neugebauer G, Matthews K. 1988b. *Ap. J. Lett.* 328:35

Sanders DB, Mazzarella JM, Surace J, Egami E, Kim DC, et al. 1996a, *Ap. J. Suppl.* submitted

Sanders DB, Young JS, Scoville NZ, Soifer BT, Danielson GE. 1987. *Ap. J. Lett.* 312:5

Saunders W, Rowan-Robinson M, Lawrence A, Efstathiou G, Kaiser N, et al. 1990. *MNRAS* 242:318

Schechter P. 1976. *Ap. J.* 203:297

Schmidt M, Green RF. 1983. *Ap. J.* 269:352

Schweizer F. 1978. In *Structure and Properties of Nearby Galaxies*, ed. EM Berkhuijsen, R Wielebinski, p. 279. Dordrecht:Reidel

Schweizer F. 1982. *Ap. J.* 252:455

Scoville NZ, Padin S, Sanders DB, Soifer BT, Yun MS. 1993. *Ap. J. Lett.* 415:75

Scoville NZ, Sanders DB. 1987. In *Interstellar Processes*, ed. H Thronson, D Hollenbach, p. 21. Dordrecht: Reidel

Scoville NZ, Sargent AZ, Sanders DB, Soifer BT. 1991. *Ap. J. Lett.* 366:5

Scoville NZ, Yun MS, Brown RL, Vanden Bout PA. 1995. *Ap. J. Lett.* 449:109

Smith BJ, Kleinmann SG, Huchra JP, Low FJ. 1987. *Ap. J.* 318:161

Soifer BT, Boehmer L, Neugebauer G, Sanders DB. 1989. *Astron. J.* 98:766

Soifer BT, Cohen JG, Armus L, Matthews K, Neugebauer G, Oke JB. 1995. *Ap. J. Lett.* 443:65

Soifer BT, Helou G, Lonsdale CJ, Neugebauer G, Hacking P, et al. 1984b. *Ap. J. Lett.* 283:1

Soifer BT, Houck JR, Neugebauer G. 1987a. *Annu. Rev. Astron. Astrophys.* 25:187

Soifer BT, Neugebauer G. 1991. *Astron. J.* 101:354

Soifer BT, Neugebauer G, Matthews K, Lawrence C, Mazzarella J. 1992. *Ap. J. Lett.* 399:55

Soifer BT, Rowan-Robinson M, Houck JR, de Jong T, Neugebauer G, et al. 1984a. *Ap. J. Lett.* 278:71

Soifer BT, Sanders DB, Madore BF, Neugebuer G, Danielson GE, et al. 1987b. *Ap. J.* 320:238
Soifer BT, Sanders DB, Neugebauer G, Danielson GE, Lonsdale CJ, et al. 1986. *Ap. J. Lett.* 303:41
Solomon PM, Downes D, Radford SJE. 1992a. *Ap. J. Lett.* 387:55
Solomon PM, Downes D, Radford SJE. 1992b. *Ap. J. Lett.* 398:29
Solomon PM, Downes D, Radford SJE, Barrett JW. 1996. *Ap. J.* In press
Solomon PM, Radford SJE, Downes D. 1992c. *Nature* 356:318
Spinoglio L, Malkan MA. 1989. *Ap. J.* 342:83
Stanford SA, Bushouse HA 1991. *Ap. J.* 371:92
Stanford SA, Sargent AI, Sanders DB, Scoville NZ. 1990. *Ap. J.* 349:492
Staveley-Smith L, Cohen RJ, Chapman JM, Pointon L, Unger SW. 1987. *MNRAS* 226:689
Stein WA. 1995. *Astron. J.* 110:1019
Stein WA, Gillett FC, Merrill KM. 1974. *Ap. J.* 187:213
Stockton A, Ridgway SE. 1991. *Astron. J.* 102:488
Strauss MA, Huchra JP. 1988. *Astron. J.* 95:1602
Strauss MA, Huchra JP, Davis M, Yahil A, Fisher KB, Tonry J. 1992. *Ap. J. Suppl.* 83:29
Sulentic JW. 1989. *Astron. J.* 98:2067
Tacconi LJ, Genzel R, Bleitz M, Cameron M, Harris AI, Madden S. 1994. *Ap. J. Lett.* 426:77
Taniguchi Y., Wada K. 1996. *Ap. J.* 469:581
Telesco CM. 1988. *Annu. Rev. Astron. Astrophys.* 26:343
Telesco CM, Harper DA. 1980. *Ap. J.* 235:392
Terlevich RJ, Boyle BJ. 1993. *MNRAS* 262:491
Tinney CG, Scoville NZ, Sanders DB, Soifer BT. 1990. *Ap. J.* 362:473
Toomre A. 1977. In *The Evolution of Galaxies and Stellar Populations*, ed. BM Tinsley, RB Larson, p. 401. New Haven, CT:Yale Univ. Obs.
Toomre A, Toomre J. 1972. *Ap. J.* 178:623
Trentham N. 1995. *MNRAS* 277:616
Ueno S, Mushotsky RF, Koyama K, Iwasawa K, Awaki H, Hayashi I. 1994 *Publ. Astron. Soc. Japan* 46:L71
Ulvestad JS, Wilson AS. 1989. *Ap. J.* 343:659
Vacca WD. 1996. In *The Interplay Between Massive Star Formation, the ISM, and Galaxy Evolution*, ed. D Knuth, B Guiderdoni, M Heydari-Malayeri, T Thuan, p. 321. Paris: Ed.Frontieres
Vader JP, Frogel JA, Terndrup DM, Heisler CA. 1993. *Astron. J.* 106:1743
Vader JP, Simon M. 1987a. *Nature* 327:304
Vader JP, Simon M. 1987b. *Astron. J.* 94:854
van den Bergh S. 1990. In *Dynamics and Interactions of Galaxies*, ed. R Wielen, p. 492. Dordrecht:Reidel
van den Bergh S. 1995a. *Nature* 374:215
van den Bergh S. 1995b. *Ap. J.* 450:27
van den Broek AC. 1990. *A study of extreme IRAS galaxies*. PhD thesis, Univ. Groningen
van der Kruit PC. 1971. *Astron. Astrophys.* 15:110
Veilleux S, Cecil G, Bland-Hawthorn J, Tully RB, Filippenko AV, Sargent WLW. 1994. *Ap. J.* 433:48
Veilleux S, Kim D-C, Sanders, DB. 1996. *Ap. J.* submitted
Veilleux S, Kim D-C, Sanders, DB, Mazzarella JM, Soifer BT. 1995. *Ap. J. Suppl.* 98:171
Veilleux S, Osterbrock DE. 1987. *Ap. J. Suppl.* 63:295
Vorontsov-Velyaminov BA. 1959. *Atlas and Catalog of Interacting Galaxies, Vol I*, Moscow: Sternberg Inst., Moscow State Univ.
Wang Z, Schweizer F, Scoville NZ. 1992. *Ap. J.* 396:510
Whitmore BC, Schweitzer F. 1995. *Astron. J.* 109:960
Whitmore BC, Schweitzer F, Leitherer C, Borne K, Robert C. 1993. *Astron. J.* 106:1354
Wright GS, James PA, Joseph RD, McLean IS. 1990. *Nature* 344:417
Wright GS, Joseph RD, Meikle WPS. 1984. *Nature* 309:430
Wunderlich E, Klein U, Wielebinski R. 1987. *Astron. Astrophys. Suppl.* 69:487
Xu C, Lisenfeld U, Völk HJ. 1994. *Astron. Astrophys.* 285:19
Young ET, Neugebauer G, Kopan EL, Benson RD, Conrow TP, et al. 1986. *A Users Guide to the IRAS Pointed Observation Products*, Pasadena: JPL
Young JS, Kenney JD, Lord S, Schloerb FP. 1984. *Ap. J. Lett.* 287:65
Young JS, Kenney JD, Tacconi L, Clausen MJ, Huang YL, et al. 1986b *Ap. J. Lett.* 311:17
Young JS, Schloerb FP, Kenney J, Lord S. 1986a. *Ap. J.* 304:443
Young JS, Scoville NZ. 1991. *Annu. Rev. Astron. Astrophys.* 29:581
Young JS, Xie S, Tacconi L, Knezek P, Viscuso P, et al. 1995. *Ap. J. Suppl.* 98:219
Zwicky F. 1956. *Ergeb. Exakten Naturwiss.* 29:34
Zwicky F, Zwicky MA. 1971. *Catalog of Selected Galaxies and Post-Eruptive Galaxies.* Zurich: Offsetdruck L. Speich

SUBJECT INDEX

A

B

C

F

G

H

I

J

K

L

M

N

Q

R

S

T

U

V

W

X

Y

CUMULATIVE INDEXES

CONTRIBUTING AUTHORS, VOLUMES 24–34

CHAPTER TITLES, VOLUMES 24–34